PEDAGOGICAL COLOR CHART

Mechanics and Thermodynamics

- Displacement and position vectors
- Displacement and position component vectors
- Linear ($\vec{\mathbf{v}}$) and angular ($\vec{\boldsymbol{\omega}}$) velocity vectors
- Velocity component vectors
- Force vectors ($\vec{\mathbf{F}}$)
- Force component vectors
- Acceleration vectors ($\vec{\mathbf{a}}$)
- Acceleration component vectors
- Energy transfer arrows (W_{eng}, Q_c, Q_h)
- Process arrow
- Linear ($\vec{\mathbf{p}}$) and angular ($\vec{\mathbf{L}}$) momentum vectors
- Linear and angular momentum component vectors
- Torque vectors ($\vec{\boldsymbol{\tau}}$)
- Torque component vectors
- Schematic linear or rotational motion directions
- Dimensional rotational arrow
- Enlargement arrow
- Springs
- Pulleys

Electricity and Magnetism

- Electric fields
- Electric field vectors
- Electric field component vectors
- Magnetic fields
- Magnetic field vectors
- Magnetic field component vectors
- Positive charges
- Negative charges
- Resistors
- Batteries and other DC power supplies
- Switches
- Capacitors
- Inductors (coils)
- Voltmeters
- Ammeters
- AC Sources
- Lightbulbs
- Ground symbol
- Current

Light and Optics

- Light ray
- Focal light ray
- Central light ray
- Converging lens
- Diverging lens
- Mirror
- Curved mirror
- Objects
- Images

CONVERSION FACTORS

Length
1 m = 39.37 in. = 3.281 ft
1 in. = 2.54 cm (exact)
1 km = 0.621 mi
1 mi = 5 280 ft = 1.609 km
1 lightyear (ly) = 9.461×10^{15} m
1 angstrom (Å) = 10^{-10} m

Mass
1 kg = 10^3 g = 6.85×10^{-2} slug
1 slug = 14.59 kg
1 u = 1.66×10^{-27} kg = 931.5 MeV/c^2

Time
1 min = 60 s
1 h = 3 600 s
1 day = 24 h = 1.44×10^3 min = 8.64×10^4 s
1 yr = 365.242 days = 3.156×10^7 s

Volume
1 L = 1 000 cm^3 = 0.035 3 ft^3
1 ft^3 = 2.832×10^{-2} m^3
1 gal = 3.786 L = 231 in.^{3}

Angle
180° = π rad
1 rad = 57.30°
1° = 60 min = 1.745×10^{-2} rad

Speed
1 km/h = 0.278 m/s = 0.621 mi/h
1 m/s = 2.237 mi/h = 3.281 ft/s
1 mi/h = 1.61 km/h = 0.447 m/s = 1.47 ft/s

Force
1 N = 0.224 8 lb = 10^5 dynes
1 lb = 4.448 N
1 dyne = 10^{-5} N = 2.248×10^{-6} lb

Work and energy
1 J = 10^7 erg = 0.738 ft · lb = 0.239 cal
1 cal = 4.186 J
1 ft · lb = 1.356 J
1 Btu = 1.054×10^3 J = 252 cal
1 J = 6.24×10^{18} eV
1 eV = 1.602×10^{-19} J
1 kWh = 3.60×10^6 J

Pressure
1 atm = 1.013×10^5 N/m^2 (or Pa) = 14.70 lb/in.^{2}
1 Pa = 1 N/m^2 = 1.45×10^{-4} lb/in.^{2}
1 lb/in.^{2} = 6.895×10^3 N/m^2

Power
1 hp = 550 ft · lb/s = 0.746 kW
1 W = 1 J/s = 0.738 ft · lb/s
1 Btu/h = 0.293 W

Tenth Edition

College Physics

Raymond A. Serway | *Emeritus, James Madison University*

Chris Vuille | *Embry-Riddle Aeronautical University*

With contributions from John Hughes | *Embry-Riddle Aeronautical University*

CENGAGE Learning

Australia • Brazil • Mexico • Singapore • United Kingdom • United States

CENGAGE Learning

College Physics, **Tenth Edition**
Raymond A. Serway and Chris Vuille

Product Director: Mary Finch
Senior Product Manager: Charlie Hartford
Development Editor: Ed Dodd
Product Assistant: Chris Robinson
Managing Developer: Peter McGahey
Senior Market Development Manager: Rebecca Berardy Schwartz
Media Developer: Andrew Coppola
Senior Marketing Development Manager: Janet del Mundo
Brand Manager: Nicole Hamm
Content Project Manager: Alison Eigel Zade
Senior Art Director: Cate Barr
Manufacturing Planner: Sandee Milewski
Rights Acquisition Specialist: Shalice Shah-Caldwell
Production Service and Compositor: Graphic World Inc.
Text & Cover Designer: Dare Porter
Cover Image: Street luge
Cover Image Credit: © Mike Chew/CORBIS
Vice President, Director of Advanced and Elective Products Program: Alison Zetterquist
Associate Content Developer, Advanced and Elective Products Program: Ashley Bargende
Senior Media Developer, Advanced and Elective Products Program: Philip Lanza

Library of Congress Control Number: 2013948341

ISBN-13: 978-1-285-76249-4

ISBN-10: 1-285-76249-5

Cengage Learning
200 First Stamford Place, 4th Floor
Stamford, CT 06902
USA

Printed in the United States of America
1 2 3 4 5 6 7 17 16 15 14 13

We dedicate this book to our wives, children, grandchildren, relatives, and friends who have provided so much love, support, and understanding through the years, and to the students for whom this book was written.

Contents Overview

PART 1 | Mechanics

PART 2 | Thermodynamics

PART 3 | Vibrations and Waves

PART 4 | Electricity and Magnetism

PART 5 | Light and Optics

PART 6 | Modern Physics

Contents

About the Authors

Raymond A. Serway received his doctorate at Illinois Institute of Technology and is Professor Emeritus at James Madison University. In 2011, he was awarded an honorary doctorate degree from his alma mater, Utica College. He received the 1990 Madison Scholar Award at James Madison University, where he taught for 17 years. Dr. Serway began his teaching career at Clarkson University, where he conducted research and taught from 1967 to 1980. He was the recipient of the Distinguished Teaching Award at Clarkson University in 1977 and the Alumni Achievement Award from Utica College in 1985. As Guest Scientist at the IBM Research Laboratory in Zurich, Switzerland, he worked with K. Alex Müller, 1987 Nobel Prize recipient. Dr. Serway was also a visiting scientist at Argonne National Laboratory, where he collaborated with his mentor and friend, the late Sam Marshall. Early in his career, he was employed as a research scientist at the Rome Air Development Center from 1961 to 1963 and at the IIT Research Institute from 1963 to 1967. Dr. Serway is also the coauthor of *Physics for Scientists and Engineers*, ninth edition; *Principles of Physics: A Calculus-Based Text*, fifth edition; *Essentials of College Physics*, *Modern Physics*, third edition; and the high school textbook *Physics*, published by Holt, Rinehart and Winston. In addition, Dr. Serway has published more than 40 research papers in the field of condensed matter physics and has given more than 60 presentations at professional meetings. Dr. Serway and his wife Elizabeth enjoy traveling, playing golf, fishing, gardening, singing in the church choir, and especially spending quality time with their four children, nine grandchildren, and a recent great grandson.

Chris Vuille is an associate professor of physics at Embry-Riddle Aeronautical University (ERAU), Daytona Beach, Florida, the world's premier institution for aviation higher education. He received his doctorate in physics from the University of Florida in 1989 and moved to Daytona after a year at ERAU's Prescott, Arizona, campus. Although he has taught courses at all levels, including postgraduate, his primary interest has been instruction at the level of introductory physics. He has received several awards for teaching excellence, including the Senior Class Appreciation Award (three times). He conducts research in general relativity and quantum theory and was a participant in the JOVE program, a special three-year NASA grant program during which he studied neutron stars. His work has appeared in a number of scientific journals, and he has been a featured science writer in Analog Science Fiction/Science Fact magazine. In addition to this textbook, he is coauthor of *Essentials of College Physics*. Dr. Vuille enjoys tennis, swimming, and playing classical piano, and he is a former chess champion of St. Petersburg and Atlanta. In his spare time he writes fiction and goes to the beach. His wife, Dianne Kowing, is Chief of Optometry for a local Veterans' Administration clinic. They have a daughter, Kira, and two sons, Christopher and James.

Preface

College Physics is written for a one-year course in introductory physics usually taken by students majoring in biology, the health professions, or other disciplines, including environmental, earth, and social sciences, and technical fields such as architecture. The mathematical techniques used in this book include algebra, geometry, and trigonometry, but not calculus. Drawing on positive feedback from users of the ninth edition, analytics gathered from both teachers and students who use Enhanced WebAssign, as well as reviewers' suggestions, we have refined the text to better meet the needs of students and teachers.

This textbook, which covers the standard topics in classical physics and twentieth-century physics, is divided into six parts. Part 1 (Chapters 1–9) deals with Newtonian mechanics and the physics of fluids; Part 2 (Chapters 10–12) is concerned with heat and thermodynamics; Part 3 (Chapters 13 and 14) covers wave motion and sound; Part 4 (Chapters 15–21) develops the concepts of electricity and magnetism; Part 5 (Chapters 22–25) treats the properties of light and the field of geometric and wave optics; and Part 6 (Chapters 26–30) provides an introduction to special relativity, quantum physics, atomic physics, and nuclear physics.

Objectives

The main objectives of this introductory textbook are twofold: to provide the student with a clear and logical presentation of the basic concepts and principles of physics and to strengthen an understanding of those concepts and principles through a broad range of interesting, real-world applications. To meet those objectives, we have emphasized sound physical arguments and problem-solving methodology. At the same time we have attempted to motivate the student through practical examples that demonstrate the role of physics in other disciplines.

Changes to the Tenth Edition

Several changes and improvements have been made in preparing the tenth edition of this text. Some of the new features are based on our experiences and on current trends in science education. Other changes have been incorporated in response to comments and suggestions offered by users of the ninth edition. The features listed here represent the major changes made for the tenth edition.

New Learning Objectives Added for Every Section

In response to a growing trend across the discipline (and the request of many users), we have added learning objectives for every section of the tenth edition. The learning objectives identify the major concepts in a given section and also identify the specific skills/outcomes students should be able to demonstrate once they have a solid understanding of those concepts. It is hoped that these learning objectives will assist those teachers who are transitioning their course to a more outcomes-based approach.

New Online Tutorials

The new online tutorials (available via Enhanced WebAssign) offer students another training tool to assist them in understanding how to apply certain key concepts presented in a given chapter. The tutorials first present a brief review of the necessary concepts from the text, together with advice on how to solve problems involving them. The student can then attempt to solve one or two such problems, guided by questions presented in the tutorial. The tutorial automatically scores student responses and presents correct solutions together with discussion. Students can then practice on several additional problems of a similar level and, in some cases go to higher level or related problems, depending on the concepts covered in the tutorial.

New Warm-Up Exercises in Every Chapter

Warm-up exercises (over 320 are included in the full book) appear at the beginning of each chapter's problems set, and were inspired by one of the author's (Vuille) classroom experiences. The idea behind warm-up exercises is to review mathematical and physical concepts that are prerequisites for a given chapter's problems set, and also to provide students with a general preview of the new physics concepts covered in a given chapter. By doing the warm-up exercises first, students will have an easier time getting comfortable with the new concepts of a chapter before tackling harder problems.

New Algorithmic Solutions in Enhanced WebAssign

All quantitative end-of-chapter problems in Enhanced WebAssign now feature *algorithmic solutions.* Fully worked-out solutions are available to students with quantitative parameters exactly matching the version of the problem assigned to individual students. As always for all "Hints" features, Enhanced WebAssign offers great flexibility to teachers regarding when to enable algorithmic solutions.

Chapter-by-Chapter Changes

The text has been carefully edited to improve clarity of presentation and precision of language. We hope that the result is a book both accurate and enjoyable to read. Although the overall content and organization of the textbook are similar to the ninth edition, a few changes were implemented. The list below highlights some of the major changes for the tenth edition.

Chapter 1 Introduction

- Nine new warm-up exercises have been added.
- A new tutorial (*Unit conversions*) has been added in Enhanced WebAssign.

Chapter 2 Motion in One Dimension

- Seven new warm-up exercises have been added.
- A new tutorial (*One-dimensional motion at constant acceleration*) has been added in Enhanced WebAssign.

Chapter 3 Vectors and Two-Dimensional Motion

- Nine new warm-up exercises have been added.
- Two new tutorials (*Applying the kinematics equations of two-dimensional motion* and *Applying the concept of relative velocity*) have been added in Enhanced WebAssign.

Chapter 4 The Laws of Motion

- Thirteen new warm-up exercises have been added.
- Five new tutorials (*Normal forces, Applying the second law to objects in equilibrium, Applying the second law to accelerating objects, Applying the static and kinetic friction forces in the second law,* and *Applying the system approach*) have been added in Enhanced WebAssign.

Chapter 5 Energy

- Ten new warm-up exercises have been added.

- Five new tutorials (*Calculating work, Applying the work–energy theorem, Applying conservation of mechanical energy, Applying the work–energy theorem with the potential energies of gravity and springs,* and *Applying average and instantaneous power*) have been added in Enhanced WebAssign.

Chapter 6 Momentum and Collisions

- Eleven new warm-up exercises have been added.
- Two new tutorials (*Collisions in one dimension* and *Inelastic collisions in two dimensions*) have been added in Enhanced WebAssign.

Chapter 7 Rotational Motion and the Law of Gravity

- Example 7.1 has been revised.
- Fifteen new warm-up exercises have been added.
- Two new tutorials (*Applying the second law to objects in uniform circular motion* and *Applying gravitational potential energy*) have been added in Enhanced WebAssign.

Chapter 8 Rotational Equilibrium and Rotational Dynamics

- Fourteen new warm-up exercises have been added.
- Four new tutorials (*Applying the conditions of mechanical equilibrium to rigid bodies, Applying the rotational second law, Applying the work-energy theorem including rotational kinetic energy,* and *Applying conservation of angular momentum*) have been added in Enhanced WebAssign.

Chapter 9 Solids and Fluids

- Eleven new warm-up exercises have been added.
- Two new tutorials (*Applying Archimedes' principle* and *Applying Bernoulli's equation*) have been added in Enhanced WebAssign.

Chapter 10 Thermal Physics

- Ten new warm-up exercises have been added.
- A new tutorial (*Applying the ideal gas law*) has been added in Enhanced WebAssign.

Chapter 11 Energy in Thermal Processes

- Example 11.11 ("Planet of Alpha Centauri B") is completely new to this edition.
- Nine new warm-up exercises have been added.
- A new tutorial (*Calorimetry*) has been added in Enhanced WebAssign.

Chapter 12 The Laws of Thermodynamics

- Fourteen new warm-up exercises have been added.
- Two new tutorials (*Thermal processes* and *Calculating changes in entropy*) have been added in Enhanced WebAssign.

Chapter 13 Vibrations and Waves

- Eleven new warm-up exercises have been added.
- A new tutorial (*Investigating simple harmonic oscillations*) has been added in Enhanced WebAssign.

Chapter 14 Sound

- Fourteen new warm-up exercises have been added.
- Two new tutorials (*Sound intensity, decibel level, and their variation with distance* and *Calculating the Doppler effect*) have been added in Enhanced WebAssign.

Chapter 15 Electric Forces and Electric Fields

- Fourteen new warm-up exercises have been added.
- Two new tutorials (*Coulomb's law and the electric field* and *Applying Gauss's law to distributions of charge*) have been added in Enhanced WebAssign.

Chapter 16 Electrical Energy and Capacitance

- Twelve new warm-up exercises have been added.
- Two new tutorials (*Applying the work-energy theorem to systems of charges* and *Evaluating the equivalent capacitance of systems of capacitors*) have been added in Enhanced WebAssign.

Chapter 17 Current and Resistance

- Ten new warm-up exercises have been added.
- One new tutorial (*Exploring electrical current, energy, and power*) has been added in Enhanced WebAssign.

Chapter 18 Direct-Current Circuits

- Thirteen new warm-up exercises have been added.
- Two new tutorials (*Simplifying circuits with both series and parallel resistors* and *Applying Kirchhoff's rules to complex DC circuits*) have been added in Enhanced WebAssign.

Chapter 19 Magnetism

- Nine new warm-up exercises have been added.
- Two new tutorials (*The motion of charged particles in a uniform magnetic field* and *Applying Ampere's law to current-carrying wires and cylinders*) have been added in Enhanced WebAssign.

Chapter 20 Induced Voltages and Inductance

- Ten new warm-up exercises have been added.
- Three new tutorials (*Using magnetic flux and Faraday's law, Calculating motional EMF,* and *RL circuits*) have been added in Enhanced WebAssign.

Chapter 21 Alternating-Current Circuits and Electromagnetic Waves

- Eleven new warm-up exercises have been added.
- Two new tutorials (*Purely resistive, capacitive, and inductive AC circuits* and *Analyzing series RLC AC circuits*) have been added in Enhanced WebAssign.

Chapter 22 Reflection and Refraction of Light

- Five new warm-up exercises have been added.
- One new tutorial (*Reflection, refraction, and Snell's law*) has been added in Enhanced WebAssign.

Chapter 23 Mirrors and Lenses

- Seven new warm-up exercises have been added.
- Two new tutorials (*Analyze the optical properties of concave and convex mirrors* and *Analyze the optical properties of convergent and divergent thin lenses*) have been added in Enhanced WebAssign.

Chapter 24 Wave Optics

- Ten new warm-up exercises have been added.
- Three new tutorials (*Solving problems involving Young's experiment, Interference in thin films,* and *Diffraction and diffraction gratings*) have been added in Enhanced WebAssign.

Chapter 25 Optical Instruments

- Twelve new warm-up exercises have been added.
- Two new tutorials (*Prescribing corrective lenses* and *Optical instruments*) have been added in Enhanced WebAssign.

Chapter 26 Relativity

- Ten new warm-up exercises have been added.
- One new tutorial (*Comparing space and time measurements by different observers in relativity*) has been added in Enhanced WebAssign.

Chapter 27 Quantum Physics

- Eleven new warm-up exercises have been added.
- One new tutorial (*Quantum physics*) has been added in Enhanced WebAssign.

Chapter 28 Atomic Physics

- Ten new warm-up exercises have been added.
- One new tutorial (*Hydrogen and hydrogen-like atoms*) has been added in Enhanced WebAssign.

Chapter 29 Nuclear Physics

- Ten new warm-up exercises have been added.
- One new tutorial (*Analyzing radioactivity*) has been added in Enhanced WebAssign.

Chapter 30 Nuclear Energy and Elementary Particles

- Ten new warm-up exercises have been added.
- One new tutorial (*Calculating the energy released in nuclear reactions*) has been added in Enhanced WebAssign.

Textbook Features

Most teachers would agree that the textbook assigned in a course should be the student's primary guide for understanding and learning the subject matter. Further, the textbook should be easily accessible and written in a style that facilitates instruction and learning. With that in mind, we have included many pedagogical features that are intended to enhance the textbook's usefulness to both students and teachers. The following features are included.

Examples For this tenth edition we have reviewed all the worked examples and made numerous improvements. Every effort has been made to ensure the collection of examples, as a whole, is comprehensive in covering all the physical concepts, physics problem types, and required mathematical techniques. The Questions usually require a conceptual response or determination, but they also include estimates requiring knowledge of the relationships between concepts. The answers for the Questions can be found at the back of the book. The examples are in a two-column format for a pedagogic purpose: students can study the example, then cover up the right column and attempt to solve the problem using the cues in the left column. Once successful in that exercise, the student can cover up both solution columns and attempt to solve the problem using only the strategy statement, and finally just the problem statement. A sample of an in-text worked example in on the next page, with an explanation of each of the example's main parts.

Integration with Enhanced WebAssign The textbook's tight integration with Enhanced WebAssign content facilitates an online learning environment that helps students improve their problem-solving skills and gives them a variety of tools to meet their individual learning styles. Extensive user data gathered by WebAssign were used to ensure that the problems most often assigned were retained for this new edition. In each chapter's problems set, the top quartile of problems that were assigned in WebAssign have cyan-shaded problem numbers for easy identification, allowing teachers to quickly and easily find the most popular problems that were assigned in Enhanced WebAssign. Master It tutorials help students solve problems by having them work through a stepped-out solution. Problems with Master It tutorials are indicated in each chapter's problem set with a M icon. In addition, Watch It solution videos (indicated by a W icon) explain fundamental problem-solving strategies to help students step through selected problems. The problems most often assigned in Enhanced WebAssign (shaded in blue) have feedback to address student misconceptions, helping students avoid common pitfalls.

The **Goal** describes the physical concepts being explored within the worked example.

The **Problem** statement presents the problem itself.

The **Strategy** section helps students analyze the problem and create a framework for working out the solution.

The **Solution** section uses a two-column format that gives the explanation for each step of the solution in the left-hand column, while giving each accompanying mathematical step in the right-hand column. This layout facilitates matching the idea with its execution and helps students learn how to organize their work. Another benefit: students can easily use this format as a training tool, covering up the solution on the right and solving the problem using the comments on the left as a guide.

EXAMPLE 13.7 Measuring the Value of g

GOAL Determine g from pendulum motion.

PROBLEM Using a small pendulum of length 0.171 m, a geophysicist counts 72.0 complete swings in a time of 60.0 s. What is the value of g in this location?

STRATEGY First calculate the period of the pendulum by dividing the total time by the number of complete swings. Solve Equation 13.15 for g and substitute values.

SOLUTION

Calculate the period by dividing the total elapsed time by the number of complete oscillations:	$T = \dfrac{\text{time}}{\text{\# of oscillations}} = \dfrac{60.0\text{ s}}{72.0} = 0.833\text{ s}$
Solve Equation 13.15 for g and substitute values:	$T = 2\pi\sqrt{\dfrac{L}{g}} \rightarrow T^2 = 4\pi^2\dfrac{L}{g}$
	$g = \dfrac{4\pi^2 L}{T^2} = \dfrac{(39.5)(0.171\text{ m})}{(0.833\text{ s})^2} = 9.73\text{ m/s}^2$

REMARKS Measuring such a vibration is a good way of determining the local value of the acceleration of gravity.

QUESTION 13.7 True or False: A simple pendulum of length 0.50 m has a larger frequency of vibration than a simple pendulum of length 1.0 m.

EXERCISE 13.7 What would be the period of the 0.171-m pendulum on the Moon, where the acceleration of gravity is 1.62 m/s²?

ANSWER 2.04 s

Remarks follow each Solution and highlight some of the underlying concepts and methodology used in arriving at a correct solution. In addition, the remarks are often used to put the problem into a larger, real-world context.

Question Each worked example features a conceptual question that promotes student understanding of the underlying concepts contained in the example.

Exercise/Answer Every Question is followed immediately by an exercise with an answer. These exercises allow students to reinforce their understanding by working a similar or related problem, with the answers giving them instant feedback. At the option of the instructor, the exercises can also be assigned as homework. Students who work through these exercises on a regular basis will find the end-of-chapter problems less intimidating.

ENHANCED WebAssign

Many Worked Examples are also available to be assigned in the Enhanced WebAssign homework management system (visit **www.cengage.com/physics/serway** for more details).

Artwork Every piece of artwork in the tenth edition is in a modern style that helps express the physics principles at work in a clearer and more precise fashion. Every piece of art is also drawn to make certain that the physical situations presented correspond exactly to the text discussion at hand.

Guidance labels are included with many figures in the text; these point out important features of the figure and guide students through figures without having to go back and forth from the figure legend to the figure itself. This format also helps those students who are visual learners. An example of this kind of figure appears on the next page.

Warm-Up Exercises As discussed earlier, these new exercises (over 320 are included in the full book) were inspired by one of the author's (Vuille) classroom experiences. Warm-up exercises review mathematical and physical concepts that are prerequisites for a given chapter's problems set and also provide students with a general preview of the new physics concepts covered in a given chapter. By doing the warm-up exercises first, students will have an easier time getting comfortable with the new concepts of a chapter before tackling harder

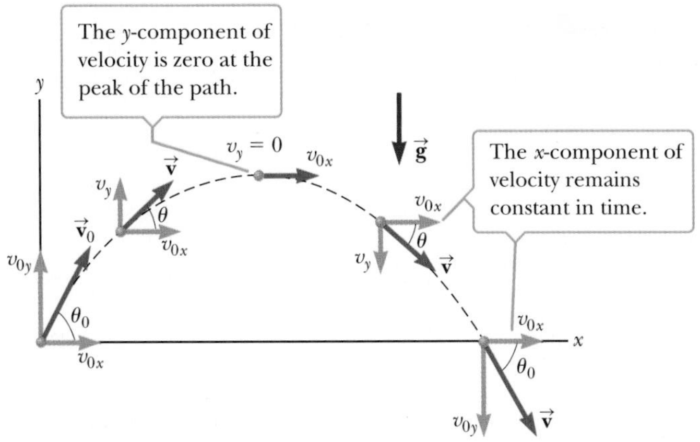

Figure 3.14
The parabolic trajectory of a particle that leaves the origin with a velocity of $\vec{v}_0$. Note that $\vec{v}$ changes with time. However, the x-component of the velocity, v_x, remains constant in time, equal to its initial velocity, v_{0x}. Also, $v_y = 0$ at the peak of the trajectory, but the acceleration is always equal to the free-fall acceleration and acts vertically downward.

problems. Answers to odd-numbered warm-up exercises are included in the Answers section at the end of the book. Answers to all warm-up exercises are in the *Instructor's Solutions Manual.*

Conceptual Questions At the end of each chapter are approximately a dozen conceptual questions. The Applying Physics examples presented in the text serve as models for students when conceptual questions are assigned and show how the concepts can be applied to understanding the physical world. The conceptual questions provide the student with a means of self-testing the concepts presented in the chapter. Some conceptual questions are appropriate for initiating classroom discussions. Answers to odd-numbered conceptual questions are included in the Answers section at the end of the book. Answers to all conceptual questions are in the *Instructor's Solutions Manual.*

Problems All questions and problems for this revision were carefully reviewed to improve their variety, interest, and pedagogical value while maintaining their clarity and quality. An extensive set of problems is included at the end of each chapter (in all, more than 2 000 problems are provided in the tenth edition). Answers to odd-numbered problems are given at the end of the book. For the convenience of both the student and teacher, about two-thirds of the problems are keyed to specific sections of the chapter. The remaining problems, labeled "Additional Problems," are not keyed to specific sections. The three levels of problems are graded according to their difficulty. Straightforward problems are numbered in **black**, intermediate level problems are numbered in **blue**, and the most challenging problems are numbered in **red**. The BIO icon identifies problems dealing with applications to the life sciences and medicine. Solutions to approximately 12 problems in each chapter are in the *Student Solutions Manual and Study Guide.*

There are three other types of problems we think teachers and students will find interesting as they work through the text:

- S **Symbolic problems** require the student to obtain an answer in terms of symbols. In general, some guidance is built into the problem statement. The goal is to better train the student to deal with mathematics at a level appropriate to this course. Most students at this level are uncomfortable with symbolic equations, which is unfortunate because symbolic equations are the most efficient vehicle for presenting relationships between physics concepts. Once students understand the physical concepts, their ability to solve problems is greatly enhanced. As soon as the numbers are substituted into an equation, however, all the concepts and their relationships to one another are lost, melded together in the student's calculator. Symbolic problems train the student to postpone substitution of values, facilitating their ability to think conceptually using the equations. An example of a symbolic problem is provided on the next page.

14\. **S** An object of mass m is dropped from the roof of a building of height h. While the object is falling, a wind blowing parallel to the face of the building exerts a constant horizontal force F on the object. (a) How long does it take the object to strike the ground? Express the time t in terms of g and h. (b) Find an expression in terms of m and F for the acceleration a_x of the object in the horizontal direction (taken as the positive x-direction). (c) How far is the object displaced horizontally before hitting the ground? Answer in terms of m, g, F, and h. (d) Find the magnitude of the object's acceleration while it is falling, using the variables F, m, and g.

- **Q|C** **Quantitative/conceptual problems** encourage the student to think conceptually about a given physics problem rather than rely solely on computational skills. Research in physics education suggests that standard physics problems requiring calculations may not be entirely adequate in training students to think conceptually. Students learn to substitute numbers for symbols in the equations without fully understanding what they are doing or what the symbols mean. Quantitative/conceptual problems combat this tendency by asking for answers that require something other than a number or a calculation. An example of a quantitative/conceptual problem is provided here:

5. **Q|C** Starting from rest, a 5.00-kg block slides 2.50 m down a rough 30.0° incline. The coefficient of kinetic friction between the block and the incline is $\mu_k = 0.436$. Determine (a) the work done by the force of gravity, (b) the work done by the friction force between block and incline, and (c) the work done by the normal force. (d) Qualitatively, how would the answers change if a shorter ramp at a steeper angle were used to span the same vertical height?

- **GP** **Guided problems** help students break problems into steps. A physics problem typically asks for one physical quantity in a given context. Often, however, several concepts must be used and a number of calculations are required to get that final answer. Many students are not accustomed to this level of complexity and often don't know where to start. A guided problem breaks a problem into smaller steps, enabling students to grasp all the concepts and strategies required to arrive at a correct solution. Unlike standard physics problems, guidance is often built into the problem statement. For example, the problem might say "Find the speed using conservation of energy" rather than asking only for the speed. In any given chapter there are usually two or three problem types that are particularly suited to this problem form. The problem must have a certain level of complexity, with a similar problem-solving strategy involved each time it appears. Guided problems are reminiscent of how a student might interact with a professor in an office visit. These problems help train students to break down complex problems into a series of simpler problems, an essential problem-solving skill. An example of a guided problem is provided here:

32\. **GP** Two blocks of masses m_1 and m_2 ($m_1 > m_2$) are placed on a frictionless table in contact with each other. A horizontal force of magnitude F is applied to the block of mass m_1 in Figure P4.32. (a) If P is the magnitude of the contact force between the blocks, draw the free-body diagrams for each block. (b) What is the net force on the system consisting of both blocks? (c) What is the net force acting on m_1? (d) What is the net force acting on m_2? (e) Write the x-component of Newton's second law for each block. (f) Solve the resulting system of two equations and two unknowns, expressing the acceleration a and contact force P in terms of the masses and force. (g) How would the answers change if the force had been applied to m_2 instead? (*Hint:* Use symmetry; don't calculate!) Is the contact force larger, smaller, or the same in this case? Why?

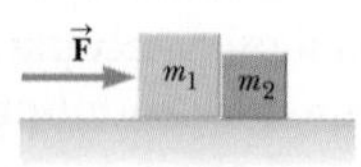

Figure P4.32

seek the advice of the teacher or other students who have taken the course. You may find it necessary to seek further instruction from experienced students. Very often, teachers offer review sessions in addition to regular class periods. It is important that you avoid the practice of delaying study until a day or two before an exam. One hour of study a day for 14 days is far more effective than 14 hours the day before the exam. "Cramming" usually produces disastrous results, especially in science. Rather than attempting an all-night study session immediately before an exam, briefly review the basic concepts and equations and get a good night's rest. If you think you need additional help in understanding the concepts, in preparing for exams, or in problem solving, we suggest you acquire a copy of the *Student Solutions Manual and Study Guide* that accompanies this textbook.

Use the Features

You should make full use of the various features of the text discussed in the preface. For example, marginal notes are useful for locating and describing important equations and concepts, and **boldfaced** type indicates important statements and definitions. Many useful tables are contained in the appendices, but most tables are incorporated in the text where they are most often referenced. Appendix A is a convenient review of mathematical techniques.

Answers to all Quick Quizzes and Example Questions, as well as odd-numbered multiple-choice questions, conceptual questions, and problems, are given at the end of the textbook. Answers to selected end-of-chapter problems are provided in the *Student Solutions Manual and Study Guide.* Problem-Solving Strategies included in selected chapters throughout the text give you additional information about how you should solve problems. The contents provide an overview of the entire text, and the index enables you to locate specific material quickly. Footnotes sometimes are used to supplement the text or to cite other references on the subject discussed.

After reading a chapter, you should be able to define any new quantities introduced in that chapter and to discuss the principles and assumptions used to arrive at certain key relations. The chapter summaries and the review sections of the *Student Solutions Manual and Study Guide* should help you in this regard. In some cases, it may be necessary for you to refer to the index of the text to locate certain topics. You should be able to correctly associate each physical quantity with the symbol used to represent that quantity and the unit in which the quantity is specified. Further, you should be able to express each important relation in a concise and accurate prose statement.

Problem Solving

R. P. Feynman, Nobel laureate in physics, once said, "You do not know anything until you have practiced." In keeping with this statement, we strongly advise that you develop the skills necessary to solve a wide range of problems. Your ability to solve problems will be one of the main tests of your knowledge of physics, so you should try to solve as many problems as possible. It is essential that you understand basic concepts and principles before attempting to solve problems. It is good practice to try to find alternate solutions to the same problem. For example, you can solve problems in mechanics using Newton's laws, but very often an alternate method that draws on energy considerations is more direct. You should not deceive yourself into thinking you understand a problem merely because you have seen it solved in class. You must be able to solve the problem and similar problems on your own. We have cast the examples in this book in a special, two-column format to help you in this regard. After studying an example, see if you can cover up the right-hand side and do it yourself, using only the written descriptions on the left as hints. Once you succeed at that, try solving the example using only the strategy statement as a guide. Finally, try to solve the problem completely on your own. At this point you are ready to answer the associated question and solve the exercise. Once you have accomplished all these steps, you will have a good mastery of the problem, its concepts, and mathematical technique. After studying all the Example Problems in this way, you are ready to tackle the problems at the end of the chapter. Of those, the guided problems provide another aid to learning how to solve some of the more complex problems.

The approach to solving problems should be carefully planned. A systematic plan is especially important when a problem involves several concepts. First, read the problem several times until you are confident you understand what is being asked. Look for any key words that will help you interpret the problem and perhaps allow you to make certain assumptions. Your ability to interpret a question properly is an integral part of problem solving. Second, you should acquire the habit of writing down the information given in a problem and those quantities that need to be found; for example, you might construct a table listing both the quantities given and the quantities to be found. This procedure is sometimes used in the worked examples of the textbook. After you have decided on the method you think is appropriate for a given problem, proceed with your solution. Finally, check your results to see if they are reasonable and consistent with your initial understanding of the problem. General problem-solving strategies of this type are included in the text and are highlighted with a surrounding box. If you follow the steps of this procedure, you will find it easier to come up with a solution and will also gain more from your efforts.

Often, students fail to recognize the limitations of certain equations or physical laws in a particular situation. It is very important that you understand and remember the assumptions underlying a particular theory or formalism. For example, certain equations in kinematics apply only to a particle moving with constant acceleration. These equations are not valid for describing motion whose acceleration is not constant, such as the motion of an object connected to a spring or the motion of an object through a fluid.

Experiments

Because physics is a science based on experimental observations, we recommend that you supplement the text by performing various types of "hands-on" experiments, either at home or in the laboratory. For example, the common Slinky™ toy is excellent for studying traveling waves, a ball swinging on the end of a long string can be used to investigate pendulum motion, various masses attached to the end of a vertical spring or rubber band can be used to determine their elastic nature, an old pair of Polaroid sunglasses and some discarded lenses and a magnifying glass are the components of various experiments in optics, and the approximate measure of the free-fall acceleration can be determined simply by measuring with a stopwatch the time it takes for a ball to drop from a known height. The list of such experiments is endless. When physical models are not available, be imaginative and try to develop models of your own.

New Media

If available, we strongly encourage you to use the **Enhanced WebAssign** product that is available with this textbook. It is far easier to understand physics if you see it in action, and the materials available in Enhanced WebAssign will enable you to become a part of that action. Enhanced WebAssign is described in the Preface.

An Invitation to Physics

It is our hope that you too will find physics an exciting and enjoyable experience and that you will profit from this experience, regardless of your chosen profession. Welcome to the exciting world of physics!

To see the World in a Grain of Sand
And a Heaven in a Wild Flower,
Hold infinity in the palm of your hand
And Eternity in an hour.

William Blake, "Auguries of Innocence"

John Van Hasselt/Sygma/Corbis

In the eighteenth century, navigators of ocean-going ships could obtain their latitude by observations of the north star, but there was no reliable way of determining longitude. The H1 clock was invented by John Harrison in 1736 in an attempt to address that need. His clock had to remain highly accurate for months at sea while withstanding constant motion, dampness, and changes of temperature. To determine longitude, navigators had only to compare local noon, when the sun was highest in the sky, with the time on the clock, which was Greenwich time. The difference in the number of hours then revealed their longitude.

Introduction 1

The goal of physics is to provide an understanding of the physical world by developing theories based on experiments. A physical theory, usually expressed mathematically, describes how a given physical system works. The theory makes certain predictions about the physical system which can then be checked by observations and experiments. If the predictions turn out to correspond closely to what is actually observed, then the theory stands, although it remains provisional. No theory to date has given a complete description of all physical phenomena, even within a given subdiscipline of physics. Every theory is a work in progress.

The basic laws of physics involve such physical quantities as force, velocity, volume, and acceleration, all of which can be described in terms of more fundamental quantities. In mechanics, it is conventional to use the quantities of **length** (L), **mass** (M), and **time** (T); all other physical quantities can be constructed from these three.

1.1 Standards of Length, Mass, and Time

LEARNING OBJECTIVES

1. State and use the SI units for length, mass, and time.
2. Give examples of the approximate magnitudes of common measurements.

To communicate the result of a measurement of a certain physical quantity, a *unit* for the quantity must be defined. If our fundamental unit of length is defined to be 1.0 meter, for example, and someone familiar with our system of measurement reports that a wall is 2.0 meters high, we know that the height of the wall is twice the fundamental unit of length. Likewise, if our fundamental unit of mass is defined as 1.0 kilogram and we are told that a person has a mass of 75 kilograms, then that person has a mass 75 times as great as the fundamental unit of mass.

In 1960 an international committee agreed on a standard system of units for the fundamental quantities of science, called **SI** (Système International). Its units of length, mass, and time are the meter, kilogram, and second, respectively.

Length

In 1799 the legal standard of length in France became the meter, defined as one ten-millionth of the distance from the equator to the North Pole. Until 1960, the official length of the meter was the distance between two lines on a specific bar of platinum-iridium alloy stored under controlled conditions. This standard was abandoned for several reasons, the principal one being that measurements of the separation between the lines were not precise enough. In 1960 the meter was defined as 1 650 763.73 wavelengths of orange-red light emitted from a krypton-86 lamp. In October 1983 this definition was abandoned also, and **the meter was redefined as the distance traveled by light in vacuum during a time interval of 1/299 792 458 second**. This latest definition establishes the speed of light at 299 792 458 meters per second.

Definition of the meter ▶

Mass

Definition of the kilogram ▶

The SI unit of mass, the kilogram, is defined as the mass of a specific platinum–iridium alloy cylinder kept at the International Bureau of Weights and Measures at Sèvres, France (similar to that shown in Fig. 1.1a). As we'll see in Chapter 4, mass is a quantity used to measure the resistance to a change in the motion of an object. It's more difficult to cause a change in the motion of an object with a large mass than an object with a small mass.

Tip 1.1 No Commas in Numbers with Many Digits

In science, numbers with more than three digits are written in groups of three digits separated by spaces rather than commas; so that 10 000 is the same as the common American notation 10,000. Similarly, $\pi = 3.14159265$ is written as 3.141 592 65.

Time

Before 1960, the time standard was defined in terms of the average length of a solar day in the year 1900. (A solar day is the time between successive appearances of the Sun at the highest point it reaches in the sky each day.) The basic unit of

Reproduced with permission of the BIPM, which retains full internationally protected copyright.

a

AP Photo/Focke Strangman

b

Figure 1.1 (a) International Prototype of the Kilogram, an accurate copy of the International Standard Kilogram kept at Sèvres, France, is housed under a double bell jar in a vault at the National Institute of Standards and Technology. (b) A cesium fountain atomic clock. The clock will neither gain nor lose a second in 20 million years.

Table 1.1 Approximate Values of Some Measured Lengths

	Length (m)
Distance from Earth to most remote known quasar	1×10^{26}
Distance from Earth to most remote known normal galaxies	4×10^{25}
Distance from Earth to nearest large galaxy (M31, the Andromeda galaxy)	2×10^{22}
Distance from Earth to nearest star (Proxima Centauri)	4×10^{16}
One light year	9×10^{15}
Mean orbit radius of Earth about Sun	2×10^{11}
Mean distance from Earth to Moon	4×10^{8}
Mean radius of Earth	6×10^{6}
Typical altitude of satellite orbiting Earth	2×10^{5}
Length of football field	9×10^{1}
Length of housefly	5×10^{-3}
Size of smallest dust particles	1×10^{-4}
Size of cells in most living organisms	1×10^{-5}
Diameter of hydrogen atom	1×10^{-10}
Diameter of atomic nucleus	1×10^{-14}
Diameter of proton	1×10^{-15}

time, the second, was defined to be $(1/60)(1/60)(1/24) = 1/86\ 400$ of the average solar day. In 1967 the second was redefined to take advantage of the high precision attainable with an atomic clock, which uses the characteristic frequency of the light emitted from the cesium-133 atom as its "reference clock." **The second is now defined as 9 192 631 700 times the period of oscillation of radiation from the cesium atom.** The newest type of cesium atomic clock is shown in Figure 1.1b.

◀ Definition of the second

Approximate Values for Length, Mass, and Time Intervals

Approximate values of some lengths, masses, and time intervals are presented in Tables 1.1, 1.2, and 1.3, respectively. Note the wide ranges of values. Study these tables to get a feel for a kilogram of mass (this book has a mass of about 2 kilograms), a time interval of 10^{10} seconds (one century is about 3×10^9 seconds), or 2 meters of length (the approximate height of a forward on a basketball team). Appendix A reviews the notation for powers of 10, such as the expression of the number 50 000 in the form 5×10^4.

Systems of units commonly used in physics are the Système International, in which the units of length, mass, and time are the meter (m), kilogram (kg), and second (s); the cgs, or Gaussian, system, in which the units of length, mass, and time

Table 1.2 Approximate Values of Some Masses

	Mass (kg)
Observable Universe	1×10^{52}
Milky Way galaxy	7×10^{41}
Sun	2×10^{30}
Earth	6×10^{24}
Moon	7×10^{22}
Shark	1×10^{2}
Human	7×10^{1}
Frog	1×10^{-1}
Mosquito	1×10^{-5}
Bacterium	1×10^{-15}
Hydrogen atom	2×10^{-27}
Electron	9×10^{-31}

Table 1.3 Approximate Values of Some Time Intervals

	Time Interval (s)
Age of Universe	5×10^{17}
Age of Earth	1×10^{17}
Average age of college student	6×10^{8}
One year	3×10^{7}
One day	9×10^{4}
Time between normal heartbeats	8×10^{-1}
Period[a] of audible sound waves	1×10^{-3}
Period[a] of typical radio waves	1×10^{-6}
Period[a] of vibration of atom in solid	1×10^{-13}
Period[a] of visible light waves	2×10^{-15}
Duration of nuclear collision	1×10^{-22}
Time required for light to travel across a proton	3×10^{-24}

[a]A *period* is defined as the time required for one complete vibration.

Table 1.4 Some Prefixes for Powers of Ten Used with "Metric" (SI and cgs) Units

Power	Prefix	Abbreviation
10^{-18}	atto-	a
10^{-15}	femto-	f
10^{-12}	pico-	p
10^{-9}	nano-	n
10^{-6}	micro-	μ
10^{-3}	milli-	m
10^{-2}	centi-	c
10^{-1}	deci-	d
10^{1}	deka-	da
10^{3}	kilo-	k
10^{6}	mega-	M
10^{9}	giga-	G
10^{12}	tera-	T
10^{15}	peta-	P
10^{18}	exa-	E

are the centimeter (cm), gram (g), and second; and the U.S. customary system, in which the units of length, mass, and time are the foot (ft), slug, and second. SI units are almost universally accepted in science and industry, and will be used throughout the book. Limited use will be made of Gaussian and U.S. customary units.

Some of the most frequently used "metric" (SI and cgs) prefixes representing powers of 10 and their abbreviations are listed in Table 1.4. For example, 10^{-3} m is equivalent to 1 millimeter (mm), and 10^3 m is 1 kilometer (km). Likewise, 1 kg is equal to 10^3 g, and 1 megavolt (MV) is 10^6 volts (V). It's a good idea to memorize the more common prefixes early on: femto- to centi-, and kilo- to giga- are used routinely by most physicists.

1.2 The Building Blocks of Matter

LEARNING OBJECTIVES

1. State the fundamental components of matter.
2. Describe qualitatively the levels of structure of matter.

A 1-kg ($\approx$ 2-lb) cube of solid gold has a length of about 3.73 cm ($\approx$ 1.5 in.) on a side. If the cube is cut in half, the two resulting pieces retain their chemical identity. But what happens if the pieces of the cube are cut again and again, indefinitely? The Greek philosophers Leucippus and Democritus couldn't accept the idea that such cutting could go on forever. They speculated that the process ultimately would end when it produced a particle that could no longer be cut. In Greek, *atomos* means "not sliceable." From this term comes our English word *atom*, once believed to be the smallest particle of matter but since found to be a composite of more elementary particles.

The atom can be naively visualized as a miniature solar system, with a dense, positively charged nucleus occupying the position of the Sun and negatively charged electrons orbiting like planets. This model of the atom, first developed by the great Danish physicist Niels Bohr nearly a century ago, led to the understanding of certain properties of the simpler atoms such as hydrogen but failed to explain many fine details of atomic structure.

Notice the size of a hydrogen atom, listed in Table 1.1, and the size of a proton—the nucleus of a hydrogen atom—one hundred thousand times smaller. If the proton were the size of a ping-pong ball, the electron would be a tiny speck about the size of a bacterium, orbiting the proton a kilometer away! Other atoms are similarly constructed. So there is a surprising amount of empty space in ordinary matter.

After the discovery of the nucleus in the early 1900s, questions arose concerning its structure. Although the structure of the nucleus remains an area of active research even today, by the early 1930s scientists determined that two basic entities—protons and neutrons—occupy the nucleus. The *proton* is nature's most common carrier of positive charge, equal in magnitude but opposite in sign to the charge on the electron. The number of protons in a nucleus determines what the element is. For instance, a nucleus containing only one proton is the nucleus of an atom of hydrogen, regardless of how many neutrons may be present. Extra neutrons correspond to different isotopes of hydrogen—deuterium and tritium—which react chemically in exactly the same way as hydrogen, but are more massive. An atom having two protons in its nucleus, similarly, is always helium, although again, differing numbers of neutrons are possible.

The existence of *neutrons* was verified conclusively in 1932. A neutron has no charge and has a mass about equal to that of a proton. Except for hydrogen, all atomic nuclei contain neutrons, which, together with the protons, interact through the strong nuclear force. That force opposes the strongly repulsive electrical force of the protons, which otherwise would cause the nucleus to disintegrate.

The division doesn't stop here; strong evidence collected over many years indicates that protons, neutrons, and a zoo of other exotic particles are composed of six particles called **quarks** (rhymes with "sharks" though some rhyme it with "forks"). These particles have been given the names *up, down, strange, charm, bottom,* and *top.* The up, charm, and top quarks each carry a charge equal to $+\frac{2}{3}$ that of the proton, whereas the down, strange, and bottom quarks each carry a charge equal to $-\frac{1}{3}$ the proton charge. The proton consists of two up quarks and one down quark (see Fig. 1.2), giving the correct charge for the proton, +1. The neutron is composed of two down quarks and one up quark and has a net charge of zero.

The up and down quarks are sufficient to describe all normal matter, so the existence of the other four quarks, indirectly observed in high-energy experiments, is something of a mystery. Despite strong indirect evidence, no isolated quark has ever been observed. Consequently, the possible existence of yet more fundamental particles remains purely speculative.

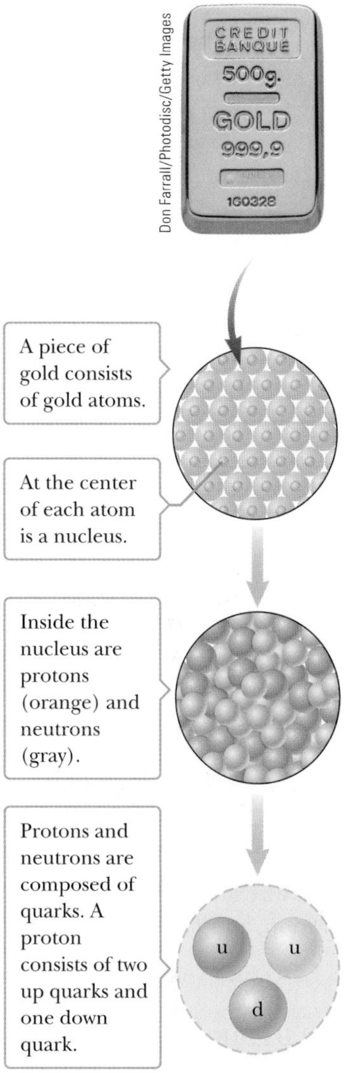

Figure 1.2 Levels of organization in matter.

1.3 Dimensional Analysis

LEARNING OBJECTIVES

1. State the definition of a dimension and give examples of the dimensions of some basic physical quantities.
2. Use dimensions to check equations for consistency.
3. Use dimensions to derive relationships between physical quantities.

In physics the word *dimension* denotes the physical nature of a quantity. The distance between two points, for example, can be measured in feet, meters, or furlongs, which are different ways of expressing the dimension of *length.*

The symbols used in this section to specify the dimensions of length, mass, and time are L, M, and T, respectively. Brackets [ ] will often be used to denote the dimensions of a physical quantity. In this notation, for example, the dimensions of velocity v are written $[v] = \mathrm{L/T}$, and the dimensions of area A are $[A] = \mathrm{L}^2$. The dimensions of area, volume, velocity, and acceleration are listed in Table 1.5, along with their units in the three common systems. The dimensions of other quantities, such as force and energy, will be described later as they are introduced.

In physics it's often necessary to deal with mathematical expressions that relate different physical quantities. One way to analyze such expressions, called **dimensional analysis,** makes use of the fact that **dimensions can be treated as algebraic quantities.** Adding masses to lengths, for example, makes no sense, so it follows that quantities can be added or subtracted only if they have the same dimensions. If the terms on the opposite sides of an equation have the same dimensions, then that equation may be correct, although correctness can't be guaranteed on the basis of dimensions alone. Nonetheless, dimensional analysis has value as a partial check of an equation and can also be used to develop insight into the relationships between physical quantities.

The procedure can be illustrated by developing some relationships between acceleration, velocity, time, and distance. Distance x has the dimension of length: $[x] = \mathrm{L}$. Time t has dimension $[t] = \mathrm{T}$. Velocity v has the dimensions length over

Table 1.5 Dimensions and Some Units of Area, Volume, Velocity, and Acceleration

System	Area (L^2)	Volume (L^3)	Velocity (L/T)	Acceleration ($\mathrm{L/T}^2$)
SI	m^2	m^3	m/s	$\mathrm{m/s}^2$
cgs	cm^2	cm^3	cm/s	$\mathrm{cm/s}^2$
U.S. customary	ft^2	ft^3	ft/s	$\mathrm{ft/s}^2$

time: $[v] = \mathrm{L/T}$, and acceleration the dimensions length divided by time squared: $[a] = \mathrm{L/T^2}$. Notice that velocity and acceleration have similar dimensions, except for an extra dimension of time in the denominator of acceleration. It follows that

$$[v] = \frac{\mathrm{L}}{\mathrm{T}} = \frac{\mathrm{L}}{\mathrm{T}^2}\mathrm{T} = [a][t]$$

From this it might be guessed that velocity equals acceleration multiplied by time, $v = at$, and that is true for the special case of motion with constant acceleration starting at rest. Noticing that velocity has dimensions of length divided by time and distance has dimensions of length, it's reasonable to guess that

$$[x] = \mathrm{L} = \mathrm{L}\frac{\mathrm{T}}{\mathrm{T}} = \frac{\mathrm{L}}{\mathrm{T}}\mathrm{T} = [v][t] = [a][t]^2$$

Here it appears that $x = at^2$ might correctly relate the distance traveled to acceleration and time; however, that equation is not even correct in the case of constant acceleration starting from rest. The correct expression in that case is $x = \frac{1}{2}at^2$. These examples serve to show the inherent limitations in using dimensional analysis to discover relationships between physical quantities. Nonetheless, such simple procedures can still be of value in developing a preliminary mathematical model for a given physical system. Further, because it's easy to make errors when solving problems, dimensional analysis can be used to check the consistency of the results. When the dimensions in an equation are not consistent, it indicates an error has been made in a prior step.

EXAMPLE 1.1 | Analysis of an Equation

GOAL Check an equation using dimensional analysis.

PROBLEM Show that the expression $v = v_0 + at$ is dimensionally correct, where v and v_0 represent velocities, a is acceleration, and t is a time interval.

STRATEGY Analyze each term, finding its dimensions, and then check to see if all the terms agree with each other.

SOLUTION

Find dimensions for v and v_0.

$$[v] = [v_0] = \frac{\mathrm{L}}{\mathrm{T}}$$

Find the dimensions of at.

$$[at] = [a][t] = \frac{\mathrm{L}}{\mathrm{T}^2}(\mathrm{T}) = \frac{\mathrm{L}}{\mathrm{T}}$$

REMARKS All the terms agree, so the equation is dimensionally correct.

QUESTION 1.1 True or False. An equation that is dimensionally correct is always physically correct, up to a constant of proportionality.

EXERCISE 1.1 Determine whether the equation $x = vt^2$ is dimensionally correct. If not, provide a correct expression, up to an overall constant of proportionality.

ANSWER Incorrect. The expression $x = vt$ is dimensionally correct.

EXAMPLE 1.2 | Find an Equation

GOAL Derive an equation by using dimensional analysis.

PROBLEM Find a relationship between an acceleration of constant magnitude a, speed v, and distance r from the origin for a particle traveling in a circle.

STRATEGY Start with the term having the most dimensionality, a. Find its dimensions, and then rewrite those dimensions in terms of the dimensions of v and r. The dimensions of time will have to be eliminated with v, because that's the only quantity (other than a, itself) in which the dimension of time appears.

SOLUTION

Write down the dimensions of a:

$$[a] = \frac{\text{L}}{\text{T}^2}$$

Solve the dimensions of speed for T:

$$[v] = \frac{\text{L}}{\text{T}} \quad \rightarrow \quad \text{T} = \frac{L}{[v]}$$

Substitute the expression for T into the equation for $[a]$:

$$[a] = \frac{\text{L}}{\text{T}^2} = \frac{\text{L}}{(\text{L}/[v])^2} = \frac{[v]^2}{\text{L}}$$

Substitute $\text{L} = [r]$, and guess at the equation:

$$[a] = \frac{[v]^2}{[r]} \quad \rightarrow \quad a = \frac{v^2}{r}$$

REMARKS This is the correct equation for the magnitude of the centripetal acceleration—acceleration toward the center of motion—to be discussed in Chapter 7. In this case it isn't necessary to introduce a numerical factor. Such a factor is often displayed explicitly as a constant k in front of the right-hand side; for example, $a = kv^2/r$. As it turns out, $k = 1$ gives the correct expression. A good technique sometimes introduced in calculus-based textbooks involves using unknown powers of the dimensions. This problem would then be set up as $[a] = [v]^b[r]^c$. Writing out the dimensions and equating powers of each dimension on both sides of the equation would result in $b = 2$ and $c = -1$.

QUESTION 1.2 True or False: Replacing v by r/t in the final answer also gives a dimensionally correct equation.

EXERCISE 1.2 In physics, energy E carries dimensions of mass times length squared divided by time squared. Use dimensional analysis to derive a relationship for energy in terms of mass m and speed v, up to a constant of proportionality. Set the speed equal to c, the speed of light, and the constant of proportionality equal to 1 to get the most famous equation in physics. (Note, however, that the first relationship is associated with energy of motion, and the second with energy of mass. See Chapter 26.)

ANSWER $E = kmv^2 \quad \rightarrow \quad E = mc^2$ when $k = 1$ and $v = c$.

1.4 Uncertainty in Measurement and Significant Figures

LEARNING OBJECTIVES

1. Identify the number of significant figures in a given physical measurement.
2. Apply significant figures to estimate the proper accuracy of a combination of physical measurements.

Physics is a science in which mathematical laws are tested by experiment. No physical quantity can be determined with complete accuracy because our senses are physically limited, even when extended with microscopes, cyclotrons, and other instruments. Consequently, it's important to develop methods of determining the accuracy of measurements.

All measurements have uncertainties associated with them, whether or not they are explicitly stated. The accuracy of a measurement depends on the sensitivity of the apparatus, the skill of the person carrying out the measurement, and the number of times the measurement is repeated. Once the measurements, along with their uncertainties, are known, it's often the case that calculations must be carried out using those measurements. Suppose two such measurements are multiplied. When a calculator is used to obtain this product, there may be eight digits in the calculator window, but often only two or three of those numbers have any significance. The rest have no value because they imply greater accuracy than was actually achieved in the original measurements. In experimental work, determining how many numbers to retain requires the application of statistics and the mathematical propagation of uncertainties. In a textbook it isn't practical to apply those sophisticated tools in the numerous calculations, so instead a simple method, called ***significant figures,*** is used to indicate the approximate number of digits that should be retained at the end of a calculation. Although that method is not mathematically rigorous, it's easy to apply and works fairly well.

Suppose that in a laboratory experiment we measure the area of a rectangular plate with a meter stick. Let's assume that the accuracy to which we can measure a particular dimension of the plate is ± 0.1 cm. If the length of the plate is measured to be 16.3 cm, we can only claim that it lies somewhere between 16.2 cm and 16.4 cm. In this case, we say the measured value has three significant figures. Likewise, if the plate's width is measured to be 4.5 cm, the actual value lies between 4.4 cm and 4.6 cm. This measured value has only two significant figures. We could write the measured values as 16.3 ± 0.1 cm and 4.5 ± 0.1 cm. In general, **a significant figure is a reliably known digit** (other than a zero used to locate a decimal point). Note that in each case, the final number has some uncertainty associated with it, and is therefore not 100% reliable. Despite the uncertainty, that number is retained and considered significant because it does convey some information.

Suppose we would like to find the area of the plate by multiplying the two measured values together. The final value can range between $(16.3 - 0.1 \text{ cm})(4.5 - 0.1 \text{ cm}) = (16.2 \text{ cm})(4.4 \text{ cm}) = 71.28 \text{ cm}^2$ and $(16.3 + 0.1 \text{ cm})(4.5 + 0.1 \text{ cm}) = (16.4 \text{ cm})(4.6 \text{ cm}) = 75.44 \text{ cm}^2$. Claiming to know anything about the hundredths place, or even the tenths place, doesn't make any sense, because it's clear we can't even be certain of the units place, whether it's the 1 in 71, the 5 in 75, or somewhere in between. The tenths and the hundredths places are clearly not significant. We have *some* information about the units place, so that number is significant. Multiplying the numbers at the middle of the uncertainty ranges gives $(16.3 \text{ cm})(4.5 \text{ cm}) = 73.35 \text{ cm}^2$, which is also in the middle of the area's uncertainty range. Because the hundredths and tenths are not significant, we drop them and take the answer to be 73 cm^2, with an uncertainty of $\pm 2 \text{ cm}^2$. Note that the answer has two significant figures, the same number of figures as the least accurately known quantity being multiplied, the 4.5-cm width.

Calculations as carried out in the preceding paragraph can indicate the proper number of significant figures, but those calculations are time-consuming. Instead, two rules of thumb can be applied. The first, concerning multiplication and division, is as follows: **In multiplying (dividing) two or more quantities, the number of significant figures in the final product (quotient) is the same as the number of significant figures in the *least accurate* of the factors being combined, where *least accurate* means *having the lowest number of significant figures.***

To get the final number of significant figures, it's usually necessary to do some rounding. If the last digit dropped is less than 5, simply drop the digit. If the last digit dropped is greater than or equal to 5, raise the last retained digit by one.[1]

Tip 1.2 Using Calculators

Calculators are designed by engineers to yield as many digits as the memory of the calculator chip permits, so be sure to round the final answer down to the correct number of significant figures.

Zeros may or may not be significant figures. Zeros used to position the decimal point in such numbers as 0.03 and 0.007 5 are not considered significant figures. Hence, 0.03 has one significant figure, and 0.007 5 has two.

When zeros are placed after other digits in a whole number, there is a possibility of misinterpretation. For example, suppose the mass of an object is given as 1 500 g. This value is ambiguous, because we don't know whether the last two zeros are being used to locate the decimal point or whether they represent significant figures in the measurement.

Using scientific notation to indicate the number of significant figures removes this ambiguity. In this case, we express the mass as 1.5×10^3 g if there are two significant figures in the measured value, 1.50×10^3 g if there are three significant figures, and 1.500×10^3 g if there are four. Likewise, 0.000 15 is expressed in scientific notation as 1.5×10^{-4} if it has two significant figures or as 1.50×10^{-4} if it has

[1]Some prefer to round to the nearest even digit when the last dropped digit is 5, which has the advantage of rounding 5 up half the time and down half the time. For example, 1.55 would round to 1.6, but 1.45 would round to 1.4. Because the final significant figure is only one representative of a range of values given by the uncertainty, this very slight refinement will not be used in this text.

three significant figures. The three zeros between the decimal point and the digit 1 in the number 0.000 15 are not counted as significant figures because they only locate the decimal point. Similarly, trailing zeros are not considered significant. However, any zeros written after a decimal point are considered significant. For example, 3.00, 30.0, and 300. have three significant figures, whereas 300 has only one. In this book, **most of the numerical examples and end-of-chapter problems will yield answers having two or three significant figures.**

For addition and subtraction, it's best to focus on the number of decimal places in the quantities involved rather than on the number of significant figures. **When numbers are added (subtracted), the number of decimal places in the result should equal the smallest number of decimal places of any term in the sum (difference).** For example, if we wish to compute 123 (zero decimal places) + 5.35 (two decimal places), the answer is 128 (zero decimal places) and not 128.35. If we compute the sum 1.000 1 (four decimal places) + 0.000 3 (four decimal places) = 1.000 4, the result has the correct number of decimal places, namely four. Observe that the rules for multiplying significant figures don't work here because the answer has five significant figures even though one of the terms in the sum, 0.000 3, has only one significant figure. Likewise, if we perform the subtraction 1.002 − 0.998 = 0.004, the result has three decimal places because each term in the subtraction has three decimal places.

To show why this rule should hold, we return to the first example in which we added 123 and 5.35, and rewrite these numbers as 123.*xxx* and 5.35*x*. Digits written with an *x* are completely unknown and can be any digit from 0 to 9. Now we line up 123.*xxx* and 5.35*x* relative to the decimal point and perform the addition, using the rule that an unknown digit added to a known or unknown digit yields an unknown:

$$\begin{array}{r} 123.xxx \\ +\quad 5.35x \\ \hline 128.xxx \end{array}$$

The answer of 128.*xxx* means that we are justified only in keeping the number 128 because everything after the decimal point in the sum is actually unknown. The example shows that the controlling uncertainty is introduced into an addition or subtraction by the term with the smallest number of decimal places.

EXAMPLE 1.3 | Carpet Calculations

GOAL Apply the rules for significant figures.

PROBLEM Several carpet installers make measurements for carpet installation in the different rooms of a restaurant, reporting their measurements with inconsistent accuracy, as compiled in Table 1.6. Compute the areas for **(a)** the banquet hall, **(b)** the meeting room, and **(c)** the dining room, taking into account significant figures. **(d)** What total area of carpet is required for these rooms?

Table 1.6 Dimensions of Rooms in Example 1.3

	Length (m)	Width (m)
Banquet hall	14.71	7.46
Meeting room	4.822	5.1
Dining room	13.8	9

STRATEGY For the multiplication problems in parts (a)–(c), count the significant figures in each number. The smaller result is the number of significant figures in the answer. Part (d) requires a sum, where the area with the least accurately known decimal place determines the overall number of significant figures in the answer.

SOLUTION

(a) Compute the area of the banquet hall.

Count significant figures:

14.71 m → 4 significant figures
7.46 m → 3 significant figures

To find the area, multiply the numbers keeping only three digits:

$14.71\text{ m} \times 7.46\text{ m} = 109.74\text{ m}^2 \rightarrow 1.10 \times 10^2\text{ m}^2$

(Continued)

(b) Compute the area of the meeting room.

Count significant figures: $4.822 \text{ m} \rightarrow$ 4 significant figures; $5.1 \text{ m} \rightarrow$ 2 significant figures

To find the area, multiply the numbers keeping only two digits: $4.822 \text{ m} \times 5.1 \text{ m} = 24.59 \text{ m}^2 \rightarrow$ 25 m^2

(c) Compute the area of the dining room.

Count significant figures: $13.8 \text{ m} \rightarrow$ 3 significant figures; $9 \text{ m} \rightarrow$ 1 significant figure

To find the area, multiply the numbers keeping only one digit: $13.8 \text{ m} \times 9 \text{ m} = 124.2 \text{ m}^2 \rightarrow$ 100 m^2

(d) Calculate the total area of carpet required, with the proper number of significant figures.

Sum all three answers without regard to significant figures: $1.10 \times 10^2 \text{ m}^2 + 25 \text{ m}^2 + 100 \text{ m}^2 = 235 \text{ m}^2$

The least accurate number is 100 m^2, with one significant figure in the hundred's decimal place: $235 \text{ m}^2 \rightarrow$ $2 \times 10^2 \text{ m}^2$

REMARKS Notice that the final answer in part (d) has only one significant figure, in the hundred's place, resulting in an answer that had to be rounded down by a sizable fraction of its total value. That's the consequence of having insufficient information. The value of 9 m, without any further information, represents a true value that could be anywhere in the interval [8.5 m, 9.5 m), all of which round to 9 when only one digit is retained.

QUESTION 1.3 How would the final answer change if the width of the dining room were given as 9.0 m?

EXERCISE 1.3 A ranch has two fenced rectangular areas. Area A has a length of 750 m and width 125 m, and area B has length 400 m and width 150 m. Find (a) area A, (b) area B, and (c) the total area, with attention to the rules of significant figures. Assume trailing zeros are not significant.

ANSWERS (a) $9.4 \times 10^4 \text{ m}^2$ (b) $6 \times 10^4 \text{ m}^2$ (c) $1.5 \times 10^5 \text{ m}^2$

In performing any calculation, especially one involving a number of steps, there will always be slight discrepancies introduced by both the rounding process and the algebraic order in which steps are carried out. For example, consider $2.35 \times 5.89/1.57$. This computation can be performed in three different orders. First, we have $2.35 \times 5.89 = 13.842$, which rounds to 13.8, followed by $13.8/1.57 = 8.789\,8$, rounding to 8.79. Second, $5.89/1.57 = 3.751\,6$, which rounds to 3.75, resulting in $2.35 \times 3.75 = 8.812\,5$, rounding to 8.81. Finally, $2.35/1.57 = 1.496\,8$ rounds to 1.50, and $1.50 \times 5.89 = 8.835$ rounds to 8.84. So three different algebraic orders, following the rules of rounding, lead to answers of 8.79, 8.81, and 8.84, respectively. Such minor discrepancies are to be expected, because the last significant digit is only one representative from a range of possible values, depending on experimental uncertainty. To avoid such discrepancies, some carry one or more extra digits during the calculation, although it isn't conceptually consistent to do so because those extra digits are not significant. As a practical matter, in the worked examples in this text, intermediate reported results will be rounded to the proper number of significant figures, and only those digits will be carried forward. In the problem sets, however, given data will usually be assumed accurate to two or three digits, even when there are trailing zeros. **In solving the problems, the student should be aware that slight differences in rounding practices can result in answers varying from the text in the last significant digit, which is normal and not cause for**

7. Use the rules for significant figures to find the answer to the addition problem 21.4 + 15 + 17.17 + 4.003. (See Section 1.4).

8. Find the polar coordinates corresponding to a point located at (−5.00, 12.00) in Cartesian coordinates. (See Section 1.7.)

9. At a horizontal distance of 45 m from the bottom of a tree, the angle of elevation to the top of the tree is 26°. How tall is the tree? (See Section 1.8.)

CONCEPTUAL QUESTIONS

ENHANCED WebAssign The conceptual questions in this chapter may be assigned online in Enhanced WebAssign.

1. Estimate the order of magnitude of the length, in meters, of each of the following: (a) a mouse, (b) a pool cue, (c) a basketball court, (d) an elephant, (e) a city block.

2. What types of natural phenomena could serve as time standards?

3. Find the order of magnitude of your age in seconds.

4. An object with a mass of 1 kg weighs approximately 2 lb. Use this information to estimate the mass of the following objects: (a) a baseball; (b) your physics textbook; (c) a pickup truck.

5. BIO (a) Estimate the number of times your heart beats in a month. (b) Estimate the number of human heartbeats in an average lifetime.

6. Estimate the number of atoms in 1 cm^3 of a solid. (Note that the diameter of an atom is about 10^{-10} m.)

7. The height of a horse is sometimes given in units of "hands." Why is this a poor standard of length?

8. How many of the lengths or time intervals given in Tables 1.2 and 1.3 could you verify, using only equipment found in a typical dormitory room?

9. (a) If an equation is dimensionally correct, does this mean that the equation must be true? (b) If an equation is not dimensionally correct, does this mean that the equation can't be true? Explain your answers.

10. Why is the metric system of units considered superior to most other systems of units?

11. How can an estimate be of value even when it is off by an order of magnitude? Explain and give an example.

12. Suppose two quantities, A and B, have different dimensions. Determine which of the following arithmetic operations *could* be physically meaningful. (a) $A + B$ (b) $B - A$ (c) $A - B$ (d) A/B (e) AB

13. Answer each question yes or no. Must two quantities have the same dimensions (a) if you are adding them? (b) If you are multiplying them? (c) If you are subtracting them? (d) If you are dividing them? (e) If you are equating them?

PROBLEMS

ENHANCED WebAssign The problems in this chapter may be assigned online in Enhanced WebAssign.

1. denotes straightforward problem; 2. denotes intermediate problem;
3. denotes challenging problem
1. denotes full solution available in *Student Solutions Manual/Study Guide*
1. denotes problems most often assigned in Enhanced WebAssign

BIO denotes biomedical problems
GP denotes guided problems
M denotes Master It tutorial available in Enhanced WebAssign
Q|C denotes asking for quantitative and conceptual reasoning
S denotes symbolic reasoning problem
W denotes Watch It video solution available in Enhanced WebAssign

1.3 Dimensional Analysis

1. The period of a simple pendulum, defined as the time necessary for one complete oscillation, is measured in time units and is given by

$$T = 2\pi\sqrt{\frac{\ell}{g}}$$

where ℓ is the length of the pendulum and g is the acceleration due to gravity, in units of length divided by time squared. Show that this equation is dimensionally consistent. (You might want to check the formula using your keys at the end of a string and a stopwatch.)

2. (a) Suppose that the displacement of an object is related to time according to the expression $x = Bt^2$. What are the dimensions of B? (b) A displacement is related to time as $x = A \sin(2\pi ft)$, where A and f are constants. Find the dimensions of A. (*Hint:* A trigonometric function appearing in an equation must be dimensionless.)

3. S A shape that covers an area A and has a uniform height h has a volume $V = Ah$. (a) Show that $V = Ah$ is dimensionally correct. (b) Show that the volumes of a cylinder and of a rectangular box can be written in the form $V = Ah$, identifying A in each case.

(Note that A, sometimes called the "footprint" of the object, can have any shape and that the height can, in general, be replaced by the average thickness of the object.)

4. Each of the following equations was given by a student during an examination: (a) $\frac{1}{2}mv^2 = \frac{1}{2}mv_0{}^2 + \sqrt{mgh}$ (b) $v = v_0 + at^2$ (c) $ma = v^2$. Do a dimensional analysis of each equation and explain why the equation can't be correct.

5. Newton's law of universal gravitation is represented by

$$F = G\frac{Mm}{r^2}$$

where F is the gravitational force, M and m are masses, and r is a length. Force has the SI units kg · m/s². What are the SI units of the proportionality constant G?

6. QIC Kinetic energy KE (Chapter 5) has dimensions kg · m²/s². It can be written in terms of the momentum p (Chapter 6) and mass m as

$$KE = \frac{p^2}{2m}$$

(a) Determine the proper units for momentum using dimensional analysis. (b) Refer to Problem 5. Given the units of force, write a simple equation relating a constant force F exerted on an object, an interval of time t during which the force is applied, and the resulting momentum of the object, p.

1.4 Uncertainty in Measurement and Significant Figures

7. W A carpet is to be installed in a room of length 9.72 m and width 5.3 m. Find the area of the room retaining the proper number of significant figures.

8. QIC Use your calculator to determine $(\sqrt{8})^3$ to three significant figures in two ways: (a) Find $\sqrt{8}$ to four significant figures; then cube this number and round to three significant figures. (b) Find $\sqrt{8}$ to three significant figures; then cube this number and round to three significant figures. (c) Which answer is more accurate? Explain.

9. How many significant figures are there in (a) 78.9 ± 0.2, (b) 3.788×10^9, (c) 2.46×10^{-6}, (d) 0.003 2

10. The speed of light is now defined to be $2.997\,924\,58 \times 10^8$ m/s. Express the speed of light to (a) three significant figures, (b) five significant figures, and (c) seven significant figures.

11. QIC A block of gold has length 5.62 cm, width 6.35 cm, and height 2.78 cm. (a) Calculate the length times the width and round the answer to the appropriate number of significant figures. (b) Now multiply the rounded result of part (a) by the height and again round, obtaining the volume. (c) Repeat the process, first finding the width times the height, rounding it, and then obtaining the volume by multiplying by the length. (d) Explain why the answers don't agree in the third significant figure.

12. The radius of a circle is measured to be (10.5 ± 0.2) m. Calculate (a) the area and (b) the circumference of the circle, and give the uncertainty in each value.

13. The edges of a shoebox are measured to be 11.4 cm, 17.8 cm, and 29 cm. Determine the volume of the box retaining the proper number of significant figures in your answer.

14. Carry out the following arithmetic operations: (a) the sum of the measured values 756, 37.2, 0.83, and 2.5; (b) the product $0.003\,2 \times 356.3$; (c) the product $5.620 \times \pi$.

1.5 Conversion of Units

15. A fathom is a unit of length, usually reserved for measuring the depth of water. A fathom is approximately 6 ft in length. Take the distance from Earth to the Moon to be 250 000 miles, and use the given approximation to find the distance in fathoms.

16. A small turtle moves at a speed of 186 furlongs per fortnight. Find the speed of the turtle in centimeters per second. Note that 1 furlong = 220 yards and 1 fortnight = 14 days.

17. A firkin is an old British unit of volume equal to 9 gallons. How many cubic meters are there in 6.00 firkins?

18. Find the height or length of these natural wonders in kilometers, meters, and centimeters: (a) The longest cave system in the world is the Mammoth Cave system in Central Kentucky, with a mapped length of 348 miles. (b) In the United States, the waterfall with the greatest single drop is Ribbon Falls in California, which drops 1 612 ft. (c) At 20 320 feet, Mount McKinley in Alaska is America's highest mountain. (d) The deepest canyon in the United States is King's Canyon in California, with a depth of 8 200 ft.

19. A car is traveling at a speed of 38.0 m/s on an interstate highway where the speed limit is 75.0 mi/h. Is the driver exceeding the speed limit? Justify your answer.

20. A certain car has a fuel efficiency of 25.0 miles per gallon (mi/gal). Express this efficiency in kilometers per liter (km/L).

21. The diameter of a sphere is measured to be 5.36 in. Find (a) the radius of the sphere in centimeters, (b) the surface area of the sphere in square centimeters, and (c) the volume of the sphere in cubic centimeters.

22. W BIO Suppose your hair grows at the rate of 1/32 inch per day. Find the rate at which it grows in nanometers per second. Because the distance between atoms in a molecule is on the order of 0.1 nm, your answer suggests how rapidly atoms are assembled in this protein synthesis.

23. The speed of light is about 3.00×10^8 m/s. Convert this figure to miles per hour.

24. M A house is 50.0 ft long and 26 ft wide and has 8.0-ft-high ceilings. What is the volume of the interior of the house in cubic meters and in cubic centimeters?

25. The amount of water in reservoirs is often measured in acre-ft. One acre-ft is a volume that covers an area of one acre to a depth of one foot. An acre is 43 560 ft^2. Find the volume in SI units of a reservoir containing 25.0 acre-ft of water.

26. The base of a pyramid covers an area of 13.0 acres (1 acre = 43 560 ft^2) and has a height of 481 ft (Fig. P1.26). If the volume of a pyramid is given by the expression $V = bh/3$, where b is the area of the base and h is the height, find the volume of this pyramid in cubic meters.

Sophie McAulay/Shutterstock.com

Figure P1.26

27. A quart container of ice cream is to be made in the form of a cube. What should be the length of a side, in centimeters? (Use the conversion 1 gallon = 3.786 liter.)

1.6 Estimates and Order-of-Magnitude Calculations

Note: In developing answers to the problems in this section, you should state your important assumptions, including the numerical values assigned to parameters used in the solution.

28. Estimate the number of steps you would have to take to walk a distance equal to the circumference of the Earth.

29. BIO Estimate the number of breaths taken by a human being during an average lifetime.

30. BIO Estimate the number of people in the world who are suffering from the common cold on any given day. (Answers may vary. Remember that a person suffers from a cold for about a week.)

31. BIO Q|C (a) About how many microorganisms are found in the human intestinal tract? (A typical bacterial length scale is one micron = 10^{-6} m. Estimate the intestinal volume and assume bacteria occupy one hundredth of it.) (b) Discuss your answer to part (a). Are these bacteria beneficial, dangerous, or neutral? What functions could they serve?

32. BIO Treat a cell in a human as a sphere of radius 1.0 μm. (a) Determine the volume of a cell. (b) Estimate the volume of your body. (c) Estimate the number of cells in your body.

33. An automobile tire is rated to last for 50 000 miles. Estimate the number of revolutions the tire will make in its lifetime.

34. BIO Q|C Bacteria and other prokaryotes are found deep underground, in water, and in the air. One micron (10^{-6} m) is a typical length scale associated with these microbes. (a) Estimate the total number of bacteria and other prokaryotes in the biosphere of the Earth. (b) Estimate the total mass of all such microbes. (c) Discuss the relative importance of humans and microbes to the ecology of planet Earth. Can *Homo sapiens* survive without them?

1.7 Coordinate Systems

35. M A point is located in a polar coordinate system by the coordinates $r = 2.5$ m and $\theta = 35°$. Find the x- and y-coordinates of this point, assuming that the two coordinate systems have the same origin.

36. A certain corner of a room is selected as the origin of a rectangular coordinate system. If a fly is crawling on an adjacent wall at a point having coordinates (2.0, 1.0), where the units are meters, what is the distance of the fly from the corner of the room?

37. Express the location of the fly in Problem 36 in polar coordinates.

38. W Two points in a rectangular coordinate system have the coordinates (5.0, 3.0) and (−3.0, 4.0), where the units are centimeters. Determine the distance between these points.

39. Two points are given in polar coordinates by $(r, \theta) = (2.00\text{ m}, 50.0°)$ and $(r, \theta) = (5.00\text{ m}, -50.0°)$, respectively. What is the distance between them?

40. S Given points (r_1, θ_1) and (r_2, θ_2) in polar coordinates, obtain a general formula for the distance between them. Simplify it as much as possible using the identity $\cos^2\theta + \sin^2\theta = 1$. *Hint:* Write the expressions for the two points in Cartesian coordinates and substitute into the usual distance formula.

1.8 Trigonometry

41. M For the triangle shown in Figure P1.41, what are (a) the length of the unknown side, (b) the tangent of θ, and (c) the sine of ϕ?

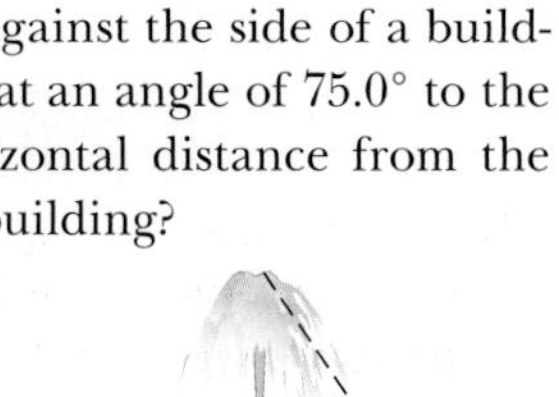

Figure P1.41

42. A ladder 9.00 m long leans against the side of a building. If the ladder is inclined at an angle of 75.0° to the horizontal, what is the horizontal distance from the bottom of the ladder to the building?

43. A high fountain of water is located at the center of a circular pool as shown in Figure P1.43. Not wishing to get his feet wet, a student walks around the pool and measures its circumference to be 15.0 m. Next, the student stands

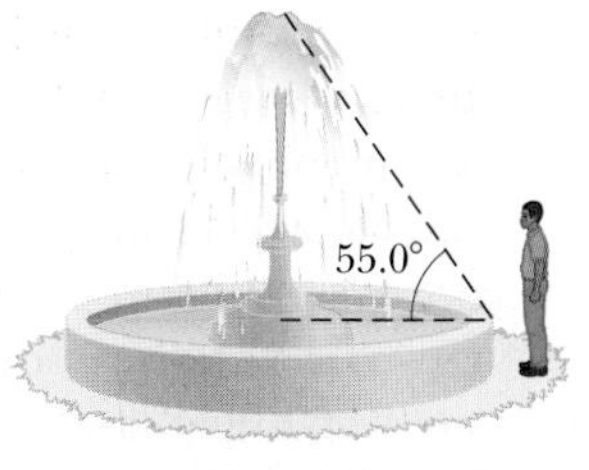

Figure P1.43

at the edge of the pool and uses a protractor to gauge the angle of elevation at the bottom of the fountain to be 55.0°. How high is the fountain?

44. W A right triangle has a hypotenuse of length 3.00 m, and one of its angles is 30.0°. What are the lengths of (a) the side opposite the 30.0° angle and (b) the side adjacent to the 30.0° angle?

45. In Figure P1.45, find (a) the side opposite θ, (b) the side adjacent to ϕ, (c) cos θ, (d) sin ϕ, and (e) tan ϕ.

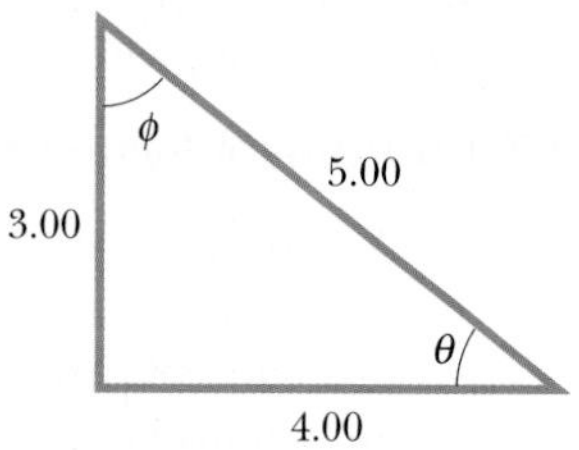

Figure P1.45

46. In a certain right triangle, the two sides that are perpendicular to each other are 5.00 m and 7.00 m long. What is the length of the third side of the triangle?

47. In Problem 46, what is the tangent of the angle for which 5.00 m is the opposite side?

48. GP S A woman measures the angle of elevation of a mountaintop as 12.0°. After walking 1.00 km closer to the mountain on level ground, she finds the angle to be 14.0°. (a) Draw a picture of the problem, neglecting the height of the woman's eyes above the ground. *Hint:* Use two triangles. (b) Select variable names for the mountain height (suggestion: y) and the woman's original distance from the mountain (suggestion: x) and label the picture. (c) Using the labeled picture and the tangent function, write two trigonometric equations relating the two selected variables. (d) Find the height y of the mountain by first solving one equation for x and substituting the result into the other equation.

49. A surveyor measures the distance across a straight river by the following method: Starting directly across from a tree on the opposite bank, he walks x = 100 m along the riverbank to establish a baseline. Then he sights across to the tree. The angle from his baseline to the tree is θ = 35.0° (Fig. P1.49). How wide is the river?

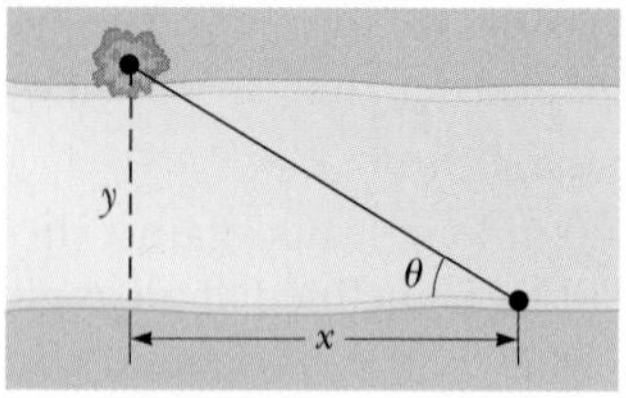

Figure P1.49

50. S Refer to Problem 48. Suppose the mountain height is y, the woman's original distance from the mountain is x, and the angle of elevation she measures from the horizontal to the top of the mountain is θ. If she moves a distance d closer to the mountain and measures an angle of elevation ϕ, find a general equation for the height of the mountain y in terms of d, ϕ, and θ, neglecting the height of her eyes above the ground.

Additional Problems

51. (a) One of the fundamental laws of motion states that the acceleration of an object is directly proportional to the resultant force on it and inversely proportional to its mass. If the proportionality constant is defined to have no dimensions, determine the dimensions of force. (b) The newton is the SI unit of force. According to the results for (a), how can you express a force having units of newtons by using the fundamental units of mass, length, and time?

52. (a) Find a conversion factor to convert from miles per hour to kilometers per hour. (b) For a while, federal law mandated that the maximum highway speed would be 55 mi/h. Use the conversion factor from part (a) to find the speed in kilometers per hour. (c) The maximum highway speed has been raised to 65 mi/h in some places. In kilometers per hour, how much of an increase is this over the 55-mi/h limit?

53. BIO One cubic centimeter (1.0 cm^3) of water has a mass of 1.0×10^{-3} kg. (a) Determine the mass of 1.0 m^3 of water. Assuming that biological substances are 98% water, estimate the masses of (b) a cell with a diameter of 1.0 μm, (c) a human kidney, and (d) a fly. Take a kidney to be roughly a sphere with a radius of 4.0 cm and a fly to be roughly a cylinder 4.0 mm long and 2.0 mm in diameter.

54. Soft drinks are commonly sold in aluminum containers. (a) To an order of magnitude, how many such containers are thrown away or recycled each year by U.S. consumers? (b) How many tons of aluminum does this represent? In your solution, state the quantities you measure or estimate and the values you take for them.

55. The displacement of an object moving under uniform acceleration is some function of time and the acceleration. Suppose we write this displacement as $s = ka^m t^n$, where k is a dimensionless constant. Show by dimensional analysis that this expression is satisfied if m = 1 and n = 2. Can the analysis give the value of k?

56. Assume that it takes 7.00 minutes to fill a 30.0-gal gasoline tank. (a) Calculate the rate at which the tank is filled in gallons per second. (b) Calculate the rate at which the tank is filled in cubic meters per second. (c) Determine the time interval, in hours, required to fill a 1.00-m^3 volume at the same rate. (1 U.S. gal = 231 $in.^3$)

57. M One gallon of paint (volume = 3.79×10^{-3} m^3) covers an area of 25.0 m^2. What is the thickness of the fresh paint on the wall?

58. S Sphere 1 has surface area A_1 and volume V_1, and sphere 2 has surface area A_2 and volume V_2. If the radius of sphere 2 is double the radius of sphere 1, what is the ratio of (a) the areas, A_2/A_1 and (b) the volumes, V_2/V_1?

59. M Assume that there are 100 million passenger cars in the United States and that the average fuel consumption is 20 mi/gal of gasoline. If the average distance traveled by each car is 10 000 mi/yr, how much gasoline would be saved per year if average fuel consumption could be increased to 25 mi/gal?

60. In 2013, the U.S. national debt was about $17 trillion. (a) If payments were made at the rate of $1 000 per second, how many years would it take to pay off the debt, assuming that no interest were charged? (b) A dollar bill is about 15.5 cm long. If seventeen trillion dollar bills were laid end to end around the Earth's equator, how many times would they encircle the planet? Take the radius of the Earth at the equator to be 6 378 km. (*Note:* Before doing any of these calculations, try to guess at the answers. You may be very surprised.)

61. (a) How many seconds are there in a year? (b) If one micrometeorite (a sphere with a diameter on the order of 10^{-6} m) struck each square meter of the Moon each second, estimate the number of years it would take to cover the Moon with micrometeorites to a depth of one meter. (*Hint:* Consider a cubic box, 1 m on a side, on the Moon, and find how long it would take to fill the box.)

62. Imagine that you are the equipment manager of a professional baseball team. One of your jobs is to keep baseballs on hand for games. Balls are sometimes lost when players hit them into the stands as either home runs or foul balls. Estimate how many baseballs you have to buy per season in order to make up for such losses. Assume that your team plays an 81-game home schedule in a season.

63. The nearest neutron star (a collapsed star made primarily of neutrons) is about 3.00×10^{18} m away from Earth. Given that the Milky Way galaxy (Fig. P1.63) is roughly a disk of diameter $\sim 10^{21}$ m and thickness $\sim 10^{19}$ m, estimate the number of neutron stars in the Milky Way to the nearest order of magnitude.

Richard Payne/NASA

Figure P1.63

The current absolute land speed record holder is the British designed ThrustSSC, a twin turbofan-powered car which achieved 763 miles per hour (1,228 km/h) for the measured mile (1.6 km), breaking the sound barrier. The car was driven by Andy Green (UK) on 10/15/1997 in the Black Rock Desert in Gerlach, Nevada.

AP Photo/Ben Margot

2 Motion in One Dimension

Life is motion. Our muscles coordinate motion microscopically to enable us to walk and jog. Our hearts pump tirelessly for decades, moving blood through our bodies. Cell wall mechanisms move select atoms and molecules in and out of cells. From the prehistoric chase of antelopes across the savanna to the pursuit of satellites in space, mastery of motion has been critical to our survival and success as a species.

The study of motion and of physical concepts such as force and mass is called **dynamics.** The part of dynamics that describes motion without regard to its causes is called **kinematics.** In this chapter the focus is on kinematics in one dimension: motion along a straight line. This kind of motion—and, indeed, *any* motion—involves the concepts of displacement, velocity, and acceleration. Here, we use these concepts to study the motion of objects undergoing constant acceleration. In Chapter 3 we will repeat this discussion for objects moving in two dimensions.

The first recorded evidence of the study of mechanics can be traced to the people of ancient Sumeria and Egypt, who were interested primarily in understanding the motions of heavenly bodies. The most systematic and detailed early studies of the heavens were conducted by the Greeks from about 300 B.C. to A.D. 300. Ancient scientists and laypeople regarded the Earth as the center of the Universe. This **geocentric model** was accepted by such notables as Aristotle (384–322 B.C.) and Claudius Ptolemy (about A.D. 140). Largely because of the authority of Aristotle, the geocentric model became the accepted theory of the Universe until the seventeenth century.

About 250 B.C., the Greek philosopher Aristarchus worked out the details of a model of the solar system based on a spherical Earth that rotated on its axis and revolved around the Sun. He proposed that the sky appeared to turn westward because the Earth was turning eastward. This model wasn't given much consideration because it was believed that a turning

Adapt these equations to phase 1, substituting $a = 29.4\ \text{m/s}^2$, $v_0 = 0$, and $y_0 = 0$:

(3) $v = (29.4\ \text{m/s}^2)t$

(4) $y = \frac{1}{2}(29.4\ \text{m/s}^2)t^2 = (14.7\ \text{m/s}^2)t^2$

Substitute $t = 4.00$ s into Equations (3) and (4) to find the rocket's velocity v and position y at the time of burnout. These will be called v_b and y_b, respectively.

$v_b = 118\ \text{m/s}$ and $y_b = 235\ \text{m}$

(b) Phase 2: Find the maximum height the rocket attains.

Adapt Equations (1) and (2) to phase 2, substituting $a = -9.8\ \text{m/s}^2$, $v_0 = v_b = 118$ m/s, and $y_0 = y_b = 235$ m:

(5) $v = (-9.8\ \text{m/s}^2)t + 118\ \text{m/s}$

(6) $y = 235\ \text{m} + (118\ \text{m/s})t - (4.90\ \text{m/s}^2)t^2$

Substitute $v = 0$ (the rocket's velocity at maximum height) in Equation (5) to get the time it takes the rocket to reach its maximum height:

$$0 = (-9.8\ \text{m/s}^2)t + 118\ \text{m/s} \quad \rightarrow \quad t = \frac{118\ \text{m/s}}{9.80\ \text{m/s}^2} = 12.0\ \text{s}$$

Substitute $t = 12.0$ s into Equation (6) to find the rocket's maximum height:

$$y_{\text{max}} = 235\ \text{m} + (118\ \text{m/s})(12.0\ \text{s}) - (4.90\ \text{m/s}^2)(12.0\ \text{s})^2 = 945\ \text{m}$$

(c) Phase 2: Find the velocity of the rocket just prior to impact.

Find the time to impact by setting $y = 0$ in Equation (6) and using the quadratic formula:

$$0 = 235\ \text{m} + (118\ \text{m/s})t - (4.90\ \text{m/s})t^2$$

$$t = 25.9\ \text{s}$$

Substitute this value of t into Equation (5):

$$v = (-9.80\ \text{m/s}^2)(25.9\ \text{s}) + 118\ \text{m/s} = -136\ \text{m/s}$$

REMARKS You may think that it is more natural to break this problem into three phases, with the second phase ending at the maximum height and the third phase a free fall from maximum height to the ground. Although this approach gives the correct answer, it's an unnecessary complication. Two phases are sufficient, one for each different acceleration.

QUESTION 2.10 If, instead, some fuel remains, at what height should the engines be fired again to brake the rocket's fall and allow a perfectly soft landing? (Assume the same acceleration as during the initial ascent.)

EXERCISE 2.10 An experimental rocket designed to land upright falls freely from a height of 2.00×10^2 m, starting at rest. At a height of 80.0 m, the rocket's engines start and provide constant upward acceleration until the rocket lands. What acceleration is required if the speed on touchdown is to be zero? (Neglect air resistance.)

ANSWER $14.7\ \text{m/s}^2$

SUMMARY

2.1 Displacement

The **displacement** of an object moving along the x-axis is defined as the change in position of the object,

$$\Delta x \equiv x_f - x_i \qquad [2.1]$$

where x_i is the initial position of the object and x_f is its final position.

A **vector** quantity is characterized by both a magnitude and a direction. A **scalar** quantity has a magnitude only.

2.2 Velocity

The **average speed** of an object is given by

$$\text{Average speed} \equiv \frac{\text{path length}}{\text{elapsed time}}$$

The **average velocity** $\bar{v}$ during a time interval Δt is the displacement Δx divided by Δt.

$$\bar{v} \equiv \frac{\Delta x}{\Delta t} = \frac{x_f - x_i}{t_f - t_i} \qquad [2.2]$$

The average velocity is equal to the slope of the straight line joining the initial and final points on a graph of the position of the object versus time.

The slope of the line tangent to the position vs. time curve at some point is equal to the **instantaneous velocity** at that time. The **instantaneous speed** of an object is defined as the magnitude of the instantaneous velocity.

2.3 Acceleration

The **average acceleration** $\overline{a}$ of an object undergoing a change in velocity Δv during a time interval Δt is

$$\overline{a} \equiv \frac{\Delta v}{\Delta t} = \frac{v_f - v_i}{t_f - t_i} \quad [2.4]$$

The **instantaneous acceleration** of an object at a certain time equals the slope of a velocity vs. time graph at that instant.

2.5 One-Dimensional Motion with Constant Acceleration

The most useful equations that describe the motion of an object moving with constant acceleration along the x-axis are as follows:

$$v = v_0 + at \quad [2.6]$$

$$\Delta x = v_0 t + \tfrac{1}{2}at^2 \quad [2.9]$$

$$v^2 = v_0{}^2 + 2a\Delta x \quad [2.10]$$

All problems can be solved with the first two equations alone, the last being convenient when time doesn't explicitly enter the problem. After the constants are properly identified, most problems reduce to one or two equations in as many unknowns.

2.6 Freely Falling Objects

An object falling in the presence of Earth's gravity exhibits a free-fall acceleration directed toward Earth's center. If air friction is neglected and if the altitude of the falling object is small compared with Earth's radius, then we can assume that the free-fall acceleration $g = 9.8 \text{ m/s}^2$ is constant over the range of motion. Equations 2.6, 2.9, and 2.10 apply, with $a = -g$.

WARM-UP EXERCISES

ENHANCED WebAssign The warm-up exercises in this chapter may be assigned online in Enhanced WebAssign.

1. **Math Review** Solve the quadratic equation $2.00t^2 - 6.00t - 9.00 = 0$ using the quadratic formula, finding both solutions.
2. **Math Review** Solve the following two equations for (a) the time t, and (b) the position, x. Assume SI units.

$$-9.8t + 49 = 0 \text{ and } x = -4.9t^2 + 49t + 16$$

3. **Math Review** Solve the following two equations for (a) the (positive) time t, and (b) the position x. Assume SI units.

$$x = 3.00t^2 \qquad x = 24.0t + 72.0$$

4. A football player runs from his own goal line to the opposing team's goal line, returning to the fifty-yard line, all in 18.0 s. Calculate (a) his average speed, and (b) the magnitude of his average velocity. (See Section 2.2.)
5. A ball is thrown downward from the top of a 40.0 m tower with an initial speed of 12.0 m/s. Assuming negligible air resistance, what is the speed of the ball just before hitting the ground? (See Section 2.6.)
6. An arrow is shot straight up in the air at an initial speed of 15.0 m/s. After how much time is the arrow heading downward at a speed of 8.00 m/s? (See Section 2.6.)
7. A red ball is dropped from rest at a height of 6.00 m. A blue ball at a height of 10.0 m is thrown down at the same instant at 4.00 m/s. How long does it take the blue ball to catch up with the red ball? (See Sections 2.5 and 2.6.)

CONCEPTUAL QUESTIONS

ENHANCED WebAssign The conceptual questions in this chapter may be assigned online in Enhanced WebAssign.

1. If the velocity of a particle is nonzero, can the particle's acceleration be zero? Explain.
2. If the velocity of a particle is zero, can the particle's acceleration be nonzero? Explain.
3. If a car is traveling eastward, can its acceleration be westward? Explain.
4. (a) Can the equations in Table 2.4 be used in a situation where the acceleration varies with time? (b) Can they be used when the acceleration is zero?
5. Two cars are moving in the same direction in parallel lanes along a highway. At some instant, the velocity of car A exceeds the velocity of car B. Does that mean that the acceleration of A is greater than that of B at that instant? Explain.
6. Figure CQ2.6 shows strobe photographs taken of a disk moving from left to right under different conditions. The time interval between images is constant. Taking the direction to the right to be positive, describe the motion of the disk in each case. For which case is (a) the acceleration positive? (b) the acceleration negative? (c) the velocity constant?

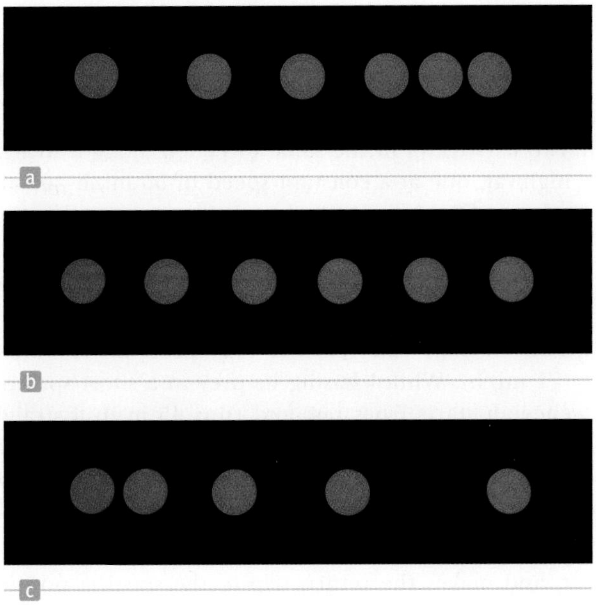

Figure CQ2.6

7. (a) Can the instantaneous velocity of an object at an instant of time ever be greater in magnitude than the average velocity over a time interval containing that instant? (b) Can it ever be less?

8. A ball is thrown vertically upward. (a) What are its velocity and acceleration when it reaches its maximum altitude? (b) What is the acceleration of the ball just before it hits the ground?

9. Consider the following combinations of signs and values for the velocity and acceleration of a particle with respect to a one-dimensional x-axis:

Velocity	Acceleration
a. Positive	Positive
b. Positive	Negative
c. Positive	Zero
d. Negative	Positive
e. Negative	Negative
f. Negative	Zero
g. Zero	Positive
h. Zero	Negative

Describe what the particle is doing in each case and give a real-life example for an automobile on an east–west one-dimensional axis, with east considered the positive direction.

10. A ball rolls in a straight line along the horizontal direction. Using motion diagrams (or multiflash photographs), describe the velocity and acceleration of the ball for each of the following situations: (a) The ball moves to the right at a constant speed. (b) The ball moves from right to left and continually slows down. (c) The ball moves from right to left and continually speeds up. (d) The ball moves to the right, first speeding up at a constant rate and then slowing down at a constant rate.

11. An object moves along the x-axis, its position measured at each instant of time. The data are organized into an accurate graph of x vs. t. Which of the following quantities cannot be obtained from this graph? (a) The velocity at any instant (b) the acceleration at any instant (c) the displacement during some time interval (d) the average velocity during some time interval (e) the speed of the particle at any instant.

12. A ball is thrown straight up in the air. For which situation are both the instantaneous velocity and the acceleration zero? (a) On the way up (b) at the top of the flight path (c) on the way down (d) halfway up and halfway down (e) none of these.

13. A juggler throws a bowling pin straight up in the air. After the pin leaves his hand and while it is in the air, which statement is true? (a) The velocity of the pin is always in the same direction as its acceleration. (b) The velocity of the pin is never in the same direction as its acceleration. (c) The acceleration of the pin is zero. (d) The velocity of the pin is opposite its acceleration on the way up. (e) The velocity of the pin is in the same direction as its acceleration on the way up.

14. A racing car starts from rest and reaches a final speed v in a time t. If the acceleration of the car is constant during this time, which of the following statements must be true? (a) The car travels a distance vt. (b) The average speed of the car is $v/2$. (c) The acceleration of the car is v/t. (d) The velocity of the car remains constant. (e) None of these.

PROBLEMS

ENHANCED WebAssign The problems in this chapter may be assigned online in Enhanced WebAssign.

1. denotes straightforward problem; **2.** denotes intermediate problem;

3. denotes challenging problem

1. (boxed) denotes full solution available in *Student Solutions Manual/ Study Guide*

1. (shaded) denotes problems most often assigned in Enhanced WebAssign

BIO denotes biomedical problems

GP denotes guided problems

M denotes Master It tutorial available in Enhanced WebAssign

Q|C denotes asking for quantitative and conceptual reasoning

S denotes symbolic reasoning problem

W denotes Watch It video solution available in Enhanced WebAssign

2.1 Displacement

2.2 Velocity

1. **BIO** The speed of a nerve impulse in the human body is about 100 m/s. If you accidentally stub your toe in the dark, estimate the time it takes the nerve impulse to travel to your brain.

2. Light travels at a speed of about 3×10^8 m/s. (a) How many miles does a pulse of light travel in a time interval of 0.1 s, which is about the blink of an eye? (b) Compare this distance to the diameter of Earth.

3. A person travels by car from one city to another with different constant speeds between pairs of cities. She drives for 30.0 min at 80.0 km/h, 12.0 min at 100 km/h, and 45.0 min at 40.0 km/h and spends 15.0 min eating lunch and buying gas. (a) Determine the average speed for the trip. (b) Determine the distance between the initial and final cities along the route.

4. The current indoor world record time in the 200-m race is 19.92 s, held by Frank Fredericks of Namibia (1996), while the indoor record time in the one-mile race is 228.5 s, held by Hicham El Guerrouj of Morroco (1997). Find the mean speed in meters per second corresponding to these record times for (a) the 200-m event and (b) the one-mile event.

5. Two boats start together and race across a 60-km-wide lake and back. Boat A goes across at 60 km/h and returns at 60 km/h. Boat B goes across at 30 km/h, and its crew, realizing how far behind it is getting, returns at 90 km/h. Turnaround times are negligible, and the boat that completes the round trip first wins. (a) Which boat wins and by how much? (Or is it a tie?) (b) What is the average velocity of the winning boat?

6. A graph of position versus time for a certain particle moving along the x-axis is shown in Figure P2.6. Find the average velocity in the time intervals from (a) 0 to 2.00 s, (b) 0 to 4.00 s, (c) 2.00 s to 4.00 s, (d) 4.00 s to 7.00 s, and (e) 0 to 8.00 s.

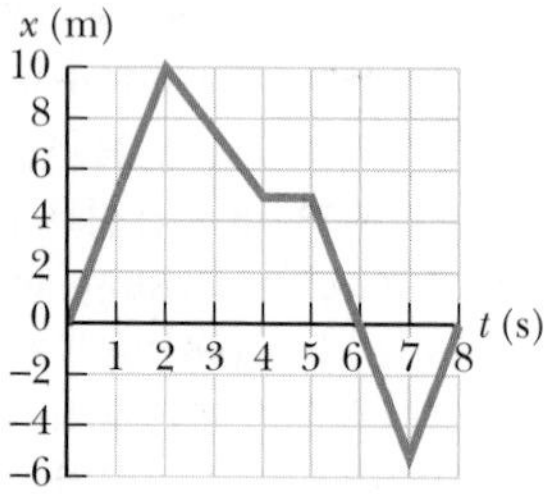

Figure P2.6 Problems 6 and 17

7. **W** A motorist drives north for 35.0 minutes at 85.0 km/h and then stops for 15.0 minutes. He then continues north, traveling 130 km in 2.00 h. (a) What is his total displacement? (b) What is his average velocity?

8. A tennis player moves in a straight-line path as shown in Figure P2.8. Find her average velocity in the time intervals from (a) 0 to 1.0 s, (b) 0 to 4.0 s, (c) 1.0 s to 5.0 s, and (d) 0 to 5.0 s.

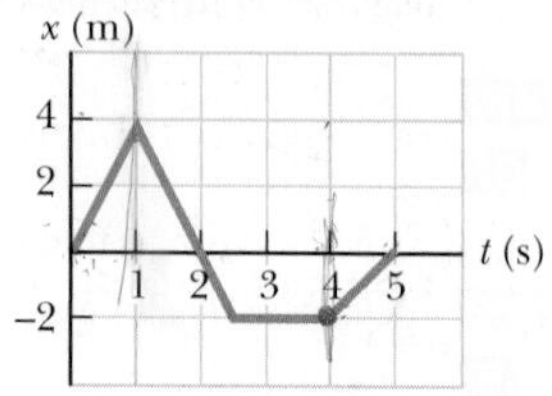

Figure P2.8

9. A jet plane has a takeoff speed of v_{to} = 75 m/s and can move along the runway at an average acceleration of 1.3 m/s^2. If the length of the runway is 2.5 km, will the plane be able to use this runway safely? Defend your answer.

10. Two cars travel in the same direction along a straight highway, one at a constant speed of 55 mi/h and the other at 70 mi/h. (a) Assuming they start at the same point, how much sooner does the faster car arrive at a destination 10 mi away? (b) How far must the faster car travel before it has a 15-min lead on the slower car?

11. The cheetah can reach a top speed of 114 km/h (71 mi/h). While chasing its prey in a short sprint, a cheetah starts from rest and runs 45 m in a straight line, reaching a final speed of 72 km/h. (a) Determine the cheetah's average acceleration during the short sprint, and (b) find its displacement at t = 3.5 s.

12. **S** An athlete swims the length L of a pool in a time t_1 and makes the return trip to the starting position in a time t_2. If she is swimming initially in the positive x-direction, determine her average velocities symbolically in (a) the first half of the swim, (b) the second half of the swim, and (c) the round trip. (d) What is her average speed for the round trip?

13. **M** A person takes a trip, driving with a constant speed of 89.5 km/h, except for a 22.0-min rest stop. If the person's average speed is 77.8 km/h, (a) how much time is spent on the trip and (b) how far does the person travel?

14. A tortoise can run with a speed of 0.10 m/s, and a hare can run 20 times as fast. In a race, they both start at the same time, but the hare stops to rest for 2.0 minutes. The tortoise wins by a shell (20 cm). (a) How long does the race take? (b) What is the length of the race?

15. To qualify for the finals in a racing event, a race car must achieve an average speed of 250 km/h on a track with a total length of 1 600 m. If a particular car covers the first half of the track at an average speed of 230 km/h, what minimum average speed must it have in the second half of the event in order to qualify?

16. **Q|C** **S** One athlete in a race running on a long, straight track with a constant speed v_1 is a distance d behind a second athlete running with a constant speed v_2. (a) Under what circumstances is the first athlete able to overtake the second athlete? (b) Find the time t it takes the first athlete to overtake the second athlete, in terms of d, v_1, and v_2. (c) At what minimum distance d_2 from the leading athlete must the finish line be located so that the trailing athlete can at least tie for first place? Express d_2 in terms of d, v_1, and v_2 by using the result of part (b).

17. A graph of position versus time for a certain particle moving along the x-axis is shown in Figure P2.6. Find the instantaneous velocity at the instants (a) t = 1.00 s, (b) t = 3.00 s, (c) t = 4.50 s, and (d) t = 7.50 s.

18. A race car moves such that its position fits the relationship

$$x = (5.0\text{ m/s})t + (0.75\text{ m/s}^3)t^3$$

where x is measured in meters and t in seconds. (a) Plot a graph of the car's position versus time. (b) Determine the instantaneous velocity of the car at $t = 4.0$ s, using time intervals of 0.40 s, 0.20 s, and 0.10 s. (c) Compare the average velocity during the first 4.0 s with the results of part (b).

19. Runner A is initially 4.0 mi west of a flagpole and is running with a constant velocity of 6.0 mi/h due east. Runner B is initially 3.0 mi east of the flagpole and is running with a constant velocity of 5.0 mi/h due west. How far are the runners from the flagpole when they meet?

2.3 Acceleration

20. A particle starts from rest and accelerates as shown in Figure P2.20. Determine (a) the particle's speed at $t = 10.0$ s and at $t = 20.0$ s, and (b) the distance traveled in the first 20.0 s.

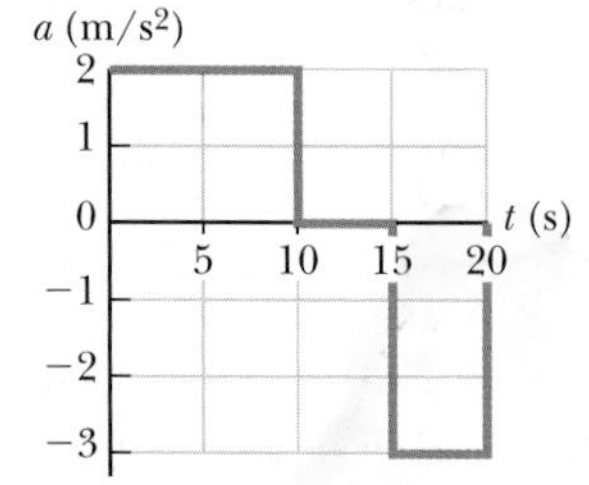

Figure P2.20

21. A 50.0-g Super Ball traveling at 25.0 m/s bounces off a brick wall and rebounds at 22.0 m/s. A high-speed camera records this event. If the ball is in contact with the wall for 3.50 ms, what is the magnitude of the average acceleration of the ball during this time interval?

22. BIO The average person passes out at an acceleration of $7g$ (that is, seven times the gravitational acceleration on Earth). Suppose a car is designed to accelerate at this rate. How much time would be required for the car to accelerate from rest to 60.0 miles per hour? (The car would need rocket boosters!)

23. W A certain car is capable of accelerating at a rate of 0.60 m/s^2. How long does it take for this car to go from a speed of 55 mi/h to a speed of 60 mi/h?

24. The velocity vs. time graph for an object moving along a straight path is shown in Figure P2.24. **(i)** Find the average acceleration of the object during the time intervals (a) 0 to 5.0 s, (b) 5.0 s to 15 s, and (c) 0 to 20 s. **(ii)** Find the instantaneous acceleration at (a) 2.0 s, (b) 10 s, and (c) 18 s.

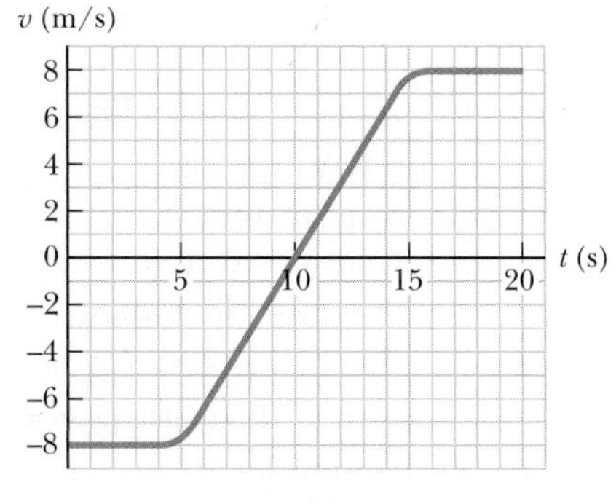

Figure P2.24

25. A steam catapult launches a jet aircraft from the aircraft carrier *John C. Stennis,* giving it a speed of 175 mi/h in 2.50 s. (a) Find the average acceleration of the plane. (b) Assuming the acceleration is constant, find the distance the plane moves.

2.5 One-Dimensional Motion with Constant Acceleration

26. Solve Example 2.5, "Car Chase" by a graphical method. On the same graph, plot position versus time for the car and the trooper. From the intersection of the two curves, read the time at which the trooper overtakes the car.

27. M An object moving with uniform acceleration has a velocity of 12.0 cm/s in the positive x-direction when its x-coordinate is 3.00 cm. If its x-coordinate 2.00 s later is −5.00 cm, what is its acceleration?

28. W In 1865 Jules Verne proposed sending men to the Moon by firing a space capsule from a 220-m-long cannon with final speed of 10.97 km/s. What would have been the unrealistically large acceleration experienced by the space travelers during their launch? (A human can stand an acceleration of $15g$ for a short time.) Compare your answer with the free-fall acceleration, 9.80 m/s^2.

29. A truck covers 40.0 m in 8.50 s while uniformly slowing down to a final velocity of 2.80 m/s. (a) Find the truck's original speed. (b) Find its acceleration.

30. GP A speedboat increases its speed uniformly from $v_i =$ 20.0 m/s to $v_f =$ 30.0 m/s in a distance of 2.00×10^2 m. (a) Draw a coordinate system for this situation and label the relevant quantities, including vectors. (b) For the given information, what single equation is most appropriate for finding the acceleration? (c) Solve the equation selected in part (b) symbolically for the boat's acceleration in terms of v_f, v_i, and Δx. (d) Substitute given values, obtaining that acceleration. (e) Find the time it takes the boat to travel the given distance.

31. A Cessna aircraft has a liftoff speed of 120 km/h. (a) What minimum constant acceleration does the aircraft require if it is to be airborne after a takeoff run of 240 m? (b) How long does it take the aircraft to become airborne?

32. An object moves with constant acceleration 4.00 m/s^2 and over a time interval reaches a final velocity of 12.0 m/s. (a) If its original velocity is 6.00 m/s, what is its displacement during the time interval? (b) What is the distance it travels during this interval? (c) If its original velocity is −6.00 m/s, what is its displacement during this interval? (d) What is the total distance it travels during the interval in part (c)?

33. Q|C In a test run, a certain car accelerates uniformly from zero to 24.0 m/s in 2.95 s. (a) What is the magnitude of the car's acceleration? (b) How long does it take the car to change its speed from 10.0 m/s to 20.0 m/s? (c) Will doubling the time always double the change in speed? Why?

34. QIC A jet plane lands with a speed of 100 m/s and can accelerate at a maximum rate of $-5.00\ m/s^2$ as it comes to rest. (a) From the instant the plane touches the runway, what is the minimum time needed before it can come to rest? (b) Can this plane land on a small tropical island airport where the runway is 0.800 km long?

35. QIC Speedy Sue, driving at 30.0 m/s, enters a one-lane tunnel. She then observes a slow-moving van 155 m ahead traveling at 5.00 m/s. Sue applies her brakes but can accelerate only at $-2.00\ m/s^2$ because the road is wet. Will there be a collision? State how you decide. If yes, determine how far into the tunnel and at what time the collision occurs. If no, determine the distance of closest approach between Sue's car and the van.

36. A record of travel along a straight path is as follows:

1. Start from rest with a constant acceleration of $2.77\ m/s^2$ for 15.0 s.
2. Maintain a constant velocity for the next 2.05 min.
3. Apply a constant negative acceleration of $-9.47\ m/s^2$ for 4.39 s.

(a) What was the total displacement for the trip? (b) What were the average speeds for legs 1, 2, and 3 of the trip, as well as for the complete trip?

37. A train is traveling down a straight track at 20 m/s when the engineer applies the brakes, resulting in an acceleration of $-1.0\ m/s^2$ as long as the train is in motion. How far does the train move during a 40-s time interval starting at the instant the brakes are applied?

38. A car accelerates uniformly from rest to a speed of 40.0 mi/h in 12.0 s. Find (a) the distance the car travels during this time and (b) the constant acceleration of the car.

39. A car starts from rest and travels for 5.0 s with a uniform acceleration of $+1.5\ m/s^2$. The driver then applies the brakes, causing a uniform acceleration of $-2.0\ m/s^2$. If the brakes are applied for 3.0 s, (a) how fast is the car going at the end of the braking period, and (b) how far has the car gone?

40. S A car starts from rest and travels for t_1 seconds with a uniform acceleration a_1. The driver then applies the brakes, causing a uniform acceleration a_2. If the brakes are applied for t_2 seconds, (a) how fast is the car going just before the beginning of the braking period? (b) How far does the car go before the driver begins to brake? (c) Using the answers to parts (a) and (b) as the initial velocity and position for the motion of the car during braking, what total distance does the car travel? Answers are in terms of the variables a_1, a_2, t_1, and t_2.

41. In the Daytona 500 auto race, a Ford Thunderbird and a Mercedes Benz are moving side by side down a straightaway at 71.5 m/s. The driver of the Thunderbird realizes that she must make a pit stop, and she smoothly slows to a stop over a distance of 250 m. She spends 5.00 s in the pit and then accelerates out, reaching her previous speed of 71.5 m/s after a distance of 350 m. At this point, how far has the Thunderbird fallen behind the Mercedes Benz, which has continued at a constant speed?

42. A certain cable car in San Francisco can stop in 10 s when traveling at maximum speed. On one occasion, the driver sees a dog a distance d in front of the car and slams on the brakes instantly. The car reaches the dog 8.0 s later, and the dog jumps off the track just in time. If the car travels 4.0 m beyond the position of the dog before coming to a stop, how far was the car from the dog? (Hint: You will need three equations.)

43. M A hockey player is standing on his skates on a frozen pond when an opposing player, moving with a uniform speed of 12 m/s, skates by with the puck. After 3.0 s, the first player makes up his mind to chase his opponent. If he accelerates uniformly at $4.0\ m/s^2$, (a) how long does it take him to catch his opponent, and (b) how far has he traveled in that time? (Assume the player with the puck remains in motion at constant speed.)

44. A train 400 m long is moving on a straight track with a speed of 82.4 km/h. The engineer applies the brakes at a crossing, and later the last car passes the crossing with a speed of 16.4 km/h. Assuming constant acceleration, determine how long the train blocked the crossing. Disregard the width of the crossing.

2.6 Freely Falling Objects

45. A ball is thrown vertically upward with a speed of 25.0 m/s. (a) How high does it rise? (b) How long does it take to reach its highest point? (c) How long does the ball take to hit the ground after it reaches its highest point? (d) What is its velocity when it returns to the level from which it started?

46. A ball is thrown directly downward with an initial speed of 8.00 m/s, from a height of 30.0 m. After what time interval does it strike the ground?

47. A certain freely falling object, released from rest, requires 1.50 s to travel the last 30.0 m before it hits the ground. (a) Find the velocity of the object when it is 30.0 m above the ground. (b) Find the total distance the object travels during the fall.

48. QIC An attacker at the base of a castle wall 3.65 m high throws a rock straight up with speed 7.40 m/s at a height of 1.55 m above the ground. (a) Will the rock reach the top of the wall? (b) If so, what is the rock's speed at the top? If not, what initial speed must the rock have to reach the top? (c) Find the change in the speed of a rock thrown straight down from the top of the wall at an initial speed of 7.40 m/s and moving between the same two points. (d) Does the change in speed of the downward-moving rock agree with the magnitude of the speed change of the rock moving upward between the same elevations? Explain physically why or why not.

49. BIO Traumatic brain injury such as concussion results when the head undergoes a very large acceleration. Generally, an acceleration less than $800\ m/s^2$ lasting for any length of time will not cause injury, whereas an acceleration greater than 1 000 m/s² lasting for at least 1 ms will cause injury. Suppose a small child rolls off a bed that is 0.40 m above the floor. If the floor is hardwood, the child's head is brought to rest in approximately 2.0 mm. If the floor is carpeted, this stopping distance is increased to about 1.0 cm. Calculate the magnitude and duration of the deceleration in both cases, to determine the risk of injury. Assume the child remains horizontal during the fall to the floor. Note that a more complicated fall could result in a head velocity greater or less than the speed you calculate.

50. A small mailbag is released from a helicopter that is descending steadily at 1.50 m/s. After 2.00 s, (a) what is the speed of the mailbag, and (b) how far is it below the helicopter? (c) What are your answers to parts (a) and (b) if the helicopter is rising steadily at 1.50 m/s?

51. A tennis player tosses a tennis ball straight up and then catches it after 2.00 s at the same height as the point of release. (a) What is the acceleration of the ball while it is in flight? (b) What is the velocity of the ball when it reaches its maximum height? Find (c) the initial velocity of the ball and (d) the maximum height it reaches.

52. **S** A package is dropped from a helicopter that is descending steadily at a speed v_0. After t seconds have elapsed, (a) what is the speed of the package in terms of v_0, g, and t? (b) What distance d is it from the helicopter in terms of g and t? (c) What are the answers to parts (a) and (b) if the helicopter is rising steadily at the same speed?

53. **Q|C** A model rocket is launched straight upward with an initial speed of 50.0 m/s. It accelerates with a constant upward acceleration of 2.00 m/s² until its engines stop at an altitude of 150 m. (a) What can you say about the motion of the rocket after its engines stop? (b) What is the maximum height reached by the rocket? (c) How long after liftoff does the rocket reach its maximum height? (d) How long is the rocket in the air?

54. **W** A baseball is hit so that it travels straight upward after being struck by the bat. A fan observes that it takes 3.00 s for the ball to reach its maximum height. Find (a) the ball's initial velocity and (b) the height it reaches.

Additional Problems

55. A truck tractor pulls two trailers, one behind the other, at a constant speed of 100 km/h. It takes 0.600 s for the big rig to completely pass onto a bridge 400 m long. For what duration of time is all or part of the truck–trailer combination on the bridge?

56. **M** **BIO** Colonel John P. Stapp, USAF, participated in studying whether a jet pilot could survive emergency ejection. On March 19, 1954, he rode a rocket-propelled sled that moved down a track at a speed of 632 mi/h (see Fig. P2.56). He and the sled were safely brought to rest in 1.40 s. Determine in SI units (a) the negative acceleration he experienced and (b) the distance he traveled during this negative acceleration.

Courtesy U.S. Air Force

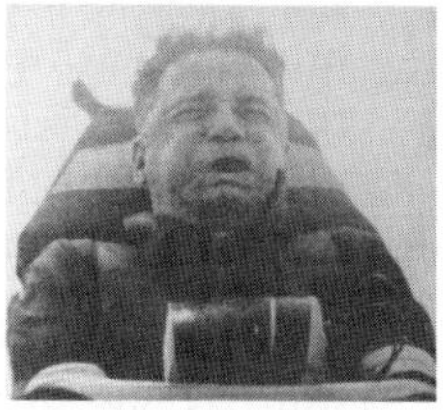
NASA/Science Source

Figure P2.56 (*left*) Col. John Stapp and his rocket sled are brought to rest in a very short time interval. (*right*) Stapp's face is contorted by the stress of rapid negative acceleration.

57. A bullet is fired through a board 10.0 cm thick in such a way that the bullet's line of motion is perpendicular to the face of the board. If the initial speed of the bullet is 400 m/s and it emerges from the other side of the board with a speed of 300 m/s, find (a) the acceleration of the bullet as it passes through the board and (b) the total time the bullet is in contact with the board.

58. A speedboat moving at 30.0 m/s approaches a no-wake buoy marker 100 m ahead. The pilot slows the boat with a constant acceleration of −3.50 m/s² by reducing the throttle. (a) How long does it take the boat to reach the buoy? (b) What is the velocity of the boat when it reaches the buoy?

59. A student throws a set of keys vertically upward to his fraternity brother, who is in a window 4.00 m above. The brother's outstretched hand catches the keys 1.50 s later. (a) With what initial velocity were the keys thrown? (b)? What was the velocity of the keys just before they were caught?

60. **S** A student throws a set of keys vertically upward to his fraternity brother, who is in a window a distance h above. The brother's outstretched hand catches the keys on their way up a time t later. (a) With what initial velocity were the keys thrown? (b) What was the velocity of the keys just before they were caught? (Answers should be in terms of h, g, and t.)

61. **BIO** It has been claimed that an insect called the froghopper (*Philaenus spumarius*) is the best jumper in the animal kingdom. This insect can accelerate at 4 000 m/s² over a distance of 2.0 mm as it straightens its specially designed "jumping legs." (a) Assuming a uniform acceleration, what is the velocity of the insect after it has accelerated through this short distance, and (b) how long did it take to reach that velocity? (c) How high would the insect jump if air resistance could be ignored? Note that the actual height obtained is about 0.7 m, so air resistance is important here.

62. **Q|C** Draw motion diagrams (see Section 2.5) for (a) an object moving to the right at constant speed, (b) an object moving to the right and speeding up at a constant rate, (c) an object moving to the right and slowing down at a constant rate, (d) an object moving to the left and speeding up at a constant rate, and (e) an object moving to the left and slowing down at a constant rate. (f) How would your drawings change if the changes in speed were not uniform; that is, if the speed were not changing at a constant rate?

63. **M** A ball is thrown upward from the ground with an initial speed of 25 m/s; at the same instant, another ball is dropped from a building 15 m high. After how long will the balls be at the same height?

64. To pass a physical education class at a university, a student must run 1.0 mi in 12 min. After running for 10 min, she still has 500 yd to go. If her maximum acceleration is $0.15\ \text{m/s}^2$, can she make it? If the answer is no, determine what acceleration she would need to be successful.

65. In Chapter 5 we will define the center of mass of an object. The center of mass moves with constant acceleration when constant forces act on the object. A gymnast jumps straight up, with her center of mass moving at 2.80 m/s as she leaves the ground. How high above this point is her center of mass (a) 0.100 s, (b) 0.200 s, (c) 0.300 s, and (d) 0.500 s thereafter?

66. **S** Two students are on a balcony a distance h above the street. One student throws a ball vertically downward at a speed v_0; at the same time, the other student throws a ball vertically upward at the same speed. Answer the following symbolically in terms of v_0, g, h, and t. (a) Write the kinematic equation for the y-coordinate of each ball. (b) Set the equations found in part (a) equal to height 0 and solve each for t symbolically using the quadratic formula. What is the difference in the two balls' time in the air? (c) Use the time-independent kinematics equation to find the velocity of each ball as it strikes the ground. (d) How far apart are the balls at a time t after they are released and before they strike the ground?

67. You drop a ball from a window on an upper floor of a building and it is caught by a friend on the ground when the ball is moving with speed v_f. You now repeat the drop, but you have a friend on the street below throw another ball upward at speed v_f exactly at the same time that you drop your ball from the window. The two balls are initially separated by 28.7 m. (a) At what time do they pass each other? (b) At what location do they pass each other relative the window?

68. The driver of a truck slams on the brakes when he sees a tree blocking the road. The truck slows down uniformly with an acceleration of $-5.60\ \text{m/s}^2$ for 4.20 s, making skid marks 62.4 m long that end at the tree. With what speed does the truck then strike the tree?

69. **QIC** Emily challenges her husband, David, to catch a \$1 bill as follows. She holds the bill vertically as in Figure P2.69, with the center of the bill between David's index finger and thumb. David must catch the bill after Emily releases it without moving his hand downward. If his reaction time is 0.2 s, will he succeed? Explain your reasoning. (This challenge is a good trick you might want to try with your friends.)

© Cengage Learning/George Semple

Figure P2.69

70. A mountain climber stands at the top of a 50.0-m cliff that overhangs a calm pool of water. She throws two stones vertically downward 1.00 s apart and observes that they cause a single splash. The first stone had an initial velocity of -2.00 m/s. (a) How long after release of the first stone did the two stones hit the water? (b) What initial velocity must the second stone have had, given that they hit the water simultaneously? (c) What was the velocity of each stone at the instant it hit the water?

71. An ice sled powered by a rocket engine starts from rest on a large frozen lake and accelerates at $+40\ \text{ft/s}^2$. After some time t_1, the rocket engine is shut down and the sled moves with constant velocity v for a time t_2. If the total distance traveled by the sled is 17 500 ft and the total time is 90 s, find (a) the times t_1 and t_2 and (b) the velocity v. At the 17 500-ft mark, the sled begins to accelerate at $-20\ \text{ft/s}^2$. (c) What is the final position of the sled when it comes to rest? (d) How long does it take to come to rest?

72. In Bosnia, the ultimate test of a young man's courage used to be to jump off a 400-year-old bridge (destroyed in 1993; rebuilt in 2004) into the River Neretva, 23 m below the bridge. (a) How long did the jump last? (b) How fast was the jumper traveling upon impact with the river? (c) If the speed of sound in air is 340 m/s, how long after the jumper took off did a spectator on the bridge hear the splash?

73. A person sees a lightning bolt pass close to an airplane that is flying in the distance. The person hears thunder 5.0 s after seeing the bolt and sees the airplane overhead 10 s after hearing the thunder. The speed of sound in air is 1 100 ft/s. (a) Find the distance of the airplane from the person at the instant of the bolt. (Neglect the time it takes the light to travel from the bolt to the eye.) (b) Assuming the plane travels with a constant speed toward the person, find the velocity of the airplane. (c) Look up the speed of light in air and defend the approximation used in part (a).

74. **QIC** A glider on an air track carries a flag of length ℓ through a stationary photogate, which measures the time interval Δt_d during which the flag blocks a beam of infrared light passing across the photogate. The ratio $v_d = \ell/\Delta t_d$ is the average velocity of the glider over this part of its motion. Suppose the glider moves with constant acceleration. (a) Is v_d necessarily equal to the instantaneous velocity of the glider when it is halfway through the photogate in space? Explain. (b) Is v_d equal to the instantaneous velocity of the glider when it is halfway through the photogate in time? Explain.

75. A stuntman sitting on a tree limb wishes to drop vertically onto a horse galloping under the tree. The constant speed of the horse is 10.0 m/s, and the man is initially 3.00 m above the level of the saddle. (a) What must be the horizontal distance between the saddle and the limb when the man makes his move? (b) How long is he in the air?

AP Photo/Denis Poroy

Dave "Human Cannonball" Smith hurtles through the air along a parabolic path, depending on the correct initial velocity and cannon angle to send him safely over the U.S.-Mexico border fence and into a net.

Vectors and Two-Dimensional Motion 3

In our discussion of one-dimensional motion in Chapter 2, we used the concept of vectors only to a limited extent. In our further study of motion, manipulating vector quantities will become increasingly important, so much of this chapter is devoted to vector techniques. We'll then apply these mathematical tools to two-dimensional motion, especially that of projectiles, and to the understanding of relative motion.

3.1 Vectors and Their Properties

LEARNING OBJECTIVES

1. Apply the definitions of scalar and vector to categorize different physical quantities.
2. Use the geometric interpretation of vector addition, subtraction and multiplication to find the resultant vectors of those operations.

Each of the physical quantities we will encounter in this book can be categorized as either a *vector quantity* or a *scalar quantity*. As noted in Chapter 2, a vector has both direction and magnitude (size). A scalar can be completely specified by its magnitude with appropriate units; it has no direction. An example of each kind of quantity is shown in Figure 3.1.

Jon Feingersh/Stone/Getty Images

Figure 3.1 A vector such as velocity has a magnitude, shown on the race car's speedometer, and a direction, straight out through the race car's front windshield. The mass of the car is a scalar quantity, as is the volume of gasoline in its fuel tank.

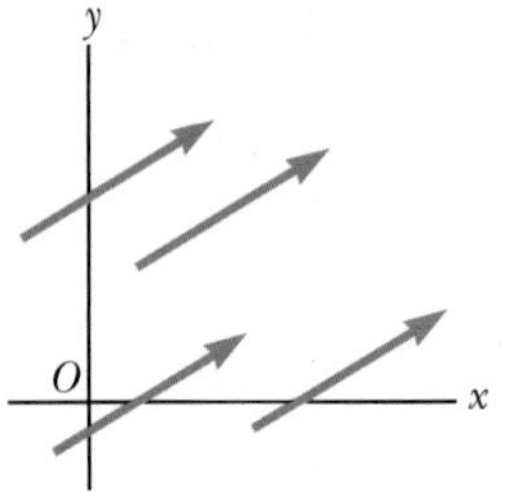

Figure 3.2 These four vectors are equal because they have equal lengths and point in the same direction.

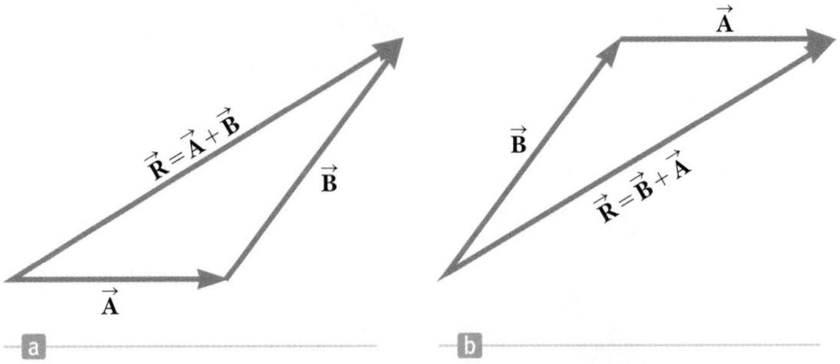

Figure 3.3
(a) When vector $\vec{\mathbf{B}}$ is added to vector $\vec{\mathbf{A}}$, the vector sum $\vec{\mathbf{R}}$ is the vector that runs from the tail of $\vec{\mathbf{A}}$ to the tip of $\vec{\mathbf{B}}$. (b) Here the resultant runs from the tail of $\vec{\mathbf{B}}$ to the tip of $\vec{\mathbf{A}}$. These constructions prove that $\vec{\mathbf{A}} + \vec{\mathbf{B}} = \vec{\mathbf{B}} + \vec{\mathbf{A}}$.

Tip 3.1 Vector Addition vs. Scalar Addition

$\vec{\mathbf{A}} + \vec{\mathbf{B}} = \vec{\mathbf{C}}$ differs significantly from $A + B = C$. The first is a vector sum, which must be handled graphically or with components, whereas the second is a simple arithmetic sum of numbers.

As described in Chapter 2, displacement, velocity, and acceleration are vector quantities. Temperature is an example of a scalar quantity. If the temperature of an object is $-5°C$, that information completely specifies the temperature of the object; no direction is required. Masses, time intervals, and volumes are scalars as well. Scalar quantities can be manipulated with the rules of ordinary arithmetic. Vectors can also be added and subtracted from each other, and multiplied, but there are a number of important differences, as will be seen in the following sections.

When a vector quantity is handwritten, it is often represented with an arrow over the letter ($\vec{A}$). As mentioned in Section 2.1, a vector quantity in this book will be represented by boldface type with an arrow on top (for example, $\vec{\mathbf{A}}$). The magnitude of the vector $\vec{\mathbf{A}}$ will be represented by italic type, as A. Italic type will also be used to represent scalars.

Equality of Two Vectors Two vectors $\vec{\mathbf{A}}$ and $\vec{\mathbf{B}}$ are equal if they have the same magnitude and the same direction. This property allows us to translate a vector parallel to itself in a diagram without affecting the vector. In fact, for most purposes, any vector can be moved parallel to itself without being affected. (See Fig. 3.2.)

Adding Vectors When two or more vectors are added, they must all have the same units. For example, it doesn't make sense to add a velocity vector, carrying units of meters per second, to a displacement vector, carrying units of meters. Scalars obey the same rule: It would be similarly meaningless to add temperatures to volumes or masses to time intervals.

Vectors can be added geometrically or algebraically. (The latter is discussed at the end of the next section.) To add vector $\vec{\mathbf{B}}$ to vector $\vec{\mathbf{A}}$ geometrically, first draw $\vec{\mathbf{A}}$ on a piece of graph paper to some scale, such as 1 cm = 1 m, so that its direction is specified relative to a coordinate system. Then draw vector $\vec{\mathbf{B}}$ to the same scale with the tail of $\vec{\mathbf{B}}$ starting at the tip of $\vec{\mathbf{A}}$, as in Figure 3.3a. Vector $\vec{\mathbf{B}}$ must be drawn along the direction that makes the proper angle relative vector $\vec{\mathbf{A}}$. The **resultant vector** $\vec{\mathbf{R}} = \vec{\mathbf{A}} + \vec{\mathbf{B}}$ is the vector drawn from the tail of $\vec{\mathbf{A}}$ to the tip of $\vec{\mathbf{B}}$. This procedure is known as the **triangle method of addition.**

When two vectors are added, their sum is independent of the order of the addition: $\vec{\mathbf{A}} + \vec{\mathbf{B}} = \vec{\mathbf{B}} + \vec{\mathbf{A}}$. This relationship can be seen from the geometric construction in Figure 3.3b, and is called the **commutative law of addition.**

This same general approach can also be used to add more than two vectors, as is done in Figure 3.4 for four vectors. The resultant vector sum $\vec{\mathbf{R}} = \vec{\mathbf{A}} + \vec{\mathbf{B}} + \vec{\mathbf{C}} + \vec{\mathbf{D}}$ is the vector drawn from the tail of the first vector to the tip of the last. Again, the order in which the vectors are added is unimportant.

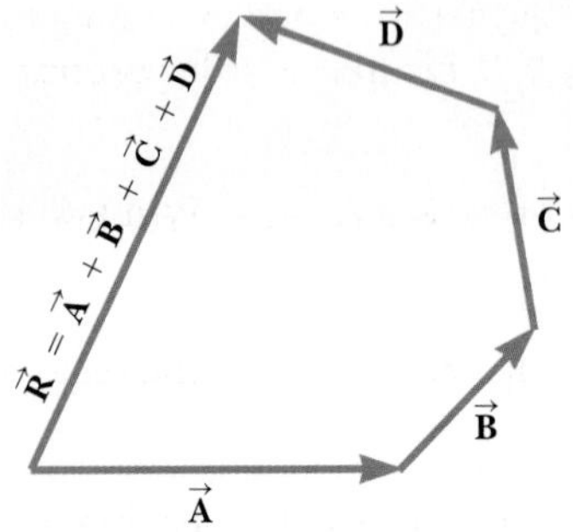

Figure 3.4 A geometric construction for summing four vectors. The resultant vector $\vec{\mathbf{R}}$ is the vector that completes the polygon.

Negative of a Vector The negative of the vector $\vec{\mathbf{A}}$ is defined as the vector that gives zero when added to $\vec{\mathbf{A}}$. This means that $\vec{\mathbf{A}}$ and $-\vec{\mathbf{A}}$ have the same magnitude but opposite directions.

Subtracting Vectors Vector subtraction makes use of the definition of the negative of a vector. We define the operation $\vec{\mathbf{A}} - \vec{\mathbf{B}}$ as the vector $-\vec{\mathbf{B}}$ added to the vector $\vec{\mathbf{A}}$:

$$\vec{\mathbf{A}} - \vec{\mathbf{B}} = \vec{\mathbf{A}} + (-\vec{\mathbf{B}}) \quad [3.1]$$

Vector subtraction is really a special case of vector addition. The geometric construction for subtracting two vectors is shown in Figure 3.5.

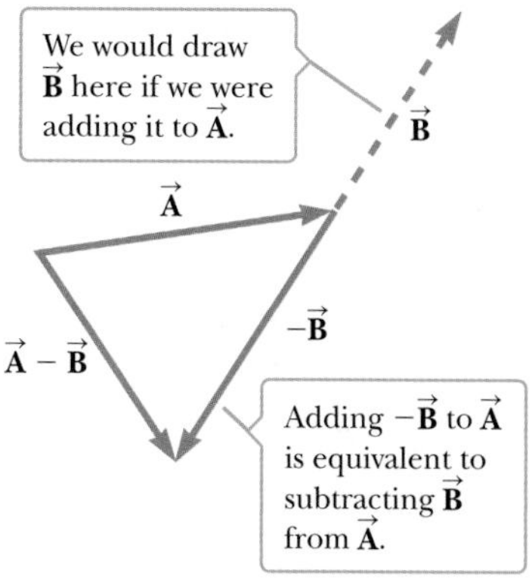

Figure 3.5 This construction shows how to subtract vector $\vec{\mathbf{B}}$ from vector $\vec{\mathbf{A}}$. The vector $-\vec{\mathbf{B}}$ has the same magnitude as the vector $\vec{\mathbf{B}}$, but points in the opposite direction.

Multiplying or Dividing a Vector by a Scalar Multiplying or dividing a vector by a scalar gives a vector. For example, if vector $\vec{\mathbf{A}}$ is multiplied by the scalar number 3, the result, written $3\vec{\mathbf{A}}$, is a vector with a magnitude three times that of $\vec{\mathbf{A}}$ and pointing in the same direction. If we multiply vector $\vec{\mathbf{A}}$ by the scalar -3, the result is $-3\vec{\mathbf{A}}$, a vector with a magnitude three times that of $\vec{\mathbf{A}}$ and pointing in the opposite direction (because of the negative sign).

Quick Quiz

3.1 The magnitudes of two vectors $\vec{\mathbf{A}}$ and $\vec{\mathbf{B}}$ are 12 units and 8 units, respectively. What are the largest and smallest possible values for the magnitude of the resultant vector $\vec{\mathbf{R}} = \vec{\mathbf{A}} + \vec{\mathbf{B}}$? (a) 14.4 and 4 (b) 12 and 8 (c) 20 and 4 (d) none of these.

EXAMPLE 3.1 Taking a Trip

GOAL Find the sum of two vectors by using a graph.

PROBLEM A car travels 20.0 km due north and then 35.0 km in a direction 60.0° west of north, as in Figure 3.6. Using a graph, find the magnitude and direction of a single vector that gives the net effect of the car's trip. This vector is called the car's *resultant displacement.*

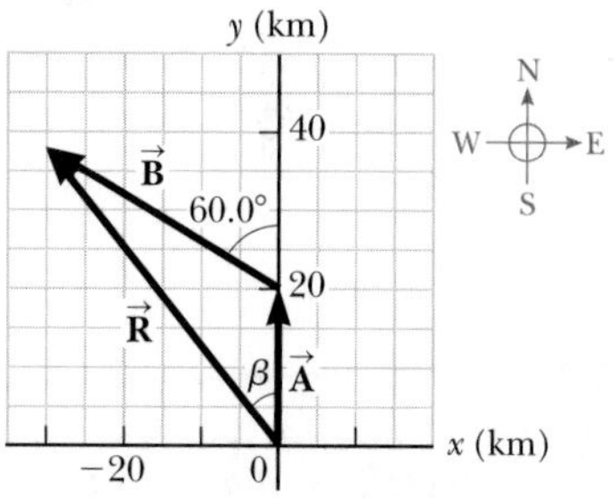

Figure 3.6 (Example 3.1) A graphical method for finding the resultant displacement vector $\vec{\mathbf{R}} = \vec{\mathbf{A}} + \vec{\mathbf{B}}$.

STRATEGY Draw a graph and represent the displacement vectors as arrows. Graphically locate the vector resulting from the sum of the two displacement vectors. Measure its length and angle with respect to the vertical.

SOLUTION

Let $\vec{\mathbf{A}}$ represent the first displacement vector, 20.0 km north, and $\vec{\mathbf{B}}$ the second displacement vector, extending west of north. Carefully graph the two vectors, drawing a resultant vector $\vec{\mathbf{R}}$ with its base touching the base of $\vec{\mathbf{A}}$ and extending to the tip of $\vec{\mathbf{B}}$. Measure the length of this vector, which turns out to be about 48 km. The angle β, measured with a protractor, is about 39° west of north.

REMARKS Notice that ordinary arithmetic doesn't work here: the correct answer of 48 km is not equal to 20.0 km + 35.0 km = 55.0 km!

QUESTION 3.1 Suppose two vectors are added. Under what conditions would the sum of the magnitudes of the vectors equal the magnitude of the resultant vector?

EXERCISE 3.1 Graphically determine the magnitude and direction of the displacement if a man walks 30.0 km 45° north of east and then walks due east 20.0 km.

ANSWER 46 km, 27° north of east

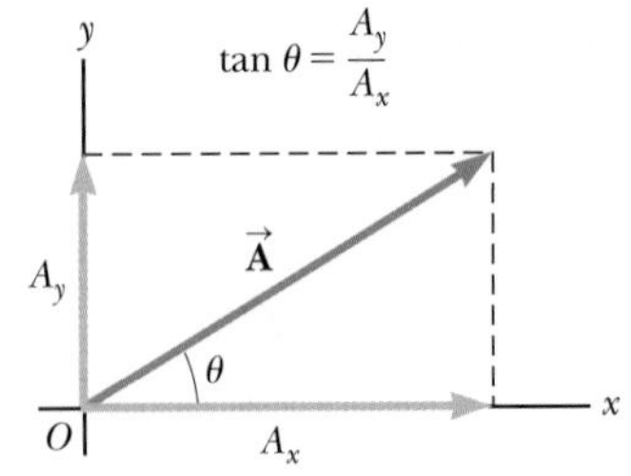

Figure 3.7 Any vector $\vec{\mathbf{A}}$ lying in the *xy*-plane can be represented by its rectangular components A_x and A_y.

3.2 Components of a Vector

LEARNING OBJECTIVES

1. Represent vectors in terms of their magnitudes and directions.
2. Represent vectors in terms of *x*- and *y*-components.
3. Perform arithmetic operations with vectors using their components.

One method of adding vectors makes use of the projections of a vector along the axes of a rectangular coordinate system. These projections are called **components.** Any vector can be completely described by its components.

Consider a vector $\vec{\mathbf{A}}$ in a rectangular coordinate system, as shown in Figure 3.7. $\vec{\mathbf{A}}$ can be expressed as the sum of two vectors: $\vec{\mathbf{A}}_x$, parallel to the *x*-axis, and $\vec{\mathbf{A}}_y$, parallel to the *y*-axis. Mathematically,

$$\vec{\mathbf{A}} = \vec{\mathbf{A}}_x + \vec{\mathbf{A}}_y$$

where $\vec{\mathbf{A}}_x$ and $\vec{\mathbf{A}}_y$ are the component vectors of $\vec{\mathbf{A}}$. The projection of $\vec{\mathbf{A}}$ along the *x*-axis, A_x, is called the *x*-component of $\vec{\mathbf{A}}$, and the projection of $\vec{\mathbf{A}}$ along the *y*-axis, A_y, is called the *y*-component of $\vec{\mathbf{A}}$. These components can be either positive or negative numbers with units. From the definitions of sine and cosine, we see that $\cos\theta = A_x/A$ and $\sin\theta = A_y/A$, so the components of $\vec{\mathbf{A}}$ are

$$A_x = A\cos\theta \qquad [3.2a]$$

$$A_y = A\sin\theta \qquad [3.2b]$$

These components form two sides of a right triangle having a hypotenuse with magnitude *A*. It follows that $\vec{\mathbf{A}}$'s magnitude and direction are related to its components through the Pythagorean theorem and the definition of the tangent:

$$A = \sqrt{A_x^{\,2} + A_y^{\,2}} \qquad [3.3]$$

$$\tan\theta = \frac{A_y}{A_x} \qquad [3.4]$$

To solve for the angle θ, which is measured counterclockwise from the positive *x*-axis by convention, the inverse tangent can be taken of both sides of Equation 3.4:

$$\theta = \tan^{-1}\left(\frac{A_y}{A_x}\right)$$

This formula gives the right answer for θ only half the time! The inverse tangent function returns values only from −90° to +90°, so the answer in your calculator window will only be correct if the vector happens to lie in the first or fourth quadrant. If it lies in the second or third quadrant, adding 180° to the number in the calculator window will always give the right answer. The angle in Equations 3.2 and 3.4 must be measured from the positive *x*-axis. Other choices of reference line are possible, but certain adjustments must then be made. (See Tip 3.2 and Fig. 3.8.)

Tip 3.2 *x*- and *y*-components

Equation 3.2 for the *x*- and *y*-components of a vector associates cosine with the *x*-component and sine with the *y*-component, as in Figure 3.8a. This association is due *solely* to the fact that we chose to measure the angle θ with respect to the positive *x*-axis. If the angle were measured with respect to the *y*-axis, as in Figure 3.8b, the components would be given by $A_x = A\sin\theta$ and $A_y = A\cos\theta$.

Tip 3.3 Inverse Tangents on Calculators: Right Half the Time

The inverse tangent function on calculators returns an angle between −90° and +90°. If the vector lies in the second or third quadrant, the angle, as measured from the positive *x*-axis, will be the angle returned by your calculator plus 180°.

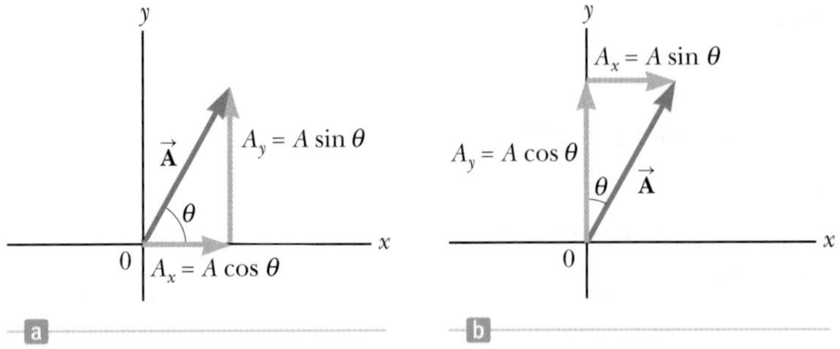

Figure 3.8 The angle θ need not always be defined from the positive *x*-axis.

Quick Quiz

3.4 Which of the following objects can't be accelerating? (a) An object moving with a constant speed; (b) an object moving with a constant velocity; (c) an object moving along a curve.

3.5 Consider the following controls in an automobile: gas pedal, brake, steering wheel. The controls in this list that can cause an acceleration of the car are (a) all three controls, (b) the gas pedal and the brake, (c) only the brake, or (d) only the gas pedal.

3.4 Motion in Two Dimensions

LEARNING OBJECTIVES

1. Describe projectile motion in two dimensions graphically.
2. Apply the two-dimensional kinematics equations to motion with constant acceleration near the surface of the Earth.

In Chapter 2 we studied objects moving along straight-line paths, such as the x-axis. In this chapter, we look at objects that move in both the x- and y-directions simultaneously under constant acceleration. An important special case of this two-dimensional motion is called **projectile motion.**

◀ Projectile motion

Anyone who has tossed any kind of object into the air has observed projectile motion. If the effects of air resistance and the rotation of Earth are neglected, the path of a projectile in Earth's gravity field is curved in the shape of a parabola, as shown in Figure 3.14.

The positive x-direction is horizontal and to the right, and the y-direction is vertical and positive upward. The most important experimental fact about projectile motion in two dimensions is that **the horizontal and vertical motions are completely independent of each other.** This means that motion in one direction has no effect on motion in the other direction. If a baseball is tossed in a parabolic path, as in Figure 3.14, the motion in the y-direction will look just like a ball tossed straight up under the influence of gravity. Figure 3.15 shows the effect of various initial angles; note that complementary angles give the same horizontal range.

Figure 3.16 is an experiment illustrating the independence of horizontal and vertical motion. The gun is aimed directly at the target ball and fired at the instant the target is released. In the absence of gravity, the projectile would hit the target because the target wouldn't move. However, the projectile still hits the target in the presence of gravity. That means

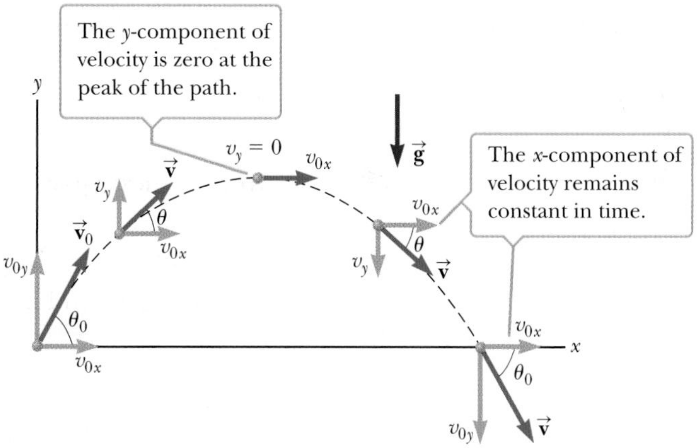

Figure 3.14
The parabolic trajectory of a particle that leaves the origin with a velocity of $\vec{v}_0$. Note that $\vec{v}$ changes with time. However, the x-component of the velocity, v_x, remains constant in time, equal to its initial velocity, v_{0x}. Also, $v_y = 0$ at the peak of the trajectory, but the acceleration is always equal to the free-fall acceleration and acts vertically downward.

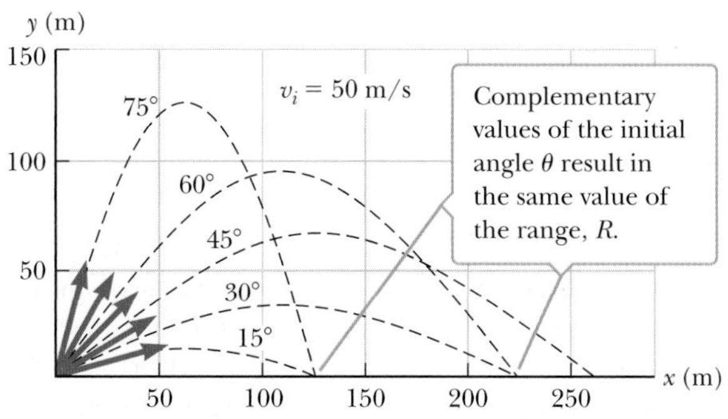

Figure 3.15
A projectile launched from the origin with an initial speed of 50 m/s at various angles of projection.

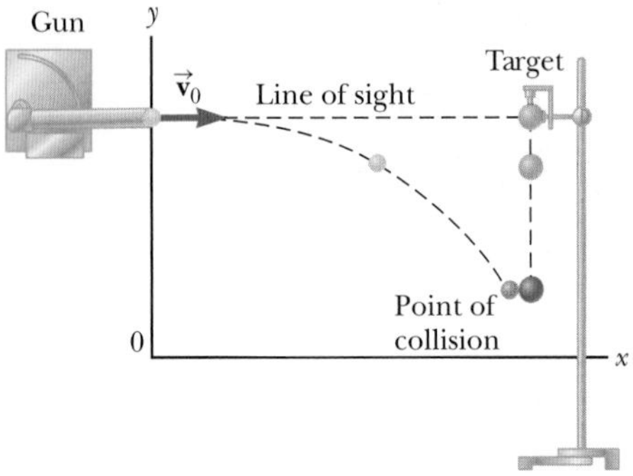

Figure 3.16 A ball is fired at a target at the same instant the target is released. Both fall vertically at the same rate and collide.

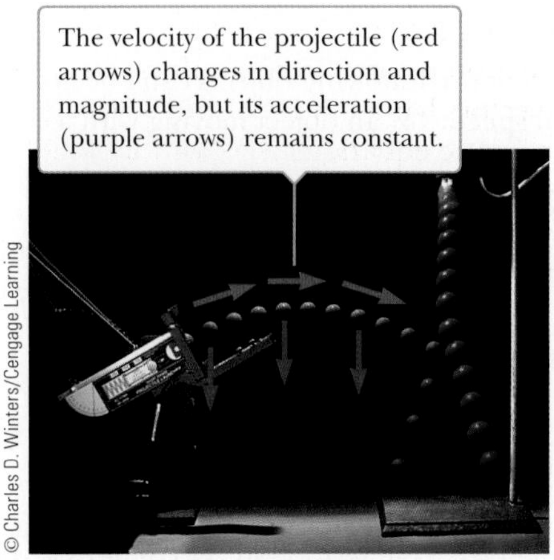

Figure 3.17 Multiflash photograph of the projectile–target demonstration. If the gun is aimed directly at the target and is fired at the same instant the target begins to fall, the projectile will hit the target.

the projectile is falling through the same vertical displacement as the target despite its horizontal motion. The experiment also works when set up as in Figure 3.17, when the initial velocity has a vertical component.

Tip 3.4 Acceleration at the Highest Point

The acceleration in the y-direction is *not* zero at the top of a projectile's trajectory. Only the y-component of the velocity is zero there. If the acceleration were zero, too, the projectile would never come down!

In general, the equations of constant acceleration developed in Chapter 2 follow separately for both the x-direction and the y-direction. An important difference is that the initial velocity now has two components, not just one as in that chapter. We assume that at $t = 0$ the projectile leaves the origin with an initial velocity $\vec{\mathbf{v}}_0$. If the velocity vector makes an angle θ_0 with the horizontal, where θ_0 is called the *projection angle*, then from the definitions of the cosine and sine functions and Figure 3.14 we have

$$v_{0x} = v_0 \cos \theta_0 \qquad \text{and} \qquad v_{0y} = v_0 \sin \theta_0$$

where v_{0x} is the initial velocity (at $t = 0$) in the x-direction and v_{0y} is the initial velocity in the y-direction.

Now, Equations 2.6, 2.9, and 2.10 developed in Chapter 2 for motion with constant acceleration in one dimension carry over to the two-dimensional case; there is one set of three equations for each direction, with the initial velocities modified as just discussed. In the x-direction, with a_x constant, we have

$$v_x = v_{0x} + a_x t \qquad [3.11a]$$

$$\Delta x = v_{0x} t + \tfrac{1}{2} a_x t^2 \qquad [3.11b]$$

$$v_x^2 = v_{0x}^2 + 2a_x \Delta x \qquad [3.11c]$$

where $v_{0x} = v_0 \cos \theta_0$. In the y-direction, we have

$$v_y = v_{0y} + a_y t \qquad [3.12a]$$

$$\Delta y = v_{0y} t + \tfrac{1}{2} a_y t^2 \qquad [3.12b]$$

$$v_y^2 = v_{0y}^2 + 2a_y \Delta y \qquad [3.12c]$$

where $v_{0y} = v_0 \sin \theta_0$ and a_y is constant. The object's speed v can be calculated from the components of the velocity using the Pythagorean theorem:

$$v = \sqrt{v_x^2 + v_y^2}$$

The angle that the velocity vector makes with the x-axis is given by

$$\theta = \tan^{-1}\left(\frac{v_y}{v_x}\right)$$

does it take the skier to reach the ground? (b) How far horizontally does the skier travel in the air before landing? (See Section 3.4.)

8. A catapult launches a large stone from ground level at a speed of 45.0 m/s at an angle of 55.0° with the horizontal. The stone returns to ground level shortly thereafter. (a) How long is it in the air? (b) What maximum height does the stone reach? (Neglect air friction.) (See Section 3.4.)

9. A cruise ship sails due north at 4.50 m/s while a coast guard patrol boat heads 45.0° north of west at 5.20 m/s. What are (a) the x- and (b) y-components of the velocity of the cruise ship relative to the patrol boat? (See Section 3.5.)

CONCEPTUAL QUESTIONS

ENHANCED WebAssign The conceptual questions in this chapter may be assigned online in Enhanced WebAssign.

1. If $\vec{\mathbf{B}}$ is added to $\vec{\mathbf{A}}$, under what conditions does the resultant vector have a magnitude equal to $A + B$? Under what conditions is the resultant vector equal to zero?

2. Under what circumstances would a vector have components that are equal in magnitude?

3. As a projectile moves in its path, is there any point along the path where the velocity and acceleration vectors are (a) perpendicular to each other? (b) Parallel to each other?

4. Construct motion diagrams showing the velocity and acceleration of a projectile at several points along its path, assuming (a) the projectile is launched horizontally and (b) the projectile is launched at an angle θ with the horizontal.

5. Explain whether the following particles do or do not have an acceleration: (a) a particle moving in a straight line with constant speed and (b) a particle moving around a curve with constant speed.

6. A ball is projected horizontally from the top of a building. One second later, another ball is projected horizontally from the same point with the same velocity. (a) At what point in the motion will the balls be closest to each other? (b) Will the first ball always be traveling faster than the second? (c) What will be the time difference between them when the balls hit the ground? (d) Can the horizontal projection velocity of the second ball be changed so that the balls arrive at the ground at the same time?

7. A spacecraft drifts through space at a constant velocity. Suddenly, a gas leak in the side of the spacecraft causes it to constantly accelerate in a direction perpendicular to the initial velocity. The orientation of the spacecraft does not change, so the acceleration remains perpendicular to the original direction of the velocity. What is the shape of the path followed by the spacecraft?

8. Determine which of the following moving objects obey the equations of projectile motion developed in this chapter. (a) A ball is thrown in an arbitrary direction. (b) A jet airplane crosses the sky with its engines thrusting the plane forward. (c) A rocket leaves the launch pad. (d) A rocket moves through the sky after its engines have failed. (e) A stone is thrown under water.

9. Two projectiles are thrown with the same initial speed, one at an angle θ with respect to the level ground and the other at angle $90° - \theta$. Both projectiles strike the ground at the same distance from the projection point. Are both projectiles in the air for the same length of time?

10. A ball is thrown upward in the air by a passenger on a train that is moving with constant velocity. (a) Describe the path of the ball as seen by the passenger. Describe the path as seen by a stationary observer outside the train. (b) How would these observations change if the train were accelerating along the track?

11. A projectile is launched at some angle to the horizontal with some initial speed v_i, and air resistance is negligible. (a) Is the projectile a freely falling body? (b) What is its acceleration in the vertical direction? (c) What is its acceleration in the horizontal direction?

12. A baseball is thrown from the outfield toward the catcher. When the ball reaches its highest point, which statement is true? (a) Its velocity and its acceleration are both zero. (b) Its velocity is not zero, but its acceleration is zero. (c) Its velocity is perpendicular to its acceleration. (d) Its acceleration depends on the angle at which the ball was thrown. (e) None of statements (a) through (d) is true.

13. A student throws a heavy red ball horizontally from a balcony of a tall building with an initial speed v_0. At the same time, a second student drops a lighter blue ball from the same balcony. Neglecting air resistance, which statement is true? (a) The blue ball reaches the ground first. (b) The balls reach the ground at the same instant. (c) The red ball reaches the ground first. (d) Both balls hit the ground with the same speed. (e) None of statements (a) through (d) is true.

14. A car moving around a circular track with constant speed (a) has zero acceleration, (b) has an acceleration component in the direction of its velocity, (c) has an acceleration directed away from the center of its path, (d) has an acceleration directed toward the center of its path, or (e) has an acceleration with

a direction that cannot be determined from the information given.

15. As an apple tree is transported by a truck moving to the right with a constant velocity, one of its apples shakes loose and falls toward the bed of the truck. Of the curves shown in Figure CQ3.15, (i) which best describes the path followed by the apple as seen by a stationary observer on the ground, who observes the truck moving from his left to his right? (ii) Which best describes the path as seen by an observer sitting in the truck?

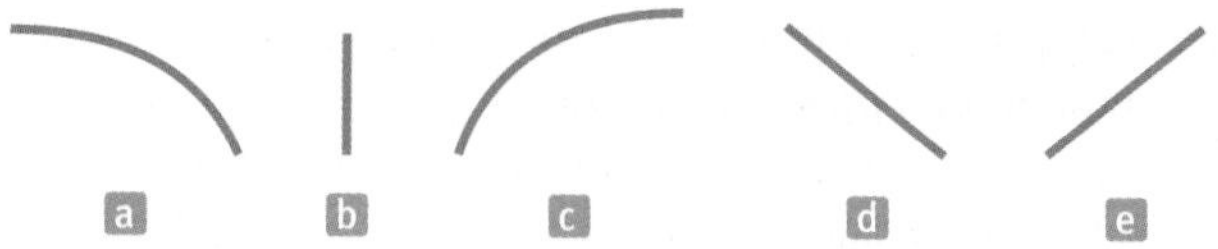

Figure CQ3.15

PROBLEMS

ENHANCED WebAssign The problems in this chapter may be assigned online in Enhanced WebAssign.

1. denotes straightforward problem; 2. denotes intermediate problem;
3. denotes challenging problem

1. denotes full solution available in *Student Solutions Manual/Study Guide*

1. denotes problems most often assigned in Enhanced WebAssign

BIO denotes biomedical problems

GP denotes guided problems

M denotes Master It tutorial available in Enhanced WebAssign

Q|C denotes asking for quantitative and conceptual reasoning

S denotes symbolic reasoning problem

W denotes Watch It video solution available in Enhanced WebAssign

3.1 Vectors and Their Properties

1. Vector $\vec{\mathbf{A}}$ has a magnitude of 29 units and points in the positive y-direction. When vector $\vec{\mathbf{B}}$ is added to $\vec{\mathbf{A}}$, the resultant vector $\vec{\mathbf{A}} + \vec{\mathbf{B}}$ points in the negative y-direction with a magnitude of 14 units. Find the magnitude and direction of $\vec{\mathbf{B}}$.

2. Vector $\vec{\mathbf{A}}$ has a magnitude of 8.00 units and makes an angle of 45.0° with the positive x-axis. Vector $\vec{\mathbf{B}}$ also has a magnitude of 8.00 units and is directed along the negative x-axis. Using graphical methods, find (a) the vector sum $\vec{\mathbf{A}} + \vec{\mathbf{B}}$ and (b) the vector difference $\vec{\mathbf{A}} - \vec{\mathbf{B}}$.

3. Vector $\vec{\mathbf{A}}$ is 3.00 units in length and points along the positive x-axis. Vector $\vec{\mathbf{B}}$ is 4.00 units in length and points along the negative y-axis. Use graphical methods to find the magnitude and direction of the vectors (a) $\vec{\mathbf{A}} + \vec{\mathbf{B}}$ and (b) $\vec{\mathbf{A}} - \vec{\mathbf{B}}$.

4. Q|C Three displacements are $\vec{\mathbf{A}}$ = 200 m due south, $\vec{\mathbf{B}}$ = 250 m due west, and $\vec{\mathbf{C}}$ = 150 m at 30.0° east of north. (a) Construct a separate diagram for each of the following possible ways of adding these vectors: $\vec{\mathbf{R}}_1 = \vec{\mathbf{A}} + \vec{\mathbf{B}} + \vec{\mathbf{C}}$; $\vec{\mathbf{R}}_2 = \vec{\mathbf{B}} + \vec{\mathbf{C}} + \vec{\mathbf{A}}$; $\vec{\mathbf{R}}_3 = \vec{\mathbf{C}} + \vec{\mathbf{B}} + \vec{\mathbf{A}}$. (b) Explain what you can conclude from comparing the diagrams.

5. A roller coaster moves 200 ft horizontally and then rises 135 ft at an angle of 30.0° above the horizontal. Next, it travels 135 ft at an angle of 40.0° below the horizontal. Use graphical techniques to find the roller coaster's displacement from its starting point to the end of this movement.

6. Q|C An airplane flies 200 km due west from city A to city B and then 300 km in the direction of 30.0° north of west from city B to city C. (a) In straight-line distance, how far is city C from city A? (b) Relative to city A, in what direction is city C? (c) Why is the answer only approximately correct?

7. A plane flies from base camp to lake A, a distance of 280 km at a direction of 20.0° north of east. After dropping off supplies, the plane flies to lake B, which is 190 km and 30.0° west of north from lake A. Graphically determine the distance and direction from lake B to the base camp.

8. A force $\vec{\mathbf{F}}_1$ of magnitude 6.00 units acts on an object at the origin in a direction θ = 30.0° above the positive x-axis (Fig. P3.8). A second force $\vec{\mathbf{F}}_2$ of magnitude 5.00 units acts on the object in the direction of the positive y-axis. Find graphically the magnitude and direction of the resultant force $\vec{\mathbf{F}}_1 + \vec{\mathbf{F}}_2$.

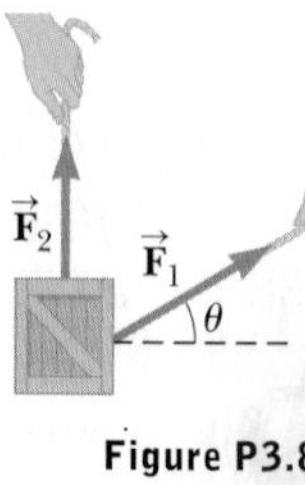

Figure P3.8

9. A man in a maze makes three consecutive displacements. His first displacement is 8.00 m westward, and the second is 13.0 m northward. At the end of his third displacement he is back to where he started. Use the graphical method to find the magnitude and direction of his third displacement.

3.2 Components of a Vector

10. A person walks 25.0° north of east for 3.10 km. How far due north and how far due east would she have to walk to arrive at the same location?

11. The magnitude of vector $\vec{\mathbf{A}}$ is 35.0 units and points in the direction 325° counterclockwise from the positive x-axis. Calculate the x- and y-components of this vector.

12. A figure skater glides along a circular path of radius 5.00 m. If she coasts around one half of the circle, find (a) the magnitude of the displacement vector and

(b) what distance she skated. (c) What is the magnitude of the displacement if she skates all the way around the circle?

13. A girl delivering newspapers covers her route by traveling 3.00 blocks west, 4.00 blocks north, and then 6.00 blocks east. (a) What is her resultant displacement? (b) What is the total distance she travels?

14. QIC A hiker starts at his camp and moves the following distances while exploring his surroundings: 75.0 m north, 2.50×10^2 m east, 125 m at an angle 30.0° north of east, and 1.50×10^2 m south. (a) Find his resultant displacement from camp. (Take east as the positive *x*-direction and north as the positive *y*-direction.) (b) Would changes in the order in which the hiker makes the given displacements alter his final position? Explain.

15. A vector has an *x*-component of −25.0 units and a *y*-component of 40.0 units. Find the magnitude and direction of the vector.

16. A quarterback takes the ball from the line of scrimmage, runs backwards for 10.0 yards, then runs sideways parallel to the line of scrimmage for 15.0 yards. At this point, he throws a 50.0-yard forward pass straight downfield, perpendicular to the line of scrimmage. What is the magnitude of the football's resultant displacement?

17. M The eye of a hurricane passes over Grand Bahama Island in a direction 60.0° north of west with a speed of 41.0 km/h. Three hours later the course of the hurricane suddenly shifts due north, and its speed slows to 25.0 km/h. How far from Grand Bahama is the hurricane 4.50 h after it passes over the island?

18. A map suggests that Atlanta is 730 miles in a direction 5.00° north of east from Dallas. The same map shows that Chicago is 560 miles in a direction 21.0° west of north from Atlanta. Figure P3.18 shows the location of these three cities. Modeling the Earth as flat, use this information to find the displacement from Dallas to Chicago.

Figure P3.18

19. A commuter airplane starts from an airport and takes the route shown in Figure P3.19. The plane first flies to city *A*, located 175 km away in a direction 30.0° north of east. Next, it flies for 150 km 20.0° west of north, to city *B*. Finally, the plane flies 190 km due west, to city *C*. Find the location of city *C* relative to the location of the starting point.

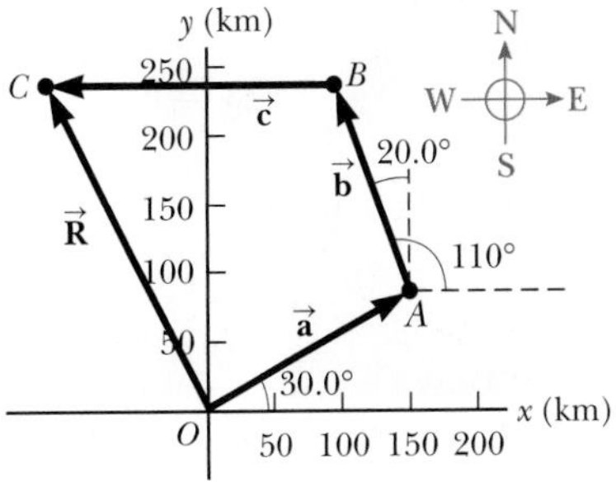

Figure P3.19

20. The helicopter view in Figure P3.20 shows two people pulling on a stubborn mule. Find (a) the single force that is equivalent to the two forces shown and (b) the force a third person would have to exert on the mule to make the net force equal to zero. The forces are measured in units of newtons (N).

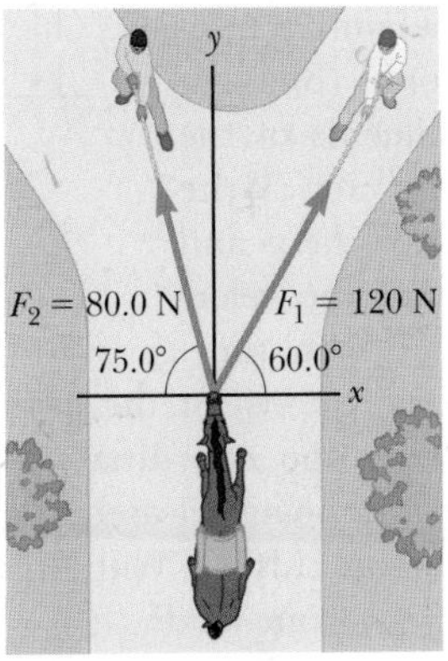

Figure P3.20

21. W A novice golfer on the green takes three strokes to sink the ball. The successive displacements of the ball are 4.00 m to the north, 2.00 m northeast, and 1.00 m at 30.0° west of south (Fig. P3.21). Starting at the same initial point, an expert golfer could make the hole in what single displacement?

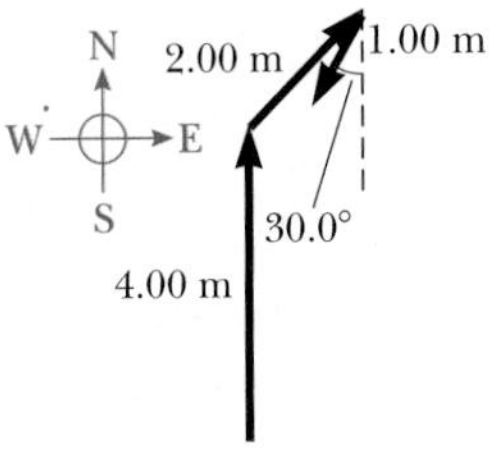

Figure P3.21

3.3 Displacement, Velocity, and Acceleration in Two Dimensions

3.4 Motion in Two Dimensions

22. One of the fastest recorded pitches in major-league baseball, thrown by Tim Lincecum in 2009, was clocked at 101.0 mi/h (Fig. P3.22). If a pitch were thrown horizontally with this velocity, how far would the ball fall vertically by the time it reached home plate, 60.5 ft away?

Figure P3.22 Tim Lincecum throws a baseball.

23. GP A student stands at the edge of a cliff and throws a stone horizontally over the edge with a speed of 18.0 m/s. The cliff is 50.0 m above a flat, horizontal beach as shown in Figure P3.23. (a) What are the coordinates of the initial position of the stone? (b) What are the components of the initial velocity? (c) Write the equations for the x- and y-components of the velocity of the stone with time. (d) Write the equations for the position of the stone with time, using the coordinates in Figure P3.23. (e) How long after being released does the stone strike the beach below the cliff? (f) With what speed and angle of impact does the stone land?

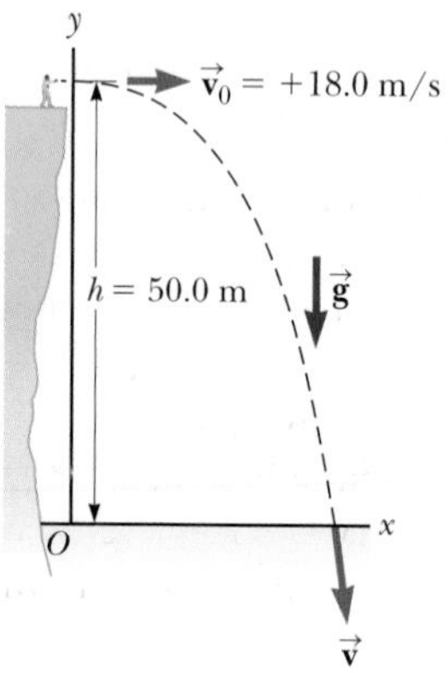

Figure P3.23

24. Q|C S A rock is thrown upward from the level ground in such a way that the maximum height of its flight is equal to its horizontal range R. (a) At what angle θ is the rock thrown? (b) In terms of the original range R, what is the range R_{max} the rock can attain if it is launched at the same speed but at the optimal angle for maximum range? (c) Would your answer to part (a) be different if the rock is thrown with the same speed on a different planet? Explain.

25. W The best leaper in the animal kingdom is the puma, which can jump to a height of 3.7 m when leaving the ground at an angle of 45°. With what speed must the animal leave the ground to reach that height?

26. Q|C The record distance in the sport of throwing cowpats is 81.1 m. This record toss was set by Steve Urner of the United States in 1981. Assuming the initial launch angle was 45° and neglecting air resistance, determine (a) the initial speed of the projectile and (b) the total time the projectile was in flight. (c) Qualitatively, how would the answers change if the launch angle were greater than 45°? Explain.

27. M A place-kicker must kick a football from a point 36.0 m (about 40 yards) from the goal. Half the crowd hopes the ball will clear the crossbar, which is 3.05 m high. When kicked, the ball leaves the ground with a speed of 20.0 m/s at an angle of 53.0° to the horizontal. (a) By how much does the ball clear or fall short of clearing the crossbar? (b) Does the ball approach the crossbar while still rising or while falling?

28. GP From the window of a building, a ball is tossed from a height y_0 above the ground with an initial velocity of 8.00 m/s and angle of 20.0° below the horizontal. It strikes the ground 3.00 s later. (a) If the base of the building is taken to be the origin of the coordinates, with upward the positive y-direction, what are the initial coordinates of the ball? (b) With the positive x-direction chosen to be out the window, find the x- and y-components of the initial velocity. (c) Find the equations for the x- and y-components of the position as functions of time. (d) How far horizontally from the base of the building does the ball strike the ground? (e) Find the height from which the ball was thrown. (f) How long does it take the ball to reach a point 10.0 m below the level of launching?

29. A brick is thrown upward from the top of a building at an angle of 25° to the horizontal and with an initial speed of 15 m/s. If the brick is in flight for 3.0 s, how tall is the building?

30. An artillery shell is fired with an initial velocity of 300 m/s at 55.0° above the horizontal. To clear an avalanche, it explodes on a mountainside 42.0 s after firing. What are the x- and y-coordinates of the shell where it explodes, relative to its firing point?

31. A car is parked on a cliff overlooking the ocean on an incline that makes an angle of 24.0° below the horizontal. The negligent driver leaves the car in neutral, and the emergency brakes are defective. The car rolls from rest down the incline with a constant acceleration of 4.00 m/s^2 for a distance of 50.0 m to the edge of the cliff, which is 30.0 m above the ocean. Find (a) the car's position relative to the base of the cliff when the car lands in the ocean and (b) the length of time the car is in the air.

32. W A fireman $d = 50.0$ m away from a burning building directs a stream of water from a ground-level fire hose at an angle of $\theta_i = 30.0°$ above the horizontal as shown in Figure P3.32. If the speed of the stream as it leaves the hose is $v_i = 40.0$ m/s, at what height will the stream of water strike the building?

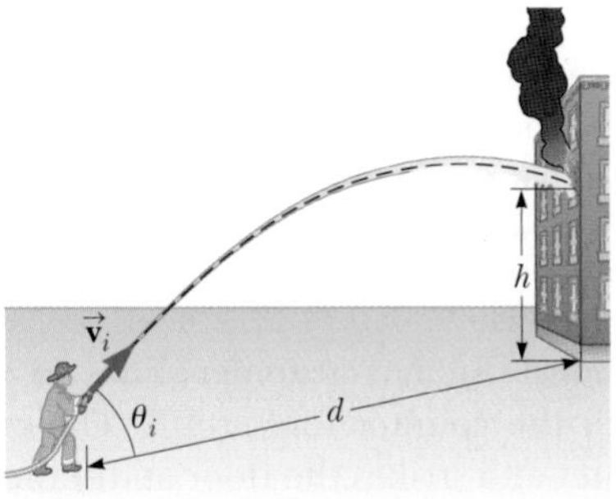

Figure P3.32

33. A projectile is launched with an initial speed of 60.0 m/s at an angle of 30.0° above the horizontal. The projectile lands on a hillside 4.00 s later. Neglect air friction. (a) What is the projectile's velocity at the highest point of its trajectory? (b) What is the straight-line distance from where the projectile was launched to where it hits its target?

34. A playground is on the flat roof of a city school, 6.00 m above the street below (Fig. P3.34). The vertical wall of the building is h = 7.00 m high, to form a 1-m-high railing around the playground. A ball has fallen to the street below, and a passerby returns it by launching it at an angle of θ = 53.0° above the horizontal at a point d = 24.0 m from the base of the building wall. The ball takes 2.20 s to reach a point vertically above the wall. (a) Find the speed at which the ball was launched. (b) Find the vertical distance by which the ball clears the wall. (c) Find the horizontal distance from the wall to the point on the roof where the ball lands.

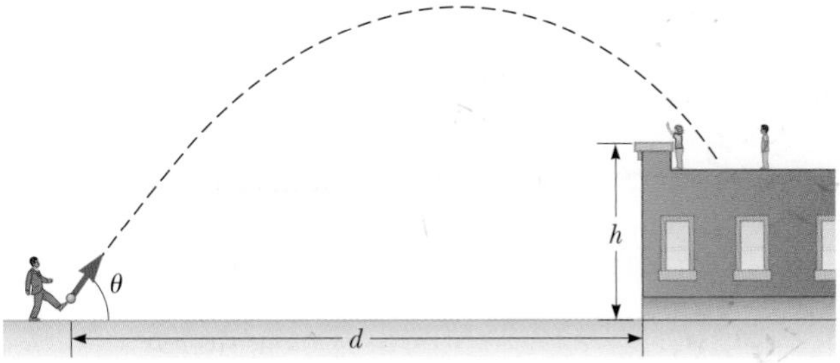

Figure P3.34

3.5 Relative Velocity

35. GP A jet airliner moving initially at 3.00×10^2 mi/h due east enters a region where the wind is blowing 1.00×10^2 mi/h in a direction 30.0° north of east. (a) Find the components of the velocity of the jet airliner relative to the air, $\vec{\mathbf{v}}_{JA}$. (b) Find the components of the velocity of the air relative to Earth, $\vec{\mathbf{v}}_{AE}$. (c) Write an equation analogous to Equation 3.16 for the velocities $\vec{\mathbf{v}}_{JA}$, $\vec{\mathbf{v}}_{AE}$, and $\vec{\mathbf{v}}_{JE}$. (d) What are the speed and direction of the aircraft relative to the ground?

36. M A car travels due east with a speed of 50.0 km/h. Raindrops are falling at a constant speed vertically with respect to the Earth. The traces of the rain on the side windows of the car make an angle of 60.0° with the vertical. Find the velocity of the rain with respect to (a) the car and (b) the Earth.

37. Q|C A bolt drops from the ceiling of a moving train car that is accelerating northward at a rate of 2.50 m/s^2. (a) What is the acceleration of the bolt relative to the train car? (b) What is the acceleration of the bolt relative to the Earth? (c) Describe the trajectory of the bolt as seen by an observer fixed on the Earth.

38. A Coast Guard cutter detects an unidentified ship at a distance of 20.0 km in the direction 15.0° east of north. The ship is traveling at 26.0 km/h on a course at 40.0° east of north. The Coast Guard wishes to send a speedboat to intercept and investigate the vessel. (a) If the speedboat travels at 50.0 km/h, in what direction should it head? Express the direction as a compass bearing with respect to due north. (b) Find the time required for the cutter to intercept the ship.

39. An airplane maintains a speed of 630 km/h relative to the air it is flying through, as it makes a trip to a city 750 km away to the north. (a) What time interval is required for the trip if the plane flies through a headwind blowing at 35.0 km/h toward the south? (b) What time interval is required if there is a tailwind with the same speed? (c) What time interval is required if there is a crosswind blowing at 35.0 km/h to the east relative to the ground?

40. BIO Suppose a chinook salmon needs to jump a waterfall that is 1.50 m high. If the fish starts from a distance 1.00 m from the base of the ledge over which the waterfall flows, (a) find the x- and y-components of the initial velocity the salmon would need to just reach the ledge at the top of its trajectory. (b) Can the fish make this jump? (Note that a chinook salmon can jump out of the water with an initial speed of 6.26 m/s.)

41. Q|C A river has a steady speed of 0.500 m/s. A student swims upstream a distance of 1.00 km and swims back to the starting point. (a) If the student can swim at a speed of 1.20 m/s in still water, how long does the trip take? (b) How much time is required in still water for the same length swim? (c) Intuitively, why does the swim take longer when there is a current?

42. Q|C S *This is a symbolic version of Problem 41.* A river has a steady speed of v_s. A student swims upstream a distance d and back to the starting point. (a) If the student can swim at a speed of v in still water, how much time t_{up} does it take the student to swim upstream a distance d? Express the answer in terms of d, v, and v_s. (b) Using the same variables, how much time t_{down} does it take to swim back downstream to the starting

point? (c) Sum the answers found in parts (a) and (b) and show that the time t_a required for the whole trip can be written as

$$t_a = \frac{2d/v}{1 - v_s^2/v^2}$$

(d) How much time t_b does the trip take in still water? (e) Which is larger, t_a or t_b? Is it always larger?

43. A bomber is flying horizontally over level terrain at a speed of 275 m/s relative to the ground and at an altitude of 3.00 km. (a) The bombardier releases one bomb. How far does the bomb travel horizontally between its release and its impact on the ground? Ignore the effects of air resistance. (b) Firing from the people on the ground suddenly incapacitates the bombardier before he can call, "Bombs away!" Consequently, the pilot maintains the plane's original course, altitude, and speed through a storm of flak. Where is the plane relative to the bomb's point of impact when the bomb hits the ground? (c) The plane has a telescopic bombsight set so that the bomb hits the target seen in the sight at the moment of release. At what angle from the vertical was the bombsight set?

Additional Problems

44. **S** A moving walkway at an airport has a speed v_1 and a length L. A woman stands on the walkway as it moves from one end to the other, while a man in a hurry to reach his flight walks on the walkway with a speed of v_2 relative to the moving walkway. (a) How long does it take the woman to travel the distance L? (b) How long does it take the man to travel this distance?

45. How long does it take an automobile traveling in the left lane of a highway at 60.0 km/h to overtake (become even with) another car that is traveling in the right lane at 40.0 km/h when the cars' front bumpers are initially 100 m apart?

46. You can use any coordinate system you like to solve a projectile motion problem. To demonstrate the truth of this statement, consider a ball thrown off the top of a building with a velocity $\vec{\mathbf{v}}$ at an angle θ with respect to the horizontal. Let the building be 50.0 m tall, the initial horizontal velocity be 9.00 m/s, and the initial vertical velocity be 12.0 m/s. Choose your coordinates such that the positive y-axis is upward, the x-axis is to the right, and the origin is at the point where the ball is released. (a) With these choices, find the ball's maximum height above the ground and the time it takes to reach the maximum height. (b) Repeat your calculations choosing the origin at the base of the building.

47. **QIC** A Nordic jumper goes off a ski jump at an angle of 10.0° below the horizontal, traveling 108 m horizontally and 55.0 m vertically before landing. (a) Ignoring friction and aerodynamic effects, calculate the speed needed by the skier on leaving the ramp. (b) Olympic Nordic jumpers can make such jumps with a jump speed of 23.0 m/s, which is considerably less than the answer found in part (a). Explain how that is possible.

48. **S** In a local diner, a customer slides an empty coffee cup down the counter for a refill. The cup slides off the counter and strikes the floor at distance d from the base of the counter. If the height of the counter is h, (a) find an expression for the time t it takes the cup to fall to the floor in terms of the variables h and g. (b) With what speed does the mug leave the counter? Answer in terms of the variables d, g, and h. (c) In the same terms, what is the speed of the cup immediately before it hits the floor? (d) In terms of h and d, what is the direction of the cup's velocity immediately before it hits the floor?

49. **M** Towns A and B in Figure P3.49 are 80.0 km apart. A couple arranges to drive from town A and meet a couple driving from town B at the lake, L. The two couples leave simultaneously and drive for 2.50 h in the directions shown. Car 1 has a speed of 90.0 km/h. If the cars arrive simultaneously at the lake, what is the speed of car 2?

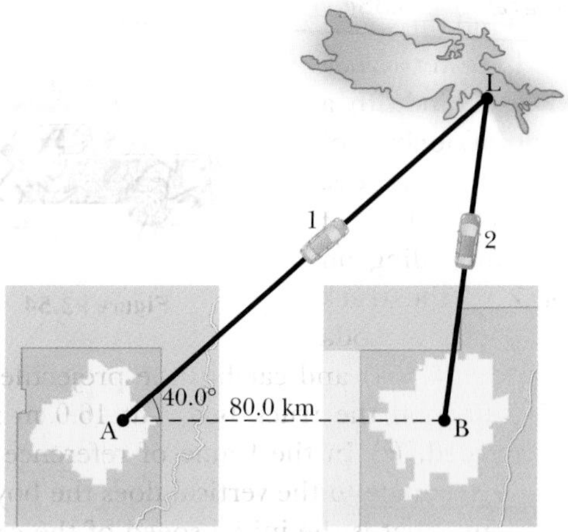

Figure P3.49

50. **BIO** A chinook salmon has a maximum underwater speed of 3.58 m/s, but it can jump out of water with a speed of 6.26 m/s. To move upstream past a waterfall, the salmon does not need to jump to the top of the fall, but only to a point in the fall where the water speed is less than 3.58 m/s; it can then swim up the fall for the remaining distance. Because the salmon must make forward progress in the water, let's assume it can swim to the top if the water speed is 3.00 m/s. If water has a speed of 1.50 m/s as it passes over a ledge, (a) how far below the ledge will the water be moving with a speed of 3.00 m/s? (Note that water undergoes projectile motion once it leaves the ledge.) (b) If the salmon

is able to jump vertically upward from the base of the fall, what is the maximum height of waterfall that the salmon can clear?

51. A rocket is launched at an angle of 53.0° above the horizontal with an initial speed of 100 m/s. The rocket moves for 3.00 s along its initial line of motion with an acceleration of 30.0 m/s^2. At this time, its engines fail and the rocket proceeds to move as a projectile. Find (a) the maximum altitude reached by the rocket, (b) its total time of flight, and (c) its horizontal range.

52. W Two canoeists in identical canoes exert the same effort paddling and hence maintain the same speed relative to the water. One paddles directly upstream (and moves upstream), whereas the other paddles directly downstream. With downstream as the positive direction, an observer on shore determines the velocities of the two canoes to be −1.2 m/s and +2.9 m/s, respectively. (a) What is the speed of the water relative to the shore? (b) What is the speed of each canoe relative to the water?

53. (a) If a person can jump a maximum horizontal distance (by using a 45° projection angle) of 3.0 m on Earth, what would be his maximum range on the Moon, where the free-fall acceleration is $g/6$ and g = 9.80 m/s^2? (b) Repeat for Mars, where the acceleration due to gravity is $0.38g$.

54. QIC A farm truck moves due east with a constant velocity of 9.50 m/s on a limitless, horizontal stretch of road. A boy riding on the back of the truck throws a can of soda upward (Fig. P3.54) and catches the projectile at the same location on the truck bed, but 16.0 m farther down the road. (a) In the frame of reference of the truck, at what angle to the vertical does the boy throw the can? (b) What is the initial speed of the can relative to the truck? (c) What is the shape of the can's trajectory as seen by the boy? An observer on the ground watches the boy throw the can and catch it. In this observer's frame of reference, (d) describe the shape of the can's path and (e) determine the initial velocity of the can.

Figure P3.54

55. M A home run is hit in such a way that the baseball just clears a wall 21 m high, located 130 m from home plate. The ball is hit at an angle of 35° to the horizontal, and air resistance is negligible. Find (a) the initial speed of the ball, (b) the time it takes the ball to reach the wall, and (c) the velocity components and the speed of the ball when it reaches the wall. (Assume the ball is hit at a height of 1.0 m above the ground.)

56. A ball is thrown straight upward and returns to the thrower's hand after 3.00 s in the air. A second ball thrown at an angle of 30.0° with the horizontal reaches the same maximum height as the first ball. (a) At what speed was the first ball thrown? (b) At what speed was the second ball thrown?

57. A quarterback throws a football toward a receiver with an initial speed of 20 m/s at an angle of 30° above the horizontal. At that instant the receiver is 20 m from the quarterback. In (a) what direction and (b) with what constant speed should the receiver run in order to catch the football at the level at which it was thrown?

58. A 2.00-m-tall basketball player is standing on the floor 10.0 m from the basket, as in Figure P3.58. If he shoots the ball at a 40.0° angle with the horizontal, at what initial speed must he throw the basketball so that it goes through the hoop without striking the backboard? The height of the basket is 3.05 m.

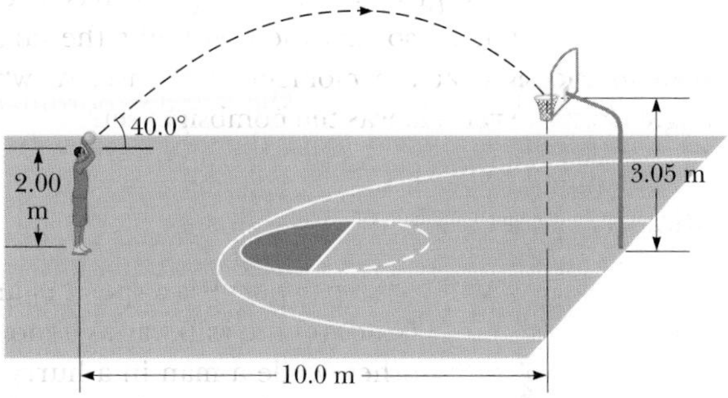

Figure P3.58

59. QIC S In a very popular lecture demonstration, a projectile is fired at a falling target as in Figure P3.59. The projectile leaves the gun at the same instant the target is dropped from rest. Assuming the gun is initially aimed at the target, show that the projectile will hit the target. (One restriction of this experiment is that the projectile must reach the target before the target strikes the floor.)

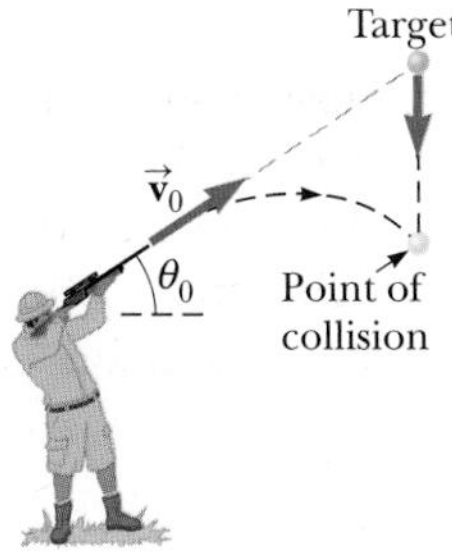

Figure P3.59

60. BIO Figure P3.60 illustrates the difference in proportions between the male (m) and female (f) anatomies. The displacements $\vec{d}_{1m}$ and $\vec{d}_{1f}$ from the bottom of the feet to the navel have magnitudes of 104 cm and 84.0 cm, respectively. The displacements $\vec{d}_{2m}$ and $\vec{d}_{2f}$ have magnitudes of 50.0 cm and 43.0 cm, respectively. (a) Find the vector sum of the displacements $\vec{d}_{d1}$ and $\vec{d}_{d2}$ in each case. (b) The male figure is 180 cm tall, the female 168 cm. Normalize the displacements of each

figure to a common height of 200 cm and re-form the vector sums as in part (a). Then find the vector difference between the two sums.

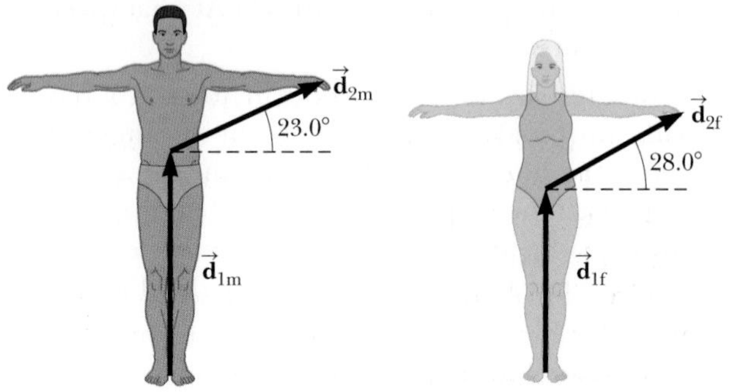

Figure P3.60

61. **Q|C** By throwing a ball at an angle of 45°, a girl can throw the ball a maximum horizontal distance R on a level field. How far can she throw the same ball vertically upward? Assume her muscles give the ball the same speed in each case. (Is this assumption valid?)

62. **S** The equation of a parabola is $y = ax^2 + bx + c$, where a, b, and c are constants. The x- and y-coordinates of a projectile launched from the origin as a function of time are given by $x = v_{0x}t$ and $y = v_{0y}t - \frac{1}{2}gt^2$, where v_{0x} and v_{0y} are the components of the initial velocity. (a) Eliminate t from these two equations and show that the path of a projectile is a parabola and has the form $y = ax + bx^2$. (b) What are the values of a, b, and c for the projectile?

63. A hunter wishes to cross a river that is 1.5 km wide and flows with a speed of 5.0 km/h parallel to its banks. The hunter uses a small powerboat that moves at a maximum speed of 12 km/h with respect to the water. What is the minimum time necessary for crossing?

64. **S** When baseball outfielders throw the ball, they usually allow it to take one bounce, on the theory that the ball arrives at its target sooner that way. Suppose that, after the bounce, the ball rebounds at the same angle θ that it had when it was released (as in Fig. P3.64), but loses half its speed. (a) Assuming that the ball is always thrown with the same initial speed, at what angle θ should the ball be thrown in order to go the same distance D with one bounce as a ball thrown upward at 45.0° with no bounce? (b) Determine the ratio of the times for the one-bounce and no-bounce throws.

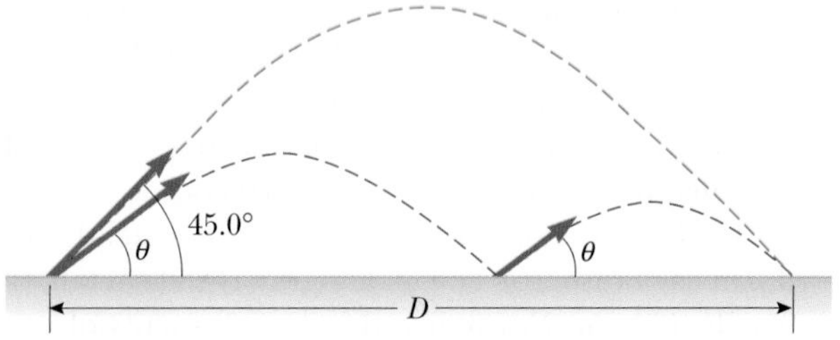

Figure P3.64

65. A daredevil is shot out of a cannon at 45.0° to the horizontal with an initial speed of 25.0 m/s. A net is positioned a horizontal distance of 50.0 m from the cannon. At what height above the cannon should the net be placed in order to catch the daredevil?

66. **BIO** Chinook salmon are able to move upstream faster by jumping out of the water periodically; this behavior is called *porpoising*. Suppose a salmon swimming in still water jumps out of the water with a speed of 6.26 m/s at an angle of 45°, sails through the air a distance L before returning to the water, and then swims a distance L underwater at a speed of 3.58 m/s before beginning another porpoising maneuver. Determine the average speed of the fish.

67. **Q|C** A student decides to measure the muzzle velocity of a pellet shot from his gun. He points the gun horizontally. He places a target on a vertical wall a distance x away from the gun. The pellet hits the target a vertical distance y below the gun. (a) Show that the position of the pellet when traveling through the air is given by $y = Ax^2$, where A is a constant. (b) Express the constant A in terms of the initial (muzzle) velocity and the free-fall acceleration. (c) If $x = 3.00$ m and $y = 0.210$ m, what is the initial speed of the pellet?

68. A sailboat is heading directly north at a speed of 20 knots (1 knot = 0.514 m/s). The wind is blowing toward the east with a speed of 17 knots. (a) Determine the magnitude and direction of the wind velocity as measured on the boat. (b) What is the component of the wind velocity in the direction parallel to the motion of the boat? (See Problem 58 in Chapter 4 for an explanation of how a sailboat can move "into the wind.")

69. A golf ball with an initial speed of 50.0 m/s lands exactly 240 m downrange on a level course. (a) Neglecting air friction, what *two* projection angles would achieve this result? (b) What is the maximum height reached by the ball, using the two angles determined in part (a)?

70. **Q|C** A landscape architect is planning an artificial waterfall in a city park. Water flowing at 0.750 m/s leaves the end of a horizontal channel at the top of a vertical wall $h = 2.35$ m high and falls into a pool (Fig. P3.70). (a) How far from the wall will the water land? Will the space behind the waterfall be wide enough for a pedestrian walkway? (b) To sell her plan to the city council, the architect wants to build a model to standard scale, one-twelfth actual size. How fast should the water flow in the channel in the model?

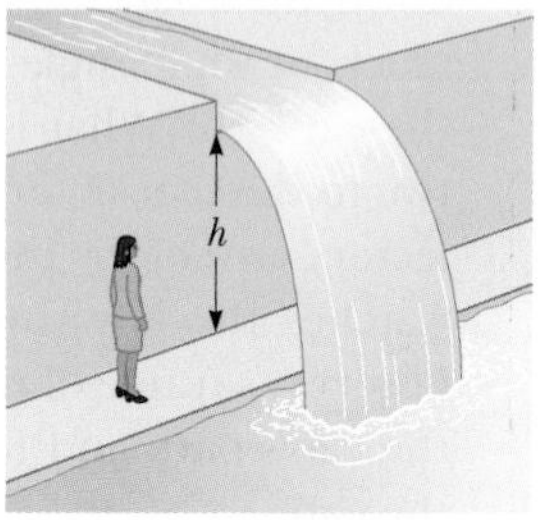

Figure P3.70

71. One strategy in a snowball fight is to throw a snowball at a high angle over level ground. Then, while your opponent is watching that snowball, you throw a second one

at a low angle timed to arrive before or at the same time as the first one. Assume both snowballs are thrown with a speed of 25.0 m/s. The first is thrown at an angle of 70.0° with respect to the horizontal. (a) At what angle should the second snowball be thrown to arrive at the same point as the first? (b) How many seconds later should the second snowball be thrown after the first in order for both to arrive at the same time?

72. A dart gun is fired while being held horizontally at a height of 1.00 m above ground level and while it is at rest relative to the ground. The dart from the gun travels a horizontal distance of 5.00 m. A college student holds the same gun in a horizontal position while sliding down a 45.0° incline at a constant speed of 2.00 m/s. How far will the dart travel if the student fires the gun when it is 1.00 m above the ground?

73. The determined Wile E. Coyote is out once more to try to capture the elusive roadrunner. The coyote wears a new pair of power roller skates, which provide a constant horizontal acceleration of 15 m/s^2, as shown in Figure P3.73. The coyote starts off at rest 70 m from the edge of a cliff at the instant the roadrunner zips by in the direction of the cliff. (a) If the roadrunner moves with constant speed, find the minimum speed the roadrunner must have to reach the cliff before the coyote. (b) If the cliff is 100 m above the base of a canyon, find where the coyote lands in the canyon. (Assume his skates are still in operation when he is in "flight" and that his horizontal component of acceleration remains constant at 15 m/s^2.)

Figure P3.73

74. A truck loaded with cannonball watermelons stops suddenly to avoid running over the edge of a washed-out bridge (Fig. P3.74). The quick stop causes a number of melons to fly off the truck. One melon rolls over the edge with an initial speed v_i = 10.0 m/s in the horizontal direction. A cross section of the bank has the shape of the bottom half of a parabola with its vertex at the edge of the road, and with the equation y^2 = (16.0 m) x, where x and y are measured in meters. What are the x- and y-coordinates of the melon when it splatters on the bank?

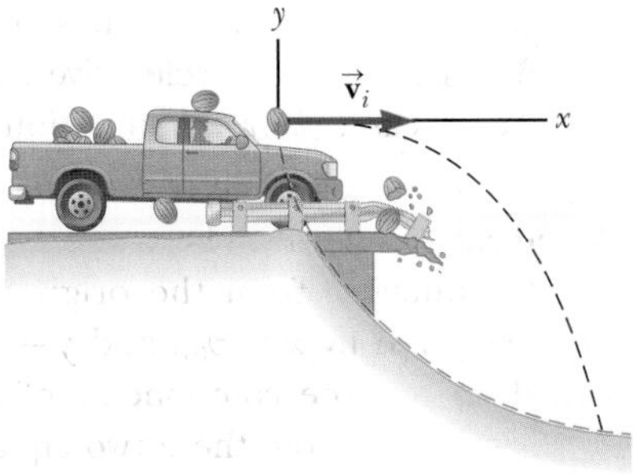

Figure P3.74 The blue dashed curve shows the parabolic shape of the bank.

A rock climber depends on a number of different forces to overcome the force of gravity and reach a summit. Her muscles apply forces to the rocks and to her own body, static friction forces helping her maintain her grip and enabling motion. Safety lines and the tension forces they can exert offer insurance against a fall. The climber uses the third law of motion continually: by exerting a downward force on the rock, the rock exerts an equal and opposite force on her, propelling her up the sheer cliff.

Greg Epperson/Shutterstock.com

4 The Laws of Motion

Classical mechanics describes the relationship between the motion of objects found in our everyday world and the forces acting on them. As long as the system under study doesn't involve objects comparable in size to an atom or traveling close to the speed of light, classical mechanics provides an excellent description of nature.

This chapter introduces Newton's three laws of motion and his law of gravity. The three laws are simple and sensible. The first law states that a force must be applied to an object in order to change its velocity. Changing an object's velocity means accelerating it, which implies a relationship between force and acceleration. This relationship, the second law, states that the net force on an object equals the object's mass times its acceleration. Finally, the third law says that whenever we push on something, it pushes back with equal force in the opposite direction. Those are the three laws in a nutshell.

Newton's three laws, together with his invention of calculus, opened avenues of inquiry and discovery that are used routinely today in virtually all areas of mathematics, science, engineering, and technology. Newton's theory of universal gravitation had a similar impact, starting a revolution in celestial mechanics and astronomy that continues to this day. With the advent of this theory, the orbits of all the planets could be calculated to high precision and the tides understood. The theory even led to the prediction of "dark stars," now called black holes, more than two centuries before any evidence for their existence was observed.[1] Newton's three laws of motion, together with his law of gravitation, are considered among the greatest achievements of the human mind.

[1] In 1783, John Michell combined Newton's theory of light and theory of gravitation, predicting the existence of "dark stars" from which light itself couldn't escape.

4.1 Forces

LEARNING OBJECTIVES

1. Describe the distinction between contact forces and field forces.
2. Identify the four field forces and describe their roles in the interactions of matter.

A **force** is commonly imagined as a push or a pull on some object, perhaps rapidly, as when we hit a tennis ball with a racket. (See Fig. 4.1.) We can hit the ball at different speeds and direct it into different parts of the opponent's court. That means we can control the magnitude of the applied force and also its direction, so force is a vector quantity, just like velocity and acceleration.

If you pull on a spring (Fig. 4.2a), the spring stretches. If you pull hard enough on a wagon (Fig. 4.2b), the wagon moves. When you kick a football (Fig. 4.2c), it deforms briefly and is set in motion. These are all examples of **contact forces,** so named because they result from physical contact between two objects.

Another class of forces doesn't involve any direct physical contact. Early scientists, including Newton, were uneasy with the concept of forces that act between two disconnected objects. Nonetheless, Newton used this "action-at-a-distance" concept in his law of gravity, whereby a mass at one location, such as the Sun, affects the motion of a distant object such as Earth despite no evident physical connection between the two objects. To overcome the conceptual difficulty associated with action at a distance, Michael Faraday (1791–1867) introduced the concept of a *field.* The corresponding forces are called **field forces.** According to this approach, an object of mass M, such as the Sun, creates an invisible influence that stretches throughout space. A second object of mass m, such as Earth, interacts with the *field* of the Sun, not directly with the Sun itself. So the force of gravitational attraction between two objects, illustrated in Figure 4.2d, is an example of a field force. The force of gravity keeps objects bound to Earth and also gives rise to what we call the *weight* of those objects.

Another common example of a field force is the electric force that one electric charge exerts on another (Fig. 4.2e). A third example is the force exerted by a bar magnet on a piece of iron (Fig. 4.2f).

The known fundamental forces in nature are all field forces. These are, in order of decreasing strength, (1) the strong nuclear force between subatomic particles, (2) the electromagnetic forces between electric charges, (3) the weak nuclear force, which arises in certain radioactive decay processes, and (4) the gravitational force between objects. The strong force keeps the nucleus of an atom from flying apart due to the repulsive electric force of the protons. The weak force is involved in most radioactive processes and plays an important role in the nuclear reactions

Figure 4.1 A tennis player applies a contact force to the ball with her racket, accelerating and directing the ball toward the open court.

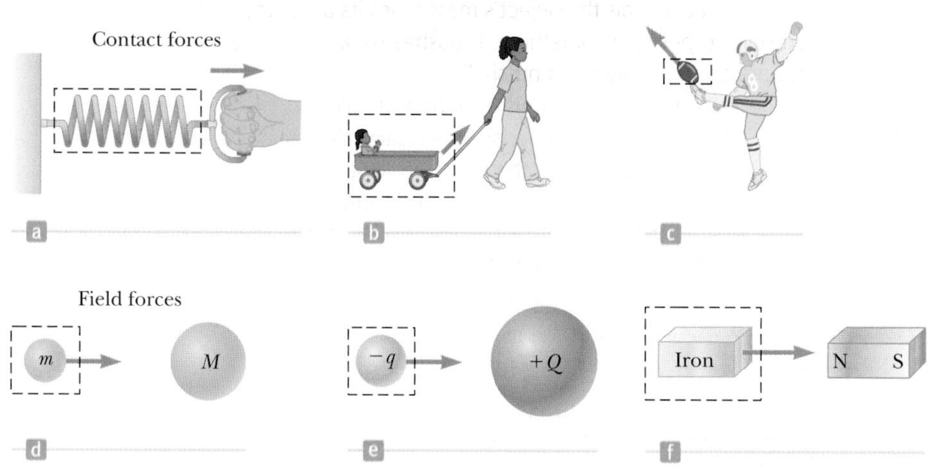

Figure 4.2 Examples of forces applied to various objects. In each case, a force acts on the object surrounded by the dashed lines. Something in the environment external to the boxed area exerts the force.

that generate the Sun's energy output. The strong and weak forces operate only on the nuclear scale, with a very short range on the order of 10^{-15} m. Outside this range they have no influence. Classical physics, however, deals only with gravitational and electromagnetic forces, which have infinite range.

Forces exerted on an object can change the object's shape. For example, striking a tennis ball with a racket, as in Figure 4.1, deforms the ball to some extent. Even objects we usually consider rigid and inflexible are deformed under the action of external forces. Often the deformations are permanent, as in the case of a collision between automobiles.

4.2 Newton's First Law

LEARNING OBJECTIVES

1. Explain what the first law of motion implies about an object's motion and the forces acting on it.
2. Explain the concepts of mass and inertia and the relationship between them.

Consider a book lying on a table. Obviously, the book remains at rest if left alone. Now imagine pushing the book with a horizontal force great enough to overcome the force of friction between the book and the table, setting the book in motion. Because the magnitude of the applied force exceeds the magnitude of the friction force, the book accelerates. When the applied force is withdrawn (Fig. 4.3a), friction soon slows the book to a stop.

Now imagine pushing the book across a smooth, waxed floor. The book again comes to rest once the force is no longer applied, but not as quickly as before. Finally, if the book is moving on a horizontal frictionless surface (Fig. 4.3b), it continues to move in a straight line with constant velocity until it hits a wall or some other obstruction.

Before about 1600, scientists felt that the natural state of matter was the state of rest. Galileo, however, devised thought experiments—such as an object moving on a frictionless surface, as just described—and concluded that **it's not the nature of an object to stop once set in motion, but rather to continue in its original state of motion.** This approach was later formalized as **Newton's first law of motion:**

Newton's first law ▶

> An object moves with a velocity that is constant in magnitude and direction unless a non-zero net force acts on it.

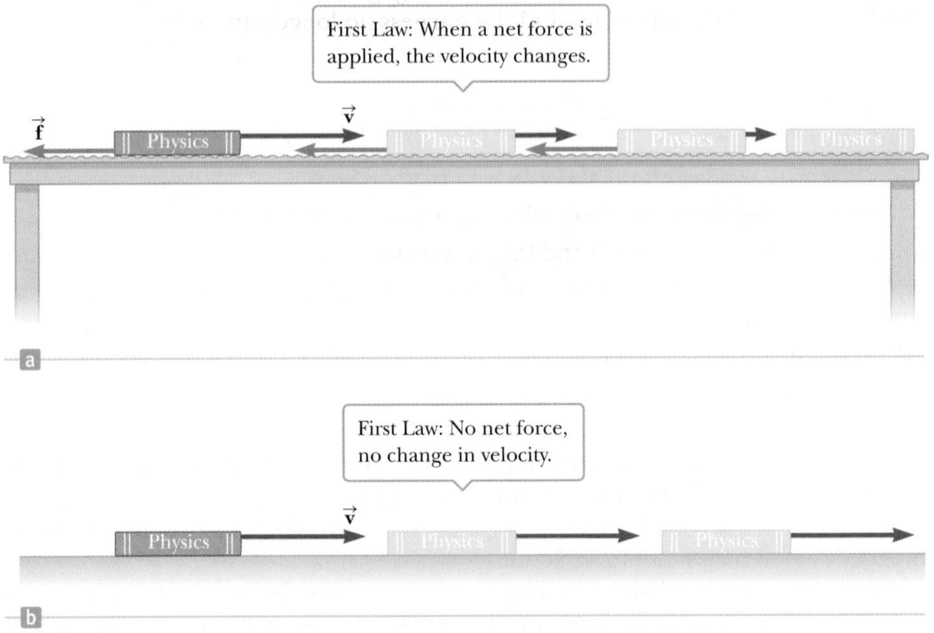

Figure 4.3 The first law of motion. (a) A book moves at an initial velocity of $\vec{v}$ on a surface with friction. Because there is a friction force acting horizontally, the book slows to rest. (b) A book moves at velocity $\vec{v}$ on a frictionless surface. In the absence of a net force, the book keeps moving at velocity v.

15. After falling from rest from a height of 30 m, a 0.50-kg ball rebounds upward, reaching a height of 20 m. If the contact between ball and ground lasted 2.0 ms, what average force was exerted on the ball?

16. M The force exerted by the wind on the sails of a sailboat is 390 N north. The water exerts a force of 180 N east. If the boat (including its crew) has a mass of 270 kg, what are the magnitude and direction of its acceleration?

4.5 Applications of Newton's Laws

17. Q|C (a) Find the tension in each cable supporting the 600-N cat burglar in Figure P4.17. (b) Suppose the horizontal cable were reattached higher up on the wall. Would the tension in the other cable increase, decrease, or stay the same? Why?

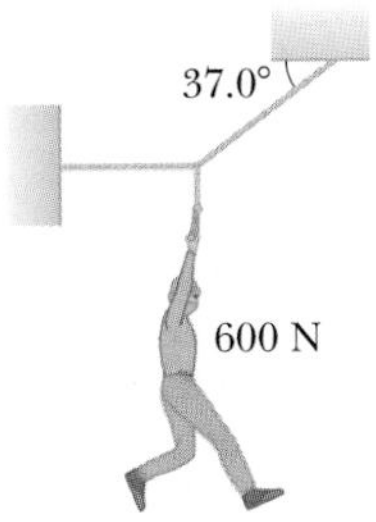

Figure P4.17

18. BIO A certain orthodontist uses a wire brace to align a patient's crooked tooth as in Figure P4.18. The tension in the wire is adjusted to have a magnitude of 18.0 N. Find the magnitude of the net force exerted by the wire on the crooked tooth.

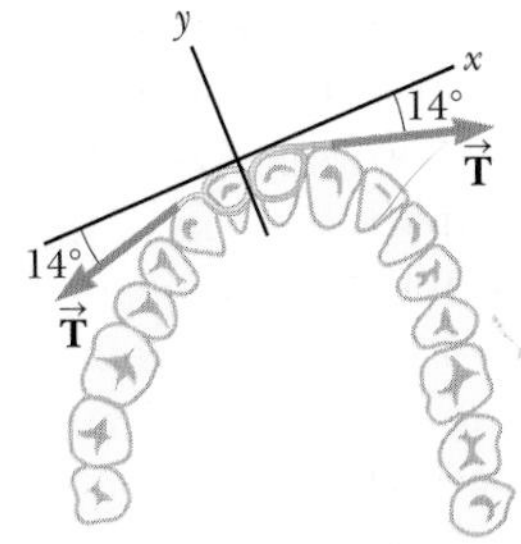

Figure P4.18

19. A 150-N bird feeder is supported by three cables as shown in Figure P4.19. Find the tension in each cable.

Figure P4.19

20. BIO The leg and cast in Figure P4.20 weigh 220 N (w_1). Determine the weight w_2 and the angle α needed so that no force is exerted on the hip joint by the leg plus the cast.

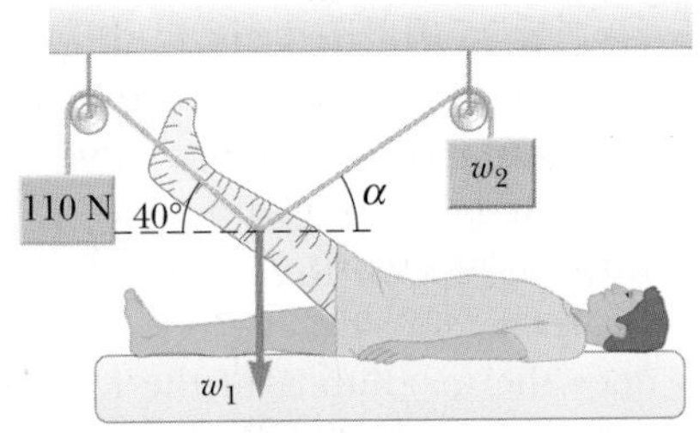

Figure P4.20

21. Two blocks each of mass m = 3.50 kg are fastened to the top of an elevator as in Figure P4.21. (a) If the elevator has an upward acceleration a = 1.60 m/s^2, find the tensions T_1 and T_2 in the upper and lower strings. (b) If the strings can withstand a maximum tension of 85.0 N, what maximum acceleration can the elevator have before the upper string breaks?

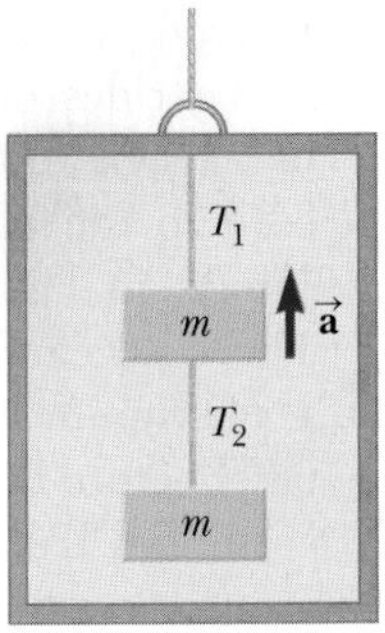

Figure P4.21
(Problems 21 and 22)

22. S Two blocks each of mass m are fastened to the top of an elevator as in Figure P4.21. The elevator has an upward acceleration a. The strings have negligible mass. (a) Find the tensions T_1 and T_2 in the upper and lower strings in terms of m, a, and g. (b) Compare the two tensions and determine which string would break first if a is made sufficiently large. (c) What are the tensions if the cable supporting the elevator breaks?

23. The distance between two telephone poles is 50.0 m. When a 1.00-kg bird lands on the telephone wire midway between the poles, the wire sags 0.200 m. Draw a free-body diagram of the bird. How much tension does the bird produce in the wire? Ignore the weight of the wire.

24. The systems shown in Figure P4.24 are in equilibrium. If the spring scales are calibrated in newtons, what do they read? Ignore the masses of the pulleys and strings and assume the pulleys and the incline in Figure P4.24d are frictionless.

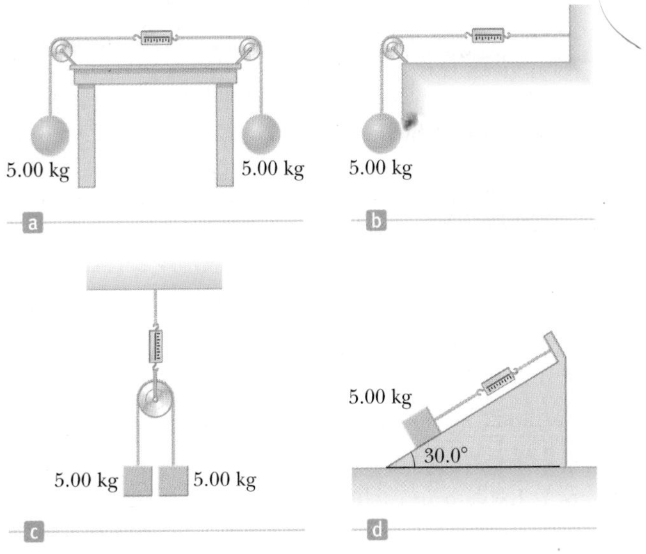

Figure P4.24

25. W A 5.0-kg bucket of water is raised from a well by a rope. If the upward acceleration of the bucket is 3.0 m/s^2, find the force exerted by the rope on the bucket.

26. A crate of mass m = 32 kg rides on the bed of a truck attached by a cord to the back of the cab as in Figure P4.26. The cord can withstand a maximum tension of 68 N before breaking. Neglecting friction between the crate and truck bed, find the maximum acceleration the truck can have before the cord breaks.

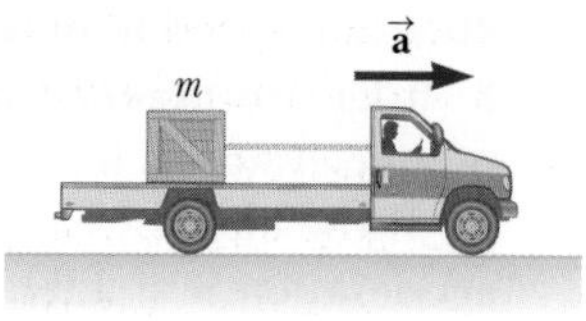

Figure P4.26

27. S Two blocks of masses m and $2m$ are held in equilibrium on a frictionless incline as in Figure P4.27. In terms of m and θ, find (a) the magnitude of the tension T_1 in the upper cord and (b) the magnitude of the tension T_2 in the lower cord connecting the two blocks.

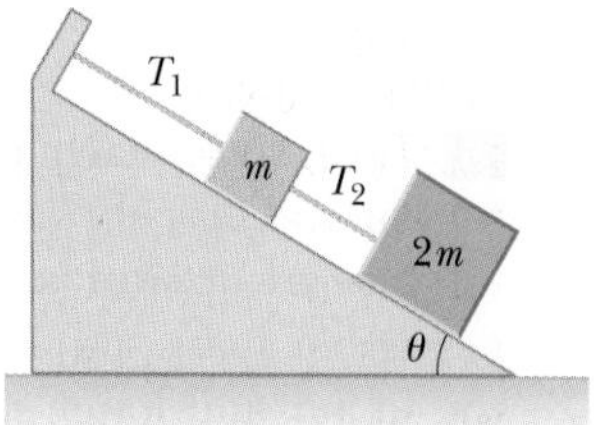

Figure P4.27

28. Two packing crates of masses 10.0 kg and 5.00 kg are connected by a light string that passes over a frictionless pulley as in Figure P4.28. The 5.00-kg crate lies on a smooth incline of angle 40.0°. Find (a) the acceleration of the 5.00-kg crate and (b) the tension in the string.

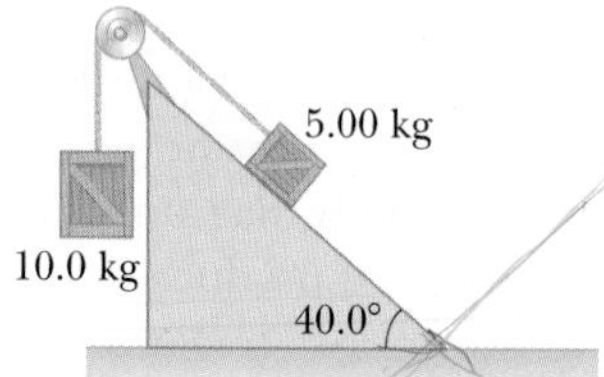

Figure P4.28

29. M Assume the three blocks portrayed in Figure P4.29 move on a frictionless surface and a 42-N force acts as shown on the 3.0-kg block. Determine (a) the acceleration given this system, (b) the tension in the cord connecting the 3.0-kg and the 1.0-kg blocks, and (c) the force exerted by the 1.0-kg block on the 2.0-kg block.

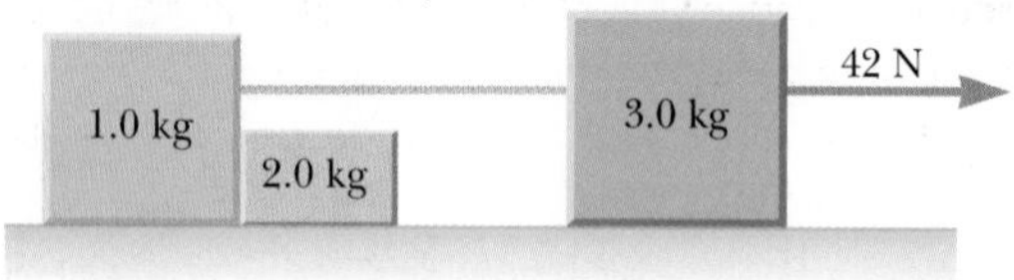

Figure P4.29

30. A block of mass m = 5.8 kg is pulled up a θ = 25° incline as in Figure P4.30 with a force of magnitude F = 32 N. (a) Find the acceleration of the block if the incline is frictionless. (b) Find the acceleration of the block if the coefficient of kinetic friction between the block and incline is 0.10.

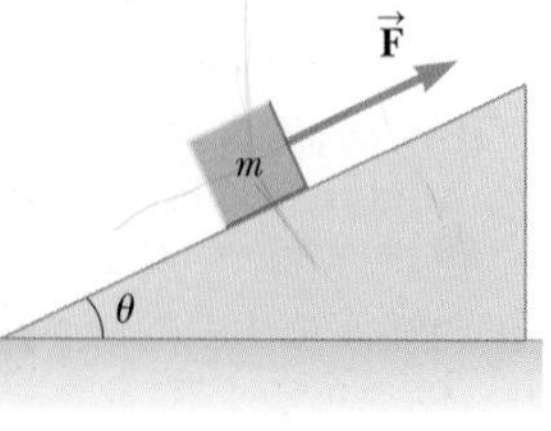

Figure P4.30

31. BIO A setup similar to the one shown in Figure P4.31 is often used in hospitals to support and apply a traction force to an injured leg. (a) Determine the force of tension in the rope supporting the leg. (b) What is the traction force exerted on the leg? Assume the traction force is horizontal.

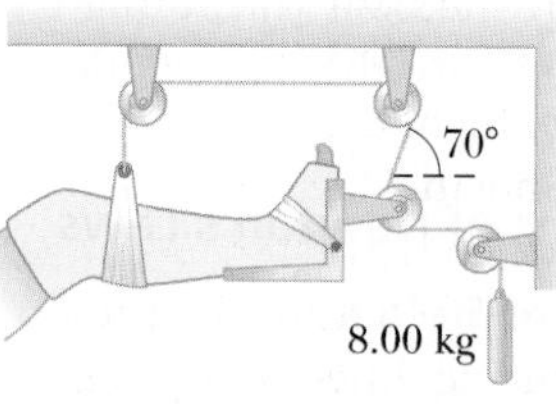

Figure P4.31

32. GP Two blocks of masses m_1 and m_2 ($m_1 > m_2$) are placed on a frictionless table in contact with each other. A horizontal force of magnitude F is applied to the block of mass m_1 in Figure P4.32. (a) If P is the magnitude of the contact force between the blocks, draw the free-body diagrams for each block. (b) What is the net force on the system consisting of both blocks? (c) What is the net force acting on m_1? (d) What is the net force acting on m_2? (e) Write the x-component of Newton's second law for each block. (f) Solve the resulting system of two equations and two unknowns, expressing the acceleration a and contact force P in terms of the masses and force. (g) How would the answers change if the force had been applied to m_2 instead? (*Hint:* use symmetry; don't calculate!) Is the contact force larger, smaller, or the same in this case? Why?

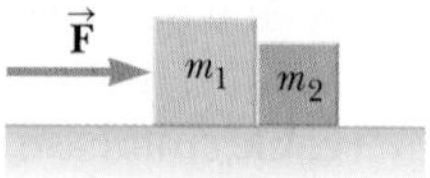

Figure P4.32

33. A 276-kg glider is being pulled by a 1 950-kg jet along a horizontal runway with an acceleration of $\vec{a}$ = 2.20 m/s^2 to the right as in Figure P4.33. Find (a) the thrust provided by the jet's engines and (b) the magnitude of the tension in the cable connecting the jet and glider.

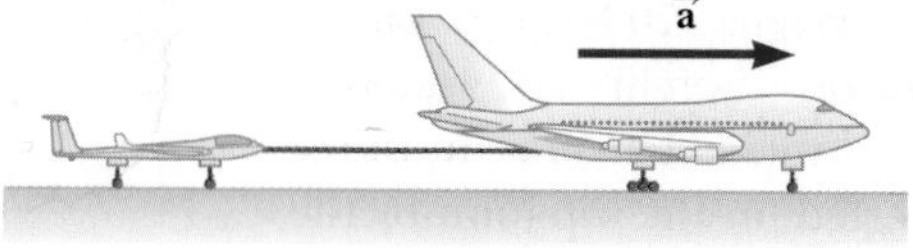

Figure P4.33

34. Q|C In Figure P4.34, the light, taut, unstretchable cord B joins block 1 and the larger-mass block 2. Cord A exerts a force on block 1 to make it accelerate forward. (a) How does the magnitude of the force exerted by cord A on block 1 compare with the magnitude of the force exerted by cord B on block 2? (b) How does the

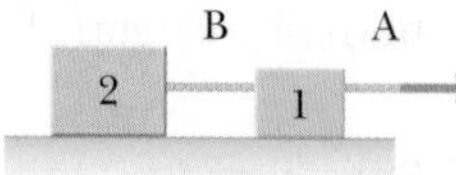

Figure P4.34

acceleration of block 1 compare with the acceleration of block 2? (c) Does cord B exert a force on block 1? Explain your answer.

35. Q|C (a) An elevator of mass m moving upward has two forces acting on it: the upward force of tension in the cable and the downward force due to gravity. When the elevator is accelerating upward, which is greater, T or w? (b) When the elevator is moving at a constant velocity upward, which is greater, T or w? (c) When the elevator is moving upward, but the acceleration is downward, which is greater, T or w? (d) Let the elevator have a mass of 1 500 kg and an upward acceleration of 2.5 m/s^2. Find T. Is your answer consistent with the answer to part (a)? (e) The elevator of part (d) now moves with a constant upward velocity of 10 m/s. Find T. Is your answer consistent with your answer to part (b)? (f) Having initially moved upward with a constant velocity, the elevator begins to accelerate downward at 1.50 m/s^2. Find T. Is your answer consistent with your answer to part (c)?

36. W An object with mass m_1 = 5.00 kg rests on a frictionless horizontal table and is connected to a cable that passes over a pulley and is then fastened to a hanging object with mass m_2 = 10.0 kg, as shown in Figure P4.36. Find (a) the acceleration of each object and (b) the tension in the cable.

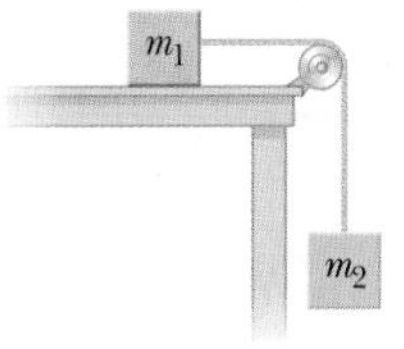

Figure P4.36
(Problems 36, 40, and 45)

37. A 1 000-kg car is pulling a 300-kg trailer. Together, the car and trailer have an acceleration of 2.15 m/s^2 in the positive x-direction. Neglecting frictional forces on the trailer, determine (a) the net force on the car, (b) the net force on the trailer, (c) the magnitude and direction of the force exerted by the trailer on the car, and (d) the resultant force exerted by the car on the road.

38. Two objects with masses of 3.00 kg and 5.00 kg are connected by a light string that passes over a frictionless pulley, as in Figure P4.38. Determine (a) the tension in the string, (b) the acceleration of each object, and (c) the distance each object will move in the first second of motion if both objects start from rest.

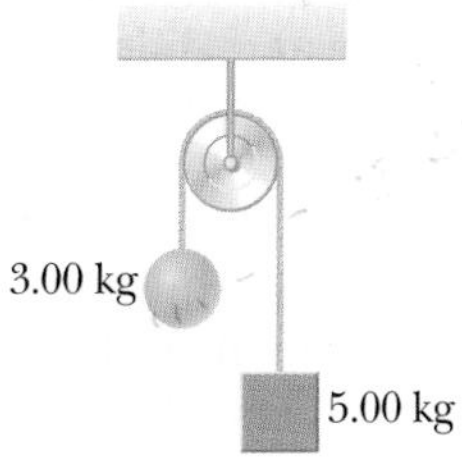

Figure P4.38

4.6 Forces of Friction

39. A dockworker loading crates on a ship finds that a 20-kg crate, initially at rest on a horizontal surface, requires a 75-N horizontal force to set it in motion. However, after the crate is in motion, a horizontal force of 60 N is required to keep it moving with a constant speed. Find the coefficients of static and kinetic friction between crate and floor.

40. In Figure P4.36, m_1 = 10 kg and m_2 = 4.0 kg. The coefficient of static friction between m_1 and the horizontal surface is 0.50, and the coefficient of kinetic friction is 0.30. (a) If the system is released from rest, what will its acceleration be? (b) If the system is set in motion with m_2 moving downward, what will be the acceleration of the system?

41. A 1 000-N crate is being pushed across a level floor at a constant speed by a force $\vec{\mathbf{F}}$ of 300 N at an angle of 20.0° below the horizontal, as shown in Figure P4.41a. (a) What is the coefficient of kinetic friction between the crate and the floor? (b) If the 300-N force is instead pulling the block at an angle of 20.0° above the horizontal, as shown in Figure P4.41b, what will be the acceleration of the crate? Assume that the coefficient of friction is the same as that found in part (a).

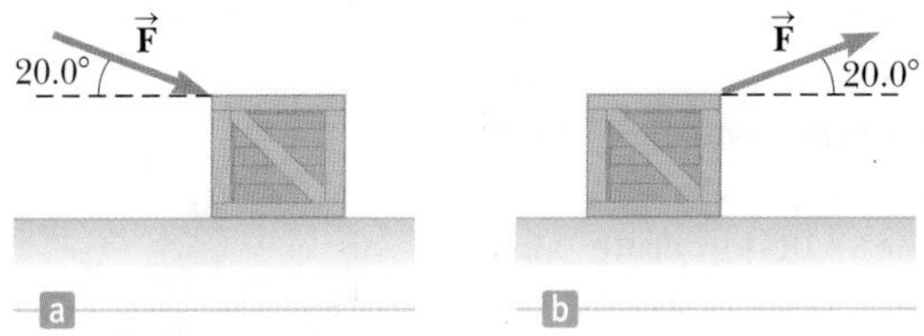

Figure P4.41

42. Q|C S A block of mass $3m$ is placed on a frictionless horizontal surface, and a second block of mass m is placed on top of the first block. The surfaces of the blocks are rough. A constant force of magnitude F is applied to the first block as in Figure P4.42. (a) Construct free-body diagrams for each block. (b) Identify the horizontal force that causes the block of mass m to accelerate. (c) Assume that the upper block does not slip on the lower block, and find the acceleration of each block in terms of m and F.

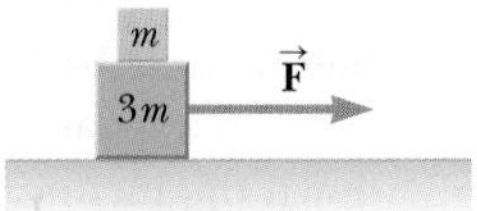

Figure P4.42

43. Consider a large truck carrying a heavy load, such as steel beams. A significant hazard for the driver is that the load may slide forward, crushing the cab, if the truck stops suddenly in an accident or even in braking. Assume, for example, a 10 000-kg load sits on the flatbed of a 20 000-kg truck moving at 12.0 m/s. Assume the load is not tied down to the truck and has a coefficient of static friction of 0.500 with the truck bed. (a) Calculate the minimum stopping distance for which the load will not slide forward relative to the truck. (b) Is any piece of data unnecessary for the solution?

44. A crate of mass 45.0 kg is being transported on the flatbed of a pickup truck. The coefficient of static friction

between the crate and the truck's flatbed is 0.350, and the coefficient of kinetic friction is 0.320. (a) The truck accelerates forward on level ground. What is the maximum acceleration the truck can have so that the crate does not slide relative to the truck's flatbed? (b) The truck barely exceeds this acceleration and then moves with constant acceleration, with the crate sliding along its bed. What is the acceleration of the crate relative to the ground?

45. M Objects with masses $m_1 = 10.0$ kg and $m_2 = 5.00$ kg are connected by a light string that passes over a frictionless pulley as in Figure P4.36. If, when the system starts from rest, m_2 falls 1.00 m in 1.20 s, determine the coefficient of kinetic friction between m_1 and the table.

46. S A hockey puck struck by a hockey stick is given an initial speed v_0 in the positive x-direction. The coefficient of kinetic friction between the ice and the puck is μ_k. (a) Obtain an expression for the acceleration of the puck. (b) Use the result of part (a) to obtain an expression for the distance d the puck slides. The answer should be in terms of the variables v_0, μ_k, and g only.

47. W The coefficient of static friction between the 3.00-kg crate and the 35.0° incline of Figure P4.47 is 0.300. What minimum force $\vec{\mathbf{F}}$ must be applied to the crate perpendicular to the incline to prevent the crate from sliding down the incline?

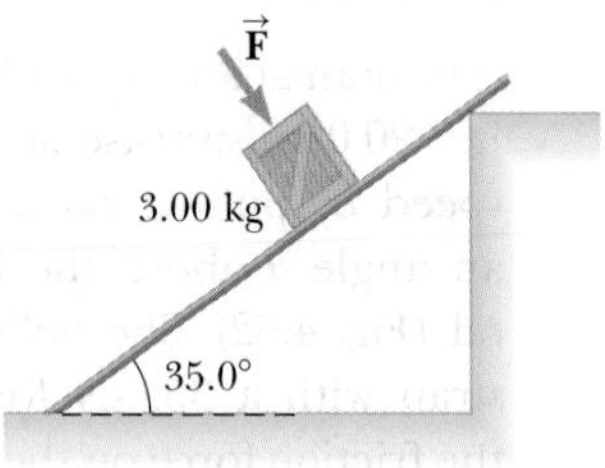

Figure P4.47

48. A student decides to move a box of books into her dormitory room by pulling on a rope attached to the box. She pulls with a force of 80.0 N at an angle of 25.0° above the horizontal. The box has a mass of 25.0 kg, and the coefficient of kinetic friction between box and floor is 0.300. (a) Find the acceleration of the box. (b) The student now starts moving the box up a 10.0° incline, keeping her 80.0 N force directed at 25.0° above the line of the incline. If the coefficient of friction is unchanged, what is the new acceleration of the box?

49. QIC An object falling under the pull of gravity is acted upon by a frictional force of air resistance. The magnitude of this force is approximately proportional to the speed of the object, which can be written as $f = bv$. Assume $b = 15$ kg/s and $m = 50$ kg. (a) What is the terminal speed the object reaches while falling? (b) Does your answer to part (a) depend on the initial speed of the object? Explain.

50. A car is traveling at 50.0 km/h on a flat highway. (a) If the coefficient of friction between road and tires on a rainy day is 0.100, what is the minimum distance in which the car will stop? (b) What is the stopping distance when the surface is dry and the coefficient of friction is 0.600?

51. A 3.00-kg block starts from rest at the top of a 30.0° incline and slides 2.00 m down the incline in 1.50 s. Find (a) the acceleration of the block, (b) the coefficient of kinetic friction between the block and the incline, (c) the frictional force acting on the block, and (d) the speed of the block after it has slid 2.00 m.

52. A 15.0-lb block rests on a horizontal floor. (a) What force does the floor exert on the block? (b) A rope is tied to the block and is run vertically over a pulley. The other end is attached to a free-hanging 10.0-lb object. What now is the force exerted by the floor on the 15.0-lb block? (c) If the 10.0-lb object in part (b) is replaced with a 20.0-lb object, what is the force exerted by the floor on the 15.0-lb block?

53. To meet a U.S. Postal Service requirement, employees' footwear must have a coefficient of static friction of 0.500 or more on a specified tile surface. A typical athletic shoe has a coefficient of 0.800. In an emergency, what is the minimum time interval in which a person starting from rest can move 3.00 m on the tile surface if she is wearing (a) footwear meeting the Postal Service minimum and (b) a typical athletic shoe?

54. Objects of masses $m_1 = 4.00$ kg and $m_2 = 9.00$ kg are connected by a light string that passes over a frictionless pulley as in Figure P4.54. The object m_1 is held at rest on the floor, and m_2 rests on a fixed incline of $\theta = 40.0°$. The objects are released from rest, and m_2 slides 1.00 m down the incline in 4.00 s. Determine (a) the acceleration of each object, (b) the tension in the string, and (c) the coefficient of kinetic friction between m_2 and the incline.

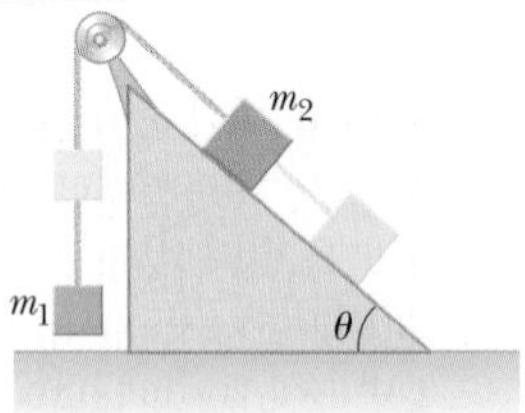

Figure P4.54

55. BIO The person in Figure P4.55 weighs 170 lb. Each crutch makes an angle of 22.0° with the vertical (as seen from the front). Half of the person's weight is supported by the crutches, the other half by the vertical forces exerted by the ground on his feet. Assuming he is at rest and the force exerted by the ground on the crutches acts along the crutches, determine (a) the smallest possible coefficient of friction between crutches and ground and (b) the magnitude of the compression force supported by each crutch.

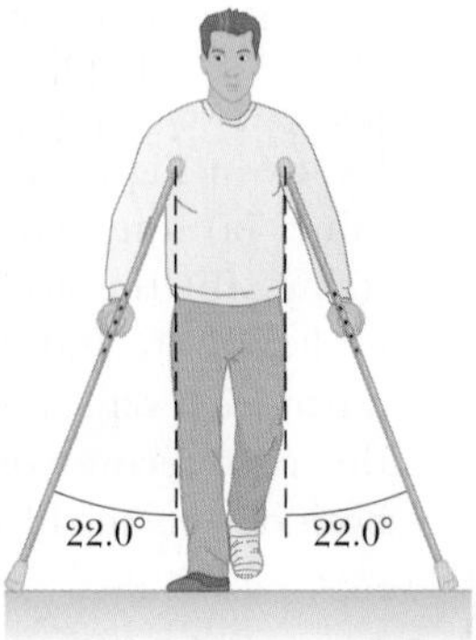

Figure P4.55

Additional Problems

56. As a protest against the umpire's calls, a baseball pitcher throws a ball straight up into the air at a speed of 20.0 m/s. In the process, he moves his hand through a distance of 1.50 m. If the ball has a mass of 0.150 kg, find the force he exerts on the ball to give it this upward speed.

57. Q|C Three objects are connected on a table as shown in Figure P4.57. The coefficient of kinetic friction between the block of mass m_2 and the table is 0.350. The objects have masses of $m_1 = 4.00$ kg, $m_2 = 1.00$ kg, and $m_3 = 2.00$ kg as shown, and the pulleys are frictionless. (a) Draw a diagram of the forces on each object. (b) Determine the acceleration of each object, including its direction. (c) Determine the tensions in the two cords. (d) If the tabletop were smooth, would the tensions increase, decrease, or remain the same? Explain.

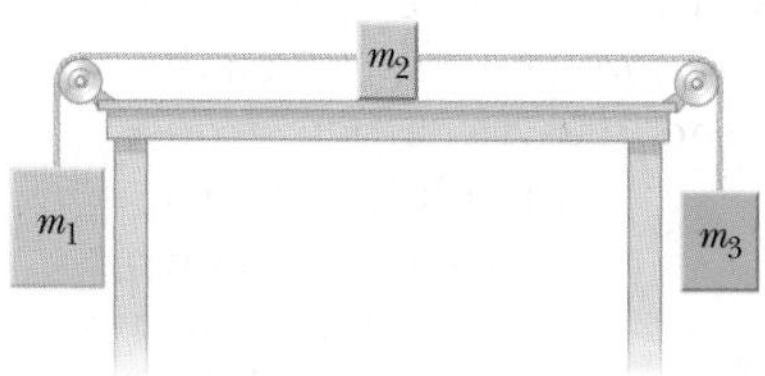

Figure P4.57

58. The force exerted by the wind on a sailboat is approximately perpendicular to the sail and proportional to the component of the wind velocity perpendicular to the sail. For the 800-kg sailboat shown in Figure P4.58, the force exerted by the wind on the sailboat is

$$F_{\text{sail}} = \left(550 \frac{\text{N}}{\text{m/s}}\right) v_{\text{wind}\perp}$$

Water exerts a force along the keel (bottom) of the boat that prevents it from moving sideways, as shown in the figure. Once the boat starts moving forward, water also exerts a drag force backwards on the boat, opposing the forward motion. If a 17-knot wind (1 knot = 0.514 m/s) is blowing to the east, what is the initial acceleration of the sailboat?

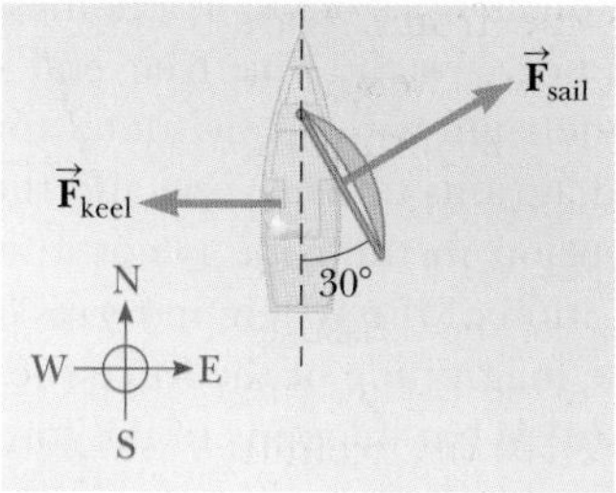

Figure P4.58

59. (a) What is the resultant force exerted by the two cables supporting the traffic light in Figure P4.59? (b) What is the weight of the light?

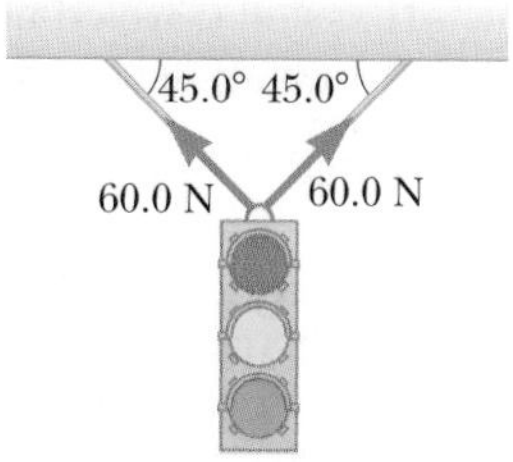

Figure P4.59

60. (a) What is the minimum force of friction required to hold the system of Figure P4.60 in equilibrium? (b) What coefficient of static friction between the 100-N block and the table ensures equilibrium? (c) If the coefficient of kinetic friction between the 100-N block and the table is 0.250, what hanging weight should replace the 50.0-N weight to allow the system to move at a constant speed once it is set in motion?

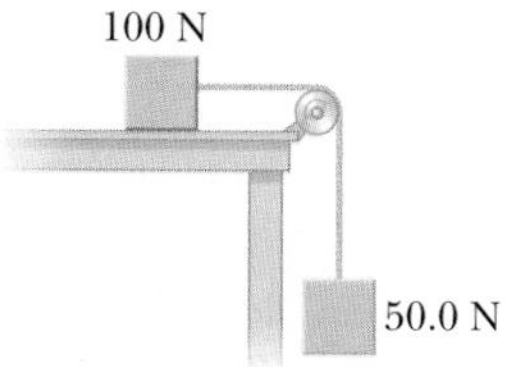

Figure P4.60

61. A boy coasts down a hill on a sled, reaching a level surface at the bottom with a speed of 7.0 m/s. If the coefficient of friction between the sled's runners and the snow is 0.050 and the boy and sled together weigh 600 N, how far does the sled travel on the level surface before coming to rest?

62. A woman at an airport is towing her 20.0-kg suitcase at constant speed by pulling on a strap at an angle θ above the horizontal (Fig. 4.62). She pulls on the strap with a 35.0-N force, and the friction force on the suitcase is 20.0 N. (a) Draw a free-body diagram of the suitcase. (b) What angle does the strap make with the horizontal? (c) What is the magnitude of the normal force that the ground exerts on the suitcase?

Figure P4.62

63. Q|C A box rests on the back of a truck. The coefficient of static friction between the box and the bed of the truck is 0.300. (a) When the truck accelerates forward, what force accelerates the box? (b) Find the maximum acceleration the truck can have before the box slides.

64. Three objects are connected by light strings as shown in Figure P4.64. The string connecting the 4.00-kg object and the 5.00-kg object passes over a light frictionless pulley. Determine (a) the acceleration of each object and (b) the tension in the two strings.

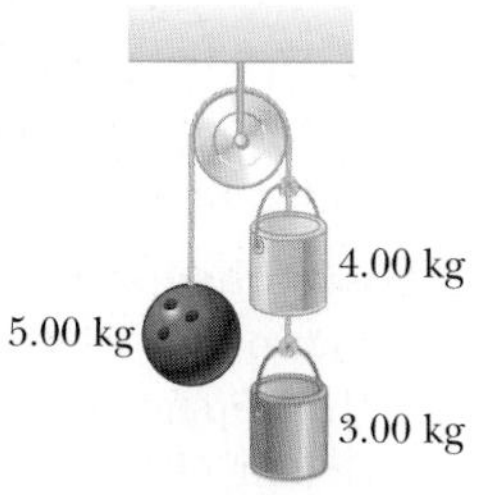

Figure P4.64

65. M A frictionless plane is 10.0 m long and inclined at 35.0°. A sled starts at the bottom with an initial speed of 5.00 m/s up the incline. When the sled reaches the point at which it momentarily stops, a second sled is released from the top of the incline with an initial speed v_i. Both sleds reach the bottom of the incline at the same moment. (a) Determine the distance that the first sled traveled up the incline. (b) Determine the initial speed of the second sled.

66. A high diver of mass 70.0 kg steps off a board 10.0 m above the water and falls vertical to the water, starting from rest. If her downward motion is stopped 2.00 s after her feet first touch the water, what average upward force did the water exert on her?

67. A 2.00-kg aluminum block and a 6.00-kg copper block are connected by a light string over a frictionless pulley. The two blocks are allowed to move on a fixed steel block wedge (of angle θ = 30.0°) as shown in Figure P4.67. Making use of Table 4.2, determine (a) the acceleration of the two blocks and (b) the tension in the string.

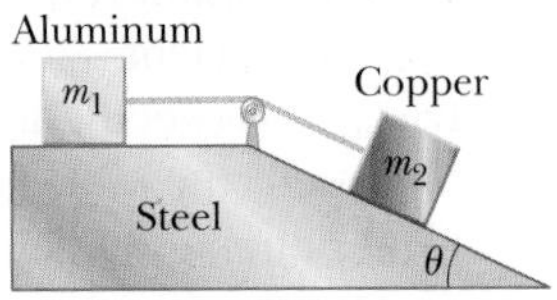

Figure P4.67

68. **S** An object of mass m_1 hangs from a string that passes over a very light fixed pulley P_1 as shown in Figure P4.68. The string connects to a second very light pulley P_2. A second string passes around this pulley with one end attached to a wall and the other to an object of mass m_2 on a frictionless, horizontal table. (a) If a_1 and a_2 are the accelerations of m_1 and m_2, respectively, what is the relation between these accelerations? Find expressions for (b) the tensions in the strings and (c) the accelerations a_1 and a_2 in terms of the masses m_1 and m_2, and g.

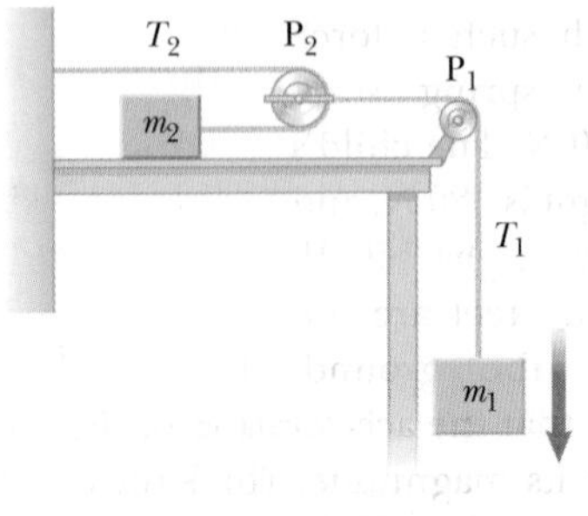

Figure P4.68

69. Two boxes of fruit on a frictionless horizontal surface are connected by a light string as in Figure P4.69, where m_1 = 10 kg and m_2 = 20 kg. A force of 50 N is applied to the 20-kg box. (a) Determine the acceleration of each box and the tension in the string. (b) Repeat the problem for the case where the coefficient of kinetic friction between each box and the surface is 0.10.

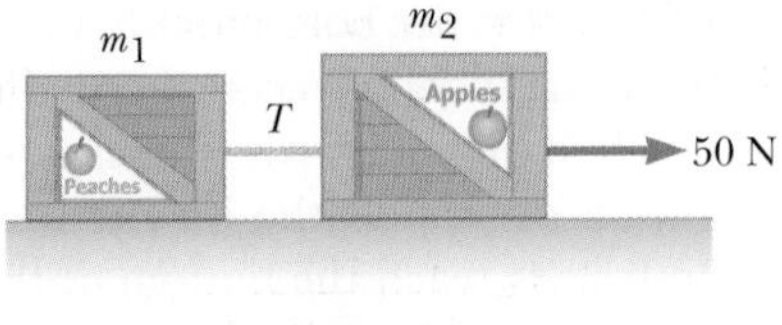

Figure P4.69

70. **QIC** *Measuring coefficients of friction* A coin is placed near one edge of a book lying on a table, and that edge of the book is lifted until the coin just slips down the incline as shown in Figure P4.70. The angle of the incline, θ_c, called the critical angle, is measured. (a) Draw a free-body diagram for the coin when it is on the verge of slipping and identify all forces acting on it. Your free-body diagram should include a force of static friction acting up the incline. (b) Is the magnitude of the friction force equal to $\mu_s n$ for angles less than θ_c? Explain. What can you definitely say about the magnitude of the friction force for any angle $\theta \leq \theta_c$? (c) Show that the coefficient of static friction is given by $\mu_s = \tan \theta_c$. (d) Once the coin starts to slide down the incline, the angle can be adjusted to a new value $\theta_c' \leq \theta_c$ such that the coin moves down the incline with constant speed. How does observation enable you to obtain the coefficient of kinetic friction?

Figure P4.70

71. **QIC** A fisherman poles a boat as he searches for his next catch. He pushes parallel to the length of the light pole, exerting a force of 240 N on the bottom of a shallow lake. The pole lies in the vertical plane containing the boat's keel. At one moment, the pole makes an angle of 35.0° with the vertical and the water exerts a horizontal drag force of 47.5 N on the boat, opposite to its forward velocity of magnitude 0.857 m/s. The mass of the boat including its cargo and the worker is 370 kg. (a) The water exerts a buoyant force vertically upward on the boat. Find the magnitude of this force. (b) Assume the forces are constant over a short interval of time. Find the velocity of the boat 0.450 s after the moment described. (c) If the angle of the pole with respect to the vertical increased but the exerted force against the bottom remained the same, what would happen to buoyant force and the acceleration of the boat?

72. **QIC** A rope with mass m_r is attached to a block with mass m_b as in Figure P4.72. Both the rope and the block rest on a horizontal, frictionless surface. The rope does not stretch. The free end of the rope is pulled to the right with a horizontal force F. (a) Draw free-body diagrams for the rope and the block, noting that the tension in the rope is not uniform. (b) Find the acceleration of the system in terms of m_b, m_r, and F. (c) Find the magnitude of the force the rope exerts on the block. (d) What happens to the force on the block as the rope's mass approaches zero? What can you state about the tension in a *light* cord joining a pair of moving objects?

73. A van accelerates down a hill (Fig. P4.73), going from rest to 30.0 m/s in 6.00 s. During the acceleration, a

toy ($m = 0.100$ kg) hangs by a string from the van's ceiling. The acceleration is such that the string remains perpendicular to the ceiling. Determine (a) the angle θ and (b) the tension in the string.

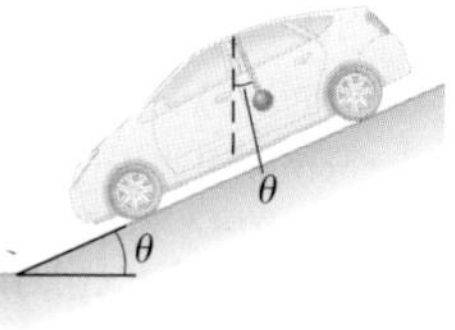

Figure P4.73

74. An inquisitive physics student, wishing to combine pleasure with scientific inquiry, rides on a roller coaster sitting on a bathroom scale. (Do not try this yourself on a roller coaster that forbids loose, heavy packages.) The bottom of the seat in the roller-coaster car is in a plane parallel to the track. The seat has a perpendicular back and a seat belt that fits around the student's chest in a plane parallel to the bottom of the seat. The student lifts his feet from the floor so that the scale reads his weight, 200 lb, when the car is horizontal. At one point during the ride, the car zooms with negligible friction down a straight slope inclined at 30.0° below the horizontal. What does the scale read at that point?

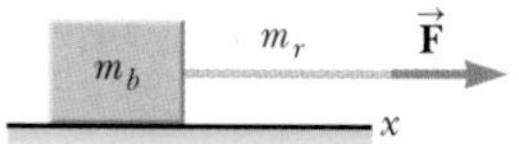

Figure P4.72

75. The parachute on a race car of weight 8 820 N opens at the end of a quarter-mile run when the car is traveling at 35 m/s. What total retarding force must be supplied by the parachute to stop the car in a distance of 1 000 m?

76. **M** On an airplane's takeoff, the combined action of the air around the engines and wings of an airplane exerts an 8 000-N force on the plane, directed upward at an angle of 65.0° above the horizontal. The plane rises with constant velocity in the vertical direction while continuing to accelerate in the horizontal direction. (a) What is the weight of the plane? (b) What is its horizontal acceleration?

77. The board sandwiched between two other boards in Figure P4.77 weighs 95.5 N. If the coefficient of friction between the boards is 0.663, what must be the magnitude of the compression forces (assumed to be horizontal) acting on both sides of the center board to keep it from slipping?

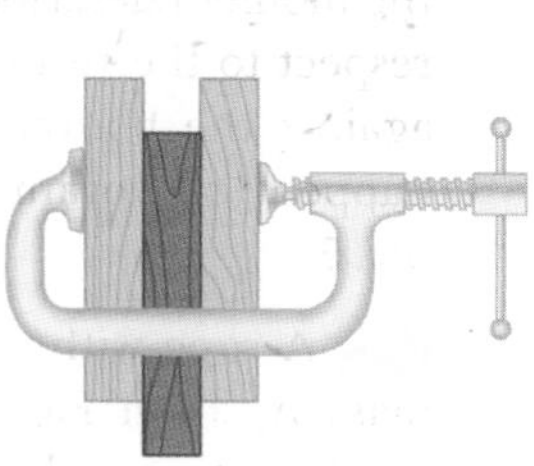
Figure P4.77

78. A sled weighing 60.0 N is pulled horizontally across snow so that the coefficient of kinetic friction between sled and snow is 0.100. A penguin weighing 70.0 N rides on the sled, as in Figure P4.78. If the coefficient of static friction between penguin and sled is 0.700, find the maximum horizontal force that can be exerted on the sled before the penguin begins to slide off.

Figure P4.78

79. A 72-kg man stands on a spring scale in an elevator. Starting from rest, the elevator ascends, attaining its maximum speed of 1.2 m/s in 0.80 s. The elevator travels with this constant speed for 5.0 s, undergoes a uniform *negative* acceleration for 1.5 s, and then comes to rest. What does the spring scale register (a) before the elevator starts to move? (b) During the first 0.80s of the elevator's ascent? (c) While the elevator is traveling at constant speed? (d) During the elevator's negative acceleration?

80. A magician pulls a tablecloth from under a 200-g mug located 30.0 cm from the edge of the cloth. The cloth exerts a friction force of 0.100 N on the mug and is pulled with a constant acceleration of 3.00 m/s². How far does the mug move relative to the horizontal tabletop before the cloth is completely out from under it? Note that the cloth must move more than 30 cm relative to the tabletop during the process.

81. An inventive child wants to reach an apple in a tree without climbing the tree. Sitting in a chair connected to a rope that passes over a frictionless pulley (Fig. P4.81), the child pulls on the loose end of the rope with such a force that the spring scale reads 250 N. The child's true weight is 320 N, and the chair weighs 160 N. The child's feet are not touching the ground. (a) Show that the acceleration of the system is *upward,* and find its magnitude. (b) Find the force the child exerts on the chair.

Figure P4.81

82. A fire helicopter carries a 620-kg bucket of water at the end of a 20.0-m-long cable. Flying back from a fire at a constant speed of 40.0 m/s, the cable makes an angle of 40.0° with respect to the vertical. Determine the force exerted by air resistance on the bucket.

83. **S** A crate of weight $\vec{F}_g$ is pushed by a force $\vec{P}$ on a horizontal floor as shown in Figure P4.83. The coefficient of static friction is μ_s, and $\vec{P}$ is directed at angle θ below the horizontal. (a) Show that the minimum value of P that will move the crate is given by

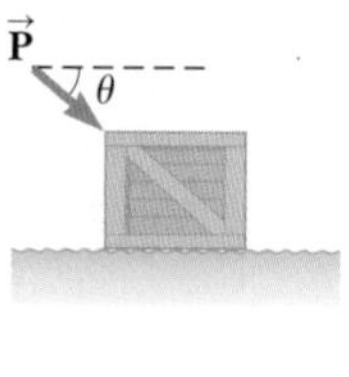

Figure P4.83

$$P = \frac{\mu_s F_g \sec\theta}{1 - \mu_s \tan\theta}$$

(b) Find the condition on θ in terms of μ_s for which motion of the crate is impossible for any value of P.

84. **Q|C** In Figure P4.84, the pulleys and the cord are light, all surfaces are frictionless, and the cord does not stretch. (a) How does the acceleration of block 1 compare with the acceleration of block 2? Explain your reasoning. (b) The mass of block 2 is $m_2 = 1.30$ kg. Derive an expression for the acceleration of the block having mass m_2 as a function of the mass of block 1, m_1. (c) What does the result of part (b) predict if m_1 is very much less than 1.30 kg? (d) What does the result of part (b) predict if m_1 approaches infinity? (e) In this last case, what is the tension in the cord? (f) Could you anticipate the answers to parts (c), (d), and (e) without first doing part (b)? Explain.

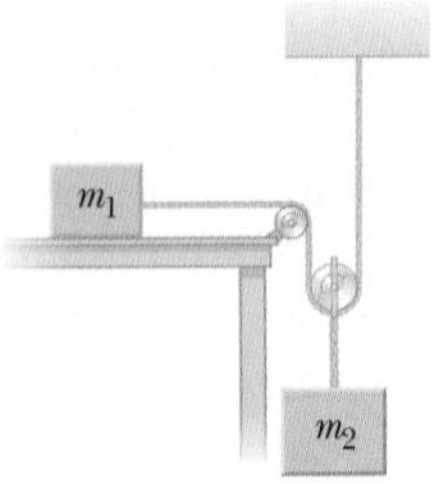

Figure P4.84

85. **S** What horizontal force must be applied to a large block of mass M shown in Figure P4.85 so that the blocks remain stationary relative to M? Assume all surfaces and the pulley are frictionless. Notice that the force exerted by the string accelerates m_2.

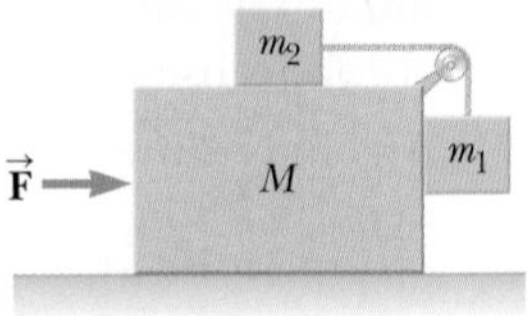

Figure P4.85

Chris Vuille

By timing his jumps, a student does work on the pogo-stick-student system, increasing his height with each jump. The work is transformed into gravitational potential energy at maximum height, converted to kinetic energy while falling, and stored as spring potential energy after contact with the ground.

Energy 5

Energy is one of the most important concepts in the world of science. In everyday use energy is associated with the fuel needed for transportation and heating, with electricity for lights and appliances, and with the foods we consume. These associations, however, don't tell us what energy *is,* only what it *does,* and that producing it requires fuel. Our goal in this chapter, therefore, is to develop a better understanding of energy and how to quantify it.

Energy is present in the Universe in a variety of forms, including mechanical, chemical, electromagnetic, and nuclear energy. Even the inert mass of everyday matter contains a very large amount of energy. Although energy can be transformed from one kind to another, all observations and experiments to date suggest that the total amount of energy in the Universe never changes. That's also true for an isolated system, which is a collection of objects that can exchange energy with each other, but not with the rest of the Universe. If one form of energy in an isolated system decreases, then another form of energy in the system must increase. For example, if the system consists of a motor connected to a battery, the battery converts chemical energy to electrical energy and the motor converts electrical energy to mechanical energy. Understanding how energy changes from one form to another is essential in all the sciences.

In this chapter the focus is mainly on *mechanical energy,* which is the sum of *kinetic energy,* the energy associated with motion, and *potential energy*—the energy associated with relative position. Using an energy approach to solve certain problems is often much easier than using forces and Newton's three laws. These two very different approaches are linked through the concept of *work.*

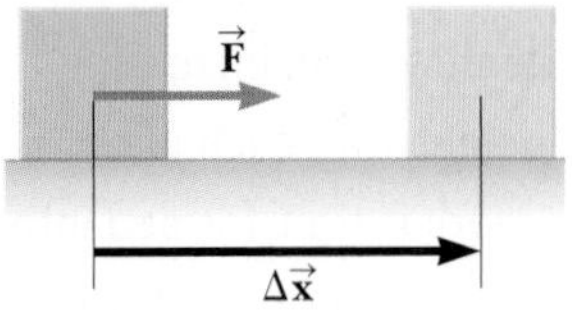

Figure 5.1 A constant force $\vec{\mathbf{F}}$ in the same direction as the displacement, $\Delta\vec{\mathbf{x}}$, does work $F\Delta x$.

5.1 Work

LEARNING OBJECTIVES

1. Contrast by example the physics concept of work with the commonly accepted concept.
2. Calculate the work done by a force on an object in basic contexts.

Work has a different meaning in physics than it does in everyday usage. In the physics definition, a physics textbook author does very little work typing away at a computer. A mason, by contrast, may do a lot of work laying concrete blocks. In physics, work is done only if an object is moved through some displacement while a force is applied to it. If either the force or displacement is doubled, the work is doubled. Double them both, and the work is quadrupled. Doing work involves applying a force to an object while moving it a given distance.

The definition for work W might be taken as

Intuitive definition of work ▶

$$W = Fd \quad [5.1]$$

where F is the magnitude of the force acting on the object and d is the magnitude of the object's displacement. That definition, however, gives only the magnitude of work done on an object when the force is constant and parallel to the displacement, which must be along a line. A more sophisticated definition is required.

Figure 5.1 shows a block undergoing a displacement $\Delta\vec{\mathbf{x}}$ along a straight line while acted on by a constant force $\vec{\mathbf{F}}$ in the same direction. We have the following definition:

The work W done on an object by a constant force $\vec{\mathbf{F}}$ during a linear displacement along the x-axis is

Work by a constant force ▶ during a linear displacement

$$W = F_x \Delta x \quad [5.2]$$

where F_x is the x-component of the force $\vec{\mathbf{F}}$ and $\Delta x = x_f - x_i$ is the object's displacement.

SI unit: joule (J) = newton · meter (N · m) = kg · m²/s²

Note that in one dimension, $\Delta x = x_f - x_i$ is a vector quantity, just as it was defined in Chapter 2, not a magnitude as might be inferred from definitions of a vector and its magnitude in Chapter 3. Therefore Δx can be either positive or negative. Work as defined in Equation 5.2 is rigorous for displacements of any object along the x-axis while a constant force acts on it and, therefore, is suitable for many one-dimensional problems. Work is a positive number if F_x and Δx are both positive or both negative, in which case, as discussed in the next section, the work increases the mechanical energy of the object. If F_x is positive and Δx is negative, or vice versa, then the work is negative, and the object loses mechanical energy. The definition in Equation 5.2 works even when the constant force $\vec{\mathbf{F}}$ is not parallel to the x-axis. Work is only done by the part of the force acting parallel to the object's direction of motion.

Tip 5.1 Work Is a Scalar Quantity

Work is a simple number—a scalar, not a vector—so there is no direction associated with it. Energy and energy transfer are also scalars.

It's easy to see the difference between the physics definition and the everyday definition of work. The author exerts very little force on the keys of a keyboard, creating only small displacements, so relatively little physics work is done. The mason must exert much larger forces on concrete blocks and move them significant distances, and so performs a much greater amount of work. Even very tiring tasks, however, may not constitute work according to the physics definition. A truck driver, for example, may drive for several hours, but if he doesn't exert a force, then $F_x = 0$ in Equation 5.2 and he doesn't do any work. Similarly, a student pressing against a wall for hours in an isometric exercise also does no work, because the displacement

zero. (b) The velocity is decreased. (c) The velocity is unchanged. (d) The speed is unchanged. (e) More information is needed.

20. A car accelerates uniformly from rest. Ignoring air friction, when does the car require the greatest power? (a) When the car first accelerates from rest, (b) just as the car reaches its maximum speed, (c) when the car reaches half its maximum speed. (d) The question is misleading because the power required is constant. (e) More information is needed.

PROBLEMS

ENHANCED WebAssign The problems in this chapter may be assigned online in Enhanced WebAssign.

1. denotes straightforward problem; **2.** denotes intermediate problem;

3. denotes challenging problem

1. denotes full solution available in *Student Solutions Manual/ Study Guide*

1. denotes problems most often assigned in Enhanced WebAssign

BIO denotes biomedical problems

GP denotes guided problems

M denotes Master It tutorial available in Enhanced WebAssign

Q|C denotes asking for quantitative and conceptual reasoning

S denotes symbolic reasoning problem

W denotes Watch It video solution available in Enhanced WebAssign

5.1 Work

1. A weight lifter lifts a 350-N set of weights from ground level to a position over his head, a vertical distance of 2.00 m. How much work does the weight lifter do, assuming he moves the weights at constant speed?

2. In 1990 Walter Arfeuille of Belgium lifted a 281.5-kg object through a distance of 17.1 cm using only his teeth. (a) How much work did Arfeuille do on the object? (b) What magnitude force did he exert on the object during the lift, assuming the force was constant?

3. The record number of boat lifts, including the boat and its ten crew members, was achieved by Sami Heinonen and Juha Räsänen of Sweden in 2000. They lifted a total mass of 653.2 kg approximately 4 in. off the ground a total of 24 times. Estimate the total mechanical work done by the two men in lifting the boat 24 times, assuming they applied the same force to the boat during each lift. (Neglect any work they may have done allowing the boat to drop back to the ground.)

4. **Q|C** A shopper in a supermarket pushes a cart with a force of 35 N directed at an angle of 25° below the horizontal. The force is just sufficient to overcome various frictional forces, so the cart moves at constant speed. (a) Find the work done by the shopper as she moves down a 50.0-m length aisle. (b) What is the *net work* done on the cart? Why? (c) The shopper goes down the next aisle, pushing horizontally and maintaining the same speed as before. If the work done by frictional forces doesn't change, would the shopper's applied force be larger, smaller, or the same? What about the work done on the cart by the shopper?

5. **Q|C** Starting from rest, a 5.00-kg block slides 2.50 m down a rough 30.0° incline. The coefficient of kinetic friction between the block and the incline is $\mu_k = 0.436$. Determine (a) the work done by the force of gravity, (b) the work done by the friction force between block and incline, and (c) the work done by the normal force. (d) Qualitatively, how would the answers change if a shorter ramp at a steeper angle were used to span the same vertical height?

6. A horizontal force of 150 N is used to push a 40.0-kg packing crate a distance of 6.00 m on a rough horizontal surface. If the crate moves at constant speed, find (a) the work done by the 150-N force and (b) the coefficient of kinetic friction between the crate and surface.

7. A sledge loaded with bricks has a total mass of 18.0 kg and is pulled at constant speed by a rope inclined at 20.0° above the horizontal. The sledge moves a distance of 20.0 m on a horizontal surface. The coefficient of kinetic friction between the sledge and surface is 0.500. (a) What is the tension in the rope? (b) How much work is done by the rope on the sledge? (c) What is the mechanical energy lost due to friction?

8. **M** A block of mass $m = 2.50$ kg is pushed a distance $d = 2.20$ m along a frictionless horizontal table by a constant applied force of magnitude $F = 16.0$ N directed at an angle $\theta = 25.0°$ below the horizontal as shown in Figure P5.8. Determine the work done by (a) the applied force, (b) the normal force exerted by the table, (c) the force of gravity, and (d) the net force on the block.

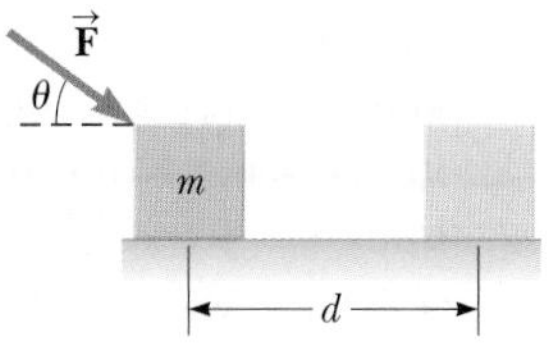

Figure P5.8

5.2 Kinetic Energy and the Work–Energy Theorem

9. A mechanic pushes a 2.50×10^3-kg car from rest to a speed of v, doing 5 000 J of work in the process. During this time, the car moves 25.0 m. Neglecting friction between car and road, find (a) v and (b) the horizontal force exerted on the car.

10. A 7.00-kg bowling ball moves at 3.00 m/s. How fast must a 2.45-g Ping-Pong ball move so that the two balls have the same kinetic energy?

11. A 65.0-kg runner has a speed of 5.20 m/s at one instant during a long-distance event. (a) What is the runner's kinetic energy at this instant? (b) If he doubles his speed to reach the finish line, by what factor does his kinetic energy change?

12. **Q|C** A worker pushing a 35.0-kg wooden crate at a constant speed for 12.0 m along a wood floor does 350 J of work by applying a constant horizontal force of magnitude F_0 on the crate. (a) Determine the value of F_0. (b) If the worker now applies a force greater than F_0, describe the subsequent motion of the crate. (c) Describe what would happen to the crate if the applied force is less than F_0.

13. **W** A 70-kg base runner begins his slide into second base when he is moving at a speed of 4.0 m/s. The coefficient of friction between his clothes and Earth is 0.70. He slides so that his speed is zero just as he reaches the base. (a) How much mechanical energy is lost due to friction acting on the runner? (b) How far does he slide?

14. A running 62-kg cheetah has a top speed of 32 m/s. (a) What is the cheetah's maximum kinetic energy? (b) Find the cheetah's speed when its kinetic energy is one half of the value found in part (a).

15. A 7.80-g bullet moving at 575 m/s penetrates a tree trunk to a depth of 5.50 cm. (a) Use work and energy considerations to find the average frictional force that stops the bullet. (b) Assuming the frictional force is constant, determine how much time elapses between the moment the bullet enters the tree and the moment it stops moving.

16. A 0.60-kg particle has a speed of 2.0 m/s at point A and a kinetic energy of 7.5 J at point B. What is (a) its kinetic energy at A? (b) Its speed at point B? (c) The total work done on the particle as it moves from A to B?

17. A large cruise ship of mass 6.50×10^7 kg has a speed of 12.0 m/s at some instant. (a) What is the ship's kinetic energy at this time? (b) How much work is required to stop it? (c) What is the magnitude of the constant force required to stop it as it undergoes a displacement of 2.50 km?

18. A man pushing a crate of mass $m = 92.0$ kg at a speed of $v = 0.850$ m/s encounters a rough horizontal surface of length $\ell = 0.65$ m as in Figure P5.18. If the coefficient of kinetic friction between the crate and rough surface is 0.358 and he exerts a constant horizontal force of 275 N on the crate, find (a) the magnitude and direction of the net force on the crate while it is on the rough surface, (b) the net work done on the crate while it is on the rough surface, and (c) the speed of the crate when it reaches the end of the rough surface.

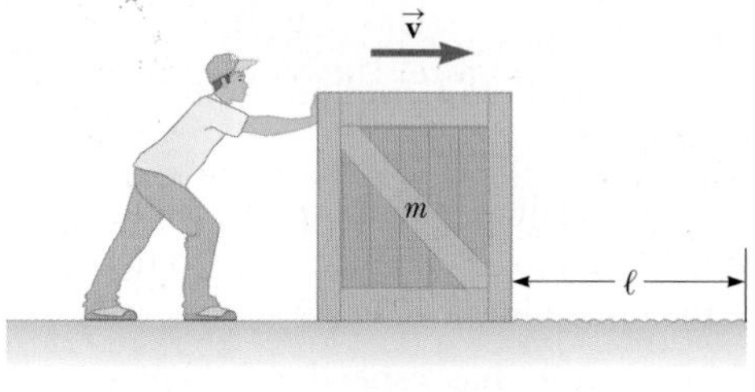

Figure P5.18

5.3 Gravitational Potential Energy

5.4 Spring Potential Energy

19. A 0.20-kg stone is held 1.3 m above the top edge of a water well and then dropped into it. The well has a depth of 5.0 m. Taking $y = 0$ at the top edge of the well, what is the gravitational potential energy of the stone–Earth system (a) before the stone is released and (b) when it reaches the bottom of the well. (c) What is the change in gravitational potential energy of the system from release to reaching the bottom of the well?

20. When a 2.50-kg object is hung vertically on a certain light spring described by Hooke's law, the spring stretches 2.76 cm. (a) What is the force constant of the spring? (b) If the 2.50-kg object is removed, how far will the spring stretch if a 1.25-kg block is hung on it? (c) How much work must an external agent do to stretch the same spring 8.00 cm from its unstretched position?

21. In a control system, an accelerometer consists of a 4.70-g object sliding on a calibrated horizontal rail. A low-mass spring attaches the object to a flange at one end of the rail. Grease on the rail makes static friction negligible, but rapidly damps out vibrations of the sliding object. When subject to a steady acceleration of $0.800g$, the object should be at a location 0.500 cm away from its equilibrium position. Find the force constant of the spring required for the calibration to be correct.

22. **GP** A 60.0-kg athlete leaps straight up into the air from a trampoline with an initial speed of 9.0 m/s. The goal of this problem is to find the maximum height she attains and her speed at half maximum height. (a) What are the interacting objects and how do they interact? (b) Select the height at which the athlete's speed is 9.0 m/s as $y = 0$. What is her kinetic energy at this point? What is the gravitational potential energy associated with the athlete? (c) What is her kinetic energy at maximum height? What is the gravitational potential energy associated with the athlete? (d) Write a general equation for energy conservation in this case and solve for the maximum height. Substitute and obtain a numerical answer. (e) Write the general equation for energy conservation and solve for the velocity at half the maximum height. Substitute and obtain a numerical answer.

23. **M** A 2 100-kg pile driver is used to drive a steel I-beam into the ground. The pile driver falls 5.00 m before coming into contact with the top of the beam, and it drives the beam 12.0 cm farther into the ground as it comes to rest. Using energy considerations, calculate the average force the beam exerts on the pile driver while the pile driver is brought to rest.

24. **S** Two blocks are connected by a light string that passes over two frictionless pulleys as in Figure P5.24.

The block of mass m_2 is attached to a spring of force constant k and $m_1 > m_2$. If the system is released from rest, and the spring is initially not stretched or compressed, find an expression for the maximum displacement d of m_2.

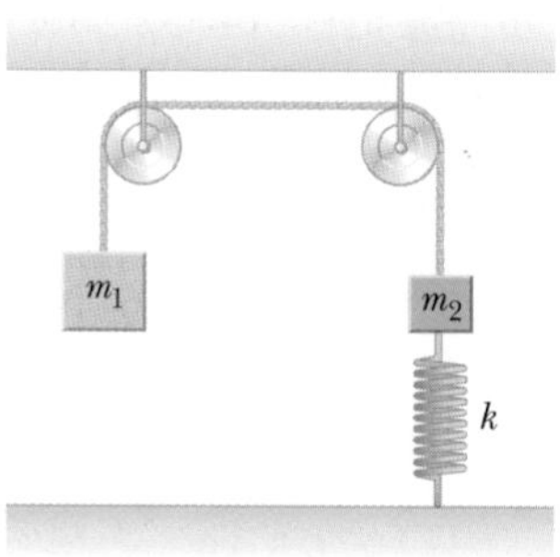

Figure P5.24

25. A daredevil on a motorcycle leaves the end of a ramp with a speed of 35.0 m/s as in Figure P5.25. If his speed is 33.0 m/s when he reaches the peak of the path, what is the maximum height that he reaches? Ignore friction and air resistance.

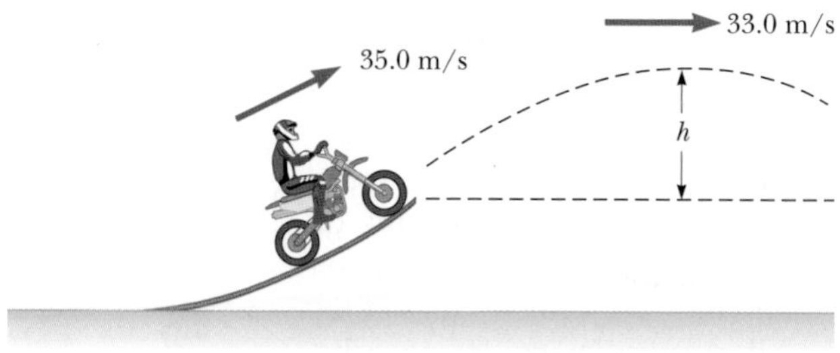

Figure P5.25

26. Truck suspensions often have "helper springs" that engage at high loads. One such arrangement is a leaf spring with a helper coil spring mounted on the axle, as shown in Figure P5.26. When the main leaf spring is compressed by distance y_0, the helper spring engages and then helps to support any additional load. Suppose the leaf spring constant is 5.25×10^5 N/m, the helper spring constant is 3.60×10^5 N/m, and $y_0 = 0.500$ m. (a) What is the compression of the leaf spring for a load of 5.00×10^5 N? (b) How much work is done in compressing the springs?

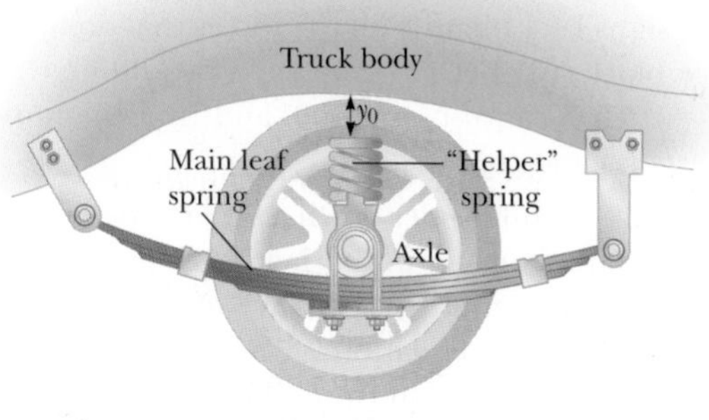

Figure P5.26

27. BIO The chin-up is one exercise that can be used to strengthen the biceps muscle. This muscle can exert a force of approximately 800 N as it contracts a distance of 7.5 cm in a 75-kg male.[3] How much work can the biceps muscles (one in each arm) perform in a single contraction? Compare this amount of work with the energy required to lift a 75-kg person 40 cm in performing a chin-up. Do you think the biceps muscle is the only muscle involved in performing a chin-up?

28. BIO A flea is able to jump about 0.5 m. It has been said that if a flea were as big as a human, it would be able to jump over a 100-story building! When an animal jumps, it converts work done in contracting muscles into gravitational potential energy (with some steps in between). The maximum force exerted by a muscle is proportional to its cross-sectional area, and the work done by the muscle is this force times the length of contraction. If we magnified a flea by a factor of 1 000, the cross section of its muscle would increase by $1\,000^2$ and the length of contraction would increase by 1 000. How high would this "superflea" be able to jump? (Don't forget that the mass of the "superflea" increases as well.)

29. A 50.0-kg projectile is fired at an angle of 30.0° above the horizontal with an initial speed of 1.20×10^2 m/s from the top of a cliff 142 m above level ground, where the ground is taken to be $y = 0$. (a) What is the initial total mechanical energy of the projectile? (b) Suppose the projectile is traveling 85.0 m/s at its maximum height of $y = 427$ m. How much work has been done on the projectile by air friction? (c) What is the speed of the projectile immediately before it hits the ground if air friction does one and a half times as much work on the projectile when it is going down as it did when it was going up?

30. S A projectile of mass m is fired horizontally with an initial speed of v_0 from a height of h above a flat, desert surface. Neglecting air friction, at the instant before the projectile hits the ground, find the following in terms of m, v_0, h, and g: (a) the work done by the force of gravity on the projectile, (b) the change in kinetic energy of the projectile since it was fired, and (c) the final kinetic energy of the projectile. (d) Are any of the answers changed if the initial angle is changed?

31. GP A horizontal spring attached to a wall has a force constant of 850 N/m. A block of mass 1.00 kg is attached to the spring and oscillates freely on a horizontal, frictionless surface as in Figure 5.20. The initial goal of this problem is to find the velocity at the equilibrium point after the block is released. (a) What objects constitute the system, and through what forces do they interact? (b) What are the two points of interest? (c) Find the energy stored in the spring when the mass is stretched 6.00 cm from equilibrium and again

[3] G. P. Pappas et al., "Nonuniform shortening in the biceps brachii during elbow flexion," *Journal of Applied Physiology* **92,** 2381, 2002.

when the mass passes through equilibrium after being released from rest. (d) Write the conservation of energy equation for this situation and solve it for the speed of the mass as it passes equilibrium. Substitute to obtain a numerical value. (e) What is the speed at the halfway point? Why isn't it half the speed at equilibrium?

5.5 Systems and Energy Conservation

32. A 50-kg pole vaulter running at 10 m/s vaults over the bar. Her speed when she is above the bar is 1.0 m/s. Neglect air resistance, as well as any energy absorbed by the pole, and determine her altitude as she crosses the bar.

33. W A child and a sled with a combined mass of 50.0 kg slide down a frictionless slope. If the sled starts from rest and has a speed of 3.00 m/s at the bottom, what is the height of the hill?

34. Hooke's law describes a certain light spring of unstretched length 35.0 cm. When one end is attached to the top of a door frame and a 7.50-kg object is hung from the other end, the length of the spring is 41.5 cm. (a) Find its spring constant. (b) The load and the spring are taken down. Two people pull in opposite directions on the ends of the spring, each with a force of 190 N. Find the length of the spring in this situation.

35. Q|C A 0.250-kg block along a horizontal track has a speed of 1.50 m/s immediately before colliding with a light spring of force constant 4.60 N/m located at the end of the track. (a) What is the spring's maximum compression if the track is frictionless? (b) If the track is *not* frictionless, would the spring's maximum compression be greater than, less than, or equal to the value obtained in part (a)?

36. A block of mass $m = 5.00$ kg is released from rest from point Ⓐ and slides on the frictionless track shown in Figure P5.36. Determine (a) the block's speed at points Ⓑ and Ⓒ and (b) the net work done by the gravitational force on the block as it moves from point from Ⓐ to Ⓒ.

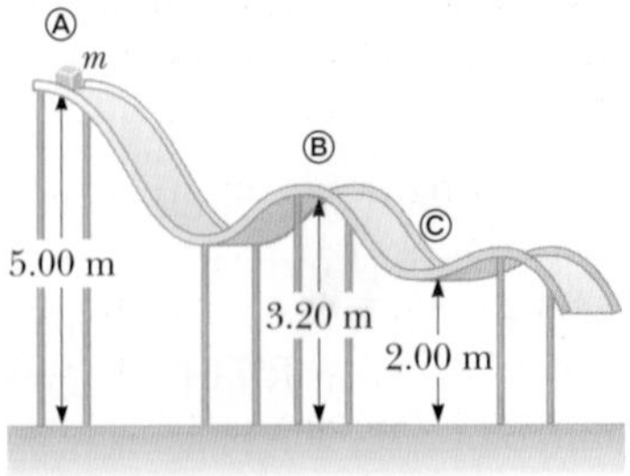

Figure P5.36

37. Tarzan swings on a 30.0-m-long vine initially inclined at an angle of 37.0° with the vertical. What is his speed at the bottom of the swing (a) if he starts from rest? (b) If he pushes off with a speed of 4.00 m/s?

38. S Two blocks are connected by a light string that passes over a frictionless pulley as in Figure P5.38. The system is released from rest while m_2 is on the floor and m_1 is a distance h above the floor. (a) Assuming $m_1 > m_2$, find an expression for the speed of m_1 just as it reaches the floor. (b) Taking $m_1 = 6.5$ kg, $m_2 = 4.2$ kg, and $h = 3.2$ m, evaluate your answer to part (a), and (c) find the speed of each block when m_1 has fallen a distance of 1.6 m.

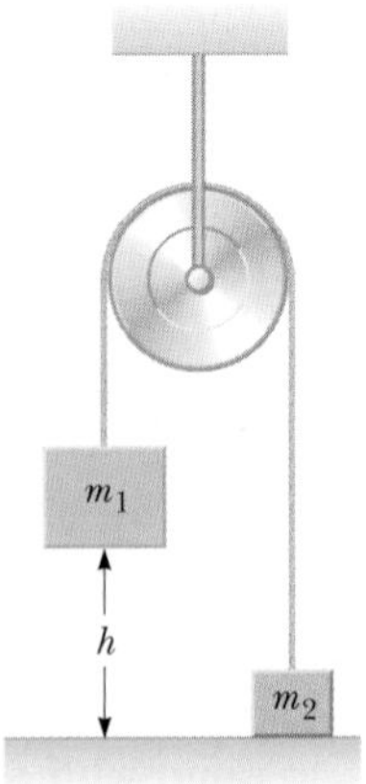

Figure P5.38

39. M The launching mechanism of a toy gun consists of a spring of unknown spring constant, as shown in Figure P5.39a. If the spring is compressed a distance of 0.120 m and the gun fired vertically as shown, the gun can launch a 20.0-g projectile from rest to a maximum height of 20.0 m above the starting point of the projectile. Neglecting all resistive forces, (a) describe the mechanical energy transformations that occur from the time the gun is fired until the projectile reaches its maximum height, (b) determine the spring constant, and (c) find the speed of the projectile as it moves through the equilibrium position of the spring (where $x = 0$), as shown in Figure P5.39b.

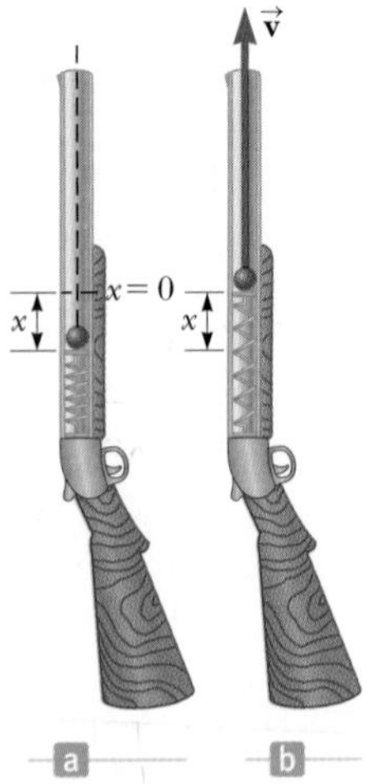

Figure P5.39

40. GP (a) A block with a mass m is pulled along a horizontal surface for a distance x by a constant force $\vec{F}$ at an angle θ with respect to the horizontal. The coefficient of kinetic friction between block and table is μ_k. Is the force exerted by friction equal to $\mu_k mg$? If not, what is the force exerted by friction? (b) How much work is done by the friction force and by $\vec{F}$? (Don't forget the signs.) (c) Identify all the forces that do no work on the block. (d) Let $m = 2.00$ kg, $x = 4.00$ m, $\theta = 37.0°$, $F = 15.0$ N, and $\mu_k = 0.400$, and find the answers to parts (a) and (b).

41. Q|C (a) A child slides down a water slide at an amusement park from an initial height h. The slide can be considered frictionless because of the water flowing down it. Can the equation for conservation of mechanical energy be used on the child? (b) Is the mass of the child a factor in determining his speed at the bottom of the slide? (c) The child drops straight down rather than following the curved ramp of the slide. In which case will he be traveling faster at ground level? (d) If friction is present, how would the conservation-of-energy equation be modified? (e) Find the maximum speed of the child when the slide is frictionless if the initial height of the slide is 12.0 m.

42. Q|C An airplane of mass 1.50×10^4 kg is moving at 60.0 m/s. The pilot then increases the engine's thrust to 7.50×10^4 N. The resistive force exerted by air on the airplane has a magnitude of 4.00×10^4 N. (a) Is the work done by the engine on the airplane equal to the change in the airplane's kinetic energy after it travels through some distance through the air? Is mechanical energy conserved? Explain. (b) Find the speed of the airplane after it has traveled 5.00×10^2 m. Assume the airplane is in level flight throughout the motion.

43. The system shown in Figure P5.43 is used to lift an object of mass $m = 76.0$ kg. A constant downward force of magnitude F is applied to the loose end of the rope such that the hanging object moves upward at constant speed. Neglecting the masses of the rope and pulleys, find (a) the required value of F, (b) the tensions T_1, T_2, and T_3, and (c) the work done by the applied force in raising the object a distance of 1.80 m.

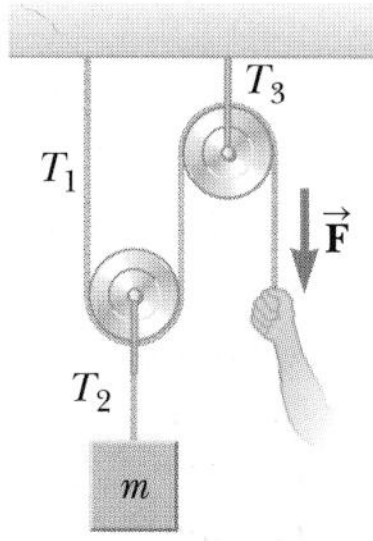

Figure P5.43

44. A 25.0-kg child on a 2.00-m-long swing is released from rest when the ropes of the swing make an angle of 30.0° with the vertical. (a) Neglecting friction, find the child's speed at the lowest position. (b) If the actual speed of the child at the lowest position is 2.00 m/s, what is the mechanical energy lost due to friction?

45. A 2.1×10^3-kg car starts from rest at the top of a 5.0-m-long driveway that is inclined at 20° with the horizontal. If an average friction force of 4.0×10^3 N impedes the motion, find the speed of the car at the bottom of the driveway.

46. GP S A child of mass m starts from rest and slides without friction from a height h along a curved waterslide (Fig. P5.46). She is launched from a height $h/5$ into the pool. (a) Is mechanical energy conserved? Why? (b) Give the gravitational potential energy associated with the child and her kinetic energy in terms of mgh at the following positions: the top of the waterslide, the launching point, and the point where she lands in the pool. (c) Determine her initial speed v_0 at the launch point in terms of g and h. (d) Determine her maximum airborne height y_{max} in terms of h, g, and the horizontal speed at that height, v_{0x}. (e) Use the x-component of the answer to part (c) to eliminate v_0 from the answer to part (d), giving the height y_{max} in terms of g, h, and the launch angle θ. (f) Would your answers be the same if the waterslide were not frictionless? Explain.

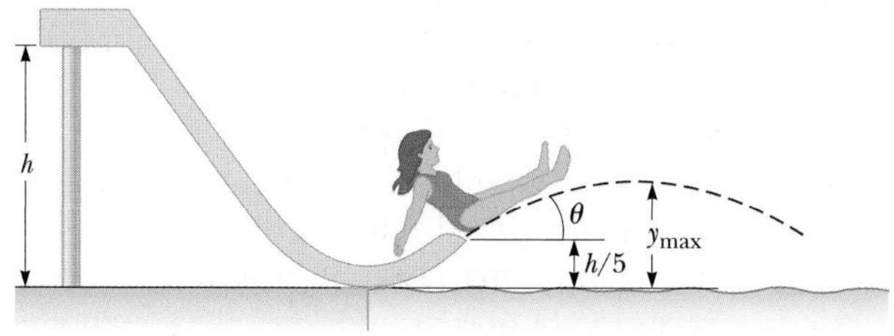

Figure P5.46

47. A skier starts from rest at the top of a hill that is inclined 10.5° with respect to the horizontal. The hillside is 200 m long, and the coefficient of friction between snow and skis is 0.075 0. At the bottom of the hill, the snow is level and the coefficient of friction is unchanged. How far does the skier glide along the horizontal portion of the snow before coming to rest?

48. In a circus performance, a monkey is strapped to a sled and both are given an initial speed of 4.0 m/s up a 20° inclined track. The combined mass of monkey and sled is 20 kg, and the coefficient of kinetic friction between sled and incline is 0.20. How far up the incline do the monkey and sled move?

49. An 80.0-kg skydiver jumps out of a balloon at an altitude of 1 000 m and opens the parachute at an altitude of 200.0 m. (a) Assuming that the total retarding force on the diver is constant at 50.0 N with the parachute closed and constant at 3 600 N with the parachute open, what is the speed of the diver when he lands on the ground? (b) Do you think the skydiver will get hurt? Explain. (c) At what height should the parachute be opened so that the final speed of the skydiver when he hits the ground is 5.00 m/s? (d) How realistic is the assumption that the total retarding force is constant? Explain.

5.6 Power

50. W A skier of mass 70 kg is pulled up a slope by a motor-driven cable. (a) How much work is required to pull him 60 m up a 30° slope (assumed frictionless) at a constant speed of 2.0 m/s? (b) What power (expressed in hp) must a motor have to perform this task?

51. A 3.50-kN piano is lifted by three workers at constant speed to an apartment 25.0 m above the street using a pulley system fastened to the roof of the building. Each worker is able to deliver 165 W of power, and the pulley system is 75.0% efficient (so that 25.0% of the mechanical energy is lost due to friction in the pulley). Neglecting the mass of the pulley, find the time required to lift the piano from the street to the apartment.

52. BIO While running, a person dissipates about 0.60 J of mechanical energy per step per kilogram of body mass. If a 60-kg person develops a power of 70 W during a race, how fast is the person running? (Assume a running step is 1.5 m long.)

53. The electric motor of a model train accelerates the train from rest to 0.620 m/s in 21.0 ms. The total mass of the train is 875 g. Find the average power delivered to the train during its acceleration.

54. When an automobile moves with constant speed down a highway, most of the power developed by the engine is used to compensate for the mechanical energy loss due to frictional forces exerted on the car by the air and the road. If the power developed by an engine is 175 hp, estimate the total frictional force acting on the car when it is moving at a speed of 29 m/s. One horsepower equals 746 W.

55. An older-model car accelerates from 0 to speed v in 10 s. A newer, more powerful sports car of the same mass accelerates from 0 to $2v$ in the same time period. Assuming the energy coming from the engine appears only as kinetic energy of the cars, compare the power of the two cars.

56. A certain rain cloud at an altitude of 1.75 km contains 3.20×10^7 kg of water vapor. How long would it take for a 2.70-kW pump to raise the same amount of water from Earth's surface to the cloud's position?

57. A 1.50×10^3-kg car starts from rest and accelerates uniformly to 18.0 m/s in 12.0 s. Assume that air resistance remains constant at 400 N during this time. Find (a) the average power developed by the engine and (b) the instantaneous power output of the engine at $t = 12.0$ s, just before the car stops accelerating.

58. A 650-kg elevator starts from rest and moves upward for 3.00 s with constant acceleration until it reaches its cruising speed, 1.75 m/s. (a) What is the average power of the elevator motor during this period? (b) How does this amount of power compare with its power during an upward trip with constant speed?

5.7 Work Done by a Varying Force

59. M The force acting on a particle varies as in Figure P5.59. Find the work done by the force as the particle moves (a) from $x = 0$ to $x = 8.00$ m, (b) from $x = 8.00$ m to $x = 10.0$ m, and (c) from $x = 0$ to $x = 10.0$ m.

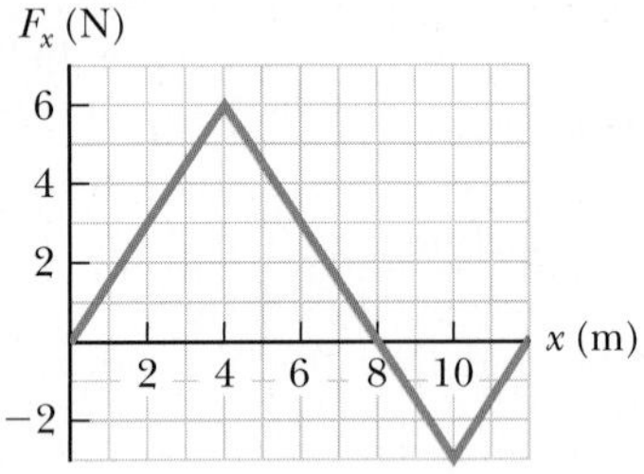

Figure P5.59

60. An object of mass 3.00 kg is subject to a force F_x that varies with position as in Figure P5.60. Find the work done by the force on the object as it moves (a) from $x = 0$ to $x = 5.00$ m, (b) from $x = 5.00$ m to $x = 10.0$ m, and (c) from $x = 10.0$ m to $x = 15.0$ m. (d) If the object has a speed of 0.500 m/s at $x = 0$, find its speed at $x = 5.00$ m and its speed at $x = 15.0$ m.

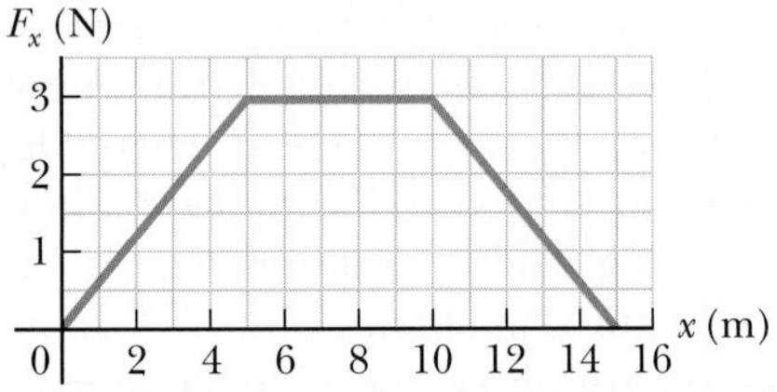

Figure P5.60

61. The force acting on an object is given by $F_x = (8x - 16)$ N, where x is in meters. (a) Make a plot of this force versus x from $x = 0$ to $x = 3.00$ m. (b) From your graph, find the net work done by the force as the object moves from $x = 0$ to $x = 3.00$ m.

Additional Problems

62. An outfielder throws a 0.150-kg baseball at a speed of 40.0 m/s and an initial angle of 30.0°. What is the kinetic energy of the ball at the highest point of its motion?

63. A person doing a chin-up weighs 700 N, exclusive of the arms. During the first 25.0 cm of the lift, each arm exerts an upward force of 355 N on the torso. If the upward movement starts from rest, what is the person's velocity at that point?

64. S A boy starts at rest and slides down a frictionless slide as in Figure P5.64. The bottom of the track is a height h above the ground. The boy then leaves the track horizontally, striking the ground a distance d as shown. Using energy methods, determine the initial height H of the boy in terms of h and d.

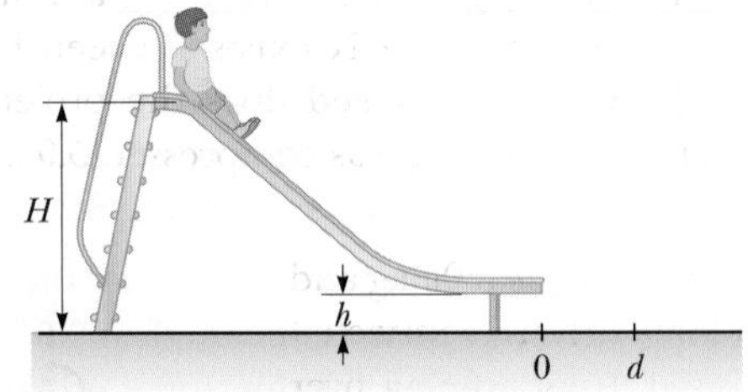

Figure P5.64

65. A roller-coaster car of mass 1.50×10^3 kg is initially at the top of a rise at point Ⓐ. It then moves 35.0 m at an angle of 50.0° below the horizontal to a lower point Ⓑ. (a) Find both the potential energy of the system when the car is at points Ⓐ and Ⓑ and the change in potential energy as the car moves from point Ⓐ to point Ⓑ, assuming $y = 0$ at point Ⓑ. (b) Repeat part (a), this time choosing $y = 0$ at point Ⓒ, which is another 15.0 m down the same slope from point Ⓑ.

66. A ball of mass $m = 1.80$ kg is released from rest at a height $h = 65.0$ cm above a light vertical spring of force constant k as in Figure P5.66a. The ball strikes the top of the spring and compresses it a distance $d = 9.00$ cm as in Figure P5.66b. Neglecting any energy losses

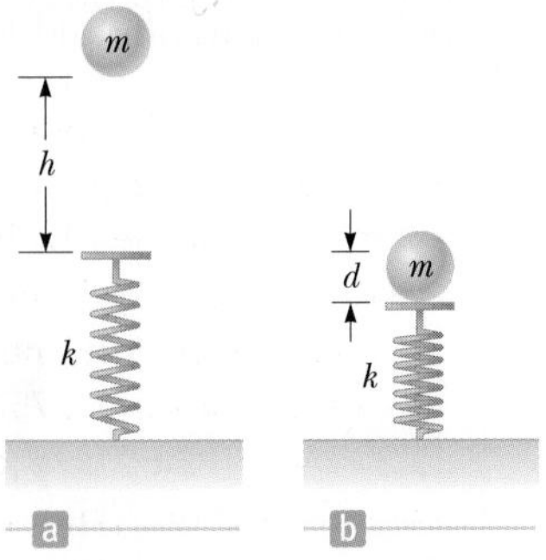

Figure P5.66

during the collision, find (a) the speed of the ball just as it touches the spring and (b) the force constant of the spring.

67. An archer pulls her bowstring back 0.400 m by exerting a force that increases uniformly from zero to 230 N. (a) What is the equivalent spring constant of the bow? (b) How much work does the archer do in pulling the bow?

68. A block of mass 12.0 kg slides from rest down a frictionless 35.0° incline and is stopped by a strong spring with $k = 3.00 \times 10^4$ N/m. The block slides 3.00 m from the point of release to the point where it comes to rest against the spring. When the block comes to rest, how far has the spring been compressed?

69. BIO (a) A 75-kg man steps out a window and falls (from rest) 1.0 m to a sidewalk. What is his speed just before his feet strike the pavement? (b) If the man falls with his knees and ankles locked, the only cushion for his fall is an approximately 0.50-cm give in the pads of his feet. Calculate the average force exerted on him by the ground during this 0.50 cm of travel. This average force is sufficient to cause damage to cartilage in the joints or to break bones.

70. A toy gun uses a spring to project a 5.3-g soft rubber sphere horizontally. The spring constant is 8.0 N/m, the barrel of the gun is 15 cm long, and a constant frictional force of 0.032 N exists between barrel and projectile. With what speed does the projectile leave the barrel if the spring was compressed 5.0 cm for this launch?

71. Two objects ($m_1 = 5.00$ kg and $m_2 = 3.00$ kg) are connected by a light string passing over a light, frictionless pulley as in Figure P5.71. The 5.00-kg object is released from rest at a point $h = 4.00$ m above the table. (a) Determine the speed of each object when the two pass each other. (b) Determine the speed of each object at the moment the 5.00-kg object hits the table. (c) How much higher does the 3.00-kg object travel after the 5.00-kg object hits the table?

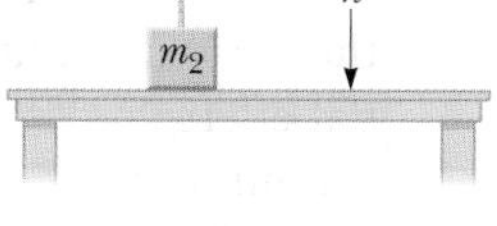

Figure P5.71

72. BIO In a needle biopsy, a narrow strip of tissue is extracted from a patient with a hollow needle. Rather than being pushed by hand, to ensure a clean cut the needle can be fired into the patient's body by a spring. Assume the needle has mass 5.60 g, the light spring has force constant 375 N/m, and the spring is originally compressed 8.10 cm to project the needle horizontally without friction. The tip of the needle then moves through 2.40 cm of skin and soft tissue, which exerts a resistive force of 7.60 N on it. Next, the needle cuts 3.50 cm into an organ, which exerts a backward force of 9.20 N on it. Find (a) the maximum speed of the needle and (b) the speed at which a flange on the back end of the needle runs into a stop, set to limit the penetration to 5.90 cm.

73. A 2.00×10^2-g particle is released from rest at point A on the inside of a smooth hemispherical bowl of radius $R = 30.0$ cm (Fig. P5.73). Calculate (a) its gravitational potential energy at A relative to B, (b) its kinetic energy at B, (c) its speed at B, (d) its potential energy at C relative to B, and (e) its kinetic energy at C.

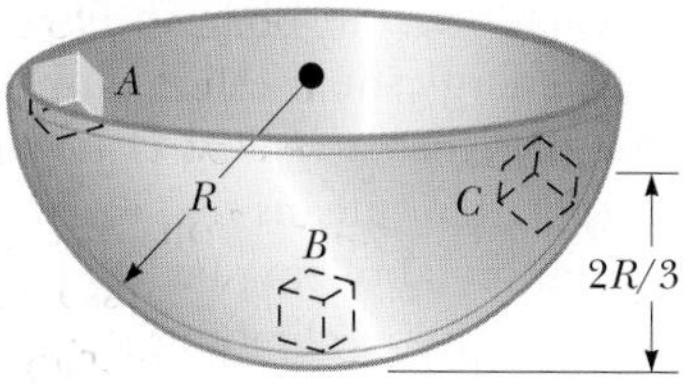

Figure P5.73 Problems 73 and 74.

74. Q|C The particle described in Problem 73 (Fig. P5.73) is released from point A at rest. Its speed at B is 1.50 m/s. (a) What is its kinetic energy at B? (b) How much mechanical energy is lost as a result of friction as the particle goes from A to B? (c) Is it possible to determine μ from these results in a simple manner? Explain.

75. A light spring with spring constant 1.20×10^3 N/m hangs from an elevated support. From its lower end hangs a second light spring, which has spring constant 1.80×10^3 N/m. A 1.50-kg object hangs at rest from the lower end of the second spring. (a) Find the total extension distance of the pair of springs. (b) Find the effective spring constant of the pair of springs as a system. We describe these springs as being *in series*. *Hint:* Consider the forces on each spring separately.

76. S *Symbolic Version of Problem 75* A light spring with spring constant k_1 hangs from an elevated support. From its lower end hangs a second light spring, which has spring constant k_2. An object of mass m hangs at rest from the lower end of the second spring. (a) Find the total extension distance x of the pair of springs in terms of the two displacements x_1 and x_2. (b) Find the effective spring constant of the pair of springs as a system. We describe these springs as being *in series*.

77. BIO In terms of saving energy, bicycling and walking are far more efficient means of transportation than is travel by automobile. For example, when riding at 10.0 mi/h, a cyclist uses food energy at a rate of about 400 kcal/h above what he would use if he were merely sitting still. (In exercise physiology, power is often measured in kcal/h rather than in watts. Here, 1 kcal = 1 nutritionist's Calorie = 4 186 J.) Walking at 3.00 mi/h requires about 220 kcal/h. It is interesting to compare these values with the energy consumption required for travel by car. Gasoline yields about

1.30×10^8 J/gal. Find the fuel economy in equivalent miles per gallon for a person (a) walking and (b) bicycling.

78. BIO Energy is conventionally measured in Calories as well as in joules. One Calorie in nutrition is 1 kilocalorie, which we define in Chapter 11 as 1 kcal = 4 186 J. Metabolizing 1 gram of fat can release 9.00 kcal. A student decides to try to lose weight by exercising. She plans to run up and down the stairs in a football stadium as fast as she can and as many times as necessary. Is this in itself a practical way to lose weight? To evaluate the program, suppose she runs up a flight of 80 steps, each 0.150 m high, in 65.0 s. For simplicity, ignore the energy she uses in coming down (which is small). Assume that a typical efficiency for human muscles is 20.0%. This means that when your body converts 100 J from metabolizing fat, 20 J goes into doing mechanical work (here, climbing stairs). The remainder goes into internal energy. Assume the student's mass is 50.0 kg. (a) How many times must she run the flight of stairs to lose 1 pound of fat? (b) What is her average power output, in watts and in horsepower, as she is running up the stairs?

79. A ski jumper starts from rest 50.0 m above the ground on a frictionless track and flies off the track at an angle of 45.0° above the horizontal and at a height of 10.0 m above the level ground. Neglect air resistance. (a) What is her speed when she leaves the track? (b) What is the maximum altitude she attains after leaving the track? (c) Where does she land relative to the end of the track?

80. A 5.0-kg block is pushed 3.0 m up a vertical wall with constant speed by a constant force of magnitude F applied at an angle of $\theta = 30°$ with the horizontal, as shown in Figure P5.80. If the coefficient of kinetic friction between block and wall is 0.30, determine the work done by (a) $\vec{\mathbf{F}}$, (b) the force of gravity, and (c) the normal force between block and wall. (d) By how much does the gravitational potential energy increase during the block's motion?

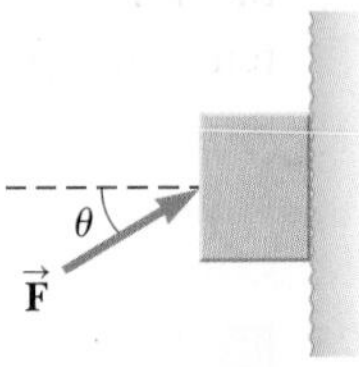

Figure P5.80

81. A child's pogo stick (Fig. P5.81) stores energy in a spring ($k = 2.50 \times 10^4$ N/m). At position Ⓐ ($x_1 = -0.100$ m), the spring compression is a maximum and the child is momentarily at rest. At position Ⓑ ($x = 0$), the spring is relaxed and the child is moving upward. At position Ⓒ, the child is again momentarily at rest at the top of the jump. Assuming that the combined mass of child and pogo stick is 25.0 kg, (a) calculate the total energy of the system if both potential energies are zero at $x = 0$, (b) determine x_2, (c) calculate the speed of the child at $x = 0$, (d) determine the value of x for which the kinetic energy of the system is a maximum, and (e) obtain the child's maximum upward speed.

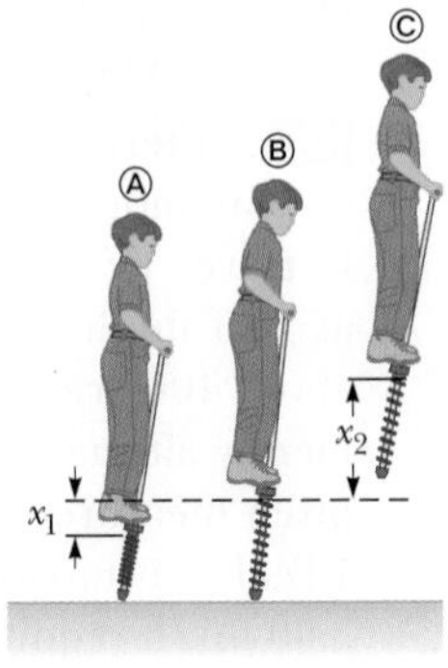

Figure P5.81

82. BIO A hummingbird is able to hover because, as the wings move downward, they exert a downward force on the air. Newton's third law tells us that the air exerts an equal and opposite force (upward) on the wings. The average of this force must be equal to the weight of the bird when it hovers. If the wings move through a distance of 3.5 cm with each stroke, and the wings beat 80 times per second, determine the work performed by the wings on the air in 1 m if the mass of the hummingbird is 3.0 g.

83. In the dangerous "sport" of bungee jumping, a daring student jumps from a hot-air balloon with a specially designed elastic cord attached to his waist. The unstretched length of the cord is 25.0 m, the student weighs 700 N, and the balloon is 36.0 m above the surface of a river below. Calculate the required force constant of the cord if the student is to stop safely 4.00 m above the river.

84. Q|C The masses of the javelin, discus, and shot are 0.80 kg, 2.0 kg, and 7.2 kg, respectively, and record throws in the corresponding track events are about 98 m, 74 m, and 23 m, respectively. Neglecting air resistance, (a) calculate the minimum initial kinetic energies that would produce these throws, and (b) estimate the average force exerted on each object during the throw, assuming the force acts over a distance of 2.0 m. (c) Do your results suggest that air resistance is an important factor?

85. A truck travels uphill with constant velocity on a highway with a 7.0° slope. A 50-kg package sits on the floor of the back of the truck and does not slide, due to a static frictional force. During an interval in which the truck travels 340 m, (a) what is the net work done on the package? What is the work done on the package by (b) the force of gravity, (c) the normal force, and (d) the friction force?

86. A daredevil wishes to bungee-jump from a hot-air balloon 65.0 m above a carnival midway. He will use a piece of uniform elastic cord tied to a harness around his body to stop his fall at a point 10.0 m above the ground. Model his body as a particle and the cord as having negligible mass and a tension force described by Hooke's force law. In a preliminary test, hanging at rest from a 5.00-m length of the cord, the jumper finds that his body weight stretches it by 1.50 m. He will drop from rest at the point where the top end of a longer section of the cord is attached to the stationary balloon. (a) What length of cord should he use? (b) What maximum acceleration will he experience?

87. M A loaded ore car has a mass of 950 kg and rolls on rails with negligible friction. It starts from rest and is pulled up a mine shaft by a cable connected to a winch. The shaft is inclined at 30.0° above the horizontal. The car accelerates uniformly to a speed of 2.20 m/s in 12.0 s and then continues at constant speed. (a) What power must the winch motor provide when the car is moving at constant speed? (b) What maximum power must the motor provide? (c) What total energy transfers out of the motor by work by the time the car moves off the end of the track, which is of length 1 250 m?

88. S An object of mass m is suspended from the top of a cart by a string of length L as in Figure P5.88a. The cart and object are initially moving to the right at a constant speed v_0. The cart comes to rest after colliding and sticking to a bumper, as in Figure P5.88b, and the suspended object swings through an angle θ. (a) Show that the initial speed is $v_0 = \sqrt{2gL(1 - \cos\theta)}$. (b) If $L = 1.20$ m and $\theta = 35.0°$, find the initial speed of the cart. (*Hint:* The force exerted by the string on the object does no work on the object.)

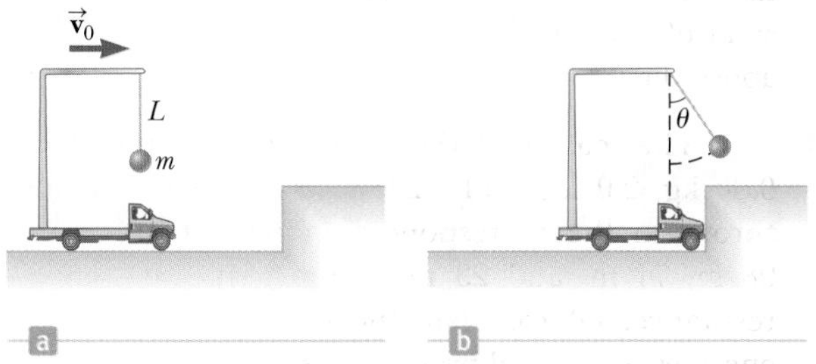

Figure P5.88

89. Three objects with masses $m_1 = 5.0$ kg, $m_2 = 10$ kg, and $m_3 = 15$ kg, respectively, are attached by strings over frictionless pulleys as indicated in Figure P5.89. The horizontal surface exerts a force of friction of 30 N on m_2. If the system is released from rest, use energy concepts to find the speed of m_3 after it moves down 4.0 m.

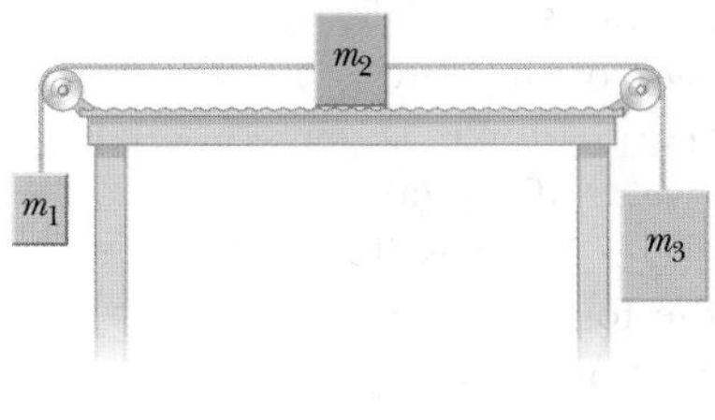

Figure P5.89

90. A light spring with force constant 3.85 N/m is compressed by 8.00 cm as it is held between a 0.250-kg block on the left and a 0.500-kg block on the right, both resting on a horizontal surface. The spring exerts a force on each block, tending to push them apart. The blocks are simultaneously released from rest. Find the acceleration with which each block starts to move, given that the coefficient of kinetic friction between each block and the surface is (a) 0, (b) 0.100, and (c) 0.462.

91. BIO In bicycling for aerobic exercise, a woman wants her heart rate to be between 136 and 166 beats per minute. Assume that her heart rate is directly proportional to her mechanical power output. Ignore all forces on the woman-plus-bicycle system, except for static friction forward on the drive wheel of the bicycle and an air resistance force proportional to the square of the bicycler's speed. When her speed is 22.0 km/h, her heart rate is 90.0 beats per minute. In what range should her speed be so that her heart rate will be in the range she wants?

92. Two blocks, A and B (with mass 50 kg and 100 kg, respectively), are connected by a string, as shown in Figure P5.92. The pulley is frictionless and of negligible mass. The coefficient of kinetic friction between block A and the incline is $\mu_k = 0.25$. Determine the change in the kinetic energy of block A as it moves from Ⓒ to Ⓓ, a distance of 20 m up the incline (and block B drops downward a distance of 20 m) if the system starts from rest.

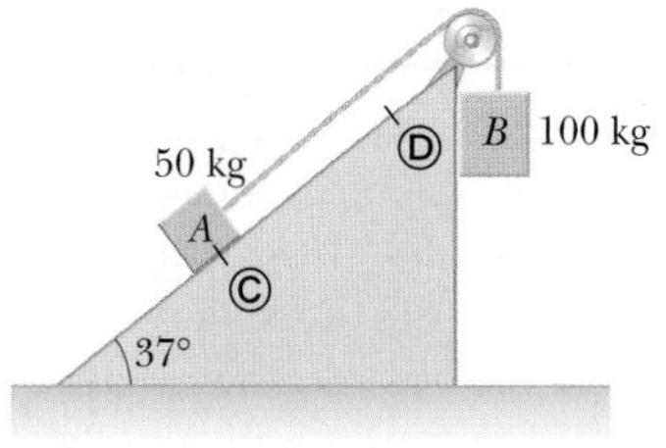

Figure P5.92

Rockets such as the Falcon 9 transform a large part of their initial mass into hot gas through chemical reactions. The energetic gas molecules collide with the reaction chamber walls, transferring momentum to the rest of the rocket before escaping out the exhaust nozzle.

NASA/Tony Gray and Robert Murray

6 Momentum and Collisions

What happens when two automobiles collide? How does the impact affect the motion of each vehicle, and what basic physical principles determine the likelihood of serious injury? How do rockets work, and what mechanisms can be used to overcome the limitations imposed by exhaust speed? Why do we have to brace ourselves when firing small projectiles at high velocity? Finally, how can we use physics to improve our golf game?

To begin answering such questions, we introduce *momentum.* Intuitively, anyone or anything that has a lot of momentum is going to be hard to stop. In politics, the term is metaphorical. Physically, the more momentum an object has, the more force has to be applied to stop it in a given time. This concept leads to one of the most powerful principles in physics: *conservation of momentum.* Using this law, complex collision problems can be solved without knowing much about the forces involved during contact. We'll also be able to derive information about the average force delivered in an impact. With conservation of momentum, we'll have a better understanding of what choices to make when designing an automobile or a moon rocket, or when addressing a golf ball on a tee.

6.1 Momentum and Impulse

LEARNING OBJECTIVES

1. Define momentum and impulse and state the impulse–momentum theorem.
2. Apply the impulse–momentum theorem to find estimates of average forces during collisions.

In physics, momentum has a precise definition. A slowly moving brontosaurus has a lot of momentum, but so does a little hot lead shot from the muzzle of a gun. We therefore expect that momentum will depend on an object's mass and velocity.

The linear momentum $\vec{\mathbf{p}}$ of an object of mass m moving with velocity $\vec{\mathbf{v}}$ is the product of its mass and velocity:

◀ Linear momentum

$$\vec{\mathbf{p}} \equiv m\vec{\mathbf{v}} \tag{6.1}$$

SI unit: kilogram-meter per second (kg · m/s)

Doubling either the mass or the velocity of an object doubles its momentum; doubling both quantities quadruples its momentum. Momentum is a vector quantity with the same direction as the object's velocity. Its components are given in two dimensions by

$$p_x = mv_x \qquad p_y = mv_y$$

where p_x is the momentum of the object in the x-direction and p_y its momentum in the y-direction.

The magnitude of the momentum p of an object of mass m can be related to its kinetic energy KE:

$$KE = \frac{p^2}{2m} \tag{6.2}$$

This relationship is easy to prove using the definitions of kinetic energy and momentum (see Problem 6) and is valid for objects traveling at speeds much less than the speed of light. Equation 6.2 is useful in grasping the interplay between the two concepts, as illustrated in Quick Quiz 6.1.

Quick Quiz

6.1 Two masses m_1 and m_2, with $m_1 < m_2$, have equal kinetic energy. How do the magnitudes of their momenta compare? (a) Not enough information is given. (b) $p_1 < p_2$ (c) $p_1 = p_2$ (d) $p_1 > p_2$.

Changing the momentum of an object requires the application of a force. This is, in fact, how Newton originally stated his second law of motion. Starting from the more common version of the second law, we have

$$\vec{\mathbf{F}}_{\text{net}} = m\vec{\mathbf{a}} = m\frac{\Delta\vec{\mathbf{v}}}{\Delta t} = \frac{\Delta(m\vec{\mathbf{v}})}{\Delta t}$$

where the mass m and the forces are assumed constant. The quantity in parentheses is just the momentum, so we have the following result:

The change in an object's momentum $\Delta\vec{\mathbf{p}}$ divided by the elapsed time Δt equals the constant net force $\vec{\mathbf{F}}_{\text{net}}$ acting on the object:

$$\frac{\Delta\vec{\mathbf{p}}}{\Delta t} = \frac{\text{change in momentum}}{\text{time interval}} = \vec{\mathbf{F}}_{\text{net}} \tag{6.3}$$

◀ Newton's second law and momentum

This equation is also valid when the forces are not constant, provided the limit is taken as Δt becomes infinitesimally small. Equation 6.3 says that if the net force on an object is zero, the object's momentum doesn't change. In other words, the linear momentum of an object is conserved when $\vec{\mathbf{F}}_{\text{net}} = 0$. Equation 6.3 also shows us that changing an object's momentum requires the

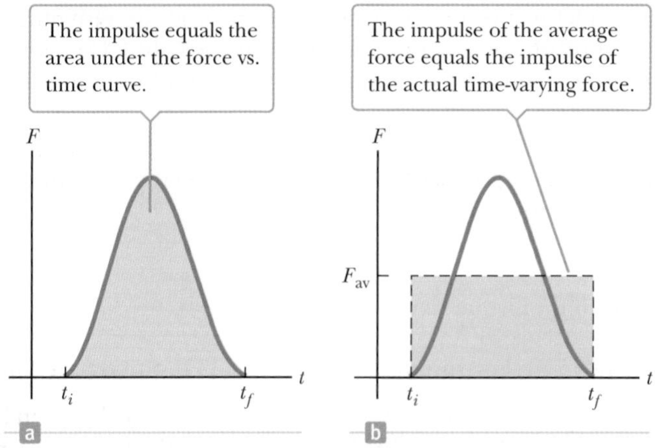

Figure 6.1 (a) A net force acting on a particle may vary in time. (b) The value of the constant force F_{av} (horizontal dashed line) is chosen so that the area of the rectangle $F_{av}\Delta t$ is the same as the area under the curve in (a).

continuous application of a force over a period of time Δt, leading to the definition of *impulse:*

If a constant force $\vec{\mathbf{F}}$ acts on an object, the **impulse** $\vec{\mathbf{I}}$ delivered to the object over a time interval Δt is given by

$$\vec{\mathbf{I}} \equiv \vec{\mathbf{F}}\,\Delta t \qquad [6.4]$$

SI unit: kilogram meter per second (kg · m/s)

Impulse is a vector quantity with the same direction as the constant force acting on the object. When a single constant force $\vec{\mathbf{F}}$ acts on an object, Equation 6.3 can be written as

Impulse–momentum theorem ▶

$$\vec{\mathbf{I}} = \vec{\mathbf{F}}\Delta t = \Delta\vec{\mathbf{p}} = m\vec{\mathbf{v}}_f - m\vec{\mathbf{v}}_i \qquad [6.5]$$

which is a special case of the **impulse–momentum theorem.** Equation 6.5 shows that **the impulse of the force acting on an object equals the change in momentum of that object.** That equality is true even if the force is not constant, as long as the time interval Δt is taken to be arbitrarily small. (The proof of the general case requires concepts from calculus.)

In real-life situations, the force on an object is only rarely constant. For example, when a bat hits a baseball, the force increases sharply, reaches some maximum value, and then decreases just as rapidly. Figure 6.1(a) shows a typical graph of force versus time for such incidents. The force starts out small as the bat comes in contact with the ball, rises to a maximum value when they are firmly in contact, and then drops off as the ball leaves the bat. In order to analyze this rather complex interaction, it's useful to define an **average force** $\vec{\mathbf{F}}_{av}$, shown as the dashed line in Figure 6.1b. The average force is the constant force delivering the same impulse to the object in the time interval Δt as the actual time-varying force. We can then write the impulse–momentum theorem as

$$\vec{\mathbf{F}}_{av}\,\Delta t = \Delta\vec{\mathbf{p}} \qquad [6.6]$$

The magnitude of the impulse delivered by a force during the time interval Δt is equal to the area under the force vs. time graph as in Figure 6.1a or, equivalently, to $F_{av}\Delta t$ as shown in Figure 6.1b. The brief collision between a bullet and an apple is illustrated in Figure 6.2.

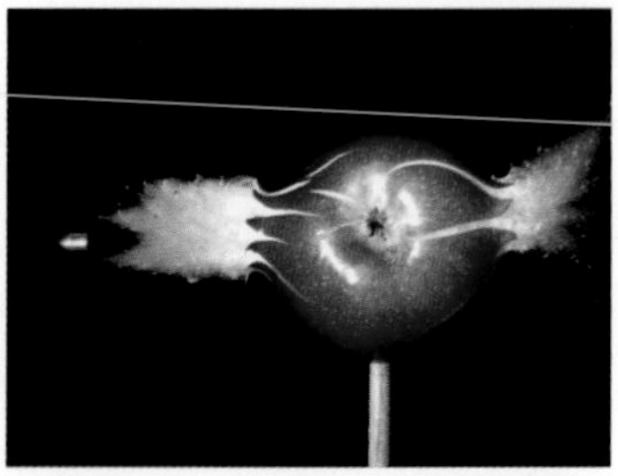

Ted Kinsman/Photo Researchers, Inc.

Figure 6.2 An apple being pierced by a 30-caliber bullet traveling at a supersonic speed of 900 m/s. This collision was photographed with a microflash stroboscope using an exposure time of 0.33 μs. Shortly after the photograph was taken, the apple disintegrated completely. Note that the points of entry and exit of the bullet are visually explosive.

APPLYING PHYSICS 6.1 | Boxing and Brain Injury BIO

Boxers in the nineteenth century used their bare fists. In modern boxing, fighters wear padded gloves. How do gloves protect the brain of the boxer from injury? Also, why do boxers often "roll with the punch"?

EXPLANATION The brain is immersed in a cushioning fluid inside the skull. If the head is struck suddenly by a bare fist, the skull accelerates rapidly. The brain matches this acceleration only because of the large impulsive force exerted by the skull on the brain. This large and sudden force (large F_{av} and small Δt) can cause severe brain injury. Padded gloves extend the time Δt over which the force is applied to the head. For a given impulse $F_{av}\Delta t$, a glove results in a longer time interval than a bare fist, decreasing the average force. Because the average force is decreased, the acceleration of the skull is decreased, reducing (but not eliminating) the chance of brain injury. The same argument can be made for "rolling with the punch": If the head is held steady while being struck, the time interval over which the force is applied is relatively short and the average force is large. If the head is allowed to move in the same direction as the punch, the time interval is lengthened and the average force reduced. ■

■ EXAMPLE 6.1 Teeing Off

GOAL Use the impulse–momentum theorem to estimate the average force exerted during an impact.

PROBLEM A golf ball with mass 5.0×10^{-2} kg is struck with a club as in Figure 6.3. The force on the ball varies from zero when contact is made up to some maximum value (when the ball is maximally deformed) and then back to zero when the ball leaves the club, as in the graph of force vs. time in Figure 6.1. Assume that the ball leaves the club face with a velocity of 44 m/s. **(a)** Find the magnitude of the impulse due to the collision. **(b)** Estimate the duration of the collision and the average force acting on the ball.

Ted Kinsman/Photo Researchers, Inc.

Figure 6.3 (Example 6.1) During impact, the club head momentarily flattens the side of the golf ball.

STRATEGY In part (a), use the fact that the impulse is equal to the change in momentum. The mass and the initial and final velocities are known, so this change can be computed. In part (b), the average force is just the change in momentum computed in part (a) divided by an estimate of the duration of the collision. Estimate the distance the ball travels on the face of the club (about 2.0 cm, roughly the same as the radius of the ball). Divide this distance by the average velocity (half the final velocity) to get an estimate of the time of contact.

SOLUTION

(a) Find the impulse delivered to the ball.

The problem is essentially one dimensional. Note that $v_i = 0$, and calculate the change in momentum, which equals the impulse:

$$I = \Delta p = p_f - p_i = (5.0 \times 10^{-2}\ \text{kg})(44\ \text{m/s}) - 0 = 2.2\ \text{kg} \cdot \text{m/s}$$

(b) Estimate the duration of the collision and the average force acting on the ball.

Estimate the time interval of the collision, Δt, using the approximate displacement (radius of the ball) and its average speed (half the maximum speed):

$$\Delta t = \frac{\Delta x}{v_{av}} = \frac{2.0 \times 10^{-2}\ \text{m}}{22\ \text{m/s}} = 9.1 \times 10^{-4}\ \text{s}$$

Estimate the average force from Equation 6.6:

$$F_{av} = \frac{\Delta p}{\Delta t} = \frac{2.2\ \text{kg} \cdot \text{m/s}}{9.1 \times 10^{-4}\ \text{s}} = 2.4 \times 10^{3}\ \text{N}$$

REMARKS This estimate shows just how large such contact forces can be. A good golfer achieves maximum momentum transfer by shifting weight from the back foot to the front foot, transmitting the body's momentum through the shaft and head of the club. This timing, involving a short movement of the hips, is more effective than a shot powered exclusively by the arms and shoulders. Following through with the swing ensures that the motion isn't slowed at the critical instant of impact.

QUESTION 6.1 What average club speed would double the average force? (Assume the final velocity is unchanged.)

EXERCISE 6.1 A 0.150-kg baseball, thrown with a speed of 40.0 m/s, is hit straight back at the pitcher with a speed of 50.0 m/s. (a) What is the magnitude of the impulse delivered by the bat to the baseball? (b) Find the magnitude of the average force exerted by the bat on the ball if the two are in contact for 2.00×10^{-3} s.

ANSWERS (a) 13.5 kg · m/s (b) 6.75 kN

EXAMPLE 6.2 How Good Are the Bumpers?

GOAL Find an impulse and estimate a force in a collision of a moving object with a stationary object.

PROBLEM In a crash test, a car of mass 1.50×10^3 kg collides with a wall and rebounds as in Figure 6.4a. The initial and final velocities of the car are $v_i = -15.0$ m/s and $v_f = 2.60$ m/s, respectively. If the collision lasts for 0.150 s, find **(a)** the impulse delivered to the car due to the collision and **(b)** the magnitude and direction of the average force exerted on the car.

STRATEGY This problem is similar to the previous example, except that the initial and final momenta are both nonzero. Find the momenta and substitute into the impulse–momentum theorem, Equation 6.6, solving for F_{av}.

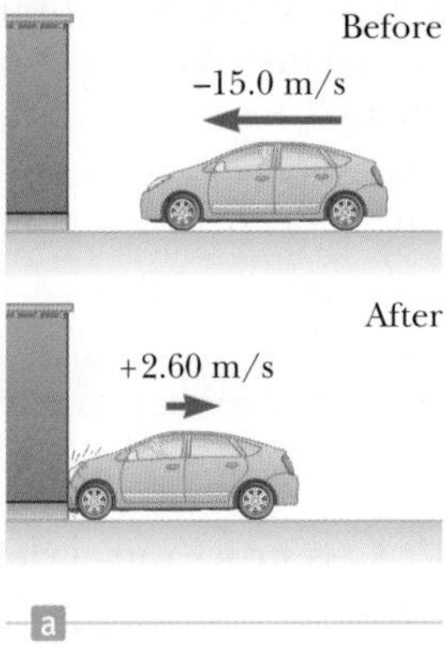

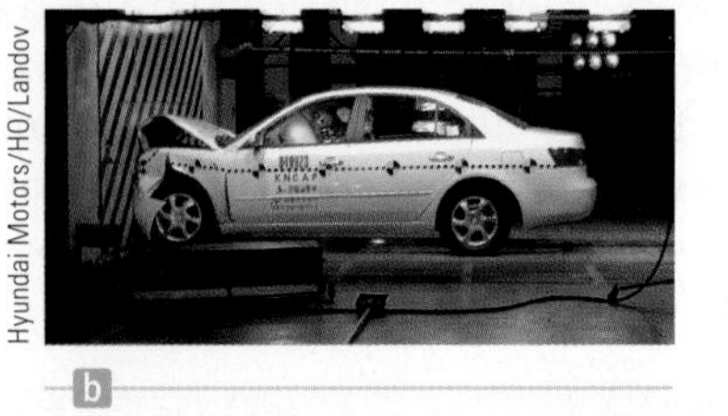

Figure 6.4 (Example 6.2) (a) This car's momentum changes as a result of its collision with the wall. (b) In a crash test (an inelastic collision), much of the car's initial kinetic energy is transformed into the energy it took to damage the vehicle.

SOLUTION

(a) Find the impulse delivered to the car.

Calculate the initial and final momenta of the car:

$$p_i = mv_i = (1.50 \times 10^3\ \text{kg})(-15.0\ \text{m/s}) = -2.25 \times 10^4\ \text{kg} \cdot \text{m/s}$$

$$p_f = mv_f = (1.50 \times 10^3\ \text{kg})(+2.60\ \text{m/s}) = +0.390 \times 10^4\ \text{kg} \cdot \text{m/s}$$

The impulse is just the difference between the final and initial momenta:

$$I = p_f - p_i = +0.390 \times 10^4\ \text{kg} \cdot \text{m/s} - (-2.25 \times 10^4\ \text{kg} \cdot \text{m/s})$$

$$I = 2.64 \times 10^4\ \text{kg} \cdot \text{m/s}$$

(b) Find the average force exerted on the car.

Apply Equation 6.6, the impulse–momentum theorem:

$$F_{av} = \frac{\Delta p}{\Delta t} = \frac{2.64 \times 10^4\ \text{kg} \cdot \text{m/s}}{0.150\ \text{s}} = +1.76 \times 10^5\ \text{N}$$

REMARKS If the car doesn't rebound off the wall, the average force exerted on the car is smaller than the value just calculated. With a final momentum of zero, the car undergoes a smaller change in momentum.

QUESTION 6.2 When a person is involved in a car accident, why is the likelihood of injury greater in a head-on collision as opposed to being hit from behind? Answer using the concepts of relative velocity, momentum, and average force.

EXERCISE 6.2 Suppose the car doesn't rebound off the wall, but the time interval of the collision remains at 0.150 s. In this case, the final velocity of the car is zero. Find the average force exerted on the car.

ANSWER $+1.50 \times 10^5$ N

Injury in Automobile Collisions

BIO APPLICATION
Injury to Passengers in Car Collisions

The main injuries that occur to a person hitting the interior of a car in a crash are brain damage, bone fracture, and trauma to the skin, blood vessels, and internal organs. Here, we compare the rather imprecisely known thresholds for human injury with typical forces and accelerations experienced in a car crash.

A force of about 90 kN (20 000 lb) compressing the tibia can cause fracture. Although the breaking force varies with the bone considered, we may take this value as the threshold force for fracture. It's well known that rapid acceleration of the head, even without skull fracture, can be fatal. Estimates show that head accelerations of 150*g* experienced for about 4 ms or 50*g* for 60 ms are fatal 50% of the time. Such injuries from rapid acceleration often result in nerve damage to the spinal cord where the nerves enter the base of the brain. The threshold for damage to skin, blood vessels, and internal organs may be estimated from whole-body impact data, where the force is uniformly distributed over the entire front surface area of 0.7 to 0.9 m^2. These data show that if the collision lasts for less than about 70 ms, a person will survive if the whole-body impact pressure (force per unit area) is less than 1.9×10^5 N/m^2 (28 lb/in.^{2}). Death results in 50% of cases in which the whole-body impact pressure reaches 3.4×10^5 N/m^2 (50 lb/in.^{2}).

Armed with the data above, we can estimate the forces and accelerations in a typical car crash and see how seat belts, air bags, and padded interiors can reduce the chance of death or serious injury in a collision. Consider a typical collision involving a 75-kg passenger not wearing a seat belt, traveling at 27 m/s (60 mi/h) who comes to rest in about 0.010 s after striking an unpadded dashboard. Using $F_{av}\Delta t = mv_f - mv_i$, we find that

$$F_{av} = \frac{mv_f - mv_i}{\Delta t} = \frac{0 - (75 \text{ kg})(27 \text{ m/s})}{0.010 \text{ s}} = -2.0 \times 10^5 \text{ N}$$

and

$$a = \left|\frac{\Delta v}{\Delta t}\right| = \frac{27 \text{ m/s}}{0.010 \text{ s}} = 2\,700 \text{ m/s}^2 = \frac{2\,700 \text{ m/s}^2}{9.8 \text{ m/s}^2}\,g = 280g$$

If we assume the passenger crashes into the dashboard and windshield so that the head and chest, with a combined surface area of 0.5 m^2, experience the force, we find a whole-body pressure of

$$\frac{F_{av}}{A} = \frac{2.0 \times 10^5 \text{ N}}{0.5 \text{ m}^2} \cong 4 \times 10^5 \text{ N/m}^2$$

We see that the force, the acceleration, and the whole-body pressure all *exceed* the threshold for fatality or broken bones and that an unprotected collision at 60 mi/h is almost certainly fatal.

What can be done to reduce or eliminate the chance of dying in a car crash? The most important factor is the collision time, or the time it takes the person to come to rest. If this time can be increased by 10 to 100 times the value of 0.01 s for a hard collision, the chances of survival in a car crash are much higher because the increase in Δt makes the contact force 10 to 100 times smaller. Seat belts restrain people so that they come to rest in about the same amount of time it takes to stop the car, typically about 0.15 s. This increases the effective collision time by an order of magnitude. Figure 6.5 shows the measured force on a car versus time for a car crash.

Air bags also increase the collision time, absorb energy from the body as they rapidly deflate, and spread the contact force over an area of the body of about 0.5 m^2, preventing penetration wounds and fractures. Air bags must deploy very rapidly (in less than 10 ms) in order to stop a human traveling at 27 m/s before he or she comes to rest against the steering column about 0.3 m away. To achieve this rapid deployment, accelerometers send a signal to discharge a bank of capacitors (devices that store electric charge), which then ignites an explosive, thereby filling the air bag with gas very quickly. The electrical charge for ignition is stored in capacitors to ensure that the air bag deploys in the event of damage to the battery or the car's electrical system in a severe collision.

The important reduction in potentially fatal forces, accelerations, and pressures to tolerable levels by the simultaneous use of seat belts and air bags is summarized as follows: If a 75-kg person traveling at 27 m/s is stopped by a seat belt in 0.15 s, the person experiences an average force of 9.8 kN, an average acceleration of 18g, and a whole-body pressure of 2.8×10^4 N/m^2 for a contact area of 0.5 m^2. These values are about one order of magnitude less than the values estimated earlier for an unprotected person and well below the thresholds for life-threatening injuries.

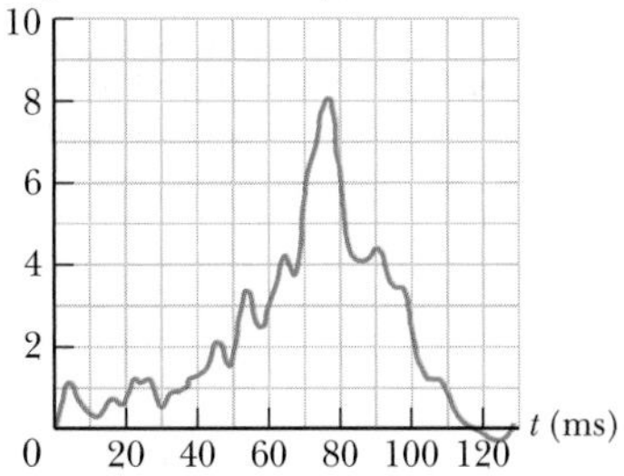

Figure 6.5 Force on a car versus time for a typical collision.

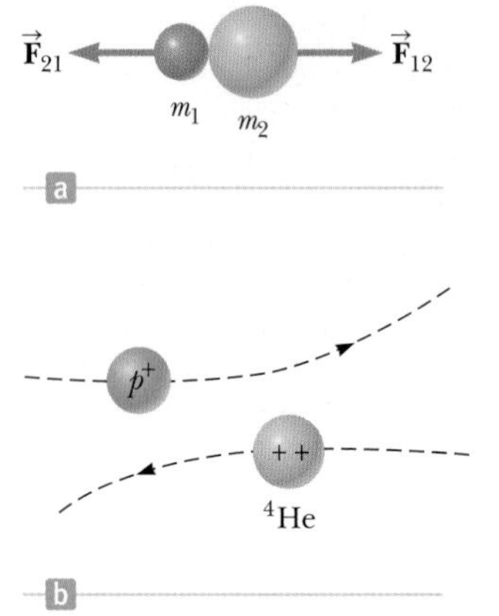

Figure 6.6 (a) A collision between two objects resulting from direct contact. (b) A collision between two charged objects (in this case, a proton and a helium nucleus).

6.2 Conservation of Momentum

LEARNING OBJECTIVES

1. Obtain the principle of conservation of momentum from the impulse–momentum theorem and the third law of motion.
2. Apply conservation of momentum to the problem of recoil.

When a collision occurs in an isolated system, the total momentum of the system doesn't change with the passage of time. Instead, it remains constant both in magnitude and in direction. The momenta of the individual objects in the system may change, but the vector sum of *all* the momenta will not change. The total momentum is therefore said to be *conserved.* In this section, we will see how the laws of motion lead us to this important conservation law.

A collision may be the result of physical contact between two objects, as illustrated in Figure 6.6a. This is a common macroscopic event, as when a pair of billiard balls or a baseball and a bat strike each other. By contrast, because contact on a submicroscopic scale is hard to define accurately, the notion of *collision* must be generalized to that scale. Forces between two objects arise from the electrostatic interaction of the electrons in the surface atoms of the objects. As will be discussed in Chapter 15, electric charges are either positive or negative. Charges with the same sign repel each other, while charges with opposite sign attract each other. To understand the distinction between macroscopic and microscopic collisions, consider the collision between two positive charges, as shown in Figure 6.6b. Because the two particles in the figure are both positively charged, they repel each other. During such a microscopic collision, particles need not touch in the normal sense in order to interact and transfer momentum.

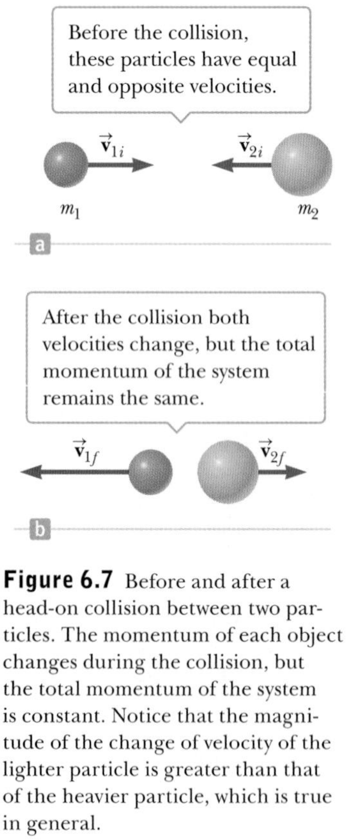

Figure 6.7 Before and after a head-on collision between two particles. The momentum of each object changes during the collision, but the total momentum of the system is constant. Notice that the magnitude of the change of velocity of the lighter particle is greater than that of the heavier particle, which is true in general.

Figure 6.7 shows an isolated system of two particles before and after they collide. By "isolated," we mean that no external forces, such as the gravitational force or friction, act on the system. Before the collision, the velocities of the two particles are $\vec{\mathbf{v}}_{1i}$ and $\vec{\mathbf{v}}_{2i}$; after the collision, the velocities are $\vec{\mathbf{v}}_{1f}$ and $\vec{\mathbf{v}}_{2f}$. The impulse–momentum theorem applied to m_1 becomes

$$\vec{\mathbf{F}}_{21}\,\Delta t = m_1\vec{\mathbf{v}}_{1f} - m_1\vec{\mathbf{v}}_{1i}$$

Likewise, for m_2, we have

$$\vec{\mathbf{F}}_{12}\,\Delta t = m_2\vec{\mathbf{v}}_{2f} - m_2\vec{\mathbf{v}}_{2i}$$

where $\vec{\mathbf{F}}_{21}$ is the average force exerted by m_2 on m_1 during the collision and $\vec{\mathbf{F}}_{12}$ is the average force exerted by m_1 on m_2 during the collision, as in Figure 6.6a.

We use average values for $\vec{\mathbf{F}}_{21}$ and $\vec{\mathbf{F}}_{12}$ even though the actual forces may vary in time in a complicated way, as is the case in Figure 6.8. Newton's third law states that at all times these two forces are equal in magnitude and opposite in direction: $\vec{\mathbf{F}}_{21} = -\vec{\mathbf{F}}_{12}$. In addition, the two forces act over the same time interval. As a result, we have

$$\vec{\mathbf{F}}_{21}\,\Delta t = -\vec{\mathbf{F}}_{12}\,\Delta t$$

or

$$m_1\vec{\mathbf{v}}_{1f} - m_1\vec{\mathbf{v}}_{1i} = -(m_2\vec{\mathbf{v}}_{2f} - m_2\vec{\mathbf{v}}_{2i})$$

after substituting the expressions obtained for $\vec{\mathbf{F}}_{21}$ and $\vec{\mathbf{F}}_{12}$. This equation can be rearranged to give the following important result:

$$m_1\vec{\mathbf{v}}_{1i} + m_2\,\vec{\mathbf{v}}_{2i} = m_1\vec{\mathbf{v}}_{1f} + m_2\vec{\mathbf{v}}_{2f} \qquad [6.7]$$

This result is a special case of the law of **conservation of momentum** and is true of isolated systems containing any number of interacting objects.

Conservation of momentum ▶

When no net external force acts on a system, the total momentum of the system remains constant in time.

aligned, both balls are released from rest at the same time, to fall through a distance of 1.20 m, as shown in Figure P6.73. (a) Find the magnitude of the downward velocity with which the basketball reaches the ground. (b) Assume that an elastic collision with the ground instantaneously reverses the velocity of the basketball while the tennis ball is still moving down. Next, the two balls meet in an elastic collision. To what height does the tennis ball rebound?

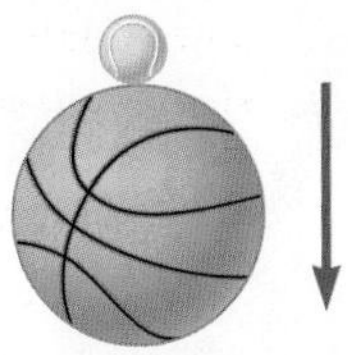

Figure P6.73

74. A 20.0-kg toboggan with 70.0-kg driver is sliding down a frictionless chute directed 30.0° below the horizontal at 8.00 m/s when a 55.0-kg woman drops from a tree limb straight down behind the driver. If she drops through a vertical displacement of 2.00 m, what is the subsequent velocity of the toboggan immediately after impact?

75. **S** *Measuring the speed of a bullet.* A bullet of mass m is fired horizontally into a wooden block of mass M lying on a table. The bullet remains in the block after the collision. The coefficient of friction between the block and table is μ, and the block slides a distance d before stopping. Find the initial speed v_0 of the bullet in terms of M, m, μ, g, and d.

76. A flying squid (family Ommastrephidae) is able to "jump" off the surface of the sea by taking water into its body cavity and then ejecting the water vertically downward. A 0.85-kg squid is able to eject 0.30 kg of water with a speed of 20 m/s. (a) What will be the speed of the squid immediately after ejecting the water? (b) How high in the air will the squid rise?

77. A 0.30-kg puck, initially at rest on a frictionless horizontal surface, is struck by a 0.20-kg puck that is initially moving along the x-axis with a velocity of 2.0 m/s. After the collision, the 0.20-kg puck has a speed of 1.0 m/s at an angle of $\theta = 53°$ to the positive x-axis. (a) Determine the velocity of the 0.30-kg puck after the collision. (b) Find the fraction of kinetic energy lost in the collision.

78. **QIC** **S** A wooden block of mass M rests on a table over a large hole as in Figure P6.78. A bullet of mass m with an initial velocity v_i is fired upward into the bottom of the block and remains in the block after the collision. The block and bullet rise to a maximum height of h. (a) Describe how you would find the initial velocity of the bullet using ideas you have learned in this chapter. (b) Find an expression for the initial velocity of the bullet.

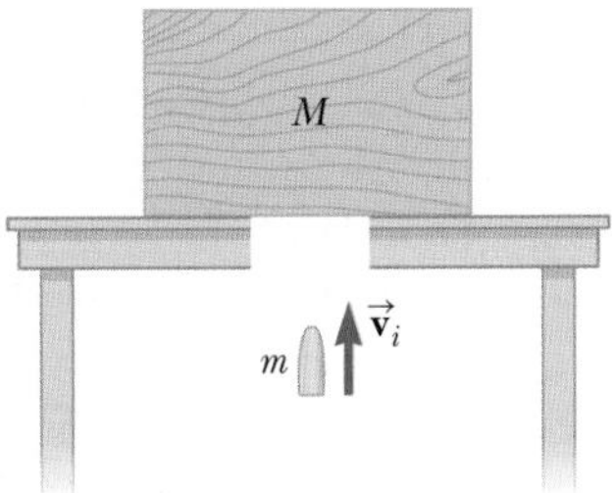

Figure P6.78 Problems 78 and 79.

79. **QIC** A 1.25-kg wooden block rests on a table over a large hole as in Figure P6.78. A 5.00-g bullet with an initial velocity v_i is fired upward into the bottom of the block and remains in the block after the collision. The block and bullet rise to a maximum height of 22.0 cm. (a) Describe how you would find the initial velocity of the bullet using ideas you have learned in this chapter. (b) Calculate the initial velocity of the bullet from the information provided.

The International Space Station falls freely around the Earth at thousands of meters per second, held in orbit by the centripetal force provided by gravity.

7 Rotational Motion and the Law of Gravity

Rotational motion is an important part of everyday life. The rotation of the Earth creates the cycle of day and night, the rotation of wheels enables easy vehicular motion, and modern technology depends on circular motion in a variety of contexts, from the tiny gears in a Swiss watch to the operation of lathes and other machinery. The concepts of *angular speed, angular acceleration,* and *centripetal acceleration* are central to understanding the motions of a diverse range of phenomena, from a car moving around a circular racetrack to clusters of galaxies orbiting a common center.

Rotational motion, when combined with Newton's law of universal gravitation and his laws of motion, can also explain certain facts about space travel and satellite motion, such as where to place a satellite so it will remain fixed in position over the same spot on the Earth. The generalization of gravitational potential energy and energy conservation offers an easy route to such results as planetary escape speed. Finally, we present Kepler's three laws of planetary motion, which formed the foundation of Newton's approach to gravity.

7.1 Angular Speed and Angular Acceleration

LEARNING OBJECTIVES

1. Define radian measure, angular position, and angular displacement.
2. Define average and instantaneous angular speed.
3. Define average and instantaneous angular acceleration.
4. Perform elementary calculations with angular variables.

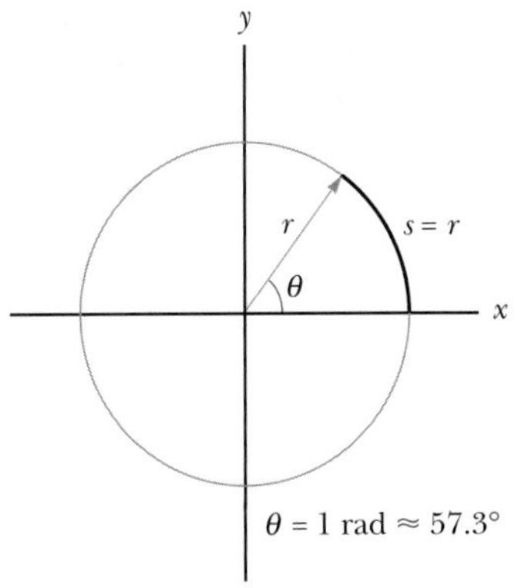

Figure 7.1 For a circle of radius r, one radian is the angle subtended by an arc length equal to r.

In the study of linear motion, the important concepts are *displacement* Δx, *velocity* v, and *acceleration* a. Each of these concepts has its analog in rotational motion: *angular displacement* $\Delta\theta$, *angular velocity* ω, and *angular acceleration* α.

The *radian,* a unit of angular measure, is essential to the understanding of these concepts. Recall that the distance s around a circle is given by $s = 2\pi r$, where r is the radius of the circle. Dividing both sides by r results in $s/r = 2\pi$. This quantity is dimensionless because both s and r have dimensions of length, but the value 2π corresponds to a displacement around a circle. A half circle would give an answer of π, a quarter circle an answer of $\pi/2$. The numbers 2π, π, and $\pi/2$ correspond to angles of 360°, 180°, and 90°, respectively, so a new unit of angular measure, the **radian,** can be introduced, with $180° = \pi$ rad relating degrees to radians.

The angle θ subtended by an arc length s along a circle of radius r, measured in radians counterclockwise from the positive x-axis, is

$$\theta = \frac{s}{r} \qquad [7.1]$$

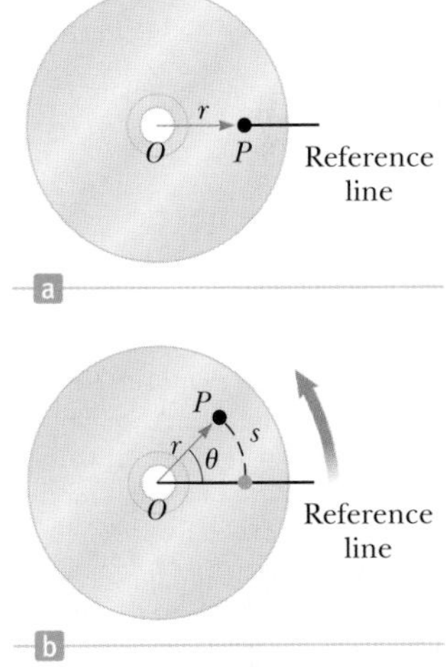

Figure 7.2 (a) The point P on a rotating compact disc at $t = 0$. (b) As the disc rotates, P moves through an arc length s.

The angle θ in Equation 7.1 is actually an angular displacement from the positive x-axis, and s the corresponding displacement along the circular arc, again measured from the positive x-axis. Figure 7.1 illustrates the size of 1 radian, which is approximately 57°. Converting from degrees to radians requires multiplying by the ratio (π rad/180°). For example, 45° (π rad/180°) = (π/4) rad.

Generally, angular quantities in physics must be expressed in radians. Be sure to set your calculator to radian mode; neglecting to do so is a common error.

Armed with the concept of radian measure, we can now discuss angular concepts in physics. Consider Figure 7.2a, a top view of a rotating compact disc. Such a disk is an example of a "rigid body," with each part of the body fixed in position relative to all other parts of the body. When a rigid body rotates through a given angle, all parts of the body rotate through the same angle at the same time. For the compact disc, the axis of rotation is at the center of the disc, O. A point P on the disc is at a distance r from the origin and moves about O in a circle of radius r. We set up a *fixed* reference line, as shown in Figure 7.2a, and assume that at time $t = 0$ the point P is on that reference line. After a time interval Δt has elapsed, P has advanced to a new position (Fig. 7.2b). In this interval, the line OP has moved through the angle θ with respect to the reference line. The angle θ, measured in radians, is called the **angular position** and is analogous to the linear position variable x. Likewise, P has moved an arc length s measured along the circumference of the circle.

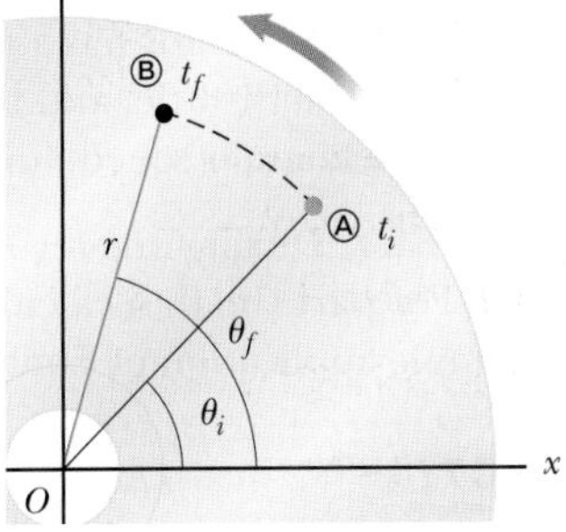

Figure 7.3 As a point on the compact disc moves from Ⓐ to Ⓑ, the disc rotates through the angle $\Delta\theta = \theta_f - \theta_i$.

In Figure 7.3, as a point on the rotating disc moves from Ⓐ to Ⓑ in a time Δt, it starts at an angle θ_i and ends at an angle θ_f. The difference $\theta_f - \theta_i$ is called the **angular displacement.**

An object's angular displacement, $\Delta\theta$, is the difference in its final and initial angles:

$$\Delta\theta = \theta_f - \theta_i \qquad [7.2]$$

SI unit: radian (rad)

For example, if a point on a disc is at $\theta_i = 4$ rad and rotates to angular position $\theta_f = 7$ rad, the angular displacement is $\Delta\theta = \theta_f - \theta_i = 7\text{ rad} - 4\text{ rad} = 3$ rad. Note that we use angular variables to describe the rotating disc because **each point on the disc undergoes the same angular displacement in any given time interval.**

Using the definition in Equation 7.2, Equation 7.1 can be written more generally as $\Delta\theta = \Delta s/r$, where Δs is a displacement along the circular arc subtended by the angular displacement. Having defined angular displacements, it's natural to define an angular speed:

The average angular speed ω_{av} of a rotating rigid object during the time interval Δt is the angular displacement $\Delta\theta$ divided by Δt:

$$\omega_{av} \equiv \frac{\theta_f - \theta_i}{t_f - t_i} = \frac{\Delta\theta}{\Delta t} \qquad [7.3]$$

SI unit: radian per second (rad/s)

For very short time intervals, the average angular speed approaches the instantaneous angular speed, just as in the linear case.

Tip 7.1 Remember the Radian

Equation 7.1 uses angles measured in *radians*. Angles expressed in terms of degrees must first be converted to radians. Also, be sure to check whether your calculator is in degree or radian mode when solving problems involving rotation.

The **instantaneous angular speed** ω of a rotating rigid object is the limit of the average speed $\Delta\theta/\Delta t$ as the time interval Δt approaches zero:

$$\omega \equiv \lim_{\Delta t \to 0} \frac{\Delta\theta}{\Delta t} \qquad [7.4]$$

SI unit: radian per second (rad/s)

We take ω to be positive when θ is increasing (counterclockwise motion) and negative when θ is decreasing (clockwise motion). When the angular speed is constant, the instantaneous angular speed is equal to the average angular speed.

EXAMPLE 7.1 Whirlybirds

GOAL Perform some elementary calculations with angular variables.

PROBLEM The rotor on a helicopter turns at an angular speed of 3.20×10^2 revolutions per minute. (In this book, we sometimes use the abbreviation rpm, but in most cases we use rev/min.) **(a)** Express this angular speed in radians per second. **(b)** If the rotor has a radius of 2.00 m, what arclength does the tip of the blade trace out in 3.00×10^2 s? **(c)** The pilot opens the throttle, and the angular speed of the blade increases while rotating twenty-six times in 3.60 s. Calculate the average angular speed during that time.

STRATEGY During one revolution, the rotor turns through an angle of 2π radians. Use this relationship as a conversion factor. For part (b), first calculate the angular displacement in radians by multiplying the angular speed by time. Part (c) is a simple application of Equation 7.3.

SOLUTION

(a) Express this angular speed in radians per second.

Apply the conversion factors 1 rev = 2π rad and 60.0 s = 1 min:

$$\omega = 3.20 \times 10^2 \frac{\text{rev}}{\text{min}}$$

$$= 3.20 \times 10^2 \frac{\cancel{\text{rev}}}{\cancel{\text{min}}}\left(\frac{2\pi\text{ rad}}{1\ \cancel{\text{rev}}}\right)\left(\frac{1.00\ \cancel{\text{min}}}{60.0\text{ s}}\right)$$

$$= \boxed{33.5\text{ rad/s}}$$

(b) Find the arclength traced out by the tip of the blade.

Multiply the angular speed by the time to obtain the angular displacement:

$$\Delta\theta = \omega t = (33.5 \text{ rad/s})(3.00 \times 10^2 \text{ s}) = 1.01 \times 10^4 \text{ rad}$$

Multiply the angular displacement by the radius to get the arc length:

$$\Delta s = r\,\Delta\theta = (2.00 \text{ m})(1.01 \times 10^4 \text{ rad}) = \boxed{2.02 \times 10^4 \text{ m}}$$

(c) Calculate the average angular speed of the blade while its angular speed increases.

Apply Equation 7.3, noticing that

$$\Delta\theta = (26 \text{ rev})(2\pi \text{ rad/rev}) = 52\pi \text{ rad}:$$

$$\omega_{\text{av}} = \frac{\Delta\theta}{\Delta t} = \frac{52\pi \text{ rad}}{3.60 \text{ s}} = 45 \text{ rad/s}$$

REMARKS It's best to express angular speeds in radians per second. Consistent use of radian measure minimizes errors.

QUESTION 7.1 Is it possible to express angular speed in degrees per second? If so, what's the conversion factor from radians per second?

EXERCISE 7.1 A Ferris wheel turns at a constant 185.0 revolutions per hour. **(a)** Express this rate of rotation in units of radians per second. **(b)** If the wheel has a radius of 12.0 m, what arclength does a passenger trace out during a ride lasting 5.00 min? **(c)** If the wheel then slows to rest in 9.72 s while making a quarter turn, calculate the magnitude of its average angular speed during that time.

ANSWERS (a) 0.323 rad/s (b) 1.16×10^3 m (c) 0.162 rad/s

■ *Quick Quiz*

7.1 A rigid body is rotating counterclockwise about a fixed axis. Each of the following pairs of quantities represents an initial angular position and a final angular position of the rigid body. Which of the sets can occur *only* if the rigid body rotates through more than 180°? (a) 3 rad, 6 rad; (b) −1 rad, 1 rad; (c) 1 rad, 5 rad.

7.2 Suppose the change in angular position for each of the pairs of values in Quick Quiz 7.1 occurred in 1 s. Which choice represents the lowest average angular speed?

Figure 7.4 shows a bicycle turned upside down so that a repair technician can work on the rear wheel. The bicycle pedals are turned so that at time t_i the wheel has angular speed ω_i (Fig. 7.4a) and at a later time t_f it has angular speed ω_f (Fig. 7.4b). Just as a changing speed leads to the concept of an acceleration, a changing angular speed leads to the concept of an angular acceleration.

An object's average angular acceleration α_{av} during the time interval Δt is the change in its angular speed $\Delta\omega$ divided by Δt:

$$\alpha_{\text{av}} \equiv \frac{\omega_f - \omega_i}{t_f - t_i} = \frac{\Delta\omega}{\Delta t} \qquad [7.5]$$

SI unit: radian per second squared (rad/s^2)

◀ Average angular acceleration

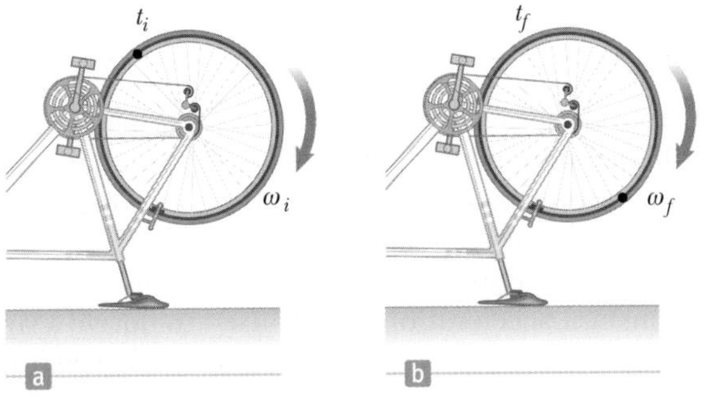

Figure 7.4 An accelerating bicycle wheel rotates with (a) angular speed ω_i at time t_i and (b) angular speed ω_f at time t_f.

As with angular velocity, positive angular accelerations are in the counterclockwise direction, negative angular accelerations in the clockwise direction. If the angular speed goes from 15 rad/s to 9.0 rad/s in 3.0 s, the average angular acceleration during that time interval is

$$\alpha_{av} = \frac{\Delta\omega}{\Delta t} = \frac{9.0 \text{ rad/s} - 15 \text{ rad/s}}{3.0 \text{ s}} = -2.0 \text{ rad/s}^2$$

The negative sign indicates that the angular acceleration is clockwise (although the angular speed, still positive but slowing down, is in the counterclockwise direction). There is also an instantaneous version of angular acceleration:

Instantaneous angular acceleration ▶

The instantaneous angular acceleration α is the limit of the average angular acceleration $\Delta\omega/\Delta t$ as the time interval Δt approaches zero:

$$\alpha \equiv \lim_{\Delta t \to 0} \frac{\Delta\omega}{\Delta t} \quad [7.6]$$

SI unit: radian per second squared (rad/s²)

When a rigid object rotates about a fixed axis, as does the bicycle wheel, every portion of the object has the same angular speed and the same angular acceleration. This fact is what makes these variables so useful for describing rotational motion. In contrast, the tangential (linear) speed and acceleration of the object take different values that depend on the distance from a given point to the axis of rotation.

7.2 Rotational Motion Under Constant Angular Acceleration

LEARNING OBJECTIVES

1. Identify the correspondence between the equations for linear motion at constant acceleration and those for angular motion.
2. Apply rotational kinematics to objects undergoing uniform angular acceleration.

A number of parallels exist between the equations for rotational motion and those for linear motion. For example, compare the defining equation for the average angular speed,

$$\omega_{av} \equiv \frac{\theta_f - \theta_i}{t_f - t_i} = \frac{\Delta\theta}{\Delta t}$$

with that of the average linear speed,

$$v_{av} \equiv \frac{x_f - x_i}{t_f - t_i} = \frac{\Delta x}{\Delta t}$$

In these equations, ω takes the place of v and θ takes the place of x, so the equations differ only in the names of the variables. In the same way, every linear quantity we have encountered so far has a corresponding "twin" in rotational motion.

The procedure used in Section 2.5 to develop the kinematic equations for linear motion under constant acceleration can be used to derive a similar set of equations for rotational motion under constant angular acceleration. The resulting equations of rotational kinematics, along with the corresponding equations for linear motion, are as follows:

Linear Motion with a Constant (Variables: x and v)	Rotational Motion About a Fixed Axis with α Constant (Variables: θ and ω)	
$v = v_i + at$	$\omega = \omega_i + \alpha t$	[7.7]
$\Delta x = v_i t + \frac{1}{2}at^2$	$\Delta\theta = \omega_i t + \frac{1}{2}\alpha t^2$	[7.8]
$v^2 = v_i^2 + 2a\Delta x$	$\omega^2 = \omega_i^2 + 2\alpha\Delta\theta$	[7.9]

Notice that every term in a given linear equation has a corresponding term in the analogous rotational equation.

Quick Quiz

7.3 Consider again the pairs of angular positions for the rigid object in Quick Quiz 7.1. If the object starts from rest at the initial angular position, moves counterclockwise with constant angular acceleration, and arrives at the final angular position with the same angular speed in all three cases, for which choice is the angular acceleration the highest?

EXAMPLE 7.2 A Rotating Wheel

GOAL Apply the rotational kinematic equations.

PROBLEM A wheel rotates with a constant angular acceleration of 3.50 rad/s^2. If the angular speed of the wheel is 2.00 rad/s at $t = 0$, **(a)** through what angle does the wheel rotate between $t = 0$ and $t = 2.00$ s? Give your answer in radians and in revolutions. **(b)** What is the angular speed of the wheel at $t = 2.00$ s? **(c)** What angular displacement (in revolutions) results while the angular speed found in part (b) doubles?

STRATEGY The angular acceleration is constant, so this problem just requires substituting given values into Equations 7.7–7.9.

SOLUTION

(a) Find the angular displacement after 2.00 s, in both radians and revolutions.

Use Equation 7.8, setting $\omega_i = 2.00$ rad/s, $\alpha = 3.5$ rad/s^2, and $t = 2.00$ s:

$$\Delta\theta = \omega_i t + \tfrac{1}{2}\alpha t^2$$
$$= (2.00 \text{ rad/s})(2.00 \text{ s}) + \tfrac{1}{2}(3.50 \text{ rad/s}^2)(2.00 \text{ s})^2$$
$$= \boxed{11.0 \text{ rad}}$$

Convert radians to revolutions.

$$\Delta\theta = (11.0 \text{ rad})(1.00 \text{ rev}/2\pi \text{ rad}) = \boxed{1.75 \text{ rev}}$$

(b) What is the angular speed of the wheel at $t = 2.00$ s?

Substitute the same values into Equation 7.7:

$$\omega = \omega_i + \alpha t = 2.00 \text{ rad/s} + (3.50 \text{ rad/s}^2)(2.00 \text{ s})$$
$$= \boxed{9.00 \text{ rad/s}}$$

(c) What angular displacement (in revolutions) results during the time in which the angular speed found in part (b) doubles?

Apply the time-independent rotational kinematics equation:

$$\omega_f^{\,2} - \omega_i^{\,2} = 2\alpha\Delta\theta$$

Substitute values, noting that $\omega_f = 2\omega_i$:

$$(2 \times 9.00 \text{ rad/s})^2 - (9.00 \text{ rad/s})^2 = 2(3.50 \text{ rad/s}^2)\Delta\theta$$

Solve for the angular displacement and convert to revolutions:

$$\Delta\theta = (34.7 \text{ rad})\left(\frac{1 \text{ rev}}{2\pi \text{ rad}}\right) = \boxed{5.52 \text{ rev}}$$

REMARKS The result of part (b) could also be obtained from Equation 7.9 and the results of part (a).

(Continued)

QUESTION 7.2 Suppose the radius of the wheel is doubled. Are the answers affected? If so, in what way?

EXERCISE 7.2 (a) Find the angle through which the wheel rotates between $t = 2.00$ s and $t = 3.00$ s. (b) Find the angular speed when $t = 3.00$ s. (c) What is the magnitude of the angular speed two revolutions following $t = 3.00$ s?

ANSWERS (a) 10.8 rad (b) 12.5 rad/s (c) 15.6 rad/s

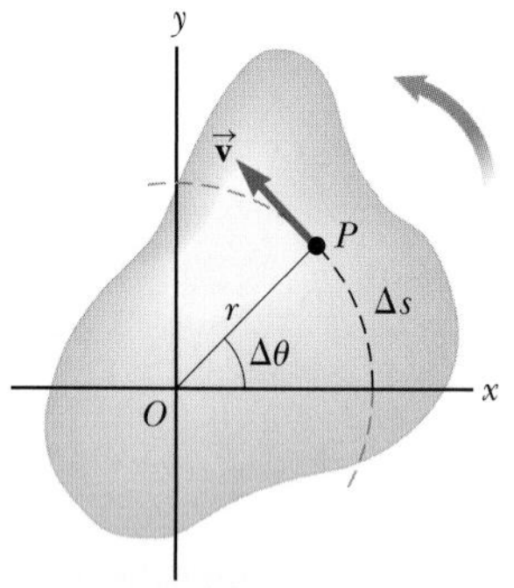

Figure 7.5 Rotation of an object about an axis through O (the z-axis) that is perpendicular to the plane of the figure. Note that a point P on the object rotates in a circle of radius r centered at O.

7.3 Relations Between Angular and Linear Quantities

LEARNING OBJECTIVE

1. Apply the relationships between angular and linear quantities.

Angular variables are closely related to linear variables. Consider the arbitrarily shaped object in Figure 7.5 rotating about the z-axis through the point O. Assume the object rotates through the angle $\Delta\theta$, and hence point P moves through the arc length Δs, in the interval Δt. We know from the defining equation for radian measure that

$$\Delta\theta = \frac{\Delta s}{r}$$

Dividing both sides of this equation by Δt, the time interval during which the rotation occurs, yields

$$\frac{\Delta\theta}{\Delta t} = \frac{1}{r}\frac{\Delta s}{\Delta t}$$

When Δt is very small, the angle $\Delta\theta$ through which the object rotates is also small and the ratio $\Delta\theta/\Delta t$ is close to the instantaneous angular speed ω. On the other side of the equation, similarly, the ratio $\Delta s/\Delta t$ approaches the instantaneous linear speed v for small values of Δt. Hence, when Δt gets arbitrarily small, the preceding equation is equivalent to

$$\omega = \frac{v}{r}$$

In Figure 7.5, the point P traverses a distance Δs along a circular arc during the time interval Δt at a linear speed of v. The direction of P's velocity vector $\vec{v}$ is *tangent to the circular path.* The magnitude of $\vec{v}$ is the linear speed $v = v_t$, called the **tangential speed** of a particle moving in a circular path, written

Tangential speed ▶

$$v_t = r\omega \qquad [7.10]$$

The tangential speed of a point on a rotating object equals the distance of that point from the axis of rotation multiplied by the angular speed. Equation 7.10 shows that the linear speed of a point on a rotating object increases as that point is moved outward from the center of rotation toward the rim, as expected; however, **every point on the rotating object has the same *angular* speed.**

Equation 7.10, derived using the defining equation for radian measure, is valid only when ω is measured in radians per unit time. Other measures of angular speed, such as degrees per second and revolutions per second, shouldn't be used.

To find a second equation relating linear and angular quantities, refer again to Figure 7.5 and suppose the rotating object changes its angular speed by $\Delta\omega$ in the time interval Δt. At the end of this interval, the speed of a point on the object, such as P, has changed by the amount Δv_t. From Equation 7.10 we have

$$\Delta v_t = r\,\Delta\omega$$

The foregoing derivations concern circular motion at constant speed. When an object moves in a circle but is speeding up or slowing down, a tangential component of acceleration, $a_t = r\alpha$, is also present. Because the tangential and centripetal components of acceleration are perpendicular to each other, we can find the magnitude of the **total acceleration** with the Pythagorean theorem:

$$a = \sqrt{a_t^2 + a_c^2} \quad [7.18]$$

◀ Total acceleration

Quick Quiz

7.6 A racetrack is constructed such that two arcs of radius 80 m at Ⓐ and 40 m at Ⓑ are joined by two stretches of straight track as in Figure 7.8. In a particular trial run, a driver travels at a constant speed of 50 m/s for one complete lap.

1. The ratio of the tangential acceleration at Ⓐ to that at Ⓑ is
   (a) $\frac{1}{2}$ (b) $\frac{1}{4}$ (c) 2 (d) 4 (e) The tangential acceleration is zero at both points.
2. The ratio of the centripetal acceleration at Ⓐ to that at Ⓑ is
   (a) $\frac{1}{2}$ (b) $\frac{1}{4}$ (c) 2 (d) 4 (e) The centripetal acceleration is zero at both points.
3. The angular speed is greatest at
   (a) Ⓐ (b) Ⓑ (c) It is equal at both Ⓐ and Ⓑ.

7.7 An object moves in a circular path with constant speed v. Which of the following statements is true concerning the object? (a) Its velocity is constant, but its acceleration is changing. (b) Its acceleration is constant, but its velocity is changing. (c) Both its velocity and acceleration are changing. (d) Its velocity and acceleration remain constant.

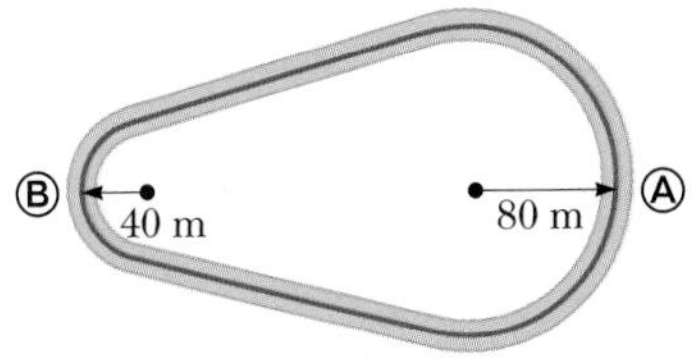

Figure 7.8 (Quick Quiz 7.6)

EXAMPLE 7.5 At the Racetrack

GOAL Apply the concepts of centripetal acceleration and tangential speed.

PROBLEM A race car accelerates uniformly from a speed of 40.0 m/s to a speed of 60.0 m/s in 5.00 s while traveling counterclockwise around a circular track of radius 4.00×10^2 m. When the car reaches a speed of 50.0 m/s, calculate **(a)** the magnitude of the car's centripetal acceleration, **(b)** the angular speed, **(c)** the magnitude of the tangential acceleration, and **(d)** the magnitude of the total acceleration.

STRATEGY Substitute values into the definitions of centripetal acceleration (Eq. 7.13), tangential speed (Eq. 7.10), and total acceleration (Eq. 7.18). Dividing the change in linear speed by the time yields the tangential acceleration.

SOLUTION

(a) Calculate the magnitude of the centripetal acceleration when $v = 50.0$ m/s.

Substitute into Equation 7.13:

$$a_c = \frac{v^2}{r} = \frac{(50.0\text{ m/s})^2}{4.00 \times 10^2\text{ m}} = \boxed{6.25\text{ m/s}^2}$$

(b) Calculate the angular speed.

Solve Equation 7.10 for ω and substitute:

$$\omega = \frac{v}{r} = \frac{50.0\text{ m/s}}{4.00 \times 10^2\text{ m}} = \boxed{0.125\text{ rad/s}}$$

(c) Calculate the magnitude of the tangential acceleration.

Divide the change in linear speed by the time:

$$a_t = \frac{v_f - v_i}{\Delta t} = \frac{60.0\text{ m/s} - 40.0\text{ m/s}}{5.00\text{ s}} = \boxed{4.00\text{ m/s}^2}$$

(d) Calculate the magnitude of the total acceleration.

Substitute into Equation 7.18:

$$a = \sqrt{a_t^2 + a_c^2} = \sqrt{(4.00\text{ m/s}^2)^2 + (6.25\text{ m/s}^2)^2}$$

$$a = \boxed{7.42\text{ m/s}^2}$$

(Continued)

REMARKS We can also find the centripetal acceleration by substituting the derived value of ω into Equation 7.17.

QUESTION 7.5 If the force causing the centripetal acceleration suddenly vanished, would the car (a) slide away along a radius, (b) proceed along a line tangent to the circular motion, or (c) proceed at an angle intermediate between the tangent and radius?

EXERCISE 7.5 Suppose the race car now slows down uniformly from 60.0 m/s to 30.0 m/s in 4.50 s to avoid an accident, while still traversing a circular path 4.00×10^2 m in radius. Calculate the car's (a) centripetal acceleration, (b) angular speed, (c) tangential acceleration, and (d) total acceleration when the speed is 40.0 m/s.

ANSWERS (a) 4.00 m/s^2 (b) 0.100 rad/s (c) -6.67 m/s^2 (d) 7.78 m/s^2

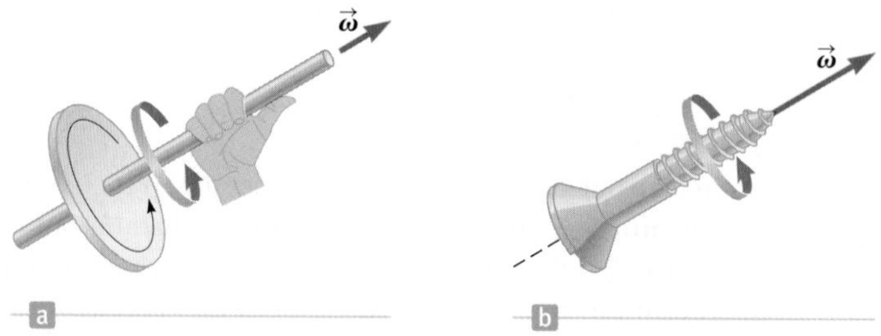

Figure 7.9 (a) The right-hand rule for determining the direction of the angular velocity vector $\vec{\omega}$. (b) The direction of $\vec{\omega}$ is in the direction of advance of a right-handed screw.

Angular Quantities Are Vectors

When we discussed linear motion in Chapter 2, we emphasized that displacement, velocity, and acceleration are all vector quantities. In describing rotational motion, angular displacement, angular velocity, and angular acceleration are also vector quantities.

The direction of the angular velocity vector $\vec{\omega}$ can be found with the **right-hand rule,** as illustrated in Figure 7.9a. Grasp the axis of rotation with your right hand so that your fingers wrap in the direction of rotation. Your extended thumb then points in the direction of $\vec{\omega}$. Figure 7.9b shows that $\vec{\omega}$ is also in the direction of advance of a rotating right-handed screw.

We can apply this rule to a disk rotating about a vertical axis through its center, as in Figure 7.10. When the disk rotates counterclockwise (Fig. 7.10a), the right-hand rule shows that the direction of $\vec{\omega}$ is upward. When the disk rotates clockwise (Fig. 7.10b), the direction of $\vec{\omega}$ is downward.

Finally, the directions of the angular acceleration $\vec{\alpha}$ and the angular velocity $\vec{\omega}$ are the same if the angular speed ω (the magnitude of $\vec{\omega}$) is increasing with time, and are opposite each other if the angular speed is decreasing with time.

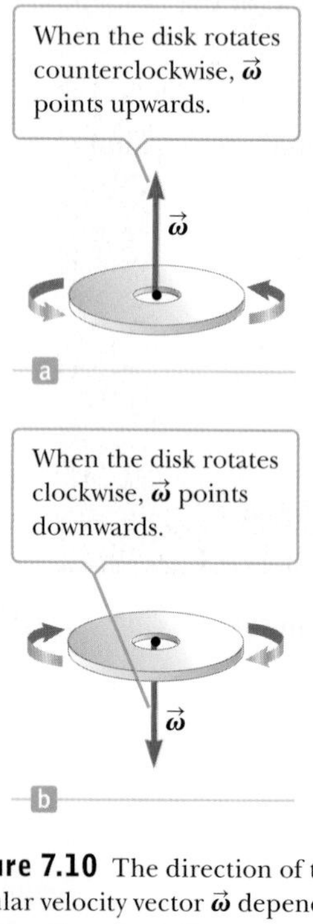

Figure 7.10 The direction of the angular velocity vector $\vec{\omega}$ depends on the direction of rotation.

Forces Causing Centripetal Acceleration

An object can have a centripetal acceleration *only* if some external force acts on it. For a ball whirling in a circle at the end of a string, that force is the tension in the string. In the case of a car moving on a flat circular track, the force is friction between the car and track. A satellite in circular orbit around Earth has a centripetal acceleration due to the gravitational force between the satellite and Earth.

Some books use the term "centripetal force," which can give the mistaken impression that it is a new force of nature. This is not the case: The adjective "centripetal" in "centripetal force" simply means that the force in question acts toward a center. The force of tension in the string of a yo-yo whirling in a vertical circle is an example of a centripetal force, as is the force of gravity on a satellite circling the Earth.

Consider a puck of mass m that is tied to a string of length r and is being whirled at constant speed in a horizontal circular path, as illustrated in Figure 7.11. Its weight is supported by a frictionless table. Why does the puck move in a circle? Because of its inertia, the tendency of the puck is to move in a straight line; however, the string prevents motion along a straight line by exerting a radial force on the puck—a tension force—that makes it follow the circular path. The tension $\vec{\mathbf{T}}$ is directed along the string toward *the center of the circle,* as shown in the figure.

In general, converting Newton's second law to polar coordinates yields an equation relating the net centripetal force, F_c, which is the sum of the radial components of all forces acting on a given object, to the centripetal acceleration. The *magnitude* of the net centripetal force equals the mass times the magnitude of the centripetal acceleration:

$$F_c = ma_c = m\frac{v^2}{r} \qquad [7.19]$$

A net force causing a centripetal acceleration acts toward the center of the circular path and effects a change in the direction of the velocity vector. If that force should vanish, the object would immediately leave its circular path and move along a straight line tangent to the circle at the point where the force vanished.

Centrifugal ('center-fleeing') forces also exist, such as the force between two particles with the same sign charge (see Chapter 15). The normal force that prevents an object from falling toward the center of the Earth is another example of a centrifugal force. Sometimes an insufficient centripetal force is mistaken for the presence of a centrifugal force (see "Fictitious Forces," page 219).

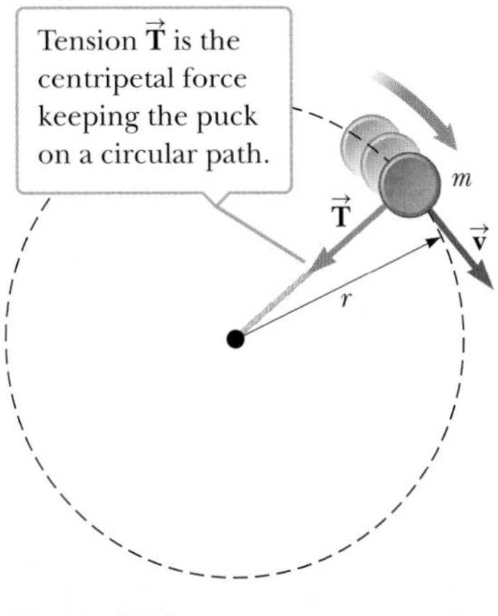

Figure 7.11 A puck attached to a string of length r rotates in a horizontal plane at constant speed.

Tip 7.2 Centripetal Force Is a *Type* of Force, Not a Force in Itself!

"Centripetal force" is a classification that includes forces acting toward a central point, like the horizontal component of the string tension on a tetherball or gravity on a satellite. A centripetal force must be *supplied* by some actual, physical force.

APPLYING PHYSICS 7.2 | Artificial Gravity

Astronauts spending lengthy periods of time in space experience a number of negative effects due to weightlessness, such as weakening of muscle tissue and loss of calcium in bones. These effects may make it very difficult for them to return to their usual environment on Earth. How could artificial gravity be generated in space to overcome such complications?

SOLUTION A rotating cylindrical space station creates an environment of artificial gravity. The normal force of the rigid walls provides the centripetal force, which keeps the astronauts moving in a circle (Fig. 7.12). To an astronaut, the normal force can't be easily distinguished from a gravitational force as long as the radius of the station is large compared with the astronaut's height. (Otherwise there are unpleasant inner ear effects.) This same principle is used in certain amusement park rides in which passengers are pressed against the inside of a rotating cylinder as it tilts in various directions. The visionary physicist Gerard O'Neill proposed creating a giant space colony a kilometer in radius that rotates slowly, creating Earth-normal artificial gravity for the inhabitants in its interior. These inside-out artificial worlds could enable safe transport on a several-thousand-year journey to another star system. ■

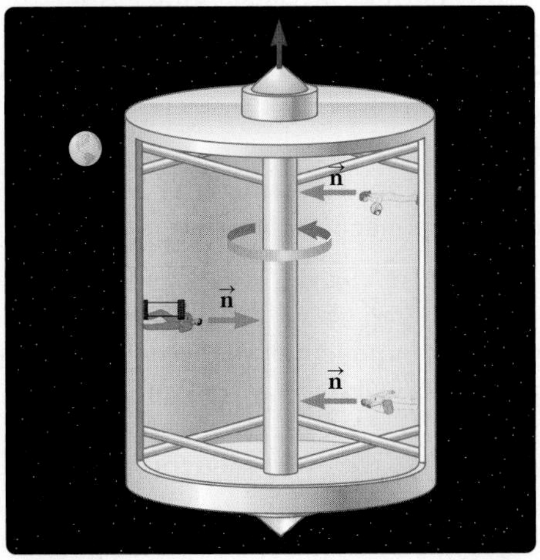

Figure 7.12 Artificial gravity inside a spinning cylinder is provided by the normal force.

PROBLEM-SOLVING STRATEGY

Forces That Cause Centripetal Acceleration

Use the following steps in dealing with centripetal accelerations and the forces that produce them:

1. **Draw a free-body diagram** of the object under consideration, labeling all forces that act on it.

2. **Choose a coordinate system** that has one axis perpendicular to the circular path followed by the object (the radial direction) and one axis tangent to the circular path (the tangential, or angular, direction). The normal direction, perpendicular to the plane of motion, is also often needed.
3. **Find the net force F_c toward the center** of the circular path, $F_c = \Sigma F_r$, where ΣF_r is the sum of the radial components of the forces. This net radial force causes the centripetal acceleration.
4. **Use Newton's second law for the radial, tangential, and normal directions,** as required, writing $\Sigma F_r = ma_c$, $\Sigma F_t = ma_t$, and $\Sigma F_n = ma_n$. Remember that the magnitude of the centripetal acceleration for uniform circular motion can always be written $a_c = v_t^2/r$.
5. **Solve** for the unknown quantities.

EXAMPLE 7.6 Buckle Up for Safety

GOAL Calculate the frictional force that causes an object to have a centripetal acceleration.

PROBLEM A car travels at a constant speed of 30.0 mi/h (13.4 m/s) on a level circular turn of radius 50.0 m, as shown in the bird's-eye view in Figure 7.13a. What minimum coefficient of static friction, μ_s, between the tires and roadway will allow the car to make the circular turn without sliding?

STRATEGY In the car's free-body diagram (Fig. 7.13b) the normal direction is vertical and the tangential direction is into the page (Step 2). Use Newton's second law. The net force acting on the car in the radial direction is the force of static friction toward the center of the circular path, which causes the car to have a centripetal acceleration. Calculating the maximum static friction force requires the normal force, obtained from the normal component of the second law.

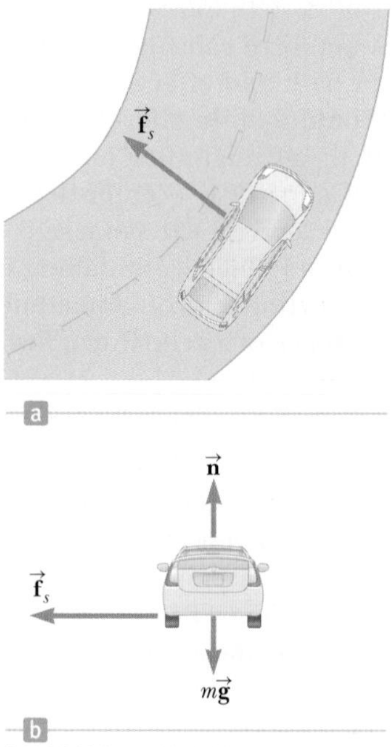

Figure 7.13 (Example 7.6) (a) The centripetal force is provided by the force of static friction, which is directed radially toward the center of the circular path. (b) Gravity, the normal force, and the static friction force act on the car.

SOLUTION

(Steps 3, 4) Write the components of Newton's second law. The radial component involves only the maximum static friction force, $f_{s,\,\max}$:

$$m\frac{v^2}{r} = f_{s,\max} = \mu_s n$$

In the vertical component of the second law, the gravity force and the normal force are in equilibrium:

$$n - mg = 0 \quad \rightarrow \quad n = mg$$

(Step 5) Substitute the expression for n into the first equation and solve for μ_s:

$$m\frac{v^2}{r} = \mu_s mg$$

$$\mu_s = \frac{v^2}{rg} = \frac{(13.4\text{ m/s})^2}{(50.0\text{ m})(9.80\text{ m/s}^2)} = 0.366$$

REMARKS The value of μ_s for rubber on dry concrete is very close to 1, so the car can negotiate the curve with ease. If the road were wet or icy, however, the value for μ_s could be 0.2 or lower. Under such conditions, the radial force provided by static friction wouldn't be great enough to keep the car on the circular path, and it would slide off on a tangent, leaving the roadway.

QUESTION 7.6 If the static friction coefficient were increased, would the maximum safe speed be reduced, increased, or remain the same?

Escape Speed

If an object is projected upward from Earth's surface with a large enough speed, it can soar off into space and never return. This speed is called Earth's **escape speed.** (It is also commonly called the *escape velocity,* but in fact is more properly a speed.)

Earth's escape speed can be found by applying conservation of energy. Suppose an object of mass m is projected vertically upward from Earth's surface with an initial speed v_i. The initial mechanical energy (kinetic plus potential energy) of the object–Earth system is given by

$$KE_i + PE_i = \tfrac{1}{2}mv_i^2 - \frac{GM_E m}{R_E}$$

We neglect air resistance and assume the initial speed is just large enough to allow the object to reach infinity with a speed of zero. This value of v_i is the escape speed v_{esc}. When the object is at an infinite distance from Earth, its kinetic energy is zero because $v_f = 0$, and the gravitational potential energy is also zero because $1/r$ goes to zero as r goes to infinity. Hence the total mechanical energy is zero, and the law of conservation of energy gives

$$\tfrac{1}{2}mv_{\text{esc}}^2 - \frac{GM_E m}{R_E} = 0$$

so that

$$v_{\text{esc}} = \sqrt{\frac{2GM_E}{R_E}} \qquad [7.22]$$

The escape speed for Earth is about 11.2 km/s, which corresponds to about 25 000 mi/h. (See Example 7.12.) Note that the expression for v_{esc} doesn't depend on the mass of the object projected from Earth, so a spacecraft has the same escape speed as a molecule. Escape speeds for the planets, the Moon, and the Sun are listed in Table 7.2. Escape speed and temperature determine to a large extent whether a world has an atmosphere and, if so, what the constituents of the atmosphere are. Planets with low escape speeds, such as Mercury, generally don't have atmospheres because the average speed of gas molecules is close to the escape speed. Venus has a very thick atmosphere, but it's almost entirely carbon dioxide, a heavy gas. The atmosphere of Earth has very little hydrogen or helium, but has retained the much heavier nitrogen and oxygen molecules.

Table 7.2 Escape Speeds for the Planets and the Moon

Planet	v_{esc} (km/s)
Mercury	4.3
Venus	10.3
Earth	11.2
Moon	2.3
Mars	5.0
Jupiter	60.0
Saturn	36.0
Uranus	22.0
Neptune	24.0
Pluto[a]	1.1

[a]In August 2006, the International Astronomical Union adopted a definition of a planet that separates Pluto from the other eight planets. Pluto is now defined as a "dwarf planet" (like the asteroid Ceres).

■ EXAMPLE 7.12 From the Earth to the Moon

GOAL Apply conservation of energy with the general form of Newton's universal law of gravity.

PROBLEM In Jules Verne's classic novel *From the Earth to the Moon,* a giant cannon dug into the Earth in Florida fired a spacecraft all the way to the Moon. **(a)** If the spacecraft leaves the cannon at escape speed, at what speed is it moving when 1.50×10^5 km from the center of Earth? Neglect any friction effects. **(b)** Approximately what constant acceleration is needed to propel the spacecraft to escape speed through a cannon bore 1.00 km long?

STRATEGY For part (a), use conservation of energy and solve for the final speed v_f. Part (b) is an application of the time-independent kinematic equation: solve for the acceleration a.

SOLUTION

(a) Find the speed at $r = 1.50 \times 10^5$ km.

Apply conservation of energy:

$$\tfrac{1}{2}mv_i^2 - \frac{GM_E m}{R_E} = \tfrac{1}{2}mv_f^2 - \frac{GM_E m}{r_f}$$

(Continued)

Multiply by $2/m$ and rearrange, solving for v_f^2. Then substitute known values and take the square root.

$$v_f^2 = v_i^2 + \frac{2GM_E}{r_f} - \frac{2GM_E}{R_E} = v_i^2 + 2GM_E\left(\frac{1}{r_f} - \frac{1}{R_E}\right)$$

$$v_f^2 = (1.12 \times 10^4 \text{ m/s})^2 + 2(6.67 \times 10^{-11} \text{ kg}^{-1}\text{m}^3\text{s}^{-2}) \times (5.98 \times 10^{24} \text{ kg})\left(\frac{1}{1.50 \times 10^8 \text{ m}} - \frac{1}{6.38 \times 10^6 \text{ m}}\right)$$

$$v_f = 2.39 \times 10^3 \text{ m/s}$$

(b) Find the acceleration through the cannon bore, assuming it's constant.

Use the time-independent kinematics equation:

$$v^2 - v_0^2 = 2a\Delta x$$

$$(1.12 \times 10^4 \text{ m/s})^2 - 0 = 2a(1.00 \times 10^3 \text{ m})$$

$$a = 6.27 \times 10^4 \text{ m/s}^2$$

REMARKS This result corresponds to an acceleration of over 6 000 times the free-fall acceleration on Earth. Such a huge acceleration is far beyond what the human body can tolerate.

QUESTION 7.12 Suppose the spacecraft managed to go into an elliptical orbit around Earth, with a nearest point (perigee) and farthest point (apogee). At which point is the kinetic energy of the spacecraft higher, and why?

EXERCISE 7.12 Using the data in Table 7.3 (see page 228), find (a) the escape speed from the surface of Mars and (b) the speed of a space vehicle when it is 1.25×10^7 m from the center of Mars if it leaves the surface at the escape speed.

ANSWERS (a) 5.04×10^3 m/s (b) 2.62×10^3 m/s

7.6 Kepler's Laws

LEARNING OBJECTIVES

1. State Kepler's three laws and explain the significance of each.
2. Apply the third law to obtain information about orbiting bodies.

The movements of the planets, stars, and other celestial bodies have been observed for thousands of years. In early history scientists regarded Earth as the center of the Universe. This **geocentric model** was developed extensively by the Greek astronomer Claudius Ptolemy in the second century A.D. and was accepted for the next 1 400 years. In 1543 Polish astronomer Nicolaus Copernicus (1473–1543) showed that Earth and the other planets revolve in circular orbits around the Sun (the **heliocentric model**).

Danish astronomer Tycho Brahe (pronounced Brah or BRAH–huh; 1546–1601) made accurate astronomical measurements over a period of 20 years, providing the data for the currently accepted model of the solar system. Brahe's precise observations of the planets and 777 stars were carried out with nothing more elaborate than a large sextant and compass; the telescope had not yet been invented.

German astronomer Johannes Kepler, who was Brahe's assistant, acquired Brahe's astronomical data and spent about 16 years trying to deduce a mathematical model for the motions of the planets. After many laborious calculations, he found that Brahe's precise data on the motion of Mars about the Sun provided the answer. Kepler's analysis first showed that the concept of circular orbits about the Sun had to be abandoned. He eventually discovered that the orbit of Mars could be accurately described by an ellipse with the Sun at one focus. He then generalized this analysis to include the motions of all planets. The complete analysis is summarized in three statements known as **Kepler's laws:**

◀ Kepler's Laws

1. All planets move in elliptical orbits with the Sun at one of the focal points.
2. A line drawn from the Sun to any planet sweeps out equal areas in equal time intervals.
3. The square of the orbital period of any planet is proportional to the cube of the average distance from the planet to the Sun.

Newton later demonstrated that these laws are consequences of the gravitational force that exists between any two objects. Newton's law of universal gravitation, together with his laws of motion, provides the basis for a full mathematical description of the motions of planets and satellites.

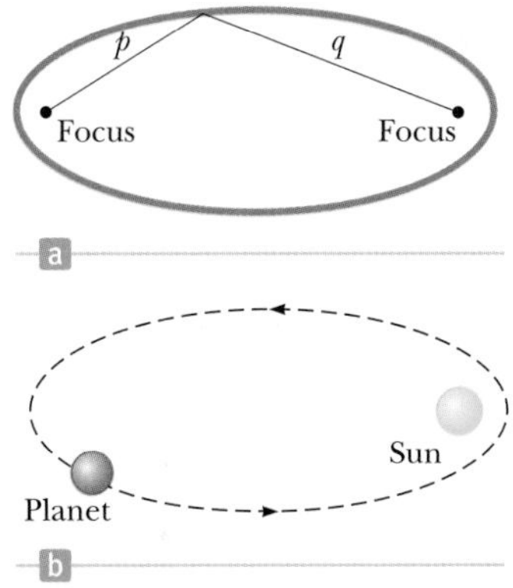

Figure 7.22 (a) The sum $p + q$ is the same for every point on the ellipse. (b) In the Solar System, the Sun is at one focus of the elliptical orbit of each planet and the other focus is empty.

Kepler's First Law

The first law arises as a natural consequence of the inverse-square nature of Newton's law of gravitation. Any object bound to another by a force that varies as $1/r^2$ will move in an elliptical orbit. As shown in Figure 7.22a, an ellipse is a curve drawn so that the sum of the distances from any point on the curve to two internal points called *focal points* or *foci* (singular, *focus*) is always the same. The semimajor axis a is half the length of the line that goes across the ellipse and contains both foci. For the Sun–planet configuration (Fig. 7.22b), the Sun is at one focus and the other focus is empty. Because the orbit is an ellipse, the distance from the Sun to the planet continuously changes.

Kepler's Second Law

Kepler's second law states that a line drawn from the Sun to any planet sweeps out equal areas in equal time intervals. Consider a planet in an elliptical orbit about the Sun, as in Figure 7.23. In a given period Δt, the planet moves from point Ⓐ to point Ⓑ. The planet moves more slowly on that side of the orbit because it's farther away from the sun. On the opposite side of its orbit, the planet moves from point Ⓒ to point Ⓓ in the same amount of time, Δt, moving faster because it's closer to the sun. Kepler's second law says that any two wedges formed as in Figure 7.23 will always have the same area. As we will see in Chapter 8, Kepler's second law is related to a physical principle known as conservation of angular momentum.

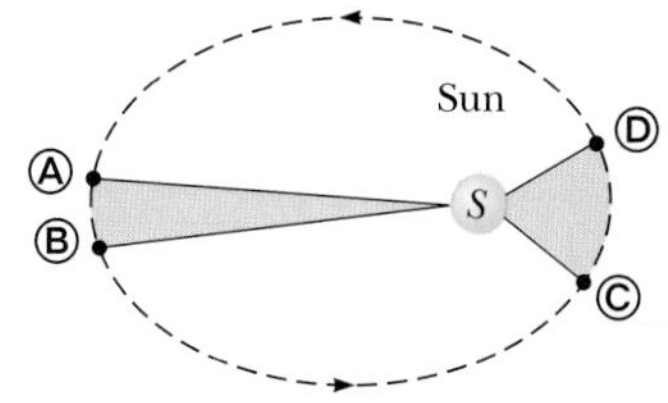

Figure 7.23 The two areas swept out by the planet in its elliptical orbit about the Sun are equal if the time interval between points Ⓐ and Ⓑ is equal to the time interval between points Ⓒ and Ⓓ.

Kepler's Third Law

The derivation of Kepler's third law is simple enough to carry out for the special case of a circular orbit. Consider a planet of mass M_p moving around the Sun, which has a mass of M_S, in a circular orbit. Because the orbit is circular, the planet moves at a constant speed v. Newton's second law, his law of gravitation, and centripetal acceleration then give the following equation:

$$M_p a_c = \frac{M_p v^2}{r} = \frac{GM_S M_p}{r^2}$$

The speed v of the planet in its orbit is equal to the circumference of the orbit divided by the time required for one revolution, T, called the **period** of the planet, so $v = 2\pi r/T$. Substituting, the preceding expression becomes

$$\frac{GM_S}{r^2} = \frac{(2\pi r/T)^2}{r}$$

$$T^2 = \left(\frac{4\pi^2}{GM_S}\right) r^3 = K_S r^3 \qquad [7.23]$$

◀ Kepler's third law

Table 7.3 Useful Planetary Data

Body	Mass (kg)	Mean Radius (m)	Period (s)	Mean Distance from Sun (m)	$\frac{T^2}{r^3}\ 10^{-19}\left(\frac{s^2}{m^3}\right)$
Mercury	3.18×10^{23}	2.43×10^{6}	7.60×10^{6}	5.79×10^{10}	2.97
Venus	4.88×10^{24}	6.06×10^{6}	1.94×10^{7}	1.08×10^{11}	2.99
Earth	5.98×10^{24}	6.38×10^{6}	3.156×10^{7}	1.496×10^{11}	2.97
Mars	6.42×10^{23}	3.37×10^{6}	5.94×10^{7}	2.28×10^{11}	2.98
Jupiter	1.90×10^{27}	6.99×10^{7}	3.74×10^{8}	7.78×10^{11}	2.97
Saturn	5.68×10^{26}	5.85×10^{7}	9.35×10^{8}	1.43×10^{12}	2.99
Uranus	8.68×10^{25}	2.33×10^{7}	2.64×10^{9}	2.87×10^{12}	2.95
Neptune	1.03×10^{26}	2.21×10^{7}	5.22×10^{9}	4.50×10^{12}	2.99
Pluto[a]	1.27×10^{23}	1.14×10^{6}	7.82×10^{9}	5.91×10^{12}	2.96
Moon	7.36×10^{22}	1.74×10^{6}	—	—	—
Sun	1.991×10^{30}	6.96×10^{8}	—	—	—

[a]In August 2006, the International Astronomical Union adopted a definition of a planet that separates Pluto from the other eight planets. Pluto is now defined as a "dwarf planet" like the asteroid Ceres.

where K_S is a constant given by

$$K_S = \frac{4\pi^2}{GM_S} = 2.97 \times 10^{-19}\ \mathrm{s^2/m^3}$$

Equation 7.23 is Kepler's third law for a circular orbit. The orbits of most of the planets are very nearly circular. Comets and asteroids, however, usually have elliptical orbits. For these orbits, the radius r must be replaced with a, the semimajor axis—half the longest distance across the elliptical orbit. (This is also the average distance of the comet or asteroid from the Sun.) A more detailed calculation shows that K_S actually depends on the sum of both the mass of a given planet and the Sun's mass. The masses of the planets, however, are negligible compared with the Sun's mass; hence can be neglected, meaning Equation 7.23 is valid for any planet in the Sun's family. If we consider the orbit of a satellite such as the Moon around Earth, then the constant has a different value, with the mass of the Sun replaced by the mass of Earth. In that case, K_E equals $4\pi^2/GM_E$.

The mass of the Sun can be determined from Kepler's third law because the constant K_S in Equation 7.23 includes the mass of the Sun and the other variables and constants can be easily measured. The value of this constant can be found by substituting the values of a planet's period and orbital radius and solving for K_S. The mass of the Sun is then

$$M_S = \frac{4\pi^2}{GK_S}$$

This same process can be used to calculate the mass of Earth (by considering the period and orbital radius of the Moon) and the mass of other planets in the solar system that have satellites.

The last column in Table 7.3 confirms that T^2/r^3 is very nearly constant. When time is measured in Earth years and the semimajor axis in astronomical units (1 AU = the distance from Earth to the Sun), Kepler's law takes the following simple form:

$$T^2 = a^3$$

This equation can be easily checked: Earth has a semimajor axis of one astronomical unit (by definition), and it takes one year to circle the Sun. This equation, of course, is valid only for the sun and its planets, asteroids, and comets.

Quick Quiz

7.10 Suppose an asteroid has a semimajor axis of 4 AU. How long does it take the asteroid to go around the Sun? (a) 2 years (b) 4 years (c) 6 years (d) 8 years

EXAMPLE 7.13 Geosynchronous Orbit and Telecommunications Satellites

GOAL Apply Kepler's third law to an Earth satellite.

PROBLEM From a telecommunications point of view, it's advantageous for satellites to remain at the same location relative to a location on Earth. This can occur only if the satellite's orbital period is the same as the Earth's period of rotation, approximately 24.0 h. **(a)** At what distance from the center of the Earth can this geosynchronous orbit be found? **(b)** What's the orbital speed of the satellite?

STRATEGY This problem can be solved with the same method that was used to derive a special case of Kepler's third law, with Earth's mass replacing the Sun's mass. There's no need to repeat the analysis; just replace the Sun's mass with Earth's mass in Kepler's third law, substitute the period T (converted to seconds), and solve for r. For part (b), find the circumference of the circular orbit and divide by the elapsed time.

SOLUTION

(a) Find the distance r to geosynchronous orbit.

Apply Kepler's third law:

$$T^2 = \left(\frac{4\pi^2}{GM_E}\right)r^3$$

Substitute the period in seconds, $T = 86\,400$ s, the gravity constant $G = 6.67 \times 10^{-11}\ \mathrm{kg^{-1}\ m^3/s^2}$, and the mass of the Earth, $M_E = 5.98 \times 10^{24}$ kg. Solve for r:

$$r = 4.23 \times 10^7\ \mathrm{m}$$

(b) Find the orbital speed.

Divide the distance traveled during one orbit by the period:

$$v = \frac{d}{T} = \frac{2\pi r}{T} = \frac{2\pi(4.23 \times 10^7\ \mathrm{m})}{8.64 \times 10^4\ \mathrm{s}} = 3.08 \times 10^3\ \mathrm{m/s}$$

REMARKS Earth's motion around the Sun was neglected; that requires using Earth's "sidereal" period (about four minutes shorter). Notice that Earth's mass could be found by substituting the Moon's distance and period into this form of Kepler's third law.

QUESTION 7.13 If the satellite was placed in an orbit three times as far away, about how long would it take to orbit the Earth once? Answer in days, rounding to one digit.

EXERCISE 7.13 Mars rotates on its axis once every 1.02 days (almost the same as Earth does). (a) Find the distance from the center of Mars at which a satellite would remain in one spot over the Martian surface. (b) Find the speed of the satellite.

ANSWERS (a) 2.03×10^7 m (b) 1.45×10^3 m/s

SUMMARY

7.1 Angular Speed and Angular Acceleration

The **average angular speed** ω_{av} of a rigid object is defined as the ratio of the angular displacement $\Delta\theta$ to the time interval Δt, or

$$\omega_{av} \equiv \frac{\theta_f - \theta_i}{t_f - t_i} = \frac{\Delta\theta}{\Delta t} \quad [7.3]$$

where ω_{av} is in radians per second (rad/s).

The **average angular acceleration** α_{av} of a rotating object is defined as the ratio of the change in angular speed $\Delta\omega$ to the time interval Δt, or

$$\alpha_{av} \equiv \frac{\omega_f - \omega_i}{t_f - t_i} = \frac{\Delta\omega}{\Delta t} \quad [7.5]$$

where α_{av} is in radians per second per second (rad/s^2).

7.2 Rotational Motion Under Constant Angular Acceleration

If an object undergoes rotational motion about a fixed axis under a constant angular acceleration α, its motion can be described with the following set of equations:

$$\omega = \omega_i + \alpha t \quad [7.7]$$

$$\Delta\theta = \omega_i t + \tfrac{1}{2}\alpha t^2 \quad [7.8]$$

$$\omega^2 = \omega_i^2 + 2\alpha\Delta\theta \quad [7.9]$$

Problems are solved as in one-dimensional kinematics.

7.3 Relations Between Angular and Linear Quantities

When an object rotates about a fixed axis, the angular speed and angular acceleration are related to the tangential speed and tangential acceleration through the relationships

$$v_t = r\omega \quad [7.10]$$

and

$$a_t = r\alpha \quad [7.11]$$

7.4 Centripetal Acceleration

Any object moving in a circular path has an acceleration directed toward the center of the circular path, called a **centripetal acceleration.** Its magnitude is given by

$$a_c = \frac{v^2}{r} = r\omega^2 \quad [7.13, 7.17]$$

Any object moving in a circular path must have a net force exerted on it that is directed toward the center of the path. Some examples of forces that cause centripetal acceleration are the force of gravity (as in the motion of a satellite) and the force of tension in a string.

7.5 Newtonian Gravitation

Newton's law of universal gravitation states that every particle in the Universe attracts every other particle with a force that is directly proportional to the product of their masses and inversely proportional to the square of the distance r between them:

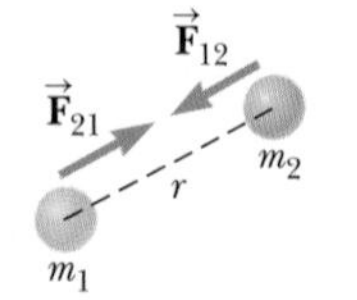

The gravitational force is attractive and acts along the line joining the particles.

$$F = G\frac{m_1 m_2}{r^2} \qquad [7.20]$$

where $G = 6.673 \times 10^{-11}$ $N \cdot m^2/kg^2$ is the **constant of universal gravitation.** A general expression for gravitational potential energy is

$$PE = -G\frac{M_E m}{r} \qquad [7.21]$$

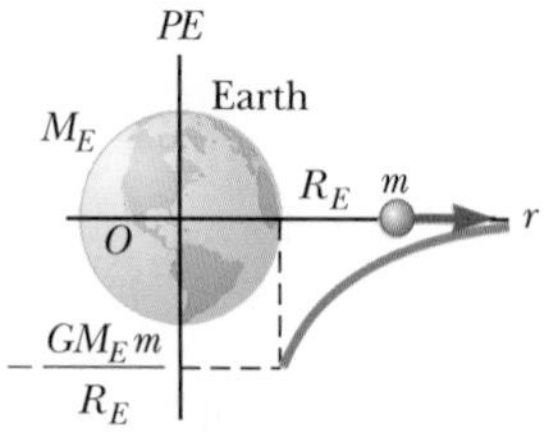

The gravitational potential energy increases towards zero as r increases.

This expression reduces to $PE = mgh$ close to the surface of Earth and holds for other worlds through replacement of the mass M_E. Problems such as finding the escape velocity from Earth can be solved by using Equation 7.21 in the conservation of energy equation.

7.6 Kepler's Laws

Kepler derived the following three laws of planetary motion:

1. All planets move in elliptical orbits with the Sun at one of the focal points.

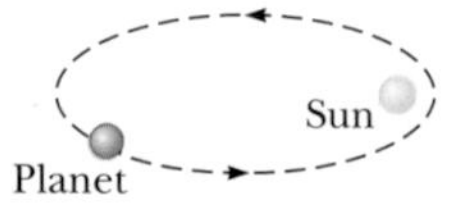

Kepler's first law.

2. A line drawn from the Sun to any planet sweeps out equal areas in equal time intervals.

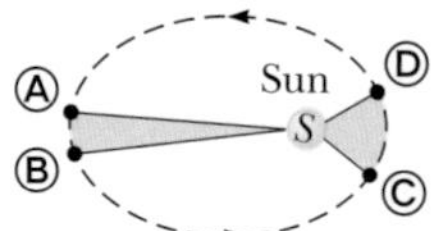

Kepler's second law.

3. The square of the orbital period of a planet is proportional to the cube of the average distance from the planet to the Sun:

$$T^2 = \left(\frac{4\pi^2}{GM_S}\right)r^3 \qquad [7.23]$$

Kepler's third law.

The third law can be applied to any large body and its system of satellites by replacing the Sun's mass with the body's mass. In particular, it can be used to determine the mass of the central body once the average distance to a satellite and its period are known.

WARM-UP EXERCISES

ENHANCED WebAssign The warm-up exercises in this chapter may be assigned online in Enhanced WebAssign.

1. **Math Review** A circular track has a radius of 125 m. (a) Calculate the distance around the track. (b) If a runner jogs 275 m along the track, through what angle has he run?

2. **Math Review** (a) Convert 47.0° to radians, using the appropriate conversion ratio. (b) Convert 2.35 rad to degrees. (c) If a circle has radius 1.70 m, what is the arc length subtended by a 47.0° angle? (See Sections 1.5 and 7.1.)

3. (a) Convert 12.0 rev/min to radians per second. (b) Convert 2.57 rad/s to rev/min. (See Sections 1.5 and 7.1.)

4. A carnival carousel accelerates nonuniformly from rest, moving through an angle of 8.60 rad in 6.00 s. If it's turning at 3.30 rad/s at that time, find (a) its average angular speed, and (b) average angular acceleration during that time interval. (See Section 7.1.)

5. Find the angular speed of a planet that circles its star in 1.00 y, in radians per second. (See Section 7.1.)

6. A grindstone increases in angular speed uniformly from 4.00 rad/s to 12.0 rad/s in 4.00 s. (a) Calculate the grindstone's angular acceleration. (b) Through what angle does it turn during that time? (See Section 7.2.)

7. A bicyclist starting at rest produces a constant angular acceleration of 1.60 rad/s^2 for wheels that are 38.0 cm in radius. (a) What is the bicyclist's linear acceleration? (b) What is the angular speed of the wheels when the bicyclist reaches 11.0 m/s? (c) How many radians have the wheels turned through in that time? (d) How far has the bicyclist traveled? (See Sections 7.2 and 7.3.)

8. A car of mass 1 230 kg travels along a circular road of radius 60.0 m at 18.0 m/s. (a) Calculate the magnitude of the car's centripetal acceleration. (b) What is the magnitude of the force of static friction acting on the car? (See Section 7.4.)

9. A man whirls a 0.20-kg piece of lead attached to the end of a string of length 0.500 m in a circular path and in a vertical plane. If the man maintains a constant speed of

4.00 m/s, what is the tension in the string when the lead is (a) at the top of the circular path? (b) at the bottom of the circular path? (See Section 7.4.)

10. (a) Find the magnitude of the gravity force between a planet with mass 5.98×10^{24} kg and its moon, with mass 7.36×10^{22} kg, if the average distance between them is 3.84×10^{8} m. (b) What is the acceleration of the moon toward the planet? (c) What is the acceleration of the planet toward the moon? (See Section 7.5.)

11. What is the gravitational acceleration close to the surface of a planet with a mass of $2M_E$ and radius of $2R_E$, where M_E and R_E are the mass and radius of Earth, respectively? Answer as a multiple of *g*, the magnitude of the gravitational acceleration near Earth's surface. (See Section 7.5.)

12. (a) Find the speed of a satellite in circular orbit 7.20×10^{6} m from the center of a world with mass 9.40×10^{23} kg. (b) How long does it take to orbit the world one time? (See Section 7.5.)

13. Calculate the escape velocity from the surface of a world with mass 9.10×10^{24} kg and radius 6.80×10^{3} km. (See Section 7.5.)

14. A space capsule of mass 645 kg is at rest 1.20×10^{7} m from the center of the Earth. When it has fallen 3.00×10^{6} m closer to the Earth, (a) what is the change in the system's gravitational potential energy? (b) Find the speed of the satellite at that point. (See Section 7.5.)

15. A comet has a period of 76.3 years and moves in an elliptical orbit in which its perihelion (closest approach to the Sun) is 0.610 AU. Find (a) the semi-major axis of the comet and (b) an estimate of the comet's maximum distance from the Sun, both in astronomical units. (See Section 7.6.)

CONCEPTUAL QUESTIONS

ENHANCED WebAssign The conceptual questions in this chapter may be assigned online in Enhanced WebAssign.

1. In a race like the Indianapolis 500, a driver circles the track counterclockwise and feels his head pulled toward one shoulder. To relieve his neck muscles from having to hold his head erect, the driver fastens a strap to one wall of the car and the other to his helmet. The length of the strap is adjusted to keep his head vertical. (a) Which shoulder does his head tend to lean toward? (b) What force or forces produce the centripetal acceleration when there is no strap? (c) What force or forces do so when there is a strap?

2. If someone told you that astronauts are weightless in Earth orbit because they are beyond the force of gravity, would you accept the statement? Explain.

3. If a car's wheels are replaced with wheels of greater diameter, will the reading of the speedometer change? Explain.

4. At night, you are farther away from the Sun than during the day. What's more, the force exerted by the Sun on you is downward into Earth at night and upward into the sky during the day. If you had a sensitive enough bathroom scale, would you appear to weigh more at night than during the day?

5. A pendulum consists of a small object called a bob hanging from a light cord of fixed length, with the top end of the cord fixed, as represented in Figure CQ7.5. The bob moves without friction, swinging equally high on both sides. It moves from its turning point *A* through point *B* and reaches its maximum speed at point *C*. (a) At what point does the bob have nonzero radial acceleration and zero tangential acceleration? What is the direction of its total acceleration at this point? (b) At what point does the bob have nonzero tangential acceleration and zero radial acceleration? What is the direction of its total acceleration at this point? (c) At what point does the bob have both nonzero tangential and radial acceleration? What is the direction of its total acceleration at this point?

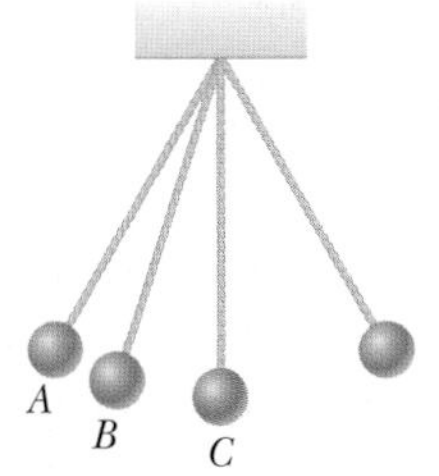

Figure CQ7.5

6. Because of Earth's rotation about its axis, you weigh slightly less at the equator than at the poles. Explain.

7. It has been suggested that rotating cylinders about 10 miles long and 5 miles in diameter be placed in space for colonies. The purpose of their rotation is to simulate gravity for the inhabitants. Explain the concept behind this proposal.

8. Describe the path of a moving object in the event that the object's acceleration is constant in magnitude at all times and (a) perpendicular to its velocity; (b) parallel to its velocity.

9. A pail of water can be whirled in a vertical circular path such that no water is spilled. Why does the water remain in the pail, even when the pail is upside down above your head?

10. Use Kepler's second law to convince yourself that Earth must move faster in its orbit during the northern hemisphere winter, when it is closest to the Sun, than during the summer, when it is farthest from the Sun.

11. Is it possible for a car to move in a circular path in such a way that it has a tangential acceleration but no centripetal acceleration?

12. A child is practicing for a BMX race. His speed remains constant as he goes counterclockwise around a level track with two nearly straight sections and two nearly semicircular sections, as shown in the aerial view of Figure CQ7.12. (a) What are the directions of his velocity at points *A*, *B*, and *C*? For each point choose one: north, south, east, west, or nonexistent? (b) What are the directions of his acceleration at points *A*, *B*, and *C*?

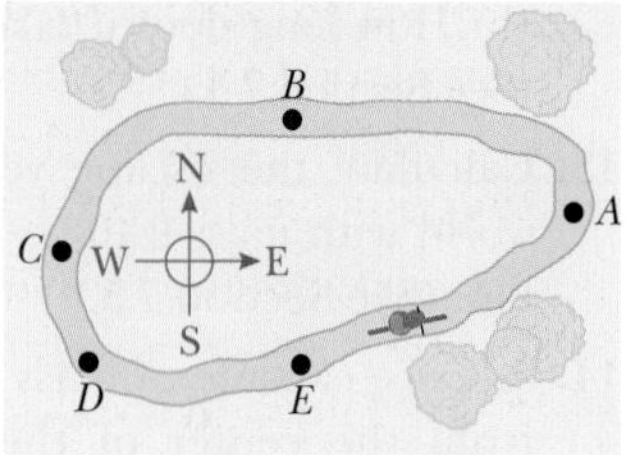

Figure CQ7.12

13. An object executes circular motion with constant speed whenever a net force of constant magnitude acts perpendicular to the velocity. What happens to the speed if the force is not perpendicular to the velocity?

PROBLEMS

ENHANCED WebAssign The problems in this chapter may be assigned online in Enhanced WebAssign.

1. denotes straightforward problem; **2.** denotes intermediate problem;
3. denotes challenging problem
1. denotes full solution available in *Student Solutions Manual/Study Guide*
1. denotes problems most often assigned in Enhanced WebAssign

BIO denotes biomedical problems
GP denotes guided problems
M denotes Master It tutorial available in Enhanced WebAssign
Q|C denotes asking for quantitative and conceptual reasoning
S denotes symbolic reasoning problem
W denotes Watch It video solution available in Enhanced WebAssign

7.1 Angular Speed and Angular Acceleration

1. Q|C (a) Find the angular speed of Earth's rotation about its axis. (b) How does this rotation affect the shape of Earth?

2. A wheel has a radius of 4.1 m. How far (path length) does a point on the circumference travel if the wheel is rotated through angles of (a) 30°, (b) 30 rad, and (c) 30 rev, respectively?

3. The tires on a new compact car have a diameter of 2.0 ft and are warranted for 60 000 miles. (a) Determine the angle (in radians) through which one of these tires will rotate during the warranty period. (b) How many revolutions of the tire are equivalent to your answer in part (a)?

4. Q|C A potter's wheel moves uniformly from rest to an angular speed of 1.00 rev/s in 30.0 s. (a) Find its angular acceleration in radians per second per second. (b) Would doubling the angular acceleration during the given period have doubled final angular speed?

7.2 Rotational Motion Under Constant Angular Acceleration

7.3 Relations Between Angular and Linear Quantities

5. A dentist's drill starts from rest. After 3.20 s of constant angular acceleration, it turns at a rate of 2.51×10^4 rev/min. (a) Find the drill's angular acceleration. (b) Determine the angle (in radians) through which the drill rotates during this period.

6. W A centrifuge in a medical laboratory rotates at an angular speed of 3 600 rev/min. When switched off, it rotates through 50.0 revolutions before coming to rest. Find the constant angular acceleration (in rad/s^2) of the centrifuge.

7. Q|C A machine part rotates at an angular speed of 0.06 rad/s; its speed is then increased to 2.2 rad/s at an angular acceleration of 0.70 rad/s^2. (a) Find the angle through which the part rotates before reaching this final speed. (b) In general, if both the initial and final angular speed are doubled at the same angular acceleration, by what factor is the angular displacement changed? Why? *Hint:* Look at the form of Equation 7.9.

8. Q|C A bicycle is turned upside down while its owner repairs a flat tire. A friend spins the other wheel and observes that drops of water fly off tangentially. She measures the heights reached by drops moving vertically (Fig. P7.8). A drop that breaks loose from the tire on one turn rises vertically 54.0 cm above the tangent point. A drop that breaks loose on the next turn rises 51.0 cm above the tangent point. The radius of the wheel is 0.381 m. (a) Why does the first drop rise higher than the second drop? (b) Neglecting air friction and using only the observed heights and the radius of the wheel, find the wheel's angular acceleration (assuming it to be constant).

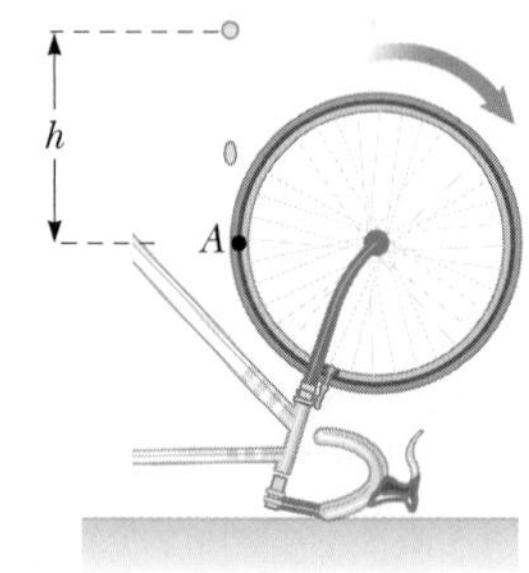

Figure P7.8 Problems 8 and 69.

9. The diameters of the main rotor and tail rotor of a single-engine helicopter are 7.60 m and 1.02 m, respectively. The respective rotational speeds are 450 rev/min and 4 138 rev/min. Calculate the speeds of the tips of both rotors. Compare these speeds with the speed of sound, 343 m/s.

10. The tub of a washer goes into its spin-dry cycle, starting from rest and reaching an angular speed of 5.0 rev/s in

8.0 s. At this point, the person doing the laundry opens the lid, and a safety switch turns off the washer. The tub slows to rest in 12.0 s. Through how many revolutions does the tub turn during the entire 20-s interval? Assume constant angular acceleration while it is starting and stopping.

11. A car initially traveling at 29.0 m/s undergoes a constant negative acceleration of magnitude 1.75 m/s^2 after its brakes are applied. (a) How many revolutions does each tire make before the car comes to a stop, assuming the car does not skid and the tires have radii of 0.330 m? (b) What is the angular speed of the wheels when the car has traveled half the total distance?

12. A 45.0-cm diameter disk rotates with a constant angular acceleration of 2.50 rad/s^2. It starts from rest at $t = 0$, and a line drawn from the center of the disk to a point P on the rim of the disk makes an angle of 57.3° with the positive x-axis at this time. At $t = 2.30$ s, find (a) the angular speed of the wheel, (b) the linear velocity and tangential acceleration of P, and (c) the position of P (in degrees, with respect to the positive x-axis).

13. M A rotating wheel requires 3.00 s to rotate 37.0 revolutions. Its angular velocity at the end of the 3.00-s interval is 98.0 rad/s. What is the constant angular acceleration (in rad/s^2) of the wheel?

14. An electric motor rotating a workshop grinding wheel at a rate of 1.00×10^2 rev/min is switched off. Assume the wheel has a constant negative angular acceleration of magnitude 2.00 rad/s^2. (a) How long does it take for the grinding wheel to stop? (b) Through how many radians has the wheel turned during the interval found in part (a)?

7.4 Centripetal Acceleration

15. A car initially traveling eastward turns north by traveling in a circular path at uniform speed as shown in Figure P7.15. The length of the arc ABC is 235 m, and the car completes the turn in 36.0 s. (a) Determine the car's speed. (b) What is the magnitude and direction of the acceleration when the car is at point B?

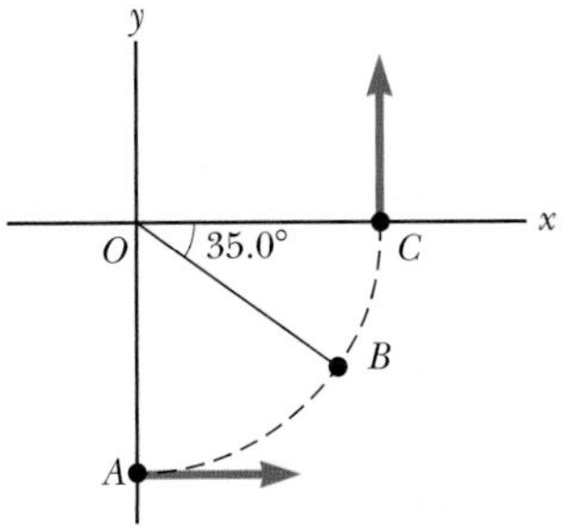

Figure P7.15

16. It has been suggested that rotating cylinders about 10 mi long and 5.0 mi in diameter be placed in space and used as colonies. What angular speed must such a cylinder have so that the centripetal acceleration at its surface equals the free-fall acceleration on Earth?

17. (a) What is the tangential acceleration of a bug on the rim of a 10.0-in.-diameter disk if the disk accelerates uniformly from rest to an angular speed of 78.0 rev/min in 3.00 s? (b) When the disk is at its final speed, what is the tangential velocity of the bug? One second after the bug starts from rest, what are its (c) tangential acceleration, (d) centripetal acceleration, and (e) total acceleration?

18. An adventurous archeologist ($m = 85.0$ kg) tries to cross a river by swinging from a vine. The vine is 10.0 m long, and his speed at the bottom of the swing is 8.00 m/s. The archeologist doesn't know that the vine has a breaking strength of 1 000 N. Does he make it across the river without falling in?

19. Q|C One end of a cord is fixed and a small 0.500-kg object is attached to the other end, where it swings in a section of a vertical circle of radius 2.00 m, as shown in Figure P7.19. When $\theta = 20.0°$, the speed of the object is 8.00 m/s. At this instant, find (a) the tension in the string, (b) the tangential and radial components of acceleration, and (c) the total acceleration. (d) Is your answer changed if the object is swinging down toward its lowest point instead of swinging up? (e) Explain your answer to part (d).

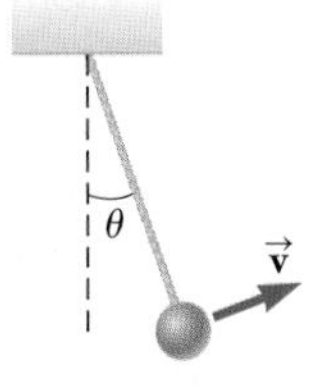

Figure P7.19

20. Q|C A coin rests 15.0 cm from the center of a turntable. The coefficient of static friction between the coin and turntable surface is 0.350. The turntable starts from rest at $t = 0$ and rotates with a constant angular acceleration of 0.730 rad/s^2. (a) Once the turntable starts to rotate, what force causes the centripetal acceleration when the coin is stationary relative to the turntable? Under what condition does the coin begin to move relative to the turntable? (b) After what period of time will the coin start to slip on the turntable?

21. A 55.0-kg ice skater is moving at 4.00 m/s when she grabs the loose end of a rope, the opposite end of which is tied to a pole. She then moves in a circle of radius 0.800 m around the pole. (a) Determine the force exerted by the horizontal rope on her arms. (b) Compare this force with her weight.

22. M A 40.0-kg child swings in a swing supported by two chains, each 3.00 m long. The tension in each chain at the lowest point is 350 N. Find (a) the child's speed at the lowest point and (b) the force exerted by the seat on the child at the lowest point. (Ignore the mass of the seat.)

23. A certain light truck can go around a flat curve having a radius of 150 m with a maximum speed of 32.0 m/s. With what maximum speed can it go around a curve having a radius of 75.0 m?

24. BIO A sample of blood is placed in a centrifuge of radius 15.0 cm. The mass of a red blood cell is 3.0×10^{-16} kg, and the magnitude of the force acting on it as it settles out of the plasma is 4.0×10^{-11} N. At how many revolutions per second should the centrifuge be operated?

25. A 50.0-kg child stands at the rim of a merry-go-round of radius 2.00 m, rotating with an angular speed of 3.00 rad/s. (a) What is the child's centripetal acceleration? (b) What is the minimum force between her feet and the floor of the carousel that is required to keep her in the circular path? (c) What minimum coefficient of static friction is required? Is the answer you found reasonable? In other words, is she likely to stay on the merry-go-round?

26. **GP** A space habitat for a long space voyage consists of two cabins each connected by a cable to a central hub as shown in Figure P7.26. The cabins are set spinning around the hub axis, which is connected to the rest of the spacecraft to generate artificial gravity. (a) What forces are acting on an astronaut in one of the cabins? (b) Write Newton's second law for an astronaut lying on the "floor" of one of the habitats, relating the astronaut's mass m, his velocity v, his radial distance from the hub r, and the normal force n. (c) What would n have to equal if the 60.0-kg astronaut is to experience half his normal Earth weight? (d) Calculate the necessary tangential speed of the habitat from Newton's second law. (e) Calculate the angular speed from the tangential speed. (f) Calculate the period of rotation from the angular speed. (g) If the astronaut stands up, will his head be moving faster, slower, or at the same speed as his feet? Why? Calculate the tangential speed at the top of his head if he is 1.80 m tall.

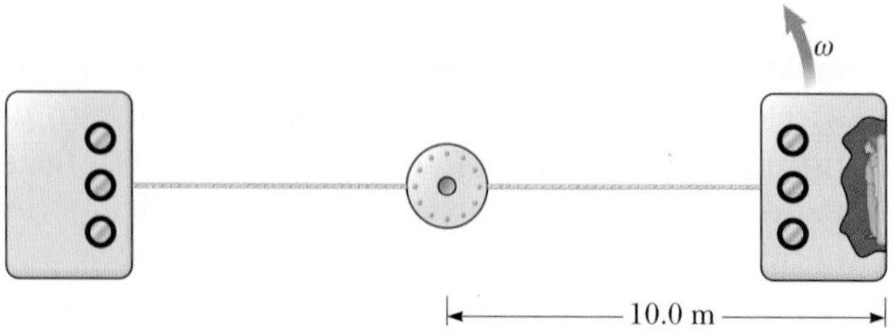

Figure P7.26

27. An air puck of mass $m_1 = 0.25$ kg is tied to a string and allowed to revolve in a circle of radius $R = 1.0$ m on a frictionless horizontal table. The other end of the string passes through a hole in the center of the table, and a mass of $m_2 = 1.0$ kg is tied to it (Fig. P7.27). The suspended mass remains in equilibrium while the puck on the tabletop revolves. (a) What is the tension in the string? (b) What is the horizontal force acting on the puck? (c) What is the speed of the puck?

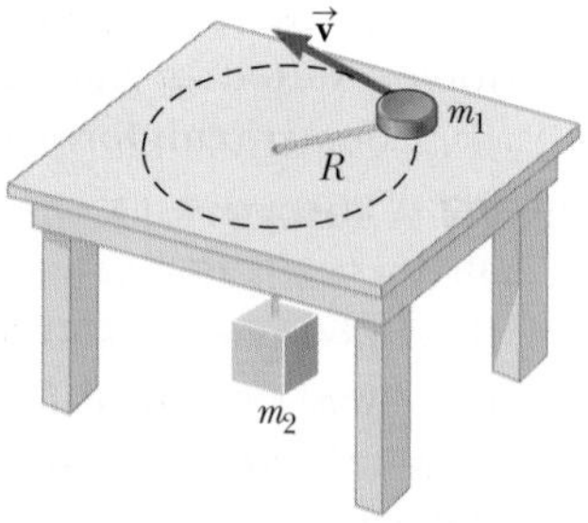

Figure P7.27 Problems 27 and 28.

28. **S** An air puck of mass m_1 is tied to a string and allowed to revolve in a circle of radius R on a horizontal, frictionless table. The other end of the string passes through a small hole in the center of the table, and an object of mass m_2 is tied to it (Fig. P7.27). The suspended object remains in equilibrium while the puck on the tabletop revolves. (a) Find a symbolic expression for the tension in the string in terms of m_2 and g. (b) Write Newton's second law for the air puck, using the variables m_1, v, R, and T. (c) Eliminate the tension T from the expressions found in parts (a) and (b) and find an expression for the speed of the puck in terms of m_1, m_2, g, and R. (d) Check your answers by substituting the values of Problem 27 and comparing the results with the answers for that problem.

29. **Q|C** A woman places her briefcase on the backseat of her car. As she drives to work, the car negotiates an unbanked curve in the road that can be regarded as an arc of a circle of radius 62.0 m. While on the curve, the speed of the car is 15.0 m/s at the instant the briefcase starts to slide across the backseat toward the side of the car. (a) What force causes the centripetal acceleration of the briefcase when it is stationary relative to the car? Under what condition does the briefcase begin to move relative to the car? (b) What is the coefficient of static friction between the briefcase and seat surface?

30. **Q|C** A pail of water is rotated in a vertical circle of radius 1.00 m. (a) What two external forces act on the water in the pail? (b) Which of the two forces is most important in causing the water to move in a circle? (c) What is the pail's minimum speed at the top of the circle if no water is to spill out? (d) If the pail with the speed found in part (c) were to suddenly disappear at the top of the circle, describe the subsequent motion of the water. Would it differ from the motion of a projectile?

31. **M** A 40.0-kg child takes a ride on a Ferris wheel that rotates four times each minute and has a diameter of 18.0 m. (a) What is the centripetal acceleration of the child? (b) What force (magnitude and direction) does the seat exert on the child at the lowest point of the ride? (c) What force does the seat exert on the child at the highest point of the ride? (d) What force does the seat exert on the child when the child is halfway between the top and bottom?

32. **W** A roller-coaster vehicle has a mass of 500 kg when fully loaded with passengers (Fig. P7.32). (a) If the

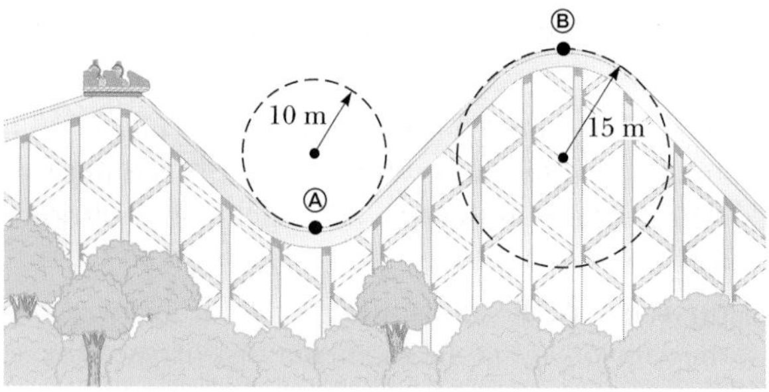

Figure P7.32

vehicle has a speed of 20.0 m/s at point Ⓐ, what is the force of the track on the vehicle at this point? (b) What is the maximum speed the vehicle can have at point Ⓑ for gravity to hold it on the track?

7.5 Newtonian Gravitation

33. The average distance separating Earth and the Moon is 384 000 km. Use the data in Table 7.3 to find the net gravitational force exerted by Earth and the Moon on a 3.00×10^4-kg spaceship located halfway between them.

34. A satellite has a mass of 100 kg and is located at 2.00×10^6 m above the surface of Earth. (a) What is the potential energy associated with the satellite at this location? (b) What is the magnitude of the gravitational force on the satellite?

35. A coordinate system (in meters) is constructed on the surface of a pool table, and three objects are placed on the table as follows: a 2.0-kg object at the origin of the coordinate system, a 3.0-kg object at (0, 2.0), and a 4.0-kg object at (4.0, 0). Find the resultant gravitational force exerted by the other two objects on the object at the origin.

36. After the Sun exhausts its nuclear fuel, its ultimate fate may be to collapse to a *white dwarf* state. In this state, it would have approximately the same mass as it has now, but its radius would be equal to the radius of Earth. Calculate (a) the average density of the white dwarf, (b) the surface free-fall acceleration, and (c) the gravitational potential energy associated with a 1.00-kg object at the surface of the white dwarf.

37. W Objects with masses of 200 kg and 500 kg are separated by 0.400 m. (a) Find the net gravitational force exerted by these objects on a 50.0-kg object placed midway between them. (b) At what position (other than infinitely remote ones) can the 50.0-kg object be placed so as to experience a net force of zero?

38. Use the data of Table 7.3 to find the point between Earth and the Sun at which an object can be placed so that the net gravitational force exerted by Earth and the Sun on that object is zero.

39. A projectile is fired straight upward from the Earth's surface at the South Pole with an initial speed equal to one third the escape speed. (a) Ignoring air resistance, determine how far from the center of the Earth the projectile travels before stopping momentarily. (b) What is the altitude of the projectile at this instant?

40. Two objects attract each other with a gravitational force of magnitude 1.00×10^{-8} N when separated by 20.0 cm. If the total mass of the objects is 5.00 kg, what is the mass of each?

7.6 Kepler's Laws

41. M A satellite is in a circular orbit around the Earth at an altitude of 2.80×10^6 m. Find (a) the period of the orbit, (b) the speed of the satellite, and (c) the acceleration of the satellite. *Hint:* Modify Equation 7.23 so it is suitable for objects orbiting the Earth rather than the Sun.

42. An artificial satellite circling the Earth completes each orbit in 110 minutes. (a) Find the altitude of the satellite. (b) What is the value of g at the location of this satellite?

43. A satellite of Mars, called Phoebus, has an orbital radius of 9.4×10^6 m and a period of 2.8×10^4 s. Assuming the orbit is circular, determine the mass of Mars.

44. A 600-kg satellite is in a circular orbit about Earth at a height above Earth equal to Earth's mean radius. Find (a) the satellite's orbital speed, (b) the period of its revolution, and (c) the gravitational force acting on it.

45. Two satellites are in circular orbits around the Earth. Satellite A is at an *altitude* equal to the Earth's radius, while satellite B is at an altitude equal to twice the Earth's radius. What is the ratio of their periods, T_B/T_A?

Additional Problems

46. W A synchronous satellite, which always remains above the same point on a planet's equator, is put in circular orbit around Jupiter to study that planet's famous red spot. Jupiter rotates once every 9.84 h. Use the data of Table 7.3 to find the altitude of the satellite.

47. (a) One of the moons of Jupiter, named Io, has an orbital radius of 4.22×10^8 m and a period of 1.77 days. Assuming the orbit is circular, calculate the mass of Jupiter. (b) The largest moon of Jupiter, named Ganymede, has an orbital radius of 1.07×10^9 m and a period of 7.16 days. Calculate the mass of Jupiter from this data. (c) Are your results to parts (a) and (b) consistent? Explain.

48. Neutron stars are extremely dense objects that are formed from the remnants of supernova explosions. Many rotate very rapidly. Suppose the mass of a certain spherical neutron star is twice the mass of the Sun and its radius is 10.0 km. Determine the greatest possible angular speed the neutron star can have so that the matter at its surface on the equator is just held in orbit by the gravitational force.

49. BIO One method of pitching a softball is called the "windmill" delivery method, in which the pitcher's arm rotates through approximately 360° in a vertical plane before the 198-gram ball is released at the lowest point of the circular motion. An experienced pitcher can throw a ball with a speed of 98.0 mi/h. Assume the angular acceleration is uniform throughout the pitching motion and take the distance between the softball and the shoulder joint to be 74.2 cm. (a) Determine the angular speed of the arm in rev/s at the instant of release. (b) Find the value of the angular acceleration in rev/s^2 and the radial and tangential acceleration of the ball just before it is released. (c) Determine the force exerted on the ball by the pitcher's hand (both radial and tangential components) just before it is released.

50. A digital audio compact disc carries data along a continuous spiral track from the inner circumference of the disc to the outside edge. Each bit occupies 0.6 μm of the track. A CD player turns the disc to carry the track counterclockwise above a lens at a constant speed of 1.30 m/s. Find the required angular speed (a) at the beginning of the recording, where the spiral has a radius of 2.30 cm, and (b) at the end of the recording, where the spiral has a radius of 5.80 cm. (c) A full-length recording lasts for 74 min, 33 s. Find the average angular acceleration of the disc. (d) Assuming the acceleration is constant, find the total angular displacement of the disc as it plays. (e) Find the total length of the track.

51. An athlete swings a 5.00-kg ball horizontally on the end of a rope. The ball moves in a circle of radius 0.800 m at an angular speed of 0.500 rev/s. What are (a) the tangential speed of the ball and (b) its centripetal acceleration? (c) If the maximum tension the rope can withstand before breaking is 100 N, what is the maximum tangential speed the ball can have?

52. A car rounds a banked curve where the radius of curvature of the road is R, the banking angle is θ, and the coefficient of static friction is μ. (a) Determine the range of speeds the car can have without slipping up or down the road. (b) What is the range of speeds possible if $R = 100$ m, $\theta = 10°$, and $\mu = 0.10$ (slippery conditions)?

53. The Solar Maximum Mission Satellite was placed in a circular orbit about 150 mi above Earth. Determine (a) the orbital speed of the satellite and (b) the time required for one complete revolution.

54. A 0.400-kg pendulum bob passes through the lowest part of its path at a speed of 3.00 m/s. (a) What is the tension in the pendulum cable at this point if the pendulum is 80.0 cm long? (b) When the pendulum reaches its highest point, what angle does the cable make with the vertical? (c) What is the tension in the pendulum cable when the pendulum reaches its highest point?

55. Q|C A car moves at speed v across a bridge made in the shape of a circular arc of radius r. (a) Find an expression for the normal force acting on the car when it is at the top of the arc. (b) At what minimum speed will the normal force become zero (causing the occupants of the car to seem weightless) if $r = 30.0$ m?

56. S Show that the escape speed from the surface of a planet of uniform density is directly proportional to the radius of the planet.

57. Q|C Because of Earth's rotation about its axis, a point on the equator has a centripetal acceleration of 0.034 0 m/s^2, whereas a point at the poles has no centripetal acceleration. (a) Show that, at the equator, the gravitational force on an object (the object's true weight) must exceed the object's apparent weight. (b) What are the apparent weights of a 75.0-kg person at the equator and at the poles? (Assume Earth is a uniform sphere and take $g = 9.800$ m/s^2.)

58. A small block of mass $m = 0.50$ kg is fired with an initial speed of $v_0 = 4.0$ m/s along a horizontal section of frictionless track, as shown in the top portion of Figure P7.58. The block then moves along the frictionless, semicircular, *vertical* tracks of radius R = 1.5 m. (a) Determine the force exerted by the track on the block at points Ⓐ and Ⓑ. (b) The bottom of the track consists of a section (L = 0.40 m) with friction. Determine the coefficient of kinetic friction between the block and that portion of the bottom track if the block just makes it to point Ⓒ on the first trip. *Hint:* If the block just makes it to point Ⓒ, the force of contact exerted by the track on the block at that point is zero.

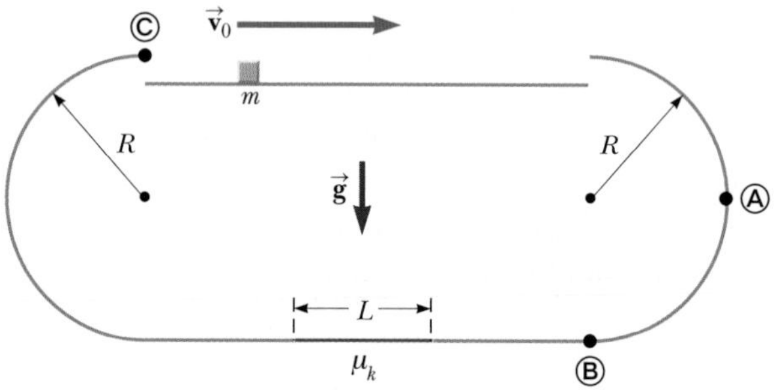

Figure P7.58

59. In Robert Heinlein's *The Moon Is a Harsh Mistress,* the colonial inhabitants of the Moon threaten to launch rocks down onto Earth if they are not given independence (or at least representation). Assuming a gun could launch a rock of mass m at twice the lunar escape speed, calculate the speed of the rock as it enters Earth's atmosphere.

60. Q|C A roller coaster travels in a circular path. (a) Identify the forces on a passenger at the top of the circular loop that cause centripetal acceleration. Show the direction of all forces in a sketch. (b) Identify the forces on the passenger at the bottom of the loop that produce centripetal acceleration. Show these in a sketch. (c) Based on your answers to parts (a) and (b), at what point, top or bottom, should the seat exert the greatest force on the passenger? (d) Assume the speed of the roller coaster is 4.00 m/s at the top of the loop of radius 8.00 m. Find the force exerted by the seat on a 70.0-kg passenger at the top of the loop. Then, assume the speed remains the same at the bottom of the loop and find the force exerted by the seat on the passenger at this point. Are your answers consistent with your choice of answers for parts (a) and (b)?

61. In a home laundry dryer, a cylindrical tub containing wet clothes is rotated steadily about a horizontal axis,

as shown in Figure P7.61. So that the clothes will dry uniformly, they are made to tumble. The rate of rotation of the smooth-walled tub is chosen so that a small piece of cloth will lose contact with the tub when the cloth is at an angle of $\theta = 68.0°$ above the horizontal. If the radius of the tub is $r = 0.330$ m, what rate of revolution is needed in revolutions per second?

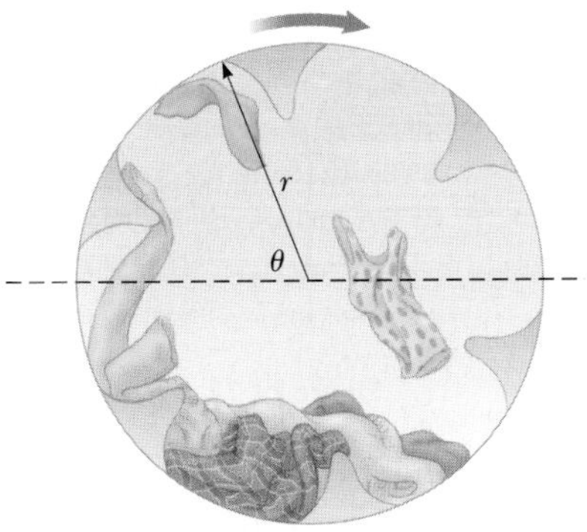

Figure P7.61

62. M A model airplane of mass 0.750 kg flies with a speed of 35.0 m/s in a horizontal circle at the end of a 60.0-m control wire as shown in Figure P7.62a. The forces exerted on the airplane are shown in Figure P7.62b; the tension in the control wire, $\theta = 20.0°$ inward from the vertical. Compute the tension in the wire, assuming the wire makes a constant angle of $\theta = 20.0°$ with the horizontal.

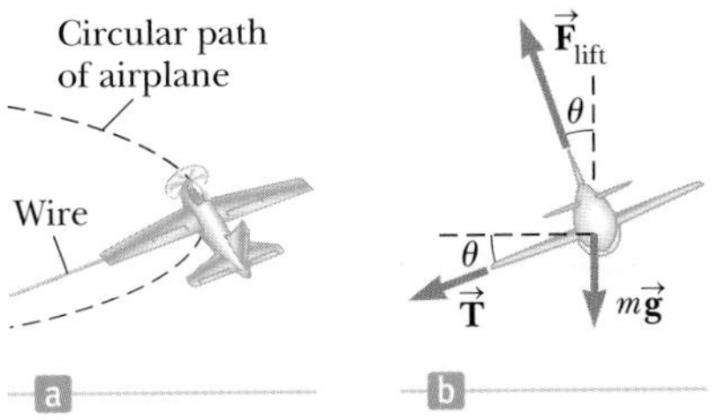

Figure P7.62

63. S A skier starts at rest at the top of a large hemispherical hill (Fig. P7.63). Neglecting friction, show that the skier will leave the hill and become airborne at a distance $h = R/3$ below the top of the hill. *Hint:* At this point, the normal force goes to zero.

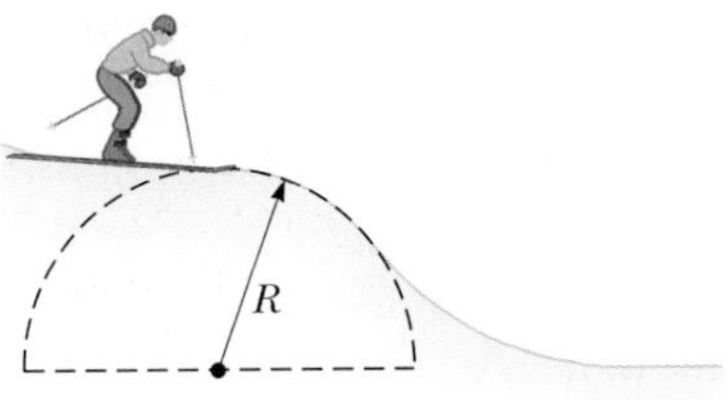

Figure P7.63

64. Casting of molten metal is important in many industrial processes. *Centrifugal casting* is used for manufacturing pipes, bearings, and many other structures. A cylindrical enclosure is rotated rapidly and steadily about a horizontal axis, as in Figure P7.64. Molten metal is poured into the rotating cylinder and then cooled, forming the finished product. Turning the cylinder at a high rotation rate forces the solidifying metal strongly to the outside. Any bubbles are displaced toward the axis so that unwanted voids will not be present in the casting.

Suppose a copper sleeve of inner radius 2.10 cm and outer radius 2.20 cm is to be cast. To eliminate bubbles and give high structural integrity, the centripetal acceleration of each bit of metal should be 100*g*. What rate of rotation is required? State the answer in revolutions per minute.

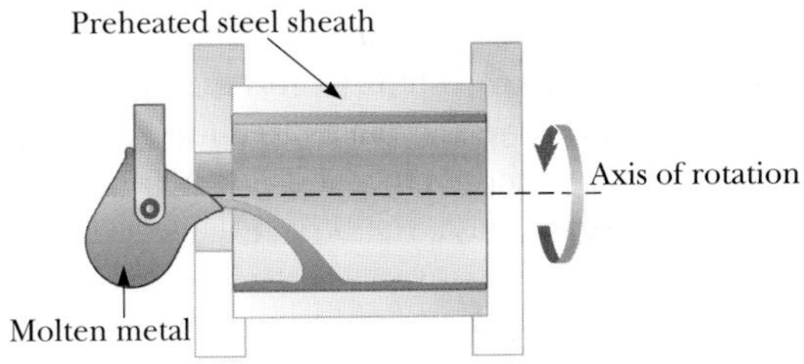

Figure P7.64

65. Suppose a 1 800-kg car passes over a bump in a roadway that follows the arc of a circle of radius 20.4 m, as in Figure P7.65. (a) What force does the road exert on the car as the car passes the highest point of the bump if the car travels at 8.94 m/s? (b) What is the maximum speed the car can have without losing contact with the road as it passes this highest point?

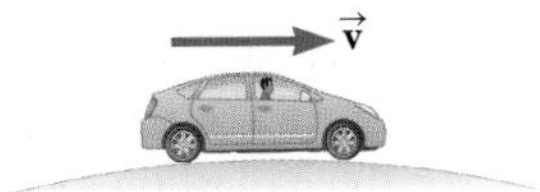

Figure P7.65

66. A stuntman whose mass is 70 kg swings from the end of a 4.0-m-long rope along the arc of a vertical circle. Assuming he starts from rest when the rope is horizontal, find the tensions in the rope that are required to make him follow his circular path (a) at the beginning of his motion, (b) at a height of 1.5 m above the bottom of the circular arc, and (c) at the bottom of the arc.

67. Q|C A minimum-energy orbit to an outer planet consists of putting a spacecraft on an elliptical trajectory with the departure planet corresponding to the perihelion of the ellipse, or closest point to the Sun, and the arrival planet corresponding to the aphelion of the ellipse, or farthest point from the Sun. (a) Use Kepler's third law to calculate how long it would take to go from Earth to Mars on such an orbit. (Answer in years.) (b) Can such an orbit be undertaken at any time? Explain.

68. Q|C The pilot of an airplane executes a constant-speed loop-the-loop maneuver in a vertical circle as in Figure 7.15b. The speed of the airplane is 2.00×10^2 m/s, and the radius of the circle is 3.20×10^3 m. (a) What is the pilot's apparent weight at the lowest

point of the circle if his true weight is 712 N? (b) What is his apparent weight at the highest point of the circle? (c) Describe how the pilot could experience weightlessness if both the radius and the speed can be varied. *Note:* His apparent weight is equal to the magnitude of the force exerted by the seat on his body. Under what conditions does this occur? (d) What speed would have resulted in the pilot experiencing weightlessness at the top of the loop?

69. **S** A piece of mud is initially at point A on the rim of a bicycle wheel of radius R rotating clockwise about a horizontal axis at a constant angular speed ω (Fig. P7.8). The mud dislodges from point A when the wheel diameter through A is horizontal. The mud then rises vertically and returns to point A. (a) Find a symbolic expression in terms of R, ω, and g for the total time the mud is in the air and returns to point A. (b) If the wheel makes one complete revolution in the time it takes the mud to return to point A, find an expression for the angular speed of the bicycle wheel ω in terms of π, g, and R.

70. **Q|C** A 0.275-kg object is swung in a *vertical* circular path on a string 0.850 m long as in Figure P7.70. (a) What are the forces acting on the ball at any point along this path? (b) Draw free-body diagrams for the ball when it is at the bottom of the circle and when it is at the top. (c) If its speed is 5.20 m/s at the top of the circle, what is the tension in the string there? (d) If the string breaks when its tension exceeds 22.5 N, what is the maximum speed the object can have at the bottom before the string breaks?

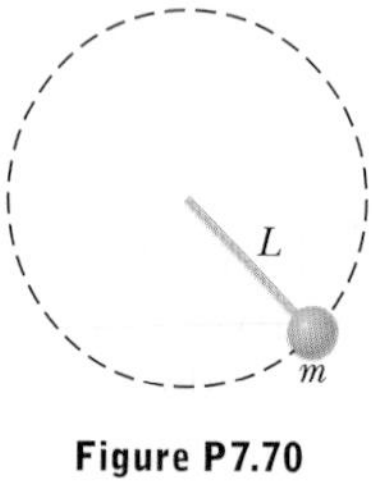

Figure P7.70

71. A 4.00-kg object is attached to a vertical rod by two strings as shown in Figure P7.71. The object rotates in a horizontal circle at constant speed 6.00 m/s. Find the tension in (a) the upper string and (b) the lower string.

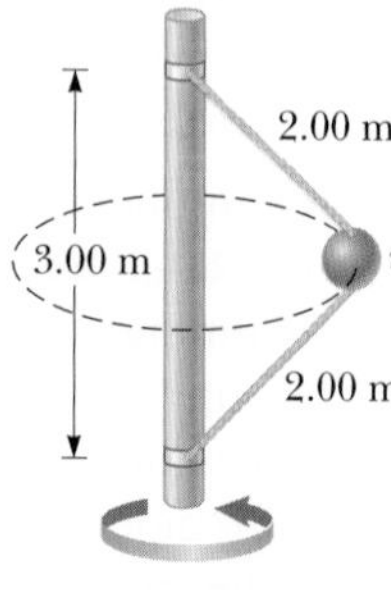

Figure P7.71

72. **BIO** The maximum lift force on a bat is proportional to the square of its flying speed v. For the hoary bat (*Lasiurus cinereus*), the magnitude of the lift force is given by

$$F_L \leq (0.018\ \text{N} \cdot \text{s}^2/\text{m}^2)v^2$$

The bat can fly in a horizontal circle by "banking" its wings at an angle θ, as shown in Figure P7.72. In this situation, the magnitude of the vertical component of the lift force must equal the bat's weight. The horizontal component of the force provides the centripetal acceleration. (a) What is the minimum speed that the bat can have if its mass is 0.031 kg? (b) If the maximum speed of the bat is 10 m/s, what is the maximum banking angle that allows the bat to stay in a horizontal plane? (c) What is the radius of the circle of its flight when the bat flies at its maximum speed? (d) Can the bat turn with a smaller radius by flying more slowly?

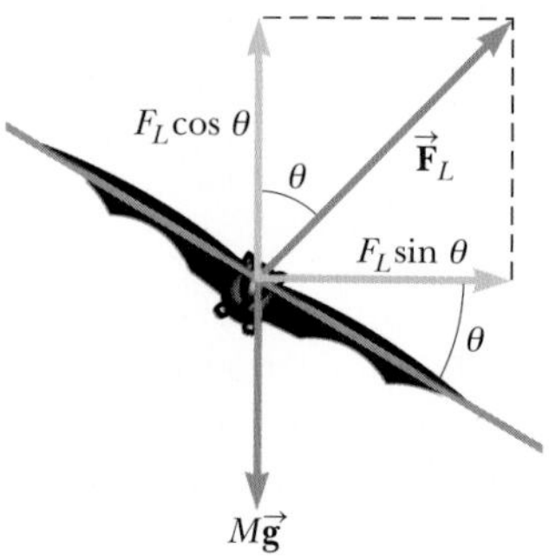

Figure P7.72

73. (a) A luggage carousel at an airport has the form of a section of a large cone, steadily rotating about its vertical axis. Its metallic surface slopes downward toward the outside, making an angle of 20.0° with the horizontal. A 30.0-kg piece of luggage is placed on the carousel, 7.46 m from the axis of rotation. The travel bag goes around once in 38.0 s. Calculate the force of static friction between the bag and the carousel. (b) The drive motor is shifted to turn the carousel at a higher constant rate of rotation, and the piece of luggage is bumped to a position 7.94 m from the axis of rotation. The bag is on the verge of slipping as it goes around once every 34.0 s. Calculate the coefficient of static friction between the bag and the carousel.

74. A 0.50-kg ball that is tied to the end of a 1.5-m light cord is revolved in a horizontal plane, with the cord making a 30° angle with the vertical. (See Fig. P7.74.) (a) Determine the ball's speed. (b) If, instead, the ball is revolved so that its speed is 4.0 m/s, what angle does the cord make with the vertical? (c) If the cord can withstand a maximum tension of 9.8 N, what is the highest speed at which the ball can move?

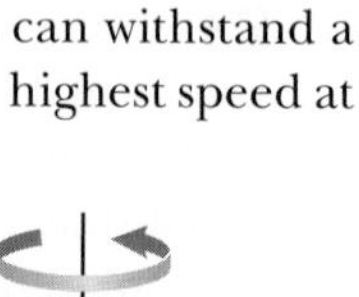

Figure P7.74

75. In a popular amusement park ride, a rotating cylinder of radius 3.00 m is set in rotation at an angular speed of 5.00 rad/s, as in Figure P7.75. The floor then drops away, leaving the riders suspended against the wall in a vertical position. What minimum coefficient of friction between a rider's clothing and the wall is needed to keep the rider from slipping? *Hint:* Recall that the magnitude of the maximum force

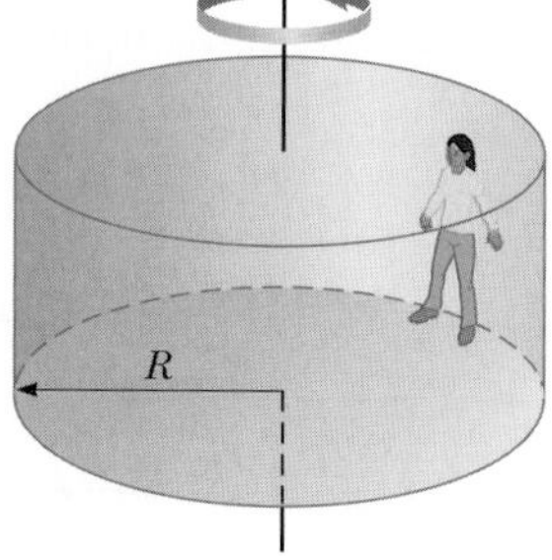

Figure P7.75

of static friction is equal to $\mu_s n$, where n is the normal force—in this case, the force causing the centripetal acceleration.

76. A massless spring of constant $k = 78.4$ N/m is fixed on the left side of a level track. A block of mass $m = 0.50$ kg is pressed against the spring and compresses it a distance d, as in Figure P7.76. The block (initially at rest) is then released and travels toward a circular loop-the-loop of radius $R = 1.5$ m. The entire track and the loop-the-loop are frictionless, except for the section of track between points A and B. Given that the coefficient of kinetic friction between the block and the track along AB is $\mu_k = 0.30$, and that the length of AB is 2.5 m, determine the minimum compression d of the spring that enables the block to just make it through the loop-the-loop at point C. *Hint:* The force exerted by the track on the block will be zero if the block barely makes it through the loop-the-loop.

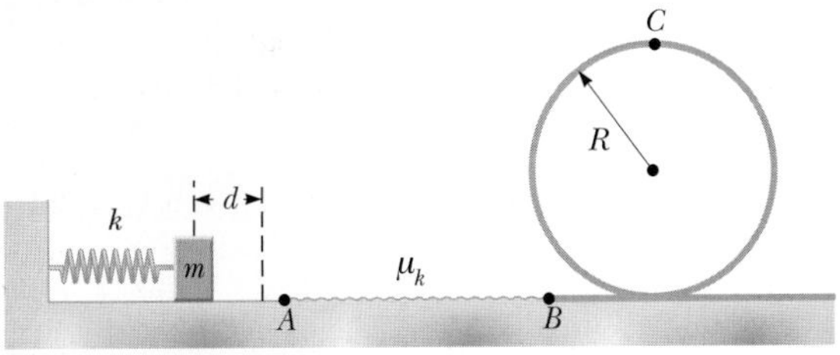

Figure P7.76

Wind exerts forces on the propellers of this wind turbine, producing a torque that causes the turbine to rotate. This process converts the kinetic energy of wind to rotational kinetic energy, which is transformed by electromagnetic induction to electrical energy.

Marnie Burkhart/Fancy/Jupiter Images

8 Rotational Equilibrium and Rotational Dynamics

In the study of linear motion, objects were treated as point particles without structure. It didn't matter *where* a force was applied, only *whether* it was applied or not.

The reality is that the point of application of a force *does* matter. In football, for example, if the ball carrier is tackled near his midriff, he might carry the tackler several yards before falling. If tackled well below the waistline, however, his center of mass rotates toward the ground, and he can be brought down immediately. Tennis provides another good example. If a tennis ball is struck with a strong horizontal force acting through its center of mass, it may travel a long distance before hitting the ground, far out of bounds. Instead, the same force applied in an upward, glancing stroke will impart topspin to the ball, which can cause it to land in the opponent's court.

The concepts of rotational equilibrium and rotational dynamics are also important in other disciplines. For example, students of architecture benefit from understanding the forces that act on buildings, and biology students should understand the forces at work in muscles and on bones and joints. These forces create torques, which tell us how the forces affect an object's equilibrium and rate of rotation.

We will find that an object remains in a state of uniform rotational motion unless acted on by a net torque. That principle is the equivalent of Newton's first law. Further, the angular acceleration of an object is proportional to the net torque acting on it, which is the analog of Newton's second law. A net torque acting on an object causes a change in its rotational energy.

Finally, torques applied to an object through a given time interval can change the object's angular momentum. In the absence of external torques, angular momentum is conserved, a property that explains some of the mysterious and formidable properties of pulsars, remnants of supernova explosions that rotate at equatorial speeds approaching that of light.

We wish to locate the point of application of the single force of magnitude $w = F_g = Mg$ (the total weight of the object), where the effect on the rotation of the object is the same as that of the individual particles. That point is called the object's **center of gravity.** Equating the torque exerted by w at the center of gravity to the sum of the torques acting on the individual particles gives

$$(m_1g + m_2g + m_3g + \cdots)x_{cg} = m_1gx_1 + m_2gx_2 + m_3gx_3 + \cdots$$

We assume that g is the same everywhere in the object (which is true for all objects we will encounter). Then the g factors in the preceding equation cancel, resulting in

$$x_{cg} = \frac{m_1x_1 + m_2x_2 + m_3x_3 + \cdots}{m_1 + m_2 + m_3 + \cdots} = \frac{\sum m_ix_i}{\sum m_i} \qquad [8.3a]$$

where x_{cg} is the x-coordinate of the center of gravity. Similarly, the y-coordinate and z-coordinate of the center of gravity of the system can be found from

$$y_{cg} = \frac{\sum m_iy_i}{\sum m_i} \qquad [8.3b]$$

and

$$z_{cg} = \frac{\sum m_iz_i}{\sum m_i} \qquad [8.3c]$$

These three equations are identical to the equations for a similar concept called **center of mass.** The center of mass and center of gravity of an object are exactly the same when g doesn't vary significantly over the object.

It's often possible to guess the location of the center of gravity. **The center of gravity of a homogeneous, symmetric body must lie on the axis of symmetry.** For example, the center of gravity of a homogeneous rod lies midway between the ends of the rod, and the center of gravity of a homogeneous sphere or a homogeneous cube lies at the geometric center of the object. The center of gravity of an irregularly shaped object, such as a wrench, can be determined experimentally by suspending the wrench from two different arbitrary points (Fig. 8.10). The wrench is first hung from point A, and a vertical line AB (which can be established with a plumb bob) is drawn when the wrench is in equilibrium. The wrench is then hung from point C, and a second vertical line CD is drawn. The center of gravity coincides with the intersection of these two lines. In fact, if the wrench is hung freely from any point, the center of gravity always lies straight below the point of support, so the vertical line through that point must pass through the center of gravity.

Several examples in Section 8.4 involve homogeneous, symmetric objects where the centers of gravity coincide with their geometric centers. A rigid object in a uniform gravitational field can be balanced by a single force equal in magnitude to the weight of the object, as long as the force is directed upward through the object's center of gravity.

Tip 8.1 Specify Your Axis

Choose the axis of rotation and use that axis exclusively throughout a given problem. The axis need not correspond to a physical axle or pivot point. Any convenient point will do.

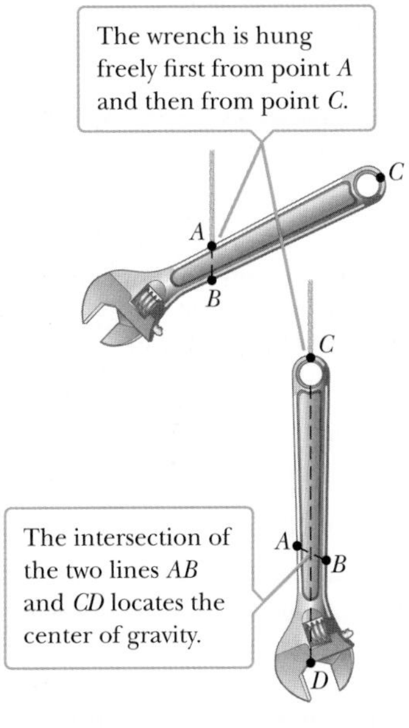

Figure 8.10 An experimental technique for determining the center of gravity of a wrench.

EXAMPLE 8.4 Where Is the Center of Gravity?

GOAL Find the center of gravity of a system of objects.

PROBLEM **(a)** Three objects are located in a coordinate system as shown in Figure 8.11a. Find the center of gravity. **(b)** How does the answer change if the object on the left is displaced upward by 1.00 m and the object on the right is displaced downward by 0.500 m (Figure 8.11b)? Treat the objects as point particles.

STRATEGY The y-coordinate and z-coordinate of the center of gravity in part (a) are both zero because all the objects are on the x-axis. We can find the x-coordinate of the center of gravity using Equation 8.3a. Part (b) requires Equation 8.3b.

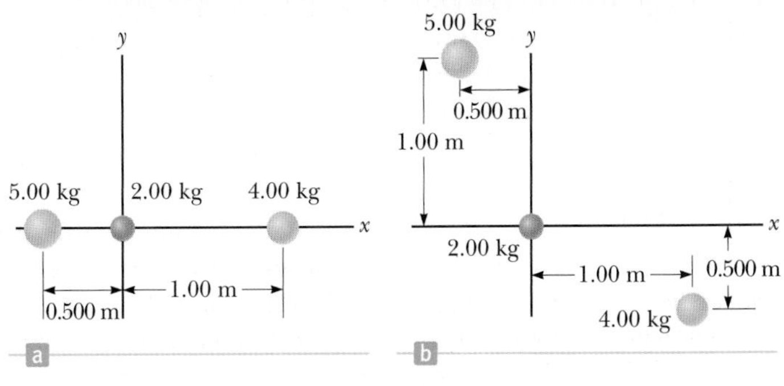

Figure 8.11 (Example 8.4) Locating the center of gravity of a system of three particles.

(Continued)

SOLUTION

(a) Find the center of gravity of the system in Figure 8.11a.

Apply Equation 8.3a to the system of three objects:

$$(1)\quad x_{cg} = \frac{\sum m_i x_i}{\sum m_i} = \frac{m_1x_1 + m_2x_2 + m_3x_3}{m_1 + m_2 + m_3}$$

Compute the numerator of Equation (1):

$$\sum m_i x_i = m_1x_1 + m_2x_2 + m_3x_3$$
$$= (5.00 \text{ kg})(-0.500 \text{ m}) + (2.00 \text{ kg})(0 \text{ m}) + (4.00 \text{ kg})(1.00 \text{ m})$$
$$= 1.50 \text{ kg} \cdot \text{m}$$

Substitute the denominator, $\Sigma m_i = 11.0$ kg, and the numerator into Equation (1).

$$x_{cg} = \frac{1.50 \text{ kg} \cdot \text{m}}{11.0 \text{ kg}} = \boxed{0.136 \text{ m}}$$

(b) How does the answer change if the positions of the objects are changed as in Figure 8.11b?

Because the x-coordinates have not been changed, the x-coordinate of the center of gravity is also unchanged:

$$x_{cg} = \boxed{0.136 \text{ m}}$$

Write Equation 8.3b:

$$y_{cg} = \frac{\sum m_i y_i}{\sum m_i} = \frac{m_1y_1 + m_2y_2 + m_3y_3}{m_1 + m_2 + m_3}$$

Substitute values:

$$y_{cg} = \frac{(5.00 \text{ kg})(1.00 \text{ m}) + (2.00 \text{ kg})(0 \text{ m}) + (4.00 \text{ kg})(-0.500 \text{ m})}{5.00 \text{ kg} + 2.00 \text{ kg} + 4.00 \text{ kg}}$$

$$y_{cg} = \boxed{0.273 \text{ m}}$$

REMARKS Notice that translating objects in the y-direction doesn't change the x-coordinate of the center of gravity. The three components of the center of gravity are each independent of the other two coordinates.

QUESTION 8.4 If 1.00 kg is added to the masses on the left and right in Figure 8.11a, does the center of mass (a) move to the left, (b) move to the right, or (c) remain in the same position?

EXERCISE 8.4 If a fourth particle of mass 2.00 kg is placed at (0, 0.25 m) in Figure 8.11a, find the x- and y-coordinates of the center of gravity for this system of four particles.

ANSWER $x_{cg} = 0.115$ m; $y_{cg} = 0.038\,5$ m

EXAMPLE 8.5 Locating Your Lab Partner's Center of Gravity BIO

GOAL Use torque to find a center of gravity.

PROBLEM In this example we show how to find the location of a person's center of gravity. Suppose your lab partner has a height L of 173 cm (5 ft, 8 in.) and a weight w of 715 N (160 lb). You can determine the position of his center of gravity by having him stretch out on a uniform board supported at one end by a scale, as shown in Figure 8.12. If the board's weight w_b is 49 N and the scale reading F is 3.50×10^2 N, find the distance of your lab partner's center of gravity from the left end of the board.

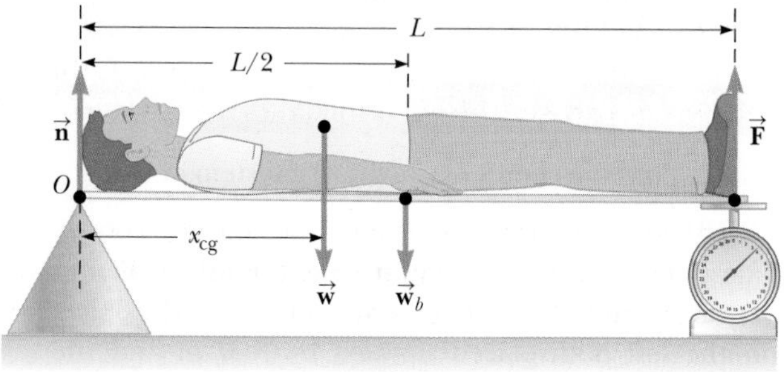

Figure 8.12 (Example 8.5) Determining your lab partner's center of gravity.

STRATEGY To find the position x_{cg} of the center of gravity, compute the torques using an axis through O. There is no torque due to the normal force $\vec{\mathbf{n}}$ because its moment arm is zero about an axis through O. Set the sum of the torques equal to zero and solve for x_{cg}.

SOLUTION

Apply the second condition of equilibrium:

$$\sum \tau_i = \tau_n + \tau_w + \tau_{w_b} + \tau_F = 0$$

Substitute expressions for the torques: $0 - wx_{cg} - w_b(L/2) + FL = 0$

Solve for x_{cg} and substitute known values:

$$x_{cg} = \frac{FL - w_b(L/2)}{w}$$

$$= \frac{(350 \text{ N})(173 \text{ cm}) - (49 \text{ N})(86.5 \text{ cm})}{715 \text{ N}} = 79 \text{ cm}$$

REMARKS The given information is sufficient only to determine the x-coordinate of the center of gravity. The other two coordinates can be estimated, based on the body's symmetry.

QUESTION 8.5 What would happen if a support is placed exactly at $x = 79$ cm followed by the removal of the supports at the subject's head and feet?

EXERCISE 8.5 Suppose a 416-kg alligator of length 3.5 m is stretched out on a board of the same length weighing 65 N. If the board is supported on the ends as in Figure 8.12, and the scale reads 1 880 N, find the x-component of the alligator's center of gravity.

ANSWER 1.59 m

8.4 Examples of Objects in Equilibrium

LEARNING OBJECTIVE

1. Apply the conditions of mechanical equilibrium to rigid bodies.

Recall from Chapter 4 that when an object is treated as a geometric point, equilibrium requires only that the net force on the object is zero. In this chapter we have shown that for extended objects a second condition for equilibrium must also be satisfied: The net torque on the object must be zero. The following general procedure is recommended for solving problems that involve objects in equilibrium.

Tip 8.2 Rotary Motion Under Zero Torque

If a net torque of zero is exerted on an object, it will continue to rotate at a constant angular speed—which need not be zero. However, zero torque *does* imply that the angular acceleration is zero.

PROBLEM-SOLVING STRATEGY

Objects in Equilibrium

1. **Diagram the system.** Include coordinates and choose a convenient rotation axis for computing the net torque on the object.
2. **Draw a force diagram** of the object of interest, showing all external forces acting on it. For systems with more than one object, draw a *separate* diagram for each object. (Most problems will have a single object of interest.)
3. **Apply $\Sigma \tau_i = 0$, the second condition of equilibrium.** This condition yields a single equation for each object of interest. If the axis of rotation has been carefully chosen, the equation often has only one unknotwn and can be solved immediately.
4. **Apply $\Sigma F_x = 0$ and $\Sigma F_y = 0$, the first condition of equilibrium.** This yields two more equations per object of interest.
5. **Solve the system of equations.** For each object, the two conditions of equilibrium yield three equations, usually with three unknowns. Solve by substitution.

EXAMPLE 8.6 A Weighted Forearm BIO

GOAL Apply the equilibrium conditions to the human body.

PROBLEM A 50.0-N (11-lb) bowling ball is held in a person's hand with the forearm horizontal, as in Figure 8.13a. The biceps muscle is attached 0.030 0 m from the joint, and the ball is 0.350 m from the joint. Find the upward

(Continued)

force $\vec{\mathbf{F}}$ exerted by the biceps on the forearm (the ulna) and the downward force $\vec{\mathbf{R}}$ exerted by the humerus on the forearm, acting at the joint. Neglect the weight of the forearm and slight deviation from the vertical of the biceps.

STRATEGY The forces acting on the forearm are equivalent to those acting on a bar of length 0.350 m, as shown in Figure 8.13b. Choose the usual *x*- and *y*-coordinates as shown and the axis at *O* on the left end. (This completes Steps 1 and 2.) Use the conditions of equilibrium to generate equations for the unknowns, and solve.

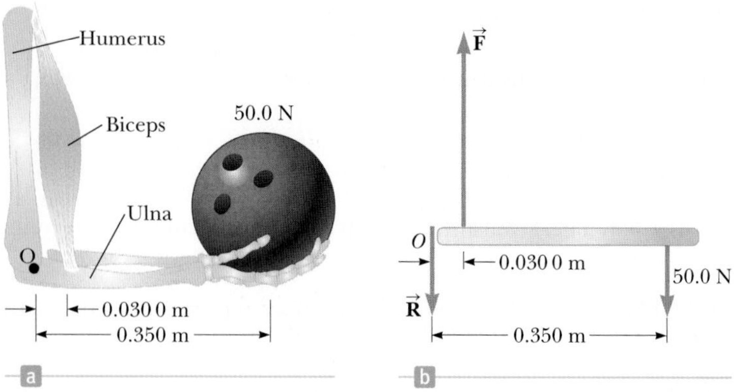

Figure 8.13 (Example 8.6) (a) A weight held with the forearm horizontal. (b) The mechanical model for the system.

SOLUTION

Apply the second condition for equilibrium (Step 3) and solve for the upward force F:

$$\sum \tau_i = \tau_R + \tau_F + \tau_{BB} = 0$$
$$R(0) + F(0.030\,0\text{ m}) - (50.0\text{ N})(0.350\text{ m}) = 0$$
$$F = 583\text{ N }(131\text{ lb})$$

Apply the first condition for equilibrium (Step 4) and solve (Step 5) for the downward force R:

$$\sum F_y = F - R - 50.0\text{ N} = 0$$
$$R = F - 50.0\text{ N} = 583\text{ N} - 50\text{ N} = 533\text{ N }(120\text{ lb})$$

REMARKS The magnitude of the force supplied by the biceps must be about ten times as large as the bowling ball it is supporting!

QUESTION 8.6 Suppose the biceps were surgically reattached three centimeters farther toward the person's hand. If the same bowling ball were again held in the person's hand, how would the force required of the biceps be affected? Explain.

EXERCISE 8.6 Suppose you wanted to limit the force acting on your joint to a maximum value of 8.00×10^2 N. (a) Under these circumstances, what maximum weight would you attempt to lift? (b) What force would your biceps apply while lifting this weight?

ANSWERS (a) 75.0 N (b) 875 N

EXAMPLE 8.7 Don't Climb the Ladder

GOAL Apply the two conditions of equilibrium.

PROBLEM A uniform ladder 10.0 m long and weighing 50.0 N rests against a frictionless vertical wall as in Figure 8.14a. If the ladder is just on the verge of slipping when it makes a 50.0° angle with the ground, find the coefficient of static friction between the ladder and ground.

STRATEGY Figure 8.14b is the force diagram for the ladder. The first condition of equilibrium, $\Sigma\vec{\mathbf{F}}_i = 0$, gives two equations for three unknowns: the magnitudes of the static friction force f and the normal force n, both acting on the base of the ladder, and the magnitude of the force of the wall, P, acting on the top of the ladder. The second condition of equilibrium, $\Sigma\tau_i = 0$, gives a third equation (for P), so all three quantities can be found. The definition of static friction then allows computation of the coefficient of static friction.

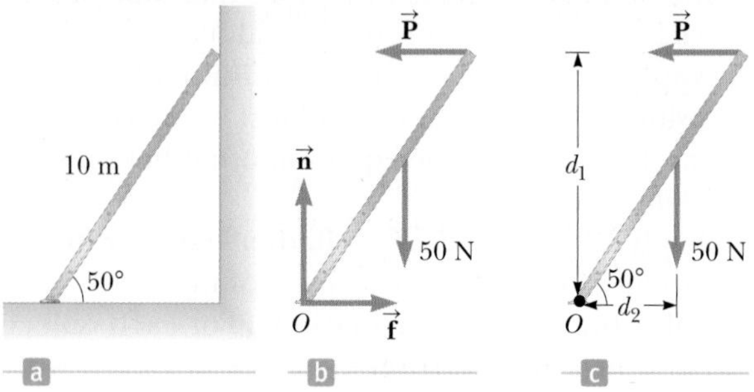

Figure 8.14 (Example 8.7) (a) A ladder leaning against a frictionless wall. (b) A force diagram of the ladder. (c) Lever arms for the force of gravity and $\vec{\mathbf{P}}$.

SOLUTION

Apply the first condition of equilibrium to the ladder:

(1) $\Sigma F_x = f - P = 0 \rightarrow f = P$

(2) $\Sigma F_y = n - 50.0\text{ N} = 0 \rightarrow n = 50.0\text{ N}$

Table 8.1 Moments of Inertia for Various Rigid Objects of Uniform Composition

Hoop or thin cylindrical shell
$I = MR^2$

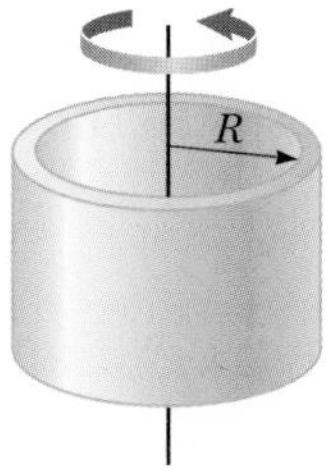

Solid sphere
$I = \frac{2}{5}MR^2$

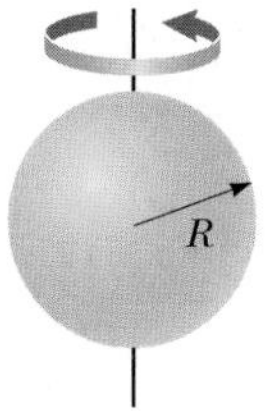

Solid cylinder or disk
$I = \frac{1}{2}MR^2$

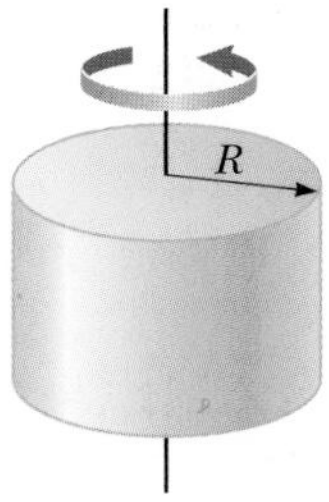

Thin spherical shell
$I = \frac{2}{3}MR^2$

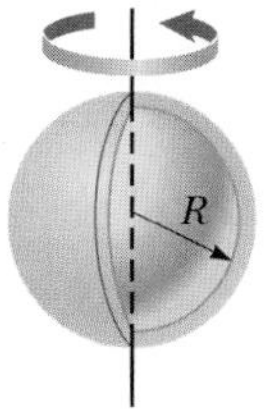

Long, thin rod with rotation axis through center
$I = \frac{1}{12}ML^2$

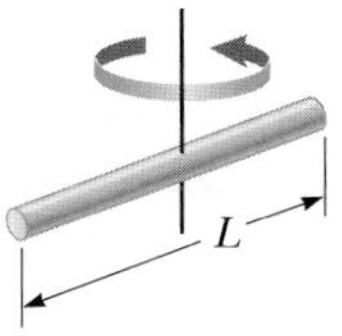

Long, thin rod with rotation axis through end
$I = \frac{1}{3}ML^2$

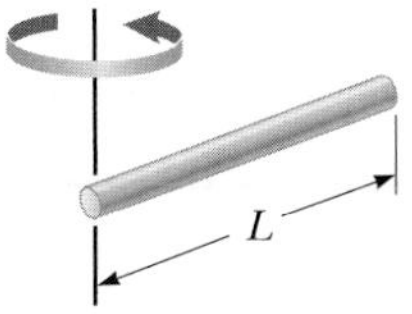

EXAMPLE 8.10 Warming Up

GOAL Find a moment of inertia and apply the rotational analog of Newton's second law.

PROBLEM A baseball player loosening up his arm before a game tosses a 0.150-kg baseball, using only the rotation of his forearm to accelerate the ball (Fig. 8.24). The forearm has a mass of 1.50 kg and the length from the elbow to the ball's center is 0.350 m. The ball starts at rest and is released with a speed of 30.0 m/s in 0.300 s. **(a)** Find the constant angular acceleration of the arm and ball. **(b)** Calculate the moment of inertia of the system consisting of the forearm and ball. **(c)** Find the torque exerted on the system that results in the angular acceleration found in part (a).

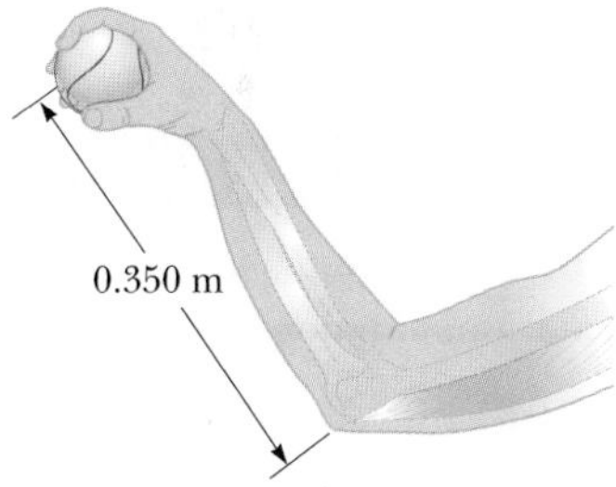

Figure 8.24 (Example 8.10) A ball being tossed by a pitcher. The forearm is used to accelerate the ball.

STRATEGY The angular acceleration can be found with rotational kinematic equations, while the moment of inertia of the system can be obtained by summing the separate moments of inertia of the ball and forearm. The ball is treated as a point particle. Multiplying these two results together gives the torque.

SOLUTION

(a) Find the angular acceleration of the ball.

The angular acceleration is constant, so use the angular velocity kinematic equation with $\omega_i = 0$:

$$\omega = \omega_i + \alpha t \quad \rightarrow \quad \alpha = \frac{\omega}{t}$$

The ball accelerates along a circular arc with radius given by the length of the forearm. Solve $v = r\omega$ for ω and substitute:

$$\alpha = \frac{\omega}{t} = \frac{v}{rt} = \frac{30.0 \text{ m/s}}{(0.350 \text{ m})(0.300 \text{ s})} = \boxed{286 \text{ rad/s}^2}$$

(Continued)

(b) Find the moment of inertia of the system (forearm plus ball).

Find the moment of inertia of the ball about an axis that passes through the elbow, perpendicular to the arm:

$$I_{\text{ball}} = mr^2 = (0.150 \text{ kg})(0.350 \text{ m})^2 = 1.84 \times 10^{-2} \text{ kg} \cdot \text{m}^2$$

Obtain the moment of inertia of the forearm, modeled as a rod rotating about an axis through one end, by consulting Table 8.1:

$$I_{\text{forearm}} = \tfrac{1}{3} ML^2 = \tfrac{1}{3}(1.50 \text{ kg})(0.350 \text{ m})^2$$
$$= 6.13 \times 10^{-2} \text{ kg} \cdot \text{m}^2$$

Sum the individual moments of inertia to obtain the moment of inertia of the system (ball plus forearm):

$$I_{\text{system}} = I_{\text{ball}} + I_{\text{forearm}} = \boxed{7.97 \times 10^{-2} \text{ kg} \cdot \text{m}^2}$$

(c) Find the torque exerted on the system.

Apply Equation 8.8, using the results of parts (a) and (b):

$$\tau = I_{\text{system}}\alpha = (7.97 \times 10^{-2} \text{ kg} \cdot \text{m}^2)(286 \text{ rad/s}^2)$$
$$= \boxed{22.8 \text{ N} \cdot \text{m}}$$

REMARKS Notice that having a long forearm can greatly increase the torque and hence the acceleration of the ball. This is one reason it's advantageous for a pitcher to be tall: the pitching arm is proportionately longer. A similar advantage holds in tennis, where taller players can usually deliver faster serves.

QUESTION 8.10 Why do pitchers step forward when delivering the pitch? Why is the timing important?

EXERCISE 8.10 A catapult with a radial arm 4.00 m long accelerates a ball of mass 20.0 kg through a quarter circle. The ball leaves the apparatus at 45.0 m/s. If the mass of the arm is 25.0 kg and the acceleration is constant, find (a) the angular acceleration, (b) the moment of inertia of the arm and ball, and (c) the net torque exerted on the ball and arm.

Hint: Use the time-independent rotational kinematics equation to find the angular acceleration, rather than the angular velocity equation.

ANSWERS (a) 40.3 rad/s^2 (b) 453 kg · m^2 (c) 1.83×10^4 N · m

EXAMPLE 8.11 The Falling Bucket

GOAL Combine Newton's second law with its rotational analog.

PROBLEM A solid, uniform, frictionless cylindrical reel of mass $M = 3.00$ kg and radius $R = 0.400$ m is used to draw water from a well (Fig. 8.25a). A bucket of mass $m = 2.00$ kg is attached to a cord that is wrapped around the cylinder. **(a)** Find the tension T in the cord and acceleration a of the bucket. **(b)** If the bucket starts from rest at the top of the well and falls for 3.00 s before hitting the water, how far does it fall?

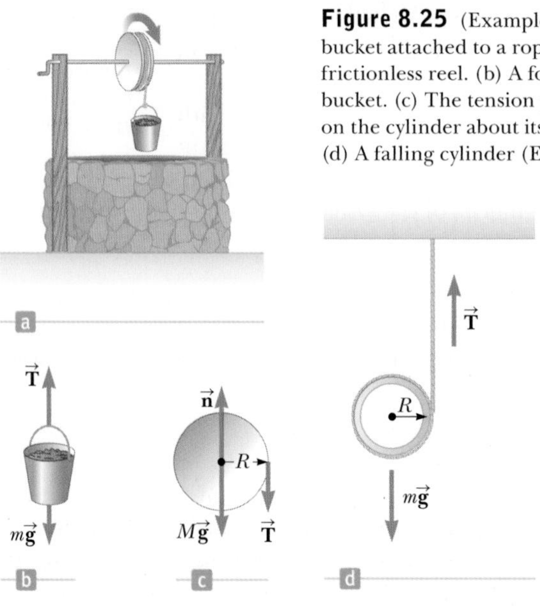

Figure 8.25 (Example 8.11) (a) A water bucket attached to a rope passing over a frictionless reel. (b) A force diagram for the bucket. (c) The tension produces a torque on the cylinder about its axis of rotation. (d) A falling cylinder (Exercise 8.11).

STRATEGY This problem involves three equations and three unknowns. The three equations are Newton's second law applied to the bucket, $ma = \Sigma F_i$; the rotational version of the second law applied to the cylinder, $I\alpha = \Sigma\tau_i$; and the relationship between linear and angular acceleration, $a = r\alpha$, which connects the dynamics of the bucket and cylinder. The three unknowns are the acceleration a of the bucket, the angular acceleration a of the cylinder, and the tension T in the rope. Assemble the terms of the three equations and solve for the three unknowns by substitution. Part (b) is a review of kinematics.

SOLUTION

(a) Find the tension in the cord and the acceleration of the bucket.

Apply Newton's second law to the bucket in Figure 8.25b. There are two forces: the tension $\vec{\mathbf{T}}$ acting upward and gravity $m\vec{\mathbf{g}}$ acting downward.

$$\textbf{(1)} \quad ma = -mg + T$$

Apply $\tau = I\alpha$ to the cylinder in Figure 8.25c:

$$\sum \tau = I\alpha = \tfrac{1}{2}MR^2\alpha \quad \text{(solid cylinder)}$$

Notice the angular acceleration is clockwise, so the torque is negative. The normal and gravity forces have zero moment arm and don't contribute any torque.

$$\textbf{(2)} \quad -TR = \tfrac{1}{2}MR^2\alpha$$

Solve for T and substitute $\alpha = a/R$ (notice that both α and a are negative):

$$\textbf{(3)} \quad T = -\tfrac{1}{2}MR\alpha = -\tfrac{1}{2}Ma$$

Substitute the expression for T in Equation (3) into Equation (1), and solve for the acceleration:

$$ma = -mg - \tfrac{1}{2}Ma \quad \rightarrow \quad a = -\frac{mg}{m + \tfrac{1}{2}M}$$

Substitute the values for m, M, and g, getting a, then substitute a into Equation (3) to get T:

$$a = \boxed{-5.60 \text{ m/s}^2} \qquad T = \boxed{8.40 \text{ N}}$$

(b) Find the distance the bucket falls in 3.00 s.

Apply the displacement kinematic equation for constant acceleration, with $t = 3.00$ s and $v_0 = 0$:

$$\Delta y = v_0 t + \tfrac{1}{2}at^2 = -\tfrac{1}{2}(5.60 \text{ m/s}^2)(3.00 \text{ s})^2 = \boxed{-25.2 \text{ m}}$$

REMARKS Proper handling of signs is very important in these problems. All such signs should be chosen initially and checked mathematically and physically. In this problem, for example, both the angular acceleration α and the acceleration a are negative, so $\alpha = a/R$ applies. If the rope had been wound the other way on the cylinder, causing counterclockwise rotation, the torque would have been positive, and the relationship would have been $\alpha = -a/R$, with the double negative making the right-hand side positive, just like the left-hand side.

QUESTION 8.11 How would the acceleration and tension change if most of the reel's mass were at its rim?

EXERCISE 8.11 A hollow cylinder of mass 0.100 kg and radius 4.00 cm has a string wrapped several times around it, as in Figure 8.25d. If the string is attached to a rigid support and the cylinder allowed to drop from rest, find (a) the acceleration of the cylinder and (b) the speed of the cylinder when a meter of string has unwound off of it.

ANSWERS (a) -4.90 m/s^2 (b) 3.13 m/s

8.6 Rotational Kinetic Energy

LEARNING OBJECTIVES

1. Define the kinetic energy of a rotating object and extend the work-energy theorem to include it.
2. Apply the extended work-energy theorem to systems involving rotation.

In Chapter 5 we defined the kinetic energy of a particle moving through space with a speed v as the quantity $\frac{1}{2}mv^2$. Analogously, **an object rotating about some axis with an angular speed ω has rotational kinetic energy given by $\frac{1}{2}I\omega^2$.** To prove this, consider an object in the shape of a thin, rigid plate rotating around some axis perpendicular to its plane, as in Figure 8.26. The plate consists of many small particles, each of mass m. All these particles rotate in circular paths around the axis. If r is the distance of one of the particles from the

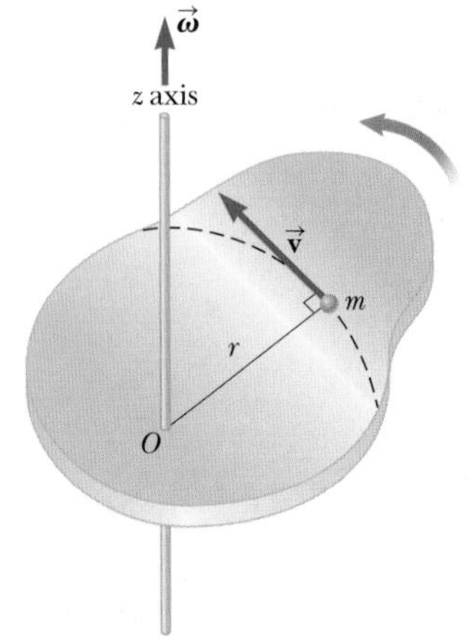

Figure 8.26 A rigid plate rotates about the z-axis with angular speed v. The kinetic energy of a particle of mass m is $\frac{1}{2}mv^2$. The total kinetic energy of the plate is $\frac{1}{2}I\omega^2$.

axis of rotation, the speed of that particle is $v = r\omega$. Because the *total* kinetic energy of the plate's rotation is the sum of all the kinetic energies associated with its particles, we have

$$KE_r = \sum\left(\tfrac{1}{2}mv^2\right) = \sum\left(\tfrac{1}{2}mr^2\omega^2\right) = \tfrac{1}{2}\left(\sum mr^2\right)\omega^2$$

In the last step, the ω^2 term is factored out because it's the same for every particle. Now, the quantity in parentheses on the right is the moment of inertia of the plate in the limit as the particles become vanishingly small, so

$$KE_r = \tfrac{1}{2}I\omega^2 \qquad [8.10]$$

where $I = \Sigma mr^2$ is the moment of inertia of the plate.

A system such as a bowling ball rolling down a ramp is described by three types of energy: **gravitational potential energy PE_g, translational kinetic energy KE_t,** and **rotational kinetic energy KE_r.** All these forms of energy, plus the potential energies of any other conservative forces, must be included in our equation for the conservation of mechanical energy of an isolated system:

Conservation of mechanical energy ▶

$$(KE_t + KE_r + PE)_i = (KE_t + KE_r + PE)_f \qquad [8.11]$$

where i and f refer to initial and final values, respectively, and PE includes the potential energies of all conservative forces in a given problem. This relation is true *only* if we ignore dissipative forces such as friction. Otherwise, it's necessary to resort to a generalization of the work–energy theorem:

Work–energy theorem including rotational energy ▶

$$W_{nc} = \Delta KE_t + \Delta KE_r + \Delta PE \qquad [8.12]$$

PROBLEM-SOLVING STRATEGY

Energy Methods and Rotation

1. **Choose two points of interest,** one where all necessary information is known, and the other where information is desired.
2. **Identify** the conservative and nonconservative forces acting on the system being analyzed.
3. **Write the general work–energy theorem,** Equation 8.12, or Equation 8.11 if all forces are conservative.
4. **Substitute general expressions** for the terms in the equation.
5. **Use $v = r\omega$** to eliminate either ω or v from the equation.
6. **Solve** for the unknown.

EXAMPLE 8.12 | A Ball Rolling Down an Incline

GOAL Combine gravitational, translational, and rotational energy.

PROBLEM A ball of mass M and radius R starts from rest at a height of $h = 2.00$ m and rolls down a $\theta = 30.0°$ slope, as in Figure 8.27. What is the linear speed of the ball when it leaves the incline? Assume that the ball rolls without slipping.

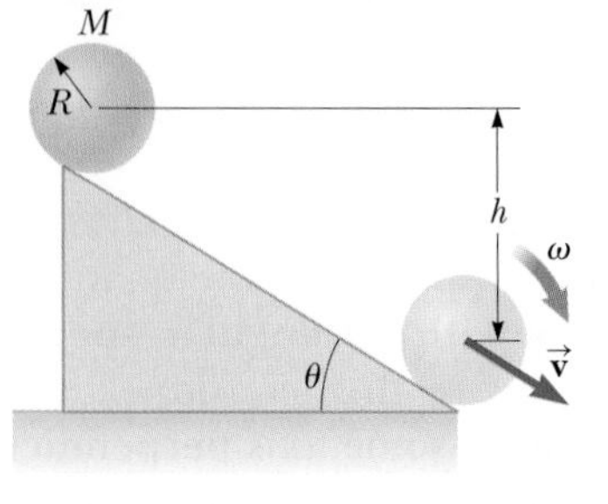

Figure 8.27 (Example 8.12) A ball starts from rest at the top of an incline and rolls to the bottom without slipping.

STRATEGY The two points of interest are the top and bottom of the incline, with the bottom acting as the zero point of gravitational potential energy. As the ball rolls down the ramp, gravitational potential energy is converted into both translational and rotational kinetic energy without dissipation, so conservation of mechanical energy can be applied with the use of Equation 8.11.

SOLUTION

Apply conservation of energy with $PE = PE_g$, the potential energy associated with gravity:

$$(KE_t + KE_r + PE_g)_i = (KE_t + KE_r + PE_g)_f$$

Substitute the appropriate general expressions, noting that $(KE_t)_i = (KE_r)_i = 0$ and $(PE_g)_f = 0$ (obtain the moment of inertia of a ball from Table 8.1):

$$0 + 0 + Mgh = \tfrac{1}{2}Mv^2 + \tfrac{1}{2}(\tfrac{2}{5}MR^2)\omega^2 + 0$$

The ball rolls without slipping, so $R\omega = v$, the "no-slip condition," can be applied:

$$Mgh = \tfrac{1}{2}Mv^2 + \tfrac{1}{5}Mv^2 = \tfrac{7}{10}Mv^2$$

Solve for v, noting that M cancels.

$$v = \sqrt{\frac{10gh}{7}} = \sqrt{\frac{10(9.80 \text{ m/s}^2)(2.00 \text{ m})}{7}} = \boxed{5.29 \text{ m/s}}$$

REMARKS Notice the translational speed is less than that of a block sliding down a frictionless slope, $v = \sqrt{2gh}$. That's because some of the original potential energy must go to increasing the rotational kinetic energy.

QUESTION 8.12 Rank from fastest to slowest: (a) a solid ball rolling down a ramp without slipping, (b) a cylinder rolling down the same ramp without slipping, (c) a block sliding down a frictionless ramp with the same height and slope.

EXERCISE 8.12 Repeat this example for a solid cylinder of the same mass and radius as the ball and released from the same height. In a race between the two objects on the incline, which one would win?

ANSWER $v = \sqrt{4gh/3} = 5.11$ m/s; the ball would win.

Quick Quiz

8.4 Two spheres, one hollow and one solid, are rotating with the same angular speed around an axis through their centers. Both spheres have the same mass and radius. Which sphere, if either, has the higher rotational kinetic energy? (a) The hollow sphere. (b) The solid sphere. (c) They have the same kinetic energy.

EXAMPLE 8.13 Blocks and Pulley

GOAL Solve a system requiring rotation concepts and the work–energy theorem.

PROBLEM Two blocks with masses $m_1 = 5.00$ kg and $m_2 = 7.00$ kg are attached by a string as in Figure 8.28a, over a pulley with mass $M = 2.00$ kg. The pulley, which turns on a frictionless axle, is a hollow cylinder with radius 0.050 0 m over

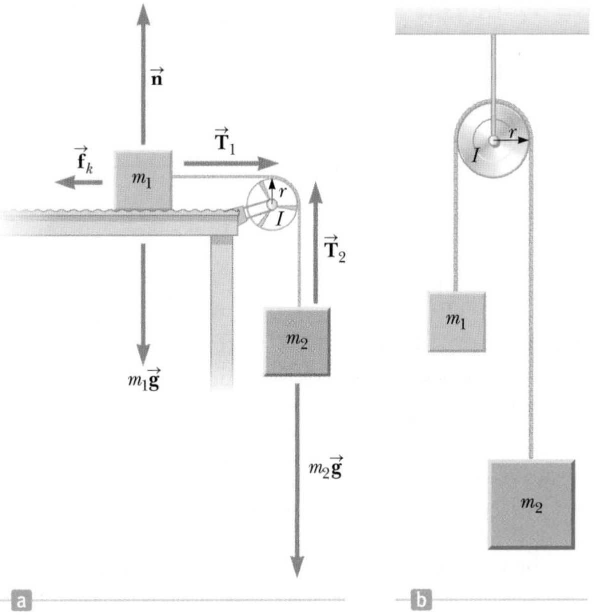

Figure 8.28 (a) (Example 8.13) $\vec{T}_1$ and $\vec{T}_2$ exert torques on the pulley. (b) (Exercise 8.13)

(Continued)

which the string moves without slipping. The horizontal surface has coefficient of kinetic friction 0.350. Find the speed of the system when the block of mass m_2 has dropped 2.00 m.

STRATEGY This problem can be solved with the extension of the work–energy theorem, Equation 8.12. If the block of mass m_2 falls from height h to 0, then the block of mass m_1 moves the same distance, $\Delta x = h$. Apply the work-energy theorem, solve for v, and substitute. Kinetic friction is the sole nonconservative force.

SOLUTION

Apply the work–energy theorem, with $PE = PE_g$, the potential energy associated with gravity:

$$W_{nc} = \Delta KE_t + \Delta KE_r + \Delta PE_g$$

Substitute the frictional work for W_{nc}, kinetic energy changes for the two blocks, the rotational kinetic energy change for the pulley, and the potential energy change for the second block:

$$-\mu_k n\,\Delta x = -\mu_k(m_1 g)\,\Delta x = (\tfrac{1}{2}m_1v^2 - 0) + (\tfrac{1}{2}m_2v^2 - 0) + (\tfrac{1}{2}I\omega^2 - 0) + (0 - m_2gh)$$

Substitute $\Delta x = h$, and write I as $(I/r^2)r^2$:

$$-\mu_k(m_1g)h = \tfrac{1}{2}m_1v^2 + \tfrac{1}{2}m_2v^2 + \tfrac{1}{2}\left(\frac{I}{r^2}\right)r^2\omega^2 - m_2gh$$

For a hoop, $I = Mr^2$ so $(I/r^2) = M$. Substitute this quantity and $v = r\omega$:

$$-\mu_k(m_1g)h = \tfrac{1}{2}m_1v^2 + \tfrac{1}{2}m_2v^2 + \tfrac{1}{2}Mv^2 - m_2gh$$

Solve for v:

$$m_2gh - \mu_k(m_1g)h = \tfrac{1}{2}m_1v^2 + \tfrac{1}{2}m_2v^2 + \tfrac{1}{2}Mv^2$$

$$= \tfrac{1}{2}(m_1 + m_2 + M)v^2$$

$$v = \sqrt{\frac{2gh(m_2 - \mu_k m_1)}{m_1 + m_2 + M}}$$

Substitute $m_1 = 5.00$ kg, $m_2 = 7.00$ kg, $M = 2.00$ kg, $g = 9.80\ \text{m/s}^2$, $h = 2.00$ m, and $\mu_k = 0.350$:

$$v = \boxed{3.83\ \text{m/s}}$$

REMARKS In the expression for the speed v, the mass m_1 of the first block and the mass M of the pulley all appear in the denominator, reducing the speed, as they should. In the numerator, m_2 is positive while the friction term is negative. Both assertions are reasonable because the force of gravity on m_2 increases the speed of the system while the force of friction on m_1 slows it down. This problem can also be solved with Newton's second law together with $\tau = I\alpha$, a good exercise.

QUESTION 8.13 How would increasing the radius of the pulley affect the final answer? Assume the angles of the cables are unchanged and the mass is the same as before.

EXERCISE 8.13 Two blocks with masses $m_1 = 2.00$ kg and $m_2 = 9.00$ kg are attached over a pulley with mass $M = 3.00$ kg, hanging straight down as in Atwood's machine (Fig. 8.28b). The pulley is a solid cylinder with radius 0.050 0 m, and there is some friction in the axle. The system is released from rest, and the string moves without slipping over the pulley. If the larger mass is traveling at a speed of 2.50 m/s when it has dropped 1.00 m, how much mechanical energy was lost due to friction in the pulley's axle?

ANSWER 29.5 J

8.7 Angular Momentum

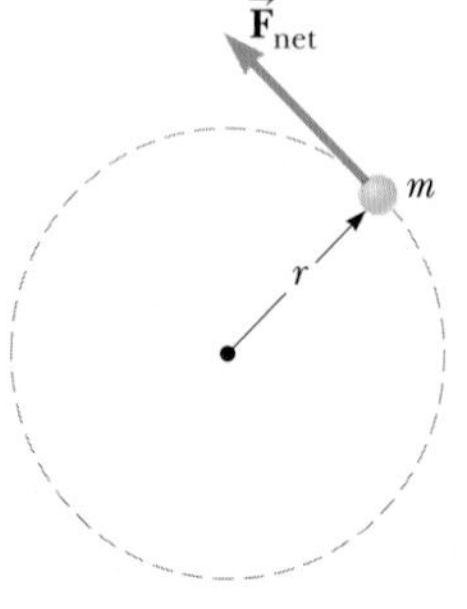

Figure 8.29 An object of mass m rotating in a circular path under the action of a constant torque.

LEARNING OBJECTIVES

1. Define angular momentum and state the rotational second law in terms of it.
2. State and apply the principle of the conservation of angular momentum.

In Figure 8.29, an object of mass m rotates in a circular path of radius r, acted on by a net force, $\vec{F}_{net}$. The resulting net torque on the object increases its angular speed from the value ω_0 to the value ω in a time interval Δt. Therefore, we can write

$$\sum \tau = I\alpha = I\frac{\Delta\omega}{\Delta t} = I\left(\frac{\omega - \omega_0}{\Delta t}\right) = \frac{I\omega - I\omega_0}{\Delta t}$$

If we define the product

$$L \equiv I\omega \qquad [8.13]$$

as the **angular momentum** of the object, then we can write

$$\sum \tau = \frac{\text{change in angular momentum}}{\text{time interval}} = \frac{\Delta L}{\Delta t} \qquad [8.14]$$

Equation 8.14 is the rotational analog of Newton's second law, which can be written in the form $F = \Delta p/\Delta t$ and states that **the net torque acting on an object is equal to the time rate of change of the object's angular momentum.** Recall that this equation also parallels the impulse–momentum theorem.

When the net external torque ($\Sigma\tau$) acting on a system is zero, Equation 8.14 gives $\Delta L/\Delta t = 0$, which says that the time rate of change of the system's angular momentum is zero. We then have the following important result:

◂ Conservation of angular momentum

Let L_i and L_f be the angular momenta of a system at two different times, and suppose there is no net external torque, so $\Sigma\tau = 0$. Then

$$L_i = L_f \qquad [8.15]$$

and angular momentum is said to be *conserved.*

Equation 8.15 gives us a third conservation law to add to our list: **conservation of angular momentum.** We can now state that **the mechanical energy, linear momentum, and angular momentum of an isolated system all remain constant.**

If the moment of inertia of an isolated rotating system changes, the system's angular speed will change. Conservation of angular momentum then requires that

$$I_i\omega_i = I_f\omega_f \quad \text{if} \quad \sum \tau = 0 \qquad [8.16]$$

APPLICATION
Figure Skating

Note that conservation of angular momentum applies to macroscopic objects such as planets and people, as well as to atoms and molecules. There are many examples of conservation of angular momentum; one of the most dramatic is that of a figure skater spinning in the finale of his act. In Figure 8.30a, the skater has pulled his arms and legs close to his body, reducing their distance from his axis of rotation and hence also reducing his moment of inertia. By conservation of angular momentum, a reduction in his moment of inertia must increase his angular

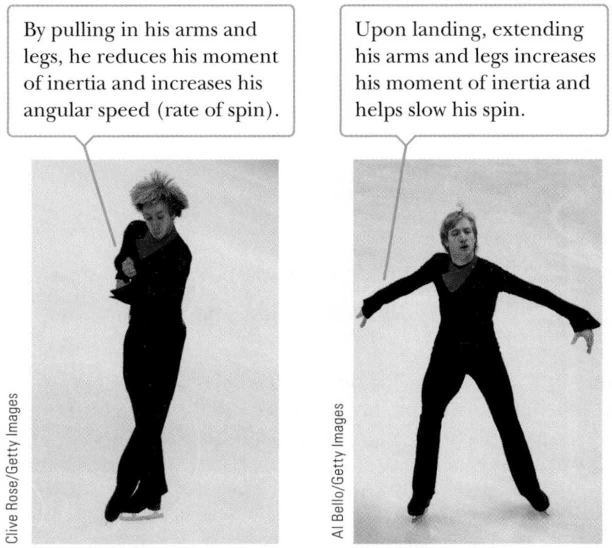

Figure 8.30 Evgeni Plushenko varies his moment of inertia to change his angular speed.

Tightly curling her body, a diver decreases her moment of inertia, increasing her angular speed.

speed. Coming out of the spin in Figure 8.30b, he needs to reduce his angular speed, so he extends his arms and legs again, increasing his moment of inertia and thereby slowing his rotation.

APPLICATION
Aerial Somersaults

Similarly, when a diver or an acrobat wishes to make several somersaults, she pulls her hands and feet close to the trunk of her body in order to rotate at a greater angular speed. In this case, the external force due to gravity acts through her center of gravity and hence exerts no torque about her axis of rotation, so the angular momentum about her center of gravity is conserved. For example, when a diver wishes to double her angular speed, she must reduce her moment of inertia to half its initial value.

APPLICATION
Rotating Neutron Stars

An interesting astrophysical example of conservation of angular momentum occurs when a massive star, at the end of its lifetime, uses up all its fuel and collapses under the influence of gravitational forces, causing a gigantic outburst of energy called a supernova. The best-studied example of a remnant of a supernova explosion is the Crab Nebula, a chaotic, expanding mass of gas (Fig. 8.31). In a supernova, part of the star's mass is ejected into space, where it eventually condenses into new stars and planets. Most of what is left behind typically collapses into a **neutron star**—an extremely dense sphere of matter with a diameter of about 10 km, greatly reduced from the 10^6-km diameter of the original star and containing a large fraction of the star's original mass. In a neutron star, pressures become so great that atomic electrons combine with protons, becoming neutrons. As the moment of inertia of the system decreases during the collapse, the star's rotational speed increases. More than 700 rapidly rotating neutron stars have been identified since their first discovery in 1967, with periods of rotation ranging from a millisecond to several seconds. The neutron star is an amazing system—an object with a mass greater than the Sun, fitting comfortably within the space of a small county and rotating so fast that the tangential speed of the surface approaches a sizable fraction of the speed of light!

Quick Quiz

8.5 A horizontal disk with moment of inertia I_1 rotates with angular speed ω_1 about a vertical frictionless axle. A second horizontal disk having moment of inertia I_2 drops onto the first, initially not rotating but sharing the same axis as the first disk. Because their surfaces are rough, the two disks eventually reach the same angular speed ω. The ratio ω/ω_1 is equal to (a) I_1/I_2 (b) I_2/I_1 (c) $I_1/(I_1 + I_2)$ (d) $I_2/(I_1 + I_2)$

8.6 If global warming continues, it's likely that some ice from the polar ice caps of the Earth will melt and the water will be distributed closer to the equator. If this occurs, would the length of the day (one rotation) (a) increase, (b) decrease, or (c) remain the same?

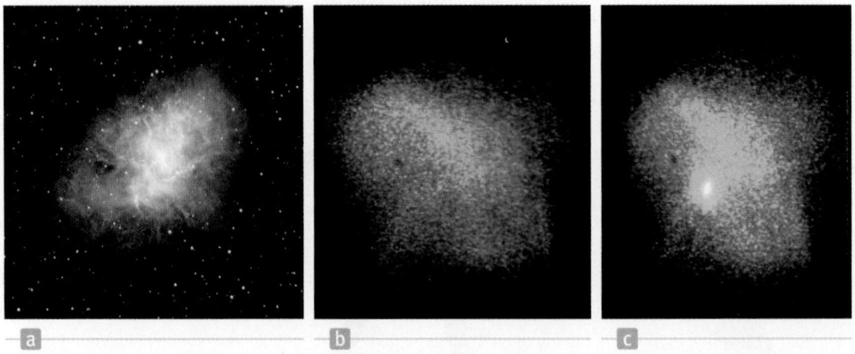

Max Planck Institute for Astronomy, Heidelberg, Germany

© Smithsonian Institute/Photo Researchers, Inc.

Figure 8.31 (a) The Crab Nebula in the constellation Taurus. This nebula is the remnant of a supernova seen on Earth in A.D. 1054. It is located some 6 300 light-years away and is approximately 6 light-years in diameter, still expanding outward. A pulsar deep inside the nebula flashes 30 times every second. (b) Pulsar off. (c) Pulsar on.

■ EXAMPLE 8.14 The Spinning Stool

GOAL Apply conservation of angular momentum to a simple system.

PROBLEM A student sits on a pivoted stool while holding a pair of weights. (See Fig. 8.32.) The stool is free to rotate about a vertical axis with negligible friction. The moment of inertia of student, weights, and stool is 2.25 kg · m^2. The student is set in rotation with arms outstretched, making one complete turn every 1.26 s, arms outstretched. **(a)** What is the initial angular speed of the system? **(b)** As he rotates, he pulls the weights inward so that the new moment of inertia of the system (student, objects, and stool) becomes 1.80 kg · m^2. What is the new angular speed of the system? **(c)** Find the work done by the student on the system while pulling in the weights. (Ignore energy lost through dissipation in his muscles.)

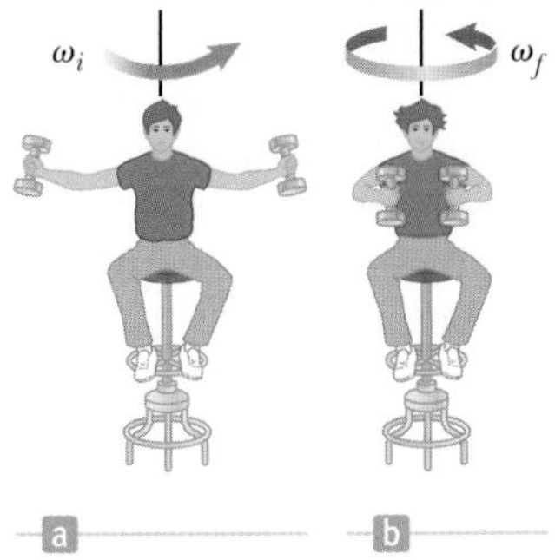

Figure 8.32 (Example 8.14) (a) The student is given an initial angular speed while holding two weights out. (b) The angular speed increases as the student draws the weights inwards.

STRATEGY (a) The angular speed can be obtained from the frequency, which is the inverse of the period. (b) There are no external torques acting on the system, so the new angular speed can be found with the principle of conservation of angular momentum. (c) The work done on the system during this process is the same as the system's change in rotational kinetic energy.

SOLUTION

(a) Find the initial angular speed of the system.

Invert the period to get the frequency, and multiply by 2π:

$$\omega_i = 2\pi f = 2\pi/T = \boxed{4.99 \text{ rad/s}}$$

(b) After he pulls the weights in, what's the system's new angular speed?

Equate the initial and final angular momenta of the system:

$$\textbf{(1)} \quad L_i = L_f \quad \rightarrow \quad I_i\omega_i = I_f\omega_f$$

Substitute and solve for the final angular speed ω_f:

$$\textbf{(2)} \quad (2.25 \text{ kg} \cdot \text{m}^2)(4.99 \text{ rad/s}) = (1.80 \text{ kg} \cdot \text{m}^2)\omega_f$$

$$\omega_f = \boxed{6.24 \text{ rad/s}}$$

(c) Find the work the student does on the system.

Apply the work–energy theorem:

$$W_{\text{student}} = \Delta K_r = \tfrac{1}{2}I_f\omega_f^2 - \tfrac{1}{2}I_i\omega_i^2$$

$$= \tfrac{1}{2}(1.80 \text{ kg} \cdot \text{m}^2)(6.24 \text{ rad/s})^2 - \tfrac{1}{2}(2.25 \text{ kg} \cdot \text{m}^2)(4.99 \text{ rad/s})^2$$

$$W_{\text{student}} = \boxed{7.03 \text{ J}}$$

REMARKS Although the angular momentum of the system is conserved, mechanical energy is not conserved because the student does work on the system.

QUESTION 8.14 If the student suddenly releases the weights, will his angular speed increase, decrease, or remain the same?

EXERCISE 8.14 A star with an initial radius of 1.0×10^8 m and period of 30.0 days collapses suddenly to a radius of 1.0×10^4 m. (a) Find the period of rotation after collapse. (b) Find the work done by gravity during the collapse if the mass of the star is 2.0×10^{30} kg. (c) What is the speed of an indestructible person standing on the equator of the collapsed star? (Neglect any relativistic or thermal effects, and assume the star is spherical before and after it collapses.)

ANSWERS (a) 2.6×10^{-2} s (b) 2.3×10^{42} J (c) 2.4×10^6 m/s

■ EXAMPLE 8.15 The Merry-Go-Round

GOAL Apply conservation of angular momentum while combining two moments of inertia.

PROBLEM A merry-go-round modeled as a disk of mass $M = 1.00 \times 10^2$ kg and radius $R = 2.00$ m is rotating in a horizontal plane about a frictionless vertical axle (Fig. 8.33 is an overhead view of the system). **(a)** After a student with mass $m = 60.0$ kg

(Continued)

jumps on the rim of the merry-go-round, the system's angular speed decreases to 2.00 rad/s. If the student walks slowly from the edge toward the center, find the angular speed of the system when she reaches a point 0.500 m from the center. **(b)** Find the change in the system's rotational kinetic energy caused by her movement to $r = 0.500$ m. **(c)** Find the work done on the student as she walks to $r = 0.500$ m.

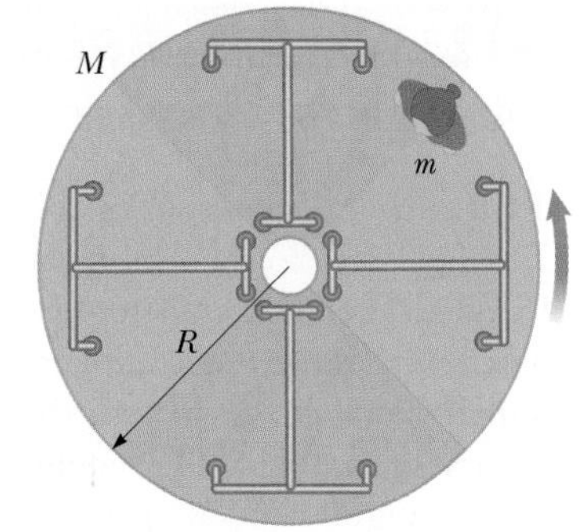

Figure 8.33 (Example 8.15) As the student walks toward the center of the merry-go-round, the moment of inertia I of the system becomes smaller. Because angular momentum is conserved and $L = I\omega$, the angular speed must increase.

STRATEGY This problem can be solved with conservation of angular momentum by equating the system's initial angular momentum when the student stands at the rim to the angular momentum when the student has reached $r = 0.500$ m. The key is to find the different moments of inertia.

SOLUTION

(a) Find the angular speed when the student reaches a point 0.500 m from the center.

Calculate the moment of inertia of the disk, I_D:

$$I_D = \tfrac{1}{2}MR^2 = \tfrac{1}{2}(1.00 \times 10^2 \text{ kg})(2.00 \text{ m})^2 = 2.00 \times 10^2 \text{ kg}\cdot\text{m}^2$$

Calculate the initial moment of inertia of the student. This is the same as the moment of inertia of a mass a distance R from the axis:

$$I_{Si} = mR^2 = (60.0 \text{ kg})(2.00 \text{ m})^2 = 2.40 \times 10^2 \text{ kg}\cdot\text{m}^2$$

Sum the two moments of inertia and multiply by the initial angular speed to find L_i, the initial angular momentum of the system:

$$L_i = (I_D + I_{Si})\omega_i = (2.00 \times 10^2 \text{ kg}\cdot\text{m}^2 + 2.40 \times 10^2 \text{ kg}\cdot\text{m}^2)(2.00 \text{ rad/s}) = 8.80 \times 10^2 \text{ kg}\cdot\text{m}^2/\text{s}$$

Calculate the student's final moment of inertia, I_{Sf}, when she is 0.500 m from the center:

$$I_{Sf} = mr_f^2 = (60.0 \text{ kg})(0.50 \text{ m})^2 = 15.0 \text{ kg}\cdot\text{m}^2$$

The moment of inertia of the platform is unchanged. Add it to the student's final moment of inertia, and multiply by the unknown final angular speed to find L_f:

$$L_f = (I_D + I_{Sf})\omega_f = (2.00 \times 10^2 \text{ kg}\cdot\text{m}^2 + 15.0 \text{ kg}\cdot\text{m}^2)\omega_f = (2.15 \times 10^2 \text{ kg}\cdot\text{m}^2)\omega_f$$

Equate the initial and final angular momenta and solve for the final angular speed of the system:

$$L_i = L_f$$

$$(8.80 \times 10^2 \text{ kg}\cdot\text{m}^2/\text{s}) = (2.15 \times 10^2 \text{ kg}\cdot\text{m}^2)\omega_f$$

$$\omega_f = \boxed{4.09 \text{ rad/s}}$$

(b) Find the change in the rotational kinetic energy of the system.

Calculate the initial kinetic energy of the system:

$$KE_i = \tfrac{1}{2}I_i\omega_i^2 = \tfrac{1}{2}(4.40 \times 10^2 \text{ kg}\cdot\text{m}^2)(2.00 \text{ rad/s})^2 = 8.80 \times 10^2 \text{ J}$$

Calculate the final kinetic energy of the system:

$$KE_f = \tfrac{1}{2}I_f\omega_f^2 = \tfrac{1}{2}(215 \text{ kg}\cdot\text{m}^2)(4.09 \text{ rad/s})^2 = 1.80 \times 10^3 \text{ J}$$

Calculate the change in kinetic energy of the system:

$$KE_f - KE_i = \boxed{920 \text{ J}}$$

(c) Find the work done on the student.

The student undergoes a change in kinetic energy that equals the work done on her. Apply the work–energy theorem:

$$W = \Delta KE_{\text{student}} = \tfrac{1}{2}I_{Sf}\omega_f^2 - \tfrac{1}{2}I_{Si}\omega_i^2 = \tfrac{1}{2}(15.0 \text{ kg}\cdot\text{m}^2)(4.09 \text{ rad/s})^2 - \tfrac{1}{2}(2.40 \times 10^2 \text{ kg}\cdot\text{m}^2)(2.00 \text{ rad/s})^2$$

$$W = \boxed{-355 \text{ J}}$$

REMARKS The angular momentum is unchanged by internal forces; however, the kinetic energy increases because the student must perform positive work in order to walk toward the center of the platform.

QUESTION 8.15 Is energy conservation violated in this example? Explain why there is a positive net change in mechanical energy. What is the origin of this energy?

EXERCISE 8.15 (a) Find the angular speed of the merry-go-round before the student jumped on, assuming the student didn't transfer any momentum or energy as she jumped on the merry-go-round. (b) By how much did the kinetic energy of the system change when the student jumped on? Notice that energy is lost in this process, as should be expected, since it is essentially a perfectly inelastic collision.

ANSWERS (a) 4.40 rad/s (b) $KE_f - KE_i = -1.06 \times 10^3$ J.

SUMMARY

8.1 Torque

Let $\vec{\mathbf{F}}$ be a force acting on an object, and let $\vec{\mathbf{r}}$ be a position vector from a chosen point O to the point of application of the force. Then the magnitude of the torque $\vec{\boldsymbol{\tau}}$ of the force $\vec{\mathbf{F}}$ is given by

$$\tau = rF\sin\theta \qquad [8.2]$$

where r is the length of the position vector, F the magnitude of the force, and θ the angle between $\vec{\mathbf{F}}$ and $\vec{\mathbf{r}}$.

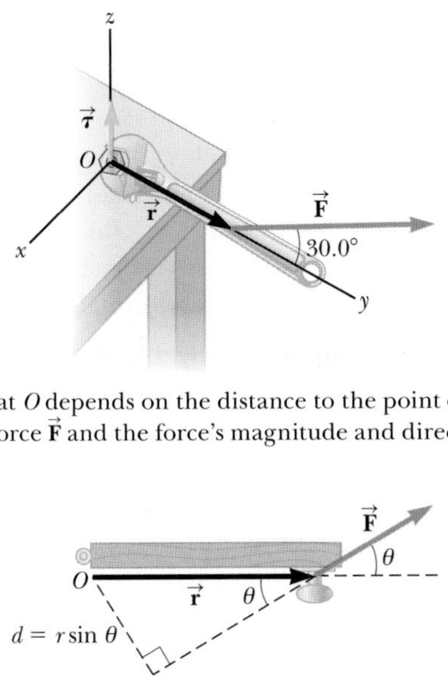

The torque at O depends on the distance to the point of application of the force $\vec{\mathbf{F}}$ and the force's magnitude and direction.

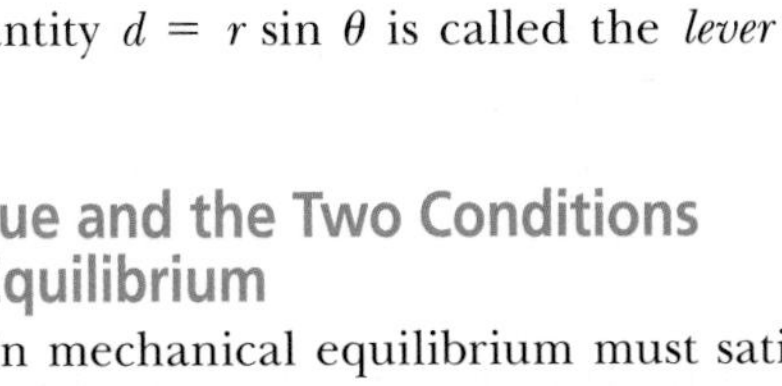

An alternate interpretation of torque involves the concept of a lever arm $d = r\sin\theta$ that is perpendicular to the force.

The quantity $d = r\sin\theta$ is called the *lever arm* of the force.

8.2 Torque and the Two Conditions for Equilibrium

An object in mechanical equilibrium must satisfy the following two conditions:

1. The net external force must be zero: $\sum\vec{\mathbf{F}} = 0$.
2. The net external torque must be zero: $\sum\vec{\boldsymbol{\tau}} = 0$.

These two conditions, used in solving problems involving rotation in a plane—result in three equations and three unknowns—two from the first condition (corresponding to the x- and y-components of the force) and one from the second condition, on torques. These equations must be solved simultaneously.

8.5 Relationship Between Torque and Angular Acceleration

The **moment of inertia** of a group of particles is

$$I \equiv \Sigma mr^2 \qquad [8.7]$$

If a rigid object free to rotate about a fixed axis has a net external torque $\Sigma\tau$ acting on it, then the object undergoes an angular acceleration a, where

$$\Sigma\tau = I\alpha \qquad [8.8]$$

This equation is the rotational equivalent of the second law of motion.

Problems are solved by using Equation 8.8 together with Newton's second law and solving the resulting equations simultaneously. The relation $a = r\alpha$ is often key in relating the translational equations to the rotational equations.

8.6 Rotational Kinetic Energy

If a rigid object rotates about a fixed axis with angular speed ω, its **rotational kinetic energy** is

$$KE_r = \tfrac{1}{2}I\omega^2 \qquad [8.10]$$

where I is the moment of inertia of the object around the axis of rotation.

A system involving rotation is described by three types of energy: potential energy PE, translational kinetic energy KE_t, and rotational kinetic energy KE_r. All these forms of energy must be included in the equation for conservation of mechanical energy for an isolated system:

$$(KE_t + KE_r + PE)_i = (KE_t + KE_r + PE)_f \qquad [8.11]$$

where i and f refer to initial and final values, respectively. When non-conservative forces are present, it's necessary to use a generalization of the work–energy theorem:

$$W_{nc} = \Delta KE_t + \Delta KE_r + \Delta PE \qquad [8.12]$$

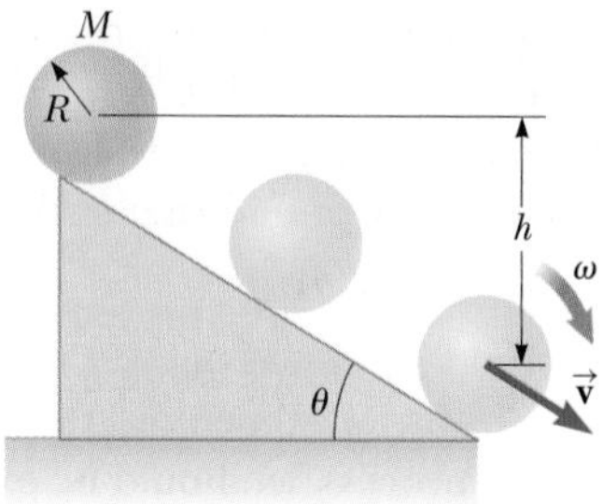

A ball rolling down an incline converts potential energy to translational and rotational kinetic energy.

8.7 Angular Momentum

The **angular momentum** of a rotating object is given by

$$L \equiv I\omega \quad [8.13]$$

Angular momentum is related to torque in the following equation:

$$\sum \tau = \frac{\text{change in angular momentum}}{\text{time interval}} = \frac{\Delta L}{\Delta t} \quad [8.14]$$

If the net external torque acting on a system is zero, the total angular momentum of the system is constant,

$$L_i = L_f \quad [8.15]$$

and is said to be conserved. Solving problems usually involves substituting into the expression

$$I_i\omega_i = I_f\omega_f \quad [8.16]$$

and solving for the unknown.

WARM-UP EXERCISES

ENHANCED WebAssign The warm-up exercises in this chapter may be assigned online in Enhanced WebAssign.

1. **Math Review** The two conditions for equilibrium (see Sections 8.2 and 8.4) often result in a system of equations such as $F_1 + F_2 = 60.0$ N and $2.00F_1 - 3.00F_2 = 0$. Find (a) F_1 and (b) F_2.
2. **Math Review** Solve the equations $\frac{1}{2}mv^2 + \frac{1}{2}I\omega^2 = mgh$ and $v = r\omega$ for the speed v using substitution, given that $I = mr^2$ and $h = 3.00$ m. (See Section 8.6. Note that mass m and radius r will both cancel, so their numerical values aren't required.)
3. **Physics Review** A spinning wheel steadily slows from an initial angular velocity of 2.00 rev/s to 0.500 rev/s in 10.0 s. (a) Calculate the wheel's angular acceleration in radians per second squared. (b) What angle does it go through during that time? (See Sections 7.1 and 7.2.)
4. **Physics Review** A construction crane's cable lifts a 50.0-kg box upward with an acceleration of 1.50 m/s^2. Find the tension in the rope. (See Section 4.5.)
5. A man opens a 1.00-m wide door by pushing on it with a force of 50.0 N directed perpendicular to its surface. What magnitude torque does he apply about an axis through the hinges if the force is applied (a) at the center of the door? (b) at the edge farthest from the hinges? (See Section 8.1.)
6. A worker applies a torque to a nut with a wrench 0.500 m long. Because of the cramped space, she must exert a force upward at an angle of 60.0° with respect to a line from the nut through the end of the wrench. If the force she exerts has magnitude 80.0 N, what magnitude torque does she apply to the nut? (See Section 8.1.)
7. A mass of 1.00 kg is at (–2.00 m, 0) and a 2.00-kg mass at (3.00 m, 3.00 m). Find the center of mass of the system. (See Section 8.3.)
8. A horizontal plank 4.00 m long and having mass 20.0 kg rests on two pivots, one at the left end and a second 1.00 m from the right end. Find the magnitude of the force exerted on the plank by the second pivot. (See Section 8.4.)
9. A student rides his bicycle at a constant speed of 3.00 m/s along a straight, level road. If the bike's tires each have a radius of 0.350 m, (a) what is the tires' angular speed? (See Section 7.3.) (b) What is the net torque on each tire? (See Section 8.5.)
10. What is the magnitude of the angular acceleration of a 25.0-kg disk of radius 0.800 m when a torque of magnitude 40.0 N · m is applied to it? (See Section 8.5.)
11. A bicycle tire has a mass of 2.70 kg and a radius of 0.350 m. (a) Treating the tire as a hoop, what is its moment of inertia about an axis passing through the hub at its center? (b) What torque is required to produce an angular acceleration of 0.750 rad/s^2? (c) What friction force applied tangentially to the edge of the tire will create a torque of that magnitude? (See Section 8.5.)
12. A bowling ball of mass 7.00 kg is rolling at 3.00 m/s along a level surface. Calculate (a) the ball's translational kinetic energy, (b) the ball's rotational kinetic energy, and (c) the ball's total kinetic energy. (d) How much work would have to be done on the ball to bring it to rest? (See Section 8.6.)
13. A basketball player entertains the crowd by spinning a basketball on his nose. The basketball has a mass of 0.600 kg and a radius of 0.121 m. If the basketball is spinning at a rate of 3.00 revolutions per second, (a) what is its rotational kinetic energy? (See Section 8.6.)

(b) What is the magnitude of its angular momentum? Treat the ball as a thin, spherical shell. (See Section 8.7.)

14. A disk of mass m is spinning freely at 6.00 rad/s when a second disk of identical mass, initially not spinning, is dropped on it so that their axes coincide. In a short time the two disks are corotating. (a) What is the angular speed of the new system? (b) If a third such disk is dropped on the first two, find the final angular speed of the system. (See Section 8.7.)

CONCEPTUAL QUESTIONS

ENHANCED WebAssign The conceptual questions in this chapter may be assigned online in Enhanced WebAssign.

1. Why can't you put your heels firmly against a wall and then bend over without falling?
2. Explain why changing the axis of rotation of an object changes its moment of inertia.
3. If you see an object rotating, is there necessarily a net torque acting on it?
4. (a) Is it possible to calculate the torque acting on a rigid object without specifying an origin? (b) Is the torque independent of the location of the origin?
5. Why does a long pole help a tightrope walker stay balanced?
6. In the movie *Jurassic Park,* there is a scene in which some members of the visiting group are trapped in the kitchen with dinosaurs outside. The paleontologist is pressing against the center of the door, trying to keep out the dinosaurs on the other side. The botanist throws herself against the door at the edge near the hinge. A pivotal point in the film is that she cannot reach a gun on the floor because she is trying to hold the door closed. If the paleontologist is pressing at the center of the door, and the botanist is pressing at the edge about 8 cm from the hinge, estimate how far the paleontologist would have to relocate in order to have a greater effect on keeping the door closed than both of them pushing together have in their original positions. (Question 6 is courtesy of Edward F. Redish. For more questions of this type, see www.physics.umd.edu/perg/.)
7. In some motorcycle races, the riders drive over small hills and the motorcycle becomes airborne for a short time. If the motorcycle racer keeps the throttle open while leaving the hill and going into the air, the motorcycle's nose tends to rise upwards. Why does this happen?
8. If you toss a textbook into the air, rotating it each time about one of the three axes perpendicular to it, you will find that it will not rotate smoothly about one of those axes. (Try placing a strong rubber band around the book before the toss so that it will stay closed.) The book's rotation is stable about those axes having the largest and smallest moments of inertia, but unstable about the axis of intermediate moment. Try this on your own to find the axis that has this intermediate moment of inertia.
9. Stars originate as large bodies of slowly rotating gas. Because of gravity, these clumps of gas slowly decrease in size. What happens to the angular speed of a star as it shrinks? Explain.
10. If a high jumper positions his body correctly when going over the bar, the center of gravity of the athlete may actually pass under the bar. (See Fig. CQ8.10.) Explain how this is possible.

Mark Dadswell/Getty Images

Figure CQ8.10

11. In a tape recorder, the tape is pulled past the read–write heads at a constant speed by the drive mechanism. Consider the reel from which the tape is pulled: As the tape is pulled off, the radius of the roll of remaining tape decreases. (a) How does the torque on the reel change with time? (b) If the tape mechanism is suddenly turned on so that the tape is quickly pulled with a large force, is the tape more likely to break when pulled from a nearly full reel or from a nearly empty reel?
12. (a) Give an example in which the net force acting on an object is zero, yet the net torque is nonzero. (b) Give an example in which the net torque acting on an object is zero, yet the net force is nonzero.
13. A ladder rests inclined against a wall. Would you feel safer climbing up the ladder if you were told that the floor was frictionless, but the wall was rough, or that the wall was frictionless, but the floor was rough? Justify your answer.
14. A cat usually lands on its feet regardless of the position from which it is dropped. A slow-motion film of a cat falling shows that the upper half of its body twists in one direction while the lower half twists in the opposite direction. (See Fig. CQ8.14.) Why does this type of rotation occur?

Agence Nature/Photo Researchers, Inc.

Figure CQ8.14

15. A solid disk and a hoop are simultaneously released from rest at the top of an incline and roll down without slipping. Which object reaches the bottom first? (a) The one that has the largest mass arrives first. (b) The one that has the largest radius arrives first. (c) The hoop arrives first. (d) The disk arrives first. (e) The hoop and the disk arrive at the same time.

16. A mouse is initially at rest on a horizontal turntable mounted on a frictionless, vertical axle. As the mouse begins to walk clockwise around the perimeter, which of the following statements *must* be true of the turntable? (a) It also turns clockwise. (b) It turns counterclockwise with the same angular velocity as the mouse. (c) It remains stationary. (d) It turns counterclockwise because angular momentum is conserved. (e) It turns clockwise because mechanical energy is conserved.

17. The cars in a soapbox derby have no engines; they simply coast downhill. Which of the following design criteria is best from a competitive point of view? The car's wheels should (a) have large moments of inertia, (b) be massive, (c) be hoop-like wheels rather than solid disks, (d) be large wheels rather than small wheels, or (e) have small moments of inertia.

PROBLEMS

ENHANCED WebAssign The problems in this chapter may be assigned online in Enhanced WebAssign.

1. denotes straightforward problem; **2.** denotes intermediate problem; **3.** denotes challenging problem

1. denotes full solution available in *Student Solutions Manual/Study Guide*

1. denotes problems most often assigned in Enhanced WebAssign

BIO denotes biomedical problems

GP denotes guided problems

M denotes Master It tutorial available in Enhanced WebAssign

Q|C denotes asking for quantitative and conceptual reasoning

S denotes symbolic reasoning problem

W denotes Watch It video solution available in Enhanced WebAssign

8.1 Torque

1. The fishing pole in Figure P8.1 makes an angle of 20.0° with the horizontal. What is the magnitude of the torque exerted by the fish about an axis perpendicular to the page and passing through the angler's hand if the fish pulls on the fishing line with a force $\vec{\mathbf{F}} = 100$ N at an angle 37.0° below the horizontal? The force is applied at a point 2.00 m from the angler's hands.

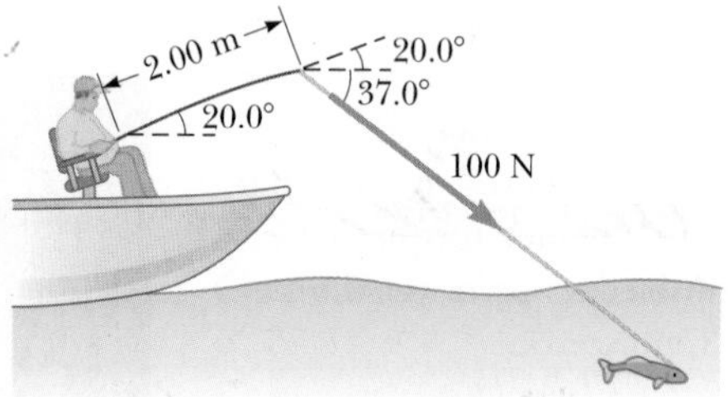

Figure P8.1

2. M Find the net torque on the wheel in Figure P8.2 about the axle through O perpendicular to the page, taking $a = 10.0$ cm and $b = 25.0$ cm.

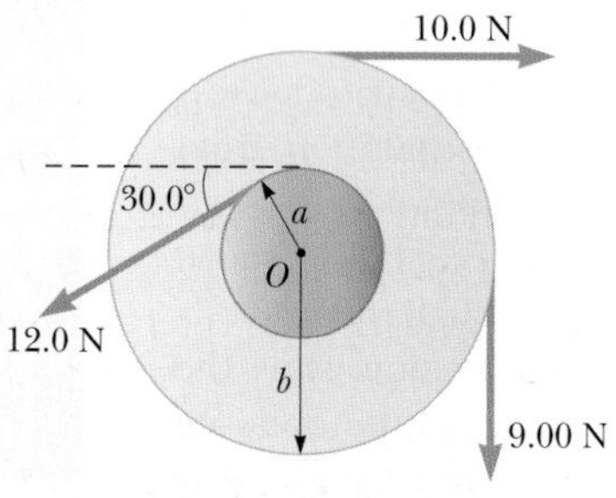

Figure P8.2

3. Calculate the net torque (magnitude and direction) on the beam in Figure P8.3 about (a) an axis through O perpendicular to the page and (b) an axis through C perpendicular to the page.

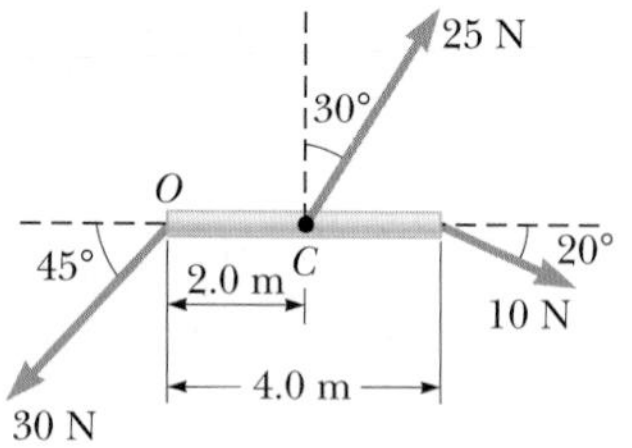

Figure P8.3

4. BIO A dental bracket exerts a horizontal force of 80.0 N on a tooth at point B in Figure P8.4. What is the torque on the root of the tooth about point A?

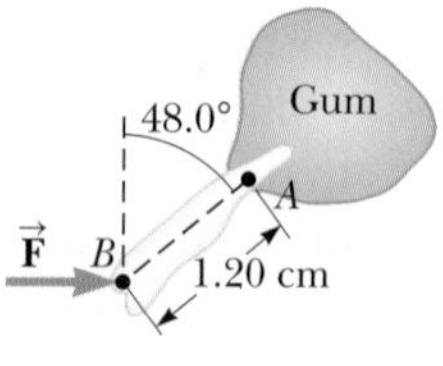

Figure P8.4

5. Q|C A simple pendulum consists of a small object of mass 3.0 kg hanging at the end of a 2.0-m-long light string that is connected to a pivot point. (a) Calculate the magnitude of the torque (due to the force of gravity) about this pivot point when the string makes a 5.0° angle with the vertical. (b) Does the torque increase or decrease as the angle increases? Explain.

6. S Write the necessary equations of equilibrium of the object shown in Figure P8.6. Take the origin of the torque equation about an axis perpendicular to the page through the point O.

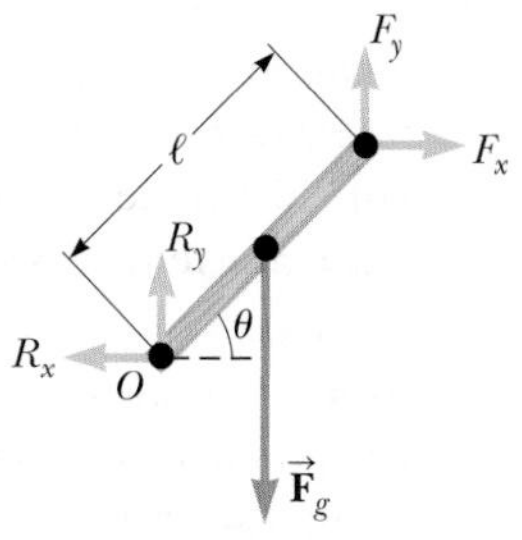

Figure P8.6

8.2 Torque and the Two Conditions for Equilibrium

8.3 The Center of Gravity

8.4 Examples of Objects in Equilibrium

7. BIO The arm in Figure P8.7 weighs 41.5 N. The force of gravity acting on the arm acts through point *A*. Determine the magnitudes of the tension force $\vec{\mathbf{F}}_t$ in the deltoid muscle and the force $\vec{\mathbf{F}}_s$ exerted by the shoulder on the humerus (upper-arm bone) to hold the arm in the position shown.

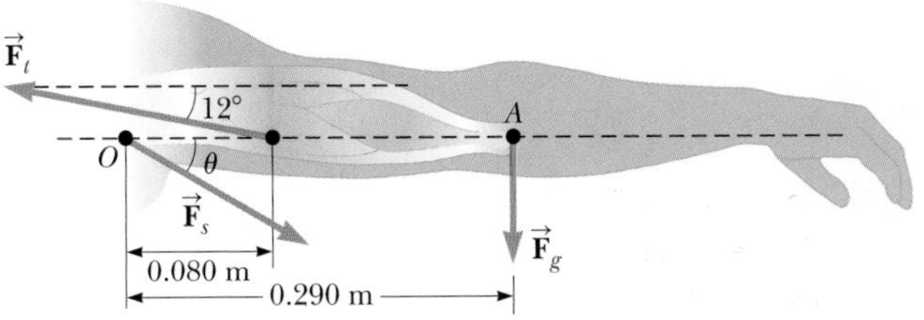

Figure P8.7

8. A uniform 35.0-kg beam of length ℓ = 5.00 m is supported by a vertical rope located *d* = 1.20 m from its left end as in Figure P8.8. The right end of the beam is supported by a vertical column. Find (a) the tension in the rope and (b) the force that the column exerts on the right end of the beam.

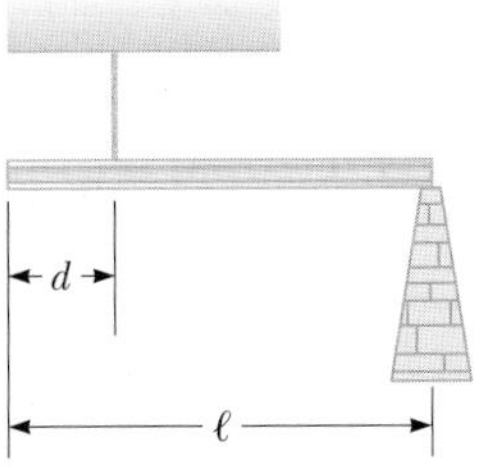

Figure P8.8

9. BIO A cook holds a 2.00-kg carton of milk at arm's length (Fig. P8.9). What force $\vec{\mathbf{F}}_B$ must be exerted by the biceps muscle? (Ignore the weight of the forearm.)

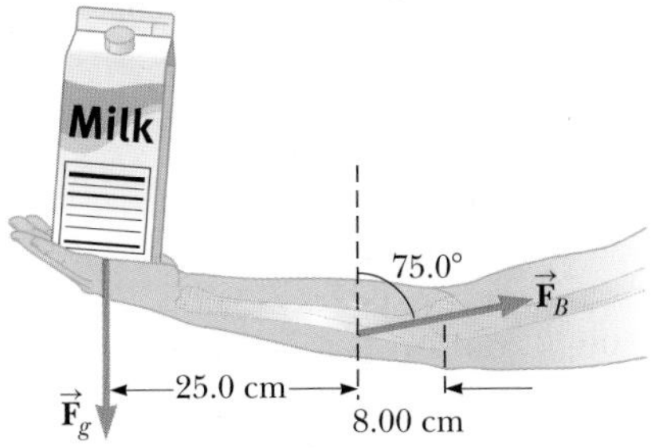

Figure P8.9

10. A meter stick is found to balance at the 49.7-cm mark when placed on a fulcrum. When a 50.0-gram mass is attached at the 10.0-cm mark, the fulcrum must be moved to the 39.2-cm mark for balance. What is the mass of the meter stick?

11. Find the *x*- and *y*-coordinates of the center of gravity of a 4.00-ft by 8.00-ft uniform sheet of plywood with the upper right quadrant removed as shown in Figure P8.11. *Hint:* The mass of any segment of the plywood sheet is proportional to the area of that segment.

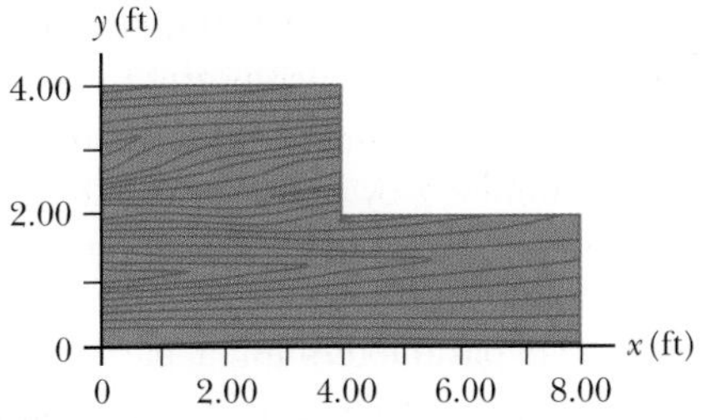

Figure P8.11

12. GP A beam resting on two pivots has a length of L = 6.00 m and mass M = 90.0 kg. The pivot under the left end exerts a normal force n_1 on the beam, and the second pivot placed a distance ℓ = 4.00 m from the left end exerts a normal force n_2. A woman of mass m = 55.0 kg steps onto the left end of the beam and begins walking to the right as in Figure P8.12. The goal is to find the woman's position when the beam begins to tip. (a) Sketch a free-body diagram, labeling the gravitational and normal forces acting on the beam and placing the woman *x* meters to the right of the first pivot, which is the origin. (b) Where is the woman when the normal force n_1 is the greatest? (c) What is n_1 when the beam is about to tip? (d) Use the force equation of equilibrium to find the value of n_2 when the beam is about to tip. (e) Using the result of part (c) and the torque equilibrium equation, with torques computed around the second pivot point, find the woman's position when the beam is about to tip. (f) Check the answer to part (e) by computing torques around the first pivot point. Except for possible slight differences due to rounding, is the answer the same?

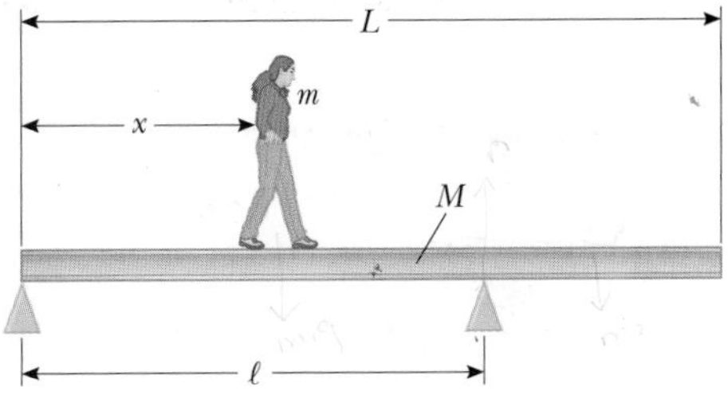

Figure P8.12 Problems 12 and 14.

13. Consider the following mass distribution, where *x*- and *y*-coordinates are given in meters: 5.0 kg at (0.0, 0.0) m, 3.0 kg at (0.0, 4.0) m, and 4.0 kg at (3.0, 0.0) m. Where should a fourth object of 8.0 kg be placed so that the center of gravity of the four-object arrangement will be at (0.0, 0.0) m?

14. S A beam of length L and mass M rests on two pivots. The first pivot is at the left end, taken as the origin, and the second pivot is at a distance ℓ from the left end. A woman of mass m starts at the left end and walks toward the right end as in Figure P8.12. When the beam is on the verge of tipping, find symbolic expressions for (a) the normal force exerted by the second pivot in terms of M, m, and g and (b) the woman's position in terms of M, m, L, and ℓ. (c) Find

the minimum value of ℓ that will allow the woman to reach the end of the beam without it tipping.

15. BIO Many of the elements in horizontal-bar exercises can be modeled by representing the gymnast by four segments consisting of arms, torso (including the head), thighs, and lower legs, as shown in Figure P8.15a. Inertial parameters for a particular gymnast are as follows:

Segment	Mass (kg)	Length (m)	r_{cg} (m)	I (kg · m^2)
Arms	6.87	0.548	0.239	0.205
Torso	33.57	0.601	0.337	1.610
Thighs	14.07	0.374	0.151	0.173
Legs	7.54	—	0.227	0.164

Note that in Figure P8.15a r_{cg} is the distance to the center of gravity measured from the joint closest to the bar and the masses for the arms, thighs, and legs include both appendages. I is the moment of inertia of each segment about its center of gravity. Determine the distance from the bar to the center of gravity of the gymnast for the two positions shown in Figures P8.15b and P8.15c.

Figure P8.15

16. BIO Using the data given in Problem 15 and the coordinate system shown in Figure P8.16b, calculate the position of the center of gravity of the gymnast shown in Figure P8.16a. Pay close attention to the definition of r_{cg} in the table.

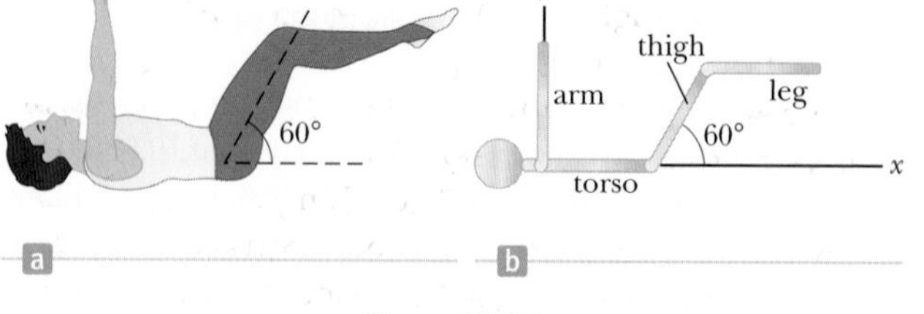

Figure P8.16

17. BIO A person bending forward to lift a load "with his back" (Fig. P8.17a) rather than "with his knees" can be injured by large forces exerted on the muscles and vertebrae. The spine pivots mainly at the fifth lumbar vertebra, with the principal supporting force provided by the erector spinalis muscle in the back. To see the magnitude of the forces involved, and to understand why back problems are common among humans, consider the model shown in Figure P8.17b of a person bending forward to lift a 200-N object. The spine and upper body are represented as a uniform horizontal rod of weight 350 N, pivoted at the base of the spine. The erector spinalis muscle, attached at a point two-thirds of the way up the spine, maintains the position of the back. The angle between the spine and this muscle is 12.0°. Find (a) the tension in the back muscle and (b) the compressional force in the spine.

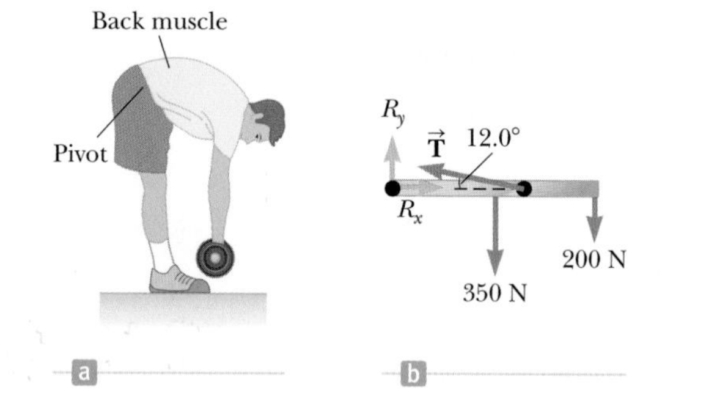

Figure P8.17

18. BIO When a person stands on tiptoe (a strenuous position), the position of the foot is as shown in Figure P8.18a. The total gravitational force on the body, $\vec{\mathbf{F}}_g$, is supported by the force $\vec{\mathbf{n}}$ exerted by the floor on the toes of one foot. A mechanical model of the situation is shown in Figure P8.18b, where $\vec{\mathbf{T}}$ is the force exerted by the Achilles tendon on the foot and $\vec{\mathbf{R}}$ is the force exerted by the tibia on the foot. Find the values of T, R, and θ when $F_g = n = 700$ N.

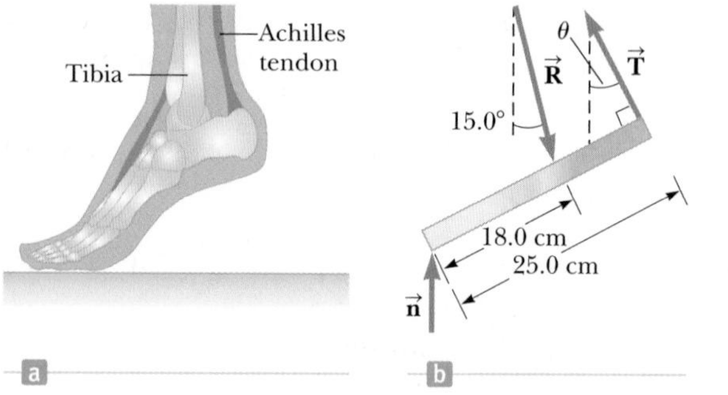

Figure P8.18

19. M A 500-N uniform rectangular sign 4.00 m wide and 3.00 m high is suspended from a horizontal, 6.00-m-long, uniform, 100-N rod as indicated in Figure P8.19. The left end of the rod is supported by a hinge, and the right end is supported by a thin cable making a 30.0° angle with the vertical. (a) Find the tension T in the cable. (b) Find the horizontal and vertical components of force exerted on the left end of the rod by the hinge.

Figure P8.19

20. A window washer is standing on a scaffold supported by a vertical rope at each end. The scaffold weighs 200 N and is 3.00 m long. What is the tension in each

rope when the 700-N worker stands 1.00 m from one end?

21. W A uniform plank of length 2.00 m and mass 30.0 kg is supported by three ropes, as indicated by the blue vectors in Figure P8.21. Find the tension in each rope when a 700-N person is $d = 0.500$ m from the left end.

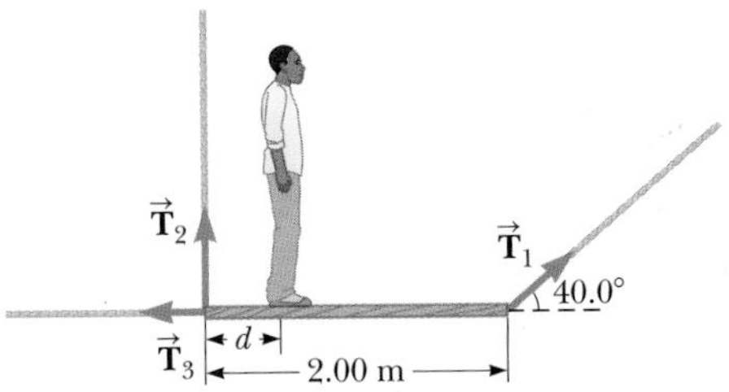

Figure P8.21

22. A hungry bear weighing 700 N walks out on a beam in an attempt to retrieve a basket of goodies hanging at the end of the beam (Fig. P8.22). The beam is uniform, weighs 200 N, and is 6.00 m long, and it is supported by a wire at an angle of $\theta = 60.0°$. The basket weighs 80.0 N. (a) Draw a force diagram for the beam. (b) When the bear is at $x = 1.00$ m, find the tension in the wire supporting the beam and the components of the force exerted by the wall on the left end of the beam. (c) If the wire can withstand a maximum tension of 900 N, what is the maximum distance the bear can walk before the wire breaks?

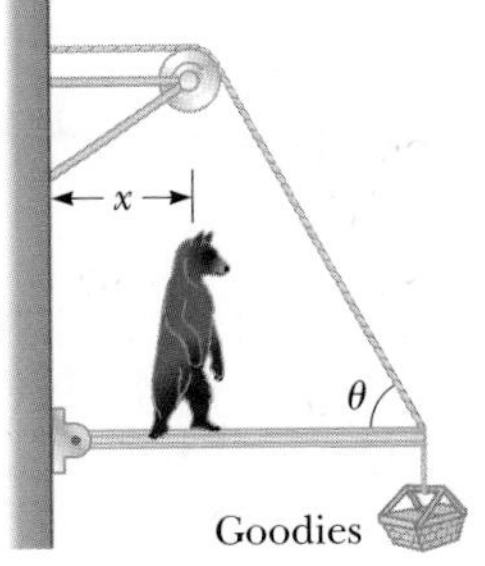

Figure P8.22

23. S Figure P8.23 shows a uniform beam of mass m pivoted at its lower end, with a horizontal spring attached between its top end and a vertical wall. The beam makes an angle θ with the horizontal. Find expressions for (a) the distance d the spring is stretched from equilibrium and (b) the components of the force exerted by the pivot on the beam.

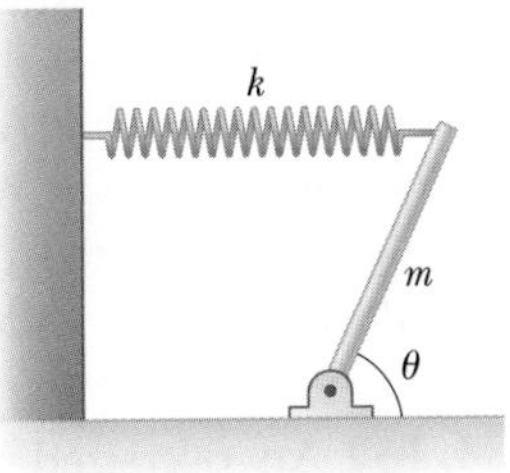

Figure P8.23

24. GP A strut of length $L = 3.00$ m and mass $m = 16.0$ kg is held by a cable at an angle of $\theta = 30.0°$ with respect to the horizontal as shown in Figure P8.24. (a) Sketch a force diagram, indicating all the forces and their placement on the strut. (b) Why is the hinge a good place to use for calculating torques? (c) Write the condition for rotational equilibrium symbolically, calculating the torques around the hinge. (d) Use the torque equation to calculate the tension in the cable. (e) Write the x- and y-components of Newton's second law for equilibrium. (f) Use the force equation to find the x- and y-components of the force on the hinge. (g) Assuming the strut position is to remain the same, would it be advantageous to attach the cable higher up on the wall? Explain the benefit in terms of the force on the hinge and cable tension.

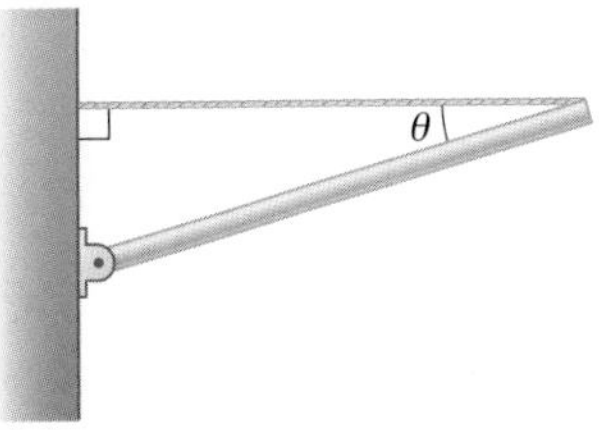

Figure P8.24

25. S A refrigerator of width w and height h rests on a rough incline as in Figure P8.25. Find an expression for the maximum value θ can have before the refrigerator tips over. Note, the contact point between the refrigerator and incline shifts as θ increases and treat the refrigerator as a uniform box.

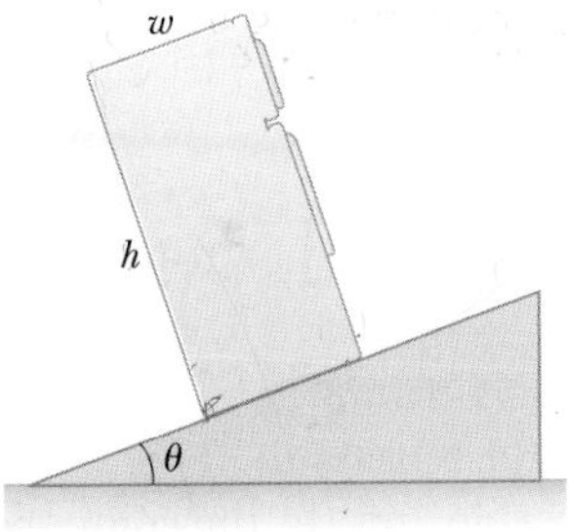

Figure P8.25

26. Q|C S A uniform beam of length L and mass m shown in Figure P8.26 is inclined at an angle θ to the horizontal. Its upper end is connected to a wall by a rope, and its lower end rests on a rough horizontal surface. The coefficient of static friction between the beam and surface is μ_s. Assume the angle is such that the static friction force is at its maximum value. (a) Draw a force diagram for the beam. (b) Using the condition of rotational equilibrium, find an expression for the tension T in the rope in terms of m, g, and θ. (c) Using Newton's second law for equilibrium, find a second expression for T in terms of μ_s, m, and g. (d) Using the foregoing results, obtain a relationship involving only μ_s and the angle θ. (e) What happens if the angle gets smaller? Is this equation valid for all values of θ? Explain.

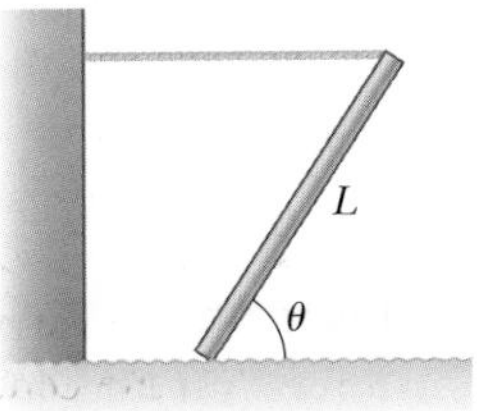

Figure P8.26

27. BIO The chewing muscle, the masseter, is one of the strongest in the human body. It is attached to the mandible (lower jawbone) as shown in Figure P8.27a. The jawbone is pivoted about a socket just in front of the auditory canal. The forces acting on the jawbone are equivalent to those acting on the curved bar in

Figure P8.27b. $\vec{\mathbf{F}}_C$ is the force exerted by the food being chewed against the jawbone, $\vec{\mathbf{T}}$ is the force of tension in the masseter, and $\vec{\mathbf{R}}$ is the force exerted by the socket on the mandible. Find $\vec{\mathbf{T}}$ and $\vec{\mathbf{R}}$ for a person who bites down on a piece of steak with a force of 50.0 N.

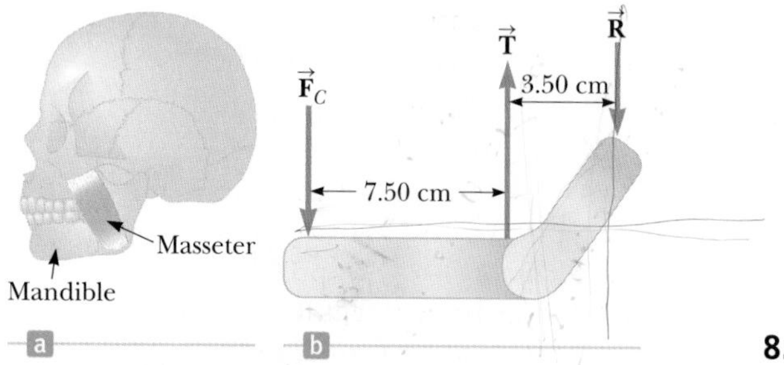

Figure P8.27

28. A 1200-N uniform boom at $\phi = 65°$ to the horizontal is supported by a cable at an angle $\theta = 25.0°$ to the horizontal as shown in Figure P8.28. The boom is pivoted at the bottom, and an object of weight $w = 2\,000$ N hangs from its top. Find (a) the tension in the support cable and (b) the components of the reaction force exerted by the pivot on the boom.

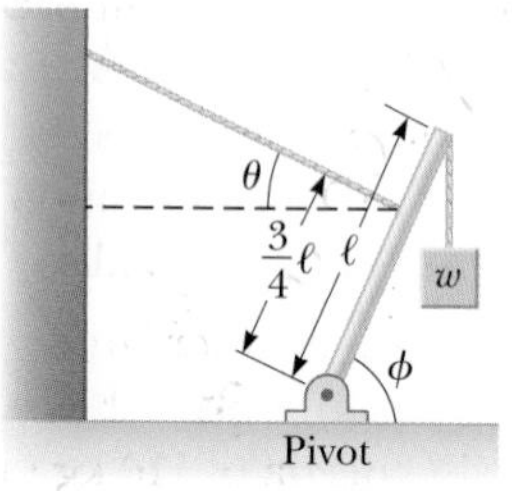

Figure P8.28

29. **BIO** The large quadriceps muscle in the upper leg terminates at its lower end in a tendon attached to the upper end of the tibia (Fig. P8.29a). The forces on the lower leg when the leg is extended are modeled as in Figure P8.29b, where $\vec{\mathbf{T}}$ is the force of tension in the tendon, $\vec{\mathbf{w}}$ is the force of gravity acting on the lower leg, and $\vec{\mathbf{F}}$ is the force of gravity acting on the foot. Find $\vec{\mathbf{T}}$ when the tendon is at an angle of 25.0° with the tibia, assuming that $w = 30.0$ N, $F = 12.5$ N, and the leg is extended at an angle θ of 40.0° with the vertical. Assume that the center of gravity of the lower leg is at its center and that the tendon attaches to the lower leg at a point one-fifth of the way down the leg.

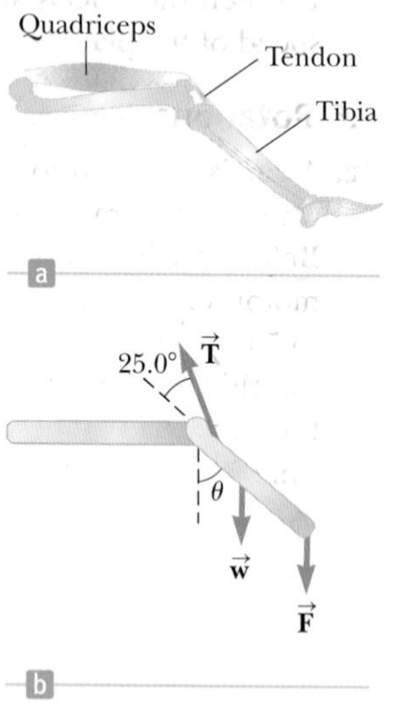

Figure P8.29

30. One end of a uniform 4.0-m-long rod of weight w is supported by a cable at an angle of $\theta = 37°$ with the rod. The other end rests against a wall, where it is held by friction. (See Fig. P8.30.) The coefficient of static friction between the wall and the rod is $\mu_s = 0.50$. Determine the minimum distance x from point A at which an additional weight w (the same as the weight of the rod) can be hung without causing the rod to slip at point A.

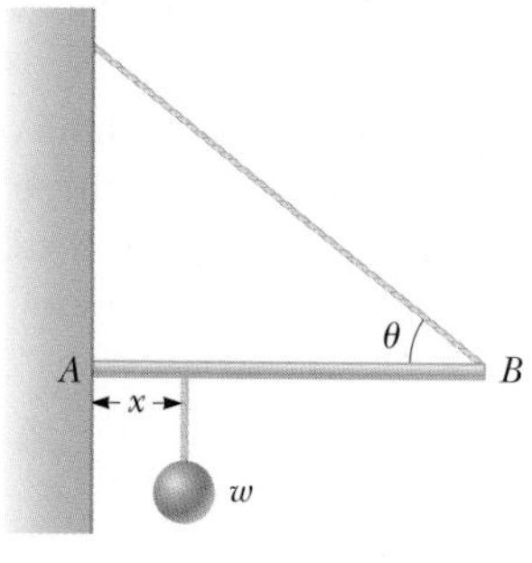

Figure P8.30

8.5 Relationship Between Torque and Angular Acceleration

31. Four objects are held in position at the corners of a rectangle by light rods as shown in Figure P8.31. Find the moment of inertia of the system about (a) the x-axis, (b) the y-axis, and (c) an axis through O and perpendicular to the page.

32. If the system shown in Figure P8.31 is set in rotation about each of the axes mentioned in Problem 31, find the torque that will produce an angular acceleration of 1.50 rad/s^2 in each case.

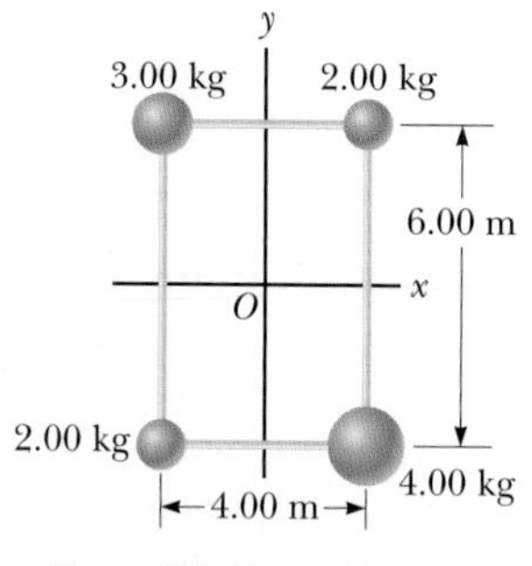

Figure P8.31 Problems 31 and 32.

33. A large grinding wheel in the shape of a solid cylinder of radius 0.330 m is free to rotate on a frictionless, vertical axle. A constant tangential force of 250 N applied to its edge causes the wheel to have an angular acceleration of 0.940 rad/s^2. (a) What is the moment of inertia of the wheel? (b) What is the mass of the wheel? (c) If the wheel starts from rest, what is its angular velocity after 5.00 s have elapsed, assuming the force is acting during that time?

34. **GP** An oversized yo-yo is made from two identical solid disks each of mass $M = 2.00$ kg and radius $R = 10.0$ cm. The two disks are joined by a solid cylinder of radius $r = 4.00$ cm and mass $m = 1.00$ kg as in Figure P8.34. Take the center of the cylinder as the axis of the system, with positive torques directed to the left along this axis. All torques and angular variables are to be calculated around this axis. Light string is wrapped around the cylinder, and the system is then allowed to drop from rest. (a) What is the moment of inertia of the system? Give a symbolic answer. (b) What torque does gravity exert on the system with respect to the given axis? (c) Take downward as the negative coordinate direction. As depicted in Figure P8.34, is the torque exerted by the tension positive or negative? Is the angular acceleration positive or negative? What about

the translational acceleration? (d) Write an equation for the angular acceleration α in terms of the translational acceleration a and radius r. (Watch the sign!) (e) Write Newton's second law for the system in terms of m, M, a, T, and g. (f) Write Newton's second law for rotation in terms of I, α, T, and r. (g) Eliminate α from the rotational second law with the expression found in part (d) and find a symbolic expression for the acceleration a in terms of m, M, g, r and R. (h) What is the numeric value for the system's acceleration? (i) What is the tension in the string? (j) How long does it take the system to drop 1.00 m from rest?

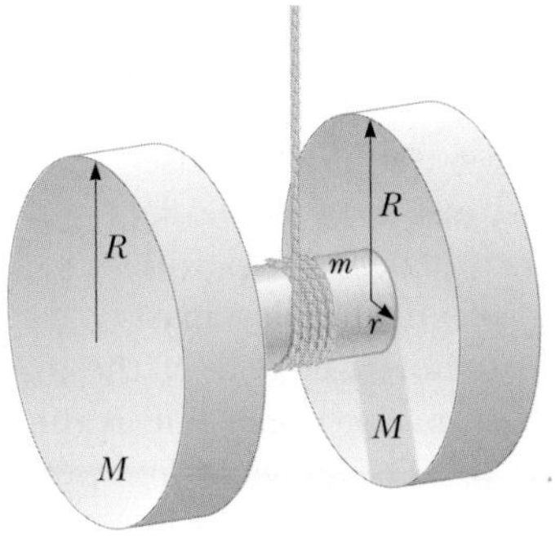

Figure P8.34

35. Q|C A rope of negligible mass is wrapped around a 225-kg solid cylinder of radius 0.400 m. The cylinder is suspended several meters off the ground with its axis oriented horizontally, and turns on that axis without friction. (a) If a 75.0-kg man takes hold of the free end of the rope and falls under the force of gravity, what is his acceleration? (b) What is the angular acceleration of the cylinder? (c) If the mass of the rope were not neglected, what would happen to the angular acceleration of the cylinder as the man falls?

36. W A potter's wheel having a radius of 0.50 m and a moment of inertia of 12 kg · m^2 is rotating freely at 50 rev/min. The potter can stop the wheel in 6.0 s by pressing a wet rag against the rim and exerting a radially inward force of 70 N. Find the effective coefficient of kinetic friction between the wheel and the wet rag.

37. A model airplane with mass 0.750 kg is tethered by a wire so that it flies in a circle 30.0 m in radius. The airplane engine provides a net thrust of 0.800 N perpendicular to the tethering wire. (a) Find the torque the net thrust produces about the center of the circle. (b) Find the angular acceleration of the airplane when it is in level flight. (c) Find the linear acceleration of the airplane tangent to its flight path.

38. A bicycle wheel has a diameter of 64.0 cm and a mass of 1.80 kg. Assume that the wheel is a hoop with all the mass concentrated on the outside radius. The bicycle is placed on a stationary stand, and a resistive force of 120 N is applied tangent to the rim of the tire. (a) What force must be applied by a chain passing over a 9.00-cm-diameter sprocket in order to give the wheel an acceleration of 4.50 rad/s^2? (b) What force is required if you shift to a 5.60-cm-diameter sprocket?

39. M A 150-kg merry-go-round in the shape of a uniform, solid, horizontal disk of radius 1.50 m is set in motion by wrapping a rope about the rim of the disk and pulling on the rope. What constant force must be exerted on the rope to bring the merry-go-round from rest to an angular speed of 0.500 rev/s in 2.00 s?

40. Q|C S An Atwood's machine consists of blocks of masses m_1 = 10.0 kg and m_2 = 20.0 kg attached by a cord running over a pulley as in Figure P8.40. The pulley is a solid cylinder with mass M = 8.00 kg and radius r = 0.200 m. The block of mass m_2 is allowed to drop, and the cord turns the pulley without slipping. (a) Why must the tension T_2 be greater than the tension T_1? (b) What is the acceleration of the system, assuming the pulley axis is frictionless? (c) Find the tensions T_1 and T_2.

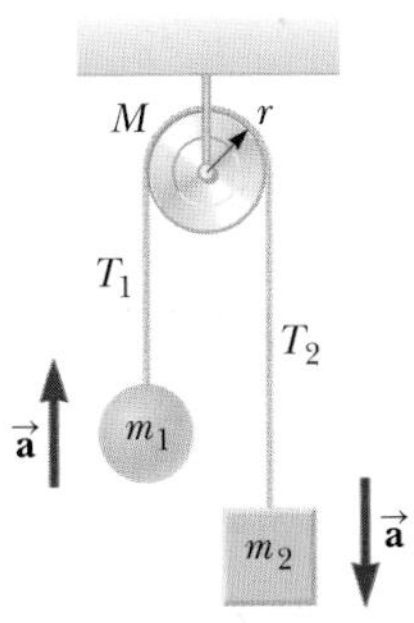

Figure P8.40

41. An airliner lands with a speed of 50.0 m/s. Each wheel of the plane has a radius of 1.25 m and a moment of inertia of 110 kg · m^2. At touchdown, the wheels begin to spin under the action of friction. Each wheel supports a weight of 1.40 × 10^4 N, and the wheels attain their angular speed in 0.480 s while rolling without slipping. What is the coefficient of kinetic friction between the wheels and the runway? Assume that the speed of the plane is constant.

8.6 Rotational Kinetic Energy

42. A car is designed to get its energy from a rotating flywheel with a radius of 2.00 m and a mass of 500 kg. Before a trip, the flywheel is attached to an electric motor, which brings the flywheel's rotational speed up to 5 000 rev/min. (a) Find the kinetic energy stored in the flywheel. (b) If the flywheel is to supply energy to the car as a 10.0-hp motor would, find the length of time the car could run before the flywheel would have to be brought back up to speed.

43. A horizontal 800-N merry-go-round of radius 1.50 m is started from rest by a constant horizontal force of 50.0 N applied tangentially to the merry-go-round. Find the kinetic energy of the merry-go-round after 3.00 s. (Assume it is a solid cylinder.)

44. Q|C Four objects—a hoop, a solid cylinder, a solid sphere, and a thin, spherical shell—each have a mass of 4.80 kg and a radius of 0.230 m. (a) Find the moment of inertia for each object as it rotates about the axes shown in Table 8.1. (b) Suppose each object is rolled down a ramp. Rank the translational speed of each object from highest to lowest. (c) Rank the objects' rotational kinetic energies from highest to lowest as the objects roll down the ramp.

45. A light rod of length $\ell = 1.00$ m rotates about an axis perpendicular to its length and passing through its center as in Figure P8.45. Two particles of masses $m_1 = 4.00$ kg and $m_2 = 3.00$ kg are connected to the ends of the rod. (a) Neglecting the mass of the rod, what is the system's kinetic energy when its angular speed is 2.50 rad/s? (b) Repeat the problem, assuming the mass of the rod is taken to be 2.00 kg.

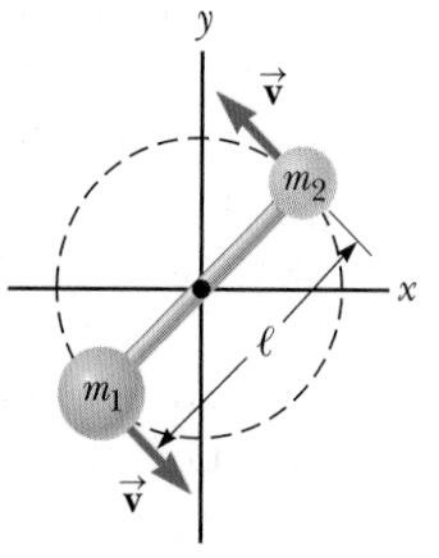

Figure P8.45 Problems 45 and 57.

46. A 240-N sphere 0.20 m in radius rolls without slipping 6.0 m down a ramp that is inclined at 37° with the horizontal. What is the angular speed of the sphere at the bottom of the slope if it starts from rest?

47. A solid, uniform disk of radius 0.250 m and mass 55.0 kg rolls down a ramp of length 4.50 m that makes an angle of 15.0° with the horizontal. The disk starts from rest from the top of the ramp. Find (a) the speed of the disk's center of mass when it reaches the bottom of the ramp and (b) the angular speed of the disk at the bottom of the ramp.

48. Q|C S A solid uniform sphere of mass m and radius R rolls without slipping down an incline of height h. (a) What forms of mechanical energy are associated with the sphere at any point along the incline when its angular speed is ω? Answer in words and symbolically in terms of the quantities m, g, y, I, ω, and v. (b) What force acting on the sphere causes it to roll rather than slip down the incline? (c) Determine the ratio of the sphere's rotational kinetic energy to its total kinetic energy at any instant.

49. The top in Figure P8.49 has a moment of inertia of 4.00×10^{-4} kg · m^2 and is initially at rest. It is free to rotate about a stationary axis AA'. A string wrapped around a peg along the axis of the top is pulled in such a manner as to maintain a constant tension of 5.57 N in the string. If the string does not slip while wound around the peg, what is the angular speed of the top after 80.0 cm of string has been pulled off the peg? *Hint:* Consider the work that is done.

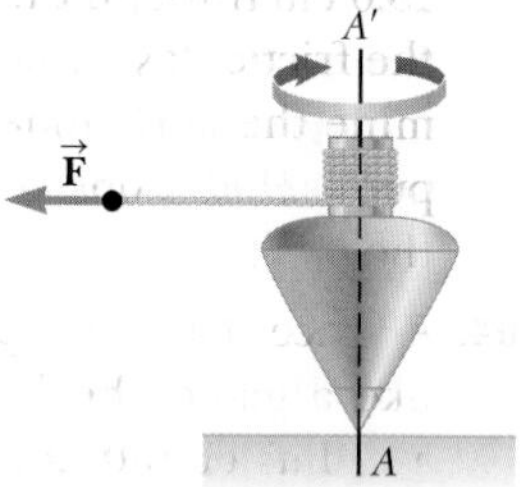

Figure P8.49

50. A constant torque of 25.0 N · m is applied to a grindstone whose moment of inertia is 0.130 kg · m^2. Using energy principles and neglecting friction, find the angular speed after the grindstone has made 15.0 revolutions. *Hint:* The angular equivalent of $W_{\text{net}} = F\Delta x = \frac{1}{2}mv_f^2 - \frac{1}{2}mv_i^2$ is $W_{\text{net}} = \tau\Delta\theta = \frac{1}{2}I\omega_f^2 - \frac{1}{2}I\omega_i^2$. You should convince yourself that this last relationship is correct.

51. A 10.0-kg cylinder rolls without slipping on a rough surface. At an instant when its center of gravity has a speed of 10.0 m/s, determine (a) the translational kinetic energy of its center of gravity, (b) the rotational kinetic energy about its center of gravity, and (c) its total kinetic energy.

52. W Use conservation of energy to determine the angular speed of the spool shown in Figure P8.52 after the 3.00-kg bucket has fallen 4.00 m, starting from rest. The light string attached to the bucket is wrapped around the spool and does not slip as it unwinds.

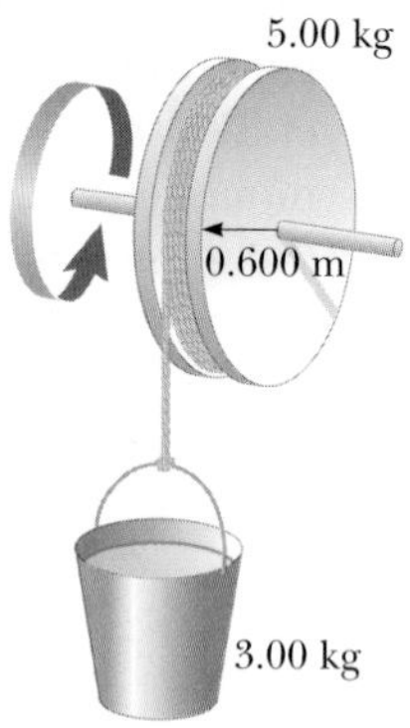

Figure P8.52

53. Q|C A giant swing at an amusement park consists of a 365-kg uniform arm 10.0 m long, with two seats of negligible mass connected at the lower end of the arm (Fig. P8.53). (a) How far from the upper end is the center of mass of the arm? (b) The gravitational potential energy of the arm is the same as if all its mass were concentrated at the center of mass. If the arm is raised through a 45.0° angle, find the gravitational potential energy, where the zero level is taken to be 10.0 m below the axis. (c) The arm drops from rest from the position described in part (b). Find the gravitational potential energy of the system when it reaches the vertical orientation. (d) Find the speed of the seats at the bottom of the swing.

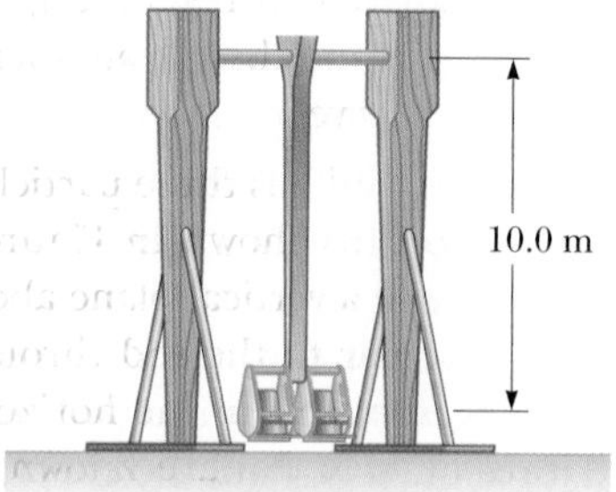

Figure P8.53

8.7 Angular Momentum

54. Each of the following objects has a radius of 0.180 m and a mass of 2.40 kg, and each rotates about an axis through its center (as in Table 8.1) with an angular speed of 35.0 rad/s. Find the magnitude of the angular momentum of each object. (a) a hoop (b) a solid cylinder (c) a solid sphere (d) a hollow spherical shell

55. (a) Calculate the angular momentum of Earth that arises from its spinning motion on its axis, treating Earth as a uniform solid sphere. (b) Calculate the angular momentum of Earth that arises from its orbital motion about the Sun, treating Earth as a point particle.

56. QIC A 0.005 00-kg bullet traveling horizontally with a speed of 1.00×10^3 m/s enters an 18.0-kg door, embedding itself 10.0 cm from the side opposite the hinges as in Figure P8.56. The 1.00-m-wide door is free to swing on its hinges. (a) Before it hits the door, does the bullet have angular momentum relative the door's axis of rotation? Explain. (b) Is mechanical energy conserved in this collision? Answer without doing a calculation. (c) At what angular speed does the door swing open immediately after the collision? (The door has the same moment of inertia as a rod with axis at one end.) (d) Calculate the energy of the door–bullet system and determine whether it is less than or equal to the kinetic energy of the bullet before the collision.

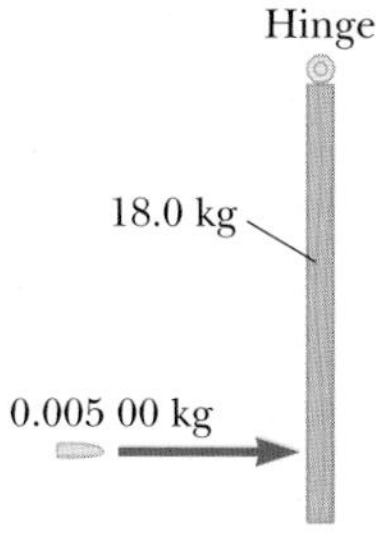

Figure P8.56. An overhead view of a bullet striking a door.

57. A light rigid rod of length $\ell = 1.00$ m rotates about an axis perpendicular to its length and through its center, as shown in Figure P8.45. Two particles of masses $m_1 = 4.00$ kg and $m_2 = 3.00$ kg are connected to the ends of the rod. What is the angular momentum of the system if the speed of each particle is 5.00 m/s? (Neglect the rod's mass.)

58. Halley's comet moves about the Sun in an elliptical orbit, with its closest approach to the Sun being 0.59 A.U. and its greatest distance being 35 A.U. (1 A.U. is the Earth–Sun distance). If the comet's speed at closest approach is 54 km/s, what is its speed when it is farthest from the Sun? You may neglect any change in the comet's mass and assume that its angular momentum about the Sun is conserved.

59. S A rigid, massless rod has three particles with equal masses attached to it as shown in Figure P8.59. The rod is free to rotate in a vertical plane about a frictionless axle perpendicular to the rod through the point P and is released from rest in the horizontal position at $t = 0$. Assuming m and d are known, find (a) the moment of inertia of the system (rod plus particles) about the pivot, (b) the torque acting on the system at $t = 0$, (c) the angular acceleration of the system at $t = 0$, (d) the linear acceleration of the particle labeled 3 at $t = 0$, (e) the maximum kinetic energy of the system, (f) the maximum angular speed reached by the rod, (g) the maximum angular momentum of the system, and (h) the maximum translational speed reached by the particle labeled 2.

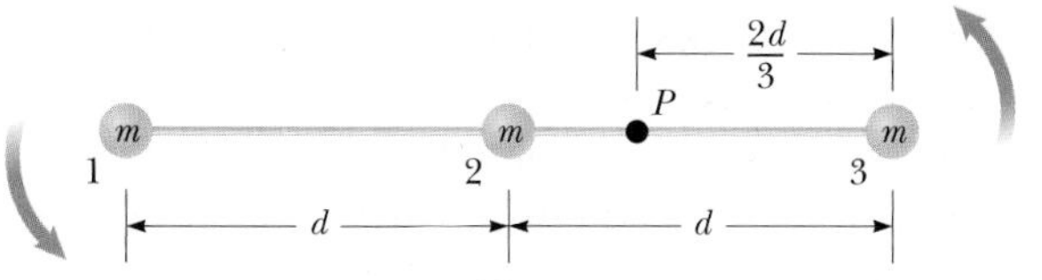

Figure P8.59

60. M A 60.0-kg woman stands at the rim of a horizontal turntable having a moment of inertia of 500 kg · m^2 and a radius of 2.00 m. The turntable is initially at rest and is free to rotate about a frictionless, vertical axle through its center. The woman then starts walking around the rim clockwise (as viewed from above the system) at a constant speed of 1.50 m/s relative to Earth. (a) In what direction and with what angular speed does the turntable rotate? (b) How much work does the woman do to set herself and the turntable into motion?

61. A solid, horizontal cylinder of mass 10.0 kg and radius 1.00 m rotates with an angular speed of 7.00 rad/s about a fixed vertical axis through its center. A 0.250-kg piece of putty is dropped vertically onto the cylinder at a point 0.900 m from the center of rotation and sticks to the cylinder. Determine the final angular speed of the system.

62. A student sits on a rotating stool holding two 3.0-kg objects. When his arms are extended horizontally, the objects are 1.0 m from the axis of rotation and he rotates with an angular speed of 0.75 rad/s. The moment of inertia of the student plus stool is 3.0 kg · m^2 and is assumed to be constant. The student then pulls in the objects horizontally to 0.30 m from the rotation axis. (a) Find the new angular speed of the student. (b) Find the kinetic energy of the student before and after the objects are pulled in.

63. W The puck in Figure P8.63 has a mass of 0.120 kg. Its original distance from the center of rotation is 40.0 cm, and it moves with a speed of 80.0 cm/s. The string is pulled downward 15.0 cm through the hole in the frictionless table. Determine the work done on the puck. *Hint:* Consider the change in kinetic energy of the puck.

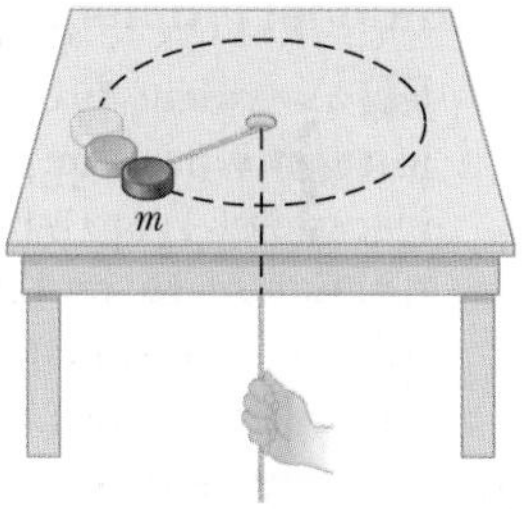

Figure P8.63

64. A space station shaped like a giant wheel has a radius of 100 m and a moment of inertia of 5.00×10^8 kg · m^2. A crew of 150 lives on the rim, and the station is rotating so that the crew experiences an apparent acceleration of 1g (Fig. P8.64). When 100 people move to the center of the station for a union meeting, the angular speed changes. What apparent acceleration is experienced by the managers remaining at the rim? Assume the average mass of a crew member is 65.0 kg.

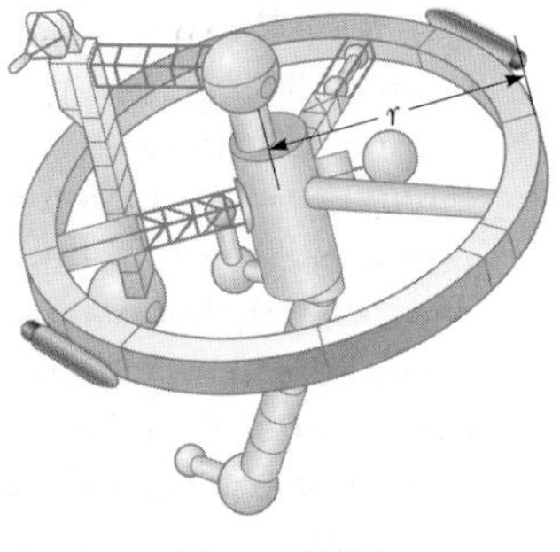

Figure P8.64

65. QIC S A cylinder with moment of inertia I_1 rotates with angular velocity ω_0 about a frictionless vertical axle. A second cylinder, with moment of inertia I_2, initially not rotating, drops onto the first cylinder (Fig. P8.65). Because the surfaces are rough, the two cylinders eventually reach the same angular speed ω. (a) Calculate ω. (b) Show that kinetic energy is lost in this situation, and calculate the ratio of the final to the initial kinetic energy.

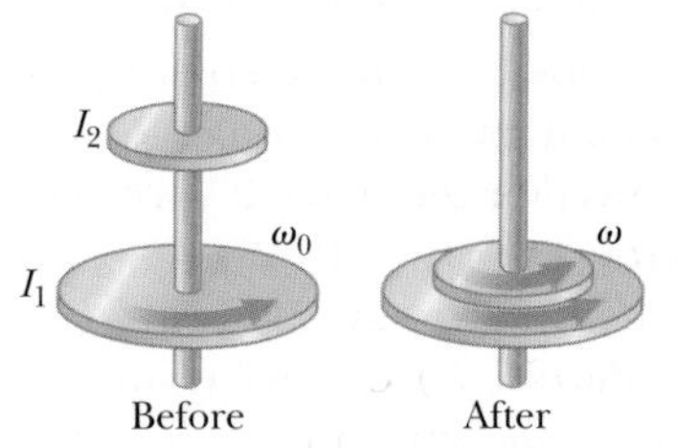

Figure P8.65

66. M A particle of mass 0.400 kg is attached to the 100-cm mark of a meter stick of mass 0.100 kg. The meter stick rotates on a horizontal, frictionless table with an angular speed of 4.00 rad/s. Calculate the angular momentum of the system when the stick is pivoted about an axis (a) perpendicular to the table through the 50.0-cm mark and (b) perpendicular to the table through the 0-cm mark.

Additional Problems

67. GP A typical propeller of a turbine used to generate electricity from the wind consists of three blades as in Figure P8.67. Each blade has a length of $L = 35$ m and a mass of $m = 420$ kg. The propeller rotates at the rate of 25 rev/min. (a) Convert the angular speed of the propeller to units of rad/s. Find (b) the moment of inertia of the propeller about the axis of rotation and (c) the total kinetic energy of the propeller.

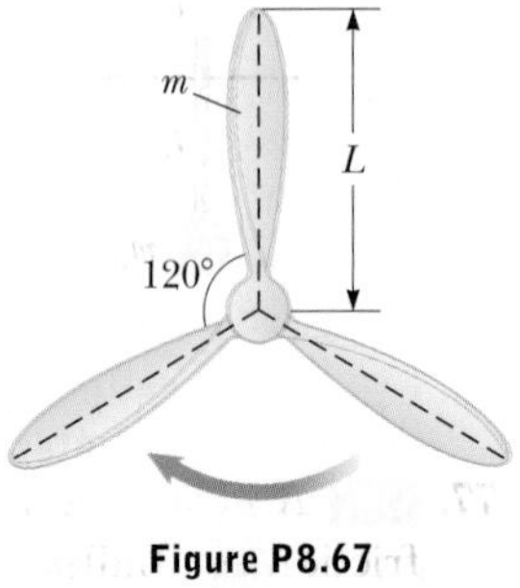

Figure P8.67

68. Figure P8.68 shows a claw-hammer as it is being used to pull a nail out of a horizontal board. If a force of magnitude 150 N is exerted horizontally as shown, find (a) the force exerted by the hammer claws on the nail and (b) the force exerted by the surface at the point of contact with the hammer head. Assume that the force the hammer exerts on the nail is parallel to the nail and perpendicular to the position vector from the point of contact.

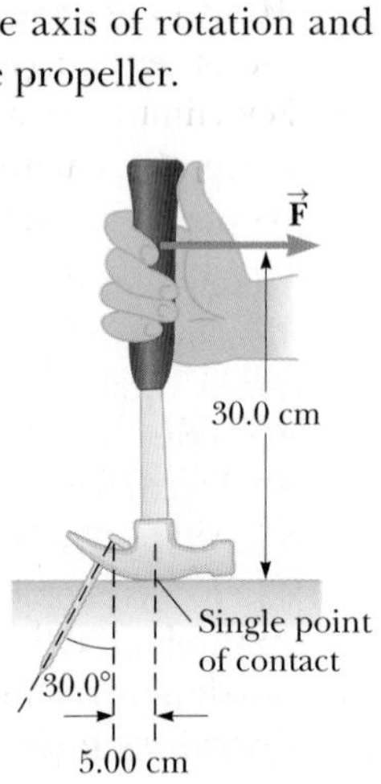

Figure P8.68

69. QIC A 40.0-kg child stands at one end of a 70.0-kg boat that is 4.00 m long (Fig. P8.69). The boat is initially 3.00 m from the pier. The child notices a turtle on a rock beyond the far end of the boat and proceeds to walk to that end to catch the turtle. (a) Neglecting friction between the boat and water, describe the motion of the system (child plus boat). (b) Where will the child be relative to the pier when he reaches the far end of the boat? (c) Will he catch the turtle? (Assume that he can reach out 1.00 m from the end of the boat.)

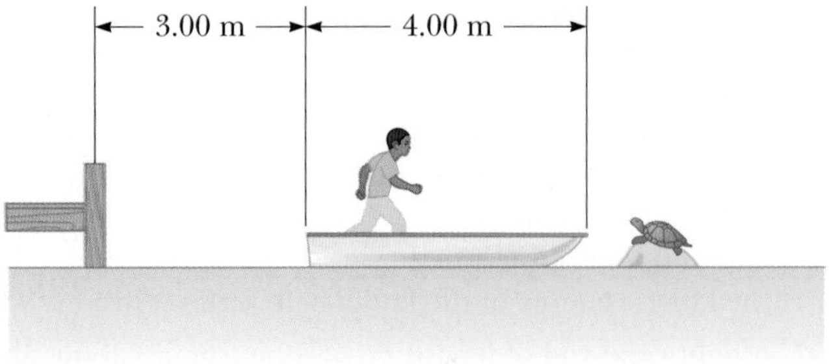

Figure P8.69

70. An object of mass $M = 12.0$ kg is attached to a cord that is wrapped around a wheel of radius $r = 10.0$ cm (Fig. P8.70). The acceleration of the object down the frictionless incline is measured to be $a = 2.00\ \text{m/s}^2$ and the incline makes an angle $\theta = 37.0°$ with the horizontal. Assuming the axle of the wheel to be frictionless, determine (a) the tension in the rope, (b) the moment of inertia of the wheel, and (c) the angular speed of the wheel 2.00 s after it begins rotating, starting from rest.

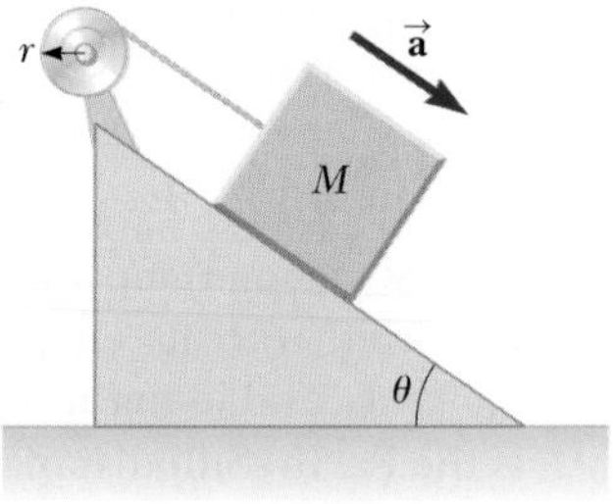

Figure P8.70

71. A uniform ladder of length L and weight w is leaning against a vertical wall. The coefficient of static friction between the ladder and the floor is the same as that between the ladder and the wall. If this coefficient of static friction is $\mu_s = 0.500$, determine the smallest angle the ladder can make with the floor without slipping.

72. Two astronauts (Fig. P8.72), each having a mass of 75.0 kg, are connected by a 10.0-m rope of negligible mass. They are isolated in space, moving in circles around the point halfway between them at a speed of 5.00 m/s. Treating the astronauts as particles, calculate (a) the magnitude of the angular momentum and (b) the rotational energy of the system. By pulling on the rope, the astronauts shorten the distance between them to 5.00 m. (c) What is the new angular momentum of the system? (d) What are their new speeds? (e) What is the new rotational energy of the system? (f) How much work is done by the astronauts in shortening the rope?

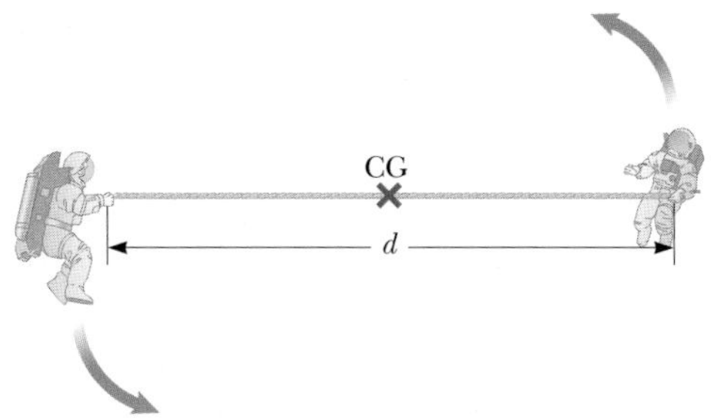

Figure P8.72 Problems 72 and 73.

73. **S** ***This is a symbolic version of problem 72.*** Two astronauts (Fig. P8.72), each having a mass M, are connected by a rope of length d having negligible mass. They are isolated in space, moving in circles around the point halfway between them at a speed v. (a) Calculate the magnitude of the angular momentum of the system by treating the astronauts as particles. (b) Calculate the rotational energy of the system. By pulling on the rope, the astronauts shorten the distance between them to $d/2$. (c) What is the new angular momentum of the system? (d) What are their new speeds? (e) What is the new rotational energy of the system? (f) How much work is done by the astronauts in shortening the rope?

74. Two window washers, Bob and Joe, are on a 3.00-m-long, 345-N scaffold supported by two cables attached to its ends. Bob weighs 750 N and stands 1.00 m from the left end, as shown in Figure P8.74. Two meters from the left end is the 500-N washing equipment. Joe is 0.500 m from the right end and weighs 1 000 N. Given that the scaffold is in rotational and translational equilibrium, what are the forces on each cable?

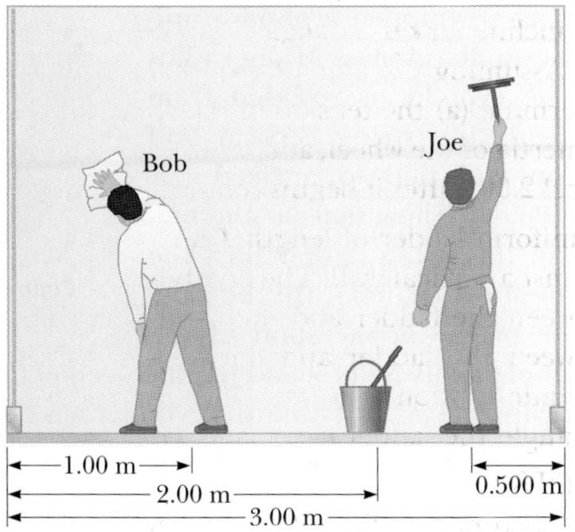

Figure P8.74

75. A 2.35-kg uniform bar of length $\ell = 1.30$ m is held in a horizontal position by three vertical springs as in Figure P8.75. The two lower springs are compressed and exert *upward* forces on the bar of magnitude $F_1 = 6.80$ N and $F_2 = 9.50$ N, respectively. Find (a) the force F_s exerted by the top spring on the bar, and (b) the location x of the upper spring that will keep the bar in equilibrium.

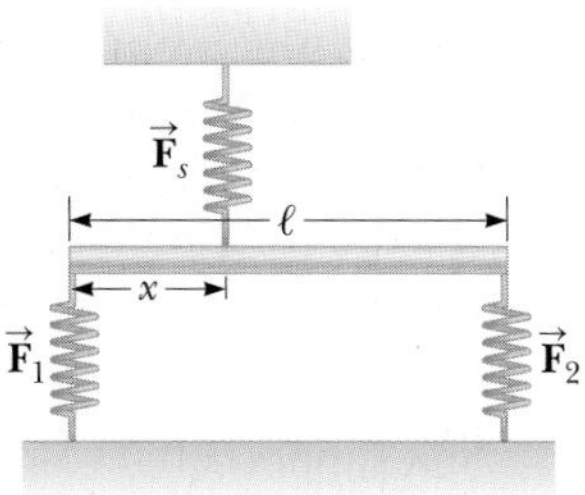

Figure P8.75

76. **Q|C** **S** A light rod of length $2L$ is free to rotate in a *vertical* plane about a frictionless pivot through its center. A particle of mass m_1 is attached at one end of the rod, and a mass m_2 is at the opposite end, where $m_1 > m_2$. The system is released from rest in the vertical position shown in Figure P8.76a, and at some later time the system is rotating in the position shown in Figure P8.76b. Take the reference point of the gravitational potential energy to be at the pivot. (a) Find an expression for the system's total mechanical energy in the vertical position. (b) Find an expression for the total mechanical energy in the rotated position shown in Figure P8.76b. (c) Using the fact that the mechanical energy of the system is conserved, how would you determine the angular speed ω of the system in the rotated position? (d) Find the magnitude of the torque on the system in the vertical position and in the rotated position. Is the torque constant? Explain what these results imply regarding the angular momentum of the system. (e) Find an expression for the magnitude of the angular acceleration of the system in the rotated position. Does your result make sense when the rod is horizontal? When it is vertical? Explain.

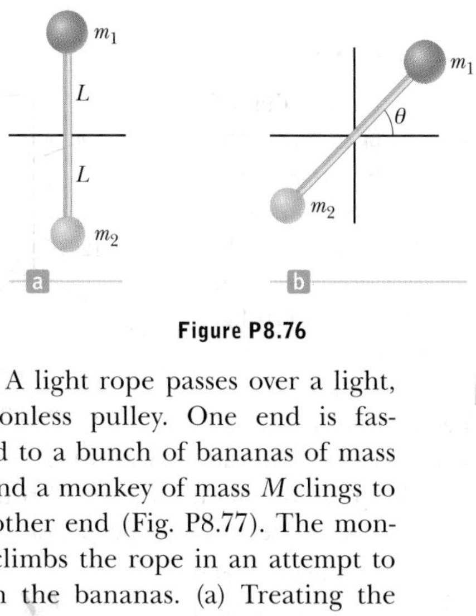

Figure P8.76

77. **Q|C** A light rope passes over a light, frictionless pulley. One end is fastened to a bunch of bananas of mass M, and a monkey of mass M clings to the other end (Fig. P8.77). The monkey climbs the rope in an attempt to reach the bananas. (a) Treating the system as consisting of the monkey, bananas, rope, and pulley, find the net torque of the system about the pulley axis. (b) Using the result of part (a), determine the total angular momentum about the pulley axis and describe the motion of the system. (c) Will the monkey reach the bananas before they get stuck in the pulley?

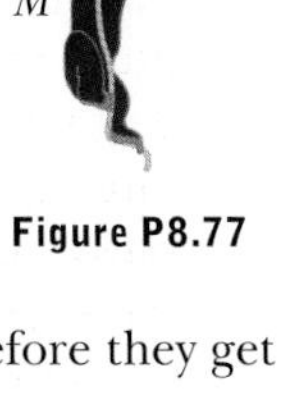

Figure P8.77

78. An electric motor turns a flywheel through a drive belt that joins a pulley on the motor and a pulley that is rigidly attached to the flywheel, as shown in Figure P8.78. The flywheel is a uniform disk with a mass of 80.0 kg and a radius of $R = 0.625$ m. It turns on a frictionless

axle. Its pulley has much smaller mass and a radius of $r = 0.230$ m. The tension T_u in the upper (taut) segment of the belt is 135 N and the flywheel has a clockwise angular acceleration of 1.67 rad/s^2. Find the tension in the lower (slack) segment of the belt.

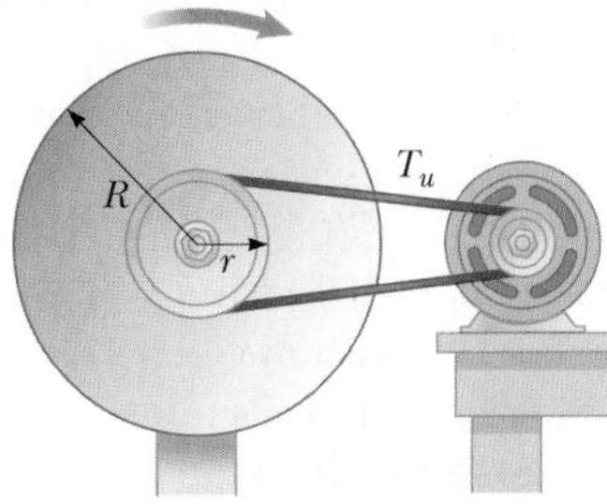

Figure P8.78

79. BIO In exercise physiology studies, it is sometimes important to determine the location of a person's center of gravity. This can be done with the arrangement shown in Figure P8.79. A light plank rests on two scales that read $F_{g1} = 380$ N and $F_{g2} = 320$ N. The scales are separated by a distance of 2.00 m. How far from the woman's feet is her center of gravity?

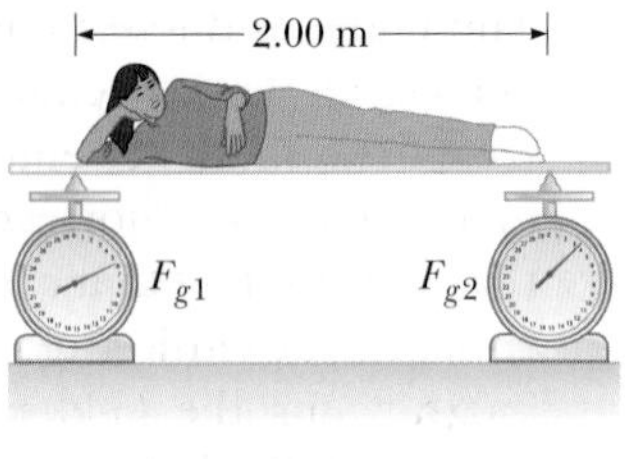

Figure P8.79

80. S A uniform thin rod of length L and mass M is free to rotate on a frictionless pin passing through one end (Fig. P8.80). The rod is released from rest in the horizontal position. (a) What is the speed of its center of gravity when the rod reaches its lowest position? (b) What is the tangential speed of the lowest point on the rod when it is in the vertical position?

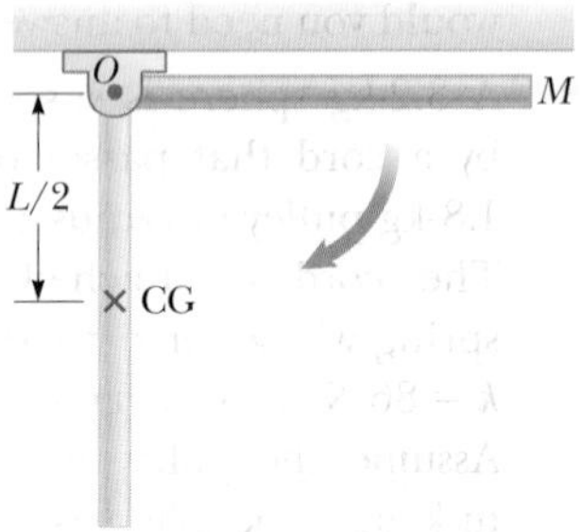

Figure P8.80

81. S A uniform solid cylinder of mass M and radius R rotates on a frictionless horizontal axle (Fig. P8.81). Two objects with equal masses m hang from light cords wrapped around the cylinder. If the system is released from rest, find (a) the tension in each cord and (b) the acceleration of each object after the objects have descended a distance h.

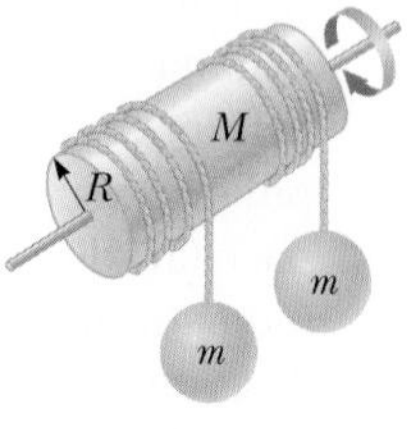

Figure P8.81

82. QIC S A painter climbs a ladder leaning against a smooth wall. At a certain height, the ladder is on the verge of slipping. (a) Explain why the force exerted by the vertical wall on the ladder is horizontal. (b) If the ladder of length L leans at an angle θ with the horizontal, what is the lever arm for this horizontal force with the axis of rotation taken at the base of the ladder? (c) If the ladder is uniform, what is the lever arm for the force of gravity acting on the ladder? (d) Let the mass of the painter be 80 kg, $L = 4.0$ m, the ladder's mass be 30 kg, $\theta = 53°$, and the coefficient of friction between ground and ladder be 0.45. Find the maximum distance the painter can climb up the ladder.

83. A *war-wolf*, or *trebuchet*, is a device used during the Middle Ages to throw rocks at castles and now sometimes used to fling pumpkins and pianos. A simple trebuchet is shown in Figure P8.83. Model it as a stiff rod of negligible mass 3.00 m long and joining particles of mass $m_1 = 0.120$ kg and $m_2 = 60.0$ kg at its ends. It can turn on a frictionless horizontal axle perpendicular to the rod and 14.0 cm from the particle of larger mass. The rod is released from rest in a horizontal orientation. Find the maximum speed that the object of smaller mass attains.

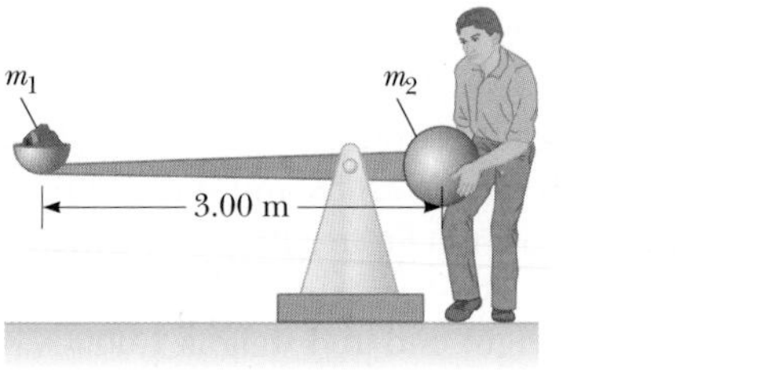

Figure P8.83

84. A string is wrapped around a uniform cylinder of mass M and radius R. The cylinder is released from rest with the string vertical and its top end tied to a fixed bar (Fig. P8.84). Show that (a) the tension in the string is one-third the weight of the cylinder, (b) the magnitude of the acceleration of the center of gravity is $2g/3$, and (c) the speed of the center of gravity is $(4gh/3)^{1/2}$ after the cylinder has descended through distance h. Verify your answer to part (c) with the energy approach.

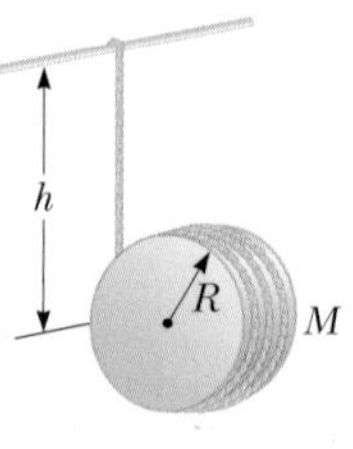

Figure P8.84

85. BIO **The Iron Cross** When a gymnast weighing 750 N executes the iron cross as in Figure P8.85a, the primary muscles involved in supporting this position are the latissimus dorsi ("lats") and the pectoralis major ("pecs"). The rings exert an upward force on the arms and support the weight of the gymnast. The force exerted by the shoulder joint on the arm is labeled $\vec{\mathbf{F}}_s$ while the two muscles exert a total force $\vec{\mathbf{F}}_m$ on the arm. Estimate the magnitude of the force $\vec{\mathbf{F}}_m$. Note that one ring supports half the weight of the gymnast, which is 375 N as indicated in Figure P8.85b. Assume that the force $\vec{\mathbf{F}}_m$ acts at an angle of 45° below the horizontal at

a distance of 4.0 cm from the shoulder joint. In your estimate, take the distance from the shoulder joint to the hand to be $L = 70$ cm and ignore the weight of the arm.

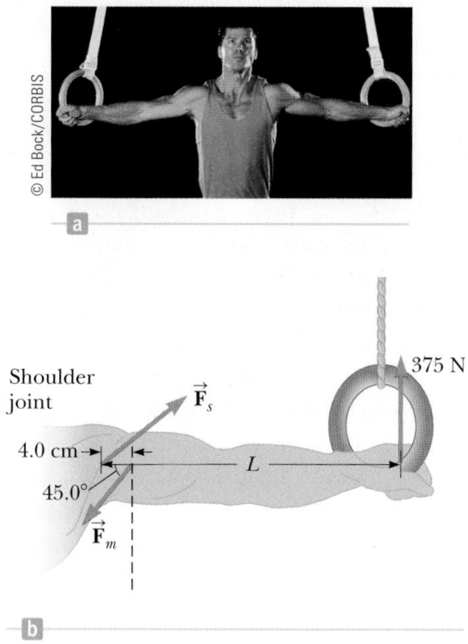

Figure P8.85

86. **BIO** In an emergency situation, a person with a broken forearm ties a strap from his hand to clip on his shoulder as in Figure P8.86. His 1.60-kg forearm remains in a horizontal position and the strap makes an angle of $\theta = 50.0°$ with the horizontal. Assume the forearm is uniform, has a length of $\ell = 0.320$ m, assume the biceps muscle is relaxed, and ignore the mass and length of the hand. Find (a) the tension in the strap and (b) the components of the reaction force exerted by the humerus on the forearm.

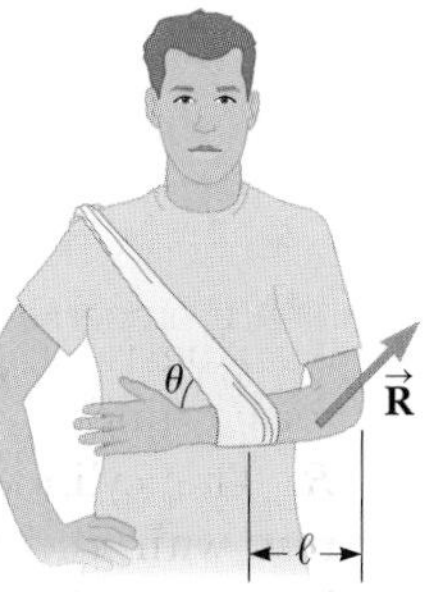

Figure P8.86

87. An object of mass $m_1 = 4.00$ kg is connected by a light cord to an object of mass $m_2 = 3.00$ kg on a frictionless surface (Fig. P8.87). The pulley rotates about a frictionless axle and has a moment of inertia of $0.500 \text{ kg} \cdot \text{m}^2$ and a radius of 0.300 m. Assuming that the cord does not slip on the pulley, find (a) the acceleration of the two masses and (b) the tensions T_1 and T_2.

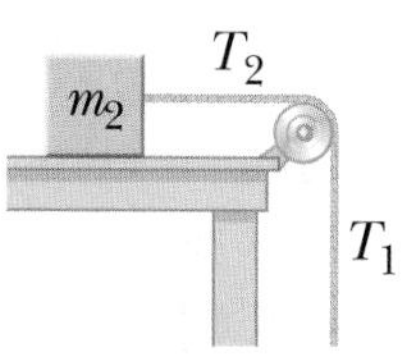

Figure P8.87

88. **Q|C** A 10.0-kg monkey climbs a uniform ladder with weight $w = 1.20 \times 10^2$ N and length $L = 3.00$ m as shown in Figure P8.88. The ladder rests against the wall at an angle of $\theta = 60.0°$. The upper and lower ends of the ladder rest on frictionless surfaces, with the lower end fastened to the wall by a horizontal rope that is frayed and that can support a maximum tension of only 80.0 N. (a) Draw a force diagram for the ladder. (b) Find the normal force exerted by the bottom of the ladder. (c) Find the tension in the rope when the monkey is two-thirds of the way up the ladder. (d) Find the maximum distance d that the monkey can climb up the ladder before the rope breaks. (e) If the horizontal surface were rough and the rope were removed, how would your analysis of the problem be changed and what other information would you need to answer parts (c) and (d)?

Figure P8.88

89. A 3.2-kg sphere is suspended by a cord that passes over a 1.8-kg pulley of radius 3.8 cm. The cord is attached to a spring whose force constant is $k = 86$ N/m as in Figure P8.89. Assume the pulley is a solid disk. (a) If the sphere is released from rest with the spring unstretched, what distance does the sphere fall through before stopping? (b) Find the speed of the sphere after it has fallen 25 cm.

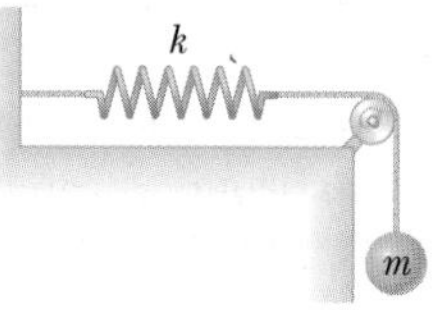

Figure P8.89

Hot air balloons exploit Archimedes' principle: the buoyant force is equal to the weight of the displaced air. The hot air expands and is less dense than the ambient air, hence lighter. When the total weight of the balloon is lighter than the air it displaces, the balloon rises.

Totophotos/Shutterstock.com

9 Solids and Fluids

There are four known states of matter: solids, liquids, gases, and plasmas. In the Universe at large, plasmas—systems of charged particles interacting electromagnetically—are the most common. In our environment on Earth, solids, liquids, and gases predominate.

An understanding of the fundamental properties of these different states of matter is important in all the sciences, in engineering, and in medicine. Forces put stresses on solids, and stresses can strain, deform, and break those solids, whether they are steel beams or bones. Fluids under pressure can perform work or carry nutrients and essential solutes, like the blood flowing through our arteries and veins. Flowing gases cause pressure differences that can lift a massive cargo plane or the roof off a house in a hurricane. High-temperature plasmas created in fusion reactors may someday allow humankind to harness the energy source of the Sun.

The study of any one of these states of matter is itself a vast discipline. Here, we'll introduce basic properties of solids and liquids, the latter including some properties of gases. In addition, we'll take a brief look at surface tension, viscosity, osmosis, and diffusion.

9.1 States of Matter

LEARNING OBJECTIVE

1. Describe and contrast the four states of matter.

Matter is normally classified as being in one of three states: **solid, liquid,** or **gas.** Often this classification system is extended to include a fourth state of matter, called a **plasma.**

Everyday experience tells us that a solid has a definite volume and shape. A brick, for example, maintains its familiar shape and size day in and day out.

A liquid has a definite volume but no definite shape. When you fill the tank on a lawn mower, the gasoline changes its shape from that of the original container to that of the tank on the mower, but the original volume is unchanged. A gas differs from solids and liquids in that it has neither definite volume nor definite shape. Because gas can flow, however, it shares many properties with liquids.

All matter consists of some distribution of atoms or molecules. The atoms in a solid, held together by forces that are mainly electrical, are located at specific positions with respect to one another and vibrate about those positions. At low temperatures, the vibrating motion is slight and the atoms can be considered essentially fixed. As energy is added to the material, the amplitude of the vibrations increases. A vibrating atom can be viewed as being bound in its equilibrium position by springs attached to neighboring atoms. A collection of such atoms and imaginary springs is shown in Figure 9.1. We can picture applied external forces as compressing these tiny internal springs. When the external forces are removed, the solid tends to return to its original shape and size. Consequently, a solid is said to have *elasticity*.

Crystals of natural quartz (SiO_2), one of the most common minerals on Earth. Quartz crystals are used to make special lenses and prisms and are employed in certain electronic applications.

Solids can be classified as either crystalline or amorphous. In a **crystalline solid** the atoms have an ordered structure. For example, in the sodium chloride crystal (common table salt), sodium and chlorine atoms occupy alternate corners of a cube, as in Figure 9.2a. In an **amorphous solid,** such as glass, the atoms are arranged almost randomly, as in Figure 9.2b.

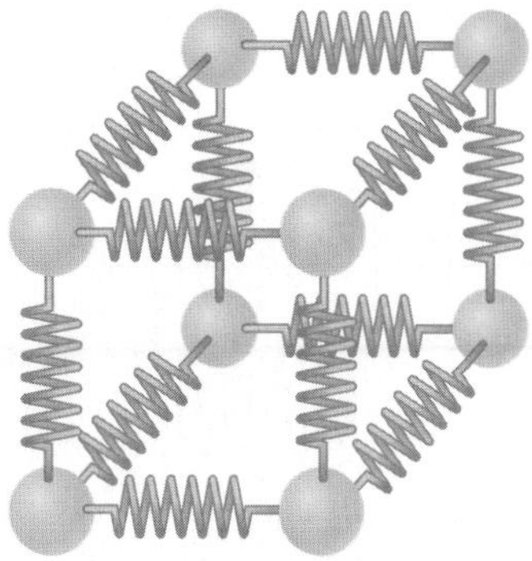

Figure 9.1 A model of a portion of a solid. The atoms (spheres) are imagined as being attached to each other by springs, which represent the elastic nature of the interatomic forces. A solid consists of trillions of segments like this, with springs connecting all of them.

For any given substance, the liquid state exists at a higher temperature than the solid state. The intermolecular forces in a liquid aren't strong enough to keep the molecules in fixed positions, and they wander through the liquid in random fashion (Fig. 9.2c). Solids and liquids both have the property that when an attempt is made to compress them, strong repulsive atomic forces act internally to resist the compression.

In the gaseous state, molecules are in constant random motion and exert only weak forces on each other. The average distance between the molecules of a gas is quite large compared with the size of the molecules. Occasionally the molecules collide with each other, but most of the time they move as nearly free, noninteracting particles. As a result, unlike solids and liquids, gases can be easily compressed. We'll say more about gases in subsequent chapters.

When a gas is heated to high temperature, many of the electrons surrounding each atom are freed from the nucleus. The resulting system is a collection of free, electrically charged particles—negatively charged electrons and positively charged ions. Such a highly ionized state of matter containing equal amounts of positive and negative charges is called a **plasma.** Unlike a neutral gas, the long-range electric and magnetic forces allow the constituents of a plasma to interact with each other. Plasmas are found inside stars and in accretion disks around black holes, for example, and are far more common than the solid, liquid, and gaseous states because there are far more stars around than any other form of celestial matter.

Normal matter, however, may constitute less than 5% of all matter in the Universe. Observations of the last several years point to the existence of an invisible **dark matter,** which affects the motion of stars orbiting the centers of galaxies. Dark matter may comprise nearly 25% of the matter in the Universe, several times larger

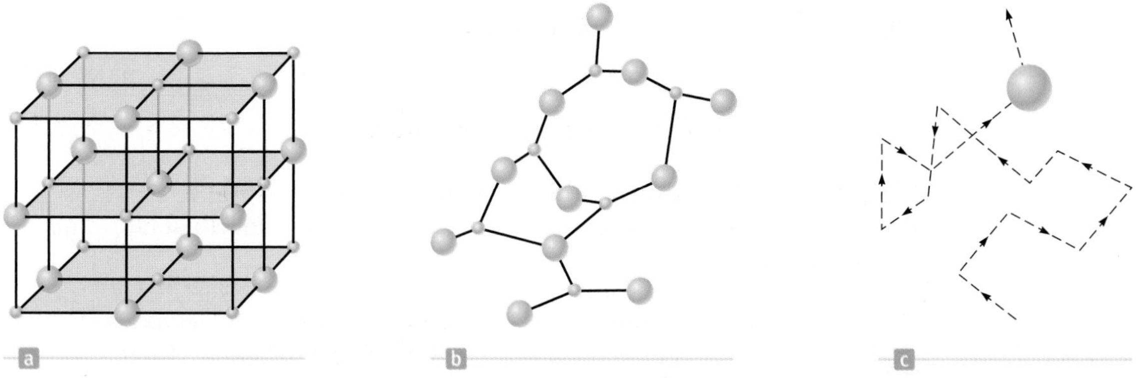

Figure 9.2 (a) The NaCl structure, with the Na^+ (gray) and Cl^- (green) ions at alternate corners of a cube. (b) In an amorphous solid, the atoms are arranged randomly. (c) Erratic motion of a molecule in a liquid.

than the amount of normal matter. Finally, the rapid acceleration of the expansion of the Universe may be driven by an even more mysterious form of matter, called **dark energy,** which may account for over 70% of all matter in the Universe.

9.2 Density and Pressure

LEARNING OBJECTIVES

1. Define the density of a uniform object.
2. Define pressure and apply it in common physical contexts.

Equal masses of aluminum and gold have an important physical difference: The aluminum takes up over seven times as much space as the gold. Although the reasons for the difference lie at the atomic and nuclear levels, a simple measure of this difference is the concept of *density*.

Density ▶

The **density** ρ of an object having uniform composition is its mass M divided by its volume V:

$$\rho \equiv \frac{M}{V} \qquad [9.1]$$

SI unit: kilogram per meter cubed (kg/m^3)

For an object with non-uniform composition, Equation 9.1 defines an average density. The most common units used for density are kilograms per cubic meter in the SI system and grams per cubic centimeter in the cgs system. Table 9.1 lists the densities of some substances. The densities of most liquids and solids vary slightly with changes in temperature and pressure; the densities of gases vary greatly with such changes. Under normal conditions, the densities of solids and liquids are about 1 000 times greater than the densities of gases. This difference implies that the average spacing between molecules in a gas under such conditions is about ten times greater than in a solid or liquid.

The **specific gravity** of a substance is the ratio of its density to the density of water at 4°C, which is 1.0×10^3 kg/m^3. (The size of the kilogram was originally defined to make the density of water 1.0×10^3 kg/m^3 at 4°C.) By definition, specific gravity is a dimensionless quantity. For example, if the specific gravity of a substance is 3.0, its density is $3.0(1.0 \times 10^3 \text{ kg/m}^3) = 3.0 \times 10^3 \text{ kg/m}^3$.

Quick Quiz

9.1 Suppose you have one cubic meter of gold, two cubic meters of silver, and six cubic meters of aluminum. Rank them by mass, from smallest to largest. (a) gold, aluminum, silver (b) gold, silver, aluminum (c) aluminum, gold, silver (d) silver, aluminum, gold

Table 9.1 Densities of Some Common Substances

Substance	ρ (kg/m^3)[a]	Substance	ρ (kg/m^3)[a]
Ice	0.917×10^3	Water	1.00×10^3
Aluminum	2.70×10^3	Glycerin	1.26×10^3
Iron	7.86×10^3	Ethyl alcohol	0.806×10^3
Copper	8.92×10^3	Benzene	0.879×10^3
Silver	10.5×10^3	Mercury	13.6×10^3
Lead	11.3×10^3	Air	1.29
Gold	19.3×10^3	Oxygen	1.43
Platinum	21.4×10^3	Hydrogen	8.99×10^{-2}
Uranium	18.7×10^3	Helium	1.79×10^{-1}

[a]All values are at standard atmospheric temperature and pressure (STP), defined as 0°C (273 K) and 1 atm (1.013×10^5 Pa). To convert to grams per cubic centimeter, multiply by 10^{-3}.

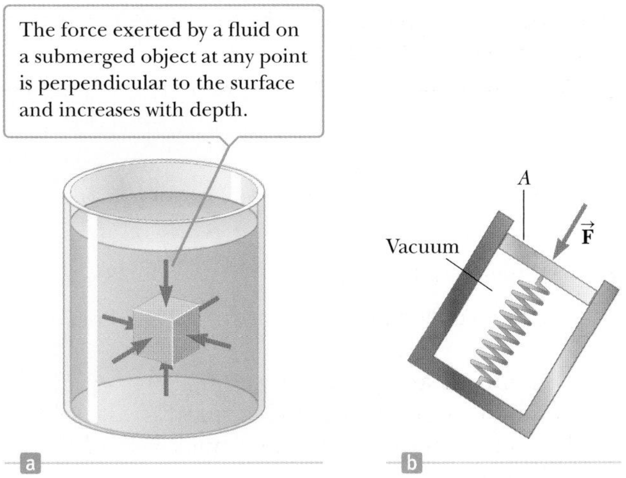

Figure 9.3 (a) The force exerted by a fluid on the surfaces of a submerged object. (b) A simple device for measuring pressure in a fluid.

The force exerted by a fluid on an object is always perpendicular to the surfaces of the object, as shown in Figure 9.3a.

The pressure at a specific point in a fluid can be measured with the device pictured in Figure 9.3b: an evacuated cylinder enclosing a light piston connected to a spring that has been previously calibrated with known weights. As the device is submerged in a fluid, the fluid presses down on the top of the piston and compresses the spring until the inward force exerted by the fluid is balanced by the outward force exerted by the spring. Let F be the magnitude of the force on the piston and A the area of the top surface of the piston. Notice that the force that compresses the spring is spread out over the entire area, motivating our formal definition of pressure:

Tip 9.1 Force and Pressure

Equation 9.2 makes a clear distinction between force and pressure. Another important distinction is that *force is a vector* and *pressure is a scalar.* There is no direction associated with pressure, but the direction of the force associated with the pressure is perpendicular to the surface of interest.

◄ Pressure

If F is the magnitude of a force exerted perpendicular to a given surface of area A, then the average pressure P is the force divided by the area:

$$P \equiv \frac{F}{A} \qquad [9.2]$$

SI unit: pascal (Pa = N/m^2)

Pressure can change from point to point, which is why the pressure in Equation 9.2 is called an average. Because pressure is defined as force per unit area, it has units of pascals (newtons per square meter). The English customary unit for pressure is the pound per inch squared. Atmospheric pressure at sea level is 14.7 lb/in.^{2}, which in SI units is 1.01×10^5 Pa.

As we see from Equation 9.2, the effect of a given force depends critically on the area to which it's applied. A 700-N man can stand on a vinyl-covered floor in regular street shoes without damaging the surface, but if he wears golf shoes, the metal cleats protruding from the soles can do considerable damage to the floor. With the cleats, the same force is concentrated into a smaller area, greatly elevating the pressure in those areas, resulting in a greater likelihood of exceeding the ultimate strength of the floor material.

Snowshoes use the same principle (Fig. 9.4). The snow exerts an upward normal force on the shoes to support the person's weight. According to Newton's third law, this upward force is accompanied by a downward force exerted by the shoes on the snow. If the person is wearing snowshoes, that force is distributed over the very large area of each snowshoe, so that the pressure at any given point is relatively low and the person doesn't penetrate very deeply into the snow.

© Royalty-Free/Corbis

Figure 9.4 Snowshoes prevent the person from sinking into the soft snow because the force on the snow is spread over a larger area, reducing the pressure on the snow's surface.

APPLYING PHYSICS 9.1 | Bed of Nails Trick

After an exciting but exhausting lecture, a physics professor stretches out for a nap on a bed of nails, as in Figure 9.5, suffering no injury and only moderate discomfort. How is that possible?

EXPLANATION If you try to support your entire weight on a single nail, the pressure on your body is your weight divided by the very small area of the end of the nail. The resulting pressure is large enough to penetrate the skin. If you distribute your weight over several hundred nails, however, as demonstrated by the professor, the pressure is considerably reduced because the area that supports your weight is the total area of all nails in contact with your body. (Why is lying on a bed of nails more comfortable than sitting on the same bed? Extend the logic to show that it would be more uncomfortable yet to stand on a bed of nails without shoes.) ■

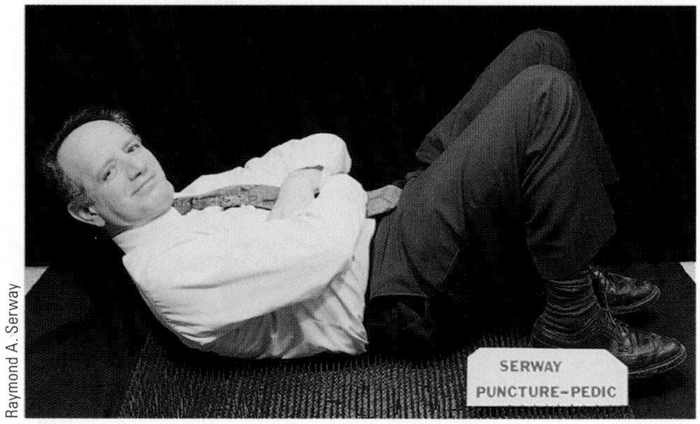

Figure 9.5 (Applying Physics 9.1) Does anyone have a pillow?

EXAMPLE 9.1 | Pressure and Weight of Water

GOAL Relate density, pressure, and weight.

PROBLEM **(a)** Calculate the weight of a cylindrical column of water with height $h = 40.0$ m and radius $r = 1.00$ m. (See Fig. 9.6.) **(b)** Calculate the force exerted by air on a disk of radius 1.00 m at the water's surface. **(c)** What pressure at a depth of 40.0 m supports the water column?

STRATEGY For part (a), calculate the volume and multiply by the density to get the mass of water, then multiply the mass by g to get the weight. Part (b) requires substitution into the definition of pressure. Adding the results of parts (a) and (b) and dividing by the area gives the pressure of water at the bottom of the column.

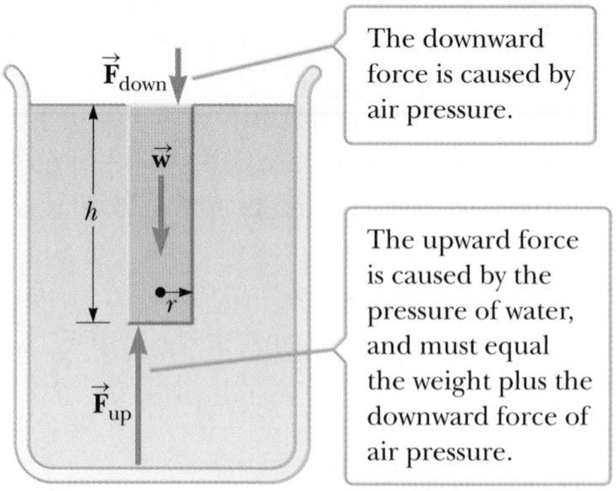

Figure 9.6 (Example 9.1)

SOLUTION

(a) Calculate the weight of a cylindrical column of water with height 40.0 m and radius 1.00 m.

Calculate the volume of the cylinder:

$$V = \pi r^2 h = \pi(1.00 \text{ m})^2(40.0 \text{ m}) = 126 \text{ m}^3$$

Multiply the volume by the density of water to obtain the mass of water in the cylinder:

$$m = \rho V = (1.00 \times 10^3 \text{ kg/m}^3)(126 \text{ m}^3) = 1.26 \times 10^5 \text{ kg}$$

Multiply the mass by the acceleration of gravity g to obtain the weight w:

$$w = mg = (1.26 \times 10^5 \text{ kg})(9.80 \text{ m/s}^2) = \boxed{1.23 \times 10^6 \text{ N}}$$

(b) Calculate the force exerted by air on a disk of radius 1.00 m at the surface of the lake.

Write the equation for pressure:

$$P = \frac{F}{A}$$

Solve the pressure equation for the force and substitute $A = \pi r^2$:

$$F = PA = P\pi r^2$$

Substitute values:

$$F = (1.01 \times 10^5 \text{ Pa})\pi\,(1.00 \text{ m})^2 = \boxed{3.17 \times 10^5 \text{ N}}$$

(c) What pressure at a depth of 40.0 m supports the water column?

Write Newton's second law for the water column: $-F_{\text{down}} - w + F_{\text{up}} = 0$

Solve for the upward force: $F_{\text{up}} = F_{\text{down}} + w = (3.17 \times 10^5 \text{ N}) + (1.23 \times 10^6 \text{ N}) = 1.55 \times 10^6 \text{ N}$

Divide the force by the area to obtain the required pressure: $P = \dfrac{F_{\text{up}}}{A} = \dfrac{1.55 \times 10^6 \text{ N}}{\pi(1.00 \text{ m})^2} = \boxed{4.93 \times 10^5 \text{ Pa}}$

REMARKS Notice that the pressure at a given depth is related to the sum of the weight of the water and the force exerted by the air pressure at the water's surface. Water at a depth of 40.0 m must push upward to maintain the column in equilibrium. Notice also the important role of density in determining the pressure at a given depth.

QUESTION 9.1 A giant oil storage facility contains oil to a depth of 40.0 m. How does the pressure at the bottom of the tank compare to the pressure at a depth of 40.0 m in water? Explain.

EXERCISE 9.1 A large rectangular tub is filled to a depth of 2.60 m with olive oil, which has density 915 kg/m^3. If the tub has length 5.00 m and width 3.00 m, calculate (a) the weight of the olive oil, (b) the force of air pressure on the surface of the oil, and (c) the pressure exerted upward by the bottom of the tub.

ANSWERS (a) 3.50×10^5 N (b) 1.52×10^6 N (c) 1.25×10^5 Pa

9.3 The Deformation of Solids

LEARNING OBJECTIVES

1. Identify the three elastic moduli related to changes in an object's length, shape and volume in response to applied stress.
2. Apply the stress–strain equations to deformation problems.

Although a solid may be thought of as having a definite shape and volume, it's possible to change its shape and volume by applying external forces. A sufficiently large force will permanently deform or break an object, but otherwise, when the external forces are removed, the object tends to return to its original shape and size. This is called *elastic behavior.*

The elastic properties of solids are discussed in terms of stress and strain. **Stress** is the force per unit area causing a deformation; **strain** is a measure of the amount of the deformation. For sufficiently small stresses, **stress is proportional to strain,** with the constant of proportionality depending on the material being deformed and on the nature of the deformation. We call this proportionality constant the **elastic modulus:**

$$\text{stress} = \text{elastic modulus} \times \text{strain} \qquad [9.3]$$

The elastic modulus is analogous to a spring constant. It can be taken as the stiffness of a material: A material having a large elastic modulus is very stiff and difficult to deform. There are three relationships having the form of Equation 9.3, corresponding to tensile, shear, and bulk deformation, and all of them satisfy an equation similar to Hooke's law for springs:

$$F = -k\,\Delta x \qquad [9.4]$$

where F is the applied force, k is the spring constant, and Δx is essentially the amount by which the spring is stretched or compressed.

Young's Modulus: Elasticity in Length

Consider a long bar of cross-sectional area A and length L_0, clamped at one end (Fig. 9.7). When an external force $\vec{\mathbf{F}}$ is applied along the bar, perpendicular to the cross section, internal forces in the bar resist the distortion ("stretching")

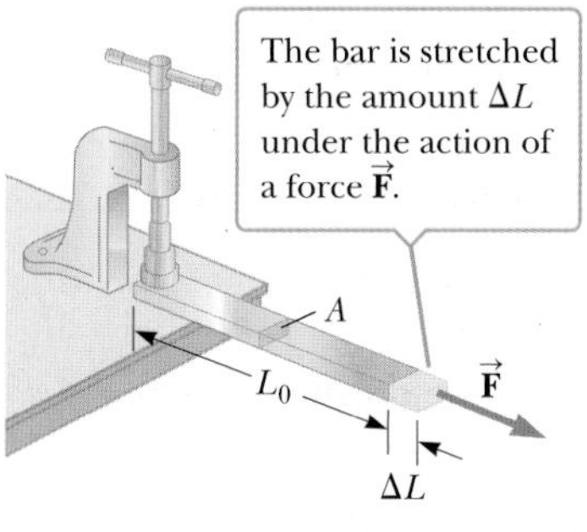

Figure 9.7 A force is applied to a long bar clamped at one end.

that $\vec{\mathbf{F}}$ tends to produce. Nevertheless, the bar attains an equilibrium in which (1) its length is greater than L_0 and (2) the external force is balanced by internal forces. Under these circumstances, the bar is said to be *stressed.* We define the **tensile stress** as the ratio of the magnitude of the external force F to the cross-sectional area A. The word "tensile" has the same root as the word "tension" and is used because the bar is under tension. The SI unit of stress is the newton per square meter (N/m^2), called the **pascal** (Pa), the same as the unit of pressure:

The pascal ▶

$$1 \text{ Pa} \equiv \text{N/m}^2$$

The **tensile strain** in this case is defined as the ratio of the change in length ΔL to the original length L_0 and is therefore a dimensionless quantity. Using Equation 9.3, we can write an equation relating tensile stress to tensile strain:

$$\frac{F}{A} = Y\frac{\Delta L}{L_0} \quad [9.5]$$

In this equation, Y is the constant of proportionality, called **Young's modulus.** Notice that Equation 9.5 could be solved for F and put in the form $F = k\,\Delta L$, where $k = YA/L_0$, making it look just like Hooke's law, Equation 9.4.

A material having a large Young's modulus is difficult to stretch or compress. This quantity is typically used to characterize a rod or wire stressed under either tension or compression. Because strain is a dimensionless quantity, Y is in pascals. Typical values are given in Table 9.2. Experiments show that (1) the change in length for a fixed external force is proportional to the original length and (2) the force necessary to produce a given strain is proportional to the cross-sectional area. The value of Young's modulus for a given material depends on whether the material is stretched or compressed. A human femur, for example, is stronger under compression than tension. For many materials, such as metals, the moduli for compression and tension differ very little from each other.

It's possible to exceed the **elastic limit** of a substance by applying a sufficiently great stress (Fig. 9.8). At the elastic limit, the stress–strain curve departs from a straight line. A material subjected to a stress beyond this limit ordinarily doesn't return to its original length when the external force is removed. As the stress is increased further, it surpasses the **ultimate strength:** the greatest stress the substance can withstand without breaking. The **breaking point** for brittle materials is just beyond the ultimate strength. For ductile metals like copper and gold, after passing the point of ultimate strength, the metal thins and stretches at a lower stress level before breaking.

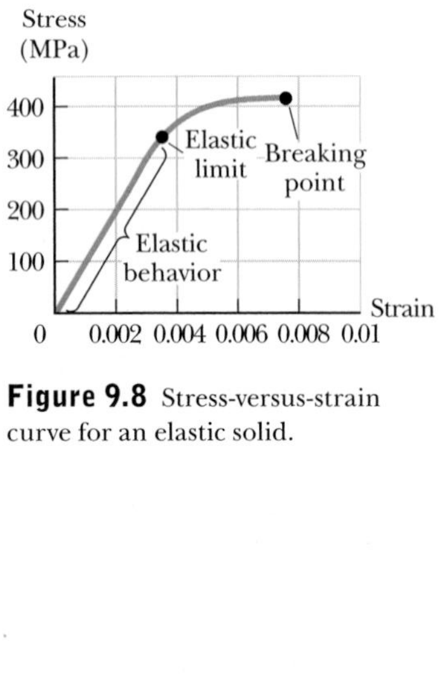

Figure 9.8 Stress-versus-strain curve for an elastic solid.

Table 9.2 Typical Values for the Elastic Modulus

Substance	Young's Modulus (Pa)	Shear Modulus (Pa)	Bulk Modulus (Pa)
Aluminum	7.0×10^{10}	2.5×10^{10}	7.0×10^{10}
Bone	1.8×10^{10}	8.0×10^{10}	—
Brass	9.1×10^{10}	3.5×10^{10}	6.1×10^{10}
Copper	11×10^{10}	4.2×10^{10}	14×10^{10}
Steel	20×10^{10}	8.4×10^{10}	16×10^{10}
Tungsten	35×10^{10}	14×10^{10}	20×10^{10}
Glass	$6.5\text{–}7.8 \times 10^{10}$	$2.6\text{–}3.2 \times 10^{10}$	$5.0\text{–}5.5 \times 10^{10}$
Quartz	5.6×10^{10}	2.6×10^{10}	2.7×10^{10}
Rib Cartilage	1.2×10^{7}	—	—
Rubber	0.1×10^{7}	—	—
Tendon	2×10^{7}	—	—
Water	—	—	0.21×10^{10}
Mercury	—	—	2.8×10^{10}

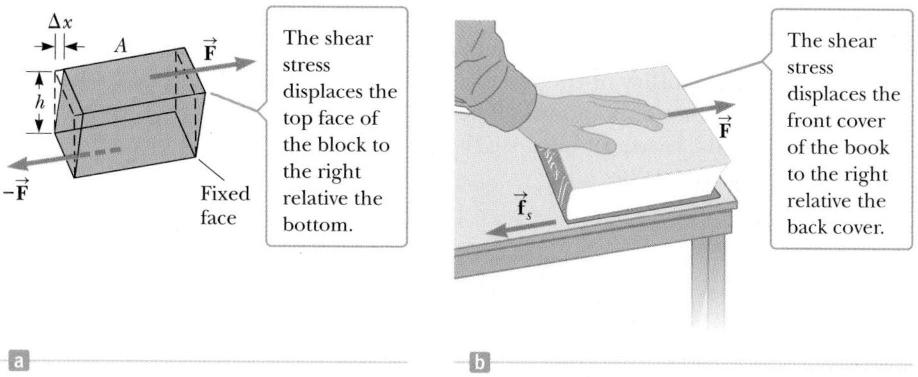

Figure 9.9 (a) A shear deformation in which a rectangular block is distorted by forces applied tangent to two of its faces. (b) A book under shear stress.

Shear Modulus: Elasticity of Shape

Another type of deformation occurs when an object is subjected to a force $\vec{\mathbf{F}}$ *parallel* to one of its faces while the opposite face is held fixed by a second force (Fig. 9.9a). If the object is originally a rectangular block, such a parallel force results in a shape with the cross section of a parallelogram. This kind of stress is called a **shear stress.** A book pushed sideways, as in Figure 9.9b, is being subjected to a shear stress. There is no change in volume with this kind of deformation. It's important to remember that in shear stress, the applied force is *parallel* to the cross-sectional area, whereas in tensile stress the force is *perpendicular* to the cross-sectional area. We define **the shear stress as F/A, the ratio of the magnitude of the parallel force to the area A of the face being sheared. The shear strain is the ratio $\Delta x/h$, where Δx is the horizontal distance the sheared face moves and h is the height of the object.** The shear stress is related to the shear strain according to

$$\frac{F}{A} = S\frac{\Delta x}{h} \qquad [9.6]$$

where S is the **shear modulus** of the material, with units of pascals (force per unit area). Once again, notice the similarity to Hooke's law.

A material having a large shear modulus is difficult to bend. Shear moduli for some representative materials are listed in Table 9.2.

Bulk Modulus: Volume Elasticity

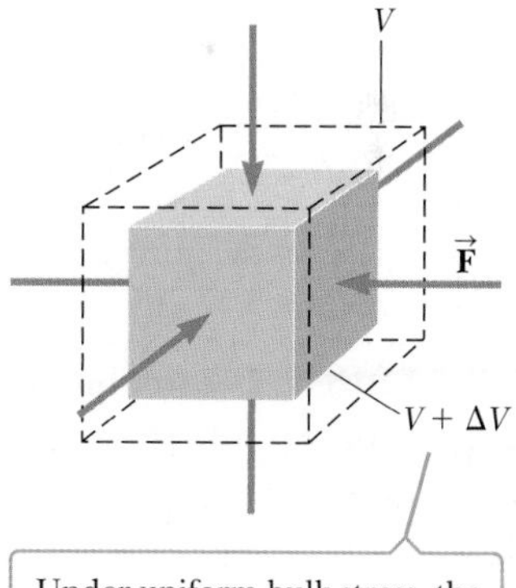

Figure 9.10 A solid cube is under uniform pressure and is therefore compressed on all sides by forces normal to its six faces. The arrowheads of force vectors on the sides of the cube that are not visible are hidden by the cube.

The bulk modulus characterizes the response of a substance to uniform squeezing. Suppose the external forces acting on an object are all perpendicular to the surface on which the force acts and are distributed uniformly over the surface of the object (Fig. 9.10). This occurs when an object is immersed in a fluid. An object subject to this type of deformation undergoes a change in volume but no change in shape. **The volume stress ΔP is defined as the ratio of the change in the magnitude of the applied force ΔF to the surface area A.** From the definition of pressure in Section 9.2, ΔP is also simply a change in pressure. The volume strain is equal to the change in volume ΔV divided by the original volume V. Again using Equation 9.3, we can relate a volume stress to a volume strain by the formula

$$\Delta P = -B\frac{\Delta V}{V} \qquad [9.7]$$

◀ Bulk modulus

A material having a large bulk modulus doesn't compress easily. Note that a negative sign is included in this defining equation so that B is always positive. An increase in pressure (positive ΔP) causes a decrease in volume (negative ΔV) and vice versa.

Table 9.2 lists bulk modulus values for some materials. If you look up such values in a different source, you may find that the reciprocal of the bulk modulus, called the **compressibility** of the material, is listed. Note from the table that both solids and liquids have bulk moduli. There is neither a Young's modulus nor shear modulus for liquids, however, because liquids simply flow when subjected to a tensile or shearing stress.

EXAMPLE 9.2 Built to Last

GOAL Calculate a compression due to tensile stress and maximum load.

PROBLEM A vertical steel beam in a building supports a load of 6.0×10^4 N. **(a)** If the length of the beam is 4.0 m and its cross-sectional area is 8.0×10^{-3} m^2, find the distance the beam is compressed along its length. **(b)** What maximum load in newtons could the steel beam support before failing?

STRATEGY Equation 9.3 pertains to compressive stress and strain and can be solved for ΔL, followed by substitution of known values. For part (b), set the compressive stress equal to the ultimate strength of steel from Table 9.3. Solve for the magnitude of the force, which is the total weight the structure can support.

SOLUTION

(a) Find the amount of compression in the beam.

Solve Equation 9.5 for ΔL and substitute, using the value of Young's modulus from Table 9.2:

$$\frac{F}{A} = Y\frac{\Delta L}{L_0}$$

$$\Delta L = \frac{FL_0}{YA} = \frac{(6.0 \times 10^4\ \text{N})(4.0\ \text{m})}{(2.0 \times 10^{11}\ \text{Pa})(8.0 \times 10^{-3}\ \text{m}^2)}$$

$$= 1.5 \times 10^{-4}\ \text{m}$$

(b) Find the maximum load that the beam can support.

Set the compressive stress equal to the ultimate compressive strength from Table 9.3, and solve for F:

$$\frac{F}{A} = \frac{F}{8.0 \times 10^{-3}\ \text{m}^2} = 5.0 \times 10^8\ \text{Pa}$$

$$F = 4.0 \times 10^6\ \text{N}$$

REMARKS In designing load-bearing structures of any kind, it's always necessary to build in a safety factor. No one would drive a car over a bridge that had been designed to supply the minimum necessary strength to keep it from collapsing.

QUESTION 9.2 Rank by the amount of fractional increase in length under increasing tensile stress, from smallest to largest: rubber, tungsten, steel, aluminum.

EXERCISE 9.2 A cable used to lift heavy materials like steel I-beams must be strong enough to resist breaking even under a load of 1.0×10^6 N. For safety, the cable must support twice that load. (a) What cross-sectional area should the cable have if it's to be made of steel? (b) By how much will an 8.0-m length of this cable stretch when subject to the 1.0×10^6-N load?

ANSWERS (a) 4.0×10^{-3} m^2 (b) 1.0×10^{-2} m

Table 9.3 Ultimate Strength of Materials

Material	Tensile Strength (N/m^2)	Compressive Strength (N/m^2)
Iron	1.7×10^8	5.5×10^8
Steel	5.0×10^8	5.0×10^8
Aluminum	2.0×10^8	2.0×10^8
Bone	1.2×10^8	1.5×10^8
Marble	—	8.0×10^7
Brick	1×10^6	3.5×10^7
Concrete	2×10^6	2×10^7

EXAMPLE 9.3 Football Injuries BIO

GOAL Obtain an estimate of shear stress.

PROBLEM A defensive lineman of mass $M = 125$ kg makes a flying tackle at $v_i = 4.00$ m/s on a stationary quarterback of mass $m = 85.0$ kg, and the lineman's helmet makes solid contact with the quarterback's femur. **(a)** What is the speed v_f of the two athletes immediately after contact? Assume a linear perfectly inelastic collision. **(b)** If the collision lasts for 0.100 s, estimate the average force exerted on the quarterback's femur. **(c)** If the cross-sectional area of the quarterback's femur is equal to 5.00×10^{-4} m^2, calculate the shear stress exerted on the bone in the collision.

STRATEGY The solution proceeds in three well-defined steps. In part (a), use conservation of linear momentum to calculate the final speed of the system consisting of the quarterback and the lineman. Second, the speed found in part (a) can be used in the impulse-momentum theorem to obtain an estimate of the average force exerted on the femur. Third, dividing the average force by the cross-sectional area of the femur gives the desired estimate of the shear stress.

SOLUTION

(a) What is the speed of the system immediately after contact?

Apply momentum conservation to the system:

$$p_{\text{initial}} = p_{\text{final}}$$

Substitute expressions for the initial and final momenta:

$$Mv_i = (M + m)\, v_f$$

Solve for the final speed v_f:

$$v_f = \frac{Mv_i}{M + m} = \frac{(125 \text{ kg})(4.00 \text{ m/s})}{125 \text{ kg} + 85.0 \text{ kg}} = \boxed{2.38 \text{ m/s}}$$

(b) Obtain an estimate for the average force delivered to the quarterback's femur.

Apply the impulse-momentum theorem:

$$F_{\text{av}}\, \Delta t = \Delta p = Mv_f - Mv_i$$

Solve for the average force exerted on the quarterback's femur:

$$F_{\text{av}} = \frac{M(v_f - v_i)}{\Delta t}$$

$$= \frac{(125 \text{ kg})(4.00 \text{ m/s} - 2.38 \text{ m/s})}{0.100 \text{ s}} = \boxed{2.03 \times 10^3 \text{ N}}$$

(c) Obtain the average shear stress exerted on the quarterback's femur.

Divide the average force found in part (b) by the cross-sectional area of the femur:

$$\text{Shear stress} = \frac{F}{A} = \frac{2.03 \times 10^3 \text{ N}}{5.00 \times 10^{-4} \text{ m}^2} = \boxed{4.06 \times 10^6 \text{ Pa}}$$

REMARKS The ultimate shear strength of a femur is approximately 7×10^7 Pa, so this collision would not be expected to break the quarterback's leg.

QUESTION 9.3 What kind of stress would be sustained by the lineman? What parts of his body would be affected?

EXERCISE 9.3 Calculate the diameter of a horizontal steel bolt if it is expected to support a maximum load hung on it having a mass of 2.00×10^3 kg but for safety reasons must be designed to support three times that load. (Assume the ultimate shear strength of steel is 2.50×10^8 Pa.)

ANSWER 1.73 cm

EXAMPLE 9.4 Lead Ballast Overboard

GOAL Apply the concepts of bulk stress and strain.

PROBLEM Ships and sailing vessels often carry lead ballast in various forms, such as bricks, to keep the ship properly oriented and upright in the water. Suppose a ship takes on cargo and the crew jettisons a total of 0.500 m^3 of lead ballast

(Continued)

into water 2.00 km deep. Calculate **(a)** the change in the pressure at that depth and **(b)** the change in volume of the lead upon reaching the bottom. Take the density of sea water to be 1.025×10^3 kg/m^3, and take the bulk modulus of lead to be 4.2×10^{10} Pa.

STRATEGY The pressure difference between the surface and a depth of 2.00 km is due to the weight of the water column. Calculate the weight of water in a column with cross section of 1.00 m^2. That number in newtons will be the same magnitude as the pressure difference in pascal. Substitute the pressure change into the bulk stress and strain equation to obtain the change in volume of the lead.

SOLUTION

(a) Calculate the pressure difference between the surface and at a depth of 2.00 km.

Use the density, volume, and acceleration of gravity g to compute the weight of water in a column having cross-sectional area of 1.00 m^2:

$$w = mg = (\rho V)g$$
$$= (1.025 \times 10^3 \text{ kg/m}^3)(2.00 \times 10^3 \text{ m}^3)(9.80 \text{ m/s}^2)$$
$$= 2.01 \times 10^7 \text{ N}$$

Divide by the area (in this case, 1.00 m^2) to obtain the pressure difference due to the column of water:

$$\Delta P = \frac{F}{A} = \frac{2.01 \times 10^7 \text{ N}}{1.00 \text{ m}^2} = 2.01 \times 10^7 \text{ Pa}$$

(b) Calculate the change in volume of the lead upon reaching the bottom.

Write the bulk stress and strain equation:

$$\Delta P = -B\frac{\Delta V}{V}$$

Solve for ΔV:

$$\Delta V = -\frac{V\Delta P}{B} = -\frac{(0.500 \text{ m}^3)(2.01 \times 10^7 \text{ Pa})}{4.2 \times 10^{10} \text{ Pa}} = -2.4 \times 10^{-4} \text{ m}^3$$

REMARKS The negative sign indicates a *decrease* in volume. The following exercise shows that even water can be compressed, although not by much.

QUESTION 9.4 Rank the following substances in order of the fractional change in volume in response to increasing pressure, from smallest to largest: copper, steel, water, mercury.

EXERCISE 9.4 (a) By what percentage does the volume of a ball of water shrink at that same depth? (b) What is the ratio of the new radius to the initial radius?

ANSWERS (a) 0.96% (b) 0.997

Arches and the Ultimate Strength of Materials

As we have seen, the ultimate strength of a material is the maximum force per unit area the material can withstand before it breaks or fractures. Such values are of great importance, particularly in the construction of buildings, bridges, and roads. Table 9.3 gives the ultimate strength of a variety of materials under both tension and compression. Note that bone and a variety of building materials (concrete, brick, and marble) are stronger under compression than under tension. The greater ability of brick and stone to resist compression is the basis of the semicircular arch, developed and used extensively by the Romans in everything from memorial arches to expansive temples and aqueduct supports.

APPLICATION
Arch Structures in Buildings

Before the development of the arch, the principal method of spanning a space was the simple post-and-beam construction (Fig. 9.11a), in which a horizontal beam is supported by two columns. This type of construction was used to build the great Greek temples. The columns of these temples were closely spaced because of the limited length of available stones and the low ultimate tensile strength of a sagging stone beam.

The semicircular arch (Fig. 9.11b) developed by the Romans was a great technological achievement in architectural design. It effectively allowed the heavy load

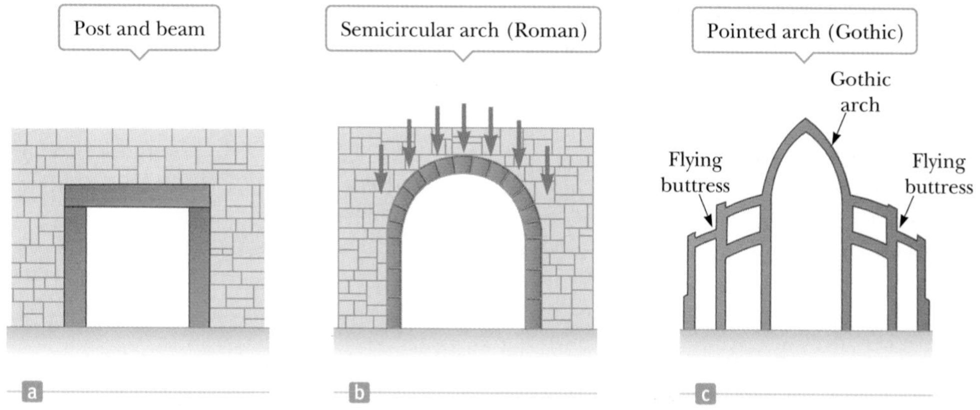

Figure 9.11 (a) A simple post-and-beam structure. (b) The semicircular arch developed by the Romans. (c) Gothic arch with flying buttresses to provide lateral support.

of a wide roof span to be channeled into horizontal and vertical forces on narrow supporting columns. The stability of this arch depends on the compression between its wedge-shaped stones. The stones are forced to squeeze against each other by the uniform loading, as shown in the figure. This compression results in horizontal outward forces at the base of the arch where it starts curving away from the vertical. These forces must then be balanced by the stone walls shown on the sides of the arch. It's common to use very heavy walls (buttresses) on either side of the arch to provide horizontal stability. If the foundation of the arch should move, the compressive forces between the wedge-shaped stones may decrease to the extent that the arch collapses. The stone surfaces used in the Roman arches were cut to make very tight joints; mortar was usually not used. The resistance to slipping between stones was provided by the compression force and the friction between the stone faces.

Another important architectural innovation was the pointed Gothic arch, shown in Figure 9.11c. This type of structure was first used in Europe beginning in the 12th century, followed by the construction of several magnificent Gothic cathedrals in France in the 13th century. One of the most striking features of these cathedrals is their extreme height. For example, the cathedral at Chartres rises to 118 ft, and the one at Reims has a height of 137 ft. Such magnificent buildings evolved over a very short time, without the benefit of any mathematical theory of structures. However, Gothic arches required flying buttresses to prevent the spreading of the arch supported by the tall, narrow columns.

9.4 Variation of Pressure with Depth

LEARNING OBJECTIVES

1. Develop the equation of hydrostatic equilibrium to explain the variation of pressure with depth in a fluid at rest.
2. Apply the equation of hydrostatic equilibrium to fluid systems.
3. Explain Pascal's principle and apply it to fluid systems.

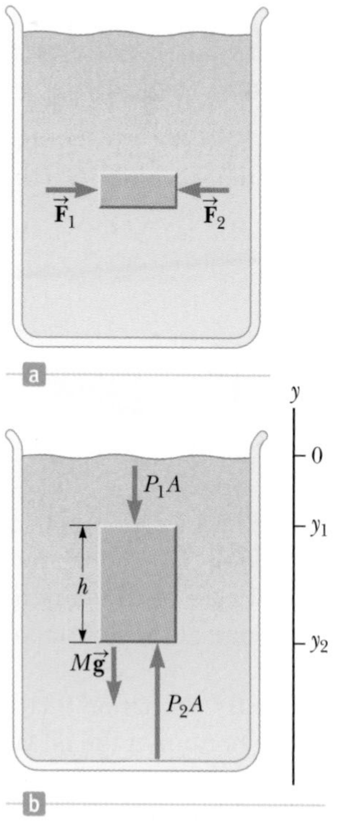

Figure 9.12 (a) In a static fluid, all points at the same depth are at the same pressure, so the force $\vec{F}_1$ must equal the force $\vec{F}_2$. (b) Because the volume of the shaded fluid isn't sinking or rising, the net force on it must equal zero.

When a fluid is at rest in a container, **all portions of the fluid must be in static equilibrium**—at rest with respect to the observer. Furthermore, **all points at the same depth must be at the same pressure.** If this were not the case, fluid would flow from the higher pressure region to the lower pressure region. For example, consider the small block of fluid shown in Figure 9.12a. If the pressure were greater on the left side of the block than on the right, $\vec{F}_1$ would be greater than $\vec{F}_2$, and the block would accelerate to the right and thus would not be in equilibrium.

Next, let's examine the fluid contained within the volume indicated by the darker region in Figure 9.12b. This region has cross-sectional area A and extends from position y_1 to position y_2 below the surface of the liquid. Three external forces act on this volume of fluid: the force of gravity, Mg; the upward force P_2A exerted by the liquid below it; and a downward force P_1A exerted by the fluid above it. Because the given volume of fluid is in equilibrium, these forces must add to zero, so we get

$$P_2A - P_1A - Mg = 0 \quad [9.8]$$

From the definition of density, we have

$$M = \rho V = \rho A(y_1 - y_2) \quad [9.9]$$

Substituting Equation 9.9 into Equation 9.8, canceling the area A, and rearranging terms, we get

$$P_2 = P_1 + \rho g(y_1 - y_2) \quad [9.10]$$

Figure 9.13 This photograph illustrates the fact that the pressure in a liquid is the same at all points lying at the same elevation. Note that the shape of the vessel does not affect the pressure.

Notice that $(y_1 - y_2)$ is positive, because $y_2 < y_1$. The force P_2A is greater than the force P_1A by exactly the weight of water between the two points. This is the same principle experienced by the person at the bottom of a pileup in football or rugby.

Atmospheric pressure is also caused by a piling up of fluid—in this case, the fluid is the gas of the atmosphere. The weight of all the air from sea level to the edge of space results in an atmospheric pressure of $P_0 = 1.013 \times 10^5$ Pa (equivalent to 14.7 lb/in.^{2}) at sea level. This result can be adapted to find the pressure P at any depth $h = (y_1 - y_2) = (0 - y_2)$ below the surface of the water:

$$P = P_0 + \rho gh \quad [9.11]$$

According to Equation 9.11, **the pressure P at a depth h below the surface of a liquid open to the atmosphere is greater than atmospheric pressure by the amount ρgh.** Moreover, the pressure isn't affected by the shape of the vessel, as shown in Figure 9.13. Equation 9.11 is often called the ***equation of hydrostatic equilibrium.*** (Similar, related equations also go by that name.)

Quick Quiz

9.2 The pressure at the bottom of a glass filled with water ($\rho = 1\,000$ kg/m^3) is P. The water is poured out and the glass is filled with ethyl alcohol ($\rho = 806$ kg/m^3). The pressure at the bottom of the glass is now (a) smaller than P (b) equal to P (c) larger than P (d) indeterminate.

EXAMPLE 9.5 Oil and Water

GOAL Calculate pressures created by layers of different fluids.

PROBLEM In a huge oil tanker, salt water has flooded an oil tank to a depth of $h_2 = 5.00$ m. On top of the water is a layer of oil $h_1 = 8.00$ m deep, as in the cross-sectional view of the tank in Figure 9.14. The oil has a density of 0.700 g/cm^3. Find the pressure at the bottom of the tank. (Take 1 025 kg/m^3 as the density of salt water.)

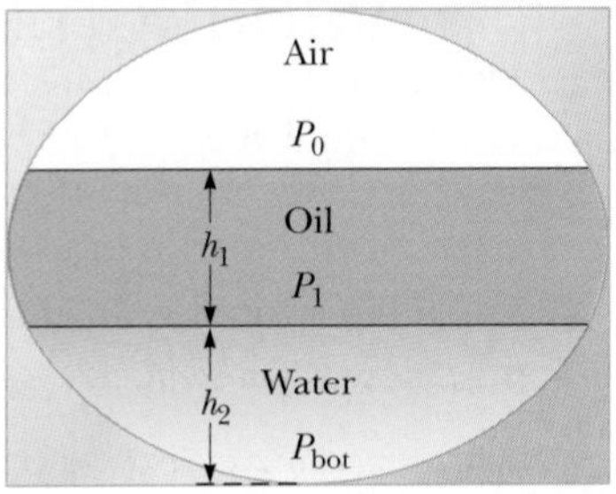

Figure 9.14 (Example 9.5)

STRATEGY Equation 9.11 must be used twice. First, use it to calculate the pressure P_1 at the bottom of the oil layer. Then use this pressure in place of P_0 in Equation 9.11 and calculate the pressure P_{bot} at the bottom of the water layer.

SOLUTION

Use Equation 9.11 to calculate the pressure at the bottom of the oil layer:

$$(1) \quad P_1 = P_0 + \rho g h_1$$
$$= 1.01 \times 10^5 \text{ Pa} + (7.00 \times 10^2 \text{ kg/m}^3)(9.80 \text{ m/s}^2)(8.00 \text{ m})$$
$$P_1 = 1.56 \times 10^5 \text{ Pa}$$

Now adapt Equation 9.11 to the new starting pressure, and use it to calculate the pressure at the bottom of the water layer:

$$(2)\quad P_{\text{bot}} = P_1 + \rho g h_2$$
$$= 1.56 \times 10^5 \text{ Pa} + (1.025 \times 10^3 \text{ kg/m}^3)(9.80 \text{ m/s}^2)(5.00 \text{ m})$$
$$P_{\text{bot}} = \boxed{2.06 \times 10^5 \text{ Pa}}$$

REMARKS The weight of the atmosphere results in P_0 at the surface of the oil layer. Then the weight of the oil and the weight of the water combine to create the pressure at the bottom.

QUESTION 9.5 Why does air pressure decrease with increasing altitude?

EXERCISE 9.5 Calculate the pressure on the top lid of a chest buried under 4.00 meters of mud with density equal to 1.75×10^3 kg/m^3 at the bottom of a 10.0-m-deep lake.

ANSWER 2.68×10^5 Pa

EXAMPLE 9.6 A Pain in the Ear BIO

GOAL Calculate a pressure difference at a given depth and estimate a force.

PROBLEM Estimate the net force exerted on your eardrum due to the water above when you are swimming at the bottom of a pool that is 5.0 m deep.

STRATEGY Use Equation 9.11 to find the pressure difference across the eardrum at the given depth. The air inside the ear is generally at atmospheric pressure. Estimate the eardrum's surface area, then use the definition of pressure to get the net force exerted on the eardrum.

SOLUTION

Use Equation 9.11 to calculate the difference between the water pressure at the depth h and the pressure inside the ear:

$$\Delta P = P - P_0 = \rho g h$$
$$= (1.00 \times 10^3 \text{ kg/m}^3)(9.80 \text{ m/s}^2)(5.0 \text{ m})$$
$$= 4.9 \times 10^4 \text{ Pa}$$

Multiply by area A to get the net force on the eardrum associated with this pressure difference, estimating the area of the eardrum as 1 cm^2.

$$F_{\text{net}} = A\Delta P \approx (1 \times 10^{-4} \text{ m}^2)(4.9 \times 10^4 \text{ Pa}) \approx \boxed{5 \text{ N}}$$

REMARKS Because a force on the eardrum of this magnitude is uncomfortable, swimmers often "pop their ears" by swallowing or expanding their jaws while underwater, an action that pushes air from the lungs into the middle ear. Using this technique equalizes the pressure on the two sides of the eardrum and relieves the discomfort.

QUESTION 9.6 Why do water containers and gas cans often have a second, smaller cap opposite the spout?

EXERCISE 9.6 An airplane takes off at sea level and climbs to a height of 425 m. Estimate the net outward force on a passenger's eardrum assuming the density of air is approximately constant at 1.3 kg/m^3 and that the inner ear pressure hasn't been equalized.

ANSWER 0.54 N

Because the pressure in a fluid depends on depth and on the value of P_0, any increase in pressure at the surface must be transmitted to every point in the fluid. This was first recognized by the French scientist Blaise Pascal (1623–1662) and is called **Pascal's principle:**

> A change in pressure applied to an enclosed fluid is transmitted undiminished to every point of the fluid and to the walls of the container.

An important application of Pascal's principle is the hydraulic press (Fig. 9.15a). A downward force $\vec{\mathbf{F}}_1$ is applied to a small piston of area A_1. The pressure is transmitted through a fluid to a larger piston of area A_2. As the pistons move

APPLICATION
Hydraulic Lifts

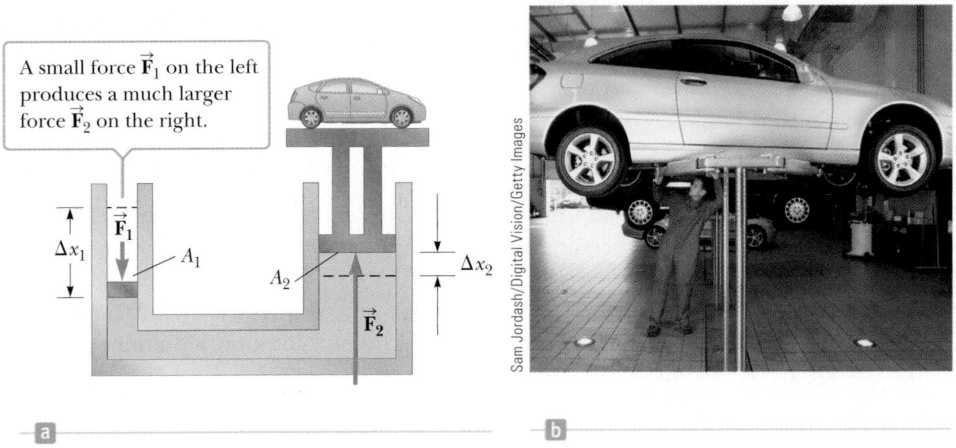

Figure 9.15 (a) In a hydraulic press, an increase of pressure in the smaller area A_1 is transmitted to the larger area A_2. Because force equals pressure times area, the force $\vec{F}_2$ is larger than $\vec{F}_1$ by a factor of A_2/A_1. (b) A vehicle under repair is supported by a hydraulic lift in a garage.

and the fluids in the left and right cylinders change their relative heights, there are slight differences in the pressures at the input and output pistons. Neglecting these small differences, the fluid pressure on each of the pistons may be taken to be the same; $P_1 = P_2$. From the definition of pressure, it then follows that $F_1/A_1 = F_2/A_2$. Therefore, the magnitude of the force $\vec{F}_2$ is larger than the magnitude of $\vec{F}_1$ by the factor A_2/A_1. That's why a large load, such as a car, can be moved on the large piston by a much smaller force on the smaller piston. Hydraulic brakes, car lifts, hydraulic jacks, forklifts, and other machines make use of this principle.

EXAMPLE 9.7 The Car Lift

GOAL Apply Pascal's principle to a car lift, and show that the input work is the same as the output work.

PROBLEM In a car lift used in a service station, compressed air exerts a force on a small piston of circular cross section having a radius of $r_1 = 5.00$ cm. This pressure is transmitted by an incompressible liquid to a second piston of radius $r_2 = 15.0$ cm. **(a)** What force must the compressed air exert on the small piston in order to lift a car weighing 13 300 N? Neglect the weights of the pistons. **(b)** What air pressure will produce a force of that magnitude? **(c)** Show that the work done by the input and output pistons is the same.

STRATEGY Substitute into Pascal's principle in part (a), while recognizing that the magnitude of the output force, F_2, must be equal to the car's weight in order to support it. Use the definition of pressure in part (b). In part (c), use $W = F\,\Delta x$ to find the ratio W_1/W_2, showing that it must equal 1. This requires combining Pascal's principle with the fact that the input and output pistons move through the same volume.

SOLUTION

(a) Find the necessary force on the small piston.

Substitute known values into Pascal's principle, using $A = \pi r^2$ for the area of each piston:

$$F_1 = \left(\frac{A_1}{A_2}\right)F_2 = \frac{\pi r_1^2}{\pi r_2^2}F_2$$

$$= \frac{\pi(5.00 \times 10^{-2}\text{ m})^2}{\pi(15.0 \times 10^{-2}\text{ m})^2}(1.33 \times 10^4\text{ N})$$

$$= \boxed{1.48 \times 10^3\text{ N}}$$

(b) Find the air pressure producing F_1.

Substitute into the definition of pressure:

$$P = \frac{F_1}{A_1} = \frac{1.48 \times 10^3\text{ N}}{\pi(5.00 \times 10^{-2}\text{ m})^2} = \boxed{1.88 \times 10^5\text{ Pa}}$$

(c) Show that the work done by the input and output pistons is the same.

First equate the volumes, and solve for the ratio of A_2 to A_1:

$$V_1 = V_2 \rightarrow A_1\Delta x_1 = A_2\Delta x_2$$

$$\frac{A_2}{A_1} = \frac{\Delta x_1}{\Delta x_2}$$

Now use Pascal's principle to get a relationship for F_1/F_2:

$$\frac{F_1}{A_1} = \frac{F_2}{A_2} \rightarrow \frac{F_1}{F_2} = \frac{A_1}{A_2}$$

Evaluate the work ratio, substituting the preceding two results:

$$\frac{W_1}{W_2} = \frac{F_1\,\Delta x_1}{F_2\,\Delta x_2} = \left(\frac{F_1}{F_2}\right)\left(\frac{\Delta x_1}{\Delta x_2}\right) = \left(\frac{A_1}{A_2}\right)\left(\frac{A_2}{A_1}\right) = 1$$

$$W_1 = W_2$$

REMARKS In this problem, we didn't address the effect of possible differences in the heights of the pistons. If the column of fluid is higher in the small piston, the fluid weight assists in supporting the car, reducing the necessary applied force. If the column of fluid is higher in the large piston, both the car and the extra fluid must be supported, so additional applied force is required.

QUESTION 9.7 True or False: If the radius of the output piston is doubled, the output force increases by a factor of 4.

EXERCISE 9.7 A hydraulic lift has pistons with diameters 8.00 cm and 36.0 cm, respectively. If a force of 825 N is exerted at the input piston, what maximum mass can be lifted by the output piston?

ANSWER 1.70×10^3 kg

APPLYING PHYSICS 9.2 | Building the Pyramids

A corollary to the statement that pressure in a fluid increases with depth is that water always seeks its own level. This means that if a vessel is filled with water, then regardless of the vessel's shape the surface of the water is perfectly flat and at the same height at all points. The ancient Egyptians used this fact to make the pyramids level. Devise a scheme showing how this could be done.

EXPLANATION There are many ways it could be done, but Figure 9.16 shows the scheme used by the Egyptians. The builders cut grooves in the base of the pyramid as in (a) and partially filled the grooves with water. The height of the water was marked as in (b), and the rock was chiseled down to the mark, as in (c). Finally, the groove was filled with crushed rock and gravel, as in (d). ■

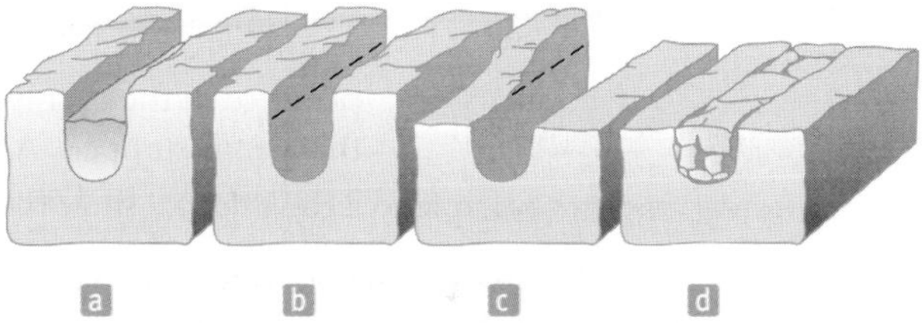

Figure 9.16 (Applying Physics 9.2)

9.5 Pressure Measurements

LEARNING OBJECTIVES

1. Define absolute pressure and gauge pressure.
2. Calculate pressures from fluid heights in a barometer.

A simple device for measuring pressure is the open-tube manometer (Fig. 9.17a). One end of a U-shaped tube containing a liquid is open to the atmosphere, and the other end is connected to a system of unknown pressure P. The pressure at point B equals $P_0 + \rho g h$, where ρ is the density of the fluid. The pressure at B, however, equals the pressure at A, which is also the unknown pressure P. We conclude that $P = P_0 + \rho g h$.

The pressure P is called the **absolute pressure,** and $P - P_0$ is called the **gauge pressure.** If P in the system is greater than atmospheric pressure, h is positive. If P is less than atmospheric pressure (a partial vacuum), h is negative, meaning that the right-hand column in Figure 9.17a is lower than the left-hand column.

Another instrument used to measure pressure is the **barometer** (Fig. 9.17b), invented by Evangelista Torricelli (1608–1647). A long tube closed at one end is filled with mercury and then inverted into a dish of mercury. The closed end of the tube is nearly a vacuum, so its pressure can be taken to be zero. It follows that $P_0 = \rho g h$, where ρ is the density of the mercury and h is the height of the mercury column. Note that the barometer measures the pressure of the atmosphere, whereas the manometer measures pressure in an enclosed fluid.

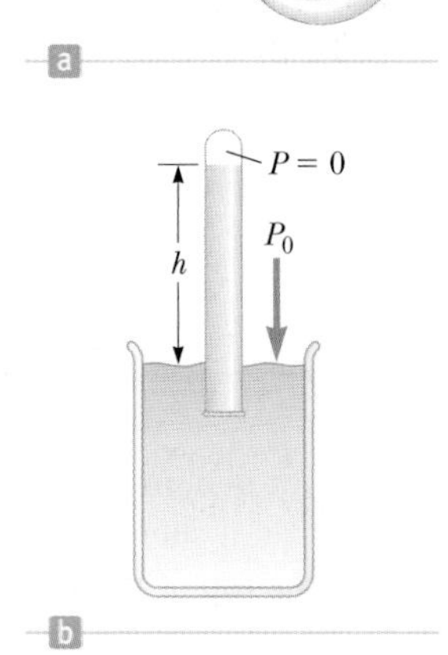

Figure 9.17 Two devices for measuring pressure: (a) an open-tube manometer and (b) a mercury barometer.

One atmosphere of pressure is defined to be the pressure equivalent of a column of mercury that is exactly 0.76 m in height at 0°C with $g = 9.806\ 65\ \text{m/s}^2$. At this temperature, mercury has a density of $13.595 \times 10^3\ \text{kg/m}^3$; therefore,

$$P_0 = \rho g h = (13.595 \times 10^3\ \text{kg/m}^3)(9.806\ 65\ \text{m/s}^2)(0.760\ 0\ \text{m})$$
$$= 1.013 \times 10^5\ \text{Pa} = 1\ \text{atm}$$

BIO APPLICATION
Decompression and Injury to the Lungs

It is interesting to note that the force of the atmosphere on our bodies (assuming a body area of 2 000 in.^{2}) is extremely large, on the order of 30 000 lb! If it were not for the fluids permeating our tissues and body cavities, our bodies would collapse. The fluids provide equal and opposite forces. In the upper atmosphere or in space, sudden decompression can lead to serious injury and death. Air retained in the lungs can damage the tiny alveolar sacs, and intestinal gas can even rupture internal organs.

Quick Quiz

9.3 Several common barometers are built using a variety of fluids. For which fluid will the column of fluid in the barometer be the highest? (Refer to Table 9.1.) (a) mercury (b) water (c) ethyl alcohol (d) benzene

BIO APPLICATION
Measuring Blood Pressure

Blood Pressure Measurements

A specialized manometer (called a sphygmomanometer) is often used to measure blood pressure. In this application, a rubber bulb forces air into a cuff wrapped tightly around the upper arm and simultaneously into a manometer, as in Figure 9.18. The pressure in the cuff is increased until the flow of blood through the brachial artery in the arm is stopped. A valve on the bulb is then opened, and the measurer listens with a stethoscope to the artery at a point just below the cuff. When the pressure in the cuff and brachial artery is just below the maximum value produced by the heart (the systolic pressure), the artery opens momentarily on each beat of the heart. At this point, the velocity of the blood is high and turbulent, and the flow is noisy and can be heard with the stethoscope. The manometer is calibrated to read the pressure in millimeters of mercury, and the value obtained is about 120 mm for a normal heart. Values of 130 mm or above are considered high, and medication to lower the blood pressure is often prescribed for such patients. As the pressure in the cuff is lowered further, intermittent sounds are still heard until the pressure falls just below the minimum heart pressure (the diastolic pressure). At this point, continuous sounds are heard. In the normal heart, this transition occurs at about 80 mm of mercury, and values above 90 require medical intervention. Blood pressure readings are usually expressed as the ratio of the systolic pressure to the diastolic pressure, which is 120/80 for a healthy heart.

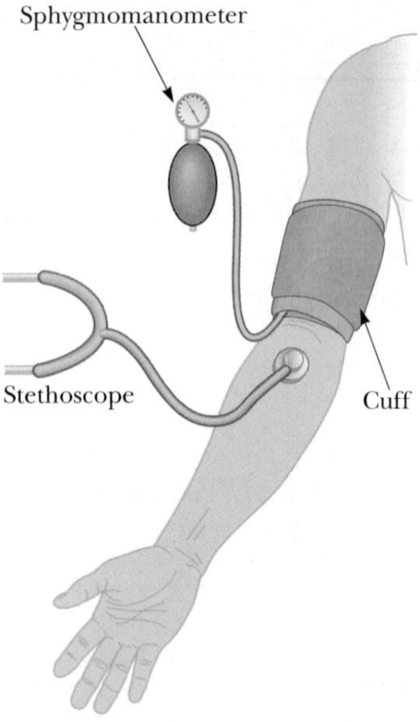

Figure 9.18 A sphygmomanometer can be used to measure blood pressure.

Quick Quiz

9.4 Blood pressure is normally measured with the cuff of the sphygmomanometer around the arm. Suppose the blood pressure is measured with the cuff around the calf of the leg of a standing person. Would the reading of the blood pressure be (a) the same here as it is for the arm, (b) greater than it is for the arm, or (c) less than it is for the arm?

APPLYING PHYSICS 9.3 | Ballpoint Pens

In a ballpoint pen, ink moves down a tube to the tip, where it is spread on a sheet of paper by a rolling stainless steel ball. Near the top of the ink cartridge, there is a small hole open to the atmosphere. If you seal this hole, you will find that the pen no longer functions. Use your knowledge of how a barometer works to explain this behavior.

EXPLANATION If the hole were sealed, or if it were not present, the pressure of the air above the ink would decrease as the ink was used. Consequently, atmospheric pressure exerted against the ink at the bottom of the cartridge would prevent some of the ink from flowing out. The hole allows the pressure above the ink to remain at atmospheric pressure. Why does a ballpoint pen seem to run out of ink when you write on a vertical surface? ■

When a fluid is in motion, its flow can be characterized in one of two ways. The flow is said to be **streamline,** or **laminar,** if every particle that passes a particular point moves along exactly the same smooth path followed by previous particles passing that point. This path is called a *streamline* (Fig. 9.27). Different streamlines can't cross each other under this steady-flow condition, and the streamline at any point coincides with the direction of the velocity of the fluid at that point.

Andy Sacks/Stone/Getty Images

Figure 9.27 An illustration of streamline flow around an automobile in a test wind tunnel. The streamlines in the airflow are made visible by smoke particles.

In contrast, the flow of a fluid becomes irregular, or **turbulent,** above a certain velocity or under any conditions that can cause abrupt changes in velocity. Irregular motions of the fluid, called *eddy currents,* are characteristic in turbulent flow, as shown in Figure 9.28.

In discussions of fluid flow, the term **viscosity** is used for the degree of internal friction in the fluid. This internal friction is associated with the resistance between two adjacent layers of the fluid moving relative to each other. A fluid such as kerosene has a lower viscosity than does crude oil or molasses.

Many features of fluid motion can be understood by considering the behavior of an **ideal fluid,** which satisfies the following conditions:

1. **The fluid is nonviscous,** which means there is no internal friction force between adjacent layers.
2. **The fluid is incompressible,** which means its density is constant.
3. **The fluid motion is steady,** meaning that the velocity, density, and pressure at each point in the fluid don't change with time.
4. **The fluid moves without turbulence.** This implies that each element of the fluid has zero angular velocity about its center, so there can't be any eddy currents present in the moving fluid. A small wheel placed in the fluid would translate but not rotate.

Zaichenko Olga/istockphoto.com

Figure 9.28 Hot gases made visible by smoke particles. The smoke first moves in laminar flow at the bottom and then in turbulent flow above.

Equation of Continuity

Figure 9.29a represents a fluid flowing through a pipe of nonuniform size. The particles in the fluid move along the streamlines in steady-state flow. In a small time interval Δt, the fluid entering the bottom end of the pipe moves a distance $\Delta x_1 = v_1 \Delta t$, where v_1 is the speed of the fluid at that location. If A_1 is the cross-sectional area in this region, then the mass contained in the bottom blue region is $\Delta M_1 = \rho_1 A_1 \Delta x_1 = \rho_1 A_1 v_1 \Delta t$, where ρ_1 is the density of the fluid at A_1. Similarly, the fluid that moves out of the upper end of the pipe in the same time interval Δt has a mass of $\Delta M_2 = \rho_2 A_2 v_2 \Delta t$. However, **because mass is conserved and because**

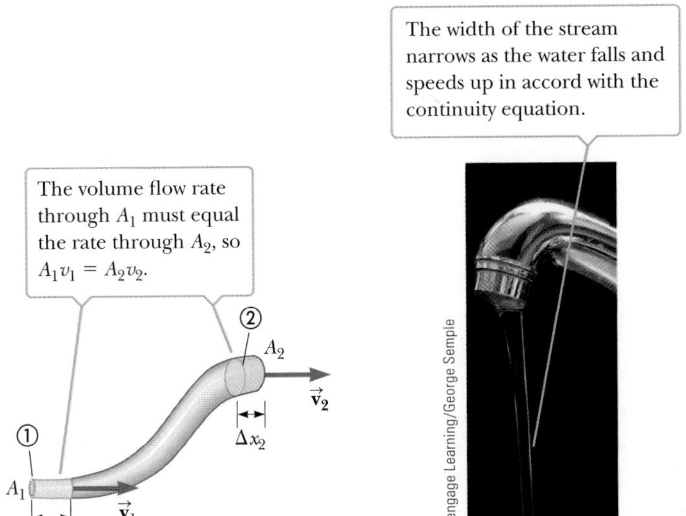

Figure 9.29 (a) A fluid moving with streamline flow through a pipe of varying cross-sectional area. The volume of fluid flowing through A_1 in a time interval Δt must equal the volume flowing through A_2 in the same time interval. (b) Water flowing slowly out of a faucet.

the flow is steady, the mass that flows into the bottom of the pipe through A_1 in the time Δt must equal the mass that flows out through A_2 in the same interval. Therefore, $\Delta M_1 = \Delta M_2$, or

$$\rho_1 A_1 v_1 = \rho_2 A_2 v_2 \quad [9.14]$$

For the case of an incompressible fluid, $\rho_1 = \rho_2$ and Equation 9.14 reduces to

Equation of continuity ▶

$$A_1 v_1 = A_2 v_2 \quad [9.15]$$

This expression is called the **equation of continuity.** From this result, we see that **the product of the cross-sectional area of the pipe and the fluid speed at that cross section is a constant.** Therefore, the speed is high where the tube is constricted and low where the tube has a larger diameter. The product Av, which has dimensions of volume per unit time, is called the **flow rate. The condition Av = constant is equivalent to the fact that the volume of fluid entering one end of the tube in a given time interval equals the volume of fluid leaving the tube in the same interval, assuming that the fluid is incompressible and there are no leaks.** Figure 9.29b is an example of an application of the equation of continuity: As the stream of water flows continuously from a faucet, the width of the stream narrows as it falls and speeds up.

Tip 9.3 Continuity Equations

The rate of flow of fluid into a system equals the rate of flow out of the system. The incoming fluid occupies a certain volume and can enter the system only if an equal volume of fluid is expelled during the same time interval.

There are many instances in everyday experience that involve the equation of continuity. Reducing the cross-sectional area of a garden hose by putting a thumb over the open end makes the water spray out with greater speed; hence the stream goes farther. Similar reasoning explains why smoke from a smoldering piece of wood first rises in a streamline pattern, getting thinner with height, eventually breaking up into a swirling, turbulent pattern. The smoke rises because it's less dense than air and the buoyant force of the air accelerates it upward. As the speed of the smoke stream increases, the cross-sectional area of the stream decreases, in accordance with the equation of continuity. The stream soon reaches a speed so great that streamline flow is not possible. We will study the relationship between speed of fluid flow and turbulence in a later discussion on the Reynolds number.

EXAMPLE 9.11 Niagara Falls

GOAL Apply the equation of continuity.

PROBLEM Each second, 5 525 m^3 of water flows over the 670-m-wide cliff of the Horseshoe Falls portion of Niagara Falls. The water is approximately 2 m deep as it reaches the cliff. Estimate its speed at that instant.

STRATEGY This is an estimate, so only one significant figure will be retained in the answer. The volume flow rate is given, and, according to the equation of continuity, is a constant equal to Av. Find the cross-sectional area, substitute, and solve for the speed.

SOLUTION

Calculate the cross-sectional area of the water as it reaches the edge of the cliff:

$$A = (670 \text{ m})(2 \text{ m}) = 1\,340 \text{ m}^2$$

Multiply this result by the speed and set it equal to the flow rate. Then solve for v:

$$Av = \text{volume flow rate}$$

$$(1\,340 \text{ m}^2)v = 5\,525 \text{ m}^3/\text{s} \quad \rightarrow \quad v \approx 4 \text{ m/s}$$

QUESTION 9.11 What happens to the speed of blood in an artery when plaque starts to build up on the artery's sides?

EXERCISE 9.11 The Garfield Thomas water tunnel at Pennsylvania State University has a circular cross section that constricts from a diameter of 3.6 m to the test section which has a diameter of 1.2 m. If the speed of flow is 3.0 m/s in the larger-diameter pipe, determine the speed of flow in the test section.

ANSWER 27 m/s

4. Figure CQ9.4 shows aerial views from directly above two dams. Both dams are equally long (the vertical dimension in the diagram) and equally deep (into the page in the diagram). The dam on the left holds back a very large lake, while the dam on the right holds back a narrow river. Which dam has to be built more strongly?

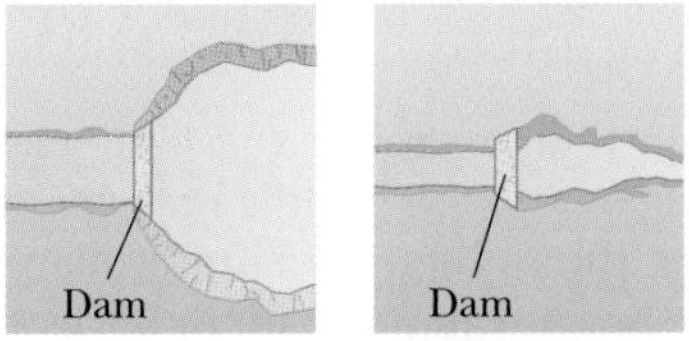

Figure CQ9.4

5. A typical silo on a farm has many bands wrapped around its perimeter, as shown in Figure CQ9.5. Why is the spacing between successive bands smaller at the lower portions of the silo?

Henry Leap and Jim Lehman

Figure CQ9.5

6. Many people believe that a vacuum created inside a vacuum cleaner causes particles of dirt to be drawn in. Actually, the dirt is pushed in. Explain.
7. Suppose a damaged ship just barely floats in the ocean after a hole in its hull has been sealed. It is pulled by a tugboat toward shore and into a river, heading toward a dry dock for repair. As the boat is pulled up the river, it sinks. Why?
8. BIO During inhalation, the pressure in the lungs is slightly less than external pressure and the muscles controlling exhalation are relaxed. Under water, the body equalizes internal and external pressures. Discuss the condition of the muscles if a person under water is breathing through a snorkel. Would a snorkel work in deep water?
9. The water supply for a city is often provided from reservoirs built on high ground. Water flows from the reservoir, through pipes, and into your home when you turn the tap on your faucet. Why is the water flow more rapid out of a faucet on the first floor of a building than in an apartment on a higher floor?
10. An ice cube is placed in a glass of water. What happens to the level of the water as the ice melts?
11. Place two cans of soft drinks, one regular and one diet, in a container of water. You will find that the diet drink floats while the regular one sinks. Use Archimedes' principle to devise an explanation. *Broad Hint:* The artificial sweetener used in diet drinks is less dense than sugar.
12. Will an ice cube float higher in water or in an alcoholic beverage?
13. Tornadoes and hurricanes often lift the roofs of houses. Use the Bernoulli effect to explain why. Why should you keep your windows open under these conditions?
14. Once ski jumpers are airborne (Fig. CQ9.14), why do they bend their bodies forward and keep their hands at their sides?

© iStockPhoto/technotr

Figure CQ9.14

15. A person in a boat floating in a small pond throws an anchor overboard. What happens to the level of the pond? (a) It rises. (b) It falls. (c) It remains the same.
16. One of the predicted problems due to global warming is that ice in the polar ice caps will melt and raise sea levels everywhere in the world. Is that more of a worry for ice (a) at the north pole, where most of the ice floats on water; (b) at the south pole, where most of the ice sits on land; (c) both at the north and south poles equally; or (d) at neither pole?

PROBLEMS

ENHANCED WebAssign The problems in this chapter may be assigned online in Enhanced WebAssign.

1. denotes straightforward problem; 2. denotes intermediate problem;
3. denotes challenging problem
[1.] denotes full solution available in *Student Solutions Manual/Study Guide*
1. denotes problems most often assigned in Enhanced WebAssign

BIO denotes biomedical problems
GP denotes guided problems
M denotes Master It tutorial available in Enhanced WebAssign
Q|C denotes asking for quantitative and conceptual reasoning
S denotes symbolic reasoning problem
W denotes Watch It video solution available in Enhanced WebAssign

9.1 States of Matter

9.2 Density and Pressure

1. S Suppose two worlds, each having mass M and radius R, coalesce into a single world. Due to gravitational contraction, the combined world has a radius of only $\frac{3}{4}R$. What is the average density of the combined world as a multiple of ρ_0, the average density of the original two worlds?

2. The British gold sovereign coin is an alloy of gold and copper having a total mass of 7.988 g, and is 22-karat gold. (a) Find the mass of gold in the sovereign in kilograms using the fact that the number of karats = 24 × (mass of gold)/(total mass). (b) Calculate the volumes of gold and copper, respectively, used to manufacture the coin. (c) Calculate the density of the British sovereign coin.

3. Q|C Four acrobats of mass 75.0 kg, 68.0 kg, 62.0 kg, and 55.0 kg form a human tower, with each acrobat standing on the shoulders of another acrobat. The 75.0-kg acrobat is at the bottom of the tower. (a) What is the normal force acting on the 75-kg acrobat? (b) If the area of each of the 75.0-kg acrobat's shoes is 425 cm^2, what average pressure (not including atmospheric pressure) does the column of acrobats exert on the floor? (c) Will the pressure be the same if a different acrobat is on the bottom?

4. M Calculate the mass of a solid gold rectangular bar that has dimensions of 4.50 cm × 11.0 cm × 26.0 cm.

5. Q|C The nucleus of an atom can be modeled as several protons and neutrons closely packed together. Each particle has a mass of 1.67×10^{-27} kg and radius on the order of 10^{-15} m. (a) Use this model and the data provided to estimate the density of the nucleus of an atom. (b) Compare your result with the density of a material such as iron. What do your result and comparison suggest about the structure of matter?

6. The four tires of an automobile are inflated to a gauge pressure of 2.0×10^5 Pa. Each tire has an area of 0.024 m^2 in contact with the ground. Determine the weight of the automobile.

7. Suppose a distant world with surface gravity of 7.44 m/s^2 has an atmospheric pressure of 8.04×10^4 Pa at the surface. (a) What force is exerted by the atmosphere on a disk-shaped region 2.00 m in radius at the surface of a methane ocean? (b) What is the weight of a 10.0-m deep cylindrical column of methane with radius 2.00 m? (c) Calculate the pressure at a depth of 10.0 m in the methane ocean. *Note:* The density of liquid methane is 415 kg/m^3.

9.3 The Deformation of Solids

8. Evaluate Young's modulus for the material whose stress–strain curve is shown in Figure 9.8.

9. M A 200-kg load is hung on a wire of length 4.00 m, cross-sectional area 0.200×10^{-4} m^2, and Young's modulus 8.00×10^{10} N/m^2. What is its increase in length?

10. Q|C Comic-book superheroes are sometimes able to punch holes through steel walls. (a) If the ultimate shear strength of steel is taken to be 2.50×10^8 Pa, what force is required to punch through a steel plate 2.00 cm thick? Assume the superhero's fist has cross-sectional area of 1.00×10^2 cm^2 and is approximately circular. (b) Qualitatively, what would happen to the superhero on delivery of the punch? What physical law applies?

11. A plank 2.00 cm thick and 15.0 cm wide is firmly attached to the railing of a ship by clamps so that the rest of the board extends 2.00 m horizontally over the sea below. A man of mass 80.0 kg is forced to stand on the very end. If the end of the board drops by 5.00 cm because of the man's weight, find the shear modulus of the wood.

12. M Assume that if the shear stress in steel exceeds about 4.00×10^8 N/m^2, the steel ruptures. Determine the shearing force necessary to (a) shear a steel bolt 1.00 cm in diameter and (b) punch a 1.00-cm-diameter hole in a steel plate 0.500 cm thick.

13. For safety in climbing, a mountaineer uses a nylon rope that is 50 m long and 1.0 cm in diameter. When supporting a 90-kg climber, the rope elongates 1.6 m. Find its Young's modulus.

14. BIO A stainless-steel orthodontic wire is applied to a tooth, as in Figure P9.14. The wire has an unstretched length of 3.1 cm and a radius of 0.11 mm. If the wire is stretched 0.10 mm, find the magnitude and direction of the force on the tooth. Disregard the width of the tooth and assume Young's modulus for stainless steel is 18×10^{10} Pa.

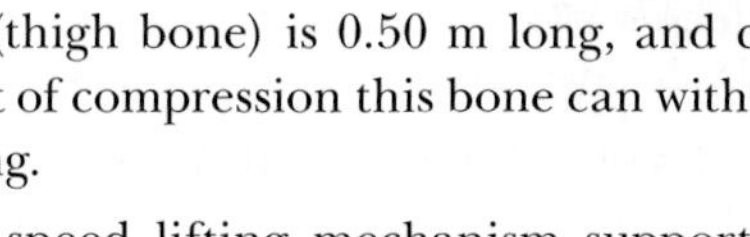

Figure P9.14

15. BIO Bone has a Young's modulus of 18×10^9 Pa. Under compression, it can withstand a stress of about 160×10^6 Pa before breaking. Assume that a femur (thigh bone) is 0.50 m long, and calculate the amount of compression this bone can withstand before breaking.

16. A high-speed lifting mechanism supports an 800-kg object with a steel cable that is 25.0 m long and 4.00 cm^2 in cross-sectional area. (a) Determine the elongation of the cable. (b) By what additional amount does the cable increase in length if the object is accelerated upward at a rate of 3.0 m/s^2? (c) What is the greatest mass that can be accelerated upward at 3.0 m/s^2 if the stress in the cable is not to exceed the elastic limit of the cable, which is 2.2×10^8 Pa?

17. A walkway suspended across a hotel lobby is supported at numerous points along its edges by a vertical cable above each point and a vertical column underneath.

The steel cable is 1.27 cm in diameter and is 5.75 m long before loading. The aluminum column is a hollow cylinder with an inside diameter of 16.14 cm, an outside diameter of 16.24 cm, and unloaded length of 3.25 m. When the walkway exerts a load force of 8 500 N on one of the support points, how much does the point move down?

18. BIO The total cross-sectional area of the load-bearing calcified portion of the two forearm bones (radius and ulna) is approximately 2.4 cm^2. During a car crash, the forearm is slammed against the dashboard. The arm comes to rest from an initial speed of 80 km/h in 5.0 ms. If the arm has an effective mass of 3.0 kg and bone material can withstand a maximum compressional stress of 16×10^7 Pa, is the arm likely to withstand the crash?

19. Determine the elongation of the rod in Figure P9.19 if it is under a tension of 5.8×10^3 N.

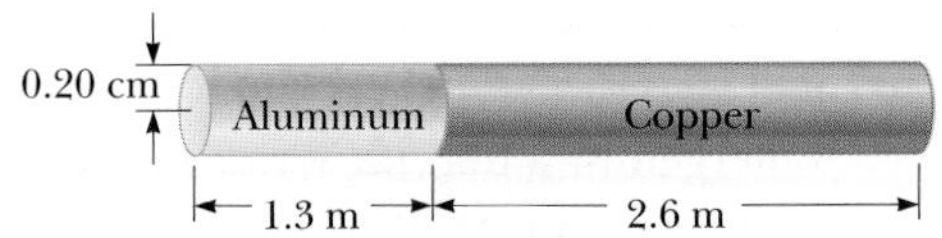

Figure P9.19

9.4 Variation of Pressure with Depth

9.5 Pressure Measurements

20. The spring of the pressure gauge shown in Figure P9.20 has a force constant of 1 250 N/m, and the piston has a radius of 1.20 cm. As the gauge is lowered into water, what change in depth causes the piston to move in by 0.750 cm?

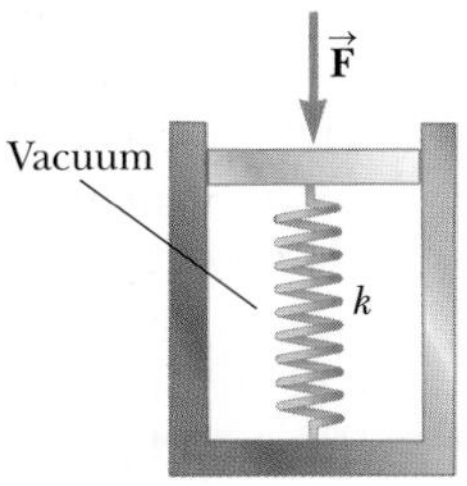

Figure P9.20

21. (a) Calculate the absolute pressure at the bottom of a fresh-water lake at a depth of 27.5 m. Assume the density of the water is 1.00×10^3 kg/m^3 and the air above is at a pressure of 101.3 kPa. (b) What force is exerted by the water on the window of an underwater vehicle at this depth if the window is circular and has a diameter of 35.0 cm?

22. Mercury is poured into a U-tube as shown in Figure P9.22a. The left arm of the tube has cross-sectional area A_1 of 10.0 cm^2, and the right arm has a cross-sectional area A_2 of 5.00 cm^2. One hundred grams of water are then poured into the right arm as shown in Figure P9.22b. (a) Determine the length of the water column in the right arm of the U-tube. (b) Given that the density of mercury is 13.6 g/cm^3, what distance h does the mercury rise in the left arm?

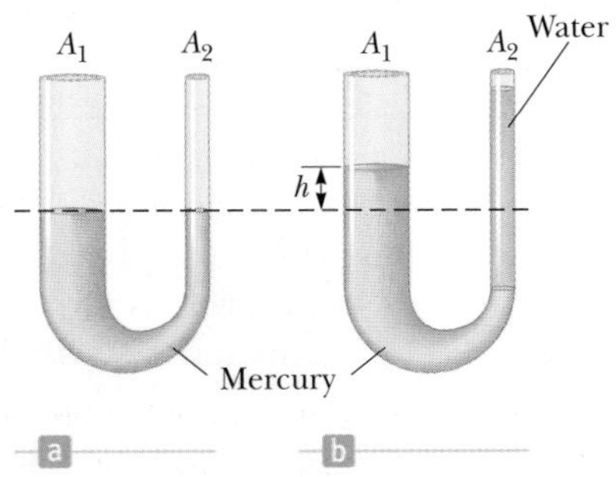

Figure P9.22

23. BIO A collapsible plastic bag (Fig. P9.23) contains a glucose solution. If the average gauge pressure in the vein is 1.33×10^3 Pa, what must be the minimum height h of the bag in order to infuse glucose into the vein? Assume the specific gravity of the solution is 1.02.

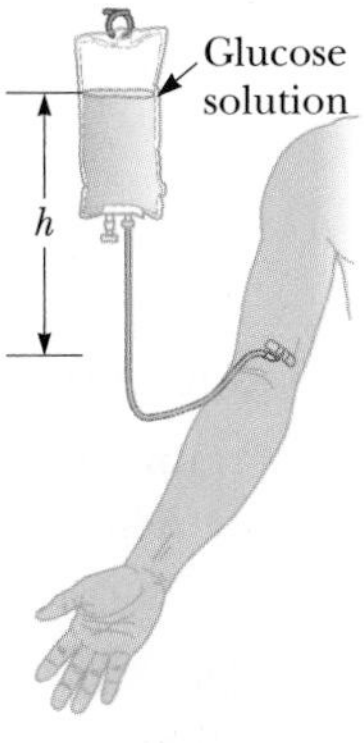

Figure P9.23

24. The deepest point in the ocean is in the Mariana Trench, about 11 km deep. The pressure at the ocean floor is huge, about 1.13×10^8 N/m^2. (a) Calculate the change in volume of 1.00 m^3 of water carried from the surface to the bottom of the Pacific. (b) The density of water at the surface is 1.03×10^3 kg/m^3. Find its density at the bottom. (c) Explain whether or when it is a good approximation to think of water as incompressible.

25. W A container is filled to a depth of 20.0 cm with water. On top of the water floats a 30.0-cm-thick layer of oil with specific gravity 0.700. What is the absolute pressure at the bottom of the container?

26. Blaise Pascal duplicated Torricelli's barometer using a red Bordeaux wine, of density 984 kg/m^3 as the working liquid (Fig. P9.26). (a) What was the height h of the wine column for normal atmospheric pressure? (b) Would you expect the vacuum above the column to be as good as for mercury?

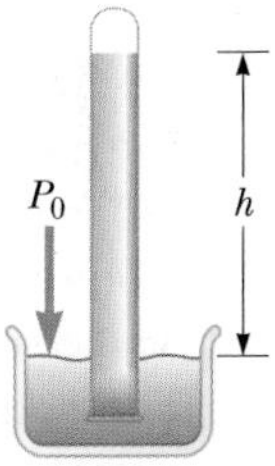

Figure P9.26

27. Figure P9.27 shows the essential parts of a hydraulic brake system. The area of the piston in the master cylinder is 1.8 cm^2 and that of the piston in the brake cylinder is 6.4 cm^2. The coefficient of friction between shoe and wheel drum is 0.50. If the wheel has a radius of 34 cm, determine the frictional torque about the axle when a force of 44 N is exerted on the brake pedal.

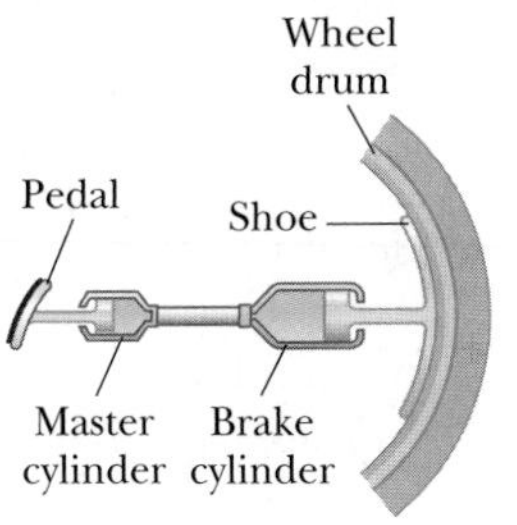

Figure P9.27

28. W Piston ① in Figure P9.28 has a diameter of 0.25 in.; piston ② has a diameter of 1.5 in. In the absence of friction, determine the force $\vec{F}$ necessary to support the 500-lb weight.

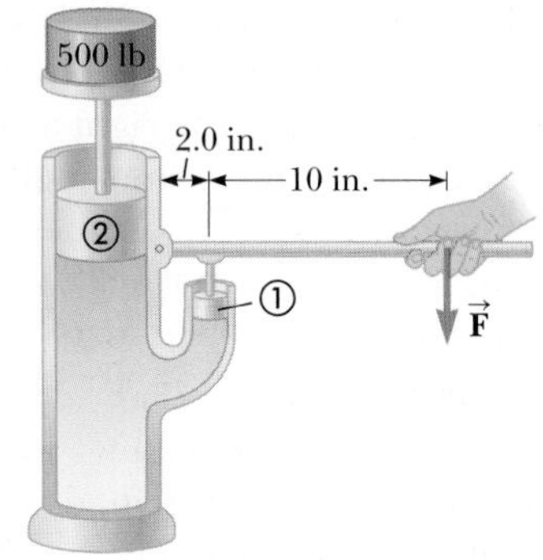

Figure P9.28

9.6 Buoyant Forces and Archimedes' Principle

29. A table-tennis ball has a diameter of 3.80 cm and average density of 0.084 0 g/cm^3. What force is required to hold it completely submerged under water?

30. BIO Q|C The average human has a density of 945 kg/m^3 after inhaling and 1 020 kg/m^3 after exhaling. (a) Without making any swimming movements, what percentage of the human body would be above the surface in the Dead Sea (a body of water with a density of about 1 230 kg/m^3) in each of these cases? (b) Given that bone and muscle are denser than fat, what physical characteristics differentiate "sinkers" (those who tend to sink in water) from "floaters" (those who readily float)?

31. A small ferryboat is 4.00 m wide and 6.00 m long. When a loaded truck pulls onto it, the boat sinks an additional 4.00 cm into the river. What is the weight of the truck?

32. GP A 62.0-kg survivor of a cruise line disaster rests atop a block of Styrofoam insulation, using it as a raft. The Styrofoam has dimensions 2.00 m × 2.00 m × 0.090 0 m. The bottom 0.024 m of the raft is submerged. (a) Draw a force diagram of the system consisting of the survivor and raft. (b) Write Newton's second law for the system in one dimension, using B for buoyancy, w for the weight of the survivor, and w_r for the weight of the raft. (Set $a = 0$.) (c) Calculate the numeric value for the buoyancy, B. (Seawater has density 1 025 kg/m^3.) (d) Using the value of B and the weight w of the survivor, calculate the weight w_r of the Styrofoam. (e) What is the density of the Styrofoam? (f) What is the maximum buoyant force, corresponding to the raft being submerged up to its top surface? (g) What total mass of survivors can the raft support?

33. Q|C A wooden block of volume 5.24×10^{-4} m^3 floats in water, and a small steel object of mass m is placed on top of the block. When $m = 0.310$ kg, the system is in equilibrium, and the top of the wooden block is at the level of the water. (a) What is the density of the wood? (b) What happens to the block when the steel object is replaced by a second steel object with a mass less than 0.310 kg? What happens to the block when the steel object is replaced by yet another steel object with a mass greater than 0.310 kg?

34. GP A large balloon of mass 226 kg is filled with helium gas until its volume is 325 m^3. Assume the density of air is 1.29 kg/m^3 and the density of helium is 0.179 kg/m^3. (a) Draw a force diagram for the balloon. (b) Calculate the buoyant force acting on the balloon. (c) Find the net force on the balloon and determine whether the balloon will rise or fall after it is released. (d) What maximum additional mass can the balloon support in equilibrium? (e) What happens to the balloon if the mass of the load is *less* than the value calculated in part (d)? (f) What limits the height to which the balloon can rise?

35. Q|C A spherical weather balloon is filled with hydrogen until its radius is 3.00 m. Its total mass including the instruments it carries is 15.0 kg. (a) Find the buoyant force acting on the balloon, assuming the density of air is 1.29 kg/m^3. (b) What is the net force acting on the balloon and its instruments after the balloon is released from the ground? (c) Why does the radius of the balloon tend to increase as it rises to higher altitude?

36. S A man of mass $m = 70.0$ kg and having a density of $\rho = 1\,050$ kg/m^3 (while holding his breath) is completely submerged in water. (a) Write Newton's second law for this situation in terms of the man's mass m, the density of water ρ_w, his volume V, and g. Neglect any viscous drag of the water. (b) Substitute $m = \rho V$ into Newton's second law and solve for the acceleration a, canceling common factors. (c) Calculate the numeric value of the man's acceleration. (d) How long does it take the man to sink 8.00 m to the bottom of the lake?

37. Q|C On October 21, 2001, Ian Ashpole of the United Kingdom achieved a record altitude of 3.35 km (11 000 ft) powered by 600 toy balloons filled with helium. Each filled balloon had a radius of about 0.50 m and an estimated mass of 0.30 kg. (a) Estimate the total buoyant force on the 600 balloons. (b) Estimate the net upward force on all 600 balloons. (c) Ashpole parachuted to Earth after the balloons began to burst at the high altitude and the system lost buoyancy. Why did the balloons burst?

38. W The gravitational force exerted on a solid object is 5.00 N as measured when the object is suspended from a spring scale as in Figure P9.38a. When the suspended object is submerged in water, the scale reads 3.50 N (Fig. P9.38b). Find the density of the object.

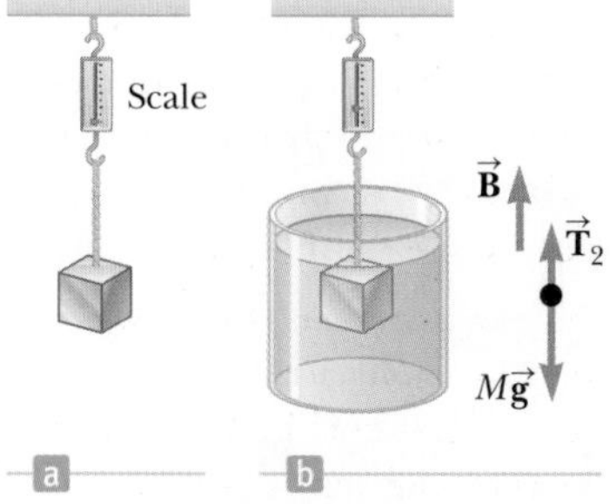

Figure P9.38

39. **M** A cube of wood having an edge dimension of 20.0 cm and a density of 650 kg/m^3 floats on water. (a) What is the distance from the horizontal top surface of the cube to the water level? (b) What mass of lead should be placed on the cube so that the top of the cube will be just level with the water surface?

40. A light spring of force constant k = 160 N/m rests vertically on the bottom of a large beaker of water (Fig. P9.40a). A 5.00-kg block of wood (density = 650 kg/m^3) is connected to the spring, and the block–spring system is allowed to come to static equilibrium (Fig. P9.40b). What is the elongation ΔL of the spring?

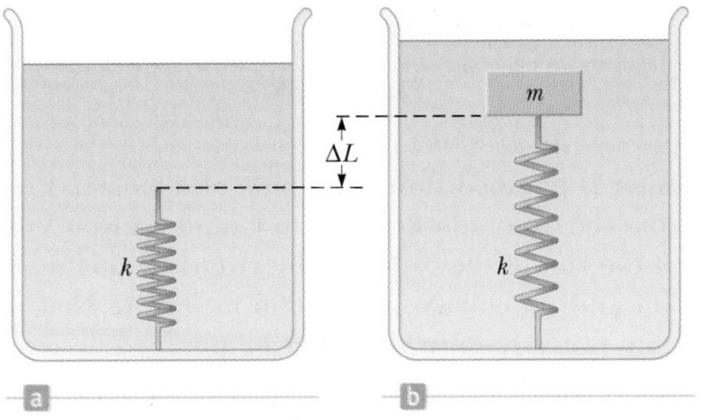

Figure P9.40

41. A sample of an unknown material appears to weigh 300 N in air and 200 N when immersed in alcohol of specific gravity 0.700. What are (a) the volume and (b) the density of the material?

42. An object weighing 300 N in air is immersed in water after being tied to a string connected to a balance. The scale now reads 265 N. Immersed in oil, the object appears to weigh 275 N. Find (a) the density of the object and (b) the density of the oil.

43. A 1.00-kg beaker containing 2.00 kg of oil (density = 916 kg/m^3) rests on a scale. A 2.00-kg block of iron is suspended from a spring scale and is completely submerged in the oil (Fig. P9.43). Find the equilibrium readings of both scales.

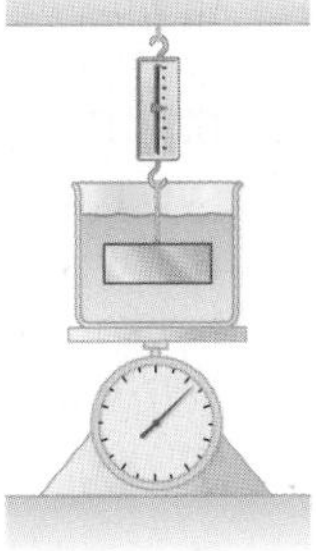

Figure P9.43

9.7 Fluids in Motion

9.8 Other Applications of Fluid Dynamics

44. Water flowing through a garden hose of diameter 2.74 cm fills a 25.0-L bucket in 1.50 min. (a) What is the speed of the water leaving the end of the hose? (b) A nozzle is now attached to the end of the hose. If the nozzle diameter is one-third the diameter of the hose, what is the speed of the water leaving the nozzle?

45. **BIO** (a) Calculate the mass flow rate (in grams per second) of blood (ρ = 1.0 g/cm^3) in an aorta with a cross-sectional area of 2.0 cm^2 if the flow speed is 40 cm/s. (b) Assume that the aorta branches to form a large number of capillaries with a combined cross-sectional area of 3.0×10^3 cm^2. What is the flow speed in the capillaries?

46. **W** A liquid (ρ = 1.65 g/cm^3) flows through a horizontal pipe of varying cross section as in Figure P9.46. In the first section, the cross-sectional area is 10.0 cm^2, the flow speed is 275 cm/s, and the pressure is 1.20×10^5 Pa. In the second section, the cross-sectional area is 2.50 cm^2. Calculate the smaller section's (a) flow speed and (b) pressure.

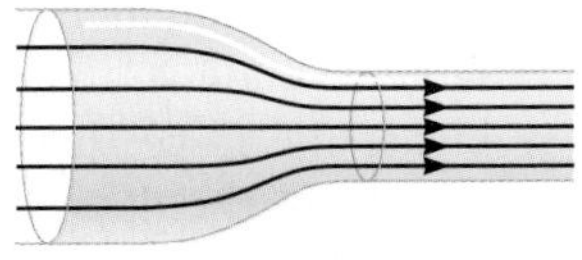

Figure P9.46

47. **BIO** A hypodermic syringe contains a medicine with the density of water (Fig. P9.47). The barrel of the syringe has a cross-sectional area of 2.50×10^{-5} m^2. In the absence of a force on the plunger, the pressure everywhere is 1.00 atm. A force $\vec{\mathbf{F}}$ of magnitude 2.00 N is exerted on the plunger, making medicine squirt from the needle. Determine the medicine's flow speed through the needle. Assume the pressure in the needle remains equal to 1.00 atm and that the syringe is horizontal.

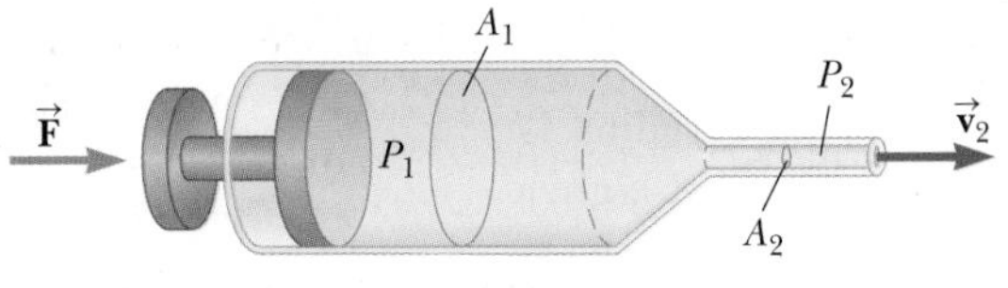

Figure P9.47

48. **BIO** When a person inhales, air moves down the bronchus (windpipe) at 15 cm/s. The average flow speed of the air doubles through a constriction in the bronchus. Assuming incompressible flow, determine the pressure drop in the constriction.

49. **QIC** A jet airplane in level flight has a mass of 8.66×10^4 kg, and the two wings have an estimated total area of 90.0 m^2. (a) What is the pressure difference between the lower and upper surfaces of the wings? (b) If the speed of air under the wings is 225 m/s, what is the speed of the air over the wings? Assume air has a density of 1.29 kg/m^3. (c) Explain why all aircraft have a "ceiling," a maximum operational altitude.

50. **S** An airplane has a mass M, and the two wings have a total area A. During level flight, the pressure on the lower wing surface is P_1. Determine the pressure P_2 on the upper wing surface.

51. **GP** In a water pistol, a piston drives water through a larger tube of radius 1.00 cm into a smaller tube of radius 1.00 mm as in Figure P9.51. (a) If the pistol is fired horizontally at a height of 1.50 m, use ballistics to determine the time it takes water to travel from the nozzle to the ground. (Neglect air resistance and assume atmospheric pressure is 1.00 atm.) (b) If the range of the stream is to be 8.00 m, with what speed must the stream leave the nozzle? (c) Given the areas of the nozzle and cylinder, use the equation of continuity to

calculate the speed at which the plunger must be moved. (d) What is the pressure at the nozzle? (e) Use Bernoulli's equation to find the pressure needed in the larger cylinder. Can gravity terms be neglected? (f) Calculate the force that must be exerted on the trigger to achieve the desired range. (The force that must be exerted is due to pressure over and above atmospheric pressure.)

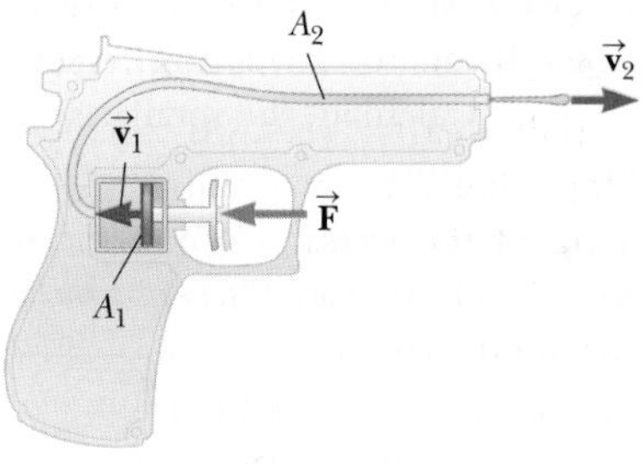

Figure P9.51

52. Water moves through a constricted pipe in steady, ideal flow. At the lower point shown in Figure P9.52, the pressure is 1.75×10^5 Pa and the pipe radius is 3.00 cm. At the higher point located at y = 2.50 m, the pressure is 1.20×10^5 Pa and the pipe radius is 1.50 cm. Find the speed of flow (a) in the lower section and (b) in the upper section. (c) Find the volume flow rate through the pipe.

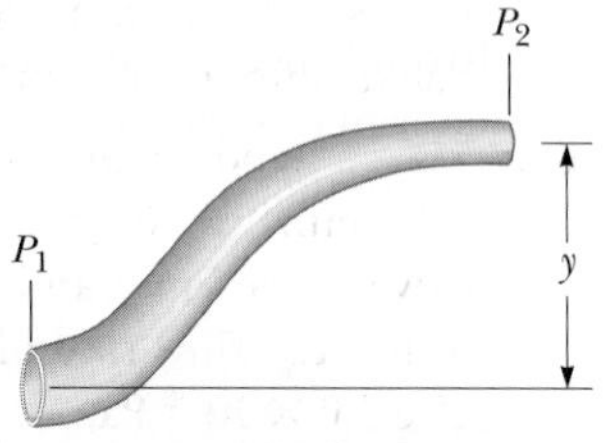

Figure P9.52

53. M A jet of water squirts out horizontally from a hole near the bottom of the tank shown in Figure P9.53. If the hole has a diameter of 3.50 mm, what is the height h of the water level in the tank?

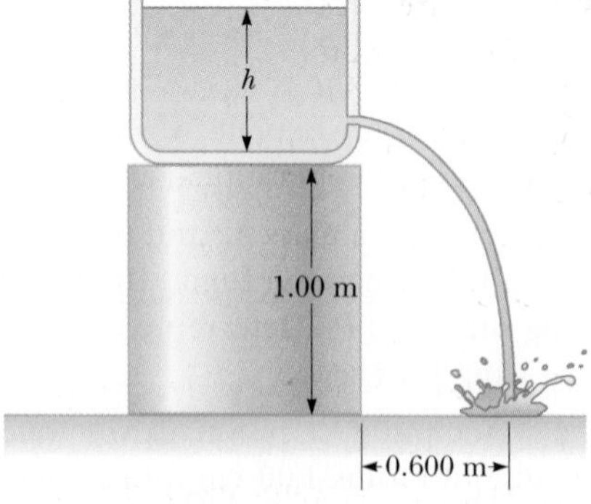

Figure P9.53

54. A large storage tank, open to the atmosphere at the top and filled with water, develops a small hole in its side at a point 16.0 m below the water level. If the rate of flow from the leak is 2.50×10^{-3} m^3/min, determine (a) the speed at which the water leaves the hole and (b) the diameter of the hole.

55. The inside diameters of the larger portions of the horizontal pipe depicted in Figure P9.55 are 2.50 cm. Water flows to the right at a rate of 1.80×10^{-4} m^3/s. Determine the inside diameter of the constriction.

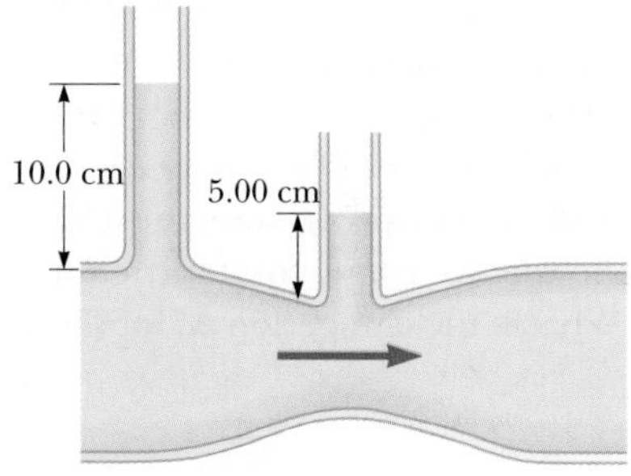

Figure P9.55

56. Water is pumped through a pipe of diameter 15.0 cm from the Colorado River up to Grand Canyon Village, on the rim of the canyon. The river is at 564 m elevation and the village is at 2 096 m. (a) At what minimum pressure must the water be pumped to arrive at the village? (b) If 4 500 m^3 are pumped per day, what is the speed of the water in the pipe? (c) What additional pressure is necessary to deliver this flow? *Note:* You may assume the free-fall acceleration and the density of air are constant over the given range of elevations.

57. Old Faithful geyser in Yellowstone Park erupts at approximately 1-hour intervals, and the height of the fountain reaches 40.0 m (Fig. P9.57). (a) Consider the rising stream as a series of separate drops. Analyze the free-fall motion of one of the drops to determine the speed at which the water leaves the ground. (b) Treat the rising stream as an ideal fluid in streamline flow. Use Bernoulli's equation to determine the speed of the water as it leaves ground level. (c) What is the pressure (above atmospheric pressure) in the heated underground chamber 175 m below the vent? You may assume the chamber is large compared with the geyser vent.

Drflet>iStockphoto.com

Figure P9.57

58. The Venturi tube shown in Figure P9.58 may be used as a fluid flowmeter. Suppose the device is used at a service station to measure the flow rate of gasoline ($\rho = 7.00 \times 10^2$ kg/m^3) through a hose having an outlet radius of 1.20 cm. If the difference in pressure is measured to be $P_1 - P_2 = 1.20$ kPa and the radius of the inlet tube to the meter is 2.40 cm, find (a) the speed of the gasoline as it leaves the hose and (b) the fluid flow rate in cubic meters per second.

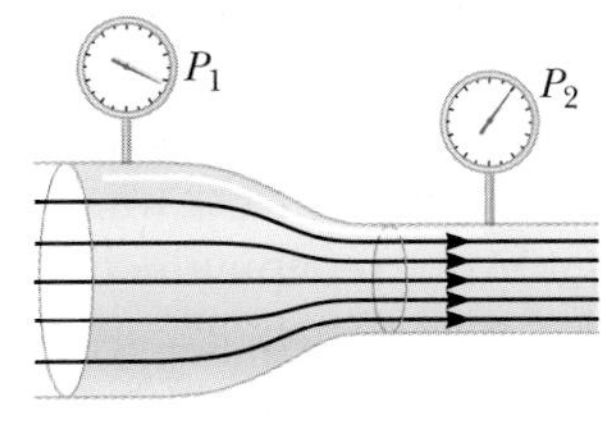

Figure P9.58

9.9 Surface Tension, Capillary Action, and Viscous Fluid Flow

59. A square metal sheet 3.0 cm on a side and of negligible thickness is attached to a balance and inserted into a container of fluid. The contact angle is found to be zero, as shown in Figure P9.59a, and the balance to which the metal sheet is attached reads 0.40 N. A thin veneer of oil is then spread over the sheet, and the contact angle becomes 180°, as shown in Figure P9.59b. The balance now reads 0.39 N. What is the surface tension of the fluid?

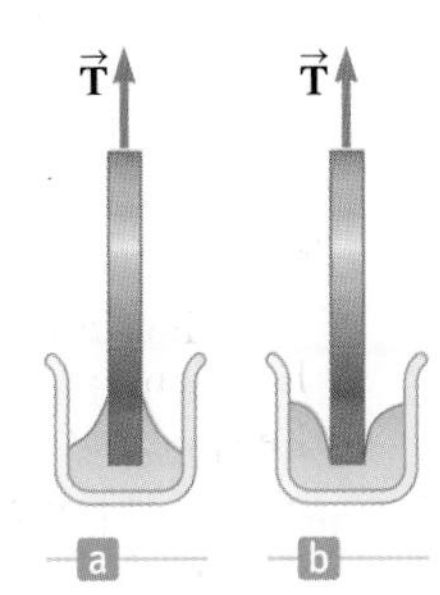

Figure P9.59

60. BIO To lift a wire ring of radius 1.75 cm from the surface of a container of blood plasma, a vertical force of 1.61×10^{-2} N greater than the weight of the ring is required. Calculate the surface tension of blood plasma from this information.

61. A certain fluid has a density of 1 080 kg/m^3 and is observed to rise to a height of 2.1 cm in a 1.0-mm-diameter tube. The contact angle between the wall and the fluid is zero. Calculate the surface tension of the fluid.

62. BIO Whole blood has a surface tension of 0.058 N/m and a density of 1 050 kg/m^3. To what height can whole blood rise in a capillary blood vessel that has a radius of 2.0×10^{-6} m if the contact angle is zero?

63. The block of ice (temperature 0°C) shown in Figure P9.63 is drawn over a level surface lubricated by a layer of water 0.10 mm thick. Determine the magnitude of the force $\vec{F}$ needed to pull the block with a constant speed of 0.50 m/s. At 0°C, the viscosity of water has the value $\eta = 1.79 \times 10^{-3}$ N · s/m^2.

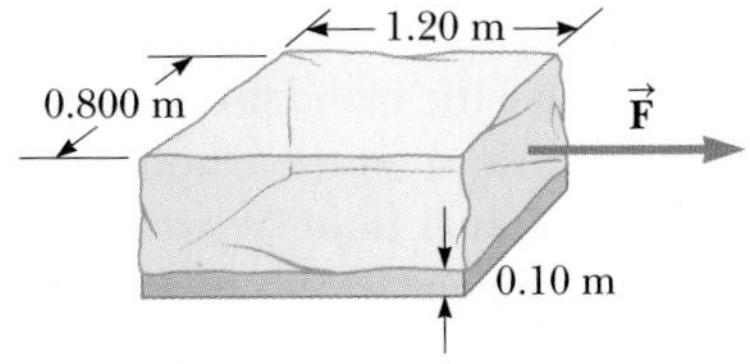

Figure P9.63

64. A thin 1.5-mm coating of glycerine has been placed between two microscope slides of width 1.0 cm and length 4.0 cm. Find the force required to pull one of the microscope slides at a constant speed of 0.30 m/s relative to the other slide.

65. A straight horizontal pipe with a diameter of 1.0 cm and a length of 50 m carries oil with a coefficient of viscosity of 0.12 N · s/m^2. At the output of the pipe, the flow rate is 8.6×10^{-5} m^3/s and the pressure is 1.0 atm. Find the *gauge* pressure at the pipe input.

66. BIO The pulmonary artery, which connects the heart to the lungs, has an inner radius of 2.6 mm and is 8.4 cm long. If the pressure drop between the heart and lungs is 400 Pa, what is the average speed of blood in the pulmonary artery?

67. Spherical particles of a protein of density 1.8 g/cm^3 are shaken up in a solution of 20°C water. The solution is allowed to stand for 1.0 h. If the depth of water in the tube is 5.0 cm, find the radius of the largest particles that remain in solution at the end of the hour.

68. BIO A hypodermic needle is 3.0 cm in length and 0.30 mm in diameter. What pressure difference between the input and output of the needle is required so that the flow rate of water through it will be 1 g/s? (Use 1.0×10^{-3} Pa · s as the viscosity of water.)

69. BIO What radius needle should be used to inject a volume of 500 cm^3 of a solution into a patient in 30 min? Assume the length of the needle is 2.5 cm and the solution is elevated 1.0 m above the point of injection. Further, assume the viscosity and density of the solution are those of pure water, and that the pressure inside the vein is atmospheric.

70. Water is forced out of a fire extinguisher by air pressure, as shown in Figure P9.70. What gauge air pressure in the tank (above atmospheric pressure) is required for the water to have a jet speed of 30.0 m/s when the water level in the tank is 0.500 m below the nozzle?

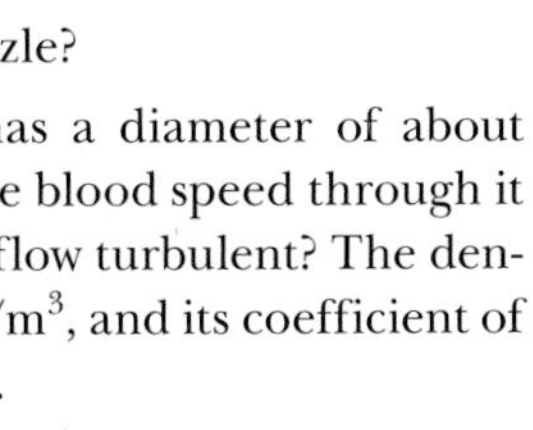

Figure P9.70

71. BIO The aorta in humans has a diameter of about 2.0 cm, and at certain times the blood speed through it is about 55 cm/s. Is the blood flow turbulent? The density of whole blood is 1 050 kg/m^3, and its coefficient of viscosity is 2.7×10^{-3} N · s/m^2.

72. A pipe carrying 20°C water has a diameter of 2.5 cm. Estimate the maximum flow speed if the flow must be streamline.

9.10 Transport Phenomena

73. BIO Sucrose is allowed to diffuse along a 10-cm length of tubing filled with water. The tube is 6.0 cm^2 in cross-sectional area. The diffusion coefficient is equal to 5.0×10^{-10} m^2/s, and 8.0×10^{-14} kg is transported along the

tube in 15 s. What is the difference in the concentration levels of sucrose at the two ends of the tube?

74. BIO Glycerin in water diffuses along a horizontal column that has a cross-sectional area of 2.0 cm^2. The concentration gradient is 3.0×10^{-2} kg/m^4, and the diffusion rate is found to be 5.7×10^{-15} kg/s. Determine the diffusion coefficient.

75. The viscous force on an oil drop is measured to be equal to 3.0×10^{-13} N when the drop is falling through air with a speed of 4.5×10^{-4} m/s. If the radius of the drop is 2.5×10^{-6} m, what is the viscosity of air?

76. Small spheres of diameter 1.00 mm fall through 20°C water with a terminal speed of 1.10 cm/s. Calculate the density of the spheres.

Additional Problems

77. An iron block of volume 0.20 m^3 is suspended from a spring scale and immersed in a flask of water. Then the iron block is removed, and an aluminum block of the same volume replaces it. (a) In which case is the buoyant force the greatest, for the iron block or the aluminum block? (b) In which case does the spring scale read the largest value? (c) Use the known densities of these materials to calculate the quantities requested in parts (a) and (b). Are your calculations consistent with your previous answers to parts (a) and (b)?

78. S The true weight of an object can be measured in a vacuum, where buoyant forces are absent. A measurement in air, however, is disturbed by buoyant forces. An object of volume V is weighed in air on an equal-arm balance with the use of counterweights of density ρ. Representing the density of air as ρ_{air} and the balance reading as F_g', show that the true weight F_g is

$$F_g = F_g' + \left(V - \frac{F_g'}{\rho g}\right)\rho_{air}\, g$$

79. As a first approximation, Earth's continents may be thought of as granite blocks floating in a denser rock (called peridotite) in the same way that ice floats in water. (a) Show that a formula describing this phenomenon is

$$\rho_g t = \rho_p d$$

where ρ_g is the density of granite (2.8×10^3 kg/m^3), ρ_p is the density of peridotite (3.3×10^3 kg/m^3), t is the thickness of a continent, and d is the depth to which a continent floats in the peridotite. (b) If a continent sinks 5.0 km into the peridotite layer (this surface may be thought of as the ocean floor), what is the thickness of the continent?

80. BIO Take the density of blood to be ρ and the distance between the feet and the heart to be h_H. Ignore the flow of blood. (a) Show that the difference in blood pressure between the feet and the heart is given by $P_F - P_H = \rho g h_H$. (b) Take the density of blood to be 1.05×10^3 kg/m^3 and the distance between the heart and the feet to be 1.20 m. Find the difference in blood pressure between these two points. This problem indicates that pumping blood from the extremities is very difficult for the heart. The veins in the legs have valves in them that open when blood is pumped toward the heart and close when blood flows away from the heart. Also, pumping action produced by physical activities such as walking and breathing assists the heart.

81. BIO The approximate diameter of the aorta is 0.50 cm; that of a capillary is 10 μm. The approximate average blood flow speed is 1.0 m/s in the aorta and 1.0 cm/s in the capillaries. If all the blood in the aorta eventually flows through the capillaries, estimate the number of capillaries in the circulatory system.

82. Superman attempts to drink water through a very long vertical straw as in Figure P9.82. With his great strength, he achieves maximum possible suction. The walls of the straw don't collapse. (a) Find the maximum height through which he can lift the water. (b) Still thirsty, the Man of Steel repeats his attempt on the Moon, which has no atmosphere. Find the difference between the water levels inside and outside the straw.

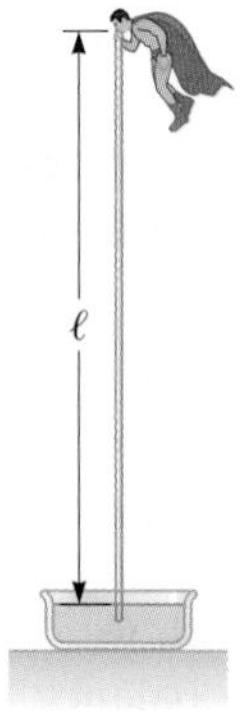

Figure P9.82

83. BIO The human brain and spinal cord are immersed in the cerebrospinal fluid. The fluid is normally continuous between the cranial and spinal cavities and exerts a pressure of 100 to 200 mm of H_2O above the prevailing atmospheric pressure. In medical work, pressures are often measured in units of mm of H_2O because body fluids, including the cerebrospinal fluid, typically have nearly the same density as water. The pressure of the cerebrospinal fluid can be measured by means of a *spinal tap*. A hollow tube is inserted into the spinal column, and the height to which the fluid rises is observed, as shown in Figure P9.83. If the fluid rises to a height of 160 mm, we write its gauge pressure as 160 mm H_2O. (a) Express this pressure in pascals, in atmospheres, and in millimeters of mercury. (b) Sometimes it is necessary to determine whether an accident victim has suffered a crushed vertebra that is blocking the flow of cerebrospinal fluid in the spinal column. In other cases, a physician may suspect that a tumor or other growth is blocking the spinal column and inhibiting the flow of cerebrospinal fluid. Such conditions can be investigated by means of the *Queckensted test*. In this procedure the veins in the patient's neck are compressed, to make the blood pressure rise in the brain. The increase in pressure in the blood vessels is transmitted to the cerebrospinal fluid. What should be the normal effect on the height of the fluid in the spinal tap? (c) Suppose compressing the veins had no effect on the level of the fluid. What might account for this phenomenon?

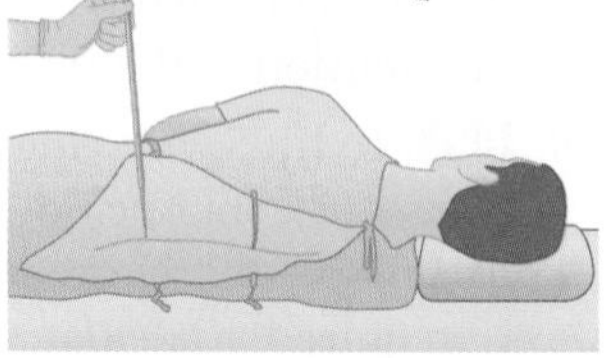

Figure P9.83

84. BIO S A *hydrometer* is an instrument used to determine liquid density. A simple one is sketched in Figure P9.84. The bulb of a syringe is squeezed and released to lift a sample of the liquid of interest into a tube containing a calibrated rod of known density. (Assume the rod is cylindrical.) The rod, of length L and average density ρ_0, floats partially immersed in the liquid of density ρ. A length h of the rod protrudes above the surface of the liquid. Show that the density of the liquid is given by

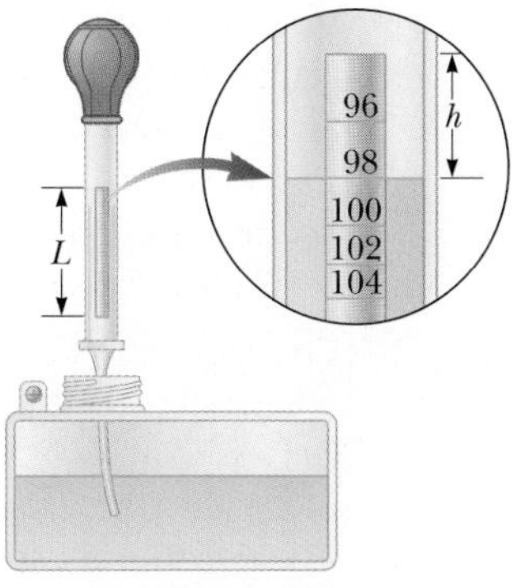

Figure P9.84

$$\rho = \frac{\rho_0 L}{L - h}$$

85. Figure P9.85 shows a water tank with a valve. If the valve is opened, what is the maximum height attained by the stream of water coming out of the right side of the tank? Assume h = 10.0 m, L = 2.00 m, and θ = 30.0°, and that the cross-sectional area at A is very large compared with that at B.

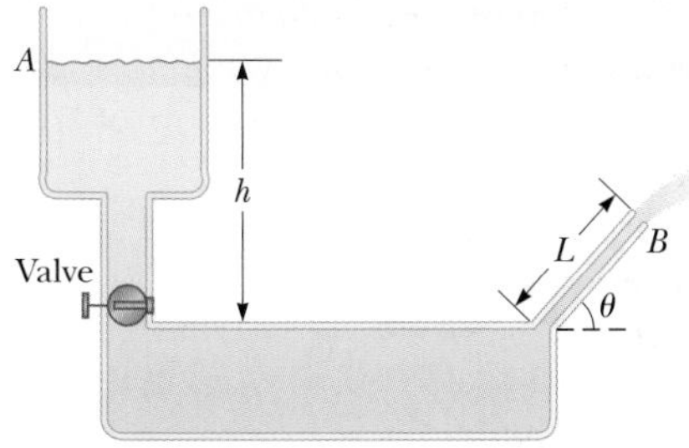

Figure P9.85

86. A helium-filled balloon, whose envelope has a mass of 0.25 kg, is tied to a 2.0-m-long, 0.050-kg string. The balloon is spherical with a radius of 0.40 m. When released, it lifts a length h of the string and then remains in equilibrium, as in Figure P9.86. Determine the value of h. *Hint:* Only that part of the string above the floor contributes to the load being supported by the balloon.

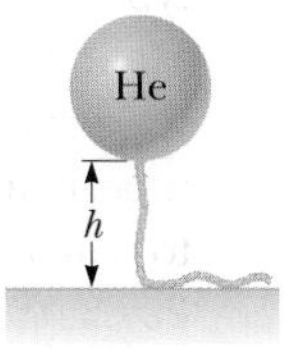

Figure P9.86

87. A light spring of constant k = 90.0 N/m is attached vertically to a table (Fig. P9.87a). A 2.00-g balloon is filled with helium (density = 0.179 kg/m^3) to a volume of 5.00 m^3 and is then connected to the spring, causing the spring to stretch as shown in Figure P89.87b. Determine the extension distance L when the balloon is in equilibrium.

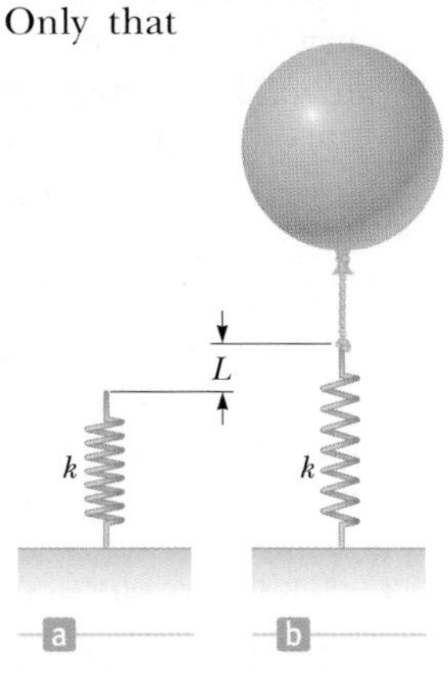

Figure P9.87

88. A U-tube open at both ends is partially filled with water (Fig. P9.88a). Oil (ρ = 750 kg/m^3) is then poured into the right arm and forms a column L = 5.00 cm high (Fig. P9.88b). (a) Determine the difference h in the heights of the two liquid surfaces. (b) The right arm is then shielded from any air motion while air is blown across the top of the left arm until the surfaces of the two liquids are at the same height (Fig. P9.88c). Determine the speed of the air being blown across the left arm. Assume the density of air is 1.29 kg/m^3.

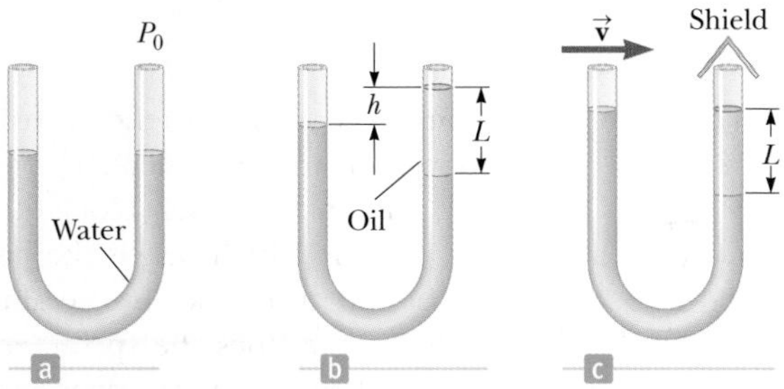

Figure P9.88

89. S In about 1657, Otto von Guericke, inventor of the air pump, evacuated a sphere made of two brass hemispheres (Fig. P9.89). Two teams of eight horses each could pull the hemispheres apart only on some trials and then "with greatest difficulty," with the resulting sound likened to a cannon firing. Find the force F required to pull the thin-walled evacuated hemispheres apart in terms of R, the radius of the hemispheres, P the pressure inside the hemispheres, and atmospheric pressure P_0.

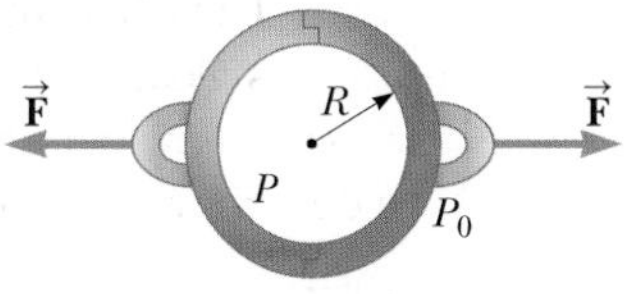

Figure P9.89

90. Oil having a density of 930 kg/m^3 floats on water. A rectangular block of wood 4.00 cm high and with a density of 960 kg/m^3 floats partly in the oil and partly in the water. The oil completely covers the block. How far below the interface between the two liquids is the bottom of the block?

91. A water tank open to the atmosphere at the top has two small holes punched in its side, one above the other. The holes are 5.00 cm and 12.0 cm above the floor. How high does water stand in the tank if the two streams of water hit the floor at the same place?

Pipelines carrying liquids often have loops to allow for expansion and contraction due to temperature changes. Without the loops, the pipes could buckle and burst.

10 Thermal Physics

How can trapped water blow off the top of a volcano in a giant explosion? What causes a sidewalk or road to fracture and buckle spontaneously when the temperature changes? How can thermal energy be harnessed to do work, running the engines that make everything in modern living possible?

Answering these and related questions is the domain of **thermal physics,** the study of temperature, heat, and how they affect matter. Quantitative descriptions of thermal phenomena require careful definitions of the concepts of temperature, heat, and internal energy. Heat leads to changes in internal energy and thus to changes in temperature, which cause the expansion or contraction of matter. Such changes can damage roadways and buildings, create stress fractures in metal, and render flexible materials stiff and brittle, the latter resulting in compromised O-rings and the *Challenger* disaster. Changes in internal energy can also be harnessed for transportation, construction, and food preservation.

Gases are critical in the harnessing of thermal energy to do work. Within normal temperature ranges, a gas acts like a large collection of non-interacting point particles, called an ideal gas. Such gases can be studied on either a macroscopic or microscopic scale. On the macroscopic scale, the pressure, volume, temperature, and number of particles associated with a gas can be related in a single equation known as the ideal gas law. On the microscopic scale, a model called the kinetic theory of gases pictures the components of a gas as small particles. That model will enable us to understand how processes on the atomic scale affect macroscopic properties like pressure, temperature, and internal energy.

10.1 Temperature and the Zeroth Law of Thermodynamics

LEARNING OBJECTIVES

1. Define thermal contact, thermal equilibrium, and heat.
2. State the zeroth law of thermodynamics and explain how it allows a definition of temperature.

Temperature is commonly associated with how hot or cold an object feels when we touch it. While our senses provide us with qualitative indications of temperature, they are unreliable and often misleading. A metal ice tray feels colder to the hand, for example, than a package of frozen vegetables at the same temperature, because metals conduct thermal energy more rapidly than a cardboard package. What we need is a reliable and reproducible method of making quantitative measurements that establish the relative "hotness" or "coldness" of objects—a method related solely to temperature. Scientists have developed a variety of thermometers for making such measurements.

When placed in contact with each other, two objects at different initial temperatures will eventually reach a common intermediate temperature. If a cup of hot coffee is cooled with an ice cube, for example, the ice rises in temperature and eventually melts while the temperature of the coffee decreases.

Understanding the concept of temperature requires understanding *thermal contact* and *thermal equilibrium.* Two objects are in **thermal contact** if energy can be exchanged between them. Two objects are in **thermal equilibrium** if they are in thermal contact and there is no net exchange of energy.

The exchange of energy between two objects due to differences in their temperatures is called **heat,** a concept examined in more detail in Chapter 11.

Using these ideas, we can develop a formal definition of temperature. Consider two objects A and B that are not in thermal contact with each other, and a third object C that acts as a **thermometer**—a device calibrated to measure the temperature of an object. We wish to determine whether A and B would be in thermal equilibrium if they were placed in thermal contact. The thermometer (object C) is first placed in thermal contact with A until thermal equilibrium is reached, as in Figure 10.1a, whereupon the reading of the thermometer is recorded. The thermometer is then placed in thermal contact with B, and its reading is again recorded at equilibrium (Fig. 10.1b). If the two readings are the same, then A and B are in thermal equilibrium with each other. If A and B are placed in thermal contact with each other, as in Figure 10.1c, there is no net transfer of energy between them.

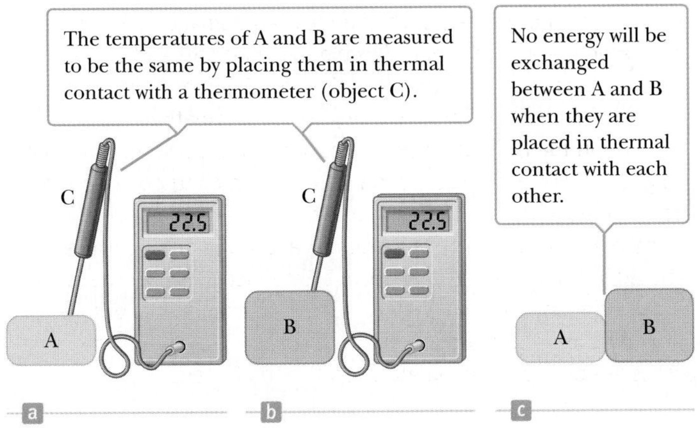

Figure 10.1 The zeroth law of thermodynamics.

We can summarize these results in a statement known as the **zeroth law of thermodynamics (the law of equilibrium):**

Zeroth law of thermodynamics ▶

> If objects A and B are separately in thermal equilibrium with a third object C, then A and B are in thermal equilibrium with each other.

This statement is important because it makes it possible to define **temperature.** We can think of temperature as the property that determines whether or not an object is in thermal equilibrium with other objects. **Two objects in thermal equilibrium with each other are at the same temperature.**

■ *Quick Quiz*

10.1 Two objects with different sizes, masses, and temperatures are placed in thermal contact. Choose the best answer: Energy travels (a) from the larger object to the smaller object (b) from the object with more mass to the one with less mass (c) from the object at higher temperature to the object at lower temperature.

10.2 Thermometers and Temperature Scales

LEARNING OBJECTIVES

1. Describe the operating principles of common thermometers.
2. Explain the origin of the absolute temperature scale.
3. Convert temperatures from one temperature scale to any other temperature scale.

Thermometers are devices used to measure the temperature of an object or a system. When a thermometer is in thermal contact with a system, energy is exchanged until the thermometer and the system are in thermal equilibrium with each other. For accurate readings, the thermometer must be much smaller than the system, so that the energy the thermometer gains or loses doesn't significantly alter the energy content of the system. All thermometers make use of some physical property that changes with temperature and can be calibrated to make the temperature measurable. Some of the physical properties used are (1) the volume of a liquid, (2) the length of a solid, (3) the pressure of a gas held at constant volume, (4) the volume of a gas held at constant pressure, (5) the electric resistance of a conductor, and (6) the color of a very hot object.

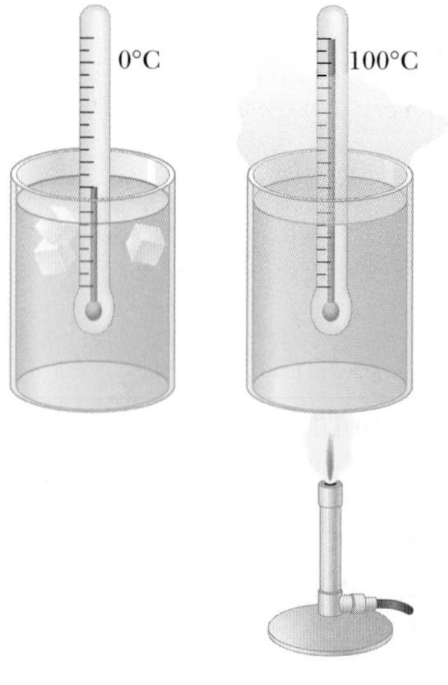

Figure 10.2 Schematic diagram of a mercury thermometer. Because of thermal expansion, the level of the mercury rises as the temperature of the mercury changes from 0°C (the ice point) to 100°C (the steam point).

One common thermometer in everyday use consists of a mass of liquid—usually mercury or alcohol—that expands into a glass capillary tube when its temperature rises (Fig. 10.2). In this case the physical property that changes is the volume of a liquid. To serve as an effective thermometer, the change in volume of the liquid with change in temperature must be very nearly constant over the temperature ranges of interest. When the cross-sectional area of the capillary tube is constant as well, the change in volume of the liquid varies linearly with its length along the tube. We can then define a temperature in terms of the length of the liquid column. The thermometer can be calibrated by placing it in thermal contact with environments that remain at constant temperature. One such environment is a mixture of water and ice in thermal equilibrium at atmospheric pressure. Another commonly used system is a mixture of water and steam in thermal equilibrium at atmospheric pressure.

Once we have marked the ends of the liquid column for our chosen environment on our thermometer, we need to define a scale of numbers associated with various temperatures. An example of such a scale is the **Celsius temperature scale,** formerly called the centigrade scale. On the Celsius scale, the temperature of the ice–water mixture is defined to be zero degrees Celsius, written 0°C and called the

ice point or **freezing point** of water. The temperature of the water–steam mixture is defined as 100°C, called the **steam point** or **boiling point** of water. Once the ends of the liquid column in the thermometer have been marked at these two points, the distance between marks is divided into 100 equal segments, each corresponding to a change in temperature of one degree Celsius.

Thermometers calibrated in this way present problems when extremely accurate readings are needed. For example, an alcohol thermometer calibrated at the ice and steam points of water might agree with a mercury thermometer only at the calibration points. Because mercury and alcohol have different thermal expansion properties, when one indicates a temperature of 50°C, say, the other may indicate a slightly different temperature. The discrepancies between different types of thermometers are especially large when the temperatures to be measured are far from the calibration points.

The Constant-Volume Gas Thermometer and the Kelvin Scale

We can construct practical thermometers such as the mercury thermometer, but these types of thermometers don't define temperature in a fundamental way. One thermometer, however, *is* more fundamental, and offers a way to define temperature and relate it directly to internal energy: the **gas thermometer.** In a gas thermometer, the temperature readings are nearly independent of the substance used in the thermometer. One type of gas thermometer is the constant-volume unit shown in Figure 10.3. The behavior observed in this device is the variation of pressure with temperature of a fixed volume of gas. When the constant-volume gas thermometer was developed, it was calibrated using the ice and steam points of water as follows (a different calibration procedure, to be discussed shortly, is now used): The gas flask is inserted into an ice–water bath, and mercury reservoir B is raised or lowered until the volume of the confined gas is at some value, indicated by the zero point on the scale. The height h, the difference between the levels in the reservoir and column A, indicates the pressure in the flask at 0°C. The flask is inserted into water at the steam point, and reservoir B is readjusted until the height in column A is again brought to zero on the scale, ensuring that the gas volume is the same as it had been in the ice bath (hence the designation "constant-volume"). A measure of the new value for h gives a value for the pressure at 100°C. These pressure and temperature values are then plotted on a graph, as in Figure 10.4. The line connecting the two points serves as a calibration curve for measuring unknown temperatures. If we want to measure the temperature of a substance, we place the gas flask in thermal contact with the substance and adjust the column of mercury until the level in column A returns to zero. The height of the mercury column tells us the pressure of the gas, and we could then find the temperature of the substance from the calibration curve.

Now suppose that temperatures are measured with various gas thermometers containing different gases. Experiments show that the thermometer readings are nearly independent of the type of gas used, as long as the gas pressure is low and the temperature is well above the point at which the gas liquefies.

We can also perform the temperature measurements with the gas in the flask at different starting pressures at 0°C. As long as the pressure is low, we will generate straight-line calibration curves for each starting pressure, as shown for three experimental trials (solid lines) in Figure 10.5 (page 340).

If the lines in Figure 10.5 are extended back toward negative temperatures, we find a startling result: In every case, regardless of the type of gas or the value of the low starting pressure, **the pressure extrapolates to zero when the temperature is −273.15°C.** This fact suggests that this particular temperature is universal in its importance, because it doesn't depend on the substance used in the thermometer. In addition, because the lowest possible pressure is $P = 0$, a perfect vacuum, the temperature −273.15°C must represent a lower bound for physical processes. We define this temperature as **absolute zero.**

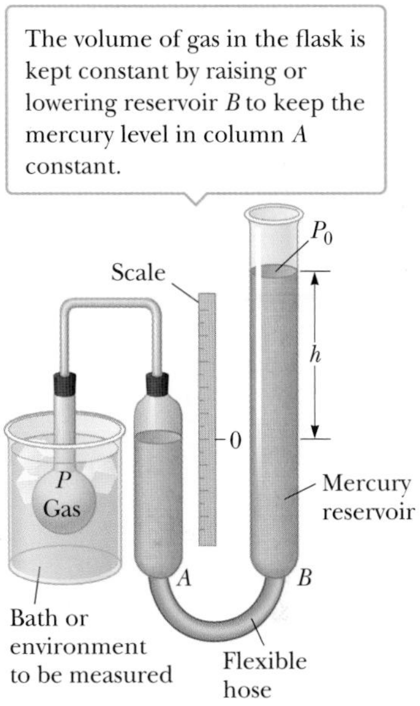

Figure 10.3 A constant-volume gas thermometer measures the pressure of the gas contained in the flask immersed in the bath.

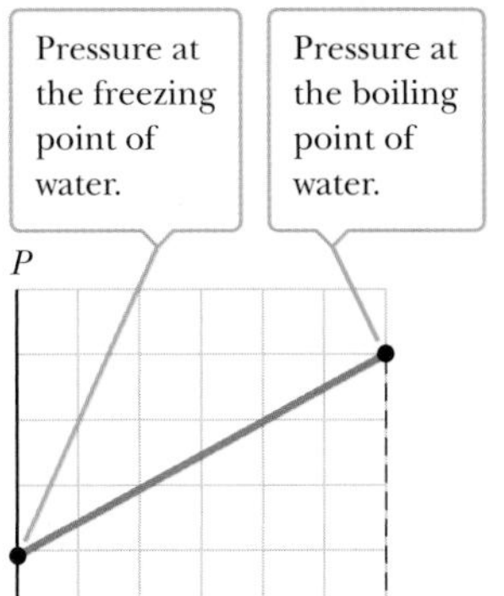

Figure 10.4 A typical graph of pressure versus temperature taken with a constant-volume gas thermometer.

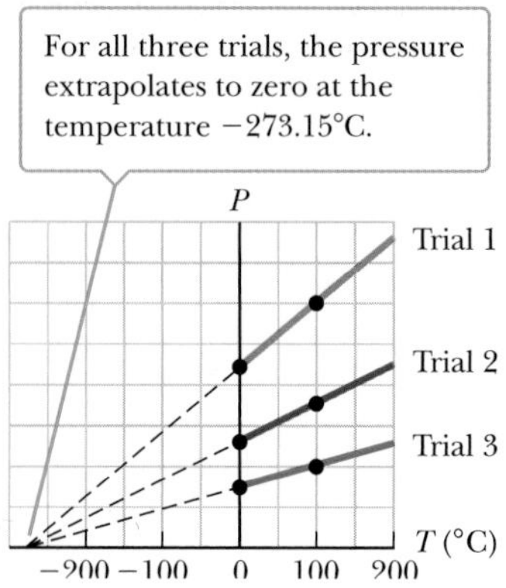

Figure 10.5 Pressure versus temperature for experimental trials in which gases have different pressures in a constant-volume gas thermometer.

Absolute zero is used as the basis for the **Kelvin temperature scale,** which sets −273.15°C as its zero point (0 K). The size of a "degree" on the Kelvin scale is chosen to be identical to the size of a degree on the Celsius scale. The relationship between these two temperature scales is

$$T_C = T - 273.15 \quad [10.1]$$

where T_C is the Celsius temperature and T is the Kelvin temperature (sometimes called the **absolute temperature**).

Technically, Equation 10.1 should have units on the right-hand side so that it reads $T_C = T\,°\mathrm{C/K} - 273.15°\mathrm{C}$. The units are rather cumbersome in this context, so we will usually suppress them in such calculations except in the final answer. (This will also be the case when discussing the Celsius and Fahrenheit scales.)

Early gas thermometers made use of ice and steam points according to the procedure just described. These points are experimentally difficult to duplicate, however, because they are pressure-sensitive. Consequently, a procedure based on two new points was adopted in 1954 by the International Committee on Weights and Measures. The first point is absolute zero. The second point is **the triple point of water, which is the single temperature and pressure at which water, water vapor, and ice can coexist in equilibrium.** This point is a convenient and reproducible reference temperature for the Kelvin scale; it occurs at a temperature of 0.01°C and a pressure of 4.58 mm of mercury. The temperature at the triple point of water on the Kelvin scale occurs at 273.16 K. Therefore, **the SI unit of temperature, the kelvin, is defined as 1/273.16 of the temperature of the triple point of water.** Figure 10.6 shows the Kelvin temperatures for various physical processes and structures. Absolute zero has been closely approached but never achieved.

What would happen to a substance if its temperature could reach 0 K? As Figure 10.5 indicates, the substance would exert zero pressure on the walls of its container (assuming the gas doesn't liquefy or solidify on the way to absolute zero). In Section 10.5 we show that the pressure of a gas is proportional to the kinetic energy of the molecules of that gas. According to classical physics, therefore, the kinetic energy of the gas would go to zero and there would be no motion at all of the individual components of the gas. According to quantum theory, however (to be discussed in Chapter 27), the gas would always retain some residual energy, called the *zero-point energy,* at that low temperature.

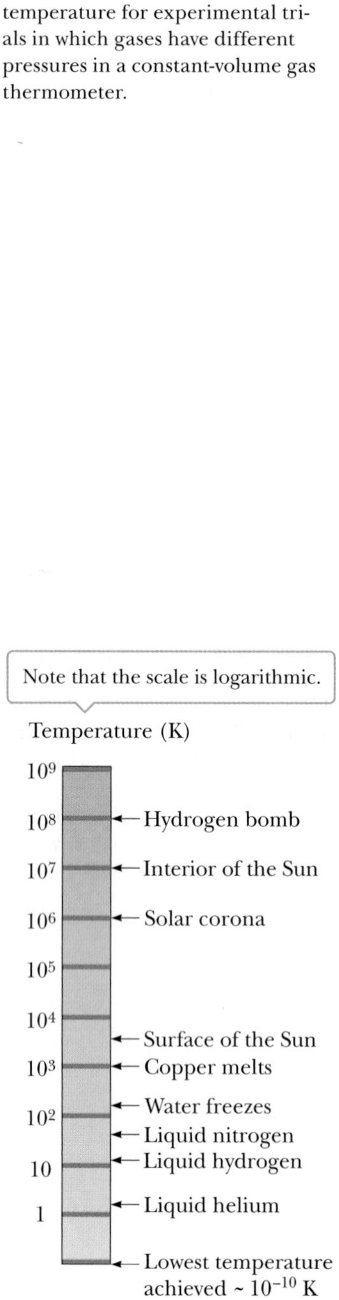

Figure 10.6 Absolute temperatures at which various selected physical processes take place.

The Celsius, Kelvin, and Fahrenheit Temperature Scales

Equation 10.1 shows that the Celsius temperature T_C is shifted from the absolute (Kelvin) temperature T by 273.15. Because the size of a Celsius degree is the same as a Kelvin, a temperature difference of 5°C is equal to a temperature difference of 5 K. The two scales differ only in the choice of zero point. The ice point (273.15 K) corresponds to 0.00°C, and the steam point (373.15 K) is equivalent to 100.00°C.

The most common temperature scale in use in the United States is the Fahrenheit scale. It sets the temperature of the ice point at 32°F and the temperature of the steam point at 212°F. The relationship between the Celsius and Fahrenheit temperature scales is

$$T_F = \tfrac{9}{5}T_C + 32 \quad [10.2a]$$

For example, a temperature of 50.0°F corresponds to a Celsius temperature of 10.0°C and an absolute temperature of 283 K.

Equation 10.2a can be inverted to give Celsius temperatures in terms of Fahrenheit temperatures:

$$T_C = \tfrac{5}{9}(T_F - 32) \quad [10.2b]$$

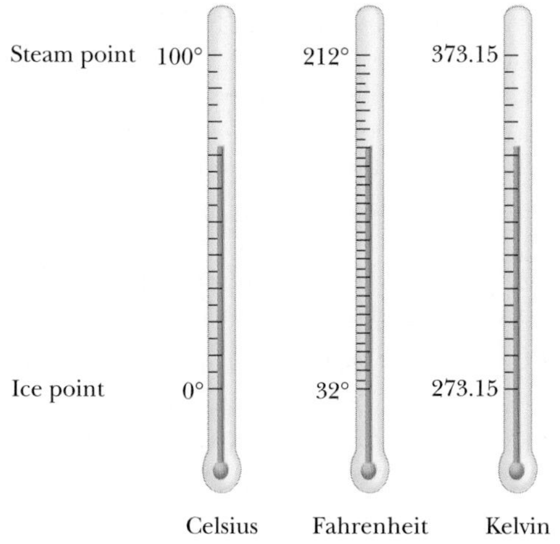

Figure 10.7 A comparison of the Celsius, Fahrenheit, and Kelvin temperature scales.

Equation 10.2 can also be used to find a relationship between changes in temperature on the Celsius and Fahrenheit scales. In a problem at the end of the chapter you will be asked to show that if the Celsius temperature changes by ΔT_C, the Fahrenheit temperature changes by the amount

$$\Delta T_F = \tfrac{9}{5}\Delta T_C \qquad [10.3]$$

Figure 10.7 compares the Celsius, Fahrenheit, and Kelvin scales. Although less commonly used, other scales do exist, such as the Rankine scale. That scale has Fahrenheit degrees and a zero point at absolute zero.

EXAMPLE 10.1 | Skin Temperature BIO

GOAL Apply the temperature conversion formulas.

PROBLEM The temperature gradient between the skin and the air is regulated by cutaneous (skin) blood flow. If the cutaneous blood vessels are constricted, the skin temperature and the temperature of the environment will be about the same. When the vessels are dilated, more blood is brought to the surface. Suppose during dilation the skin warms from 72.0°F to 84.0°F. **(a)** Convert these temperatures to Celsius and find the difference. **(b)** Convert the temperatures to Kelvin, again finding the difference.

STRATEGY This is a matter of applying the conversion formulas, Equations 10.1 and 10.2. For part (b) it's easiest to use the answers for Celsius rather than develop another set of conversion equations.

SOLUTION

(a) Convert the temperatures from Fahrenheit to Celsius and find the difference.

Convert the lower temperature, using Equation 10.2b:

$$T_C = \tfrac{5}{9}(T_F - 32.0) = \tfrac{5}{9}(72.0 - 32.0) = \boxed{22.2°\text{C}}$$

Convert the upper temperature:

$$T_C = \tfrac{5}{9}(T_F - 32.0) = \tfrac{5}{9}(84.0 - 32.0) = \boxed{28.9°\text{C}}$$

Find the difference of the two temperatures:

$$\Delta T_C = 28.9°\text{C} - 22.2°\text{C} = \boxed{6.7°\text{C}}$$

(b) Convert the temperatures from Fahrenheit to Kelvin and find their difference.

Convert the lower temperature, using the answers for Celsius found in part (a):

$$T_C = T - 273.15 \quad \rightarrow \quad T = T_C + 273.15$$

$$T = 22.2 + 273.15 = \boxed{295.4\text{ K}}$$

(Continued)

Convert the upper temperature: $T = 28.9 + 273.15 = \boxed{302.1\ \text{K}}$

Find the difference of the two temperatures: $\Delta T = 302.1\ \text{K} - 295.4\ \text{K} = \boxed{6.7\ \text{K}}$

REMARKS The change in temperature in Kelvin and Celsius is the same, as it should be.

QUESTION 10.1 Which represents a larger temperature change, a Celsius degree or a Fahrenheit degree?

EXERCISE 10.1 Core body temperature can rise from 98.6°F to 107°F during extreme exercise, such as a marathon run. Such elevated temperatures can also be caused by viral or bacterial infections or tumors and are dangerous if sustained. (a) Convert the given temperatures to Celsius and find the difference. (b) Convert the temperatures to Kelvin, again finding the difference.

ANSWERS (a) 37.0°C, 41.7°C, 4.7°C (b) 310.2 K, 314.9 K, 4.7 K

■ EXAMPLE 10.2 | Extraterrestrial Temperature Scale

GOAL Understand how to relate different temperature scales.

PROBLEM An extraterrestrial scientist invents a temperature scale such that water freezes at −75°E and boils at 325°E, where E stands for an extraterrestrial scale. Find an equation that relates temperature in °E to temperature in °C.

STRATEGY Using the given data, find the ratio of the number of °E between the two temperatures to the number of °C. This ratio will be the same as a similar ratio for any other such process—say, from the freezing point to an unknown temperature—corresponding to T_E and T_C. Setting the two ratios equal and solving for T_E in terms of T_C yields the desired relationship. For clarity, the rules of significant figures will not be applied here.

SOLUTION

Find the change in temperature in °E between the freezing and boiling points of water:

$$\Delta T_E = 325°\text{E} - (-75°\text{E}) = 400°\text{E}$$

Find the change in temperature in °C between the freezing and boiling points of water:

$$\Delta T_C = 100°\text{C} - 0°\text{C} = 100°\text{C}$$

Form the ratio of these two quantities.

$$\frac{\Delta T_E}{\Delta T_C} = \frac{400°\text{E}}{100°\text{C}} = 4\,\frac{°\text{E}}{°\text{C}}$$

This ratio is the same between any other two temperatures—say, from the freezing point to an unknown final temperature. Set the two ratios equal to each other:

$$\frac{\Delta T_E}{\Delta T_C} = \frac{T_E - (-75°\text{E})}{T_C - 0°\text{C}} = 4\,\frac{°\text{E}}{°\text{C}}$$

Solve for T_E:

$$T_E - (-75°\text{E}) = 4(°\text{E}/°\text{C})(T_C - 0°\text{C})$$

$$T_E = \boxed{4T_C - 75}$$

REMARKS The relationship between any other two temperatures scales can be derived in the same way.

QUESTION 10.2 True or False: Finding the relationship between two temperature scales using knowledge of the freezing and boiling point of water in each system is equivalent to finding the equation of a straight line.

EXERCISE 10.2 Find the equation converting °F to °E.

ANSWER $T_E = \frac{20}{9}T_F - 146$

10.3 Thermal Expansion of Solids and Liquids

LEARNING OBJECTIVES

1. Explain the physical origins of thermal expansion.
2. Apply the equations of thermal expansion to physical systems.

Our discussion of the liquid thermometer made use of one of the best-known changes that occur in most substances: As temperature of the substance increases, its volume increases. This phenomenon, known as **thermal expansion,** plays an important role in numerous applications. Thermal expansion joints, for example, must be included in buildings, concrete highways, and bridges to compensate for changes in dimensions with variations in temperature (Fig. 10.8).

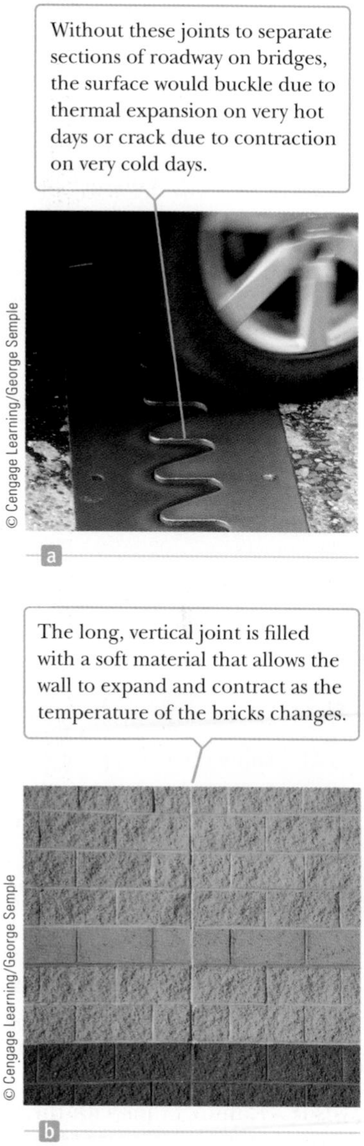

Figure 10.8 Thermal expansion joints in (a) bridges and (b) walls.

The overall thermal expansion of an object is a consequence of the change in the average separation between its constituent atoms or molecules. To understand this idea, consider how the atoms in a solid substance behave. These atoms are located at fixed equilibrium positions; if an atom is pulled away from its position, a restoring force pulls it back. We can imagine that the atoms are particles connected by springs to their neighboring atoms. (See Fig. 9.1 in the previous chapter.) If an atom is pulled away from its equilibrium position, the distortion of the springs provides a restoring force.

At ordinary temperatures, the atoms vibrate around their equilibrium positions with an amplitude (maximum distance from the center of vibration) of about 10^{-11} m, with an average spacing between the atoms of about 10^{-10} m. As the temperature of the solid increases, the atoms vibrate with greater amplitudes and the average separation between them increases. Consequently, the solid as a whole expands.

If the thermal expansion of an object is sufficiently small compared with the object's initial dimensions, then the change in any dimension is, to a good approximation, proportional to the first power of the temperature change. Suppose an object has an initial length L_0 along some direction at some temperature T_0. Then the length increases by ΔL for a change in temperature ΔT. So for small changes in temperature,

$$\Delta L = \alpha L_0 \, \Delta T \qquad [10.4]$$

or

$$L - L_0 = \alpha L_0(T - T_0)$$

where L is the object's final length, T is its final temperature, and the proportionality constant α is called the **coefficient of linear expansion** for a given material and has units of $(°\text{C})^{-1}$.

Table 10.1 lists the coefficients of linear expansion for various materials. Note that for these materials α is positive, indicating an increase in length with increasing temperature.

Thermal expansion affects the choice of glassware used in kitchens and laboratories. If hot liquid is poured into a cold container made of ordinary glass, the container may well break due to thermal stress. The inside surface of the glass becomes hot and expands, while the outside surface is at room temperature, and ordinary glass may not withstand the difference in expansion without breaking. Pyrex® glass has a coefficient of linear expansion of about one-third that of ordinary glass, so the thermal stresses are smaller. Kitchen measuring cups and laboratory beakers are often made of Pyrex so they can be used with hot liquids.

APPLICATION
Pyrex Glass

Table 10.1 Average Coefficients of Expansion for Some Materials Near Room Temperature

Material	Average Coefficient of Linear Expansion $[(°C)^{-1}]$	Material	Average Coefficient of Volume Expansion $[(°C)^{-1}]$
Aluminum	24×10^{-6}	Acetone	1.5×10^{-4}
Brass and bronze	19×10^{-6}	Benzene	1.24×10^{-4}
Concrete	12×10^{-6}	Ethyl alcohol	1.12×10^{-4}
Copper	17×10^{-6}	Gasoline	9.6×10^{-4}
Glass (ordinary)	9×10^{-6}	Glycerin	4.85×10^{-4}
Glass (Pyrex®)	3.2×10^{-6}	Mercury	1.82×10^{-4}
Invar (Ni–Fe alloy)	0.9×10^{-6}	Turpentine	9.0×10^{-4}
Lead	29×10^{-6}	Air[a] at 0°C	3.67×10^{-3}
Steel	11×10^{-6}	Helium	3.665×10^{-3}

[a]Gases do not have a specific value for the volume expansion coefficient because the amount of expansion depends on the type of process through which the gas is taken. The values given here assume the gas undergoes an expansion at constant pressure.

Tip 10.1 Coefficients of Expansion Are Not Constants

The coefficients of expansion can vary somewhat with temperature, so the given coefficients are actually averages.

EXAMPLE 10.3 Expansion of a Railroad Track

GOAL Apply the concept of linear expansion and relate it to stress.

PROBLEM **(a)** A steel railroad track has a length of 30.000 m when the temperature is 0°C. What is its length on a hot day when the temperature is 40.0°C? **(b)** Suppose the track is nailed down so that it can't expand. What stress results in the track due to the temperature change?

(Example 10.3) Thermal expansion: The extreme heat of a July day in Asbury Park, New Jersey, caused these railroad tracks to buckle.

STRATEGY **(a)** Apply the linear expansion equation, using Table 10.1 and Equation 10.4. **(b)** A track that cannot expand by ΔL due to external constraints is equivalent to compressing the track by ΔL, creating a stress in the track. Using the equation relating tensile stress to tensile strain together with the linear expansion equation, the amount of (compressional) stress can be calculated using Equation 9.5.

SOLUTION

(a) Find the length of the track at 40.0°C.

Substitute given quantities into Equation 10.4, finding the change in length:

$$\Delta L = \alpha L_0 \Delta T = [11 \times 10^{-6} (°C)^{-1}](30.000 \text{ m})(40.0°C)$$
$$= 0.013 \text{ m}$$

Add the change to the original length to find the final length:

$$L = L_0 + \Delta L = \boxed{30.013 \text{ m}}$$

(b) Find the stress if the track cannot expand.

Substitute into Equation 9.5 to find the stress:

$$\frac{F}{A} = Y\frac{\Delta L}{L} = (2.00 \times 10^{11} \text{ Pa})\left(\frac{0.013 \text{ m}}{30.0 \text{ m}}\right)$$
$$= \boxed{8.7 \times 10^{7} \text{ Pa}}$$

REMARKS Repeated heating and cooling is an important part of the weathering process that gradually wears things out, weakening structures over time.

QUESTION 10.3 What happens to the tension of wires in a piano when the temperature decreases?

EXERCISE 10.3 What is the length of the same railroad track on a cold winter day when the temperature is 0°F?

ANSWER 29.994 m

APPLYING PHYSICS 10.1 | Bimetallic Strips and Thermostats

How can different coefficients of expansion for metals be used as a temperature gauge and control electronic devices such as air conditioners?

EXPLANATION When the temperatures of a brass rod and a steel rod of equal length are raised by the same amount from some common initial value, the brass rod expands more than the steel rod because brass has a larger coefficient of expansion than steel. A simple device that uses this principle is a **bimetallic strip.** Such strips can be found in the thermostats of certain home heating systems. The strip is made by securely bonding two different metals together. As the temperature of the strip increases, the two metals expand by different amounts and the strip bends, as in Figure 10.9. The change in shape can make or break an electrical connection. ■

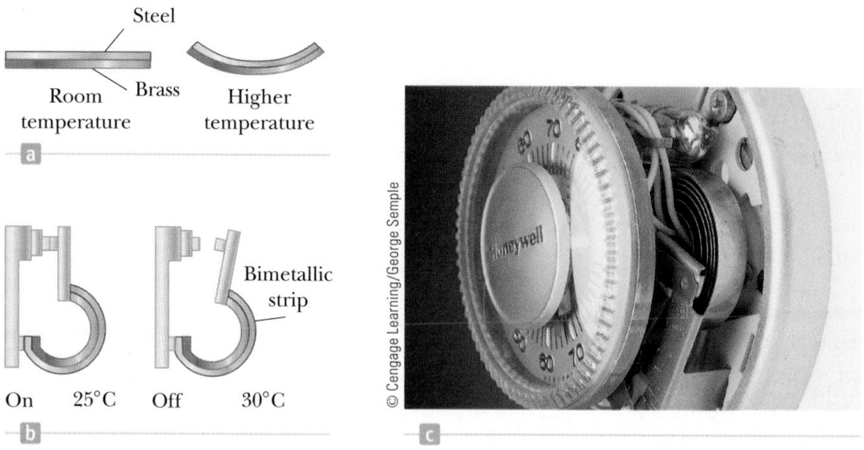

Figure 10.9 (Applying Physics 10.1) (a) A bimetallic strip bends as the temperature changes because the two metals have different coefficients of expansion. (b) A bimetallic strip used in a thermostat to break or make electrical contact. (c) The interior of a thermostat, showing the coiled bimetallic strip. Why do you suppose the strip is coiled?

It may be helpful to picture a thermal expansion as a magnification or a photographic enlargement. For example, as the temperature of a metal washer increases (Fig. 10.10), all dimensions, including the radius of the hole, increase according to Equation 10.4.

One practical application of thermal expansion is the common technique of using hot water to loosen a metal lid stuck on a glass jar. This works because the circumference of the lid expands more than the rim of the jar.

Because the linear dimensions of an object change due to variations in temperature, it follows that surface area and volume of the object also change. Consider a square of material having an initial length L_0 on a side and therefore an initial area $A_0 = L_0^2$. As the temperature is increased, the length of each side increases to

$$L = L_0 + \alpha L_0\,\Delta T$$

The new area A is

$$A = L^2 = (L_0 + \alpha L_0\,\Delta T)(L_0 + \alpha L_0\,\Delta T) = L_0^2 + 2\alpha L_0^2\,\Delta T + \alpha^2 L_0^2(\Delta T)^2$$

The last term in this expression contains the quantity $\alpha\Delta T$ raised to the second power. Because $\alpha\Delta T$ is much less than one, squaring it makes it even smaller. Consequently, we can neglect this term to get a simpler expression:

$$A = L_0^2 + 2\alpha L_0^2\,\Delta T$$

$$A = A_0 + 2\alpha A_0\,\Delta T$$

so that

$$\Delta A = A - A_0 = \gamma A_0\,\Delta T \qquad [10.5]$$

where $\gamma = 2\alpha$. The quantity γ (Greek letter gamma) is called the **coefficient of area expansion.**

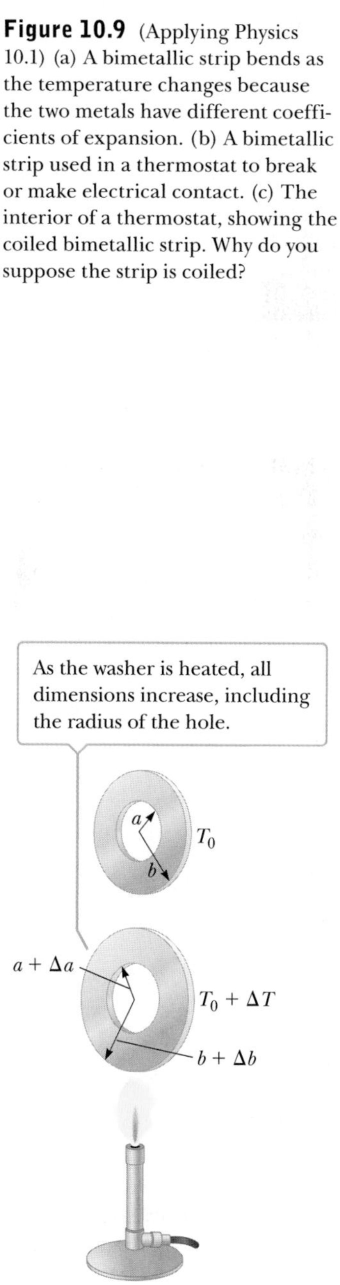

Figure 10.10 Thermal expansion of a homogeneous metal washer. (Note that the expansion is exaggerated in this figure.)

EXAMPLE 10.4 Rings and Rods

GOAL Apply the equation of area expansion.

PROBLEM **(a)** A circular copper ring at 20.0°C has a hole with an area of 9.980 cm^2. What minimum temperature must it have so that it can be slipped onto a steel metal rod having a cross-sectional area of 10.000 cm^2? **(b)** Suppose the ring and the rod are heated simultaneously. What minimum change in temperature of both will allow the ring to be slipped onto the end of the rod? (Assume no significant change in the coefficients of linear expansion over this temperature range.)

STRATEGY In part (a), finding the necessary temperature change is just a matter of substituting given values into Equation 10.5, the equation of area expansion. Remember that $\gamma = 2\alpha$. Part (b) is a little harder because now the rod is also expanding. If the ring is to slip onto the rod, however, the final cross-sectional areas of both ring and rod must be equal. Write this condition in mathematical terms, using Equation 10.5 on both sides of the equation, and solve for ΔT.

SOLUTION

(a) Find the temperature of the ring that will allow it to slip onto the rod.

Write Equation 10.5 and substitute known values, leaving ΔT as the sole unknown:

$$\Delta A = \gamma A_0 \, \Delta T$$

$$0.020 \text{ cm}^2 = [34 \times 10^{-6} \, (°\text{C})^{-1}](9.980 \text{ cm}^2)(\Delta T)$$

Solve for ΔT, then add this change to the initial temperature to get the final temperature:

$$\Delta T = 59°\text{C}$$

$$T = T_0 + \Delta T = 20.0°\text{C} + 59°\text{C} = 79°\text{C}$$

(b) If both ring and rod are heated, find the minimum change in temperature that will allow the ring to be slipped onto the rod.

Set the final areas of the copper ring and steel rod equal to each other:

$$A_C + \Delta A_C = A_S + \Delta A_S$$

Substitute for each change in area, ΔA:

$$A_C + \gamma_C A_C \, \Delta T = A_S + \gamma A_S \, \Delta T$$

Rearrange terms to get ΔT on one side only, factor it out, and solve:

$$\gamma_C A_C \, \Delta T - \gamma_S A_S \, \Delta T = A_S - A_C$$

$$(\gamma_C A_C - \gamma_S A_S) \, \Delta T = A_S - A_C$$

$$\Delta T = \frac{A_S - A_C}{\gamma_C A_C - \gamma_S A_S}$$

$$= \frac{10.000 \text{ cm}^2 - 9.980 \text{ cm}^2}{(34 \times 10^{-6} \, °\text{C}^{-1})(9.980 \text{ cm}^2) - (22 \times 10^{-6} \, °\text{C}^{-1})(10.000 \text{ cm}^2)}$$

$$\Delta T = 170°\text{C}$$

REMARKS Warming and cooling strategies are sometimes useful for separating glass parts in a chemistry lab, such as the glass stopper in a bottle of reagent.

QUESTION 10.4 If instead of heating the copper ring in part (a) the steel rod is cooled, would the magnitude of the required temperature change be larger, smaller, or the same? Why? (Don't calculate it!)

EXERCISE 10.4 A steel ring with a hole having area of 3.990 cm^2 is to be placed on an aluminum rod with cross-sectional area of 4.000 cm^2. Both rod and ring are initially at a temperature of 35.0°C. At what common temperature can the steel ring be slipped onto one end of the aluminum rod?

ANSWER −61°C

We can also show that the *increase in volume* of an object accompanying a change in temperature is

$$\Delta V = \beta V_0 \, \Delta T \qquad [10.6]$$

where β, the **coefficient of volume expansion,** is equal to 3α. (Note that $\gamma = 2\alpha$ and $\beta = 3\alpha$ only if the coefficient of linear expansion of the object is the same in all directions.) The proof of Equation 10.6 is similar to the proof of Equation 10.5.

Table 11.2 Latent Heats of Fusion and Vaporization

Substance	Melting Point (°C)	Latent Heat of Fusion (J/kg)	Latent Heat of Fusion (cal/g)	Boiling Point (°C)	Latent Heat of Vaporization (J/kg)	Latent Heat of Vaporization (cal/g)
Helium	−269.65	5.23×10^3	1.25	−268.93	2.09×10^4	4.99
Nitrogen	−209.97	2.55×10^4	6.09	−195.81	2.01×10^5	48.0
Oxygen	−218.79	1.38×10^4	3.30	−182.97	2.13×10^5	50.9
Ethyl alcohol	−114	1.04×10^5	24.9	78	8.54×10^5	204
Water	0.00	3.33×10^5	79.7	100.00	2.26×10^6	540
Sulfur	119	3.81×10^4	9.10	444.60	3.26×10^5	77.9
Lead	327.3	2.45×10^4	5.85	1 750	8.70×10^5	208
Aluminum	660	3.97×10^5	94.8	2 450	1.14×10^7	2 720
Silver	960.80	8.82×10^4	21.1	2 193	2.33×10^6	558
Gold	1 063.00	6.44×10^4	15.4	2 660	1.58×10^6	377
Copper	1 083	1.34×10^5	32.0	1 187	5.06×10^6	1 210

The unit of latent heat is the joule per kilogram (J/kg). The word *latent* means "lying hidden within a person or thing." The positive sign in Equation 11.6 is chosen when energy is absorbed by a substance, as when ice is melting. The negative sign is chosen when energy is removed from a substance, as when steam condenses to water.

Tip 11.3 Signs Are Critical

For phase changes, use the correct explicit sign in Equation 11.6, positive if you are adding energy to the substance, negative if you're taking it away.

The **latent heat of fusion** L_f is used when a phase change occurs during melting or freezing, whereas the **latent heat of vaporization** L_v is used when a phase change occurs during boiling or condensing.[1] For example, at atmospheric pressure the latent heat of fusion for water is 3.33×10^5 J/kg and the latent heat of vaporization for water is 2.26×10^6 J/kg. The latent heats of different substances vary considerably, as can be seen in Table 11.2.

Another process, sublimation, is the passage from the solid to the gaseous phase without going through a liquid phase. The fuming of dry ice (frozen carbon dioxide) illustrates this process, which has its own latent heat associated with it, the heat of sublimation.

To better understand the physics of phase changes, consider the addition of energy to a 1.00-g cube of ice at −30.0°C in a container held at constant pressure. Suppose this input of energy turns the ice to steam (water vapor) at 120.0°C. Figure 11.3 (page 376) is a plot of the experimental measurement of temperature as energy is added to the system. We examine each portion of the curve separately.

Part A During this portion of the curve, the temperature of the ice changes from −30.0°C to 0.0°C. Because the specific heat of ice is 2 090 J/kg · °C, we can calculate the amount of energy added from Equation 11.3:

$$Q = mc_{\text{ice}}\,\Delta T = (1.00 \times 10^{-3}\text{ kg})(2\,090\text{ J/kg}\cdot{}^\circ\text{C})(30.0^\circ\text{C}) = 62.7\text{ J}$$

Part B When the ice reaches 0°C, the ice–water mixture remains at that temperature—even though energy is being added—until all the ice melts to become water at 0°C. According to Equation 11.6, the energy required to melt 1.00 g of ice at 0°C is

$$Q = mL_f = (1.00 \times 10^{-3}\text{ kg})(3.33 \times 10^5\text{ J/kg}) = 333\text{ J}$$

[1]When a gas cools, it eventually returns to the liquid phase, or *condenses*. The energy per unit mass given up during the process is called the *heat of condensation*, and it equals the heat of vaporization. When a liquid cools, it eventually solidifies, and the *heat of solidification* equals the heat of fusion.

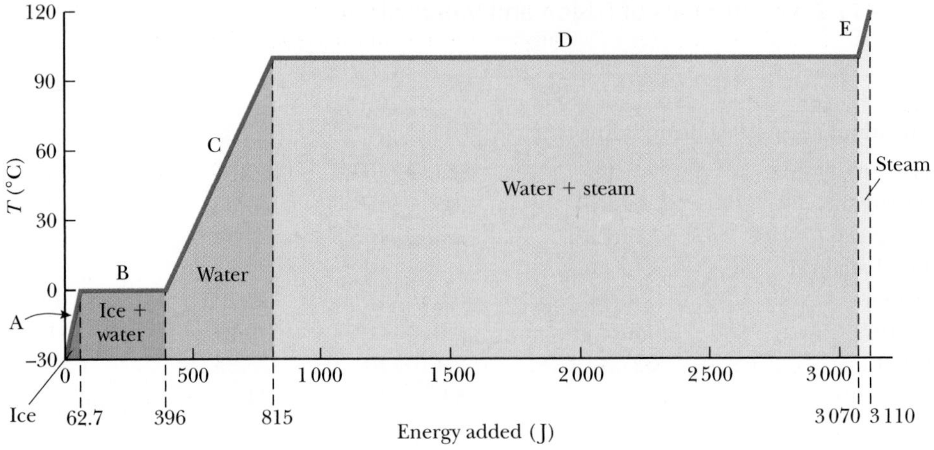

Figure 11.3 A plot of temperature versus energy added when 1.00 g of ice, initially at −30.0°C, is converted to steam at 120°C.

Part C Between 0°C and 100°C, no phase change occurs. The energy added to the water is used to increase its temperature, as in part A. The amount of energy necessary to increase the temperature from 0°C to 100°C is

$$Q = mc_{\text{water}}\,\Delta T = (1.00 \times 10^{-3}\text{ kg})(4.19 \times 10^3\text{ J/kg}\cdot{}^\circ\text{C})(1.00 \times 10^2\ {}^\circ\text{C})$$

$$Q = 4.19 \times 10^2\text{ J}$$

Part D At 100°C, another phase change occurs as the water changes to steam at 100°C. As in Part B, the water–steam mixture remains at constant temperature, this time at 100°C—even though energy is being added—until all the liquid has been converted to steam. The energy required to convert 1.00 g of water at 100°C to steam at 100°C is

$$Q = mL_v = (1.00 \times 10^{-3}\text{ kg})(2.26 \times 10^6\text{ J/kg}) = 2.26 \times 10^3\text{ J}$$

Part E During this portion of the curve, as in parts A and C, no phase change occurs, so all the added energy goes into increasing the temperature of the steam. The energy that must be added to raise the temperature of the steam to 120.0°C is

$$Q = mc_{\text{steam}}\,\Delta T = (1.00 \times 10^{-3}\text{ kg})(2.01 \times 10^3\text{ J/kg}\cdot{}^\circ\text{C})(20.0^\circ\text{C}) = 40.2\text{ J}$$

The total amount of energy that must be added to change 1.00 g of ice at −30.0°C to steam at 120.0°C is the sum of the results from all five parts of the curve, 3.11×10^3 J. Conversely, to cool 1.00 g of steam at 120.0°C to the point at which it becomes ice cooled to −30.0°C, 3.11×10^3 J of energy must be removed.

Phase changes can be described in terms of rearrangements of molecules when energy is added to or removed from a substance. Consider first the liquid-to-gas phase change. The molecules in a liquid are close together, and the forces between them are stronger than the forces between the more widely separated molecules of a gas. Work must therefore be done on the liquid against these attractive molecular forces so as to separate the molecules. The latent heat of vaporization is the amount of energy that must be added to the one kilogram of liquid to accomplish this separation.

Similarly, at the melting point of a solid, the amplitude of vibration of the atoms about their equilibrium positions becomes great enough to allow the atoms to pass the barriers of adjacent atoms and move to their new positions. On average, these new positions are less symmetrical than the old ones and therefore have higher energy. The latent heat of fusion is equal to the work required at the molecular level to transform the mass from the ordered solid phase to the disordered liquid phase.

The average distance between atoms is much greater in the gas phase than in either the liquid or the solid phase. Each atom or molecule is removed from its neighbors, overcoming the attractive forces of nearby neighbors. Therefore, more work is increase the amplitude of their vibrations, until eventually those in the handle are affected. This increased vibration represents an increase in temperature of the metal (and possibly a burned hand!).

Although the transfer of energy through a substance can be partly explained by atomic vibrations, the rate of conduction depends on the properties of the substance. For example, it's possible to hold a piece of asbestos in a flame indefinitely, which implies that very little energy is conducted through the asbestos. In general, metals are good thermal conductors because they contain large numbers of electrons that are relatively free to move through the metal and can transport energy from one region to another. In a good conductor such as copper, conduction takes place via the vibration of atoms and the motion of free electrons. Materials such as asbestos, cork, paper, and fiberglass are poor thermal conductors. Gases are also poor thermal conductors because of the large distance between their molecules.

Conduction occurs only if there is a difference in temperature between two parts of the conducting medium. The temperature difference drives the flow of energy. Consider a slab of material of thickness Δx and cross-sectional area A with its opposite faces at different temperatures T_c and T_h, where $T_h > T_c$ (Fig. 11.5). The slab allows energy to transfer from the region of higher temperature to the region of lower temperature by thermal conduction. The rate of energy transfer, $P = Q/\Delta t$, is proportional to the cross-sectional area of the slab and the temperature difference and is inversely proportional to the thickness of the slab:

$$P = \frac{Q}{\Delta t} \propto A\frac{\Delta T}{\Delta x}$$

Note that P has units of watts when Q is in joules and Δt is in seconds.

Suppose a substance is in the shape of a long, uniform rod of length L, as in Figure 11.6. We assume the rod is insulated, so thermal energy can't escape by conduction from its surface except at the ends. One end is in thermal contact with an energy reservoir at temperature T_c and the other end is in thermal contact with a reservoir at temperature $T_h > T_c$. When a steady state is reached, the temperature at each point along the rod is constant in time. In this case $\Delta T = T_h - T_c$ and $\Delta x = L$, so

$$\frac{\Delta T}{\Delta x} = \frac{T_h - T_c}{L}$$

The rate of energy transfer by conduction through the rod is given by

$$P = kA\frac{(T_h - T_c)}{L} \quad [11.7]$$

where k, a proportionality constant that depends on the material, is called the **thermal conductivity.** Substances that are good conductors have large thermal conductivities, whereas good insulators have low thermal conductivities. Table 11.3 lists the thermal conductivities for various substances.

Quick Quiz

11.3 Will an ice cube wrapped in a wool blanket remain frozen for (a) less time, (b) the same length of time, or (c) a longer time than an identical ice cube exposed to air at room temperature?

11.4 Two rods of the same length and diameter are made from different materials. The rods are to connect two regions of different temperature so that energy will transfer through the rods by heat. They can be connected in series, as in Figure 11.7a (page 382), or in parallel, as in Figure 11.7b. In which case is the rate of energy transfer by heat larger? (a) When the rods are in series (b) When the rods are in parallel (c) The rate is the same in both cases.

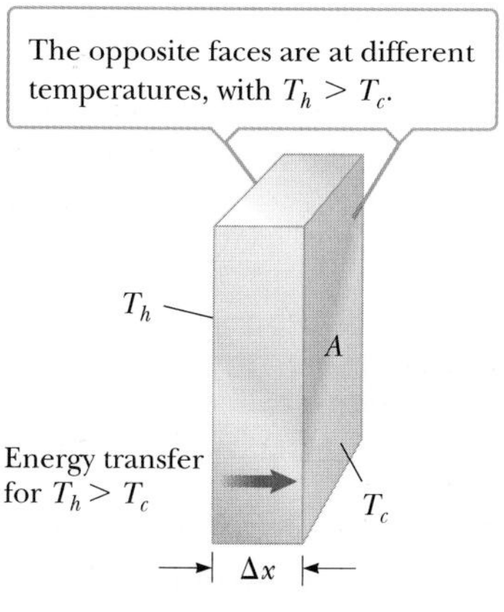

Figure 11.5 Energy transfer through a conducting slab of cross-sectional area A and thickness Δx.

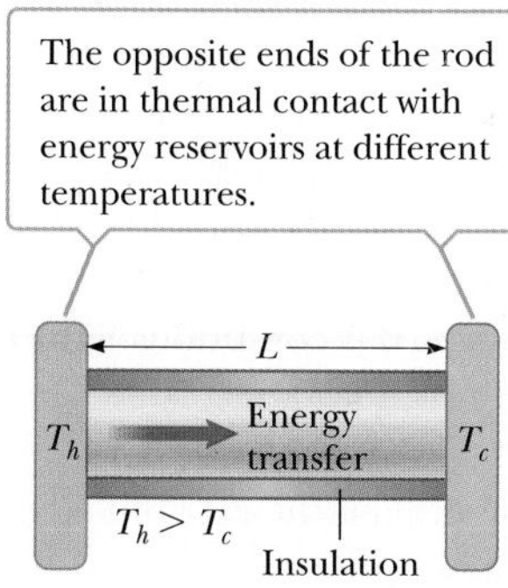

Figure 11.6 Conduction of energy through a uniform, insulated rod of length L.

Table 11.3 Thermal Conductivities

Substance	Thermal Conductivity (J/s · m · °C)
Metals (at 25°C)	
Aluminum	238
Copper	397
Gold	314
Iron	79.5
Lead	34.7
Silver	427
Gases (at 20°C)	
Air	0.023 4
Helium	0.138
Hydrogen	0.172
Nitrogen	0.023 4
Oxygen	0.023 8
Nonmetals	
(approximate values)	
Asbestos	0.08
Concrete	0.8
Glass	0.8
Ice	2
Rubber	0.2
Water	0.6
Wood	0.08

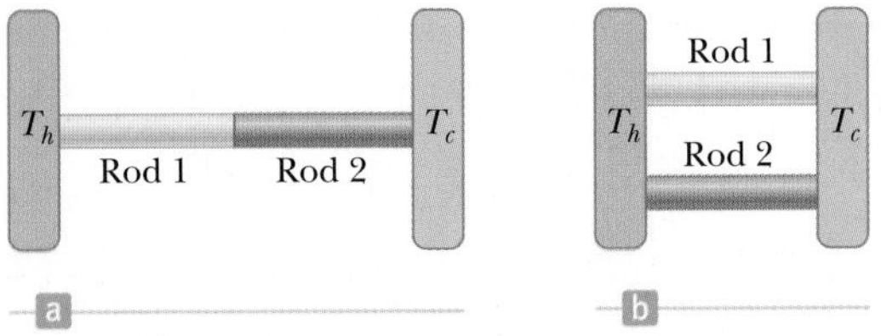

Figure 11.7 (Quick Quiz 11.4) In which case is the rate of energy transfer larger?

EXAMPLE 11.8 Conductive Losses from the Human Body BIO

GOAL Apply the conduction equation to a human being.

PROBLEM In a human being, a layer of fat and muscle lies under the skin having various thicknesses depending on location. In response to a cold environment, capillaries near the surface of the body constrict, reducing blood flow and thereby reducing the conductivity of the tissues. These tissues form a shell up to an inch thick having a thermal conductivity of about 0.21 W/m · K, the same as skin or fat. **(a)** Estimate the rate of loss of thermal energy due to conduction from the human core region to the skin surface, assuming a shell thickness of 2.0 cm and a skin temperature of 33.0°C. (Skin temperature varies, depending on external conditions.) **(b)** Calculate the thermal energy lost due to conduction in 1.0 h. **(c)** Estimate the change in body temperature in 1.0 h if the energy is not replenished. Assume a body mass of 75 kg and a skin surface area of 1.73 m^2.

STRATEGY The solution to part (a) requires applying Equation 11.7 for the rate of energy transfer due to conduction. Multiplying the power found in part (a) by the elapsed time yields the total thermal energy transfer in the given time. In part (c), an estimate for the change in temperature if the energy is not replenished can be developed using Equation 11.3, $Q = mc\Delta T$.

SOLUTION

(a) Estimate the rate of loss of thermal energy due to conduction.

Write the thermal conductivity equation:

$$P = \frac{kA(T_h - T_c)}{L}$$

Substitute values:

$$P = \frac{(0.21\ \text{J/m}\cdot\text{K})(1.73\ \text{m}^2)(37.0°\text{C} - 33.0°\text{C})}{2.0 \times 10^{-2}\ \text{m}} = \boxed{73\ \text{W}}$$

(b) Calculate the thermal energy lost due to conduction in 1.0 h.

Multiply the power P by the time Δt:

$$Q = P\Delta t = (73\ \text{W})(3\,600\ \text{s}) = \boxed{2.6 \times 10^5\ \text{J}}$$

(c) Estimate the change in body temperature in 1.0 h if the energy is not replenished.

Write Equation 11.3 and solve it for ΔT:

$$Q = mc\Delta T$$

$$\Delta T = \frac{Q}{mc} = \frac{2.6 \times 10^5\ \text{J}}{(75\ \text{kg})(3\,470\ \text{J/kg}\cdot\text{K})} = \boxed{1.0°\text{C}}$$

REMARKS The calculation doesn't take into account the thermal gradient, which further reduces the rate of conduction through the shell. Whereas thermal energy transfers through the shell by conduction, other mechanisms remove that energy from the body's surface because air is a poor conductor of thermal energy. Convection, radiation, and evaporation of sweat are the primary mechanisms that remove thermal energy from the skin. The calculation shows that even under mild conditions the body must constantly replenish its internal energy. It's possible to die of exposure even in temperatures well above freezing.

QUESTION 11.8 Why does a long distance runner require very little in the way of warm clothing when running in cold weather, but puts on a sweater after finishing the run?

EXERCISE 11.8 BIO A female minke whale has a core body temperature of 35°C and a core/blubber interface temperature of 29°C, with an average blubber thickness of 4.0 cm and thermal conductivity of 0.25 W/m·K. (a) At what rate is energy lost from the whale's core by conduction from the core/blubber interface through the blubber to the skin? Assume a skin temperature of 12°C and a total body area of 22 m^2. (b) What percent of the daily energy budget is this number? (The average female minke whale requires 8.0 × 10^8 J of energy per day—that's a lot of plankton and krill.)

ANSWERS (a) 2.3 × 10^3 W (b) 25%

Home Insulation

To determine whether to add insulation to a ceiling or some other part of a building, the preceding discussion of conduction must be extended for two reasons:

1. The insulating properties of materials used in buildings are usually expressed in engineering (U.S. customary) rather than SI units. Measurements stamped on a package of fiberglass insulating board will be in units such as British thermal units, feet, and degrees Fahrenheit.
2. In dealing with the insulation of a building, conduction through a compound slab must be considered, with each portion of the slab having a certain thickness and a specific thermal conductivity. A typical wall in a house consists of an array of materials, such as wood paneling, drywall, insulation, sheathing, and wood siding.

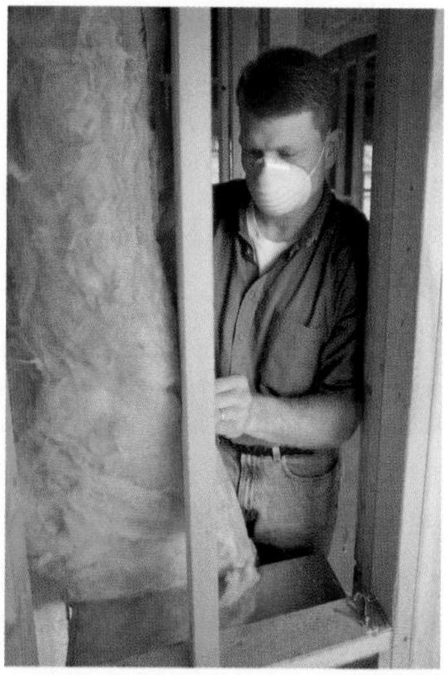

A worker installing fiberglass insulation in a home. The mask protects the worker against the inhalation of microscopic fibers, which could be hazardous to his health.

The rate of energy transfer by conduction through a compound slab is

$$\frac{Q}{\Delta t} = \frac{A(T_h - T_c)}{\sum_i L_i/k_i} \qquad [11.8]$$

where T_h and T_c are the temperatures of the *outer extremities* of the slab and the summation is over all portions of the slab. This formula can be derived algebraically, using the facts that the temperature at the interface between two insulating materials must be the same and that the rate of energy transfer through one insulator must be the same as through all the other insulators. If the slab consists of three different materials, the denominator is the sum of three terms. In engineering practice, the term L/k for a particular substance is referred to as the *R*-value of the material, so Equation 11.8 reduces to

$$\frac{Q}{\Delta t} = \frac{A(T_h - T_c)}{\sum_i R_i} \qquad [11.9]$$

The *R*-values for a few common building materials are listed in Table 11.4. Note the unit of *R* and the fact that the *R*-values are defined for specific thicknesses.

Table 11.4 *R*-Values for Some Common Building Materials

Material	*R* value[a] (ft² · °F · h/Btu)
Hardwood siding (1.0 in. thick)	0.91
Wood shingles (lapped)	0.87
Brick (4.0 in. thick)	4.00
Concrete block (filled cores)	1.93
Styrofoam (1.0 in. thick)	5.0
Fiberglass batting (3.5 in. thick)	10.90
Fiberglass batting (6.0 in. thick)	18.80
Fiberglass board (1.0 in. thick)	4.35
Cellulose fiber (1.0 in. thick)	3.70
Flat glass (0.125 in. thick)	0.89
Insulating glass (0.25-in. space)	1.54
Vertical air space (3.5 in. thick)	1.01
Stagnant layer of air	0.17
Drywall (0.50 in. thick)	0.45
Sheathing (0.50 in. thick)	1.32

[a]The values in this table can be converted to SI units by multiplying the values by 0.1761.

Next to any vertical outside surface is a very thin, stagnant layer of air that must be considered when the total *R*-value for a wall is computed. The thickness of this stagnant layer depends on the speed of the wind. As a result, energy loss by conduction from a house on a day when the wind is blowing is greater than energy loss on a day when the wind speed is zero. A representative *R*-value for a stagnant air layer is given in Table 11.4. The values are typically given in British units, but they can be converted to the equivalent metric units by multiplying the values in the table by 0.176 1.)

EXAMPLE 11.9 Construction and Thermal Insulation

GOAL Calculate the *R*-value of several layers of insulating material and its effect on thermal energy transfer.

PROBLEM **(a)** Find the energy transferred in 1.00 h by conduction through a concrete wall 2.0 m high, 3.65 m long, and 0.20 m thick if one side of the wall is held at 5.00°C and the other side is at 20.0°C (Fig. 11.8). Assume the concrete has a thermal conductivity of 0.80 J/s · m · °C. **(b)** The owner of the home decides to increase the insulation, so he installs 0.50 in of thick sheathing, 3.5 in of fiberglass batting, and a drywall 0.50 in thick. Calculate the *R*-factor. **(c)** Calculate the energy transferred in 1.00 h by conduction. **(d)** What is the temperature between the concrete wall and the sheathing? Assume there is an air layer on the exterior of the concrete wall but not between the concrete and the sheathing.

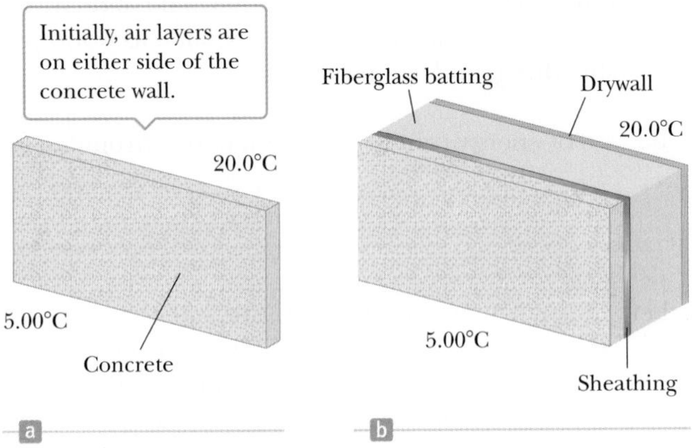

Figure 11.8 (Example 11.9) A cross-sectional view of (a) a concrete wall with two air spaces and (b) the same wall with sheathing, fiberglass batting, drywall, and two air layers.

STRATEGY The *R*-value of the concrete wall is given by L/k. Add this to the *R*-value of two air layers and then substitute into Equation 11.8, multiplying by the seconds in an hour to get the total energy transferred through the wall in an hour. Repeat this process, with different materials, for parts (b) and (c). Part (d) requires finding the *R*-value for an air layer and the concrete wall and then substituting into the thermal conductivity equation. In this problem metric units are used, so be sure to convert the *R*-values in the table. (Converting to SI requires multiplication of the British units by 0.176 1.)

SOLUTION

(a) Find the energy transferred in 1.00 h by conduction through a concrete wall.

Calculate the *R*-value of concrete plus two air layers:

$$\sum R = \frac{L}{k} + 2R_{\text{air layer}} = \frac{0.20\text{ m}}{0.80\text{ J/s}\cdot\text{m}\cdot{}^\circ\text{C}} + 2\left(0.030\,\frac{\text{m}^2}{\text{J/s}\cdot{}^\circ\text{C}}\right)$$

$$= 0.31\,\frac{\text{m}^2}{\text{J/s}\cdot{}^\circ\text{C}}$$

Write the thermal conduction equation:

$$P = \frac{A(T_h - T_c)}{\sum R}$$

Substitute values:

$$P = \frac{(7.3\text{ m}^2)(20.0^\circ\text{C} - 5.00^\circ\text{C})}{0.31\text{ m}^2\cdot\text{s}\cdot{}^\circ\text{C/J}} = 353\text{ W} \quad\rightarrow\quad 350\text{ W}$$

Multiply the power in watts times the seconds in an hour:

$$Q = P\Delta t = (350\text{ W})(3\,600\text{ s}) = \boxed{1.3\times10^6\text{ J}}$$

(b) Calculate the *R*-factor of the newly insulated wall.

Refer to Table 11.4 and sum the appropriate quantities after converting them to SI units:

$$R_{\text{total}} = R_{\text{outside air layer}} + R_{\text{concrete}} + R_{\text{sheath}} + R_{\text{fiberglass}} + R_{\text{drywall}} + R_{\text{inside air layer}}$$

$$= (0.030 + 0.25 + 0.232 + 1.92 + 0.079 + 0.030)$$

$$= \boxed{2.5\text{ m}^2\cdot{}^\circ\text{C}\cdot\text{s/J}}$$

(Continued)

(c) Calculate the energy transferred in 1.00 h by conduction.

Write the thermal conduction equation:

$$P = \frac{A(T_h - T_c)}{\sum R}$$

Substitute values:

$$P = \frac{(7.3 \text{ m}^2)(20.0°\text{C} - 5.00°\text{C})}{2.5 \text{ m}^2 \cdot \text{s} \cdot °\text{C/J}} = 44 \text{ W}$$

Multiply the power in watts times the seconds in an hour:

$$Q = P\Delta t = (44 \text{ W})(3\,600 \text{ s}) = \boxed{1.6 \times 10^5 \text{ J}}$$

(d) Calculate the temperature between the concrete and the sheathing.

Write the thermal conduction equation:

$$P = \frac{A(T_h - T_c)}{\sum R}$$

Solve algebraically for T_h by multiplying both sides by ΣR and dividing both sides by area A:

$$P\sum R = A(T_h - T_c) \quad \rightarrow \quad (T_h - T_c) = \frac{P\sum R}{A}$$

Add T_c to both sides:

$$T_h = \frac{P\sum R}{A} + T_c$$

Substitute the R-value for the concrete wall from part (a), but subtract the R-value of one air layer from that calculated in part (a):

$$T_h = \frac{(44 \text{ W})(0.31 \text{ m}^2 \cdot \text{s} \cdot °\text{C/J} - 0.03 \text{ m}^2 \cdot \text{s} \cdot °\text{C/J})}{7.3 \text{ m}^2} + 5.00°\text{C}$$

$$= \boxed{6.7°\text{C}}$$

REMARKS Notice the enormous energy savings that can be realized with good insulation!

QUESTION 11.9 Which of the following choices results in the best possible R-value? (a) Use material with a small thermal conductivity and large thickness. (b) Use thin material with a large thermal conductivity. (c) Use material with a small thermal conductivity and small thickness.

EXERCISE 11.9 Instead of the layers of insulation, the owner installs a brick wall on the exterior of the concrete wall. (a) Calculate the R-factor, including the two stagnant air layers on the inside and outside of the wall. (b) Calculate the energy transferred in 1.00 h by conduction, under the same conditions as in the example. (c) What is the temperature between the concrete and the brick?

ANSWERS (a) 1.01 $\text{m}^2 \cdot °\text{C} \cdot \text{s/J}$ (b) 3.9×10^5 J (c) 16°C

Convection

When you warm your hands over an open flame, as illustrated in Figure 11.9, the air directly above the flame, being warmed, expands. As a result, the density of this air decreases and the air rises, warming your hands as it flows by. **The transfer of energy by the movement of a substance is called convection.** When the movement results from differences in density, as with air around a fire, it's referred to as *natural convection*. Airflow at a beach is an example of natural convection, as is the mixing that occurs as surface water in a lake cools and sinks. When the substance is forced to move by a fan or pump, as in some hot air and hot water heating systems, the process is called *forced convection*.

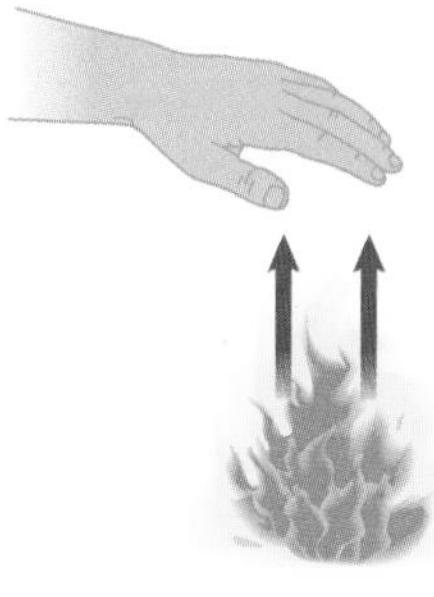

Figure 11.9 Warming a hand by convection.

Convection currents assist in the boiling of water. In a teakettle on a hot stovetop, the lower layers of water are warmed first. The warmed water has a lower density and rises to the top, while the denser, cool water at the surface sinks to the bottom of the kettle and is warmed.

The same process occurs when a radiator raises the temperature of a room. The hot radiator warms the air in the lower regions of the room. The warm air expands and, because of its lower density, rises to the ceiling. The denser, cooler air from above sinks, setting up the continuous air current pattern shown in Figure 11.10.

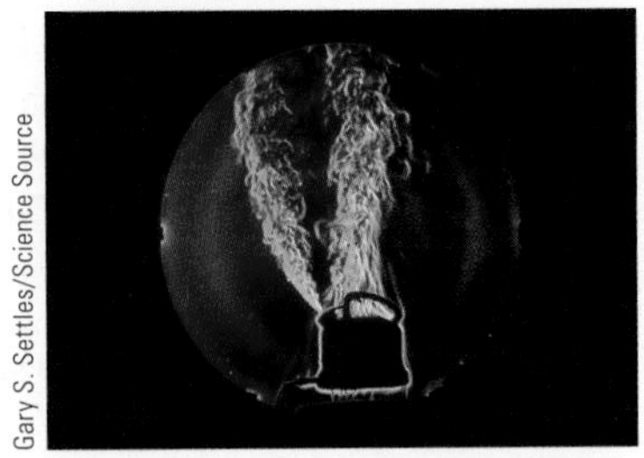

Photograph of a teakettle, showing steam and turbulent convection air currents.

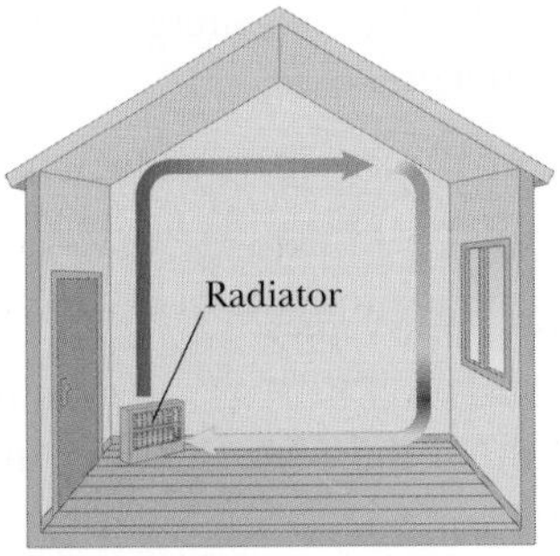

Figure 11.10 Convection currents are set up in a room warmed by a radiator.

APPLICATION
Cooling Automobile Engines

An automobile engine is maintained at a safe operating temperature by a combination of conduction and forced convection. Water (actually, a mixture of water and antifreeze) circulates in the interior of the engine. As the metal of the engine block increases in temperature, energy passes from the hot metal to the cooler water by thermal conduction. The water pump forces water out of the engine and into the radiator, carrying energy along with it (by forced convection). In the radiator the hot water passes through metal pipes that are in contact with the cooler outside air, and energy passes into the air by conduction. The cooled water is then returned to the engine by the water pump to absorb more energy. The process of air being pulled past the radiator by the fan is also forced convection.

BIO APPLICATION
Algal Blooms in Ponds and Lakes

The algal blooms often seen in temperate lakes and ponds during the spring or fall are caused by convection currents in the water. To understand this process, consider Figure 11.11. During the summer, bodies of water develop temperature gradients, with a warm upper layer of water separated from a cold lower layer by a buffer zone called a thermocline. In the spring and fall temperature changes in the water break down this thermocline, setting up convection currents that mix the water. The mixing process transports nutrients from the bottom to the surface. The nutrient-rich water forming at the surface can cause a rapid, temporary increase in the algae population.

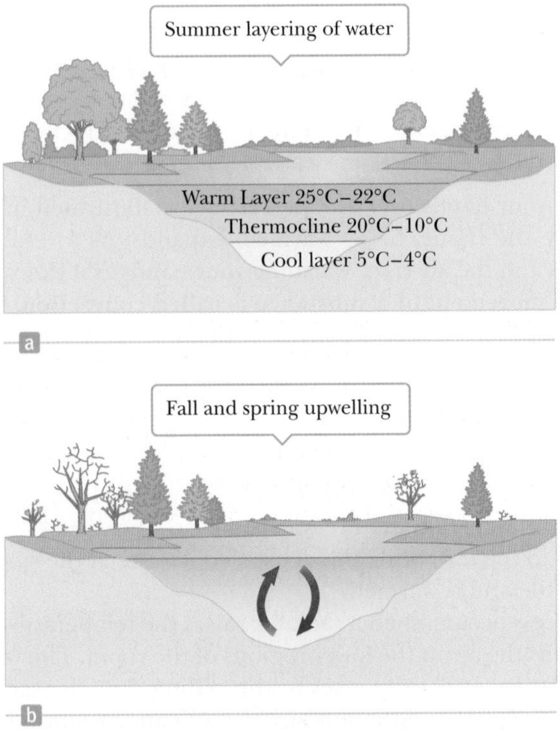

Figure 11.11 (a) During the summer, a warm upper layer of water is separated from a cooler lower layer by a thermocline. (b) Convection currents during the spring and fall mix the water and can cause algal blooms.

APPLYING PHYSICS 11.1 | Body Temperature BIO

The body temperature of mammals ranges from about 35°C to 38°C, whereas that of birds ranges from about 40°C to 43°C. How can these narrow ranges of body temperature be maintained in cold weather?

EXPLANATION A natural method of maintaining body temperature is via layers of fat beneath the skin. Fat protects against both conduction and convection because of its low thermal conductivity and because there are few blood vessels in fat to carry blood to the surface, where energy losses by convection can occur. Birds ruffle their feathers in cold weather to trap a layer of air with a low thermal conductivity between the feathers and the skin. Bristling the fur produces the same effect in fur-bearing animals.

Humans keep warm with wool sweaters and down jackets that trap the warmer air in regions close to their bodies, reducing energy loss by convection and conduction. ■

Radiation

Another process of transferring energy is through **radiation.** Figure 11.12 shows how your hands can be warmed by a lamp through radiation. Because your hands aren't in physical contact with the lamp and the conductivity of air is very low, conduction can't account for the energy transfer. Nor can convection be responsible for any transfer of energy because your hands aren't above the lamp in the path of convection currents. The warmth felt in your hands must therefore come from the transfer of energy by radiation.

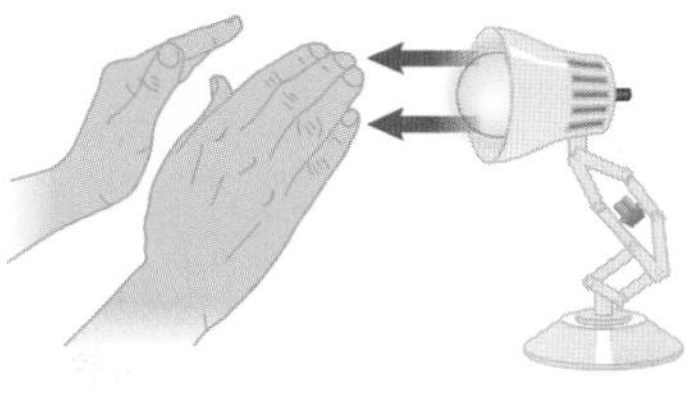

Figure 11.12 Warming hands by radiation.

All objects radiate energy continuously in the form of electromagnetic waves due to thermal vibrations of their molecules. These vibrations create the orange glow of an electric stove burner, an electric space heater, and the coils of a toaster.

The rate at which an object radiates energy is proportional to the fourth power of its absolute temperature. This is known as **Stefan's law,** expressed in equation form as

$$P = \sigma A e T^4 \qquad [11.10]$$

◀ Stefan's law

where P is the power in watts (or joules per second) radiated by the object, σ is the Stefan–Boltzmann constant, equal to $5.669\,6 \times 10^{-8}$ W/m$^2 \cdot$ K^4, A is the surface area of the object in square meters, e is a constant called the **emissivity** of the object, and T is the object's Kelvin temperature. The value of e can vary between zero and one, depending on the properties of the object's surface.

Approximately 1 370 J of electromagnetic radiation from the Sun passes through each square meter at the top of the Earth's atmosphere every second. This radiation is primarily visible light, accompanied by significant amounts of infrared and ultraviolet light. We will study these types of radiation in detail in Chapter 21. Some of this energy is reflected back into space, and some is absorbed by the atmosphere, but enough arrives at the surface of the Earth each day to supply all our energy needs hundreds of times over, if it could be captured and used efficiently. The growth in the number of solar houses in the United States is one example of an attempt to make use of this abundant energy. Radiant energy from the Sun affects our day-to-day existence in a number of ways, influencing Earth's average temperature, ocean currents, agriculture, and rain patterns. It can also affect behavior.

As another example of the effects of energy transfer by radiation, consider what happens to the atmospheric temperature at night. If there is a cloud cover above the Earth, the water vapor in the clouds absorbs part of the infrared radiation emitted by the Earth and re-emits it back to the surface. Consequently, the temperature at the surface remains at moderate levels. In the absence of cloud cover, there is nothing to prevent the radiation from escaping into space, so the temperature drops more on a clear night than on a cloudy night.

As an object radiates energy at a rate given by Equation 11.10, it also absorbs radiation. If it didn't, the object would eventually radiate all its energy and its temperature would reach absolute zero. The energy an object absorbs comes from its environment, which consists of other bodies that radiate energy. If an object is at a temperature T and its surroundings are at a temperature T_0, the net energy gained or lost each second by the object as a result of radiation is

$$P_{\text{net}} = \sigma A e(T^4 - T_0^{\,4}) \qquad [11.11]$$

When an object is in equilibrium with its surroundings, it radiates and absorbs energy at the same rate, so its temperature remains constant. When an object is hotter than its surroundings, it radiates more energy than it absorbs and therefore cools.

An **ideal absorber** is an object that absorbs all the light radiation incident on it, including invisible infrared and ultraviolet light. Such an object is called a **black body** because a room-temperature black body would look black. Because a black body doesn't reflect radiation at any wavelength, any light coming from it is due to atomic and molecular vibrations alone. A perfect black body has emissivity $e = 1$. An ideal absorber is also an ideal radiator of energy. The Sun, for example, is nearly a perfect black body. This statement may seem contradictory because the Sun is bright, not dark; the light that comes from the Sun, however, is emitted, not reflected. Black bodies are perfect absorbers that look black at room temperature because they don't reflect any light. All black bodies, except those at absolute zero, emit light that has a characteristic spectrum, discussed in Chapter 27. In contrast to black bodies, an object for which $e = 0$ absorbs none of the energy incident on it, reflecting it all. Such a body is an **ideal reflector.**

White clothing is more comfortable to wear in the summer than black clothing. Black fabric acts as a good absorber of incoming sunlight and as a good emitter of this absorbed energy. About half of the emitted energy, however, travels toward the body, causing the person wearing the garment to feel uncomfortably warm. White or light-colored clothing reflects away much of the incoming energy.

APPLICATION
Light-Colored Summer Clothing

The amount of energy radiated by an object can be measured with temperature-sensitive recording equipment via a technique called **thermography.** An image of the pattern formed by varying radiation levels, called a **thermogram,** is brightest in the warmest areas. Figure 11.13 reproduces a thermogram of a house. More energy escapes in the lighter regions, such as the door and windows. The owners of this house could conserve energy and reduce their heating costs by adding insulation to the attic area and by installing thermal draperies over the windows. Thermograms have also been used to image injured or diseased tissue in medicine, because such areas are often at a different temperature than surrounding healthy tissue, although many radiologists consider thermograms inadequate as a diagnostic tool.

BIO APPLICATION
Thermography

BIO APPLICATION
Radiation Thermometers for Measuring Body Temperature

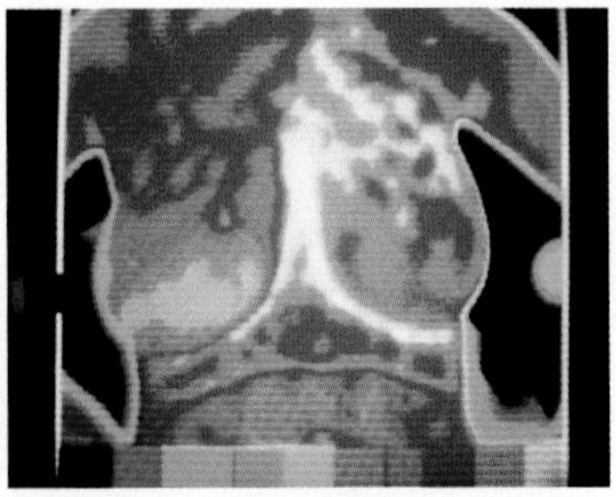

SPL/Science Source

Thermogram of a woman's breasts. Her left breast is diseased (red and orange) and her right breast (blue) is healthy.

Science Source

Figure 11.13 This thermogram of a house, made during cold weather.

Figure 11.14 shows a recently developed radiation thermometer that has removed most of the risk of taking the temperature of young children or the aged with a rectal thermometer, such as bowel perforation or bacterial contamination. The instrument measures the intensity of the infrared radiation leaving the eardrum and surrounding tissues and converts this information to a standard numerical reading. The eardrum is a particularly good location to measure body temperature because it's near the hypothalamus, the body's temperature control center.

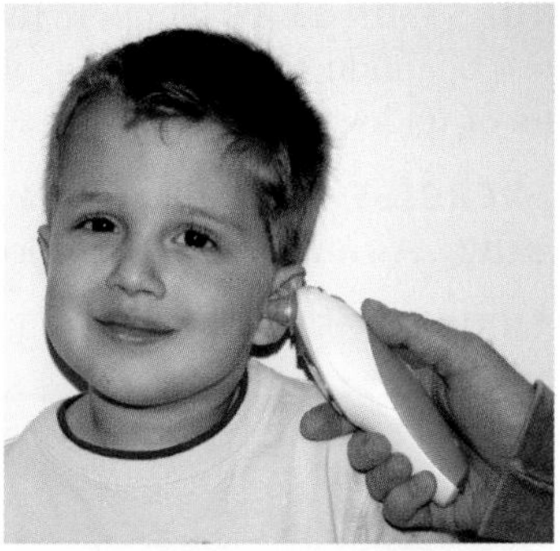

Figure 11.14 A radiation thermometer measures a patient's temperature by monitoring the intensity of infrared radiation leaving the ear.

Quick Quiz

11.5 Stars A and B have the same temperature, but star A has twice the radius of star B. (a) What is the ratio of star A's power output to star B's output due to electromagnetic radiation? The emissivity of both stars can be assumed to be 1. (b) Repeat the question if the stars have the same radius, but star A has twice the absolute temperature of star B. (c) What's the ratio if star A has both twice the radius and twice the absolute temperature of star B?

APPLYING PHYSICS 11.2 | Thermal Radiation and Night Vision

How can thermal radiation be used to see objects in near total darkness?

EXPLANATION There are two methods of night vision, one enhancing a combination of very faint visible light and infrared light, and another using infrared light only. The latter is valuable for creating images in absolute darkness. Because all objects above absolute zero emit thermal radiation due to the vibrations of their atoms, the infrared (invisible) light can be focused by a special lens and scanned by an array of infrared detector elements. These elements create a thermogram. The information from thousands of separate points in the field of view is converted to electrical impulses and translated by a microchip into a form suitable for display. Different temperature areas are assigned different colors, which can then be easily discerned on the display. ■

EXAMPLE 11.10 | Polar Bear Club BIO

GOAL Apply Stefan's law.

PROBLEM A member of the Polar Bear Club, dressed only in bathing trunks of negligible size, prepares to plunge into the Gulf of Finland from the beach in St. Petersburg, Russia. The air is calm, with a temperature of 5°C. If the swimmer's surface body temperature is 25°C, compute the net rate of energy loss from his skin due to radiation. How much energy is lost in 10.0 min? Assume his emissivity is 0.900 and his surface area is 1.50 m^2.

STRATEGY Use Equation 11.11, the thermal radiation equation, substituting the given information. Remember to convert temperatures to Kelvin by adding 273 to each value in degrees Celsius!

SOLUTION

Convert temperatures from Celsius to Kelvin:

$$T_{5^\circ\text{C}} = T_C + 273 = 5 + 273 = 278\text{ K}$$

$$T_{25^\circ\text{C}} = T_C + 273 = 25 + 273 = 298\text{ K}$$

Compute the net rate of energy loss, using Equation 11.11:

$$P_{\text{net}} = \sigma A e(T^4 - T_0{}^4)$$
$$= (5.67 \times 10^{-8}\text{ W/m}^2\cdot\text{K}^4)(1.50\text{ m}^2) \times (0.900)[(298\text{ K})^4 - (278\text{ K})^4]$$
$$P_{\text{net}} = \boxed{146\text{ W}}$$

Multiply the preceding result by the time, 10 minutes, to get the energy lost in that time due to radiation:

$$Q = P_{\text{net}} \times \Delta t = (146\text{ J/s})(6.00 \times 10^2\text{ s}) = \boxed{8.76 \times 10^4\text{ J}}$$

REMARKS Energy is also lost from the body through convection and conduction. Clothing traps layers of air next to the skin, which are warmed by radiation and conduction. In still air these warm layers are more readily retained. Even a Polar Bear Club member enjoys some benefit from the still air, better retaining a stagnant air layer next to the surface of his skin.

(Continued)

QUESTION 11.10 Suppose that at a given temperature the rate of an object's energy loss due to radiation is equal to its loss by conduction. When the object's temperature is raised, is the energy loss due to radiation (a) greater than, (b) equal to, or (c) less than the rate of energy loss due to conduction? (Assume the temperature of the environment is constant.)

EXERCISE 11.10 Repeat the calculation when the man is standing in his bedroom, with an ambient temperature of 20.0°C. Assume his body surface temperature is 27.0°C, with emissivity of 0.900.

ANSWER 55.9 W, 3.35×10^4 J

EXAMPLE 11.11 Planet of Alpha Centauri B

GOAL Apply Stefan's law to stars and their planets.

PROBLEM The star Alpha Centauri B is one member of the triple star system, Alpha Centauri AB-C, the closest star system to Earth. **(a)** Calculate the power output P of Alpha Centauri B, given its surface temperature of 5 790 K and radius $R = 6.02 \times 10^8$ m. **(b)** Calculate the power P_I intercepted by a possible Earth-sized planet, Alpha Centauri Bb, with radius $r = 6.64 \times 10^6$ m, orbiting its star at a distance of $r_O = 6.00 \times 10^9$ m. **(c)** Estimate the temperature of the planet using Stefan's equation. Assume all worlds are black bodies, with $e = 1$.

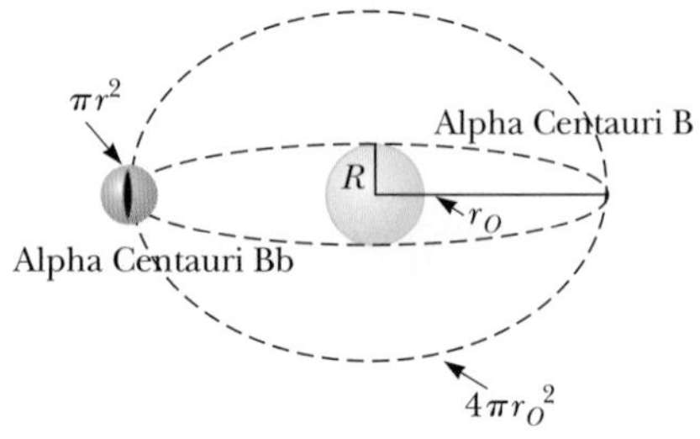

Figure 11.15 (Example 11.11) The power emitted by Alpha Centauri B travels radially outward, crossing a sphere with the same radius, r_O, as Alpha Centauri Bb's orbital radius. The cross-sectional area of the planet intercepts a small part of that radiation. (*Note:* Figure not drawn to scale)

STRATEGY Calculating the power output in part (a) is a matter of substitution. To solve part (b), it's necessary to find the fraction of the star's power intercepted by the planet. The star's energy crosses a sphere of area $A_O = 4\pi r_O^2$, where radius r_O is the planet's distance from Alpha Centauri B. The cross-sectional area of the planet's disk, $A_{pd} = \pi r^2$, intercepts a fraction of this energy given by A_{pd}/A_O. (See Figure 11.15.) Multiplying the star's power output by the fraction gives the amount of power the planet must both absorb and emit if in equilibrium, which is the answer to part (b). Substitute it into Stefan's equation and solve for the planet's temperature, the answer for part (c).

SOLUTION

(a) Calculate the power output of Alpha Centauri B.

Compute the surface area of Alpha Centauri B:

$$A = 4\pi R^2 = 4\pi(6.02 \times 10^8 \text{ m})^2 = 4.55 \times 10^{18} \text{ m}^2$$

Write Stefan's equation and substitute values:

$$P = \sigma A e T^4 = (5.67 \times 10^{-8} \text{ W/m}^2 \cdot \text{K}^4)(4.55 \times 10^{18} \text{ m}^2)(1.00)(5\,790 \text{ K})^4 = \boxed{2.90 \times 10^{26} \text{ W}}$$

(b) Calculate the power P_I intercepted by a possible Earth-sized planet, Alpha Centauri Bb.

Calculate the area of the planet's disk, A_{pd}, and the area of a sphere, A_O, with the same radius as the planet's orbital radius:

$$A_{pd} = \pi r^2 = \pi(6.64 \times 10^6 \text{ m})^2 = 1.39 \times 10^{14} \text{ m}^2$$
$$A_O = 4\pi r_O^2 = 4\pi(6.00 \times 10^9 \text{ m})^2 = 4.52 \times 10^{20} \text{ m}^2$$

Find the fraction of the star's power intercepted by the planet:

$$P_I = \left(\frac{A_{pd}}{A_O}\right)P = \left(\frac{1.39 \times 10^{14} \text{ m}^2}{4.52 \times 10^{20} \text{ m}^2}\right)(2.90 \times 10^{26} \text{ W})$$
$$= \boxed{8.92 \times 10^{19} \text{ W}}$$

(c) Estimate the temperature of the planet using Stefan's equation.

Write Stefan's equation, set it equal to the intercepted power, P_I, and solve for the temperature. Note that the full planetary area, $4\pi r^2$, not just the disk area, must be used:

$$P_I = \sigma A e T^4 = (5.67 \times 10^{-8} \text{ W/m}^2 \cdot \text{K}^4)(5.54 \times 10^{14} \text{ m}^2)(1.00)T^4$$
$$= (3.15 \times 10^7 \text{ W/K}^4)T^4 = 8.92 \times 10^{19} \text{ W}$$
$$T = \boxed{1.30 \times 10^3 \text{ K}}$$

REMARKS This calculation is only an estimate because the planet may not be a perfect black body, and the effects of an atmosphere—unlikely in this case—can greatly affect the typical average temperature on a given world.

5. On a clear, cold night, why does frost tend to form on the tops, rather than the sides, of mailboxes and cars?

6. The U.S. penny is now made of copper-coated zinc. Can a calorimetric experiment be devised to test for the metal content in a collection of pennies? If so, describe the procedure.

7. Cups of water for coffee or tea can be warmed with a coil that is immersed in the water and raised to a high temperature by means of electricity. (a) Why do the instructions warn users not to operate the coils in the absence of water? (b) Can the immersion coil be used to warm up a cup of stew?

8. The air temperature above coastal areas is profoundly influenced by the large specific heat of water. One reason is that the energy released when 1 cubic meter of water cools by 1.0°C will raise the temperature of an enormously larger volume of air by 1.0°C. Estimate that volume of air. The specific heat of air is approximately 1.0 kJ/kg · °C. Take the density of air to be 1.3 kg/m^3.

9. A tile floor may feel uncomfortably cold to your bare feet, but a carpeted floor in an adjoining room at the same temperature feels warm. Why?

10. On a very hot day, it's possible to cook an egg on the hood of a car. Would you select a black car or a white car on which to cook your egg? Why?

11. Concrete has a higher specific heat than does soil. Use this fact to explain (partially) why a city has a higher average temperature than the surrounding countryside. Would you expect evening breezes to blow from city to country or from country to city? Explain.

12. You need to pick up a very hot cooking pot in your kitchen. You have a pair of hot pads. Should you soak them in cold water or keep them dry in order to pick up the pot most comfortably?

13. A poker is a stiff, nonflammable rod used to push burning logs around in a fireplace. Suppose it is to be made of a single material. For best functionality and safety, should the poker be made from a material with (a) high specific heat and high thermal conductivity, (b) low specific heat and low thermal conductivity, (c) low specific heat and high thermal conductivity, (d) high specific heat and low thermal conductivity, or (e) low specific heat and low density?

14. Star A has twice the radius and twice the absolute temperature of star B. What is the ratio of the power output of star A to that of star B? The emissivity of both stars can be assumed to be 1. (a) 4 (b) 8 (c) 16 (d) 32 (e) 64

15. A person shakes a sealed, insulated bottle containing coffee for a few minutes. What is the change in the temperature of the coffee? (a) a large decrease (b) a slight decrease (c) no change (d) a slight increase (e) a large increase

PROBLEMS

ENHANCED WebAssign The problems in this chapter may be assigned online in Enhanced WebAssign.

1. denotes straightforward problem; 2. denotes intermediate problem;
3. denotes challenging problem
[1.] denotes full solution available in *Student Solutions Manual/ Study Guide*
1. denotes problems most often assigned in Enhanced WebAssign

BIO denotes biomedical problems
GP denotes guided problems
M denotes Master It tutorial available in Enhanced WebAssign
Q|C denotes asking for quantitative and conceptual reasoning
S denotes symbolic reasoning problem
W denotes Watch It video solution available in Enhanced WebAssign

11.1 Heat and Internal Energy

11.2 Specific Heat

1. The highest recorded waterfall in the world is found at Angel Falls in Venezuela. Its longest single waterfall has a height of 807 m. If water at the top of the falls is at 15.0°C, what is the maximum temperature of the water at the bottom of the falls? Assume all the kinetic energy of the water as it reaches the bottom goes into raising the water's temperature.

2. M The temperature of a silver bar rises by 10.0°C when it absorbs 1.23 kJ of energy by heat. The mass of the bar is 525 g. Determine the specific heat of silver from these data.

3. Lake Erie contains roughly 4.00×10^{11} m^3 of water. (a) How much energy is required to raise the temperature of that volume of water from 11.0°C to 12.0°C? (b) How many years would it take to supply this amount of energy by using the 1 000-MW exhaust energy of an electric power plant?

4. An aluminum rod is 20.0 cm long at 20°C and has a mass of 350 g. If 10 000 J of energy is added to the rod by heat, what is the change in length of the rod?

[5.] Q|C A 3.00-g copper coin at 25.0°C drops 50.0 m to the ground. (a) Assuming 60.0% of the change in gravitational potential energy of the coin–Earth system goes into increasing the internal energy of the coin, determine the coin's final temperature. (b) Does the result depend on the mass of the coin? Explain.

6. BIO A 55-kg woman cheats on her diet and eats a 540-Calorie (540 kcal) jelly doughnut for breakfast. (a) How many joules of energy are the equivalent of

one jelly doughnut? (b) How many stairs must the woman climb to perform an amount of mechanical work equivalent to the food energy in one jelly doughnut? Assume the height of a single stair is 15 cm. (c) If the human body is only 25% efficient in converting chemical energy to mechanical energy, how many stairs must the woman climb to work off her breakfast?

7. BIO Q|C A 75-kg sprinter accelerates from rest to a speed of 11.0 m/s in 5.0 s. (a) Calculate the mechanical work done by the sprinter during this time. (b) Calculate the average power the sprinter must generate. (c) If the sprinter converts food energy to mechanical energy with an efficiency of 25%, at what average rate is he burning Calories? (d) What happens to the other 75% of the food energy being used?

8. BIO S A sprinter of mass m accelerates uniformly from rest to velocity v in t seconds. (a) Write a symbolic expression for the instantaneous mechanical power P required by the sprinter in terms of force F and velocity v. (b) Use Newton's second law and a kinematics equation for the velocity at any time to obtain an expression for the instantaneous power in terms of m, a, and t only. (c) If a 75.0-kg sprinter reaches a speed of 11.0 m/s in 5.00 s, calculate the sprinter's acceleration, assuming it to be constant. (d) Calculate the 75.0-kg sprinter's instantaneous mechanical power as a function of time t and (e) give the maximum rate at which he burns Calories during the sprint, assuming 25% efficiency of conversion form food energy to mechanical energy.

9. W A 5.00-g lead bullet traveling at 300 m/s is stopped by a large tree. If half the kinetic energy of the bullet is transformed into internal energy and remains with the bullet while the other half is transmitted to the tree, what is the increase in temperature of the bullet?

10. The apparatus shown in Figure P11.10 was used by Joule to measure the mechanical equivalent of heat. Work is done on the water by a rotating paddle wheel, which is driven by two blocks falling at a constant speed. The temperature of the stirred water increases due to the friction between the water and the paddles. If the energy lost in the bearings and through the walls is neglected, then the loss in potential energy associated with the blocks equals the work done by the paddle wheel on the water. If each block has a mass of 1.50 kg and the insulated tank is filled with 200 g of water, what is the increase in temperature of the water after the blocks fall through a distance of 3.00 m?

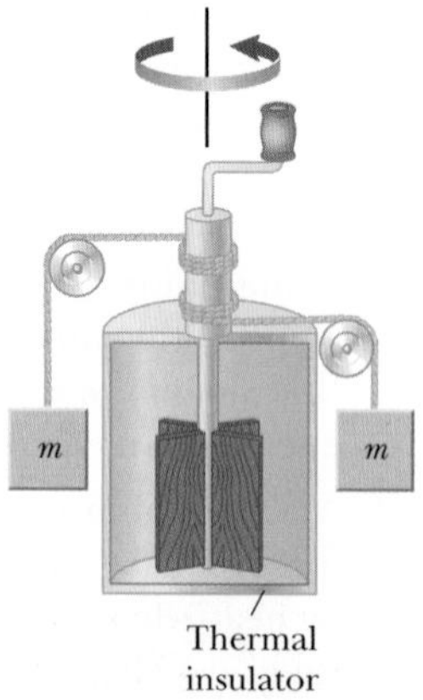

Figure P11.10 The falling weights rotate the paddles, causing the temperature of the water to increase.

11. A 200-g aluminum cup contains 800 g of water in thermal equilibrium with the cup at 80°C. The combination of cup and water is cooled uniformly so that the temperature decreases by 1.5°C per minute. At what rate is energy being removed? Express your answer in watts.

12. A 1.5-kg copper block is given an initial speed of 3.0 m/s on a rough horizontal surface. Because of friction, the block finally comes to rest. (a) If the block absorbs 85% of its initial kinetic energy as internal energy, calculate its increase in temperature. (b) What happens to the remaining energy?

13. A certain steel railroad rail is 13 yd in length and weighs 70.0 lb/yd. How much thermal energy is required to increase the length of such a rail by 3.0 mm? *Note:* Assume the steel has the same specific heat as iron.

14. GP In the summer of 1958 in St. Petersburg, Florida, a new sidewalk was poured near the childhood home of one of the authors. No expansion joints were supplied, and by mid-July the sidewalk had been completely destroyed by thermal expansion and had to be replaced, this time with the important addition of expansion joints! This event is modeled here.

A slab of concrete 4.00 cm thick, 1.00 m long, and 1.00 m wide is poured for a sidewalk at an ambient temperature of 25.0°C and allowed to set. The slab is exposed to direct sunlight and placed in a series of such slabs without proper expansion joints, so linear expansion is prevented. (a) Using the linear expansion equation (Eq. 10.4), eliminate ΔL from the equation for compressive stress and strain (Eq. 9.3). (b) Use the expression found in part (a) to eliminate ΔT from Equation 11.3, obtaining a symbolic equation for thermal energy transfer Q. (c) Compute the mass of the concrete slab given that its density is 2.40×10^3 kg/m^3. (d) Concrete has an ultimate compressive strength of 2.00×10^7 Pa, specific heat of 880 J/kg · °C, and Young's modulus of 2.1×10^{10} Pa. How much thermal energy must be transferred to the slab to reach this compressive stress? (e) What temperature change is required? (f) If the Sun delivers 1.00×10^3 W of power to the top surface of the slab and if half the energy, on the average, is absorbed and retained, how long does it take the slab to reach the point at which it is in danger of cracking due to compressive stress?

11.3 Calorimetry

15. What mass of water at 25.0°C must be allowed to come to thermal equilibrium with a 1.85-kg cube of aluminum initially at 1.50×10^2°C to lower the temperature of the aluminum to 65.0°C? Assume any water turned to steam subsequently recondenses.

16. Lead pellets, each of mass 1.00 g, are heated to 200°C. How many pellets must be added to 500 g of water that is initially at 20.0°C to make the equilibrium temperature 25.0°C? Neglect any energy transfer to or from the container.

17. W An aluminum cup contains 225 g of water and a 40-g copper stirrer, all at 27°C. A 400-g sample of silver at an initial temperature of 87°C is placed in the water. The stirrer is used to stir the mixture until it reaches its final equilibrium temperature of 32°C. Calculate the mass of the aluminum cup.

18. In a showdown on the streets of Laredo, the good guy drops a 5.00-g silver bullet at a temperature of 20.0°C into a 100-cm^3 cup of water at 90.0°C. Simultaneously, the bad guy drops a 5.00-g copper bullet at the same initial temperature into an identical cup of water. Which one ends the showdown with the coolest cup of water in the West? Neglect any energy transfer into or away from the container.

19. Q|C An aluminum calorimeter with a mass of 100 g contains 250 g of water. The calorimeter and water are in thermal equilibrium at 10.0°C. Two metallic blocks are placed into the water. One is a 50.0-g piece of copper at 80.0°C. The other has a mass of 70.0 g and is originally at a temperature of 100°C. The entire system stabilizes at a final temperature of 20.0°C. (a) Determine the specific heat of the unknown sample. (b) Using the data in Table 11.1, can you make a positive identification of the unknown material? Can you identify a possible material? (c) Explain your answers for part (b).

20. M A 1.50-kg iron horseshoe initially at 600°C is dropped into a bucket containing 20.0 kg of water at 25.0°C. What is the final temperature of the water–horseshoe system? Ignore the heat capacity of the container and assume a negligible amount of water boils away.

21. A student drops two metallic objects into a 120-g steel container holding 150 g of water at 25°C. One object is a 200-g cube of copper that is initially at 85°C, and the other is a chunk of aluminum that is initially at 5.0°C. To the surprise of the student, the water reaches a final temperature of 25°C, precisely where it started. What is the mass of the aluminum chunk?

22. When a driver brakes an automobile, the friction between the brake drums and the brake shoes converts the car's kinetic energy to thermal energy. If a 1 500-kg automobile traveling at 30 m/s comes to a halt, how much does the temperature rise in each of the four 8.0-kg iron brake drums? (The specific heat of iron is 448 J/kg · °C.)

23. Q|C Equal 0.400-kg masses of lead and tin at 60.0°C are placed in 1.00 kg of water at 20.0°C. (a) What is the equilibrium temperature of the system? (b) If an alloy is half lead and half tin by mass, what specific heat would you anticipate for the alloy? (c) How many atoms of tin N_{Sn} are in 0.400 kg of tin, and how many atoms of lead N_{Pb} are in 0.400 kg of lead? (d) Divide the number N_{Sn} of tin atoms by the number N_{Pb} of lead atoms and compare this ratio with the specific heat of tin divided by the specific heat of lead. What conclusion can be drawn?

24. An unknown substance has a mass of 0.125 kg and an initial temperature of 95.0°C. The substance is then dropped into a calorimeter made of aluminum containing 0.285 kg of water initially at 25.0°C. The mass of the aluminum container is 0.150 kg, and the temperature of the calorimeter increases to a final equilibrium temperature of 32.0°C. Assuming no thermal energy is transferred to the environment, calculate the specific heat of the unknown substance.

11.4 Latent Heat and Phase Change

25. A 75-g ice cube at 0°C is placed in 825 g of water at 25°C. What is the final temperature of the mixture?

26. A 50-g ice cube at 0°C is heated until 45 g has become water at 100°C and 5.0 g has become steam at 100°C. How much energy was added to accomplish the transformation?

27. W A 100-g cube of ice at 0°C is dropped into 1.0 kg of water that was originally at 80°C. What is the final temperature of the water after the ice has melted?

28. How much energy is required to change a 40-g ice cube from ice at −10°C to steam at 110°C?

29. M A 75-kg cross-country skier glides over snow as in Figure P11.29. The coefficient of friction between skis and snow is 0.20. Assume all the snow beneath her skis is at 0°C and that all the internal energy generated by friction is added to snow, which sticks to her skis until it melts. How far would she have to ski to melt 1.0 kg of snow?

technotr/iStockphoto.com

Figure P11.29

30. GP Into a 0.500-kg aluminum container at 20.0°C is placed 6.00 kg of ethyl alcohol at 30.0°C and 1.00 kg ice at −10.0°C. Assume the system is insulated from its environment. (a) Identify all five thermal energy transfers that occur as the system goes to a final equilibrium temperature *T*. Use the form "substance at *X*°C to substance at *Y*°C." (b) Construct a table similar to the table in Example 11.5. (c) Sum all terms in the right-most column of the table and set the sum equal to zero. (d) Substitute information from the table into the equation found in part (c) and solve for the final equilibrium temperature, *T*.

31. A 40-g block of ice is cooled to −78°C and is then added to 560 g of water in an 80-g copper calorimeter at a temperature of 25°C. Determine the final temperature of the system consisting of the ice, water, and calorimeter. (If not all the ice melts, determine how much ice is left.) Remember that the ice must first warm to 0°C, melt, and then continue warming as water. (The specific heat of ice is 0.500 cal/g · °C = 2 090 J/kg · °C.)

32. BIO When you jog, most of the food energy you burn above your basal metabolic rate (BMR) ends up as internal energy that would raise your body temperature if it were not eliminated. The evaporation of perspiration is the primary mechanism for eliminating this energy. Determine the amount of water you lose to evaporation when running for 30 minutes at a rate that uses 400 kcal/h above your BMR. (That amount is often considered to be the "maximum fat-burning" energy output.) The metabolism of 1 gram of fat generates approximately 9.0 kcal of energy and produces approximately 1 gram of water. (The hydrogen atoms in the fat molecule are transferred to oxygen to form water.) What fraction of your need for water will be provided by fat metabolism? (The latent heat of vaporization of water at room temperature is 2.5×10^6 J/kg.)

33. A high-end gas stove usually has at least one burner rated at 14 000 Btu/h. (a) If you place a 0.25-kg aluminum pot containing 2.0 liters of water at 20°C on this burner, how long will it take to bring the water to a boil, assuming all the heat from the burner goes into the pot? (b) Once boiling begins, how much time is required to boil all the water out of the pot?

34. BIO A 60.0-kg runner expends 300 W of power while running a marathon. Assuming 10.0% of the energy is delivered to the muscle tissue and that the excess energy is removed from the body primarily by sweating, determine the volume of bodily fluid (assume it is water) lost per hour. (At 37.0°C, the latent heat of vaporization of water is 2.41×10^6 J/kg.)

35. Steam at 100°C is added to ice at 0°C. (a) Find the amount of ice melted and the final temperature when the mass of steam is 10 g and the mass of ice is 50 g. (b) Repeat with steam of mass 1.0 g and ice of mass 50 g.

36. BIO The excess internal energy of metabolism is exhausted through a variety of channels, such as through radiation and evaporation of perspiration. Consider another pathway for energy loss: moisture in exhaled breath. Suppose you breathe out 22.0 breaths per minute, each with a volume of 0.600 L. Suppose also that you inhale dry air and exhale air at 37°C containing water vapor with a vapor pressure of 3.20 kPa. The vapor comes from the evaporation of liquid water in your body. Model the water vapor as an ideal gas. Assume its latent heat of evaporation at 37°C is the same as its heat of vaporization at 100°C. Calculate the rate at which you lose energy by exhaling humid air.

37. Q|C A 3.00-g lead bullet at 30.0°C is fired at a speed of 2.40×10^2 m/s into a large, fixed block of ice at 0°C, in which it becomes embedded. (a) Describe the energy transformations that occur as the bullet is cooled. What is the final temperature of the bullet? (b) What quantity of ice melts?

11.5 Energy Transfer

38. A glass windowpane in a home is 0.62 cm thick and has dimensions of 1.0 m × 2.0 m. On a certain day, the indoor temperature is 25°C and the outdoor temperature is 0°C. (a) What is the rate at which energy is transferred by heat through the glass? (b) How much energy is lost through the window in one day, assuming the temperatures inside and outside remain constant?

39. A pond with a flat bottom has a surface area of 820 m^2 and a depth of 2.0 m. On a warm day, the surface water is at a temperature of 25°C, while the bottom of the pond is at 12°C. Find the rate at which energy is transferred by conduction from the surface to the bottom of the pond.

40. BIO The thermal conductivities of human tissues vary greatly. Fat and skin have conductivities of about 0.20 W/m · K and 0.020 W/m · K, respectively, while other tissues inside the body have conductivities of about 0.50 W/m · K. Assume that between the core region of the body and the skin surface lies a skin layer of 1.0 mm, fat layer of 0.50 cm, and 3.2 cm of other tissues. (a) Find the *R*-factor for each of these layers, and the equivalent *R*-factor for all layers taken together, retaining two digits. (b) Find the rate of energy loss when the core temperature is 37°C and the exterior temperature is 0°C. Assume that both a protective layer of clothing and an insulating layer of unmoving air are absent, and a body area of 2.0 m^2.

41. A steam pipe is covered with 1.50-cm-thick insulating material of thermal conductivity 0.200 cal/cm · °C · s. How much energy is lost every second when the steam is at 200°C and the surrounding air is at 20.0°C? The pipe has a circumference of 800 cm and a length of 50.0 m. Neglect losses through the ends of the pipe.

42. The average thermal conductivity of the walls (including windows) and roof of a house in Figure P11.42 is 4.8×10^{-4} kW/m · °C, and their average thickness is 21.0 cm. The house is heated with natural gas, with a heat of combustion (energy released per cubic meter of gas burned) of 9 300 kcal/m^3. How many cubic meters of gas must be burned each day to maintain an inside temperature of 25.0°C if the outside temperature is 0.0°C? Disregard surface air layers, radiation, and energy loss by heat through the ground.

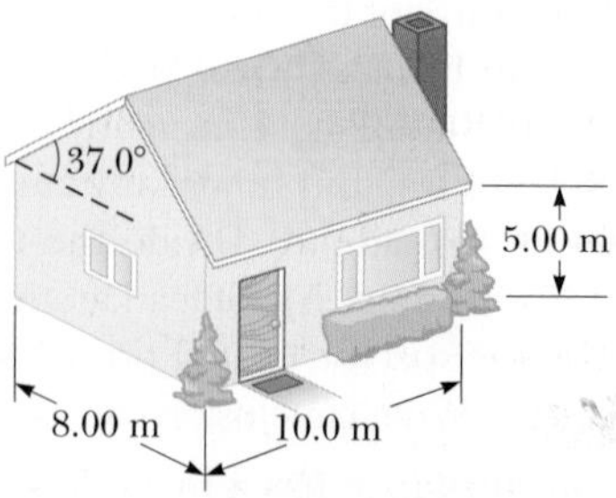

Figure P11.42

43. Consider two cooking pots of the same dimensions, each containing the same amount of water at the same initial temperature. The bottom of the first pot is made of copper, while the bottom of the second pot is made of aluminum. Both pots are placed on a hot surface having a temperature of 145°C. The water in the copper-bottomed pot boils away completely in 425 s. How long does it take the water in the aluminum-bottomed pot to boil away completely?

44. A thermopane window consists of two glass panes, each 0.50 cm thick, with a 1.0-cm-thick sealed layer of air in between. (a) If the inside surface temperature is 23°C and the outside surface temperature is 0.0°C, determine the rate of energy transfer through 1.0 m^2 of the window. (b) Compare your answer to (a) with the rate of energy transfer through 1.0 m^2 of a single 1.0-cm-thick pane of glass. Disregard surface air layers.

45. M A copper rod and an aluminum rod of equal diameter are joined end to end in good thermal contact. The temperature of the free end of the copper rod is held constant at 100°C and that of the far end of the aluminum rod is held at 0°C. If the copper rod is 0.15 m long, what must be the length of the aluminum rod so that the temperature at the junction is 50°C?

46. A Styrofoam box has a surface area of 0.80 m^2 and a wall thickness of 2.0 cm. The temperature of the inner surface is 5.0°C, and the outside temperature is 25°C. If it takes 8.0 h for 5.0 kg of ice to melt in the container, determine the thermal conductivity of the Styrofoam.

47. W A rectangular glass window pane on a house has a width of 1.0 m, a height of 2.0 m, and a thickness of 0.40 cm. Find the energy transferred through the window by conduction in 12 hours on a day when the inside temperature of the house is 22°C and the outside temperature is 2.0°C. Take surface air layers into consideration.

48. A solar sail is made of aluminized Mylar having an emissivity of 0.03 and reflecting 97% of the light that falls on it. Suppose a sail with area 1.00 km^2 is oriented so that sunlight falls perpendicular to its surface with an intensity of 1.40×10^3 W/m^2. To what temperature will it warm before it emits as much energy (from both sides) by radiation as it absorbs on the sunny side? Assume the sail is so thin that the temperature is uniform and no energy is emitted from the edges. Take the environment temperature to be 0 K.

49. Measurements on two stars indicate that Star X has a surface temperature of 5 727°C and Star Y has a surface temperature of 11 727°C. If both stars have the same radius, what is the ratio of the luminosity (total power output) of Star Y to the luminosity of Star X? Both stars can be considered to have an emissivity of 1.0.

50. The filament of a 75-W light bulb is at a temperature of 3 300 K. Assuming the filament has an emissivity $e = 1.0$, find its surface area.

Additional Problems

51. The bottom of a copper kettle has a 10-cm radius and is 2.0 mm thick. The temperature of the outside surface is 102°C, and the water inside the kettle is boiling at 1 atm of pressure. Find the rate at which energy is being transferred through the bottom of the kettle.

52. A family comes home from a long vacation with laundry to do and showers to take. The water heater has been turned off during the vacation. If the heater has a capacity of 50.0 gallons and a 4 800-W heating element, how much time is required to raise the temperature of the water from 20.0°C to 60.0°C? Assume the heater is well insulated and no water is withdrawn from the tank during that time.

53. M A 40-g ice cube floats in 200 g of water in a 100-g copper cup; all are at a temperature of 0°C. A piece of lead at 98°C is dropped into the cup, and the final equilibrium temperature is 12°C. What is the mass of the lead?

54. BIO The surface area of an unclothed person is 1.50 m^2, and his skin temperature is 33.0°C. The person is located in a dark room with a temperature of 20.0°C, and the emissivity of the skin is $e = 0.95$. (a) At what rate is energy radiated by the body? (b) What is the significance of the sign of your answer?

55. A 200-g block of copper at a temperature of 90°C is dropped into 400 g of water at 27°C. The water is contained in a 300-g glass container. What is the final temperature of the mixture?

56. Q|C Liquid nitrogen has a boiling point of 77 K and a latent heat of vaporization of 2.01×10^5 J/kg. A 25-W electric heating element is immersed in an insulated vessel containing 25 L of liquid nitrogen at its boiling point. (a) Describe the energy transformations that occur as power is supplied to the heating element. (b) How many kilograms of nitrogen are boiled away in a period of 4.0 hours?

57. Q|C A student measures the following data in a calorimetry experiment designed to determine the specific heat of aluminum:

Initial temperature of water and calorimeter:	70.0°C
Mass of water:	0.400 kg
Mass of calorimeter:	0.040 kg
Specific heat of calorimeter:	0.63 kJ/kg · °C
Initial temperature of aluminum:	27.0°C
Mass of aluminum:	0.200 kg
Final temperature of mixture:	66.3°C

Use these data to determine the specific heat of aluminum. Explain whether your result is within 15% of the value listed in Table 11.1.

58. BIO Overall, 80% of the energy used by the body must be eliminated as excess thermal energy and needs to

be dissipated. The mechanisms of elimination are radiation, evaporation of sweat (2 430 kJ/kg), evaporation from the lungs (38 kJ/h), conduction, and convection.

A person working out in a gym has a metabolic rate of 2 500 kJ/h. His body temperature is 37°C, and the outside temperature 24°C. Assume the skin has an area of 2.0 m^2 and emissivity of 0.97. (a) At what rate is his excess thermal energy dissipated by radiation? (b) If he eliminates 0.40 kg of perspiration during that hour, at what rate is thermal energy dissipated by evaporation of sweat? (c) At what rate is energy eliminated by evaporation from the lungs? (d) At what rate must the remaining excess energy be eliminated through conduction and convection?

59. Liquid helium has a very low boiling point, 4.2 K, as well as a very low latent heat of vaporization, 2.00×10^4 J/kg. If energy is transferred to a container of liquid helium at the boiling point from an immersed electric heater at a rate of 10.0 W, how long does it take to boil away 2.00 kg of the liquid?

60. A class of 10 students taking an exam has a power output per student of about 200 W. Assume the initial temperature of the room is 20°C and that its dimensions are 6.0 m by 15.0 m by 3.0 m. What is the temperature of the room at the end of 1.0 h if all the energy remains in the air in the room and none is added by an outside source? The specific heat of air is 837 J/kg · °C, and its density is about 1.3×10^{-3} g/cm^3.

61. A bar of gold (Au) is in thermal contact with a bar of silver (Ag) of the same length and area (Fig. P11.61). One end of the compound bar is maintained at 80.0°C, and the opposite end is at 30.0°C. Find the temperature at the junction when the energy flow reaches a steady state.

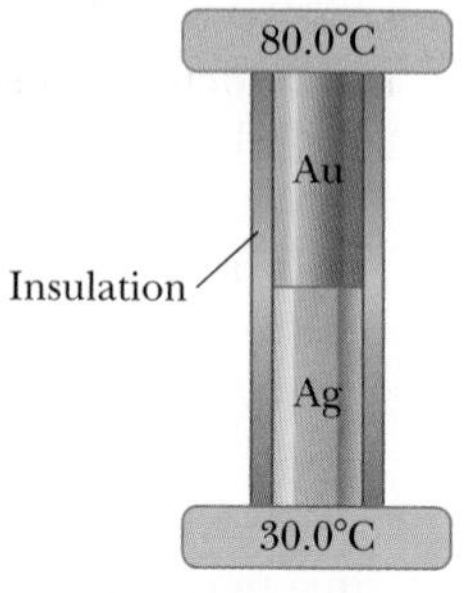

Figure P11.61

62. An iron plate is held against an iron wheel so that a sliding frictional force of 50 N acts between the two pieces of metal. The relative speed at which the two surfaces slide over each other is 40 m/s. (a) Calculate the rate at which mechanical energy is converted to internal energy. (b) The plate and the wheel have masses of 5.0 kg each, and each receives 50% of the internal energy. If the system is run as described for 10 s and each object is then allowed to reach a uniform internal temperature, what is the resultant temperature increase?

63. An automobile has a mass of 1 500 kg, and its aluminum brakes have an overall mass of 6.00 kg. (a) Assuming all the internal energy transformed by friction when the car stops is deposited in the brakes and neglecting energy transfer, how many times could the car be braked to rest starting from 25.0 m/s before the brakes would begin to melt? (Assume an initial temperature of 20.0°C.) (b) Identify some effects that are neglected in part (a), but are likely to be important in a more realistic assessment of the temperature increase of the brakes.

64. Three liquids are at temperatures of 10°C, 20°C, and 30°C, respectively. Equal masses of the first two liquids are mixed, and the equilibrium temperature is 17°C. Equal masses of the second and third are then mixed, and the equilibrium temperature is 28°C. Find the equilibrium temperature when equal masses of the first and third are mixed.

65. QIC A *flow calorimeter* is an apparatus used to measure the specific heat of a liquid. The technique is to measure the temperature difference between the input and output points of a flowing stream of the liquid while adding energy at a known rate. (a) Start with the equations $Q = mc(\Delta T)$ and $m = \rho V$, and show that the rate at which energy is added to the liquid is given by the expression $\Delta Q/\Delta t = \rho c(\Delta T)(\Delta V/\Delta t)$. (b) In a particular experiment, a liquid of density 0.72 g/cm^3 flows through the calorimeter at the rate of 3.5 cm^3/s. At steady state, a temperature difference of 5.8°C is established between the input and output points when energy is supplied at the rate of 40 J/s. What is the specific heat of the liquid?

66. A wood stove is used to heat a single room. The stove is cylindrical in shape, with a diameter of 40.0 cm and a length of 50.0 cm, and operates at a temperature of 400°F. (a) If the temperature of the room is 70.0°F, determine the amount of radiant energy delivered to the room by the stove each second if the emissivity is 0.920. (b) If the room is a square with walls that are 8.00 ft high and 25.0 ft wide, determine the *R*-value needed in the walls and ceiling to maintain the inside temperature at 70.0°F if the outside temperature is 32.0°F. Note that we are ignoring any heat conveyed by the stove via convection and any energy lost through the walls (and windows!) via convection or radiation.

67. A "solar cooker" consists of a curved reflecting mirror that focuses sunlight onto the object to be heated (Fig. P11.67). The solar power per unit area reaching the Earth at the location of a 0.50-m-diameter solar cooker is 600 W/m^2. Assuming 50% of the incident energy is converted to thermal energy, how long would it take to boil away 1.0 L of water initially at 20°C? (Neglect the specific heat of the container.)

Figure P11.67

68. BIO For bacteriological testing of water supplies and in medical clinics, samples must routinely be incubated for 24 h at 37°C. A standard constant-temperature bath

with electric heating and thermostatic control is not suitable in developing nations without continuously operating electric power lines. Peace Corps volunteer and MIT engineer Amy Smith invented a low-cost, low-maintenance incubator to fill the need. The device consists of a foam-insulated box containing several packets of a waxy material that melts at 37.0°C, interspersed among tubes, dishes, or bottles containing the test samples and growth medium (food for bacteria). Outside the box, the waxy material is first melted by a stove or solar energy collector. Then it is put into the box to keep the test samples warm as it solidifies. The heat of fusion of the phase-change material is 205 kJ/kg. Model the insulation as a panel with surface area 0.490 m^2, thickness 9.50 cm, and conductivity 0.012 0 W/m°C. Assume the exterior temperature is 23.0°C for 12.0 h and 16.0°C for 12.0 h. (a) What mass of the waxy material is required to conduct the bacteriological test? (b) Explain why your calculation can be done without knowing the mass of the test samples or of the insulation.

69. The surface of the Sun has a temperature of about 5 800 K. The radius of the Sun is 6.96×10^8 m. Calculate the total energy radiated by the Sun each second. Assume the emissivity of the Sun is 0.986.

70. BIO The evaporation of perspiration is the primary mechanism for cooling the human body. Estimate the amount of water you will lose when you bake in the sun on the beach for an hour. Use a value of 1 000 W/m^2 for the intensity of sunlight and note that the energy required to evaporate a liquid at a particular temperature is approximately equal to the sum of the energy required to raise its temperature to the boiling point and the latent heat of vaporization (determined at the boiling point).

71. At time $t = 0$, a vessel contains a mixture of 10 kg of water and an unknown mass of ice in equilibrium at 0°C. The temperature of the mixture is measured over a period of an hour, with the following results: During the first 50 min, the mixture remains at 0°C; from 50 min to 60 min, the temperature increases steadily from 0°C to 2°C. Neglecting the heat capacity of the vessel, determine the mass of ice that was initially placed in it. Assume a constant power input to the container.

72. Q|C An ice-cube tray is filled with 75.0 g of water. After the filled tray reaches an equilibrium temperature 20.0°C, it is placed in a freezer set at −8.00°C to make ice cubes. (a) Describe the processes that occur as energy is being removed from the water to make ice. (b) Calculate the energy that must be removed from the water to make ice cubes at −8.00°C.

73. An aluminum rod and an iron rod are joined end to end in good thermal contact. The two rods have equal lengths and radii. The free end of the aluminum rod is maintained at a temperature of 100°C, and the free end of the iron rod is maintained at 0°C. (a) Determine the temperature of the interface between the two rods. (b) If each rod is 15 cm long and each has a cross-sectional area of 5.0 cm^2, what quantity of energy is conducted across the combination in 30 min?

A cyclist is an engine: she requires fuel and oxygen to burn it, and the result is work that drives her forward as her excess waste energy is expelled in her evaporating sweat.

Erik Isakson/Getty Images

12 The Laws of Thermodynamics

According to the first law of thermodynamics, the internal energy of a system can be increased either by adding energy to the system or by doing work on it. That means the internal energy of a system, which is just the sum of the molecular kinetic and potential energies, can change as a result of two separate types of energy transfer across the boundary of the system. Although the first law imposes conservation of energy for both energy added by heat and work done on a system, it doesn't predict which of several possible energy-conserving processes actually occur in nature.

The second law of thermodynamics constrains the first law by establishing which processes allowed by the first law actually occur. For example, the second law tells us that energy never flows by heat spontaneously from a cold object to a hot object. One important application of this law is in the study of heat engines (such as the internal combustion engine) and the principles that limit their efficiency.

12.1 Work in Thermodynamic Processes

LEARNING OBJECTIVES

1. Define the work done **on** an ideal gas in an isobaric (constant pressure) process, and relate it to the work done **by** a gas on its environment.
2. Calculate the work done on a gas at constant pressure.
3. Evaluate the work done on a gas using a graph of the gas pressure versus its volume.

Energy can be transferred to a system by heat and by work done on the system. In most cases of interest treated here, the system is a volume of gas, which is important in understanding engines. All such systems of gas will be assumed to be in thermodynamic equilibrium, so that every part of the gas is at the same temperature and pressure. If that were not the case, the ideal gas law wouldn't apply and most of the results presented here wouldn't be valid. Consider a gas contained by a cylinder fitted with a movable piston (Fig. 12.1a) and in equilibrium. The gas occupies a volume V and exerts a uniform pressure P on the cylinder walls and the piston. The gas is compressed slowly enough so the system remains essentially in thermodynamic equilibrium at all times. As the piston is pushed downward by an external force F through a displacement Δy, the work done on the gas is

$$W = -F\Delta y = -PA\Delta y$$

where we have set the magnitude F of the external force equal to PA, possible because the pressure is the same everywhere in the system (by the assumption of equilibrium). Note that if the piston is pushed downward, $\Delta y = y_f - y_i$ is negative, so we need an explicit negative sign in the expression for W to make the work positive. The change in volume of the gas is $\Delta V = A\Delta y$, which leads to the following definition:

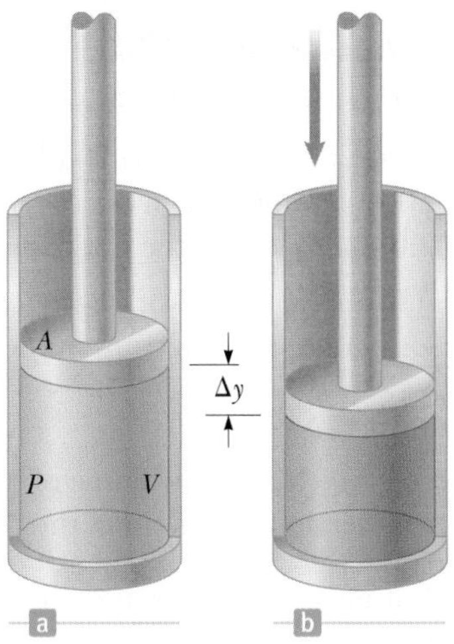

Figure 12.1 (a) A gas in a cylinder occupying a volume V at a pressure P. (b) Pushing the piston down compresses the gas.

The **work W done on a gas** at constant pressure is given by

$$W = -P\Delta V \qquad [12.1]$$

where P is the pressure throughout the gas and ΔV is the change in volume of the gas during the process.

If the gas is compressed as in Figure 12.1b, ΔV is negative and the work done on the gas is positive. If the gas expands, ΔV is positive and the work done on the gas is negative. The work done by the gas on its environment, W_{env}, is simply the negative of the work done on the gas. In the absence of a change in volume, the work is zero.

The definition of work W in Equation 12.1 specifies **work done on** a gas. In many texts, work W is defined as **work done by** a gas. In this text, work done by a gas is denoted by W_{env}. In every case, $W = -W_{env}$, so the two definitions differ by a minus sign. The reason it's important to define work W as work done *on* a gas is to make the concept of work in thermodynamics consistent with the concept of work in mechanics. In mechanics, the system is some object, and when positive work is done on that object, its energy increases. When work W done on a gas as defined in Equation 12.1 is positive, the internal energy of the gas increases, which is consistent with the mechanics definition.

Tip 12.1 Work Done *on* Versus Work Done *by*

Work done ***on*** the gas is labeled W. That definition focuses on the internal energy of the system. Work done ***by*** the gas, say on a piston, is labeled W_{env}, where the focus is on harnessing a system's internal energy to do work on something external to the gas. W and W_{env} are two different ways of looking at the same thing. It's always true that $W = -W_{env}$.

In Figure 12.2a the man pushes a crate, doing positive work on it, so the crate's speed and therefore its kinetic energy both increase. In Figure 12.2b a man pushes a piston to the right, compressing the gas in the container and doing positive work on the gas. The average speed of the molecules of gas increases, so the temperature and therefore the internal energy of the gas increase. Consequently, just as doing work on a crate increases its kinetic energy, doing work on a system of gas increases its internal energy.

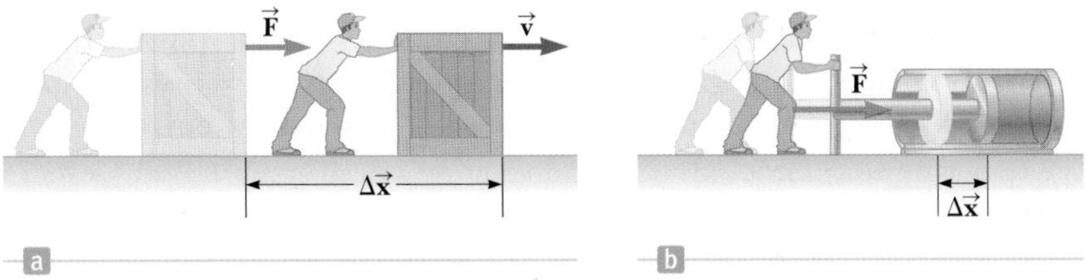

Figure 12.2 (a) When a force is exerted on a crate, the work done by that force increases the crate's mechanical energy. (b) When a piston is pushed, the gas in the container is compressed, increasing the gas's thermal energy.

EXAMPLE 12.1 Work Done by an Expanding Gas

GOAL Apply the definition of work at constant pressure.

PROBLEM In a system similar to that shown in Figure 12.1, the gas in the cylinder is at a pressure equal to 1.01×10^5 Pa and the piston has an area of 0.100 m². As energy is slowly added to the gas by heat, the piston is pushed up a distance of 4.00 cm. Calculate the work done by the expanding gas on the surroundings, W_{env}, assuming the pressure remains constant.

STRATEGY The work done on the environment is the negative of the work done on the gas given in Equation 12.1. Compute the change in volume and multiply by the pressure.

SOLUTION

Find the change in volume of the gas, ΔV, which is the cross-sectional area times the displacement:

$$\Delta V = A\,\Delta y = (0.100\ \text{m}^2)(4.00 \times 10^{-2}\ \text{m})$$
$$= 4.00 \times 10^{-3}\ \text{m}^3$$

Multiply this result by the pressure, getting the work the gas does on the environment, W_{env}:

$$W_{env} = P\Delta V = (1.01 \times 10^5\ \text{Pa})(4.00 \times 10^{-3}\ \text{m}^3)$$
$$= \boxed{404\ \text{J}}$$

REMARKS The volume of the gas increases, so the work done on the environment is positive. The work done on the system during this process is $W = -404$ J. The energy required to perform positive work on the environment must come from the energy of the gas.

QUESTION 12.1 If no energy were added to the gas during the expansion, could the pressure remain constant?

EXERCISE 12.1 Gas in a cylinder similar to Figure 12.1 moves a piston with area 0.200 m² as energy is slowly added to the system. If 2.00×10^3 J of work is done on the environment and the pressure of the gas in the cylinder remains constant at 1.01×10^5 Pa, find the displacement of the piston.

ANSWER 9.90×10^{-2} m

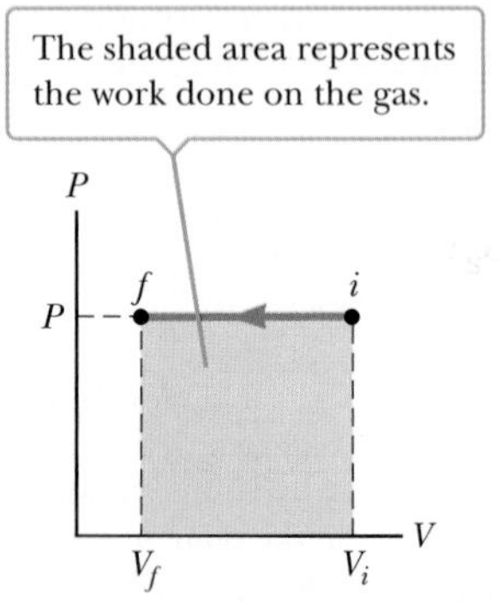

Figure 12.3 The *PV* diagram for a gas being compressed at constant pressure.

Equation 12.1 can be used to calculate the work done on the system *only* when the pressure of the gas remains constant during the expansion or compression. A process in which the pressure remains constant is called an **isobaric process.** The pressure vs. volume graph, or ***PV* diagram,** of an isobaric process is shown in Figure 12.3. The curve on such a graph is called the *path* taken between the initial and final states, with the arrow indicating the direction the process is going, in this case from larger to smaller volume. The area under the graph is

$$\text{Area} = P(V_f - V_i) = P\Delta V$$

The area under the graph in a *PV* diagram is equal in magnitude to the work done on the gas.

That statement is true in general, whether or not the process proceeds at constant pressure. Just draw the *PV* diagram of the process, find the area underneath the graph (and above the horizontal axis), and that area will be the equal to the magnitude of the work done on the gas. If the arrow on the graph points toward larger volumes, the work done on the gas is negative. If the arrow on the graph points toward smaller volumes, the work done on the gas is positive.

Whenever negative work is done on a system, positive work is done by the system on its environment. The negative work done on the system represents a loss of energy from the system—the cost of doing positive work on the environment.

Quick Quiz

12.1 By visual inspection, order the *PV* diagrams shown in Figure 12.4 from the most negative work done on the system to the most positive work done on the system. (a) a, b, c, d (b) a, c, b, d (c) d, b, c, a (d) d, a, c, b

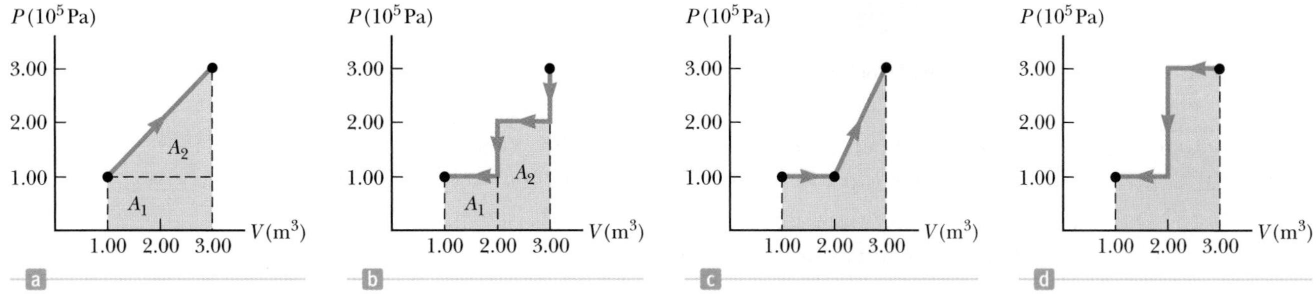

Figure 12.4 (Quick Quiz 12.1 and Example 12.2)

Notice that the graphs in Figure 12.4 all have the same endpoints, but the areas beneath the curves are different. The work done on a system depends on the path taken in the *PV* diagram.

EXAMPLE 12.2 Work and *PV* Diagrams

GOAL Calculate work from a *PV* diagram.

PROBLEM Find the numeric value of the work done on the gas in **(a)** Figure 12.4a and **(b)** Figure 12.4b.

STRATEGY The regions in question are composed of rectangles and triangles. Use basic geometric formulas to find the area underneath each curve. Check the direction of the arrow to determine signs.

SOLUTION

(a) Find the work done on the gas in Figure 12.4a.

Compute the areas A_1 and A_2 in Figure 12.4a. A_1 is a rectangle and A_2 is a triangle.

$$A_1 = \text{height} \times \text{width} = (1.00 \times 10^5 \text{ Pa})(2.00 \text{ m}^3)$$
$$= 2.00 \times 10^5 \text{ J}$$
$$A_2 = \tfrac{1}{2}\text{base} \times \text{height}$$
$$= \tfrac{1}{2}(2.00 \text{ m}^3)(2.00 \times 10^5 \text{ Pa}) = 2.00 \times 10^5 \text{ J}$$

Sum the areas (the arrows point to increasing volume, so the work done on the gas is negative):

$$\text{Area} = A_1 + A_2 = 4.00 \times 10^5 \text{ J}$$
$$W = -4.00 \times 10^5 \text{ J}$$

(b) Find the work done on the gas in Figure 12.4b.

Compute the areas of the two rectangular regions:

$$A_1 = \text{height} \times \text{width} = (1.00 \times 10^5 \text{ Pa})(1.00 \text{ m}^3)$$
$$= 1.00 \times 10^5 \text{ J}$$
$$A_2 = \text{height} \times \text{width} = (2.00 \times 10^5 \text{ Pa})(1.00 \text{ m}^3)$$
$$= 2.00 \times 10^5 \text{ J}$$

Sum the areas (the arrows point to decreasing volume, so the work done on the gas is positive):

$$\text{Area} = A_1 + A_2 = 3.00 \times 10^5 \text{ J}$$
$$W = +3.00 \times 10^5 \text{ J}$$

REMARKS Notice that in both cases the paths in the *PV* diagrams start and end at the same points, but the answers are different.

QUESTION 12.2 Is work done on a system during a process in which its volume remains constant?

EXERCISE 12.2 Compute the work done on the system in Figures 12.4c and 12.4d.

ANSWERS -3.00×10^5 J, $+4.00 \times 10^5$ J

12.2 The First Law of Thermodynamics

LEARNING OBJECTIVES

1. State the first law of thermodynamics and discuss its physical origins.
2. Apply the first law of thermodynamics to simple systems and processes.
3. Define molar specific heat and the change of internal energy of an ideal gas.
4. Discuss the concept of degrees of freedom and their physical effect on the molar specific heat of a gas.

The **first law of thermodynamics** is another energy conservation law that relates changes in internal energy—the energy associated with the position and jiggling of all the molecules of a system—to energy transfers due to heat and work. The first law is universally valid, applicable to all kinds of processes, providing a connection between the microscopic and macroscopic worlds.

There are two ways energy can be transferred between a system and its surrounding environment: by doing work, which requires a macroscopic displacement of an object through the application of a force, and by a direct exchange of energy across the system boundary, often by heat. Heat is the transfer of energy between a system and its environment due to a temperature difference and usually occurs through one or more of the mechanisms of radiation, conduction, and convection. For example, in Figure 12.5 hot gases and radiation impinge on the cylinder, raising its temperature, and energy Q is transferred by conduction to the gas, where it is distributed mainly through convection. Other processes for transferring energy into a system are possible, such as a chemical reaction or an electrical discharge. Any energy Q exchanged between the system and the environment and any work done through the expansion or compression of the system results in a ***change in the internal energy,*** ΔU, of the system. A change in internal energy results in measurable changes in the macroscopic variables of the system such as the pressure, temperature, and volume. The relationship between the change in internal energy, ΔU, energy Q, and the work W done on the system is given by the **first law of thermodynamics:**

First law of thermodynamics ▶

If a system undergoes a change from an initial state to a final state, then the change in the internal energy ΔU is given by

$$\Delta U = U_f - U_i = Q + W \qquad [12.2]$$

where Q is the energy exchanged between the system and the environment, and W is the work done on the system.

The quantity Q is positive when energy is transferred into the system and negative when energy is removed from the system.

Figure 12.5 illustrates the first law for a cylinder of gas and how the system interacts with the environment. The gas cylinder contains a frictionless piston, and the block is initially at rest. Energy Q is introduced into the gas as the gas expands

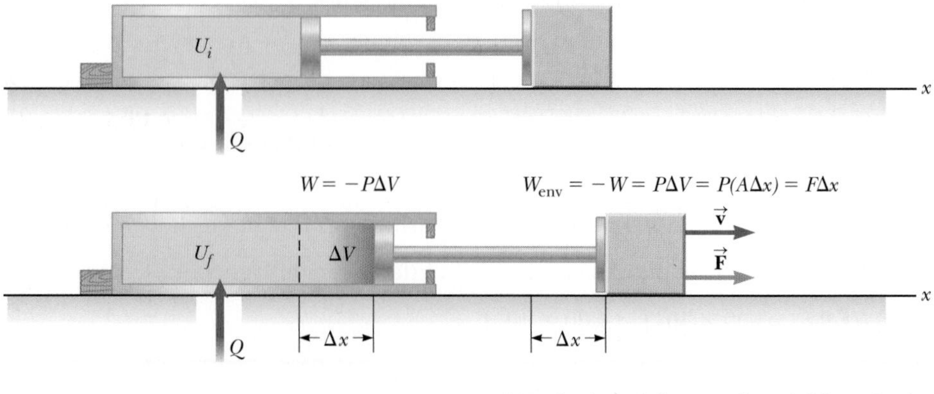

Figure 12.5 Thermal energy Q is transferred to the gas, increasing its internal energy. The gas presses against the piston, displacing it and performing mechanical work on the environment or, equivalently, doing negative work on the gas, reducing the internal energy.

against the piston with constant pressure P. Until the piston hits the stops, it exerts a force on the block, which accelerates on a frictionless surface. Negative work W is done on the gas, and at the same time positive work $W_{env} = -W$ is done by the gas on the block. Adding the work done on the environment, W_{env}, to the work done on the gas, W, gives zero net work, as it must because energy must be conserved.

From Equation 12.2 we also see that the internal energy of any isolated system must remain constant, so that $\Delta U = 0$. Even when a system isn't isolated, the change in internal energy will be zero if the system goes through a cyclic process in which all the thermodynamic variables—pressure, volume, temperature, and moles of gas—return to their original values.

It's important to remember that the quantities in Equation 12.2 concern a *system*, not the effect on the system's environment through work. If the system is hot gas expanding against a piston, as in Figure 12.5, the system work W is *negative* because the piston can only expand at the expense of the internal energy of the gas. The work W_{env} done by the hot gas on the *environment*—in this case, moving a piston which moves the block—is positive, but that's not the work W in Equation 12.2. This way of defining work in the first law makes it consistent with the concept of work defined in Chapter 5. In both the mechanical and thermal cases, the effect on the system is the same: positive work increases the system's energy, and negative work decreases the system's energy.

Some textbooks identify W as the work done by the gas on its environment. This is an equivalent formulation, but it means that W must carry a minus sign in the first law. That convention isn't consistent with previous discussions of the energy of a system, because when W is positive the system *loses* energy, whereas in Chapter 5 positive W means that the system *gains* energy. For that reason, the old convention is not used in this book.

Tip 12.2 Dual Sign Conventions

Many physics and engineering textbooks present the first law as $\Delta U = Q - W$, with a minus sign between the heat and the work. The reason is that work is defined in these treatments as the work done *by* the system rather than *on* the system, as in our treatment. Using our notation, this equivalent first law would read $\Delta U = Q - W_{env}$.

EXAMPLE 12.3 Heating a Gas

GOAL Combine the first law of thermodynamics with work done during a constant pressure process.

PROBLEM An ideal gas absorbs 5.00×10^3 J of energy while doing 2.00×10^3 J of work on the environment during a constant pressure process. **(a)** Compute the change in the internal energy of the gas. **(b)** If the internal energy now drops by 4.50×10^3 J and 7.50×10^3 J is expelled from the system, find the change in volume, assuming a constant pressure process at 1.01×10^5 Pa.

STRATEGY Part (a) requires substitution of the given information into the first law, Equation 12.2. Notice, however, that the given work is done on the *environment.* The negative of this amount is the work done on the *system,* representing a loss of internal energy. Part (b) is a matter of substituting the equation for work at constant pressure into the first law and solving for the change in volume.

SOLUTION

(a) Compute the change in internal energy of the gas.

Substitute values into the first law, noting that the work done on the gas is negative:

$$\Delta U = Q + W = 5.00 \times 10^3\text{ J} - 2.00 \times 10^3\text{ J}$$
$$= \boxed{3.00 \times 10^3\text{ J}}$$

(b) Find the change in volume, noting that ΔU and Q are both negative in this case.

Substitute the equation for work done at constant pressure into the first law:

$$\Delta U = Q + W = Q - P\Delta V$$
$$-4.50 \times 10^3\text{ J} = -7.50 \times 10^3\text{ J} - (1.01 \times 10^5\text{ Pa})\Delta V$$

Solve for the change in volume, ΔV:

$$\Delta V = \boxed{-2.97 \times 10^{-2}\text{ m}^3}$$

REMARKS The change in volume is negative, so the system contracts, doing negative work on the environment, whereas the work W on the system is positive.

(Continued)

QUESTION 12.3 True or False: When an ideal gas expands at constant pressure, the change in the internal energy must be positive.

EXERCISE 12.3 Suppose the internal energy of an ideal gas rises by 3.00×10^3 J at a constant pressure of 1.00×10^5 Pa, while the system gains 4.20×10^3 J of energy by heat. Find the change in volume of the system.

ANSWER 1.20×10^{-2} m^3

Recall that an expression for the internal energy of an ideal gas is

$$U = \tfrac{3}{2}nRT \qquad [12.3a]$$

This expression is valid only for a *monatomic* ideal gas, which means the particles of the gas consist of single atoms. The change in the internal energy, ΔU, for such a gas is given by

$$\Delta U = \tfrac{3}{2}nR\Delta T \qquad [12.3b]$$

The **molar specific heat at constant volume** of a monatomic ideal gas, C_v, is defined by

$$C_v \equiv \tfrac{3}{2}R \qquad [12.4]$$

The change in internal energy of an ideal gas can then be written

$$\Delta U = nC_v\,\Delta T \qquad [12.5]$$

For ideal gases, this expression is always valid, even when the volume isn't constant. The value of the molar specific heat, however, depends on the gas and can vary under different conditions of temperature and pressure.

A gas with a larger molar specific heat requires more energy to realize a given temperature change. The size of the molar specific heat depends on the structure of the gas molecule and how many different ways it can store energy. A monatomic gas such as helium can store energy as motion in three different directions. A gas such as hydrogen, on the other hand, is diatomic in normal temperature ranges, and aside from moving in three directions, it can also tumble, rotating in two different directions. So hydrogen molecules can store energy in the form of translational motion and in addition can store energy through tumbling. Further, molecules can also store energy in the vibrations of their constituent atoms. A gas composed of molecules with more ways to store energy will have a larger molar specific heat.

Each different way a gas molecule can store energy is called a *degree of freedom*. Each degree of freedom contributes $\frac{1}{2}R$ to the molar specific heat. Because an atomic ideal gas can move in three directions, it has a molar specific heat capacity $C_v = 3(\frac{1}{2}R) = \frac{3}{2}R$. A diatomic gas like molecular oxygen, O_2, can also tumble in two different directions. This adds $2 \times \frac{1}{2}R = R$ to the molar heat specific heat, so $C_v = \frac{5}{2}R$ for diatomic gases. The spinning about the long axis connecting the two atoms is generally negligible. Vibration of the atoms in a molecule can also contribute to the heat capacity. A full analysis of a given system is often complex, so in general, molar specific heats must be determined by experiment. Some representative values of C_v can be found in Table 12.1.

12.3 Thermal Processes

LEARNING OBJECTIVES

1. Identify, define and discuss in physical terms the four most common thermal processes.
2. Evaluate thermodynamic quantities for isobaric, adiabatic, isovolumetric, isothermal and general processes.

Table 12.1 Molar Specific Heats of Various Gases

Gas	Molar Specific Heat (J/mol · K)[a]			
	C_p	C_v	$C_p - C_v$	$\gamma = C_p/C_v$
Monatomic Gases				
He	20.8	12.5	8.33	1.67
Ar	20.8	12.5	8.33	1.67
Ne	20.8	12.7	8.12	1.64
Kr	20.8	12.3	8.49	1.69
Diatomic Gases				
H_2	28.8	20.4	8.33	1.41
N_2	29.1	20.8	8.33	1.40
O_2	29.4	21.1	8.33	1.40
CO	29.3	21.0	8.33	1.40
Cl_2	34.7	25.7	8.96	1.35
Polyatomic Gases				
CO_2	37.0	28.5	8.50	1.30
SO_2	40.4	31.4	9.00	1.29
H_2O	35.4	27.0	8.37	1.30
CH_4	35.5	27.1	8.41	1.31

[a]All values except that for water were obtained at 300 K.

Engine cycles can be complex. Fortunately, they can often be broken down into a series of simple processes. In this section the four most common processes will be studied and illustrated by their effect on an ideal gas. Each process corresponds to making one of the variables in the ideal gas law a constant or assuming one of the three quantities in the first law of thermodynamics is zero. The four processes are called isobaric (constant pressure), adiabatic (no thermal energy transfer, or $Q = 0$), isovolumetric (constant volume, corresponding to $W = 0$) and isothermal (constant temperature, corresponding to $\Delta U = 0$). Naturally, many other processes don't fall into one of these four categories, so they will be covered in a fifth category, called a general process. What is essential in each case is to be able to calculate the three thermodynamic quantities from the first law: the work W, the thermal energy transfer Q, and the change in the internal energy ΔU.

Isobaric Processes

Recall from Section 12.1 that in an isobaric process the pressure remains constant as the gas expands or is compressed. An expanding gas does work on its environment, given by $W_{env} = P\,\Delta V$. The PV diagram of an isobaric expansion is given in Figure 12.3. As previously discussed, the magnitude of the work done on the gas is just the area under the path in its PV diagram: height times length, or $P\,\Delta V$. The negative of this quantity, $W = -P\,\Delta V$, is the energy lost by the gas because the gas does work as it expands. This is the quantity that should be substituted into the first law.

The work done by the gas on its environment must come at the expense of the change in its internal energy, ΔU. Because the change in the internal energy of an ideal gas is given by $\Delta U = nC_v\,\Delta T$, the temperature of an expanding gas must decrease as the internal energy decreases. Expanding volume and decreasing temperature means the pressure must also decrease, in conformity with the ideal gas law, $PV = nRT$. Consequently, the only way such a process can remain at constant pressure is if thermal energy Q is transferred into the gas by heat. Rearranging the first law, we obtain

$$Q = \Delta U - W = \Delta U + P\,\Delta V$$

Now we can substitute the expression in Equation 12.3b for ΔU and use the ideal gas law to substitute $P\Delta V = nR\Delta T$:

$$Q = \tfrac{3}{2}nR\Delta T + nR\Delta T = \tfrac{5}{2}nR\Delta T$$

Another way to express this transfer by heat is

$$Q = nC_p \Delta T \qquad [12.6]$$

where $C_p = \tfrac{5}{2}R$. For ideal gases, the molar heat capacity at constant pressure, C_p, is the sum of the molar heat capacity at constant volume, C_v, and the gas constant R:

$$C_p = C_v + R \qquad [12.7]$$

This can be seen in the fourth column of Table 12.1, where $C_p - C_v$ is calculated for a number of different gases. The difference works out to be approximately R in virtually every case.

EXAMPLE 12.4 | Expanding Gas

GOAL Use molar specific heats and the first law in a constant pressure process.

PROBLEM Suppose a system of monatomic ideal gas at 2.00×10^5 Pa and an initial temperature of 293 K slowly expands at constant pressure from a volume of 1.00 L to 2.50 L. **(a)** Find the work done on the environment. **(b)** Find the change in internal energy of the gas. **(c)** Use the first law of thermodynamics to obtain the thermal energy absorbed by the gas during the process. **(d)** Use the molar heat capacity at constant pressure to find the thermal energy absorbed. **(e)** How would the answers change for a diatomic ideal gas?

STRATEGY This problem mainly involves substituting values into the appropriate equations. Substitute into the equation for work at constant pressure to obtain the answer to part (a). In part (b) use the ideal gas law twice: to find the temperature when $V = 2.00$ L and to find the number of moles of the gas. These quantities can then be used to obtain the change in internal energy, ΔU. Part (c) can then be solved by substituting into the first law, yielding Q, the answer checked in part (d) with Equation 12.6. Repeat these steps for part (e) after increasing the molar specific heats by R because of the extra two degrees of freedom associated with a diatomic gas.

SOLUTION

(a) Find the work done on the environment.

Apply the definition of work at constant pressure:

$$W_{env} = P\Delta V = (2.00 \times 10^5 \text{ Pa})(2.50 \times 10^{-3} \text{ m}^3 - 1.00 \times 10^{-3} \text{ m}^3)$$

$$W_{env} = \boxed{3.00 \times 10^2 \text{ J}}$$

(b) Find the change in the internal energy of the gas.

First, obtain the final temperature, using the ideal gas law, noting that $P_i = P_f$:

$$\frac{P_f V_f}{P_i V_i} = \frac{T_f}{T_i} \quad \rightarrow \quad T_f = T_i \frac{V_f}{V_i} = (293 \text{ K})\frac{(2.50 \times 10^{-3} \text{ m}^3)}{(1.00 \times 10^{-3} \text{ m}^3)}$$

$$T_f = 733 \text{ K}$$

Again using the ideal gas law, obtain the number of moles of gas:

$$n = \frac{P_i V_i}{RT_i} = \frac{(2.00 \times 10^5 \text{ Pa})(1.00 \times 10^{-3} \text{ m}^3)}{(8.31 \text{ J/K}\cdot\text{mol})(293 \text{ K})}$$

$$= 8.21 \times 10^{-2} \text{ mol}$$

Use these results and given quantities to calculate the change in internal energy, ΔU:

$$\Delta U = nC_v\Delta T = \tfrac{3}{2}nR\Delta T$$

$$= \tfrac{3}{2}(8.21 \times 10^{-2} \text{ mol})(8.31 \text{ J/K}\cdot\text{mol})(733 \text{ K} - 293 \text{ K})$$

$$\Delta U = \boxed{4.50 \times 10^2 \text{ J}}$$

(c) Use the first law to obtain the energy transferred by heat.

Solve the first law for Q, and substitute ΔU and $W = -W_{env} = -3.00 \times 10^2$ J:

$$\Delta U = Q + W \quad \rightarrow \quad Q = \Delta U - W$$

$$Q = 4.50 \times 10^2 \text{ J} - (-3.00 \times 10^2 \text{ J}) = \boxed{7.50 \times 10^2 \text{ J}}$$

(d) Use the molar heat capacity at constant pressure to obtain Q.

Substitute values into Equation 12.6:

$$Q = nC_p\Delta T = \tfrac{5}{2}nR\Delta T$$
$$= \tfrac{5}{2}(8.21 \times 10^{-2}\ \text{mol})(8.31\ \text{J/K}\cdot\text{mol})(733\ \text{K} - 293\ \text{K})$$
$$= 7.50 \times 10^2\ \text{J}$$

(e) How would the answers change for a diatomic gas?

Obtain the new change in internal energy, ΔU, noting that $C_v = \tfrac{5}{2}R$ for a diatomic gas:

$$\Delta U = nC_v\Delta T = (\tfrac{3}{2} + 1)nR\,\Delta T$$
$$= \tfrac{5}{2}(8.21 \times 10^{-2}\ \text{mol})(8.31\ \text{J/K}\cdot\text{mol})(733\ \text{K} - 293\ \text{K})$$
$$\Delta U = 7.50 \times 10^2\ \text{J}$$

Obtain the new energy transferred by heat, Q:

$$Q = nC_p\Delta T = (\tfrac{5}{2} + 1)nR\Delta T$$
$$= \tfrac{7}{2}(8.21 \times 10^{-2}\ \text{mol})(8.31\ \text{J/K}\cdot\text{mol})(733\ \text{K} - 293\ \text{K})$$
$$Q = 1.05 \times 10^3\ \text{J}$$

REMARKS Part (b) could also be solved with fewer steps by using the ideal gas equation $PV = nRT$ once the work is known. The pressure and number of moles are constant, and the gas is ideal, so $P\Delta V = nR\Delta T$. Given that $C_v = \tfrac{3}{2}R$, the change in the internal energy ΔU can then be calculated in terms of the expression for work:

$$\Delta U = nC_v\Delta T = \tfrac{3}{2}nR\Delta T = \tfrac{3}{2}P\Delta V = \tfrac{3}{2}W$$

Similar methods can be used in other processes.

QUESTION 12.4 True or False: During a constant pressure compression, the temperature of an ideal gas must always decrease, and the gas must always exhaust thermal energy ($Q < 0$).

EXERCISE 12.4 Suppose an ideal monatomic gas at an initial temperature of 475 K is compressed from 3.00 L to 2.00 L while its pressure remains constant at 1.00×10^5 Pa. Find (a) the work done on the gas, (b) the change in internal energy, and (c) the energy transferred by heat, Q.

ANSWERS (a) 1.00×10^2 J (b) -1.50×10^2 J (c) -2.50×10^2 J

Adiabatic Processes

In an adiabatic process, no energy enters or leaves the system by heat. Such a system is insulated, thermally isolated from its environment. In general, however, the system isn't mechanically isolated, so it can still do work. A sufficiently rapid process may be considered approximately adiabatic because there isn't time for any significant transfer of energy by heat.

For adiabatic processes $Q = 0$, so the first law becomes

$$\Delta U = W \quad \text{(adiabatic processes)}$$

The work done during an adiabatic process can be calculated by finding the change in the internal energy. Alternately, the work can be computed from a PV diagram. For an ideal gas undergoing an adiabatic process, it can be shown that

$$PV^{\gamma} = \text{constant} \qquad \textbf{[12.8a]}$$

where

$$\gamma = \frac{C_p}{C_v} \qquad \textbf{[12.8b]}$$

is called the *adiabatic index* of the gas. Values of the adiabatic index for several different gases are given in Table 12.1. After computing the constant on the right-hand side of Equation 12.8a and solving for the pressure P, the area under the curve in the PV diagram can be found by counting boxes, yielding the work.

If a hot gas is allowed to expand so quickly that there is no time for energy to enter or leave the system by heat, the work done on the gas is negative and the internal energy decreases. This decrease occurs because kinetic energy is transferred from the gas molecules to the moving piston. Such an adiabatic expansion is of practical importance and is nearly realized in an internal combustion engine when a gasoline–air mixture is ignited and expands rapidly against a piston. The following example illustrates this process.

EXAMPLE 12.5 Work and an Engine Cylinder

GOAL Use the first law to find the work done in an adiabatic expansion.

PROBLEM In a car engine operating at a frequency of 1.80×10^3 rev/min, the expansion of hot, high-pressure gas against a piston occurs in about 10 ms. Because energy transfer by heat typically takes a time on the order of minutes or hours, it's safe to assume little energy leaves the hot gas during the expansion. Find the work done by the gas on the piston during this adiabatic expansion by assuming the engine cylinder contains 0.100 moles of an ideal monatomic gas that goes from 1.200×10^3 K to 4.00×10^2 K, typical engine temperatures, during the expansion.

STRATEGY Find the change in internal energy using the given temperatures. For an adiabatic process, this equals the work done on the gas, which is the negative of the work done on the environment—in this case, the piston.

SOLUTION

Start with the first law, taking $Q = 0$:

$$W = \Delta U - Q = \Delta U - 0 = \Delta U$$

Find ΔU from the expression for the internal energy of an ideal monatomic gas:

$$\Delta U = U_f - U_i = \tfrac{3}{2}nR(T_f - T_i)$$

$$= \tfrac{3}{2}(0.100\text{ mol})(8.31\text{ J/mol}\cdot\text{K})(4.00 \times 10^2\text{ K} - 1.20 \times 10^3\text{ K})$$

$$\Delta U = -9.97 \times 10^2\text{ J}$$

The change in internal energy equals the work done on the system, which is the negative of the work done on the piston:

$$W_{\text{piston}} = -W = -\Delta U = 9.97 \times 10^2\text{ J}$$

REMARKS The work done on the piston comes at the expense of the internal energy of the gas. In an ideal adiabatic expansion, the loss of internal energy is completely converted into useful work. In a real engine, there are always losses.

QUESTION 12.5 In an adiabatic expansion of an ideal gas, why must the change in temperature always be negative?

EXERCISE 12.5 A monatomic ideal gas with volume 0.200 L is rapidly compressed, so the process can be considered adiabatic. If the gas is initially at 1.01×10^5 Pa and 3.00×10^2 K and the final temperature is 477 K, find the work done by the gas on the environment, W_{env}.

ANSWER −17.9 J

EXAMPLE 12.6 An Adiabatic Expansion

GOAL Use the adiabatic pressure vs. volume relation to find a change in pressure and the work done on a gas.

PROBLEM A monatomic ideal gas at an initial pressure of 1.01×10^5 Pa expands adiabatically from an initial volume of 1.50 m^3, doubling its volume (Fig. 12.6). **(a)** Find the new pressure. **(b)** Sketch the *PV* diagram and estimate the work done on the gas.

STRATEGY There isn't enough information to solve this problem with the ideal gas law. Instead, use Equation 12.8a,b and the given information to find the adiabatic index and the constant C for the process. For part (b), sketch the *PV* diagram and count boxes to estimate the area under the graph, which gives the work.

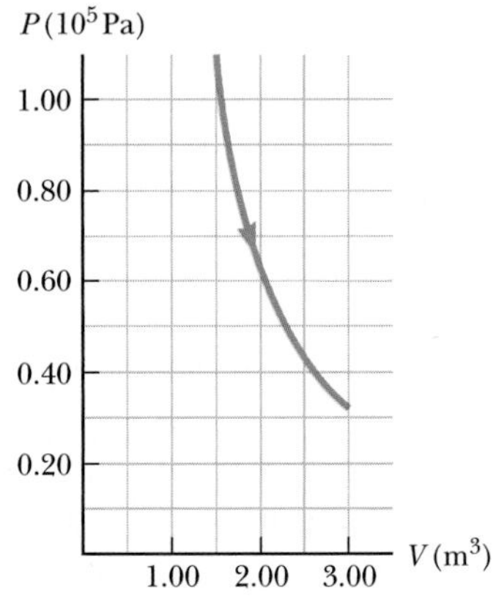

Figure 12.6 (Example 12.6) The *PV* diagram of an adiabatic expansion: the graph of $P = CV^{-\gamma}$, where C is a constant and $\gamma = C_p/C_v$.

SOLUTION

(a) Find the new pressure.

First, calculate the adiabatic index:

$$\gamma = \frac{C_p}{C_v} = \frac{\frac{5}{2}R}{\frac{3}{2}R} = \frac{5}{3}$$

Use Equation 12.8a to find the constant C:

$$C = P_1V_1^{\gamma} = (1.01 \times 10^5 \text{ Pa})(1.50 \text{ m}^3)^{5/3}$$
$$= 1.99 \times 10^5 \text{ Pa} \cdot \text{m}^5$$

The constant C is fixed for the entire process and can be used to find P_2:

$$C = P_2V_2^{\gamma} = P_2(3.00 \text{ m}^3)^{5/3}$$
$$1.99 \times 10^5 \text{ Pa} \cdot \text{m}^5 = P_2\,(6.24 \text{ m}^5)$$
$$P_2 = \boxed{3.19 \times 10^4 \text{ Pa}}$$

(b) Estimate the work done on the gas from a *PV* diagram.

Count the boxes between $V_1 = 1.50 \text{ m}^3$ and $V_2 = 3.00 \text{ m}^3$ in the graph of $P = (1.99 \times 10^5 \text{ Pa} \cdot \text{m}^5)V^{-5/3}$ in the *PV* diagram shown in Figure 12.6:

Number of boxes ≈ 17

Each box has an 'area' of 5.00×10^3 J.

$$W \approx -17 \cdot 5.00 \times 10^3 \text{ J} = \boxed{-8.5 \times 10^4 \text{ J}}$$

REMARKS The exact answer, obtained with calculus, is -8.43×10^4 J, so our result is a very good estimate. The answer is negative because the gas is expanding, doing positive work on the environment, thereby reducing its own internal energy.

QUESTION 12.6 For an adiabatic expansion between two given volumes and an initial pressure, which gas does more work, a monatomic gas or a diatomic gas?

EXERCISE 12.6 Repeat the preceding calculations for an ideal diatomic gas expanding adiabatically from an initial volume of 0.500 m^3 to a final volume of 1.25 m^3, starting at a pressure of $P_1 = 1.01 \times 10^5$ Pa. Use the same techniques as in the example.

ANSWERS $P_2 = 2.80 \times 10^4$ Pa, $W \approx -4 \times 10^4$ J

Isovolumetric Processes

An **isovolumetric process,** sometimes called an *isochoric* process (which is harder to remember), proceeds at constant volume, corresponding to vertical lines in a *PV* diagram. If the volume doesn't change, no work is done on or by the system, so $W = 0$ and the first law of thermodynamics reads

$$\Delta U = Q \quad \text{(isovolumetric process)}$$

This result tells us that **in an isovolumetric process, the change in internal energy of a system equals the energy transferred to the system by heat.** From Equation 12.5, the energy transferred by heat in constant volume processes is given by

$$Q = nC_v\,\Delta T \qquad [12.9]$$

EXAMPLE 12.7 An Isovolumetric Process

GOAL Apply the first law to a constant-volume process.

PROBLEM A monatomic ideal gas has a temperature $T = 3.00 \times 10^2$ K and a constant volume of 1.50 L. If there are 5.00 moles of gas, **(a)** how much thermal energy must be added in order to raise the temperature of the gas to 3.80×10^2 K? **(b)** Calculate the change in pressure of the gas, ΔP. **(c)** How much thermal energy would be required if the gas were ideal and diatomic? **(d)** Calculate the change in the pressure for the diatomic gas.

(Continued)

SOLUTION

(a) How much thermal energy must be added in order to raise the temperature of the gas to 3.80×10^2 K?

Apply Equation 12.9, using the fact that $C_v = 3R/2$ for an ideal monatomic gas:

$$\text{(1)}\quad Q = \Delta U = nC_v\Delta T = \tfrac{3}{2}nR\Delta T$$
$$= \tfrac{3}{2}(5.00\text{ mol})(8.31\text{ J/K}\cdot\text{mol})(80.0\text{ K})$$
$$Q = \boxed{4.99 \times 10^3\text{ J}}$$

(b) Calculate the change in pressure, ΔP.

Use the ideal gas equation $PV = nRT$ and Equation (1) to relate ΔP to Q:

$$\Delta(PV) = (\Delta P)V = nR\Delta T = \tfrac{2}{3}Q$$

Solve for ΔP:

$$\Delta P = \frac{2}{3}\frac{Q}{V} = \frac{2}{3}\frac{4.99 \times 10^3\text{ J}}{1.50 \times 10^{-3}\text{ m}^3}$$
$$= \boxed{2.22 \times 10^6\text{ Pa}}$$

(c) How much thermal energy would be required if the gas were ideal and diatomic?

Repeat the calculation with $C_v = 5R/2$:

$$Q = \Delta U = nC_v\Delta T = \tfrac{5}{2}\,nR\Delta T = 8.31 \times 10^3\text{ J}$$

(d) Calculate the change in the pressure for the diatomic gas.

Use the result of part (c) and repeat the calculation of part (b), with 2/3 replaced by 2/5 because the gas is diatomic:

$$\Delta P = \frac{2}{5}\frac{Q}{V} = \frac{2}{5}\frac{8.31 \times 10^3\text{ J}}{1.50 \times 10^{-3}\text{ m}^3}$$
$$= \boxed{2.22 \times 10^6\text{ Pa}}$$

REMARKS The constant volume diatomic gas, under the same conditions, requires more thermal energy per degree of temperature change because there are more ways for the diatomic molecules to store energy. Despite the extra energy added, the diatomic gas reaches the same final pressure as the monatomic gas.

QUESTION 12.7 If the same amount of energy as found in part (a) were transferred to 5.00 moles of carbon dioxide at the same initial temperature, would the final temperature be lower, higher, or unchanged?

EXERCISE 12.7 (a) Find the change in temperature ΔT of 22.0 mol of a monatomic ideal gas if it absorbs 9 750 J at a constant volume of 2.40 L. (b) What is the change in pressure, ΔP? (c) If the system is an ideal diatomic gas, find the change in its temperature. (d) Find the change in pressure of the diatomic gas.

ANSWERS (a) 35.6 K (b) 2.71×10^6 Pa (c) 21.3 K (d) 1.63×10^6 Pa

Isothermal Processes

During an isothermal process, the temperature of a system doesn't change. In an ideal gas the internal energy U depends only on the temperature, so it follows that $\Delta U = 0$ because $\Delta T = 0$. In this case, the first law of thermodynamics gives

$$W = -Q \quad \text{(isothermal process)}$$

We see that if the system is an ideal gas undergoing an isothermal process, the work done on the system is equal to the negative of the thermal energy transferred to the system. Such a process can be visualized in Figure 12.7. A cylinder filled with gas is in contact with a large energy reservoir that can exchange energy with the gas without changing its temperature. For a constant temperature ideal gas,

$$P = \frac{nRT}{V}$$

where the numerator on the right-hand side is constant. The PV diagram of a typical isothermal process is graphed in Figure 12.8, contrasted with an adiabatic

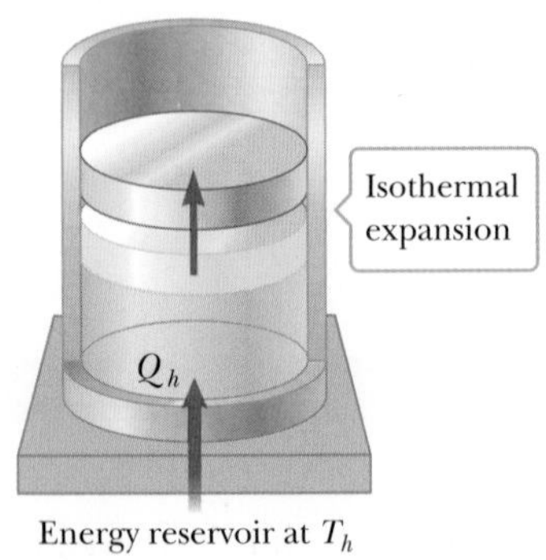

Figure 12.7 The gas in the cylinder expands isothermally while in contact with a reservoir at temperature T_h.

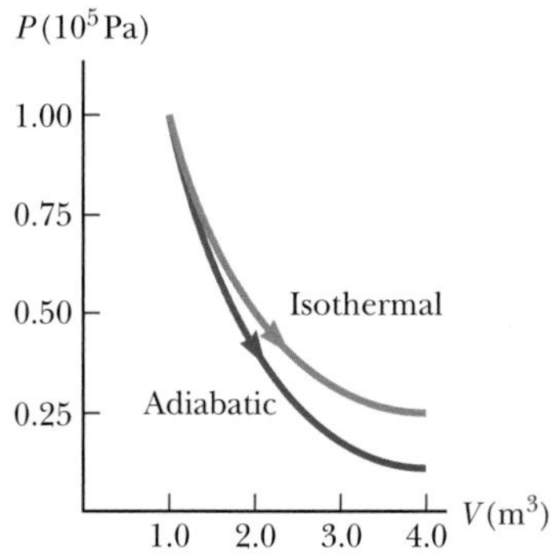

Figure 12.8 The PV diagram of an isothermal expansion, graph of $P = CV^{-1}$, where C is a constant, compared to an adiabatic expansion, $P = C_A V^{-\gamma}$. C_A is a constant equal in magnitude to C in this case but carrying different units.

process. The pressure falls off more rapidly for an adiabatic expansion because thermal energy can't be transferred into the system. In an isothermal expansion, the system loses energy by doing work on the environment but regains an equal amount of energy across the boundary.

Using methods of calculus, it can be shown that the work done on the environment during an isothermal process is given by

$$W_{\text{env}} = nRT \ln\left(\frac{V_f}{V_i}\right) \qquad [12.10]$$

The symbol "ln" in Equation 12.10 is an abbreviation for the natural logarithm, discussed in Appendix A. The work W done on the gas is just the negative of W_{env}.

EXAMPLE 12.8 | An Isothermally Expanding Balloon

GOAL Find the work done during an isothermal expansion.

PROBLEM A balloon contains 5.00 moles of a monatomic ideal gas. As energy is added to the system by heat (say, by absorption from the Sun), the volume increases by 25% at a constant temperature of 27.0°C. Find the work W_{env} done by the gas in expanding the balloon, the thermal energy Q transferred to the gas, and the work W done on the gas.

STRATEGY Be sure to convert temperatures to kelvins. Use Equation 12.10 for isothermal work W_{env} done on the environment to find the work W done on the balloon, which satisfies $W = -W_{\text{env}}$. Further, for an isothermal process, the thermal energy Q transferred to the system equals the work W_{env} done by the system on the environment.

SOLUTION

Substitute into Equation 12.10, finding the work done during the isothermal expansion. Note that $T = 27.0°\text{C} = 3.00 \times 10^2$ K.

$$W_{\text{env}} = nRT \ln\left(\frac{V_f}{V_i}\right)$$

$$= (5.00 \text{ mol})(8.31 \text{ J/K} \cdot \text{mol})(3.00 \times 10^2 \text{ K}) \times \ln\left(\frac{1.25 V_0}{V_0}\right)$$

$$W_{\text{env}} = 2.78 \times 10^3 \text{ J}$$

$$Q = W_{\text{env}} = 2.78 \times 10^3 \text{ J}$$

The negative of this amount is the work done on the gas:

$$W = -W_{\text{env}} = -2.78 \times 10^3 \text{ J}$$

REMARKS Notice the relationship between the work done on the gas, the work done on the environment, and the energy transferred. These relationships are true of all isothermal processes.

QUESTION 12.8 True or False: In an isothermal process no thermal energy transfer takes place.

(Continued)

EXERCISE 12.8 Suppose that subsequent to this heating, 1.50×10^4 J of thermal energy is removed from the gas isothermally. Find the final volume in terms of the initial volume of the example, V_0. (*Hint:* Follow the same steps as in the example, but in reverse. Also note that the initial volume in this exercise is $1.25V_0$.)

ANSWER $0.375V_0$

General Case

When a process follows none of the four given models, it's still possible to use the first law to get information about it. The work can be computed from the area under the curve of the *PV* diagram, and if the temperatures at the endpoints can be found, ΔU follows from Equation 12.5, as illustrated in the following example.

EXAMPLE 12.9 A General Process

GOAL Find thermodynamic quantities for a process that doesn't fall into any of the four previously discussed categories.

PROBLEM A quantity of 4.00 moles of a monatomic ideal gas expands from an initial volume of 0.100 m^3 to a final volume of 0.300 m^3 and pressure of 2.5×10^5 Pa (Fig. 12.9a). Compute **(a)** the work done on the gas, **(b)** the change in internal energy of the gas, and **(c)** the thermal energy transferred to the gas.

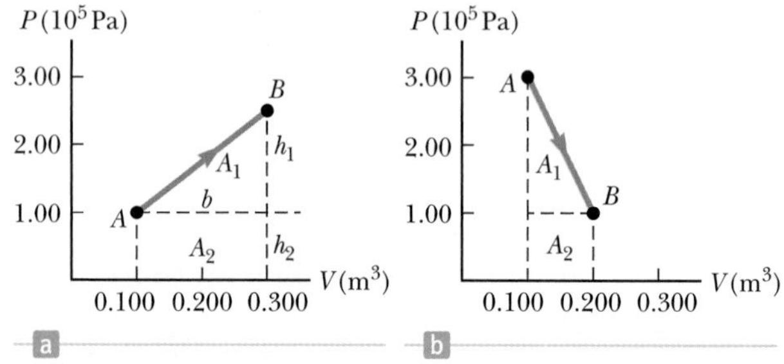

Figure 12.9 (a) (Example 12.9) (b) (Exercise 12.9)

STRATEGY The work done on the gas is equal to the negative of the area under the curve in the *PV* diagram. Use the ideal gas law to get the temperature change and, subsequently, the change in internal energy. Finally, the first law gives the thermal energy transferred by heat.

SOLUTION

(a) Find the work done on the gas by computing the area under the curve in Figure 12.9a.

Find A_1, the area of the triangle:

$$A_1 = \tfrac{1}{2}bh_1 = \tfrac{1}{2}(0.200\text{ m}^3)(1.50 \times 10^5\text{ Pa}) = 1.50 \times 10^4\text{ J}$$

Find A_2, the area of the rectangle:

$$A_2 = bh_2 = (0.200\text{ m}^3)(1.00 \times 10^5\text{ Pa}) = 2.00 \times 10^4\text{ J}$$

Sum the two areas (the gas is expanding, so the work done on the gas is negative and a minus sign must be supplied):

$$W = -(A_1 + A_2) = \boxed{-3.50 \times 10^4\text{ J}}$$

(b) Find the change in the internal energy during the process.

Compute the temperature at points *A* and *B* with the ideal gas law:

$$T_A = \frac{P_A V_A}{nR} = \frac{(1.00 \times 10^5\text{ Pa})(0.100\text{ m}^3)}{(4.00\text{ mol})(8.31\text{ J/K}\cdot\text{mol})} = 301\text{ K}$$

$$T_B = \frac{P_B V_B}{nR} = \frac{(2.50 \times 10^5\text{ Pa})(0.300\text{ m}^3)}{(4.00\text{ mol})(8.31\text{ J/K}\cdot\text{mol})} = 2.26 \times 10^3\text{ K}$$

Compute the change in internal energy:

$$\Delta U = \tfrac{3}{2}nR\,\Delta T$$

$$= \tfrac{3}{2}(4.00\text{ mol})(8.31\text{ J/K}\cdot\text{mol})(2.26 \times 10^3\text{ K} - 301\text{ K})$$

$$\Delta U = \boxed{9.77 \times 10^4\text{ J}}$$

(c) Compute *Q* with the first law:

$$Q = \Delta U - W = 9.77 \times 10^4\text{ J} - (-3.50 \times 10^4\text{ J})$$

$$= \boxed{1.33 \times 10^5\text{ J}}$$

REMARKS As long as it's possible to compute the work, cycles involving these more exotic processes can be completely analyzed. Usually, however, it's necessary to use calculus. Note that the solution to part (b) could have been facilitated by yet another application of $PV = nRT$:

$$\Delta U = \tfrac{3}{2}nR\Delta T = \tfrac{3}{2}\Delta(PV) = \tfrac{3}{2}(P_BV_B - P_AV_A)$$

This result means that even in the absence of information about the number of moles or the temperatures, the problem could be solved knowing the initial and final pressures and volumes.

QUESTION 12.9 For a curve with lower pressures but the same endpoints as in Figure 12.9a, would the thermal energy transferred be (a) smaller than, (b) equal to, or (c) greater than the thermal energy transfer of the straight-line path?

EXERCISE 12.9 Figure 12.9b represents a process involving 3.00 moles of a monatomic ideal gas expanding from 0.100 m^3 to 0.200 m^3. Find the work done on the system, the change in the internal energy of the system, and the thermal energy transferred in the process.

ANSWERS $W = -2.00 \times 10^4$ J, $\Delta U = -1.50 \times 10^4$ J, $Q = 5.00 \times 10^3$ J

Given all the different processes and formulas, it's easy to become confused when approaching one of these ideal gas problems, although most of the time only substitution into the correct formula is required. The essential facts and formulas are compiled in Table 12.2, both for easy reference and also to display the similarities and differences between the processes.

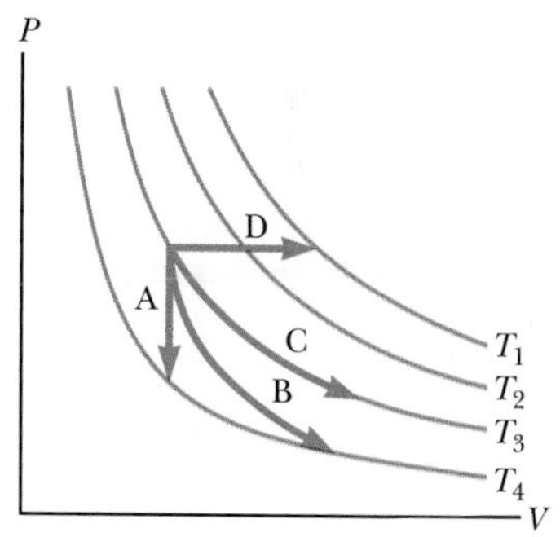

Figure 12.10 (Quick Quiz 12.2) Identify the nature of paths A, B, C, and D.

Table 12.2 The First Law and Thermodynamic Processes (Ideal Gases)

Process	ΔU	Q	W
Isobaric	$nC_v\,\Delta T$	$nC_p\,\Delta T$	$-P\,\Delta V$
Adiabatic	$nC_v\,\Delta T$	0	ΔU
Isovolumetric	$nC_v\,\Delta T$	ΔU	0
Isothermal	0	$-W$	$-nRT\ln\left(\dfrac{V_f}{V_i}\right)$
General	$nC_v\,\Delta T$	$\Delta U - W$	(*PV* Area)

Quick Quiz

12.2 Identify the paths A, B, C, and D in Figure 12.10 as isobaric, isothermal, isovolumetric, or adiabatic. For path B, $Q = 0$.

12.4 Heat Engines and the Second Law of Thermodynamics

LEARNING OBJECTIVES

1. Analyze heat engines using the first law of thermodynamics, and calculate their efficiency at converting thermal energy into work.
2. Analyze refrigerators and heat pumps and calculate their coefficients of performance.
3. State the second law of thermodynamics in two formulations, and interpret it and the first law in terms of the output of engines.
4. Explain the idea of a reversible process and how it relates to real processes.
5. Define and analyze an ideal (Carnot) engine, calculate its efficiency, and state the implications for real engines.
6. Explain the relationship between the ideal engine's efficiency and the third law of thermodynamics.

A **heat engine** takes in energy by heat and partially converts it to other forms, such as electrical and mechanical energy. In a typical process for producing electricity in a power plant, for instance, coal or some other fuel is burned, and the resulting internal energy is used to convert water to steam. The steam is then directed at the blades of a turbine, setting it rotating. Finally, the mechanical energy associated with this rotation is used to drive an electric generator. In another heat engine—the internal combustion engine in an automobile—energy enters the engine as fuel is injected into the cylinder and combusted, and a fraction of this energy is converted to mechanical energy.

Cyclic process ▶

In general, a heat engine carries some working substance through a **cyclic process**[1] during which (1) energy is transferred by heat from a source at a high temperature, (2) work is done by the engine, and (3) energy is expelled from the engine by heat to a source at lower temperature. As an example, consider the operation of a steam engine in which the working substance is water. The water in the engine is carried through a cycle in which it first evaporates into steam in a boiler and then expands against a piston. After the steam is condensed with cooling water, it returns to the boiler, and the process is repeated.

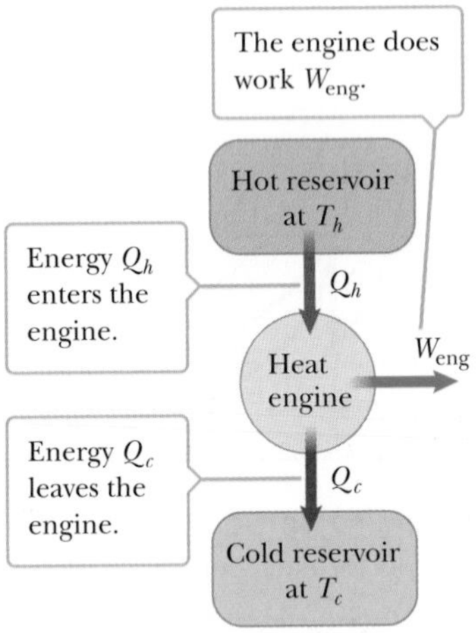

Figure 12.11 In this schematic representation of a heat engine, part of the thermal energy from the hot reservoir is turned into work while the rest is expelled to the cold reservoir.

It's useful to draw a heat engine schematically, as in Figure 12.11. The engine absorbs energy Q_h from the hot reservoir, does work W_{eng}, then gives up energy Q_c to the cold reservoir. (Note that *negative* work is done *on* the engine, so that $W = -W_{eng}$.) Because the working substance goes through a cycle, always returning to its initial thermodynamic state, its initial and final internal energies are equal, so $\Delta U = 0$. From the first law of thermodynamics, therefore,

$$\Delta U = 0 = Q + W \rightarrow Q_{net} = -W = W_{eng}$$

The last equation shows that **the work W_{eng} done by a heat engine equals the net energy absorbed by the engine.** As we can see from Figure 12.11, $Q_{net} = |Q_h| - |Q_c|$. Therefore,

$$W_{eng} = |Q_h| - |Q_c| \qquad [12.11]$$

Ordinarily, a transfer of thermal energy Q can be either positive or negative, so the use of absolute value signs makes the signs of Q_h and Q_c explicit.

If the working substance is a gas, then **the work done by the engine for a cyclic process is the area enclosed by the curve representing the process on a *PV* diagram.** This area is shown for an arbitrary cyclic process in Figure 12.12.

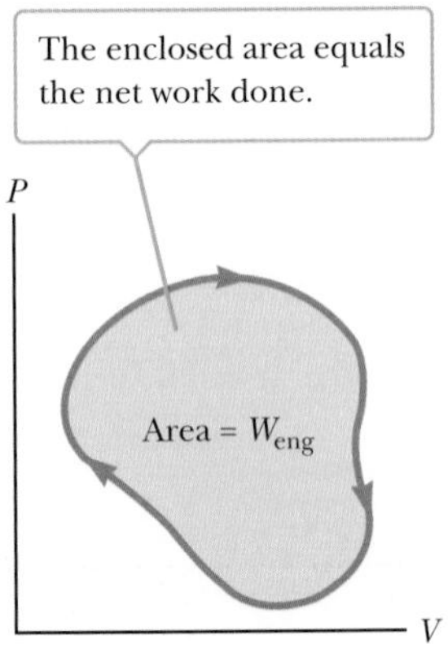

Figure 12.12 The *PV* diagram for an arbitrary cyclic process.

The **thermal efficiency** e of a heat engine is defined as the work done by the engine, W_{eng}, divided by the energy absorbed during one cycle:

$$e \equiv \frac{W_{eng}}{|Q_h|} = \frac{|Q_h| - |Q_c|}{|Q_h|} = 1 - \frac{|Q_c|}{|Q_h|} \qquad [12.12]$$

We can think of thermal efficiency as the ratio of the benefit received (work) to the cost incurred (energy transfer at the higher temperature). Equation 12.12 shows that a heat engine has 100% efficiency ($e = 1$) only if $Q_c = 0$, meaning no energy is expelled to the cold reservoir. In other words, a heat engine with perfect efficiency would have to use all the input energy for doing mechanical work. That isn't possible, as will be seen in Section 12.5.

[1]Strictly speaking, the internal combustion engine is not a heat engine according to the description of the cyclic process, because the air–fuel mixture undergoes only one cycle and is then expelled through the exhaust system.

EXAMPLE 12.10 The Efficiency of an Engine

GOAL Apply the efficiency formula to a heat engine.

PROBLEM During one cycle, an engine extracts 2.00×10^3 J of energy from a hot reservoir and transfers 1.50×10^3 J to a cold reservoir. **(a)** Find the thermal efficiency of the engine. **(b)** How much work does this engine do in one cycle? **(c)** What average power does the engine generate if it goes through four cycles in 2.50 s?

STRATEGY Apply Equation 12.12 to obtain the thermal efficiency, then use the first law, adapted to engines (Eq. 12.11), to find the work done in one cycle. To obtain the power generated, divide the work done in four cycles by the time it takes to run those cycles.

SOLUTION

(a) Find the engine's thermal efficiency.

Substitute Q_c and Q_h into Equation 12.12:

$$e = 1 - \frac{|Q_c|}{|Q_h|} = 1 - \frac{1.50 \times 10^3 \text{ J}}{2.00 \times 10^3 \text{ J}} = \boxed{0.250, \text{ or } 25.0\%}$$

(b) How much work does this engine do in one cycle?

Apply the first law in the form of Equation 12.11 to find the work done by the engine:

$$W_{\text{eng}} = |Q_h| - |Q_c| = 2.00 \times 10^3 \text{ J} - 1.50 \times 10^3 \text{ J} = \boxed{5.00 \times 10^2 \text{ J}}$$

(c) Find the average power output of the engine.

Multiply the answer of part (b) by four and divide by time:

$$P = \frac{W}{\Delta t} = \frac{4.00 \times (5.00 \times 10^2 \text{ J})}{2.50 \text{ s}} = \boxed{8.00 \times 10^2 \text{ W}}$$

REMARKS Problems like this usually reduce to solving two equations and two unknowns, as here, where the two equations are the efficiency equation and the first law and the unknowns are the efficiency and the work done by the engine.

QUESTION 12.10 Can the efficiency of an engine always be improved by increasing the thermal energy put into the system during a cycle? Explain.

EXERCISE 12.10 The energy absorbed by an engine is three times as great as the work it performs. (a) What is its thermal efficiency? (b) What fraction of the energy absorbed is expelled to the cold reservoir? (c) What is the average power output of the engine if the energy input is 1 650 J each cycle and it goes through two cycles every 3 seconds?

ANSWERS (a) $\frac{1}{3}$ (b) $\frac{2}{3}$ (c) 367 W

EXAMPLE 12.11 Analyzing an Engine Cycle

GOAL Combine several concepts to analyze an engine cycle.

PROBLEM A heat engine contains an ideal monatomic gas confined to a cylinder by a movable piston. The gas starts at A, where $T = 3.00 \times 10^2$ K. (See Fig. 12.13a.) The process $B \rightarrow C$ is an isothermal expansion. **(a)** Find the number n of moles of gas and the temperature at B. **(b)** Find ΔU, Q, and W for the isovolumetric process $A \rightarrow B$. **(c)** Repeat for the isothermal process $B \rightarrow C$. **(d)** Repeat for the isobaric process $C \rightarrow A$. **(e)** Find the net change in the internal energy for the complete cycle. **(f)** Find the thermal energy Q_h transferred into the system, the thermal energy rejected, Q_c, the thermal efficiency, and net work on the environment performed by the engine.

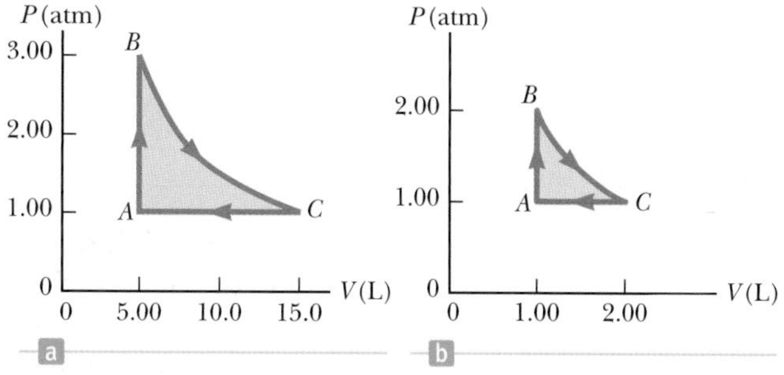

Figure 12.13 (a) (Example 12.11) (b) (Exercise 12.11)

STRATEGY In part (a) n and T can be found from the ideal gas law, which connects the equilibrium values of P, V, and T. Once the temperature T is known at the points A, B, and C, the change in internal energy, ΔU, can be computed from

(Continued)

the formula in Table 12.2 for each process. Q and W can be similarly computed, or deduced from the first law, using the techniques applied in the single-process examples.

SOLUTION

(a) Find n and T_B with the ideal gas law:

$$n = \frac{P_A V_A}{RT_A} = \frac{(1.00\text{ atm})(5.00\text{ L})}{(0.082\,1\text{ L}\cdot\text{atm/mol}\cdot\text{K})(3.00\times 10^2\text{ K})}$$
$$= \boxed{0.203\text{ mol}}$$

$$T_B = \frac{P_B V_B}{nR} = \frac{(3.00\text{ atm})(5.00\text{ L})}{(0.203\text{ mol})(0.082\,1\text{ L}\cdot\text{atm/mol}\cdot\text{K})}$$
$$= \boxed{9.00\times 10^2\text{ K}}$$

(b) Find ΔU_{AB}, Q_{AB}, and W_{AB} for the constant volume process $A \rightarrow B$.

Compute ΔU_{AB}, noting that $C_v = \frac{3}{2}R = 12.5\text{ J/mol}\cdot\text{K}$:

$$\Delta U_{AB} = nC_v\,\Delta T = (0.203\text{ mol})(12.5\text{ J/mol}\cdot\text{K}) \times (9.00\times 10^2\text{ K} - 3.00\times 10^2\text{ K})$$
$$\Delta U_{AB} = \boxed{1.52\times 10^3\text{ J}}$$

$\Delta V = 0$ for isovolumetric processes, so no work is done:

$$W_{AB} = \boxed{0}$$

We can find Q_{AB} from the first law:

$$Q_{AB} = \Delta U_{AB} = \boxed{1.52\times 10^3\text{ J}}$$

(c) Find ΔU_{BC}, Q_{BC}, and W_{BC} for the isothermal process $B \rightarrow C$.

This process is isothermal, so the temperature doesn't change, and the change in internal energy is zero:

$$\Delta U_{BC} = nC_v\,\Delta T = \boxed{0}$$

Compute the work done on the system, using the negative of Equation 12.10:

$$W_{BC} = -nRT\ln\left(\frac{V_C}{V_B}\right)$$
$$= -(0.203\text{ mol})(8.31\text{ J/mol}\cdot\text{K})(9.00\times 10^2\text{ K}) \times \ln\left(\frac{1.50\times 10^{-2}\text{ m}^3}{5.00\times 10^{-3}\text{ m}^3}\right)$$
$$W_{BC} = \boxed{-1.67\times 10^3\text{ J}}$$

Compute Q_{BC} from the first law:

$$0 = Q_{BC} + W_{BC} \rightarrow Q_{BC} = -W_{BC} = \boxed{1.67\times 10^3\text{ J}}$$

(d) Find ΔU_{CA}, Q_{CA}, and W_{CA} for the isobaric process $C \rightarrow A$.

Compute the work on the system, with pressure constant:

$$W_{CA} = -P\Delta V = -(1.01\times 10^5\text{ Pa})(5.00\times 10^{-3}\text{ m}^3 - 1.50\times 10^{-2}\text{ m}^3)$$
$$W_{CA} = \boxed{1.01\times 10^3\text{ J}}$$

Find the change in internal energy, ΔU_{CA}:

$$\Delta U_{CA} = \tfrac{3}{2}nR\,\Delta T = \tfrac{3}{2}(0.203\text{ mol})(8.31\text{ J/K}\cdot\text{mol}) \times (3.00\times 10^2\text{ K} - 9.00\times 10^2\text{ K})$$
$$\Delta U_{CA} = \boxed{-1.52\times 10^3\text{ J}}$$

Compute the thermal energy, Q_{CA}, from the first law:

$$Q_{CA} = \Delta U_{CA} - W_{CA} = -1.52\times 10^3\text{ J} - 1.01\times 10^3\text{ J}$$
$$= \boxed{-2.53\times 10^3\text{ J}}$$

(e) Find the net change in internal energy, ΔU_{net}, for the cycle:

$$\Delta U_{\text{net}} = \Delta U_{AB} + \Delta U_{BC} + \Delta U_{CA}$$
$$= 1.52\times 10^3\text{ J} + 0 - 1.52\times 10^3\text{ J} = \boxed{0}$$

(f) Find the energy input, Q_h; the energy rejected, Q_c; the thermal efficiency; and the net work performed by the engine:

Sum all the positive contributions to find Q_h:

$$Q_h = Q_{AB} + Q_{BC} = 1.52 \times 10^3\text{ J} + 1.67 \times 10^3\text{ J} = \boxed{3.19 \times 10^3\text{ J}}$$

Sum any negative contributions (in this case, there is only one):

$$Q_c = \boxed{-2.53 \times 10^3\text{ J}}$$

Find the engine efficiency and the net work done by the engine:

$$e = 1 - \frac{|Q_c|}{|Q_h|} = 1 - \frac{2.53 \times 10^3\text{ J}}{3.19 \times 10^3\text{ J}} = \boxed{0.207}$$

$$W_{\text{eng}} = -(W_{AB} + W_{BC} + W_{CA}) = -(0 - 1.67 \times 10^3\text{ J} + 1.01 \times 10^3\text{ J}) = \boxed{6.60 \times 10^2\text{ J}}$$

REMARKS Cyclic problems are rather lengthy, but the individual steps are often short substitutions. Notice that the change in internal energy for the cycle is zero and that the net work done on the environment is identical to the net thermal energy transferred, both as they should be.

QUESTION 12.11 If *BC* were a straight-line path, would the work done by the cycle be affected? How?

EXERCISE 12.11 4.05×10^{-2} mol of monatomic ideal gas goes through the process shown in Figure 12.13b. The temperature at point A is 3.00×10^2 K and is 6.00×10^2 K during the isothermal process $B \rightarrow C$. (a) Find Q, ΔU, and W for the constant volume process $A \rightarrow B$. (b) Do the same for the isothermal process $B \rightarrow C$. (c) Repeat, for the constant pressure process $C \rightarrow A$. (d) Find Q_h, Q_c, and the efficiency. (e) Find W_{eng}.

ANSWERS (a) $Q_{AB} = \Delta U_{AB} = 151$ J, $W_{AB} = 0$ (b) $\Delta U_{BC} = 0$, $Q_{BC} = -W_{BC} = 1.40 \times 10^2$ J (c) $Q_{CA} = -252$ J, $\Delta U_{CA} = -151$ J, $W_{CA} = 101$ J (d) $Q_h = 291$ J, $Q_c = -252$ J, $e = 0.134$ (e) $W_{\text{eng}} = 39$ J

Refrigerators and Heat Pumps

Heat engines can operate in reverse. In this case, energy is injected into the engine, modeled as work W in Figure 12.14, resulting in energy being extracted from the cold reservoir and transferred to the hot reservoir. The system now operates as a heat pump, a common example being a refrigerator (Fig. 12.15 on page 422). Energy Q_c is extracted from the interior of the refrigerator and delivered as energy Q_h to the warmer air in the kitchen. The work is done in the compressor unit of the refrigerator, compressing a refrigerant such as freon, causing its temperature to increase.

A household air conditioner is another example of a heat pump. Some homes are both heated and cooled by heat pumps. In winter, the heat pump extracts energy Q_c from the cool outside air and delivers energy Q_h to the warmer air inside. In summer, energy Q_c is removed from the cool inside air, while energy Q_h is ejected to the warm air outside.

For a refrigerator or an air conditioner—a heat pump operating in cooling mode—work W is what you pay for, in terms of electrical energy running the compressor, whereas Q_c is the desired benefit. The most efficient refrigerator or air conditioner is one that removes the greatest amount of energy from the cold reservoir in exchange for the least amount of work.

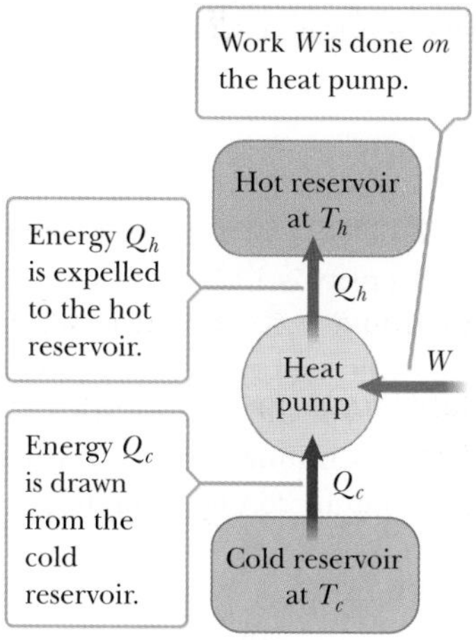

Figure 12.14
In this schematic representation of a heat pump, thermal energy is extracted from the cold reservoir and "pumped" to the hot reservoir.

The coefficient of performance (COP) for a refrigerator or an air conditioner is the magnitude of the energy extracted from the cold reservoir, $|Q_c|$, divided by the work W performed by the device:

$$\text{COP(cooling mode)} = \frac{|Q_c|}{W} \qquad [12.13]$$

SI unit: dimensionless

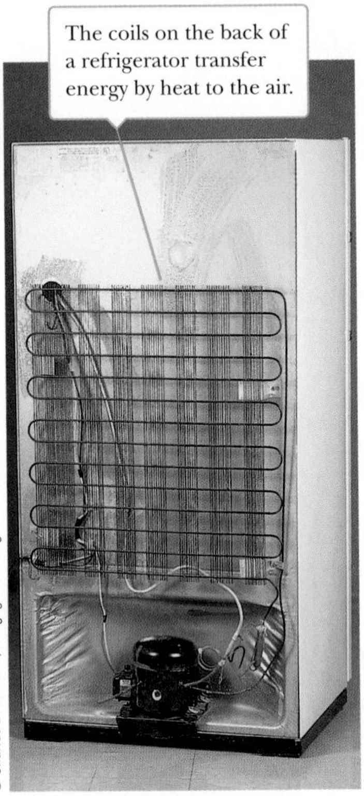

© Charles D. Winters/Cengage Learning

Figure 12.15 The back of a household refrigerator. The air surrounding the coils is the hot reservoir.

The larger this ratio, the better the performance, because more energy is being removed for a given amount of work. A good refrigerator or air conditioner will have a COP of 5 or 6.

A heat pump operating in heating mode warms the inside of a house in winter by extracting energy from the colder outdoor air. This statement may seem paradoxical, but recall that this process is equivalent to a refrigerator removing energy from its interior and ejecting it into the kitchen.

The coefficient of performance of a heat pump operating in the heating mode is the magnitude of the energy rejected to the hot reservoir, $|Q_h|$, divided by the work W done by the pump:

$$\text{COP(heating mode)} = \frac{|Q_h|}{W} \qquad [12.14]$$

SI unit: dimensionless

In effect, the COP of a heat pump in the heating mode is the ratio of what you gain (energy delivered to the interior of your home) to what you give (work input). Typical values for this COP are greater than 1, because $|Q_h|$ is usually greater than W.

In a groundwater heat pump, energy is extracted in the winter from water deep in the ground rather than from the outside air, while energy is delivered to that water in the summer. This strategy increases the year-round efficiency of the heating and cooling unit because the groundwater is at a higher temperature than the air in winter and at a cooler temperature than the air in summer.

EXAMPLE 12.12 Cooling the Leftovers

GOAL Apply the coefficient of performance of a refrigerator.

PROBLEM A 2.00-L container of leftover soup at a temperature of 323 K is placed in a refrigerator. Assume the specific heat of the soup is the same as that of water and the density is 1.25×10^3 kg/m^3. The refrigerator cools the soup to 283 K. **(a)** If the COP of the refrigerator is 5.00, find the energy needed, in the form of work, to cool the soup. **(b)** If the compressor has a power rating of 0.250 hp, for what minimum length of time must it operate to cool the soup to 283 K? (The minimum time assumes the soup cools at the same rate that the heat pump ejects thermal energy from the refrigerator.)

STRATEGY The solution to this problem requires three steps. First, find the total mass m of the soup. Second, using $Q = mc\Delta T$, where $Q = Q_c$, find the energy transfer required to cool the soup. Third, substitute Q_c and the COP into Equation 12.13, solving for W. Divide the work by the power to get an estimate of the time required to cool the soup.

SOLUTION

(a) Find the work needed to cool the soup.

Calculate the mass of the soup:

$$m = \rho V = (1.25 \times 10^3 \text{ kg/m}^3)(2.00 \times 10^{-3} \text{ m}^3) = 2.50 \text{ kg}$$

Find the energy transfer required to cool the soup:

$$Q_c = Q = mc\Delta T$$
$$= (2.50 \text{ kg})(4\,190 \text{ J/kg}\cdot\text{K})(283 \text{ K} - 323 \text{ K})$$
$$= -4.19 \times 10^5 \text{ J}$$

Substitute Q_c and the COP into Equation 12.13:

$$\text{COP} = \frac{|Q_c|}{W} = \frac{4.19 \times 10^5 \text{ J}}{W} = 5.00$$

$$W = \boxed{8.38 \times 10^4 \text{ J}}$$

(b) Find the time needed to cool the soup.

Convert horsepower to watts:

$$P = (0.250 \text{ hp})(746 \text{ W/1 hp}) = 187 \text{ W}$$

Divide the work by the power to find the elapsed time: $\Delta t = \frac{W}{P} = \frac{8.38 \times 10^4\ \text{J}}{187\ \text{W}} = 448\ \text{s}$

REMARKS This example illustrates how cooling different substances requires differing amounts of work due to differences in specific heats. The problem doesn't take into account the insulating properties of the soup container and of the soup itself, which retard the cooling process.

QUESTION 12.12 If the refrigerator door is left open, does the kitchen become cooler? Why or why not?

EXERCISE 12.12 (a) How much work must a heat pump with a COP of 2.50 do to extract 1.00 MJ of thermal energy from the outdoors (the cold reservoir)? (b) If the unit operates at 0.500 hp, how long will the process take? (Be sure to use the correct COP!)

ANSWERS (a) 6.67×10^5 J (b) 1.79×10^3 s

The Second Law of Thermodynamics

There are limits to the efficiency of heat engines. The ideal engine would convert all input energy into useful work, but it turns out that such an engine is impossible to construct. The Kelvin-Planck formulation of the **second law of thermodynamics** can be stated as follows:

> No heat engine operating in a cycle can absorb energy from a reservoir and use it entirely for the performance of an equal amount of work.

This form of the second law means that the efficiency $e = W_{\text{eng}}/|Q_h|$ of engines must always be less than 1. Some energy $|Q_c|$ must always be lost to the environment. In other words, it's theoretically impossible to construct a heat engine with an efficiency of 100%.

To summarize, the first law says **we can't get a greater amount of energy out of a cyclic process than we put in,** and the second law says **we can't break even.** No matter what engine is used, some energy must be transferred by heat to the cold reservoir. In Equation 12.11, the second law simply means $|Q_c|$ is always greater than zero.

There is another equivalent statement of the second law:

> If two systems are in thermal contact, net thermal energy transfers spontaneously by heat from the hotter system to the colder system.

Here, spontaneous means the energy transfer occurs naturally, with no work being done. Thermal energy naturally transfers from hotter systems to colder systems. Work must be done to transfer thermal energy from a colder system to a hotter system, however. An example is the refrigerator, which transfers thermal energy from inside the refrigerator to the warmer kitchen.

Lord Kelvin
British Physicist and Mathematician (1824–1907)
Born William Thomson in Belfast, Kelvin was the first to propose the use of an absolute scale of temperature. His study of Carnot's theory led to the idea that energy cannot pass spontaneously from a colder object to a hotter object; this principle is known as the second law of thermodynamics.

Reversible and Irreversible Processes

No engine can operate with 100% efficiency, but different designs yield different efficiencies, and it turns out that one design in particular delivers the maximum possible efficiency. This design is the Carnot cycle, discussed in the next subsection. Understanding it requires the concepts of reversible and irreversible processes. In a **reversible** process, every state along the path is an equilibrium state, so the system can return to its initial conditions by going along the same path in the reverse direction. A process that doesn't satisfy this requirement is **irreversible.**

Most natural processes are known to be irreversible; the reversible process is an idealization. Although real processes are always irreversible, some are *almost* reversible. If a real process occurs so slowly that the system is virtually always in equilibrium, the process can be considered reversible. Imagine compressing a gas very slowly by

Figure 12.16 A method for compressing a gas in a reversible isothermal process.

Sadi Carnot
French Engineer (1796–1832)
Carnot is considered to be the founder of the science of thermodynamics. Some of his notes found after his death indicate that he was the first to recognize the relationship between work and heat.

Tip 12.3 Don't Shop for a Carnot Engine

The Carnot engine is only an idealization. If a Carnot engine were developed in an effort to maximize efficiency, it would have zero power output because for all of the processes to be reversible, the engine would have to run infinitely slowly.

dropping grains of sand onto a frictionless piston, as in Figure 12.16. The temperature can be kept constant by placing the gas in thermal contact with an energy reservoir. The pressure, volume, and temperature of the gas are well defined during this isothermal compression. Each added grain of sand represents a change to a new equilibrium state. The process can be reversed by slowly removing grains of sand from the piston.

The Carnot Engine

In 1824, in an effort to understand the efficiency of real engines, a French engineer named Sadi Carnot (1796–1832) described a theoretical engine now called a *Carnot engine* that is of great importance from both a practical and a theoretical viewpoint. He showed that a heat engine operating in an ideal, reversible cycle—now called a **Carnot cycle**—between two energy reservoirs is the most efficient engine possible. Such an engine establishes an upper limit on the efficiencies of all real engines. **Carnot's theorem** can be stated as follows:

> No real engine operating between two energy reservoirs can be more efficient than a Carnot engine operating between the same two reservoirs.

In a Carnot cycle, an ideal gas is contained in a cylinder with a movable piston at one end. The temperature of the gas varies between T_c and T_h. The cylinder walls and the piston are thermally nonconducting. Figure 12.17 shows the four stages of the Carnot cycle, and Figure 12.18 is the *PV* diagram for the cycle. The cycle consists of two adiabatic and two isothermal processes, all reversible:

1. The process $A \rightarrow B$ is an isothermal expansion at temperature T_h in which the gas is placed in thermal contact with a hot reservoir (a large oven, for example) at temperature T_h (Fig. 12.17a). During the process, the gas absorbs energy Q_h from the reservoir and does work W_{AB} in raising the piston.
2. In the process $B \rightarrow C$, the base of the cylinder is replaced by a thermally nonconducting wall and the gas expands adiabatically, so no energy enters or leaves the system by heat (Fig. 12.17b). During the process, the temperature falls from T_h to T_c and the gas does work W_{BC} in raising the piston.
3. In the process $C \rightarrow D$, the gas is placed in thermal contact with a cold reservoir at temperature T_c (Fig. 12.17c) and is compressed isothermally at temperature T_c. During this time, the gas expels energy Q_c to the reservoir and the work done on the gas is W_{CD}.
4. In the final process, $D \rightarrow A$, the base of the cylinder is again replaced by a thermally nonconducting wall (Fig. 12.17d), and the gas is compressed adiabatically. The temperature of the gas increases to T_h, and the work done on the gas is W_{DA}.

For a Carnot engine, the following relationship between the thermal energy transfers and the absolute temperatures can be derived:

$$\frac{|Q_c|}{|Q_h|} = \frac{T_c}{T_h} \qquad [12.15]$$

Substituting this expression into Equation 12.12, we find that the thermal efficiency of a Carnot engine is

$$e_C = 1 - \frac{T_c}{T_h} \qquad [12.16]$$

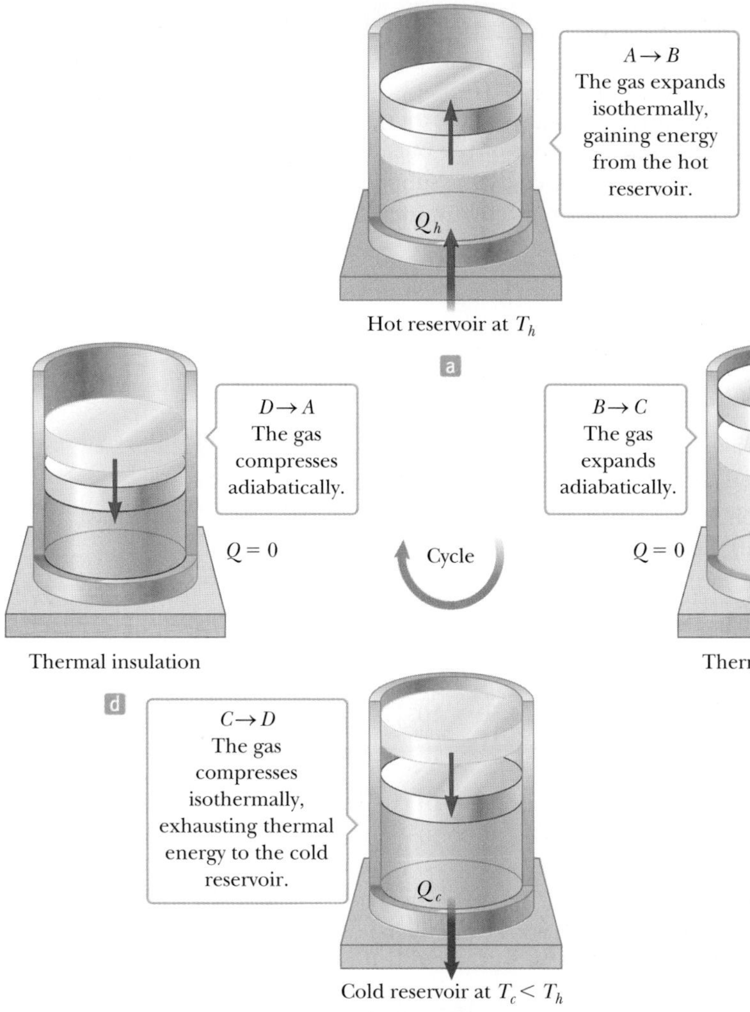

Figure 12.17
The Carnot cycle. The letters A, B, C, and D refer to the states of the gas shown in Figure 12.18. The arrows on the piston indicate the direction of its motion during each process.

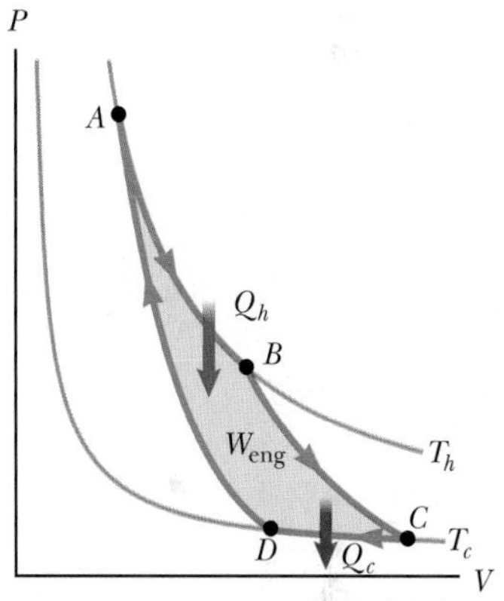

Figure 12.18
The PV diagram for the Carnot cycle. The net work done, W_{eng}, equals the net energy transferred into the Carnot engine in one cycle, $|Q_h| - |Q_c|$.

where T must be in kelvins. From this result, we see that **all Carnot engines operating reversibly between the same two temperatures have the same efficiency.**

Equation 12.16 can be applied to any working substance operating in a Carnot cycle between two energy reservoirs. According to that equation, the efficiency is zero if $T_c = T_h$. The efficiency increases as T_c is lowered and as T_h is increased. The efficiency can be one (100%), however, only if $T_c = 0$ K. According to the **third law of thermodynamics,** it's impossible to lower the temperature of a system to absolute zero in a finite number of steps, so such reservoirs are not available and the maximum efficiency is always less than 1. In most practical cases, the cold reservoir is near room temperature, about 300 K, so increasing the efficiency requires raising the temperature of the hot reservoir. **All real engines operate irreversibly, due to friction and the brevity of their cycles, and are therefore *less* efficient than the Carnot engine.**

◀ Third law of thermodynamics

■ *Quick Quiz*

12.3 Three engines operate between reservoirs separated in temperature by 300 K. The reservoir temperatures are as follows:

Engine A: $T_h = 1\,000$ K, $T_c = 700$ K
Engine B: $T_h = 800$ K, $T_c = 500$ K
Engine C: $T_h = 600$ K, $T_c = 300$ K

Rank the engines in order of their theoretically possible efficiency, from highest to lowest. (a) A, B, C (b) B, C, A (c) C, B, A (d) C, A, B

EXAMPLE 12.13 | The Steam Engine

GOAL Apply the equations of an ideal (Carnot) engine.

PROBLEM A steam engine has a boiler that operates at 5.00×10^2 K. The energy from the boiler changes water to steam, which drives the piston. The temperature of the exhaust is that of the outside air, 3.00×10^2 K. **(a)** What is the engine's efficiency if it's an ideal engine? **(b)** If the 3.50×10^3 J of energy is supplied from the boiler, find the energy transferred to the cold reservoir and the work done by the engine on its environment.

STRATEGY This problem requires substitution into Equations 12.15 and 12.16, both applicable to a Carnot engine. The first equation relates the ratio Q_c/Q_h to the ratio T_c/T_h, and the second gives the Carnot engine efficiency.

SOLUTION

(a) Find the engine's efficiency, assuming it's ideal.

Substitute into Equation 12.16, the equation for the efficiency of a Carnot engine:

$$e_C = 1 - \frac{T_c}{T_h} = 1 - \frac{3.00 \times 10^2 \text{ K}}{5.00 \times 10^2 \text{ K}} = \boxed{0.400, \text{ or } 40.0\%}$$

(b) Find the energy transferred to the cold reservoir and the work done on the environment if 3.50×10^3 J is delivered to the engine during one cycle.

Equation 12.15 shows that the ratio of energies equals the ratio of temperatures:

$$\frac{|Q_c|}{|Q_h|} = \frac{T_c}{T_h} \quad \rightarrow \quad |Q_c| = |Q_h| \frac{T_c}{T_h}$$

Substitute, finding the energy transferred to the cold reservoir:

$$|Q_c| = (3.50 \times 10^3 \text{ J})\left(\frac{3.00 \times 10^2 \text{ K}}{5.00 \times 10^2 \text{ K}}\right) = \boxed{2.10 \times 10^3 \text{ J}}$$

Use Equation 12.11 to find the work done by the engine:

$$W_{\text{eng}} = |Q_h| - |Q_c| = 3.50 \times 10^3 \text{ J} - 2.10 \times 10^3 \text{ J}$$
$$= \boxed{1.40 \times 10^3 \text{ J}}$$

REMARKS This problem differs from the earlier examples on work and efficiency because we used the special Carnot relationships, Equations 12.15 and 12.16. Remember that these equations can only be used when the cycle is identified as ideal or a Carnot.

QUESTION 12.13 True or False: A nonideal engine operating between the same temperature extremes as a Carnot engine and having the same input thermal energy will perform the same amount of work as the Carnot engine.

EXERCISE 12.13 The highest theoretical efficiency of a gasoline engine based on the Carnot cycle is 0.300, or 30.0%. (a) If this engine expels its gases into the atmosphere, which has a temperature of 3.00×10^2 K, what is the temperature in the cylinder immediately after combustion? (b) If the heat engine absorbs 837 J of energy from the hot reservoir during each cycle, how much work can it perform in each cycle?

ANSWERS (a) 429 K (b) 251 J

12.5 Entropy

LEARNING OBJECTIVES

1. State the thermodynamic and the statistical definitions of entropy.
2. Calculate the change in entropy of various physical systems.
3. Relate entropy to disorder.
4. Discuss the degradation of energy and the entropy of the Universe.

Temperature and internal energy, associated with the zeroth and first laws of thermodynamics, respectively, are both state variables, meaning they can be used to describe the thermodynamic state of a system. A state variable called the **entropy** S is related to the second law of thermodynamics. We define entropy on a macroscopic scale as German physicist Rudolf Clausius (1822–1888) first expressed it in 1865:

> Let Q_r be the energy absorbed or expelled during a reversible, constant temperature process between two equilibrium states. Then the change in entropy during any constant temperature process connecting the two equilibrium states is defined as
>
> $$\Delta S \equiv \frac{Q_r}{T} \qquad [12.17]$$
>
> **SI unit: joules/kelvin (J/K)**

A similar formula holds when the temperature isn't constant, but its derivation entails calculus and won't be considered here. Calculating the change in entropy, ΔS, during a transition between two equilibrium states requires finding a reversible path that connects the states. The entropy change calculated on that reversible path is taken to be ΔS for the *actual* path. This approach is necessary because quantities such as the temperature of a system can be defined only for systems in equilibrium, and a reversible path consists of a sequence of equilibrium states. The subscript r on the term Q_r emphasizes that the path chosen for the calculation must be reversible. The change in entropy ΔS, like changes in internal energy ΔU and changes in potential energy, depends only on the endpoints, and not on the path connecting them.

The concept of entropy gained wide acceptance in part because it provided another variable to describe the state of a system, along with pressure, volume, and temperature. Its significance was enhanced when it was found that **the entropy of the Universe increases in all natural processes.** This is yet another way of stating the second law of thermodynamics.

Although the entropy of the *Universe* increases in all natural processes, the entropy of a *system* can decrease. For example, if system A transfers energy Q to system B by heat, the entropy of system A decreases. This transfer, however, can only occur if the temperature of system B is less than the temperature of system A. Because temperature appears in the denominator in the definition of entropy, system B's increase in entropy will be greater than system A's decrease, so taken together, the entropy of the Universe increases.

For centuries, individuals have attempted to build perpetual motion machines that operate continuously without any input of energy or increase in entropy. The laws of thermodynamics preclude the invention of any such machines.

The concept of entropy is satisfying because it enables us to present the second law of thermodynamics in the form of a mathematical statement. In the next section we find that entropy can also be interpreted in terms of probabilities, a relationship that has profound implications.

Library of Congress

Rudolf Clausius
German Physicist (1822–1888)
Born with the name Rudolf Gottlieb, he adopted the classical name of Clausius, which was a popular thing to do in his time. "I propose . . . to call S the entropy of a body, after the Greek word 'transformation.' I have designedly coined the word 'entropy' to be similar to energy, for these two quantities are so analogous in their physical significance, that an analogy of denominations seems to be helpful."

Tip 12.4 Entropy ≠ Energy
Don't confuse energy and entropy. Although the names sound similar the concepts are different.

Quick Quiz

12.4 Which of the following is true for the entropy change of a system that undergoes a reversible, adiabatic process? (a) $\Delta S < 0$ (b) $\Delta S = 0$ (c) $\Delta S > 0$

EXAMPLE 12.14 Melting a Piece of Lead

GOAL Calculate the change in entropy due to a phase change.

PROBLEM **(a)** Find the change in entropy of 3.00×10^2 g of lead when it melts at 327°C. Lead has a latent heat of fusion of 2.45×10^4 J/kg. **(b)** Suppose the same amount of energy is used to melt part of a piece of silver, which is already at its melting point of 961°C. Find the change in the entropy of the silver.

STRATEGY This problem can be solved by substitution into Equation 12.17. Be sure to use the Kelvin temperature scale.

(Continued)

SOLUTION

(a) Find the entropy change of the lead.

Find the energy necessary to melt the lead:

$$Q = mL_f = (0.300\text{ kg})(2.45 \times 10^4\text{ J/kg}) = 7.35 \times 10^3\text{ J}$$

Convert the temperature in degrees Celsius to Kelvins:

$$T = T_C + 273 = 327 + 273 = 6.00 \times 10^2\text{ K}$$

Substitute the quantities found into the entropy equation:

$$\Delta S = \frac{Q}{T} = \frac{7.35 \times 10^3\text{ J}}{6.00 \times 10^2\text{ K}} = \boxed{12.3\text{ J/K}}$$

(b) Find the entropy change of the silver.

The added energy is the same as in part (a), by supposition. Substitute into the entropy equation, after first converting the melting point of silver to kelvins:

$$T = T_C + 273 = 961 + 273 = 1.234 \times 10^3\text{ K}$$

$$\Delta S = \frac{Q}{T} = \frac{7.35 \times 10^3\text{ J}}{1.234 \times 10^3\text{ K}} = \boxed{5.96\text{ J/K}}$$

REMARKS This example shows that adding a given amount of energy to a system increases its entropy, but adding the same amount of energy to another system at higher temperature results in a smaller increase in entropy. This is because the change in entropy is inversely proportional to the temperature.

QUESTION 12.14 If the same amount of energy were used to melt ice at 0°C to water at 0°C, rank the entropy changes for ice, silver, and lead, from smallest to largest.

EXERCISE 12.14 Find the change in entropy of a 2.00-kg block of gold at 1 063°C when it melts to become liquid gold at 1 063°C. (The latent heat of fusion for gold is 6.44×10^4 J/kg.)

ANSWER 96.4 J/K

EXAMPLE 12.15 Ice, Steam, and the Entropy of the Universe

GOAL Calculate the change in entropy for a system and its environment.

PROBLEM A block of ice at 273 K is put in thermal contact with a container of steam at 373 K, converting 25.0 g of ice to water at 273 K while condensing some of the steam to water at 373 K. Find **(a)** the change in entropy of the ice, **(b)** the change in entropy of the steam, and **(c)** the change in entropy of the Universe.

STRATEGY First, calculate the energy transfer necessary to melt the ice. The amount of energy gained by the ice is lost by the steam. Compute the entropy change for each process and sum to get the entropy change of the Universe.

SOLUTION

(a) Find the change in entropy of the ice.

Use the latent heat of fusion, L_f, to compute the thermal energy needed to melt 25.0 g of ice:

$$Q_{\text{ice}} = mL_f = (0.025\text{ kg})(3.33 \times 10^5\text{ J}) = 8.33 \times 10^3\text{ J}$$

Calculate the change in entropy of the ice:

$$\Delta S_{\text{ice}} = \frac{Q_{\text{ice}}}{T_{\text{ice}}} = \frac{8.33 \times 10^3\text{ J}}{273\text{ K}} = \boxed{30.5\text{ J/K}}$$

(b) Find the change in entropy of the steam.

By supposition, the thermal energy lost by the steam is equal to the thermal energy gained by the ice:

$$\Delta S_{\text{steam}} = \frac{Q_{\text{steam}}}{T_{\text{steam}}} = \frac{-8.33 \times 10^3\text{ J}}{373\text{ K}} = \boxed{-22.3\text{ J/K}}$$

(c) Find the change in entropy of the Universe.

Sum the two changes in entropy:

$$\Delta S_{\text{universe}} = \Delta S_{\text{ice}} + \Delta S_{\text{steam}} = 30.5\text{ J/k} - 22.3\text{ J/K}$$

$$= \boxed{+\,8.2\text{ J/K}}$$

REMARKS Notice that the entropy of the Universe increases, as it must in all natural processes.

QUESTION 12.15 True or False: For a given magnitude of thermal energy transfer, the change in entropy is smaller for processes that proceed at lower temperature.

EXERCISE 12.15 A 4.00-kg block of ice at 273 K encased in a thin plastic shell of negligible mass melts in a large lake at 293 K. At the instant the ice has completely melted in the shell and is still at 273 K, calculate the change in entropy of (a) the ice, (b) the lake (which essentially remains at 293 K), and (c) the Universe.

ANSWERS (a) 4.88×10^3 J/K (b) -4.55×10^3 J/K (c) $+3.3 \times 10^2$ J/K

■ EXAMPLE 12.16 | A Falling Boulder

GOAL Combine mechanical energy and entropy.

PROBLEM A chunk of rock of mass 1.00×10^3 kg at 293 K falls from a cliff of height 125 m into a large lake, also at 293 K. Find the change in entropy of the lake, assuming all the rock's kinetic energy upon entering the lake converts to thermal energy absorbed by the lake.

STRATEGY Gravitational potential energy when the rock is at the top of the cliff converts to kinetic energy of the rock before it enters the lake, then is transferred to the lake as thermal energy. The change in the lake's temperature is negligible (due to its mass). Divide the mechanical energy of the rock by the temperature of the lake to estimate the lake's change in entropy.

SOLUTION

Calculate the gravitational potential energy associated with the rock at the top of the cliff:

$$PE = mgh = (1.00 \times 10^3 \text{ kg})(9.80 \text{ m/s}^2)(125 \text{ m})$$
$$= 1.23 \times 10^6 \text{ J}$$

This energy is transferred to the lake as thermal energy, resulting in an entropy increase of the lake:

$$\Delta S = \frac{Q}{T} = \frac{1.23 \times 10^6 \text{ J}}{293 \text{ K}} = \boxed{4.20 \times 10^3 \text{ J/K}}$$

REMARKS This example shows how even simple mechanical processes can bring about increases in the Universe's entropy.

QUESTION 12.16 If you carefully remove your physics book from a shelf and place it on the ground, what happens to the entropy of the Universe? Does it increase, decrease, or remain the same? Explain.

EXERCISE 12.16 Estimate the change in entropy of a tree trunk at 15.0°C when a bullet of mass 5.00 g traveling at 1.00×10^3 m/s embeds itself in it. (Assume the kinetic energy of the bullet transforms to thermal energy, all of which is absorbed by the tree.)

ANSWER 8.68 J/K

Entropy and Disorder

A large element of chance is inherent in natural processes. The spacing between trees in a natural forest, for example, is random; if you discovered a forest where all the trees were equally spaced, you would conclude that it had been planted. Likewise, leaves fall to the ground with random arrangements. It would be highly unlikely to find the leaves laid out in perfectly straight rows. We can express the results of such observations by saying that **a disorderly arrangement is much more probable than an orderly one if the laws of nature are allowed to act without interference.**

Entropy originally found its place in thermodynamics, but its importance grew tremendously as the field of statistical mechanics developed. This analytical approach employs an alternate interpretation of entropy. In statistical mechanics, the behavior of a substance is described by the statistical behavior of the atoms and molecules contained in it. One of the main conclusions of the statistical mechanical approach is that **isolated systems tend toward greater disorder, and entropy is a measure of that disorder.**

Tip 12.5 Don't Confuse the *W*'s

The symbol *W* used here is a *probability,* not to be confused with the same symbol used for work.

In light of this new view of entropy, Boltzmann found another method for calculating entropy through use of the relation

$$S = k_B \ln W \qquad [12.18]$$

where $k_B = 1.38 \times 10^{-23}$ J/K is Boltzmann's constant and W is a number proportional to the probability that the system has a particular configuration. The symbol "ln" again stands for natural logarithm, discussed in Appendix A.

Equation 12.18 could be applied to a bag of marbles. Imagine that you have 100 marbles—50 red and 50 green—stored in a bag. You are allowed to draw four marbles from the bag according to the following rules: Draw one marble, record its color, return it to the bag, and draw again. Continue this process until four marbles have been drawn. Note that because each marble is returned to the bag before the next one is drawn, the probability of drawing a red marble is always the same as the probability of drawing a green one.

The results of all possible drawing sequences are shown in Table 12.3. For example, the result RRGR means that a red marble was drawn first, a red one second, a green one third, and a red one fourth. The table indicates that there is only one possible way to draw four red marbles. There are four possible sequences that produce one green and three red marbles, six sequences that produce two green and two red, four sequences that produce three green and one red, and one sequence that produces all green. From Equation 12.18, we see that the state with the greatest disorder (two red and two green marbles) has the highest entropy because it is most probable. In contrast, the most ordered states (all red marbles and all green marbles) are least likely to occur and are states of lowest entropy.

The outcome of the draw can range between these highly ordered (lowest-entropy) and highly disordered (highest-entropy) states. Entropy can be regarded as an index of how far a system has progressed from an ordered to a disordered state.

The second law of thermodynamics is really a statement of what is most probable rather than of what must be. Imagine placing an ice cube in contact with a hot piece of pizza. There is nothing in nature that absolutely forbids the transfer of energy by heat from the ice to the much warmer pizza. Statistically, it's possible for a slow-moving molecule in the ice to collide with a faster-moving molecule in the pizza so that the slow one transfers some of its energy to the faster one. When the great number of molecules present in the ice and pizza are considered, however, the odds are overwhelmingly in favor of the transfer of energy from the faster-moving molecules to the slower-moving molecules. Furthermore, this example demonstrates that a system naturally tends to move from a state of order to a state of disorder. The initial state, in which all the pizza molecules have high kinetic energy and all the ice molecules have lower kinetic energy, is much more ordered than the final state after energy transfer has taken place and the ice has melted.

APPLICATION
The Direction of Time

Even more generally, the second law of thermodynamics defines the direction of time for all events as the direction in which the entropy of the universe increases. Although conservation of energy isn't violated if energy flows spontaneously from a cold object (the ice cube) to a hot object (the pizza slice), that event

Table 12.3 Possible Results of Drawing Four Marbles from a Bag

End Result	Possible Draws	Total Number of Same Results
All R	RRRR	1
1G, 3R	RRRG, RRGR, RGRR, GRRR	4
2G, 2R	RRGG, RGRG, GRRG, RGGR, GRGR, GGRR	6
3G, 1R	GGGR, GGRG, GRGG, RGGG	4
All G	GGGG	1

violates the second law because it represents a spontaneous increase in order. Of course, such an event also violates everyday experience. If the melting ice cube is filmed and the film speeded up, the difference between running the film in forward and reverse directions would be obvious to an audience. The same would be true of filming any event involving a large number of particles, such as a dish dropping to the floor and shattering.

As another example, suppose you were able to measure the velocities of all the air molecules in a room at some instant. It's very unlikely that you would find all molecules moving in the same direction with the same speed; that would be a highly ordered state, indeed. The most probable situation is a system of molecules moving haphazardly in all directions with a wide distribution of speeds, a highly disordered state. This physical situation can be compared to the drawing of marbles from a bag: If a container held 10^{23} molecules of a gas, the probability of finding all the molecules moving in the same direction with the same speed at some instant would be similar to that of drawing a marble from the bag 10^{23} times and getting a red marble on every draw, clearly an unlikely set of events.

a

b

(a) A royal flush is a highly ordered poker hand with a low probability of occurrence. (b) A disordered and worthless poker hand. The probability of this *particular* hand occurring is the same as that of the royal flush. There are so many worthless hands, however, that the probability of being dealt a worthless hand is much higher than that of being dealt a royal flush. Can you calculate the probability of being dealt a full house (a pair and three of a kind) from a standard deck of 52 cards?

The tendency of nature to move toward a state of disorder affects the ability of a system to do work. Consider a ball thrown toward a wall. The ball has kinetic energy, and its state is an ordered one, which means that all the atoms and molecules of the ball move in unison at the same speed and in the same direction (apart from their random internal motions). When the ball hits the wall, however, part of the ball's kinetic energy is transformed into the random, disordered, internal motion of the molecules in the ball and the wall, and the temperatures of the ball and the wall both increase slightly. Before the collision, the ball was capable of doing work. It could drive a nail into the wall, for example. With the transformation of part of the ordered energy into disordered internal energy, this capability of doing work is reduced. The ball rebounds with less kinetic energy than it originally had, because the collision is inelastic.

Various forms of energy can be converted to internal energy, as in the collision between the ball and the wall, but the reverse transformation is never complete. In general, given two kinds of energy, A and B, if A can be completely converted to B and vice versa, we say that A and B are of the *same grade*. However, if A can be completely converted to B and the reverse is never complete, then A is of a *higher grade* of energy than B. In the case of a ball hitting a wall, the kinetic energy of the ball is of a higher grade than the internal energy contained in the ball and the wall after the collision. When high-grade energy is converted to internal energy, it can never be fully recovered as high-grade energy.

This conversion of high-grade energy to internal energy is referred to as the **degradation of energy.** The energy is said to be degraded because it takes on a form that is less useful for doing work. In other words, **in all real processes, the energy available for doing work decreases.**

Finally, note once again that the statement that entropy must increase in all natural processes is true only for isolated systems. There are instances in which the entropy of some system decreases, but with a corresponding net increase in entropy for some other system. When all systems are taken together to form the Universe, **the entropy of the Universe always increases.**

Ultimately, the entropy of the Universe should reach a maximum. When it does, the Universe will be in a state of uniform temperature and density. All physical, chemical, and biological processes will cease, because a state of perfect disorder implies no available energy for doing work. This gloomy state of affairs is sometimes referred to as the ultimate "heat death" of the Universe.

Quick Quiz

12.5 Suppose you are throwing two dice in a friendly game of craps. For any given throw, the two numbers that are faceup can have a sum of 2, 3, 4, 5, 6, 7, 8, 9, 10, 11, or 12. Which outcome is most probable? Which is least probable?

12.6 Human Metabolism BIO

LEARNING OBJECTIVES

1. Define metabolic rate and its relation to the physical rate of oxygen consumption.
2. Analyze the biological impacts of metabolic rate, physical activity and weight gain using thermodynamics.
3. Describe how the laws of thermodynamics can quantify physical fitness and the human body's efficiency.

Animals do work and give off energy by heat, leading us to believe the first law of thermodynamics can be applied to living organisms to describe them in a general way. The internal energy stored in humans goes into other forms needed for maintaining and repairing the major body organs and is transferred out of the body by work as a person walks or lifts a heavy object, and by heat when the body is warmer than its surroundings. Because the rates of change of internal energy, energy loss by heat, and energy loss by work vary widely with the intensity and duration of human activity, it's best to measure the time rates of change of ΔU, Q, and W. Rewriting the first law, these time rates of change are related by

$$\frac{\Delta U}{\Delta t} = \frac{Q}{\Delta t} + \frac{W}{\Delta t} \qquad [12.19]$$

On average, energy Q flows *out* of the body, and work is done *by* the body on its surroundings, so both $Q/\Delta t$ and $W/\Delta t$ are negative. This means that $\Delta U/\Delta t$ would be negative and the internal energy and body temperature would decrease with time if a human were a closed system with no way of ingesting matter or replenishing internal energy stores. Because all animals are actually open systems, they acquire internal energy (chemical potential energy) by eating and breathing, so their internal energy and temperature are kept constant. Overall, the energy from the oxidation of food ultimately supplies the work done by the body and energy lost from the body by heat, and this is the interpretation we give Equation 12.19. That is, $\Delta U/\Delta t$ is the rate at which internal energy is added to our bodies by food, and this term just balances the rate of energy loss by heat, $Q/\Delta t$, and by work, $W/\Delta t$. Finally, if we have a way of measuring $\Delta U/\Delta t$ and $W/\Delta t$ for a human, we can calculate $Q/\Delta t$ from Equation 12.19 and gain useful information on the efficiency of the body as a machine.

Measuring the Metabolic Rate $\Delta U/\Delta t$

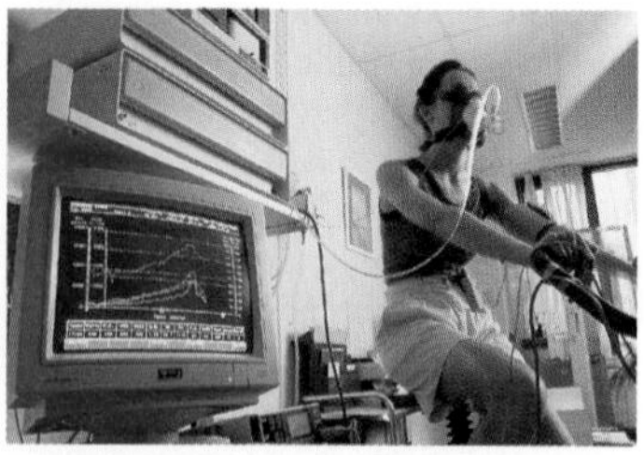

Figure 12.19 This bike rider is being monitored for oxygen consumption.

The value of $W/\Delta t$, the work done by a person per unit time, can easily be determined by measuring the power output supplied by the person (in pedaling a bike, for example). **The metabolic rate $\Delta U/\Delta t$ is the rate at which chemical potential energy in food and oxygen are transformed into internal energy to just balance the body losses of internal energy by work and heat.** Although the mechanisms of food oxidation and energy release in the body are complicated, involving many intermediate reactions and enzymes (organic compounds that speed up the chemical reactions taking place at "low" body temperatures), an amazingly simple rule summarizes these processes: **The metabolic rate is directly proportional to the rate of oxygen consumption by volume.** It is found that for an average diet, the consumption of one liter of oxygen releases 4.8 kcal, or 20 kJ, of energy. We may write this important summary rule as

Metabolic rate equation ▶

$$\frac{\Delta U}{\Delta t} = 4.8\frac{\Delta V_{O_2}}{\Delta t} \qquad [12.20]$$

where the metabolic rate $\Delta U/\Delta t$ is measured in kcal/s and $\Delta V_{O_2}/\Delta t$, the volume rate of oxygen consumption, is in L/s. Measuring the rate of oxygen consumption during

different activities for several hours. These values were obtained by measuring the power output and simultaneous oxygen consumption of mine workers and calculating the metabolic rate from their oxygen consumption. The table shows that a person can steadily supply mechanical power for several hours at about 100 W with an efficiency of about 17%. It also shows the dependence of efficiency on activity, and that e can drop to values as low as 3% for highly inefficient activities like shoveling, which involves many starts and stops. Finally, it is interesting in comparison to the average results of Table 12.6 that a superbly conditioned athlete, efficiently coupled to a mechanical device for extracting power (a bike!), can supply a power of around 300 W for about 30 minutes at a peak efficiency of 22%.

SUMMARY

12.1 Work in Thermodynamic Processes

The work done on a gas at a constant pressure is

$$W = -P\,\Delta V \quad [12.1]$$

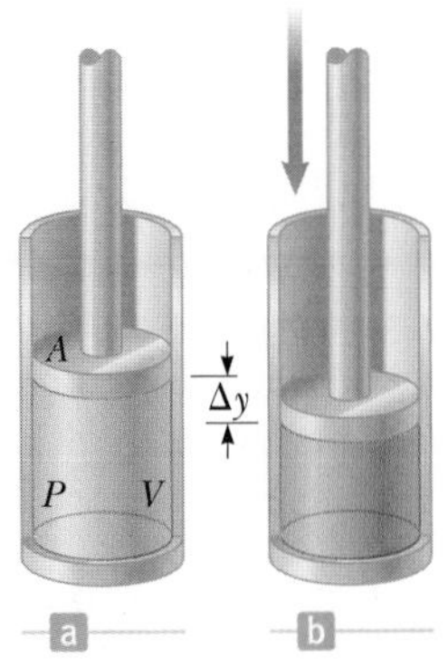

Positive work is done on a gas by compressing it.

The work done on the gas is positive if the gas is compressed (ΔV is negative) and negative if the gas expands (ΔV is positive). In general, the work done on a gas that takes it from some initial state to some final state is the negative of the area under the curve on a PV diagram.

12.2 The First Law of Thermodynamics

According to the first law of thermodynamics, when a system undergoes a change from one state to another, the **change in its internal energy** ΔU is

$$\Delta U = U_f - U_i = Q + W \quad [12.2]$$

where Q is the energy exchanged across the boundary between the system and the environment and W is the work done on the system. The quantity Q is positive when energy is transferred into the system by heating and negative when energy is removed from the system by cooling. W is positive when work is done on the system (for example, by compression) and negative when the system does positive work on its environment.

The change of the internal energy, ΔU, of an ideal gas is given by

$$\Delta U = nC_v\Delta T \quad [12.5]$$

where C_v is the molar specific heat at constant volume.

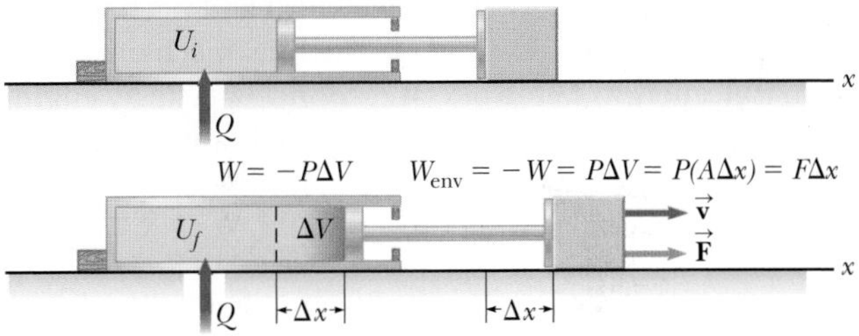

Illustration of the first law of thermodynamics.

12.3 Thermal Processes

An **isobaric process** is one that occurs at constant pressure. The work done on the system in such a process is $-P\,\Delta V$, whereas the thermal energy transferred by heat is given by

$$Q = nC_p\,\Delta T \quad [12.6]$$

with the molar heat capacity at constant pressure given by $C_p = C_v + R$.

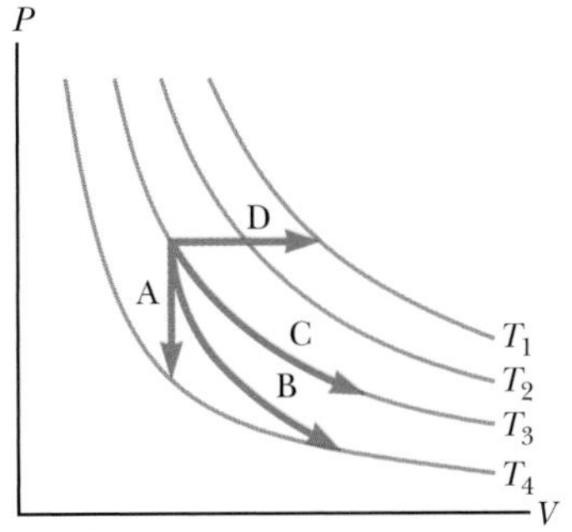

Four gas processes: A is an isochoric process (constant volume); B is an adiabatic expansion (no thermal energy transfer); C is an isothermal process (constant temperature); D is an isobaric process (constant pressure).

In an **adiabatic process** no energy is transferred by heat between the system and its surroundings ($Q = 0$). In this case the first law gives $\Delta U = W$, which means the internal energy changes solely as a consequence of work being done on the system. The pressure and volume in adiabatic processes are related by

$$PV^{\gamma} = \text{constant} \quad [12.8a]$$

where $\gamma = C_p/C_v$ is the adiabatic index.

In an **isovolumetric process** the volume doesn't change and no work is done. For such processes, the first law gives $\Delta U = Q$.

An **isothermal process** occurs at constant temperature. The work done by an ideal gas on the environment is

$$W_{\text{env}} = nRT \ln\left(\frac{V_f}{V_i}\right) \qquad [12.10]$$

12.4 Heat Engines and the Second Law of Thermodynamics

In a **cyclic process** (in which the system returns to its initial state), $\Delta U = 0$ and therefore $Q = W_{\text{eng}}$, meaning the energy transferred into the system by heat equals the work done on the system during the cycle.

A **heat engine** takes in energy by heat and partially converts it to other forms of energy, such as mechanical and electrical energy. The work W_{eng} done by a heat engine in carrying a working substance through a cyclic process ($\Delta U = 0$) is

$$W_{\text{eng}} = |Q_h| - |Q_c| \qquad [12.11]$$

where Q_h is the energy absorbed from a hot reservoir and Q_c is the energy expelled to a cold reservoir.

The **thermal efficiency** of a heat engine is defined as the ratio of the work done by the engine to the energy transferred into the engine per cycle:

$$e \equiv \frac{W_{\text{eng}}}{|Q_h|} = 1 - \frac{|Q_c|}{|Q_h|} \qquad [12.12]$$

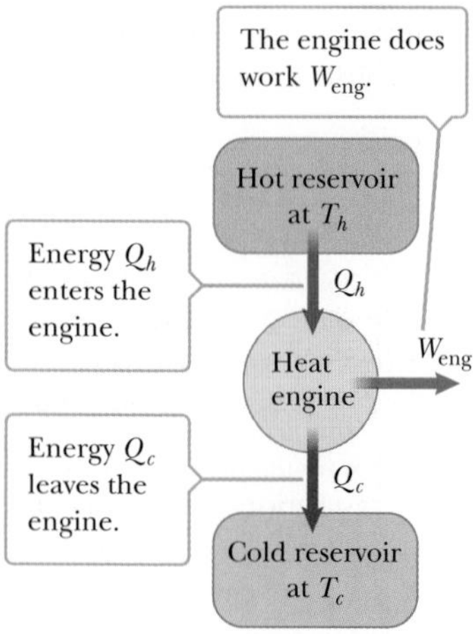

Schematic diagram of a heat engine.

Heat pumps are heat engines in reverse. In a refrigerator the heat pump removes thermal energy from inside the refrigerator. Heat pumps operating in cooling mode have coefficient of performance given by

$$\text{COP(cooling mode)} = \frac{|Q_c|}{W} \qquad [12.13]$$

A heat pump in heating mode has coefficient of performance

$$\text{COP(heating mode)} = \frac{|Q_h|}{W} \qquad [12.14]$$

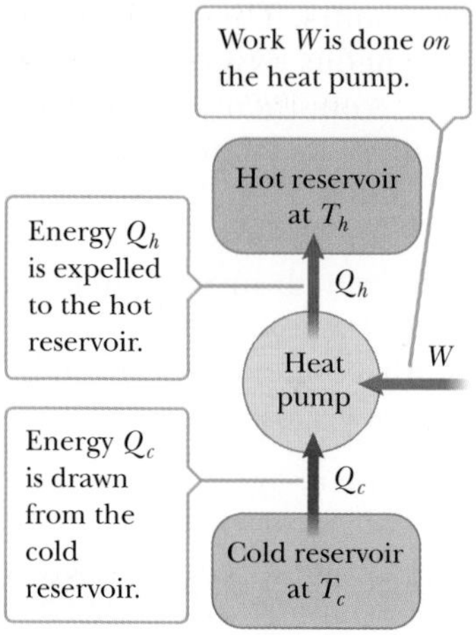

Schematic diagram of a heat pump.

Real processes proceed in an order governed by the **second law of thermodynamics,** which can be stated in two ways:

1. Energy will not flow spontaneously by heat from a cold object to a hot object.
2. No heat engine operating in a cycle can absorb energy from a reservoir and perform an equal amount of work.

No real heat engine operating between the Kelvin temperatures T_h and T_c can exceed the efficiency of an engine operating between the same two temperatures in a **Carnot cycle,** given by

$$e_C = 1 - \frac{T_c}{T_h} \qquad [12.16]$$

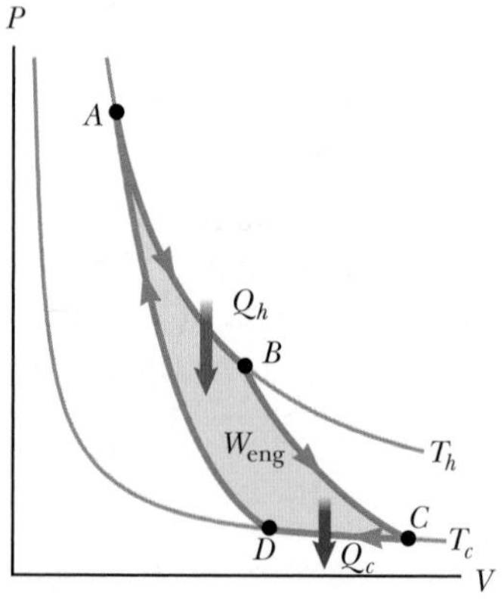

PV diagram of a Carnot cycle.

Perfect efficiency of a Carnot engine requires a cold reservoir of 0 K, absolute zero. According to the **third law of thermodynamics,** however, it is impossible to lower the temperature of a system to absolute zero in a finite number of steps.

12.5 Entropy

The second law can also be stated in terms of a quantity called **entropy** (S). The **change in entropy** of a system is equal to the energy Q_r flowing by heat into (or out of) the system as the system changes from one state to another by a reversible process, divided by the absolute temperature:

$$\Delta S \equiv \frac{Q_r}{T} \qquad [12.17]$$

One of the primary findings of statistical mechanics is that systems tend toward disorder, and entropy is a measure of that disorder. An alternate statement of the second law is that the entropy of the Universe increases in all natural processes.

WARM-UP EXERCISES

ENHANCED WebAssign The warm-up exercises in this chapter may be assigned online in Enhanced WebAssign.

1. **Math Review** For each of the following functions, graph pressure (P) versus volume (V) for $1.00\ \text{m}^3 \leq V \leq 3.00\ \text{m}^3$ on the same set of axes. Use units of 10^5 Pa for pressure and m^3 for volume. (See also Section 12.3.) (a) $P = (3.00 \times 10^5\ \text{Pa})$ (b) $P = (3.00 \times 10^5\ \text{Pa} \cdot \text{m}^3)\ V^{-1}$ (c) $P = (3.00 \times 10^5\ \text{Pa} \cdot \text{m}^{5/3})\ V^{-5/3}$

2. **Math Review** Let $W = (2.30 \times 10^5\ \text{J}) \ln (V_f/V_i)$. Solve for the unknown quantity in each of the following cases: (a) the final volume V_f is twice the initial volume V_i (b) the initial volume is $1.00\ \text{m}^3$ and the work W is 2.50×10^5 J. (See also Section 12.3.)

3. **Physics Review** The specific heat of steam at atmospheric pressure is 2 010 J/kg · °C. Evaluate the energy required to raise the temperature of 2.50 kg of steam from 105°C to 120°C. (See Section 11.2.)

4. **Physics Review** An ideal gas has initial volume of $0.400\ \text{m}^3$ and pressure of 9.60×10^4 Pa. (a) If the initial temperature is 282 K, find the number of moles of gas in the system. (b) If the gas is heated at constant volume to 382 K, what is the final pressure? (See Section 10.4.)

5. **Physics Review** (a) Calculate the internal energy of a 2.70 moles of a monatomic gas at a temperature of 0°C. (b) By how much does the internal energy change if the gas is heated to 425 K? (See Section 10.5.)

6. A monatomic ideal gas expands from $1.00\ \text{m}^3$ to $2.50\ \text{m}^3$ at a constant pressure of 2.00×10^5 Pa. Find (a) the work done on the gas, (see Section 12.1), (b) the thermal energy Q transferred into the gas by heat (see Section 12.3, subsection "Isobaric Processes"), and (c) the change in the internal energy of the gas. (See Section 12.2.)

7. A 2.00-mole ideal gas system is maintained at a constant volume of 4.00 liters. If 1.00×10^2 J of thermal energy is transferred to the system, find (a) the work done on the gas, (b) the change in the internal energy of the system, (c) the change in temperature of the gas in kelvin, if the gas is monatomic, and (d) the change in temperature if the gas is diatomic. (See Section 12.3, subsection "Isovolumetric Processes".)

8. How much net work is done by the gas undergoing the cyclic process illustrated in (a) Figure WU12.8a, (b) Figure WU12.8b, and (c) Figure WU12.8c? Round your answer to two significant figures. (See Sections 12.1 and 12.4.)

9. A diatomic ideal gas expands adiabatically from a volume of $1.00\ \text{m}^3$ to a final volume of $3.50\ \text{m}^3$. If the initial pressure is 1.00×10^5 Pa, find (a) the adiabatic index of the gas (see Sections 12.2 and 12.3), and (b) the final pressure. (See Section 12.3, subsection "Adiabatic Processes".)

10. An ideal gas drives a piston as it expands isothermally from $1.00\ \text{m}^3$ to $2.00\ \text{m}^3$ at 850.0 K. If there are 3.90×10^2 moles of gas in the piston, (a) what is the change in the internal energy of the gas? (b) How much work does the gas do in displacing the piston? (c) How much thermal energy is transferred by heat? (See Section 12.3, subsection "Isothermal Processes".)

11. An engine does 15.0 kJ of work while absorbing 75.0 kJ from the hot reservoir. Calculate (a) the efficiency of the engine and (b) the energy it transfers to the cold reservoir. (See Section 12.4.)

12. A refrigerator does 18.0 kJ of work while moving 115 kJ of thermal energy from inside the refrigerator. Calculate (a) the refrigerator's coefficient of performance and (b) the energy it transfers to its environment. (See Section 12.4.)

13. A steam turbine operates at a boiler temperature of 450.0 K and an exhaust temperature of 300.0 K. (a) What is the maximum theoretical efficiency of this system? (b) If the system operates at maximum efficiency

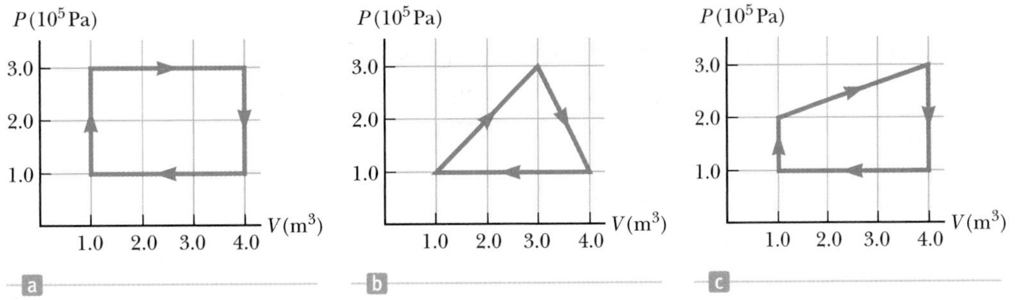

Figure WU12.8

and the boiler expels 10.0 kJ of energy to the cold reservoir, how much energy does it absorb from the hot reservoir? (See Section 12.4.)

14. Distillers purify water by boiling contaminated water into purified steam and condensing the steam into a separate container. Suppose a distiller boils 10.0 kg of liquid water at 373.15 K into steam at the same temperature. (a) Determine the amount of thermal energy added to the liquid water. (See Section 11.4.) (b) Evaluate the water's change in entropy. (See Section 12.5.)

CONCEPTUAL QUESTIONS

ENHANCED WebAssign The conceptual questions in this chapter may be assigned online in Enhanced WebAssign.

1. What are some factors that affect the efficiency of automobile engines?

2. If you shake a jar full of jelly beans of different sizes, the larger beans tend to appear near the top and the smaller ones tend to fall to the bottom. (a) Why does that occur? (b) Does this process violate the second law of thermodynamics?

3. BIO Consider the human body performing a strenuous exercise, such as lifting weights or riding a bicycle. Work is being done by the body, and energy is leaving by conduction from the skin into the surrounding air. According to the first law of thermodynamics, the temperature of the body should be steadily decreasing during the exercise. That isn't what happens, however. Is the first law invalid for this situation? Explain.

4. Clearly distinguish among temperature, heat, and internal energy.

5. For an ideal gas in an isothermal process, there is no change in internal energy. Suppose the gas does work W during such a process. How much energy is transferred by heat?

6. A steam-driven turbine is one major component of an electric power plant. Why is it advantageous to increase the temperature of the steam as much as possible?

7. Is it possible to construct a heat engine that creates no thermal pollution?

8. In solar ponds constructed in Israel, the Sun's energy is concentrated near the bottom of a salty pond. With the proper layering of salt in the water, convection is prevented and temperatures of 100°C may be reached. Can you guess the maximum efficiency with which useful mechanical work can be extracted from the pond?

9. When a sealed Thermos bottle full of hot coffee is shaken, what changes, if any, take place in (a) the temperature of the coffee and (b) its internal energy?

10. Give some examples of irreversible processes that occur in nature. Give an example of a process in nature that is nearly reversible.

11. The first law of thermodynamics says we can't get more out of a process than we put in, but the second law says that we can't break even. Explain this statement.

12. If a supersaturated sugar solution is allowed to evaporate slowly, sugar crystals form in the container. Hence, sugar molecules go from a disordered form (in solution) to a highly ordered, crystalline form. Does this process violate the second law of thermodynamics? Explain.

13. Using the first law of thermodynamics, explain why the *total* energy of an isolated system is always constant.

14. What is wrong with the following statement: "Given any two bodies, the one with the higher temperature contains more heat."

15. An ideal gas is compressed to half its initial volume by means of several possible processes. Which of the following processes results in the most work done on the gas? (a) isothermal (b) adiabatic (c) isobaric (d) The work done is independent of the process.

16. A thermodynamic process occurs in which the entropy of a system changes by -6 J/K. According to the second law of thermodynamics, what can you conclude about the entropy change of the environment? (a) It must be $+6$ J/K or less. (b) It must be equal to 6 J/K. (c) It must be between $+6$ J/K and 0. (d) It must be 0. (e) It must be $+6$ J/K or more.

17. A window air conditioner is placed on a table inside a well-insulated apartment, plugged in and turned on. What happens to the average temperature of the apartment? (a) It increases. (b) It decreases. (c) It remains constant. (d) It increases until the unit warms up and then decreases. (e) The answer depends on the initial temperature of the apartment.

PROBLEMS

ENHANCED WebAssign The problems in this chapter may be assigned online in Enhanced WebAssign.

1. denotes straightforward problem; **2.** denotes intermediate problem;
3. denotes challenging problem
1. denotes full solution available in *Student Solutions Manual/Study Guide*
1. denotes problems most often assigned in Enhanced WebAssign

BIO denotes biomedical problems
GP denotes guided problems
M denotes Master It tutorial available in Enhanced WebAssign
Q|C denotes asking for quantitative and conceptual reasoning
S denotes symbolic reasoning problem
W denotes Watch It video solution available in Enhanced WebAssign

12.1 Work in Thermodynamic Processes

1. QIC An ideal gas is enclosed in a cylinder with a movable piston on top of it. The piston has a mass of 8 000 g and an area of 5.00 cm^2 and is free to slide up and down, keeping the pressure of the gas constant. (a) How much work is done on the gas as the temperature of 0.200 mol of the gas is raised from 20.0°C to 300°C? (b) What does the sign of your answer to part (a) indicate?

2. Sketch a *PV* diagram and find the work done by the gas during the following stages. (a) A gas is expanded from a volume of 1.0 L to 3.0 L at a constant pressure of 3.0 atm. (b) The gas is then cooled at constant volume until the pressure falls to 2.0 atm. (c) The gas is then compressed at a constant pressure of 2.0 atm from a volume of 3.0 L to 1.0 L. *Note:* Be careful of signs. (d) The gas is heated until its pressure increases from 2.0 atm to 3.0 atm at a constant volume. (e) Find the net work done during the complete cycle.

3. M Gas in a container is at a pressure of 1.5 atm and a volume of 4.0 m^3. What is the work done *on* the gas (a) if it expands at constant pressure to twice its initial volume, and (b) if it is compressed at constant pressure to one-quarter its initial volume?

4. A 40.0-g projectile is launched by the expansion of hot gas in an arrangement shown in Figure P12.4a. The cross-sectional area of the launch tube is 1.0 cm^2, and the length that the projectile travels down the tube after starting from rest is 32 cm. As the gas expands, the pressure varies as shown in Figure P12.4b. The values for the initial pressure and volume are $P_i = 11 \times 10^5$ Pa and $V_i = 8.0$ cm^3 while the final values are $P_f = 1.0 \times 10^5$ Pa and $V_f = 40.0$ cm^3. Friction between the projectile and the launch tube is negligible. (a) If the projectile is launched into a vacuum, what is the speed of the projectile as it leaves the launch tube? (b) If instead the projectile is launched into air at a pressure of 1.0×10^5 Pa, what fraction of the work done by the expanding gas in the tube is spent by the projectile pushing air out of the way as it proceeds down the tube?

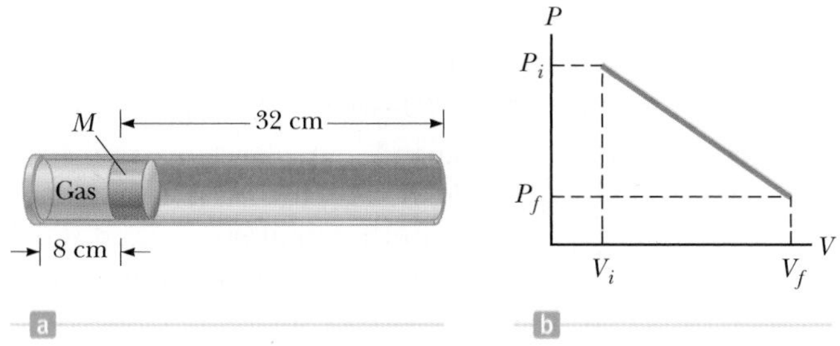

Figure P12.4

5. A gas expands from *I* to *F* along the three paths indicated in Figure P12.5. Calculate the work done *on* the gas along paths (a) *IAF*, (b) *IF*, and (c) *IBF*.

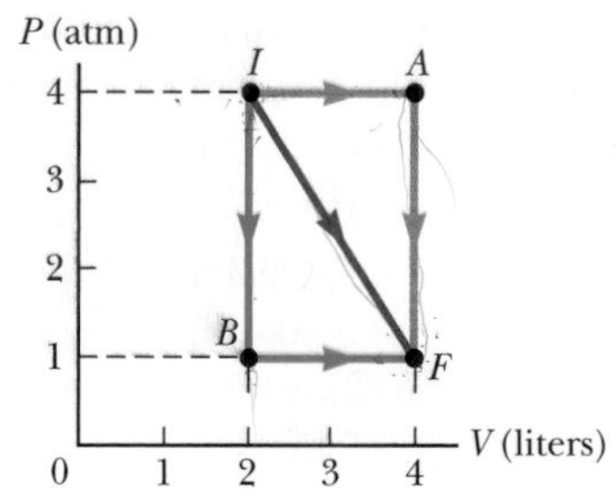

Figure P12.5 Problems 5 and 13.

6. Sketch a *PV* diagram of the following processes: (a) A gas expands at constant pressure P_1 from volume V_1 to volume V_2. It is then kept at constant volume while the pressure is reduced to P_2. (b) A gas is reduced in pressure from P_1 to P_2 while its volume is held constant at V_1. It is then expanded at constant pressure P_2 to a final volume V_2. (c) In which of the processes is more work done *by* the gas? Why?

7. A sample of helium behaves as an ideal gas as it is heated at constant pressure from 273 K to 373 K. If 20.0 J of work is done by the gas during this process, what is the mass of helium present?

8. (a) Find the work done *by* an ideal gas as it expands from point *A* to point *B* along the path shown in Figure P12.8. (b) How much work is done by the gas if it compressed from *B* to *A* along the same path?

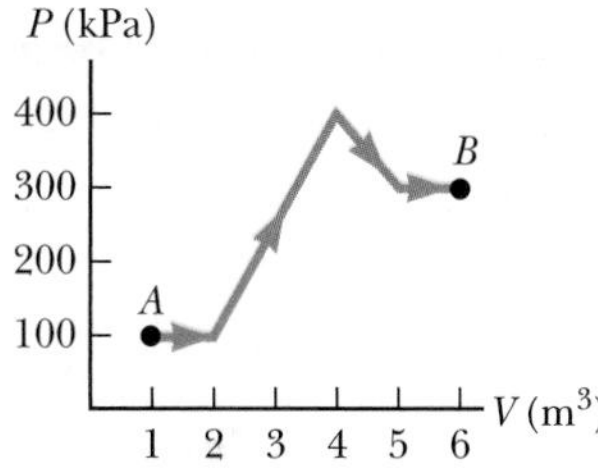

Figure P12.8

9. W One mole of an ideal gas initially at a temperature of $T_i = 0$°C undergoes an expansion at a constant pressure of 1.00 atm to four times its original volume. (a) Calculate the new temperature T_f of the gas. (b) Calculate the work done *on* the gas during the expansion.

10. (a) Determine the work done *on* a fluid that expands from *i* to *f* as indicated in Figure P12.10. (b) How much work is done *on* the fluid if it is compressed from *f* to *i* along the same path?

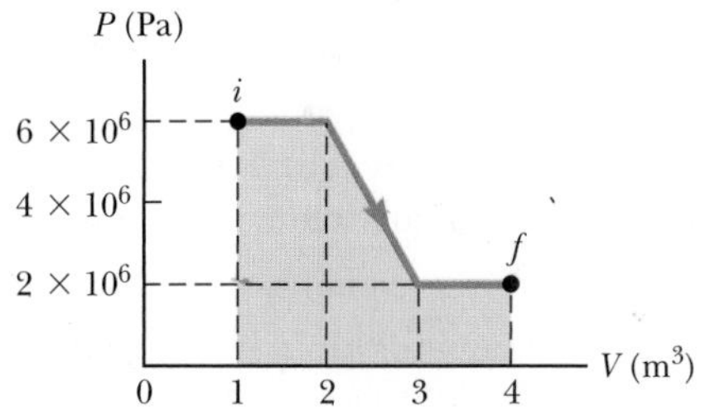

Figure P12.10

12.2 The First Law of Thermodynamics

12.3 Thermal Processes

11. S The only form of energy possessed by molecules of a monatomic ideal gas is translational kinetic energy.

Using the results from the discussion of kinetic theory in Section 10.5, show that the internal energy of a monatomic ideal gas at pressure P and occupying volume V may be written as $U = \frac{3}{2}PV$.

12. GP A cylinder of volume 0.300 m^3 contains 10.0 mol of neon gas at 20.0°C. Assume neon behaves as an ideal gas. (a) What is the pressure of the gas? (b) Find the internal energy of the gas. (c) Suppose the gas expands at constant pressure to a volume of 1.000 m^3. How much work is done on the gas? (d) What is the temperature of the gas at the new volume? (e) Find the internal energy of the gas when its volume is 1.000 m^3. (f) Compute the change in the internal energy during the expansion. (g) Compute $\Delta U - W$. (h) Must thermal energy be transferred to the gas during the constant pressure expansion or be taken away? (i) Compute Q, the thermal energy transfer. (j) What symbolic relationship between Q, ΔU, and W is suggested by the values obtained?

13. M A gas expands from I to F in Figure P12.5. The energy added to the gas by heat is 418 J when the gas goes from I to F along the diagonal path. (a) What is the change in internal energy of the gas? (b) How much energy must be added to the gas by heat for the indirect path IAF to give the same change in internal energy?

14. Q|C In a running event, a sprinter does 4.8×10^5 J of work and her internal energy decreases by 7.5×10^5 J. (a) Determine the heat transferred between her body and surroundings during this event. (b) What does the sign of your answer to part (a) indicate?

15. W A gas is compressed at a constant pressure of 0.800 atm from 9.00 L to 2.00 L. In the process, 400 J of energy leaves the gas by heat. (a) What is the work done *on* the gas? (b) What is the change in its internal energy?

16. A quantity of a monatomic ideal gas undergoes a process in which both its pressure and volume are doubled as shown in Figure P12.16. What is the energy absorbed by heat into the gas during this process? *Hint:* The internal energy of a monatomic ideal gas at pressure P and occupying volume V is given by $U = \frac{3}{2}PV$.

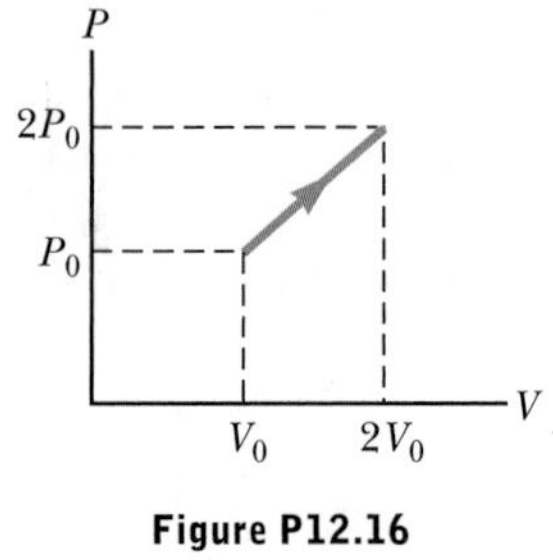

Figure P12.16

17. A gas is enclosed in a container fitted with a piston of cross-sectional area 0.150 m^2. The pressure of the gas is maintained at 6 000 Pa as the piston moves inward 20.0 cm. (a) Calculate the work done by the gas. (b) If the internal energy of the gas decreases by 8.00 J, find the amount of energy removed from the system by heat during the compression.

18. A monatomic ideal gas undergoes the thermodynamic process shown in the PV diagram of Figure P12.18. Determine whether each of the values ΔU, Q, and W for the gas is positive, negative, or zero. *Hint:* The internal energy of a monatomic ideal gas at pressure P and occupying volume V is given by $U = \frac{3}{2}PV$.

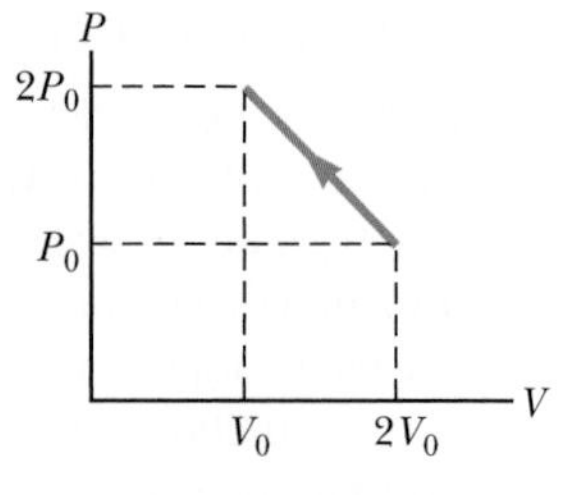

Figure P12.18

19. Q|C An ideal gas is compressed from a volume of V_i = 5.00 L to a volume of V_f = 3.00 L while in thermal contact with a heat reservoir at T = 295 K as in Figure P12.19. During the compression process, the piston moves down a distance of d = 0.130 m under the action of an average external force of F = 25.0 kN. Find (a) the work done on the gas, (b) the change in internal energy of the gas, and (c) the thermal energy exchanged between the gas and the reservoir. (d) If the gas is thermally insulated so no thermal energy could be exchanged, what would happen to the temperature of the gas during the compression?

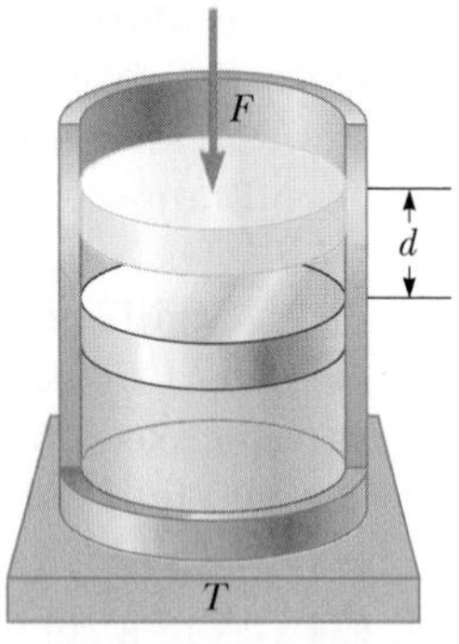

Figure P12.19

20. A system consisting of 0.025 6 moles of a diatomic ideal gas is taken from state A to state C along the path in Figure P12.20. (a) How much work is done *on* the gas during this process? (b) What is the lowest temperature of the gas during this process, and where does it occur? (c) Find the change in internal energy of the gas and (d) the energy delivered to the gas in going from A to C. *Hint:* For part (c), adapt the equation in the remarks of Example 12.9 to a diatomic ideal gas.

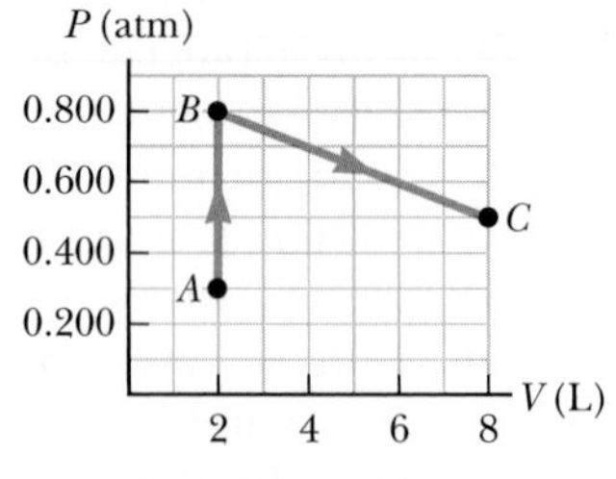

Figure P12.20

21. An ideal monatomic gas expands isothermally from 0.500 m^3 to 1.25 m^3 at a constant temperature of 675 K. If the initial pressure is 1.00×10^5 Pa, find (a) the work done on the gas, (b) the thermal energy transfer Q, and (c) the change in the internal energy.

22. S An ideal gas expands at constant pressure. (a) Show that $P\Delta V = nR\Delta T$. (b) If the gas is monatomic, start from the definition of internal energy and show that $\Delta U = \frac{3}{2}W_{env}$, where W_{env} is the work done by the gas on its environment. (c) For the same monatomic ideal gas, show with the first law that $Q = \frac{5}{2}W_{env}$. (d) Is it possible for an ideal gas to expand at constant pressure while exhausting thermal energy? Explain.

23. GP An ideal monatomic gas is contained in a vessel of *constant* volume 0.200 m^3. The initial temperature

and pressure of the gas are 300 K and 5.00 atm, respectively. The goal of this problem is to find the temperature and pressure of the gas after 16.0 kJ of thermal energy is supplied to the gas. (a) Use the ideal gas law and initial conditions to calculate the number of moles of gas in the vessel. (b) Find the specific heat of the gas. (c) What is the work done by the gas during this process? (d) Use the first law of thermodynamics to find the change in internal energy of the gas. (e) Find the change in temperature of the gas. (f) Calculate the final temperature of the gas. (g) Use the ideal gas expression to find the final pressure of the gas.

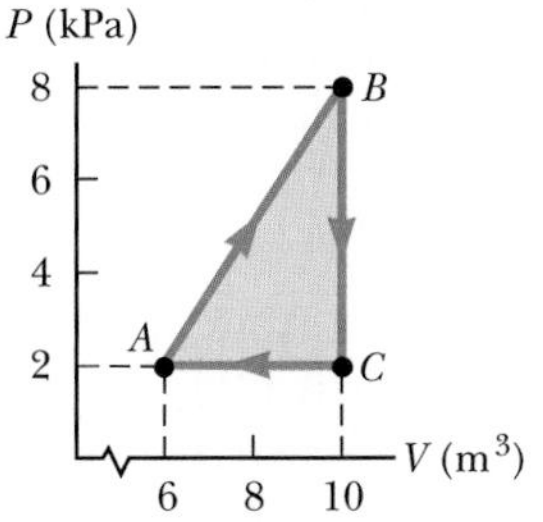

Figure P12.24

24. Consider the cyclic process described by Figure P12.24. If Q is negative for the process BC and ΔU is negative for the process CA, determine the signs of Q, W, and ΔU associated with each process.

25. A 5.0-kg block of aluminum is heated from 20°C to 90°C at atmospheric pressure. Find (a) the work done by the aluminum, (b) the amount of energy transferred to it by heat, and (c) the increase in its internal energy.

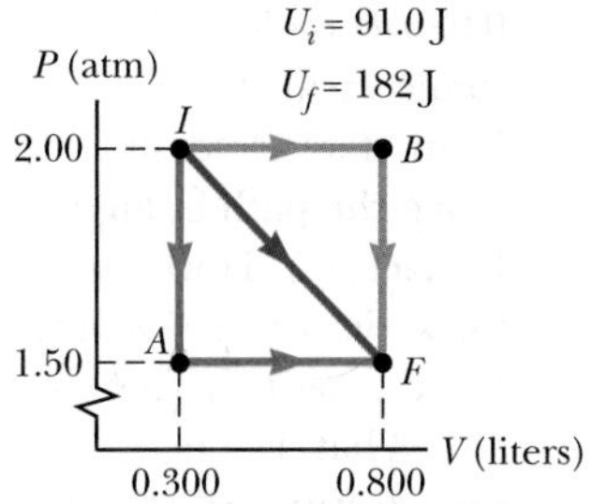

Figure P12.26

26. One mole of gas initially at a pressure of 2.00 atm and a volume of 0.300 L has an internal energy equal to 91.0 J. In its final state, the gas is at a pressure of 1.50 atm and a volume of 0.800 L, and its internal energy equals 182 J. For the paths IAF, IBF, and IF in Figure P12.26, calculate (a) the work done *on* the gas and (b) the net energy transferred to the gas by heat in the process.

27. **S** Consider the Universe to be an adiabatic expansion of atomic hydrogen gas. (a) Use the ideal gas law and Equation 12.8a to show that $TV^{\gamma-1} = C$, where C is a constant. (b) The current Universe extends at least 15 billion light-years in all directions (1.4×10^{26} m), and the current temperature of the Universe is 2.7 K. Estimate the temperature of the Universe when it was the size of a nutshell, with a radius of 2 cm. (For this calculation, assume the Universe is spherical.)

28. Suppose the Universe is considered to be an ideal gas of hydrogen atoms expanding adiabatically. (a) If the density of the gas in the Universe is one hydrogen atom per cubic meter, calculate the number of moles per unit volume (n/V). (b) Calculate the pressure of the Universe, taking the temperature of the Universe as 2.7 K. (c) If the current radius of the Universe is 15 billion light-years (1.4×10^{26} m), find the pressure of the Universe when it was the size of a nutshell, with radius 2.0×10^{-2} m. (*Be careful:* Calculator overflow can occur.)

12.4 Heat Engines and the Second Law of Thermodynamics

29. A gas increases in pressure from 2.00 atm to 6.00 atm at a constant volume of 1.00 m³ and then expands at constant pressure to a volume of 3.00 m³ before returning to its initial state as shown in Figure P12.29. How much work is done in one cycle?

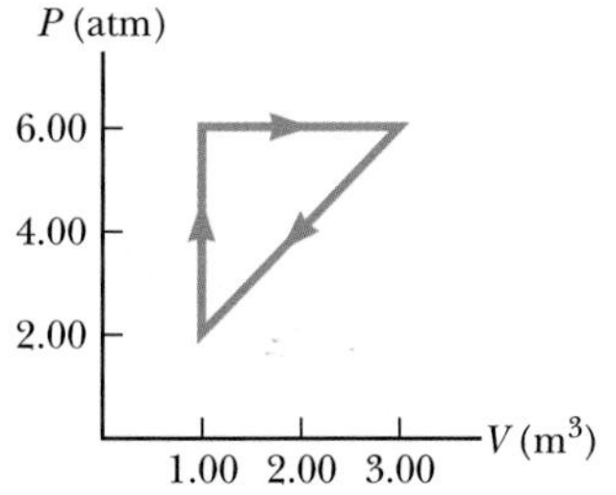

Figure P12.29

30. An ideal gas expands at a constant pressure of 6.00×10^5 Pa from a volume of 1.00 m³ to a volume of 4.00 m³ and then is compressed to one-third that pressure and a volume of 2.50 m³ as shown in Figure P12.30 before returning to its initial state. How much work is done in taking a gas through one cycle of the process shown in the figure?

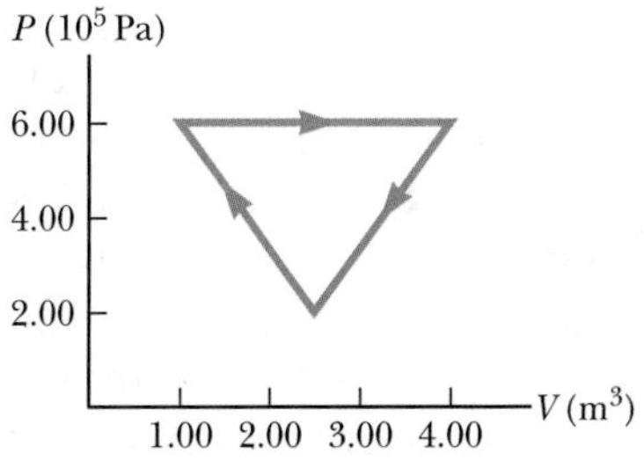

Figure P12.30

31. A heat engine operates between a reservoir at 25°C and one at 375°C. What is the maximum efficiency possible for this engine?

32. **QIC** A heat engine is being designed to have a Carnot efficiency of 65% when operating between two heat reservoirs. (a) If the temperature of the cold reservoir is 20°C, what must be the temperature of the hot reservoir? (b) Can the actual efficiency of the engine be equal to 65%? Explain.

33. The work done by an engine equals one-fourth the energy it absorbs from a reservoir. (a) What is its thermal efficiency? (b) What fraction of the energy absorbed is expelled to the cold reservoir?

34. In each cycle of its operation, a heat engine expels 2 400 J of energy and performs 1 800 J of mechanical work. (a) How much thermal energy must be added to the engine in each cycle? (b) Find the thermal efficiency of the engine.

35. One of the most efficient engines ever built is a coal-fired steam turbine engine in the Ohio River valley,

driving an electric generator as it operates between 1 870°C and 430°C. (a) What is its maximum theoretical efficiency? (b) Its actual efficiency is 42.0%. How much mechanical power does the engine deliver if it absorbs 1.40×10^5 J of energy each second from the hot reservoir.

36. A gun is a heat engine. In particular, it is an internal combustion piston engine that does not operate in a cycle, but comes apart during its adiabatic expansion process. A certain gun consists of 1.80 kg of iron. It fires one 2.40-g bullet at 320 m/s with an energy efficiency of 1.10%. Assume the body of the gun absorbs all the energy exhaust and increases uniformly in temperature for a short time before it loses any energy by heat into the environment. Find its temperature increase.

37. An engine absorbs 1.70 kJ from a hot reservoir at 277°C and expels 1.20 kJ to a cold reservoir at 27°C in each cycle. (a) What is the engine's efficiency? (b) How much work is done by the engine in each cycle? (c) What is the power output of the engine if each cycle lasts 0.300 s?

38. A heat pump has a coefficient of performance of 3.80 and operates with a power consumption of 7.03×10^3 W. (a) How much energy does the heat pump deliver into a home during 8.00 h of continuous operation? (b) How much energy does it extract from the outside air in 8.00 h?

39. A freezer has a coefficient of performance of 6.30. The freezer is advertised as using 457 kW-h/y. (a) On average, how much energy does the freezer use in a single day? (b) On average, how much thermal energy is removed from the freezer each day? (c) What maximum mass of water at 20.0°C could the freezer freeze in a single day? *Note:* One kilowatt-hour (kW-h) is an amount of energy equal to operating a 1-kW appliance for one hour.

40. Q|C Suppose an ideal (Carnot) heat pump could be constructed. (a) Using Equation 12.15, obtain an expression for the coefficient of performance for such a heat pump in terms of T_h and T_c. (b) Would such a heat pump work better if the difference in the operating temperatures were greater or smaller? (c) Compute the coefficient of performance for such a heat pump if the cold reservoir is 50.0°C and indoor temperature is 70.0°C.

41. In one cycle a heat engine absorbs 500 J from a high-temperature reservoir and expels 300 J to a low-temperature reservoir. If the efficiency of this engine is 60% of the efficiency of a Carnot engine, what is the ratio of the low temperature to the high temperature in the Carnot engine?

42. Q|C A power plant has been proposed that would make use of the temperature gradient in the ocean. The system is to operate between 20.0°C (surface water temperature) and 5.00°C (water temperature at a depth of about 1 km). (a) What is the maximum efficiency of such a system? (b) If the useful power output of the plant is 75.0 MW, how much energy is absorbed per hour? (c) In view of your answer to part (a), do you think such a system is worthwhile (considering that there is no charge for fuel)?

43. A certain nuclear power plant has an electrical power output of 435 MW. The rate at which energy must be supplied to the plant is 1 420 MW. (a) What is the thermal efficiency of the power plant? (b) At what rate is thermal energy expelled by the plant?

44. M A heat engine operates in a Carnot cycle between 80.0°C and 350°C. It absorbs 21 000 J of energy per cycle from the hot reservoir. The duration of each cycle is 1.00 s. (a) What is the mechanical power output of this engine? (b) How much energy does it expel in each cycle by heat?

12.5 Entropy

45. A Styrofoam cup holding 125 g of hot water at 1.00×10^2°C cools to room temperature, 20.0°C. What is the change in entropy of the room? (Neglect the specific heat of the cup and any change in temperature of the room.)

46. A 65-g ice cube is initially at 0.0°C. (a) Find the change in entropy of the cube after it melts completely at 0.0°C. (b) What is the change in entropy of the environment in this process? *Hint:* The latent heat of fusion for water is 3.33×10^5 J/kg.

47. A freezer is used to freeze 1.0 L of water completely into ice. The water and the freezer remain at a constant temperature of $T = 0$°C. Determine (a) the change in the entropy of the water and (b) the change in the entropy of the freezer.

48. W What is the change in entropy of 1.00 kg of liquid water at 100°C as it changes to steam at 100°C?

49. A 70.0-kg log falls from a height of 25.0 m into a lake. If the log, the lake, and the air are all at 300 K, find the change in entropy of the Universe during this process.

50. If you roll a pair of dice, what is the total number of ways in which you can obtain (a) a 12? (b) a 7?

51. M The surface of the Sun is approximately at 5 700 K, and the temperature of the Earth's surface is approximately 290 K. What entropy change occurs when 1 000 J of energy is transferred by heat from the Sun to the Earth?

52. Q|C When an aluminum bar is temporarily connected between a hot reservoir at 725 K and a cold reservoir at 310 K, 2.50 kJ of energy is transferred by heat from the hot reservoir to the cold reservoir. In this irreversible process, calculate the change in entropy of (a) the hot reservoir, (b) the cold reservoir, and (c) the Universe, neglecting any change in entropy of the aluminum rod. (d) Mathematically, why did the result for the Universe in part (c) have to be positive?

53. Prepare a table like Table 12.3 for the following occurrence: You toss four coins into the air simultaneously and record all the possible results of the toss in terms of the numbers of heads and tails that can result. (For example, HHTH and HTHH are two possible ways in which three heads and one tail can be achieved.) (a) On the basis of your table, what is the most probable result of a toss? In terms of entropy, (b) what is the most ordered state, and (c) what is the most disordered?

54. S *This is a symbolic version of Problem 52.* When a metal bar is temporarily connected between a hot reservoir at T_h and a cold reservoir at T_c, the energy transferred by heat from the hot reservoir to the cold reservoir is Q_h. In this irreversible process, find expressions for the change in entropy of (a) the hot reservoir, (b) the cold reservoir, and (c) the Universe.

12.6 Human Metabolism

55. On a typical day, a 65-kg man sleeps for 8.0 h, does light chores for 3.0 h, walks slowly for 1.0 h, and jogs at moderate pace for 0.5 h. What is the change in his internal energy for all these activities?

56. BIO Q|C A weightlifter has a basal metabolic rate of 80.0 W. As he is working out, his metabolic rate increases by about 650 W. (a) How many hours does it take him to work off a 450-Calorie bagel if he stays in bed all day? (b) How long does it take him if he's working out? (c) Calculate the amount of mechanical work necessary to lift a 120-kg barbell 2.00 m. (d) He drops the barbell to the floor and lifts it repeatedly. How many times per minute must he repeat this process to do an amount of mechanical work equivalent to his metabolic rate increase of 650 W during exercise? (e) Could he actually do repetitions at the rate found in part (d) at the given metabolic level? Explain.

57. BIO Sweating is one of the main mechanisms with which the body dissipates heat. Sweat evaporates with a latent heat of 2 430 kJ/kg at body temperature, and the body can produce as much as 1.5 kg of sweat per hour. If sweating were the only heat dissipation mechanism, what would be the maximum sustainable metabolic rate, in watts, if 80% of the energy used by the body goes into waste heat?

Additional Problems

58. A Carnot engine operates between the temperatures $T_h = 100°C$ and $T_c = 20°C$. By what factor does the theoretical efficiency increase if the temperature of the hot reservoir is increased to 550°C?

59. A 1 500-kW heat engine operates at 25% efficiency. The heat energy expelled at the low temperature is absorbed by a stream of water that enters the cooling coils at 20°C. If 60 L flows across the coils per second, determine the increase in temperature of the water.

60. W A Carnot engine operates between 100°C and 20°C. How much ice can the engine melt from its exhaust after it has done 5.0×10^4 J of work?

61. A substance undergoes the cyclic process shown in Figure P12.61. Work output occurs along path *AB* while work input is required along path *BC*, and no work is involved in the constant volume process *CA*. Energy transfers by heat occur during each process involved in the cycle. (a) What is the work output during process *AB*? (b) How much work input is required during process *BC*? (c) What is the net energy input *Q* during this cycle?

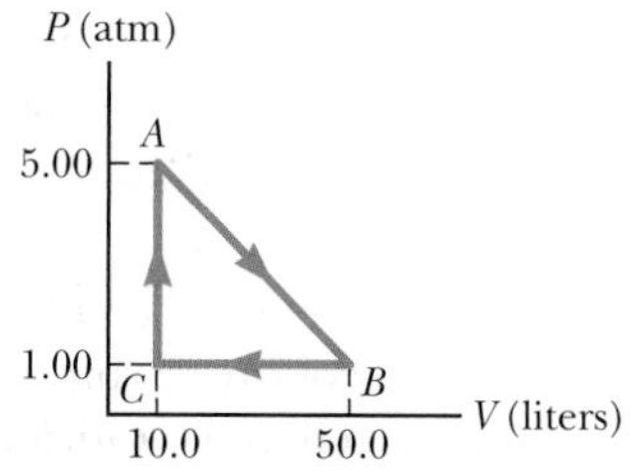

Figure P12.61

62. When a gas follows path 123 on the *PV* diagram in Figure P12.62, 418 J of energy flows into the system by heat and −167 J of work is done *on* the gas. (a) What is the change in the internal energy of the system? (b) How much energy *Q* flows into the system if the gas follows path 143? The work done on the gas along this path is −63.0 J. What net work would be done on or by the system if the system followed (c) path 12341 and (d) path 14321? (e) What is the change in internal energy of the system in the processes described in parts (c) and (d)?

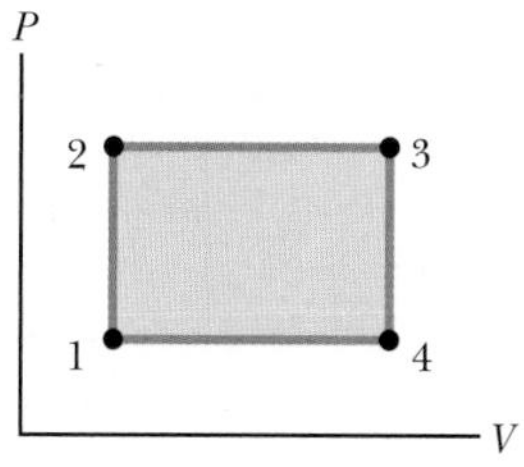

Figure P12.62

63. A 100-kg steel support rod in a building has a length of 2.0 m at a temperature of 20°C. The rod supports a hanging load of 6 000 kg. Find (a) the work done *on* the rod as the temperature increases to 40°C, (b) the energy *Q* added to the rod (assume the specific heat of steel is the same as that for iron), and (c) the change in internal energy of the rod.

64. S An ideal gas initially at pressure P_0, volume V_0, and temperature T_0 is taken through the cycle described in Figure P12.64. (a) Find the net work done *by* the gas per cycle in terms of P_0 and V_0. (b) What is the net energy *Q* added to the system per cycle? (c) Obtain a numerical value for the net work done per cycle for 1.00 mol of gas initially at 0°C. *Hint:* Recall that the work done by the system equals the area under a *PV* curve.

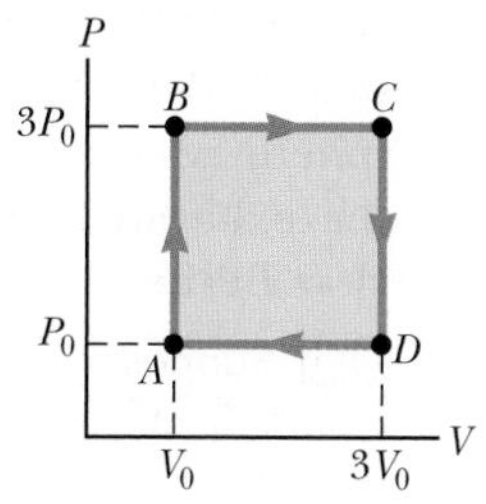

Figure P12.64

65. M One mole of neon gas is heated from 300 K to 420 K at constant pressure. Calculate (a) the energy Q transferred to the gas, (b) the change in the internal energy of the gas, and (c) the work done *on* the gas. Note that neon has a molar specific heat of $c = 20.79$ J/mol · K for a constant-pressure process.

66. Every second at Niagara Falls, approximately 5 000 m^3 of water falls a distance of 50.0 m. What is the increase in entropy per second due to the falling water? Assume the mass of the surroundings is so great that its temperature and that of the water stay nearly constant at 20.0°C. Also assume a negligible amount of water evaporates.

67. A cylinder containing 10.0 moles of a monatomic ideal gas expands from Ⓐ to Ⓑ along the path shown in Figure P12.67. (a) Find the temperature of the gas at point A and the temperature at point Ⓑ. (b) How much work is done by the gas during this expansion? (c) What is the change in internal energy of the gas? (d) Find the energy transferred to the gas by heat in this process.

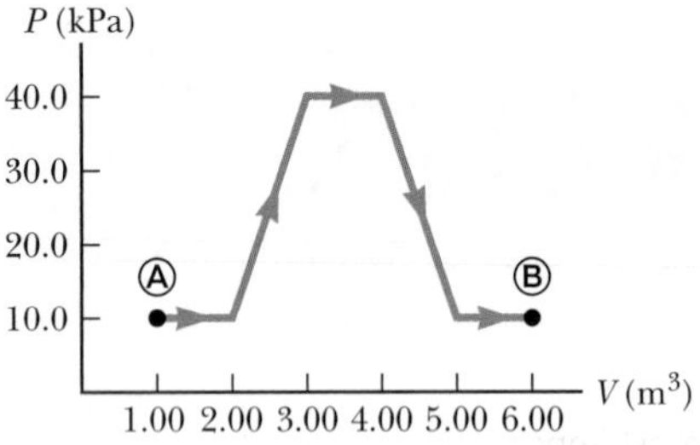

Figure P12.67

68. GP Two moles of molecular hydrogen (H_2) react with 1 mole of molecular oxygen (O_2) to produce 2 moles of water (H_2O) together with an energy release of 241.8 kJ/mole of water. Suppose a spherical vessel of radius 0.500 m contains 14.4 moles of H_2 and 7.2 moles of O_2 at 20.0°C. (a) What is the initial pressure in the vessel? (b) What is the initial internal energy of the gas? (c) Suppose a spark ignites the mixture and the gases burn completely into water vapor. How much energy is produced? (d) Find the temperature and pressure of the steam, assuming it's an ideal gas. (e) Find the mass of steam and then calculate the steam's density. (f) If a small hole were put in the sphere, what would be the initial exhaust velocity of the exhausted steam if spewed out into a vacuum? (Use Bernoulli's equation.)

69. BIO Suppose you spend 30.0 minutes on a stair-climbing machine, climbing at a rate of 90.0 steps per minute, with each step 8.00 inches high. If you weigh 150 lb and the machine reports that 600 kcal have been burned at the end of the workout, what efficiency is the machine using in obtaining this result? If your actual efficiency is 0.18, how many kcal did you actually burn?

70. BIO Hydrothermal vents deep on the ocean floor spout water at temperatures as high as 570°C. This temperature is below the boiling point of water because of the immense pressure at that depth. Because the surrounding ocean temperature is at 4.0°C, an organism could use the temperature gradient as a source of energy. (a) Assuming the specific heat of water under these conditions is 1.0 cal/g · °C, how much energy is released when 1.0 liter of water is cooled from 570°C to 4.0°C? (b) What is the maximum usable energy an organism can extract from this energy source? (Assume the organism has some internal type of heat engine acting between the two temperature extremes.) (c) Water from these vents contains hydrogen sulfide (H_2S) at a concentration of 0.90 mmole/liter. Oxidation of 1.0 mole of H_2S produces 310 kJ of energy. How much energy is available through H_2S oxidation of 1.0 L of water?

71. An electrical power plant has an overall efficiency of 15%. The plant is to deliver 150 MW of electrical power to a city, and its turbines use coal as fuel. The burning coal produces steam at 190°C, which drives the turbines. The steam is condensed into water at 25°C by passing through coils that are in contact with river water. (a) How many metric tons of coal does the plant consume each day (1 metric ton $= 1 \times 10^3$ kg)? (b) What is the total cost of the fuel per year if the delivery price is \$8 per metric ton? (c) If the river water is delivered at 20°C, at what minimum rate must it flow over the cooling coils so that its temperature doesn't exceed 25°C? *Note:* The heat of combustion of coal is 7.8×10^3 cal/g.

72. A diatomic ideal gas expands from a volume of $V_A = 1.00$ m^3 to $V_B = 3.00$ m^3 along the path shown in Figure P12.72. If the initial pressure is $P_A = 2.00 \times 10^5$ Pa and there are 87.5 mol of gas, calculate (a) the work done on the gas during this process, (b) the change in temperature of the gas, and (c) the change in internal energy of the gas. (d) How much thermal energy is transferred to the system?

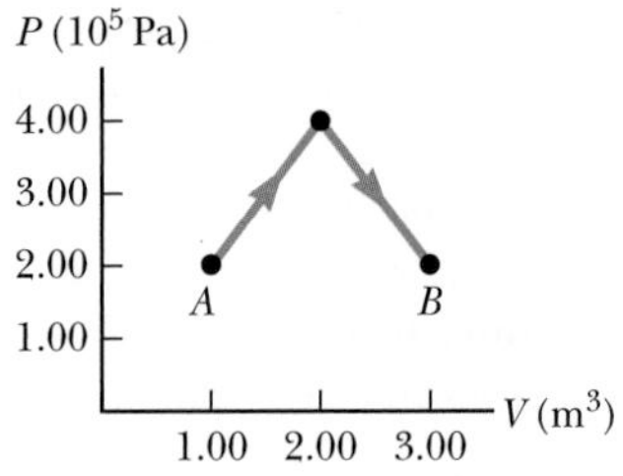

Figure P12.72

Ocean waves combine properties of both transverse and longitudinal waves. With proper balance and timing, a surfer can capture some of the wave's energy and take it for a ride.

Vibrations and Waves 13

Periodic motion, from masses on springs to vibrations of atoms, is one of the most important kinds of physical behavior. In this chapter we take a more detailed look at Hooke's law, where the force is proportional to the displacement, tending to restore objects to some equilibrium position. A large number of physical systems can be successfully modeled with this simple idea, including the vibrations of strings, the swinging of a pendulum, and the propagation of waves of all kinds. All these physical phenomena involve periodic motion.

Periodic vibrations can cause disturbances that move through a medium in the form of waves. Many kinds of waves occur in nature, such as sound waves, water waves, seismic waves, and electromagnetic waves. These very different physical phenomena are described by common terms and concepts introduced here.

13.1 Hooke's Law

LEARNING OBJECTIVES

1. Define Hooke's force law for springs and describe the elements of simple harmonic motion that arise from it.
2. Apply Hooke's law and the second law of motion to spring systems.

One of the simplest types of vibrational motion is that of an object attached to a spring, previously discussed in the context of energy in Chapter 5. We assume the object moves on a frictionless horizontal surface. If the spring is stretched or compressed a small distance x from its unstretched or equilibrium position and then released, it exerts a force on the object as shown in Figure 13.1. From experiment, the spring force F_s is found to obey the equation

Hooke's law ▶

$$F_s = -kx \qquad [13.1]$$

where x is the displacement of the object from its equilibrium position ($x = 0$) and k is a positive constant called the **spring constant.** This force law for springs was discovered by Robert Hooke in 1678 and is known as **Hooke's law.** The value of k is a measure of the stiffness of the spring. Stiff springs have large k values, and soft springs have small k values.

The negative sign in Equation 13.1 means that the force exerted by the spring is always directed *opposite* the displacement of the object. When the object is to the right of the equilibrium position, as in Figure 13.1a, x is positive and F_s is negative. This means that the force is in the negative direction, to the left. When the object is to the left of the equilibrium position, as in Figure 13.1c, x is negative and F_s is positive, indicating that the direction of the force is to the right. Of course, when $x = 0$, as in Figure 13.1b, the spring is unstretched and $F_s = 0$. Because the spring force always acts toward the equilibrium position, it is sometimes called a restoring force. **A restoring force always pushes or pulls the object toward the equilibrium position.**

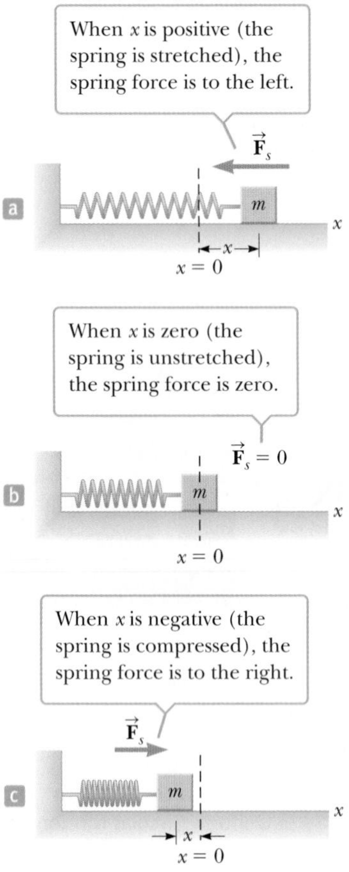

Figure 13.1 The force exerted by a spring on an object varies with the displacement of the object from the equilibrium position, $x = 0$.

Suppose the object is initially pulled a distance A to the right and released from rest. The force exerted by the spring on the object pulls it back toward the equilibrium position. As the object moves toward $x = 0$, the magnitude of the force decreases (because x decreases) and reaches zero at $x = 0$. The object gains speed as it moves toward the equilibrium position, however, reaching its maximum speed when $x = 0$. The momentum gained by the object causes it to overshoot the equilibrium position and compress the spring. As the object moves to the left of the equilibrium position (negative x-values), the spring force acts on it to the right, steadily increasing in strength, and the speed of the object decreases. The object finally comes briefly to rest at $x = -A$ before accelerating back towards $x = 0$ and ultimately returning to the original position at $x = A$. The process is then repeated, and the object continues to oscillate back and forth over the same path. This type of motion is called **simple harmonic motion. Simple harmonic motion occurs when the net force along the direction of motion obeys Hooke's law—when the net force is proportional to the displacement from the equilibrium point and is always directed toward the equilibrium point.**

Not all periodic motions over the same path can be classified as simple harmonic motion. A ball being tossed back and forth between a parent and a child moves repetitively, but the motion isn't simple harmonic motion because the force acting on the ball doesn't take the form of Hooke's law, Equation 13.1.

The motion of an object suspended from a vertical spring is also simple harmonic. In this case the force of gravity acting on the attached object stretches the spring until equilibrium is reached and the object is suspended at rest. By definition, the equilibrium position of the object is $x = 0$. When the object is moved away from equilibrium by a distance x and released, a net force acts toward the equilibrium position. Because the net force is proportional to x, the motion is simple harmonic.

The following three concepts are important in discussing any kind of periodic motion:

- The **amplitude** A is the maximum distance of the object from its equilibrium position. In the absence of friction, an object in simple harmonic motion oscillates between the positions $x = -A$ and $x = +A$.
- The **period** T is the time it takes the object to move through one complete cycle of motion, from $x = A$ to $x = -A$ and back to $x = A$.
- The **frequency** f is the number of complete cycles or vibrations per unit of time, and is the reciprocal of the period ($f = 1/T$).

The acceleration of an object moving with simple harmonic motion can be found by using Hooke's law in the equation for Newton's second law, $F = ma$. This gives

$$ma = F = -kx$$

$$a = -\frac{k}{m}x \qquad [13.2]$$

◀ Acceleration in simple harmonic motion

Equation 13.2, an example of a *harmonic oscillator equation,* gives the acceleration as a function of position. Because the maximum value of x is defined to be the amplitude A, the acceleration ranges over the values $-kA/m$ to $+kA/m$. In the next section we will find equations for velocity as a function of position and for position as a function of time. Springs satisfying Hooke's law are also called *ideal* springs. In real springs, spring mass, internal friction, and varying elasticity affect the force law and motion.

Tip 13.1 Constant-Acceleration Equations Don't Apply

The acceleration a of a particle in simple harmonic motion is *not* constant; it changes, varying with x, so we can't apply the constant acceleration kinematic equations of Chapter 2.

Quick Quiz

13.1 A block on the end of a horizontal spring is pulled from equilibrium at $x = 0$ to $x = A$ and released. Through what total distance does it travel in one full cycle of its motion? (a) $A/2$ (b) A (c) $2A$ (d) $4A$

13.2 For a simple harmonic oscillator, which of the following pairs of vector quantities can't both point in the same direction? (The position vector is the displacement from equilibrium.) (a) position and velocity (b) velocity and acceleration (c) position and acceleration

EXAMPLE 13.1 Simple Harmonic Motion on a Frictionless Surface

GOAL Calculate forces and accelerations for a horizontal spring system.

PROBLEM A 0.350-kg object attached to a spring of force constant 1.30×10^2 N/m is free to move on a frictionless horizontal surface, as in Figure 13.1. If the object is released from rest at $x = 0.100$ m, find the force on it and its acceleration at $x = 0.100$ m, $x = 0.050\,0$ m, $x = 0$ m, $x = -0.050\,0$ m, and $x = -0.100$ m.

STRATEGY Substitute given quantities into Hooke's law to find the forces, then calculate the accelerations with Newton's second law. The amplitude A is the same as the point of release from rest, $x = 0.100$ m.

SOLUTION

Write Hooke's force law:

$$F_s = -kx$$

Substitute the value for k and take $x = A = 0.100$ m, finding the spring force at that point:

$$F_{max} = -kA = -(1.30 \times 10^2 \text{ N/m})(0.100 \text{ m})$$
$$= -13.0 \text{ N}$$

Solve Newton's second law for a and substitute to find the acceleration at $x = A$:

$$ma = F_{max}$$
$$a = \frac{F_{max}}{m} = \frac{-13.0 \text{ N}}{0.350 \text{ kg}} = -37.1 \text{ m/s}^2$$

Repeat the same process for the other four points, assembling a table:

Position (m)	Force (N)	Acceleration (m/s^2)
0.100	−13.0	−37.1
0.050	−6.50	−18.6
0	0	0
−0.050	+6.50	+18.6
−0.100	+13.0	+37.1

REMARKS The table above shows that when the initial position is halved, the force and acceleration are also halved. Further, positive values of x give negative values of the force and acceleration, whereas negative values of x give positive values of the force and acceleration. As the object moves to the left and passes the equilibrium point, the spring force becomes positive (for negative values of x), slowing the object down.

(Continued)

QUESTION 13.1 Will doubling a given displacement always result in doubling the magnitude of the spring force? Explain.

EXERCISE 13.1 For the same spring and mass system, find the force exerted by the spring and the position x when the object's acceleration is $+9.00\ \text{m/s}^2$.

ANSWERS 3.15 N, −2.42 cm

EXAMPLE 13.2 | Mass on a Vertical Spring

GOAL Apply Newton's second law together with the force of gravity and Hooke's law.

PROBLEM A spring is hung vertically (Fig. 13.2a), and an object of mass m attached to the lower end is then slowly lowered a distance d to the equilibrium point (Fig. 13.2b). **(a)** Find the value of the spring constant if the magnitude of the displacement d is 2.0 cm and the mass is 0.55 kg. **(b)** If a second identical spring is attached to the object in parallel with the first spring (Fig. 13.2d), where is the new equilibrium point of the system? **(c)** What is the effective spring constant of the two springs acting as one?

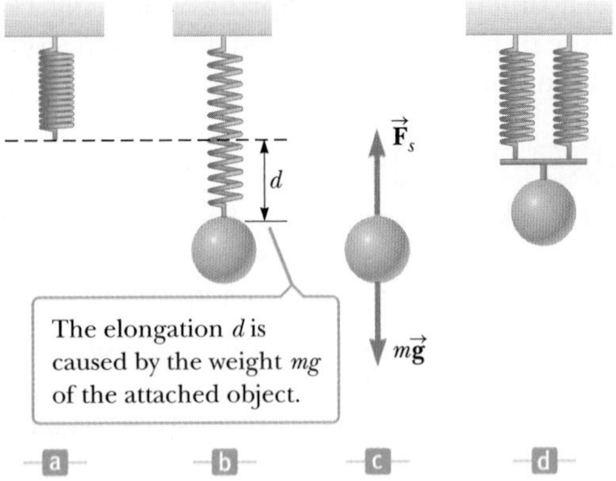

Figure 13.2 (Example 13.2) (a)–(c) Determining the spring constant. Because the upward spring force balances the weight when the system is in equilibrium, it follows that $k = mg/d$. (d) A system involving two springs in parallel.

STRATEGY This example is an application of Newton's second law. The spring force is upward, balancing the downward force of gravity mg when the system is in equilibrium. (See Fig. 13.2c.) Because the suspended object is in equilibrium, the forces on the object sum to zero, and it's possible to solve for the spring constant k. Part (b) is solved the same way, but has two spring forces balancing the force of gravity. The spring constants are known, so the second law for equilibrium can be solved for the displacement of the spring. Part (c) involves using the displacement found in part (b). Treating the two springs as one equivalent spring, the second law then leads to the effective spring constant of the two-spring system.

SOLUTION

(a) Find the value of the spring constant if the spring is displaced by 2.0 cm and the mass of the object is 0.55 kg.

Apply Newton's second law to the object (with $a = 0$) and solve for the spring constant k:

$$\sum F = F_g + F_s = -mg + kd = 0$$

$$k = \frac{mg}{d} = \frac{(0.55\ \text{kg})(9.80\ \text{m/s}^2)}{2.0 \times 10^{-2}\ \text{m}} = \boxed{2.7 \times 10^2\ \text{N/m}}$$

(b) If a second identical spring is attached to the object in parallel with the first spring (Fig. 13.2d), find the new equilibrium point of the system.

Apply Newton's second law, but with two springs acting on the object:

$$\sum F = F_g + F_{s1} + F_{s2} = -mg + kd_2 + kd_2 = 0$$

Solve for d_2:

$$d_2 = \frac{mg}{2k} = \frac{(0.55\ \text{kg})(9.80\ \text{m/s}^2)}{2(2.7 \times 10^2\ \text{N/m})} = \boxed{1.0 \times 10^{-2}\ \text{m}}$$

(c) What is the effective spring constant of the two springs acting as one?

Write the second law for the system, with an effective spring constant k_{eff}:

$$\sum F = F_g + F_s = -mg + k_{\text{eff}} d_2 = 0$$

Solve for k_{eff}:

$$k_{\text{eff}} = \frac{mg}{d_2} = \frac{(0.55\ \text{kg})(9.80\ \text{m/s}^2)}{1.0 \times 10^{-2}\ \text{m}} = 5.4 \times 10^2\ \text{N/m}$$

REMARKS In this example, the spring force is positive because it's directed upward. If the object were displaced from the equilibrium position and released, it would oscillate around that point, just like a horizontal spring. Notice that attaching an extra identical spring in parallel is equivalent to having a single spring with twice the force constant. With springs attached end to end in series, however, the exercise illustrates that, all other things being equal, longer springs have smaller force constants than shorter springs.

QUESTION 13.2 Generalize: When two springs with force constants k_1 and k_2 act in parallel on an object, what is the spring constant k_{eff} of the single spring that would be equivalent to the two springs, in terms of k_1 and k_2?

EXERCISE 13.2 When a 75.0-kg man slowly adds his weight to a vertical spring attached to the ceiling, he reaches equilibrium when the spring is stretched by 6.50 cm. (a) Find the spring constant. (b) If a second, identical spring is hung on the first in series, and the man again adds his weight to the system, by how much does the system of springs stretch? (c) What would be the spring constant of a single, equivalent spring?

ANSWERS (a) 1.13×10^4 N/m (b) 13.0 cm (c) 5.65×10^3 N/m

13.2 Elastic Potential Energy

LEARNING OBJECTIVE

1. Review the application of the work-energy theorem to systems involving spring potential energy.

In this section we review the material covered in Section 4 of Chapter 5.

A system of interacting objects has potential energy associated with the configuration of the system. A compressed spring has potential energy that, when allowed to expand, can do work on an object, transforming spring potential energy into the object's kinetic energy. As an example, Figure 13.3 shows a ball being projected from a spring-loaded toy gun, where the spring is compressed a distance x. As the gun is fired, the compressed spring does work on the ball and imparts kinetic energy to it.

Recall that the energy stored in a stretched or compressed spring or some other elastic material is called elastic potential energy, PE_s, given by

$$PE_s \equiv \tfrac{1}{2}kx^2 \qquad [13.3]$$

◀ Elastic potential energy

Recall also that the law of conservation of energy, including both gravitational and spring potential energy, is given by

$$(KE + PE_g + PE_s)_i = (KE + PE_g + PE_s)_f \qquad [13.4]$$

If nonconservative forces such as friction are present, then the change in mechanical energy must equal the work done by the nonconservative forces:

$$W_{nc} = (KE + PE_g + PE_s)_f - (KE + PE_g + PE_s)_i \qquad [13.5]$$

Rotational kinetic energy must be included in both Equation 13.4 and Equation 13.5 for systems involving torques.

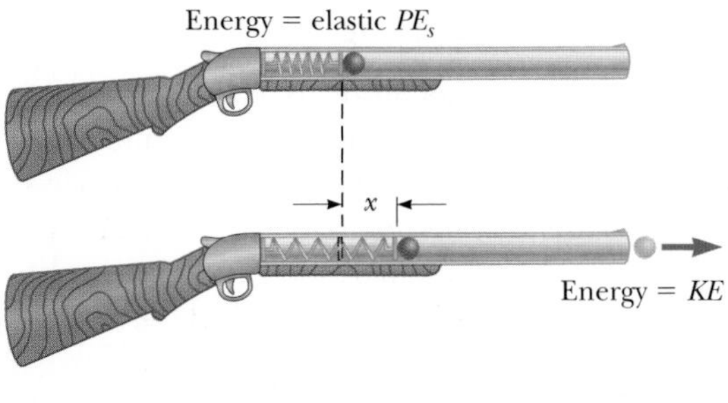

Figure 13.3 A ball projected from a spring-loaded gun. The elastic potential energy stored in the spring is transformed into the kinetic energy of the ball.

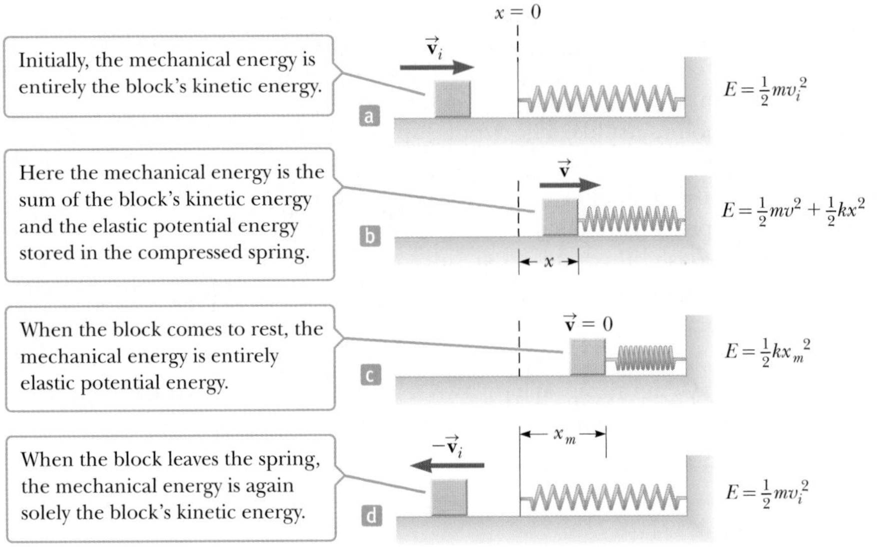

Figure 13.4 A block sliding on a frictionless horizontal surface collides with a light spring. In the absence of friction, the mechanical energy in this process remains constant.

APPLICATION
Archery

As an example of the energy conversions that take place when a spring is included in a system, consider Figure 13.4. A block of mass m slides on a frictionless horizontal surface with constant velocity $\vec{v}_i$ and collides with a coiled spring. The description that follows is greatly simplified by assuming the spring is very light (an ideal spring) and therefore has negligible kinetic energy. As the spring is compressed, it exerts a force to the left on the block. At maximum compression, the block comes to rest for just an instant (Fig. 13.4c). The initial total energy in the system (block plus spring) before the collision is the kinetic energy of the block. After the block collides with the spring and the spring is partially compressed, as in Figure 13.4b, the block has kinetic energy $\frac{1}{2}mv^2$ (where $v < v_i$) and the spring has potential energy $\frac{1}{2}kx^2$. When the block stops for an instant at the point of maximum compression, the kinetic energy is zero. Because the spring force is conservative and because there are no external forces that can do work on the system, **the total mechanical energy of the system consisting of the block and spring remains constant.** Energy is transformed from the kinetic energy of the block to the potential energy stored in the spring. As the spring expands, the block moves in the opposite direction and regains all its initial kinetic energy, as in Figure 13.4d.

When an archer pulls back on a bowstring, elastic potential energy is stored in both the bent bow and stretched bowstring (Fig. 13.5). When the arrow is released, the potential energy stored in the system is transformed into the kinetic energy of the arrow. Devices such as crossbows and slingshots work the same way.

Figure 13.5 Elastic potential energy is stored in this drawn bow.

Quick Quiz

13.3 When an object moving in simple harmonic motion is at its maximum displacement from equilibrium, which of the following is at a maximum? (a) velocity, (b) acceleration, or (c) kinetic energy

EXAMPLE 13.3 | Stop That Car!

GOAL Apply conservation of energy and the work–energy theorem with spring and gravitational potential energy.

PROBLEM A 13 000-N car starts at rest and rolls down a hill from a height of 10.0 m (Fig. 13.6). It then moves across a level surface and collides with a light spring-loaded guardrail. **(a)** Neglecting any losses due to friction, and ignoring the rotational kinetic energy of the wheels, find the maximum distance the spring is compressed. Assume a spring constant of 1.0×10^6 N/m. **(b)** Calculate the magnitude of the car's maximum acceleration after contact with the spring, assuming no frictional losses. **(c)** If the spring is compressed by only 0.30 m, find the change in the mechanical energy due to friction.

STRATEGY Because friction losses are neglected, use conservation of energy in the form of Equation 13.4 to solve for the spring displacement in part (a). The initial and final values of the car's kinetic energy are zero, so the initial potential energy of the car–spring–Earth system is completely converted to elastic potential energy in the spring at the end of the ride. In part (b) apply Newton's second law, substituting the answer to part (a) for x because the maximum compression will give the maximum acceleration. In part (c) friction is no longer neglected, so use the work–energy theorem, Equation 13.5. The change in mechanical energy must equal the mechanical energy lost due to friction.

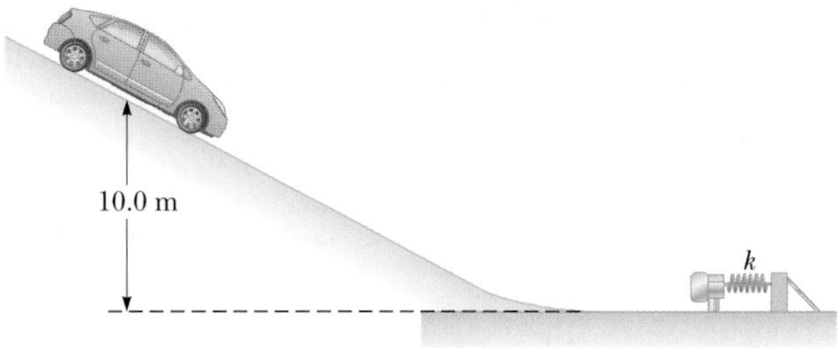

Figure 13.6 (Example 13.3) A car starts from rest on a hill at the position shown. When the car reaches the bottom of the hill, it collides with a spring-loaded guardrail.

SOLUTION

(a) Find the maximum spring compression, assuming no energy losses due to friction.

Apply conservation of mechanical energy. Initially, there is only gravitational potential energy, and at maximum compression of the guardrail, there is only spring potential energy.

$$(KE + PE_g + PE_s)_i = (KE + PE_g + PE_s)_f$$
$$0 + mgh + 0 = 0 + 0 + \tfrac{1}{2}kx^2$$

Solve for x:

$$x = \sqrt{\frac{2mgh}{k}} = \sqrt{\frac{2(13\,000\text{ N})(10.0\text{ m})}{1.0 \times 10^6\text{ N/m}}} = \boxed{0.51\text{ m}}$$

(b) Calculate the magnitude of the car's maximum acceleration by the spring, neglecting friction.

Apply Newton's second law to the car:

$$ma = -kx \quad \rightarrow \quad a = -\frac{kx}{m} = -\frac{kxg}{mg} = -\frac{kxg}{w}$$

Substitute values:

$$a = -\frac{(1.0 \times 10^6\text{ N/m})(0.51\text{ m})(9.8\text{ m/s}^2)}{13\,000\text{ N}}$$
$$= -380\text{ m/s}^2 \rightarrow |a| = \boxed{380\text{ m/s}^2}$$

(c) If the compression of the guardrail is only 0.30 m, find the change in the mechanical energy due to friction.

Use the work–energy theorem:

$$W_{nc} = (KE + PE_g + PE_s)_f - (KE + PE_g + PE_s)_i$$
$$= (0 + 0 + \tfrac{1}{2}kx^2) - (0 + mgh + 0)$$
$$= \tfrac{1}{2}(1.0 \times 10^6\text{ N/m})(0.30)^2 - (13\,000\text{ N})(10.0\text{ m})$$
$$W_{nc} = \boxed{-8.5 \times 10^4\text{ J}}$$

REMARKS The answer to part (b) is about 40 times greater than the acceleration of gravity, so we'd better be wearing our seat belts. Note that the solution didn't require calculation of the velocity of the car.

QUESTION 13.3 True or False: In the absence of energy losses due to friction, doubling the height of the hill doubles the maximum acceleration delivered by the spring.

EXERCISE 13.3 A spring-loaded gun fires a 0.100-kg puck along a tabletop. The puck slides up a curved ramp and flies straight up into the air. If the spring is displaced 12.0 cm from equilibrium and the spring constant is 875 N/m, how high does the puck rise, neglecting friction? (b) If instead it only rises to a height of 5.00 m because of friction, what is the change in mechanical energy?

ANSWERS (a) 6.43 m (b) −1.40 J

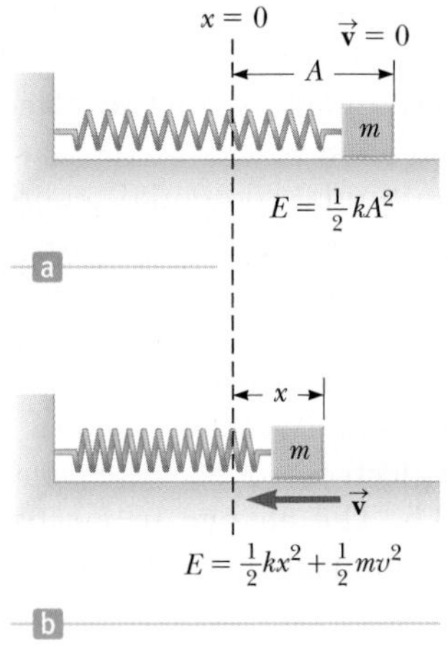

Figure 13.7 (a) An object attached to a spring on a frictionless surface is released from rest with the spring extended a distance A. Just before the object is released, the total energy is the elastic potential energy $\frac{1}{2}kA^2$. (b) When the object reaches position x, it has kinetic energy $\frac{1}{2}mv^2$ and the elastic potential energy has decreased to $\frac{1}{2}kx^2$.

In addition to studying the preceding example, it's a good idea to review those given in Section 5.4.

Velocity as a Function of Position

Conservation of energy provides a simple method of deriving an expression for the velocity of an object undergoing periodic motion as a function of position. The object in question is initially at its maximum extension A (Fig. 13.7a) and is then released from rest. The initial energy of the system is entirely elastic potential energy stored in the spring, $\frac{1}{2}kA^2$. As the object moves toward the origin to some new position x (Fig. 13.7b), part of this energy is transformed into kinetic energy, and the potential energy stored in the spring is reduced to $\frac{1}{2}kx^2$. Because the total energy of the system is equal to $\frac{1}{2}kA^2$ (the initial energy stored in the spring), we can equate this quantity to the sum of the kinetic and potential energies at the position x:

$$\tfrac{1}{2}kA^2 = \tfrac{1}{2}mv^2 + \tfrac{1}{2}kx^2$$

Solving for v, we get

$$v = \pm\sqrt{\frac{k}{m}(A^2 - x^2)} \qquad [13.6]$$

This expression shows that the object's speed is a maximum at $x = 0$ and is zero at the extreme positions $x = \pm A$.

The right side of Equation 13.6 is preceded by the $\pm$ sign because the square root of a number can be either positive or negative. If the object in Figure 13.7 is moving to the right, v is positive; if the object is moving to the left, v is negative.

EXAMPLE 13.4 | The Object–Spring System Revisited

GOAL Apply the time-independent velocity expression, Equation 13.6, to an object–spring system.

PROBLEM A 0.500-kg object connected to a light spring with a spring constant of 20.0 N/m oscillates on a frictionless horizontal surface. **(a)** Calculate the total energy of the system and the maximum speed of the object if the amplitude of the motion is 3.00 cm. **(b)** What is the velocity of the object when the displacement is 2.00 cm? **(c)** Compute the kinetic and potential energies of the system when the displacement is 2.00 cm.

STRATEGY The total energy of the system can be found most easily at $x = A$, where the kinetic energy is zero. There, the potential energy alone is equal to the total energy. Conservation of energy then yields the speed at $x = 0$. For part (b), obtain the velocity by substituting the given value of x into the time-independent velocity equation. Using this result, the kinetic energy asked for in part (c) can be found by substitution, and the potential energy can be found by substitution into Equation 13.3.

SOLUTION

(a) Calculate the total energy and maximum speed if the amplitude is 3.00 cm.

Substitute $x = A = 3.00$ cm and $k = 20.0$ N/m into the equation for the total mechanical energy E:

$$E = KE + PE_g + PE_s$$
$$= 0 + 0 + \tfrac{1}{2}kA^2 = \tfrac{1}{2}(20.0 \text{ N/m})(3.00 \times 10^{-2} \text{ m})^2$$
$$= \boxed{9.00 \times 10^{-3} \text{ J}}$$

Use conservation of energy with $x_i = A$ and $x_f = 0$ to compute the speed of the object at the origin:

$$(KE + PE_g + PE_s)_i = (KE + PE_g + PE_s)_f$$
$$0 + 0 + \tfrac{1}{2}kA^2 = \tfrac{1}{2}mv_{\text{max}}^2 + 0 + 0$$
$$\tfrac{1}{2}mv_{\text{max}}^2 = 9.00 \times 10^{-3} \text{ J}$$
$$v_{\text{max}} = \sqrt{\frac{18.0 \times 10^{-3} \text{ J}}{0.500 \text{ kg}}} = \boxed{0.190 \text{ m/s}}$$

(b) Compute the velocity of the object when the displacement is 2.00 cm.

Substitute known values directly into Equation 13.6:

$$v = \pm\sqrt{\frac{k}{m}(A^2 - x^2)}$$
$$= \pm\sqrt{\frac{20.0 \text{ N/m}}{0.500 \text{ kg}}[(0.030\,0 \text{ m})^2 - (0.020\,0 \text{ m})^2]}$$
$$= \boxed{\pm 0.141 \text{ m/s}}$$

(c) Compute the kinetic and potential energies when the displacement is 2.00 cm.

Substitute into the equation for kinetic energy:

$$KE = \tfrac{1}{2}mv^2 = \tfrac{1}{2}(0.500 \text{ kg})(0.141 \text{ m/s})^2 = \boxed{4.97 \times 10^{-3} \text{ J}}$$

Substitute into the equation for spring potential energy:

$$PE_s = \tfrac{1}{2}kx^2 = \tfrac{1}{2}(20.0 \text{ N/m})(2.00 \times 10^{-2} \text{ m})^2$$
$$= \boxed{4.00 \times 10^{-3} \text{ J}}$$

REMARKS With the given information, it is impossible to choose between the positive and negative solutions in part (b). Notice that the sum $KE + PE_s$ in part (c) equals the total energy E found in part (a), as it should (except for a small discrepancy due to rounding).

QUESTION 13.4 True or False: Doubling the initial displacement doubles the speed of the object at the equilibrium point.

EXERCISE 13.4 For what values of x is the speed of the object 0.10 m/s?

ANSWER ±2.55 cm

13.3 Comparing Simple Harmonic Motion with Uniform Circular Motion

LEARNING OBJECTIVES

1. Describe the relationship between simple harmonic motion and uniform circular motion.
2. Define and apply the related concepts of the period, frequency, and angular frequency of a spring harmonic oscillator.

We can better understand and visualize many aspects of simple harmonic motion along a straight line by looking at its relationship to uniform circular motion. Figure 13.8 is a top view of an experimental arrangement that is useful for this purpose. A ball is attached to the rim of a turntable of radius A, illuminated from

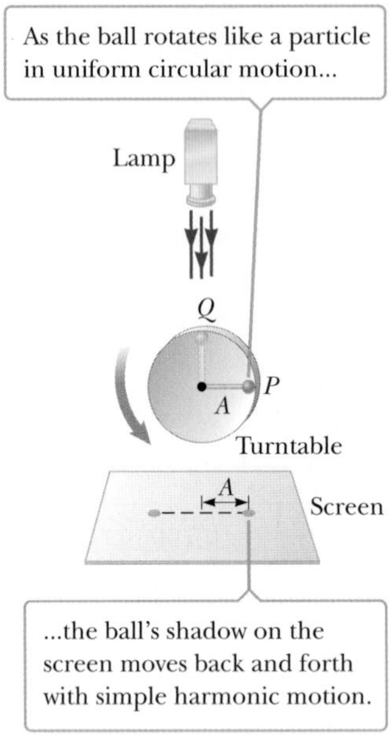

Figure 13.8 An experimental setup for demonstrating the connection between simple harmonic motion and uniform circular motion.

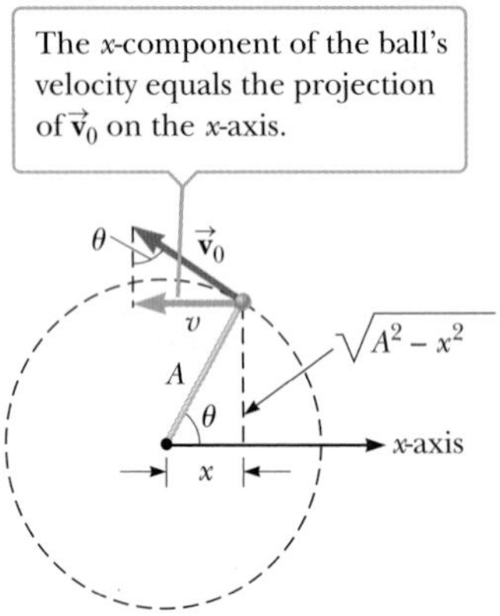

Figure 13.9 The ball rotates with constant speed v_0.

the side by a lamp. We find that **as the turntable rotates with constant angular speed, the shadow of the ball moves back and forth with simple harmonic motion.**

This fact can be understood from Equation 13.6, which says that the velocity of an object moving with simple harmonic motion is related to the displacement by

$$v = C\sqrt{A^2 - x^2}$$

where C is a constant. To see that the shadow also obeys this relation, consider Figure 13.9, which shows the ball moving with a constant speed v_0 in a direction tangent to the circular path. At this instant, the velocity of the ball in the x-direction is given by $v = v_0 \sin\theta$, or

$$\sin\theta = \frac{v}{v_0}$$

From the larger triangle in Figure 13.9 we can obtain a second expression for $\sin\theta$:

$$\sin\theta = \frac{\sqrt{A^2 - x^2}}{A}$$

Equating the right-hand sides of the two expressions for $\sin\theta$, we find the following relationship between the velocity v and the displacement x:

$$\frac{v}{v_0} = \frac{\sqrt{A^2 - x^2}}{A}$$

or

$$v = \frac{v_0}{A}\sqrt{A^2 - x^2} = C\sqrt{A^2 - x^2}$$

The velocity of the ball in the x-direction is related to the displacement x in exactly the same way as the velocity of an object undergoing simple harmonic motion. The shadow therefore moves with simple harmonic motion.

A valuable example of the relationship between simple harmonic motion and circular motion can be seen in vehicles and machines that use the back-and-forth motion of a piston to create rotational motion in a wheel. Consider the drive wheel of a locomotive. In Figure 13.10, the rods are connected to a piston that moves back and forth in simple harmonic motion. The rods transform the back-and-forth motion of the piston into rotational motion of the wheels. A similar mechanism in an automobile engine transforms the back-and-forth motion of the pistons to rotational motion of the crankshaft.

APPLICATION
Pistons and Drive Wheels

Figure 13.10 The drive wheel mechanism of an old locomotive.

Period and Frequency

The period T of the shadow in Figure 13.8, which represents the time required for one complete trip back and forth, is also the time it takes the ball to make one complete circular trip on the turntable. Because the ball moves through the distance $2\pi A$ (the circumference of the circle) in the time T, the speed v_0 of the ball around the circular path is

$$v_0 = \frac{2\pi A}{T}$$

and the period is

$$T = \frac{2\pi A}{v_0} \qquad [13.7]$$

Imagine that the ball moves from P to Q, a quarter of a revolution, in Figure 13.8. The motion of the shadow is equivalent to the horizontal motion of an object on the end of a spring. For this reason, the radius A of the circular motion is the same as the amplitude A of the simple harmonic motion of the shadow. During the quarter of a cycle shown, the shadow moves from a point where the energy of

the wave speed depends on an elastic property of the medium (B or F) and on an inertial property of the medium (ρ or μ). In fact, the speed of all mechanical waves follows an expression of the general form

$$v = \sqrt{\frac{\text{elastic property}}{\text{inertial property}}}$$

Another example of this general form is the **speed of a longitudinal wave in a solid rod,** which is

$$v = \sqrt{\frac{Y}{\rho}} \qquad [14.3]$$

where Y is the Young's modulus of the solid (see Eq. 9.3) and ρ is its density. This expression is valid only for a thin, solid rod.

Table 14.1 lists the speeds of sound in various media. The speed of sound is much higher in solids than in gases because the molecules in a solid interact more strongly with each other than do molecules in a gas. Striking a long steel rail with a hammer, for example, produces two sound waves, one moving through the rail and a slower wave moving through the air. A person with an ear pressed against the rail first hears the faster sound moving through the rail, then the sound moving through air. In general, sound travels faster through solids than liquids and faster through liquids than gases, although there are exceptions.

The speed of sound also depends on the temperature of the medium. For sound traveling through air, the relationship between the speed of sound and temperature is

$$v = (331 \text{ m/s})\sqrt{\frac{T}{273 \text{ K}}} \qquad [14.4]$$

where 331 m/s is the speed of sound in air at 0°C and T is the absolute (Kelvin) temperature. Using this equation, the speed of sound in air at 293 K (a typical room temperature) is approximately 343 m/s.

Table 14.1 Speeds of Sound in Various Media

Medium	v (m/s)
Gases	
Air (0°C)	331
Air (100°C)	386
Hydrogen (0°C)	1 286
Oxygen (0°C)	317
Helium (0°C)	972
Liquids at 25°C	
Water	1 493
Methyl alcohol	1 143
Sea water	1 533
Solids[a]	
Aluminum	6 420
Copper (rolled)	5 010
Steel	5 950
Lead (rolled)	1 960
Synthetic rubber	1 600

[a]Values given are for propagation of longitudinal waves in bulk media. Speeds for longitudinal waves in thin rods are smaller, and speeds of transverse waves in bulk are smaller yet.

Quick Quiz

14.1 Which of the following actions will increase the speed of sound in air? (a) decreasing the air temperature (b) increasing the frequency of the sound (c) increasing the air temperature (d) increasing the amplitude of the sound wave (e) reducing the pressure of the air

APPLYING PHYSICS 14.1 | The Sounds Heard During a Storm

How does lightning produce thunder, and what causes the extended rumble?

EXPLANATION Assume you're at ground level, and neglect ground reflections. When lightning strikes, a channel of ionized air carries a large electric current from a cloud to the ground. This results in a rapid temperature increase of the air in the channel as the current moves through it, causing a similarly rapid expansion of the air. The expansion is so sudden and so intense that a tremendous disturbance—thunder—is produced in the air. The entire length of the channel produces the sound at essentially the same instant of time. Sound produced at the bottom of the channel reaches you first because that's the point closest to you. Sounds from progressively higher portions of the channel reach you at later times, resulting in an extended roar. If the lightning channel were a perfectly straight line, the roar might be steady, but the zigzag shape of the path results in the rumbling variation in loudness, with different quantities of sound energy from different segments arriving at any given instant. ■

EXAMPLE 14.1 Explosion over an Ice Sheet

GOAL Calculate time of travel for sound through various media.

PROBLEM An explosion occurs 275 m above an 867-m-thick ice sheet that lies over ocean water. If the air temperature is −7.00°C, how long does it take the sound to reach a research vessel 1 250 m below the ice? Neglect any changes in the bulk modulus and density with temperature and depth. (Use $B_{\text{ice}} = 9.2 \times 10^9$ Pa.)

STRATEGY Calculate the speed of sound in air with Equation 14.4, and use $d = vt$ to find the time needed for the sound to reach the surface of the ice. Use Equation 14.1 to compute the speed of sound in ice, again finding the time with the distance equation. Finally, use the speed of sound in salt water to find the time needed to traverse the water and then sum the three times.

SOLUTION

Calculate the speed of sound in air at −7.00°C, which is equivalent to 266 K:

$$v_{\text{air}} = (331 \text{ m/s})\sqrt{\frac{T}{273 \text{ K}}} = (331 \text{ m/s})\sqrt{\frac{266 \text{ K}}{273 \text{ K}}} = 327 \text{ m/s}$$

Calculate the travel time through the air:

$$t_{\text{air}} = \frac{d}{v_{\text{air}}} = \frac{275 \text{ m}}{327 \text{ m/s}} = 0.841 \text{ s}$$

Compute the speed of sound in ice, using the bulk modulus and density of ice:

$$v_{\text{ice}} = \sqrt{\frac{B}{\rho}} = \sqrt{\frac{9.2 \times 10^9 \text{ Pa}}{917 \text{ kg/m}^3}} = 3.2 \times 10^3 \text{ m/s}$$

Compute the travel time through the ice:

$$t_{\text{ice}} = \frac{d}{v_{\text{ice}}} = \frac{867 \text{ m}}{3\,200 \text{ m/s}} = 0.27 \text{ s}$$

Compute the travel time through the ocean water:

$$t_{\text{water}} = \frac{d}{v_{\text{water}}} = \frac{1\,250 \text{ m}}{1\,533 \text{ m/s}} = 0.815 \text{ s}$$

Sum the three times to obtain the total time of propagation:

$$t_{\text{tot}} = t_{\text{air}} + t_{\text{ice}} + t_{\text{water}} = 0.841 \text{ s} + 0.27 \text{ s} + 0.815 \text{ s}$$

$$= 1.93 \text{ s}$$

REMARKS Notice that the speed of sound is highest in solid ice, second highest in liquid water, and slowest in air. The speed of sound depends on temperature, so the answer would have to be modified if the actual temperatures of ice and the sea water were known. At 0°C, for example, the speed of sound in sea water falls to 1 449 m/s.

QUESTION 14.1 Is the speed of sound in rubber higher or lower than the speed of sound in aluminum? Explain.

EXERCISE 14.1 Compute the speed of sound in the following substances at 273 K: (a) a thin lead rod ($Y = 1.6 \times 10^{10}$ Pa), (b) mercury ($B = 2.8 \times 10^{10}$ Pa), and (c) air at −15.0°C.

ANSWERS (a) 1.2×10^3 m/s (b) 1.4×10^3 m/s (c) 322 m/s

14.4 Energy and Intensity of Sound Waves

LEARNING OBJECTIVES

1. Define the average intensity of a wave, the threshold of hearing, and the threshold of pain.
2. Define sound intensity level (decibel scale) and discuss the reason it's required.
3. Apply the equations for sound intensity and decibel level to multiple sources of sound waves.

As the tines of a tuning fork move back and forth through the air, they exert a force on a layer of air and cause it to move. In other words, the tines do work on the layer of air. That the fork pours sound energy into the air is one reason the vibration of the fork slowly dies out. (Other factors, such as the energy lost to friction as the tines bend, are also responsible for the lessening of movement.)

The average **intensity** I of a wave on a given surface is defined as the rate at which energy flows through the surface, $\Delta E/\Delta t$, divided by the surface area A:

$$I \equiv \frac{1}{A}\frac{\Delta E}{\Delta t} \qquad [14.5]$$

where the direction of energy flow is perpendicular to the surface at every point.

SI unit: watt per meter squared (W/m^2)

A rate of energy transfer is power, so Equation 14.5 can be written in the alternate form

$$I \equiv \frac{\text{power}}{\text{area}} = \frac{P}{A} \qquad [14.6]$$

◀ Intensity of a wave

where P is the sound power passing through the surface, measured in watts, and the intensity again has units of watts per square meter.

The faintest sounds the human ear can detect at a frequency of 1 000 Hz have an intensity of about 1×10^{-12} W/m². This intensity is called the **threshold of hearing.** The loudest sounds the ear can tolerate have an intensity of about 1 W/m² (the **threshold of pain**). At the threshold of hearing, the increase in pressure in the ear is approximately 3×10^{-5} Pa over normal atmospheric pressure. Because atmospheric pressure is about 1×10^{5} Pa, this means the ear can detect pressure fluctuations as small as about 3 parts in 10^{10}! The maximum displacement of an air molecule at the threshold of hearing is about 1×10^{-11} m, a remarkably small number! If we compare this displacement with the diameter of a molecule (about 10^{-10} m), we see that the ear is an extremely sensitive detector of sound waves.

The loudest sounds the human ear can tolerate at 1 kHz correspond to a pressure variation of about 29 Pa away from normal atmospheric pressure, with a maximum displacement of air molecules of 1×10^{-5} m.

Intensity Level in Decibels

The loudest tolerable sounds have intensities about 1.0×10^{12} times greater than the faintest detectable sounds. The most intense sound, however, isn't perceived as being 1.0×10^{12} times louder than the faintest sound because the sensation of loudness is approximately logarithmic in the human ear. (For a review of logarithms, see Section A.3, heading G, in Appendix A.) The relative intensity of a sound is called the **intensity level** or **decibel level,** defined by

$$\beta \equiv 10 \log\left(\frac{I}{I_0}\right) \qquad [14.7]$$

◀ Intensity level

The constant $I_0 = 1.0 \times 10^{-12}$ W/m² is the reference intensity, the sound intensity at the threshold of hearing, I is the intensity, and β is the corresponding intensity level measured in decibels (dB). (The word *decibel*, which is one-tenth of a *bel*, comes from the name of the inventor of the telephone, Alexander Graham Bell (1847–1922).)

To get a feel for various decibel levels, we can substitute a few representative numbers into Equation 14.7, starting with $I = 1.0 \times 10^{-12}$ W/m²:

$$\beta = 10 \log\left(\frac{1.0 \times 10^{-12}\ \text{W/m}^2}{1.0 \times 10^{-12}\ \text{W/m}^2}\right) = 10 \log(1) = 0\ \text{dB}$$

Tip 14.1 Intensity Versus Intensity Level

Don't confuse intensity with intensity level. Intensity is a physical quantity with units of watts per meter squared; intensity level, or decibel level, is a convenient mathematical transformation of intensity to a logarithmic scale.

Table 14.2 Intensity Levels in Decibels for Different Sources

Source of Sound	β(dB)
Nearby jet airplane	150
Jackhammer, machine gun	130
Siren, rock concert	120
Subway, power mower	100
Busy traffic	80
Vacuum cleaner	70
Normal conversation	50
Mosquito buzzing	40
Whisper	30
Rustling leaves	10
Threshold of hearing	0

From this result, we see that the lower threshold of human hearing has been chosen to be zero on the decibel scale. Progressing upward by powers of ten yields

$$\beta = 10 \log \left(\frac{1.0 \times 10^{-11}\ \mathrm{W/m^2}}{1.0 \times 10^{-12}\ \mathrm{W/m^2}} \right) = 10 \log (10) = 10\ \mathrm{dB}$$

$$\beta = 10 \log \left(\frac{1.0 \times 10^{-10}\ \mathrm{W/m^2}}{1.0 \times 10^{-12}\ \mathrm{W/m^2}} \right) = 10 \log (100) = 20\ \mathrm{dB}$$

Notice the pattern: *Multiplying* a given intensity by ten *adds* 10 db to the intensity level. This pattern holds throughout the decibel scale. For example, a 50-dB sound is 10 times as intense as a 40-dB sound, whereas a 60-dB sound is 100 times as intense as a 40-dB sound.

On this scale, the threshold of pain ($I = 1.0\ \mathrm{W/m^2}$) corresponds to an intensity level of $\beta = 10 \log (1/1 \times 10^{-12}) = 10 \log (10^{12}) = 120$ dB. Nearby jet airplanes can create intensity levels of 150 dB, and subways and riveting machines have levels of 90-100 dB. The electronically amplified sound heard at rock concerts can attain levels of up to 120 dB, the threshold of pain. Exposure to such high intensity levels can seriously damage the ear. Earplugs are recommended whenever prolonged intensity levels exceed 90 dB. Recent evidence suggests that noise pollution, which is common in most large cities and in some industrial environments, may be a contributing factor to high blood pressure, anxiety, and nervousness. Table 14.2 gives the approximate intensity levels of various sounds.

EXAMPLE 14.2 | A Noisy Grinding Machine

GOAL Working with watts and decibels.

PROBLEM A noisy grinding machine in a factory produces a sound intensity of $1.00 \times 10^{-5}\ \mathrm{W/m^2}$ at a certain location. Calculate **(a)** the decibel level of this machine at that point and **(b)** the new intensity level when a second, identical machine is added to the factory. **(c)** A certain number of additional such machines are put into operation alongside these two machines. When all the machines are running at the same time the decibel level is 77.0 dB. Find the sound intensity. (Assume, in each part, that the sound intensity is measured at the same point, equidistant from all the machines.)

STRATEGY Parts (a) and (b) require substituting into the decibel formula, Equation 14.7, with the intensity in part (b) twice the intensity in part (a). In part (c), the intensity level in decibels is given, and it's necessary to work backwards, using the inverse of the logarithm function, to get the intensity in watts per meter squared.

SOLUTION

(a) Calculate the intensity level of the single grinder.

Substitute the intensity into the decibel formula:

$$\beta = 10 \log \left(\frac{1.00 \times 10^{-5}\ \mathrm{W/m^2}}{1.00 \times 10^{-12}\ \mathrm{W/m^2}} \right) = 10 \log (10^7)$$

$$= 70.0\ \mathrm{dB}$$

(b) Calculate the new intensity level when an additional machine is added.

Substitute twice the intensity of part (a) into the decibel formula:

$$\beta = 10 \log \left(\frac{2.00 \times 10^{-5}\ \mathrm{W/m^2}}{1.00 \times 10^{-12}\ \mathrm{W/m^2}} \right) = 73.0\ \mathrm{dB}$$

(c) Find the intensity corresponding to an intensity level of 77.0 dB.

Substitute 77.0 dB into the decibel formula and divide both sides by 10:

$$\beta = 77.0\ \mathrm{dB} = 10 \log \left(\frac{I}{I_0} \right)$$

$$7.70 = \log \left(\frac{I}{10^{-12}\ \mathrm{W/m^2}} \right)$$

Make each side the exponent of 10. On the right-hand side, $10^{\log u} = u$, by definition of base 10 logarithms.

$$10^{7.70} = 5.01 \times 10^7 = \frac{I}{1.00 \times 10^{-12}\ \mathrm{W/m^2}}$$

$$I = 5.01 \times 10^{-5}\ \mathrm{W/m^2}$$

REMARKS The answer is five times the intensity of the single grinder, so in part (c) there are five such machines operating simultaneously. Because of the logarithmic definition of intensity level, large changes in intensity correspond to small changes in intensity level.

QUESTION 14.2 By how many decibels is the sound intensity level raised when the sound intensity is doubled?

EXERCISE 14.2 Suppose a manufacturing plant has an average sound intensity level of 97.0 dB created by 25 identical machines. (a) Find the total intensity created by all the machines. (b) Find the sound intensity created by one such machine. (c) What's the sound intensity level if five such machines are running?

ANSWERS (a) 5.01×10^{-3} W/m^2 (b) 2.00×10^{-4} W/m^2 (c) 90.0 dB

BIO APPLICATION
OSHA Noise-Level Regulations

Federal OSHA regulations now demand that no office or factory worker be exposed to noise levels that average more than 85 dB over an 8-h day. From a management point of view, here's the good news: one machine in the factory may produce a noise level of 70 dB, but a second machine, though doubling the total intensity, increases the noise level by only 3 dB. Because of the logarithmic nature of intensity levels, doubling the intensity doesn't double the intensity level; in fact, it alters it by a surprisingly small amount. This means that equipment can be added to the factory without appreciably altering the intensity level of the environment.

Now here's the bad news: as you remove noisy machinery, the intensity level isn't lowered appreciably. In Exercise 14.2, reducing the intensity level by 7 dB would require the removal of 20 of the 25 machines! To lower the level another 7 dB would require removing 80% of the remaining machines, in which case only one machine would remain.

14.5 Spherical and Plane Waves

LEARNING OBJECTIVES

1. Discuss wave fronts and rays and apply them to both spherical waves and plane waves.
2. Relate the intensity of a spherical sound wave to the distance from its point source.

If a small spherical object oscillates so that its radius changes periodically with time, a spherical sound wave is produced (Fig. 14.4). The wave moves outward from the source at a constant speed.

Because all points on the vibrating sphere behave in the same way, we conclude that the energy in a spherical wave propagates equally in all directions. This means that no one direction is preferred over any other. If P_{av} is the average power emitted by the source, then at any distance r from the source, this power must be distributed over a spherical surface of area $4\pi r^2$, assuming no absorption in the medium. (Recall that $4\pi r^2$ is the surface area of a sphere.) Hence, the **intensity** of the sound at a distance r from the source is

$$I = \frac{\text{average power}}{\text{area}} = \frac{P_{av}}{A} = \frac{P_{av}}{4\pi r^2} \qquad [14.8]$$

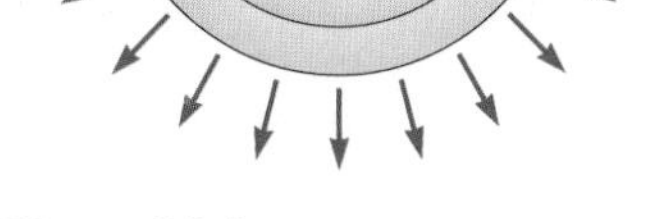

Figure 14.4 A spherical wave propagating radially outward from an oscillating sphere. The intensity of the wave varies as $1/r^2$.

This equation shows that the intensity of a wave decreases with increasing distance from its source, as you might expect. The fact that I varies as $1/r^2$ is a result of the assumption that the small source (sometimes called a **point source**) emits a spherical wave. (In fact, light waves also obey this so-called inverse-square relationship.) Because the average power is the same through any spherical surface centered at

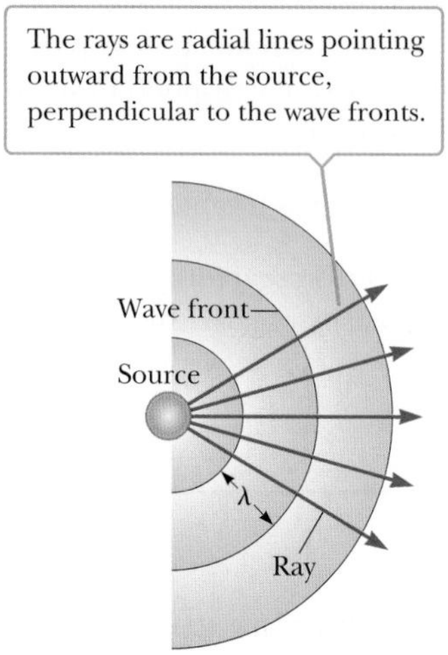

Figure 14.5 Spherical waves emitted by a point source. The circular arcs represent the spherical wave fronts concentric with the source.

the source, we see that the intensities at distances r_1 and r_2 (Fig. 14.4) from the center of the source are

$$I_1 = \frac{P_{av}}{4\pi r_1^2} \qquad I_2 = \frac{P_{av}}{4\pi r_2^2}$$

The ratio of the intensities at these two spherical surfaces is

$$\frac{I_1}{I_2} = \frac{r_2^2}{r_1^2} \qquad [14.9]$$

It's useful to represent spherical waves graphically with a series of circular arcs (lines of maximum intensity) concentric with the source representing part of a spherical surface, as in Figure 14.5. We call such an arc a **wave front.** The distance between adjacent wave fronts equals the wavelength λ. The radial lines pointing outward from the source and perpendicular to the arcs are called **rays.**

Now consider a small portion of a wave front that is at a *great* distance (relative to λ) from the source, as in Figure 14.6. In this case the rays are nearly parallel to each other and the wave fronts are very close to being planes. At distances from the source that are great relative to the wavelength, therefore, we can approximate the wave front with parallel planes, called **plane waves.** Any small portion of a spherical wave that is far from the source can be considered a plane wave. Figure 14.7 illustrates a plane wave propagating along the *x*-axis. If the positive *x*-direction is taken to be the direction of the wave motion (or ray) in this figure, then the wave fronts are parallel to the plane containing the *y*- and *z*-axes.

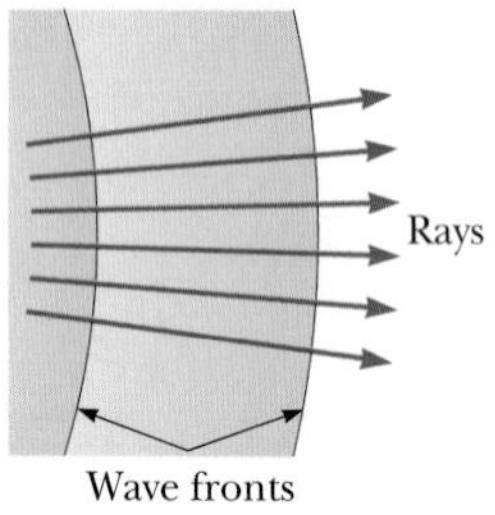

Figure 14.6 Far away from a point source, the wave fronts are nearly parallel planes and the rays are nearly parallel lines perpendicular to the planes. A small segment of a spherical wave front is approximately a plane wave.

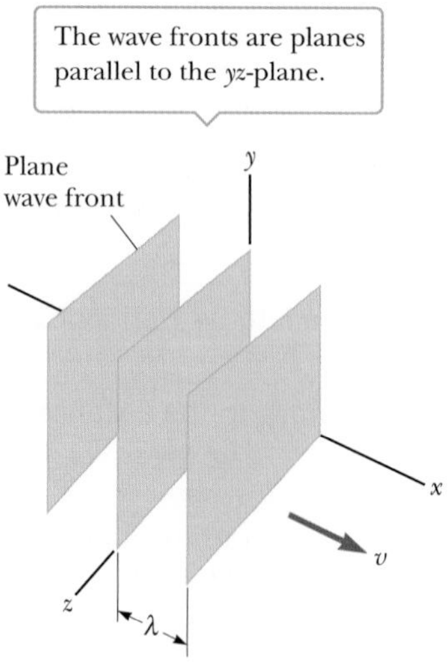

Figure 14.7 A representation of a plane wave moving in the positive *x*-direction with a speed v.

EXAMPLE 14.3 Intensity Variations of a Point Source

GOAL Relate sound intensities and their distances from a point source.

PROBLEM A small source emits sound waves with a power output of 80.0 W. **(a)** Find the intensity 3.00 m from the source. **(b)** At what distance would the intensity be one-fourth as much as it is at $r = 3.00$ m? **(c)** Find the distance at which the sound intensity level is 40.0 dB.

STRATEGY The source is small, so the emitted waves are spherical and the intensity in part (a) can be found by substituting values into Equation 14.8. Part (b) involves solving for r in Equation 14.8 followed by substitution (although Eq. 14.9 can be used instead). In part (c), convert from the sound intensity level to the intensity in W/m^2, using Equation 14.7. Then substitute into Equation 14.9 (although Eq. 14.8 could be used instead) and solve for r_2.

SOLUTION

(a) Find the intensity 3.00 m from the source.

Substitute $P_{av} = 80.0$ W and $r = 3.00$ m into Equation 14.8:

$$I = \frac{P_{av}}{4\pi r^2} = \frac{80.0\text{ W}}{4\pi(3.00\text{ m})^2} = 0.707\text{ W/m}^2$$

(b) At what distance would the intensity be one-fourth as much as it is at $r = 3.00$ m?

Take $I = (0.707\text{ W/m}^2)/4$, and solve for r in Equation 14.8:

$$r = \left(\frac{P_{av}}{4\pi I}\right)^{1/2} = \left[\frac{80.0\text{ W}}{4\pi(0.707\text{ W/m}^2)/4.0}\right]^{1/2} = 6.00\text{ m}$$

(c) Find the distance at which the intensity level is 40.0 dB.

Convert the intensity level of 40.0 dB to an intensity in W/m² by solving Equation 14.7 for I:

$$40.0 = 10\log\left(\frac{I}{I_0}\right) \rightarrow 4.00 = \log\left(\frac{I}{I_0}\right)$$

$$10^{4.00} = \frac{I}{I_0} \rightarrow I = 10^{4.00}I_0 = 1.00 \times 10^{-8}\text{ W/m}^2$$

Solve Equation 14.9 for $r_2^{\,2}$, substitute the intensity and the result of part (a), and take the square root:

$$\frac{I_1}{I_2} = \frac{r_2^{\,2}}{r_1^{\,2}} \rightarrow r_2^{\,2} = r_1^{\,2}\frac{I_1}{I_2}$$

$$r_2^{\,2} = (3.00\text{ m})^2\left(\frac{0.707\text{ W/m}^2}{1.00 \times 10^{-8}\text{ W/m}^2}\right)$$

$$r_2 = 2.52 \times 10^4\text{ m}$$

REMARKS Once the intensity is known at one position a certain distance away from the source, it's easier to use Equation 14.9 rather than Equation 14.8 to find the intensity at any other location. This is particularly true for part (b), where, using Equation 14.9, we can see right away that doubling the distance reduces the intensity to one- fourth its previous value.

QUESTION 14.3 The power output of a sound system is increased by a factor of 25. By what factor should you adjust your distance from the speakers so the sound intensity is the same?

EXERCISE 14.3 Suppose a certain jet plane creates an intensity level of 125 dB at a distance of 5.00 m. What intensity level does it create on the ground directly underneath it when flying at an altitude of 2.00 km?

ANSWER 73.0 dB

14.6 The Doppler Effect

LEARNING OBJECTIVES

1. Discuss the physical origins of the Doppler effect.
2. Apply the equations for Doppler-shifted frequencies to moving sources and observers of sound.
3. Discuss the physical conditions that result in a shock wave and define the Mach number.

If a car or truck is moving while its horn is blowing, the frequency of the sound you hear is higher as the vehicle approaches you and lower as it moves away from you. This phenomenon is one example of the *Doppler effect,* named for Austrian physicist Christian Doppler (1803–1853), who discovered it. The same effect is heard if you're on a motorcycle and the horn is stationary: the frequency is higher as you approach the source and lower as you move away.

Although the Doppler effect is most often associated with sound, it's common to all waves, including light.

In deriving the Doppler effect, we assume the air is stationary and that all speed measurements are made relative to this stationary medium. In the general case, the speed of the observer v_O, the speed of the source v_S, and the speed of sound v are all measured relative to the medium in which the sound is propagated.

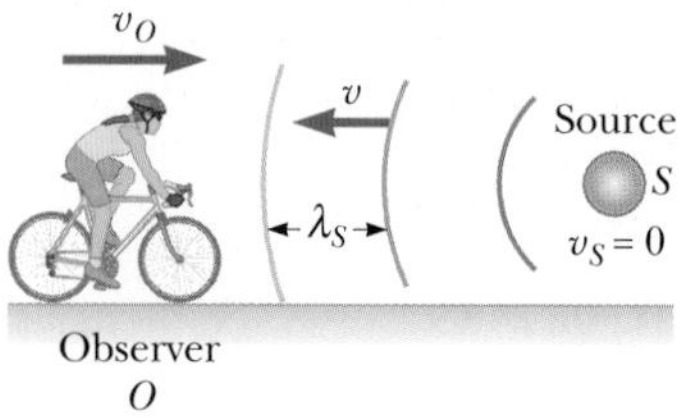

Figure 14.8 An observer moving with a speed v_O *toward* a stationary point source (S) hears a frequency f_O that is *greater* than the source frequency f_S.

Case 1: The Observer Is Moving Relative to a Stationary Source

In Figure 14.8 an observer is moving with a speed of v_O toward the source (considered a point source), which is at rest ($v_S = 0$).

We take the frequency of the source to be f_S, the wavelength of the source to be λ_S, and the speed of sound in air to be v. If both observer and source are stationary, the observer detects f_S wave fronts per second. (That is, when $v_O = 0$ and $v_S = 0$, the observed frequency f_O equals the source frequency f_S.) When moving toward the source, the observer moves a distance of $v_O t$ in t seconds. During this interval, **the observer detects an additional number of wave fronts.** The number of extra wave fronts is equal to the distance traveled, $v_O t$, divided by the wavelength λ_S:

$$\text{Additional wave fronts detected} = \frac{v_O t}{\lambda_S}$$

Divide this equation by the time t to get the number of additional wave fronts detected *per second*, v_O/λ_S. Hence, the frequency heard by the observer is *increased* to

$$f_O = f_S + \frac{v_O}{\lambda_S}$$

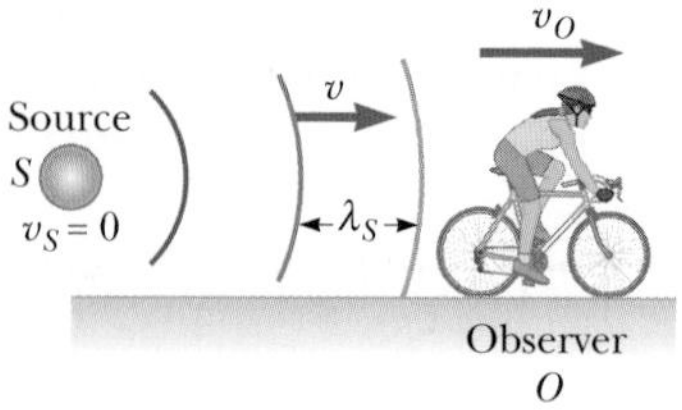

Figure 14.9 An observer moving with a speed of v_O *away* from a stationary source hears a frequency f_O that is *lower* than the source frequency f_S.

Substituting $\lambda_S = v/f_S$ into this expression for f_O, we obtain

$$f_O = f_S\left(\frac{v + v_O}{v}\right) \qquad [14.10]$$

When the observer is *moving away* from a stationary source (Fig. 14.9), the observed frequency decreases. A derivation yields the same result as Equation 14.10, but with $v - v_O$ in the numerator. Therefore, when the observer is moving away from the source, substitute $-v_O$ for v_O in Equation 14.10.

Case 2: The Source Is Moving Relative to a Stationary Observer

Now consider a source moving toward an observer at rest, as in Figure 14.10. Here, the wave fronts passing observer A are closer together because the source is moving in the direction of the outgoing wave. As a result, the wavelength λ_O measured

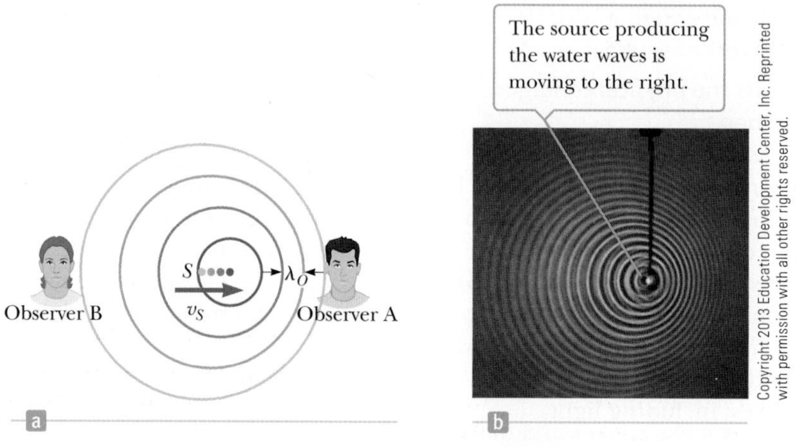

Figure 14.10 (a) A source S moving with speed v_S toward stationary observer A and away from stationary observer B. Observer A hears an *increased* frequency, and observer B hears a *decreased* frequency. (b) The Doppler effect in water, observed in a ripple tank.

by observer A is shorter than the wavelength λ_S of the source at rest. During each vibration, which lasts for an interval T (the period), the source moves a distance $v_S T = v_S/f_S$ and **the wavelength is shortened by that amount.** The observed wavelength is therefore given by

$$\lambda_O = \lambda_S - \frac{v_S}{f_S}$$

Because $\lambda_S = v/f_S$, the frequency observed by A is

$$f_O = \frac{v}{\lambda_O} = \frac{v}{\lambda_S - \frac{v_S}{f_S}} = \frac{v}{\frac{v}{f_S} - \frac{v_S}{f_S}}$$

or

$$f_O = f_S\left(\frac{v}{v - v_S}\right) \qquad [14.11]$$

As expected, **the observed frequency increases when the source is moving toward the observer.** When the source is *moving away* from an observer at rest, the minus sign in the denominator must be replaced with a plus sign, so the factor becomes $(v + v_S)$.

Tip 14.2 Doppler Effect Doesn't Depend on Distance

The sound from a source approaching at constant speed will increase in intensity, but the observed (elevated) frequency will remain unchanged. The Doppler effect doesn't depend on distance.

General Case

When both the source and the observer are in motion relative to Earth, Equations 14.10 and 14.11 can be combined to give

$$f_O = f_S\left(\frac{v + v_O}{v - v_S}\right) \qquad [14.12]$$

◀ Doppler shift: observer and source in motion

In this expression, the signs for the values substituted for v_O and v_S depend on the direction of the velocity. When the observer moves *toward* the source, a *positive* speed is substituted for v_O; when the observer moves *away from* the source, a *negative* speed is substituted for v_O. Similarly, a *positive* speed is substituted for v_S when the source moves *toward* the observer, a *negative* speed when the source moves away from the observer.

Choosing incorrect signs is the most common mistake made in working a Doppler effect problem. The following rules may be helpful: The word *toward* is associated with an *increase* in the observed frequency; the words *away from* are associated with a *decrease* in the observed frequency.

These two rules derive from the physical insight that when the observer is moving toward the source (or the source toward the observer), there is a smaller observed period between wave crests, hence a larger frequency, with the reverse holding—a smaller observed frequency—when the observer is moving away from the source (or the source away from the observer). Keep the physical insight in mind whenever you're in doubt about the signs in Equation 14.12: Adjust them as necessary to get the correct physical result.

The second most common mistake made in applying Equation 14.12 is to accidentally reverse numerator and denominator. Some find it helpful to remember the equation in the following form:

$$\frac{f_O}{v + v_O} = \frac{f_S}{v - v_S}$$

The advantage of this form is its symmetry: both sides are very nearly the same, with O's on the left and S's on the right. Forgetting which side has the plus sign and which has the minus sign is not a serious problem as long as physical insight is used to check the answer and make adjustments as necessary.

Quick Quiz

14.2 Suppose you're on a hot air balloon ride, carrying a buzzer that emits a sound of frequency f. If you accidentally drop the buzzer over the side while the balloon is rising at constant speed, what can you conclude about the sound you hear as the buzzer falls toward the ground? (a) The frequency and intensity increase. (b) The frequency decreases and the intensity increases. (c) The frequency decreases and the intensity decreases. (d) The frequency remains the same, but the intensity decreases.

APPLYING PHYSICS 14.2 | Out-of-Tune Speakers

Suppose you place your stereo speakers far apart and run past them from right to left or left to right. If you run rapidly enough and have excellent pitch discrimination, you may notice that the music playing seems to be out of tune when you're between the speakers. Why?

EXPLANATION When you are between the speakers, you are running away from one of them and toward the other, so there is a Doppler shift downward for the sound from the speaker behind you and a Doppler shift upward for the sound from the speaker ahead of you. As a result, the sound from the two speakers will not be in tune. A simple calculation shows that a world-class sprinter could run fast enough to generate about a semitone difference in the sound from the two speakers. ■

EXAMPLE 14.4 | Listen, but Don't Stand on the Track

GOAL Solve a Doppler shift problem when only the source is moving.

PROBLEM A train moving at a speed of 40.0 m/s sounds its whistle, which has a frequency of 5.00×10^2 Hz. Determine the frequency heard by a stationary observer as the train *approaches* the observer. The ambient temperature is 24.0°C.

STRATEGY Use Equation 14.4 to get the speed of sound at the ambient temperature, then substitute values into Equation 14.12 for the Doppler shift. Because the train approaches the observer, the observed frequency will be larger. Choose the sign of v_S to reflect this fact.

SOLUTION

Use Equation 14.4 to calculate the speed of sound in air at $T = 24.0$°C:

$$v = (331 \text{ m/s})\sqrt{\frac{T}{273 \text{ K}}}$$

$$= (331 \text{ m/s})\sqrt{\frac{(273 + 24.0) \text{ K}}{273 \text{ K}}} = \boxed{345 \text{ m/s}}$$

The observer is stationary, so $v_O = 0$. The train is moving *toward* the observer, so $v_S = +40.0$ m/s (*positive*). Substitute these values and the speed of sound into the Doppler shift equation:

$$f_O = f_S\left(\frac{v + v_O}{v - v_S}\right)$$

$$= (5.00 \times 10^2 \text{ Hz})\left(\frac{345 \text{ m/s}}{345 \text{ m/s} - 40.0 \text{ m/s}}\right)$$

$$= \boxed{566 \text{ Hz}}$$

REMARKS If the train were going away from the observer, $v_S = -40.0$ m/s would have been chosen instead.

QUESTION 14.4 Does the Doppler shift change due to temperature variations? If so, why? For typical daily variations in temperature in a moderate climate, would any change in the Doppler shift be best characterized as (a) nonexistent, (b) small, or (c) large?

EXERCISE 14.4 Determine the frequency heard by the stationary observer as the train *recedes* from the observer.

ANSWER 448 Hz

on the string are called nodes, labeled N in Figure 14.17a. From the figure, observe that the distance between adjacent nodes is one-half the wavelength of the wave:

$$d_{NN} = \frac{1}{2}\lambda$$

Consider a string of length L that is fixed at both ends, as in Figure 14.18. For a string, we can set up standing-wave patterns at many frequencies—the more loops, the higher the frequency. Figure 14.19 is a multiflash photograph of a standing wave on a string.

First, **the ends of the string must be nodes, because these points are fixed.** If the string is displaced at its midpoint and released, the vibration shown in Figure 14.18b can be produced, in which case the center of the string is an antinode, labeled A. Note that from end to end, the pattern is N–A–N. The distance from a node to its adjacent antinode, N–A, is always equal to a quarter wavelength, $\lambda_1/4$.

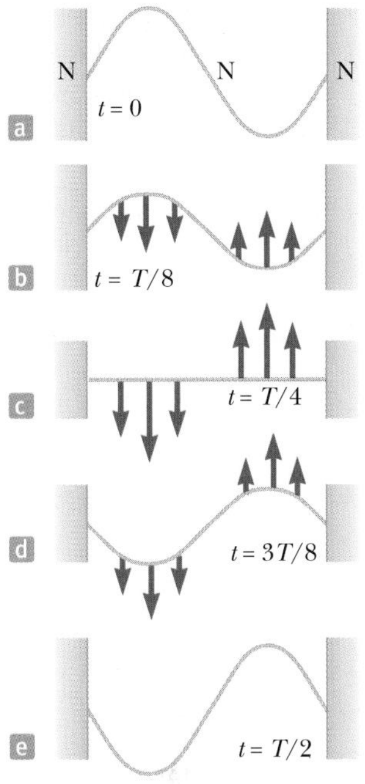

Figure 14.17 A standing-wave pattern in a stretched string, shown by snapshots of the string during one-half of a cycle. In part (a) N denotes a node.

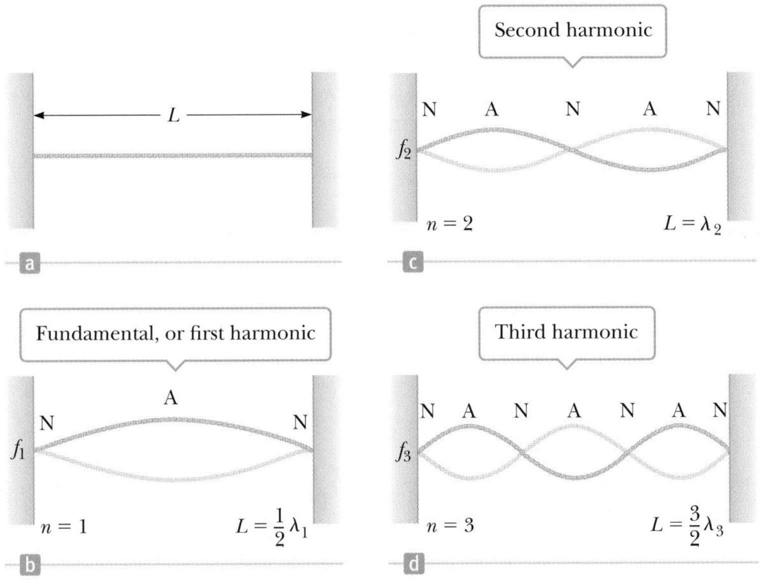

Figure 14.18 (a) Standing waves in a stretched string of length L fixed at both ends. The characteristic frequencies of vibration form a harmonic series: (b) the fundamental frequency, or first harmonic; (c) the second harmonic; and (d) the third harmonic. Note that N denotes a node, A an antinode.

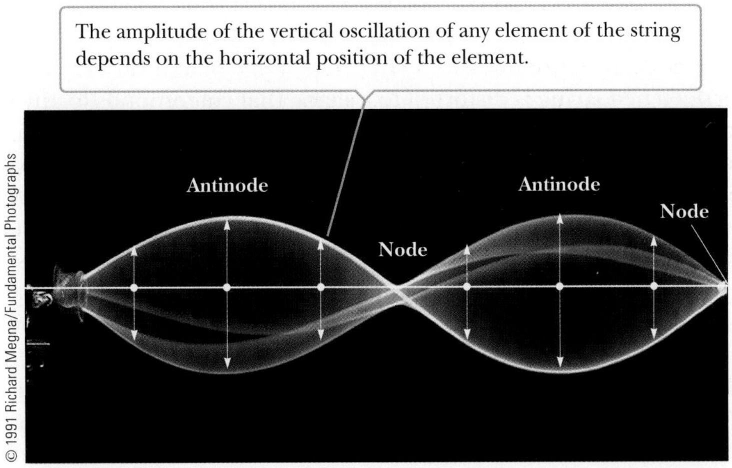

Figure 14.19 Multiflash photograph of a standing-wave two-loop pattern in a second harmonic ($n = 2$), using a cord driven by a vibrator at the left end.

There are two such segments, N–A and A–N, so $L = 2(\lambda_1/4) = \lambda_1/2$, and $\lambda_1 = 2L$. The frequency of this vibration is therefore

$$f_1 = \frac{v}{\lambda_1} = \frac{v}{2L} \qquad [14.15]$$

Recall that the speed of a wave on a string is $v = \sqrt{F/\mu}$, where F is the tension in the string and μ is its mass per unit length (Chapter 13). Substituting into Equation 14.15, we obtain

$$f_1 = \frac{1}{2L}\sqrt{\frac{F}{\mu}} \qquad [14.16]$$

This lowest frequency of vibration is called the **fundamental frequency** of the vibrating string, or the **first harmonic.**

The first harmonic has nodes only at the ends: the points of attachment, with node-antinode pattern of N–A–N. The next harmonic, called the **second harmonic** (also called the **first overtone**), can be constructed by inserting an additional node–antinode segment between the endpoints. This makes the pattern N–A–N–A–N, as in Figure 14.18c. We count the node–antinode pairs: N–A, A–N, N–A, and A–N, four segments in all, each representing a quarter wavelength. We then have $L = 4(\lambda_2/4) = \lambda_2$, and the second harmonic (first overtone) is

$$f_2 = \frac{v}{\lambda_2} = \frac{v}{L} = 2\left(\frac{v}{2L}\right) = 2f_1$$

This frequency is equal to *twice* the fundamental frequency. The **third harmonic (second overtone)** is constructed similarly. Inserting one more N–A segment, we obtain Figure 14.18d, the pattern of nodes reading N–A–N–A–N–A–N. There are six node–antinode segments, so $L = 6(\lambda_3/4) = 3(\lambda_3/2)$, which means that $\lambda_3 = 2L/3$, giving

$$f_3 = \frac{v}{\lambda_3} = \frac{3v}{2L} = 3f_1$$

All the higher harmonics, it turns out, are positive integer multiples of the fundamental:

Natural frequencies of a string fixed at both ends ▶

$$f_n = nf_1 = \frac{n}{2L}\sqrt{\frac{F}{\mu}} \qquad n = 1, 2, 3, \ldots \qquad [14.17]$$

The frequencies f_1, $2f_1$, $3f_1$, and so on form a **harmonic series.**

Quick Quiz

14.4 Which of the following frequencies are higher harmonics of a string with fundamental frequency of 150 Hz? (a) 200 Hz (b) 300 Hz (c) 400 Hz (d) 500 Hz (e) 600 Hz

When a stretched string is distorted to a shape that corresponds to any one of its harmonics, after being released it vibrates only at the frequency of that harmonic. If the string is struck or bowed, however, the resulting vibration includes different amounts of various harmonics, including the fundamental frequency. Waves not in the harmonic series are quickly damped out on a string fixed at both ends. In effect, when disturbed, the string "selects" the standing-wave frequencies. As we'll see later, the presence of several harmonics on a string gives stringed instruments their characteristic sound, which enables us to distinguish one from another even when they are producing identical fundamental frequencies.

APPLICATION
Tuning a Musical Instrument

The frequency of a string on a musical instrument can be changed by varying either the tension or the length. The tension in guitar and violin strings is varied by turning pegs on the neck of the instrument. As the tension is increased, the frequency of the harmonic series increases according to Equation 14.17. Once the instrument is tuned, the musician varies the frequency by pressing the strings against the neck at a variety of positions, thereby changing the effective lengths of the vibrating portions of the strings. As the length is reduced, the frequency again increases, as follows from Equation 14.17.

Finally, Equation 14.17 shows that a string of fixed length can be made to vibrate at a lower fundamental frequency by increasing its mass per unit length. This increase is achieved in the bass strings of guitars and pianos by wrapping the strings with metal windings.

EXAMPLE 14.7 | Guitar Fundamentals

GOAL Apply standing-wave concepts to a stringed instrument.

PROBLEM The high E string on a certain guitar measures 64.0 cm in length and has a fundamental frequency of 329 Hz. When a guitarist presses down so that the string is in contact with the first fret (Fig. 14.20a), the string is shortened so that it plays an F note that has a frequency of 349 Hz. **(a)** How far is the fret from the nut? **(b)** Overtones can be produced on a guitar string by gently placing the index finger in the location of a node of a higher harmonic. The string should be touched, but not depressed against a fret. (Given the width of a finger, pressing too hard will damp out higher harmonics as well.) The fundamental frequency is thereby suppressed, making it possible to hear overtones. Where on the guitar string relative to the nut should the finger be lightly placed so as to hear the second harmonic of the high E string? The fourth harmonic? (This is equivalent to finding the location of the nodes in each case.)

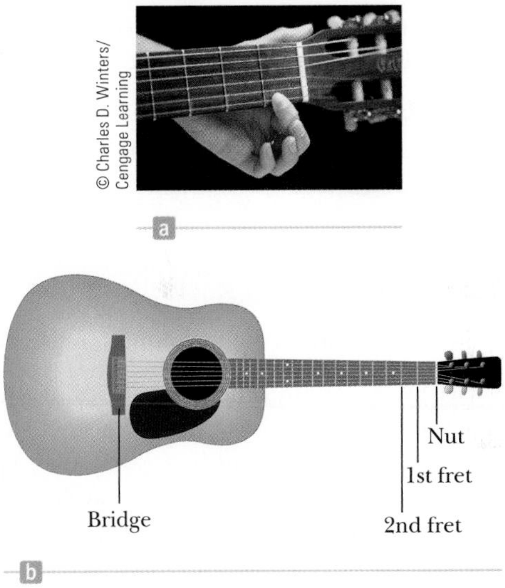

Figure 14.20 (Example 14.7) (a) Playing an F note on a guitar. (b) Some parts of a guitar.

STRATEGY For part (a) use Equation 14.15, corresponding to the fundamental frequency, to find the speed of waves on the string. Shortening the string by playing a higher note doesn't affect the wave speed, which depends only on the tension and linear density of the string (which are unchanged). Solve Equation 14.15 for the new length L, using the new fundamental frequency, and subtract this length from the original length to find the distance from the nut to the first fret. In part (b) remember that the distance from node to node is half a wavelength. Calculate the wavelength, divide it in two, and locate the nodes, which are integral numbers of half-wavelengths from the nut. *Note:* The nut is a small piece of wood or ebony at the top of the fret board. The distance from the nut to the bridge (below the sound hole) is the length of the string. (See Fig. 14.20b.)

SOLUTION

(a) Find the distance from the nut to the first fret.

Substitute $L_0 = 0.640$ m and $f_1 = 329$ Hz into Equation 14.15, finding the wave speed on the string:

$$f_1 = \frac{v}{2L_0}$$

$$v = 2L_0 f_1 = 2(0.640\text{ m})(329\text{ Hz}) = 421\text{ m/s}$$

Solve Equation 14.15 for the length L, and substitute the wave speed and the frequency of an F note.

$$L = \frac{v}{2f} = \frac{421\text{ m/s}}{2(349\text{ Hz})} = 0.603\text{ m} = 60.3\text{ cm}$$

Subtract this length from the original length L_0 to find the distance from the nut to the first fret:

$$\Delta x = L_0 - L = 64.0\text{ cm} - 60.3\text{ cm} = \boxed{3.7\text{ cm}}$$

(b) Find the locations of nodes for the second and fourth harmonics.

The second harmonic has a wavelength $\lambda_2 = L_0 =$ 64.0 cm. The distance from nut to node corresponds to half a wavelength.

$$\Delta x = \tfrac{1}{2}\lambda_2 = \tfrac{1}{2}L_0 = 32.0\text{ cm}$$

(Continued)

The fourth harmonic, of wavelength $\lambda_4 = \frac{1}{2}L_O = 32.0$ cm, has three nodes between the endpoints:

$$\Delta x = \frac{1}{2}\lambda_4 = \boxed{16.0 \text{ cm}}, \Delta x = 2(\lambda_4/2) = \boxed{32.0 \text{ cm}},$$

$$\Delta x = 3(\lambda_4/2) = \boxed{48.0 \text{ cm}}$$

REMARKS Placing a finger at the position $\Delta x = 32.0$ cm damps out the fundamental and odd harmonics, but not all the higher even harmonics. The second harmonic dominates, however, because the rest of the string is free to vibrate. Placing the finger at $\Delta x = 16.0$ cm or 48.0 cm damps out the first through third harmonics, allowing the fourth harmonic to be heard.

QUESTION 14.7 True or False: If a guitar string has length L, gently placing a thin object at the position $(\frac{1}{2})^n L$ will always result in the sounding of a higher harmonic, where n is a positive integer.

EXERCISE 14.7 Pressing the E string down on the fret board just above the second fret pinches the string firmly against the fret, giving an F-sharp, which has frequency 3.70×10^2 Hz. (a) Where should the second fret be located? (b) Find two locations where you could touch the open E string and hear the third harmonic.

ANSWERS (a) 7.1 cm from the nut and 3.4 cm from the first fret. Note that the distance from the first to the second fret isn't the same as from the nut to the first fret. (b) 21.3 cm and 42.7 cm from the nut

EXAMPLE 14.8 Harmonics of a Stretched Wire

GOAL Calculate string harmonics, relate them to sound, and combine them with tensile stress.

PROBLEM **(a)** Find the frequencies of the fundamental, second, and third harmonics of a steel wire 1.00 m long with a mass per unit length of 2.00×10^{-3} kg/m and under a tension of 80.0 N. **(b)** Find the wavelengths of the sound waves created by the vibrating wire for all three modes. Assume the speed of sound in air is 345 m/s. **(c)** Suppose the wire is carbon steel with a density of 7.80×10^3 kg/m^3, a cross-sectional area $A = 2.56 \times 10^{-7}$ m^2, and an elastic limit of 2.80×10^8 Pa. Find the fundamental frequency if the wire is tightened to the elastic limit. Neglect any stretching of the wire (which would slightly reduce the mass per unit length).

STRATEGY (a) It's easiest to find the speed of waves on the wire then substitute into Equation 14.15 to find the first harmonic. The next two are multiples of the first, given by Equation 14.17. (b) The frequencies of the sound waves are the same as the frequencies of the vibrating wire, but the wavelengths are different. Use $v_s = f\lambda$, where v_s is the speed of sound in air, to find the wavelengths in air. (c) Find the force corresponding to the elastic limit and substitute it into Equation 14.16.

SOLUTION

(a) Find the first three harmonics at the given tension.

Use Equation 13.18 to calculate the speed of the wave on the wire:

$$v = \sqrt{\frac{F}{\mu}} = \sqrt{\frac{80.0 \text{ N}}{2.00 \times 10^{-3} \text{ kg/m}}} = 2.00 \times 10^2 \text{ m/s}$$

Find the wire's fundamental frequency from Equation 14.15:

$$f_1 = \frac{v}{2L} = \frac{2.00 \times 10^2 \text{ m/s}}{2(1.00 \text{ m})} = \boxed{1.00 \times 10^2 \text{ Hz}}$$

Find the next two harmonics by multiplication:

$$f_2 = 2f_1 = \boxed{2.00 \times 10^2 \text{ Hz}}, f_3 = 3f_1 = \boxed{3.00 \times 10^2 \text{ Hz}}$$

(b) Find the wavelength of the sound waves produced.

Solve $v_s = f\lambda$ for the wavelength and substitute the frequencies:

$$\lambda_1 = v_s/f_1 = (345 \text{ m/s})/(1.00 \times 10^2 \text{ Hz}) = \boxed{3.45 \text{ m}}$$

$$\lambda_2 = v_s/f_2 = (345 \text{ m/s})/(2.00 \times 10^2 \text{ Hz}) = \boxed{1.73 \text{ m}}$$

$$\lambda_3 = v_s/f_3 = (345 \text{ m/s})/(3.00 \times 10^2 \text{ Hz}) = \boxed{1.15 \text{ m}}$$

(c) Find the fundamental frequency corresponding to the elastic limit.

Calculate the tension in the wire from the elastic limit:

$$\frac{F}{A} = \text{elastic limit} \quad \rightarrow \quad F = (\text{elastic limit})A$$

$$F = (2.80 \times 10^8 \text{ Pa})(2.56 \times 10^{-7} \text{ m}^2) = 71.7 \text{ N}$$

Substitute the values of F, μ, and L into Equation 14.16:

$$f_1 = \frac{1}{2L}\sqrt{\frac{F}{\mu}}$$

$$f_1 = \frac{1}{2(1.00 \text{ m})}\sqrt{\frac{71.7 \text{ N}}{2.00 \times 10^{-3} \text{ kg/m}}} = \boxed{94.7 \text{ Hz}}$$

REMARKS From the answer to part (c), it appears we need to choose a thicker wire or use a better grade of steel with a higher elastic limit. The frequency corresponding to the elastic limit is smaller than the fundamental!

QUESTION 14.8 A string on a guitar is replaced with one of lower linear density. To obtain the same frequency sound as previously, must the tension of the new string be (a) greater than, (b) less than, or (c) equal to the tension in the old string?

EXERCISE 14.8 (a) Find the fundamental frequency and second harmonic if the tension in the wire is increased to 115 N. (Assume the wire doesn't stretch or break.) (b) Using a sound speed of 345 m/s, find the wavelengths of the sound waves produced.

ANSWERS (a) 1.20×10^2 Hz, 2.40×10^2 Hz (b) 2.88 m, 1.44 m

14.9 Forced Vibrations and Resonance

LEARNING OBJECTIVES

1. Explain the concept of resonance as applied to forced vibrations.
2. Discuss physical examples of resonance.

In Chapter 13 we learned that the energy of a damped oscillator decreases over time because of friction. It's possible to compensate for this energy loss by applying an external force that does positive work on the system.

For example, suppose an object–spring system having some natural frequency of vibration f_0 is pushed back and forth by a periodic force with frequency f. The system vibrates at the frequency f of the driving force. This type of motion is referred to as a **forced vibration.** Its amplitude reaches a maximum when the frequency of the driving force equals the natural frequency of the system f_0, called the **resonant frequency** of the system. Under this condition, the system is said to be in **resonance.**

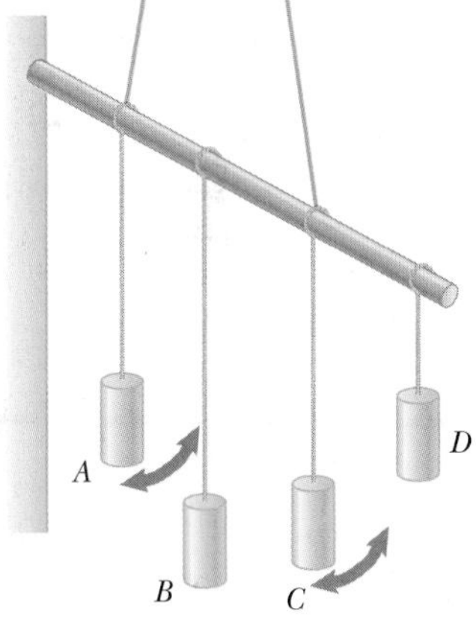

Figure 14.21 A demonstration of resonance.

In Section 14.8 we learned that a stretched string can vibrate in one or more of its natural modes. Here again, if a periodic force is applied to the string, the amplitude of vibration increases as the frequency of the applied force approaches one of the string's natural frequencies of vibration.

Resonance vibrations occur in a wide variety of circumstances. Figure 14.21 illustrates one experiment that demonstrates a resonance condition. Several pendulums of different lengths are suspended from a flexible beam. If one of them, such as *A*, is set in motion, the others begin to oscillate because of vibrations in the flexible beam. Pendulum *C*, the same length as *A*, oscillates with the greatest amplitude because its natural frequency matches that of pendulum *A* (the driving force).

Another simple example of resonance is a child being pushed on a swing, which is essentially a pendulum with a natural frequency that depends on its length. The swing is kept in motion by a series of appropriately timed pushes. For its amplitude to increase, the swing must be pushed each time it returns to the person's hands. This corresponds to a frequency equal to the natural frequency of the swing. If the energy put into the system per cycle of motion equals the energy lost due to friction, the amplitude remains constant.

APPLICATION
Shattering Goblets with the Voice

Opera singers have been known to make audible vibrations in crystal goblets with their powerful voices. This is yet another example of resonance: The sound

Figure 14.22 (a) In 1940, turbulent winds set up torsional vibrations in the Tacoma Narrows Bridge, causing it to oscillate at a frequency near one of the natural frequencies of the bridge structure. (b) Once established, this resonance condition led to the bridge's collapse. A number of scientists, however, have challenged the resonance interpretation.

waves emitted by the singer can set up large-amplitude vibrations in the glass. If a highly amplified sound wave has the right frequency, the amplitude of forced vibrations in the glass increases to the point where the glass becomes heavily strained and shatters.

APPLICATION
Structural Integrity and Resonance

The classic example of structural resonance occurred in 1940, when the Tacoma Narrows bridge in the state of Washington was put into oscillation by the wind (Fig. 14.22). The amplitude of the oscillations increased rapidly and reached a high value until the bridge ultimately collapsed (probably because of metal fatigue). In recent years, however, a number of researchers have called this explanation into question. Gusts of wind, in general, don't provide the periodic force necessary for a sustained resonance condition, and the bridge exhibited large twisting oscillations, rather than the simple up-and-down oscillations expected of resonance.

A more recent example of destruction by structural resonance occurred during the Loma Prieta earthquake near Oakland, California, in 1989. In a mile-long section of the double-decker Nimitz Freeway, the upper deck collapsed onto the lower deck, killing several people. The collapse occurred because that particular section was built on mud fill, whereas other parts were built on bedrock. As seismic waves pass through mud fill or other loose soil, their speed decreases and their amplitude increases. The section of the freeway that collapsed oscillated at the same frequency as other sections, but at a much larger amplitude.

14.10 Standing Waves in Air Columns

LEARNING OBJECTIVES

1. Contrast standing waves in air columns open at one end and at both ends.
2. Apply the equations for the frequencies of open and closed pipes to harmonic systems.
3. Discuss applications of sound waves in closed and open columns.

Standing longitudinal waves can be set up in a tube of air, such as an organ pipe, as the result of interference between sound waves traveling in opposite directions. The relationship between the incident wave and the reflected wave depends on whether the reflecting end of the tube is open or closed. A portion of the sound wave is reflected back into the tube even at an open end. **If one end is closed, a node must exist at that end because the movement of air is restricted. If the end is open, the elements of air have complete freedom of motion, and an antinode exists.**

Figure 14.23a shows the first three modes of vibration of a pipe open at both ends. When air is directed against an edge at the left, longitudinal standing waves are formed and the pipe vibrates at its natural frequencies. Note that, from end to

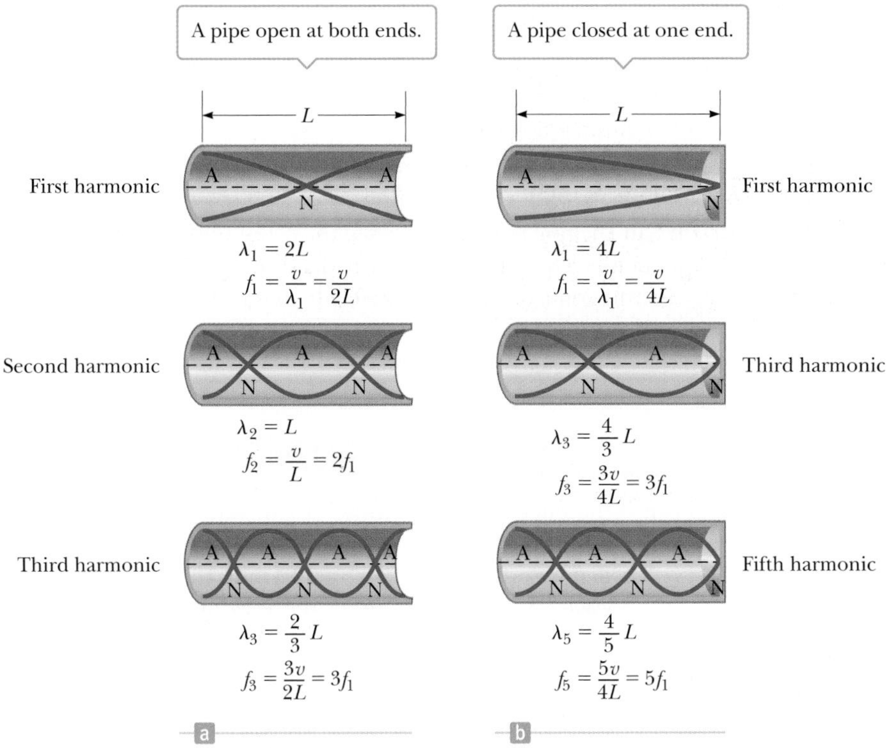

Figure 14.23 (a) Standing longitudinal waves in an organ pipe open at both ends. The natural frequencies $f_1, 2f_1, 3f_1 \ldots$ form a harmonic series. (b) Standing longitudinal waves in an organ pipe closed at one end. Only *odd* harmonics are present, and the natural frequencies are f_1, $3f_1$, $5f_1$, and so on.

end, the pattern is A–N–A, the same pattern as in the vibrating string, except node and antinode have exchanged positions. As before, an antinode and its adjacent node, A–N, represent a quarter-wavelength, and there are two, A–N and N–A, so $L = 2(\lambda_1/4) = \lambda_1/2$ and $\lambda_1 = 2L$. The fundamental frequency of the pipe open at both ends is then $f_1 = v/\lambda_1 = v/2L$. The next harmonic has an additional node and antinode between the ends, creating the pattern A–N–A–N–A. We count the pairs: A–N, N–A, A–N, and N–A, making four segments, each with length $\lambda_2/4$. We have $L = 4(\lambda_2/4) = \lambda_2$, and the second harmonic (first overtone) is $f_2 = v/\lambda_2 = v/L = 2(v/2L) = 2f_1$. All higher harmonics, it turns out, are positive integer multiples of the fundamental:

Tip 14.4 Sound Waves Are Not Transverse

The standing longitudinal waves in Figure 14.23 are drawn as transverse waves only because it's difficult to draw longitudinal displacements: they're in the same direction as the wave propagation. In the figure, the vertical axis represents either pressure or horizontal displacement of the elements of the medium.

$$f_n = n\frac{v}{2L} = nf_1 \qquad n = 1, 2, 3, \ldots \qquad [14.18]$$

◄ Pipe open at both ends; all harmonics are present

where v is the speed of sound in air. Notice the similarity to Equation 14.17, which also involves multiples of the fundamental.

If a pipe is open at one end and closed at the other, the open end is an antinode and the closed end is a node (Fig. 14.23b). In such a pipe, the fundamental frequency consists of a single antinode–node pair, A–N, so $L = \lambda_1/4$ and $\lambda_1 = 4L$. The fundamental harmonic for a pipe closed at one end is then $f_1 = v/\lambda_1 = v/4L$. The first overtone has another node and antinode between the open end and closed end, making the pattern A–N–A–N. There are three antinode–node segments in this pattern (A–N, N–A, and A–N), so $L = 3(\lambda_3/4)$ and $\lambda_3 = 4L/3$. The first overtone therefore has frequency $f_3 = v/\lambda_3 = 3v/4L = 3f_1$. Similarly, $f_5 = 5f_1$. In contrast to the pipe open at both ends, **there are no even multiples of the fundamental harmonic.** The odd harmonics for a pipe open at one end only are given by

$$f_n = n\frac{v}{4L} = nf_1 \qquad n = 1, 3, 5, \ldots \qquad [14.19]$$

◄ Pipe closed at one end; only odd harmonics are present

Quick Quiz

14.5 A pipe open at both ends resonates at a fundamental frequency f_{open}. When one end is covered and the pipe is again made to resonate, the fundamental frequency is f_{closed}. Which of the following expressions describes how these two resonant frequencies compare? (a) $f_{\text{closed}} = f_{\text{open}}$ (b) $f_{\text{closed}} = \frac{3}{2} f_{\text{open}}$ (c) $f_{\text{closed}} = 2 f_{\text{open}}$ (d) $f_{\text{closed}} = \frac{1}{2} f_{\text{open}}$ (e) none of these

14.6 Balboa Park in San Diego has an outdoor organ. When the air temperature increases, the fundamental frequency of one of the organ pipes (a) increases, (b) decreases, (c) stays the same, or (d) is impossible to determine. (The thermal expansion of the pipe is negligible.)

APPLYING PHYSICS 14.3 | Oscillations in a Harbor

Why do passing ocean waves sometimes cause the water in a harbor to undergo very large oscillations, called a *seiche* (pronounced *saysh*)?

EXPLANATION Water in a harbor is enclosed and possesses a natural frequency based on the size of the harbor. This is similar to the natural frequency of the enclosed air in a bottle, which can be excited by blowing across the edge of the opening. Ocean waves pass by the opening of the harbor at a certain frequency. If this frequency matches that of the enclosed harbor, then a large standing wave can be set up in the water by resonance. This situation can be simulated by carrying a fish tank filled with water. If your walking frequency matches the natural frequency of the water as it sloshes back and forth, a large standing wave develops in the fish tank. ■

APPLYING PHYSICS 14.4 | Why are Instruments Warmed Up?

Why do the strings go flat and the wind instruments go sharp during a performance if an orchestra doesn't warm up beforehand?

EXPLANATION Without warming up, all the instruments will be at room temperature at the beginning of the concert. As the wind instruments are played, they fill with warm air from the player's exhalation. The increase in temperature of the air in the instruments causes an increase in the speed of sound, which raises the resonance frequencies of the air columns. As a result, the instruments go sharp. The strings on the stringed instruments also increase in temperature due to the friction of rubbing with the bow. This results in thermal expansion, which causes a decrease in tension in the strings. With the decrease in tension, the wave speed on the strings drops and the fundamental frequencies decrease, so the stringed instruments go flat. ■

APPLYING PHYSICS 14.5 | How Do Bugles Work?

A bugle has no valves, keys, slides, or finger holes. How can it be used to play a song?

EXPLANATION Songs for the bugle are limited to harmonics of the fundamental frequency because there is no control over frequencies without valves, keys, slides, or finger holes. The player obtains different notes by changing the tension in the lips as the bugle is played, exciting different harmonics. The normal playing range of a bugle is among the third, fourth, fifth, and sixth harmonics of the fundamental. "Reveille," for example, is played with just the three notes G, C, and F, and "Taps" is played with these three notes and the G one octave above the lower G. ■

EXAMPLE 14.9 | Harmonics of a Pipe

GOAL Find frequencies of open and closed pipes.

PROBLEM A pipe is 2.46 m long. **(a)** Determine the frequencies of the first three harmonics if the pipe is open at both ends. Take 343 m/s as the speed of sound in air. **(b)** How many harmonic frequencies of this pipe lie in the audible range, from 20 Hz to 20 000 Hz? **(c)** What are the three lowest possible frequencies if the pipe is closed at one end and open at the other?

STRATEGY Substitute into Equation 14.18 for part (a) and Equation 14.19 for part (c). All harmonics, $n = 1, 2, 3 \ldots$ are available for the pipe open at both ends, but only the harmonics with $n = 1, 3, 5, \ldots$ for the pipe closed at one end. For part (b), set the frequency in Equation 14.18 equal to 2.00×10^4 Hz.

SOLUTION

(a) Find the frequencies if the pipe is open at both ends.

Substitute into Equation 14.18, with $n = 1$:

$$f_1 = \frac{v}{2L} = \frac{343 \text{ m/s}}{2(2.46 \text{ m})} = \boxed{69.7 \text{ Hz}}$$

Multiply to find the second and third harmonics:

$$f_2 = 2f_1 = \boxed{139 \text{ Hz}} \qquad f_3 = 3f_1 = \boxed{209 \text{ Hz}}$$

(b) How many harmonics lie between 20 Hz and 20 000 Hz for this pipe?

Set the frequency in Equation 14.18 equal to 2.00×10^4 Hz and solve for n:

$$f_n = n\frac{v}{2L} = n\frac{343 \text{ m/s}}{2 \cdot 2.46 \text{ m}} = 2.00 \times 10^4 \text{ Hz}$$

This works out to $n = 286.88$, which must be truncated down ($n = 287$ gives a frequency over 2.00×10^4 Hz):

$$n = \boxed{286}$$

(c) Find the frequencies for the pipe closed at one end.

Apply Equation 14.19 with $n = 1$:

$$f_1 = \frac{v}{4L} = \frac{343 \text{ m/s}}{4(2.46 \text{ m})} = \boxed{34.9 \text{ Hz}}$$

The next two harmonics are odd multiples of the first:

$$f_3 = 3f_1 = \boxed{105 \text{ Hz}} \qquad f_5 = 5f_1 = \boxed{175 \text{ Hz}}$$

QUESTION 14.9 True or False: The fundamental wavelength of a longer pipe is greater than the fundamental wavelength of a shorter pipe.

EXERCISE 14.9 (a) What length pipe open at both ends has a fundamental frequency of 3.70×10^2 Hz? Find the first overtone. (b) If the one end of this pipe is now closed, what is the new fundamental frequency? Find the first overtone. (c) If the pipe is open at one end only, how many harmonics are possible in the normal hearing range from 20 to 20 000 Hz?

ANSWERS (a) 0.464 m, 7.40×10^2 Hz (b) 185 Hz, 555 Hz (c) 54

EXAMPLE 14.10 | Resonance in a Tube of Variable Length

GOAL Understand resonance in tubes and perform elementary calculations.

PROBLEM Figure 14.24a shows a simple apparatus for demonstrating resonance in a tube. A long tube open at both ends is partially submerged in a beaker of water, and a vibrating tuning fork of unknown frequency is placed near the top of the tube. The length of the air column, L, is adjusted by moving the tube vertically. The sound waves generated by the fork are reinforced when the length of the air column corresponds to one of the resonant frequencies of the tube. Suppose the smallest value of L for which a peak occurs in the sound intensity is 9.00 cm. **(a)** With this measurement, determine the frequency of the tuning fork. **(b)** Find the wavelength and the next two air-column lengths giving resonance. Take the speed of sound to be 343 m/s.

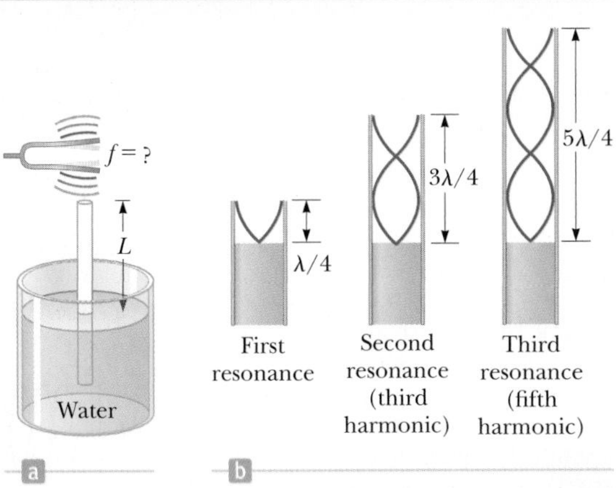

Figure 14.24 (Example 14.10) (a) Apparatus for demonstrating the resonance of sound waves in a tube closed at one end. The length L of the air column is varied by moving the tube vertically while it is partially submerged in water. (b) The first three resonances of the system.

STRATEGY Once the tube is in the water, the setup is the same as a pipe closed at one end. For part (a), substitute values for v and L into Equation 14.19 with $n = 1$, and find the frequency of the tuning fork. (b) The next resonance maximum occurs when the water

(Continued)

level is low enough in the straw to allow a second node (see Fig. 14.24b), which is another half-wavelength in distance. The third resonance occurs when the third node is reached, requiring yet another half-wavelength of distance. The frequency in each case is the same because it's generated by the tuning fork.

SOLUTION

(a) Find the frequency of the tuning fork.

Substitute $n = 1$, $v = 343$ m/s, and $L_1 = 9.00 \times 10^{-2}$ m into Equation 14.19:

$$f_1 = \frac{v}{4L_1} = \frac{343 \text{ m/s}}{4(9.00 \times 10^{-2} \text{ m})} = \boxed{953 \text{ Hz}}$$

(b) Find the wavelength and the next two water levels giving resonance.

Calculate the wavelength, using the fact that, for a tube open at one end, $\lambda = 4L$ for the fundamental.

$$\lambda = 4L_1 = 4(9.00 \times 10^{-2} \text{ m}) = \boxed{0.360 \text{ m}}$$

Add a half-wavelength of distance to L_1 to get the next resonance position:

$$L_2 = L_1 + \lambda/2 = 0.090\,0 \text{ m} + 0.180 \text{ m} = \boxed{0.270 \text{ m}}$$

Add another half-wavelength to L_2 to obtain the third resonance position:

$$L_3 = L_2 + \lambda/2 = 0.270 \text{ m} + 0.180 \text{ m} = \boxed{0.450 \text{ m}}$$

REMARKS This experimental arrangement is often used to measure the speed of sound, in which case the frequency of the tuning fork must be known in advance.

QUESTION 14.10 True or False: The resonant frequency of an air column depends on the length of the column and the speed of sound.

EXERCISE 14.10 An unknown gas is introduced into the aforementioned apparatus using the same tuning fork, and the first resonance occurs when the air column is 5.84 cm long. Find the speed of sound in the gas.

ANSWER 223 m/s

14.11 Beats

LEARNING OBJECTIVES

1. Discuss the interference phenomenon of beats.
2. Apply the concept of beats to situations involving two frequencies.

The interference phenomena we have been discussing so far have involved the superposition of two or more waves with the same frequency, traveling in opposite directions. Another type of interference effect results from the superposition of two waves with slightly different frequencies. In such a situation, the waves at some fixed point are periodically in and out of phase, corresponding to an alternation in time between constructive and destructive interference. To understand this phenomenon, consider Figure 14.25. The two waves shown

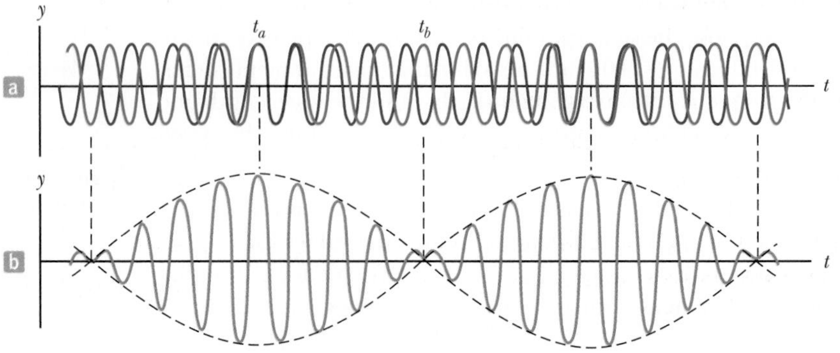

Figure 14.25 Beats are formed by the combination of two waves of slightly different frequencies traveling in the same direction. (a) The individual waves heard by an observer at a fixed point in space. (b) The combined wave has an amplitude (dashed line) that oscillates in time.

WARM-UP EXERCISES

WebAssign The warm-up exercises in this chapter may be assigned online in Enhanced WebAssign.

1. **Math Review** Determine the value of x in each of the following equations: (a) $10^x = 5$ (b) $\log(3x) = 2$. (See also Section 14.4.)

2. **Math Review** Determine the value of I if 75 dB = $10 \log \left(\dfrac{I}{10^{-12}\ \text{W/m}^2} \right)$. (See also Section 14.4.)

3. **Physics Review** A piano tuning fork typically vibrates at 440 Hz. If the speed of sound is 343 m/s, what is the wavelength of the sound produced by the fork? (See Section 13.8.)

4. **Physics Review** A string has a mass of 12.0 g and length of 1.50 m. (a) Calculate the string's linear density, and (b) the speed of waves on the string, if it's put under a tension of 85.0 N. (See Section 13.9.)

5. The temperature at Furnace Creek in Death Valley reached 136°F (331 K) on July 10, 1913. What is the speed of sound in air at this temperature? (See Section 14.3.)

6. (a) Ethyl alcohol has a density of 0.806×10^3 kg/m^3. Compute the speed of sound in ethyl alcohol, which has a bulk modulus of 1.0×10^{-9} Pa. (b) Calculate the speed of sound in aluminum. (See Section 14.3.)

7. The sound intensity level of a jet plane going down the runway as observed from a certain location is 105 dB. Determine the physical intensity of the sound at this point. (See Section 14.4.)

8. A rock band creates a sound intensity level of 118 dB at a distance of 32.0 m. (a) Calculate the sound intensity. (See Section 14.4.) (b) Assuming sound from the amplifiers travels as a spherical wave, what average power do the amplifiers generate? (See Section 14.5.)

9. On a hot 95.0°F (308 K) day at a racetrack, a Formula One racecar is traveling at a speed of 1.50×10^2 mph (67.1 m/s) away from a stationary siren emitting sound waves at a frequency of 3.30×10^2 Hz. (a) Determine the sound speed for the given temperature. (See Section 14.3.) (b) What frequency did the racecar driver hear? (See Section 14.6.)

10. The driver of a car traveling 30.0 m/s sounds his horn as he approaches an intersection. If the horn has a frequency of 675 Hz, what frequency does a pedestrian hear, if she is at rest at the intersection's crosswalk? Assume a sound speed of 343 m/s. (See Section 14.6.)

11. Two speakers, several meters apart and facing each other, emit identical sound waves at a frequency of 225 Hz. Suppose the speed of sound in air is 331 m/s and a man is standing exactly half way between the two speakers, where the interference is constructive. (a) What is the wavelength for sound waves coming from each speaker? (See Section 13.8.) (b) What minimum distance along the line between the speakers would the listener find the next point of constructive interference? (See Section 14.7.)

12. A 2.00-m long string has a mass of 0.025 0 kg. If this string is pulled to a tension of 50.0 N and tied between two fixed supports, determine (a) the mass per unit length of the string, (b) the speed of the waves on the string, (c) the fundamental frequency for vibrations, and (d) the frequency of the second harmonic. (See Section 14.8.)

13. The pipe of a flute has a length of 58.0 cm, is closed at one end and is open at the other. If the speed of sound in air is 343 m/s, what is (a) the fundamental frequency of the flute? (b) the frequency of the next higher harmonic? (c) What is the fundamental frequency of the same pipe if it's open at both ends? (See Section 14.10.)

14. When two tuning forks are sounded at the same time, a beat frequency of 3 Hz occurs. If the first tuning fork has a frequency of 4.40×10^2 Hz, what are the two possible frequencies of the second tuning fork? (See Section 14.11.)

CONCEPTUAL QUESTIONS

WebAssign The conceptual questions in this chapter may be assigned online in Enhanced WebAssign.

1. (a) You are driving down the highway in your car when a police car sounding its siren overtakes you and passes you. If its frequency at rest is f_0, is the frequency you hear while the car is catching up to you higher or lower than f_0? (b) What about the frequency you hear after the car has passed you?

2. A crude model of the human throat is that of a pipe open at both ends with a vibrating source to introduce the sound into the pipe at one end. Assuming the vibrating source produces a range of frequencies, discuss the effect of changing the pipe's length.

3. Older auto-focus cameras sent out a pulse of sound and measured the time interval required for the pulse to reach an object, reflect off of it, and return to be detected. Can air temperature affect the camera's focus? New cameras use a more reliable infrared system.

4. Explain how the distance to a lightning bolt (Fig. CQ14.4) can be determined by counting the seconds between the flash and the sound of thunder.

Figure CQ14.4

5. Secret agents in the movies always want to get to a secure phone with a voice scrambler. How do these devices work?

6. Why does a vibrating guitar string sound louder when placed on the instrument than it would if allowed to vibrate in the air while off the instrument?

7. You are driving toward the base of a cliff and you honk your horn. (a) Is there a Doppler shift of the sound when you hear the echo? If so, is it like a moving source or moving observer? (b) What if the reflection occurs not from a cliff, but from the forward edge of a huge alien spacecraft moving toward you as you drive?

8. The radar systems used by police to detect speeders are sensitive to the Doppler shift of a pulse of radio waves. Discuss how this sensitivity can be used to measure the speed of a car.

9. An archer shoots an arrow from a bow. Does the string of the bow exhibit standing waves after the arrow leaves? If so, and if the bow is perfectly symmetric so that the arrow leaves from the center of the string, what harmonics are excited?

10. A soft drink bottle resonates as air is blown across its top. What happens to the resonant frequency as the level of fluid in the bottle decreases?

11. An airplane mechanic notices that the sound from a twin-engine aircraft varies rapidly in loudness when both engines are running. What could be causing this variation from loud to soft?

PROBLEMS

ENHANCED WebAssign The problems in this chapter may be assigned online in Enhanced WebAssign.

1. denotes straightforward problem; 2. denotes intermediate problem;
3. denotes challenging problem
1. denotes full solution available in *Student Solutions Manual/Study Guide*
1. denotes problems most often assigned in Enhanced WebAssign

BIO denotes biomedical problems
GP denotes guided problems
M denotes Master It tutorial available in Enhanced WebAssign
Q|C denotes asking for quantitative and conceptual reasoning
S denotes symbolic reasoning problem
W denotes Watch It video solution available in Enhanced WebAssign

14.2 Characteristics of Sound Waves

14.3 The Speed of Sound

Note: Unless otherwise specified, assume the speed of sound in air is 343 m/s, its value at an air temperature of 20.0°C. At any other Celsius temperature T_C, the speed of sound in air is described by Equation 14.4:

$$v = 331\sqrt{1 + \frac{T_C}{273}}$$

where v is in m/s and T is in °C. Use Table 14.1 to find speeds of sound in other media.

1. Q|C Suppose you hear a clap of thunder 16.2 s after seeing the associated lightning stroke. The speed of light in air is 3.00×10^8 m/s. (a) How far are you from the lightning stroke? (b) Do you need to know the value of the speed of light to answer? Explain.

2. Earthquakes at fault lines in Earth's crust create seismic waves, which are longitudinal (P-waves) or transverse (S-waves). The P-waves have a speed of about 7 km/s. Estimate the average bulk modulus of Earth's crust given that the density of rock is about 2 500 kg/m^3.

3. On a hot summer day, the temperature of air in Arizona reaches 114°F. What is the speed of sound in air at this temperature?

4. BIO A dolphin located in seawater at a temperature of 25°C emits a sound directed toward the bottom of the ocean 150 m below. How much time passes before it hears an echo?

5. A group of hikers hears an echo 3.00 s after shouting. How far away is the mountain that reflected the sound wave?

6. BIO The range of human hearing extends from approximately 20 Hz to 20 000 Hz. Find the wavelengths of these extremes at a temperature of 27°C.

7. A sound wave propagating in air has a frequency of 4.00 kHz. Calculate the change in wavelength when the wave, initially traveling in a region where T = 27.0°C, enters a region where T = 10.0°C.

8. A stone is dropped from rest into a well. The sound of the splash is heard exactly 2.00 s later. Find the depth of the well if the air temperature is 10.0°C.

9. QIC A hammer strikes one end of a thick steel rail of length 8.50 m. A microphone located at the opposite end of the rail detects two pulses of sound, one that travels through the air and a longitudinal wave that travels through the rail. (a) Which pulse reaches the microphone first? (b) Find the separation in time between the arrivals of the two pulses.

14.4 Energy and Intensity of Sound Waves

14.5 Spherical and Plane Waves

10. QIC The intensity level produced by a jet airplane at a certain location is 150 dB. (a) Calculate the intensity of the sound wave generated by the jet at the given location. (b) Compare the answer to part (a) to the threshold of pain and explain why employees directing jet airplanes at airports must wear hearing protection equipment.

11. W One of the loudest sounds in recent history was that made by the explosion of Krakatoa on August 26–27, 1883. According to barometric measurements, the sound had a decibel level of 180 dB at a distance of 161 km. Assuming the intensity falls off as the inverse of the distance squared, what was the decibel level on Rodriguez Island, 4 800 km away?

12. A sound wave from a siren has an intensity of 100.0 W/m^2 at a certain point, and a second sound wave from a nearby ambulance has an intensity level 10 dB greater than the siren's sound wave at the same point. What is the intensity level of the sound wave due to the ambulance?

13. BIO A person wears a hearing aid that uniformly increases the intensity level of all audible frequencies of sound by 30.0 dB. The hearing aid picks up sound having a frequency of 250 Hz at an intensity of 3.0×10^{-11} W/m^2. What is the intensity delivered to the eardrum?

14. BIO The area of a typical eardrum is about 5.0×10^{-5} m^2. Calculate the sound power (the energy per second) incident on an eardrum at (a) the threshold of hearing and (b) the threshold of pain.

15. BIO The toadfish makes use of resonance in a closed tube to produce very loud sounds. The tube is its swim bladder, used as an amplifier. The sound level of this creature has been measured as high as 100 dB. (a) Calculate the intensity of the sound wave emitted. (b) What is the intensity level if three of these toadfish try to make a sound at the same time?

16. GP A trumpet creates a sound intensity level of 1.15×10^2 dB at a distance of 1.00 m. (a) What is the sound intensity of a trumpet at this distance? (b) What is the sound intensity of five trumpets at this distance? (c) Find the sound intensity of five trumpets at the location of the first row of an audience, 8.00 m away, assuming, for simplicity, the sound energy propagates uniformly in all directions. (d) Calculate the decibel level of the five trumpets in the first row. (e) If the trumpets are being played in an outdoor auditorium, how far away, in theory, can their combined sound be heard? (f) In practice such a sound could not be heard once the listener was 2–3 km away. Why can't the sound be heard at the distance found in part (e) *Hint:* In a very quiet room the ambient sound intensity level is about 30 dB.

17. There is evidence that elephants communicate via infrasound, generating rumbling vocalizations as low as 14 Hz that can travel up to 10 km. The intensity level of these sounds can reach 103 dB, measured a distance of 5.0 m from the source. Determine the intensity level of the infrasound 10 km from the source, assuming the sound energy radiates uniformly in all directions.

18. A family ice show is held at an enclosed arena. The skaters perform to music playing at a level of 80.0 dB. This intensity level is too loud for your baby, who yells at 75.0 dB. (a) What total sound intensity engulfs you? (b) What is the combined sound level?

19. M A train sounds its horn as it approaches an intersection. The horn can just be heard at a level of 50 dB by an observer 10 km away. (a) What is the average power generated by the horn? (b) What intensity level of the horn's sound is observed by someone waiting at an intersection 50 m from the train? Treat the horn as a point source and neglect any absorption of sound by the air.

20. An outside loudspeaker (considered a small source) emits sound waves with a power output of 100 W. (a) Find the intensity 10.0 m from the source. (b) Find the intensity level in decibels at that distance. (c) At what distance would you experience the sound at the threshold of pain, 120 dB?

21. S Show that the difference in decibel levels β_1 and β_2 of a sound source is related to the ratio of its distances r_1 and r_2 from the receivers by the formula

$$\beta_2 - \beta_1 = 20 \log\left(\frac{r_1}{r_2}\right)$$

22. A skyrocket explodes 100 m above the ground (Fig. P14.22). Three observers are spaced 100 m apart, with the first (A) directly under the explosion. (a) What

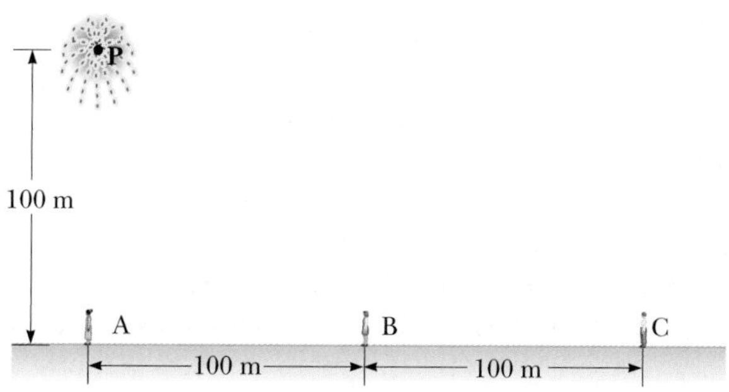

Figure P14.22

is the ratio of the sound intensity heard by observer A to that heard by observer B? (b) What is the ratio of the intensity heard by observer A to that heard by observer C?

14.6 The Doppler Effect

23. A commuter train passes a passenger platform at a constant speed of 40.0 m/s. The train horn is sounded at its characteristic frequency of 320 Hz. (a) What overall change in frequency is detected by a person on the platform as the train moves from approaching to receding? (b) What wavelength is detected by a person on the platform as the train approaches?

24. An airplane traveling at half the speed of sound emits a sound of frequency 5.00 kHz. At what frequency does a stationary listener hear the sound (a) as the plane approaches? (b) After it passes?

25. W Two trains on separate tracks move toward each other. Train 1 has a speed of 130 km/h; train 2, a speed of 90.0 km/h. Train 2 blows its horn, emitting a frequency of 500 Hz. What is the frequency heard by the engineer on train 1?

26. At rest, a car's horn sounds the note A (440 Hz). The horn is sounded while the car is moving down the street. A bicyclist moving in the same direction with one-third the car's speed hears a frequency of 415 Hz. (a) Is the cyclist ahead of or behind the car? (b) What is the speed of the car?

27. An alert physics student stands beside the tracks as a train rolls slowly past. He notes that the frequency of the train whistle is 465 Hz when the train is approaching him and 441 Hz when the train is receding from him. Using these frequencies, he calculates the speed of the train. What value does he find?

28. BIO Q|C A bat flying at 5.00 m/s is chasing an insect flying in the same direction. If the bat emits a 40.0-kHz chirp and receives back an echo at 40.4 kHz, (a) what is the speed of the insect? (b) Will the bat be able to catch the insect? Explain.

29. A tuning fork vibrating at 512 Hz falls from rest and accelerates at 9.80 m/s^2. How far below the point of release is the tuning fork when waves of frequency 485 Hz reach the release point?

30. BIO Expectant parents are thrilled to hear their unborn baby's heartbeat, revealed by an ultrasonic motion detector. Suppose the fetus's ventricular wall moves in simple harmonic motion with amplitude 1.80 mm and frequency 115 beats per minute. The motion detector in contact with the mother's abdomen produces sound at precisely 2 MHz, which travels through tissue at 1.50 km/s. (a) Find the maximum linear speed of the heart wall. (b) Find the maximum frequency at which sound arrives at the wall of the baby's heart. (c) Find the maximum frequency at which reflected sound is received by the motion detector. (By electronically "listening" for echoes at a frequency different from the broadcast frequency, the motion detector can produce beeps of audible sound in synchrony with the fetal heartbeat.)

31. M A supersonic jet traveling at Mach 3.00 at an altitude of $h = 20\ 000$ m is directly over a person at time $t = 0$ as shown in Figure P14.31. Assume the average speed of sound in air is 335 m/s over the path of the sound. (a) At what time will the person encounter the shock wave due to the sound emitted at $t = 0$? (b) Where will the plane be when this shock wave is heard?

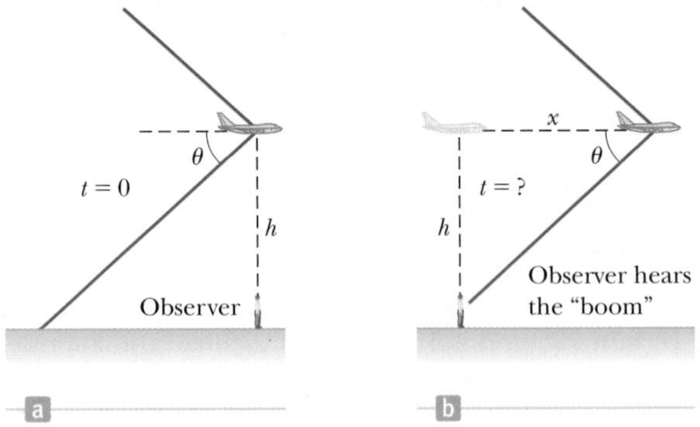

Figure P14.31

32. GP A yellow submarine traveling horizontally at 11.0 m/s uses sonar with a frequency of 5.27×10^3 Hz. A red submarine is in front of the yellow submarine and moving 3.00 m/s relative to the water in the same direction. A crewman in the red submarine observes sound waves ("pings") from the yellow submarine. Take the speed of sound in seawater as 1 533 m/s. (a) Write Equation 14.12. (b) Which submarine is the source of the sound? (c) Which submarine carries the observer? (d) Does the motion of the observer's submarine increase or decrease the time between the pressure maxima of the incoming sound waves? How does that affect the observed period? The observed frequency? (e) Should the sign of v_0 be positive or negative? (f) Does the motion of the source submarine increase or decrease the time observed between the pressure maxima? How does this motion affect the observed period? The observed frequency? (g) What sign should be chosen for v_s? (h) Substitute the appropriate numbers and obtain the frequency observed by the crewman on the red submarine.

14.7 Interference of Sound Waves

33. Two small speakers are driven by a common oscillator at 8.00×10^2 Hz. The speakers face each other and are separated by 1.25 m. Locate the points along a line joining the two speakers where relative minima would be expected. (Use $v = 343$ m/s.)

34. The acoustical system shown in Figure P14.34 is driven by a speaker emitting sound of frequency 756 Hz. (a) If constructive interference occurs at a particular instant, by what minimum amount should the path length in the upper U-shaped tube be increased so that destructive interference occurs instead? (b) What minimum increase in the original length of the upper tube will again result in constructive interference?

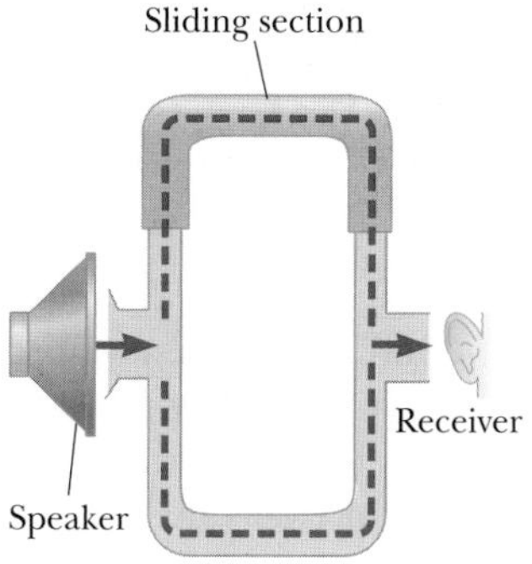

Figure P14.34

35. The ship in Figure P14.35 travels along a straight line parallel to the shore and a distance $d = 600$ m from it. The ship's radio receives simultaneous signals of the same frequency from antennas A and B, separated by a distance $L = 800$ m. The signals interfere constructively at point C, which is equidistant from A and B. The signal goes through the first minimum at point D, which is directly outward from the shore from point B. Determine the wavelength of the radio waves.

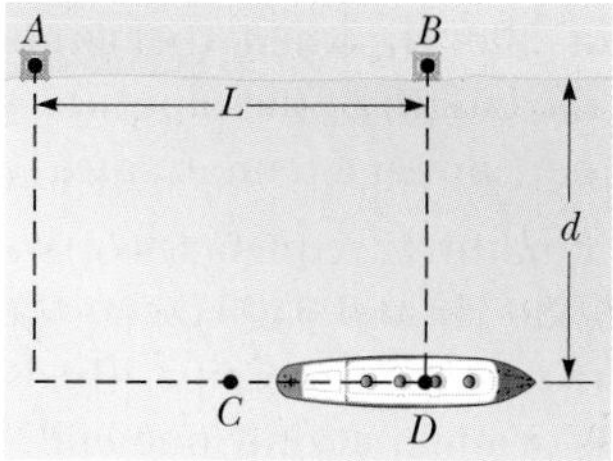

Figure P14.35

36. Two loudspeakers are placed above and below each other, as in Figure P14.36 and driven by the same source at a frequency of 4.50×10^2 Hz. An observer is in front of the speakers (to the right) at point O, at the same distance from each speaker. What minimum vertical distance upward should the top speaker be moved to create destructive interference at point O?

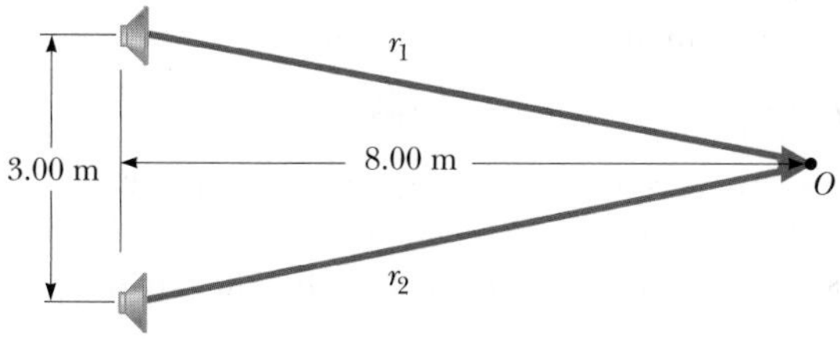

Figure P14.36

37. A pair of speakers separated by a distance $d = 0.700$ m are driven by the same oscillator at a frequency of 686 Hz. An observer originally positioned at one of the speakers begins to walk along a line perpendicular to the line joining the speakers as in Figure P14.37. (a) How far must the observer walk before reaching a relative maximum in intensity? (b) How far will the observer be from the speaker when the first relative minimum is detected in the intensity?

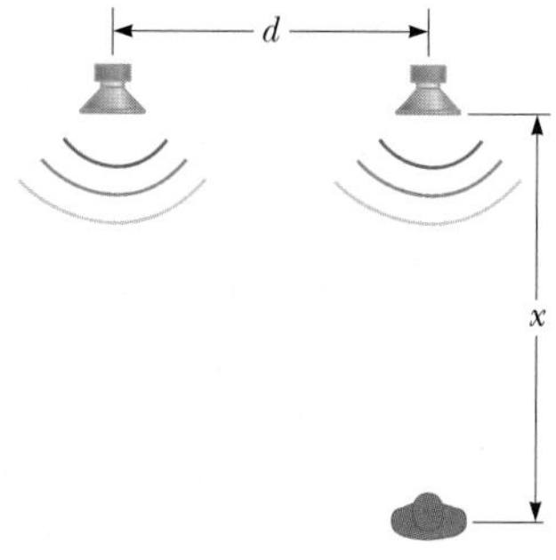

Figure P14.37

14.8 Standing Waves

38. A steel wire in a piano has a length of 0.700 0 m and a mass of 4.300×10^{-3} kg. To what tension must this wire be stretched so that the fundamental vibration corresponds to middle C ($f_C = 261.6$ Hz on the chromatic musical scale)?

39. W A stretched string fixed at each end has a mass of 40.0 g and a length of 8.00 m. The tension in the string is 49.0 N. (a) Determine the positions of the nodes and antinodes for the third harmonic. (b) What is the vibration frequency for this harmonic?

40. How far, and in what direction, should a cellist move her finger to adjust a string's tone from an out-of-tune 449 Hz to an in-tune 440 Hz? The string is 68.0 cm long, and the finger is 20.0 cm from the nut for the 449-Hz tone.

41. A stretched string of length L is observed to vibrate in five equal segments when driven by a 630-Hz oscillator. What oscillator frequency will set up a standing wave so that the string vibrates in three segments?

42. Two pieces of steel wire with identical cross sections have lengths of L and $2L$. The wires are each fixed at both ends and stretched so that the tension in the longer wire is four times greater than in the shorter wire. If the fundamental frequency in the shorter wire is 60 Hz, what is the frequency of the second harmonic in the longer wire?

43. A steel wire with mass 25.0 g and length 1.35 m is strung on a bass so that the distance from the nut to the bridge is 1.10 m. (a) Compute the linear density of the string. (b) What velocity wave on the string will produce the desired fundamental frequency of the E_1 string, 41.2 Hz? (c) Calculate the tension required to obtain the proper frequency. (d) Calculate the wavelength of the string's vibration. (e) What is the wavelength of the sound produced in air? (Assume the speed of sound in air is 343 m/s.)

44. S A standing wave is set up in a string of variable length and tension by a vibrator of variable frequency. Both ends of the string are fixed. When the vibrator has a frequency f_A, in a string of length L_A and under tension T_A, n_A antinodes are set up in the string. (a) Write an expression for the frequency f_A of a standing wave in terms of the number n_A, length L_A, tension

T_A, and linear density μ_A. (b) If the length of the string is doubled to $L_B = 2L_A$, what frequency f_B (written as a multiple of f_A) will result in the same number of antinodes? Assume the tension and linear density are unchanged. *Hint:* Make a ratio of expressions for f_B and f_A. (c) If the frequency and length are held constant, what tension T_B will produce $n_A + 1$ antinodes? (d) If the frequency is tripled and the length of the string is halved, by what factor should the tension be changed so that twice as many antinodes are produced?

45. A 12.0-kg object hangs in equilibrium from a string with total length of $L = 5.00$ m and linear mass density of $\mu = 0.001\ 00$ kg/m. The string is wrapped around two light, frictionless pulleys that are separated by the distance $d = 2.00$ m (Fig. P14.45a). (a) Determine the tension in the string. (b) At what frequency must the string between the pulleys vibrate in order to form the standing-wave pattern shown in Figure P14.45b?

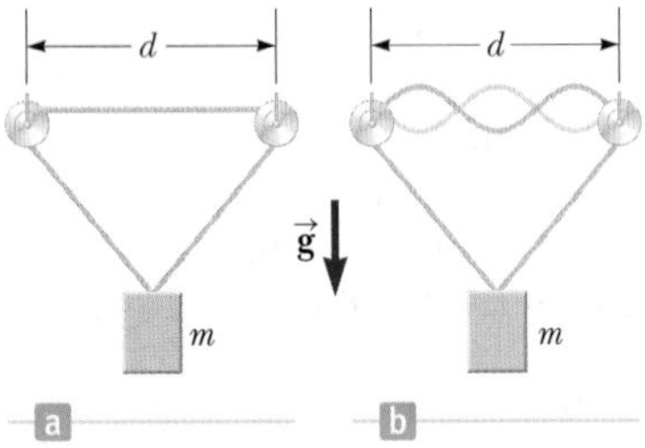

Figure P14.45

46. M In the arrangement shown in Figure P14.46, an object of mass $m = 5.0$ kg hangs from a cord around a light pulley. The length of the cord between point P and the pulley is $L = 2.0$ m. (a) When the vibrator is set to a frequency of 150 Hz, a standing wave with six loops is formed. What must be the linear mass density of the cord? (b) How many loops (if any) will result if m is changed to 45 kg? (c) How many loops (if any) will result if m is changed to 10 kg?

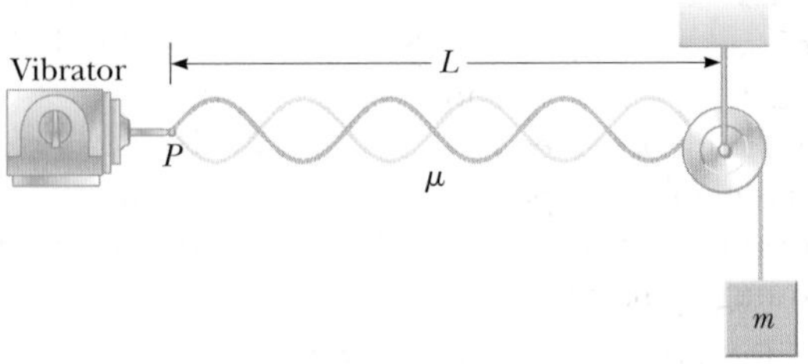

Figure P14.46

47. BIO A 60.00-cm guitar string under a tension of 50.000 N has a mass per unit length of 0.100 00 g/cm. What is the highest resonant frequency that can be heard by a person capable of hearing frequencies up to 20 000 Hz?

14.9 Forced Vibrations and Resonance

48. Standing-wave vibrations are set up in a crystal goblet with four nodes and four antinodes equally spaced around the 20.0-cm circumference of its rim. If transverse waves move around the glass at 900 m/s, an opera singer would have to produce a high harmonic with what frequency in order to shatter the glass with a resonant vibration?

14.10 Standing Waves in Air Columns

49. The windpipe of a typical whooping crane is about 5.0 ft. long. What is the lowest resonant frequency of this pipe, assuming it is closed at one end? Assume a temperature of 37°C.

50. The overall length of a piccolo is 32.0 cm. The resonating air column vibrates as in a pipe that is open at both ends. (a) Find the frequency of the lowest note a piccolo can play. (b) Opening holes in the side effectively shortens the length of the resonant column. If the highest note a piccolo can sound is 4 000 Hz, find the distance between adjacent antinodes for this mode of vibration.

51. BIO W The human ear canal is about 2.8 cm long. If it is regarded as a tube that is open at one end and closed at the eardrum, what is the fundamental frequency around which we would expect hearing to be most sensitive?

52. Q|C A tunnel under a river is 2.00 km long. (a) At what frequencies can the air in the tunnel resonate? (b) Explain whether it would be good to make a rule against blowing your car horn when you are in the tunnel.

53. M A pipe open at both ends has a fundamental frequency of 300 Hz when the temperature is 0°C. (a) What is the length of the pipe? (b) What is the fundamental frequency at a temperature of 30.0°C?

54. Two adjacent natural frequencies of an organ pipe are found to be 550 Hz and 650 Hz. (a) Calculate the fundamental frequency of the pipe. (b) Is the pipe open at both ends or open at only one end? (c) What is the length of the pipe?

14.11 Beats

55. In certain ranges of a piano keyboard, more than one string is tuned to the same note to provide extra loudness. For example, the note at 1.10×10^2 Hz has two strings at this frequency. If one string slips from its normal tension of 6.00×10^2 N to 5.40×10^2 N, what beat frequency is heard when the hammer strikes the two strings simultaneously?

56. The G string on a violin has a fundamental frequency of 196 Hz. It is 30.0 cm long and has a mass of 0.500 g. While this string is sounding, a nearby violinist effectively shortens the G string on her identical violin (by sliding her finger down the string) until a beat frequency of 2.00 Hz is heard between the two strings. When that occurs, what is the effective length of her string?

57. Two train whistles have identical frequencies of 1.80×10^2 Hz. When one train is at rest in the station and the other is moving nearby, a commuter standing on the station platform hears beats with a frequency of 2.00 beats/s when the whistles operate together. What are the two possible speeds and directions that the moving train can have?

58. Two pipes of equal length are each open at one end. Each has a fundamental frequency of 480 Hz at 300 K. In one pipe the air temperature is increased to 305 K. If the two pipes are sounded together, what beat frequency results?

59. A student holds a tuning fork oscillating at 256 Hz. He walks toward a wall at a constant speed of 1.33 m/s. (a) What beat frequency does he observe between the tuning fork and its echo? (b) How fast must he walk away from the wall to observe a beat frequency of 5.00 Hz?

14.13 The Ear

60. BIO If a human ear canal can be thought of as resembling an organ pipe, closed at one end, that resonates at a fundamental frequency of 3 000 Hz, what is the length of the canal? Use a normal body temperature of 37°C for your determination of the speed of sound in the canal.

61. BIO Some studies suggest that the upper frequency limit of hearing is determined by the diameter of the eardrum. The wavelength of the sound wave and the diameter of the eardrum are approximately equal at this upper limit. If the relationship holds exactly, what is the diameter of the eardrum of a person capable of hearing 20 000 Hz? (Assume a body temperature of 37.0°C.)

Additional Problems

62. A typical sound level for a buzzing mosquito is 40 dB, and that of a vacuum cleaner is approximately 70 dB. Approximately how many buzzing mosquitoes will produce a sound intensity equal to that of a vacuum cleaner?

63. Assume a 150-W loudspeaker broadcasts sound equally in all directions and produces sound with a level of 103 dB at a distance of 1.60 m from its center. (a) Find its sound power output. If a salesperson claims the speaker is rated at 150 W, he is referring to the maximum electrical power input to the speaker. (b) Find the efficiency of the speaker, that is, the fraction of input power that is converted into useful output power.

64. Two small loudspeakers emit sound waves of different frequencies equally in all directions. Speaker *A* has an output of 1.00 mW, and speaker *B* has an output of 1.50 mW. Determine the sound level (in decibels) at point *C* in Figure P14.64 assuming (a) only speaker *A* emits sound, (b) only speaker *B* emits sound, and (c) both speakers emit sound.

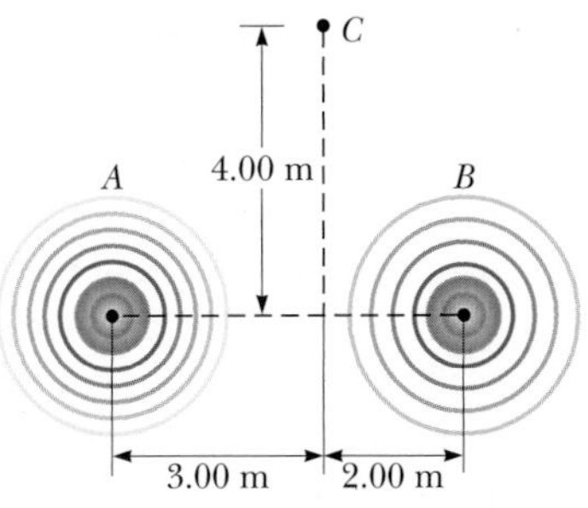

Figure P14.64

65. An interstate highway has been built through a neighborhood in a city. In the afternoon, the sound level in an apartment in the neighborhood is 80.0 dB as 100 cars pass outside the window every minute. Late at night, the traffic flow is only five cars per minute. What is the average late-night sound level?

66. A student uses an audio oscillator of adjustable frequency to measure the depth of a water well. He reports hearing two successive resonances at 52.0 Hz and 60.0 Hz. How deep is the well?

67. A stereo speaker is placed between two observers who are 36 m apart, along the line connecting them. If one observer records an intensity level of 60 dB, and the other records an intensity level of 80 dB, how far is the speaker from each observer?

68. M Two ships are moving along a line due east (Fig. P14.68). The trailing vessel has a speed relative to a land-based observation point of $v_1 = 64.0$ km/h, and the leading ship has a speed of $v_2 = 45.0$ km/h relative to that point. The two ships are in a region of the ocean where the current is moving uniformly due west at $v_{\text{current}} =$ 10.0 km/h. The trailing ship transmits a sonar signal at a frequency of 1 200.0 Hz through the water. What frequency is monitored by the leading ship?

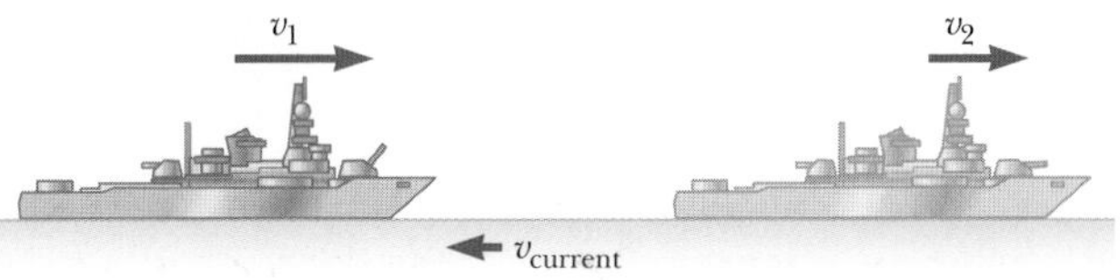

Figure P14.68

69. A quartz watch contains a crystal oscillator in the form of a block of quartz that vibrates by contracting and expanding. Two opposite faces of the block, 7.05 mm apart, are antinodes, moving alternately toward and away from each other. The plane halfway between these two faces is a node of the vibration. The speed of sound in quartz is 3.70×10^3 m/s. Find the frequency of the vibration. An oscillating electric voltage accompanies the mechanical oscillation, so the quartz is described as *piezoelectric*. An electric circuit feeds in energy to maintain the oscillation and also counts the voltage pulses to keep time.

70. A flowerpot is knocked off a window ledge from a height $d = 20.0$ m above the sidewalk as shown in Figure P14.70. It falls toward an unsuspecting man of height $h = 1.75$ m who is standing below. Assume the man below requires 0.300 s to respond to a warning. How close to the sidewalk can the flowerpot fall before it is too late for a warning shouted from the window to reach the man in time?

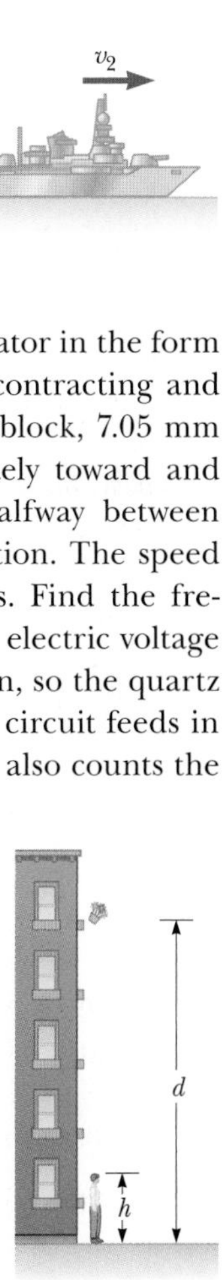

Figure P14.70

71. On a workday, the average decibel level of a busy street is 70 dB, with 100 cars passing

a given point every minute. If the number of cars is reduced to 25 every minute on a weekend, what is the decibel level of the street?

72. A flute is designed so that it plays a frequency of 261.6 Hz, middle C, when all the holes are covered and the temperature is 20.0°C. (a) Consider the flute to be a pipe open at both ends and find its length, assuming the middle-C frequency is the fundamental frequency. (b) A second player, nearby in a colder room, also attempts to play middle C on an identical flute. A beat frequency of 3.00 beats/s is heard. What is the temperature of the room?

73. A block with a speaker bolted to it is connected to a spring having spring constant $k = 20.0$ N/m, as shown in Figure P14.73. The total mass of the block and speaker is 5.00 kg, and the amplitude of the unit's motion is 0.500 m. If the speaker emits sound waves of frequency 440 Hz, determine the (a) lowest and (b) highest frequencies heard by the person to the right of the speaker.

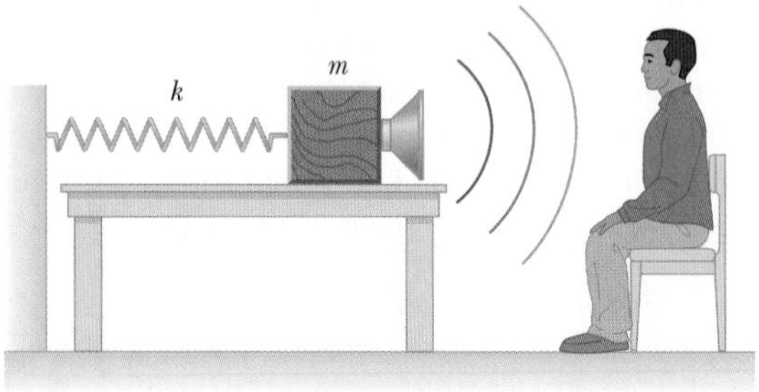

Figure P14.73

74. A student stands several meters in front of a smooth reflecting wall, holding a board on which a wire is fixed at each end. The wire, vibrating in its third harmonic, is 75.0 cm long, has a mass of 2.25 g, and is under a tension of 400 N. A second student, moving toward the wall, hears 8.30 beats per second. What is the speed of the student approaching the wall?

75. By proper excitation, it is possible to produce both longitudinal and transverse waves in a long metal rod. In a particular case, the rod is 150 cm long and 0.200 cm in radius and has a mass of 50.9 g. Young's modulus for the material is 6.80×10^{10} Pa. Determine the required tension in the rod so that the ratio of the speed of longitudinal waves to the speed of transverse waves is 8.

76. A 0.500-m-long brass pipe open at both ends has a fundamental frequency of 350 Hz. (a) Determine the temperature of the air in the pipe. (b) If the temperature is increased by 20.0°C, what is the new fundamental frequency of the pipe? Be sure to include the effects of temperature on both the speed of sound in air and the length of the pipe.

Shvaygert Ekaterina/Shutterstock.com

View of lightning over a city at night. During a thunderstorm, a high concentration of electrical charge in a thundercloud creates a higher-than-normal electric field between the thundercloud and the negatively charged Earth's surface. This strong electric field creates an electric discharge—an enormous spark—between the charged cloud and the ground. Other discharges observed in the sky include cloud-to-cloud discharges and the more frequent intracloud discharges.

Electric Forces and Electric Fields 15

Electricity is the lifeblood of technological civilization and modern society. Without it, we revert to the mid-nineteenth century: no telephones, no television, none of the household appliances that we take for granted. Modern medicine would be a fantasy, and due to the lack of sophisticated experimental equipment and fast computers—and especially the slow dissemination of information—science and technology would grow at a glacial pace.

Instead, with the discovery and harnessing of electric forces and fields, we can view arrangements of atoms, probe the inner workings of the cell, and send spacecraft beyond the limits of the solar system. All this has become possible in just the last few generations of human life, a blink of the eye compared to the million years our kind spent foraging the savannahs of Africa.

Around 700 B.C. the ancient Greeks conducted the earliest known study of electricity. It all began when someone noticed that a fossil material called amber would attract small objects after being rubbed with wool. Since then we have learned that this phenomenon is not restricted to amber and wool, but occurs (to some degree) when almost any two nonconducting substances are rubbed together.

In this chapter we use the effect of charging by friction to begin an investigation of electric forces. We then discuss Coulomb's law, which is the fundamental law of force between any two stationary charged particles. The concept of an electric field associated with charges is introduced and its effects on other charged particles described. We end with discussions of the Van de Graaff generator and Gauss's law.

Benjamin Franklin
(1706–1790)
Franklin was a printer, author, physical scientist, inventor, diplomat, and a founding father of the United States. His work on electricity in the late 1740s changed a jumbled, unrelated set of observations into a coherent science.

15.1 Properties of Electric Charges

LEARNING OBJECTIVES

1. Define the SI unit of electric charge and identify the basic carriers of positive and negative charge.
2. Discuss the concept of charge conservation and the role of experiments in early studies of electricity.

After running a plastic comb through your hair, you will find that the comb attracts bits of paper. The attractive force is often strong enough to suspend the paper from the comb, defying the gravitational pull of the entire Earth. The same effect occurs with other rubbed materials, such as glass and hard rubber.

Another simple experiment is to rub an inflated balloon against wool (or across your hair). On a dry day, the rubbed balloon will then stick to the wall of a room, often for hours. These materials have become electrically charged. You can give your body an electric charge by vigorously rubbing your shoes on a wool rug or by sliding across a car seat. You can then surprise and annoy a friend or coworker with a light touch on the arm, delivering a slight shock to both yourself and your victim. (If the coworker is your boss, don't expect a promotion!) These experiments work best on a dry day because excessive moisture can facilitate a leaking away of the charge.

Experiments also demonstrate that there are two kinds of electric charge, which Benjamin Franklin (1706–1790) named **positive** and **negative.** Figure 15.1 illustrates the interaction of the two charges. A hard rubber (or plastic) rod that has been rubbed with fur is suspended by a piece of string. When a glass rod that has been rubbed with silk is brought near the rubber rod, the rubber rod is attracted toward the glass rod (Fig. 15.1a). If two charged rubber rods (or two charged glass rods) are brought near each other, as in Figure 15.1b, the force between them is repulsive. These observations may be explained by assuming the rubber and glass rods have acquired different kinds of excess charge. We use the convention suggested by Franklin, where the excess electric charge on the glass rod is called positive and that on the rubber rod is called negative. On the basis of such observations, we conclude that **like charges repel one another and unlike charges attract one another.** Objects usually contain equal amounts of positive and negative charge; electrical forces between objects arise when those objects have net negative or positive charges.

Like charges repel; unlike charges attract. ▶

Nature's basic carriers of positive charge are protons, which, along with neutrons, are located in the nuclei of atoms. The nucleus, about 10^{-15} m in radius, is surrounded by a cloud of negatively charged electrons with a radius about ten

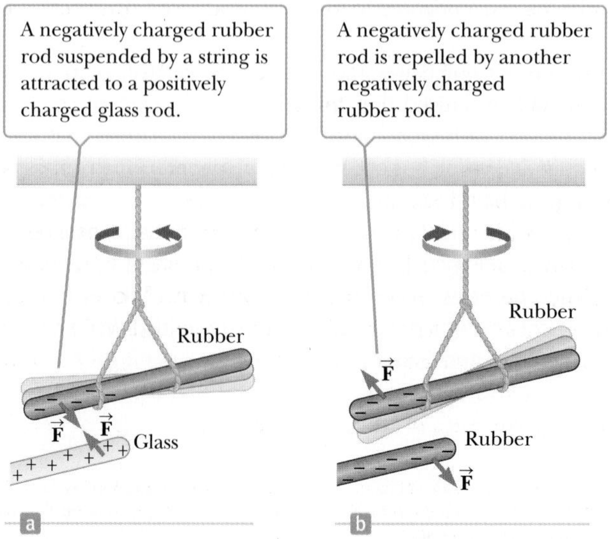

Figure 15.1 An experimental setup for observing the electrical force between two charged objects.

thousand times larger. An electron has the same magnitude charge as a proton, but the opposite sign. In a gram of matter there are approximately 10^{23} positively charged protons and just as many negatively charged electrons, so the net charge is zero. Because the nucleus of an atom is held firmly in place inside a solid, protons never move from one material to another. Electrons are far lighter than protons and hence more easily accelerated by forces. Further, they occupy the outer regions of the atom. Consequently, objects become charged by gaining or losing electrons.

Charge transfers readily from one type of material to another. Rubbing the two materials together serves to increase the area of contact, facilitating the transfer process.

◀ Charge is conserved

An important characteristic of charge is that **electric charge is always conserved.** Charge isn't *created* when two neutral objects are rubbed together; rather, the objects become charged because **negative charge is transferred from one object to the other.** One object gains a negative charge while the other loses an equal amount of negative charge and hence is left with a net positive charge. When a glass rod is rubbed with silk, as in Figure 15.2, electrons are transferred from the rod to the silk. As a result, the glass rod carries a net positive charge, the silk a net negative charge. Likewise, when rubber is rubbed with fur, electrons are transferred from the fur to the rubber.

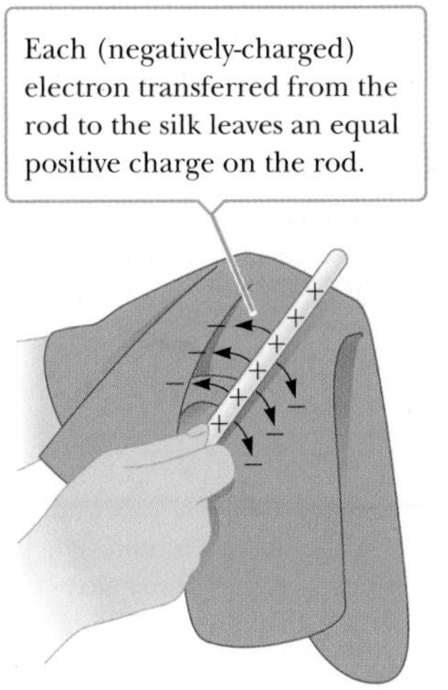

Figure 15.2 When a glass rod is rubbed with silk, electrons are transferred from the glass to the silk. Because the charges are transferred in discrete bundles, the charges on the two objects are $\pm e$, $\pm 2e$, $\pm 3e$, and so on.

In 1909 Robert Millikan (1886–1953) discovered that if an object is charged, its charge is always a multiple of a fundamental unit of charge, designated by the symbol e. In modern terms, the charge is said to be **quantized,** meaning that charge occurs in discrete chunks that can't be further subdivided. An object may have a charge of $\pm e$, $\pm 2e$, $\pm 3e$, and so on, but never[1] a fractional charge of $\pm 0.5e$ or $\pm 0.22e$. Other experiments in Millikan's time showed that the electron has a charge of $-e$ and the proton has an equal and opposite charge of $+e$. Some particles, such as a neutron, have no net charge. A neutral atom (an atom with no net charge) contains as many protons as electrons. The value of e is now known to be $1.602\ 19 \times 10^{-19}$ C. (The SI unit of electric charge is the **coulomb,** or C.)

15.2 Insulators and Conductors

LEARNING OBJECTIVES

1. Describe conductors, insulators, and semi-conductors on the basis of their relative abilities to conduct electric charge.
2. Describe the physical processes of polarization and charging by conduction or induction.

Substances can be classified in terms of their ability to conduct electric charge.

> In **conductors,** electric charges move freely in response to an electric force. All other materials are called **insulators.**

Glass and rubber are insulators. When such materials are charged by rubbing, only the rubbed area becomes charged, and there is no tendency for the charge to move into other regions of the material. In contrast, materials such as copper, aluminum, and silver are good conductors. When such materials are charged in some small region, the charge readily distributes itself over the entire surface of the material. If you hold a copper rod in your bare hand and rub the rod with wool or fur, it will not attract a piece of paper. This might suggest that a metal can't be charged. However, if you hold the copper rod with an insulator and then rub it with wool or fur, the rod remains charged and attracts the paper. In the first case,

[1]There is strong evidence for the existence of fundamental particles called **quarks** that have charges of $\pm e/3$ or $\pm 2e/3$. The charge is *still* quantized, but in units of $\pm e/3$ rather than $\pm e$. A more complete discussion of quarks and their properties is presented in Chapter 30.

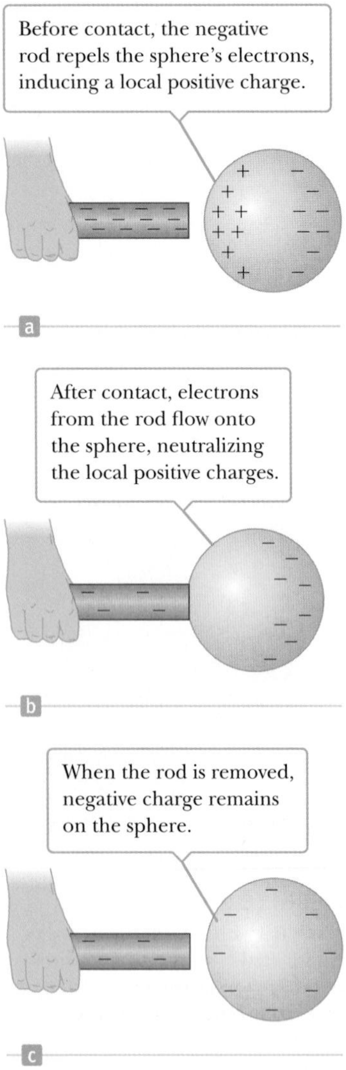

Figure 15.3 Charging a metallic object by conduction.

the electric charges produced by rubbing readily move from the copper through your body and finally to ground. In the second case, the insulating handle prevents the flow of charge to ground.

Semiconductors are a third class of materials, and their electrical properties are somewhere between those of insulators and those of conductors. Silicon and germanium are well-known semiconductors that are widely used in the fabrication of a variety of electronic devices.

Charging by Conduction

Consider a negatively charged rubber rod brought into contact with an insulated neutral conducting sphere. The excess electrons on the rod repel electrons on the sphere, creating local positive charges on the neutral sphere. On contact, some electrons on the rod are now able to move onto the sphere, as in Figure 15.3, neutralizing the positive charges. When the rod is removed, the sphere is left with a net negative charge. This process is referred to as charging by **conduction.** The object being charged in such a process (the sphere) is always left with a charge having the same sign as the object doing the charging (the rubber rod).

Charging by Induction

An object connected to a conducting wire or copper pipe buried in the Earth is said to be **grounded.** The Earth can be considered an infinite reservoir for electrons; in effect, it can accept or supply an unlimited number of electrons. With this idea in mind, we can understand the charging of a conductor by a process known as **induction.**

Consider a negatively charged rubber rod brought near a neutral (uncharged) conducting sphere that is insulated, so there is no conducting path to ground (Fig. 15.4). Initially the sphere is electrically neutral (Fig. 15.4a). When the negatively charged rod is brought close to the sphere, the repulsive force between the electrons in the rod and those in the sphere causes some electrons to move to the side of the sphere farthest away from the rod (Fig. 15.4b). The region of the sphere nearest the negatively charged rod has an excess of positive charge because of the migration of electrons away from that location. If a grounded conducting wire is then connected to the sphere, as in Figure 15.4c, some of the electrons leave the sphere and travel to ground. If the wire to ground is then removed (Fig. 15.4d), the conducting sphere is left with an excess of induced positive charge. Finally, when the rubber rod is removed from the vicinity of the sphere (Fig. 15.4e), the induced positive charge remains on the ungrounded sphere. Even though the positively charged atomic nuclei remain fixed, this excess positive charge becomes uniformly distributed over the surface of the ungrounded sphere because of the repulsive forces among the like charges and the high mobility of electrons in a metal.

In the process of inducing a charge on the sphere, the charged rubber rod doesn't lose any of its negative charge because it never comes in contact with the sphere. Further, the sphere is left with a charge opposite that of the rubber rod. **Charging an object by induction requires no contact with the object inducing the charge.**

A process similar to charging by induction in conductors also takes place in insulators. In most neutral atoms or molecules, the center of positive charge coincides with the center of negative charge. In the presence of a charged object, however, these centers may separate slightly, resulting in more positive charge on one side of the molecule than on the other. This effect is known as **polarization.** The realignment of charge within individual molecules produces an induced charge on the surface of the insulator, as shown in Figure 15.5a. This property explains why a balloon charged through rubbing will stick to an electrically neutral wall or why the comb you just used on your hair attracts tiny bits of neutral paper.

SUMMARY

15.1 Properties of Electric Charges

Electric charges have the following properties:

1. Unlike charges attract one another and like charges repel one another.
2. Electric charge is always conserved.
3. Charge comes in discrete packets that are integral multiples of the basic electric charge $e = 1.6 \times 10^{-19}$ C.
4. The force between two charged particles is proportional to the inverse square of the distance between them.

15.2 Insulators and Conductors

Conductors are materials in which charges move freely in response to an electric field. All other materials are called **insulators.**

15.3 Coulomb's Law

Coulomb's law states that the electric force between two stationary charged particles separated by a distance r has the magnitude

$$F = k_e \frac{|q_1||q_2|}{r^2} \quad [15.1]$$

where $|q_1|$ and $|q_2|$ are the magnitudes of the charges on the particles in coulombs and

$$k_e \approx 8.99 \times 10^9 \text{ N} \cdot \text{m}^2/\text{C}^2 \quad [15.2]$$

is the **Coulomb constant.**

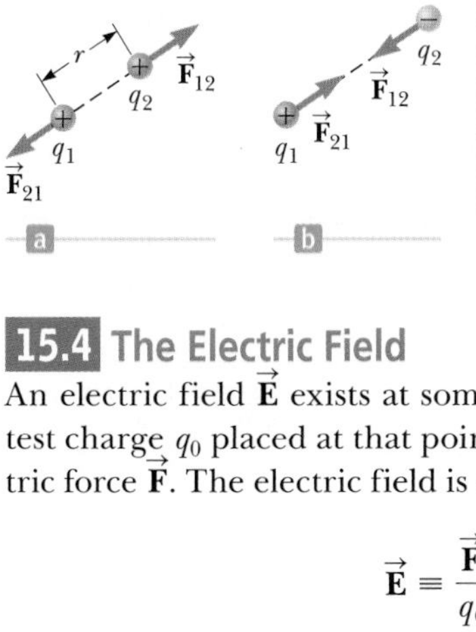

(a) The electric force between two charges with the same sign is repulsive, and (b) it is attractive when the charges have opposite signs.

15.4 The Electric Field

An electric field $\vec{\mathbf{E}}$ exists at some point in space if a small test charge q_0 placed at that point is acted upon by an electric force $\vec{\mathbf{F}}$. The electric field is defined as

$$\vec{\mathbf{E}} \equiv \frac{\vec{\mathbf{F}}}{q_0} \quad [15.3]$$

The **direction** of the electric field at a point in space is defined to be the direction of the electric force that would be exerted on a small positive charge placed at that point.

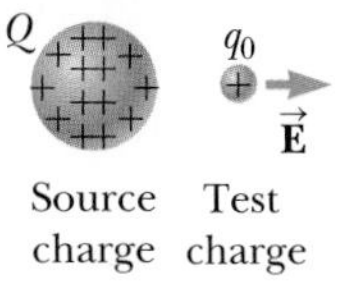

The electric force of charge Q on a test charge q_0 divided by q_0 gives the electric field $\vec{\mathbf{E}}$ of Q at that point.

The magnitude of the electric field due to a *point charge* q at a distance r from the point charge is

$$E = k_e \frac{|q|}{r^2} \quad [15.6]$$

15.5 Electric Field Lines

Electric field lines are useful for visualizing the electric field in any region of space. The electric field vector $\vec{\mathbf{E}}$ is tangent to the electric field lines at every point. Further, the number of electric field lines per unit area through a surface perpendicular to the lines is proportional to the strength of the electric field at that surface.

15.6 Conductors in Electrostatic Equilibrium

A **conductor in electrostatic equilibrium** has the following properties:

1. The electric field is zero everywhere inside the conducting material.
2. Any excess charge on an isolated conductor must reside entirely on its surface.
3. The electric field just outside a charged conductor is perpendicular to the conductor's surface.
4. On an irregularly shaped conductor, charge accumulates where the radius of curvature of the surface is smallest, at sharp points.

15.9 Electric Flux and Gauss's Law

Gauss's law states that the electric flux through any closed surface is equal to the net charge Q inside the surface divided by the permittivity of free space, ϵ_0:

$$EA = \Phi_E = \frac{Q_{\text{inside}}}{\epsilon_0} \quad [15.12]$$

For highly symmetric distributions of charge, Gauss's law can be used to calculate electric fields.

The electric flux Φ through any arbitrary surface surrounding a charge q is q/ϵ_0.

WARM-UP EXERCISES

ENHANCED WebAssign The warm-up exercises in this chapter may be assigned online in Enhanced WebAssign.

1. **Math Review** A force vector has components given by $F_x = -7.00$ N and $F_y = 4.50$ N. Find (a) the magnitude and (b) the direction of the force, measured counterclockwise from the positive x-axis.

2. **Math Review** The force acting on a particle has a magnitude of 125 N and is directed 30.0° above the positive x-axis. Determine (a) the x-component and (b) y-component of the force.

3. Math Review Two electric force vectors act on a particle. Their x-components are 15.0 N and −7.50 N and their y-components are −11.5 N and −4.50 N, respectively. For the resultant electric force, find (a) the x-component, (b) the y-component, and the (c) magnitude and (d) direction of the resultant electric force, measured counterclockwise from the positive x-axis. (See also Section 15.4.)

4. Physics Review A large firecracker explodes under a soda can, sending it straight upward with an initial speed of 7.27 m/s. Neglecting air friction, find (a) the can's maximum height, (b) the first time at which it reaches a height of 1.00 m, and (c) the velocity at that time. (See Section 2.5.)

5. Physics Review A simple pendulum with mass 0.250 kg hangs from string 1.40 m long. (See Fig. WU15.5.) A leaf blower blows air at it, exerting a constant horizontal drag force $\vec{F}_{drag}$ of magnitude 1.67 N. Find (a) the tension in the string and (b) the angle the pendulum makes with respect to the vertical. (See Section 4.5.)

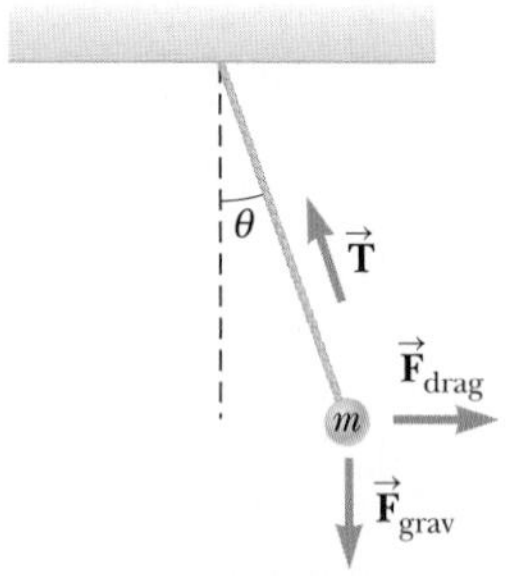

Figure WU15.5

6. Physics Review The Falcon 9 rocket, manufactured by the commercial SpaceX Corporation, has a takeoff mass of 4.80×10^5 kg and an initial thrust of 5.88×10^6 N. At takeoff, determine (a) the net upward force acting on the rocket, and (b) the upward acceleration of the rocket. (See Section 4.5.)

7. Physics Review A horizontal spring with force constant k = 57.0 N/m is used to accelerate a 0.500-kg mass on a frictionless surface. If the spring is compressed by 5.00×10^{-2} m and released, find (a) the initial magnitude of the spring force when the mass is released and (b) the initial acceleration of the mass. (See Section 5.4.)

8. Two protons are separated by a distance of 0.100 m. Given the proton charge of 1.60×10^{-19} C, determine the magnitude of the electric force that one proton exerts on the other. (See Section 15.3.)

9. A charge of 16.0 nC is located at the origin. Find (a) the magnitude and (b) the direction of the displacement vector pointing from the origin to (3.00, 4.00) m. Calculate (c) the magnitude, and (d) the x- and (e) y-components of the electric field at that point. (See Section 15.4.)

10. Consider a system of two charges where the first charge is on the positive x-axis and the second charge is on the positive y-axis. If these two charges produce electric fields at the origin given by (2.50, 0) N/C and (0, −2.50) N/C, respectively, determine (a) the magnitude and (b) the direction of the resultant electric field at the origin, measured counterclockwise from the positive x-axis. (c) What would be the magnitude of the force on a 1.41×10^{-9} C charge placed at the origin? (See Sections 15.3 and 15.4.)

11. A tiny droplet of oil of mass 1.50×10^{-15} kg having a charge of -2.56×10^{-18} C accelerates straight down inside a vacuum chamber. If the electric field is −5 625 N/C, find (a) the force of gravity on the droplet, (b) the electric force, and (c) the droplet's acceleration. (Give the correct signs, with down the negative direction.)

12. A constant electric field of magnitude 5.00 N/C and pointing in the positive z-direction passes through a square with sides of length 0.500 m, located in the xy-plane. (a) Calculate the magnitude of the electric flux through the square. (b) Repeat the calculation if the electric field is oriented at an angle of 30.0° with respect to the positive z-direction. (See Section 15.9.)

13. Three charges are placed inside a basketball. If the charges are 2.50 nC, −1.25 nC and 0.500 nC, respectively, what is the electric flux through the basketball? (See Section 15.9.)

14. A thin, hollow sphere has a charge q = −2.00 nC evenly distributed over its surface. The radius of the charged sphere is 1.00 m and the radii of two spherical Gaussian surfaces are a = 0.500 m and b = 1.50 m, respectively. (See Figure WU15.14.) A charge of Q = 5.00 nC is placed at the center of the charged sphere. (a) How much charge is inside the inner Gaussian surface? (b) Find the electric flux through the inner Gaussian surface, and (c) the electric field there, by dividing by the sphere's area. (d) How much charge is inside the outer dashed sphere? Find (e) the electric flux through the outer Gaussian surface and (f) the electric field strength on that surface. (See Section 15.9.)

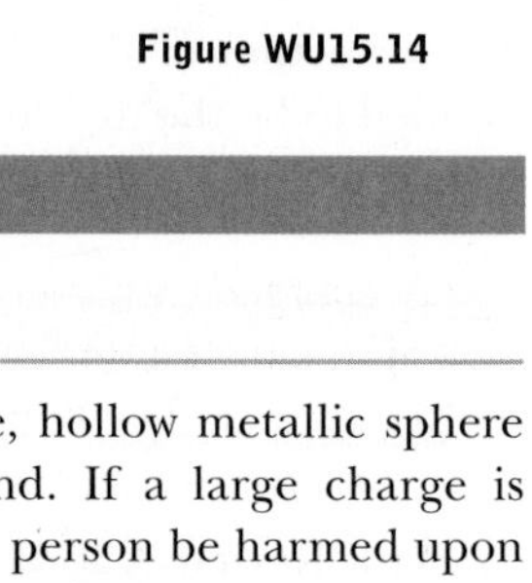

Figure WU15.14

CONCEPTUAL QUESTIONS

ENHANCED WebAssign The conceptual questions in this chapter may be assigned online in Enhanced WebAssign.

1. A glass object receives a positive charge of +3 nC by rubbing it with a silk cloth. In the rubbing process, have protons been added to the object or have electrons been removed from it?

2. Explain from an atomic viewpoint why charge is usually transferred by electrons.

3. A person is placed in a large, hollow metallic sphere that is insulated from ground. If a large charge is placed on the sphere, will the person be harmed upon touching the inside of the sphere?

4. Why must hospital personnel wear special conducting shoes while working around oxygen in an operating room? What might happen if the personnel wore shoes with rubber soles?

5. (a) Would life be different if the electron were positively charged and the proton were negatively charged? (b) Does the choice of signs have any bearing on physical and chemical interactions? Explain your answers.

6. If a suspended object A is attracted to a charged object B, can we conclude that A is charged? Explain.

7. Explain how a positively charged object can be used to leave another metallic object with a net negative charge. Discuss the motion of charges during the process.

8. Consider point A in Figure CQ15.8 located an arbitrary distance from two point charges in otherwise empty space. (a) Is it possible for an electric field to exist at point A in empty space? (b) Does charge exist at this point? (c) Does a force exist at this point?

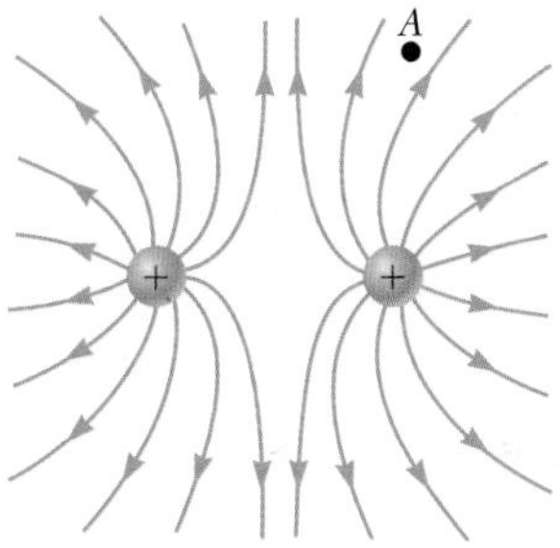

Figure CQ15.8

9. A student stands on a thick piece of insulating material, places her hand on top of a Van de Graaff generator, and then turns on the generator. Does she receive a shock?

10. In fair weather, there is an electric field at the surface of the Earth, pointing down into the ground. What is the sign of the electric charge on the ground in this situation?

11. A charged comb often attracts small bits of dry paper that then fly away when they touch the comb. Explain why that occurs.

12. Why should a ground wire be connected to the metal support rod for a television antenna?

13. There are great similarities between electric and gravitational fields. A room can be electrically shielded so that there are no electric fields in the room by surrounding it with a conductor. Can a room be gravitationally shielded? Explain.

14. A spherical surface surrounds a point charge q. Describe what happens to the total flux through the surface if (a) the charge is tripled, (b) the volume of the sphere is doubled, (c) the surface is changed to a cube, (d) the charge is moved to another location inside the surface, and (e) the charge is moved outside the surface.

15. If more electric field lines leave a Gaussian surface than enter it, what can you conclude about the net charge enclosed by that surface?

16. A student who grew up in a tropical country and is studying in the United States may have no experience with static electricity sparks and shocks until his or her first American winter. Explain.

17. What happens when a charged insulator is placed near an uncharged metallic object? (a) They repel each other. (b) They attract each other. (c) They may attract or repel each other, depending on whether the charge on the insulator is positive or negative. (d) They exert no electrostatic force on each other. (e) The charged insulator always spontaneously discharges.

PROBLEMS

ENHANCED WebAssign The problems in this chapter may be assigned online in Enhanced WebAssign.

1. denotes straightforward problem; 2. denotes intermediate problem;
3. denotes challenging problem
[1.] denotes full solution available in *Student Solutions Manual/ Study Guide*
1. denotes problems most often assigned in Enhanced WebAssign

BIO denotes biomedical problems
GP denotes guided problems
M denotes Master It tutorial available in Enhanced WebAssign
Q|C denotes asking for quantitative and conceptual reasoning
S denotes symbolic reasoning problem
W denotes Watch It video solution available in Enhanced WebAssign

15.3 Coulomb's Law

1. Q|C A 7.50-nC charge is located 1.80 m from a 4.20-nC charge. (a) Find the magnitude of the electrostatic force that one particle exerts on the other. (b) Is the force attractive or repulsive?

[2.] A charged particle A exerts a force of 2.62 N to the right on charged particle B when the particles are 13.7 mm apart. Particle B moves straight away from A to make the distance between them 17.7 mm. What vector force does particle B then exert on A?

3. Q|C Two metal balls A and B of negligible radius are floating at rest on Space Station Freedom between two metal bulkheads, connected by a taut nonconducting thread of length 2.00 m. Ball A carries charge q, and ball B carries charge $2q$. Each ball is 1.00 m away from a bulkhead. (a) If the tension in the string is 2.50 N, what is the magnitude of q? (b) What happens to the system as time passes? Explain.

4. A small sphere of mass m = 7.50 g and charge q_1 = 32.0 nC is attached to the end of a string and hangs vertically as in Figure P15.4. A second charge of equal mass and charge q_2 = −58.0 nC is located below the first charge a

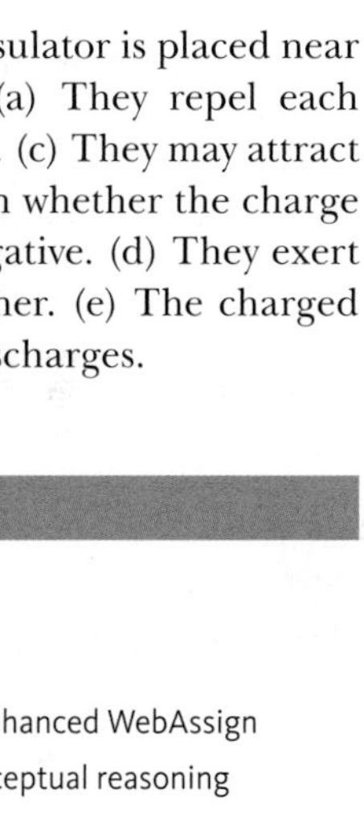

Figure P15.4

distance $d = 2.00$ cm below the first charge as in Figure P15.4. (a) Find the tension in the string. (b) If the string can withstand a maximum tension of 0.180 N, what is the smallest value d can have before the string breaks?

5. The nucleus of ^{8}Be, which consists of 4 protons and 4 neutrons, is very unstable and spontaneously breaks into two alpha particles (helium nuclei, each consisting of 2 protons and 2 neutrons). (a) What is the force between the two alpha particles when they are 5.00×10^{-15} m apart, and (b) what is the initial magnitude of the acceleration of the alpha particles due to this force? Note that the mass of an alpha particle is 4.002 6 u.

6. BIO A molecule of DNA (deoxyribonucleic acid) is 2.17 μm long. The ends of the molecule become singly ionized: negative on one end, positive on the other. The helical molecule acts like a spring and compresses 1.00% upon becoming charged. Determine the effective spring constant of the molecule.

7. A small sphere of charge 0.800 μC hangs from the end of a spring as in Figure P15.7a. When another small sphere of charge -0.600 μC is held beneath the first sphere as in Figure P15.7b, the spring stretches by $d =$ 3.50 cm from its original length and reaches a new equilibrium position with a separation between the charges of $r = 5.00$ cm. What is the force constant of the spring?

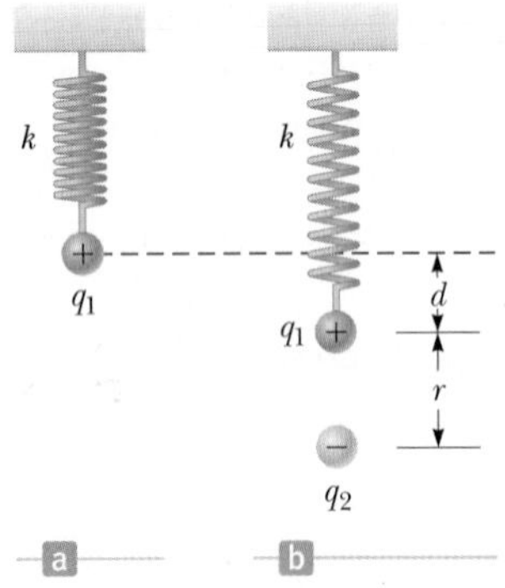

Figure P15.7

8. S Four point charges are at the corners of a square of side a as shown in Figure P15.8. Determine the magnitude and direction of the resultant electric force on q, with k_e, q, and a left in symbolic form.

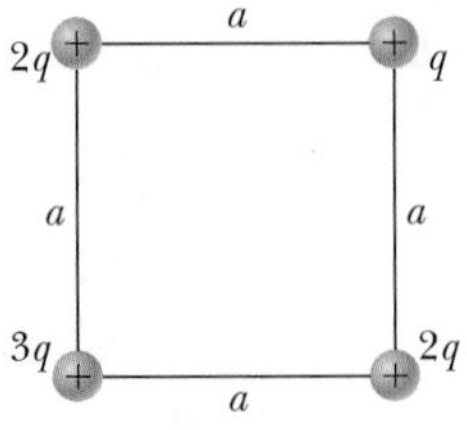

Figure P15.8

9. Two small identical conducting spheres are placed with their centers 0.30 m apart. One is given a charge of 12×10^{-9} C, the other a charge of -18×10^{-9} C. (a) Find the electrostatic force exerted on one sphere by the other. (b) The spheres are connected by a conducting wire. Find the electrostatic force between the two after equilibrium is reached, where both spheres have the same charge.

10. Calculate the magnitude and direction of the Coulomb force on each of the three charges shown in Figure P15.10.

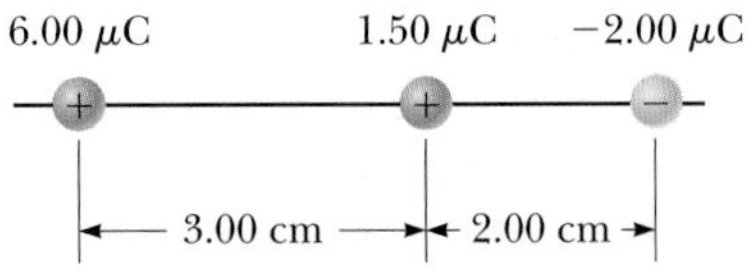

Figure P15.10 Problems 10 and 18.

11. M Three charges are arranged as shown in Figure P15.11. Find the magnitude and direction of the electrostatic force on the charge at the origin.

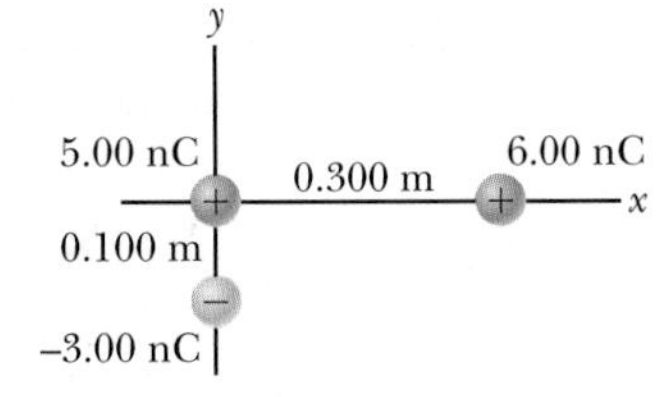

Figure P15.11

12. A positive charge $q_1 = 2.70$ μC on a frictionless horizontal surface is attached to a spring of force constant k as in Figure P15.12. When a charge of $q_2 = -8.60$ μC is placed 9.50 cm away from the positive charge, the spring stretches by 5.00 mm, reducing the distance between charges to $d = 9.00$ cm. Find the value of k.

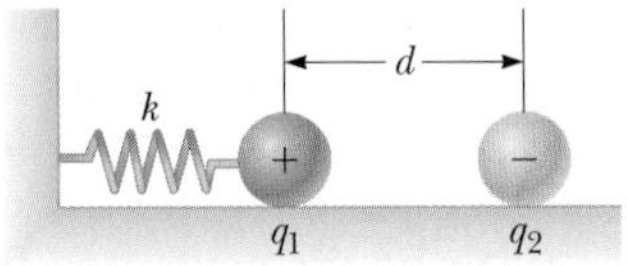

Figure P15.12

13. Three point charges are located at the corners of an equilateral triangle as in Figure P15.13. Find the magnitude and direction of the net electric force on the 2.00 μC charge.

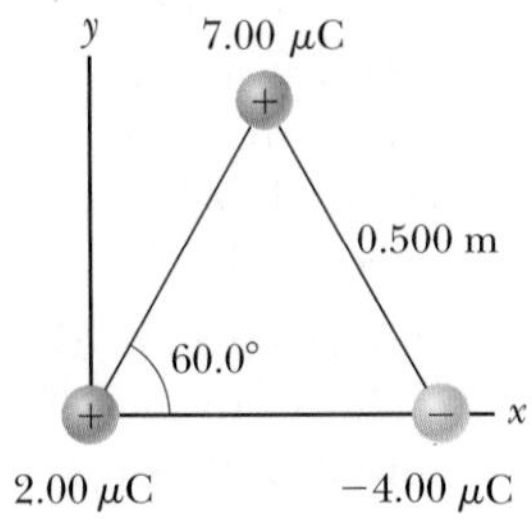

Figure P15.13 Problems 13 and 24.

14. Two identical metal blocks resting on a frictionless horizontal surface are connected by a light metal spring having constant $k = 100$ N/m and unstretched length $L_i = 0.400$ m as in Figure P15.14a. A charge Q is slowly placed on each block causing the spring to stretch to an equilibrium length $L = 0.500$ m as in Figure P15.14b. Determine the value of Q, modeling the blocks as charged particles.

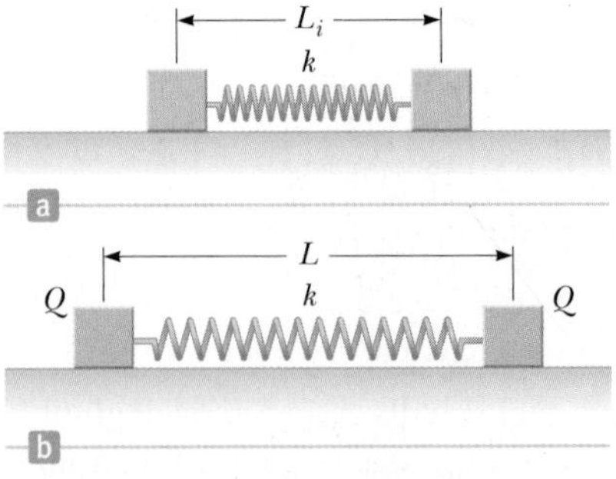

Figure P15.14

15. W Two small metallic spheres, each of mass $m = 0.20$ g, are suspended as pendulums by light strings from a common point as shown in Figure P15.15. The spheres are given the same electric charge, and it is found that they come to equilibrium when each string is at an angle of $\theta = 5.0°$ with the vertical. If each string has length $L = 30.0$ cm, what is the magnitude of the charge on each sphere?

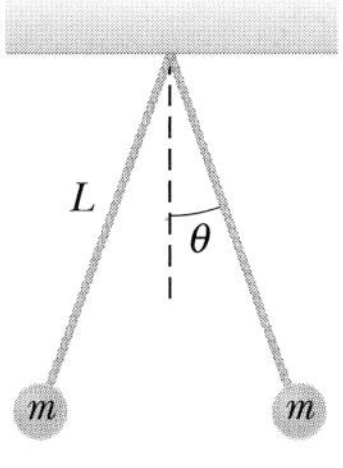

Figure P15.15

16. GP Particle A of charge 3.00×10^{-4} C is at the origin, particle B of charge -6.00×10^{-4} C is at (4.00 m, 0) and particle C of charge 1.00×10^{-4} C is at (0, 3.00 m). (a) What is the x-component of the electric force exerted by A on C? (b) What is the y-component of the force exerted by A on C? (c) Find the magnitude of the force exerted by B on C. (d) Calculate the x-component of the force exerted by B on C. (e) Calculate the y-component of the force exerted by B on C. (f) Sum the two x-components to obtain the resultant x-component of the electric force acting on C. (g) Repeat part (f) for the y-component. (h) Find the magnitude and direction of the resultant electric force acting on C.

15.4 The Electric Field

17. A small object of mass 3.80 g and charge $-18\ \mu$C is suspended motionless above the ground when immersed in a uniform electric field perpendicular to the ground. What is the magnitude and direction of the electric field?

18. (a) Determine the electric field strength at a point 1.00 cm to the left of the middle charge shown in Figure P15.10. (b) If a charge of $-2.00\ \mu$C is placed at this point, what are the magnitude and direction of the force on it?

19. An electric field of magnitude 5.25×10^5 N/C points due south at a certain location. Find the magnitude and direction of the force on a $-6.00\ \mu$C charge at this location.

20. W An electron is accelerated by a constant electric field of magnitude 300 N/C. (a) Find the acceleration of the electron. (b) Use the equations of motion with constant acceleration to find the electron's speed after 1.00×10^{-8} s, assuming it starts from rest.

21. A small block of mass m and charge Q is placed on an insulated, frictionless, inclined plane of angle θ as in Figure P15.21. An electric field is applied parallel to the incline. (a) Find an expression for the magnitude of the electric field that enables the block to remain at rest. (b) If $m = 5.40$ g, $Q = -7.00\ \mu$C, and $\theta = 25.0°$, determine the magnitude and direction of the electric field that enables the block to remain at rest on the incline.

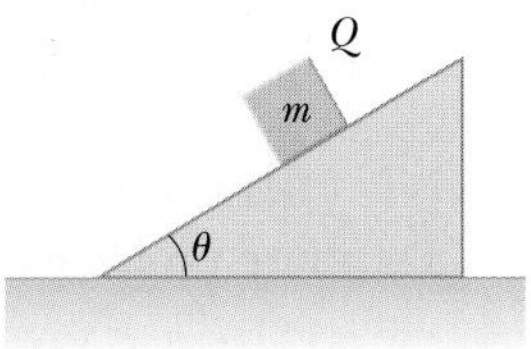

Figure P15.21

22. A small sphere of charge $q = +68\ \mu$C and mass $m = 5.8$ g is attached to a light string and placed in a uniform electric field $\vec{\mathbf{E}}$ that makes an angle $\theta = 37°$ with the horizontal. The opposite end of the string is attached to a wall and the sphere is in static equilibrium when the string is horizontal as in Figure P15.22. (a) Construct a free body diagram for the sphere. Find (b) the magnitude of the electric field and (c) the tension in the string.

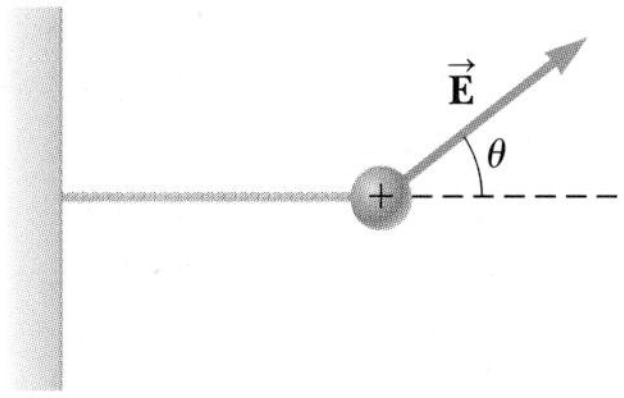

Figure P15.22

23. M A proton accelerates from rest in a uniform electric field of 640 N/C. At some later time, its speed is 1.20×10^6 m/s. (a) Find the magnitude of the acceleration of the proton. (b) How long does it take the proton to reach this speed? (c) How far has it moved in that interval? (d) What is its kinetic energy at the later time?

24. Q|C (a) Find the magnitude and direction of the electric field at the position of the 2.00 μC charge in Figure P15.13. (b) How would the electric field at that point be affected if the charge there were doubled? Would the magnitude of the electric force be affected?

25. S Two equal positive charges are at opposite corners of a trapezoid as in Figure P15.25. Find symbolic expressions for the components of the electric field at the point P.

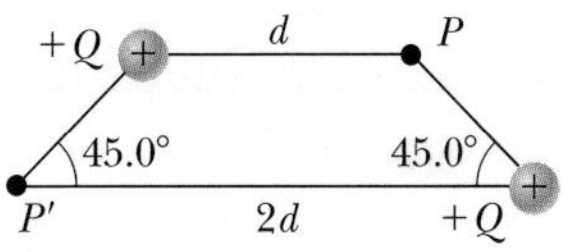

Figure P15.25

26. Three point charges are located on a circular arc as shown in Figure P15.26. (a) What is the total electric field at P, the center of the arc? (b) Find the electric force that would be exerted on a -5.00-nC charge placed at P.

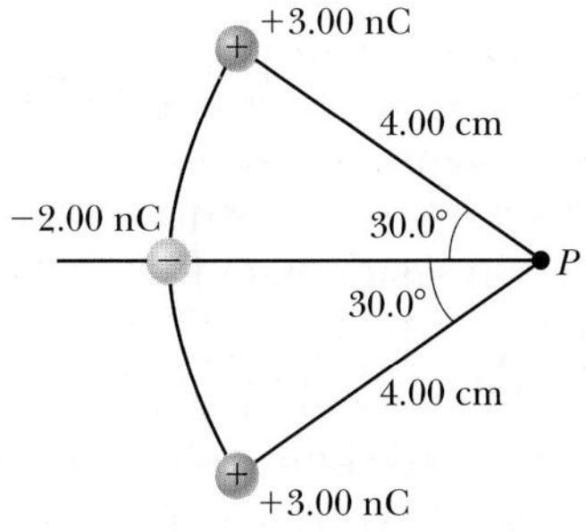

Figure P15.26

27. In Figure P15.27 determine the point (other than infinity) at which the total electric field is zero.

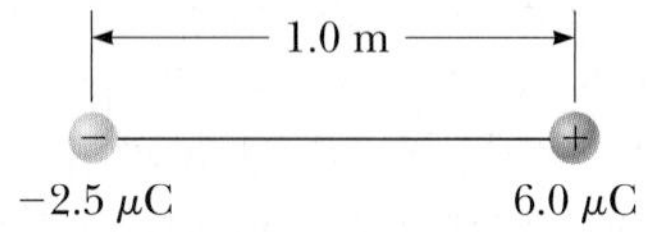

Figure P15.27

28. Three charges are at the corners of an equilateral triangle, as shown in Figure P15.28. Calculate the electric field at a point midway between the two charges on the x-axis.

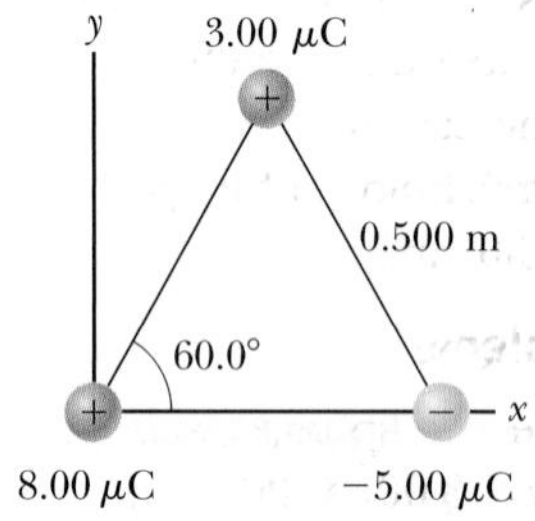

Figure P15.28

29. Three identical charges ($q = -5.0\ \mu C$) lie along a circle of radius 2.0 m at angles of 30°, 150°, and 270°, as shown in Figure P15.29. What is the resultant electric field at the center of the circle?

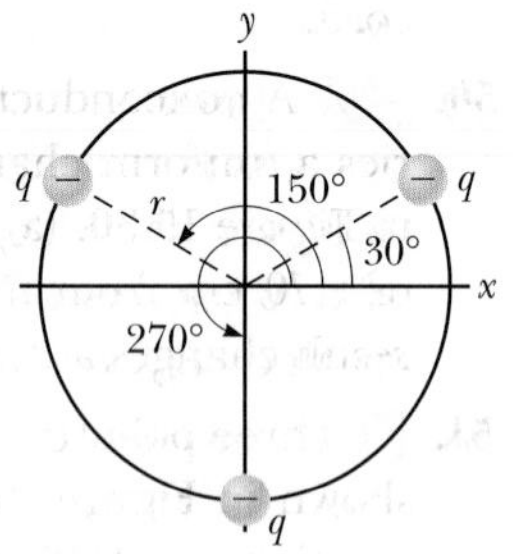

Figure P15.29

15.5 Electric Field Lines

15.6 Conductors in Electrostatic Equilibrium

30. Figure P15.30 shows the electric field lines for two point charges separated by a small distance. (a) Determine the ratio q_1/q_2. (b) What are the signs of q_1 and q_2?

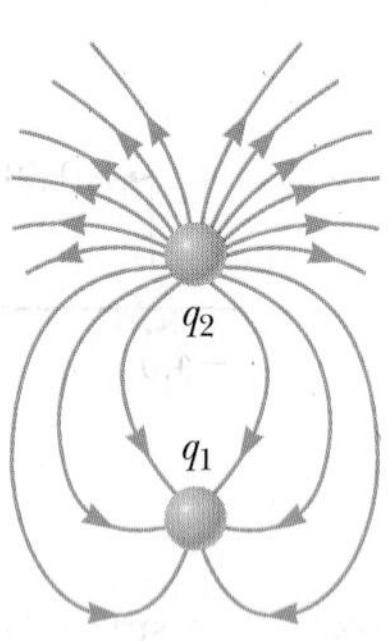

Figure P15.30

31. (a) Sketch the electric field lines around an isolated point charge $q > 0$. (b) Sketch the electric field pattern around an isolated negative point charge of magnitude $-2q$.

32. (a) Sketch the electric field pattern around two positive point charges of magnitude 1 μC placed close together. (b) Sketch the electric field pattern around two negative point charges of $-2\ \mu$C, placed close together. (c) Sketch the pattern around two point charges of $+1\ \mu$C and $-2\ \mu$C, placed close together.

33. Two point charges are a small distance apart. (a) Sketch the electric field lines for the two if one has a charge four times that of the other and both charges are positive. (b) Repeat for the case in which both charges are negative.

34. S Three equal positive charges are at the corners of an equilateral triangle of side a as in Figure P15.34. Assume the three charges together create an electric field. (a) Sketch the electric field lines in the plane of the charges. (b) Find the location of one point (other than ∞) where the electric field is zero. What are (c) the magnitude and (d) the direction of the electric field at P due to the two charges at the base?

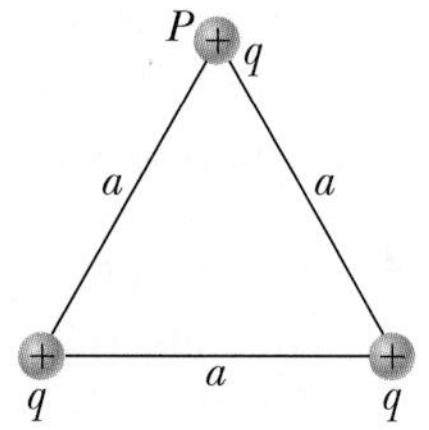

Figure P15.34

35. Refer to Figure 15.20. The charge lowered into the center of the hollow conductor has a magnitude of 5 μC. Find the magnitude and sign of the charge on the inside and outside of the hollow conductor when the charge is as shown in (a) Figure 15.20a, (b) Figure 15.20b, (c) Figure 15.20c, and (d) Figure 15.20d.

15.8 The Van de Graaff Generator

36. The dome of a Van de Graaff generator receives a charge of 2.0×10^{-4} C. Find the strength of the electric field (a) inside the dome, (b) at the surface of the dome, assuming it has a radius of 1.0 m, and (c) 4.0 m from the center of the dome. *Hint:* See Section 15.6 to review properties of conductors in electrostatic equilibrium. Also, note that the points on the surface are outside a spherically symmetric charge distribution; the total charge may be considered to be located at the center of the sphere.

37. If the electric field strength in air exceeds 3.0×10^6 N/C, the air becomes a conductor. Using this fact, determine the maximum amount of charge that can be carried by a metal sphere 2.0 m in radius. (See the hint in Problem 36.)

38. In the Millikan oil-drop experiment illustrated in Figure 15.21, an atomizer (a sprayer with a fine nozzle) is used to introduce many tiny droplets of oil between two oppositely charged parallel metal plates. Some of the droplets pick up one or more excess electrons. The charge on the plates is adjusted so that the electric force on the excess electrons exactly balances the weight of the droplet. The idea is to look for a droplet that has the smallest electric force and assume it has only one excess electron. This strategy lets the observer measure the charge on the electron. Suppose we are using an electric field of 3×10^4 N/C. The charge on one electron is about 1.6×10^{-19} C. Estimate the radius of an oil drop of density 858 kg/m^3 for which its weight could be balanced by the electric force of this field on one electron. (Problem 38 is courtesy of E. F. Redish. For more problems of this type, visit www.physics.umd.edu/perg/.)

39. W A Van de Graaff generator is charged so that a proton at its surface accelerates radially outward at 1.52×10^{12} m/s^2. Find (a) the magnitude of the electric force on the proton at that instant and (b) the

magnitude and direction of the electric field at the surface of the generator.

15.9 Electric Flux and Gauss's Law

40. A uniform electric field of magnitude E = 435 N/C makes an angle of θ = 65.0° with a plane surface of area A = 3.50 m^2 as in Figure P15.40. Find the electric flux through this surface.

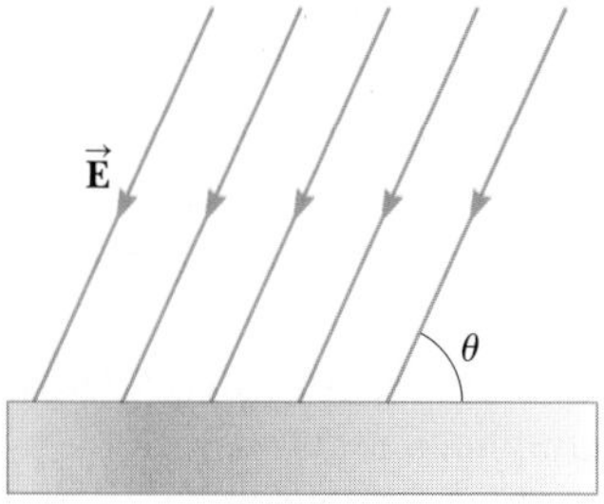

Figure P15.40

41. **M** An electric field of intensity 3.50 kN/C is applied along the x-axis. Calculate the electric flux through a rectangular plane 0.350 m wide and 0.700 m long if (a) the plane is parallel to the yz-plane, (b) the plane is parallel to the xy-plane, and (c) the plane contains the y-axis and its normal makes an angle of 40.0° with the x-axis.

42. **Q|C** The electric field everywhere on the surface of a charged sphere of radius 0.230 m has a magnitude of 575 N/C and points radially outward from the center of the sphere. (a) What is the net charge on the sphere? (b) What can you conclude about the nature and distribution of charge inside the sphere?

43. **S** Four closed surfaces, S_1 through S_4, together with the charges $-2Q$, Q, and $-Q$, are sketched in Figure P15.43. (The colored lines are the intersections of the surfaces with the page.) Find the electric flux through each surface.

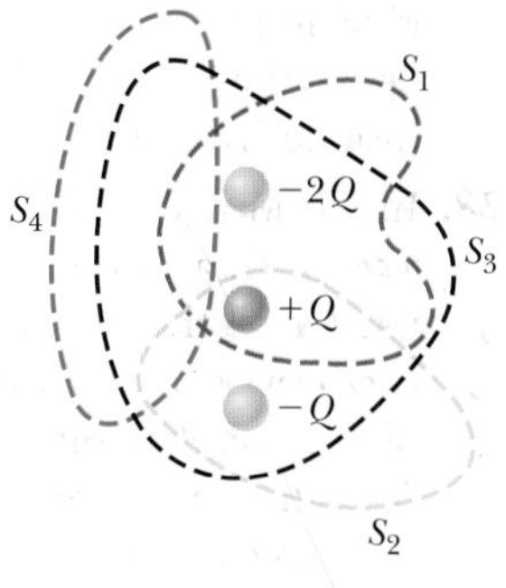

Figure P15.43

44. A charge q = +5.80 μC is located at the center of a regular tetrahedron (a four-sided surface) as in Figure P15.44. Find (a) the total electric flux through the tetrahedron and (b) the electric flux through one face of the tetrahedron.

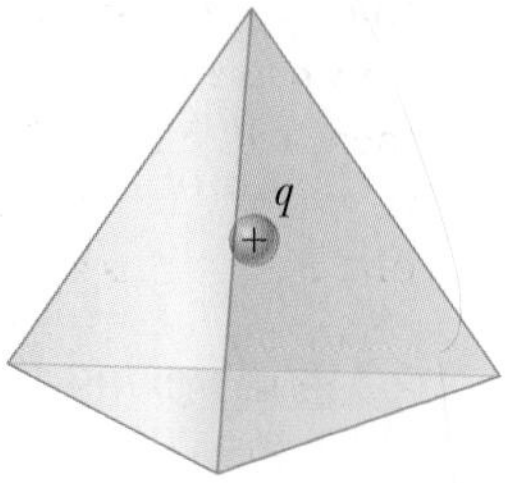

Figure P15.44

45. A point charge q is located at the center of a spherical shell of radius a that has a charge $-q$ uniformly distributed on its surface. Find the electric field (a) for all points outside the spherical shell and (b) for a point inside the shell a distance r from the center.

46. **Q|C** A charge of 1.70 × 10^2 μC is at the center of a cube of edge 80.0 cm. No other charges are nearby. (a) Find the flux through the whole surface of the cube. (b) Find the flux through each face of the cube. (c) Would your answers to parts (a) or (b) change if the charge were not at the center? Explain.

47. Suppose the conducting spherical shell of Figure 15.29 carries a charge of 3.00 nC and that a charge of −2.00 nC is at the center of the sphere. If a = 2.00 m and b = 2.40 m, find the electric field at (a) r = 1.50 m, (b) r = 2.20 m, and (c) r = 2.50 m. (d) What is the charge distribution on the sphere?

48. **S** A very large nonconducting plate lying in the xy-plane carries a charge per unit area of σ. A second such plate located at z = 2.00 cm and oriented parallel to the xy-plane carries a charge per unit area of -2σ. Find the electric field (a) for $z < 0$, (b) $0 < z < 2.00$ cm, and (c) $z > 2.00$ cm.

Additional Problems

49. In deep space two spheres each of radius 5.00 m are connected by a 3.00 × 10^2 m nonconducting cord. If a uniformly distributed charge of 35.0 mC resides on the surface of each sphere, calculate the tension in the cord.

50. **Q|C** A nonconducting, thin plane sheet of charge carries a uniform charge per unit area of 5.20 μC/m^2 as in Figure 15.30. (a) Find the electric field at a distance of 8.70 cm from the plate. (b) Explain whether your result changes as the distance from the sheet is varied.

51. **W** Three point charges are aligned along the x-axis as shown in Figure P15.51. Find the electric field at the position x = +2.0 m, y = 0.

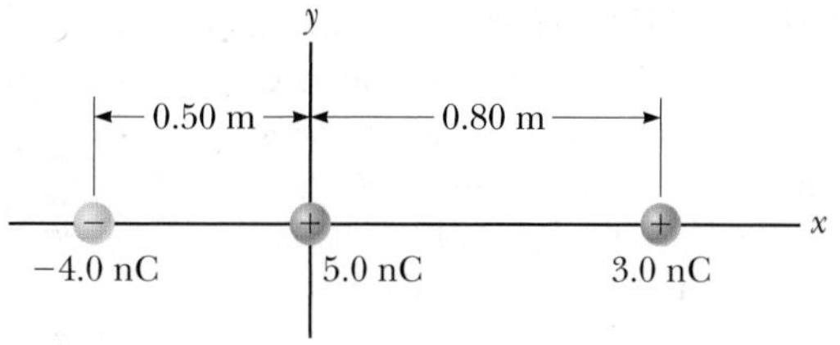

Figure P15.51

52. **M** A small plastic ball of mass m = 2.00 g is suspended by a string of length L = 20.0 cm in a uniform electric field, as shown in Figure P15.52. If the ball is in

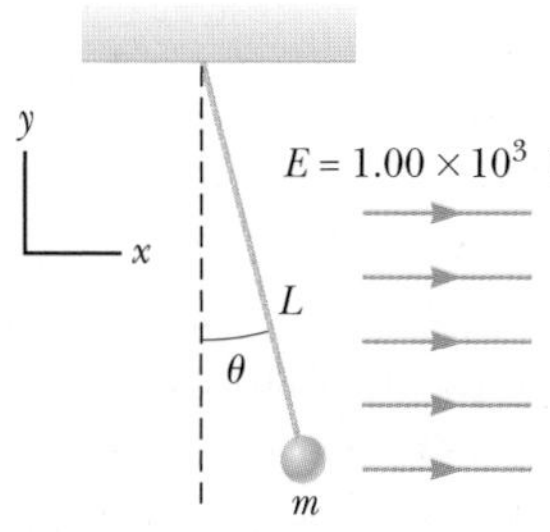

Figure P15.52

equilibrium when the string makes a $\theta = 15.0°$ angle with the vertical as indicated, what is the net charge on the ball?

53. (a) Two identical point charges $+q$ are located on the y-axis at $y = +a$ and $y = -a$. What is the electric field along the x-axis at $x = b$? (b) A circular ring of charge of radius a has a total positive charge Q distributed uniformly around it. The ring is in the $x = 0$ plane with its center at the origin. What is the electric field along the x-axis at $x = b$ due to the ring of charge? *Hint:* Consider the charge Q to consist of many pairs of identical point charges positioned at the ends of diameters of the ring.

54. **S** The electrons in a particle beam each have a kinetic energy K. Find the magnitude of the electric field that will stop these electrons in a distance d, expressing the answer symbolically in terms of K, e, and d. Should the electric field point in the direction of the motion of the electron, or should it point in the opposite direction?

55. **S** A point charge $+2Q$ is at the origin and a point charge $-Q$ is located along the x-axis at $x = d$ as in Figure P15.55. Find symbolic expressions for the components of the net force on a third point charge $+Q$ located along the y-axis at $y = d$.

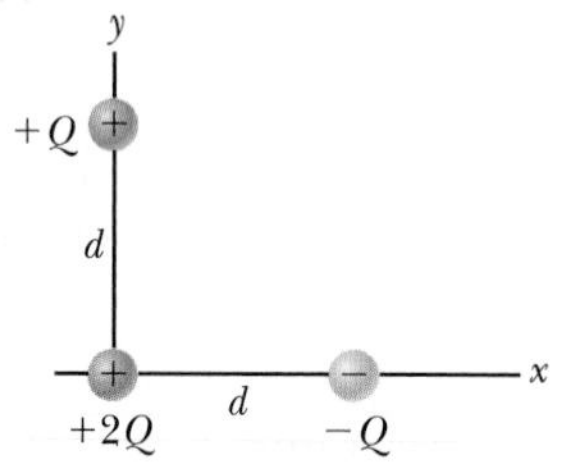

Figure P15.55

56. **Q|C** A 1.00-g cork ball having a positive charge of 2.00 mC is suspended vertically on a 0.500-m-long light string in the presence of a uniform downward-directed electric field of magnitude $E = 1.00 \times 10^5$ N/C as in Figure P15.56. If the ball is displaced slightly from the vertical, it oscillates like a simple pendulum. (a) Determine the period of the ball's oscillation. (b) Should gravity be included in the calculation for part (a)? Explain.

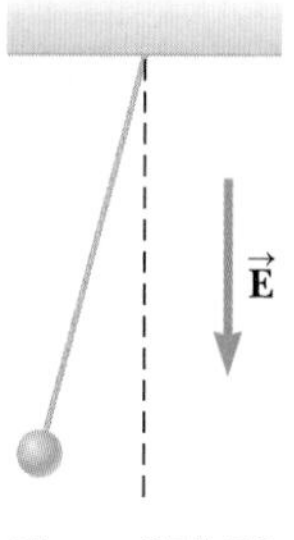

Figure P15.56

57. Two 2.0-g spheres are suspended by 10.0-cm-long light strings (Fig. P15.57). A uniform electric field is applied in the x-direction. If the spheres have charges of -5.0×10^{-8} C and $+5.0 \times 10^{-8}$ C, determine the electric field intensity that enables the spheres to be in equilibrium at $\theta = 10°$.

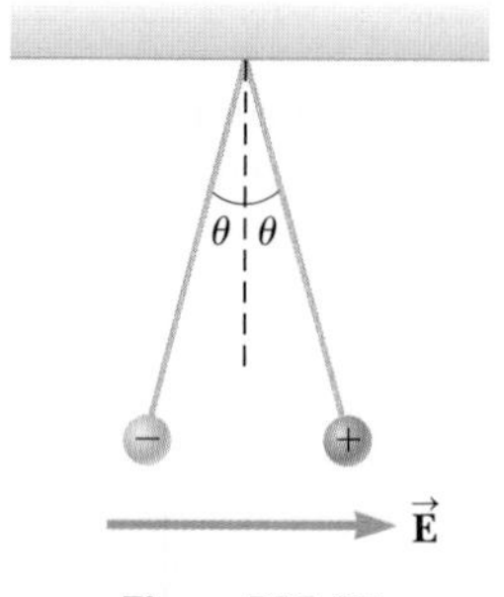

Figure P15.57

58. **Q|C** A point charge of magnitude 5.00 μC is at the origin of a coordinate system, and a charge of -4.00 μC is at the point $x = 1.00$ m. There is a point on the x-axis, at x less than infinity, where the electric field goes to zero. (a) Show by conceptual arguments that this point cannot be located between the charges. (b) Show by conceptual arguments that the point cannot be at any location between $x = 0$ and negative infinity. (c) Show by conceptual arguments that the point must be between $x = 1.00$ m and $x =$ positive infinity. (d) Use the values given to find the point and show that it is consistent with your conceptual argument.

59. Two hard rubber spheres, each of mass $m = 15.0$ g, are rubbed with fur on a dry day and are then suspended with two insulating strings of length $L = 5.00$ cm whose support points are a distance $d = 3.00$ cm from each other as shown in Figure P15.59. During the rubbing process, one sphere receives exactly twice the charge of the other. They are observed to hang at equilibrium, each at an angle of $\theta = 10.0°$ with the vertical. Find the amount of charge on each sphere.

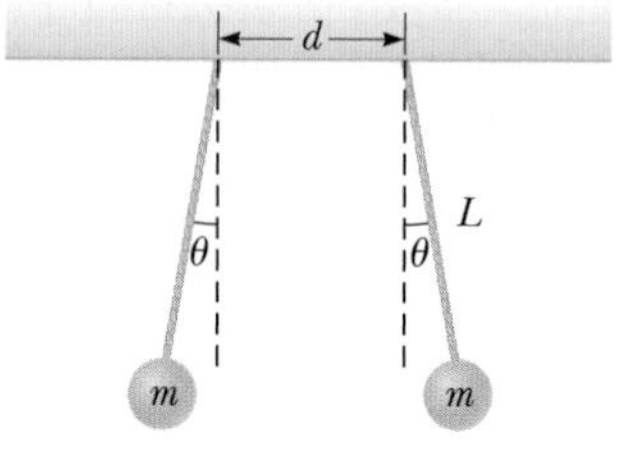

Figure P15.59

60. Two small beads having positive charges $q_1 = 3q$ and $q_2 = q$ are fixed at the opposite ends of a horizontal insulating rod of length $d = 1.50$ m. The bead with charge q_1 is at the origin. As shown in Figure P15.60, a third small charged bead is free to slide on the rod. At what position x is the third bead in equilibrium?

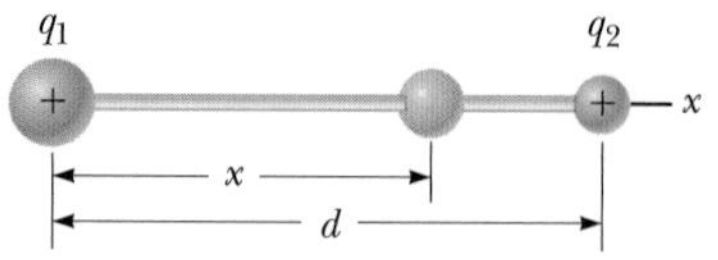

Figure P15.60

61. **M** A solid conducting sphere of radius 2.00 cm has a charge of 8.00 μC. A conducting spherical shell of inner radius 4.00 cm and outer radius 5.00 cm is concentric with the solid sphere and has a charge of -4.00 μC. Find the electric field at (a) $r = 1.00$ cm, (b) $r = 3.00$ cm, (c) $r = 4.50$ cm, and (d) $r = 7.00$ cm from the center of this charge configuration.

62. Three identical point charges, each of mass $m = 0.100$ kg, hang from three strings, as shown in Figure P15.62. If the lengths of the left and right strings are

each $L = 30.0$ cm and if the angle θ is 45.0°, determine the value of q.

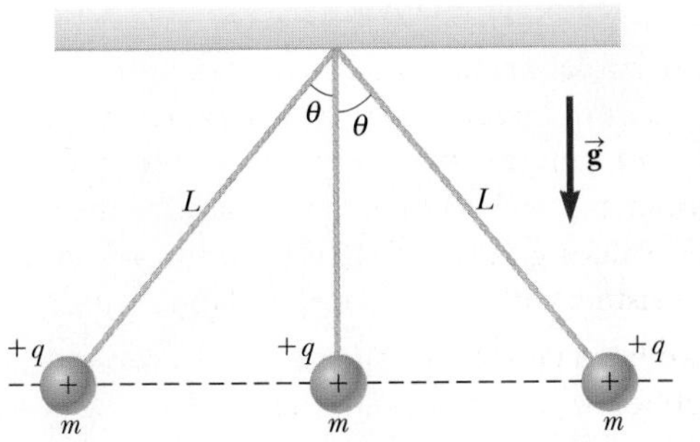

Figure P15.62

63. Each of the electrons in a particle beam has a kinetic energy of 1.60×10^{-17} J. (a) What is the magnitude of the uniform electric field (pointing in the direction of the electrons' movement) that will stop these electrons in a distance of 10.0 cm? (b) How long will it take to stop the electrons? (c) After the electrons stop, what will they do? Explain.

64. Protons are projected with an initial speed $v_0 =$ 9 550 m/s into a region where a uniform electric field of magnitude $E = 720$ N/C is present (Fig. P15.64). The protons are to hit a target that lies a horizontal distance of 1.27 mm from the point where the protons are launched. Find (a) the two projection angles θ that will result in a hit and (b) the total duration of flight for each of the two trajectories.

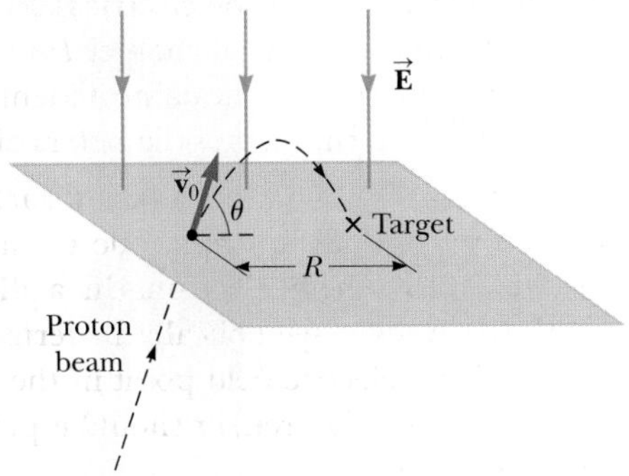

Figure P15.64

The only effective treatment for a patient with a heart in ventricular fibrillation, a spastic quivering of the heart muscle that is fatal in minutes, is an electrical shock delivered by a defibrillator. A capacitor in the defibrillator stores a large charge at high voltage and delivers it rapidly, shocking the heart and restoring a normal heart beat.

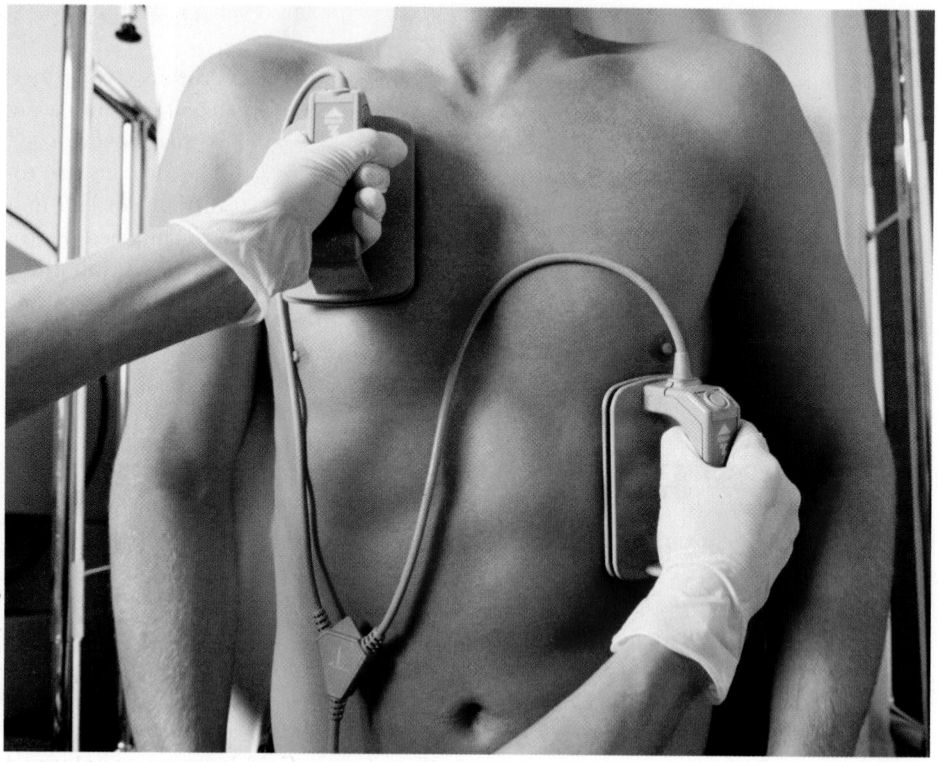

Andrew Olney/Getty Images

16 Electrical Energy and Capacitance

The concept of potential energy was first introduced in Chapter 5 in connection with the conservative forces of gravity and springs. By using the principle of conservation of energy, we were often able to avoid working directly with forces when solving problems. Here we learn that the potential energy concept is also useful in the study of electricity. Because the Coulomb force is conservative, we can define an electric potential energy corresponding to that force. In addition, we define an electric potential—the potential energy per unit charge—corresponding to the electric field.

With the concept of electric potential in hand, we can begin to understand electric circuits, starting with an investigation of common circuit elements called capacitors. These simple devices store electrical energy and have found uses virtually everywhere, from etched circuits on a microchip to the creation of enormous bursts of power in fusion experiments.

16.1 Electric Potential Energy and Electric Potential

LEARNING OBJECTIVES

1. Define electric potential energy difference for a constant electric field in terms of the work done by the field.
2. Contrast the concept of electric potential energy with that of electric potential.
3. Apply the work-energy theorem to systems involving electric potential energy and electric potential.

Electric potential energy and electric potential are closely related concepts. The electric potential turns out to be just the electric potential energy per unit charge. This relationship is similar to that between electric force and the electric field, which is the electric force per unit charge.

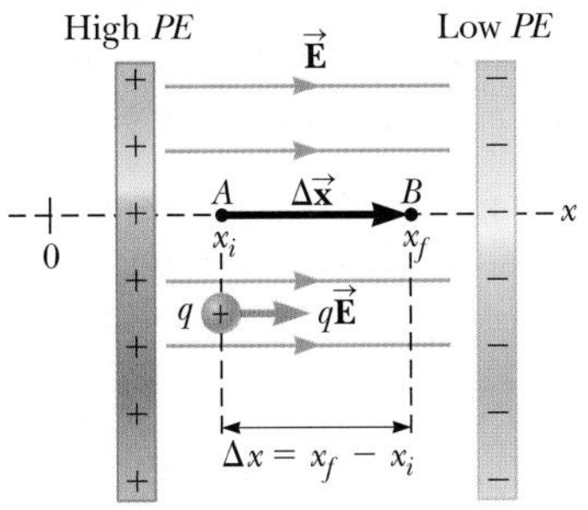

Figure 16.1 When a charge q moves in a uniform electric field $\vec{\mathbf{E}}$ from point A to point B, the work done on the charge by the electric force is $qE_x\,\Delta x$.

Work and Electric Potential Energy

Recall from Chapter 5 that the work done by a conservative force $\vec{\mathbf{F}}$ on an object depends only on the initial and final positions of the object and not on the path taken between those two points. This, in turn, means that a potential energy function PE exists. As we have seen, potential energy is a scalar quantity with the change in potential energy equal by definition to the negative of the work done by the conservative force: $\Delta PE = PE_f - PE_i = -W_F$.

Both the Coulomb force law and the universal law of gravity are proportional to $1/r^2$. Because they have the same mathematical form and because the gravity force is conservative, it follows that **the Coulomb force is also conservative.** As with gravity, an electrical potential energy function can be associated with this force.

To make these ideas more quantitative, imagine a small positive charge placed at point A in a *uniform* electric field $\vec{\mathbf{E}}$, as in Figure 16.1. For simplicity, we first consider only constant electric fields and charges that move parallel to that field in one dimension (taken to be the x-axis). The electric field between equally and oppositely charged parallel plates is an example of a field that is approximately constant. (See Chapter 15.) As the charge moves from point A to point B under the influence of the electric field $\vec{\mathbf{E}}$, the work done on the charge by the electric field is equal to the part of the electric force $q\vec{\mathbf{E}}$ acting parallel to the displacement times the displacement $\Delta x = x_f - x_i$:

$$W_{AB} = F_x\,\Delta x = qE_x(x_f - x_i)$$

In this expression q is the charge and E_x is the vector component of $\vec{\mathbf{E}}$ in the x-direction (*not* the magnitude of $\vec{\mathbf{E}}$). Unlike the magnitude of $\vec{\mathbf{E}}$, the component E_x can be positive or negative, depending on the direction of $\vec{\mathbf{E}}$, although in Figure 16.1 E_x is positive. Finally, note that the displacement, like q and E_x, can also be either positive or negative, depending on the direction of the displacement.

The preceding expression for the work done by an electric field on a charge moving in one dimension is valid for both positive and negative charges and for constant electric fields pointing in *any* direction. When numbers are substituted with correct signs, the overall correct sign automatically results. In some books the expression $W = qEd$ is used, instead, where E is the magnitude of the electric field and d is the distance the particle travels. The weakness of this formulation is that it doesn't allow, mathematically, for negative electric work on positive charges, nor for positive electric work on negative charges! Nonetheless, the expression is easy to remember and useful for finding magnitudes: the magnitude of the work done by a constant electric field on a charge moving parallel to the field is always given by $|W| = |q|Ed$.

We can substitute our definition of electric work into the work–energy theorem (assume other forces are absent):

$$W = qE_x\,\Delta x = \Delta KE$$

The electric force is conservative, so the electric work depends only on the endpoints of the path, A and B, not on the path taken. Therefore, as the charge accelerates to the right in Figure 16.1, it gains kinetic energy and loses an equal amount of potential energy. Recall from Chapter 5 that **the work done by a conservative force can be reinterpreted as the negative of the change in a potential energy**

associated with that force. This interpretation motivates the definition of the change in electric potential energy:

Change in electric potential energy ▶

The change in the electric potential energy, ΔPE, of a system consisting of an object of charge q moving through a displacement Δx in a constant electric field $\vec{\mathbf{E}}$ is given by

$$\Delta PE = -W_{AB} = -qE_x \Delta x \qquad [16.1]$$

where E_x is the x-component of the electric field and $\Delta x = x_f - x_i$ is the displacement of the charge along the x-axis.

SI unit: joule (J)

Although potential energy can be defined for any electric field, **Equation 16.1 is valid only for the case of a uniform (i.e., constant) electric field, for a particle that undergoes a displacement along a given axis (here called the *x*-axis)**. Because the electric field is conservative, the change in potential energy doesn't depend on the path. Consequently, it's unimportant whether or not the charge remains on the axis at all times during the displacement: the change in potential energy will be the same. In subsequent sections we will examine situations in which the electric field is not uniform.

Electric and gravitational potential energy can be compared in Figure 16.2. In this figure the electric and gravitational fields are both directed downwards. We see that positive charge in an electric field acts very much like mass in a gravity field: a positive charge at point A falls in the direction of the electric field, just as a positive mass falls in the direction of the gravity field. Let point B be the zero point for potential energy in both Figures 16.2a and 16.2b. From conservation of energy, in falling from point A to point B the positive charge gains kinetic energy equal in magnitude to the loss of electric potential energy:

$$\Delta KE + \Delta PE_{\text{el}} = \Delta KE + (0 - |q|Ed) = 0 \quad \rightarrow \quad \Delta KE = |q|Ed$$

The absolute-value signs on q are there only to make explicit that the charge is positive in this case. Similarly, the object in Figure 16.2b gains kinetic energy equal in magnitude to the loss of gravitational potential energy:

$$\Delta KE + \Delta PE_g = \Delta KE + (0 - mgd) = 0 \quad \rightarrow \quad \Delta KE = mgd$$

So for positive charges, electric potential energy works very much like gravitational potential energy. In both cases moving an object opposite the direction of the field results in a gain of potential energy, and upon release, the potential energy is converted to the object's kinetic energy.

Electric potential energy differs significantly from gravitational potential energy, however, in that there are two kinds of electrical charge—positive and negative—whereas gravity has only positive "gravitational charge" (i.e., mass). A negatively charged particle at rest at point A in Figure 16.2a would have to be *pushed* down to point B. To see why, apply the work–energy theorem to a negative charge at rest at point A and assumed to have some speed v on arriving at point B:

$$W = \Delta KE + \Delta PE_{\text{el}} = \left(\tfrac{1}{2}mv^2 - 0\right) + [0 - (-|q|Ed)]$$

$$W = \tfrac{1}{2}mv^2 + |q|Ed$$

Notice that the negative charge, $-|q|$, unlike the positive charge, had a positive change in electric potential energy in moving from point A to point B. If the negative charge has any speed at point B, the kinetic energy corresponding to that speed is also positive. Because both terms on the right-hand side of the work–energy equation are positive, there is no way of getting the negative charge from point A to point B without doing positive work W on it. In fact, if the negative charge is simply released at point A, it will "fall" upwards against the direction of the field!

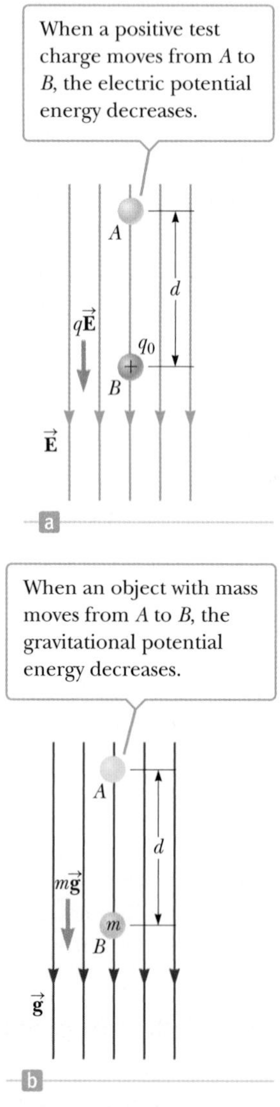

Figure 16.2 (a) When the electric field $\vec{\mathbf{E}}$ is directed downward, point B is at a lower electric potential than point A. (b) An object of mass m moves in the direction of the gravitational field $\vec{\mathbf{g}}$.

■ Quick Quiz

16.1 If an electron is released from rest in a uniform electric field, does the electric potential energy of the charge–field system (a) increase, (b) decrease, or (c) remain the same?

■ EXAMPLE 16.1 Potential Energy Differences in an Electric Field

GOAL Illustrate the concept of electric potential energy.

PROBLEM A proton is released from rest at $x = -2.00$ cm in a constant electric field with magnitude 1.50×10^3 N/C, pointing in the positive x-direction. **(a)** Calculate the change in the electric potential energy associated with the proton when it reaches $x = 5.00$ cm. **(b)** An electron is now fired in the same direction from the same position. What is the change in electric potential energy associated with the electron if it reaches $x = 12.0$ cm? **(c)** If the direction of the electric field is reversed and an electron is released from rest at $x = 3.00$ cm, by how much has the electric potential energy changed when the electron reaches $x = 7.00$ cm?

STRATEGY This problem requires a straightforward substitution of given values into the definition of electric potential energy, Equation 16.1.

SOLUTION

(a) Calculate the change in the electric potential energy associated with the proton.

Apply Equation 16.1:

$$\begin{aligned}\Delta PE &= -qE_x\,\Delta x = -qE_x(x_f - x_i)\\ &= -(1.60 \times 10^{-19}\text{ C})(1.50 \times 10^3\text{ N/C})\\ &\quad \times [0.050\,0\text{ m} - (-0.020\,0\text{ m})]\\ &= \boxed{-1.68 \times 10^{-17}\text{ J}}\end{aligned}$$

(b) Find the change in electric potential energy associated with an electron fired from $x = -0.020\,0$ m and reaching $x = 0.120$ m.

Apply Equation 16.1, but in this case note that the electric charge q is negative:

$$\begin{aligned}\Delta PE &= -qE_x\,\Delta x = -qE_x(x_f - x_i)\\ &= -(-1.60 \times 10^{-19}\text{ C})(1.50 \times 10^3\text{ N/C})\\ &\quad \times [(0.120\text{ m} - (-0.020\,0\text{ m})]\\ &= \boxed{+3.36 \times 10^{-17}\text{ J}}\end{aligned}$$

(c) Find the change in potential energy associated with an electron traveling from $x = 3.00$ cm to $x = 7.00$ cm if the direction of the electric field is reversed.

Substitute, but now the electric field points in the negative x-direction, hence carries a minus sign:

$$\begin{aligned}\Delta PE &= -qE_x\,\Delta x = -qE_x\,(x_f - x_i)\\ &= -(-1.60 \times 10^{-19}\text{ C})(-1.50 \times 10^3\text{ N/C})\\ &\quad \times (0.070\text{ m} - 0.030\text{ m})\\ &= \boxed{-9.60 \times 10^{-18}\text{ J}}\end{aligned}$$

REMARKS Notice that the proton (actually the proton–field system) lost potential energy when it moved in the positive x-direction, whereas the electron gained potential energy when it moved in the same direction. Finding changes in potential energy with the field reversed was only a matter of supplying a minus sign, bringing the total number in this case to three! It's important not to drop any of the signs.

QUESTION 16.1 True or False: When an electron is released from rest in a constant electric field, the change in the electric potential energy associated with the electron becomes more negative with time.

EXERCISE 16.1 Find the change in electric potential energy associated with the electron in part (b) as it goes on from $x = 0.120$ m to $x = -0.180$ m. (Note that the electron must turn around and go back at some point. The location of the turning point is unimportant because changes in potential energy depend only on the endpoints of the path.)

ANSWER -7.20×10^{-17} J

EXAMPLE 16.2 | Dynamics of Charged Particles

GOAL Use electric potential energy in conservation of energy problems.

PROBLEM **(a)** Find the speed of the proton at $x = 0.050\,0$ m in part (a) of Example 16.1. **(b)** Find the initial speed of the electron (at $x = -2.00$ cm) in part (b) of Example 16.1 given that its speed has fallen by half when it reaches $x = 0.120$ m.

STRATEGY Apply conservation of energy, solving for the unknown speeds. Part (b) involves two equations: the conservation of energy equation and the condition $v_f = \frac{1}{2}v_i$ for the unknown initial and final speeds. The changes in electric potential energy have already been calculated in Example 16.1.

SOLUTION

(a) Calculate the proton's speed at $x = 0.050$ m.

Use conservation of energy, with an initial speed of zero:

$$\Delta KE + \Delta PE = 0 \quad \rightarrow \quad (\tfrac{1}{2}m_p v^2 - 0) + \Delta PE = 0$$

Solve for v and substitute the change in potential energy found in Example 16.1a:

$$v^2 = -\frac{2}{m_p}\Delta PE$$

$$v = \sqrt{-\frac{2}{m_p}\Delta PE}$$

$$= \sqrt{-\frac{2}{(1.67 \times 10^{-27}\text{ kg})}(-1.68 \times 10^{-17}\text{ J})}$$

$$= \boxed{1.42 \times 10^5 \text{ m/s}}$$

(b) Find the electron's initial speed (at $x = -2.00$ cm) given that its speed has fallen by half at $x = 0.120$ m.

Apply conservation of energy once again, substituting expressions for the initial and final kinetic energies:

$$\Delta KE + \Delta PE = 0$$

$$(\tfrac{1}{2}m_e v_f^2 - \tfrac{1}{2}m_e v_i^2) + \Delta PE = 0$$

Substitute the condition $v_f = \frac{1}{2}v_i$ and subtract the change in potential energy from both sides:

$$\tfrac{1}{2}m_e(\tfrac{1}{2}v_i)^2 - \tfrac{1}{2}m_e v_i^2 = -\Delta PE$$

Combine terms and solve for v_i, the initial speed, and substitute the change in potential energy found in Example 16.1b:

$$-\tfrac{3}{8}m_e v_i^2 = -\Delta PE$$

$$v_i = \sqrt{\frac{8\Delta PE}{3m_e}} = \sqrt{\frac{8(3.36 \times 10^{-17}\text{ J})}{3(9.11 \times 10^{-31}\text{ kg})}}$$

$$= \boxed{9.92 \times 10^6 \text{ m/s}}$$

REMARKS Although the changes in potential energy associated with the proton and electron were similar in magnitude, the effect on their speeds differed dramatically. The change in potential energy had a far greater effect on the much lighter electron than on the proton.

QUESTION 16.2 True or False: If a proton and electron both move through the same displacement in an electric field, the change in potential energy associated with the proton must be equal in magnitude and opposite in sign to the change in potential energy associated with the electron.

EXERCISE 16.2 Refer to Exercise 16.1. Find the electron's speed at $x = -0.180$ m. *Note:* Use the initial velocity from part (b) of Example 16.2.

ANSWER 1.35×10^7 m/s The answer is 4.5% of the speed of light.

Electric Potential

In Chapter 15 it was convenient to define an electric field $\vec{\mathbf{E}}$ related to the electric force $\vec{\mathbf{F}} = q\vec{\mathbf{E}}$. In this way the properties of fixed collections of charges could be easily studied, and the force on any particle in the electric field could be obtained simply by multiplying by the particle's charge q. For the same reasons, it's useful to define an *electric potential difference* ΔV related to the potential energy by $\Delta PE = q\Delta V$:

> The electric potential difference ΔV between points A and B is the change in electric potential energy as a charge q moves from A to B divided by the charge q:
>
> $$\Delta V = V_B - V_A = \frac{\Delta PE}{q} \qquad [16.2]$$
>
> **SI unit: joule per coulomb, or volt (J/C, or V)**

◀ Potential difference between two points

This definition is completely general, although in many cases calculus would be required to compute the change in potential energy of the system. Because electric potential energy is a scalar quantity, **electric potential is also a scalar quantity**. From Equation 16.2, we see that electric potential difference is a measure of the change in electric potential energy per unit charge. Alternately, the electric potential difference is the work per unit charge that would have to be done by some force to move a charge from point A to point B in the electric field. The SI unit of electric potential is the joule per coulomb, called the volt (V). From the definition of that unit, 1 J of work must be done to move a 1-C charge between two points that are at a potential difference of 1 V. In the process of moving through a potential difference of 1 V, the 1-C charge gains 1 J of energy.

For the special case of a uniform electric field such as that between charged parallel plates, dividing Equation 16.1 by q gives

$$\frac{\Delta PE}{q} = -E_x \Delta x$$

Comparing this equation with Equation 16.2, we find that

$$\Delta V = -E_x \Delta x \qquad [16.3]$$

Equation 16.3 shows that potential difference also has units of electric field times distance. It then follows that the SI unit of the electric field, the newton per coulomb, can also be expressed as volts per meter:

$$1 \text{ N/C} = 1 \text{ V/m}$$

Because Equation 16.3 is directly related to Equation 16.1, remember that it's valid only for the system consisting of a uniform electric field and a charge moving in one dimension.

Released from rest, positive charges accelerate spontaneously from regions of high potential to low potential. If a positive charge is given some initial velocity in the direction of high potential, it can move in that direction, but will slow and finally turn around, just like a ball tossed upwards in a gravity field. Negative charges do exactly the opposite: released from rest, they accelerate from regions of low potential toward regions of high potential. Work must be done on negative charges to make them go in the direction of lower electric potential.

Tip 16.1 Potential and Potential Energy

Electric potential is characteristic of the field only, independent of a test charge that may be placed in that field. On the other hand, potential energy is a characteristic of the charge-field system due to an interaction between the field and a charge placed in the field.

Quick Quiz

16.2 If a negatively charged particle is placed at rest in an electric potential field that increases in the positive x-direction, will the particle (a) accelerate in the positive x-direction, (b) accelerate in the negative x-direction, or (c) remain at rest?

16.3 Figure 16.3 is a graph of an electric potential as a function of position. If a positively charged particle is placed at point A, what will its subsequent motion be? Will it (a) go to the right, (b) go to the left, (c) remain at point A, or (d) oscillate around point B?

(Continued)

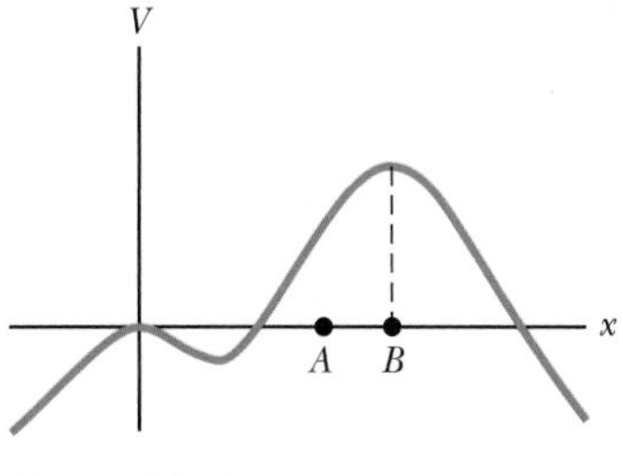

Figure 16.3 (Quick Quizzes 16.3 and 16.4)

16.4 If a negatively charged particle is placed at point *B* in Figure 16.3 and given a very small kick to the right, what will its subsequent motion be? Will it (a) go to the right and not return, (b) go to the left, (c) remain at point *B*, or (d) oscillate around point *B*?

APPLICATION
Automobile Batteries

An application of potential difference is the 12-V battery found in an automobile. Such a battery maintains a potential difference across its terminals, with the positive terminal 12 V higher in potential than the negative terminal. In practice the negative terminal is usually connected to the metal body of the car, which can be considered to be at a potential of zero volts. The battery provides the electrical current necessary to operate headlights, a radio, power windows, motors, and so forth. Now consider a charge of +1 C, to be moved around a circuit that contains the battery and some of these external devices. As the charge is moved inside the battery from the negative terminal (at 0 V) to the positive terminal (at 12 V), the work done on the charge by the battery is 12 J. Every coulomb of positive charge that leaves the positive terminal of the battery carries an energy of 12 J. As the charge moves through the external circuit toward the negative terminal, it gives up its 12 J of electrical energy to the external devices. When the charge reaches the negative terminal, its electrical energy is zero again. At this point, the battery takes over and restores 12 J of energy to the charge as it is moved from the negative to the positive terminal, enabling it to make another transit of the circuit. The actual amount of charge that leaves the battery each second and traverses the circuit depends on the properties of the external devices, as seen in the next chapter.

EXAMPLE 16.3 | TV Tubes and Atom Smashers

GOAL Relate electric potential to an electric field and conservation of energy.

PROBLEM In atom smashers (also known as cyclotrons and linear accelerators) charged particles are accelerated in much the same way they are accelerated in TV tubes: through potential differences. Suppose a proton is injected at a speed of 1.00×10^6 m/s between two plates 5.00 cm apart, as shown in Figure 16.4. The proton subsequently accelerates across the gap and exits through the opening. **(a)** What must the electric potential difference be if the exit speed is to be 3.00×10^6 m/s? **(b)** What is the electric field between the plates, assuming it's constant? The positive *x*-direction is to the right.

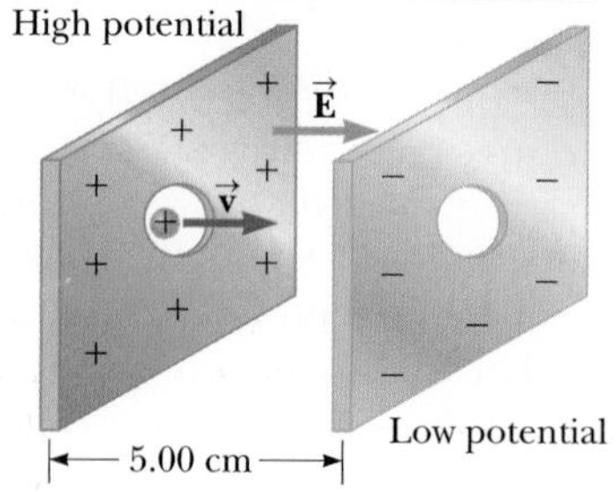

Figure 16.4 (Example 16.3) A proton enters a cavity and accelerates from one charged plate toward the other in an electric field $\vec{\mathbf{E}}$.

STRATEGY Use conservation of energy, writing the change in potential energy in terms of the change in electric potential, ΔV, and solve for ΔV. For part (b), solve Equation 16.3 for the electric field.

SOLUTION

(a) Find the electric potential yielding the desired exit speed of the proton.

Apply conservation of energy, writing the potential energy in terms of the electric potential:

$$\Delta KE + \Delta PE = \Delta KE + q\,\Delta V = 0$$

Solve the energy equation for the change in potential:

$$\Delta V = -\frac{\Delta KE}{q} = -\frac{\frac{1}{2}m_p{v_f}^2 - \frac{1}{2}m_p{v_i}^2}{q} = -\frac{m_p}{2q}({v_f}^2 - {v_i}^2)$$

Substitute the given values, obtaining the necessary potential difference:

$$\Delta V = -\frac{(1.67 \times 10^{-27}\text{ kg})}{2(1.60 \times 10^{-19}\text{ C})}[(3.00 \times 10^6\text{ m/s})^2 - (1.00 \times 10^6\text{ m/s})^2]$$

$$\Delta V = -4.18 \times 10^4\text{ V}$$

(b) What electric field must exist between the plates?

Solve Equation 16.3 for the electric field and substitute:

$$E = -\frac{\Delta V}{\Delta x} = \frac{4.18 \times 10^4\text{ V}}{0.050\,0\text{ m}} = 8.36 \times 10^5\text{ N/C}$$

REMARKS Systems of such cavities, consisting of alternating positive and negative plates, are used to accelerate charged particles to high speed before smashing them into targets. To prevent a slowing of, say, a positively charged particle after it passes through the negative plate of one cavity and enters the next, the charges on the plates are reversed. Otherwise, the particle would be traveling from the negative plate to a positive plate in the second cavity, and the kinetic energy gained in the previous cavity would be lost in the second.

QUESTION 16.3 True or False: A more massive particle gains less energy in traversing a given potential difference than does a lighter particle having the same charge.

EXERCISE 16.3 Suppose electrons in a TV tube are accelerated through a potential difference of 2.00×10^4 V from the heated cathode (negative electrode), where they are produced, toward the screen, which also serves as the anode (positive electrode), 25.0 cm away. (a) At what speed would the electrons impact the phosphors on the screen? Assume they accelerate from rest and ignore relativistic effects (Chapter 26). (b) What's the magnitude of the electric field, if it is assumed constant?

ANSWERS (a) 8.38×10^7 m/s (b) 8.00×10^4 V/m

16.2 Electric Potential and Potential Energy Due to Point Charges

LEARNING OBJECTIVES

1. Define the electric potential of a point charge and the potential energy of a pair of point charges.
2. Apply electric potential and electric potential energy to systems of charged particles.

In electric circuits a point of zero electric potential is often defined by grounding (connecting to the Earth) some point in the circuit. For example, if the negative terminal of a 12-V battery were connected to ground, it would be considered to have a potential of zero, whereas the positive terminal would have a potential of +12 V. The potential difference created by the battery, however, is only locally defined. In this section we describe the electric potential of a point charge, which is defined throughout space.

The electric field of a point charge extends throughout space, so its electric potential does, also. The zero point of electric potential could be taken anywhere, but is usually taken to be an infinite distance from the charge, far from its influence and the influence of any other charges. With this choice, the methods of calculus can be used to show that the electric potential created by a point charge q at any distance r from the charge is given by

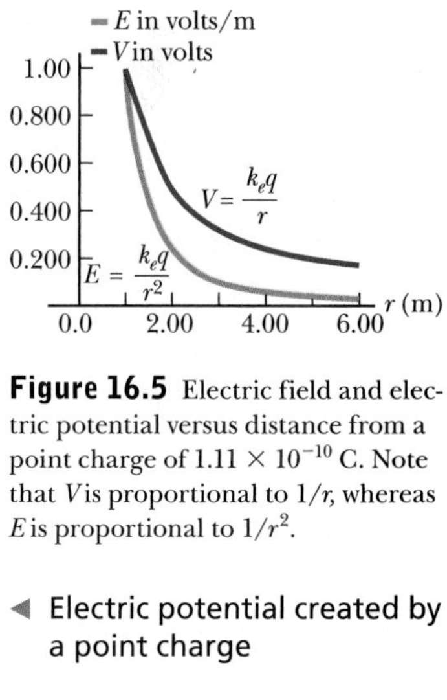

Figure 16.5 Electric field and electric potential versus distance from a point charge of 1.11×10^{-10} C. Note that V is proportional to $1/r$, whereas E is proportional to $1/r^2$.

$$V = k_e \frac{q}{r} \qquad [16.4]$$

◀ Electric potential created by a point charge

Equation 16.4 shows that the electric potential, or work per unit charge, required to move a positive test charge in from infinity to a distance r from a positive point charge q increases as the test charge moves closer to q. A plot of Equation 16.4 in Figure 16.5 shows that the potential associated with a point charge decreases as $1/r$ with increasing r, in contrast to the magnitude of the charge's electric field, which decreases as $1/r^2$.

◀ Superposition principle

The electric potential of two or more charges is obtained by applying the **superposition principle: the total electric potential at some point P due to several point charges is the algebraic sum of the electric potentials due to the individual charges.** This method is similar to the one used in Chapter 15 to find the resultant electric field at a point in space. Unlike electric field superposition, which involves a sum of vectors, the superposition of electric potentials requires evaluating a sum

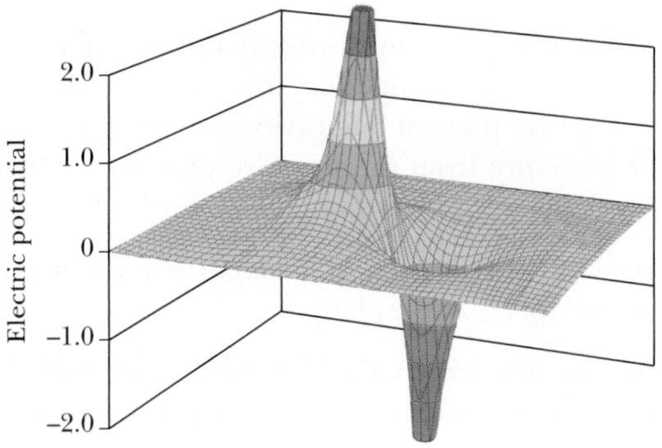

Figure 16.6 The electric potential (in arbitrary units) in the plane containing an electric dipole. Potential is plotted in the vertical dimension.

of scalars. As a result, it's much easier to evaluate the electric potential at some point due to several charges than to evaluate the electric field, which is a vector quantity.

Figure 16.6 is a computer-generated plot of the electric potential associated with an electric dipole, which consists of two charges of equal magnitude but opposite in sign. The charges lie in a horizontal plane at the center of the potential spikes. The value of the potential is plotted in the vertical dimension. The computer program has added the potential of each charge to arrive at total values of the potential.

Just as in the case of constant electric fields, there is a relationship between electric potential and electric potential energy. If V_1 is the electric potential due to charge q_1 at a point P (Fig. 16.7a) the work required to bring charge q_2 from infinity to P without acceleration is q_2V_1. By definition, this work equals the potential energy PE of the two-particle system when the particles are separated by a distance r (Fig. 16.7b).

We can therefore express the electrical potential energy of the *pair* of charges as

Potential energy of a pair of charges ▶

$$PE = q_2V_1 = k_e\frac{q_1q_2}{r} \quad [16.5]$$

If the charges are of the *same* sign, PE is positive. Because like charges repel, positive work must be done on the system by an external agent to force the two charges near each other. Conversely, if the charges are of *opposite* sign, the force is attractive and PE is negative. This means that negative work must be done to prevent unlike charges from accelerating toward each other as they are brought close together.

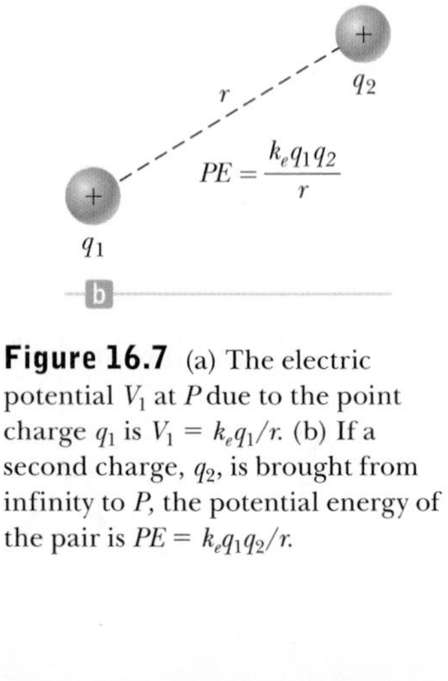

Figure 16.7 (a) The electric potential V_1 at P due to the point charge q_1 is $V_1 = k_eq_1/r$. (b) If a second charge, q_2, is brought from infinity to P, the potential energy of the pair is $PE = k_eq_1q_2/r$.

■ Quick Quiz

16.5 Consider a collection of charges in a given region and suppose all other charges are distant and have a negligible effect. Further, the electric potential is taken to be zero at infinity. If the electric potential at a given point in the region is zero, which of the following statements must be true? (a) The electric field is zero at that point. (b) The electric potential energy is a minimum at that point. (c) There is no net charge in the region. (d) Some charges in the region are positive, and some are negative. (e) The charges have the same sign and are symmetrically arranged around the given point.

16.6 A spherical balloon contains a positively charged particle at its center. As the balloon is inflated to a larger volume while the charged particle remains at the center, which of the following are true? (a) The electric potential at the surface of the balloon increases. (b) The magnitude of the electric field at the surface of the balloon increases. (c) The electric flux through the balloon remains the same. (d) None of these.

of spheres centered on the point charge. Figure 16.11b shows the equipotentials associated with two charges of equal magnitude but opposite sign.

16.5 Applications

LEARNING OBJECTIVE

1. Describe several important applications of electrically charged systems.

The Electrostatic Precipitator

APPLICATION
The Electrostatic Precipitator

One important application of electric discharge in gases is a device called an *electrostatic precipitator*. This device removes particulate matter from combustion gases, thereby reducing air pollution. It's especially useful in coal-burning power plants and in industrial operations that generate large quantities of smoke. Systems currently in use can eliminate approximately 90% by mass of the ash and dust from smoke. Unfortunately, a very high percentage of the lighter particles still escape, and they contribute significantly to smog and haze.

Figure 16.12 illustrates the basic idea of the electrostatic precipitator. A high voltage (typically 40 kV to 100 kV) is maintained between a wire running down the center of a duct and the outer wall, which is grounded. The wire is maintained at a negative electric potential with respect to the wall, so the electric field is directed toward the wire. The electric field near the wire reaches a high enough value to cause a discharge around the wire and the formation of positive ions, electrons, and negative ions, such as O_2^-. As the electrons and negative ions are accelerated toward the outer wall by the nonuniform electric field, the dirt particles in the streaming gas become charged by collisions and ion capture. Because most of the charged dirt particles are negative, they are also drawn to the outer wall by the electric field. When the duct is shaken, the particles fall loose and are collected at the bottom.

In addition to reducing the amounts of harmful gases and particulate matter in the atmosphere, the electrostatic precipitator recovers valuable metal oxides from the stack.

Figure 16.12 (a) A schematic diagram of an electrostatic precipitator. Compare the air pollution when the precipitator is (b) operating and (c) turned off.

APPLICATION
The Electrostatic Air Cleaner

A similar device called an *electrostatic air cleaner* is used in homes to relieve the discomfort of allergy sufferers. Air laden with dust and pollen is drawn into the device across a positively charged mesh screen. The airborne particles become positively charged when they make contact with the screen, and then they pass through a second, negatively charged mesh screen. The electrostatic force of attraction between the positively charged particles in the air and the negatively charged screen causes the particles to precipitate out on the surface of the screen, removing a very high percentage of contaminants from the air stream.

APPLICATION
Xerographic Copiers

Xerography and Laser Printers

Xerography is widely used to make photocopies of printed materials. The basic idea behind the process was developed by Chester Carlson, who was granted a patent for his invention in 1940. In 1947 the Xerox Corporation launched a full-scale program to develop automated duplicating machines using Carlson's process. The huge success of that development is evident: today, practically all offices and libraries have one or more duplicating machines, and the capabilities of these machines continue to evolve.

Some features of the xerographic process involve simple concepts from electrostatics and optics. The one idea that makes the process unique, however, is the use of photoconductive material to form an image. A photoconductor is a material that is a poor conductor of electricity in the dark, but a reasonably good conductor when exposed to light.

Figure 16.13 illustrates the steps in the xerographic process. First, the surface of a plate or drum is coated with a thin film of the photoconductive material (usually selenium or some compound of selenium), and the photoconductive surface is given a positive electrostatic charge in the dark (Fig. 16.13a). The page to be copied is then projected onto the charged surface (Fig. 16.13b). The photoconducting surface becomes conducting only in areas where light strikes; there the light produces charge carriers in the photoconductor that neutralize the positively charged surface. The charges remain on those areas of the photoconductor not exposed to light, however, leaving a hidden image of the object in the form of a positive distribution of surface charge.

APPLICATION
Laser Printers

Next, a negatively charged powder called a *toner* is dusted onto the photoconducting surface (Fig. 16.13c). The charged powder adheres only to the areas that contain the positively charged image. At this point, the image becomes visible. It is then transferred to the surface of a sheet of positively charged paper. Finally,

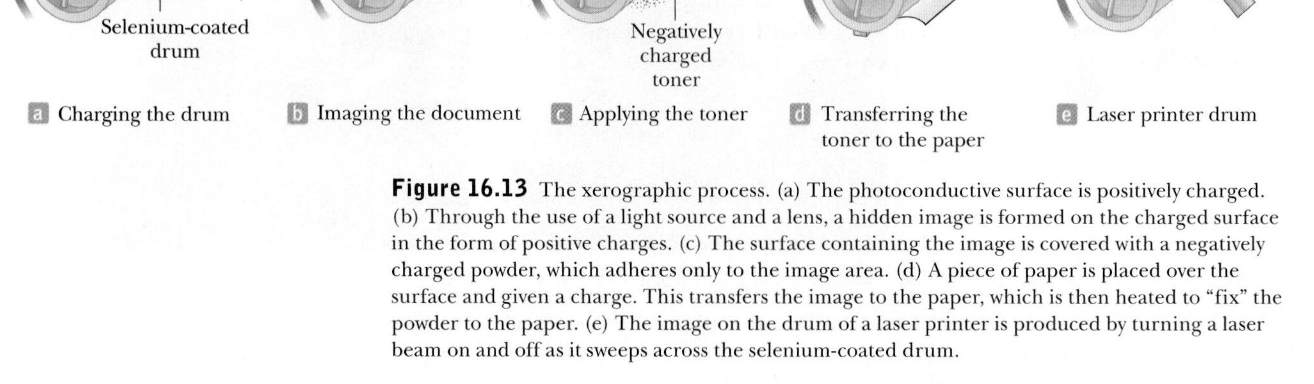

Figure 16.13 The xerographic process. (a) The photoconductive surface is positively charged. (b) Through the use of a light source and a lens, a hidden image is formed on the charged surface in the form of positive charges. (c) The surface containing the image is covered with a negatively charged powder, which adheres only to the image area. (d) A piece of paper is placed over the surface and given a charge. This transfers the image to the paper, which is then heated to "fix" the powder to the paper. (e) The image on the drum of a laser printer is produced by turning a laser beam on and off as it sweeps across the selenium-coated drum.

the toner is "fixed" to the surface of the paper by heat (Fig. 16.13d), resulting in a permanent copy of the original.

The steps for producing a document on a laser printer are similar to those used in a photocopy machine in that parts (a), (c), and (d) of Figure 16.13 remain essentially the same. The difference between the two techniques lies in the way the image is formed on the selenium-coated drum. In a laser printer the command to print the letter *O*, for instance, is sent to a laser from the memory of a computer. A rotating mirror inside the printer causes the beam of the laser to sweep across the selenium-coated drum in an interlaced pattern (Fig. 16.13e). Electrical signals generated by the printer turn the laser beam on and off in a pattern that traces out the letter *O* in the form of positive charges on the selenium. Toner is then applied to the drum, and the transfer to paper is accomplished as in a photocopy machine.

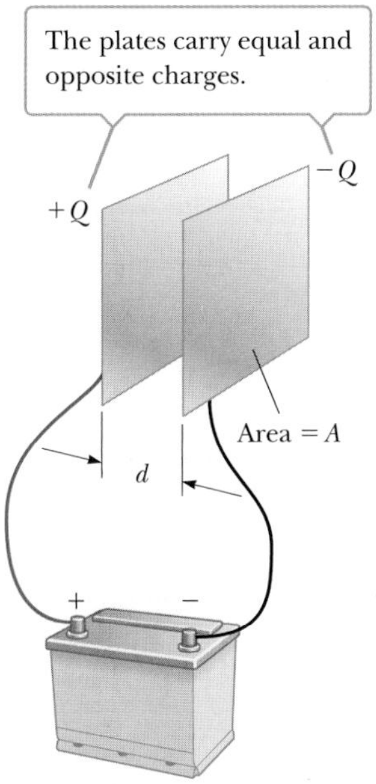

Figure 16.14 A parallel-plate capacitor consists of two parallel plates, each of area *A*, separated by a distance *d*.

16.6 Capacitance

LEARNING OBJECTIVE

1. Describe a capacitor and define capacitance.

A **capacitor** is a device used in a variety of electric circuits, such as to tune the frequency of radio receivers, eliminate sparking in automobile ignition systems, or store short-term energy for rapid release in electronic flash units. Figure 16.14 shows a typical design for a capacitor. It consists of two parallel metal plates separated by a distance *d*. Used in an electric circuit, the plates are connected to the positive and negative terminals of a battery or some other voltage source. When this connection is made, electrons are pulled off one of the plates, leaving it with a charge of $+Q$, and are transferred through the battery to the other plate, leaving it with a charge of $-Q$, as shown in the figure. The transfer of charge stops when the potential difference across the plates equals the potential difference of the battery. A charged capacitor is a device that stores energy that can be reclaimed when needed for a specific application.

The capacitance C of a capacitor is the ratio of the magnitude of the charge on either conductor (plate) to the magnitude of the potential difference between the conductors (plates):

$$C \equiv \frac{Q}{\Delta V} \qquad [16.8]$$

◀ Capacitance of a pair of conductors

SI unit: farad (F) = coulomb per volt (C/V)

The quantities Q and ΔV are always taken to be positive when used in Equation 16.8. For example, if a 3.0-μF capacitor is connected to a 12-V battery, the magnitude of the charge on each plate of the capacitor is

$$Q = C\Delta V = (3.0 \times 10^{-6}\ \text{F})(12\ \text{V}) = 36\ \mu\text{C}$$

From Equation 16.8, we see that a large capacitance is needed to store a large amount of charge for a given applied voltage. The farad is a very large unit of capacitance. In practice, most typical capacitors have capacitances ranging from microfarads ($1\ \mu\text{F} = 1 \times 10^{-6}\ \text{F}$) to picofarads ($1\ \text{pF} = 1 \times 10^{-12}\ \text{F}$).

16.7 The Parallel-Plate Capacitor

LEARNING OBJECTIVES

1. Derive an expression for the capacitance of a parallel-plate capacitor.
2. Calculate the fundamental physical properties of a parallel-plate capacitor.

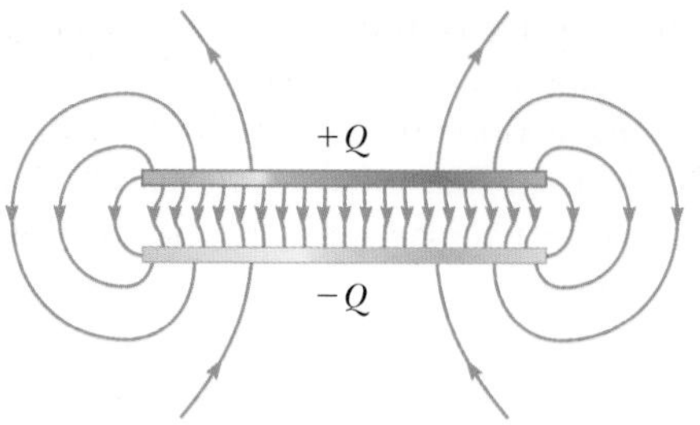

Figure 16.15 The electric field between the plates of a parallel-plate capacitor is uniform near the center, but nonuniform near the edges.

Tip 16.2 Potential Difference Is ΔV, Not V

Use the symbol ΔV for the potential difference across a circuit element or a device (many other books use simply V for potential difference). The dual use of V to represent potential in one place and a potential difference in another can lead to unnecessary confusion.

The capacitance of a device depends on the geometric arrangement of the conductors. The capacitance of a parallel-plate capacitor with plates separated by air (see Fig. 16.14) can be easily calculated from three facts. First, recall from Chapter 15 that the magnitude of the electric field between two plates is given by $E = \sigma/\epsilon_0$, where σ is the magnitude of the charge per unit area on each plate. Second, we found earlier in this chapter that the potential difference between two plates is $\Delta V = Ed$, where d is the distance between the plates. Third, the charge on one plate is given by $Q = \sigma A$, where A is the area of the plate. Substituting these three facts into the definition of capacitance gives the desired result:

$$C = \frac{Q}{\Delta V} = \frac{\sigma A}{Ed} = \frac{\sigma A}{(\sigma/\epsilon_0)d}$$

Canceling the charge per unit area, σ, yields

Capacitance of a parallel-plate capacitor ▶

$$C = \epsilon_0 \frac{A}{d} \qquad [16.9]$$

where A is the area of one of the plates, d is the distance between the plates, and ϵ_0 is the permittivity of free space.

From Equation 16.9, we see that plates with larger area can store more charge. The same is true for a small plate separation d because then the positive charges on one plate exert a stronger force on the negative charges on the other plate, allowing more charge to be held on the plates.

Figure 16.15 shows the electric field lines of a more realistic parallel-plate capacitor. The electric field is very nearly constant in the center between the plates, but becomes less so when approaching the edges. For most purposes, however, the field may be taken as constant throughout the region between the plates.

APPLICATION
Camera Flash Attachments

One practical device that uses a capacitor is the flash attachment on a camera. A battery is used to charge the capacitor, and the stored charge is then released when the shutter-release button is pressed to take a picture. The stored charge is delivered to a flash tube very quickly, illuminating the subject at the instant more light is needed.

APPLICATION
Computer Keyboards

Computers make use of capacitors in many ways. For example, one type of computer keyboard has capacitors at the bases of its keys, as in Figure 16.16. Each key is connected to a movable plate, which represents one side of the capacitor; the fixed plate on the bottom of the keyboard represents the other side of the capacitor. When a key is pressed, the capacitor spacing decreases, causing an increase in capacitance. External electronic circuits recognize each key by the *change* in its capacitance when it is pressed.

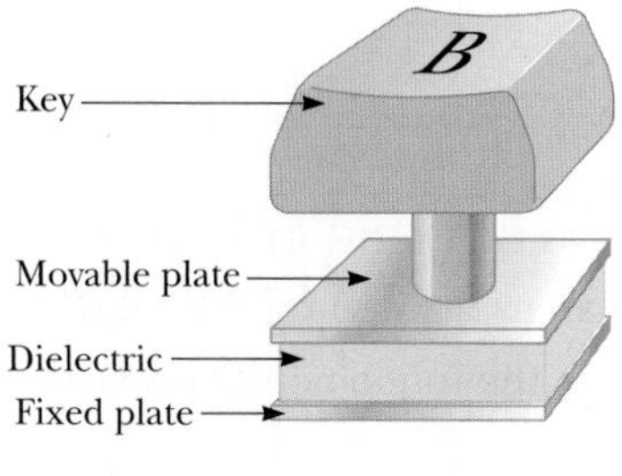

Figure 16.16 When the key of one type of keyboard is pressed, the capacitance of a parallel-plate capacitor increases as the plate spacing decreases. The substance labeled "dielectric" is an insulating material, as described in Section 16.10.

Quick Quiz

16.10 A fully charged parallel-plate capacitor remains connected to a battery while a dielectric is slid between the plates. Do the following quantities increase, decrease, or stay the same? (a) C (b) Q (c) E between the plates (d) ΔV (e) energy stored in the capacitor

EXAMPLE 16.11 | A Paper-Filled Capacitor

GOAL Calculate fundamental physical properties of a parallel-plate capacitor with a dielectric.

PROBLEM A parallel-plate capacitor has plates 2.0 cm by 3.0 cm. The plates are separated by a 1.0-mm thickness of paper. Find **(a)** the capacitance of this device and **(b)** the maximum charge that can be placed on the capacitor. **(c)** After the fully charged capacitor is disconnected from the battery, the dielectric is subsequently removed. Find the new electric field across the capacitor. Does the capacitor discharge?

STRATEGY For part (a), obtain the dielectric constant for paper from Table 16.1 and substitute, with other given quantities, into Equation 16.19. For part (b), note that Table 16.1 also gives the dielectric strength of paper, which is the maximum electric field that can be applied before electrical breakdown occurs. Use Equation 16.3, $\Delta V = Ed$, to obtain the maximum voltage and substitute into the basic capacitance equation. For part (c), remember that disconnecting the battery traps the extra charge on the plates, which must remain even after the dielectric is removed. Find the charge density on the plates and use Gauss's law to find the new electric field between the plates.

SOLUTION

(a) Find the capacitance of this device.

Substitute into Equation 16.19:

$$C = \kappa \epsilon_0 \frac{A}{d}$$

$$= 3.7\left(8.85 \times 10^{-12}\,\frac{\mathrm{C^2}}{\mathrm{N \cdot m^2}}\right)\left(\frac{6.0 \times 10^{-4}\ \mathrm{m^2}}{1.0 \times 10^{-3}\ \mathrm{m}}\right)$$

$$= \boxed{2.0 \times 10^{-11}\ \mathrm{F}}$$

(b) Find the maximum charge that can be placed on the capacitor.

Calculate the maximum applied voltage, using the dielectric strength of paper, E_{max}:

$$\Delta V_{max} = E_{max}d = (16 \times 10^6\ \mathrm{V/m})(1.0 \times 10^{-3}\ \mathrm{m})$$

$$= 1.6 \times 10^4\ \mathrm{V}$$

Solve the basic capacitance equation for Q_{max} and substitute ΔV_{max} and C:

$$Q_{max} = C\,\Delta V_{max} = (2.0 \times 10^{-11}\ \mathrm{F})(1.6 \times 10^4\ \mathrm{V})$$

$$= \boxed{0.32\ \mu\mathrm{C}}$$

(c) Suppose the fully charged capacitor is disconnected from the battery and the dielectric is subsequently removed. Find the new electric field between the plates of the capacitor. Does the capacitor discharge?

Compute the charge density on the plates:

$$\sigma = \frac{Q_{max}}{A} = \frac{3.2 \times 10^{-7}\ \mathrm{C}}{6.0 \times 10^{-4}\ \mathrm{m^2}} = 5.3 \times 10^{-4}\ \mathrm{C/m^2}$$

Calculate the electric field from the charge density:

$$E = \frac{\sigma}{\epsilon_0} = \frac{5.3 \times 10^{-4}\ \mathrm{C/m^2}}{8.85 \times 10^{-12}\ \mathrm{C^2/m^2 \cdot N}} = \boxed{6.0 \times 10^7\ \mathrm{N/C}}$$

Because the electric field without the dielectric exceeds the value of the dielectric strength of air, the capacitor discharges across the gap.

REMARKS Dielectrics allow κ times as much charge to be stored on a capacitor for a given voltage. They also allow an increase in the applied voltage by increasing the threshold of electrical breakdown.

QUESTION 16.11 Without the paper dielectric, is the maximum charge that can be stored on this capacitor (a) larger than, (b) smaller than, or (c) the same as found in part (b)?

(Continued)

EXERCISE 16.11 A parallel-plate capacitor has plate area of 2.50×10^{-3} m^2 and distance between the plates of 2.00 mm. (a) Find the maximum charge that can be placed on the capacitor if air is between the plates. (b) Find the maximum charge if the air is replaced by polystyrene.

ANSWERS (a) 7×10^{-8} C (b) 1.4×10^{-6} C

EXAMPLE 16.12 Capacitors with Two Dielectrics

GOAL Derive a symbolic expression for a parallel-plate capacitor with two dielectrics.

PROBLEM A parallel-plate capacitor has dielectrics with constants κ_1 and κ_2 between the two plates, as shown in Figure 16.29. Each dielectric fills exactly half the volume between the plates. Derive expressions for **(a)** the potential difference between the two plates and **(b)** the resulting capacitance of the system.

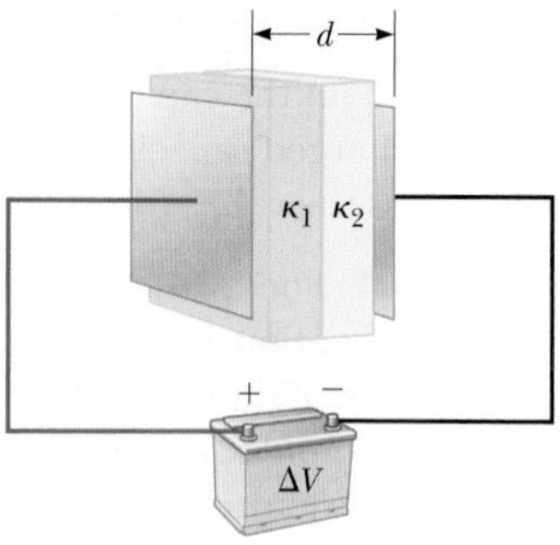

Figure 16.29 (Exercise 16.12)

STRATEGY The magnitude of the potential difference between the two plates of a capacitor is equal to the electric field multiplied by the plate separation. The electric field in a region is reduced by a factor of $1/\kappa$ when a dielectric is introduced, so $E = \sigma/\epsilon = \sigma/\kappa\epsilon_0$. Add the potential difference across each dielectric to find the total potential difference ΔV between the plates. The voltage difference across each dielectric is given by $\Delta V = Ed$, where E is the electric field and d the displacement. Obtain the capacitance from the relationship $C = Q/\Delta V$.

SOLUTION

(a) Derive an expression for the potential difference between the two plates.

Write a general expression for the potential difference across both slabs:

$$\Delta V = \Delta V_1 + \Delta V_2 = E_1 d_1 + E_2 d_2$$

Substitute expressions for the electric fields and dielectric thicknesses, $d_1 = d_2 = d/2$:

$$\Delta V = \frac{\sigma}{\kappa_1\epsilon_0}\frac{d}{2} + \frac{\sigma}{\kappa_2\epsilon_0}\frac{d}{2} = \frac{\sigma d}{2\epsilon_0}\left(\frac{1}{\kappa_1} + \frac{1}{\kappa_2}\right)$$

(b) Derive an expression for the resulting capacitance of the system.

Write the general expression for capacitance:

$$C = \frac{Q}{\Delta V}$$

Substitute $Q = \sigma A$ and the expression for the potential difference from part (a):

$$C = \frac{\sigma A}{\frac{\sigma d}{2\epsilon_0}\left(\frac{1}{\kappa_1} + \frac{1}{\kappa_2}\right)} = \frac{2\epsilon_0 A}{d}\frac{\kappa_1\kappa_2}{\kappa_1 + \kappa_2}$$

REMARKS The answer is the same as if there had been two capacitors in series with the respective dielectrics. When a capacitor consists of two dielectrics as shown in Figure 16.30, however, it's equivalent to two different capacitors in parallel.

QUESTION 16.12 What answer is obtained when the two dielectrics are removed so there is vacuum between the plates?

EXERCISE 16.12 Suppose a capacitor has two dielectrics arranged as shown in Figure 16.30, each dielectric filling exactly half of the volume between the two plates. Derive an expression for the capacitance if each dielectric fills exactly half the volume between the plates.

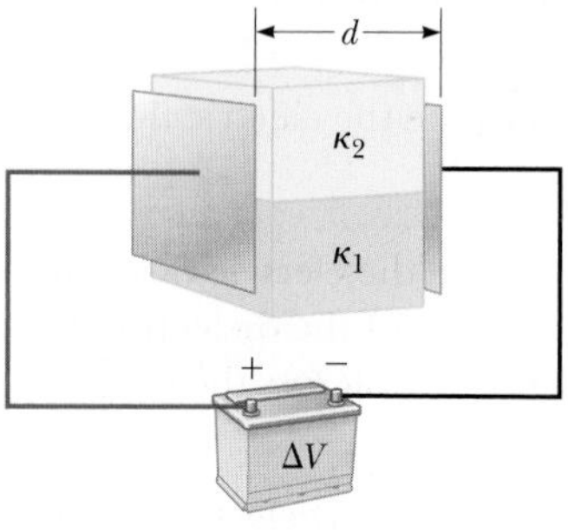

Figure 16.30 (Exercise 16.12)

ANSWER $C = \frac{\kappa_1 + \kappa_2}{2}\frac{\epsilon_0 A}{d}$

An Atomic Description of Dielectrics

The explanation of why a dielectric increases the capacitance of a capacitor is based on an atomic description of the material, which in turn involves a property of some molecules called **polarization**. A molecule is said to be polarized when there is a separation between the average positions of its negative charge and its positive charge. In some molecules, such as water, this condition is always present. To see why, consider the geometry of a water molecule (Fig. 16.31).

The molecule is arranged so that the negative oxygen atom is bonded to the positively charged hydrogen atoms with a 105° angle between the two bonds. The center of negative charge is at the oxygen atom, and the center of positive charge lies at a point midway along the line joining the hydrogen atoms (point x in the diagram). Materials composed of molecules that are permanently polarized in this way have large dielectric constants, and indeed, Table 16.1 shows that the dielectric constant of water is large ($\kappa = 80$) compared with other common substances.

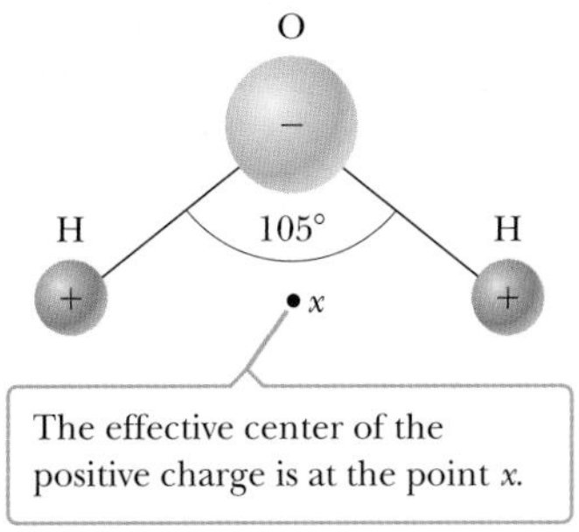

Figure 16.31 The water molecule, H_2O, has a permanent polarization resulting from its bent geometry.

A symmetric molecule (Fig. 16.32a) can have no permanent polarization, but a polarization can be induced in it by an external electric field. A field directed to the left, as in Figure 16.32b, would cause the center of positive charge to shift to the left from its initial position and the center of negative charge to shift to the right. This *induced polarization* is the effect that predominates in most materials used as dielectrics in capacitors.

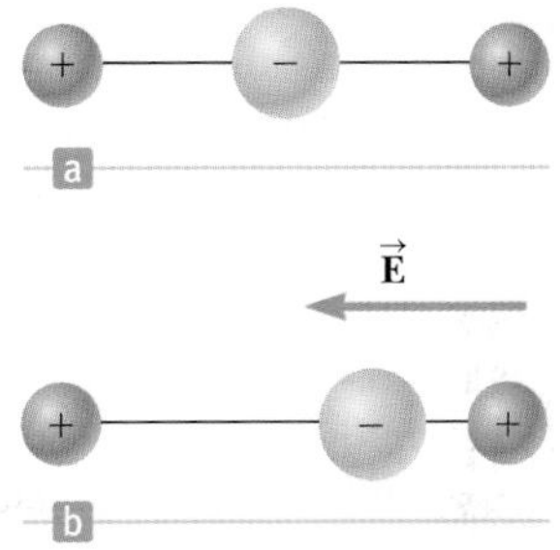

Figure 16.32 (a) A symmetric molecule has no permanent polarization. (b) An external electric field induces a polarization in the molecule.

To understand why the polarization of a dielectric can affect capacitance, consider the slab of dielectric shown in Figure 16.33. Before placing the slab between the plates of the capacitor, the polar molecules are randomly oriented (Fig. 16.33a). The polar molecules are dipoles, and each creates a dipole electric field, but because of their random orientation, this field averages to zero.

After insertion of the dielectric slab into the electric field $\vec{\mathbf{E}}_0$ between the plates (Fig. 16.33b), the positive plate attracts the negative ends of the dipoles and the negative plate attracts the positive ends of the dipoles. These forces exert a torque on the molecules making up the dielectric, reorienting them so that on average the negative pole is more inclined toward the positive plate and the positive pole is more aligned toward the negative plate. The positive and negative charges in the middle still cancel each other, but there is a net accumulation of negative charge in the dielectric next to the positive plate and a net accumulation of positive charge next to the negative plate. This configuration can be modeled as an additional pair of charged plates, as in Figure 16.33c, creating an induced electric field $\vec{\mathbf{E}}_{\text{ind}}$ that partly cancels the original electric field $\vec{\mathbf{E}}_0$. If the battery is not connected when the dielectric is inserted, the potential difference ΔV_0 across the plates is reduced to $\Delta V_0/\kappa$.

If the capacitor is still connected to the battery, however, the negative poles push more electrons off the positive plate, making it more positive. Meanwhile, the

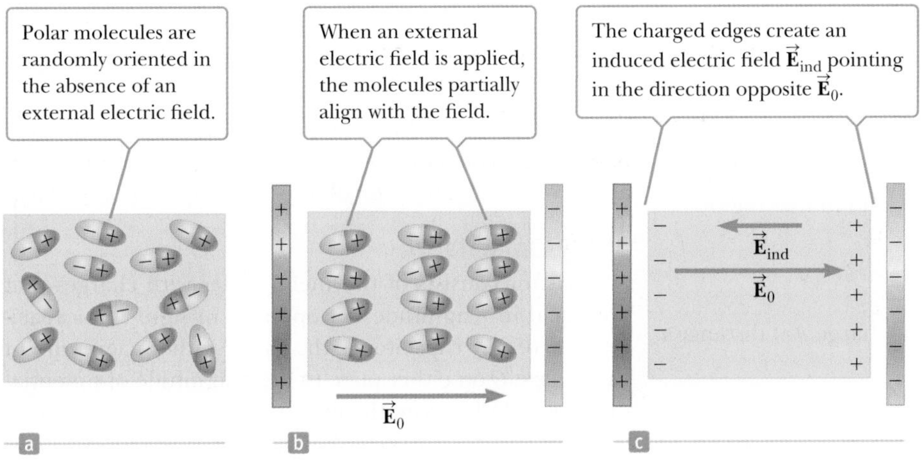

Figure 16.33 (a) Polar molecules are randomly oriented in a dielectric. (b) An electric field is applied to the dielectric. (c) The charged edges of the dielectric act like an additional pair of parallel plates, reducing the overall field between the actual plates. The interior of the dielectric is still neutral.

positive poles attract more electrons onto the negative plate. This situation continues until the potential difference across the battery reaches its original magnitude, equal to the potential gain across the battery. The net effect is an increase in the amount of charge stored on the capacitor. Because the plates can store more charge for a given voltage, it follows from $C = Q\Delta V$ that the capacitance must increase.

Quick Quiz

16.11 Consider a parallel-plate capacitor with a dielectric material between the plates. If the temperature of the dielectric increases, does the capacitance (a) decrease, (b) increase, or (c) remain the same?

SUMMARY

16.1 Potential Difference and Electric Potential

The change in the electric potential energy of a system consisting of an object of charge q moving through a displacement Δx in a constant electric field $\vec{\mathbf{E}}$ is given by

$$\Delta PE = -W_{AB} = -qE_x\,\Delta x \qquad [16.1]$$

When a charge q moves in a uniform electric field $\vec{\mathbf{E}}$ from point A to point B, the work done on the charge by the electric force is $qE_x\,\Delta x$.

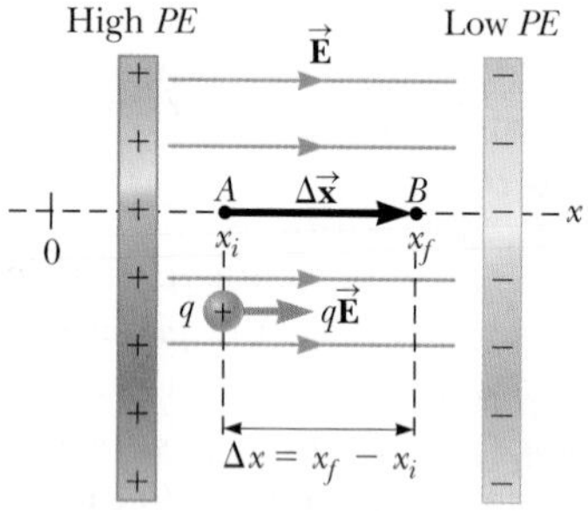

where E_x is the component of the electric field in the x-direction and $\Delta x = x_f - x_i$. The **difference in electric potential** between two points A and B is

$$\Delta V = V_B - V_A = \frac{\Delta PE}{q} \qquad [16.2]$$

where ΔPE is the *change* in electrical potential energy as a charge q moves between A and B. The units of potential difference are joules per coulomb, or **volts;** 1 J/C = 1 V.

The **electric potential difference** between two points A and B in a *uniform* electric field $\vec{\mathbf{E}}$ is

$$\Delta V = -E_x\,\Delta x \qquad [16.3]$$

where $\Delta x = x_f - x_i$ is the displacement between A and B and E_x is the x-component of the electric field in that region.

16.2 Electric Potential and Potential Energy Due to Point Charges

The **electric potential** due to a point charge q at distance r from the point charge is

$$V = k_e\frac{q}{r} \qquad [16.4]$$

Electric field and electric potential versus distance from a point charge of 1.11×10^{-10} C. Note that V is proportional to $1/r$, whereas E is proportional to $1/r^2$.

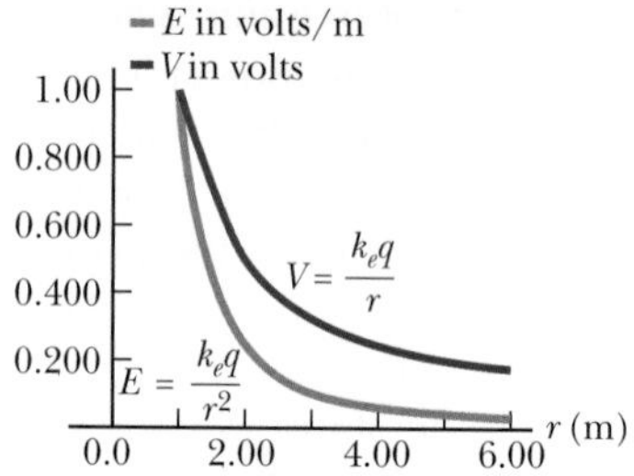

The **electric potential energy** of a pair of point charges separated by distance r is

$$PE = k_e\frac{q_1 q_2}{r} \qquad [16.5]$$

These equations can be used in the solution of conservation of energy problems and in the work–energy theorem.

16.3 Potentials and Charged Conductors

16.4 Equipotential Surfaces

Every point on the surface of a charged conductor in electrostatic equilibrium is at the same potential. Further, the potential is constant everywhere inside the conductor and equals its value on the surface.

The **electron volt** is defined as the energy that an electron (or proton) gains when accelerated through a potential difference of 1 V. The conversion between electron volts and joules is

$$1\text{ eV} = 1.60 \times 10^{-19}\text{ C}\cdot\text{V} = 1.60 \times 10^{-19}\text{ J} \qquad [16.7]$$

Any surface on which the potential is the same at every point is called an equipotential surface. The electric field is always oriented perpendicular to an equipotential surface.

16.6 Capacitance

A capacitor consists of two metal plates with charges that are equal in magnitude but opposite in sign. The capacitance C of any capacitor is the ratio of the magnitude of the charge Q on either plate to the magnitude of potential difference ΔV between them:

$$C \equiv \frac{Q}{\Delta V} \qquad [16.8]$$

Capacitance has the units coulombs per volt, or farads; 1 C/V = 1 F.

16.7 The Parallel-Plate Capacitor

The capacitance of two parallel metal plates of area A separated by distance d is

$$C = \epsilon_0 \frac{A}{d} \qquad [16.9]$$

where $\epsilon_0 = 8.85 \times 10^{-12}\ \mathrm{C^2/N \cdot m^2}$ is a constant called the **permittivity of free space.**

A parallel-plate capacitor consists of two parallel plates, each of area A, separated by a distance d.

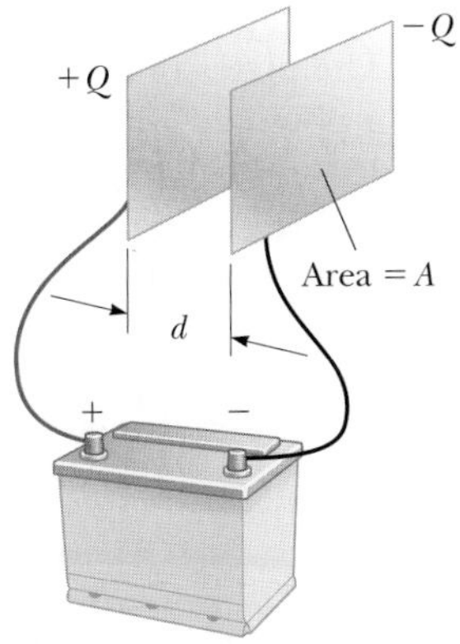

16.8 Combinations of Capacitors

The **equivalent capacitance of a parallel combination** of capacitors is

$$C_{eq} = C_1 + C_2 + C_3 + \cdots \qquad [16.12]$$

If two or more capacitors are connected in series, the **equivalent capacitance of the series combination** is

$$\frac{1}{C_{eq}} = \frac{1}{C_1} + \frac{1}{C_2} + \frac{1}{C_3} + \cdots \qquad [16.15]$$

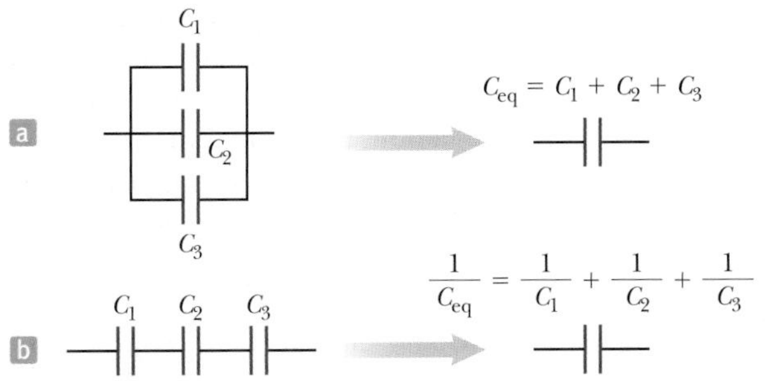

Capacitors in (a) parallel or in (b) series can be written as a single equivalent capacitor.

Problems involving a combination of capacitors can be solved by applying Equations 16.12 and 16.15 repeatedly to a circuit diagram, simplifying it as much as possible. This step is followed by working backwards to the original diagram, applying $C = Q/\Delta V$, that parallel capacitors have the same voltage drop, and that series capacitors have the same charge.

16.9 Energy Stored in a Charged Capacitor

Three equivalent expressions for calculating the **energy stored** in a charged capacitor are

$$\text{Energy stored} = \tfrac{1}{2}Q\,\Delta V = \tfrac{1}{2}C(\Delta V)^2 = \frac{Q^2}{2C} \qquad [16.17]$$

16.10 Capacitors with Dielectrics

When a nonconducting material, called a **dielectric,** is placed between the plates of a capacitor, the capacitance is multiplied by the factor κ, which is called the **dielectric constant**, a property of the dielectric material. The capacitance of a parallel-plate capacitor filled with a dielectric is

$$C = \kappa\epsilon_0 \frac{A}{d} \qquad [16.19]$$

WARM-UP EXERCISES

ENHANCED WebAssign The warm-up exercises in this chapter may be assigned online in Enhanced WebAssign.

1. **Math Review** Determine the value of C_{eq} in the following equation if $C_1 = 5.00\ \Omega$, $C_2 = 6.00\ \Omega$, and $C_3 = 7.00\ \Omega$: $\frac{1}{C_{eq}} = \frac{1}{C_1} + \frac{1}{C_2} + \frac{1}{C_3}$. (See also Section 16.8.)

2. **Physics Review** A 2.50-kg object initially at rest has a gravitational potential energy of 72.0 J. Determine the speed of the object when it has moved under the influence of gravity to a location where its gravitational potential energy is 45.0 J. (See Section 5.3.)

3. A uniform electric field of magnitude 3.00 N/C is directed along the $+x$-axis. If a 2.00 μC charge moves from (1.00, 0) m to (2.50, 0) m in this field, determine (a) the work done by the electric force, (b) the change in the electric potential energy of the particle, and (c) the electric potential difference between the particle's initial and final points. (See Section 16.1.)

4. A +4.00 μC charge is located at the origin. Determine (a) the electric potential at a distance of 2.00 m from the charge, and (b) the electric potential energy of a −2.03 μC charge located at (0, 2.00) m. (See Section 16.2.)

5. Two protons are located at (1.00, 0) m and (0,1.50) m, respectively. Determine (a) the electric potential at the origin, and (b) the electric potential energy of a third proton located at the origin. (See Section 16.2.)

6. (a) A hydrogen atom can be ionized by a photon having an energy of 13.6 eV. Convert that quantity to joules. (b) It requires 4.186 J of thermal energy to warm a gram of water by 1.00 K. Convert that energy to electron volts. (See Section 16.3.)

7. A capacitor with capacitance 3.00 μF is connected to a 9.00-V battery. (a) Find the charge on the capacitor in coulombs. (b) What voltage battery would be required to store 7.20×10^{-5} C on the capacitor? (See Section 16.6.)

8. A parallel-plate capacitor with no dielectric is made from two plates separated by 1.15×10^{-3} m. (a) Determine the device's capacitance if each plate has an area of 0.250 m^2. (See Section 16.6.) (b) What magnitude of charge is stored on each plate when the capacitor is connected to a 12.0-V battery? (See Section 16.6.) (c) If a material of dielectric constant 12.5 is inserted between the plates, what are the new values of the capacitance and (d) the magnitude of charge stored on each plate? (See Section 16.10.)

9. Two capacitors have capacitance 2.00 μF and 3.00 μF, respectively. Calculate the equivalent capacitance in microfarads if the capacitors are put (a) in parallel with each other, and (b) in series with each other. (See Section 16.8.)

10. Two capacitors are connected in parallel across a 12.0-V battery. If their capacitances are 12.0 μF and 25.0 μF, determine (a) the voltage across each capacitor, (b) the magnitude of charge stored on each plate of the 12.0 μF capacitor, (c) the magnitude of charge stored on each plate of the 25.0 μF capacitor, and (d) the equivalent capacitance of the system. (See Section 16.8.)

11. Two capacitors are connected in series between the terminals of a 12.0-V battery. If their capacitances are 12.0 μF and 25.0 μF, determine (a) the equivalent capacitance of the system, (b) the magnitude of charge stored on each plate of either capacitor, (c) the voltage across the 12.0 μF capacitor, and (d) the voltage across the 25.0 μF capacitor. (See Section 16.8.)

12. The plates of a parallel-plate capacitor are charged to a potential difference of 12.0 V. If the capacitance is 15.0 μF, calculate (a) the energy stored in the capacitor and (b) the magnitude of charge stored on each plate of the capacitor. (See Section 16.9.)

CONCEPTUAL QUESTIONS

ENHANCED WebAssign The conceptual questions in this chapter may be assigned online in Enhanced WebAssign.

1. (a) Describe the motion of a proton after it is released from rest in a uniform electric field. (b) Describe the changes (if any) in its kinetic energy and the electric potential energy associated with the proton.

2. Rank the potential energies of the four systems of particles shown in Figure CQ16.2 from largest to smallest. Include equalities if appropriate.

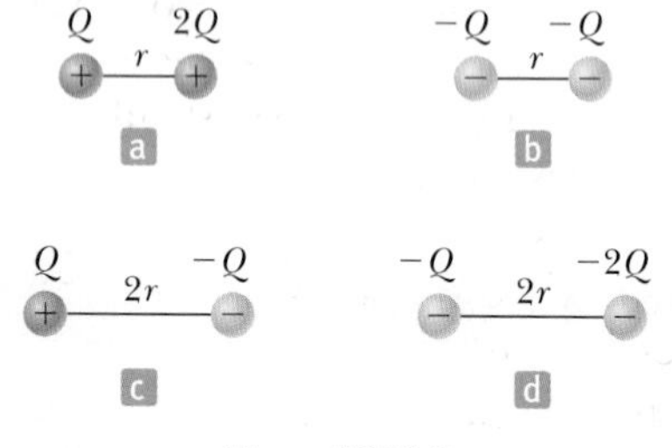

Figure CQ16.2

3. A parallel-plate capacitor is charged by a battery, and the battery is then disconnected from the capacitor. Because the charges on the capacitor plates are opposite in sign, they attract each other. Hence, it takes positive work to increase the plate separation. Show that the external work done when the plate separation is increased leads to an increase in the energy stored in the capacitor.

4. When charged particles are separated by an infinite distance, the electric potential energy of the pair is zero. When the particles are brought close, the electric potential energy of a pair with the same sign is positive, whereas the electric potential energy of a pair with opposite signs is negative. Explain.

5. Suppose you are sitting in a car and a 20-kV power line drops across the car. Should you stay in the car or get out? The power line potential is 20 kV compared to the potential of the ground.

6. Why is it important to avoid sharp edges or points on conductors used in high-voltage equipment?

7. Explain why, under static conditions, all points in a conductor must be at the same electric potential.

8. If you are given three different capacitors C_1, C_2, and C_3, how many different combinations of capacitance can you produce, using all capacitors in your circuits?

9. (a) Why is it dangerous to touch the terminals of a high-voltage capacitor even after the voltage source that charged the battery is disconnected from the capacitor? (b) What can be done to make the capacitor safe to handle after the voltage source has been removed?

10. The plates of a capacitor are connected to a battery. (a) What happens to the charge on the plates if the connecting wires are removed from the battery? (b) What happens to the charge if the wires are removed from the battery and connected to each other?

11. Rank the electric potentials at the four points shown in Figure CQ16.11 from largest to smallest.

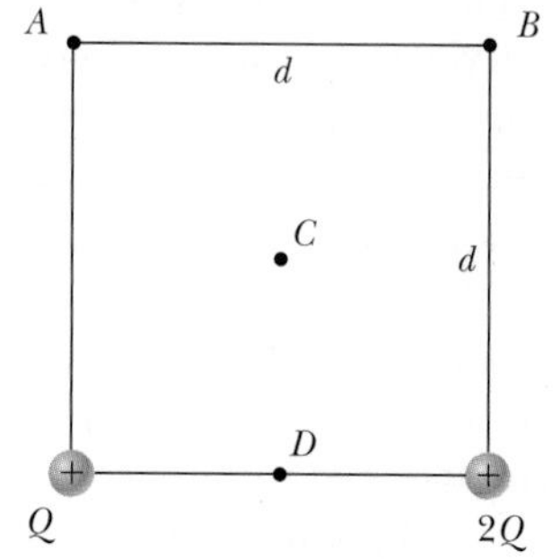

Figure CQ16.11

12. If you were asked to design a capacitor in which small size and large capacitance were required, what would be the two most important factors in your design?

13. Is it always possible to reduce a combination of capacitors to one equivalent capacitor with the rules developed in this chapter? Explain.

14. Explain why a dielectric increases the maximum operating voltage of a capacitor even though the physical size of the capacitor doesn't change.

PROBLEMS

ENHANCED WebAssign The problems in this chapter may be assigned online in Enhanced WebAssign.

1. denotes straightforward problem; 2. denotes intermediate problem;

3. denotes challenging problem

1. denotes full solution available in *Student Solutions Manual/ Study Guide*

1. denotes problems most often assigned in Enhanced WebAssign

BIO denotes biomedical problems

GP denotes guided problems

M denotes Master It tutorial available in Enhanced WebAssign

Q|C denotes asking for quantitative and conceptual reasoning

S denotes symbolic reasoning problem

W denotes Watch It video solution available in Enhanced WebAssign

16.1 Potential Difference and Electric Potential

1. A uniform electric field of magnitude 375 N/C pointing in the positive *x*-direction acts on an electron, which is initially at rest. After the electron has moved 3.20 cm, what is (a) the work done by the field on the electron, (b) the change in potential energy associated with the electron, and (c) the velocity of the electron?

2. A proton is released from rest in a uniform electric field of magnitude 385 N/C. Find (a) the electric force on the proton, (b) the acceleration of the proton, and (c) the distance it travels in 2.00 μs.

3. BIO W A potential difference of 90 mV exists between the inner and outer surfaces of a cell membrane. The inner surface is negative relative to the outer surface. How much work is required to eject a positive sodium ion (Na^+) from the interior of the cell?

4. A metal sphere of radius 5.00 cm is initially uncharged. How many electrons would have to be placed on the sphere to produce an electric field of magnitude 1.50×10^5 N/C at a point 8.00 cm from the center of the sphere?

5. The potential difference between the accelerating plates of a TV set is about 25 kV. If the distance between the plates is 1.5 cm, find the magnitude of the uniform electric field in the region between the plates.

6. A point charge $q = +40.0\ \mu$C moves from A to B separated by a distance $d = 0.180$ m in the presence of an external electric field $\vec{\mathbf{E}}$ of magnitude 275 N/C directed toward the right as in Figure P16.6. Find (a) the electric force exerted on the charge, (b) the work done by the electric force, (c) the change in the electric potential energy of the charge, and (d) the potential difference between A and B.

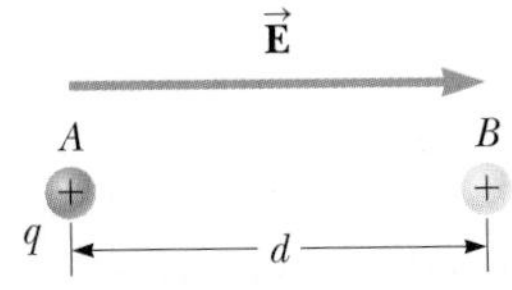

Figure P16.6

7. M Oppositely charged parallel plates are separated by 5.33 mm. A potential difference of 600 V exists between the plates. (a) What is the magnitude of the electric field between the plates? (b) What is the magnitude of the force on an electron between the plates? (c) How much work must be done on the electron to move it to the negative plate if it is initially positioned 2.90 mm from the positive plate?

8. Q|C S (a) Find the potential difference ΔV_e required to stop an electron (called a "stopping potential") moving with an initial speed of 2.85×10^7 m/s. (b) Would a proton traveling at the same speed require a greater or lesser magnitude potential difference? Explain. (c) Find a symbolic expression for the ratio of the proton stopping potential and the electron stopping potential, $\Delta V_p/\Delta V_e$. The answer should be in terms of the proton mass m_p and electron mass m_e.

9. Q|C A 74.0-g block carrying a charge $Q = 35.0\ \mu$C is connected to a spring for which $k = 78.0$ N/m. The block lies on a frictionless, horizontal surface and is immersed in a uniform electric field of magnitude $E = 4.86 \times 10^4$ N/C directed as shown in Figure P16.9. If the block is released from rest when the spring is unstretched ($x = 0$), (a) by what maximum distance does the block move from its initial position? (b) Find the subsequent equilibrium position of the block and the amplitude of its motion. (c) Using conservation of energy, find a symbolic relationship giving the potential difference between its initial position and the point of maximum extension in terms of the spring constant k, the amplitude A, and the charge Q.

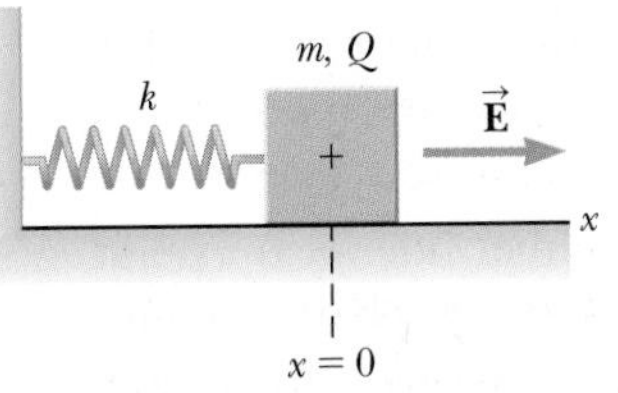

Figure P16.9

10. On planet Tehar, the free-fall acceleration is the same as that on the Earth, but there is also a strong downward electric field that is uniform close to the planet's surface. A 2.00-kg ball having a charge of 5.00 μC is thrown upward at a speed of 20.1 m/s. It hits the ground after an interval of 4.10 s. What is the potential difference between the starting point and the top point of the trajectory?

16.2 Electric Potential and Potential Energy Due to Point Charges

16.3 Potentials and Charged Conductors

16.4 Equipotential Surfaces

11. QIC An electron is at the origin. (a) Calculate the electric potential V_A at point A, $x = 0.250$ cm. (b) Calculate the electric potential V_B at point B, $x = 0.750$ cm. What is the potential difference $V_B - V_A$? (c) Would a negatively charged particle placed at point A necessarily go through this same potential difference upon reaching point B? Explain.

12. The two charges in Figure P16.12 are separated by d = 2.00 cm. Find the electric potential at (a) point A and (b) point B, which is halfway between the charges.

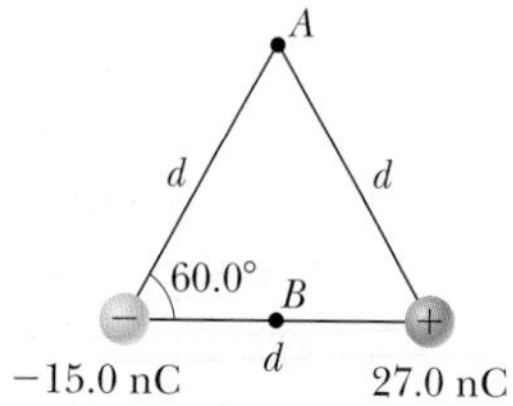

Figure P16.12

13. (a) Find the electric potential, taking zero at infinity, at the upper right corner (the corner without a charge) of the rectangle in Figure P16.13. (b) Repeat if the 2.00-μC charge is replaced with a charge of -2.00 μC.

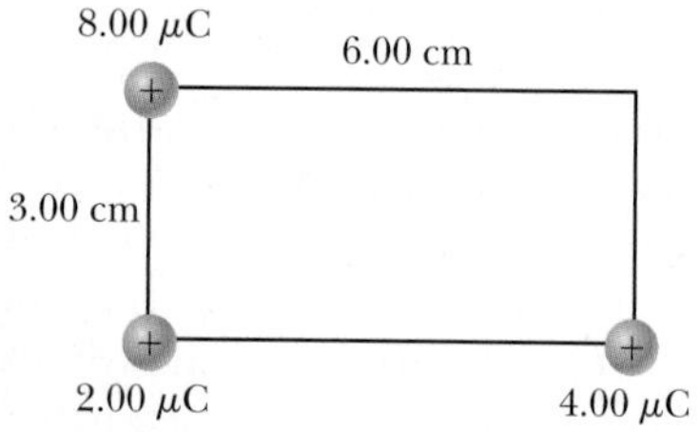

Figure P16.13 Problems 13 and 14.

14. Three charges are situated at corners of a rectangle as in Figure P16.13. How much work must an external agent do to move the 8.00-μC charge to infinity?

15. M QIC Two point charges $Q_1 = +5.00$ nC and $Q_2 = -3.00$ nC are separated by 35.0 cm. (a) What is the electric potential at a point midway between the charges? (b) What is the potential energy of the pair of charges? What is the significance of the algebraic sign of your answer?

16. QIC S Three identical point charges each of charge q are located at the vertices of an equilateral triangle as in Figure P16.16. The distance from the center of the triangle to each vertex is a. (a) Show that the electric field at the center of the triangle is zero. (b) Find a symbolic expression for the electric potential at the center of the triangle. (c) Give a physical explanation of the fact that the electric potential is not zero, yet the electric field is zero at the center.

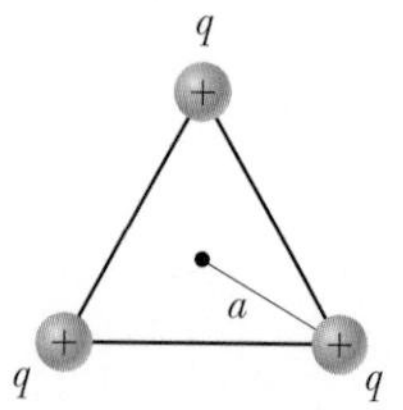

Figure P16.16

17. The three charges in Figure P16.17 are at the vertices of an isosceles triangle. Let q = 7.00 nC and calculate the electric potential at the midpoint of the base.

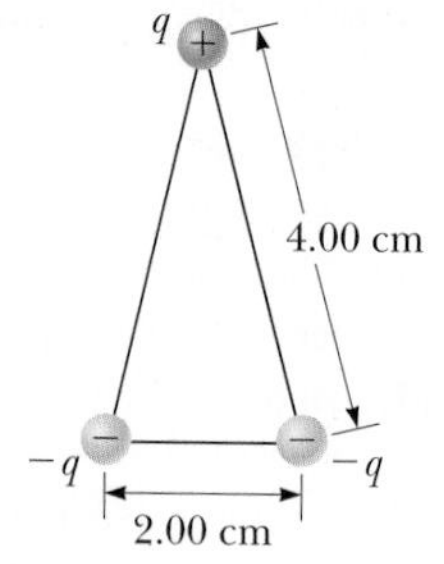

Figure P16.17

18. S A positive point charge q = +2.50 nC is located at x = 1.20 m and a negative charge of $-2q$ = -5.00 nC is located at the origin as in Figure P16.18. (a) Sketch the electric potential versus x for points along the x-axis in the range -1.50 m $< x <$ 1.50 m. (b) Find a symbolic expression for the potential on the x-axis at an arbitrary point P between the two charges. (c) Find the electric potential at x = 0.600 m. (d) Find the point along the x-axis between the two charges where the electric potential is zero.

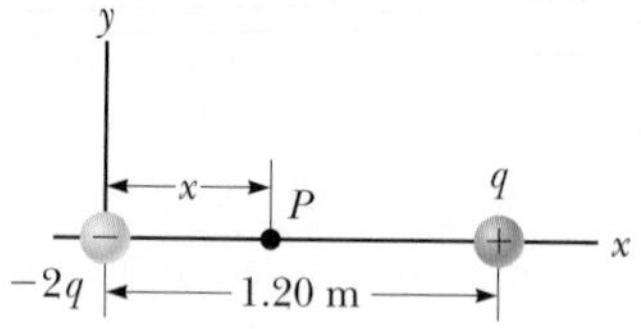

Figure P16.18

19. GP A proton is located at the origin, and a second proton is located on the x-axis at x = 6.00 fm (1 fm = 10^{-15} m). (a) Calculate the electric potential energy associated with this configuration. (b) An alpha particle (charge = $2e$, mass = 6.64×10^{-27} kg) is now placed at (x, y) = (3.00, 3.00) fm. Calculate the electric potential energy associated with this configuration. (c) Starting with the three-particle system, find the change in electric potential energy if the alpha particle is allowed to escape to infinity while the two protons remain fixed in place. (Throughout, neglect any radiation effects.) (d) Use conservation of energy to calculate the speed of the alpha particle at infinity. (e) If the two protons are released from rest and the alpha particle remains fixed, calculate the speed of the protons at infinity.

20. QIC A proton and an alpha particle (charge = $2e$, mass = 6.64×10^{-27} kg) are initially at rest, separated by 4.00×10^{-15} m. (a) If they are both released simultaneously, explain why you can't find their velocities at infinity using only conservation of energy. (b) What other conservation law can be applied in this case? (c) Find the speeds of the proton and alpha particle, respectively, at infinity.

21. A tiny sphere of mass 8.00 μg and charge -2.80 nC is initially at a distance of 1.60 μm from a fixed charge of +8.50 nC. If the 8.00-mg sphere is released from rest, find (a) its kinetic energy when it is 0.500 μm from the fixed charge and (b) its speed when it is 0.500 μm from the fixed charge.

22. The metal sphere of a small Van de Graaff generator illustrated in Figure 15.23 has a radius of 18 cm. When the electric field at the surface of the sphere reaches 3.0×10^6 V/m, the air breaks down, and the generator discharges. What is the maximum potential the sphere can have before breakdown occurs?

23. W In Rutherford's famous scattering experiments that led to the planetary model of the atom, alpha particles (having charges of $+2e$ and masses of 6.64×10^{-27} kg) were fired toward a gold nucleus with charge $+79e$. An alpha particle, initially very far from the gold nucleus, is fired at 2.00×10^7 m/s directly toward the nucleus, as in Figure P16.23. How close does the alpha particle get to the gold nucleus before turning around? Assume the gold nucleus remains stationary.

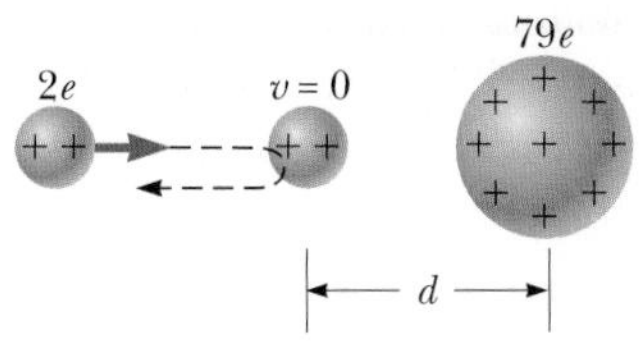

Figure P16.23

24. S Four point charges each having charge Q are located at the corners of a square having sides of length a. Find symbolic expressions for (a) the total electric potential at the center of the square due to the four charges and (b) the work required to bring a fifth charge q from infinity to the center of the square.

16.6 Capacitance

16.7 The Parallel-Plate Capacitor

25. W Consider the Earth and a cloud layer 800 m above the planet to be the plates of a parallel-plate capacitor. (a) If the cloud layer has an area of 1.0 km^2 = 1.0×10^6 m^2, what is the capacitance? (b) If an electric field strength greater than 3.0×10^6 N/C causes the air to break down and conduct charge (lightning), what is the maximum charge the cloud can hold?

26. (a) When a 9.00-V battery is connected to the plates of a capacitor, it stores a charge of 27.0 μC. What is the value of the capacitance? (b) If the same capacitor is connected to a 12.0-V battery, what charge is stored?

27. An air-filled parallel-plate capacitor has plates of area 2.30 cm^2 separated by 1.50 mm. The capacitor is connected to a 12.0-V battery. (a) Find the value of its capacitance. (b) What is the charge on the capacitor? (c) What is the magnitude of the uniform electric field between the plates?

28. Two conductors having net charges of +10.0 μC and -10.0 μC have a potential difference of 10.0 V between them. (a) Determine the capacitance of the system. (b) What is the potential difference between the two conductors if the charges on each are increased to +100 μC and -100 μC?

29. M An air-filled capacitor consists of two parallel plates, each with an area of 7.60 cm^2 and separated by a distance of 1.80 mm. If a 20.0-V potential difference is applied to these plates, calculate (a) the electric field between the plates, (b) the capacitance, and (c) the charge on each plate.

30. A 1-megabit computer memory chip contains many 60.0×10^{-15}-F capacitors. Each capacitor has a plate area of 21.0×10^{-12} m^2. Determine the plate separation of such a capacitor. (Assume a parallel-plate configuration.) The diameter of an atom is on the order of 10^{-10} m = 1 Å. Express the plate separation in angstroms.

31. QIC A parallel-plate capacitor with area 0.200 m^2 and plate separation of 3.00 mm is connected to a 6.00-V battery. (a) What is the capacitance? (b) How much charge is stored on the plates? (c) What is the electric field between the plates? (d) Find the magnitude of the charge density on each plate. (e) Without disconnecting the battery, the plates are moved farther apart. Qualitatively, what happens to each of the previous answers?

32. A small object with a mass of 350 μg carries a charge of 30.0 nC and is suspended by a thread between the vertical plates of a parallel-plate capacitor. The plates are separated by 4.00 cm. If the thread makes an angle of 15.0° with the vertical, what is the potential difference between the plates?

16.8 Combinations of Capacitors

33. Given a 2.50-μF capacitor, a 6.25-μF capacitor, and a 6.00-V battery, find the charge on each capacitor if you connect them (a) in series across the battery and (b) in parallel across the battery.

34. Two capacitors, C_1 = 5.00 μF and C_2 = 12.0 μF, are connected in parallel, and the resulting combination is connected to a 9.00-V battery. Find (a) the equivalent capacitance of the combination, (b) the potential difference across each capacitor, and (c) the charge stored on each capacitor.

35. Find (a) the equivalent capacitance of the capacitors in Figure P16.35, (b) the charge on each capacitor, and (c) the potential difference across each capacitor.

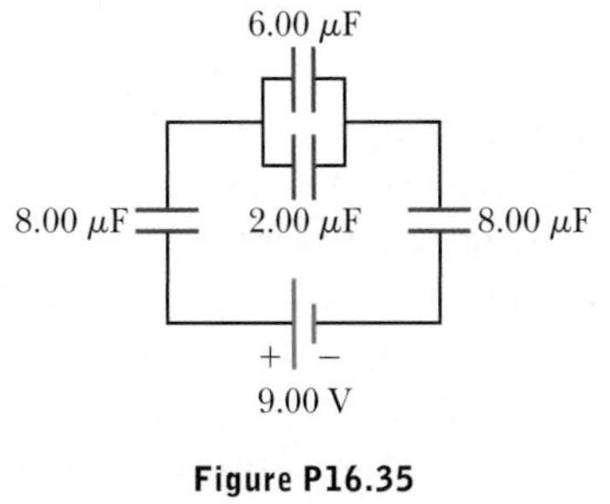

Figure P16.35

36. W Two capacitors give an equivalent capacitance of 9.00 pF when connected in parallel and an equivalent capacitance of 2.00 pF when connected in series. What is the capacitance of each capacitor?

37. For the system of capacitors shown in Figure P16.37, find (a) the equivalent capacitance of the system, (b) the charge on each capacitor, and (c) the potential difference across each capacitor.

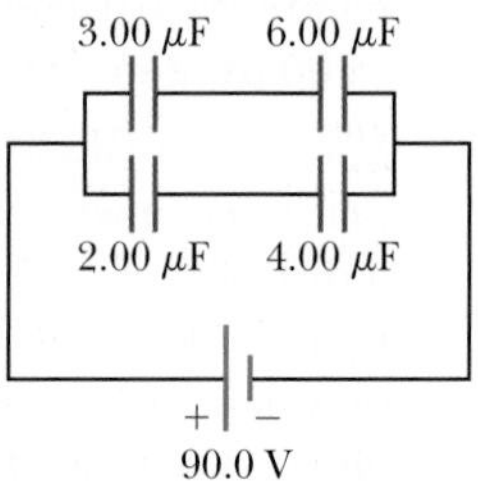

Figure P16.37 Problems 37 and 56.

38. GP Consider the combination of capacitors in Figure P16.38. (a) Find the equivalent single capacitance of the two capacitors in series and redraw the diagram (called diagram 1) with this equivalent capacitance. (b) In diagram 1 find the equivalent capacitance of the three capacitors in parallel and redraw the diagram as a single battery and single capacitor in a loop. (c) Compute the charge on the single equivalent capacitor. (d) Returning to diagram 1, compute the charge on each individual capacitor. Does the sum agree with the value found in part (c)? (e) What is the charge on the 24.0-μF capacitor and on the 8.00-μF capacitor? Compute the voltage drop across (f) the 24.0-μF capacitor and (g) the 8.00-μF capacitor.

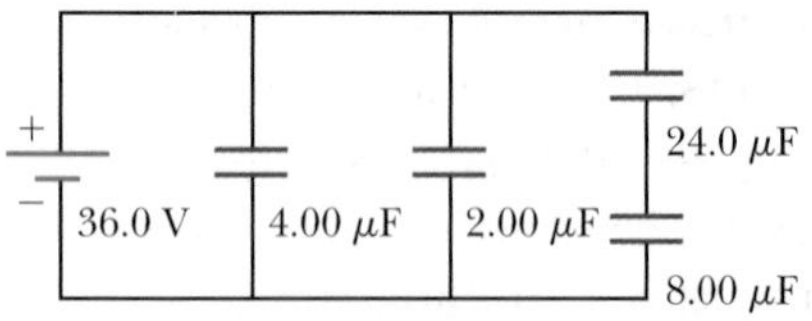

Figure P16.38

39. Find the charge on each of the capacitors in Figure P16.39.

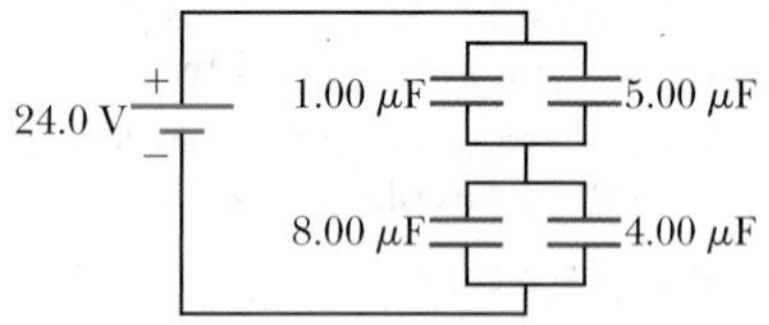

Figure P16.39

40. Q|C S Three capacitors are connected to a battery as shown in Figure P16.40. Their capacitances are $C_1 = 3C$, $C_2 = C$, and $C_3 = 5C$. (a) What is the equivalent capacitance of this set of capacitors? (b) State the ranking of the capacitors according to the charge they store from largest to smallest. (c) Rank the capacitors according to the potential differences across them from largest to smallest. (d) Assume C_3 is increased. Explain what happens to the charge stored by each capacitor.

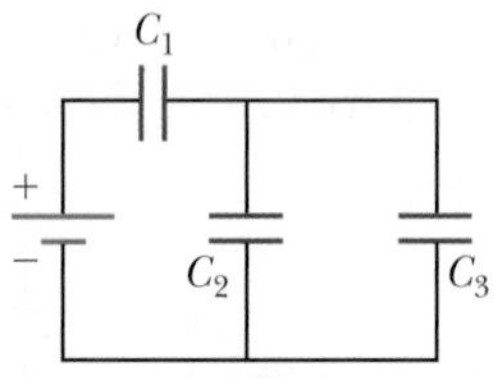

Figure P16.40

41. A 25.0-μF capacitor and a 40.0-μF capacitor are charged by being connected across separate 50.0-V batteries. (a) Determine the resulting charge on each capacitor. (b) The capacitors are then disconnected from their batteries and connected to each other, with each negative plate connected to the other positive plate. What is the final charge of each capacitor? (c) What is the final potential difference across the 40.0-μF capacitor?

42. (a) Find the equivalent capacitance between points a and b for the group of capacitors connected as shown in Figure P16.42 if $C_1 = 5.00\ \mu$F, $C_2 = 10.00\ \mu$F, and $C_3 = 2.00\ \mu$F. (b) If the potential between points a and b is 60.0 V, what charge is stored on C_3?

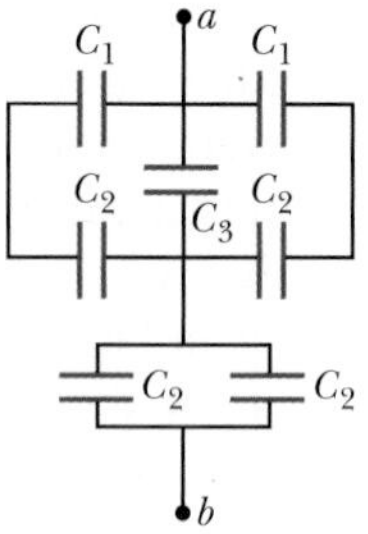

Figure P16.42

43. A 1.00-μF capacitor is charged by being connected across a 10.0-V battery. It is then disconnected from the battery and connected across an uncharged 2.00-μF capacitor. Determine the resulting charge on each capacitor.

44. M Four capacitors are connected as shown in Figure P16.44. (a) Find the equivalent capacitance between points a and b. (b) Calculate the charge on each capacitor, taking $\Delta V_{ab} = 15.0$ V.

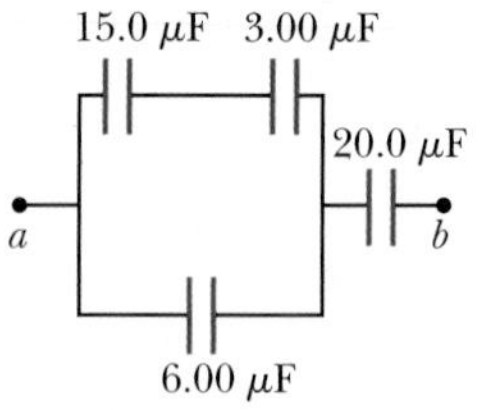

Figure P16.44

16.9 Energy Stored in a Charged Capacitor

45. A 12.0-V battery is connected to a 4.50-μF capacitor. How much energy is stored in the capacitor?

46. Q|C Two capacitors, $C_1 = 18.0\ \mu$F and $C_2 = 36.0\ \mu$F, are connected in series, and a 12.0-V battery is connected across them. (a) Find the equivalent capacitance, and the energy contained in this equivalent capacitor. (b) Find the energy stored in each individual capacitor. Show that the sum of these two energies is the same as the energy found in part (a). Will this equality always be true, or does it depend on the number of capacitors and their capacitances? (c) If the same capacitors were connected in parallel, what potential difference would be required across them so that the combination stores the same energy as in part (a)? Which capacitor stores more energy in this situation, C_1 or C_2?

47. A parallel-plate capacitor has capacitance 3.00 μF. (a) How much energy is stored in the capacitor if it is connected to a 6.00-V battery? (b) If the battery is disconnected and the distance between the charged

plates doubled, what is the energy stored? (c) The battery is subsequently reattached to the capacitor, but the plate separation remains as in part (b). How much energy is stored? (Answer each part in microjoules.)

48. A certain storm cloud has a potential difference of 1.00×10^8 V relative to a tree. If, during a lightning storm, 50.0 C of charge is transferred through this potential difference and 1.00% of the energy is absorbed by the tree, how much water (sap in the tree) initially at 30.0°C can be boiled away? Water has a specific heat of 4 186 J/kg·°C, a boiling point of 100°C, and a heat of vaporization of 2.26×10^6 J/kg.

16.10 Capacitors with Dielectrics

49. QIC The voltage across an air-filled parallel-plate capacitor is measured to be 85.0 V. When a dielectric is inserted and completely fills the space between the plates as in Figure P16.49, the voltage drops to 25.0 V. (a) What is the dielectric constant of the inserted material? Can you identify the dielectric? (b) If the dielectric doesn't completely fill the space between the plates, what could you conclude about the voltage across the plates?

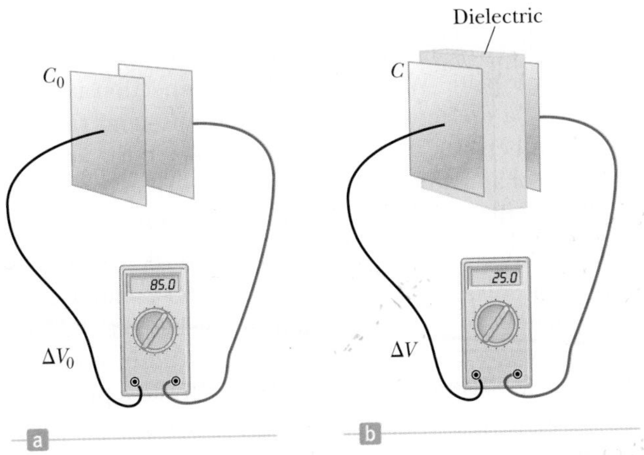

Figure P16.49

50. (a) How much charge can be placed on a capacitor with air between the plates before it breaks down if the area of each plate is 5.00 cm^2? (b) Find the maximum charge if polystyrene is used between the plates instead of air. Assume the dielectric strength of air is 3.00×10^6 V/m and that of polystyrene is 24.0×10^6 V/m.

51. Determine (a) the capacitance and (b) the maximum voltage that can be applied to a Teflon-filled parallel-plate capacitor having a plate area of 175 cm^2 and an insulation thickness of 0.040 0 mm.

52. Consider a plane parallel-plate capacitor made of two strips of aluminum foil separated by a layer of paraffin-coated paper. Each strip of foil and paper is 7.00 cm wide. The foil is 0.004 00 mm thick, and the paper is 0.025 0 mm thick and has a dielectric constant of 3.70. What length should the strips be if a capacitance of 9.50×10^{-8} F is desired? (If, after this plane capacitor is formed, a second paper strip can be added below the foil-paper-foil stack and the resulting assembly rolled into a cylindrical form—similar to that shown in Figure 16.26—the capacitance can be doubled because both surfaces of each foil strip would then store charge. Without the second strip of paper, however, rolling the layers would result in a short circuit.)

53. BIO A model of a red blood cell portrays the cell as a spherical capacitor, a positively charged liquid sphere of surface area A separated from the surrounding negatively charged fluid by a membrane of thickness t. Tiny electrodes introduced into the interior of the cell show a potential difference of 100 mV across the membrane. The membrane's thickness is estimated to be 100 nm and has a dielectric constant of 5.00. (a) If an average red blood cell has a mass of 1.00×10^{-12} kg, estimate the volume of the cell and thus find its surface area. The density of blood is 1 100 kg/m^3. (b) Estimate the capacitance of the cell by assuming the membrane surfaces act as parallel plates. (c) Calculate the charge on the surface of the membrane. How many electronic charges does the surface charge represent?

Additional Problems

54. When a potential difference of 150 V is applied to the plates of an air-filled parallel-plate capacitor, the plates carry a surface charge density of 3.00×10^{-10} C/cm^2. What is the spacing between the plates?

55. S Three parallel-plate capacitors are constructed, each having the same plate area A and with C_1 having plate spacing d_1, C_2 having plate spacing d_2, and C_3 having plate spacing d_3. Show that the total capacitance C of the three capacitors connected in series is the same as a capacitor of plate area A and with plate spacing $d = d_1 + d_2 + d_3$.

56. QIC For the system of four capacitors shown in Figure P16.37, find (a) the total energy stored in the system and (b) the energy stored by each capacitor. (c) Compare the sum of the answers in part (b) with your result to part (a) and explain your observation.

57. S A parallel-plate capacitor with a plate separation d has a capacitance C_0 in the absence of a dielectric. A slab of dielectric material of dielectric constant κ and thickness $d/3$ is then inserted between the plates as in Figure P16.57a. Show that the capacitance of this partially filled capacitor is given by

$$C = \left(\frac{3\kappa}{2\kappa + 1}\right)C_0$$

Hint: Treat the system as two capacitors connected in series as in Figure P16.57b, one with dielectric in it and the other one empty.

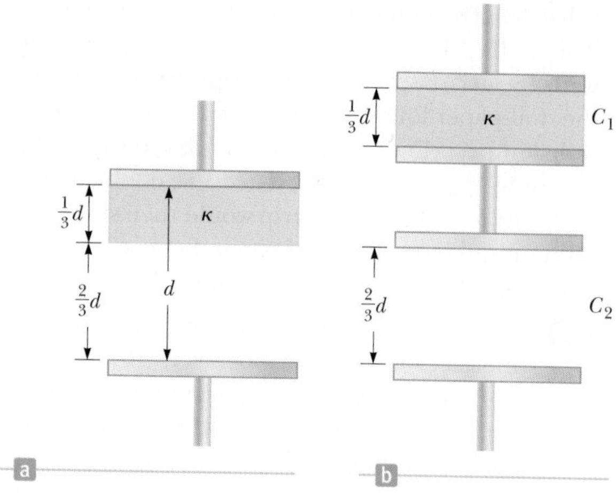

Figure P16.57

58. **S** Two capacitors give an equivalent capacitance of C_p when connected in parallel and an equivalent capacitance of C_s when connected in series. What is the capacitance of each capacitor?

59. **M** A parallel-plate capacitor is constructed using a dielectric material whose dielectric constant is 3.00 and whose dielectric strength is 2.00×10^8 V/m. The desired capacitance is 0.250 μF, and the capacitor must withstand a maximum potential difference of 4.00 kV. Find the minimum area of the capacitor plates.

60. Two charges of 1.0 μC and -2.0 μC are 0.50 m apart at two vertices of an equilateral triangle as in Figure P16.60. (a) What is the electric potential due to the 1.0-μC charge at the third vertex, point P? (b) What is the electric potential due to the -2.0-μC charge at P? (c) Find the total electric potential at P. (d) What is the work required to move a 3.0-μC charge from infinity to P?

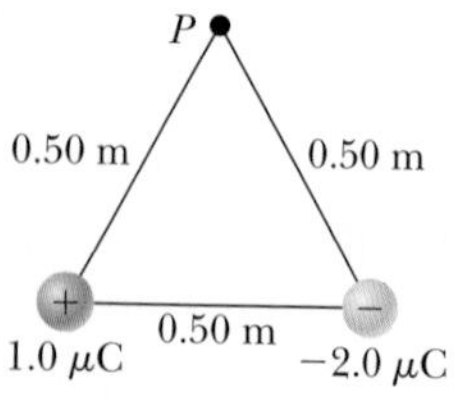

Figure P16.60

61. Find the equivalent capacitance of the group of capacitors shown in Figure P16.61.

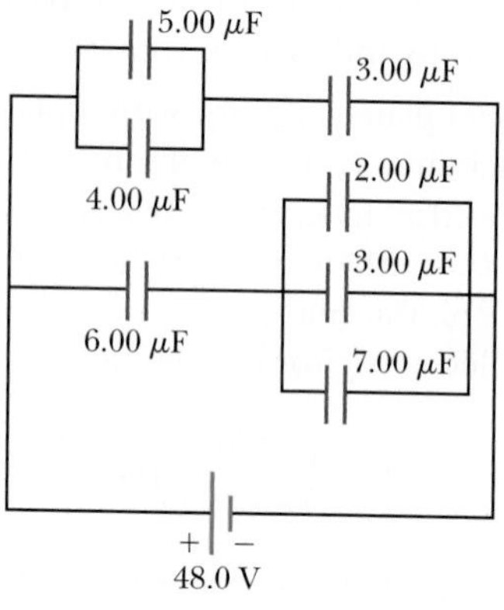

Figure P16.61

62. A spherical capacitor consists of a spherical conducting shell of radius b and charge $-Q$ concentric with a smaller conducting sphere of radius a and charge Q. (a) Find the capacitance of this device. (b) Show that as the radius b of the outer sphere approaches infinity, the capacitance approaches the value $a/k_e = 4\pi\epsilon_0 a$.

63. **BIO** The immediate cause of many deaths is ventricular fibrillation, an uncoordinated quivering of the heart, as opposed to proper beating. An electric shock to the chest can cause momentary paralysis of the heart muscle, after which the heart will sometimes start organized beating again. A *defibrillator* is a device that applies a strong electric shock to the chest over a time of a few milliseconds. The device contains a capacitor of a few microfarads, charged to several thousand volts. Electrodes called paddles, about 8 cm across and coated with conducting paste, are held against the chest on both sides of the heart. Their handles are insulated to prevent injury to the operator, who calls "Clear!" and pushes a button on one paddle to discharge the capacitor through the patient's chest. Assume an energy of 300 W · s is to be delivered from a 30.0-μF capacitor. To what potential difference must it be charged?

64. When a certain air-filled parallel-plate capacitor is connected across a battery, it acquires a charge of 150 μC on each plate. While the battery connection is maintained, a dielectric slab is inserted into, and fills, the region between the plates. This results in the accumulation of an additional charge of 200 μC on each plate. What is the dielectric constant of the slab?

65. Capacitors $C_1 = 6.0$ μF and $C_2 = 2.0$ μF are charged as a parallel combination across a 250-V battery. The capacitors are disconnected from the battery and from each other. They are then connected positive plate to negative plate and negative plate to positive plate. Calculate the resulting charge on each capacitor.

66. **S** Two positive charges each of charge q are fixed on the y-axis, one at $y = d$ and the other at $y = -d$ as in Figure P16.66. A third positive charge $2q$ located on the x-axis at $x = 2d$ is released from rest. Find symbolic expressions for (a) the total electric potential due to the first two charges at the location of the charge $2q$, (b) the electric potential energy of the charge $2q$, (c) the kinetic energy of the charge $2q$ after it has moved infinitely far from the other charges, and (d) the speed of the charge $2q$ after it has moved infinitely far from the other charges if its mass is m.

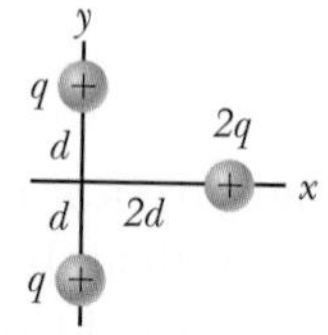

Figure P16.66

67. Metal sphere A of radius 12.0 cm carries 6.00 μC of charge, and metal sphere B of radius 18.0 cm carries -4.00 μC of charge. If the two spheres are attached by

a very long conducting thread, what is the final distribution of charge on the two spheres?

68. An electron is fired at a speed $v_0 = 5.6 \times 10^6$ m/s and at an angle $\theta_0 = -45°$ between two parallel conducting plates that are $D = 2.0$ mm apart, as in Figure P16.68. If the voltage difference between the plates is $\Delta V = 100$ V, determine (a) how close, d, the electron will get to the bottom plate and (b) where the electron will strike the top plate.

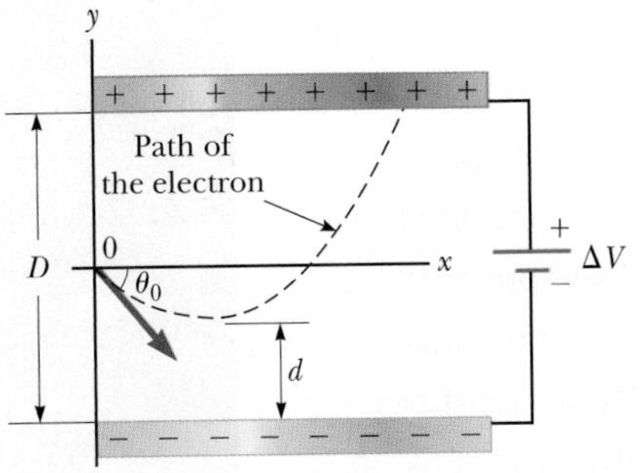

Figure P16.68

The blue glow comes from positively-charged xenon atoms that are electrostatically accelerated, then expelled from an ion engine prototype. The current of ions produces ninety millinewtons of thrust continuously for months at a time. Electrons must be fed back into the exhaust to prevent a buildup of negative charge. Such engines are highly efficient and suitable for extended deep space missions.

17 Current and Resistance

Many practical applications and devices are based on the principles of static electricity, but electricity was destined to become an inseparable part of our daily lives when scientists learned how to produce a continuous flow of charge for relatively long periods of time using batteries. The battery or voltaic cell was invented in 1800 by Italian physicist Alessandro Volta. Batteries supplied a continuous flow of charge at low potential, in contrast to earlier electrostatic devices that produced a tiny flow of charge at high potential for brief periods. This steady source of electric current allowed scientists to perform experiments to learn how to control the flow of electric charges in circuits. Today, electric currents power our lights, radios, television sets, air conditioners, computers, and refrigerators. They ignite the gasoline in automobile engines, travel through miniature components making up the chips of microcomputers, and provide the power for countless other invaluable tasks.

In this chapter we define current and discuss some of the factors that contribute to the resistance to the flow of charge in conductors. We also discuss energy transformations in electric circuits. These topics will be the foundation for additional work with circuits in later chapters.

17.1 Electric Current

LEARNING OBJECTIVES

1. Define both average and instantaneous electrical current and discuss their physical meaning.
2. Apply the concept of current to simple electrical systems.

Tip 17.1 *Current Flow* Is Redundant

The phrases *flow of current* and *current flow* are commonly used, but here the word *flow* is redundant because current is already defined as a flow (of charge). Avoid this construction!

In Figure 17.1 charges move in a direction perpendicular to a surface of area A. (That area could be the cross-sectional area of a wire, for example.) **The current is the rate at which charge flows through this surface.**

Suppose ΔQ is the amount of charge that flows through an area A in a time interval Δt and that the direction of flow is perpendicular to the area. Then the **average current I_{av}** is equal to the amount of charge divided by the time interval:

$$I_{av} \equiv \frac{\Delta Q}{\Delta t} \qquad [17.1a]$$

SI unit: coulomb/second (C/s), or the ampere (A)

Current is composed of individual moving charges, so for an extremely low current, it is conceivable that a single charge could pass through area A in one instant and no charge in the next instant. All currents, then, are essentially averages over time. Given the very large number of charges usually involved, however, it makes sense to define an instantaneous current.

The **instantaneous current I** is the limit of the average current as the time interval goes to zero:

$$I = \lim_{\Delta t \to 0} I_{av} = \lim_{\Delta t \to 0} \frac{\Delta Q}{\Delta t} \qquad [17.1b]$$

SI unit: coulomb/second (C/s), or the ampere (A)

When the current is steady, the average and instantaneous currents are the same. Note that one ampere of current is equivalent to one coulomb of charge passing through a given area A in a time interval of 1 s.

When charges flow through a surface as in Figure 17.1, they can be positive, negative, or both. **The direction of conventional current used in this book is the direction positive charges flow.** (This historical convention originated about 200 years ago, when the ideas of positive and negative charges were introduced.) In a common conductor such as copper, the current is due to the motion of negatively charged electrons, so the direction of the current is opposite the direction of motion of the electrons. On the other hand, for a beam of positively charged protons in an accelerator, the current is in the same direction as the motion of the protons. In some cases—gases and electrolytes, for example—the current is the result of the flows of both positive and negative charges. Moving charges, whether positive or negative, are referred to as *charge carriers*. In a metal, for example, the charge carriers are electrons.

In electrostatics, where charges are stationary, the electric potential is the same everywhere in a conductor. That is no longer true for conductors carrying current: as charges move along a wire, the electric potential is continually decreasing (except in the special case of superconductors). The decreasing electric potential means that the moving charges lose energy according to the relationship $\Delta U_{charges} = q\Delta V$, while

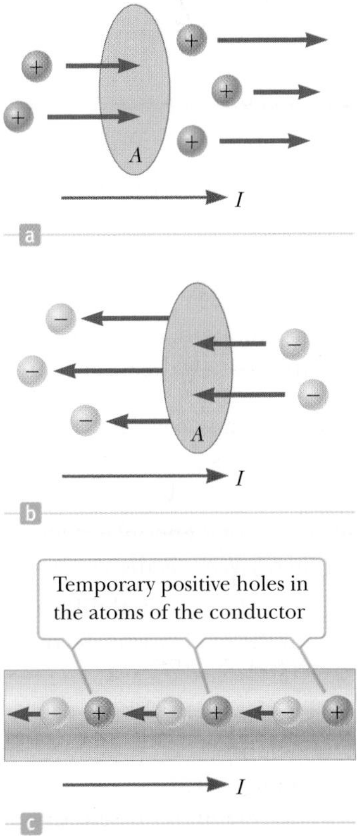

Figure 17.1 The time rate of flow of charge through area A is the current I. (a) The direction of current is the same as the flow of positive charge. (b) Negative charge flowing to the left is equivalent to an equal amount of positive charge flowing to the right. (c) In a conductor, positive holes open in the lattice of the conductor's atoms as electrons move in response to a potential. Negative electrons moving actively to the left are equivalent to positive holes migrating to the right.

an energy $\Delta U_{wire} = -q\Delta V$ is deposited in the current-carrying wire. (Those expressions derive from Equation 16.2.) If q is taken to be positive, corresponding to the convention of positive current, then $\Delta V = V_f - V_i$ is negative because, in a circuit, positive charges move from regions of high potential to regions of low potential. That in turn means $\Delta U_{charges} = q\Delta V$ is negative, as it should be, because the moving charges lose energy. Often only the magnitude is desired, however, in which case absolute values are substituted into q and ΔV. If the current is constant, then dividing the energy by the elapsed time yields the power delivered to the circuit element, such as a lightbulb filament.

EXAMPLE 17.1 Turn On the Light

GOAL Apply the concept of current.

PROBLEM The amount of charge that passes through the filament of a certain lightbulb in 2.00 s is 1.67 C. Find **(a)** the average current in the lightbulb and **(b)** the number of electrons that pass through the filament in 5.00 s. **(c)** If the current is supplied by a 12.0-V battery, what total energy is delivered to the lightbulb filament during 2.00 s? What is the average power?

STRATEGY Substitute into Equation 17.1a for part (a), then multiply the answer by the time given in part (b) to get the total charge that passes in that time. The total charge equals the number N of electrons going through the circuit times the charge per electron. To obtain the energy delivered to the filament, multiply the potential difference, ΔV, by the total charge. Dividing the energy by time yields the average power.

SOLUTION

(a) Compute the average current in the lightbulb.

Substitute the charge and time into Equation 17.1a:

$$I_{av} = \frac{\Delta Q}{\Delta t} = \frac{1.67\text{ C}}{2.00\text{ s}} = \boxed{0.835\text{ A}}$$

(b) Find the number of electrons passing through the filament in 5.00 s.

The total number N of electrons times the charge per electron equals the total charge, $I_{av}\,\Delta t$:

$$\text{(1)}\quad Nq = I_{av}\,\Delta t$$

Substitute and solve for N:

$$N(1.60 \times 10^{-19}\text{ C/electron}) = (0.835\text{ A})(5.00\text{ s})$$

$$N = \boxed{2.61 \times 10^{19}\text{ electrons}}$$

(c) What total energy is delivered to the lightbulb filament? What is the average power?

Multiply the potential difference by the total charge to obtain the energy transferred to the filament:

$$\text{(2)}\quad \Delta U = q\Delta V = (1.67\text{ C})(12.0\text{ V}) = \boxed{20.0\text{ J}}$$

Divide the energy by the elapsed time to calculate the average power:

$$P_{av} = \frac{\Delta U}{\Delta t} = \frac{20.0\text{ J}}{2.00\text{ s}} = \boxed{10.0\text{ W}}$$

REMARKS It's important to use units to ensure the correctness of equations such as Equation (1). Notice the enormous number of electrons that pass through a given point in a typical circuit. Magnitudes were used in calculating the energies in Equation (2). Technically, the charge carriers are electrons with negative charge moving from a lower potential to a higher potential, so the change in their energy is $\Delta U_{charge} = q\Delta V = (-1.67\text{ C})(+12.0\text{ V}) = -20.0\text{ J}$, a loss of energy that is delivered to the filament, $\Delta U_{fil} = -\Delta U_{charge} = +20.0\text{ J}$. The energy and power, calculated here using the definitions of Chapter 16, will be further addressed in Section 17.6.

QUESTION 17.1 Is it possible to have an instantaneous current of $e/2$ per second? Explain. Can the average current take this value?

EXERCISE 17.1 A 9.00-V battery delivers a current of 1.34 A to the lightbulb filament of a pocket flashlight. (a) How much charge passes through the filament in 2.00 min? (b) How many electrons pass through the filament? Calculate (c) the energy delivered to the filament during that time and (d) the power delivered by the battery.

ANSWERS (a) 161 C (b) 1.01×10^{21} electrons (c) 1.45×10^{3} J (d) 12.1 W

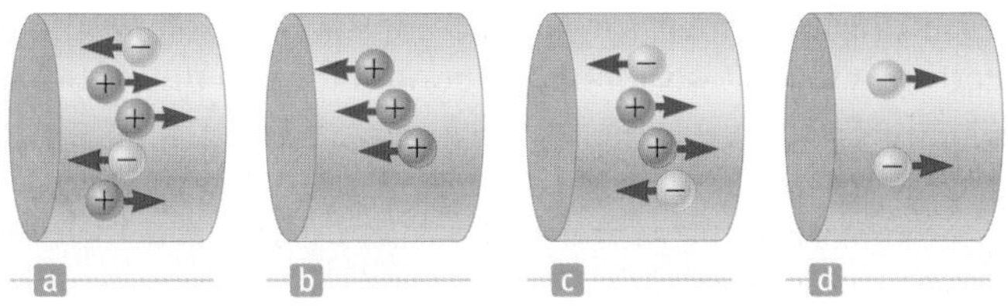

Figure 17.2
(Quick Quiz 17.1)

Quick Quiz

17.1 Consider positive and negative charges all moving horizontally with the same speed through the four regions in Figure 17.2. Rank the magnitudes of the currents in these four regions from lowest to highest. (I_a is the current in Figure 17.2a, I_b the current in Figure 17.2b, etc.) (a) I_d, I_a, I_c, I_b (b) I_a, I_c, I_b, I_d (c) I_c, I_a, I_d, I_b (d) I_d, I_b, I_c, I_a (e) I_a, I_b, I_c, I_d (f) None of these

17.2 A Microscopic View: Current and Drift Speed

LEARNING OBJECTIVES

1. Relate electrical current to the drift speed of charge carriers.
2. Evaluate the drift speed in typical electrical conductors.

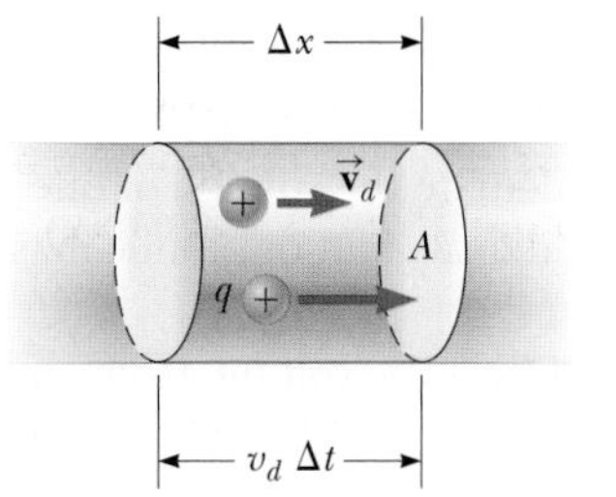

Figure 17.3 A section of a uniform conductor of cross-sectional area A. The charge carriers move with a speed v_d, and the distance they travel in time Δt is given by $\Delta x = v_d \Delta t$. The number of mobile charge carriers in the section of length Δx is given by $nAv_d\Delta t$, where n is the number of mobile carriers per unit volume.

Macroscopic currents can be related to the motion of the microscopic charge carriers making up the current. It turns out that current depends on the average speed of the charge carriers in the direction of the current, the number of charge carriers per unit volume, and the of the charge carried by each charge carrier.

Consider identically charged particles moving in a conductor of cross-sectional area A (Fig. 17.3). The volume of an element of length Δx of the conductor is $A\Delta x$. If n represents the number of mobile charge carriers per unit volume, the number of carriers in the volume element is $nA\Delta x$. The mobile charge ΔQ in this element is therefore

$$\Delta Q = \text{number of carriers} \times \text{charge per carrier} = (nA\Delta x)q$$

where q is the charge on each carrier. If the carriers move with a constant average speed called the **drift speed** v_d, the distance they move in the time interval Δt is $\Delta x = v_d\,\Delta t$. We can therefore write

$$\Delta Q = (nAv_d\,\Delta t)q$$

If we divide both sides of this equation by Δt and take the limit as Δt goes to zero, we see that the current in the conductor is

$$I = \lim_{\Delta t \to 0} \frac{\Delta Q}{\Delta t} = nqv_dA \qquad [17.2]$$

To understand the meaning of drift speed, consider a conductor in which the charge carriers are free electrons. If the conductor is isolated, these electrons undergo random motion similar to the motion of the molecules in a gas. The drift speed is normally much smaller than the free electrons' average speed between collisions with the fixed atoms of the conductor. When a potential difference is applied between the ends of the conductor (say, with a battery), an electric field is set up in the conductor, creating an electric force on the electrons and hence a current. In reality, the electrons don't simply move in straight lines along the conductor. Instead, they undergo repeated collisions with the atoms of the metal, and the result is a complicated zigzag motion with only a small average drift speed along the wire (Fig. 17.4). The energy transferred from the electrons to the metal atoms during a collision increases the vibrational energy of the atoms and causes

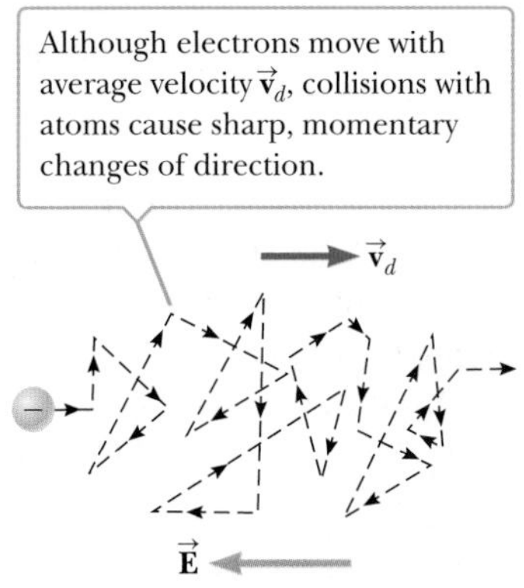

Figure 17.4 A schematic representation of the zigzag motion of a charge carrier in a conductor. Notice that the drift velocity $\vec{v}_d$ is opposite the direction of the electric field.

a corresponding increase in the temperature of the conductor. Despite the collisions, however, the electrons move slowly along the conductor in a direction opposite $\vec{\mathbf{E}}$ with the drift velocity $\vec{\mathbf{v}}_d$.

EXAMPLE 17.2 Drift Speed of Electrons

GOAL Calculate a drift speed and compare it with the rms speed of an electron gas.

PROBLEM A copper wire of cross-sectional area 3.00×10^{-6} m^2 carries a current of 10.0 A. **(a)** Assuming each copper atom contributes one free electron to the metal, find the drift speed of the electrons in this wire. **(b)** Use the ideal gas model to compare the drift speed with the random rms speed an electron would have at 20.0°C. The density of copper is 8.92 g/cm^3, and its atomic mass is 63.5 u.

STRATEGY All the variables in Equation 17.2 are known except for n, the number of free charge carriers per unit volume. We can find n by recalling that one mole of copper contains an Avogadro's number (6.02×10^{23}) of atoms and each atom contributes one charge carrier to the metal. The volume of one mole can be found from copper's known density and atomic mass. The atomic mass is the same, numerically, as the number of grams in a mole of the substance.

SOLUTION

(a) Find the drift speed of the electrons.

Calculate the volume of one mole of copper from its density and its atomic mass:

$$V = \frac{m}{\rho} = \frac{63.5 \text{ g/mol}}{8.92 \text{ g/cm}^3} = 7.12 \text{ cm}^3/\text{mol}$$

Convert the volume from cm^3 to m^3:

$$7.12 \text{ cm}^3/\text{mol}\left(\frac{1 \text{ m}}{10^2 \text{ cm}}\right)^3 = 7.12 \times 10^{-6} \text{ m}$$

Divide Avogadro's number (the number of electrons in one mole) by the volume per mole to obtain the number density:

$$n = \frac{6.02 \times 10^{23} \text{ electrons/mole}}{7.12 \times 10^{-6} \text{ m}^3/\text{mole}}$$
$$= 8.46 \times 10^{28} \text{ electrons/m}^3$$

Solve Equation 17.2 for the drift speed and substitute:

$$v_d = \frac{I}{nqA}$$
$$= \frac{10.0 \text{ C/s}}{(8.46 \times 10^{28} \text{ electrons/m}^3)(1.60 \times 10^{-19} \text{ C})(3.00 \times 10^{-6} \text{ m}^2)}$$
$$v_d = \boxed{2.46 \times 10^{-4} \text{ m/s}}$$

(b) Find the rms speed of a gas of electrons at 20.0°C.

Apply Equation 10.18:

$$v_{\text{rms}} = \sqrt{\frac{3k_BT}{m_e}}$$

Convert the temperature to the Kelvin scale and substitute values:

$$v_{\text{rms}} = \sqrt{\frac{3(1.38 \times 10^{-23} \text{ J/K})(293 \text{ K})}{9.11 \times 10^{-31} \text{ kg}}}$$
$$= \boxed{1.15 \times 10^5 \text{ m/s}}$$

REMARKS The drift speed of an electron in a wire is very small, only about one-billionth of its random thermal speed.

QUESTION 17.2 True or False: The drift velocity in a wire of a given composition is inversely proportional to the number density of charge carriers.

EXERCISE 17.2 What current in a copper wire with a cross-sectional area of 7.50×10^{-7} m^2 would result in a drift speed equal to 5.00×10^{-4} m/s?

ANSWER 5.08 A

Example 17.2 shows that drift speeds are typically very small. In fact, the drift speed is much smaller than the average speed between collisions. Electrons traveling at 2.46×10^{-4} m/s, as in the example, would take about 68 min to travel 1 m! In view of this low speed, why does a lightbulb turn on almost instantaneously when a switch is thrown? Think of the flow of water through a pipe. If a drop of water is forced into one end of a pipe that is already filled with water, a drop must be pushed out the other end of the pipe. Although it may take an individual drop a long time to make it through the pipe, a flow initiated at one end produces a similar flow at the other end very quickly. Another familiar analogy is the motion of a bicycle chain. When the sprocket moves one link, the other links all move more or less immediately, even though it takes a given link some time to make a complete rotation. In a conductor, the change in the electric field that drives the free electrons travels at a speed close to that of light, so when you flip a light switch, the message for the electrons to start moving through the wire (the electric field) reaches them at a speed on the order of 10^8 m/s!

Tip 17.2 Electrons Are Everywhere in the Circuit

Electrons don't have to travel from the light switch to the lightbulb for the lightbulb to operate. Electrons already in the filament of the lightbulb move in response to the electric field set up by the battery. Also, the battery does *not* provide electrons to the circuit; it provides *energy* to the existing electrons.

Quick Quiz

17.2 Suppose a current-carrying wire has a cross-sectional area that gradually becomes smaller along the wire so that the wire has the shape of a very long, truncated cone. How does the drift speed vary along the wire? (a) It slows down as the cross section becomes smaller. (b) It speeds up as the cross section becomes smaller. (c) It doesn't change. (d) More information is needed.

17.3 Current and Voltage Measurements In Circuits

LEARNING OBJECTIVES

1. Discuss the concept of an electrical circuit.
2. Discuss ammeters and voltmeters, instruments used to measure currents and potential differences in circuits.

To study electric current in circuits, we need to understand how to measure currents and voltages.

The circuit shown in Figure 17.5a is a drawing of the actual circuit necessary for measuring the current in Example 17.1. Figure 17.5b shows a stylized figure called a circuit diagram that represents the actual circuit of Figure 17.5a. This circuit consists of only a battery and a lightbulb. The word *circuit* means "a closed loop of some sort around which current circulates." The battery pumps charge through the bulb and around the loop. No charge would flow without a complete conducting path from the positive terminal of the battery into one side of the bulb, out the other side, and

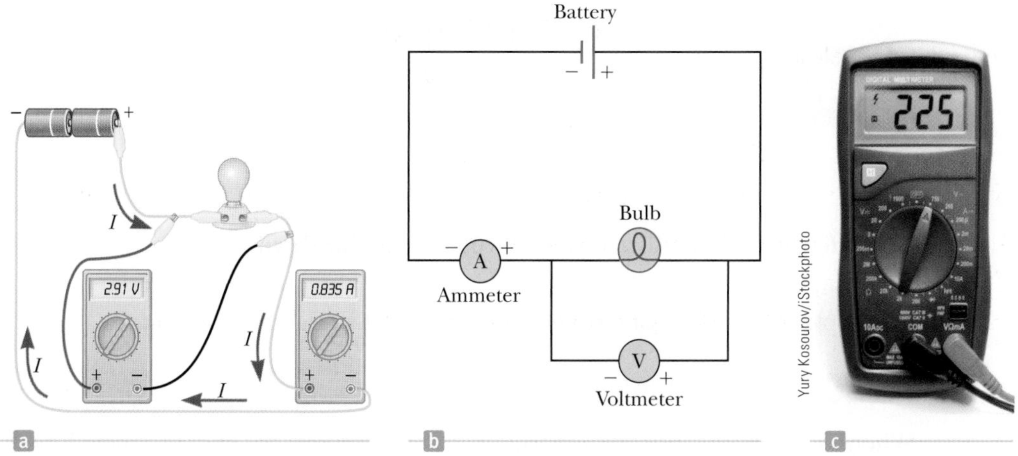

Figure 17.5 (a) A sketch of an actual circuit used to measure the current in a flashlight bulb and the potential difference across it. (b) A schematic diagram of the circuit shown in (a). (c) A digital multimeter can be used to measure both current and potential difference.

through the copper conducting wires back to the negative terminal of the battery. The most important quantities that characterize how the bulb works in different situations are the current I in the bulb and the potential difference ΔV across the bulb. To measure the current in the bulb, we place an ammeter, the device for measuring current, in line with the bulb so there is no path for the current to bypass the meter; all the charge passing through the bulb must also pass through the ammeter. The voltmeter measures the potential difference, or voltage, between the two ends of the bulb's filament. If we use two meters simultaneously as in Figure 17.5a, we can remove the voltmeter and see if its presence affects the current reading. Figure 17.5c shows a digital multimeter, a convenient device, with a digital readout, that can be used to measure voltage, current, or resistance. An advantage of using a digital multimeter as a voltmeter is that it will usually not affect the current because a digital meter has enormous resistance to the flow of charge in the voltmeter mode.

At this point, you can measure the current as a function of voltage (an I–ΔV curve) of various devices in the lab. All you need is a variable voltage supply (an adjustable battery) capable of supplying potential differences from about −5 V to +5 V, a bulb, a resistor, some wires and alligator clips, and a couple of multimeters. Be sure to always start your measurements using the highest multimeter scales (say, 10 A and 1 000 V), and increase the sensitivity one scale at a time to obtain the highest accuracy without overloading the meters. (Increasing the sensitivity means lowering the maximum current or voltage that the scale reads.) Note that the meters must be connected with the proper polarity with respect to the voltage supply, as shown in Figure 17.5b. Finally, follow your instructor's directions carefully to avoid damaging the meters and incurring a soaring lab fee.

Quick Quiz

17.3 Look at the four "circuits" shown in Figure 17.6 and select those that will light the bulb.

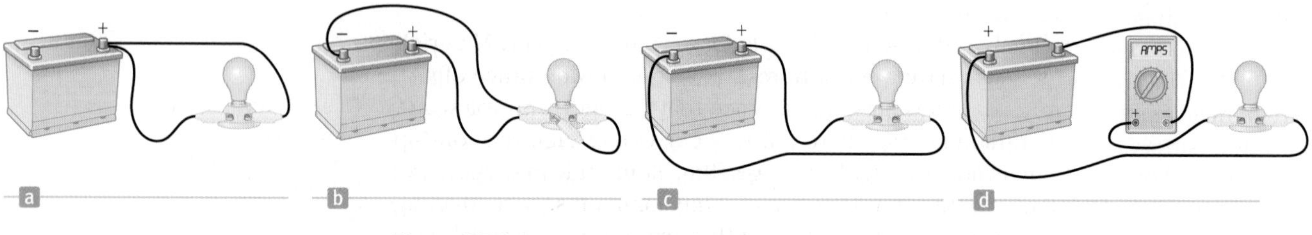

Figure 17.6 (Quick Quiz 17.3)

17.4 Resistance, Resistivity, and Ohm's Law

LEARNING OBJECTIVES

1. Define electrical resistance and discuss its physical origins.
2. Relate resistance to resistivity.
3. Apply the concepts of resistivity and resistance to electrical systems.

Resistance and Ohm's Law

When a voltage (potential difference) ΔV is applied across the ends of a metallic conductor as in Figure 17.7, the current in the conductor is found to be proportional to the applied voltage; $I \propto \Delta V$. If the proportionality holds, we can write $\Delta V = IR$, where the proportionality constant R is called the *resistance* of the conductor. In fact, we define the **resistance** as the ratio of the voltage across the conductor to the current it carries:

$$R \equiv \frac{\Delta V}{I} \qquad [17.3]$$

◄ Resistance

Resistance has SI units of volts per ampere, called **ohms** (Ω). If a potential difference of 1 V across a conductor produces a current of 1 A, the resistance of the conductor is 1 Ω. For example, if an electrical appliance connected to a 120-V source carries a current of 6 A, its resistance is 20 Ω.

The concepts of electric current, voltage, and resistance can be compared to the flow of water in a river. As water flows downhill in a river of constant width and depth, the flow rate (water current) depends on the steepness of descent of the river and the effects of rocks, the riverbank, and other obstructions. The voltage difference is analogous to the steepness, and the resistance to the obstructions. Based on this analogy, it seems reasonable that increasing the voltage applied to a circuit should increase the current in the circuit, just as increasing the steepness of descent increases the water current. Also, increasing the obstructions in the river's path will reduce the water current, just as increasing the resistance in a circuit will lower the electric current. Resistance in a circuit arises due to collisions between the electrons carrying the current with fixed atoms inside the conductor. These collisions inhibit the movement of charges in much the same way as would a force of friction. For many materials, including most metals, experiments show that **the resistance remains constant over a wide range of applied voltages or currents**. This statement is known as **Ohm's law**, after Georg Simon Ohm (1789–1854), who was the first to conduct a systematic study of electrical resistance.

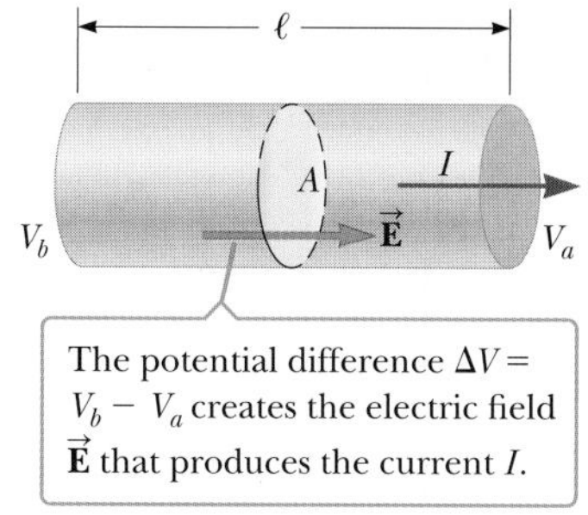

Figure 17.7 A uniform conductor of length ℓ and cross-sectional area A. The current I is proportional to the potential difference or, equivalently, to the electric field and length.

Ohm's law is given by

$$\Delta V = IR \qquad [17.4]$$

where R is understood to be independent of ΔV, the potential drop across the resistor, and I, the current in the resistor. We will continue to use this traditional form of Ohm's law when discussing electrical circuits. A **resistor** is a conductor that provides a specified resistance in an electric circuit. The symbol for a resistor in circuit diagrams is a zigzag line: —\/\/\/—.

Ohm's law is an empirical relationship valid only for certain materials. Materials that obey Ohm's law, and hence have a constant resistance over a wide range of voltages, are said to be **ohmic**. Materials having resistance that changes with voltage or current are **nonohmic**. Ohmic materials have a linear current–voltage relationship over a large range of applied voltages (Fig. 17.8a page 608). Nonohmic materials have a nonlinear current–voltage relationship (Fig. 17.8b). One common semiconducting device that is nonohmic is the *diode*, a circuit element that acts like a one-way valve for current. Its resistance is small for currents in one direction (positive ΔV) and large for currents in the reverse direction (negative ΔV). Most modern electronic devices, such as transistors, have nonlinear current–voltage relationships; their operation depends on the particular ways in which they violate Ohm's law.

© Bettmann/CORBIS

Georg Simon Ohm
(1787–1854)
A high school teacher in Cologne and later a professor at Munich, Ohm formulated the concept of resistance and discovered the proportionalities expressed in Equation 17.5.

■ Quick Quiz

17.4 In Figure 17.8b does the resistance of the diode (a) increase or (b) decrease as the positive voltage ΔV increases?

17.5 All electric devices are required to have identifying plates that specify their electrical characteristics. The plate on a certain steam iron states that the iron carries a current of 6.00 A when connected to a source of 1.20×10^2 V. What is the resistance of the steam iron? (a) 0.050 0 Ω (b) 20.0 Ω (c) 36.0 Ω

Courtesy of Henry Leap and Jim Lehman

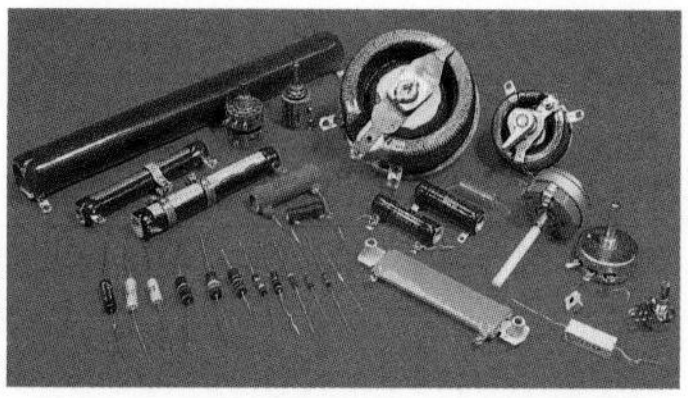

An assortment of resistors used for a variety of applications in electronic circuits.

Resistivity

Electrons don't move in straight-line paths through a conductor. Instead, they undergo repeated collisions with the metal atoms. Consider a conductor with a voltage applied across its ends. An electron gains speed as the electric force associated with the internal electric field accelerates it, giving it a velocity in the direction

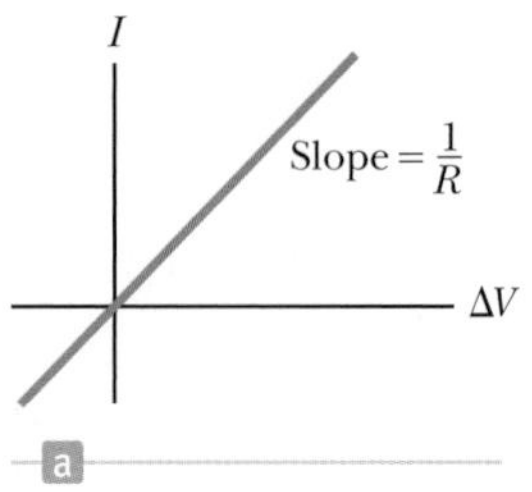

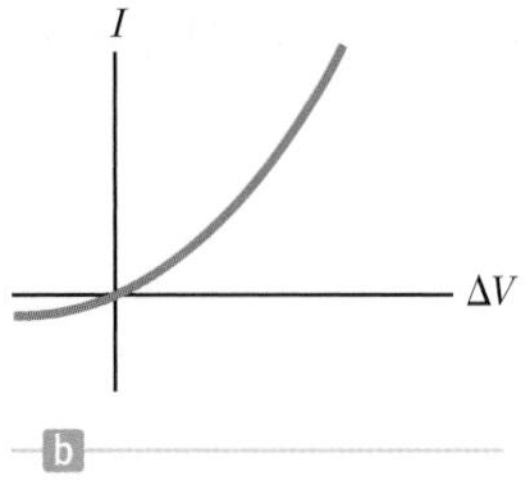

Figure 17.8 (a) The current–voltage curve for an ohmic material. The curve is linear, and the slope gives the resistance of the conductor. (b) A nonlinear current–voltage curve for a semiconducting diode. This device doesn't obey Ohm's law.

opposite that of the electric field. A collision with an atom randomizes the electron's velocity, reducing it in the direction opposite the field. The process then repeats itself. Together, these collisions affect the electron somewhat as a force of internal friction would. This step is the origin of a material's resistance.

The resistance of an ohmic conductor increases with length, which makes sense because the electrons going through it must undergo more collisions in a longer conductor. A smaller cross-sectional area also increases the resistance of a conductor, just as a smaller pipe slows the fluid moving through it. The resistance, then, is proportional to the conductor's length ℓ and inversely proportional to its cross-sectional area A,

$$R = \rho \frac{\ell}{A} \qquad [17.5]$$

where the constant of proportionality, ρ, is called the **resistivity** of the material. Every material has a characteristic resistivity that depends on its electronic structure and on temperature. Good electric conductors have very low resistivities, and good insulators have very high resistivities. Table 17.1 lists the resistivities of various materials at 20°C. Because resistance values are in ohms, resistivity values must be in ohm-meters ($\Omega \cdot$ m).

Table 17.1 Resistivities and Temperature Coefficients of Resistivity for Various Materials (at 20°C)

Material	Resistivity ($\Omega \cdot$ m)	Temperature Coefficient of Resistivity [(°C)$^{-1}$]
Silver	1.59×10^{-8}	3.8×10^{-3}
Copper	1.7×10^{-8}	3.9×10^{-3}
Gold	2.44×10^{-8}	3.4×10^{-3}
Aluminum	2.82×10^{-8}	3.9×10^{-3}
Tungsten	5.6×10^{-8}	4.5×10^{-3}
Iron	10.0×10^{-8}	5.0×10^{-3}
Platinum	11×10^{-8}	3.92×10^{-3}
Lead	22×10^{-8}	3.9×10^{-3}
Nichrome[a]	150×10^{-8}	0.4×10^{-3}
Carbon	3.5×10^{-5}	-0.5×10^{-3}
Germanium	0.46	-48×10^{-3}
Silicon	640	-75×10^{-3}
Glass	10^{10}–10^{14}	
Hard rubber	$\approx 10^{13}$	
Sulfur	10^{15}	
Quartz (fused)	75×10^{16}	

[a]A nickel-chromium alloy commonly used in heating elements.

APPLYING PHYSICS 17.1 | Dimming of Aging Lightbulbs

As a lightbulb ages, why does it gives off less light than when new?

EXPLANATION There are two reasons for the lightbulb's behavior, one electrical and one optical, but both are related to the same phenomenon occurring within the bulb. The filament of an old lightbulb is made of a tungsten wire that has been kept at a high temperature for many hours. High temperatures evaporate tungsten from the filament, decreasing its radius. From $R = \rho\ell/A$, we see that a decreased cross-sectional area leads to an increase in the resistance of the filament. This increasing resistance with age means that the filament will carry less current for the same applied voltage. With less current in the filament, there is less light output, and the filament glows more dimly.

At the high operating temperature of the filament, tungsten atoms leave its surface, much as water molecules evaporate from a puddle of water. The atoms are carried away by convection currents in the gas in the bulb and are deposited on the inner surface of the glass. In time, the glass becomes less transparent because of the tungsten coating, which decreases the amount of light that passes through the glass. ■

EXAMPLE 17.3 The Resistance of Nichrome Wire

GOAL Combine the concept of resistivity with Ohm's law.

PROBLEM **(a)** Calculate the resistance per unit length of a 22-gauge Nichrome wire of radius 0.321 mm. **(b)** If a potential difference of 10.0 V is maintained across a 1.00-m length of the Nichrome wire, what is the current in the wire? **(c)** The wire is melted down and recast with twice its original length. Find the new resistance R_N as a multiple of the old resistance R_O.

STRATEGY Part (a) requires substitution into Equation 17.5, after calculating the cross-sectional area, whereas part (b) is a matter of substitution into Ohm's law. Part (c) requires some algebra. The idea is to take the expression for the new resistance and substitute expressions for ℓ_N and A_N, the new length and cross-sectional area, in terms of the old length and cross section. For the area substitution, remember that the volumes of the old and new wires are the same.

SOLUTION

(a) Calculate the resistance per unit length.

Find the cross-sectional area of the wire:

$$A = \pi r^2 = \pi(0.321 \times 10^{-3}\ \text{m})^2 = 3.24 \times 10^{-7}\ \text{m}^2$$

Obtain the resistivity of Nichrome from Table 17.1, solve Equation 17.5 for R/ℓ, and substitute:

$$\frac{R}{\ell} = \frac{\rho}{A} = \frac{1.5 \times 10^{-6}\ \Omega \cdot \text{m}}{3.24 \times 10^{-7}\ \text{m}^2} = \boxed{4.6\ \Omega/\text{m}}$$

(b) Find the current in a 1.00-m segment of the wire if the potential difference across it is 10.0 V.

Substitute given values into Ohm's law:

$$I = \frac{\Delta V}{R} = \frac{10.0\ \text{V}}{4.6\ \Omega} = \boxed{2.2\ \text{A}}$$

(c) If the wire is melted down and recast with twice its original length, find the new resistance as a multiple of the old.

Find the new area A_N in terms of the old area A_O, using the fact the volume doesn't change and $\ell_N = 2\ell_O$:

$$V_N = V_O \;\rightarrow\; A_N\ell_N = A_O\ell_O \;\rightarrow\; A_N = A_O(\ell_O/\ell_N)$$

$$A_N = A_O(\ell_O/2\ell_O) = A_O/2$$

Substitute into Equation 17.5:

$$R_N = \frac{\rho\ell_N}{A_N} = \frac{\rho(2\ell_O)}{(A_O/2)} = 4\frac{\rho\ell_O}{A_O} = \boxed{4R_O}$$

REMARKS From Table 17.1, the resistivity of Nichrome is about 100 times that of copper, a typical good conductor. Therefore, a copper wire of the same radius would have a resistance per unit length of only 0.052 Ω/m, and a 1.00-m length of copper wire of the same radius would carry the same current (2.2 A) with an applied voltage of only 0.115 V.

Because of its resistance to oxidation, Nichrome is often used for heating elements in toasters, irons, and electric heaters.

QUESTION 17.3 Would replacing the Nichrome with copper result in a higher current or lower current?

EXERCISE 17.3 What is the resistance of a 6.0-m length of Nichrome wire that has a radius 0.321 mm? How much current does it carry when connected to a 120-V source?

ANSWERS 28 Ω; 4.3 A

Quick Quiz

17.6 Suppose an electrical wire is replaced with one having every linear dimension doubled (i.e., the length and radius have twice their original values). Does the wire now have (a) more resistance than before, (b) less resistance, or (c) the same resistance?

17.5 Temperature Variation of Resistance

In an old-fashioned carbon filament incandescent lamp, the electrical resistance is typically 10 Ω, but changes with temperature.

LEARNING OBJECTIVES

1. Explain the physical reasons for the increase in resistivity and resistance with increasing temperature.
2. State the dependence of resistivity on temperature for limited temperature ranges, and extend it to the temperature dependence of resistance.
3. Apply the temperature dependence of resistance to electrical systems.

The resistivity ρ, and hence the resistance, of a conductor depends on a number of factors. One of the most important is the temperature of the metal. For most metals, resistivity increases with increasing temperature. This correlation can be understood as follows: as the temperature of the material increases, its constituent atoms vibrate with greater amplitudes. As a result, the electrons find it more difficult to get by those atoms, just as it is more difficult to weave through a crowded room when the people are in motion than when they are standing still. The increased electron scattering with increasing temperature results in increased resistivity. Technically, thermal expansion also affects resistance; however, this is a very small effect.

Over a limited temperature range, the resistivity of most metals increases linearly with increasing temperature according to the expression

$$\rho = \rho_0[1 + \alpha(T - T_0)] \tag{17.6}$$

where ρ is the resistivity at some temperature T (in Celsius degrees), ρ_0 is the resistivity at some reference temperature T_0 (usually taken to be 20°C), and α is a parameter called the **temperature coefficient of resistivity**. Temperature coefficients for various materials are provided in Table 17.1. The interesting negative values of α for semiconductors arise because these materials possess weakly bound charge carriers that become free to move and contribute to the current as the temperature rises.

Because the resistance of a conductor with a uniform cross section is proportional to the resistivity according to Equation 17.5 ($R = \rho l/A$), the temperature variation of resistance can be written

$$R = R_0[1 + \alpha(T - T_0)] \tag{17.7}$$

Precise temperature measurements are often made using this property, as shown by the following example.

■ EXAMPLE 17.4 | A Platinum Resistance Thermometer

GOAL Apply the temperature dependence of resistance.

PROBLEM A resistance thermometer, which measures temperature by measuring the change in resistance of a conductor, is made of platinum and has a resistance of 50.0 Ω at 20.0°C. **(a)** When the device is immersed in a vessel containing indium at its melting point, its resistance increases to 76.8 Ω. From this information, find the melting point of indium. **(b)** The indium is heated further until it reaches a temperature of 235°C. Estimate the ratio of the new current in the platinum to the current I_{mp} at the melting point, assuming the coefficient of resistivity for platinum doesn't change significantly with temperature.

STRATEGY For part (a), solve Equation 17.7 for $T - T_0$ and get α for platinum from Table 17.1, substituting known quantities. For part (b), use Ohm's law in Equation 17.7.

QUESTION 17.6 If the heater wire is replaced by a wire with lower resistance, is the time required to heat the gas (a) unchanged, (b) increased, or (c) decreased?

EXERCISE 17.6 A hot-water heater is rated at 4.50×10^3 W and operates at 2.40×10^2 V. (a) Find the resistance in the heating element and the current. (b) How long does it take to heat 125 L of water from 20.0°C to 50.0°C, neglecting conduction and other losses? (c) How much does it cost at \$0.12/kWh?

ANSWERS (a) 12.8 Ω, 18.8 A (b) 3.49×10^3 s (c) \$0.52

17.7 Superconductors

LEARNING OBJECTIVE

1. Discuss superconductivity and describe important applications.

There is a class of metals and compounds with resistances that fall virtually to *zero* below a certain temperature T_c called the *critical temperature.* These materials are known as **superconductors**. The resistance vs. temperature graph for a superconductor follows that of a normal metal at temperatures above T_c (Fig. 17.11). When the temperature is at or below T_c, however, the resistance suddenly drops to zero. This phenomenon was discovered in 1911 by Dutch physicist H. Kamerlingh Onnes as he and a graduate student worked with mercury, which is a superconductor below 4.1 K. Recent measurements have shown that the resistivities of superconductors below T_c are less than 4×10^{-25} Ω · m, around 10^{17} times smaller than the resistivity of copper and in practice considered to be zero.

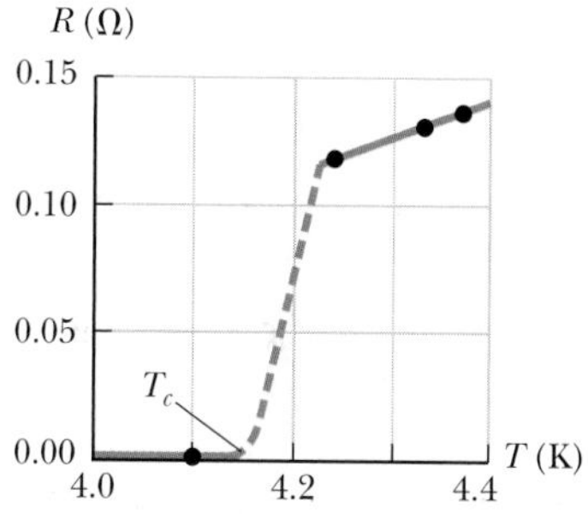

Figure 17.11 Resistance versus temperature for a sample of mercury (Hg). The graph follows that of a normal metal above the critical temperature T_c. The resistance drops to zero at the critical temperature, which is 4.2 K for mercury, and remains at zero for lower temperatures.

Today, thousands of superconductors are known, including such common metals as aluminum, tin, lead, zinc, and indium. Table 17.2 lists the critical temperatures of several superconductors. The value of T_c is sensitive to chemical composition, pressure, and crystalline structure. Interestingly, copper, silver, and gold, which are excellent conductors, don't exhibit superconductivity.

Table 17.2 Critical Temperatures for Various Superconductors

Material	T_c(K)
Zn	0.88
Al	1.19
Sn	3.72
Hg	4.15
Pb	7.18
Nb	9.46
Nb_3Sn	18.05
Nb_3Ge	23.2
$YBa_2Cu_3O_7$	92
Bi–Sr–Ca–Cu–O	105
Tl–Ba–Ca–Cu–O	125
$HgBa_2Ca_2Cu_3O_8$	134

A truly remarkable feature of superconductors is that once a current is set up in them, it persists *without any applied voltage* (because $R = 0$). In fact, steady currents in superconducting loops have been observed to persist for years with no apparent decay!

An important development in physics that created much excitement in the scientific community was the discovery of high-temperature copper-oxide-based superconductors. The excitement began with a 1986 publication by J. Georg Bednorz and K. Alex Müller, scientists at the IBM Zurich Research Laboratory in Switzerland, in which they reported evidence for superconductivity at a temperature near 30 K in an oxide of barium, lanthanum, and copper. Bednorz and Müller were awarded the Nobel Prize in Physics in 1987 for their important discovery. The discovery was remarkable because the critical temperature was significantly higher than that of any previously known superconductor. Shortly thereafter a new family of compounds was investigated, and research activity in the field of superconductivity proceeded vigorously. In early 1987 groups at the University of Alabama at Huntsville and the University of Houston announced the discovery of superconductivity at about 92 K in an oxide of yttrium, barium, and copper ($YBa_2Cu_3O_7$). Late in 1987, teams of scientists from Japan and the United States reported superconductivity at 105 K in an oxide of bismuth, strontium, calcium, and copper. More recently, scientists have reported superconductivity at temperatures as high as 150 K in an oxide containing mercury. The search for novel superconducting materials continues, with the hope of someday obtaining a room-temperature superconducting material. This research is important both for scientific reasons and for practical applications. A superconducting ceramic levitates a permanent magnet in Figure 17.12.

An important and useful application is the construction of superconducting magnets in which the magnetic field intensities are about ten times greater than those of the best normal electromagnets. Such magnets are being considered as a means of storing energy. The idea of using superconducting power lines to transmit power efficiently is also receiving serious consideration. Modern superconducting electronic devices consisting of two thin-film superconductors separated by a thin insulator have been constructed. Among these devices are magnetometers (magnetic-field measuring devices) and various microwave devices.

Yoav Levy/PhotoTake

Figure 17.12 A small permanent magnet floats freely above a nitrogen-cooled ceramic superconductor. The superconductor has zero resistance and expels any magnetic field from its interior by creating a mirror image of the magnetic poles of the permanent magnet. This "Meissner effect" results in magnetic levitation of the permanent magnet.

17.8 Electrical Activity in the Heart BIO

LEARNING OBJECTIVES

1. Discuss the critical role of electrical activity in the human body.
2. Discuss medical instruments used to monitor and regulate the body's electrical activity.

Electrocardiograms

APPLICATION
Electrocardiograms

Every action involving the body's muscles is initiated by electrical activity. The voltages produced by muscular action in the heart are particularly important to physicians. Voltage pulses cause the heart to beat, and the waves of electrical excitation that sweep across the heart associated with the heartbeat are conducted through the body via the body fluids. These voltage pulses are large enough to be detected by suitable monitoring equipment attached to the skin. A sensitive voltmeter making good electrical contact with the skin by means of contacts attached with conducting paste can be used to measure heart pulses, which are typically of the order of 1 mV at the surface of the body. The voltage pulses can be recorded on an instrument called an **electrocardiograph**, and the pattern recorded by this instrument is called an **electrocardiogram** (EKG). To understand the information contained in an EKG pattern, it is necessary first to describe the underlying principles concerning electrical activity in the heart.

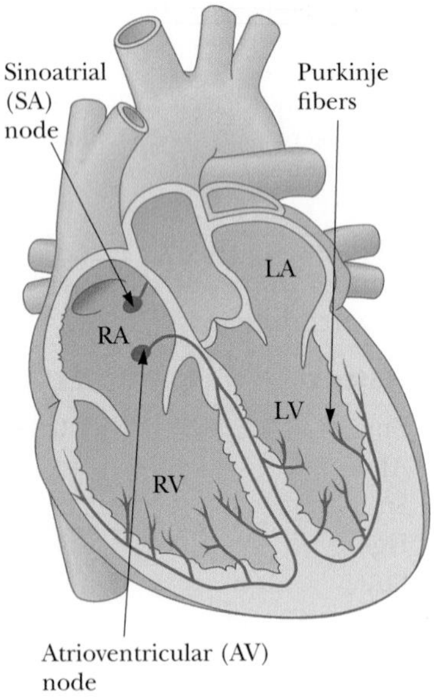

Figure 17.13 The electrical conduction system of the human heart. (RA: right atrium; LA: left atrium; RV: right ventricle; LV: left ventricle.)

The right atrium of the heart contains a specialized set of muscle fibers called the SA (sinoatrial) node that initiates the heartbeat (Fig. 17.13). Electric impulses that originate in these fibers gradually spread from cell to cell throughout the right and left atrial muscles, causing them to contract. The pulse that passes through the muscle cells is often called a *depolarization wave* because of its effect on individual cells. If an individual muscle cell were examined in its resting state, a double-layer electric charge distribution would be found on its surface, as shown in Figure 17.14a. The impulse generated by the SA node momentarily and locally allows positive charge on the outside of the

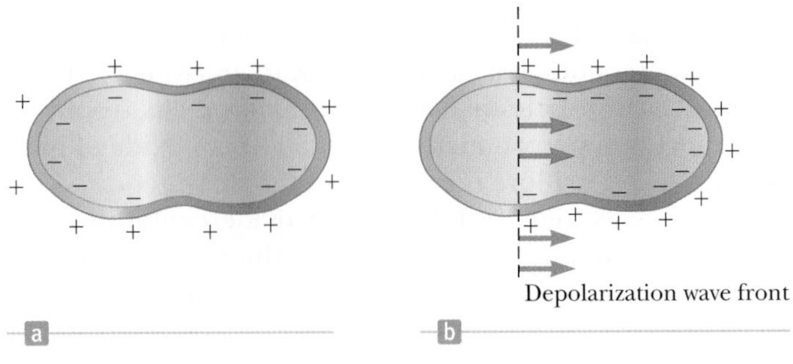

Figure 17.14 (a) Charge distribution of a muscle cell in the atrium before a depolarization wave has passed through the cell. (b) Charge distribution as the wave passes.

cell to flow in and neutralize the negative charge on the inside layer. This effect changes the cell's charge distribution to that shown in Figure 17.14b. Once the depolarization wave has passed through an individual heart muscle cell, the cell recovers the resting-state charge distribution (positive out, negative in) shown in Figure 17.14a in about 250 ms. When the impulse reaches the atrioventricular (AV) node (Fig. 17.13), the muscles of the atria begin to relax, and the pulse is directed to the ventricular muscles by the AV node. The muscles of the ventricles contract as the depolarization wave spreads through the ventricles along a group of fibers called the *Purkinje fibers*. The ventricles then relax after the pulse has passed through. At this point, the SA node is triggered again and the cycle is repeated.

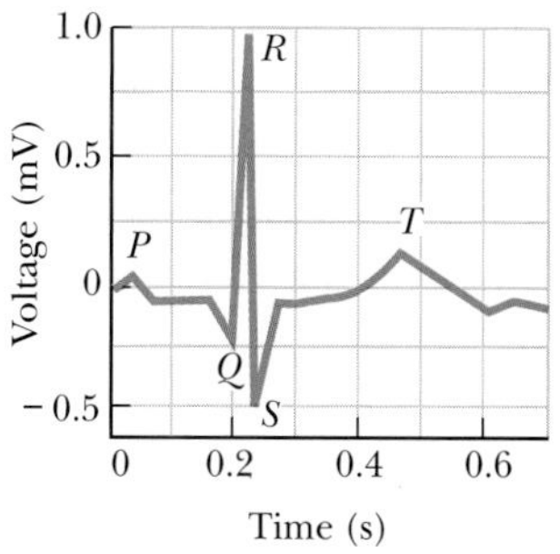

Figure 17.15 An EKG response for a normal heart.

A sketch of the electrical activity registered on an EKG for one beat of a normal heart is shown in Figure 17.15. The pulse indicated by *P* occurs just before the atria begin to contract. The *QRS* pulse occurs in the ventricles just before they contract, and the *T* pulse occurs when the cells in the ventricles begin to recover. EKGs for an abnormal heart are shown in Figure 17.16. The *QRS* portion of the pattern shown in Figure 17.16a is wider than normal, indicating that the patient may have an enlarged heart. (Why?) Figure 17.16b indicates that there is no constant relationship between the *P* pulse and the *QRS* pulse. This suggests a blockage in the electrical conduction path between the SA and AV nodes that results in the atria and ventricles beating independently and inefficient heart pumping. Finally, Figure 17.16c shows a situation in which there is no *P* pulse and an irregular spacing between the *QRS* pulses. This is symptomatic of irregular atrial contraction, which is called *fibrillation*. In this condition, the atrial and ventricular contractions are irregular.

BIO APPLICATION
Cardiac Pacemakers

As noted previously, the sinoatrial node directs the heart to beat at the appropriate rate, usually about 72 beats per minute. Disease or the aging process, however, can damage the heart and slow its beating, and a medical assist may be necessary in the form of a *cardiac pacemaker* attached to the heart. This matchbox-sized electrical device implanted under the skin has a lead that is connected to the wall of the right ventricle. Pulses from this lead stimulate the heart to maintain its proper rhythm. In general, a pacemaker is designed to produce pulses at a rate of about 60 per minute, slightly slower than the normal number of beats per minute, but sufficient to maintain life. The circuitry

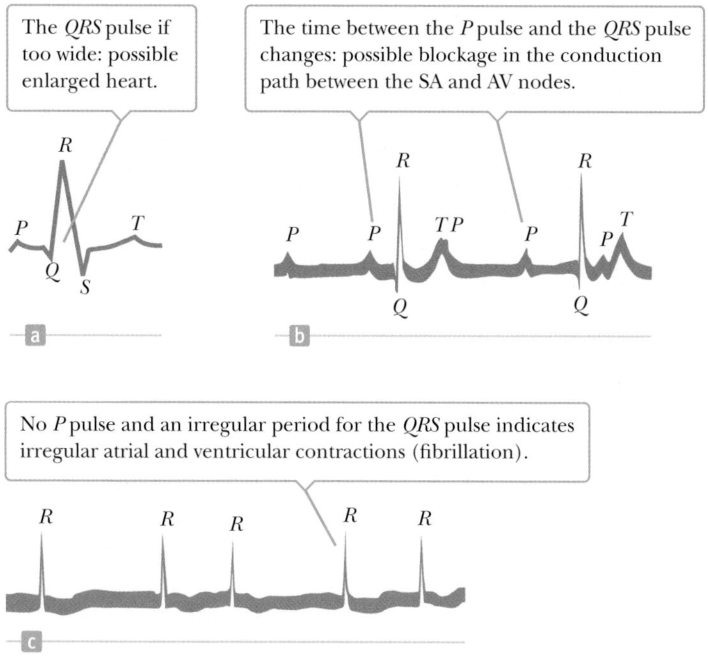

Figure 17.16 Abnormal EKGs.

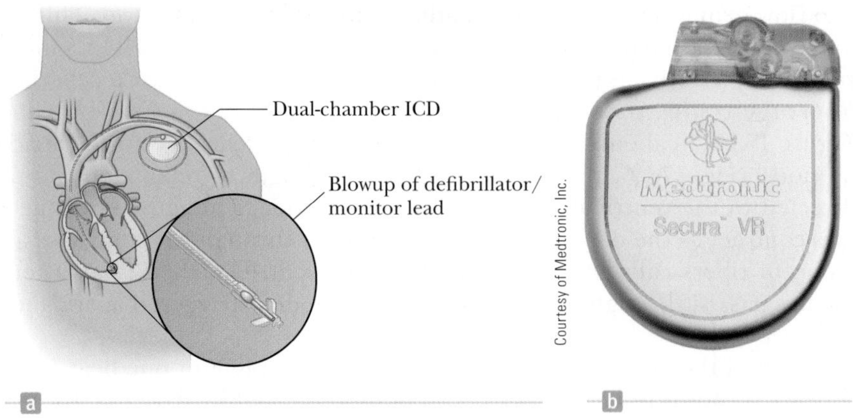

Figure 17.17 (a) A dual-chamber ICD with leads in the heart. One lead monitors and stimulates the right atrium, and the other monitors and stimulates the right ventricle. (b) Medtronic Dual Chamber ICD.

consists of a capacitor charging up to a certain voltage from a lithium battery and then discharging. The design of the circuit is such that if the heart is beating normally, the capacitor is not allowed to charge completely and send pulses to the heart.

An Emergency Room in Your Chest

BIO APPLICATION
Implanted Cardioverter Defibrillators

In June 2001 an operation on then–Vice President Dick Cheney focused attention on the progress in treating heart problems with tiny implanted electrical devices. Aptly called "an emergency room in your chest" by Cheney's attending physician, devices called *I*mplanted *C*ardioverter *D*efibrillators (**ICDs**) can monitor, record, and logically process heart signals and then supply different corrective signals to hearts beating too slowly, too rapidly, or irregularly. ICDs can even monitor and send signals to the atria and ventricles independently! Figure 17.17a shows a sketch of an ICD with conducting leads that are implanted in the heart. Figure 17.17b shows an actual titanium-encapsulated dual-chamber ICD.

The latest ICDs are sophisticated devices capable of a number of functions:

1. monitoring both atrial and ventricular chambers to differentiate between atrial and potentially fatal ventricular arrhythmias, which require prompt regulation
2. storing about a half hour of heart signals that can easily be read out by a physician
3. being easily reprogrammed with an external magnetic wand
4. performing complicated signal analysis and comparison
5. supplying 0.25- to 10-V repetitive pacing signals to speed up or slow down a malfunctioning heart, or a high-voltage pulse of about 800 V to halt the potentially fatal condition of ventricular fibrillation, in which the heart quivers rapidly rather than beats (people who have experienced such a high-voltage jolt say that it feels like a kick or a bomb going off in the chest)
6. automatically adjusting the number of pacing pulses per minute to match the patient's activity

ICDs are powered by lithium batteries and have implanted lifetimes of 4 to 6 years. Some basic properties of these adjustable ICDs are given in Table 17.3. In the table *tachycardia* means "rapid heartbeat" and *bradycardia* means "slow heartbeat." A key factor in developing tiny electrical implants that serve as defibrillators is the development of capacitors with relatively large capacitance (125 μF) and small physical size.

59. A 120-V motor has mechanical power output of 2.50 hp. It is 90.0% efficient in converting power that it takes in by electrical transmission into mechanical power. (a) Find the current in the motor. (b) Find the energy delivered to the motor by electrical transmission in 3.00 h of operation. (c) If the electric company charges \$0.110/kWh, what does it cost to run the motor for 3.00 h?

60. Q|C A Nichrome heating element in an oven has a resistance of 8.0 Ω at 20°C. (a) What is its resistance at 350°C? (b) What assumption did you have to make to obtain your answer to part (a)?

61. A length of metal wire has a radius of 5.00×10^{-3} m and a resistance of 0.100 Ω. When the potential difference across the wire is 15.0 V, the electron drift speed is found to be 3.17×10^{-4} m/s. On the basis of these data, calculate the density of free electrons in the wire.

62. In a certain stereo system, each speaker has a resistance of 4.00 Ω. The system is rated at 60.0 W in each channel. Each speaker circuit includes a fuse rated at a maximum current of 4.00 A. Is this system adequately protected against overload?

63. A resistor is constructed by forming a material of resistivity $3.5 \times 10^5\ \Omega \cdot \text{m}$ into the shape of a hollow cylinder of length 4.0 cm and inner and outer radii 0.50 cm and 1.2 cm, respectively. In use, a potential difference is applied between the ends of the cylinder, producing a current parallel to the length of the cylinder. Find the resistance of the cylinder.

64. S When a straight wire is heated, its resistance changes according to the equation

$$R = R_0[1 + \alpha(T - T_0)]$$

(Eq. 17.7), where α is the temperature coefficient of resistivity. (a) Show that a more precise result, which includes the length and area of a wire change when it is heated, is

$$R = \frac{R_0[1 + \alpha(T - T_0)][1 + \alpha'(T - T_0)]}{[1 + 2\alpha'(T - T_{0)})]}$$

where α' is the coefficient of linear expansion. (See Chapter 10.) (b) Compare the two results for a 2.00-m-long copper wire of radius 0.100 mm, starting at 20.0°C and heated to 100.0°C.

65. BIO An x-ray tube used for cancer therapy operates at 4.0 MV, with a beam current of 25 mA striking the metal target. Nearly all the power in the beam is transferred to a stream of water flowing through holes drilled in the target. What rate of flow, in kilograms per second, is needed if the rise in temperature (ΔT) of the water is not to exceed 50°C?

66. A man wishes to vacuum his car with a canister vacuum cleaner marked 535 W at 120 V. The car is parked far from the building, so he uses an extension cord 15.0 m long to plug the cleaner into a 120-V source. Assume the cleaner has constant resistance. (a) If the resistance of each of the two conductors of the extension cord is 0.900 Ω, what is the actual power delivered to the cleaner? (b) If, instead, the power is to be at least 525 W, what must be the diameter of each of two identical copper conductors in the cord the young man buys? (c) Repeat part (b) if the power is to be at least 532 W. *Suggestion:* A symbolic solution can simplify the calculations.

This vision-impaired patient is wearing an artificial vision device. A camera on the lens sends video to a portable computer. The computer then converts the video to signals that stimulate implants in the patient's visual cortex, giving him enough sight to move around on his own.

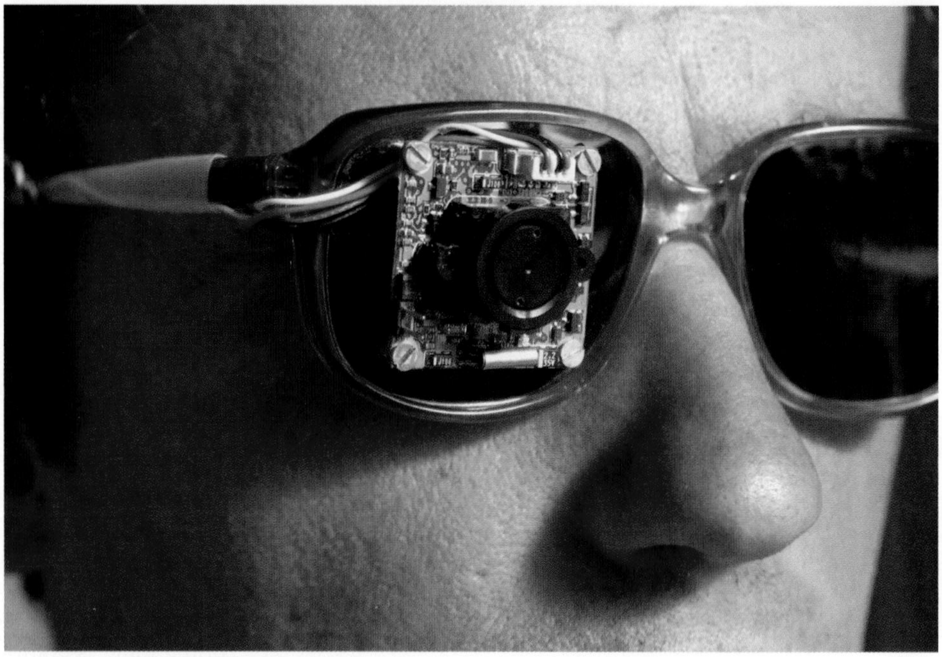

18 Direct-Current Circuits

Batteries, resistors, and capacitors can be used in various combinations to construct electric circuits, which direct and control the flow of electricity and the energy it conveys. Such circuits make possible all the modern conveniences in a home: electric lights, electric stove tops and ovens, washing machines, and a host of other appliances and tools. Electric circuits are also found in our cars, in tractors that increase farming productivity, and in all types of medical equipment that saves so many lives every day.

In this chapter we study and analyze a number of simple direct-current circuits. The analysis is simplified by the use of two rules known as Kirchhoff's rules, which follow from the principle of conservation of energy and the law of conservation of charge. Most of the circuits are assumed to be in *steady state,* which means that the currents are constant in magnitude and direction. We close the chapter with a discussion of circuits containing resistors and capacitors, in which current varies with time.

18.1 Sources of emf

LEARNING OBJECTIVES

1. Describe in physical terms the concepts of emf, terminal voltage, internal resistance, and load resistance.
2. Evaluate the current and total power output of an emf source.

A current is maintained in a closed circuit by a source of emf.[1] Among such sources are any devices (for example, batteries and generators) that increase the potential

[1]The term was originally an abbreviation for *electromotive force,* but emf is not really a force, so the long form is discouraged.

energy of the circulating charges. A source of emf can be thought of as a "charge pump" that forces electrons to move in a direction opposite the electrostatic field inside the source. The emf $\mathcal{E}$ of a source is the work done per unit charge; hence the SI unit of emf is the volt.

Consider the circuit in Figure 18.1a consisting of a battery connected to a resistor. We assume the connecting wires have no resistance. If we neglect the internal resistance of the battery, the potential drop across the battery (the terminal voltage) equals the emf of the battery. Because a real battery always has some internal resistance r, however, the terminal voltage is not equal to the emf. The circuit of Figure 18.1a can be described schematically by the diagram in Figure 18.1b. The battery, represented by the dashed rectangle, consists of a source of emf $\mathcal{E}$ in series with an internal resistance r. Now imagine a positive charge moving through the battery from a to b in the figure. As the charge passes from the negative to the positive terminal of the battery, the potential of the charge increases by $\mathcal{E}$. As the charge moves through the resistance r, however, its potential decreases by the amount Ir, where I is the current in the circuit. The terminal voltage of the battery, $\Delta V = V_b - V_a$, is therefore given by

$$\Delta V = \mathcal{E} - Ir \quad [18.1]$$

From this expression, we see that **$\mathcal{E}$ is equal to the terminal voltage when the current is zero**, called the **open-circuit voltage**. By inspecting Figure 18.1b, we find that the terminal voltage ΔV must also equal the potential difference across the external resistance R, often called the **load resistance**; that is, $\Delta V = IR$. Combining this relationship with Equation 18.1, we arrive at

$$\mathcal{E} = IR + Ir \quad [18.2]$$

Solving for the current gives

$$I = \frac{\mathcal{E}}{R + r}$$

The preceding equation shows that the current in this simple circuit depends on both the resistance external to the battery and the internal resistance of the battery. If R is much greater than r, we can neglect r in our analysis (an option we usually select).

If we multiply Equation 18.2 by the current I, we get

$$I\mathcal{E} = I^2R + I^2r$$

This equation tells us that the total power output $I\mathcal{E}$ of the source of emf is converted at the rate I^2R at which energy is delivered to the load resistance, *plus* the rate I^2r at which energy is delivered to the internal resistance. Again, if $r << R$, most of the power delivered by the battery is transferred to the load resistance.

Unless otherwise stated, in our examples and end-of-chapter problems we will assume the internal resistance of a battery in a circuit is negligible.

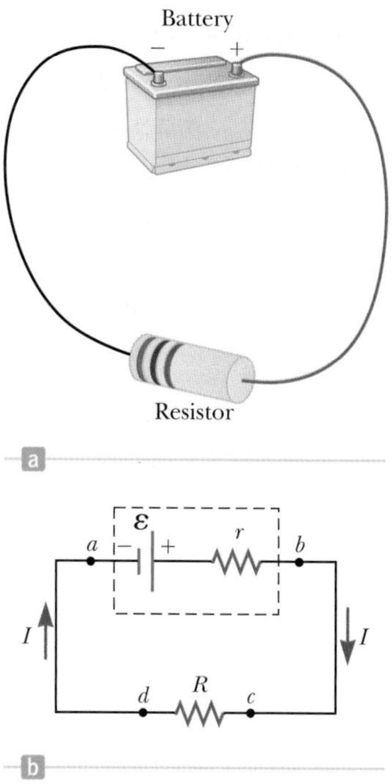

Figure 18.1 (a) A circuit consisting of a resistor connected to the terminals of a battery. (b) A circuit diagram of a source of emf $\mathcal{E}$ having internal resistance r connected to an external resistor R.

© Cengage Learning/George Semple

An assortment of batteries.

Quick Quiz

18.1 True or False: While discharging, the terminal voltage of a battery can never be greater than the emf of the battery.

18.2 Why does a battery get warm while in use?

Tip 18.1 What's Constant in a Battery?

Equation 18.2 shows that the current in a circuit depends on the resistance of the battery, so a battery can't be considered a source of constant current. Even the terminal voltage of a battery given by Equation 18.1 can't be considered constant because the internal resistance can change (due to warming, for example, during the operation of the battery). A battery is, however, a source of constant emf.

18.2 Resistors in Series

LEARNING OBJECTIVES

1. Derive an expression for the equivalent resistance of a series of resistors.
2. Analyze circuits having resistors in series.

When two or more resistors are connected end to end as in Figure 18.2, they are said to be in *series*. The resistors could be simple devices, such as lightbulbs or heating elements. When two resistors R_1 and R_2 are connected to a battery as in Figure 18.2, **the current is the same in the two resistors because any charge that flows through R_1 must also flow through R_2**. This is analogous to water flowing through a pipe with two constrictions, corresponding to R_1 and R_2. Whatever volume of water flows in one end in a given time interval must exit the opposite end.

Because the potential difference between a and b in Figure 18.2b equals IR_1 and the potential difference between b and c equals IR_2, the potential difference between a and c is

$$\Delta V = IR_1 + IR_2 = I(R_1 + R_2)$$

Regardless of how many resistors we have in series, the sum of the potential differences across the resistors is equal to the total potential difference across the combination. As we will show later, this result is a consequence of the conservation of energy. Figure 18.2c shows an equivalent resistor R_{eq} that can replace the two resistors of the original circuit. The equivalent resistor has the same effect on the circuit because it results in the same current in the circuit as the two resistors. Applying Ohm's law to this equivalent resistor, we have

$$\Delta V = IR_{eq}$$

Equating the preceding two expressions, we have

$$IR_{eq} = I(R_1 + R_2)$$

or

$$R_{eq} = R_1 + R_2 \text{ (series combination)} \qquad [18.3]$$

An extension of the preceding analysis shows that the equivalent resistance of three or more resistors connected in series is

Equivalent resistance of a series combination of resistors ▶

$$R_{eq} = R_1 + R_2 + R_3 + \cdots \qquad [18.4]$$

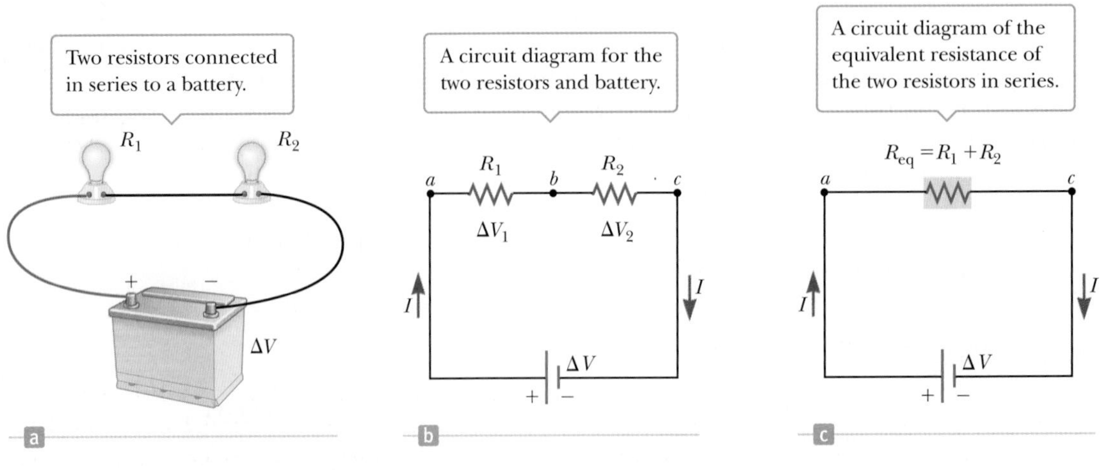

Figure 18.2 Two resistors, R_1 and R_2, in the form of incandescent light bulbs, in series with a battery. The currents in the resistors are the same, and the equivalent resistance of the combination is $R_{eq} = R_1 + R_2$.

Therefore, **the equivalent resistance of a series combination of resistors is the algebraic sum of the individual resistances and is always greater than any individual resistance.**

Note that if the filament of one lightbulb in Figure 18.2 were to fail, the circuit would no longer be complete (an open-circuit condition would exist) and the second bulb would also go out.

APPLYING PHYSICS 18.1 | Christmas Lights in Series

A new design for Christmas lights allows them to be connected in series. A failed bulb in such a string would result in an open circuit, and all the bulbs would go out. How can the bulbs be redesigned to prevent that from happening?

EXPLANATION If the string of lights contained the usual kind of bulbs, a failed bulb would be hard to locate. Each bulb would have to be replaced with a good bulb, one by one, until the failed bulb was found. If there happened to be two or more failed bulbs in the string of lights, finding them would be a lengthy and annoying task.

Christmas lights use special bulbs that have an insulated loop of wire (a jumper) across the conducting supports to the bulb filaments (Fig. 18.3). If the filament breaks and the bulb fails, the bulb's resistance increases dramatically. As a result, most of the applied voltage appears across the loop of wire. This voltage causes the insulation around the loop of wire to burn, causing the metal wire to make electrical contact with the supports. This produces a conducting path through the bulb, so the other bulbs remain lit. ■

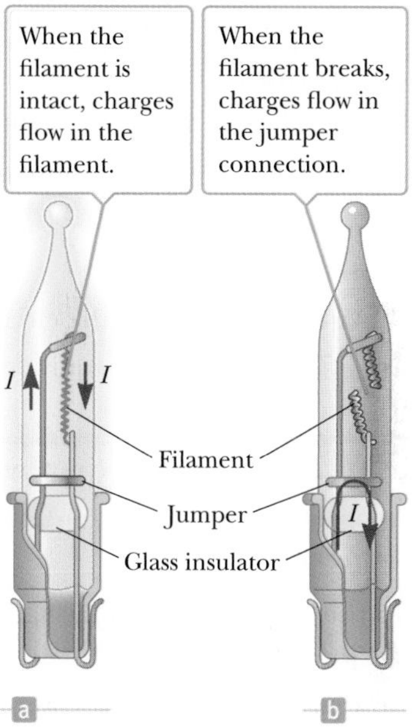

Figure 18.3 (Applying Physics 18.1) (a) Schematic diagram of a modern "miniature" holiday lightbulb, with a jumper connection to provide a current path if the filament breaks. (b) A holiday lightbulb with a broken filament.

Quick Quiz

18.3 In Figure 18.4 the current is measured with the ammeter at the bottom of the circuit. When the switch is opened, does the reading on the ammeter (a) increase, (b) decrease, or (c) not change?

18.4 The circuit in Figure 18.4 consists of two resistors, a switch, an ammeter, and a battery. When the switch is closed, power P_c is delivered to resistor R_1. When the switch is opened, which of the following statements is true about the power P_o delivered to R_1? (a) $P_o < P_c$ (b) $P_o = P_c$ (c) $P_o > P_c$

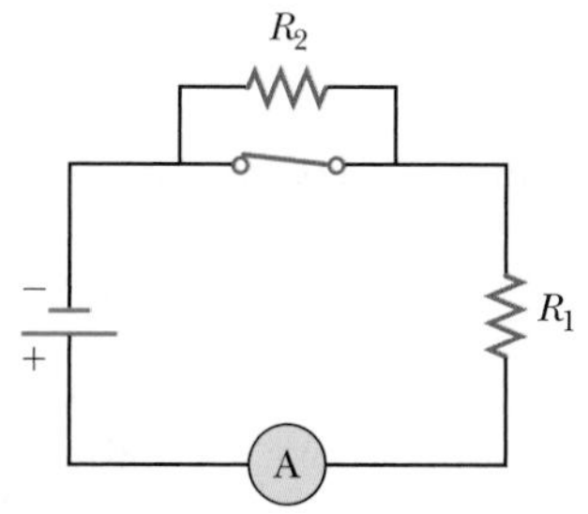

Figure 18.4 (Quick Quizzes 18.3 and 18.4)

EXAMPLE 18.1 | Four Resistors in Series

GOAL Analyze several resistors connected in series.

PROBLEM Four resistors are arranged as shown in Figure 18.5a. Find **(a)** the equivalent resistance of the circuit and **(b)** the current in the circuit if the closed-circuit terminal voltage of the battery is 6.0 V. **(c)** Calculate the electric potential at point A if the potential at the positive terminal is 6.0 V. **(d)** Suppose the open circuit voltage, or emf $\mathcal{E}$, is 6.2 V. Calculate the battery's internal resistance. **(e)** What fraction f of the battery's power is delivered to the load resistors?

(Continued)

STRATEGY Because the resistors are connected in series, summing their resistances gives the equivalent resistance. Ohm's law can then be used to find the current. To find the electric potential at point A, calculate the voltage drop ΔV across the 2.0-Ω resistor and subtract the result from 6.0 V. In part (d) use Equation 18.1 to find the internal resistance of the battery. The fraction of the power delivered to the load resistance is just the power delivered to load, $I\Delta V$, divided by the total power, $I\mathcal{E}$.

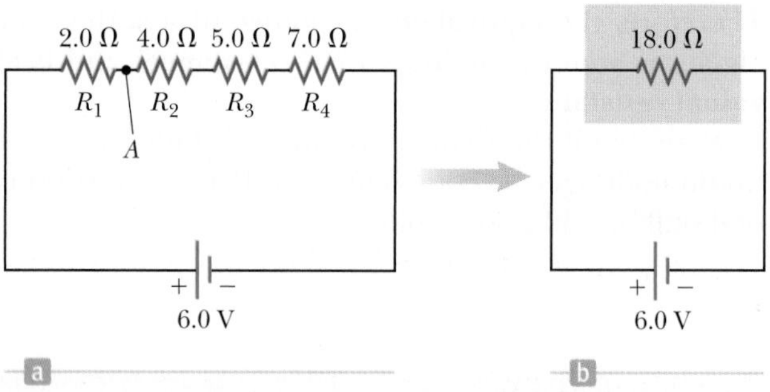

Figure 18.5 (Example 18.1) (a) Four resistors connected in series. (b) The equivalent resistance of the circuit in (a).

SOLUTION

(a) Find the equivalent resistance of the circuit.

Apply Equation 18.4, summing the resistances:

$$R_{eq} = R_1 + R_2 + R_3 + R_4 = 2.0\ \Omega + 4.0\ \Omega + 5.0\ \Omega + 7.0\ \Omega = \boxed{18.0\ \Omega}$$

(b) Find the current in the circuit.

Apply Ohm's law to the equivalent resistor in Figure 18.5b, solving for the current:

$$I = \frac{\Delta V}{R_{eq}} = \frac{6.0\ \text{V}}{18.0\ \Omega} = \boxed{0.33\ \text{A}}$$

(c) Calculate the electric potential at point A.

Apply Ohm's law to the 2.0-Ω resistor to find the voltage drop across it:

$$\Delta V = IR = (0.33\ \text{A})(2.0\ \Omega) = 0.66\ \text{V}$$

To find the potential at A, subtract the voltage drop from the potential at the positive terminal:

$$V_A = 6.0\ \text{V} - 0.66\ \text{V} = \boxed{5.3\ \text{V}}$$

(d) Calculate the battery's internal resistance if the battery's emf is 6.2 V.

Write Equation 18.1:

$$\Delta V = \mathcal{E} - Ir$$

Solve for the internal resistance r and substitute values:

$$r = \frac{\mathcal{E} - \Delta V}{I} = \frac{6.2\ \text{V} - 6.0\ \text{V}}{0.33\ \text{A}} = \boxed{0.6\ \Omega}$$

(e) What fraction f of the battery's power is delivered to the load resistors?

Divide the power delivered to the load by the total power output:

$$f = \frac{I\Delta V}{I\mathcal{E}} = \frac{\Delta V}{\mathcal{E}} = \frac{6.0\ \text{V}}{6.2\ \text{V}} = \boxed{0.97}$$

REMARKS A common misconception is that the current is "used up" and steadily declines as it progresses through a series of resistors. That would be a violation of the conservation of charge. What is actually used up is the electric potential energy of the charge carriers, some of which is delivered to each resistor.

QUESTION 18.1 Explain why the current in a real circuit very slowly decreases with time compared to its initial value.

EXERCISE 18.1 A closed circuit consists of a battery with a terminal voltage of 12.0 V, and 3.0-Ω, 6.0-Ω, 8.0-Ω, and 9.0-Ω resistors connected in series, oriented as in Figure 18.5a, with the battery in the bottom of the loop, positive terminal on the left, and resistors in increasing order, left to right, in the top of the loop. Calculate (a) the equivalent resistance of the circuit, (b) the current, and (c) the total power dissipated by the load resistors. (d) What is the electric potential at a point between the 6.0-Ω and 8.0-Ω resistors, if the electric potential at the positive terminal is 12.0 V? (e) If the battery has an emf of 12.1 V, find the battery's internal resistance.

ANSWERS (a) 26.0 Ω (b) 0.462 A (c) 5.54 W (d) 7.84 V (e) 0.2 Ω

approaches its maximum value, Q, as t approaches infinity. The voltage ΔV across the capacitor at any time is obtained by dividing the charge by the capacitance: $\Delta V = q/C$.

As you can see from Equation 18.7, it would take an infinite amount of time, in this model, for the capacitor to become fully charged. The reason is mathematical: in obtaining that equation, charges are assumed to be infinitely small, whereas in reality the smallest charge is that of an electron, with a magnitude equal to 1.60×10^{-19} C. For all practical purposes, the capacitor is fully charged after a finite amount of time. The term RC that appears in Equation 18.7 is called the **time constant** τ (Greek letter tau), so

$$\tau = RC \qquad [18.8]$$

The time constant represents the time required for the charge to increase from zero to 63.2% of its maximum equilibrium value. This means that in a period of time equal to one time constant, the charge on the capacitor increases from zero to $0.632Q$. This can be seen by substituting $t = \tau = RC$ into Equation 18.7 and solving for q. (Note that $1 - e^{-1} = 0.632$.) It's important to note that a capacitor charges very slowly in a circuit with a long time constant, whereas it charges very rapidly in a circuit with a short time constant. After a time equal to ten time constants, the capacitor is more than 99.99% charged.

Now consider the circuit in Figure 18.17a, consisting of a capacitor with an initial charge Q, a resistor, and a switch. Before the switch is closed, the potential difference across the charged capacitor is Q/C. Once the switch is closed, the charge begins to flow through the resistor from one capacitor plate to the other until the capacitor is fully discharged. If the switch is closed at $t = 0$, it can be shown that the charge q on the capacitor varies with time according to the equation

$$q = Qe^{-t/RC} \qquad [18.9]$$

The charge decreases exponentially with time, as shown in Figure 18.17b. In the interval $t = \tau = RC$, the charge decreases from its initial value Q to $0.368Q$. In other words, in a time equal to one time constant, the capacitor loses 63.2% of its initial charge. Because $\Delta V = q/C$, the voltage across the capacitor also decreases exponentially with time according to the equation $\Delta V = \mathcal{E}e^{-t/RC}$, where $\mathcal{E}$ (which equals Q/C) is the initial voltage across the fully charged capacitor.

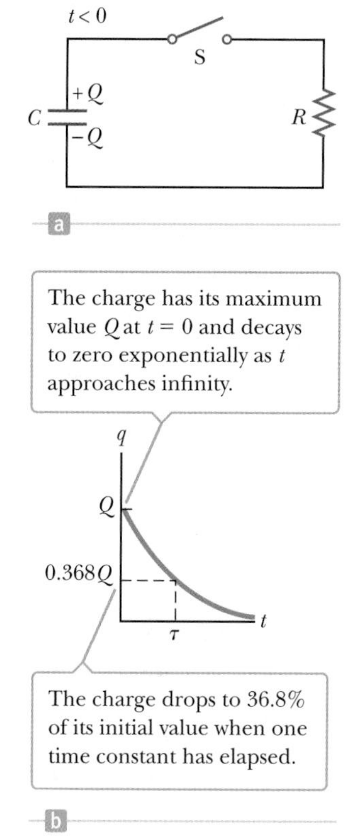

Figure 18.17 (a) A charged capacitor connected to a resistor and a switch. (b) A graph of the charge on the capacitor versus time after the switch is closed.

APPLYING PHYSICS 18.4 | Timed Windshield Wipers

Many automobiles are equipped with windshield wipers that can be used intermittently during a light rainfall. How does the operation of this feature depend on the charging and discharging of a capacitor?

EXPLANATION The wipers are part of an RC circuit with time constant that can be varied by selecting different values of R through a multiposition switch. The brief time that the wipers remain on and the time they are off are determined by the value of the time constant of the circuit. ■

APPLYING PHYSICS 18.5 | Bacterial Growth

In biological research concerning population growth, an equation is used that is similar to the exponential equations encountered in the analysis of RC circuits. Applied to a number of bacteria, this equation is

$$N_f = N_i 2^n$$

where N_f is the number of bacteria present after n doubling times, N_i is the number present initially, and n is the number of growth cycles or doubling times. Doubling times vary according to the organism. The doubling time is about 30 days for the bacteria responsible for leprosy, and about 20 minutes for the salmonella bacteria responsible for food poisoning. Suppose only 10 salmonella bacteria find their way onto a turkey leg after your Thanksgiving meal. Four hours later you come back for a midnight snack. How many bacteria are present now?

EXPLANATION The number of doubling times is 240 min/20 min = 12. Thus,

$$N_f = N_i 2^n = (10 \text{ bacteria})(2^{12}) = 40\,960 \text{ bacteria}$$

So your system will have to deal with an invading host of about 41 000 bacteria, which are going to continue to double in a very promising environment. ■

APPLYING PHYSICS 18.6 | Roadway Flashers

Many roadway construction sites have flashing yellow lights to warn motorists of possible dangers. What causes the lights to flash?

EXPLANATION A typical circuit for such a flasher is shown in Figure 18.18. The lamp L is a gas-filled lamp that acts as an open circuit until a large potential difference causes a discharge, which gives off a bright light. During this discharge, charge flows through the gas between the electrodes of the lamp. When the switch is closed, the battery charges the capacitor. At the beginning, the current is high and the charge on the capacitor is low, so most of the potential difference appears across the resistance *R*. As the capacitor charges, more potential difference appears across it, reflecting the lower current and lower potential difference across the resistor. Eventually, the potential difference across the capacitor reaches a value at which the lamp will conduct, causing a flash. This flash discharges the capacitor through the lamp, and the process of charging begins again. The period between flashes can be adjusted by changing the time constant of the *RC* circuit. ■

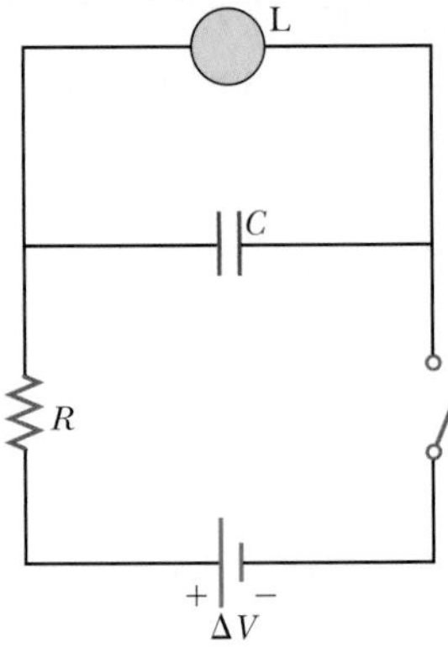

Figure 18.18 (Applying Physics 18.6)

Quick Quiz

18.9 The switch is closed in Figure 18.19. After a long time compared with the time constant of the circuit, what will the current be in the 2-Ω resistor? (a) 4 A (b) 3 A (c) 2 A (d) 1 A (e) More information is needed.

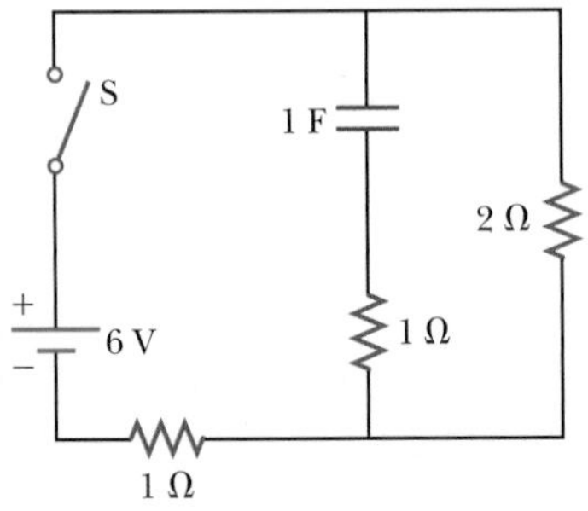

Figure 18.19 (Quick Quiz 18.9)

EXAMPLE 18.6 | Charging a Capacitor in an *RC* Circuit

GOAL Calculate elementary properties of a simple *RC* circuit.

PROBLEM An uncharged capacitor and a resistor are connected in series to a battery, as in Figure 18.16a. If $\mathcal{E} =$ 12.0 V, $C = 5.00\ \mu$F, and $R = 8.00 \times 10^5\ \Omega$, find **(a)** the time constant of the circuit, **(b)** the maximum charge on the capacitor, **(c)** the charge on the capacitor after 6.00 s, **(d)** the potential difference across the resistor after 6.00 s, and **(e)** the current in the resistor at that time.

STRATEGY Finding the time constant in part (a) requires substitution into Equation 18.8. For part (b), the maximum charge occurs after a long time, when the current has dropped to zero. By Ohm's law, $\Delta V = IR$, the potential difference across the resistor is also zero at that time, and Kirchhoff's loop rule then gives the maximum charge. Finding the charge at some particular time, as in part (c), is a matter of substituting into Equation 18.7. Kirchhoff's loop rule and the capacitance equation can be used to indirectly find the potential drop across the resistor in part (d), and then Ohm's law yields the current.

SOLUTION

(a) Find the time constant of the circuit.

Use the definition of the time constant, Equation 18.8:

$$\tau = RC = (8.00 \times 10^5\ \Omega)(5.00 \times 10^{-6}\ \text{F}) = 4.00\ \text{s}$$

(b) Calculate the maximum charge on the capacitor.

Apply Kirchhoff's loop rule to the *RC* circuit, going clockwise, which means that the voltage difference across the battery is positive and the differences across the capacitor and resistor are negative:

$$\text{(1)}\quad \Delta V_{\text{bat}} + \Delta V_C + \Delta V_R = 0$$

From the definition of capacitance (Eq. 16.8) and Ohm's law, we have $\Delta V_C = -q/C$ and $\Delta V_R = -IR$. These are voltage drops, so they're negative. Also, $\Delta V_{\text{bat}} = +\mathcal{E}$.

$$\text{(2)}\quad \mathcal{E} - \frac{q}{C} - IR = 0$$

When the maximum charge $q = Q$ is reached, $I = 0$. Solve Equation (2) for the maximum charge:

$$\mathcal{E} - \frac{Q}{C} = 0 \quad \rightarrow \quad Q = C\mathcal{E}$$

Substitute to find the maximum charge:

$$Q = (5.00 \times 10^{-6}\ \text{F})(12.0\ \text{V}) = \boxed{60.0\ \mu\text{C}}$$

(c) Find the charge on the capacitor after 6.00 s.

Substitute into Equation 18.7:

$$q = Q(1 - e^{-t/\tau}) = (60.0\ \mu\text{C})(1 - e^{-6.00\ \text{s}/4.00\ \text{s}})$$

$$= \boxed{46.6\ \mu\text{C}}$$

(d) Compute the potential difference across the resistor after 6.00 s.

Compute the voltage drop ΔV_C across the capacitor at that time:

$$\Delta V_C = -\frac{q}{C} = -\frac{46.6\ \mu\text{C}}{5.00\ \mu\text{F}} = -9.32\ \text{V}$$

Solve Equation (1) for ΔV_R and substitute:

$$\Delta V_R = -\Delta V_{\text{bat}} - \Delta V_C = -12.0\ \text{V} - (-9.32\ \text{V})$$

$$= \boxed{-2.68\ \text{V}}$$

(e) Find the current in the resistor after 6.00 s.

Apply Ohm's law, using the results of part (d) (remember that $\Delta V_R = -IR$ here):

$$I = \frac{-\Delta V_R}{R} = \frac{-(-2.68\ \text{V})}{(8.00 \times 10^5\ \Omega)}$$

$$= \boxed{3.35 \times 10^{-6}\ \text{A}}$$

REMARKS In solving this problem, we paid scrupulous attention to signs. These signs must always be chosen when applying Kirchhoff's loop rule and must remain consistent throughout the problem. Alternately, magnitudes can be used and the signs chosen by physical intuition. For example, the magnitude of the potential difference across the resistor must equal the magnitude of the potential difference across the battery minus the magnitude of the potential difference across the capacitor.

QUESTION 18.6 In an *RC* circuit as depicted in Figure 18.16a, what happens to the time required for the capacitor to be charged to half its maximum value if either the resistance or capacitance is increased? (a) It increases. (b) It decreases. (c) It remains the same.

EXERCISE 18.6 Find (a) the charge on the capacitor after 2.00 s have elapsed, (b) the magnitude of the potential difference across the capacitor after 2.00 s, and (c) the magnitude of the potential difference across the resistor at that same time.

ANSWERS (a) 23.6 μC (b) 4.72 V (c) 7.28 V

■ EXAMPLE 18.7 | Discharging a Capacitor in an *RC* Circuit

GOAL Calculate some elementary properties of a discharging capacitor in an *RC* circuit.

PROBLEM Consider a capacitor C being discharged through a resistor R as in Figure 18.17a (page 641). **(a)** How long does it take the charge on the capacitor to drop to one-fourth its initial value? Answer as a multiple of τ. **(b)** Compute the initial charge and time constant, and **(c)** the time it takes to discharge all but the last quantum of charge, 1.60×10^{-19} C, if the initial potential difference across the capacitor is 12.0 V, the capacitance is equal to 3.50×10^{-6} F, and the resistance is 2.00 Ω. (Assume an exponential decrease during the entire discharge process.)

STRATEGY This problem requires substituting given values into various equations, as well as a few algebraic manipulations involving the natural logarithm. In part (a) set $q = \frac{1}{4}Q$ in Equation 18.9 for a discharging capacitor, where Q is the initial charge, and solve for time t. In part (b) substitute into Equations 16.8 and 18.8 to find the initial capacitor charge and time constant, respectively. In part (c) substitute the results of part (b) and the final charge $q = 1.60 \times 10^{-19}$ C into the discharging-capacitor equation, again solving for time.

SOLUTION

(a) How long does it take the charge on the capacitor to reduce to one-fourth its initial value?

Apply Equation 18.9:

$$q(t) = Qe^{-t/RC}$$

(Continued)

Substitute $q(t) = Q/4$ into the preceding equation and cancel Q:

$$\tfrac{1}{4}Q = Qe^{-t/RC} \rightarrow \tfrac{1}{4} = e^{-t/RC}$$

Take natural logarithms of both sides and solve for the time t:

$$\ln\left(\tfrac{1}{4}\right) = -t/RC$$

$$t = -RC\ln\left(\tfrac{1}{4}\right) = 1.39RC = \boxed{1.39\tau}$$

(b) Compute the initial charge and time constant from the given data.

Use the capacitance equation to find the initial charge:

$$C = \frac{Q}{\Delta V} \rightarrow Q = C\,\Delta V = (3.50 \times 10^{-6}\ \text{F})(12.0\ \text{V})$$

$$Q = \boxed{4.20 \times 10^{-5}\ \text{C}}$$

Now calculate the time constant:

$$\tau = RC = (2.00\ \Omega)(3.50 \times 10^{-6}\ \text{F}) = \boxed{7.00 \times 10^{-6}\ \text{s}}$$

(c) How long does it take to drain all but the last quantum of charge?

Apply Equation 18.9, divide by Q, and take natural logarithms of both sides:

$$q(t) = Qe^{-t/\tau} \rightarrow e^{-t/\tau} = \frac{q}{Q}$$

Take the natural logs of both sides:

$$-t/\tau = \ln\left(\frac{q}{Q}\right) \rightarrow t = -\tau\ln\left(\frac{q}{Q}\right)$$

Substitute $q = 1.60 \times 10^{-19}$ C and the values for Q and τ found in part (b):

$$t = -(7.00 \times 10^{-6}\ \text{s})\ln\left(\frac{1.60 \times 10^{-19}\ \text{C}}{4.20 \times 10^{-5}\ \text{C}}\right)$$

$$= \boxed{2.32 \times 10^{-4}\ \text{s}}$$

REMARKS Part (a) shows how useful information can often be obtained even when no details concerning capacitances, resistances, or voltages are known. Part (c) demonstrates that capacitors can be rapidly discharged (or conversely, charged), despite the mathematical form of Equations 18.7 and 18.9, which indicate an infinite time would be required.

QUESTION 18.7 Suppose the initial voltage used to charge the capacitor were doubled. Would the time required for discharging all but the last quantum of charge (a) increase, (b) decrease, (c) remain the same?

EXERCISE 18.7 Suppose the same type of series circuit has $R = 8.00 \times 10^4\ \Omega$, $C = 5.00\ \mu\text{F}$, and an initial voltage across the capacitor of 6.00 V. (a) How long does it take the capacitor to lose half its initial charge? (b) How long does it take to lose all but the last 10 electrons on the negative plate?

ANSWERS (a) 0.277 s (b) 12.2 s

18.6 Household Circuits

LEARNING OBJECTIVE

1. Describe fundamental properties of household circuits.

Household circuits are a practical application of some of the ideas presented in this chapter. In a typical installation the utility company distributes electric power to individual houses with a pair of wires, or power lines. Electrical devices in a house are then connected in parallel to these lines, as shown in Figure 18.20. The potential difference between the two wires is about 120 V. (These currents and voltages are actually alternating currents and voltages, but for the present discussion we will assume they are direct currents and voltages.) One of the wires is connected to ground, and the other wire, sometimes called the "hot" wire, is at a potential of 120 V. A meter and a circuit breaker (or a fuse) are connected in series with the wire entering the house, as indicated in the figure.

In modern homes, circuit breakers are used in place of fuses. When the current in a circuit exceeds some value (typically 15 A or 20 A), the circuit breaker acts as a switch and opens the circuit. Figure 18.21 shows one design for a circuit breaker. Current passes through a bimetallic strip, the top of which bends to the left when excessive current heats it. If the strip bends far enough to the left, it settles into a groove in the spring-loaded metal bar. When this settling occurs, the bar drops enough to open the circuit at the contact point. The bar also flips a switch that indicates that the circuit breaker is not operational. (After the overload is removed, the switch can be flipped back on.) Circuit breakers based on this design have the disadvantage that some time is required for the heating of the strip, so the circuit may not be opened rapidly enough when it is overloaded. Therefore, many circuit breakers are now designed to use electromagnets (discussed in Chapter 19).

The wire and circuit breaker are carefully selected to meet the current demands of a circuit. If the circuit is to carry currents as large as 30 A, a heavy-duty wire and an appropriate circuit breaker must be used. Household circuits that are normally used to power lamps and small appliances often require only 20 A. Each circuit has its own circuit breaker to accommodate its maximum safe load.

As an example, consider a circuit that powers a toaster, a microwave oven, and a heater (represented by R_1, R_2, and R_3 in Fig. 18.20). Using the equation $P = I\Delta V$, we can calculate the current carried by each appliance. The toaster, rated at 1 000 W, draws a current of $1\ 000/120 = 8.33$ A. The microwave oven, rated at 800 W, draws a current of 6.67 A, and the heater, rated at 1 300 W, draws a current of 10.8 A. If the three appliances are operated simultaneously, they draw a total current of 25.8 A. Therefore, the breaker should be able to handle at least this much current, or else it will be tripped. As an alternative, the toaster and microwave oven could operate on one 20-A circuit with the heater on a separate 20-A circuit.

Many heavy-duty appliances, such as electric ranges and clothes dryers, require 240 V to operate. The power company supplies this voltage by providing, in addition to a live wire that is 120 V above ground potential, another wire, also considered live, that is 120 V below ground potential (Fig. 18.22, page 646). Therefore, the potential drop across the two live wires is 240 V. An appliance operating from a 240-V line requires half the current of one operating from a 120-V line; consequently, smaller wires can be used in the higher-voltage circuit without becoming overheated.

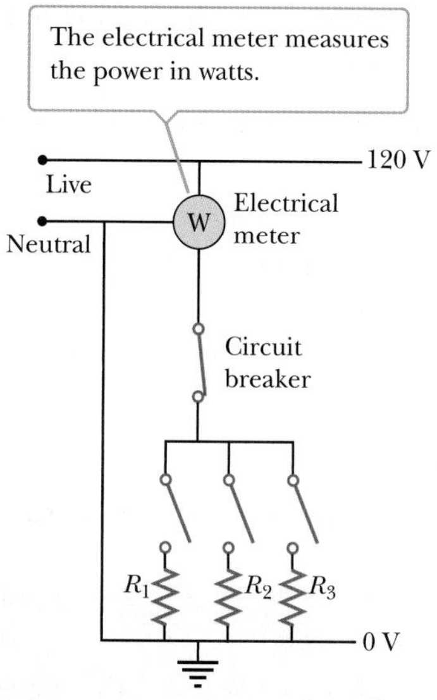

Figure 18.20 A wiring diagram for a household circuit. The resistances R_1, R_2, and R_3 represent appliances or other electrical devices that operate at an applied voltage of 120 V.

APPLICATION

Fuses and Circuit Breakers

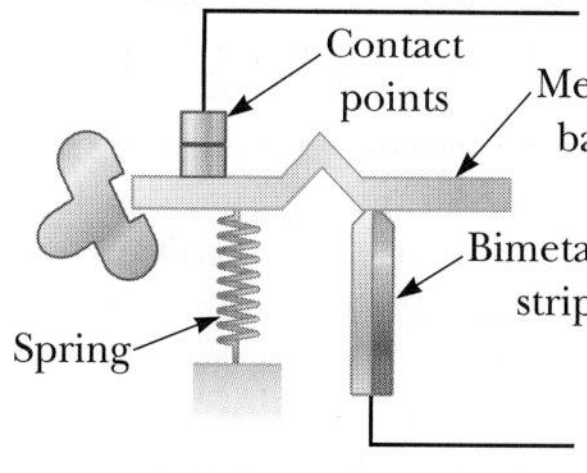

Figure 18.21 A circuit breaker that uses a bimetallic strip for its operation.

18.7 Electrical Safety

LEARNING OBJECTIVE

1. Describe several important aspects of electrical safety.

A person can be electrocuted by touching a live wire while in contact with ground. Such a hazard is often due to frayed insulation that exposes the conducting wire. The ground contact might be made by touching a water pipe (which is normally at ground potential) or by standing on the ground with wet feet because impure water is a good conductor. Obviously, such situations should be avoided at all costs.

Electric shock can result in fatal burns or cause the muscles of vital organs, such as the heart, to malfunction. The degree of damage to the body depends on the magnitude of the current, the length of time it acts, and the part of the body through which it passes. Currents of 5 mA or less can cause a sensation of shock, but ordinarily do little or no damage. If the current is larger than about 10 mA, the hand muscles contract and the person may be unable to let go of the live wire. If a current of about 100 mA passes through the body for just a few seconds, it

APPLICATION
Third Wire on Consumer Appliances

can be fatal. Such large currents paralyze the respiratory muscles. In some cases, currents of about 1 A through the body produce serious (and sometimes fatal) burns.

As an additional safety feature for consumers, electrical equipment manufacturers now use electrical cords that have a third wire, called a *case ground.* To understand how this works, consider the drill being used in Figure 18.23. A two-wire device has one wire, called the "hot" wire, connected to the high-potential (120-V) side of the input power line, while the second wire is connected to ground (0 V). If the high-voltage wire comes in contact with the case of the drill (Fig. 18.23a), a short circuit occurs. In this undesirable circumstance, the pathway for the current is from the high-voltage wire through the person holding the drill and to Earth, a pathway that can be fatal. Protection is provided by a third wire, connected to the case of the drill (Fig. 18.23b). In this case, if a short occurs, the path of least resistance for the current is from the high-voltage wire through the case and back to ground through the third wire. The resulting high current produced will blow a fuse or trip a circuit breaker before the consumer is injured.

Special power outlets called ground-fault interrupters (GFIs) are now being used in kitchens, bathrooms, basements, and other hazardous areas of new homes. They are designed to protect people from electrical shock by sensing small currents—approximately 5 mA and greater—leaking to ground. When current above this level is detected, the device shuts off (interrupts) the current in less than a millisecond. (Ground-fault interrupters will be discussed in Chapter 20.)

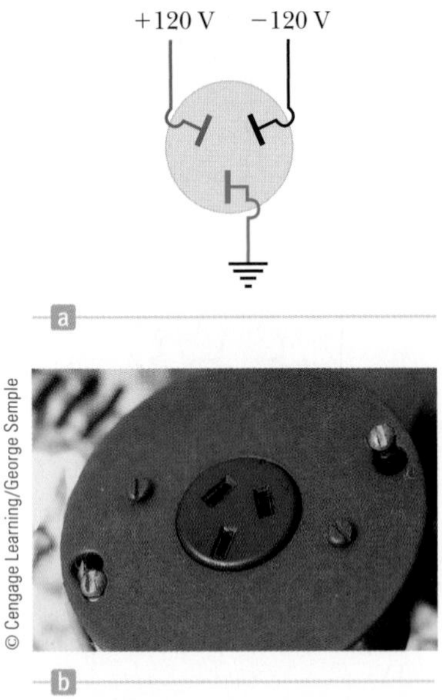

Figure 18.22 (a) The connections for each of the openings in a 240-V outlet. (b) An outlet for connection to a 240-V supply.

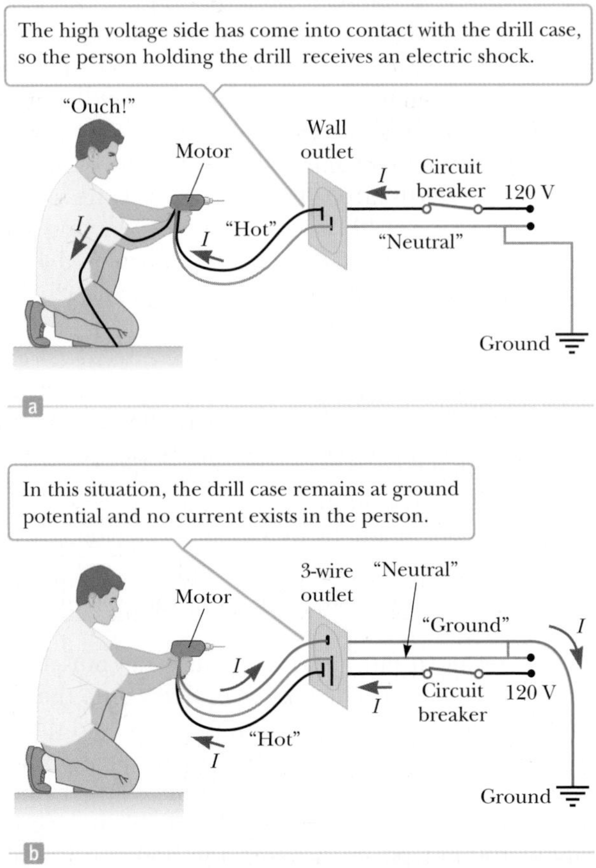

Figure 18.23 The "hot" (or "live") wire, at 120 V, always includes a circuit breaker for safety. (a) When the drill is operated with two wires, the normal current path is from the "hot" wire, through the motor connections, and back to ground through the "neutral" wire. (b) Shock can be prevented by a third wire running from the drill case to the ground. The wire colors represent electrical standards in the United States: the "hot" wire is black, the ground wire is green, and the neutral wire is white (shown as gray in the figure).

18.8 Conduction of Electrical Signals by Neurons[2] BIO

LEARNING OBJECTIVES

1. Describe the conduction of electrical signals by neurons.
2. Describe the structure and operational principles of neurons.

The most remarkable use of electrical phenomena in living organisms is found in the nervous system of animals. Specialized cells in the body called **neurons** form a complex network that receives, processes, and transmits information from one part of the body to another. The center of this network is located in the brain, which has the ability to store and analyze information. On the basis of this information, the nervous system controls parts of the body.

The nervous system is highly complex and consists of about 10^{10} interconnected neurons. Some aspects of the nervous system are well known. Over the past several decades the method of signal propagation through the nervous system has been established. The messages transmitted by neurons are voltage pulses called *action potentials.* When a neuron receives a strong enough stimulus, it produces identical voltage pulses that are actively propagated along its structure. The strength of the stimulus is conveyed by the number of pulses produced. When the pulses reach the end of the neuron, they activate either muscle cells or other neurons. There is a "firing threshold" for neurons: action potentials propagate along a neuron only if the stimulus is sufficiently strong.

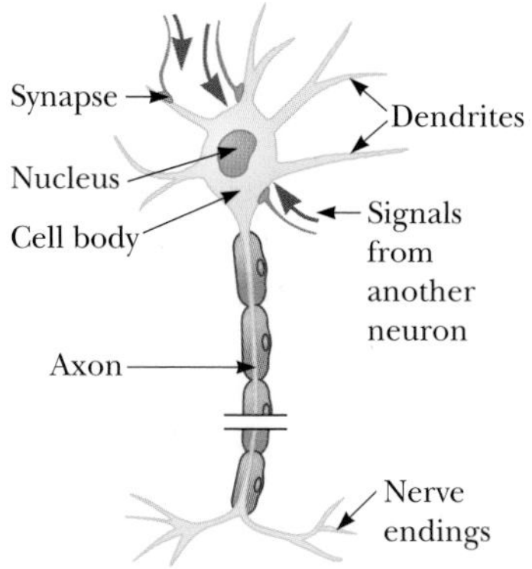

Figure 18.24 Diagram of a neuron.

Neurons can be divided into three classes: sensory neurons, motor neurons, and interneurons. The sensory neurons receive stimuli from sensory organs that monitor the external and internal environment of the body. Depending on their specialized functions, the sensory neurons convey messages about factors such as light, temperature, pressure, muscle tension, and odor to higher centers in the nervous system. The motor neurons carry messages that control the muscle cells. Those messages are based on the information provided by the sensory neurons and by the brain. The interneurons transmit information from one neuron to another.

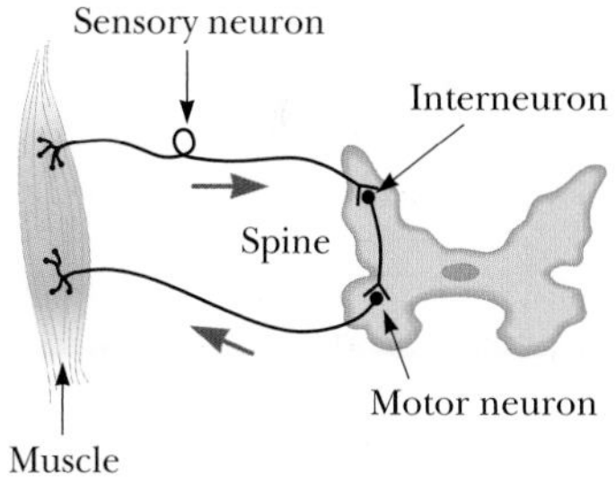

Figure 18.25 A simple neural circuit.

Each neuron consists of a cell body to which are attached input ends called **dendrites** and a long tail called the **axon**, which transmits signals away from the cell (Fig. 18.24). The far end of the axon branches into nerve endings that transmit signals across small gaps to other neurons or to muscle cells. A simple sensorimotor neuron circuit is shown in Figure 18.25. A stimulus from a muscle produces nerve impulses that travel to the spine. Here the signal is transmitted to a motor neuron, which in turn sends impulses to control the muscle. Figure 18.26 shows an electron microscope image of neurons in the brain.

The axon, which is an extension of the neuron cell, conducts electric impulses away from the cell body. Some axons are extremely long. In humans, for example, the axons connecting the spine with the fingers and toes are more than 1 m long. The neuron can transmit messages because of the special active electrical characteristics of the axon. (The axon acts as an **active** source of energy like a battery, rather than like a **passive** stretch of resistive wire.) Much of the information about the electrical and chemical properties of the axon is obtained by inserting small needlelike probes into it. Figure 18.27 (page 648) shows an experimental setup.

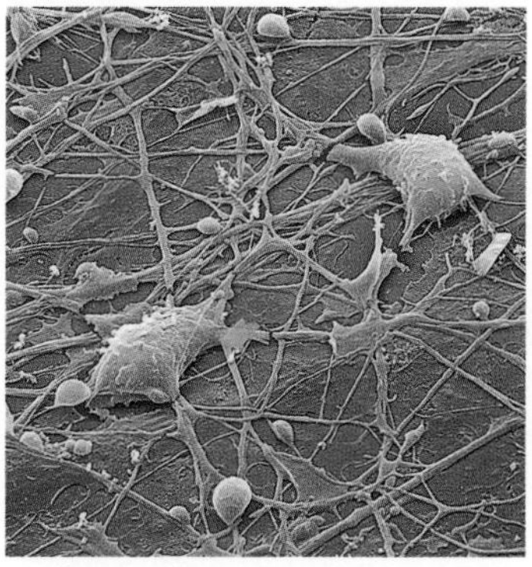

Juergen Berger, Max-Planck Institute/Science Photo Library/Science Source

Figure 18.26 Stellate neuron from human cortex.

Note that the outside of the axon is grounded, so all measured voltages are with respect to a zero potential on the outside. With these probes it is possible to inject current into the axon, measure the resulting action potential as a function of time at a fixed point, and sample the cell's chemical composition. Such experiments are usually difficult to run because the diameter of most axons is very small. Even the largest axons in the human nervous system have a diameter of only about 20×10^{-4} cm. The

[2]This section is based on an essay by Paul Davidovits of Boston College.

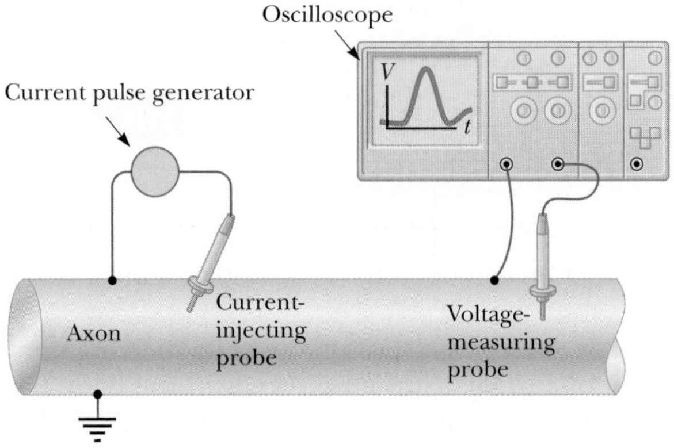

Figure 18.27 An axon stimulated electrically. The left probe injects a short pulse of current, and the right probe measures the resulting action potential as a function of time.

giant squid, however, has an axon with a diameter of about 0.5 mm, which is large enough for the convenient insertion of probes. Much of the information about signal transmission in the nervous system has come from experiments with the squid axon.

In the aqueous environment of the body, salts and other molecules dissociate into positive and negative ions. As a result, body fluids are relatively good conductors of electricity. The inside of the axon is filled with an ionic fluid that is separated from the surrounding body fluid by a thin membrane that is only about 5 nm to 10 nm thick. The resistivities of the internal and external fluids are about the same, but their chemical compositions are substantially different. The external fluid is similar to seawater: its ionic solutes are mostly positive sodium ions and negative chloride ions. Inside the axon, the positive ions are mostly potassium ions and the negative ions are mostly large organic ions.

Ordinarily, the concentrations of sodium and potassium ions inside and outside the axon would be equalized by diffusion. The axon, however, is a living cell with an energy supply and can change the permeability of its membranes on a time scale of milliseconds.

When the axon is not conducting an electric pulse, the axon membrane is highly permeable to potassium ions, slightly permeable to sodium ions, and impermeable to large organic ions. Consequently, although sodium ions cannot easily enter the axon, potassium ions can leave it. As the potassium ions leave the axon, however, they leave behind large negative organic ions, which cannot follow them through the membrane. As a result, a negative potential builds up inside the axon with respect to the outside. The final negative potential reached, which has been measured at about −70 mV, holds back the outflow of potassium ions so that at equilibrium the concentration of ions remains as stated above.

The mechanism for the production of an electric signal by the neuron is conceptually simple, but was experimentally difficult to sort out. When a neuron changes its resting potential because of an appropriate stimulus, the properties of its membrane change locally. As a result, there is a sudden flow of sodium ions into the cell that lasts for about 2 ms. This flow produces the +30 mV peak in the action potential shown in Figure 18.28a. Immediately afterward, there is an increase in potassium ion flow out of the cell that restores the resting action potential of −70 mV in an additional 3 ms. Both the Na^+ and K^+ ion flows have been measured by using radioactive Na and K tracers. The nerve signal has been measured to travel along the axon at speeds from 50 m/s to about 150 m/s. This flow of charged particles (or signal transmission) in a nerve axon is *unlike* signal transmission in a metal wire. In an axon, charges move perpendicular to the direction of travel of the nerve signal, and the signal moves much more slowly than a voltage pulse traveling along a metallic wire.

Although the axon is a highly complex structure and much of how Na^+ and K^+ ion channels open and close is not understood, standard electric circuit concepts of current and capacitance can be used to analyze axons. It is left as a problem (Problem 41) to show that the axon, having equal and opposite charges separated by a thin dielectric membrane, acts like a capacitor.

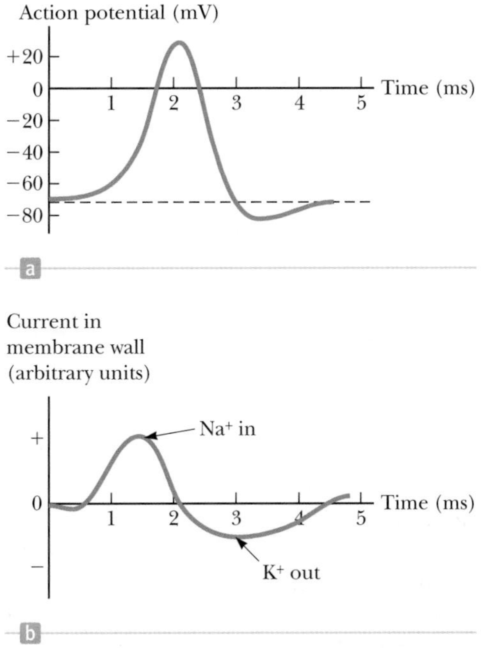

Figure 18.28 (a) Typical action potential as a function of time. (b) Current in the axon membrane wall as a function of time.

SUMMARY

18.1 Sources of emf

Any device, such as a battery or generator, that increases the electric potential energy of charges in an electric circuit is called a **source of emf**. Batteries convert chemical energy into electrical potential energy, and generators convert mechanical energy into electrical potential energy.

The terminal voltage ΔV of a battery is given by

$$\Delta V = \mathcal{E} - Ir \qquad [18.1]$$

where ε is the emf of the battery, I is the current, and r is the internal resistance of the battery. Generally, the internal resistance is small enough to be neglected.

18.2 Resistors in Series

The **equivalent resistance** of a set of resistors connected in **series** is

$$R_{eq} = R_1 + R_2 + R_3 + \cdots \qquad [18.4]$$

Several resistors in series can be replaced with a single equivalent resistor.

The current remains at a constant value as it passes through a series of resistors. The potential difference across any two resistors in series is different, unless the resistors have the same resistance.

18.3 Resistors in Parallel

The **equivalent resistance** of a set of resistors connected in **parallel** is

$$\frac{1}{R_{eq}} = \frac{1}{R_1} + \frac{1}{R_2} + \frac{1}{R_3} + \cdots \qquad [18.6]$$

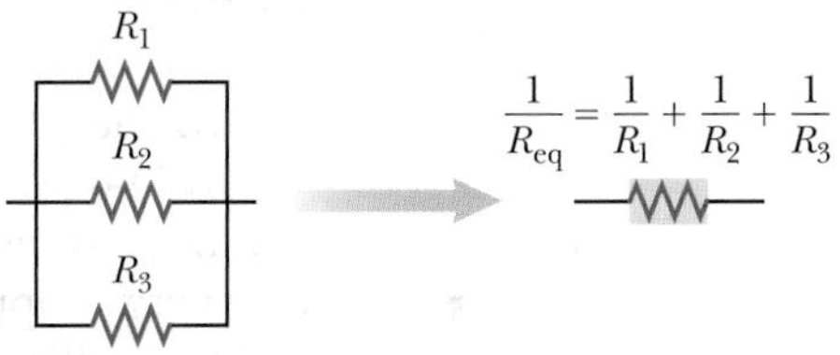

Several resistors in parallel can be replaced with a single equivalent resistor.

The potential difference across any two parallel resistors is the same; the current in each resistor, however, will be different unless the two resistances are equal.

18.4 Kirchhoff's Rules and Complex DC Circuits

Complex circuits can be analyzed using **Kirchhoff's rules**:

1. The sum of the currents entering any junction must equal the sum of the currents leaving that junction.
2. The sum of the potential differences across all the elements around any closed circuit loop must be zero.

The first rule, called the junction rule, is a statement of **conservation of charge**. The second rule, called the loop rule, is a statement of **conservation of energy**. Solving problems involves using these rules to generate as many equations as there are unknown currents. The equations can then be solved simultaneously.

18.5 *RC* Circuits

In a simple *RC* circuit with a battery, a resistor, and a capacitor in series, the charge on the capacitor increases according to the equation

$$q = Q(1 - e^{-t/RC}) \qquad [18.7]$$

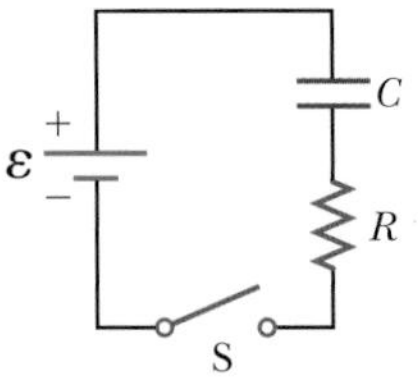

An *RC* circuit with a battery and a resistor charges a capacitor when the switch is closed.

The term *RC* in Equation 18.7 is called the **time constant** τ (Greek letter tau), so

$$\tau = RC \qquad [18.8]$$

The time constant represents the time required for the charge to increase from zero to 63.2% of its maximum equilibrium value.

A simple *RC* circuit consisting of a charged capacitor in series with a resistor discharges according to the expression

$$q = Qe^{-t/RC} \qquad [18.9]$$

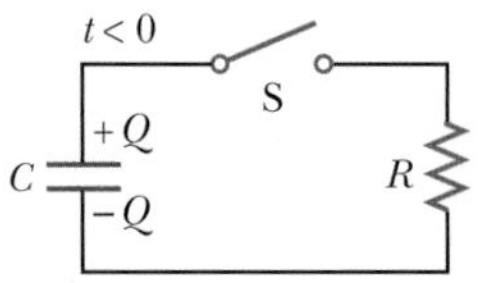

An *RC* circuit with charged capacitor discharges across a resistor when the switch is closed.

Problems can be solved by substituting values for q, Q, t or τ into these equations. The voltage ΔV across the capacitor at any time is obtained by dividing the charge by the capacitance: $\Delta V = q/C$. Using Kirchhoff's loop rule yields the potential difference across the resistor. Ohm's law applied to the resistor then gives the current.

WARM-UP EXERCISES

ENHANCED WebAssign The warm-up exercises in this chapter may be assigned online in Enhanced WebAssign.

1. **Math Review** Determine the values of (a) I_1 and (b) I_2 that satisfy the following two equations: $9.00 + 3.00\,I_1 - 2.00\,I_2 = 0$ and $-2.00\,I_1 + I_2 = 0$. (See also Section 18.4.)

2. **Math Review** Determine the values of (a) I_1, (b) I_2, and (c) I_3 that satisfy the following three equations: $-I_1 + I_2 + I_3 = 5.00$, $-3.00\,I_1 + I_2 = 0$, and $I_1 + I_3 = 0$.

3. **Math Review** Suppose the charge q on a capacitor is given by $q = (3.00\text{ C})\,e^{-t/(0.200\text{ s})}$. Find (a) the value of q when $t = 0.040\,0$ s, and (b) the value of t when $q = 1.20$ C. (See also Section 18.5.)

4. **Physics Review** Find the charge on a 25.0 μF capacitor when connected to a 12.0-V battery. (See Section 16.6.)

5. **Physics Review** Two capacitors have capacitances of 2.00 μF and 0.500 μF. Determine the equivalent capacitance if they are connected (a) in parallel, and (b) in series. (See Section 16.8.)

6. **Physics Review** A resistor with resistance 30.0 Ω is connected to the terminals of a voltage source. Calculate (a) the current through the resistor, and (b) the power dissipated by the resistor when the voltage is set at 6.00 V. (c) If the voltage is doubled, recalculate the power. (See Sections 17.4 and 17.6.)

7. A battery has an open-circuit voltage (emf) of 9.20 V and an internal resistance of 0.400 Ω. Determine (a) the current when the terminals are connected across a 3.00-Ω resistor, (b) the terminal voltage ΔV, and (c) the total power supplied by the battery. (See Section 18.1.)

8. Two resistors have resistances of 2.00 Ω and 0.500 Ω. Determine the equivalent resistance if they are connected (a) in series, (b) in parallel. (See Sections 18.2 and 18.3.)

9. Three resistors have resistances of 2.00 Ω, 10.0 Ω and 25.0 Ω. Determine the equivalent resistance if they are connected (a) in parallel, and (b) in series. (See Sections 18.2 and 18.3.)

10. A simple circuit is constructed from a 12.0-V battery, a 3.00-Ω resistor and a 5.00-Ω resistor. Ignoring the internal resistance of the battery, determine the current supplied by the battery if the resistors are connected across the battery terminals (a) in series, and (b) in parallel. (See Sections 18.2 and 18.3.)

11. Consider the circuit shown in Figure WU18.11. Determine (a) the equivalent resistance of the three resistors in parallel, (b) the equivalent resistance of all four resistors in the circuit, (c) the current supplied by the battery, and (d) the magnitude of the potential difference across the 2.00-Ω resistor, (e) the magnitude of the potential difference across the 10.0-Ω resistor, and (f) the current through the 10.0-Ω resistor. (See Sections 18.2, 18.3 and 18.4)

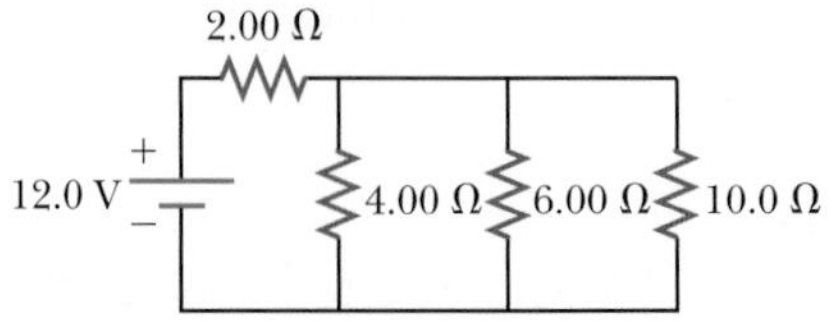

Figure WU18.11

12. A flashing light blub is driven by a simple series *RC* circuit with a time constant of 1.30 s. If the circuit's equivalent resistance is $5.00 \times 10^3\ \Omega$, determine its equivalent capacitance. (See Section 18.5.)

13. A 275-Ω resistor is in series with a 36.0 μF capacitor and a 12.0-V voltage source with the circuit switch open and the capacitor uncharged. (a) What is the time constant of this *RC* circuit? Calculate (b) the maximum charge the capacitor can accumulate, and (c) the charge on the capacitor 5.00 ms after the switch is closed. (See Section 18.5.)

CONCEPTUAL QUESTIONS

ENHANCED WebAssign The conceptual questions in this chapter may be assigned online in Enhanced WebAssign.

1. Is the direction of current in a battery always from the negative terminal to the positive one? Explain.
2. Given three lightbulbs and a battery, sketch as many different circuits as you can.
3. Suppose the energy transferred to a dead battery during charging is *W*. The recharged battery is then used until fully discharged again. Is the total energy transferred out of the battery during use also *W*?
4. A short circuit is a circuit containing a path of very low resistance in parallel with some other part of the circuit. Discuss the effect of a short circuit on the portion of the circuit it parallels. Use a lamp with a frayed line cord as an example.
5. Connecting batteries in series increases the emf applied to a circuit. What advantage might there be to connecting them in parallel?
6. If electrical power is transmitted over long distances, the resistance of the wires becomes significant. Why? Which mode of transmission would result in less energy loss, high current and low voltage or low current and high voltage? Discuss.
7. If you have your headlights on while you start your car, why do they dim while the car is starting?
8. Two sets of Christmas lights are available. For set A, when one bulb is removed, the remaining bulbs remain illuminated. For set B, when one bulb is removed, the remaining bulbs do not operate. Explain the difference in wiring for the two sets.
9. Why is it possible for a bird to sit on a high-voltage wire without being electrocuted? (See Fig. CQ18.9.)

Superstock

Figure CQ18.9

10. (a) Two resistors are connected in series across a battery. Is the power delivered to each resistor (i) the same or (ii) not necessarily the same? (b) Two resistors are connected in parallel across a battery. Is the power delivered to each resistor (i) the same or (ii) not necessarily the same?
11. Suppose a parachutist lands on a high-voltage wire and grabs the wire as she prepares to be rescued. Will she be electrocuted? If the wire then breaks, should she continue to hold onto the wire as she falls to the ground?
12. A ski resort consists of a few chairlifts and several interconnected downhill runs on the side of a mountain, with a lodge at the bottom. The lifts are analogous to batteries, and the runs are analogous to resistors. Describe how two runs can be in series. Describe how three runs can be in parallel. Sketch a junction of one lift and two runs. One of the skiers is carrying an altimeter. State Kirchhoff's junction rule and Kirchhoff's loop rule for ski resorts.
13. Embodied in Kirchhoff's rules are two conservation laws. What are they?
14. Why is it dangerous to turn on a light when you are in a bathtub?

PROBLEMS

ENHANCED WebAssign The problems in this chapter may be assigned online in Enhanced WebAssign.

1. denotes straightforward problem; 2. denotes intermediate problem;
3. denotes challenging problem
[1.] denotes full solution available in *Student Solutions Manual/Study Guide*
1. denotes problems most often assigned in Enhanced WebAssign

BIO denotes biomedical problems
GP denotes guided problems
M denotes Master It tutorial available in Enhanced WebAssign
Q|C denotes asking for quantitative and conceptual reasoning
S denotes symbolic reasoning problem
W denotes Watch It video solution available in Enhanced WebAssign

18.1 Sources of emf

18.2 Resistors in Series

18.3 Resistors in Parallel

1. A battery having an emf of 9.00 V delivers 117 mA when connected to a 72.0-Ω load. Determine the internal resistance of the battery.

2. Three 9.0-Ω resistors are connected in series with a 12-V battery. Find (a) the equivalent resistance of the circuit and (b) the current in each resistor. (c) Repeat for the case in which all three resistors are connected in parallel across the battery.

3. A lightbulb marked "75 W [at] 120 V" is screwed into a socket at one end of a long extension cord in which each of the two conductors has a resistance of 0.800 Ω. The other end of the extension cord is plugged into a 120-V outlet. (a) Draw a circuit diagram, and (b) find the actual power of the bulb in the circuit described.

4. (a) Find the current in an 8.00-Ω resistor connected to a battery that has an internal resistance of 0.15 Ω if the voltage across the battery (the terminal voltage) is 9.00 V. (b) What is the emf of the battery?

5. M (a) Find the equivalent resistance between points *a* and *b* in Figure P18.5. (b) Calculate the current in each resistor if a potential difference of 34.0 V is applied between points *a* and *b*.

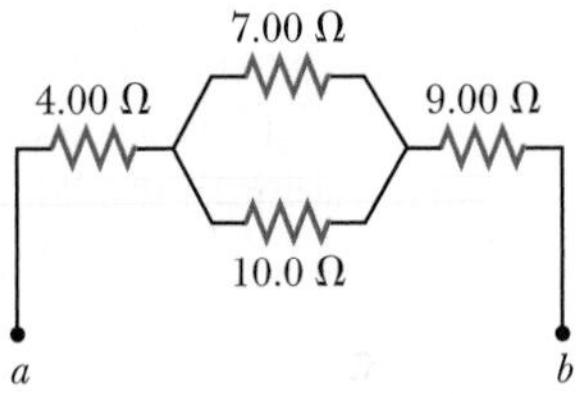

Figure P18.5

6. Consider the combination of resistors shown in Figure P18.6. (a) Find the equivalent resistance between point *a* and *b*. (b) If a voltage of 35.0 V is applied between points *a* and *b*, find the current in each resistor.

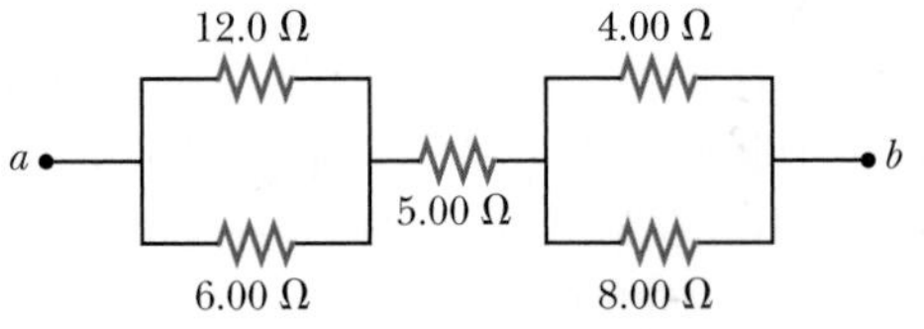

Figure P18.6

7. Two resistors connected in series have an equivalent resistance of 690 Ω. When they are connected in parallel, their equivalent resistance is 150 Ω. Find the resistance of each resistor.

8. GP Consider the circuit shown in Figure P18.8. (a) Calculate the equivalent resistance of the 10.0-Ω and 5.00-Ω resistors connected in parallel. (b) Using the result of part (a), calculate the combined resistance of the 10.0-Ω, 5.00-Ω, and 4.00-Ω resistors. (c) Calculate the equivalent resistance of the combined resistance found in part (b) and the parallel 3.00-Ω resistor. (d) Combine the equivalent resistance found in part (c) with the 2.00-Ω resistor. (e) Calculate the total current in the circuit. (f) What is the voltage drop across the 2.00-Ω resistor? (g) Subtracting the result of part (f) from the battery voltage, find the voltage across the 3.00-Ω resistor. (h) Calculate the current in the 3.00-Ω resistor.

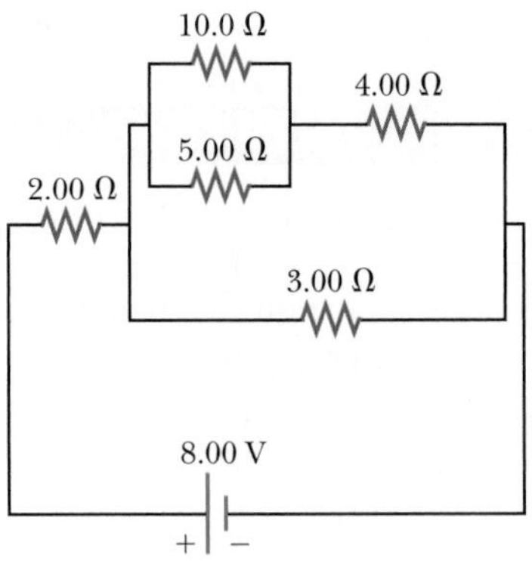

Figure P18.8

9. Consider the circuit shown in Figure P18.9. Find (a) the potential difference between points *a* and *b* and (b) the current in the 20.0-Ω resistor.

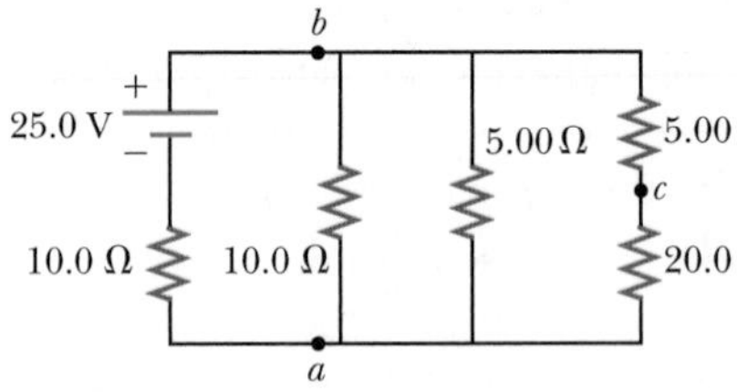

Figure P18.9

10. S Four resistors are connected to a battery as shown in Figure P18.10. (a) Determine the potential difference across each resistor in terms of $\mathcal{E}$. (b) Determine the current in each resistor in terms of *I*.

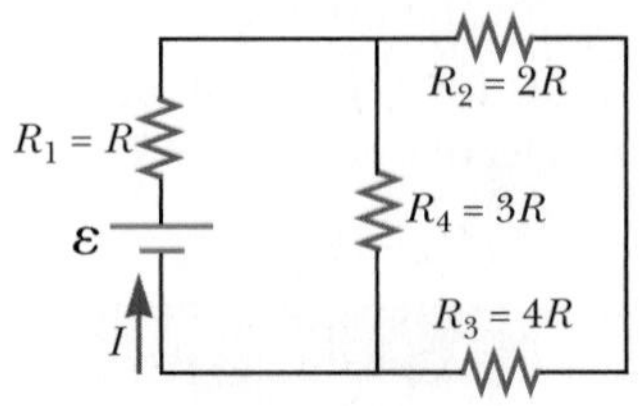

Figure P18.10

11. The resistance between terminals *a* and *b* in Figure P18.11 is 75 Ω. If the resistors labeled *R* have the same value, determine *R*.

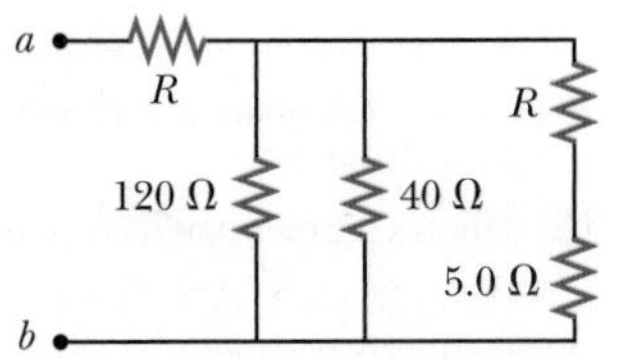

Figure P18.11

12. A battery with $\mathcal{E} = 6.00$ V and no internal resistance supplies current to the circuit shown in Figure P18.12. When the double-throw switch S is open as shown in the figure, the current in the battery is 1.00 mA. When the switch is closed in position a, the current in the battery is 1.20 mA. When the switch is closed in position b, the current in the battery is 2.00 mA. Find the resistances (a) R_1, (b) R_2, and (c) R_3.

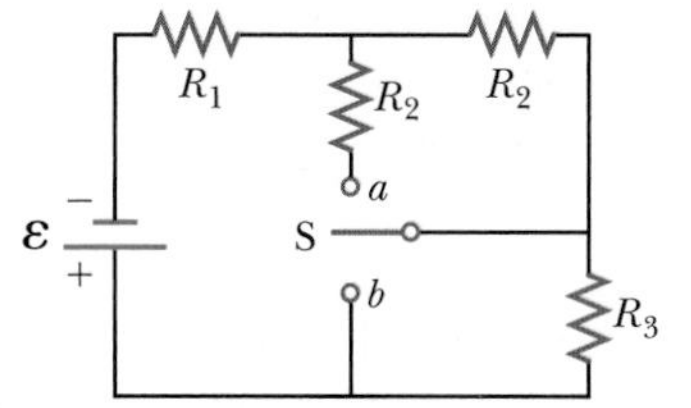

Figure P18.12

13. Find the current in the 12-Ω resistor in Figure P18.13.

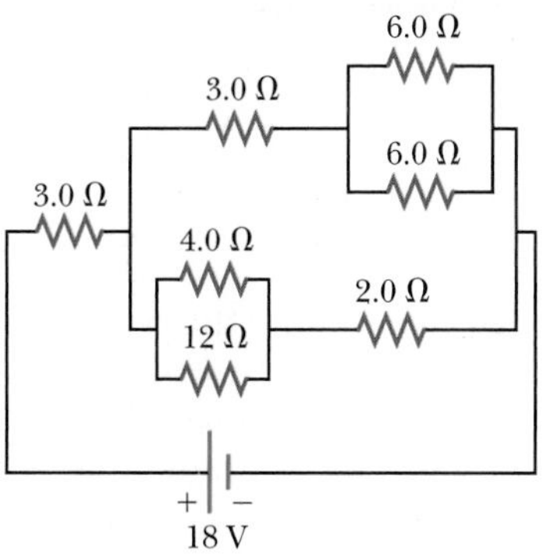

Figure P18.13

14. QIC (a) Is it possible to reduce the circuit shown in Figure P18.14 to a single equivalent resistor connected across the battery? Explain. (b) Find the current in the 2.00-Ω resistor. (c) Calculate the power delivered by the battery to the circuit.

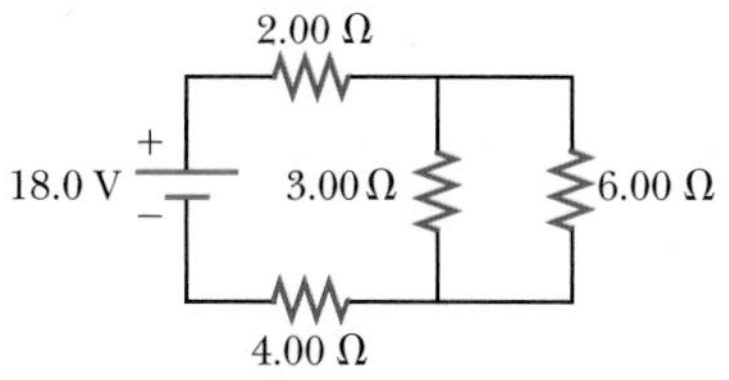

Figure P18.14

15. W (a) You need a 45-Ω resistor, but the stockroom has only 20-Ω and 50-Ω resistors. How can the desired resistance be achieved under these circumstances? (b) What can you do if you need a 35-Ω resistor?

18.4 Kirchhoff's Rules and Complex DC Circuits

Note: For some circuits, the currents are not necessarily in the direction shown.

16. S (a) Find the current in each resistor of Figure P18.16 by using the rules for resistors in series and parallel. (b) Write three independent equations for the three currents using Kirchhoff's laws: one with the node rule; a second using the loop rule through the battery, the 6.0-Ω resistor, and the 24.0-Ω resistor; and the third using the loop rule through the 12.0-Ω and 24.0-Ω resistors. Solve to check the answers found in part (a).

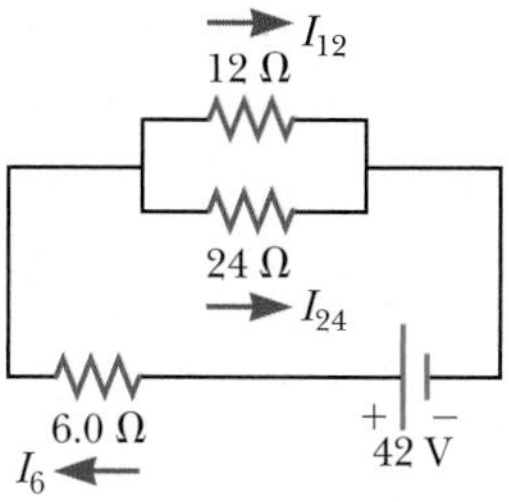

Figure P18.16

17. The ammeter shown in Figure P18.17 reads 2.00 A. Find I_1, I_2, and $\mathcal{E}$.

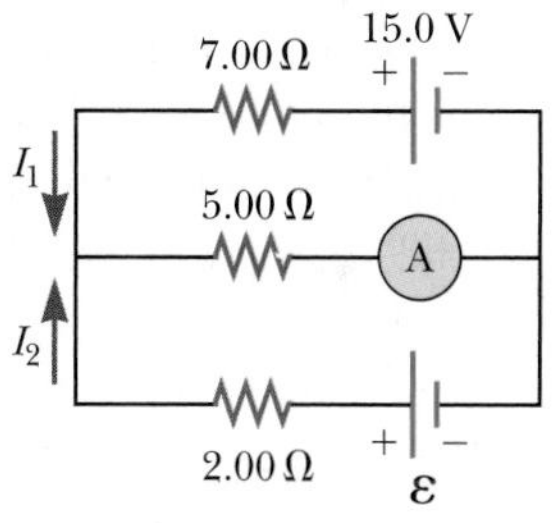

Figure P18.17

18. W For the circuit shown in Figure P18.18, calculate (a) the current in the 2.00-Ω resistor and (b) the potential difference between points a and b, $\Delta V = V_b - V_a$.

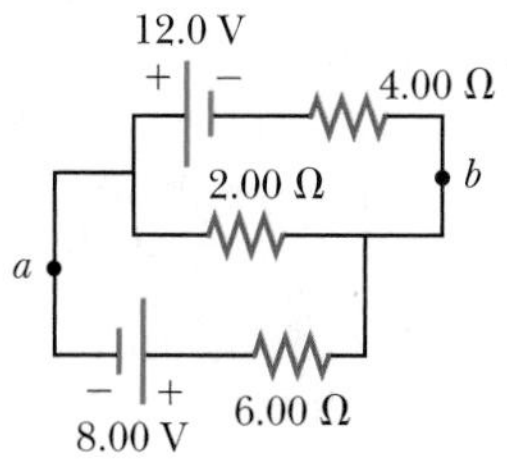

Figure P18.18

19. Taking $R = 1.00$ kΩ and $\mathcal{E} = 250$ V in Figure P18.19, determine the direction and magnitude of the current in the horizontal wire between a and e.

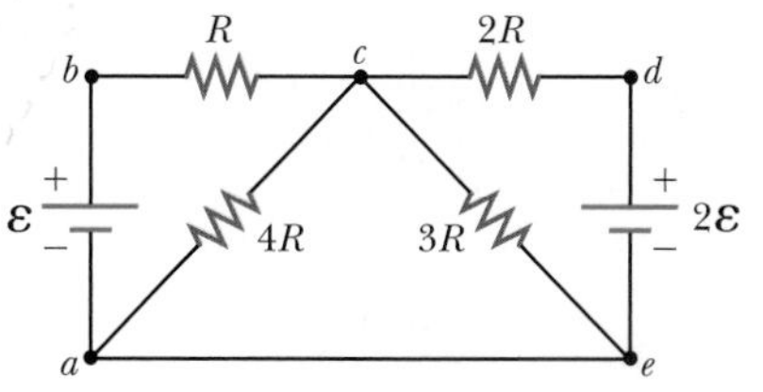

Figure P18.19

20. W In the circuit of Figure P18.20, the current I_1 is 3.0 A and the values of $\mathcal{E}$ and R are unknown. What are the currents I_2 and I_3?

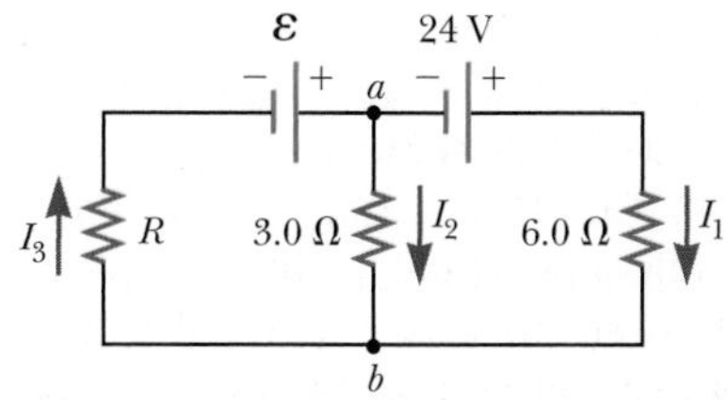

Figure P18.20

21. In the circuit of Figure P18.21, determine (a) the current in each resistor and (b) the potential difference across the 200-Ω resistor.

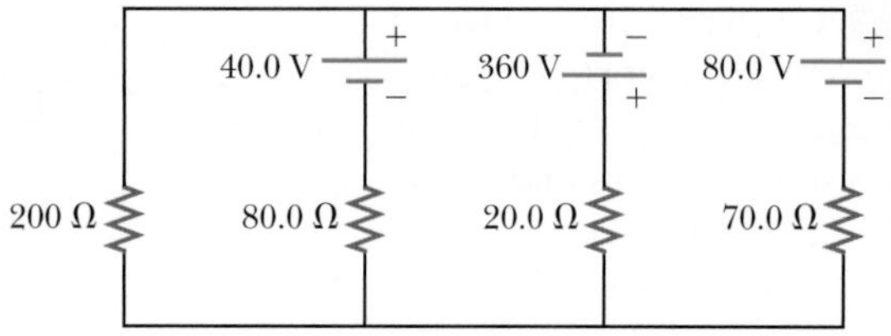

Figure P18.21

22. QIC Four resistors are connected to a battery with a terminal voltage of 12 V, as shown in Figure P18.22. (a) How would you reduce the circuit to an equivalent single resistor connected to the battery? Use this procedure to find the equivalent resistance of the circuit. (b) Find the current delivered by the battery to this equivalent resistance. (c) Determine the power delivered by the battery. (d) Determine the power delivered to the 50.0-Ω resistor.

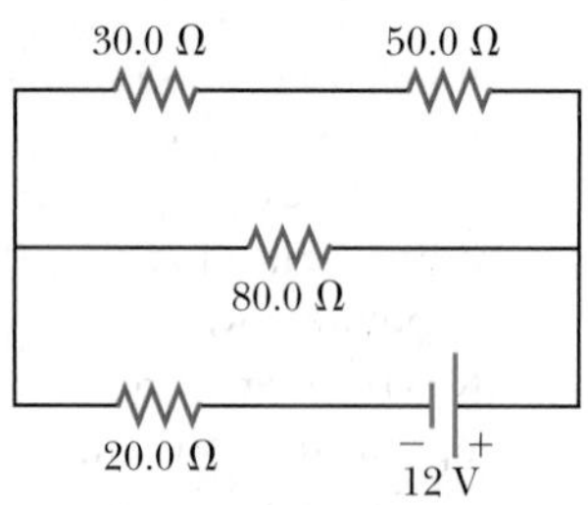

Figure P18.22

23. M Using Kirchhoff's rules, (a) find the current in each resistor shown in Figure P18.23 and (b) find the potential difference between points c and f.

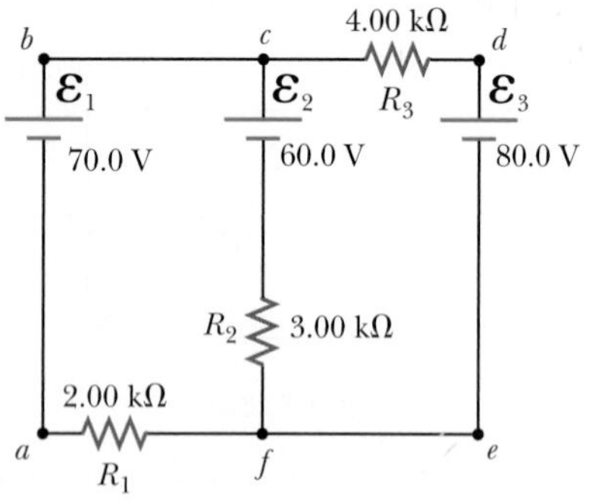

Figure P18.23

24. Two 1.50-V batteries—with their positive terminals in the same direction—are inserted in series into the barrel of a flashlight. One battery has an internal resistance of 0.255 Ω, the other an internal resistance of 0.153 Ω. When the switch is closed, a current of 0.600 A passes through the lamp. (a) What is the lamp's resistance? (b) What fraction of the power dissipated is dissipated in the batteries?

25. QIC (a) Can the circuit shown in Figure P18.25 be reduced to a single resistor connected to the batteries? Explain. (b) Calculate each of the unknown currents I_1, I_2, and I_3 for the circuit.

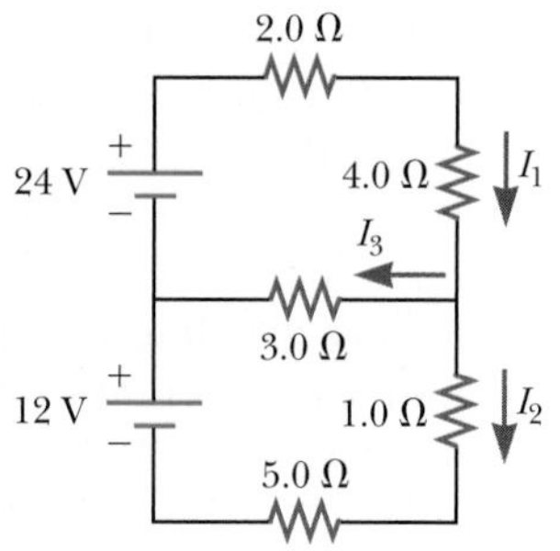

Figure P18.25

26. A dead battery is charged by connecting it to the live battery of another car with jumper cables (Fig. P18.26). Determine the current in (a) the starter and in (b) the dead battery.

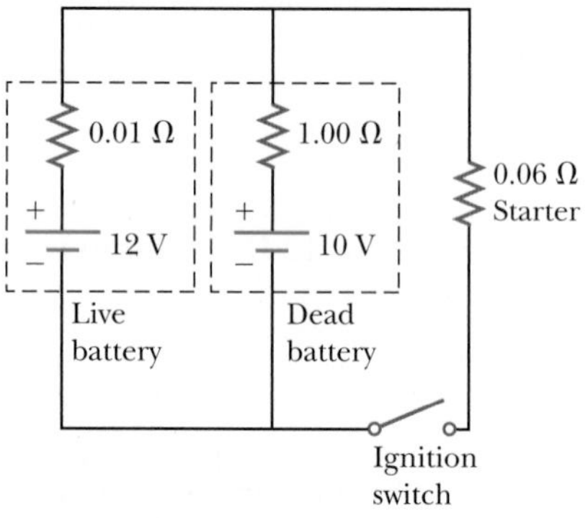

Figure P18.26

27. QIC (a) Can the circuit shown in Figure P18.27 be reduced to a single resistor connected to the batteries? Explain. (b) Find the magnitude of the current and its direction in each resistor.

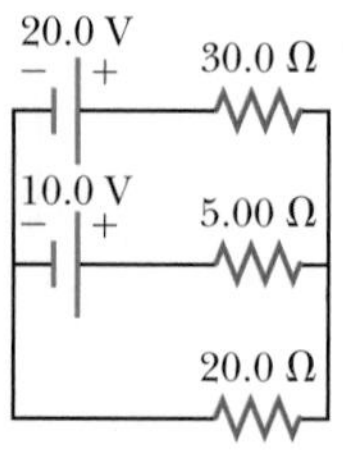

Figure P18.27

28. GP S For the circuit shown in Figure P18.28, use Kirchhoff's rules to obtain equations for (a) the upper loop, (b) the lower loop, and (c) the node on the left side. In each case suppress units for clarity and simplify, combining like terms. (d) Solve the node equation for I_{36}. (e) Using the equation found in (d), eliminate I_{36} from the equation found in part (b). (f) Solve the equations found in part (a) and part (e) simultaneously for the two

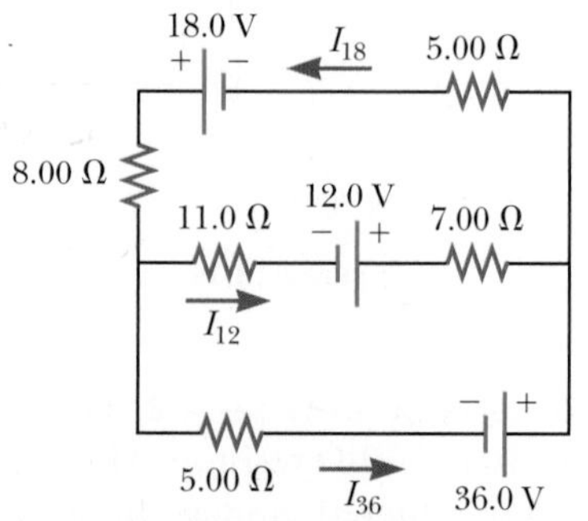

Figure P18.28

unknowns for I_{18} and I_{12}, respectively. (g) Substitute the answers found in part (f) into the node equation found in part (d), solving for I_{36}. (h) What is the significance of the negative answer for I_{12}?

29. M Find the potential difference across each resistor in Figure P18.29.

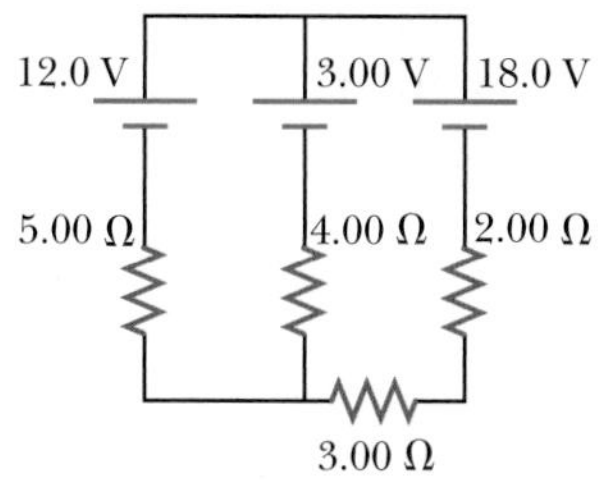

Figure P18.29

18.5 *RC* Circuits

30. Show that $\tau = RC$ has units of time.

31. Consider the series *RC*-circuit shown in Figure 18.16 for which $R = 75.0\ \text{k}\Omega$, $C = 25.0\ \mu\text{F}$, and $\mathcal{E} = 12.0$ V. Find (a) the time constant of the circuit and (b) the charge on the capacitor one time constant after the switch is closed.

32. An uncharged capacitor and a resistor are connected in series to a source of emf. If $\mathcal{E} = 9.00$ V, $C = 20.0\ \mu\text{F}$, and $R = 100\ \Omega$, find (a) the time constant of the circuit, (b) the maximum charge on the capacitor, and (c) the charge on the capacitor after one time constant.

33. Consider a series *RC* circuit as in Figure P18.33 for which $R = 1.00\ \text{M}\Omega$, $C = 5.00\ \mu\text{F}$, and $\mathcal{E} = 30.0$ V. Find (a) the time constant of the circuit and (b) the maximum charge on the capacitor after the switch is thrown closed. (c) Find the current in the resistor 10.0 s after the switch is closed.

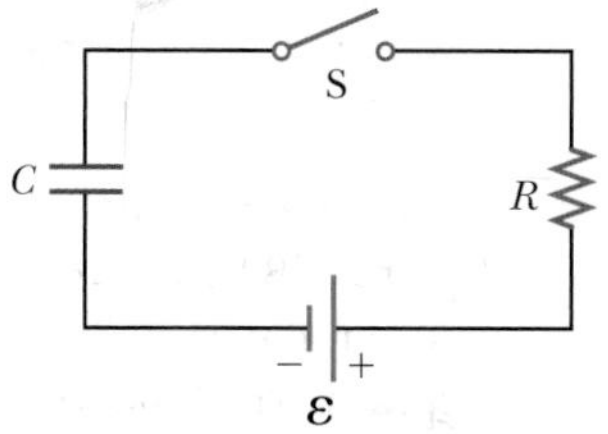

Figure P18.33
Problems 33 and 34.

34. W Consider a series *RC* circuit for which $R = 1.0\ \text{M}\Omega$, $C = 5.0\ \mu\text{F}$, and $\mathcal{E} = 30$ V as in Figure P18.33. Find the charge on the capacitor 10 s after the switch is closed.

35. A charged capacitor is connected to a resistor and a switch as in Figure P18.35. The circuit has a time constant of 1.50 s. Soon after the switch is closed, the charge on the capacitor is 75.0% of its initial charge. (a) Find the time interval required for the capacitor to reach this charge. (b) If $R = 250\ \text{k}\Omega$, what is the value of *C*?

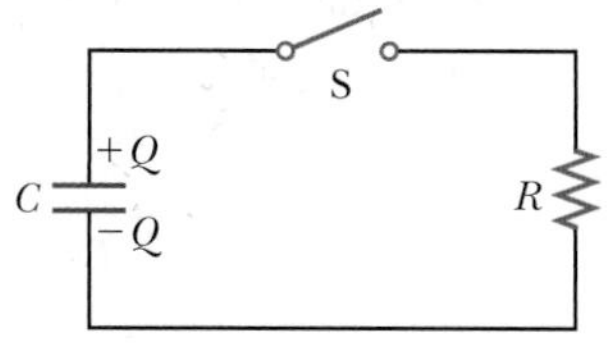

Figure P18.35

36. A 10.0-μF capacitor is charged by a 10.0-V battery through a resistance *R*. The capacitor reaches a potential difference of 4.00 V in a time interval of 3.00 s after charging begins. Find *R*.

18.6 Household Circuits

37. What minimum number of 75-W lightbulbs must be connected in parallel to a single 120-V household circuit to trip a 30.0-A circuit breaker?

38. A lamp ($R = 150\ \Omega$), an electric heater ($R = 25\ \Omega$), and a fan ($R = 50\ \Omega$) are connected in parallel across a 120-V line. (a) What total current is supplied to the circuit? (b) What is the voltage across the fan? (c) What is the current in the lamp? (d) What power is expended in the heater?

39. M A heating element in a stove is designed to dissipate 3 000 W when connected to 240 V. (a) Assuming the resistance is constant, calculate the current in the heating element if it is connected to 120 V. (b) Calculate the power it dissipates at that voltage.

40. Q|C A coffee maker is rated at 1 200 W, a toaster at 1 100 W, and a waffle maker at 1 400 W. The three appliances are connected in parallel to a common 120-V household circuit. (a) What is the current in each appliance when operating independently? (b) What total current is delivered to the appliances when all are operating simultaneously? (c) Is a 15-A circuit breaker sufficient in this situation? Explain.

18.8 Conduction of Electrical Signals by Neurons

41. BIO Assume a length of axon membrane of about 0.10 m is excited by an action potential (length excited = nerve speed × pulse duration = 50.0 m/s × 2.0×10^{-3} s = 0.10 m). In the resting state, the outer surface of the axon wall is charged positively with K^+ ions and the inner wall has an equal and opposite charge of negative organic ions, as shown in Figure P18.41. Model the axon as a parallel-plate capacitor and take $C = \kappa\epsilon_0 A/d$ and $Q = C\Delta V$ to investigate the charge as follows. Use typical values for a cylindrical axon of cell wall thickness $d = 1.0 \times 10^{-8}$ m, axon radius $r = 1.0 \times 10^{1}\ \mu\text{m}$, and cell-wall dielectric constant

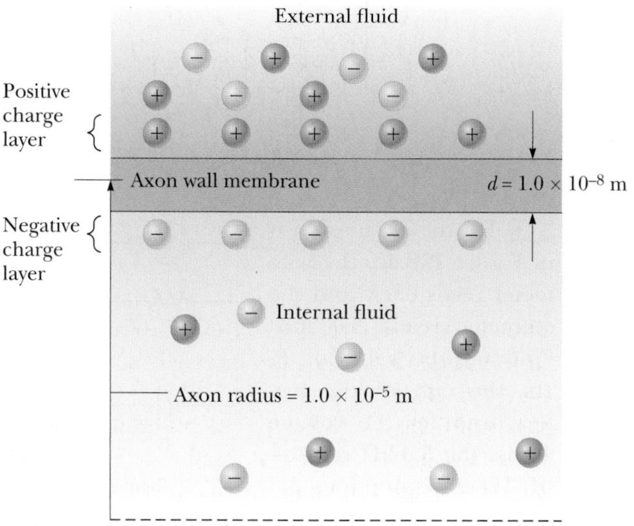

Figure P18.41 Problems 41 and 42.

$\kappa = 3.0$. (a) Calculate the positive charge on the outside of a 0.10-m piece of axon when it is not conducting an electric pulse. How many K^+ ions are on the outside of the axon assuming an initial potential difference of 7.0×10^{-2} V? Is this a large charge per unit area? *Hint:* Calculate the charge per unit area in terms of electronic charge *e* per squared (Å²). An atom has a cross section of about 1 Å² (1 Å = 10^{-10} m). (b) How much positive charge must flow through the cell membrane to reach the excited state of $+3.0 \times 10^{-2}$ V from the resting state of -7.0×10^{-2} V? How many sodium ions (Na^+) is this? (c) If it takes 2.0 ms for the Na^+ ions to enter the axon, what is the average current in the axon wall in this process? (d) How much energy does it take to raise the potential of the inner axon wall to $+3.0 \times 10^{-2}$ V, starting from the resting potential of -7.0×10^{-2} V?

42. BIO Consider the model of the axon as a capacitor from Problem 41 and Figure P18.41. (a) How much energy does it take to restore the inner wall of the axon to -7.0×10^{-2} V, starting from $+3.0 \times 10^{-2}$ V? (b) Find the average current in the axon wall during this process.

43. BIO Using Figure 18.28b and the results of Problems 18.41d and 18.42a, find the power supplied by the axon per action potential.

Additional Problems

44. How many different resistance values can be constructed from a 2.0-Ω, a 4.0-Ω, and a 6.0-Ω resistor? Show how you would get each resistance value either individually or by combining them.

45. (a) Calculate the potential difference between points *a* and *b* in Figure P18.45 and (b) identify which point is at the higher potential.

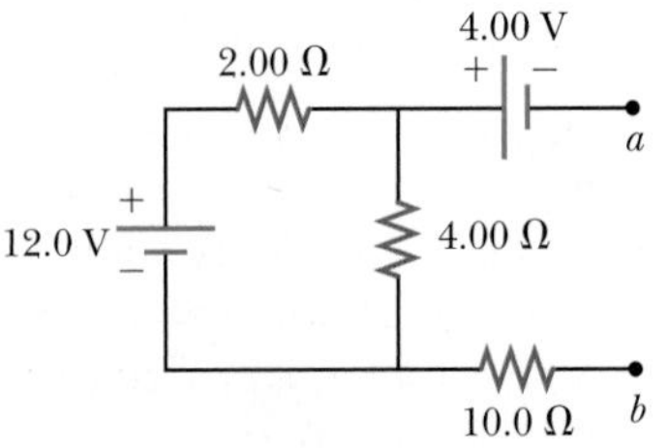

Figure P18.45

46. Q|C For the circuit shown in Figure P18.46, the voltmeter reads 6.0 V and the ammeter reads 3.0 mA. Find (a) the value of *R*, (b) the emf of the battery, and (c) the voltage across the 3.0-kΩ resistor. (d) What assumptions did you have to make to solve this problem?

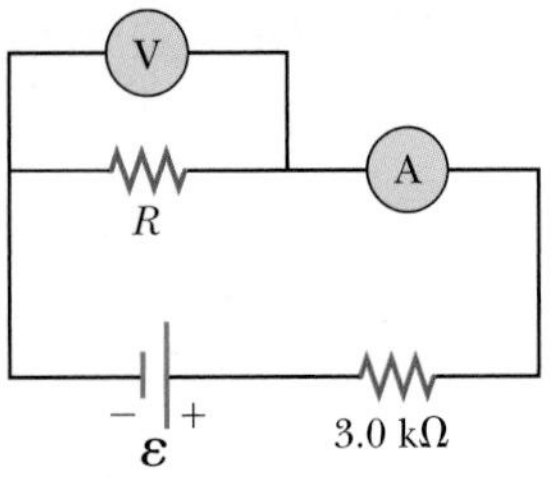

Figure P18.46

47. Find (a) the equivalent resistance of the circuit in Figure P18.47, (b) each current in the circuit, (c) the potential difference across each resistor, and (d) the power dissipated by each resistor.

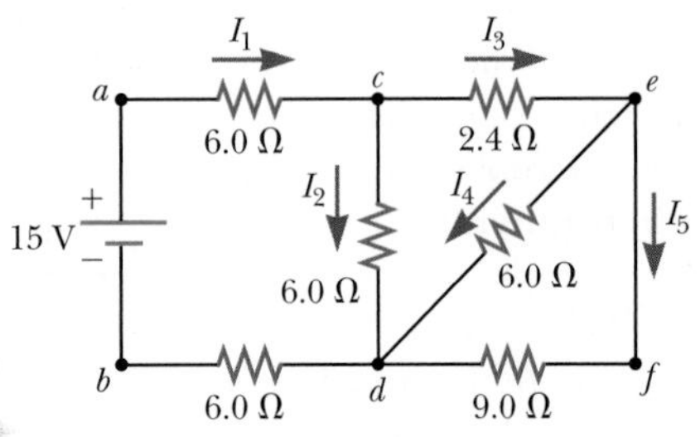

Figure P18.47

48. Three 60.0-W, 120-V lightbulbs are connected across a 120-V power source, as shown in Figure P18.48. Find (a) the total power delivered to the three bulbs and (b) the potential difference across each. Assume the resistance of each bulb is constant (even though, in reality, the resistance increases markedly with current).

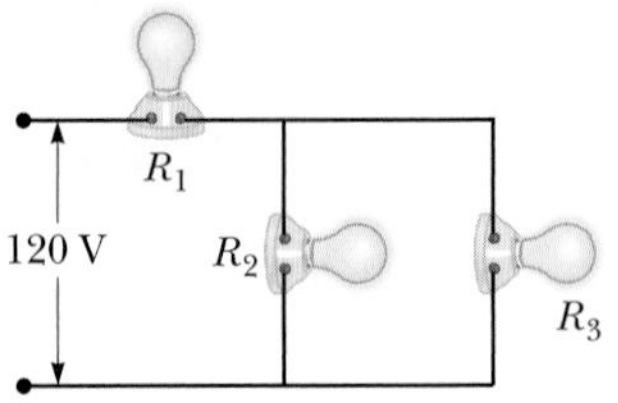

Figure P18.48

49. When two unknown resistors are connected in series with a battery, the battery delivers 225 W and carries a total current of 5.00 A. For the same total current, 50.0 W is delivered when the resistors are connected in parallel. Determine the value of each resistor.

50. Q|C The circuit in Figure P18.50a consists of three resistors and one battery with no internal resistance. (a) Find the current in the 5.00-Ω resistor. (b) Find the power delivered to the 5.00-Ω resistor. (c) In each of the circuits in Figures P18.50b, P18.50c, and P18.50d, an additional 15.0-V battery has been inserted into the circuit. Which diagram or diagrams represent a circuit that requires the use of Kirchhoff's rules to find the

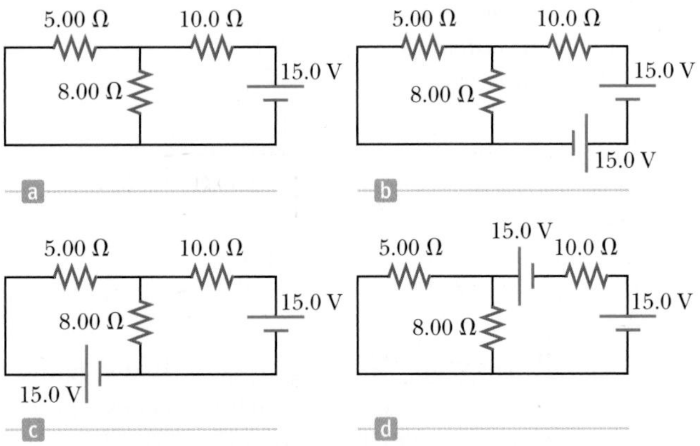

Figure P18.50

currents? Explain why. (d) In which of these three new circuits is the smallest amount of power delivered to the 10.0-Ω resistor? (You need not calculate the power in each circuit if you explain your answer.)

51. **QIC** **S** A circuit consists of three identical lamps, each of resistance R, connected to a battery as in Figure P18.51. (a) Calculate an expression for the equivalent resistance of the circuit when the switch is open. Repeat the calculation when the switch is closed. (b) Write an expression for the power supplied by the battery when the switch is open. Repeat the calculation when the switch is closed. (c) Using the results already obtained, explain what happens to the brightness of the lamps when the switch is closed.

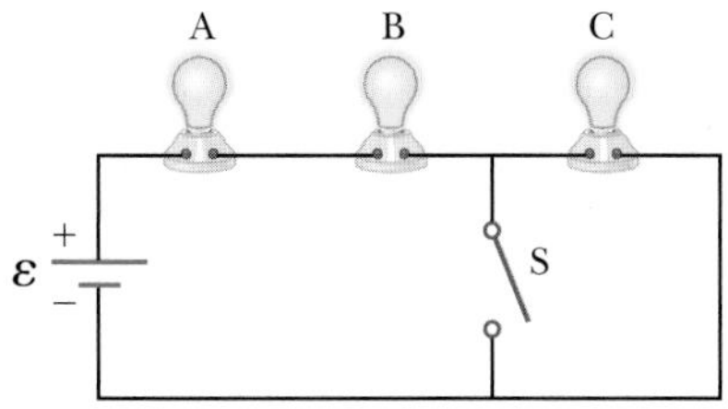

Figure P18.51

52. The resistance between points a and b in Figure P18.52 drops to one-half its original value when switch S is closed. Determine the value of R.

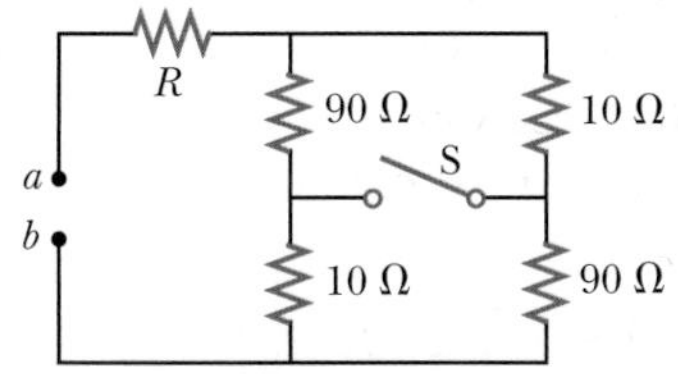

Figure P18.52

53. The circuit in Figure P18.53 has been connected for several seconds. Find the current (a) in the 4.00-V battery, (b) in the 3.00-Ω resistor, (c) in the 8.00-V battery, and (d) in the 3.00-V battery. (e) Find the charge on the capacitor.

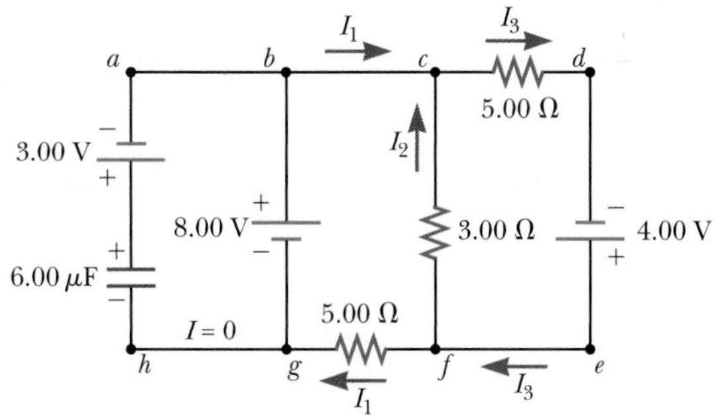

Figure P18.53

54. An emf of 10 V is connected to a series RC circuit consisting of a resistor of 2.0×10^6 Ω and an initially uncharged capacitor of 3.0 μF. Find the time required for the charge on the capacitor to reach 90% of its final value.

55. The student engineer of a campus radio station wishes to verify the effectiveness of the lightning rod on the antenna mast (Fig. P18.55). The unknown resistance R_x is between points C and E. Point E is a "true ground," but is inaccessible for direct measurement because the stratum in which it is located is several meters below Earth's surface. Two identical rods are driven into the ground at A and B, introducing an unknown resistance R_y. The procedure for finding the unknown resistance R_x is as follows. Measure resistance R_1 between points A and B. Then connect A and B with a heavy conducting wire and measure resistance R_2 between points A and C. (a) Derive a formula for R_x in terms of the observable resistances R_1 and R_2. (b) A satisfactory ground resistance would be $R_x < 2.0$ Ω. Is the grounding of the station adequate if measurements give $R_1 = 13$ Ω and $R_2 = 6.0$ Ω?

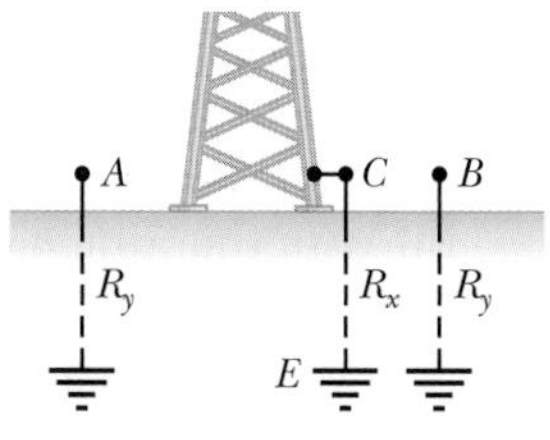

Figure P18.55

56. The resistor R in Figure P18.56 dissipates 20 W of power. Determine the value of R.

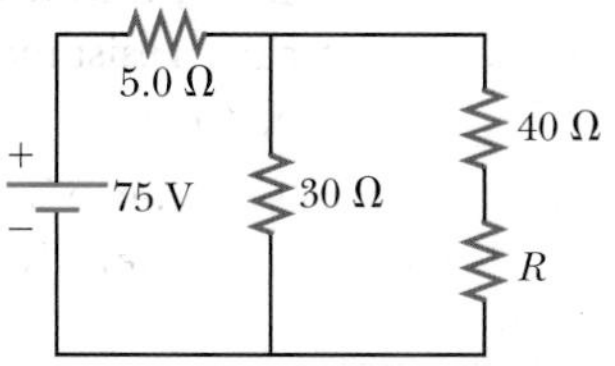

Figure P18.56

57. A voltage ΔV is applied to a series configuration of n resistors, each of resistance R. The circuit components are reconnected in a parallel configuration, and voltage ΔV is again applied. Show that the power consumed by the series configuration is $1/n^2$ times the power consumed by the parallel configuration.

58. For the network in Figure P18.58, show that the resistance between points a and b is $R_{ab} = \frac{27}{17}$ Ω. (*Hint:* Connect a battery with emf $\mathcal{E}$ across points a and b and determine $\mathcal{E}/I$, where I is the current in the battery.)

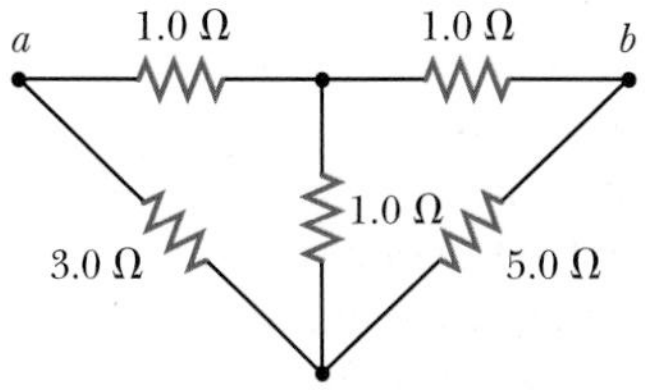

Figure P18.58

59. A battery with an internal resistance of 10.0 Ω produces an open circuit voltage of 12.0 V. A variable load resistance with a range from 0 to 30.0 Ω is connected across the battery. (*Note:* A battery has a resistance that depends on the condition of its chemicals and that increases as the battery ages. This internal resistance can be represented in a simple circuit diagram as a resistor in series with the battery.) (a) Graph the power dissipated in the load resistor as a function of the load resistance. (b) With your graph, demonstrate the following important theorem: *The power delivered to a load is a maximum if the load resistance equals the internal resistance of the source.*

60. The circuit in Figure P18.60 contains two resistors, $R_1 = 2.0\ \text{k}\Omega$ and $R_2 = 3.0\ \text{k}\Omega$, and two capacitors, $C_1 = 2.0\ \mu\text{F}$ and $C_2 = 3.0\ \mu\text{F}$, connected to a battery with emf $\mathcal{E} = 120$ V. If there are no charges on the capacitors before switch S is closed, determine the charges q_1 and q_2 on capacitors C_1 and C_2, respectively, as functions of time, after the switch is closed. *Hint:* First reconstruct the circuit so that it becomes a simple *RC* circuit containing a single resistor and single capacitor in series, connected to the battery, and then determine the total charge q stored in the circuit.

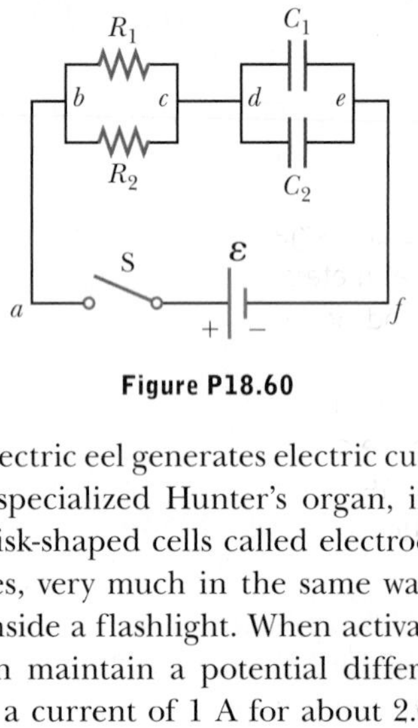

Figure P18.60

61. BIO An electric eel generates electric currents through its highly specialized Hunter's organ, in which thousands of disk-shaped cells called electrocytes are lined up in series, very much in the same way batteries are lined up inside a flashlight. When activated, each electrocyte can maintain a potential difference of about 150 mV at a current of 1 A for about 2.0 ms. Suppose a grown electric eel has 4.0×10^3 electrocytes and can deliver up to 300 shocks in rapid series over about 1 s. (a) What maximum electrical power can an electric eel generate? (b) Approximately how much energy does it release in one shock? (c) How high would a mass of 1 kg have to be lifted so that its gravitational potential energy equals the energy released in 300 such shocks?

62. In Figure P18.62, $R_1 = 0.100\ \Omega$, $R_2 = 1.00\ \Omega$, and $R_3 = 10.0\ \Omega$. Find the equivalent resistance of the circuit and the current in each resistor when a 5.00-V power supply is connected between (a) points *A* and *B*, (b) points *A* and *C*, and (c) points *A* and *D*.

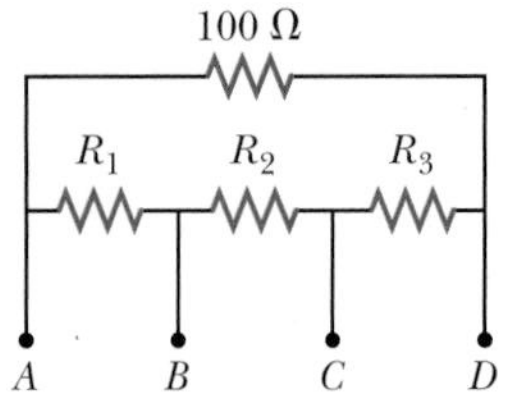

Figure P18.62

63. M What are the expected readings of the ammeter and voltmeter for the circuit in Figure P18.63?

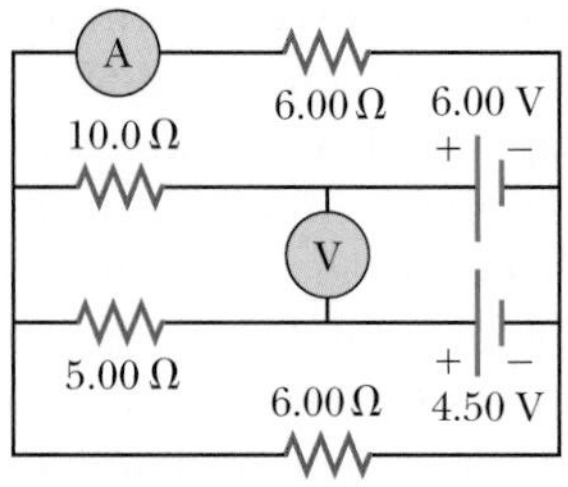

Figure P18.63

64. Consider the two arrangements of batteries and bulbs shown in Figure P18.64. The two bulbs are identical and have resistance *R*, and the two batteries are identical with output voltage Δ*V*. (a) In case 1, with the two bulbs in series, compare the brightness of each bulb, the current in each bulb, and the power delivered to each bulb. (b) In case 2, with the two bulbs in parallel, compare the brightness of each bulb, the current in each bulb, and the power supplied to each bulb. (c) Which bulbs are brighter, those in case 1 or those in case 2? (d) In each case, if one bulb fails, will the other go out as well? If the other bulb doesn't fail, will it get brighter or stay the same? (Problem 64 is courtesy of E. F. Redish. For other problems of this type, visit http://www.physics.umd.edu/perg/.)

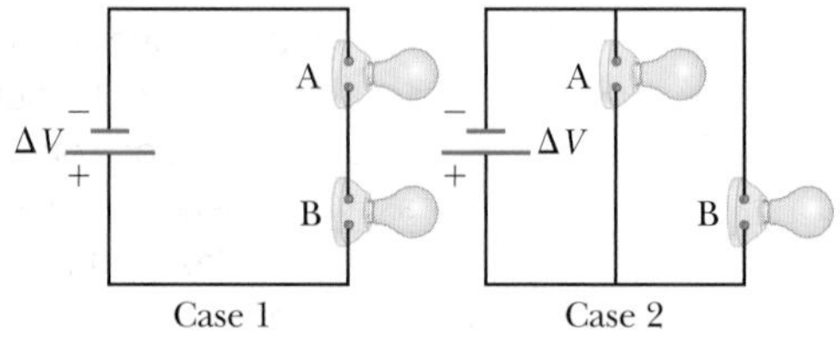

Figure P18.64

65. The given pair of capacitors in Figure P18.65 is fully charged by a 12.0-V battery. The battery is disconnected and the circuit closed. After 1.00 ms, how much charge remains on (a) the 3.00-μF capacitor? (b) The 2.00-μF capacitor? (c) What is the current in the resistor?

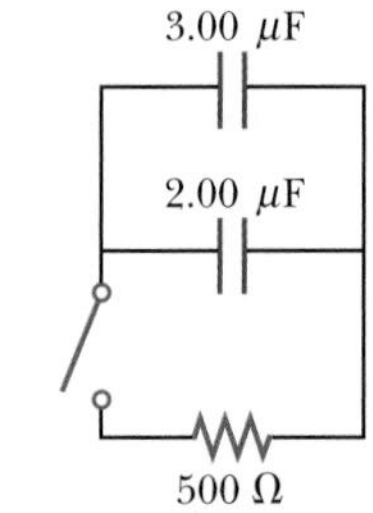

Figure P18.65

66. A 2.00-nF capacitor with an initial charge of 5.10 μC is discharged through a 1.30-kΩ resistor. (a) Calculate the magnitude of the current in the resistor 9.00 μs after the resistor is connected across the terminals of the capacitor. (b) What charge remains on the capacitor after 8.00 μs? (c) What is the maximum current in the resistor?

26. A current of 17.0 mA is maintained in a single circular loop with a circumference of 2.00 m. A magnetic field of 0.800 T is directed parallel to the plane of the loop. What is the magnitude of the torque exerted by the magnetic field on the loop?

27. **M** An eight-turn coil encloses an elliptical area having a major axis of 40.0 cm and a minor axis of 30.0 cm (Fig. P19.27). The coil lies in the plane of the page and carries a clockwise current of 6.00 A. If the coil is in a uniform magnetic field of 2.00×10^{-4} T directed toward the left of the page, what is the magnitude of the torque on the coil? *Hint:* The area of an ellipse is $A = \pi ab$, where a and b are, respectively, the semimajor and semiminor axes of the ellipse.

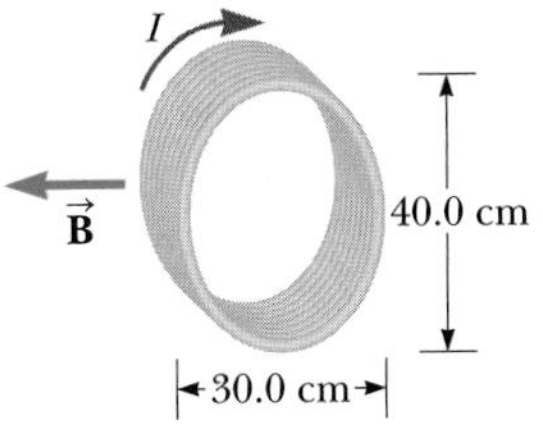

Figure P19.27

28. **W** A rectangular loop consists of 100 closely wrapped turns and has dimensions 0.400 m by 0.300 m. The loop is hinged along the y-axis, and the plane of the coil makes an angle of 30.0° with the x-axis (Fig. P19.28). What is the magnitude of the torque exerted on the loop by a uniform magnetic field of 0.800 T directed along the x-axis when the current in the windings has a value of 1.20 A in the direction shown? What is the expected direction of rotation of the loop?

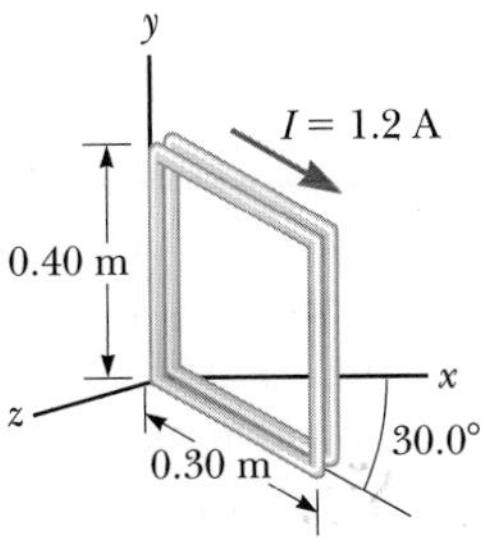

Figure P19.28

29. A 200-turn rectangular coil having dimensions of 3.0 cm by 5.0 cm is placed in a uniform magnetic field of magnitude 0.90 T. (a) Find the current in the coil if the maximum torque exerted on it by the magnetic field is 0.15 N · m. (b) Find the magnitude of the torque on the coil when the magnetic field makes an angle of 25° with the normal to the plane of the coil.

30. A copper wire is 8.00 m long and has a cross-sectional area of 1.00×10^{-4} m^2. The wire forms a one-turn loop in the shape of a square and is then connected to a battery that applies a potential difference of 0.100 V. If the loop is placed in a uniform magnetic field of magnitude 0.400 T, what is the maximum torque that can act on it? The resistivity of copper is $1.70 \times 10^{-8}\ \Omega \cdot$ m.

31. A long piece of wire with a mass of 0.100 kg and a total length of 4.00 m is used to make a square coil with a side of 0.100 m. The coil is hinged along a horizontal side, carries a 3.40-A current, and is placed in a vertical magnetic field with a magnitude of 0.010 0 T. (a) Determine the angle that the plane of the coil makes with the vertical when the coil is in equilibrium. (b) Find the torque acting on the coil due to the magnetic force at equilibrium.

32. **GP** A rectangular loop has dimensions 0.500 m by 0.300 m. The loop is hinged along the x-axis and lies in the xy-plane (Fig. P19.32). A uniform magnetic field of 1.50 T is directed at an angle of 40.0° with respect to the positive y-axis and lies parallel everywhere to the yz-plane. The loop carries a current of 0.900 A in the direction shown. (Ignore gravitation.) (a) In what direction is magnetic force exerted on wire segment ab? What is the direction of the magnetic torque associated with this force, as computed with respect to the x-axis? (b) What is the direction of the magnetic force exerted on segment cd? What is the direction of the magnetic torque associated with this force, again computed with respect to the x-axis? (c) Can the forces examined in parts (a) and (b) combine to cause the loop to rotate around the x-axis? Can they affect the motion of the loop in any way? Explain. (d) What is the direction (in the yz-plane) of the magnetic force exerted on segment bc? Measuring torques with respect to the x-axis, what is the direction of the torque exerted by the force on segment bc? (e) Looking toward the origin along the positive x-axis, will the loop rotate clockwise or counterclockwise? (f) Compute the magnitude of the magnetic moment of the loop. (g) What is the angle between the magnetic moment vector and the magnetic field? (h) Compute the torque on the loop using the values found for the magnetic moment and magnetic field.

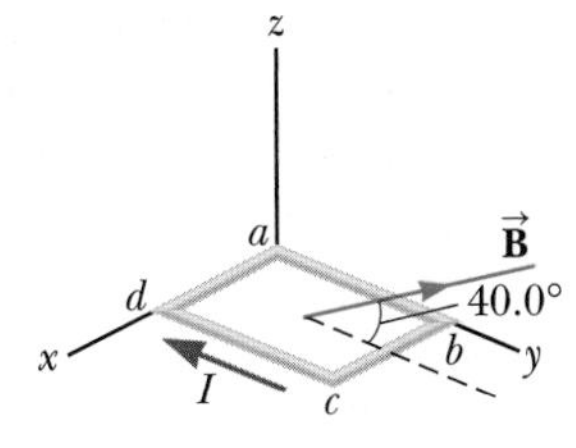

Figure P19.32

19.6 Motion of a Charged Particle in a Magnetic Field

33. An electron moves in a circular path perpendicular to a magnetic field of magnitude 0.235 T. If the kinetic energy of the electron is 3.30×10^{-19} J, find (a) the speed of the electron and (b) the radius of the circular path.

34. A proton travels with a speed of 5.02×10^6 m/s at an angle of 60° with the direction of a magnetic field of magnitude 0.180 T in the positive x-direction. What are (a) the magnitude of the magnetic force on the proton and (b) the proton's acceleration?

35. Figure P19.35a is a diagram of a device called a velocity selector, in which particles of a specific velocity pass through undeflected while those with greater or lesser velocities are deflected either upwards or downwards. An electric field is directed perpendicular to a magnetic field, producing an electric force and a magnetic force on the charged particle that can be equal in magnitude and opposite in direction (Fig. P19.35b) and hence cancel. Show that particles with a speed of $v = E/B$ will pass through the velocity selector undeflected.

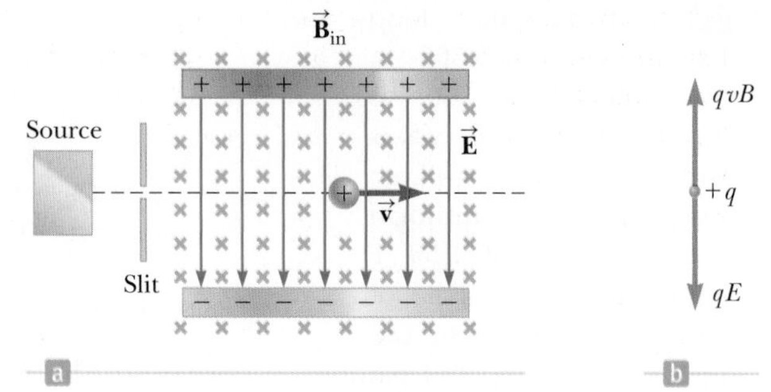

Figure P19.35

36\. **W** Consider the mass spectrometer shown schematically in Figure P19.36. The electric field between the plates of the velocity selector is 950 V/m, and the magnetic fields in both the velocity selector and the deflection chamber have magnitudes of 0.930 T. Calculate the radius of the path in the system for a singly charged ion with mass $m = 2.18 \times 10^{-26}$ kg. *Hint:* See Problem 35.

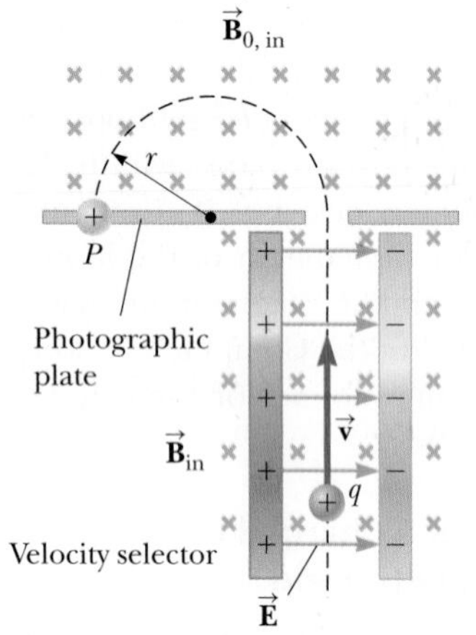

Figure P19.36 (Problems 36 and 41) A mass spectrometer. Charged particles are first sent through a velocity selector. They then enter a region where a magnetic field $\vec{B}_0$ (directed inward) causes positive ions to move in a semicircular path and strike a photographic film at *P*.

37\. A singly charged positive ion has a mass of 2.50×10^{-26} kg. After being accelerated through a potential difference of 250 V, the ion enters a magnetic field of 0.500 T, in a direction perpendicular to the field. Calculate the radius of the path of the ion in the field.

38\. A mass spectrometer is used to examine the isotopes of uranium. Ions in the beam emerge from the velocity selector at a speed of 3.00×10^5 m/s and enter a uniform magnetic field of 0.600 T directed perpendicularly to the velocity of the ions. What is the distance between the impact points formed on the photographic plate by singly charged ions of ^{235}U and ^{238}U?

39\. A proton is at rest at the plane vertical boundary of a region containing a uniform vertical magnetic field *B* (Fig. P19.39). An alpha particle moving horizontally makes a head-on elastic collision with the proton. Immediately after the collision, both particles enter the magnetic field, moving perpendicular to the direction of the field. The radius of the proton's trajectory is *R*. The mass of the alpha particle is four times that of the proton, and its charge is twice that of the proton. Find the radius of the alpha particle's trajectory.

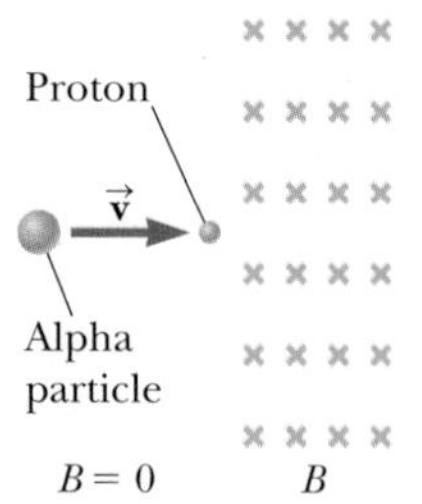

Figure P19.39

40\. **S** A particle with charge *q* and kinetic energy *KE* travels in a uniform magnetic field of magnitude *B*. If the particle moves in a circular path of radius *R*, find expressions for (a) its speed and (b) its mass.

41\. A particle passes through a mass spectrometer as illustrated in Figure P19.36. The electric field between the plates of the velocity selector has a magnitude of 8 250 V/m, and the magnetic fields in both the velocity selector and the deflection chamber have magnitudes of 0.093 1 T. In the deflection chamber the particle strikes a photographic plate 39.6 cm removed from its exit point after traveling in a semicircle. (a) What is the mass-to-charge ratio of the particle? (b) What is the mass of the particle if it is doubly ionized? (c) What is its identity, assuming it's an element?

42\. **S** A proton (charge $+e$, mass m_p), a deuteron (charge $+e$, mass $2m_p$), and an alpha particle (charge $+2e$, mass $4m_p$) are accelerated from rest through a common potential difference ΔV. Each of the particles enters a uniform magnetic field $\vec{B}$, with its velocity in a direction perpendicular to $\vec{B}$. The proton moves in a circular path of radius r_p. In terms of r_p, determine (a) the radius r_d of the circular orbit for the deuteron and (b) the radius r_α for the alpha particle.

19.7 Magnetic Field of a Long, Straight Wire and Ampère's Law

43\. A lightning bolt may carry a current of 1.00×10^4 A for a short time. What is the resulting magnetic field 100 m from the bolt? Suppose the bolt extends far above and below the point of observation.

44\. In each of parts (a), (b), and (c) of Figure P19.44, find the direction of the current in the wire that would produce a magnetic field directed as shown.

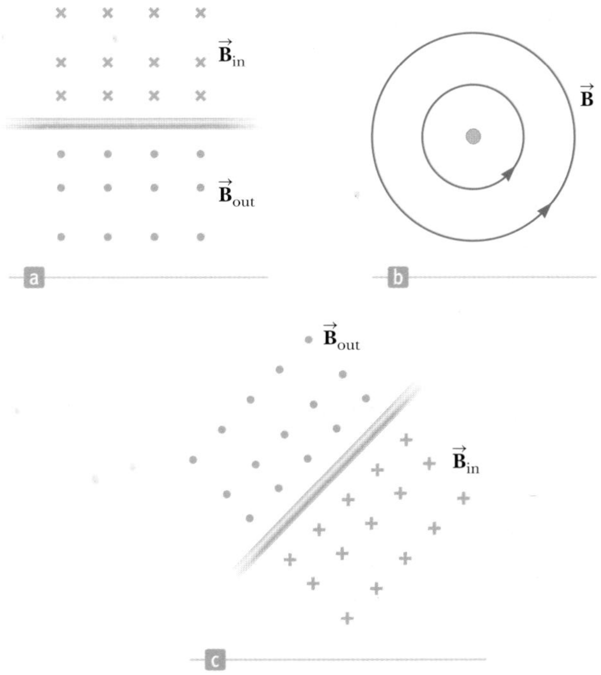

Figure P19.44

45. BIO Neurons in our bodies carry weak currents that produce detectable magnetic fields. A technique called *magnetoencephalography*, or MEG, is used to study electrical activity in the brain using this concept. This technique is capable of detecting magnetic fields as weak as 1.0×10^{-15} T. Model the neuron as a long wire carrying a current and find the current it must carry to produce a field of this magnitude at a distance of 4.0 cm from the neuron.

46. In 1962 measurements of the magnetic field of a large tornado were made at the Geophysical Observatory in Tulsa, Oklahoma. If the magnitude of the tornado's field was $B = 1.50 \times 10^{-8}$ T pointing north when the tornado was 9.00 km east of the observatory, what current was carried up or down the funnel of the tornado? Model the vortex as a long, straight wire carrying a current.

47. BIO A cardiac pacemaker can be affected by a static magnetic field as small as 1.7 mT. How close can a pacemaker wearer come to a long, straight wire carrying 20 A?

48. The two wires shown in Figure P19.48 are separated by $d = 10.0$ cm and carry currents of $I = 5.00$ A in opposite directions. Find the magnitude and direction of the net magnetic field (a) at a point midway between the wires; (b) at point P_1, 10.0 cm to the right of the wire on the right; and (c) at point P_2, $2d = 20.0$ cm to the left of the wire on the left.

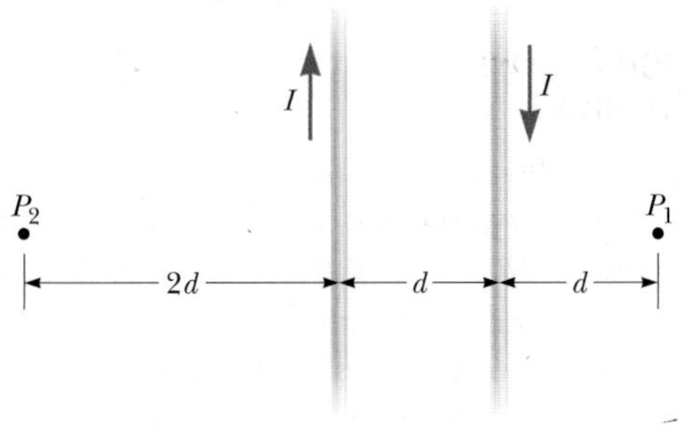

Figure P19.48

49. Four long, parallel conductors carry equal currents of $I = 5.00$ A. Figure P19.49 is an end view of the conductors. The direction of the current is into the page at points A and B (indicated by the crosses) and out of the page at C and D (indicated by the dots). Calculate the magnitude and direction of the magnetic field at point P, located at the center of the square with edge of length 0.200 m.

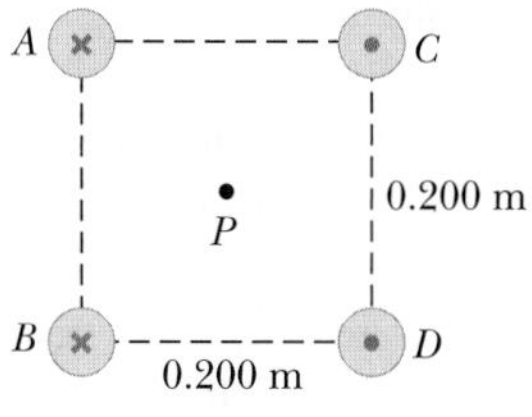

Figure P19.49

50. Two long, parallel wires carry currents of $I_1 = 3.00$ A and $I_2 = 5.00$ A in the direction indicated in Figure P19.50. (a) Find the magnitude and direction of the magnetic field at a point midway between the wires ($d = 20.0$ cm). (b) Find the magnitude and direction of the magnetic field at point P, located $d = 20.0$ cm above the wire carrying the 5.00-A current.

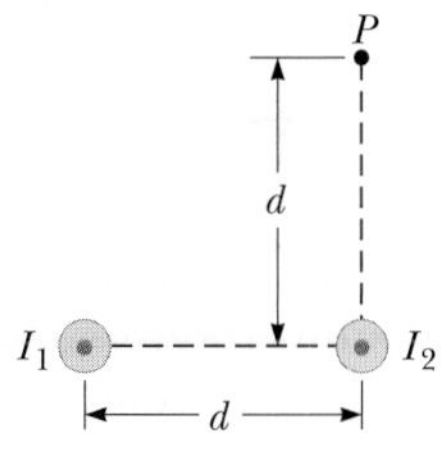

Figure P19.50

51. M A wire carries a 7.00-A current along the x-axis, and another wire carries a 6.00-A current along the y-axis, as shown in Figure P19.51. What is the magnetic field at point P, located at $x = 4.00$ m, $y = 3.00$ m?

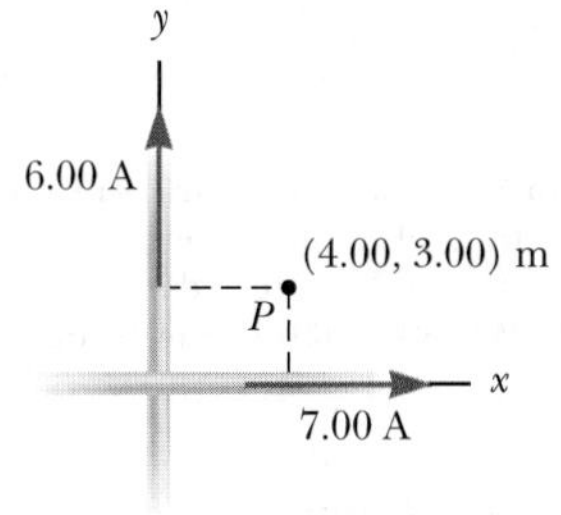

Figure P19.51

52. GP A long, straight wire lies on a horizontal table in the xy-plane and carries a current of 1.20 μA in the positive x-direction along the x-axis. A proton is traveling in the negative x-direction at speed 2.30×10^4 m/s a distance d above the wire (i.e., $z = d$). (a) What is the direction of the magnetic field of the wire at the position of the proton? (b) What is the direction of the magnetic force acting on the proton? (c) Explain why the direction of the proton's motion doesn't change. (d) Using Newton's second law, find a symbolic expression for d in terms of the acceleration of gravity g, the proton mass m, its speed v, charge q, and the current I.

(e) Find the numeric answer for the distance d using the results of part (d).

53. The magnetic field 40.0 cm away from a long, straight wire carrying current 2.00 A is 1.00 μT. (a) At what distance is it 0.100 μT? (b) At one instant, the two conductors in a long household extension cord carry equal 2.00-A currents in opposite directions. The two wires are 3.00 mm apart. Find the magnetic field 40.0 cm away from the middle of the straight cord, in the plane of the two wires. (c) At what distance is it one-tenth as large? (d) The center wire in a coaxial cable carries current 2.00 A in one direction, and the sheath around it carries current 2.00 A in the opposite direction. What magnetic field does the cable create at points outside?

54. QIC S Two long, parallel wires separated by a distance $2d$ carry equal currents in the same direction. An end view of the two wires is shown in Figure P19.54, where the currents are out of the page. (a) What is the direction of the magnetic field at P on the x-axis set up by the two wires? (b) Find an expression for the magnitude of the field at P. (c) From your result to part (b), determine the field at a point midway between the two wires. Does your result meet with your expectation? Explain.

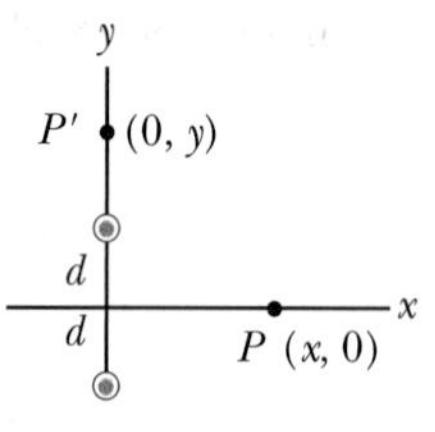

Figure P19.54

19.8 Magnetic Force Between Two Parallel Conductors

55. Two long, parallel wires separated by 2.50 cm carry currents in opposite directions. The current in one wire is 1.25 A, and the current in the other is 3.50 A. (a) Find the magnitude of the force per unit length that one wire exerts on the other. (b) Is the force attractive or repulsive?

56. QIC Two parallel wires separated by 4.0 cm repel each other with a force per unit length of 2.0×10^{-4} N/m. The current in one wire is 5.0 A. (a) Find the current in the other wire. (b) Are the currents in the same direction or in opposite directions? (c) What would happen if the direction of one current were reversed and doubled?

57. A wire with a weight per unit length of 0.080 N/m is suspended directly above a second wire. The top wire carries a current of 30.0 A, and the bottom wire carries a current of 60.0 A. Find the distance of separation between the wires so that the top wire will be held in place by magnetic repulsion.

58. In Figure P19.58 the current in the long, straight wire is $I_1 = 5.00$ A, and the wire lies in the plane of the rectangular loop, which carries 10.0 A. The dimensions shown are $c = 0.100$ m, $a = 0.150$ m, and $\ell = 0.450$ m. Find the magnitude and direction of the net force exerted by the magnetic field due to the straight wire on the loop.

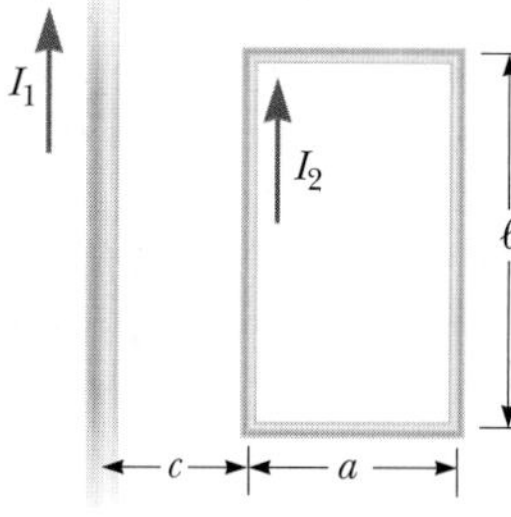

Figure P19.58

19.9 Magnetic Fields of Current Loops and Solenoids

59. M A long solenoid that has 1 000 turns uniformly distributed over a length of 0.400 m produces a magnetic field of magnitude 1.00×10^{-4} T at its center. What current is required in the windings for that to occur?

60. A certain superconducting magnet in the form of a solenoid of length 0.50 m can generate a magnetic field of 9.0 T in its core when its coils carry a current of 75 A. The windings, made of a niobium–titanium alloy, must be cooled to 4.2 K. Find the number of turns in the solenoid.

61. It is desired to construct a solenoid that will have a resistance of 5.00 Ω (at 20°C) and produce a magnetic field of 4.00×10^{-2} T at its center when it carries a current of 4.00 A. The solenoid is to be constructed from copper wire having a diameter of 0.500 mm. If the radius of the solenoid is to be 1.00 cm, determine (a) the number of turns of wire needed and (b) the length the solenoid should have.

62. A solenoid 10.0 cm in diameter and 75.0 cm long is made from copper wire of diameter 0.100 cm, with very thin insulation. The wire is wound onto a cardboard tube in a single layer, with adjacent turns touching each other. What power must be delivered to the solenoid if it is to produce a field of 8.00 mT at its center?

63. An electron is moving at a speed of 1.0×10^4 m/s in a circular path of radius 2.0 cm inside a solenoid. The magnetic field of the solenoid is perpendicular to the plane of the electron's path. Find (a) the strength of the magnetic field inside the solenoid and (b) the current in the solenoid if it has 25 turns per centimeter.

Additional Problems

64. QIC Figure P19.64 is a setup that can be used to measure magnetic fields. A rectangular coil of wire contains N turns has a width w. The coil is attached to one arm of a balance and is suspended between the poles of a magnet. The field is uniform and perpendicular to the plane of the coil. The system is first balanced when the current in the coil is zero. When the switch is closed and the coil carries a current I, a mass m must be added to the right side to balance the system. (a) Find an expression for the magnitude of the magnetic field

and determine its direction. (b) Why is the result independent of the vertical dimension of the coil? (c) Suppose the coil has 50 turns and width of 5.0 cm. When the switch is closed, the coil carries a current of 0.30 A, and a mass of 20.0 g must be added to the right side to balance the system. What is the magnitude of the magnetic field?

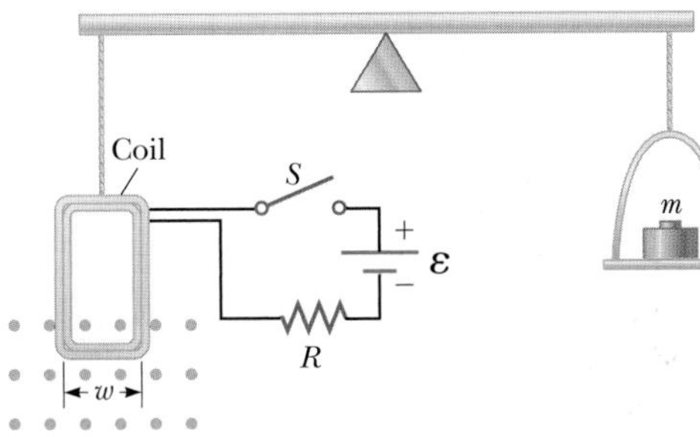

Figure P19.64

65. Two coplanar and concentric circular loops of wire carry currents of $I_1 = 5.00$ A and $I_2 = 3.00$ A in opposite directions as in Figure P19.65. (a) If $r_1 = 12.0$ cm and $r_2 = 9.00$ cm, what are (a) the magnitude and (b) the direction of the net magnetic field at the center of the two loops? (c) Let r_1 remain fixed at 12.0 cm and let r_2 be a variable. Determine the value of r_2 such that the net field at the center of the loop is zero.

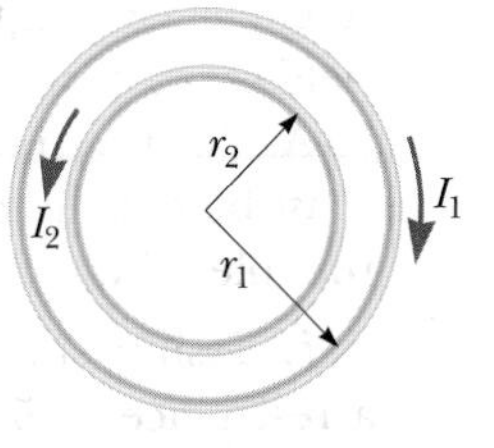

Figure P19.65

66. An electron moves in a circular path perpendicular to a constant magnetic field of magnitude 1.00 mT. The angular momentum of the electron about the center of the circle is 4.00×10^{-25} kg · m²/s. Determine (a) the radius of the circular path and (b) the speed of the electron.

67. Two long, straight wires cross each other at right angles, as shown in Figure P19.67. (a) Find the direction and magnitude of the magnetic field at point *P*, which is in the same plane as the two wires. (b) Find the magnetic field at a point 30.0 cm above the point of intersection (30.0 cm out of the page, toward you).

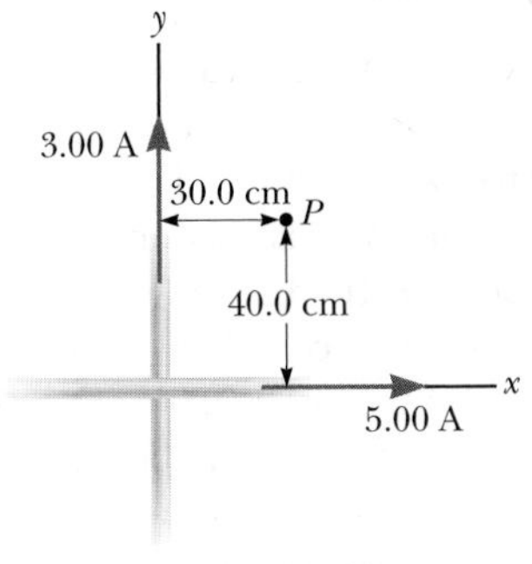

Figure P19.67

68. A 0.200-kg metal rod carrying a current of 10.0 A glides on two horizontal rails 0.500 m apart. What vertical magnetic field is required to keep the rod moving at a constant speed if the coefficient of kinetic friction between the rod and rails is 0.100?

69. BIO Using an electromagnetic flowmeter (Fig. P19.69), a heart surgeon monitors the flow rate of blood through an artery. Electrodes *A* and *B* make contact with the outer surface of the blood vessel, which has interior diameter 3.00 mm. (a) For a magnetic field magnitude of 0.040 0 T, a potential difference of 160 μV appears between the electrodes. Calculate the speed of the blood. (b) Verify that electrode *A* is positive, as shown. Does the sign of the emf depend on whether the mobile ions in the blood are predominantly positively or negatively charged? Explain.

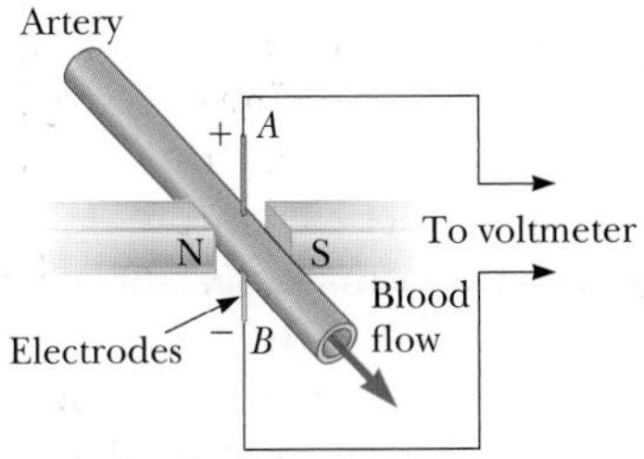

Figure P19.69

70. A uniform horizontal wire with a linear mass density of 0.50 g/m carries a 2.0-A current. It is placed in a constant magnetic field with a strength of 4.0×10^{-3} T. The field is horizontal and perpendicular to the wire. As the wire moves upward starting from rest, (a) what is its acceleration and (b) how long does it take to rise 50 cm? Neglect the magnetic field of Earth.

71. Three long, parallel conductors carry currents of $I =$ 2.0 A. Figure P19.71 is an end view of the conductors, with each current coming out of the page. Given that $a = 1.0$ cm, determine the magnitude and direction of the magnetic field at points *A*, *B*, and *C*.

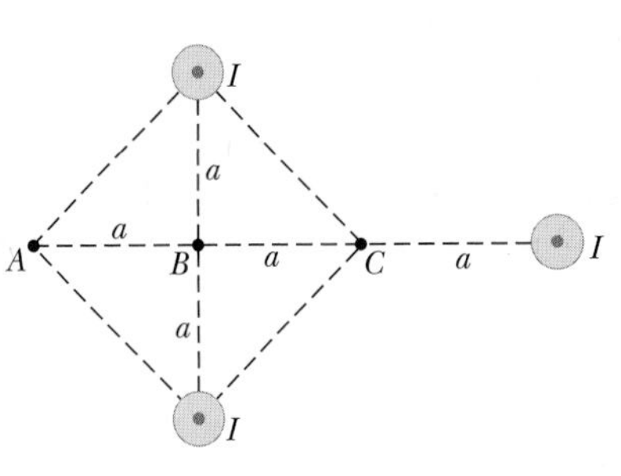

Figure P19.71

72. Two long, parallel wires, each with a mass per unit length of 40 g/m, are supported in a horizontal plane by 6.0-cm-long strings, as shown in Figure P19.72. Each wire carries the same current *I*, causing the wires to

repel each other so that the angle θ between the supporting strings is 16°. (a) Are the currents in the same or opposite directions? (b) Determine the magnitude of each current.

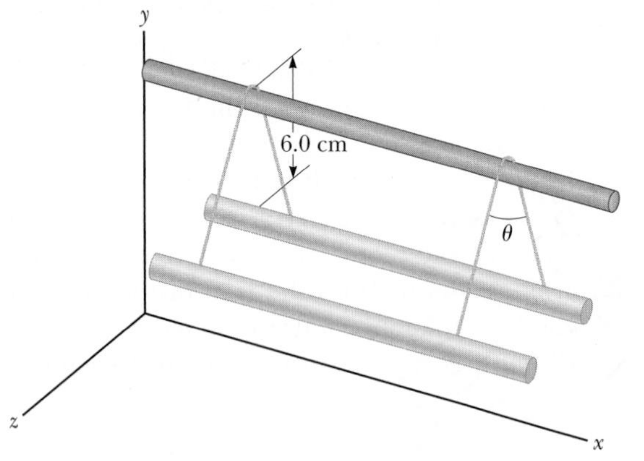

Figure P19.72

73. Protons having a kinetic energy of 5.00 MeV are moving in the positive x-direction and enter a magnetic field of 0.050 0 T in the z-direction, out of the plane of the page, and extending from $x = 0$ to $x = 1.00$ m as in Figure P19.73. (a) Calculate the y-component of the protons' momentum as they leave the magnetic field. (b) Find the angle α between the initial velocity vector of the proton beam and the velocity vector after the beam emerges from the field. *Hint:* Neglect relativistic effects and note that 1 eV = 1.60×10^{-19} J.

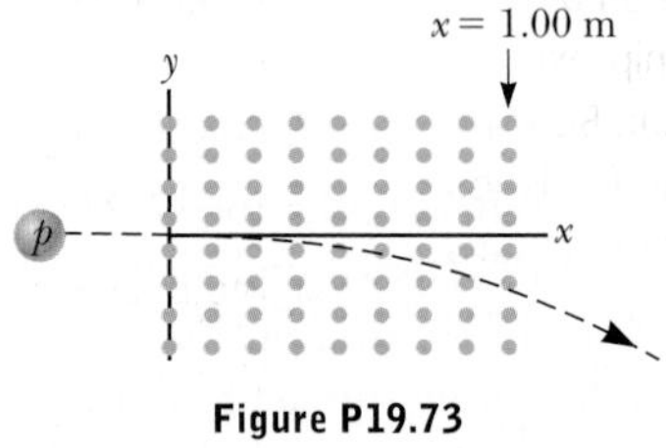

Figure P19.73

74. A straight wire of mass 10.0 g and length 5.0 cm is suspended from two identical springs that, in turn, form a closed circuit (Fig. P19.74). The springs stretch a distance of 0.50 cm under the weight of the wire. The circuit has a total resistance of 12 Ω. When a magnetic field directed out of the page (indicated by the dots in the figure) is turned on, the springs are observed to stretch an additional 0.30 cm. What is the strength of the magnetic field? (The upper portion of the circuit is fixed.)

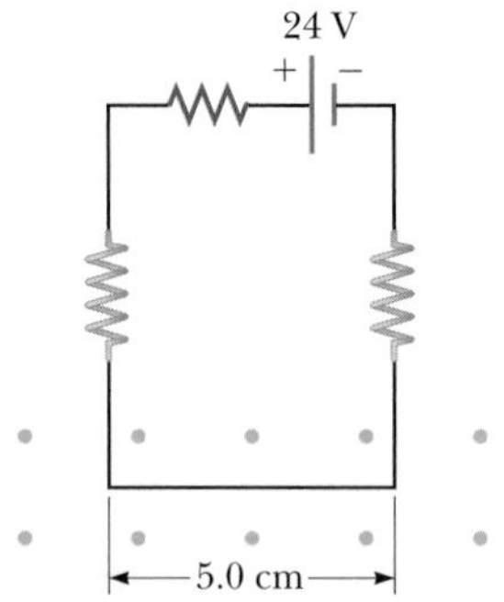

Figure P19.74

75. A 1.00-kg ball having net charge $Q = 5.00\ \mu$C is thrown out of a window horizontally at a speed $v = 20.0$ m/s. The window is at a height $h = 20.0$ m above the ground. A uniform horizontal magnetic field of magnitude $B = 0.010\ 0$ T is perpendicular to the plane of the ball's trajectory. Find the magnitude of the magnetic force acting on the ball just before it hits the ground. *Hint:* Ignore magnetic forces in finding the ball's final velocity.

76. Two long, parallel conductors separated by 10.0 cm carry currents in the same direction. The first wire carries a current $I_1 = 5.00$ A, and the second carries $I_2 = 8.00$ A. (a) What is the magnitude of the magnetic field created by I_1 at the location of I_2? (b) What is the force per unit length exerted by I_1 on I_2? (c) What is the magnitude of the magnetic field created by I_2 at the location of I_1? (d) What is the force per length exerted by I_2 on I_1?

South West News Service

A forest of fluorescent lights, not wired to any power source, are lighted by electromagnetic induction. Changing currents in the power wires overhead create time-dependent magnetic flux in the vicinity of the tubes, inducing a voltage across them.

Induced Voltages and Inductance 20

In 1819 Hans Christian Oersted discovered that an electric current exerted a force on a magnetic compass. Although there had long been speculation that such a relationship existed, Oersted's finding was the first evidence of a link between electricity and magnetism. Because nature is often symmetric, the discovery that electric currents produce magnetic fields led scientists to suspect that magnetic fields could produce electric currents. Indeed, experiments conducted by Michael Faraday in England and independently by Joseph Henry in the United States in 1831 showed that a changing magnetic field could induce an electric current in a circuit. The results of these experiments led to a basic and important law known as Faraday's law. In this chapter we discuss Faraday's law and several practical applications, one of which is the production of electrical energy in power plants throughout the world.

20.1 Induced emf and Magnetic Flux

LEARNING OBJECTIVES

1. Define magnetic flux and discuss its role in producing an induced emf.
2. Evaluate the magnitude and change in the magnetic flux through a given area.

An experiment first conducted by Faraday demonstrated that a current can be produced by a changing magnetic field. The apparatus shown in Figure 20.1 consists of a coil connected to a switch and a battery. We call this coil the *primary coil* and the corresponding circuit the primary circuit. The coil is wrapped around an iron ring to intensify the magnetic field produced by the current in the coil. A

Figure 20.1 Faraday's experiment.

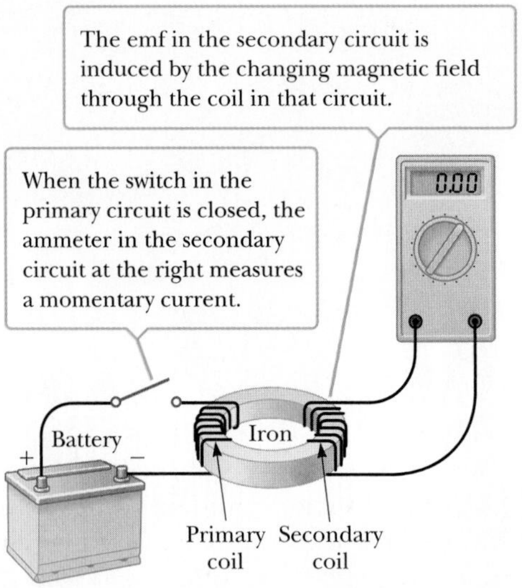

© iStockphoto.com/Steven Wynn Photography

Michael Faraday
British physicist and chemist (1791–1867)
Faraday is often regarded as the greatest experimental scientist of the 1800s. His many contributions to the study of electricity include the inventions of the electric motor, electric generator, and transformer, as well as the discovery of electromagnetic induction and the laws of electrolysis. Greatly influenced by religion, he refused to work on military poison gas for the British government.

secondary coil, at the right, is wrapped around the iron ring and is connected to an ammeter. The corresponding circuit is called the secondary circuit. It's important to notice that **there is no battery in the secondary circuit**.

At first glance, you might guess that no current would ever be detected in the secondary circuit. When the switch in the primary circuit in Figure 20.1 is suddenly closed, however, something amazing happens: the ammeter measures a current in the secondary circuit and then returns to zero! When the switch is opened again, the ammeter reads a current in the opposite direction and again returns to zero. Finally, whenever there is a steady current in the primary circuit, the ammeter reads zero.

From such observations, Faraday concluded that an electric current could be produced by a *changing* magnetic field. (A steady magnetic field doesn't produce a current unless the coil is moving, as explained below.) The current produced in the secondary circuit occurs for only an instant while the magnetic field through the secondary coil is changing. In effect, the secondary circuit behaves as though a source of emf were connected to it for a short time. It's customary to say that **an induced emf is produced in the secondary circuit by the changing magnetic field**.

Magnetic Flux

To evaluate induced emfs quantitatively, we need to understand what factors affect the phenomenon. Although changing magnetic fields always induce electric fields, in other situations the magnetic field remains constant, yet an induced electric field is still produced. The best example of this is an electric generator: A loop of conductor rotating in a constant magnetic field creates an electric current.

The physical quantity associated with magnetism that creates an electric field is a **changing magnetic flux**. Magnetic flux is defined in the same way as electric flux (Section 15.9) and is proportional to both the strength of the magnetic field passing through the plane of a loop of wire and the area of the loop.

Magnetic flux ▶

The **magnetic flux $\boldsymbol{\Phi_B}$** through a loop of wire with area A is defined by

$$\Phi_B \equiv B_\perp A = BA\cos\theta \qquad [20.1]$$

where $B_\perp$ is the component of a uniform magnetic field $\vec{\mathbf{B}}$ perpendicular to the plane of the loop, as in Figure 20.2a, and θ is the angle between $\vec{\mathbf{B}}$ and the normal (perpendicular) to the plane of the loop.

SI unit: weber (Wb)

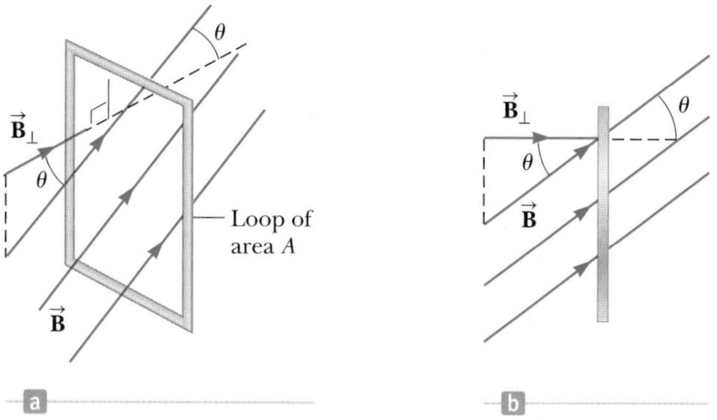

Figure 20.2 (a) A uniform magnetic field $\vec{B}$ making an angle θ with a direction normal to the plane of a wire loop of area A. (b) An edge view of the loop.

Note that there are always two directions normal to a given plane surface. In Figure 20.2, for example, that direction could be chosen to be to the right, resulting in positive flux. The normal direction could also be chosen to point to the left, which would result in a negative flux of the same magnitude. The choice of normal direction is called the orientation of the surface. Once chosen in a given problem, the normal direction remains fixed. A good default is to choose the normal direction so that the initial angle between the magnetic field and the normal direction is less than 90°.

From Equation 20.1, it follows that $B_\perp = B\cos\theta$. The magnetic flux, in other words, is the magnitude of the part of $\vec{B}$ that is perpendicular to the plane of the loop times the area of the loop. Figure 20.2b is an edge view of the loop and the penetrating magnetic field lines. When the field is perpendicular to the plane of the loop as in Figure 20.3a, $\theta = 0$ and Φ_B has a maximum value, $\Phi_{B,\max} = BA$. When the plane of the loop is parallel to $\vec{B}$ as in Figure 20.3b, $\theta = 90°$ and $\Phi_B = 0$. The flux can also be negative. For example, when $\theta = 180°$, the flux is equal to $-BA$. Because the SI unit of B is the tesla, or weber per square meter, the unit of flux is $\mathrm{T \cdot m^2}$, or weber (Wb).

We can emphasize the qualitative meaning of Equation 20.1 by first drawing magnetic field lines, as in Figure 20.3. The number of lines per unit area increases as the field strength increases. **The value of the magnetic flux is proportional to the total number of lines passing through the loop.** We see that the most lines pass through the loop when its plane is perpendicular to the field, as in Figure 20.3a, so the flux has its maximum value at that time. As Figure 20.3b shows, no lines pass through the loop when its plane is parallel to the field, so in that case $\Phi_B = 0$.

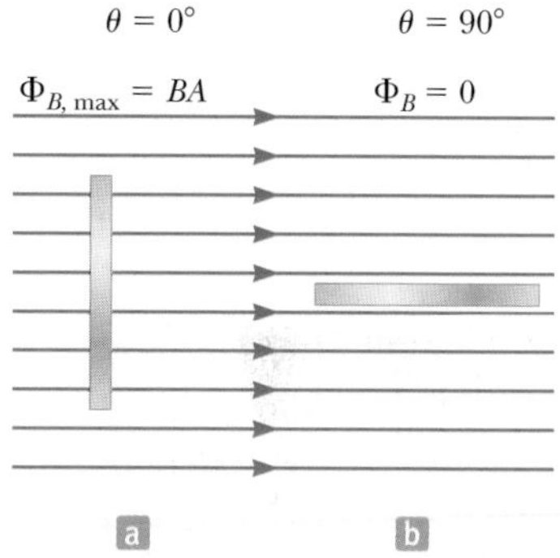

Figure 20.3 An edge view of a loop in a uniform magnetic field. (a) When the field lines are perpendicular to the plane of the loop, the magnetic flux through the loop is a maximum and equal to $\Phi_B = BA$. (b) When the field lines are parallel to the plane of the loop, the magnetic flux through the loop is zero.

APPLYING PHYSICS 20.1 | Flux Compared

Argentina has more land area ($2.8 \times 10^6\ \mathrm{km^2}$) than Greenland ($2.2 \times 10^6\ \mathrm{km^2}$). Why is the magnetic flux of Earth's magnetic field larger through Greenland than through Argentina?

EXPLANATION Greenland (latitude 60° north to 80° north) is closer to a magnetic pole than Argentina (latitude 20° south to 50° south), so the magnetic field is stronger there. That in itself isn't sufficient to conclude that the magnetic flux is greater, but Greenland's proximity to a pole also means that the angle the magnetic field lines make with the vertical is smaller than in Argentina. As a result, more field lines penetrate the surface in Greenland, despite Argentina's slightly larger area. ■

EXAMPLE 20.1 | Magnetic Flux

GOAL Calculate magnetic flux and a change in flux.

PROBLEM A conducting circular loop of radius 0.250 m is placed in the *xy*-plane in a uniform magnetic field of 0.360 T that points in the positive *z*-direction, the same direction as the normal to the plane. **(a)** Calculate the magnetic flux through the loop. **(b)** Suppose the loop is rotated clockwise around the *x*-axis, so the normal direction now points at a

(Continued)

45.0° angle with respect to the z-axis. Recalculate the magnetic flux through the loop. **(c)** What is the change in flux due to the rotation of the loop?

STRATEGY After finding the area, substitute values into the equation for magnetic flux for each part. Because the normal direction was chosen to be the same direction as the magnetic field, the angle between the magnetic field and the normal is initially 0°. After the rotation, that angle becomes 45°.

SOLUTION

(a) Calculate the initial magnetic flux through the loop.

First, calculate the area of the loop:

$$A = \pi r^2 = \pi(0.250\text{ m})^2 = 0.196\text{ m}^2$$

Substitute A, B, and $\theta = 0°$ into Equation 20.1 to find the initial magnetic flux:

$$\Phi_B = AB\cos\theta = (0.196\text{ m}^2)(0.360\text{ T})\cos(0°)$$
$$= 0.070\,6\text{ T}\cdot\text{m}^2 = \boxed{0.070\,6\text{ Wb}}$$

(b) Calculate the magnetic flux through the loop after it has rotated 45.0° around the x-axis.

Make the same substitutions as in part (a), except the angle between $\vec{\mathbf{B}}$ and the normal is now $\theta = 45.0°$:

$$\Phi_B = AB\cos\theta = (0.196\text{ m}^2)(0.360\text{ T})\cos(45.0°)$$
$$= 0.049\,9\text{ T}\cdot\text{m}^2 = \boxed{0.049\,9\text{ Wb}}$$

(c) Find the change in the magnetic flux due to the rotation of the loop.

Subtract the result of part (a) from the result of part (b):

$$\Delta\Phi_B = 0.049\,9\text{ Wb} - 0.070\,6\text{ Wb} = \boxed{-0.020\,7\text{ Wb}}$$

REMARKS Notice that the rotation of the loop, not any change in the magnetic field, is responsible for the change in flux. This changing magnetic flux is essential in the functioning of electric motors and generators.

QUESTION 20.1 True or False: If the loop is rotated in the opposite direction by the same amount, the change in magnetic flux has the same magnitude but opposite sign.

EXERCISE 20.1 The loop, having rotated by 45°, rotates clockwise another 30°, so the normal to the plane points at an angle of 75° with respect to the direction of the magnetic field. Find (a) the magnetic flux through the loop when $\theta = 75°$ and (b) the change in magnetic flux during the rotation from 45° to 75°.

ANSWERS (a) 0.018 3 Wb (b) −0.031 6 Wb

20.2 Faraday's Law of Induction and Lenz's Law

LEARNING OBJECTIVES

1. State Faraday's and Lenz's laws and describe applications based on them.
2. Apply Faraday's and Lenz's laws to systems with changing magnetic flux.

The usefulness of the concept of magnetic flux can be made obvious by another simple experiment that demonstrates the basic idea of electromagnetic induction. Consider a wire loop connected to an ammeter as in Figure 20.4. If a magnet is moved toward the loop, the ammeter reads a current in one direction, as in Figure 20.4a. When the magnet is held stationary, as in Figure 20.4b, the ammeter reads zero current. If the magnet is moved away from the loop, the ammeter reads a current in the opposite direction, as in Figure 20.4c. If the magnet is held stationary and the loop is moved either toward or away from the magnet, the ammeter also reads a current. From these observations, it can be concluded that **a current is established in the circuit as long as there is relative motion between the magnet and the loop**. The same experimental results are found whether the loop moves or the magnet moves. We call such a current an **induced current** because it is produced by an **induced emf**.

Tip 20.1 Induced Current Requires a Change in Magnetic Flux

The existence of magnetic flux through an area is not sufficient to create an induced emf. A *change* in the magnetic flux over some time interval Δt must occur for an emf to be induced.

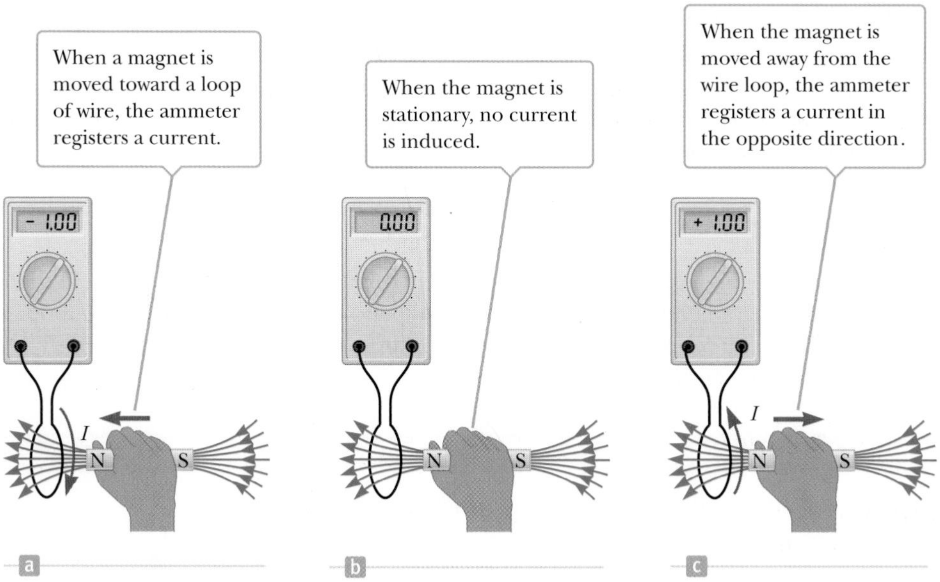

Figure 20.4 A simple experiment showing that a current is induced in a loop when a magnet is moved toward or away from the loop.

Tip 20.2 There Are Two Magnetic Fields to Consider

When applying Lenz's law, there are *two* magnetic fields to consider. The first is the external changing magnetic field that induces the current in a conducting loop. The second is the magnetic field produced by the induced current in the loop.

This experiment is similar to the Faraday experiment discussed in Section 20.1. In each case, an emf is induced in a circuit when the magnetic flux through the circuit changes with time. It turns out that the instantaneous emf induced in a circuit equals the negative of the rate of change of magnetic flux with respect to time through the circuit. This is **Faraday's law of magnetic induction**.

◀ Faraday's law

If a circuit contains N tightly wound loops and the magnetic flux through each loop changes by the amount $\Delta\Phi_B$ during the interval Δt, the average emf induced in the circuit during time Δt is

$$\mathcal{E} = -N\frac{\Delta\Phi_B}{\Delta t} \qquad [20.2]$$

Because $\Phi_B = BA\cos\theta$, a change of any of the factors B, A, or θ with time produces an emf. We explore the effect of a change in each of these factors in the following sections. The minus sign in Equation 20.2 is included to indicate the polarity of the induced emf. This polarity determines the direction of the current in the loop, and is given by **Lenz's law**:

The current caused by the induced emf travels in the direction that creates a magnetic field with flux opposing the change in the original flux through the circuit.

Lenz's law says that if the magnetic flux through a loop is becoming more positive, say, then the induced emf creates a current and associated magnetic field that produces negative magnetic flux. Some mistakenly think this "counter magnetic field" created by the induced current, called $\vec{\mathbf{B}}_{\text{ind}}$ ("ind" for induced), will always point in a direction opposite the applied magnetic field $\vec{\mathbf{B}}$, but that is only true half the time! Figure 20.5a shows a field penetrating a loop. The graph in Figure 20.5b shows that the magnitude of the magnetic field $\vec{\mathbf{B}}$ shrinks with time, which means that the flux of $\vec{\mathbf{B}}$ is shrinking with time, so the induced field $\vec{\mathbf{B}}_{\text{ind}}$ will actually be in the same direction as $\vec{\mathbf{B}}$. In effect, $\vec{\mathbf{B}}_{\text{ind}}$ "shores up" the field $\vec{\mathbf{B}}$, slowing the loss of flux through the loop.

The direction of the current in Figure 20.5a can be determined by right-hand rule number 2: Point your right thumb in the direction that will cause the fingers

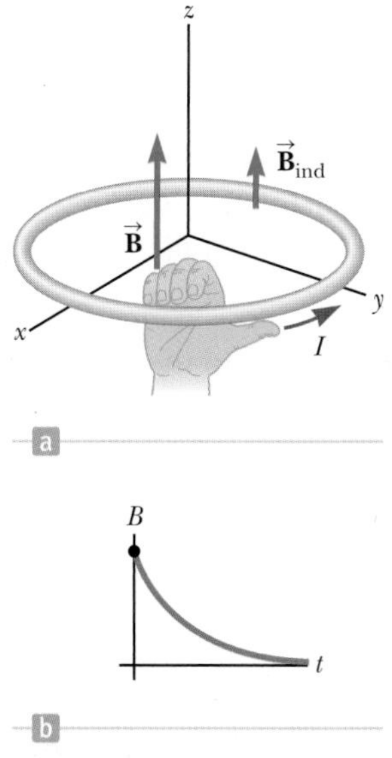

Figure 20.5 (a) The magnetic field $\vec{\mathbf{B}}$ becomes smaller with time, reducing the flux, so current is induced in a direction that creates an induced magnetic field $\vec{\mathbf{B}}_{\text{ind}}$ opposing the change in magnetic flux. (b) Graph of the magnitude of the magnetic field as a function of time.

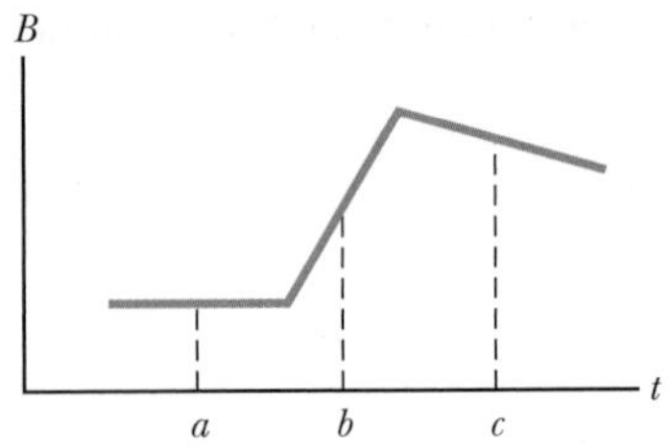

Figure 20.6 (Quick Quiz 20.1)

on your right hand to curl in the direction of the induced field $\vec{\mathbf{B}}_{\text{ind}}$. In this case, that direction is counterclockwise: with the right thumb pointed in the direction of the current, your fingers curl down outside the loop and around and ***up through the inside of the loop***. Remember, inside the loop is where it's important for the induced magnetic field to be pointing up.

Quick Quiz

20.1 Figure 20.6 is a graph of the magnitude B versus time for a magnetic field that passes through a fixed loop and is oriented perpendicular to the plane of the loop. Rank the magnitudes of the emf generated in the loop from largest to smallest at the three instants indicated.

EXAMPLE 20.2 Faraday and Lenz to the Rescue

GOAL Calculate an induced emf and current with Faraday's law and apply Lenz's law when the magnetic field changes with time.

PROBLEM A coil with 25 turns of wire is wrapped on a frame with a square cross section 1.80 cm on a side. Each turn has the same area, equal to that of the frame, and the total resistance of the coil is 0.350 Ω. An applied uniform magnetic field is perpendicular to the plane of the coil, as in Figure 20.7. **(a)** If the field changes uniformly from 0.00 T to 0.500 T in 0.800 s, what is the induced emf in the coil while the field is changing? Find **(b)** the magnitude and **(c)** the direction of the induced current in the coil while the field is changing.

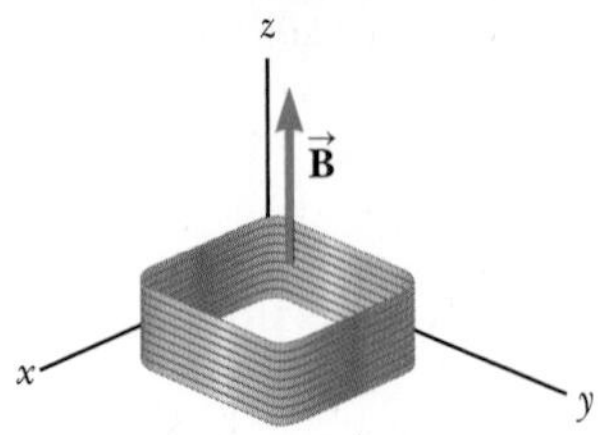

Figure 20.7 (Example 20.2)

STRATEGY Part (a) requires substituting into Faraday's law, Equation 20.2. The necessary information is given, except for $\Delta\Phi_B$, the change in the magnetic flux during the elapsed time. Using the normal direction to coincide with the positive z-axis, compute the initial and final magnetic fluxes with Equation 20.1, find the difference, and assemble all terms in Faraday's law. The current can then be found with Ohm's law, and its direction with Lenz's law.

SOLUTION

(a) Find the induced emf in the coil.

To compute the flux, the area of the coil is needed:

$$A = L^2 = (0.018\,0 \text{ m})^2 = 3.24 \times 10^{-4} \text{ m}^2$$

The magnetic flux $\Phi_{B,i}$ through the coil at $t = 0$ is zero because $B = 0$. Calculate the flux at $t = 0.800$ s:

$$\Phi_{B,f} = BA\cos\theta = (0.500 \text{ T})(3.24 \times 10^{-4} \text{ m}^2)\cos(0°) = 1.62 \times 10^{-4} \text{ Wb}$$

Compute the change in the magnetic flux through the cross section of the coil over the 0.800-s interval:

$$\Delta\Phi_B = \Phi_{B,f} - \Phi_{B,i} = 1.62 \times 10^{-4} \text{ Wb}$$

Substitute into Faraday's law of induction to find the induced emf in the coil:

$$\mathcal{E} = -N\frac{\Delta\Phi_B}{\Delta t} = -(25 \text{ turns})\left(\frac{1.62 \times 10^{-4} \text{ Wb}}{0.800 \text{ s}}\right) = -5.06 \times 10^{-3} \text{ V}$$

(b) Find the magnitude of the induced current in the coil.

Substitute the voltage difference and the resistance into Ohm's law, where $\Delta V = \mathcal{E}$:

$$I = \frac{\Delta V}{R} = \frac{5.06 \times 10^{-3} \text{ V}}{0.350\ \Omega} = 1.45 \times 10^{-2} \text{ A}$$

(c) Find the direction of the induced current in the coil.

The magnetic field is increasing up through the loop, in the same direction as the normal to the plane; hence, the flux is positive and is also increasing. A downward-pointing induced magnetic field will create negative flux, opposing the change. If you point your right thumb in the clockwise direction along the loop as viewed from above, your fingers curl down through the loop, which is the correct direction for the counter magnetic field. Hence the current must proceed in a clockwise direction as viewed from above the coil.

REMARKS Lenz's law can best be handled by first sketching a diagram.

QUESTION 20.2 What average emf is induced in the loop if, instead, the magnetic field changes uniformly from 0.500 T to 0 in 0.800 s? How would that affect the induced current?

EXERCISE 20.2 Suppose the magnetic field changes uniformly from 0.500 T to 0.200 T in the next 0.600 s. Compute (a) the induced emf in the coil and (b) the magnitude and direction of the induced current.

ANSWERS (a) 4.05×10^{-3} V (b) 1.16×10^{-2} A (counterclockwise as viewed from above the coil)

Finding the Direction of the Induced Current

Finding the direction of the induced current can be tricky. The following three examples illustrate how the direction is found using Lenz's law.

Lenz's Law Example 1 The current in the wire of Figure 20.8 is steadily increasing in the direction indicated. Let's choose the normal direction to be out of the page so that magnetic field vectors coming out of the page will produce positive magnetic flux. The magnetic field created by the current I circulates around the wire, going into the page on the right side of the long wire in the region of the rectangular coil and coming out of the page on the left side of the long straight wire. Therefore, the magnetic flux through the rectangular coil due to the current I is negative. Because the current is increasing up the page, the magnetic field is becoming stronger, increasing the magnitude of the negative flux through the rectangular coil. By Lenz's law, the induced current in the coil must produce positive flux, countering the increasing negative flux. That requires an induced magnetic field pointing out of the page through the coil. Mentally curl the fingers of the right hand around the right branch of the rectangular coil, and note that the fingers come straight up out of the page through the coil, as required. The right thumb, meanwhile, points up the page, indicating the current direction in that part of the coil. Therefore the induced current in the coil is counterclockwise.

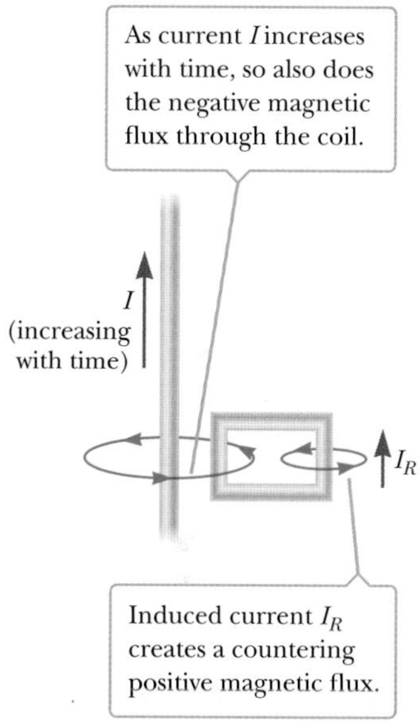

Figure 20.8 (Lenz's Law Example 1) Current I increases in magnitude with time, strengthening the magnetic field that circulates around the wire.

Lenz's Law Example 2 In Figure 20.9a, the north pole of the magnet moves toward the coil. If the normal direction is chosen to the right, then the magnetic flux through the coil due to the magnet is positive and increases with time. A negative flux to the left must therefore be created by the induced current in the coil, so the induced magnetic field must also point to the left as indicated in Figure 20.9b. Imagine curling the fingers of the right hand around the coil so they point through the coil to the left. The right thumb then points upward, indicating the induced current is counterclockwise as viewed from the left side of the loop.

Lenz's Law Example 3 Consider a coil of wire placed near a solenoid in Figure 20.10a (page 706). The wire is wrapped in such a way as to create a south magnetic pole at the right end when the switch is closed in Figure 20.10b. Choose left as the normal direction. When the switch is closed, the current in the solenoid begins to increase, and the magnetic flux through the coil is positive and increasing with time. Therefore the induced current in the coil must create negative magnetic flux to counteract the increasing positive flux created by the current in the

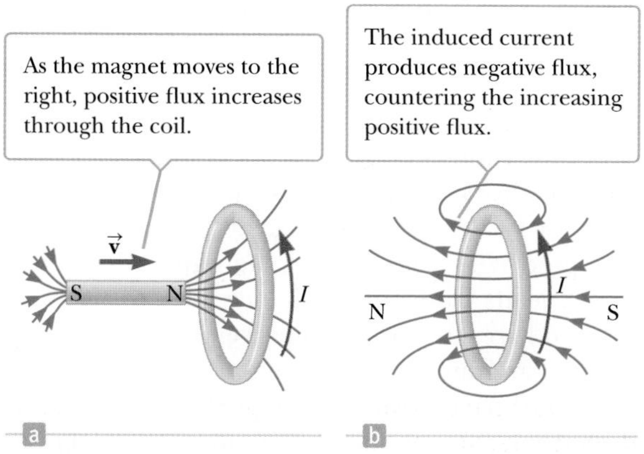

Figure 20.9 (Lenz's Law Example 2) (a) The north pole of the magnet approaches the coil from the left, with the normal direction taken to the right. (b) A current is induced in the coil.

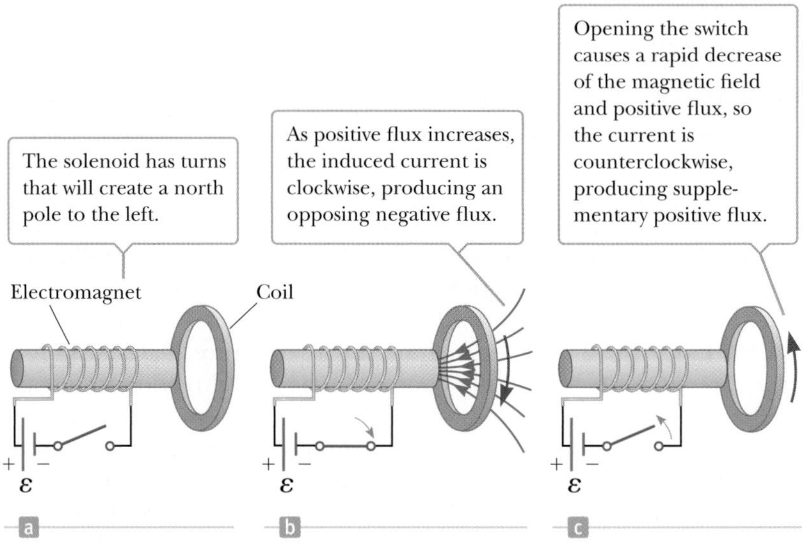

Figure 20.10 (Lenz's Law Example 3) (a) The turns of the solenoid create a magnetic field with north pole pointing left, which is also taken as the normal direction. (b) When the switch is closed, positive flux begins increasing through the coil as field lines converge on the solenoid's south pole. (c) Opening the switch causes the solenoid's field to rapidly decrease.

solenoid. That requires an induced magnetic field directed to the right through the coil. Turning the right hand so the thumb is pointed downward, the right fingers can curl through the coil and to the right. The induced current in the coil follows the direction of the right thumb, which is clockwise as viewed from the left end of the coil. When the switch is opened again in Figure 20.10c the current in the solenoid changes direction because the magnetic field and positive flux begin to decrease. Moving counterclockwise, the induced current creates positive flux through the coil, opposing the decrease in positive flux.

In all three of these examples, the critical idea is that a changing flux causes an induced current, and the associated induced magnetic field produces flux opposing the change in flux in accordance with Lenz's law. When the flux stops changing, the induced current stops. Although in each case the magnetic flux changed because of a changing magnetic field, induced currents can result even when the magnetic field is constant provided the flux through the loop changes. That fact will become clear when discussing motional emf in Section 20.3 and generators in Section 20.4.

Quick Quiz

20.2 A bar magnet is falling toward the center of a loop of wire, with the north pole oriented downward. Viewed from the same side of the loop as the magnet, as the north pole approaches the loop, what is the direction of the induced current? (a) clockwise (b) zero (c) counterclockwise (d) along the length of the magnet

20.3 Two circular loops are side by side and lie in the xy-plane. A switch is closed, starting a counterclockwise current in the left-hand loop, as viewed from a point on the positive z-axis passing through the center of the loop. Which of the following statements is true of the right-hand loop? (a) The current remains zero. (b) An induced current moves counterclockwise. (c) An induced current moves clockwise.

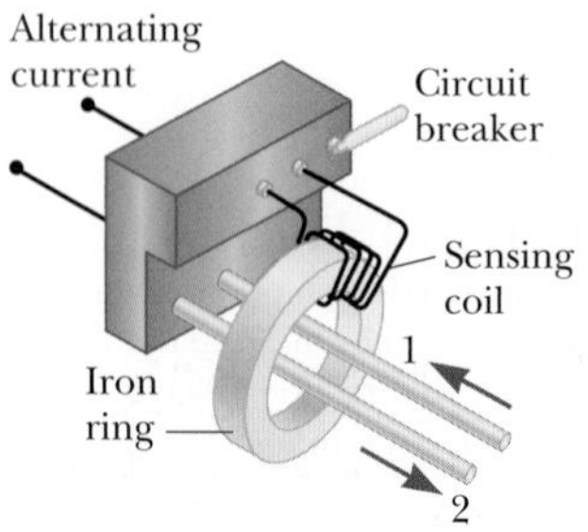

Figure 20.11 Essential components of a ground fault interrupter (contents of the gray box in Fig. 20.12a). In newer homes such devices are built directly into wall outlets. The purpose of the sensing coil and circuit breaker is to cut off the current before damage is done.

The ground fault interrupter (GFI) is an interesting safety device that protects people against electric shock when they touch appliances and power tools. Its operation makes use of Faraday's law. Figure 20.11 shows the essential parts of a ground fault interrupter. Wire 1 leads from the wall outlet to the appliance to be protected, and wire 2 leads from the appliance back to the wall outlet. An iron ring surrounds the two wires to confine the magnetic field set up by each wire. A sensing coil, which can activate a circuit breaker when changes in magnetic flux occur, is wrapped around part of the iron ring. Because the currents in the wires are in opposite directions, the net magnetic field through the sensing coil due to

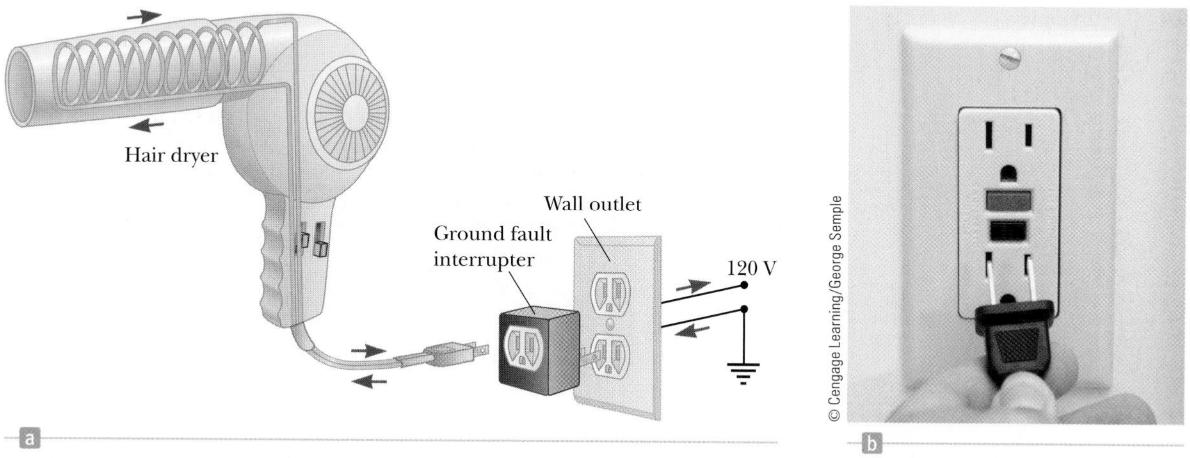

Figure 20.12 (a) This hair dryer has been plugged into a ground fault interrupter that is in turn plugged into an unprotected wall outlet. (b) You likely have seen this kind of ground fault interrupter in a hotel bathroom, where hair dryers and electric shavers are often used by people just out of the shower or who might touch a water pipe, providing a ready path to ground in the event of a short circuit.

the currents is zero. If a short circuit occurs in the appliance so that there is no returning current, however, the net magnetic field through the sensing coil is no longer zero. A short circuit can happen if, for example, one of the wires loses its insulation, providing a path through you to ground if you happen to be touching the appliance and are grounded as in Figure 18.23a. Because the current is alternating, the magnetic flux through the sensing coil changes with time, producing an induced voltage in the coil. This induced voltage is used to trigger a circuit breaker, stopping the current quickly (in about 1 ms) before it reaches a level that might be harmful to the person using the appliance. A ground fault interrupter provides faster and more complete protection than even the case-ground-and-circuit-breaker combination shown in Figure 18.23b. For this reason, ground fault interrupters are commonly found in bathrooms, where electricity poses a hazard to people. (See Fig. 20.12.)

APPLICATION
Ground fault interrupters

Another interesting application of Faraday's law is the production of sound in an electric guitar. A vibrating string induces an emf in a coil (Fig. 20.13). The pickup coil is placed near the vibrating guitar string, which is made of a metal that can be magnetized. The permanent magnet inside the coil magnetizes the portion of the string nearest the coil. When the guitar string vibrates at some frequency, its magnetized segment produces a changing magnetic flux through the pickup coil. The changing flux induces a voltage in the coil, which is fed to an amplifier. The output of the amplifier is sent to the loudspeakers, producing the sound waves that we hear.

APPLICATION
Electric guitar pickups

Sudden infant death syndrome, or SIDS, is a devastating affliction in which a baby suddenly stops breathing during sleep without an apparent cause. One type of monitoring device, called an apnea monitor, is sometimes used to alert caregivers

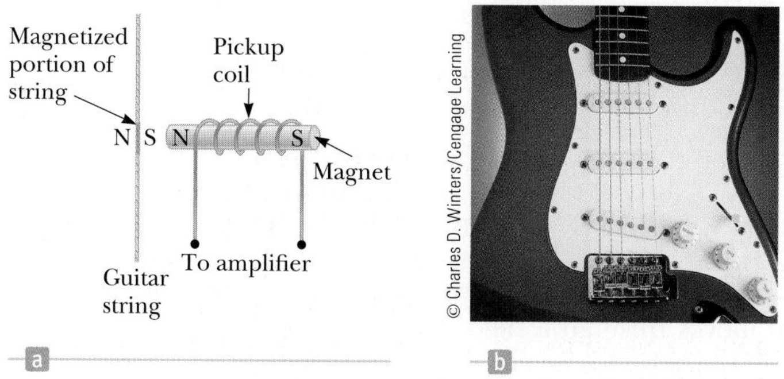

Figure 20.13 (a) In an electric guitar a vibrating string induces a voltage in the pickup coil. (b) Several pickups allow the vibration to be detected from different portions of the string.

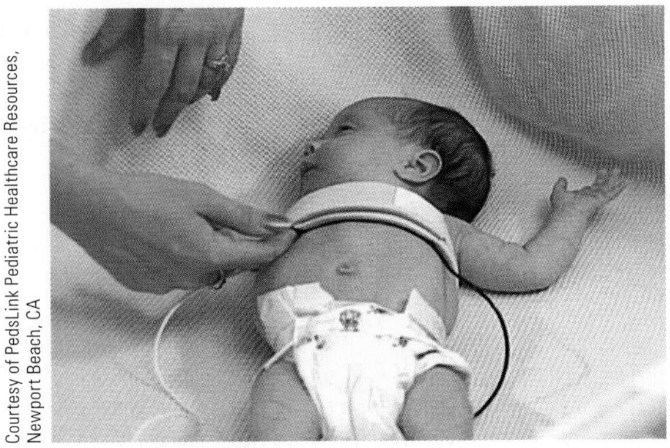

Figure 20.14 This infant is wearing a monitor designed to alert care-givers if breathing stops. Notice the two wires attached to opposite sides of the chest.

BIO APPLICATION
Apnea monitors

of the cessation of breathing. The device uses induced currents, as shown in Figure 20.14. A coil of wire attached to one side of the chest carries an alternating current. The varying magnetic flux produced by this current passes through a pickup coil attached to the opposite side of the chest. Expansion and contraction of the chest caused by breathing or movement change the strength of the voltage induced in the pickup coil. If breathing stops, however, the pattern of the induced voltage stabilizes, and external circuits monitoring the voltage sound an alarm to the caregivers after a momentary pause to ensure that a problem actually does exist.

20.3 Motional emf

LEARNING OBJECTIVES

1. Define motional emf and use Faraday's law to discuss the potential difference across a moving conductor.
2. Apply Faraday's law to systems involving a conductor moving through a magnetic field.

In Section 20.2 we considered emfs induced in a circuit when the magnetic field changes with time. In this section we describe a particular application of Faraday's law in which a so-called **motional emf** is produced. It is the emf induced in a conductor moving through a magnetic field.

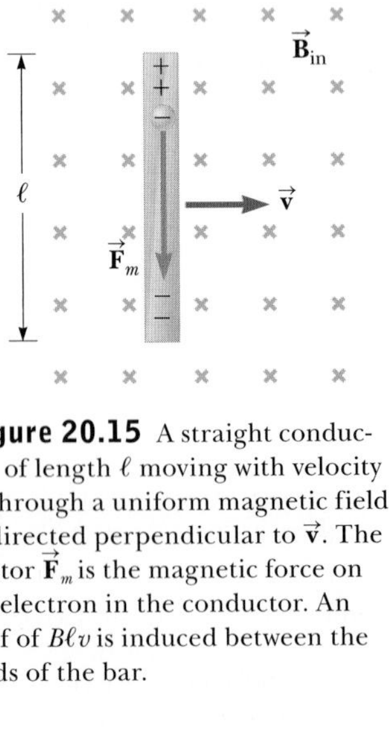

Figure 20.15 A straight conductor of length ℓ moving with velocity $\vec{\mathbf{v}}$ through a uniform magnetic field $\vec{\mathbf{B}}$ directed perpendicular to $\vec{\mathbf{v}}$. The vector $\vec{\mathbf{F}}_m$ is the magnetic force on an electron in the conductor. An emf of $B\ell v$ is induced between the ends of the bar.

First consider a straight conductor of length ℓ moving with constant velocity through a uniform magnetic field directed into the paper, as in Figure 20.15. For simplicity, we assume the conductor moves in a direction perpendicular to the field. A magnetic force of magnitude $F_m = qvB$, directed downward, acts on the electrons in the conductor. Because of this magnetic force, the free electrons move to the lower end of the conductor and accumulate there, leaving a net positive charge at the upper end. As a result of this charge separation, an electric field is produced in the conductor. The charge at the ends builds up until the downward magnetic force qvB is balanced by the upward electric force qE. At this point, charge stops flowing and the condition for equilibrium requires that

$$qE = qvB \qquad \text{or} \qquad E = vB$$

Because the electric field is uniform, the field produced in the conductor is related to the potential difference across the ends by $\Delta V = E\ell$, giving

$$\Delta V = E\ell = B\ell v \qquad [20.3]$$

Because there is an excess of positive charge at the upper end of the conductor and an excess of negative charge at the lower end, the upper end is at a higher potential than the lower end. There is a potential difference across a conductor as long as it moves through a field. If the motion is reversed, the polarity of the potential difference is also reversed.

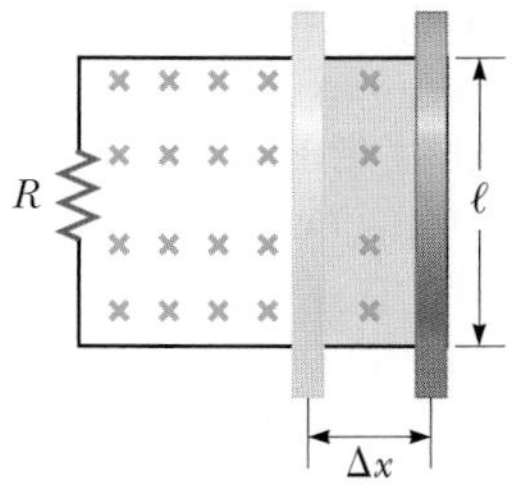

Figure 20.17 As the bar moves to the right, the area of the loop increases by the amount $\ell\Delta x$ and the magnetic flux through the loop increases by $B\ell\Delta x$.

A more interesting situation occurs if the moving conductor is part of a closed conducting path. This situation is particularly useful for illustrating how a changing loop area induces a current in a closed circuit described by Faraday's law. Consider a circuit consisting of a conducting bar of length ℓ, sliding along two fixed, parallel conducting rails, as in Figure 20.16a. For simplicity, assume the moving bar has zero resistance and the stationary part of the circuit has constant resistance R. Take the normal direction to coincide with the z-axis, out of the page. A uniform and constant magnetic field $\vec{\mathbf{B}}$ is applied perpendicular to the plane of the circuit. As the bar is pulled to the right in the positive x-direction with velocity $\vec{\mathbf{v}}$ under the influence of an applied force $\vec{\mathbf{F}}_{\text{app}}$, a magnetic force along the length of the bar acts on the free charges in the bar. This force in turn sets up an induced current because the charges are free to move in a closed conducting path. In this case, the changing magnetic flux through the loop and the corresponding induced emf across the moving bar arise from the *change in area of the loop* as the bar moves through the magnetic field. Because the flux into the page increases, by Lenz's law the induced current circulates counterclockwise, producing flux out of the page that opposes the change.

Assume the bar moves a distance Δx in time Δt, as shown in Figure 20.17. The increase in flux $\Delta\Phi_B$ through the loop in that time is the amount of flux that now passes through the portion of the circuit that has area $\ell\ \Delta x$:

$$\Delta\Phi_B = BA = B\ell\ \Delta x$$

Using Faraday's law and noting that there is one loop ($N = 1$), we find that the magnitude of the induced emf is

$$|\mathcal{E}| = \frac{\Delta\Phi_B}{\Delta t} = B\ell\frac{\Delta x}{\Delta t} = B\ell v \qquad [20.4]$$

This induced emf is often called a **motional emf** because it arises from the motion of a conductor through a magnetic field.

Further, if the resistance of the circuit is R, the magnitude of the induced current in the circuit is

$$I = \frac{|\mathcal{E}|}{R} = \frac{B\ell v}{R} \qquad [20.5]$$

Figure 20.16b shows the equivalent circuit diagram for this example.

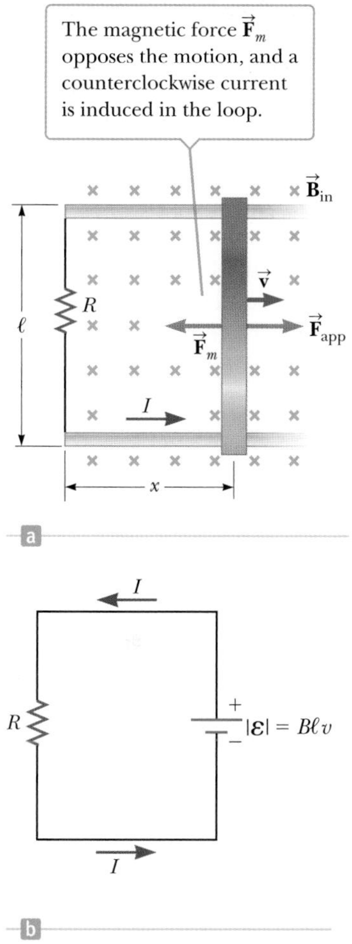

Figure 20.16 (a) A conducting bar sliding with velocity $\vec{\mathbf{v}}$ along two conducting rails under the action of an applied force $\vec{\mathbf{F}}_{\text{app}}$. (b) The equivalent circuit of that in (a).

APPLYING PHYSICS 20.2 | Space Catapult

Applying a force on the bar will result in an induced emf in the circuit shown in Figure 20.16. Suppose we remove the external magnetic field in the diagram and replace the resistor with a high-voltage source and a switch, as in Figure 20.18. What will happen when the switch is closed? Will the bar move, and does it matter which way we connect the high-voltage source?

EXPLANATION Suppose the source is capable of establishing high current. Then the two horizontal conducting rods will create a strong magnetic field in the area between

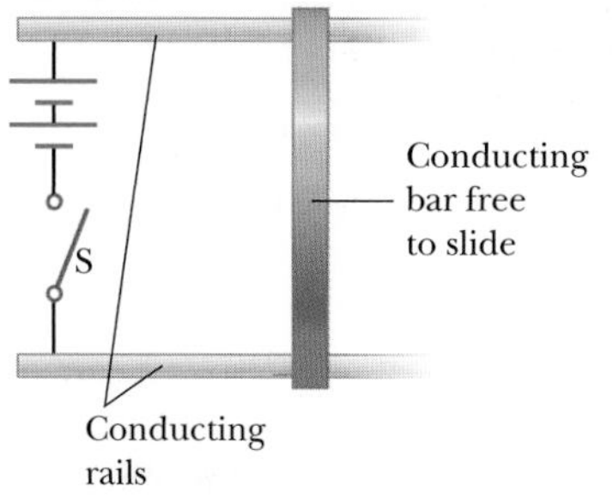

Figure 20.18 (Applying Physics 20.2)

(Continued)

them, directed into the page. (The movable bar also creates a magnetic field, but this field can't exert force on the bar itself.) Because the moving bar carries a downward current, a magnetic force is exerted on the bar, directed to the right. Hence, the bar accelerates along the rails away from the power supply. If the polarity of the power were reversed, the magnetic field would be out of the page, the current in the bar would be upward, and the force on the bar would still be directed to the right. The $BI\ell$ force exerted by a magnetic field according to Equation 19.6 causes the bar to accelerate away from the voltage source. Studies have shown that it's possible to launch payloads into space with this technology. (This is the working principle of a rail gun.) Very large accelerations can be obtained with currently available technology, with payloads being accelerated to a speed of several kilometers per second in a fraction of a second. This acceleration is larger than humans can tolerate.

Rail guns have been proposed as propulsion systems for moving asteroids into more useful orbits. The material of an asteroid could be mined and launched off the surface by a rail gun, which would act like a rocket engine, modifying the velocity and hence the orbit of the asteroid. Some asteroids contain trillions of dollars worth of valuable metals. ■

Quick Quiz

20.4 A horizontal metal bar oriented east-west drops straight down in a location where Earth's magnetic field is due north. As a result, an emf develops between the ends. Which end is positively charged? (a) the east end (b) the west end (c) neither end carries a charge

20.5 You intend to move a rectangular loop of wire into a region of uniform magnetic field at a given speed so as to induce an emf in the loop. The plane of the loop must remain perpendicular to the magnetic field lines. In which orientation should you hold the loop while you move it into the region with the magnetic field to generate the largest emf? (a) with the long dimension of the loop parallel to the velocity vector (b) with the short dimension of the loop parallel to the velocity vector (c) either way because the emf is the same regardless of orientation

EXAMPLE 20.3 | A Potential Difference Induced Across Airplane Wings

GOAL Find the emf induced by motion through a magnetic field.

PROBLEM An airplane with a wingspan of 30.0 m flies due north at a location where the downward component of Earth's magnetic field is 0.600×10^{-4} T. There is also a component pointing due north that has a magnitude of 0.470×10^{-4} T. **(a)** Find the difference in potential between the wingtips when the speed of the plane is 2.50×10^2 m/s. **(b)** Which wingtip is positive?

STRATEGY Because the plane is flying north, the northern component of magnetic field won't have any effect on the induced emf. The induced emf across the wing is caused solely by the downward component of the Earth's magnetic field. Substitute the given quantities into Equation 20.4. Use right-hand rule number 1 to find the direction positive charges would be propelled by the magnetic force.

SOLUTION

(a) Calculate the difference in potential across the wingtips.

Write the motional emf equation and substitute the given quantities:

$$\mathcal{E} = B\ell v = (0.600 \times 10^{-4}\ \text{T})(30.0\ \text{m})(2.50 \times 10^2\ \text{m/s})$$
$$= 0.450\ \text{V}$$

(b) Which wingtip is positive?

Apply right-hand rule number 1:

Point the fingers of your right hand north, in the direction of the velocity, and curl them down, in the direction of the magnetic field. Your thumb points west. Therefore the west wingtip is positive.

REMARKS An induced emf such as this one can cause problems on an aircraft.

QUESTION 20.3 In what directions are magnetic forces exerted on electrons in the metal aircraft if it is flying due west? (a) north (b) south (c) east (d) west (e) up (f) down

EXERCISE 20.3 Suppose a space station is in orbit where the magnetic field is parallel to Earth's surface, points north, and has magnitude 1.80×10^{-4} T. A metal cable attached to the space station stretches radially outwards 2.50 km. (a) Estimate the potential difference that develops between the ends of the cable if it's traveling eastward around Earth at a speed of 7.70×10^{3} m/s. (b) Which end of the cable is positive, the lower end or the upper end?

ANSWERS (a) 3.47×10^{3} V (b) The upper end is positive.

EXAMPLE 20.4 Where Is the Energy Source?

GOAL Use motional emf to find an induced emf and a current.

PROBLEM **(a)** The sliding bar in Figure 20.16a has a length of 0.500 m and moves at 2.00 m/s in a magnetic field of magnitude 0.250 T. Using the concept of motional emf, find the induced voltage in the moving rod. **(b)** If the resistance in the circuit is 0.500 Ω, find the current in the circuit and the power delivered to the resistor. (*Note:* The current in this case goes counterclockwise around the loop.) **(c)** Calculate the magnetic force on the bar. **(d)** Use the concepts of work and power to calculate the applied force.

STRATEGY For part (a), substitute into Equation 20.4 for the motional emf. Once the emf is found, substitution into Ohm's law gives the current. In part (c), use Equation 19.6 for the magnetic force on a current-carrying conductor. In part (d), use the fact that the power dissipated by the resistor multiplied by the elapsed time must equal the work done by the applied force.

SOLUTION

(a) Find the induced emf with the concept of motional emf.

Substitute into Equation 20.4 to find the induced emf:

$$\mathcal{E} = B\ell v = (0.250\ \text{T})(0.500\ \text{m})(2.00\ \text{m/s}) = \boxed{0.250\ \text{V}}$$

(b) Find the induced current in the circuit and the power dissipated by the resistor.

Substitute the emf and the resistance into Ohm's law to find the induced current:

$$I = \frac{\mathcal{E}}{R} = \frac{0.250\ \text{V}}{0.500\ \Omega} = \boxed{0.500\ \text{A}}$$

Substitute $I = 0.500$ A and $\mathcal{E} = 0.250$ V into Equation 17.8 to find the power dissipated by the 0.500-Ω resistor:

$$P = I\Delta V = (0.500\ \text{A})(0.250\ \text{V}) = \boxed{0.125\ \text{W}}$$

(c) Calculate the magnitude and direction of the magnetic force on the bar.

Substitute values for I, B, and ℓ into Equation 19.6, with $\sin\theta = \sin(90°) = 1$, to find the magnitude of the force:

$$F_m = IB\ell = (0.500\ \text{A})(0.250\ \text{T})(0.500\ \text{m}) = \boxed{6.25 \times 10^{-2}\ \text{N}}$$

Apply right-hand rule number 1 to find the direction of the force:

Point the fingers of your right hand in the direction of the positive current, then curl them in the direction of the magnetic field. Your thumb points in the negative x-direction.

(d) Find the value of F_{app}, the applied force.

Set the work done by the applied force equal to the dissipated power times the elapsed time:

$$W_{\text{app}} = F_{\text{app}}d = P\Delta t$$

Solve for F_{app} and substitute $d = v\,\Delta t$:

$$F_{\text{app}} = \frac{P\Delta t}{d} = \frac{P\Delta t}{v\Delta t} = \frac{P}{v} = \frac{0.125\ \text{W}}{2.00\ \text{m/s}} = \boxed{6.25 \times 10^{-2}\ \text{N}}$$

REMARKS Part (d) could be solved by using Newton's second law for an object in equilibrium: Two forces act horizontally on the bar and the acceleration of the bar is zero, so the forces must be equal in magnitude and opposite in direction. Notice the agreement between the answers for F_m and F_{app}, despite the very different concepts used.

QUESTION 20.4 Suppose the applied force and magnetic field in Figure 20.16a are removed, but a battery creates a current in the same direction as indicated. What happens to the bar?

(Continued)

EXERCISE 20.4 Suppose the current suddenly increases to 1.25 A in the same direction as before due to an increase in speed of the bar. Find (a) the emf induced in the rod and (b) the new speed of the rod.

ANSWERS (a) 0.625 V (b) 5.00 m/s

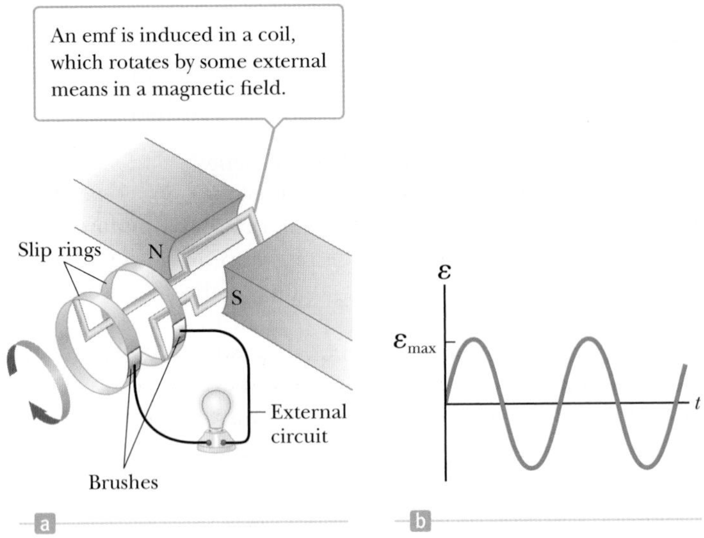

Figure 20.19 (a) A schematic diagram of an AC generator. An emf is induced in a coil, which rotates by some external means in a magnetic field. (b) A plot of the alternating emf induced in the loop versus time.

20.4 Generators

LEARNING OBJECTIVES

1. Contrast the operating principles of AC and DC generators.
2. Describe the operating principles of motors and the phenomenon of back emf.
3. Apply Faraday's law to generators and motors.

APPLICATION
Alternating-current generators

Generators and motors are important practical devices that operate on the principle of electromagnetic induction. First, consider the **alternating-current** (AC) **generator**, a device that converts mechanical energy to electrical energy. In its simplest form, the AC generator consists of a wire loop rotated in a magnetic field by some external means (Fig. 20.19a). In commercial power plants, the energy required to rotate the loop can be derived from a variety of sources. In a hydroelectric plant, for example, falling water directed against the blades of a turbine produces the rotary motion; in a coal-fired plant, heat produced by burning coal is used to convert water to steam, and this steam is directed against the turbine blades. As the loop rotates, the magnetic flux through it changes with time, inducing an emf and a current in an external circuit. The ends of the loop are connected to slip rings that rotate with the loop. Connections to the external circuit are made by stationary brushes in contact with the slip rings.

We can derive an expression for the emf generated in the rotating loop by making use of the equation for motional emf, $\mathcal{E} = B\ell v$. Figure 20.20a shows a loop of wire rotating clockwise in a uniform magnetic field directed to the right. The magnetic force (qvB) on the charges in wires AB and CD is not along the lengths of the wires. (The force on the electrons in these wires is perpendicular to the wires.) Hence, an emf is generated only in wires BC and AD. At any instant, wire BC has velocity $\vec{\mathbf{v}}$ at an angle θ with the magnetic field, as shown in Figure 20.20b. (Note that the component of velocity parallel to the field has no effect on the charges in the wire, whereas the component of velocity perpendicular to the field produces a magnetic force on the charges that moves electrons from C to B.) The emf generated in wire BC equals $B\ell v_{\perp}$,

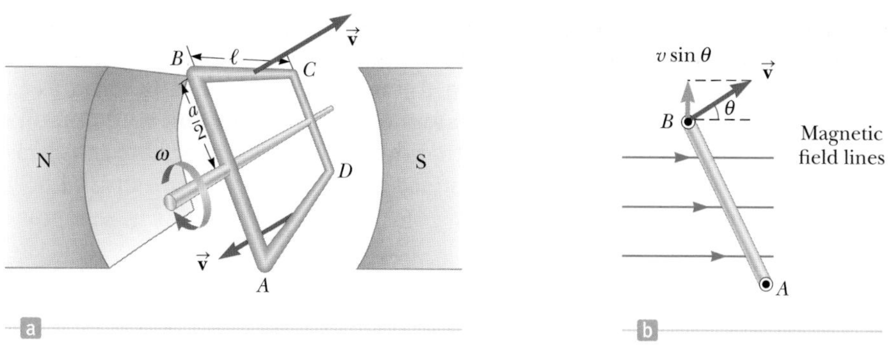

Figure 20.20 (a) A loop rotating at a constant angular velocity in an external magnetic field. The emf induced in the loop varies sinusoidally with time. (b) An edge view of the rotating loop.

where ℓ is the length of the wire and $v_\perp$ is the component of velocity perpendicular to the field. An emf of $B\ell v_\perp$ is also generated in wire DA, and the sense of this emf is the same as that in wire BC. Because $v_\perp = v \sin \theta$, the total induced emf is

$$\mathcal{E} = 2B\ell v_\perp = 2B\ell v \sin \theta \qquad [20.6]$$

If the loop rotates with a constant angular speed ω, we can use the relation $\theta = \omega t$ in Equation 20.6. Furthermore, because every point on the wires BC and DA rotates in a circle about the axis of rotation with the same angular speed ω, we have $v = r\omega = (a/2)\omega$, where a is the length of sides AB and CD. Equation 20.6 therefore reduces to

$$\mathcal{E} = 2B\ell\left(\frac{a}{2}\right)\omega \sin \omega t = B\ell a\omega \sin \omega t$$

If a coil has N turns, the emf is N times as large because each loop has the same emf induced in it. Further, because the area of the loop is $A = \ell a$, the total emf is

$$\mathcal{E} = NBA\omega \sin \omega t \qquad [20.7]$$

Turbines turn electric generators at a hydroelectric power plant.

This result shows that the emf varies sinusoidally with time, as plotted in Figure 20.19b. Note that the maximum emf has the value

$$\mathcal{E}_{max} = NBA\omega \qquad [20.8]$$

which occurs when $\omega t = 90°$ or $270°$. In other words, $\mathcal{E} = \mathcal{E}_{max}$ when the plane of the loop is parallel to the magnetic field. Further, the emf is zero when $\omega t = 0$ or $180°$, which happens whenever the magnetic field is perpendicular to the plane of the loop. In the United States and Canada the frequency of rotation for commercial generators is 60 Hz, whereas in some European countries 50 Hz is used. (Recall that $\omega = 2\pi f$, where f is the frequency in hertz.)

APPLICATION
Direct-current generators

The **direct-current** (DC) **generator** is illustrated in Figure 20.21a. The components are essentially the same as those of the AC generator except that the contacts to the rotating loop are made by a split ring, or commutator. In this design the output voltage always has the same polarity and the current is a pulsating direct current, as in Figure 20.21b. Note that the contacts to the split ring reverse their

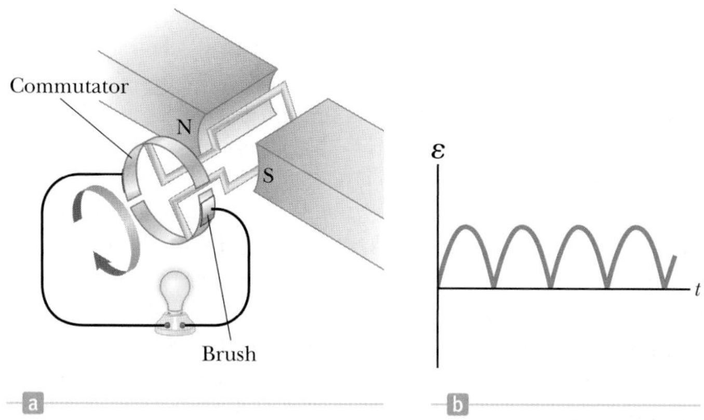

Figure 20.21 (a) A schematic diagram of a DC generator. (b) The emf fluctuates in magnitude, but always has the same polarity.

roles every half cycle. At the same time, the polarity of the induced emf reverses. Hence, the polarity of the split ring remains the same.

A pulsating DC current is not suitable for most applications. To produce a steady DC current, commercial DC generators use many loops and commutators distributed around the axis of rotation so that the sinusoidal pulses from the loops overlap in phase. When these pulses are superimposed, the DC output is almost free of fluctuations.

■ EXAMPLE 20.5 | emf Induced in an AC Generator

GOAL Understand physical aspects of an AC generator.

PROBLEM An AC generator consists of eight turns of wire, each having area $A = 0.090\ 0\ \text{m}^2$, with a total resistance of 12.0 Ω. The coil rotates in a magnetic field of 0.500 T at a constant frequency of 60.0 Hz, with axis of rotation perpendicular to the direction of the magnetic field. **(a)** Find the maximum induced emf. **(b)** What is the maximum induced current? **(c)** Determine the induced emf and current as functions of time. **(d)** What maximum torque must be applied to keep the coil turning?

STRATEGY From the given frequency, calculate the angular frequency ω and substitute it, together with given quantities, into Equation 20.8. As functions of time, the emf and current have the form $A \sin \omega t$, where A is the maximum emf or current, respectively. For part (d), calculate the magnetic torque on the coil when the current is at a maximum. (See Chapter 19.) The applied torque must do work against this magnetic torque to keep the coil turning.

SOLUTION

(a) Find the maximum induced emf.

First, calculate the angular frequency of the rotational motion:

$$\omega = 2\pi f = 2\pi(60.0\ \text{Hz}) = 377\ \text{rad/s}$$

Substitute the values for N, A, B, and ω into Equation 20.8, obtaining the maximum induced emf:

$$\mathcal{E}_{\text{max}} = NAB\omega = 8(0.090\ 0\ \text{m}^2)(0.500\ \text{T})(377\ \text{rad/s})$$
$$= \boxed{136\ \text{V}}$$

(b) What is the maximum induced current?

Substitute the maximum induced emf $\mathcal{E}_{\text{max}}$ and the resistance R into Ohm's law to find the maximum induced current:

$$I_{\text{max}} = \frac{\mathcal{E}_{\text{max}}}{R} = \frac{136\ \text{V}}{12.0\ \Omega} = \boxed{11.3\ \text{A}}$$

(c) Determine the induced emf and the current as functions of time.

Substitute $\mathcal{E}_{\text{max}}$ and ω into Equation 20.7 to obtain the variation of $\mathcal{E}$ with time t in seconds:

$$\mathcal{E} = \mathcal{E}_{\text{max}} \sin \omega t = \boxed{(136\ \text{V}) \sin 377t}$$

The time variation of the current looks just like this expression, except with the maximum current out in front:

$$I = \boxed{(11.3\ \text{A}) \sin 377t}$$

(d) Calculate the maximum applied torque necessary to keep the coil turning.

Write the equation for magnetic torque:

$$\tau = \mu B \sin \theta$$

Calculate the maximum magnetic moment of the coil, μ:

$$\mu = I_{\text{max}}AN = (11.3\ \text{A})(0.090\ \text{m}^2)(8) = 8.14\ \text{A} \cdot \text{m}^2$$

Substitute into the magnetic torque equation, with $\theta = 90°$ to find the maximum applied torque:

$$\tau_{\text{max}} = (8.14\ \text{A} \cdot \text{m}^2)(0.500\ \text{T}) \sin 90° = \boxed{4.07\ \text{N} \cdot \text{m}}$$

REMARKS The number of loops, N, can't be arbitrary because there must be a force strong enough to turn the coil.

QUESTION 20.5 What effect does doubling the frequency have on the maximum induced emf?

EXERCISE 20.5 An AC generator is to have a maximum output of 301 V. Each circular turn of wire has an area of 0.100 m^2 and a resistance of 0.80 Ω. The coil rotates in a magnetic field of 0.600 T with a frequency of 40.0 Hz, with the axis of rotation perpendicular to the direction of the magnetic field. (a) How many turns of wire should the coil have to produce the desired emf? (b) Find the maximum current induced in the coil. (c) Determine the induced emf as a function of time.

ANSWERS (a) 20 turns (b) 18.8 A (c) $\mathcal{E} = (301 \text{ V}) \sin 251t$

Motors and Back emf

Motors are devices that convert electrical energy to mechanical energy. Essentially, **a motor is a generator run in reverse**: instead of a current being generated by a rotating loop, a current is supplied to the loop by a source of emf, and the magnetic torque on the current-carrying loop causes it to rotate.

A motor can perform useful mechanical work when a shaft connected to its rotating coil is attached to some external device. As the coil in the motor rotates, however, the changing magnetic flux through it induces an emf that acts to reduce the current in the coil. If it *increased* the current, Lenz's law would be violated. The phrase **back emf** is used for an emf that tends to reduce the applied current. The back emf increases in magnitude as the rotational speed of the coil increases. We can picture this state of affairs as the equivalent circuit in Figure 20.22. For illustrative purposes, assume the external power source supplying current in the coil of the motor has a voltage of 120 V, the coil has a resistance of 10 Ω, and the back emf induced in the coil at this instant is 70 V. The voltage available to supply current equals the difference between the applied voltage and the back emf, or 50 V in this case. The current is always reduced by the back emf.

When a motor is turned on, there is no back emf initially and the current is very large because it's limited only by the resistance of the coil. As the coil begins to rotate, the induced back emf opposes the applied voltage and the current in the coil is reduced. If the mechanical load increases, the motor slows down, which decreases the back emf. This reduction in the back emf increases the current in the coil and therefore also increases the power needed from the external voltage source. As a result, the power requirements for starting a motor and for running it under heavy loads are greater than those for running the motor under average loads. If the motor is allowed to run under no mechanical load, the back emf reduces the current to a value just large enough to balance energy losses by heat and friction.

APPLICATION
Motors

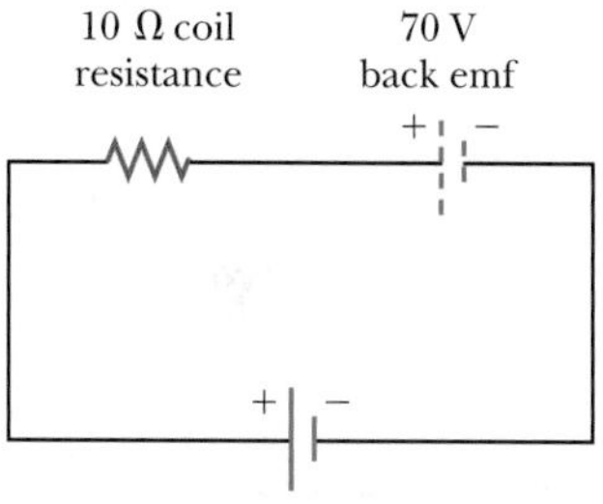

Figure 20.22 A motor can be represented as a resistance plus a back emf.

EXAMPLE 20.6 Induced Current in a Motor

GOAL Apply the concept of a back emf in calculating the induced current in a motor.

PROBLEM A motor has coils with a resistance of 10.0 Ω and is supplied by a voltage of $\Delta V = 1.20 \times 10^2$ V. When the motor is running at its maximum speed, the back emf is 70.0 V. Find the current in the coils **(a)** when the motor is first turned on and **(b)** when the motor has reached its maximum rotation rate.

STRATEGY For each part, find the net voltage, which is the applied voltage minus the induced emf. Divide the net voltage by the resistance to get the current.

SOLUTION

(a) Find the initial current, when the motor is first turned on.

If the coil isn't rotating, the back emf is zero and the current has its maximum value. Calculate the difference between the emf and the initial back emf and divide by the resistance R, obtaining the initial current:

$$I = \frac{\mathcal{E} - \mathcal{E}_{\text{back}}}{R} = \frac{1.20 \times 10^2 \text{ V} - 0}{10.0 \ \Omega} = \boxed{12.0 \text{ A}}$$

(Continued)

(b) Find the current when the motor is rotating at its maximum rate.

Repeat the calculation, using the maximum value of the back emf:

$$I = \frac{\mathcal{E} - \mathcal{E}_{\text{back}}}{R} = \frac{1.20 \times 10^2\ \text{V} - 70.0\ \text{V}}{10.0\ \Omega} = \frac{50.0\ \text{V}}{10.0\ \Omega}$$

$$= 5.00\ \text{A}$$

REMARKS The phenomenon of back emf is one way in which the rotation rate of electric motors is limited.

QUESTION 20.6 As a motor speeds up, what happens to the magnitude of the magnetic torque? (a) It increases. (b) It decreases. (c) It remains constant.

EXERCISE 20.6 If the current in the motor is 8.00 A at some instant, what is the back emf at that time?

ANSWER 40.0 V

20.5 Self-Inductance

LEARNING OBJECTIVES

1. Describe the concepts of self-induction and self-induced emf.
2. Evaluate the self-inductance and self-induced emf in simple electrical systems.

Consider a circuit consisting of a switch, a resistor, and a source of emf, as in Figure 20.23. When the switch is closed, the current doesn't immediately change from zero to its maximum value, $\mathcal{E}/R$. The law of electromagnetic induction, Faraday's law, prevents this change. What happens instead is the following: as the current increases with time, the magnetic flux through the loop due to this current also increases. The increasing flux induces an emf in the circuit that opposes the change in magnetic flux. By Lenz's law, the induced emf is in the direction indicated by the dashed battery in the figure. The net potential difference across the resistor is the emf of the battery minus the opposing induced emf. As the magnitude of the current increases, the *rate* of increase lessens and hence the induced emf decreases. This opposing emf results in a gradual increase in the current. For the same reason, when the switch is opened, the current doesn't immediately fall to zero. This effect is called **self-induction** because the changing flux through the circuit arises from the circuit itself. The emf that is set up in the circuit is called a **self-induced emf**.

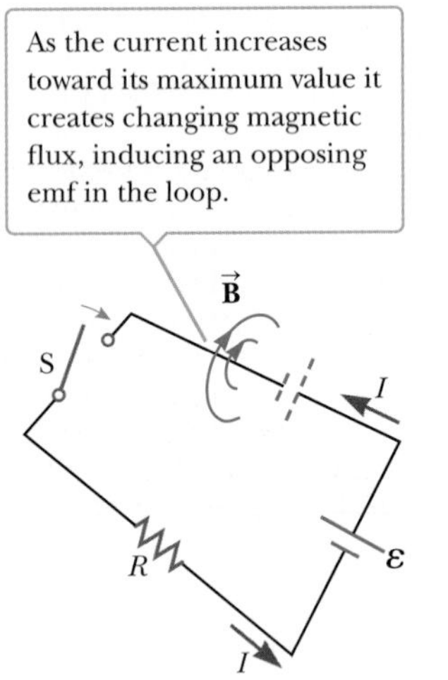

Figure 20.23 After the switch in the circuit is closed, the current produces its own magnetic flux through the loop. An opposing emf is therefore induced, meaning the current relatively slowly increases toward its maximum value rather than jumping to that value right away. The battery with the dashed lines is a symbol for the self-induced emf.

As a second example of self-inductance, consider Figure 20.24, which shows a coil wound on a cylindrical iron core. (A practical device would have several hundred turns.) Assume the current changes with time. When the current is in the direction shown, a magnetic field is set up inside the coil, directed from right to left. As a result, some lines of magnetic flux pass through the cross-sectional area of the coil. As the current changes with time, the flux through the coil changes and induces an emf in the coil. Lenz's law shows that this induced emf has a direction that opposes the change in the current. If the current is increasing, the induced emf is as pictured in Figure 20.24b, and if the current is decreasing, the induced emf is as shown in Figure 20.24c.

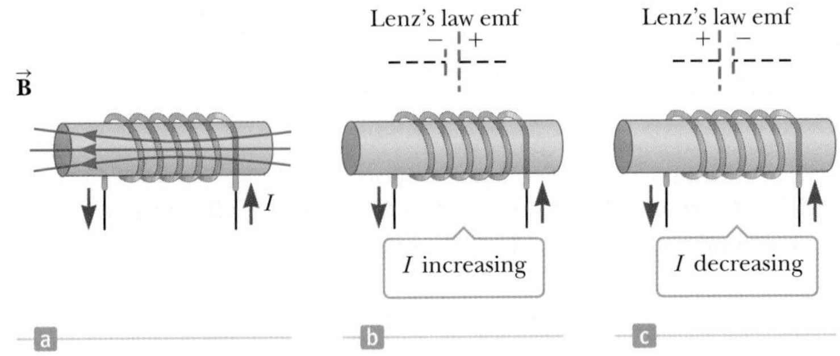

Figure 20.24 (a) A current in the coil produces a magnetic field directed to the left. (b) If the current increases, the coil acts as a source of emf directed as shown by the dashed battery. (c) The induced emf in the coil changes its polarity if the current decreases. The battery symbols drawn with dashed lines represent the included emf in the coil.

To evaluate self-inductance quantitatively, first note that, according to Faraday's law, the induced emf is given by Equation 20.2:

$$\mathcal{E} = -N\frac{\Delta\Phi_B}{\Delta t}$$

The magnetic flux is proportional to the magnetic field, which is proportional to the current in the coil. Therefore, **the self-induced emf must be proportional to the rate of change of the current with time,** or

$$\mathcal{E} \equiv -L\frac{\Delta I}{\Delta t} \qquad [20.9]$$

where L is a proportionality constant called the **inductance** of the device. The negative sign indicates that a changing current induces an emf in opposition to the change. In other words, if the current is increasing (ΔI positive), the induced emf is negative, indicating opposition to the increase in current. Likewise, if the current is decreasing (ΔI negative), the sign of the induced emf is positive, indicating that the emf is acting to oppose the decrease.

The inductance of a coil depends on the cross-sectional area of the coil and other quantities, which can all be grouped under the general heading of geometric factors. The SI unit of inductance is the **henry** (H), which, from Equation 20.9, is equal to 1 volt-second per ampere:

$$1\text{ H} = 1\text{ V}\cdot\text{s/A}$$

In the process of calculating self-inductance, it is often convenient to equate Equations 20.2 and 20.9 to find an expression for L:

$$N\frac{\Delta\Phi_B}{\Delta t} = L\frac{\Delta I}{\Delta t}$$

$$L = N\frac{\Delta\Phi_B}{\Delta I} = \frac{N\Phi_B}{I} \qquad [20.10]$$

◀ Inductance

Joseph Henry
American physicist (1797–1878)
Henry became the first director of the Smithsonian Institution and first president of the Academy of Natural Science. He was the first to produce an electric current with a magnetic field, but he failed to publish his results as early as Faraday because of his heavy teaching duties at the Albany Academy in New York State. He improved the design of the electromagnet and constructed one of the first motors. He also discovered the phenomenon of self-induction. The unit of inductance, the henry, is named in his honor.

APPLYING PHYSICS 20.3 | Making Sparks Fly

In some circuits a spark occurs between the poles of a switch when the switch is opened. Why isn't there a spark when the switch for this circuit is closed?

EXPLANATION According to Lenz's law, the direction of induced emfs is such that the induced magnetic field opposes change in the original magnetic flux. When the switch is opened, the sudden drop in the magnetic field in the circuit induces an emf in a direction that opposes change in the original current. This induced emf can cause a spark as the current bridges the air gap between the poles of the switch. The spark doesn't occur when the switch is closed, because the original current is zero and the induced emf opposes any change in that current. ■

In general, determining the inductance of a given current element can be challenging. Finding an expression for the inductance of a common solenoid, however, is straightforward. Let the solenoid have N turns and length ℓ. Assume ℓ is large compared with the radius and the core of the solenoid is air. We take the interior magnetic field to be uniform and given by Equation 19.16,

$$B = \mu_0 n I = \mu_0\frac{N}{\ell}I$$

where $n = N/\ell$ is the number of turns per unit length. The magnetic flux through each turn is therefore

$$\Phi_B = BA = \mu_0\frac{N}{\ell}AI$$

where A is the cross-sectional area of the solenoid. From this expression and Equation 20.10, we find that

$$L = \frac{N\Phi_B}{I} = \frac{\mu_0 N^2 A}{\ell} \qquad [20.11a]$$

This equation shows that L depends on the geometric factors ℓ and A and on μ_0 and is proportional to the square of the number of turns. Because $N = n\ell$, we can also express the result in the form

$$L = \mu_0 \frac{(n\ell)^2}{\ell} A = \mu_0 n^2 A\ell = \mu_0 n^2 V \qquad [20.11b]$$

where $V = A\ell$ is the volume of the solenoid.

■ EXAMPLE 20.7 | Inductance, Self-Induced emf, and Solenoids

GOAL Calculate the inductance and self-induced emf of a solenoid.

PROBLEM **(a)** Calculate the inductance of a solenoid containing 300 turns if the length of the solenoid is 25.0 cm and its cross-sectional area is 4.00×10^{-4} m². **(b)** Calculate the self-induced emf in the solenoid described in part (a) if the current in the solenoid decreases at the rate of 50.0 A/s.

STRATEGY Substituting given quantities into Equation 20.11a gives the inductance L. For part (b), substitute the result of part (a) and $\Delta I/\Delta t = -50.0$ A/s into Equation 20.9 to get the self-induced emf.

SOLUTION

(a) Calculate the inductance of the solenoid.

Substitute the number N of turns, the area A, and the length ℓ into Equation 20.11a to find the inductance:

$$L = \frac{\mu_0 N^2 A}{\ell}$$

$$= (4\pi \times 10^{-7}\,\text{T}\cdot\text{m/A})\frac{(300)^2(4.00 \times 10^{-4}\,\text{m}^2)}{25.0 \times 10^{-2}\,\text{m}}$$

$$= 1.81 \times 10^{-4}\,\text{T}\cdot\text{m}^2/\text{A} = 0.181\,\text{mH}$$

(b) Calculate the self-induced emf in the solenoid.

Substitute L and $\Delta I/\Delta t = -50.0$ A/s into Equation 20.9, finding the self-induced emf:

$$\mathcal{E} = -L\frac{\Delta I}{\Delta t} = -(1.81 \times 10^{-4}\,\text{H})(-50.0\,\text{A/s})$$

$$= 9.05\,\text{mV}$$

REMARKS Notice that $\Delta I/\Delta t$ is negative because the current is decreasing with time. The expression for the inductance in part (a) relies on the assumption that the radius of the solenoid is small compared to its length.

QUESTION 20.7 If the solenoid were wrapped into a circle so as to become a toroidal solenoid, what would be true of its self-inductance? (a) It would be the same. (b) It would be larger. (c) It would be smaller.

EXERCISE 20.7 A solenoid is to have an inductance of 0.285 mH, a cross-sectional area of 6.00×10^{-4} m², and a length of 36.0 cm. (a) How many turns per unit length should it have? (b) If the self-induced emf is −12.5 mV at a given time, at what rate is the current changing at that instant?

ANSWERS (a) 1.02×10^3 turns/m (b) 43.9 A/s

20.6 *RL* Circuits

LEARNING OBJECTIVES

1. Describe the physical effect of introducing inductance into a circuit with changing current.
2. State the time constant for an *RL* circuit.
3. Apply the *RL* circuit time constant to circuits containing resistors and inductors.

A circuit element that has a large inductance, such as a closely wrapped coil of many turns, is called an **inductor**. The circuit symbol for an inductor is . We will always assume the self-inductance of the remainder of the circuit is negligible compared with that of the inductor in the circuit.

To gain some insight into the effect of an inductor in a circuit, consider the two circuits in Figure 20.25. Figure 20.25a shows a resistor connected to the terminals of a battery. For this circuit, Kirchhoff's loop rule is $\mathcal{E} - IR = 0$. The voltage drop across the resistor is

$$\Delta V_R = -IR \qquad [20.12]$$

In this case, **we interpret resistance as a measure of opposition to the current**. Now consider the circuit in Figure 20.25b, consisting of an inductor connected to the terminals of a battery. At the instant the switch in this circuit is closed, because $IR = 0$, the emf of the battery equals the back emf generated in the coil. Hence, we have

$$\mathcal{E}_L = -L\frac{\Delta I}{\Delta t} \qquad [20.13]$$

From this expression, **we can interpret *L* as a measure of opposition to the rate of change of current.**

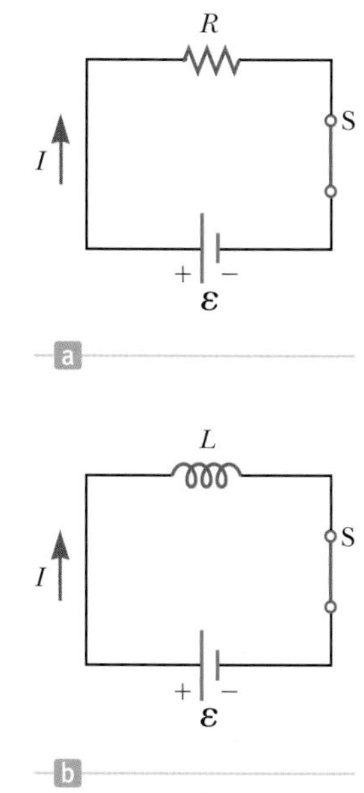

Figure 20.25 A comparison of the effect of a resistor with that of an inductor in a simple circuit.

Figure 20.26 shows a circuit consisting of a resistor, an inductor, and a battery. Suppose the switch is closed at $t = 0$. The current begins to increase, but the inductor produces an emf that opposes the increasing current. As a result, the current can't change from zero to its maximum value of $\mathcal{E}/R$ instantaneously. Equation 20.13 shows that the induced emf is a maximum when the current is changing most rapidly, which occurs when the switch is first closed. As the current approaches its steady-state value, the back emf of the coil falls off because the current is changing more slowly. Finally, when the current reaches its steady-state value, the rate of change is zero and the back emf is also zero. Figure 20.27 plots current in the circuit as a function of time. This plot is similar to that of the charge on a capacitor as a function of time, discussed in Chapter 18, section 5, in connection with RC circuits. In that case, we found it convenient to introduce a quantity called the *time constant of the circuit*, which told us something about the time required for the capacitor to approach its steady-state charge. In the same way, time constants are defined for circuits containing resistors and inductors. The **time constant** τ for an *RL* circuit is the time required for the current in the circuit to reach 63.2% of its final value $\mathcal{E}/R$; the time constant of an *RL* circuit is given by

$$\tau = \frac{L}{R} \qquad [20.14]$$

◀ Time constant for an *RL* circuit

Using methods of calculus, it can be shown that the current in such a circuit is given by

$$I = \frac{\mathcal{E}}{R}(1 - e^{-t/\tau}) \qquad [20.15]$$

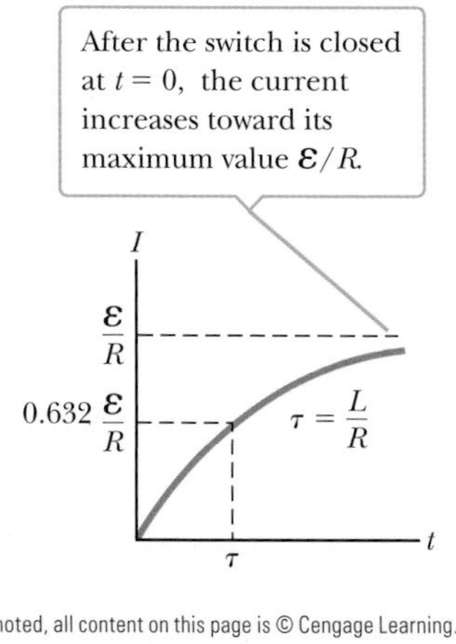

Figure 20.27 A plot of current versus time for the *RL* circuit shown in Figure 20.26. The switch is closed at $t = 0$, and the current increases toward its maximum value $\mathcal{E}/R$. The time constant τ is the time it takes the current to reach 63.2% of its maximum value.

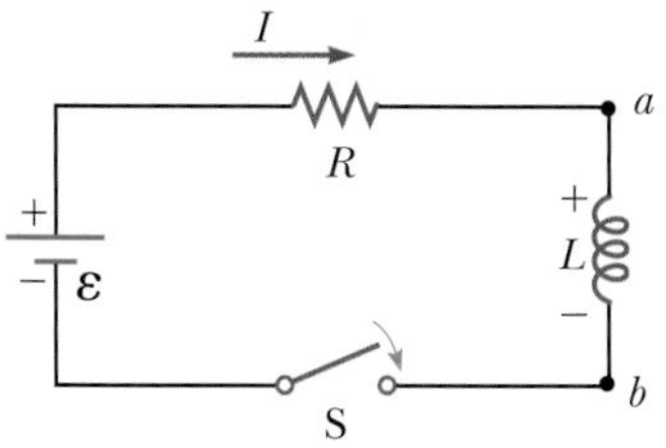

Figure 20.26 A series *RL* circuit. As the current increases toward its maximum value, the inductor produces an emf that opposes the increasing current.

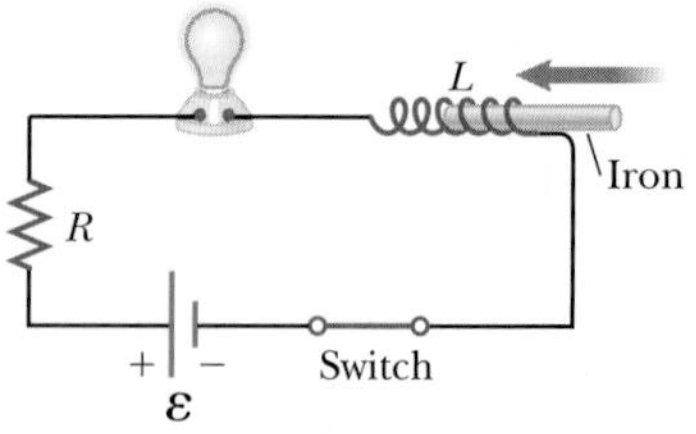

Figure 20.28 (Quick Quiz 20.6)

This equation is consistent with our intuition: when the switch is closed at $t = 0$, the current is initially zero, rising with time to some maximum value. Notice the mathematical similarity between Equation 20.15 and Equation 18.7, which features a capacitor instead of an inductor. As in the case of a capacitor, the equation's form suggests an infinite amount of time is required for the current in the inductor to reach its maximum value. This is an artifact of assuming current is composed of moving charges that are infinitesimal, as will be demonstrated in Example 20.9.

Quick Quiz

20.6 The switch in the circuit shown in Figure 20.28 is closed, and the lightbulb glows steadily. The inductor is a simple air-core solenoid. An iron rod is inserted into the interior of the solenoid, increasing the magnitude of the magnetic field in the solenoid. As the rod is inserted, the brightness of the lightbulb (a) increases, (b) decreases, or (c) remains the same.

EXAMPLE 20.8 An *RL* Circuit

GOAL Calculate a time constant and relate it to current in an *RL* circuit.

PROBLEM A 12.6-V battery is in a circuit with a 30.0-mH inductor and a 0.150-Ω resistor, as in Figure 20.26. The switch is closed at $t = 0$. **(a)** Find the time constant of the circuit. **(b)** Find the current after one time constant has elapsed. **(c)** Find the voltage drops across the resistor when $t = 0$ and $t =$ one time constant, τ. **(d)** What's the rate of change of the current after one time constant?

STRATEGY Part (a) requires only substitution into the definition of time constant. With this value and Ohm's law, the current after one time constant can be found, and multiplying this current by the resistance yields the voltage drop across the resistor after one time constant. With the voltage drop and Kirchhoff's loop law, the voltage across the inductor can be found. This value can be substituted into Equation 20.13 to obtain the rate of change of the current.

SOLUTION

(a) What's the time constant of the circuit?

Substitute the inductance L and resistance R into Equation 20.14, finding the time constant:

$$\tau = \frac{L}{R} = \frac{30.0 \times 10^{-3}\,\mathrm{H}}{0.150\,\Omega} = \boxed{0.200\,\mathrm{s}}$$

(b) Find the current after one time constant has elapsed.

First, use Ohm's law to compute the final value of the current after many time constants have elapsed:

$$I_{\max} = \frac{\mathcal{E}}{R} = \frac{12.6\,\mathrm{V}}{0.150\,\Omega} = 84.0\,\mathrm{A}$$

After one time constant, the current rises to 63.2% of its final value:

$$I_{1\tau} = (0.632)I_{\max} = (0.632)(84.0\,\mathrm{A}) = \boxed{53.1\,\mathrm{A}}$$

(c) Find the voltage drops across the resistance when $t = 0$ and $t =$ one time constant.

Initially, the current in the circuit is zero, so, from Ohm's law, the voltage across the resistor is zero:

$$\Delta V_R = IR$$

$$\Delta V_R\,(t = 0\,\mathrm{s}) = (0\,\mathrm{A})(0.150\,\Omega) = \boxed{0}$$

Next, using Ohm's law, find the magnitude of the voltage drop across the resistor after one time constant:

$$\Delta V_R\,(t = 0.200\,\mathrm{s}) = (53.1\,\mathrm{A})(0.150\,\Omega) = \boxed{7.97\,\mathrm{V}}$$

(d) What's the rate of change of the current after one time constant?

Using Kirchhoff's voltage rule, calculate the voltage drop across the inductor at that time:

$$\mathcal{E} + \Delta V_R + \Delta V_L = 0$$

Solve for ΔV_L:

$$\Delta V_L = -\mathcal{E} - \Delta V_R = -12.6\,\mathrm{V} - (-7.97\,\mathrm{V}) = -4.6\,\mathrm{V}$$

Now solve Equation 20.13 for $\Delta I/\Delta t$ and substitute:

$$\Delta V_L = -L\frac{\Delta I}{\Delta t}$$

$$\frac{\Delta I}{\Delta t} = -\frac{\Delta V_L}{L} = -\frac{-4.6\text{ V}}{30.0 \times 10^{-3}\text{ H}} = \boxed{150\text{ A/s}}$$

REMARKS The values used in this problem were taken from actual components salvaged from the starter system of a car. Because the current in such an *RL* circuit is initially zero, inductors are sometimes referred to as "chokes" because they temporarily choke off the current. In solving part (d), we traversed the circuit in the direction of positive current, so the voltage difference across the battery was positive and the differences across the resistor and inductor were negative.

QUESTION 20.8 Find the current in the circuit after two time constants.

EXERCISE 20.8 A 12.6-V battery is in series with a resistance of 0.350 Ω and an inductor. (a) After a long time, what is the current in the circuit? (b) What is the current after one time constant? (c) What's the voltage drop across the inductor at this time? (d) Find the inductance if the time constant is 0.130 s.

ANSWERS (a) 36.0 A (b) 22.8 A (c) 4.62 V (d) 4.55×10^{-2} H

EXAMPLE 20.9 | Formation of a Magnetic Field

GOAL Understand the role of time in setting up an inductor's magnetic field.

PROBLEM Given the *RL* circuit of Example 20.8, find the time required for the current to reach 99.9% of its maximum value after the switch is closed.

STRATEGY The solution requires solving Equation 20.15 for time followed by substitution. Notice that the maximum current is $I_{\text{max}} = \mathcal{E}/R$.

SOLUTION

Write Equation 20.15, with I_f substituted for the current:

$$I_f = \frac{\mathcal{E}}{R}(1 - e^{-t/\tau}) = I_{\text{max}}(1 - e^{-t/\tau})$$

Divide both sides by I_{max}:

$$\frac{I_f}{I_{\text{max}}} = 1 - e^{-t/\tau}$$

Subtract 1 from both sides and then multiply both sides by −1:

$$1 - \frac{I_f}{I_{\text{max}}} = e^{-t/\tau}$$

Take the natural log of both sides:

$$\ln\left(1 - \frac{I_f}{I_{\text{max}}}\right) = \ln\left(e^{-t/\tau}\right) = -t/\tau$$

Solve for t and substitute the expression for τ from Equation 20.14:

$$t = -\frac{L}{R}\ln\left(1 - \frac{I_f}{I_{\text{max}}}\right)$$

Substitute values, obtaining the desired time:

$$t = -\frac{30.0 \times 10^{-3}\text{ H}}{0.150\ \Omega}\ln(1 - 0.999) = \boxed{1.38\text{ s}}$$

REMARKS From this calculation, it's found that forming the magnetic field in an inductor and approaching the maximum current occurs relatively rapidly. Contrary to what might be expected from the mathematical form of Equation 20.15, an infinite amount of time is not actually required.

QUESTION 20.9 If the inductance were doubled, by what factor would the length of time found be changed? (a) 1 (i.e., no change) (b) 2 (c) $\frac{1}{2}$

EXERCISE 20.9 Suppose a series *RL* circuit is composed of a 2.00-Ω resistor, a 15.0 H inductor, and a 6.00 V battery. (a) What is the time constant for this circuit? (b) Once the switch is closed, how long does it take the current to reach half its maximum value?

ANSWERS (a) 7.50 s (b) 5.20 s

20.7 Energy Stored in a Magnetic Field

LEARNING OBJECTIVES

1. Define the energy stored in an inductor and contrast it with the energy stored in a capacitor.
2. Apply the energy stored by an inductor to *RL* circuits.

The emf induced by an inductor prevents a battery from establishing an instantaneous current in a circuit. The battery has to do work to produce a current. We can think of this needed work as energy stored in the inductor in its magnetic field. In a manner similar to that used in Section 16.9 to find the energy stored in a capacitor, we find that the energy stored by an inductor is

Energy stored in an inductor ▶

$$PE_L = \tfrac{1}{2}LI^2 \qquad [20.16]$$

Notice that the result is similar in form to the expression for the energy stored in a charged capacitor (Eq. 16.18):

Energy stored in a capacitor ▶

$$PE_C = \tfrac{1}{2}C(\Delta V)^2$$

■ EXAMPLE 20.10 | Magnetic Energy

GOAL Relate the storage of magnetic energy to currents in an *RL* circuit.

PROBLEM A 12.0-V battery is connected in series to a 25.0-Ω resistor and a 5.00-H inductor. **(a)** Find the maximum current in the circuit. **(b)** Find the energy stored in the inductor at this time. **(c)** How much energy is stored in the inductor when the current is changing at a rate of 1.50 A/s?

STRATEGY In part (a) Ohm's law and Kirchhoff's voltage rule yield the maximum current because the voltage across the inductor is zero when the current is maximal. Substituting the current into Equation 20.16 gives the energy stored in the inductor. In part (c) the given rate of change of the current can be used to calculate the voltage drop across the inductor at the specified time. Kirchhoff's voltage rule and Ohm's law then give the current I at that time, which can be used to find the energy stored in the inductor.

SOLUTION

(a) Find the maximum current in the circuit.

Apply Kirchhoff's voltage rule to the circuit:

$$\Delta V_{\text{batt}} + \Delta V_R + \Delta V_L = 0$$

$$\mathcal{E} - IR - L\frac{\Delta I}{\Delta t} = 0$$

When the maximum current is reached, $\Delta I/\Delta t$ is zero, so the voltage drop across the inductor is zero. Solve for the maximum current I_{max}:

$$I_{\text{max}} = \frac{\mathcal{E}}{R} = \frac{12.0\text{ V}}{25.0\ \Omega} = \boxed{0.480\text{ A}}$$

(b) Find the energy stored in the inductor at this time.

Substitute known values into Equation 20.16:

$$PE_L = \tfrac{1}{2}LI_{\text{max}}^2 = \tfrac{1}{2}(5.00\text{ H})(0.480\text{ A})^2 = \boxed{0.576\text{ J}}$$

(c) Find the energy in the inductor when the current changes at a rate of 1.50 A/s.

Apply Kirchhoff's voltage rule to the circuit once again:

$$\mathcal{E} - IR - L\frac{\Delta I}{\Delta t} = 0$$

Solve this equation for the current I and substitute:

$$I = \frac{1}{R}\left(\mathcal{E} - L\frac{\Delta I}{\Delta t}\right)$$

$$= \frac{1}{25.0\ \Omega}[12.0\text{ V} - (5.00\text{ H})(1.50\text{ A/s})] = 0.180\text{ A}$$

12. A circular loop of wire of radius 12.0 cm is placed in a magnetic field directed perpendicular to the plane of the loop, as shown in Figure P20.10. If the field decreases at the rate of 0.050 0 T/s in some time interval, what is the magnitude of the emf induced in the loop during this interval?

13. A technician wearing a circular metal band on his wrist moves his hand into a uniform magnetic field of magnitude 2.5 T in a time of 0.18 s. If the diameter of the band is 6.5 cm and the field is at an angle of 45° with the plane of the metal band while the hand is in the field, find the magnitude of the average emf induced in the band.

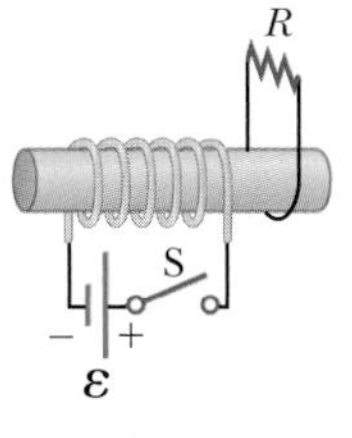

Figure P20.14

14. In Figure P20.14 what is the direction of the current induced in the resistor at the instant the switch is closed?

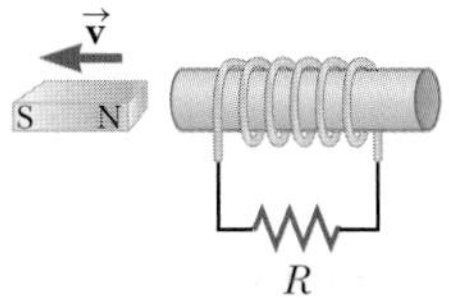

Figure P20.15

15. A bar magnet is positioned near a coil of wire, as shown in Figure P20.15. What is the direction of the current in the resistor when the magnet is moved (a) to the left and (b) to the right?

16. Find the direction of the current in the resistor shown in Figure P20.16 (a) at the instant the switch is closed, (b) after the switch has been closed for several minutes, and (c) at the instant the switch is opened.

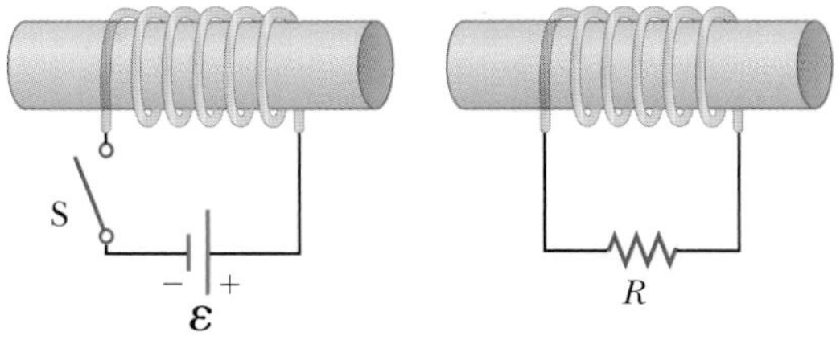

Figure P20.16

17. A circular loop of wire lies below a long wire carrying a current that is increasing as in Figure P20.17a. (a) What is the direction of the induced current in the loop, if any? (b) Now suppose the loop is next to the same wire as in P20.17b. What is the direction of the induced current in the loop, if any? Explain your answers.

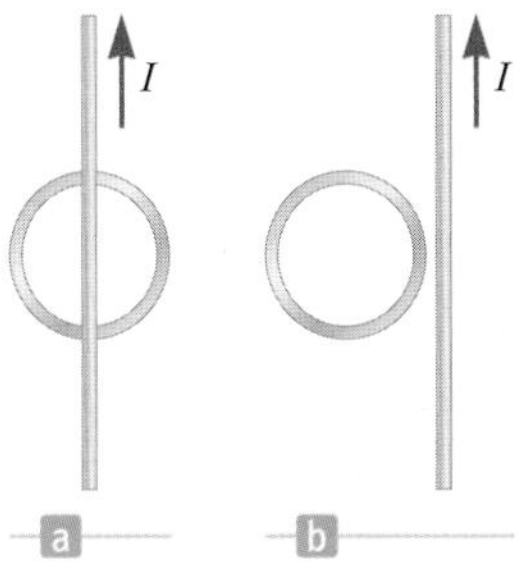

Figure P20.17

18. A square, single-turn wire loop $\ell = 1.00$ cm on a side is placed inside a solenoid that has a circular cross section of radius $r = 3.00$ cm, as shown in the end view of Figure P20.18. The solenoid is 20.0 cm long and wound with 100 turns of wire. (a) If the current in the solenoid is 3.00 A, what is the flux through the square loop? (b) If the current in the solenoid is reduced to zero in 3.00 s, what is the magnitude of the average induced emf in the square loop?

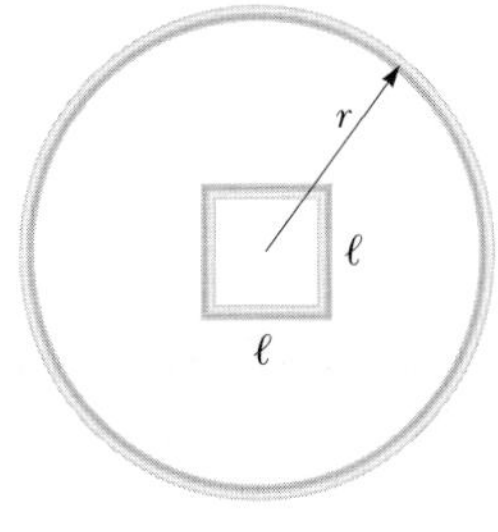

Figure P20.18

19. GP A 300-turn solenoid with a length of 20.0 cm and a radius of 1.50 cm carries a current of 2.00 A. A second coil of four turns is wrapped tightly around this solenoid, so it can be considered to have the same radius as the solenoid. The current in the 300-turn solenoid increases steadily to 5.00 A in 0.900 s. (a) Use Ampere's law to calculate the initial magnetic field in the middle of the 300-turn solenoid. (b) Calculate the magnetic field of the 300-turn solenoid after 0.900 s. (c) Calculate the area of the 4-turn coil. (d) Calculate the change in the magnetic flux through the 4-turn coil during the same period. (e) Calculate the average induced emf in the 4-turn coil. Is it equal to the instantaneous induced emf? Explain. (f) Why could contributions to the magnetic field by the current in the 4-turn coil be neglected in this calculation?

20. M A circular coil enclosing an area of 100 cm^2 is made of 200 turns of copper wire. The wire making up the coil has resistance of 5.0 Ω, and the ends of the wire are connected to form a closed circuit. Initially, a 1.1-T uniform magnetic field points perpendicularly upward through the plane of the coil. The direction of the field then reverses so that the final magnetic field has a magnitude of 1.1 T and points downward through the coil. If the time required for the field to reverse directions is 0.10 s, what is the average current in the coil during that time?

21. BIO To monitor the breathing of a hospital patient, a thin belt is girded around the patient's chest as in Figure P20.21. The belt is a 200-turn coil. When the patient inhales, the area encircled by the coil increases

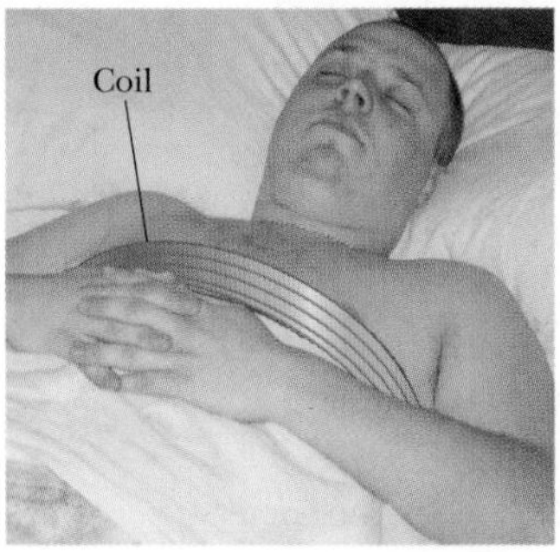

Figure P20.21

by 39.0 cm^2. The magnitude of Earth's magnetic field is 50.0 μT and makes an angle of 28.0° with the plane of the coil. Assuming a patient takes 1.80 s to inhale, find the magnitude of the average induced emf in the coil during that time.

22. Q|C S An N-turn circular wire coil of radius r lies in the xy-plane (the plane of the page), as in Figure P20.10. A uniform magnetic field is turned on, increasing steadily from 0 to B_0 in the positive z-direction in t seconds. (a) Find a symbolic expression for the emf, $\mathcal{E}$, induced in the coil in terms of the variables given. (b) Looking down on at the xy-plane from the positive z-axis, is the direction of the induced current clockwise or counterclockwise? (c) If each loop has resistance R, find an expression for the magnitude of the induced current, I.

20.3 Motional emf

23. A truck is carrying a steel beam of length 15.0 m on a freeway. An accident causes the beam to be dumped off the truck and slide horizontally along the ground at a speed of 25.0 m/s. The velocity of the center of mass of the beam is northward while the length of the beam maintains an east–west orientation. The vertical component of the Earth's magnetic field at this location has a magnitude of 35.0 μT. What is the magnitude of the induced emf between the ends of the beam?

24. A 2.00-m length of wire is held in an east–west direction and moves horizontally to the north with a speed of 15.0 m/s. The vertical component of Earth's magnetic field in this region is 40.0 μT directed downward. Calculate the induced emf between the ends of the wire and determine which end is positive.

25. A pickup truck has a width of 79.8 in. If it is traveling north at 37 m/s through a magnetic field with vertical component of 35 μT, what magnitude emf is induced between the driver and passenger sides of the truck?

26. Q|C A bar magnet is held stationary while a circular loop of wire is moved toward the magnet at constant velocity at position A as in Figure P20.26. The loop passes over the magnet at position B and moves away from the magnet at position C. Find the direction of the induced current in the loop using Lenz's law (a) at position A and (b) at position C. (c) What is the induced current in the loop at position B? Explain.

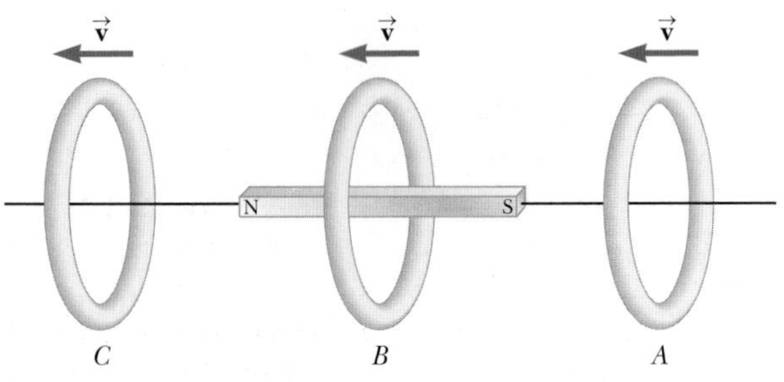

Figure P20.26 Problems 26 and 43.

27. M An automobile has a vertical radio antenna 1.20 m long. The automobile travels at 65.0 km/h on a horizontal road where Earth's magnetic field is 50.0 μT, directed toward the north and downward at an angle of 65.0° below the horizontal. (a) Specify the direction the automobile should move so as to generate the maximum motional emf in the antenna, with the top of the antenna positive relative to the bottom. (b) Calculate the magnitude of this induced emf.

28. Q|C An astronaut is connected to her spacecraft by a 25-m-long tether cord as she and the spacecraft orbit Earth in a circular path at a speed of 3.0×10^3 m/s. At one instant, the voltage measured between the ends of a wire embedded in the cord is measured to be 0.45 V. Assume the long dimension of the cord is perpendicular to the vertical component of Earth's magnetic field at that instant. (a) What is the magnitude of the vertical component of Earth's field at this location? (b) Does the measured voltage change as the system moves from one location to another? Explain.

29. Figure P20.29 shows a bar of mass m = 0.200 kg that can slide without friction on a pair of rails separated by a distance ℓ = 1.20 m and located on an inclined plane that makes an angle θ = 25.0° with respect to the ground. The resistance of the resistor is R = 1.00 Ω, and a uniform magnetic field of magnitude B = 0.500 T is directed downward, perpendicular to the ground, over the entire region through which the bar moves. With what constant speed v does the bar slide along the rails?

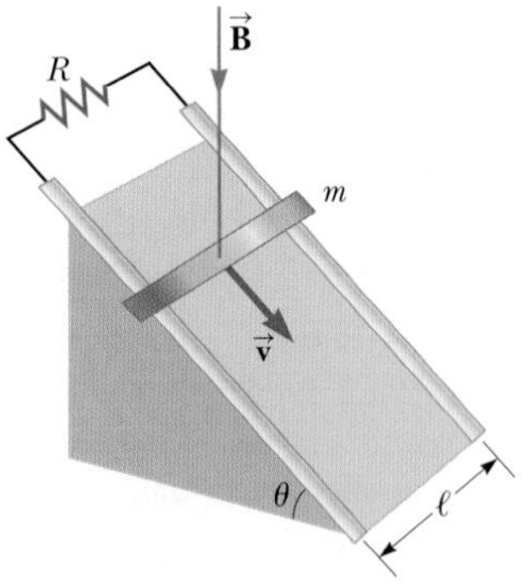

Figure P20.29

30. W Consider the arrangement shown in Figure P20.30. Assume R = 6.00 Ω, ℓ = 1.20 m, and a uniform 2.50-T magnetic field is directed *into* the page. At what speed should the bar be moved to produce a current of 0.500 A in the resistor?

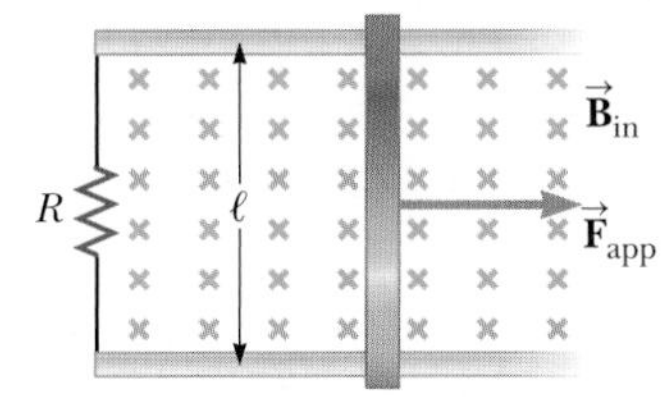

Figure P20.30 Problems 30, 56, and 59.

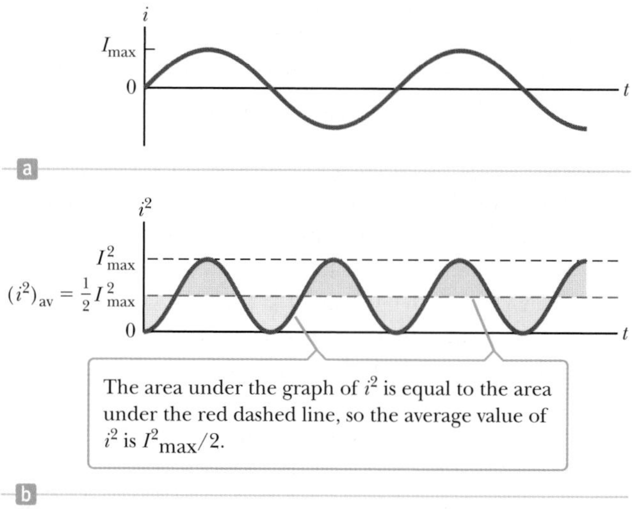

Figure 21.3 (a) Plot of the current in a resistor as a function of time. (b) Plot of the square of the current in a resistor as a function of time. The area under the red-dashed line omits the tan-shaded regions, but that is made up by the inclusion of the gray-shaded regions with the same area. That shows the area under the curve of i^2 is the same as the area under the red dashed line, so $(i^2)_{av} = I^2_{max}/2$.

same amount of energy in a resistor as the actual alternating current. Finding the average of the alternating current i, depicted in Figure 21.3a, would not be useful because that average is zero, whereas the rms current is always positive. To find the rms current, we first square the current, then find its average value, and finally take the square root of this average value. Hence, the rms current is the square *root* of the average (*mean*) of the *square* of the current. Because i^2 varies as $\sin^2 2\pi ft$, the average value of i^2 is $\frac{1}{2}I^2_{max}$ (Fig. 21.3b).[1] Therefore, the rms current I_{rms} is related to the maximum value of the alternating current I_{max} by

$$I_{rms} = \frac{I_{max}}{\sqrt{2}} = 0.707 I_{max} \qquad [21.2]$$

This equation says that an alternating current with a maximum value of 3 A produces the same heating effect in a resistor as a direct current of $(3/\sqrt{2})$ A. We can therefore say that the average power dissipated in a resistor that carries alternating current I is

$$P_{av} = I^2_{rms}R$$

Alternating voltages are also best discussed in terms of rms voltages, with a relationship identical to the preceding one,

$$\Delta V_{rms} = \frac{\Delta V_{max}}{\sqrt{2}} = 0.707\,\Delta V_{max} \qquad [21.3]$$

◀ rms voltage

where ΔV_{rms} is the rms voltage and ΔV_{max} is the maximum value of the alternating voltage.

[1]We can show that $(i^2)_{av} = I^2_{max}/2$ as follows: The current in the circuit varies with time according to the expression $i = I_{max} \sin 2\pi ft$, so $i^2 = I^2_{max} \sin^2 2\pi ft$. Therefore, we can find the average value of i^2 by calculating the average value of $\sin^2 2\pi ft$. Note that a graph of $\cos^2 2\pi ft$ versus time is identical to a graph of $\sin^2 2\pi ft$ versus time, except that the points are shifted on the time axis. Thus, the time average of $\sin^2 2\pi ft$ is equal to the time average of $\cos^2 2\pi ft$, taken over one or more cycles. That is,

$$(\sin^2 2\pi ft)_{av} = (\cos^2 2\pi ft)_{av}$$

With this fact and the trigonometric identity $\sin^2\theta + \cos^2\theta = 1$, we get

$$(\sin^2 2\pi ft)_{av} + (\cos^2 2\pi ft)_{av} = 2(\sin^2 2\pi ft)_{av} = 1$$

$$(\sin^2 2\pi ft)_{av} = \tfrac{1}{2}$$

When this result is substituted into the expression $i^2 = I^2_{max} \sin^2 2\pi ft$, we get $(i^2)_{av} = I^2_{rms} = I^2_{max}/2$, or $I_{rms} = I_{max}/\sqrt{2}$, where I_{rms} is the rms current.

Table 21.1 Notation Used in This Chapter

	Voltage	Current
Instantaneous value	Δv	i
Maximum value	$\Delta V_{\max}$	$I_{\max}$
rms value	ΔV_{rms}	I_{rms}

When we speak of measuring an AC voltage of 120 V from an electric outlet, we actually mean an rms voltage of 120 V. A quick calculation using Equation 21.3 shows that such an AC voltage actually has a peak value of about 170 V. In this chapter we use rms values when discussing alternating currents and voltages. One reason is that AC ammeters and voltmeters are designed to read rms values. Further, if we use rms values, many of the equations for alternating current will have the same form as those used in the study of direct-current (DC) circuits. Table 21.1 summarizes the notations used throughout this chapter.

Consider the series circuit in Figure 21.1 (page 734), consisting of a resistor connected to an AC generator. A resistor impedes the current in an AC circuit, just as it does in a DC circuit. Ohm's law is therefore valid for an AC circuit, and we have

$$\Delta V_{R,\text{rms}} = I_{\text{rms}} R \qquad [21.4a]$$

The rms voltage across a resistor is equal to the rms current in the circuit times the resistance. This equation is also true if maximum values of current and voltage are used:

$$\Delta V_{R,\max} = I_{\max} R \qquad [21.4b]$$

Quick Quiz

21.1 Which of the following statements can be true for a resistor connected in a simple series circuit to an operating AC generator? (a) $P_{\text{av}} = 0$ and $i_{\text{av}} = 0$ (b) $P_{\text{av}} = 0$ and $i_{\text{av}} > 0$ (c) $P_{\text{av}} > 0$ and $i_{\text{av}} = 0$ (d) $P_{\text{av}} > 0$ and $i_{\text{av}} > 0$

EXAMPLE 21.1 What Is the rms Current?

GOAL Perform basic AC circuit calculations for a purely resistive circuit.

PROBLEM An AC voltage source has an output of $\Delta v = (2.00 \times 10^2\text{ V}) \sin 2\pi ft$. This source is connected to a $1.00 \times 10^2\ \Omega$ resistor as in Figure 21.1. Find the rms voltage and rms current in the resistor.

STRATEGY Compare the expression for the voltage output just given with the general form, $\Delta v = \Delta V_{\max} \sin 2\pi ft$, finding the maximum voltage. Substitute this result into the expression for the rms voltage.

SOLUTION

Obtain the maximum voltage by comparison of the given expression for the output with the general expression:

$$\Delta v = (2.00 \times 10^2\text{ V}) \sin 2\pi ft \qquad \Delta v = \Delta V_{\max} \sin 2\pi ft$$
$$\to \quad \Delta V_{\max} = 2.00 \times 10^2\text{ V}$$

Next, substitute into Equation 21.3 to find the rms voltage of the source:

$$\Delta V_{\text{rms}} = \frac{\Delta V_{\max}}{\sqrt{2}} = \frac{2.00 \times 10^2\text{ V}}{\sqrt{2}} = 141\text{ V}$$

Substitute this result into Ohm's law to find the rms current:

$$I_{\text{rms}} = \frac{\Delta V_{\text{rms}}}{R} = \frac{141\text{ V}}{1.00 \times 10^2\ \Omega} = 1.41\text{ A}$$

REMARKS Notice how the concept of rms values allows the handling of an AC circuit quantitatively in much the same way as a DC circuit.

QUESTION 21.1 True or False: The rms current in an AC circuit oscillates sinusoidally with time.

EXERCISE 21.1 Find the maximum current in the circuit and the average power delivered to the circuit.

ANSWER 2.00 A; 2.00×10^2 W

APPLYING PHYSICS 21.1 | Electric Fields and Cancer Treatment BIO

Cancer cells multiply far more frequently than most normal cells, spreading throughout the body, using its resources and interfering with normal functioning. Most therapies damage both cancerous and healthy cells, so finding methods that target cancer cells is important in developing better treatments for the disease.

Because cancer cells multiply so rapidly, it's natural to consider treatments that prevent or disrupt cell division. Treatments such as chemotherapy interfere with the cell division cycle, but can also damage healthy cells. It has recently been found that alternating electric fields produced by alternating currents in the range of 100 kHz can disrupt the cell division cycle, either by slowing the division or by causing a dividing cell to disintegrate. Healthy cells that divide at only a very slow rate are less vulnerable than the rapidly dividing cancer cells, so such therapy holds out promise for certain types of cancer.

The alternating electric fields are thought to affect the process of mitosis, which is the dividing of the cell nucleus into two sets of identical chromosomes. Near the end of the first phase of mitosis, called the prophase, the mitotic spindle forms, a structure of fine filaments that guides the two replicated sets of chromosomes into separate daughter cells. The mitotic spindle is made up of a polymerization of dimers of tubulin, a protein with a large electric dipole moment. The alternating electric field exerts forces on these dipoles, disrupting their proper functioning.

Electric field therapy is especially promising for the treatment of brain tumors because healthy brain cells don't divide and therefore would be unharmed by the alternating electric fields. Research on such therapies is ongoing. ■

21.2 Capacitors in an AC Circuit

LEARNING OBJECTIVES

1. Define capacitive reactance.
2. Discuss the phase relationship between current and voltage difference in an AC circuit capacitor.
3. Calculate the capacitive reactance and rms current of capacitive AC circuits.

To understand the effect of a capacitor on the behavior of a circuit containing an AC voltage source, we first review what happens when a capacitor is placed in a circuit containing a DC source, such as a battery. When the switch is closed in a series circuit containing a battery, a resistor, and a capacitor, the initial charge on the plates of the capacitor is zero. The motion of charge through the circuit is therefore relatively free, and there is a large current in the circuit. As more charge accumulates on the capacitor, the voltage across it increases, opposing the current. After some time interval, which depends on the time constant RC, the current approaches zero. Consequently, a capacitor in a DC circuit limits or impedes the current so that it approaches zero after a brief time.

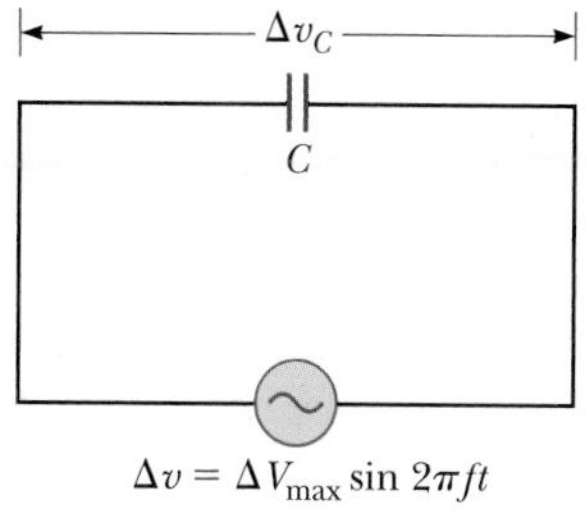

Figure 21.4 A series circuit consisting of a capacitor C connected to an AC generator.

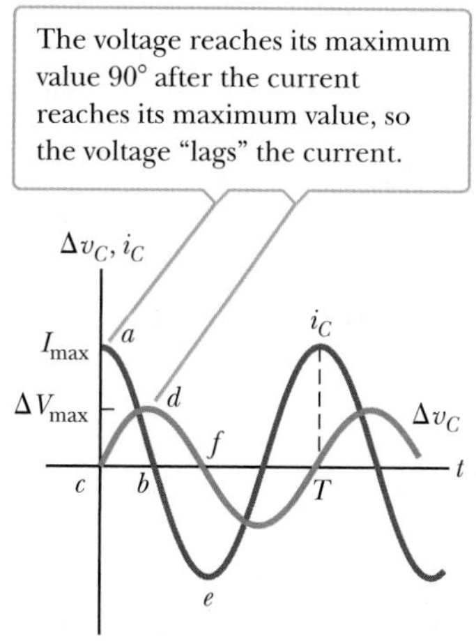

Figure 21.5 Plots of current and voltage across a capacitor versus time in an AC circuit.

Now consider the simple series circuit in Figure 21.4, consisting of a capacitor connected to an AC generator. We sketch curves of current versus time and voltage versus time, and then attempt to make the graphs seem reasonable. The curves are shown in Figure 21.5. First, notice that the segment of the current curve from a to b indicates that the current starts out at a rather large value. This large value can be understood by recognizing that there is no charge on the capacitor at $t = 0$; as a consequence, there is nothing in the circuit except the resistance of the wires to hinder the flow of charge at this instant. The current decreases, however, as the voltage across the capacitor increases from c to d on the voltage curve. When the voltage is at point d, the current reverses and begins to increase in the opposite direction (from b to e on the current curve). During this time, the voltage across the capacitor decreases from d to f because the plates are now losing the charge they accumulated earlier. The remainder of the cycle for both voltage and current is a repeat of what happened during the first half of the cycle. The current reaches a maximum value in the opposite direction at point e on the current curve and then decreases as the voltage across the capacitor builds up.

In a purely resistive circuit, the current and voltage are always in step with each other. That isn't the case when a capacitor is in the circuit. In Figure 21.5, when an alternating voltage is applied across a capacitor, the voltage reaches its maximum value one-quarter of a cycle after the current reaches its maximum value. We say that **the voltage across a capacitor always lags the current by 90°.**

The voltage across a capacitor lags the current by 90° ▶

The impeding effect of a capacitor on the current in an AC circuit is expressed in terms of a factor called the **capacitive reactance** X_C, defined as

Capacitive reactance ▶

$$X_C \equiv \frac{1}{2\pi f C} \quad [21.5]$$

When C is in farads and f is in hertz, the unit of X_C is the ohm. Notice that $2\pi f = \omega$, the angular frequency.

From Equation 21.5, as the frequency f of the voltage source increases, the capacitive reactance X_C (the impeding effect of the capacitor) decreases, so the current increases. At high frequency, there is less time available to charge the capacitor, so less charge and voltage accumulate on the capacitor, which translates into less opposition to the flow of charge and, consequently, a higher current. The analogy between capacitive reactance and resistance means that we can write an equation of the same form as Ohm's law to describe AC circuits containing capacitors. This equation relates the rms voltage and rms current in the circuit to the capacitive reactance:

$$\Delta V_{C,\text{rms}} = I_{\text{rms}} X_C \quad [21.6]$$

EXAMPLE 21.2 | A Purely Capacitive AC Circuit

GOAL Perform basic AC circuit calculations for a capacitive circuit.

PROBLEM An 8.00-μF capacitor is connected to the terminals of an AC generator with an rms voltage of 1.50×10^2 V and a frequency of 60.0 Hz. Find the capacitive reactance and the rms current in the circuit.

STRATEGY Substitute values into Equations 21.5 and 21.6.

SOLUTION

Substitute the values of f and C into Equation 21.5:

$$X_C = \frac{1}{2\pi f C} = \frac{1}{2\pi(60.0\text{ Hz})(8.00 \times 10^{-6}\text{ F})} = 332\ \Omega$$

Solve Equation 21.6 for the current and substitute the values for X_C and the rms voltage to find the rms current:

$$I_{\text{rms}} = \frac{\Delta V_{C,\text{rms}}}{X_C} = \frac{1.50 \times 10^2\text{ V}}{332\ \Omega} = 0.452\text{ A}$$

REMARKS Again, notice how similar the technique is to that of analyzing a DC circuit with a resistor.

QUESTION 21.2 True or False: The larger the capacitance of a capacitor, the larger the capacitive reactance.

EXERCISE 21.2 If the frequency is doubled, what happens to the capacitive reactance and the rms current?

ANSWER X_C is halved, and I_{rms} is doubled.

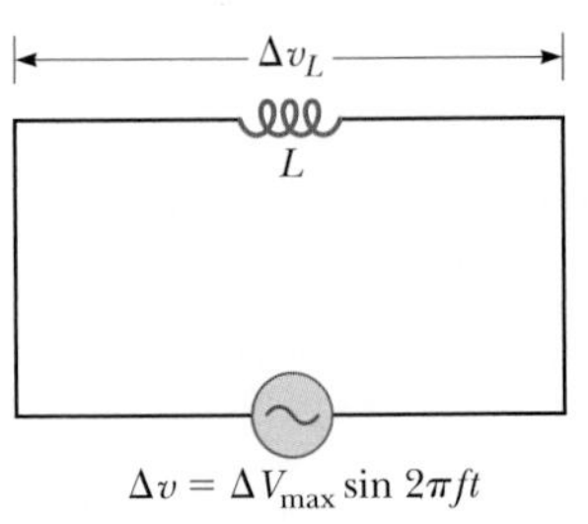

Figure 21.6 A series circuit consisting of an inductor L connected to an AC generator.

21.3 Inductors in an AC Circuit

LEARNING OBJECTIVES

1. Define inductive reactance.
2. Discuss the phase relationship between current and voltage difference in an AC circuit inductor.
3. Calculate the inductive reactance and rms current in inductive AC circuits.

Now consider an AC circuit consisting only of an inductor connected to the terminals of an AC source, as in Figure 21.6. (In any real circuit there is some resistance in the wire forming the inductive coil, but we ignore this consideration for now.) The changing current output of the generator produces a back emf that impedes the current in the circuit. The magnitude of this back emf is

$$\Delta v_L = L\frac{\Delta I}{\Delta t} \qquad [21.7]$$

The effective resistance of the coil in an AC circuit is measured by a quantity called the **inductive reactance,** X_L:

$$X_L \equiv 2\pi f L \qquad [21.8]$$

When f is in hertz and L is in henries, the unit of X_L is the ohm. The inductive reactance *increases* with increasing frequency and increasing inductance. Contrast this with capacitors, where increasing frequency or capacitance *decreases* the capacitive reactance.

To understand the meaning of inductive reactance, compare Equation 21.8 with Equation 21.7. First, note from Equation 21.8 that the inductive reactance depends on the inductance L, which is reasonable because the back emf (Eq. 21.7) is large for large values of L. Second, note that the inductive reactance depends on the frequency f. This dependence, too, is reasonable because the back emf depends on $\Delta I/\Delta t$, a quantity that is large when the current changes rapidly, as it would for high frequencies.

With inductive reactance defined in this way, we can write an equation of the same form as Ohm's law for the voltage across the coil or inductor:

$$\Delta V_{L,\text{rms}} = I_{\text{rms}} X_L \qquad [21.9]$$

where $\Delta V_{L,\text{rms}}$ is the rms voltage across the coil and I_{rms} is the rms current in the coil.

Figure 21.7 shows the instantaneous voltage and instantaneous current across the coil as functions of time. When a sinusoidal voltage is applied across an inductor, the voltage reaches its maximum value one-quarter of an oscillation period before the current reaches its maximum value. In this situation we say that **the voltage across an inductor always leads the current by 90°.**

To see why there is a phase relationship between voltage and current, we examine a few points on the curves of Figure 21.7. At point a on the current curve, the current is beginning to increase in the positive direction. At this instant the rate of change of current, $\Delta I/\Delta t$ (the slope of the current curve), is at a maximum, and we see from Equation 21.7 that the voltage across the inductor is consequently also at a maximum. As the current rises between points a and b on the curve, $\Delta I/\Delta t$ gradually decreases until it reaches zero at point b. As a result, the voltage across the inductor is decreasing during this same time interval, as the segment between c and d on the voltage curve indicates. Immediately after point b, the current begins to decrease, although it still has the same direction it had during the previous quarter cycle. As the current decreases to zero (from b to e on the curve), a voltage is again induced in the coil (from d to f), but the polarity of this voltage is opposite the polarity of the voltage induced between c and d. This occurs because back emfs always oppose the change in the current.

We could continue to examine other segments of the curves, but no new information would be gained because the current and voltage variations are repetitive.

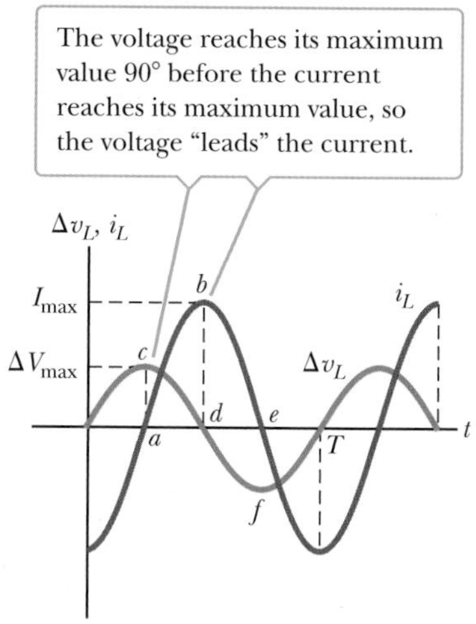

Figure 21.7 Plots of current and voltage across an inductor versus time in an AC circuit.

■ EXAMPLE 21.3 | A Purely Inductive AC Circuit

GOAL Perform basic AC circuit calculations for an inductive circuit.

PROBLEM In a purely inductive AC circuit (see Fig. 21.6), $L = 25.0$ mH and the rms voltage is 1.50×10^2 V. Find the inductive reactance and rms current in the circuit if the frequency is 60.0 Hz.

(Continued)

SOLUTION

Substitute L and f into Equation 21.8 to get the inductive reactance:

$$X_L = 2\pi f L = 2\pi(60.0\ \text{s}^{-1})(25.0 \times 10^{-3}\ \text{H}) = 9.42\ \Omega$$

Solve Equation 21.9 for the rms current and substitute:

$$I_{\text{rms}} = \frac{\Delta V_{L,\text{rms}}}{X_L} = \frac{1.50 \times 10^2\ \text{V}}{9.42\ \Omega} = 15.9\ \text{A}$$

REMARKS The analogy with DC circuits is even closer than in the capacitive case because in the inductive equivalent of Ohm's law, the voltage across an inductor is *proportional* to the inductance L, just as the voltage across a resistor is proportional to R in Ohm's law.

QUESTION 21.3 True or False: A larger inductance or frequency results in a larger inductive reactance.

EXERCISE 21.3 Calculate the inductive reactance and rms current in a similar circuit if the frequency is again 60.0 Hz, but the rms voltage is 85.0 V and the inductance is 47.0 mH.

ANSWER $X_L = 17.7\ \Omega$, $I_{\text{rms}} = 4.80\ \text{A}$

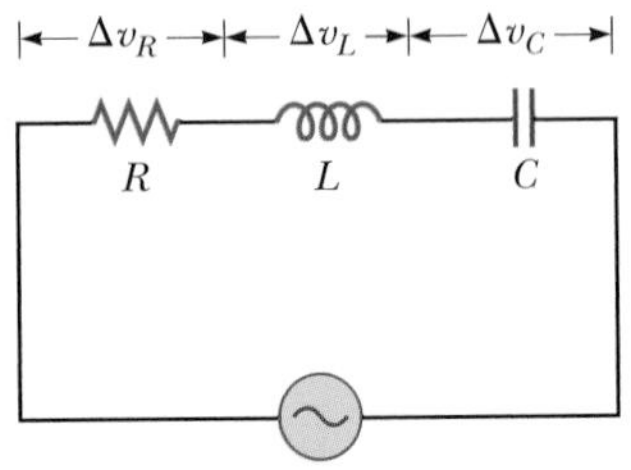

Figure 21.8 A series circuit consisting of a resistor, an inductor, and a capacitor connected to an AC generator.

21.4 The *RLC* Series Circuit

LEARNING OBJECTIVES

1. Define impedance in an AC circuit.
2. Discuss the AC relationship between current and voltage differences for elements in a series *RLC* circuit.
3. Apply the concepts of impedance, phasors and phase diagrams to series *RLC* circuits.

In the foregoing sections we examined the effects of an inductor, a capacitor, and a resistor when they are connected separately across an AC voltage source. We now consider what happens when these elements are combined.

Figure 21.8 shows a circuit containing a resistor, an inductor, and a capacitor connected in series across an AC source that supplies a total voltage Δv at some instant. The current in the circuit is the same at all points in the circuit at any instant and varies sinusoidally with time, as indicated in Figure 21.9a. This fact can be expressed mathematically as

$$i = I_{\text{max}} \sin 2\pi f t$$

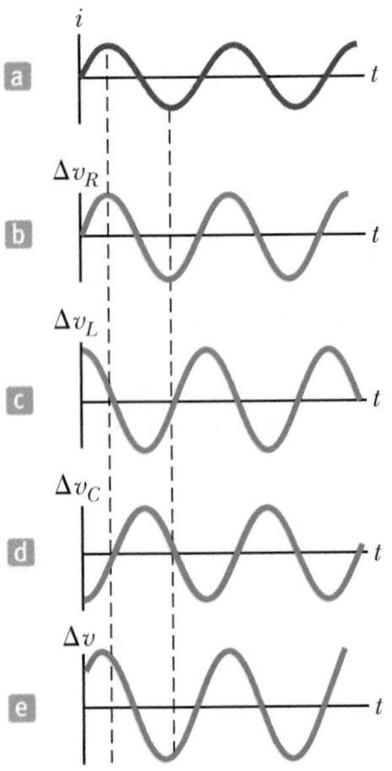

Figure 21.9 Phase relations in the series *RLC* circuit shown in Figure 21.8.

Earlier, we learned that the voltage across each element may or may not be in phase with the current. The instantaneous voltages across the three elements, shown in Figure 21.9, have the following phase relations to the instantaneous current:

1. The instantaneous voltage Δv_R across the resistor is *in phase* with the instantaneous current. (See Fig. 21.9b.)
2. The instantaneous voltage Δv_L across the inductor *leads* the current by 90°. (See Fig. 21.9c.)
3. The instantaneous voltage Δv_C across the capacitor *lags* the current by 90°. (See Fig. 21.9d.)

The net instantaneous voltage Δv supplied by the AC source equals the sum of the instantaneous voltages across the separate elements: $\Delta v = \Delta v_R + \Delta v_C + \Delta v_L$. This doesn't mean, however, that the voltages measured with an AC voltmeter across R, C, and L sum to the measured source voltage! In fact, the measured voltages *don't* sum to the measured source voltage because the voltages across R, C, and L all have different phases.

To account for the different phases of the voltage drops, we use a technique involving vectors. We represent the voltage across each element with a rotating vector, as in

Figure 21.10. The rotating vectors are referred to as **phasors**, and the diagram is called a **phasor diagram**. This particular diagram represents the circuit voltage given by the expression $\Delta v = \Delta V_{\max} \sin (2\pi ft + \phi)$, where $\Delta V_{\max}$ is the maximum voltage (the magnitude or length of the rotating vector or phasor) and ϕ is the angle between the phasor and the positive x-axis when $t = 0$. The phasor can be viewed as a vector of magnitude $\Delta V_{\max}$ rotating at a constant frequency f so that its projection along the y-axis is the instantaneous voltage in the circuit. Because ϕ is the phase angle between the voltage and current in the circuit, the phasor for the current (not shown in Fig. 21.10) lies along the positive x-axis when $t = 0$ and is expressed by the relation $i = I_{\max} \sin (2\pi ft)$.

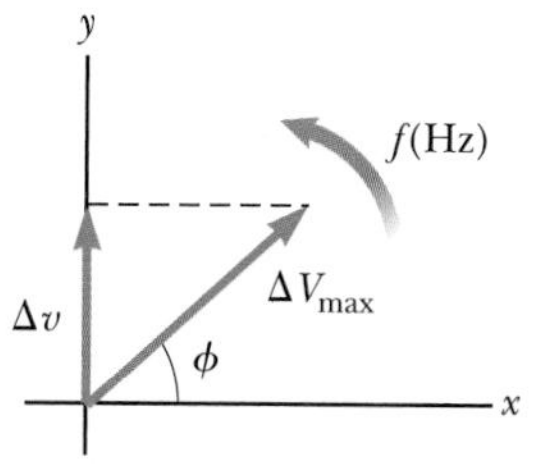

Figure 21.10 A phasor diagram for the voltage in an AC circuit, where ϕ is the phase angle between the voltage and the current and Δv is the instantaneous voltage.

The phasor diagrams in Figure 21.11 are useful for analyzing the *series RLC* circuit. Voltages in phase with the current are represented by vectors along the positive x-axis, and voltages out of phase with the current lie along other directions. ΔV_R is horizontal and to the right because it's in phase with the current. Likewise, ΔV_L is represented by a phasor along the positive y-axis because it leads the current by 90°. Finally, ΔV_C is along the negative y-axis because it lags the current[2] by 90°. If the phasors are added as vector quantities so as to account for the different phases of the voltages across R, L, and C, Figure 21.11a shows that the only x-component for the voltages is ΔV_R and the net y-component is $\Delta V_L - \Delta V_C$. We now add the phasors vectorially to find the phasor $\Delta V_{\max}$ (Fig. 21.11b), which represents the maximum voltage. The right triangle in Figure 21.11b gives the following equations for the maximum voltage and the phase angle ϕ between the maximum voltage and the current:

$$\Delta V_{\max} = \sqrt{\Delta V_R{}^2 + (\Delta V_L - \Delta V_C)^2} \qquad [21.10]$$

$$\tan \phi = \frac{\Delta V_L - \Delta V_C}{\Delta V_R} \qquad [21.11]$$

In these equations, all voltages are maximum values. Although we choose to use maximum voltages in our analysis, the preceding equations apply equally well to rms voltages because the two quantities are related to each other by the same factor for all circuit elements. The result for the maximum voltage $\Delta V_{\max}$ given by Equation 21.10 reinforces the fact that **the voltages across the resistor, capacitor, and inductor are not in phase, so one cannot simply add them to get the voltage across the combination of element or to get the source voltage.**

Quick Quiz

21.2 For the circuit in Figure 21.8, is the instantaneous voltage of the source equal to (a) the sum of the maximum voltages across the elements, (b) the sum of the instantaneous voltages across the elements, or (c) the sum of the rms voltages across the elements?

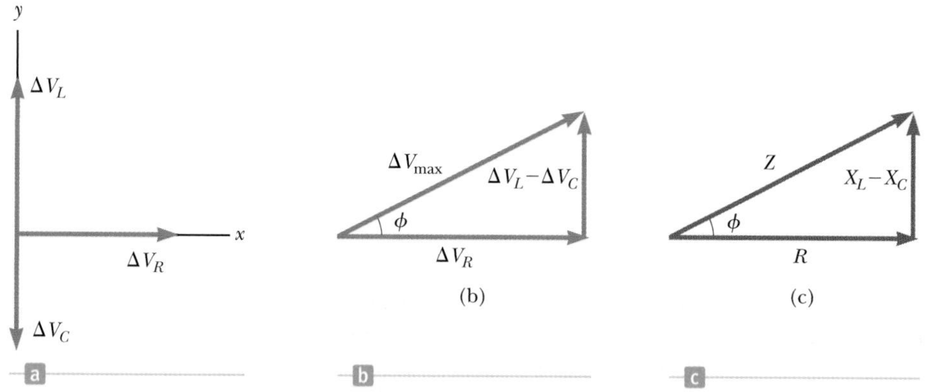

Figure 21.11 (a) A phasor diagram for the *RLC* circuit. (b) Addition of the phasors as vectors gives $\Delta V_{\max} = \sqrt{\Delta V_R{}^2 + (\Delta V_L - \Delta V_C)^2}$. (c) The reactance triangle that gives the impedance relation $Z = \sqrt{R^2 + (X_L - X_C)^2}$.

[2] A mnemonic to help you remember the phase relationships in *RLC* circuits is "*ELI* the *ICE* man." *E* represents the voltage $\mathcal{E}$, *I* the current, *L* the inductance, and *C* the capacitance. Thus, the name *ELI* means that in an inductive circuit, the voltage $\mathcal{E}$ leads the current *I*. In a capacitive circuit *ICE* means that the current leads the voltage.

We can write Equation 21.10 in the form of Ohm's law, using the relations $\Delta V_R = I_{\text{max}}R$, $\Delta V_L = I_{\text{max}}X_L$, and $\Delta V_C = I_{\text{max}}X_C$, where I_{max} is the maximum current in the circuit:

$$\Delta V_{\text{max}} = I_{\text{max}}\sqrt{R^2 + (X_L - X_C)^2} \qquad [21.12]$$

It's convenient to define a parameter called the **impedance** Z of the circuit as

Impedance ▶

$$Z \equiv \sqrt{R^2 + (X_L - X_C)^2} \qquad [21.13]$$

so that Equation 21.12 becomes

$$\Delta V_{\text{max}} = I_{\text{max}}Z \qquad [21.14]$$

Equation 21.14 is in the form of Ohm's law, $\Delta V = IR$, with R replaced by the impedance in ohms. Indeed, Equation 21.14 can be regarded as a generalized form of Ohm's law applied to a series AC circuit. Both the impedance and therefore the current in an AC circuit depend on the resistance, the inductance, the capacitance, *and* the frequency (because the reactances are frequency dependent).

It's useful to represent the impedance Z with a vector diagram such as the one depicted in Figure 21.11c. A right triangle is constructed with right side $X_L - X_C$, base R, and hypotenuse Z. Applying the Pythagorean theorem to this triangle, we see that

$$Z = \sqrt{R^2 + (X_L - X_C)^2}$$

which is Equation 21.13. Furthermore, we see from the vector diagram in Figure 21.11c that the phase angle ϕ between the current and the voltage obeys the relationship

Phase angle ϕ ▶

$$\tan\phi = \frac{X_L - X_C}{R} \qquad [21.15]$$

The physical significance of the phase angle will become apparent in Section 21.5.

Table 21.2 provides impedance values and phase angles for some series circuits containing different combinations of circuit elements.

Bettmann/Corbis

Nikola Tesla
(1856–1943)
Tesla was born in Croatia, but spent most of his professional life as an inventor in the United States. He was a key figure in the development of alternating-current electricity, high-voltage transformers, and the transport of electrical power via AC transmission lines. Tesla's viewpoint was at odds with the ideas of Edison, who committed himself to the use of direct current in power transmission. Tesla's AC approach won out.

Table 21.2 Impedance Values and Phase Angles for Various Combinations of Circuit Elements

Circuit Elements	Impedance Z	Phase Angle ϕ
R	R	0°
C	X_C	−90°
L	X_L	+90°
R C	$\sqrt{R^2 + X_C^2}$	Negative, between −90° and 0°
R L	$\sqrt{R^2 + X_L^2}$	Positive, between 0° and 90°
R L C	$\sqrt{R^2 + (X_L - X_C)^2}$	Negative if $X_C > X_L$ Positive if $X_C < X_L$

Note: In each case an AC voltage (not shown) is applied across the combination of elements (that is, across the dots).

Parallel alternating current circuits are also useful in everyday applications. We won't discuss them here, however, because their analysis is beyond the scope of this book.

■ Quick Quiz

21.3 If switch A is closed in Figure 21.12, what happens to the impedance of the circuit? (a) It increases. (b) It decreases. (c) It doesn't change.

21.4 Suppose $X_L > X_C$ in Figure 21.12. If switch A is closed, what happens to the phase angle? (a) It increases. (b) It decreases. (c) It doesn't change.

21.5 Suppose $X_L > X_C$ in Figure 21.12. If switch A is left open and switch B is closed, what happens to the phase angle? (a) It increases. (b) It decreases. (c) It doesn't change.

21.6 Suppose $X_L > X_C$ in Figure 21.12 and, with both switches open, a piece of iron is slipped into the inductor. During this process, what happens to the brightness of the bulb? (a) It increases. (b) It decreases. (c) It doesn't change.

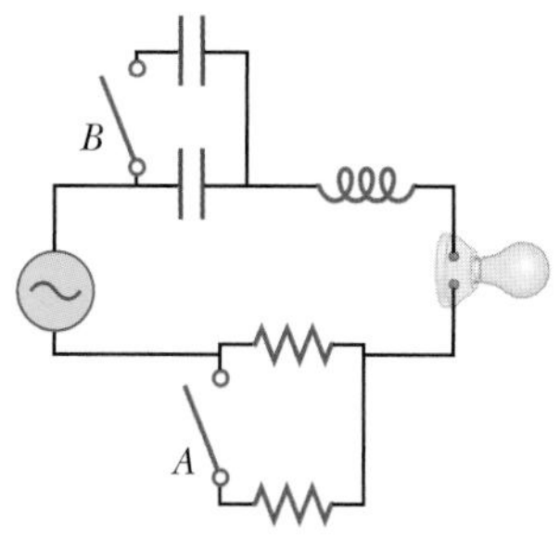

Figure 21.12 (Quick Quizzes 21.3–21.6)

■ PROBLEM-SOLVING STRATEGY

RLC Circuits

The following procedure is recommended for solving series RLC circuit problems:

1. Calculate the inductive and capacitive reactances, X_L and X_C.
2. Use X_L and X_C together with the resistance R to calculate the impedance Z of the circuit.
3. Find the maximum current or maximum voltage drop with the equivalent of Ohm's law, $\Delta V_{\text{max}} = I_{\text{max}} Z$.
4. Calculate the voltage drops across the individual elements with the appropriate variations of Ohm's law: $\Delta V_{R,\text{max}} = I_{\text{max}} R$, $\Delta V_{L,\text{max}} = I_{\text{max}} X_L$, and $\Delta V_{C,\text{max}} = I_{\text{max}} X_C$.
5. Obtain the phase angle using $\tan \phi = (X_L - X_C)/R$.

■ EXAMPLE 21.4 | An *RLC* Circuit

GOAL Analyze a series *RLC* AC circuit and find the phase angle.

PROBLEM A series *RLC* AC circuit has resistance $R = 2.50 \times 10^2\ \Omega$, inductance $L = 0.600$ H, capacitance $C = 3.50\ \mu\text{F}$, frequency $f = 60.0$ Hz, and maximum voltage $\Delta V_{\text{max}} = 1.50 \times 10^2$ V. Find **(a)** the impedance of the circuit, **(b)** the maximum current in the circuit, **(c)** the phase angle, and **(d)** the maximum voltages across the elements.

STRATEGY Calculate the inductive and capacitive reactances, which can be used with the resistance to calculate the impedance and phase angle. The impedance and Ohm's law yield the maximum current.

SOLUTION

(a) Find the impedance of the circuit.

First, calculate the inductive and capacitive reactances:

$$X_L = 2\pi f L = 226\ \Omega \qquad X_C = 1/2\pi f C = 758\ \Omega$$

Substitute these results and the resistance R into Equation 21.13 to obtain the impedance of the circuit:

$$Z = \sqrt{R^2 + (X_L - X_C)^2}$$

$$= \sqrt{(2.50 \times 10^2\ \Omega)^2 + (226\ \Omega - 758\ \Omega)^2} = \boxed{588\ \Omega}$$

(b) Find the maximum current in the circuit.

Use Equation 21.12, the equivalent of Ohm's law, to find the maximum current:

$$I_{\text{max}} = \frac{\Delta V_{\text{max}}}{Z} = \frac{1.50 \times 10^2\ \text{V}}{588\ \Omega} = \boxed{0.255\ \text{A}}$$

(Continued)

(c) Find the phase angle.

Calculate the phase angle between the current and the voltage with Equation 21.15:

$$\phi = \tan^{-1}\frac{X_L - X_C}{R} = \tan^{-1}\left(\frac{226\ \Omega - 758\ \Omega}{2.50 \times 10^2\ \Omega}\right) = \boxed{-64.8^\circ}$$

(d) Find the maximum voltages across the elements.

Use the "Ohm's law" expressions for each individual type of current element:

$$\Delta V_{R,\max} = I_{\max}R = (0.255\text{ A})(2.50 \times 10^2\ \Omega) = \boxed{63.8\text{ V}}$$

$$\Delta V_{L,\max} = I_{\max}X_L = (0.255\text{ A})(2.26 \times 10^2\ \Omega) = \boxed{57.6\text{ V}}$$

$$\Delta V_{C,\max} = I_{\max}X_C = (0.255\text{ A})(7.58 \times 10^2\ \Omega) = \boxed{193\text{ V}}$$

REMARKS Because the circuit is more capacitive than inductive ($X_C > X_L$), ϕ is negative. A negative phase angle means that the current leads the applied voltage. Notice also that the sum of the maximum voltages across the elements is $\Delta V_R + \Delta V_L + \Delta V_C = 314$ V, which is much greater than the maximum voltage of the generator, 150 V. As we saw in Quick Quiz 21.2, the sum of the maximum voltages is a meaningless quantity because when alternating voltages are added, *both their amplitudes and their phases* must be taken into account. We know that the maximum voltages across the various elements occur at different times, so it doesn't make sense to add all the maximum values. The correct way to "add" the voltages is through Equation 21.10.

QUESTION 21.4 True or False: In an *RLC* circuit, the impedance must always be greater than or equal to the resistance.

EXERCISE 21.4 Analyze a series *RLC* AC circuit for which $R = 175\ \Omega$, $L = 0.500$ H, $C = 22.5\ \mu$F, $f = 60.0$ Hz, and $\Delta V_{\max} = 325$ V. Find (a) the impedance, (b) the maximum current, (c) the phase angle, and (d) the maximum voltages across the elements.

ANSWERS (a) 189 Ω (b) 1.72 A (c) 22.0° (d) $\Delta V_{R,\max} = 301$ V, $\Delta V_{L,\max} = 324$ V, $\Delta V_{C,\max} = 203$ V

21.5 Power in an AC Circuit

LEARNING OBJECTIVES

1. Discuss the conservation of energy in purely capacitive and purely inductive AC circuits.
2. Discuss the energy dissipated by resistive *RLC* circuits.
3. Evaluate the average power dissipated by an *RLC* circuit.

No power losses are associated with pure capacitors and pure inductors in an AC circuit. A pure capacitor, by definition, has no resistance or inductance, whereas a pure inductor has no resistance or capacitance. (These definitions are idealizations: in a real capacitor, for example, inductive effects could become important at high frequencies.) We begin by analyzing the power dissipated in an AC circuit that contains only a generator and a capacitor.

When the current increases in one direction in an AC circuit, charge accumulates on the capacitor and a voltage drop appears across it. When the voltage reaches its maximum value, the energy stored in the capacitor is

$$PE_C = \tfrac{1}{2}C(\Delta V_{\max})^2$$

This energy storage is only momentary, however: When the current reverses direction, the charge leaves the capacitor plates and returns to the voltage source. During one-half of each cycle the capacitor is being charged, and during the other half the charge is being returned to the voltage source. Therefore, the average power supplied by the source is zero. In other words, **no power losses occur in a capacitor in an AC circuit.**

Similarly, the source must do work against the back emf of an inductor that is carrying a current. When the current reaches its maximum value, the energy stored in the inductor is a maximum and is given by

$$PE_L = \tfrac{1}{2}LI_{\max}^2$$

SOLUTION

(a) Determine the percentage of power lost in the line.

Substitute into Equation 21.23 to find the current in the transmission line:

$$I_2 = \frac{I_1 \Delta V_1}{\Delta V_2} = \frac{(1.00 \times 10^2 \text{ A})(4.00 \times 10^3 \text{ V})}{2.40 \times 10^5 \text{ V}} = 1.67 \text{ A}$$

Now use Equation 21.16 to find the power lost in the transmission line:

$$\textbf{(1)}\ P_{\text{lost}} = I_2^{\,2} R = (1.67 \text{ A})^2 (30.0\ \Omega) = 83.7 \text{ W}$$

Calculate the power output of the generator:

$$P = I_1 \Delta V_1 = (1.00 \times 10^2 \text{ A})(4.00 \times 10^3 \text{ V}) = 4.00 \times 10^5 \text{ W}$$

Finally, divide P_{lost} by the power output and multiply by 100 to find the percentage of power lost:

$$\% \text{ power lost} = \left(\frac{83.7 \text{ W}}{4.00 \times 10^5 \text{ W}}\right) \times 100 = 0.020\,9\%$$

(b) What percentage of the original power would be lost in the transmission line if the voltage were not stepped up?

Replace the stepped-up current in Equation (1) by the original current of 1.00×10^2 A:

$$P_{\text{lost}} = I^2 R = (1.00 \times 10^2 \text{ A})^2 (30.0\ \Omega) = 3.00 \times 10^5 \text{ W}$$

Calculate the percentage loss, as before:

$$\% \text{ power lost} = \left(\frac{3.00 \times 10^5 \text{ W}}{4.00 \times 10^5 \text{ W}}\right) \times 100 = 75\%$$

REMARKS This example illustrates the advantage of high-voltage transmission lines. In the city, a transformer at a substation steps the voltage back down to about 4000 V, and this voltage is maintained across utility lines throughout the city. When the power is to be used at a home or business, a transformer on a utility pole near the establishment reduces the voltage to 240 V or 120 V.

QUESTION 21.7 If the voltage is stepped up to double the amount in this problem, by what factor is the power loss changed? (a) 2 (b) no change (c) $\frac{1}{2}$ (d) $\frac{1}{4}$

© George Semple/Cengage Learning

This cylindrical step-down transformer drops the voltage from 4 000 V to 220 V for delivery to a group of residences.

EXERCISE 21.7 Suppose the same generator has the voltage stepped up to only 7.50×10^4 V and the resistance of the line is 85.0 Ω. Find the percentage of power lost in this case.

ANSWER 0.604%

21.8 Maxwell's Predictions

LEARNING OBJECTIVES

1. State and discuss the four fundamental ideas underlying Maxwell's unifying theory of electricity and magnetism.
2. Discuss the concept of electromagnetic waves.

During the early stages of their study and development, electric and magnetic phenomena were thought to be unrelated. In 1865, however, James Clerk Maxwell (1831–1879) provided a mathematical theory that showed a close relationship between all electric and magnetic phenomena. In addition to unifying the formerly separate fields of electricity and magnetism, his brilliant theory predicted

James Clerk Maxwell
Scottish Theoretical Physicist (1831–1879)
Maxwell developed the electromagnetic theory of light and the kinetic theory of gases, and he explained the nature of Saturn's rings and color vision. Maxwell's successful interpretation of the electromagnetic field resulted in the equations that bear his name. Formidable mathematical ability combined with great insight enabled him to lead the way in the study of electromagnetism and kinetic theory.

that electric and magnetic fields can move through space as waves. The theory he developed is based on the following four pieces of information:

1. Electric field lines originate on positive charges and terminate on negative charges.
2. Magnetic field lines always form closed loops; they don't begin or end anywhere.
3. A varying magnetic field induces an emf and hence an electric field. This fact is a statement of Faraday's law (Chapter 20).
4. Magnetic fields are generated by moving charges (or currents), as summarized in Ampère's law (Chapter 19).

The first statement is a consequence of the nature of the electrostatic force between charged particles, given by Coulomb's law. It embodies the fact that **free charges (electric monopoles) exist in nature**.

The second statement—that magnetic fields form continuous loops—is exemplified by the magnetic field lines around a long, straight wire, which are closed circles, and the magnetic field lines of a bar magnet, which form closed loops. It says, in contrast to the first statement, that **free magnetic charges (magnetic monopoles) don't exist in nature**.

The third statement is equivalent to Faraday's law of induction, and the fourth is equivalent to Ampère's law.

In one of the greatest theoretical developments of the 19th century, Maxwell used these four statements within a corresponding mathematical framework to prove that electric and magnetic fields play symmetric roles in nature. It was already known from experiments that a changing magnetic field produced an electric field according to Faraday's law. Maxwell believed that nature was symmetric, and he therefore hypothesized that a changing electric field should produce a magnetic field. This hypothesis could not be proven experimentally at the time it was developed because the magnetic fields generated by changing electric fields are generally very weak and therefore difficult to detect.

To justify his hypothesis, Maxwell searched for other phenomena that might be explained by it. He turned his attention to the motion of rapidly oscillating (accelerating) charges, such as those in a conducting rod connected to an alternating voltage. Such charges are accelerated and, according to Maxwell's predictions, generate changing electric and magnetic fields. The changing fields cause electromagnetic disturbances that travel through space as waves, similar to the spreading water waves created by a pebble thrown into a pool. The waves sent out by the oscillating charges are fluctuating electric and magnetic fields, so they are called *electromagnetic waves*. From Faraday's law and from Maxwell's own generalization of Ampère's law, Maxwell calculated the speed of the waves to be equal to the speed of light, $c = 3 \times 10^8$ m/s. He concluded that visible light and other electromagnetic waves consist of fluctuating electric and magnetic fields traveling through empty space, with each varying field inducing the other! His was truly one of the greatest discoveries of science, on a par with Newton's discovery of the laws of motion. Like Newton's laws, it had a profound influence on later scientific developments.

21.9 Hertz's Confirmation of Maxwell's Predictions

LEARNING OBJECTIVE

1. Discuss Heinrich Hertz's experimental confirmation of Maxwell's predictions about electromagnetic waves.

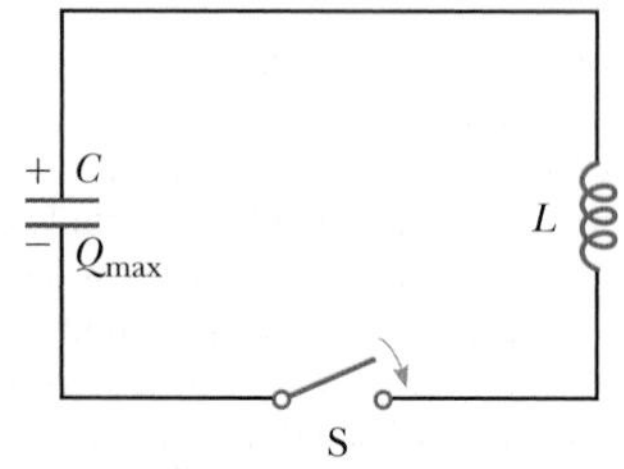

Figure 21.16 A simple *LC* circuit. The capacitor has an initial charge of Q_{max}, and the switch is closed at $t = 0$.

In 1887, after Maxwell's death, Heinrich Hertz (1857–1894) was the first to generate and detect electromagnetic waves in a laboratory setting, using *LC* circuits. In such a circuit a charged capacitor is connected to an inductor, as in Figure 21.16.

When the switch is closed, oscillations occur in the current in the circuit and in the charge on the capacitor. If the resistance of the circuit is neglected, no energy is dissipated and the oscillations continue.

In the following analysis, we neglect the resistance in the circuit. We assume the capacitor has an initial charge of $Q_{\max}$ and the switch is closed at $t = 0$. When the capacitor is fully charged, the total energy in the circuit is stored in the electric field of the capacitor and is equal to $Q_{\max}^2/2C$. At this time, the current is zero, so no energy is stored in the inductor. As the capacitor begins to discharge, the energy stored in its electric field decreases. At the same time, the current increases and energy equal to $LI^2/2$ is now stored in the magnetic field of the inductor. Thus, energy is transferred from the electric field of the capacitor to the magnetic field of the inductor. When the capacitor is fully discharged, it stores no energy. At this time, the current reaches its maximum value and all the energy is stored in the inductor. The process then repeats in the reverse direction. The energy continues to transfer between the inductor and the capacitor, corresponding to oscillations in the current and charge.

As we saw in Section 21.6, the frequency of oscillation of an LC circuit is called the *resonance frequency* of the circuit and is given by

$$f_0 = \frac{1}{2\pi\sqrt{LC}}$$

The circuit Hertz used in his investigations of electromagnetic waves is similar to that just discussed and is shown schematically in Figure 21.17. An induction coil (a large coil of wire) is connected to two metal spheres with a narrow gap between them to form a capacitor. Oscillations are initiated in the circuit by short voltage pulses sent via the coil to the spheres, charging one positive, the other negative. Because L and C are quite small in this circuit, the frequency of oscillation is quite high, $f \approx 100$ MHz. This circuit is called a transmitter because it produces electromagnetic waves.

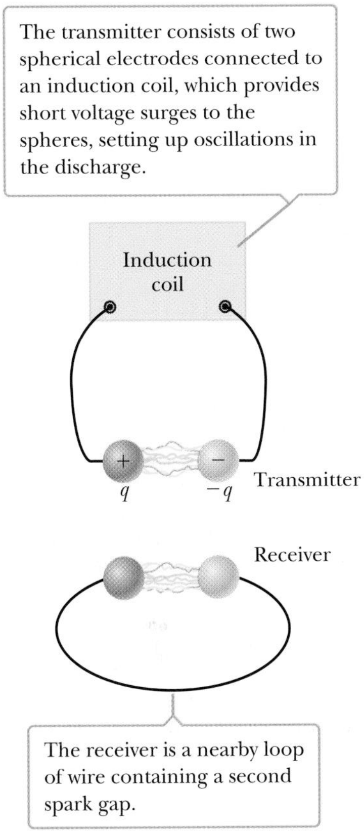

Figure 21.17 A schematic diagram of Hertz's apparatus for generating and detecting electromagnetic waves.

Several meters from the transmitter circuit, Hertz placed a second circuit, the receiver, which consisted of a single loop of wire connected to two spheres. It had its own effective inductance, capacitance, and natural frequency of oscillation. Hertz found that energy was being sent from the transmitter to the receiver when the resonance frequency of the receiver was adjusted to match that of the transmitter. The energy transfer was detected when the voltage across the spheres in the receiver circuit became high enough to produce ionization in the air, which caused sparks to appear in the air gap separating the spheres. Hertz's experiment is analogous to the mechanical phenomenon in which a tuning fork picks up the vibrations from another, identical tuning fork.

Hertz hypothesized that the energy transferred from the transmitter to the receiver is carried in the form of waves, now recognized as electromagnetic waves. In a series of experiments, he also showed that the radiation generated by the transmitter exhibits wave properties: interference, diffraction, reflection, refraction, and polarization. As you will see shortly, all these properties are exhibited by light. It became evident that Hertz's electromagnetic waves had the same known properties of light waves and differed only in frequency and wavelength. Hertz effectively confirmed Maxwell's theory by showing that Maxwell's mysterious electromagnetic waves existed and had all the properties of light waves.

Perhaps the most convincing experiment Hertz performed was the measurement of the speed of waves from the transmitter, accomplished as follows: waves of known frequency from the transmitter were reflected from a metal sheet so that an interference pattern was set up, much like the standing-wave pattern on a stretched string. As we learned in our discussion of standing waves, the distance between nodes is $\lambda/2$, so Hertz was able to determine the wavelength λ. Using the relationship $v = \lambda f$, he found that v was close to 3×10^8 m/s, the known speed of visible light. Hertz's experiments thus provided the first evidence in support of Maxwell's theory.

Heinrich Rudolf Hertz
German Physicist (1857–1894)
Hertz made his most important discovery of radio waves in 1887. After finding that the speed of a radio wave was the same as that of light, Hertz showed that radio waves, like light waves, could be reflected, refracted, and diffracted. Hertz died of blood poisoning at the age of 36. During his short life, he made many contributions to science. The hertz, equal to one complete vibration or cycle per second, is named after him.

21.10 Production of Electromagnetic Waves by an Antenna

LEARNING OBJECTIVES

1. Discuss the radiation of energy by the accelerated charges in an AC circuit.
2. Discuss the transverse nature of electromagnetic waves.

In the previous section we found that the energy stored in an *LC* circuit is continually transferred between the electric field of the capacitor and the magnetic field of the inductor. This energy transfer, however, continues for prolonged periods of time only when the changes occur slowly. If the current alternates rapidly, the circuit loses some of its energy in the form of electromagnetic waves. In fact, electromagnetic waves are radiated by *any* circuit carrying an alternating current. The fundamental mechanism responsible for this radiation is the acceleration of a charged particle. **Whenever a charged particle accelerates, it radiates energy.**

APPLICATION
Radio-Wave Transmission

An alternating voltage applied to the wires of an antenna forces electric charges in the antenna to oscillate. This common technique for accelerating charged particles is the source of the radio waves emitted by the broadcast antenna of a radio station.

Figure 21.18 illustrates the production of an electromagnetic wave by oscillating electric charges in an antenna. Two metal rods are connected to an AC source, which causes charges to oscillate between the rods. The output voltage of the generator is sinusoidal. At $t = 0$, the upper rod is given a maximum positive charge and the bottom rod an equal negative charge, as in Figure 21.18a. The electric field near the antenna at this instant is also shown in the figure. As the charges oscillate, the rods become less charged, the field near the rods decreases in strength, and the downward-directed maximum electric field produced at $t = 0$ moves away from the rod. When the charges are neutralized, as in Figure 21.18b, the electric field has dropped to zero, after an interval equal to one-quarter of the period of oscillation. Continuing in this fashion, the upper rod soon obtains a maximum negative charge and the lower rod becomes positive, as in Figure 21.18c, resulting in an electric field directed upward. This occurs after an interval equal to one-half the period of oscillation. The oscillations continue as indicated in Figure 21.18d. Note that the electric field near the antenna oscillates in phase with the charge distribution: the field points down when the upper rod is positive and up when the upper rod is negative. Further, the magnitude of the field at any instant depends on the amount of charge on the rods at that instant.

As the charges continue to oscillate (and accelerate) between the rods, the electric field set up by the charges moves away from the antenna in all directions at the speed of light. Figure 21.18 shows the electric field pattern on one side of the antenna at certain times during the oscillation cycle. As you can see, one cycle of charge oscillation produces one full wavelength in the electric field pattern.

Tip 21.1 Accelerated Charges Produce Electromagnetic Waves

Stationary charges produce only electric fields, whereas charges in uniform motion (i.e., constant velocity) produce electric and magnetic fields, but no electromagnetic waves. In contrast, accelerated charges produce electromagnetic waves as well as electric and magnetic fields. An accelerating charge also radiates energy.

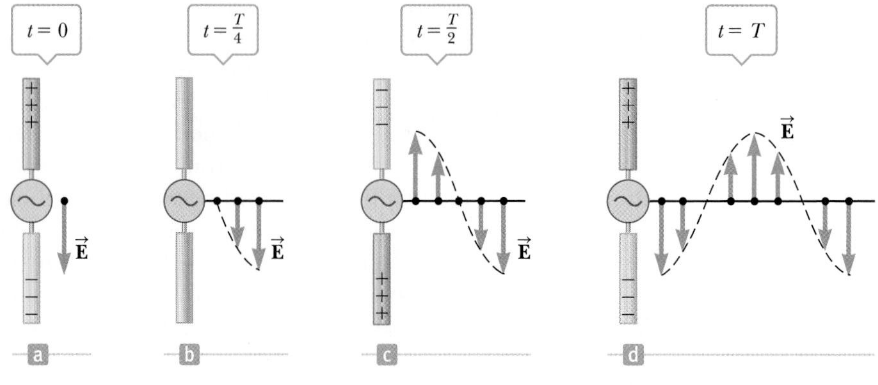

Figure 21.18 An electric field set up by oscillating charges in an antenna. The field moves away from the antenna at the speed of light.

Because the oscillating charges create a current in the rods, a magnetic field is also generated when the current in the rods is upward, as shown in Figure 21.19. The magnetic field lines circle the antenna (recall right-hand rule number 2) and are perpendicular to the electric field at all points. As the current changes with time, the magnetic field lines spread out from the antenna. At great distances from the antenna, the strengths of the electric and magnetic fields become very weak. At these distances, however, it is necessary to take into account the facts that (1) a changing magnetic field produces an electric field and (2) a changing electric field produces a magnetic field, as predicted by Maxwell. These induced electric and magnetic fields are in phase: at any point, the two fields reach their maximum values at the same instant. This synchrony is illustrated at one instant of time in Figure 21.20. Note that (1) the $\vec{\mathbf{E}}$ and $\vec{\mathbf{B}}$ fields are perpendicular to each other and (2) both fields are perpendicular to the direction of motion of the wave. This second property is characteristic of transverse waves. Hence, we see that **an electromagnetic wave is a transverse wave**.

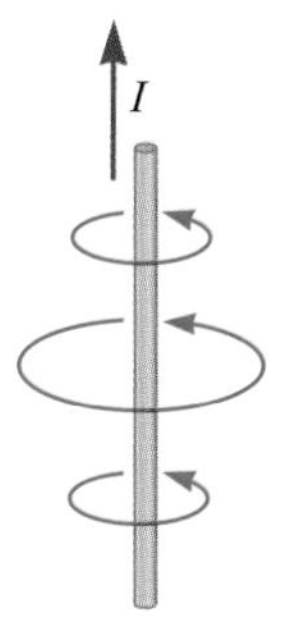

Figure 21.19 Magnetic field lines around an antenna carrying a changing current.

21.11 Properties of Electromagnetic Waves

LEARNING OBJECTIVES

1. Discuss the concept of a plane wave.
2. Relate the speed of light to the permeability and permittivity of the propagation medium.
3. Relate the speed of light to the ratio of the electric and magnetic field magnitudes in an electromagnetic wave.
4. Discuss the fundamental properties of electromagnetic waves.
5. Evaluate the average intensity of an electromagnetic wave and the momentum delivered by the wave to perfectly absorbing or reflecting surfaces.
6. Calculate thermal and mechanical effects due to the interaction of electromagnetic waves and matter.

We have seen that Maxwell's detailed analysis predicted the existence and properties of electromagnetic waves. In this section we summarize what we know about electromagnetic waves thus far and consider some additional properties. In our discussion here and in future sections, we will often make reference to a type of wave called a **plane wave**. A plane electromagnetic wave is a wave traveling from a very distant source. Figure 21.20 pictures such a wave at a given instant of time. In this case the oscillations of the electric and magnetic fields take place in planes perpendicular

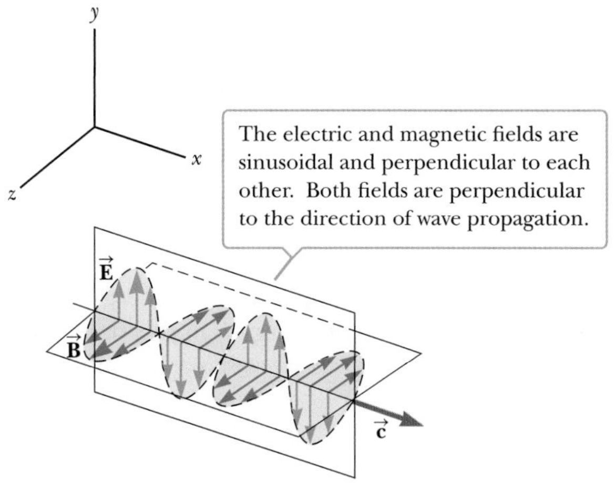

Figure 21.20 An electromagnetic wave sent out by oscillating charges in an antenna, represented at one instant of time and far from the antenna, moving in the positive x-direction with speed c.

to the x-axis and are therefore perpendicular to the direction of travel of the wave. Because of the latter property, electromagnetic waves are transverse waves. In the figure the electric field $\vec{\mathbf{E}}$ is in the y-direction and the magnetic field $\vec{\mathbf{B}}$ is in the z-direction. Light propagates in a direction perpendicular to these two fields. That direction is determined by yet another right-hand rule: (1) point the fingers of your right hand in the direction of $\vec{\mathbf{E}}$, (2) curl them in the direction of $\vec{\mathbf{B}}$, and (3) the right thumb then points in the direction of propagation of the wave.

Electromagnetic waves travel with the speed of light. In fact, it can be shown that the speed of an electromagnetic wave is related to the permeability and permittivity of the medium through which it travels. Maxwell found this relationship for free space to be

Speed of light ▶

$$c = \frac{1}{\sqrt{\mu_0 \epsilon_0}} \qquad [21.24]$$

where c is the speed of light, $\mu_0 = 4\pi \times 10^{-7}\ \mathrm{N{\cdot}s^2/C^2}$ is the permeability constant of vacuum, and $\epsilon_0 = 8.854\ 19 \times 10^{-12}\ \mathrm{C^2/N{\cdot}m^2}$ is the permittivity of free space. Substituting these values into Equation 21.24, we find that

$$c = 2.997\ 92 \times 10^8\ \mathrm{m/s} \qquad [21.25]$$

Tip 21.2 *E* Stronger than *B*?

The relationship $E = Bc$ makes it appear that the electric fields associated with light are much larger than the magnetic fields. That is not the case: The units are different, so the quantities can't be directly compared. The two fields contribute equally to the energy of a light wave.

Because electromagnetic waves travel at the same speed as light in vacuum, scientists concluded (correctly) that **light is an electromagnetic wave**.

Maxwell also proved the following relationship for electromagnetic waves:

$$\frac{E}{B} = c \qquad [21.26]$$

which states that the ratio of the magnitude of the electric field to the magnitude of the magnetic field equals the speed of light.

Electromagnetic waves carry energy as they travel through space, and this energy can be transferred to objects placed in their paths. The average rate at which energy passes through an area perpendicular to the direction of travel of a wave, or the average power per unit area, is called the **intensity *I*** of the wave and is given by

$$I = \frac{E_{\max} B_{\max}}{2\mu_0} \qquad [21.27]$$

where $E_{\max}$ and $B_{\max}$ are the *maximum* values of E and B. The quantity I is analogous to the intensity of sound waves introduced in Chapter 14. From Equation 21.26, we see that $E_{\max} = cB_{\max} = B_{\max}/\sqrt{\mu_0 \epsilon_0}$. Equation 21.27 can therefore also be expressed as

$$I = \frac{E_{\max}^2}{2\mu_0 c} = \frac{c}{2\mu_0} B_{\max}^2 \qquad [21.28]$$

Note that in these expressions we use the *average* power per unit area. A detailed analysis would show that the energy carried by an electromagnetic wave is shared equally by the electric and magnetic fields.

Light is an electromagnetic wave and transports energy and momentum ▶

Electromagnetic waves have an average intensity given by Equation 21.28. When the waves strike an area A of an object's surface for a given time Δt, energy $U = IA\Delta t$ is transferred to the surface. Momentum is transferred, as well. Hence, pressure is exerted on a surface when an electromagnetic wave impinges on it. In what follows, we assume the electromagnetic wave transports a total energy U to a surface in a time Δt. If the surface absorbs all the incident energy U in this time, Maxwell showed that the total momentum $\vec{\mathbf{p}}$ delivered to this surface has a magnitude

$$p = \frac{U}{c} \quad \text{(complete absorption)} \qquad [21.29]$$

If the surface is a perfect reflector, then the momentum transferred in a time Δt for normal incidence is twice that given by Equation 21.29. This is analogous to a molecule of gas bouncing off the wall of a container in a perfectly elastic collision. If the molecule is initially traveling in the positive x-direction at velocity v and after the collision is traveling in the negative x-direction at velocity $-v$, its change in momentum is given by $\Delta p = mv - (-mv) = 2mv$. Light bouncing off a perfect reflector is a similar process, so for complete reflection,

$$p = \frac{2U}{c} \quad \text{(complete reflection)} \qquad [21.30]$$

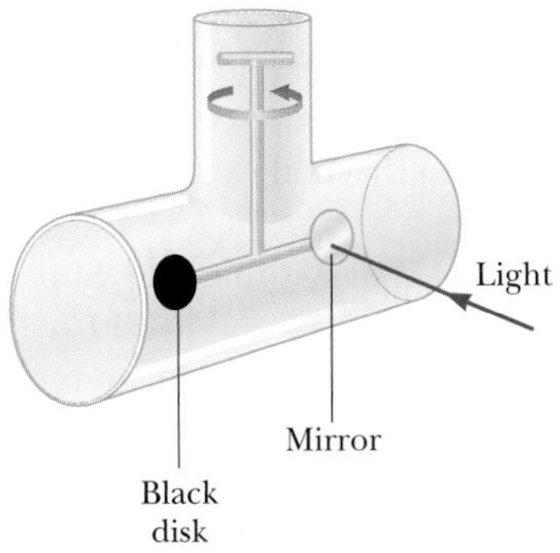

Figure 21.21 An apparatus for measuring the radiation pressure of light. In practice, the system is contained in a high vacuum.

Although radiation pressures are very small (about 5×10^{-6} N/m^2 for direct sunlight), they have been measured with a device such as the one shown in Figure 21.21. Light is allowed to strike a mirror and a black disk that are connected to each other by a horizontal bar suspended from a fine fiber. Light striking the black disk is completely absorbed, so *all* the momentum of the light is transferred to the disk. Light striking the mirror head-on is totally reflected; hence, the momentum transfer to the mirror is twice that transmitted to the disk. As a result, the horizontal bar supporting the disks twists counterclockwise as seen from above. The bar comes to equilibrium at some angle under the action of the torques caused by radiation pressure and the twisting of the fiber. The radiation pressure can be determined by measuring the angle at which equilibrium occurs. The apparatus must be placed in a high vacuum to eliminate the effects of air currents. It's interesting that similar experiments demonstrate that electromagnetic waves carry angular momentum, as well.

In summary, electromagnetic waves traveling through free space have the following properties:

◀ Some properties of electromagnetic waves

1. Electromagnetic waves travel at the speed of light.
2. Electromagnetic waves are transverse waves because the electric and magnetic fields are perpendicular to the direction of propagation of the wave and to each other.
3. The ratio of the electric field to the magnetic field in an electromagnetic wave equals the speed of light.
4. Electromagnetic waves carry both energy and momentum, which can be delivered to a surface.

APPLYING PHYSICS 21.3 | Solar System Dust

In the interplanetary space in the solar system, there is a large amount of dust. Although interplanetary dust can in theory have a variety of sizes—from molecular size upward—why are there very few dust particles smaller than about 0.2 μm in the solar system? *Hint:* The solar system originally contained dust particles of all sizes.

EXPLANATION Dust particles in the solar system are subject to two forces: the gravitational force toward the Sun and the force from radiation pressure, which is directed away from the Sun. The gravitational force is proportional to the cube of the radius of a spherical dust particle because it is proportional to the mass (ρV) of the particle. The radiation pressure is proportional to the square of the radius because it depends on the cross-sectional area of the particle. For large particles, the gravitational force is larger than the force of radiation pressure, and the weak attraction to the Sun causes such particles to move slowly toward it. For small particles, less than about 0.2 μm, the larger force from radiation pressure sweeps them out of the solar system. ■

Quick Quiz

21.7 In an apparatus such as the one in Figure 21.21, suppose the black disk is replaced by one with half the radius. Which of the following are different after the disk is replaced? (a) radiation pressure on the disk (b) radiation force on the disk (c) radiation momentum delivered to the disk in a given time interval

EXAMPLE 21.8 A Hot Tin Roof (Solar-Powered Homes)

GOAL Calculate some basic properties of light and relate them to thermal radiation.

PROBLEM Assume the Sun delivers an average power per unit area of about 1.00×10^3 W/m^2 to Earth's surface. **(a)** Calculate the total power incident on a flat tin roof 8.00 m by 20.0 m. Assume the radiation is incident *normal* (perpendicular) to the roof. **(b)** The tin roof reflects some light, and convection, conduction, and radiation transport the rest of the thermal energy away until some equilibrium temperature is established. If the roof is a perfect blackbody and rids itself of one-half of the incident radiation through thermal radiation, what's its equilibrium temperature? Assume the ambient temperature is 298 K.

(Example 21.8) A solar home.

SOLUTION

(a) Calculate the power delivered to the roof.

Multiply the intensity by the area to get the power:

$$P = IA = (1.00 \times 10^3 \text{ W/m}^2)(8.00 \text{ m} \times 20.0 \text{ m})$$

$$= 1.60 \times 10^5 \text{ W}$$

(b) Find the equilibrium temperature of the roof.

Substitute into Stefan's law. Only one-half the incident power should be substituted, and twice the area of the roof (both the top and the underside of the roof count).

$$P = \sigma e A(T^4 - T_0^{\,4})$$

$$T^4 = T_0^{\,4} + \frac{P}{\sigma e A}$$

$$= (298 \text{ K})^4 + \frac{(0.500)(1.60 \times 10^5 \text{ W/m}^2)}{(5.67 \times 10^{-8} \text{ W/m}^2 \cdot \text{K}^4)(1)(3.20 \times 10^2 \text{ m}^2)}$$

$$T = 333 \text{ K} = 6.0 \times 10^1 \text{ °C}$$

REMARKS If the incident power could *all* be converted to electric power, it would be more than enough for the average home. Unfortunately, solar energy isn't easily harnessed, and the prospects for large-scale conversion are not as bright as they may appear from this simple calculation. For example, the conversion efficiency from solar to electrical energy is far less than 100%; 10–20% is typical for photovoltaic cells. Roof systems for using solar energy to raise the temperature of water with efficiencies of around 50% have been built. Other practical problems must be considered, however, such as overcast days, geographic location, and energy storage.

QUESTION 21.8 Does the angle the roof makes with respect to the horizontal affect the amount of power absorbed by the roof? Explain.

EXERCISE 21.8 A spherical satellite orbiting Earth is lighted on one side by the Sun, with intensity 1 340 W/m^2. If the radius of the satellite is 1.00 m, what power is incident upon it? *Note:* The satellite effectively intercepts radiation only over a cross section, an area equal to that of a disk, πr^2.

ANSWERS 4.21×10^3 W

EXAMPLE 21.9 Clipper Ships of Space

GOAL Relate the intensity of light to its mechanical effect on matter.

PROBLEM Aluminized Mylar film is a highly reflective, lightweight material that could be used to make sails for spacecraft driven by the light of the Sun. Suppose a sail with area 1.00 km^2 is orbiting the Sun at a distance of 1.50×10^{11} m. The sail has a mass of 5.00×10^3 kg and is tethered to a payload of mass 2.00×10^4 kg. **(a)** If the intensity of sunlight is 1.34×10^3 W/m^2 and the sail is oriented perpendicular to the incident light, what radial force is exerted on the sail? **(b)** About how long would it take to change the radial speed of the sail by 1.00 km/s? Assume the sail is perfectly reflecting. **(c)** Suppose the light were supplied by a large, powerful laser beam instead of the Sun. (Such systems have been proposed.) Calculate the peak electric and magnetic fields of the laser light.

STRATEGY Equation 21.30 gives the momentum imparted when light strikes an object and is totally reflected. The change in this momentum with time is a force.

For part (b), use Newton's second law to obtain the acceleration. The velocity kinematics equation then yields the necessary time to achieve the desired change in speed. Part (c) follows from Equation 21.27 and $E = Bc$.

SOLUTION

(a) Find the force exerted on the sail.

Write Equation 21.30 and substitute $U = P\Delta t = IA\,\Delta t$ for the energy delivered to the sail:

$$\Delta p = \frac{2U}{c} = \frac{2P\Delta t}{c} = \frac{2IA\Delta t}{c}$$

Divide both sides by Δt, obtaining the force $\Delta p/\Delta t$ exerted by the light on the sail:

$$F = \frac{\Delta p}{\Delta t} = \frac{2IA}{c} = \frac{2(1340\ \mathrm{W/m^2})(1.00 \times 10^6\ \mathrm{m^2})}{3.00 \times 10^8\ \mathrm{m/s}}$$

$$= \boxed{8.93\ \mathrm{N}}$$

(b) Find the time it takes to change the radial speed by 1.00 km/s.

Substitute the force into Newton's second law and solve for the acceleration of the sail:

$$a = \frac{F}{m} = \frac{8.93\ \mathrm{N}}{2.50 \times 10^4\ \mathrm{kg}} = 3.57 \times 10^{-4}\ \mathrm{m/s^2}$$

Apply the kinematics velocity equation:

$$v = at + v_0$$

Solve for t:

$$t = \frac{v - v_0}{a} = \frac{1.00 \times 10^3\ \mathrm{m/s}}{3.57 \times 10^{-4}\ \mathrm{m/s^2}} = \boxed{2.80 \times 10^6\ \mathrm{s}}$$

(c) Calculate the peak electric and magnetic fields if the light is supplied by a laser.

Solve Equation 21.28 for E_{max}:

$$I = \frac{E_{\mathrm{max}}^2}{2\mu_0 c} \quad \rightarrow \quad E_{\mathrm{max}} = \sqrt{2\mu_0 c I}$$

$$E_{\mathrm{max}} = \sqrt{2(4\pi \times 10^{-7}\ \mathrm{N \cdot s^2/C^2})(3.00 \times 10^8\ \mathrm{m/s})(1.34 \times 10^3\ \mathrm{W/m^2})}$$

$$= \boxed{1.01 \times 10^3\ \mathrm{N/C}}$$

Obtain B_{max} using $E_{\mathrm{max}} = B_{\mathrm{max}}c$:

$$B_{\mathrm{max}} = \frac{E_{\mathrm{max}}}{c} = \frac{1.01 \times 10^3\ \mathrm{N/C}}{3.00 \times 10^8\ \mathrm{m/s}} = \boxed{3.37 \times 10^{-6}\ \mathrm{T}}$$

REMARKS The answer to part (b) is a little over a month. While the acceleration is very low, there are no fuel costs, and within a few months the velocity can change sufficiently to allow the spacecraft to reach any planet in the solar system. Such spacecraft may be useful for certain purposes and are highly economical, but require a considerable amount of patience.

QUESTION 21.9 By what factor will the force exerted by the Sun's light be changed when the spacecraft is twice as far from the Sun? (a) no change (b) $\frac{1}{2}$ (c) $\frac{1}{4}$ (d) $\frac{1}{8}$

EXERCISE 21.9 A laser has a power of 22.0 W and a beam radius of 0.500 mm. (a) Find the intensity of the laser. (b) Suppose you were floating in space and pointed the laser beam away from you. What would your acceleration be? Assume your total mass, including equipment, is 72.0 kg and the force is directed through your center of mass. *Hint:* The change in momentum is the same as in the nonreflective case. (c) Calculate your acceleration if it were due to the gravity of a space station with mass 1.00×10^6 kg and center of mass 100.0 m away. (d) Calculate the peak electric and magnetic fields of the laser.

ANSWERS (a) $2.80 \times 10^7\ \mathrm{W/m^2}$ (b) $1.02 \times 10^{-9}\ \mathrm{m/s^2}$ (c) $6.67 \times 10^{-9}\ \mathrm{m/s^2}$ (d) 1.45×10^5 N/C, 4.84×10^{-4} T. **Remark** If you were planning to use your laser welding torch as a thruster to get you back to the station, don't bother, because the force of gravity is stronger. Better yet, get somebody to toss you a line.

21.12 The Spectrum of Electromagnetic Waves

LEARNING OBJECTIVE

1. Describe and discuss the electromagnetic spectrum.

All electromagnetic waves travel in a vacuum with the speed of light, c. These waves transport energy and momentum from some source to a receiver. In 1887 Hertz successfully generated and detected the radio-frequency electromagnetic

Raymond A. Serway

Wearing sunglasses lacking ultraviolet (UV) protection is worse for your eyes than wearing no sunglasses at all. Sunglasses without protection absorb some visible light, causing the pupils to dilate. This allows more UV light to enter the eye, increasing the damage to the lens of the eye over time. Without the sunglasses, the pupils constrict, reducing both visible and dangerous UV radiation. Be cool: wear sunglasses with UV protection.

waves predicted by Maxwell. Maxwell himself had recognized as electromagnetic waves both visible light and the infrared radiation discovered in 1800 by William Herschel. It is now known that other forms of electromagnetic waves exist that are distinguished by their frequencies and wavelengths.

Because all electromagnetic waves travel through free space with a speed c, their frequency f and wavelength λ are related by the important expression

$$c = f\lambda \qquad [21.31]$$

The various types of electromagnetic waves are presented in Figure 21.22. Notice the wide and overlapping range of frequencies and wavelengths. For instance, an AM radio wave with a frequency of 1.50 MHz (a typical value) has a wavelength of

$$\lambda = \frac{c}{f} = \frac{3.00 \times 10^8\ \text{m/s}}{1.50 \times 10^6\ \text{s}^{-1}} = 2.00 \times 10^2\ \text{m}$$

The following abbreviations are often used to designate short wavelengths and distances:

$$1\ \text{micrometer}\ (\mu\text{m}) = 10^{-6}\ \text{m}$$

$$1\ \text{nanometer}\ (\text{nm}) = 10^{-9}\ \text{m}$$

$$1\ \text{angstrom}\ (\text{Å}) = 10^{-10}\ \text{m}$$

The wavelengths of visible light, for example, range from 0.4 μm to 0.7 μm, or 400 nm to 700 nm, or 4 000 Å to 7 000 Å.

Quick Quiz

21.8 Which of the following statements are true about light waves? (a) The higher the frequency, the longer the wavelength. (b) The lower the frequency, the longer the wavelength. (c) Higher-frequency light travels faster than lower-frequency light. (d) The shorter the wavelength, the higher the frequency. (e) The lower the frequency, the shorter the wavelength.

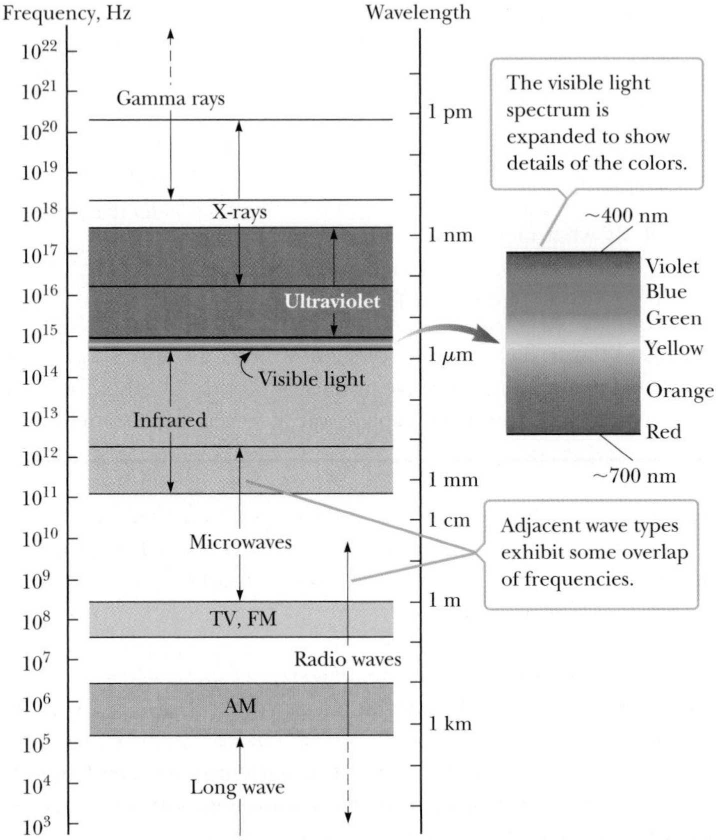

Figure 21.22 The electromagnetic spectrum.

Brief descriptions of the wave types follow, in order of decreasing wavelength. There is no sharp division between one kind of electromagnetic wave and the next. All forms of electromagnetic radiation are produced by accelerating charges.

Radio waves, which were discussed in Section 21.10, are the result of charges accelerating through conducting wires. They are, of course, used in radio and television communication systems.

Microwaves (short-wavelength radio waves) have wavelengths ranging between about 1 mm and 30 cm and are generated by electronic devices. Their short wavelengths make them well suited for the radar systems used in aircraft navigation and for the study of atomic and molecular properties of matter. Microwave ovens are an interesting domestic application of these waves. It has been suggested that solar energy might be harnessed by beaming microwaves to Earth from a solar collector in space.

Infrared waves (sometimes incorrectly called "heat waves"), produced by hot objects and molecules, have wavelengths ranging from about 1 mm to the longest wavelength of visible light, 7×10^{-7} m. They are readily absorbed by most materials. The infrared energy absorbed by a substance causes it to get warmer because the energy agitates the atoms of the object, increasing their vibrational or translational motion. The result is a rise in temperature. Infrared radiation has many practical and scientific applications, including physical therapy, infrared photography, and the study of the vibrations of atoms.

Visible light, the most familiar form of electromagnetic waves, may be defined as the part of the spectrum that is detected by the human eye. Light is produced by the rearrangement of electrons in atoms and molecules. The wavelengths of visible light are classified as colors ranging from violet ($\lambda \approx 4 \times 10^{-7}$ m) to red ($\lambda \approx 7 \times 10^{-7}$ m). The eye's sensitivity is a function of wavelength and is greatest at a wavelength of about 5.6×10^{-7} m (yellow green).

Ultraviolet (UV) light covers wavelengths ranging from about 4×10^{-7} m (400 nm) down to 6×10^{-10} m (0.6 nm). The Sun is an important source of ultraviolet light (which is the main cause of suntans). Most of the ultraviolet light from the Sun is absorbed by atoms in the upper atmosphere, or stratosphere, which is fortunate, because UV light in large quantities has harmful effects on humans. One important constituent of the stratosphere is ozone (O_3), produced from reactions of oxygen with ultraviolet radiation. The resulting ozone shield causes lethal high-energy ultraviolet radiation to warm the stratosphere.

X-rays are electromagnetic waves with wavelengths from about 10^{-8} m (10 nm) down to 10^{-13} m (10^{-4} nm). The most common source of x-rays is the acceleration of high-energy electrons bombarding a metal target. X-rays are used as a diagnostic tool in medicine and as a treatment for certain forms of cancer. Because x-rays easily penetrate and damage or destroy living tissues and organisms, care must be taken to avoid unnecessary exposure and overexposure.

Gamma rays—electromagnetic waves emitted by radioactive nuclei—have wavelengths ranging from about 10^{-10} m to less than 10^{-14} m. They are highly penetrating and cause serious damage when absorbed by living tissues. Accordingly, those working near such radiation must be protected by garments containing heavily absorbing materials, such as layers of lead.

When astronomers observe the same celestial object using detectors sensitive to different regions of the electromagnetic spectrum, striking variations in the object's features can be seen. Figure 21.23 (page 760) shows images of the Crab Nebula made in three different wavelength ranges. The Crab Nebula is the remnant of a supernova explosion that was seen on Earth in 1054 A.D. (Compare with Fig. 8.31.)

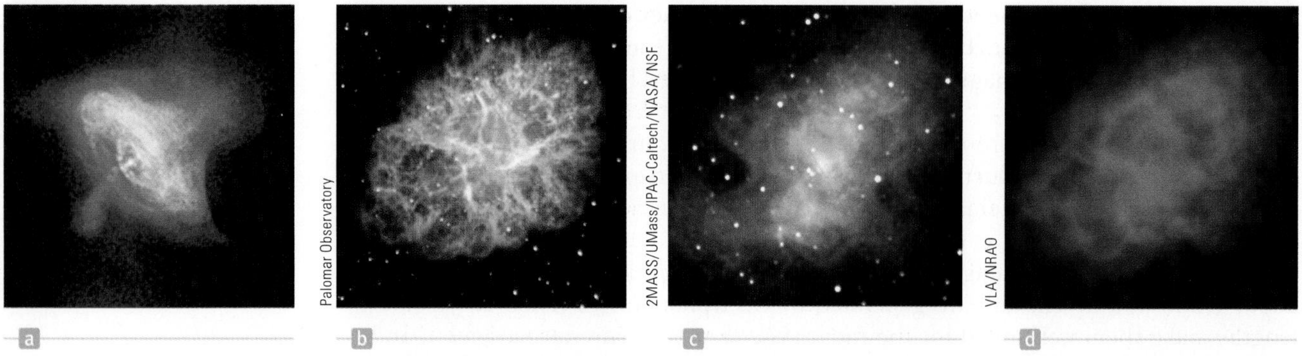

Figure 21.23 Observations in different parts of the electromagnetic spectrum show different features of the Crab Nebula. (a) X-ray image. (b) Optical image. (c) Infrared. (d) Radio image.

APPLYING PHYSICS 21.4 | Light and Wound Treatment BIO

An important issue in human health is wound management. Chronic wounds affect five million to seven million people in the United States at an annual cost of more than twenty billion dollars. Low-level laser therapy has been shown to facilitate the healing and closure of wounds.

Infrared light increases the generation of adenosine triphosphate (ATP) in mitochondria and may stimulate the activation of genes and enzymes associated with cellular respiration. (ATP molecules provide energy for a variety of important cell functions.) Infrared light may also increase the concentration of reactive oxygen molecules, which could increase communication between the nucleus, cytosol, and the mitochondria. This mechanism may enhance and accelerate the healing process.

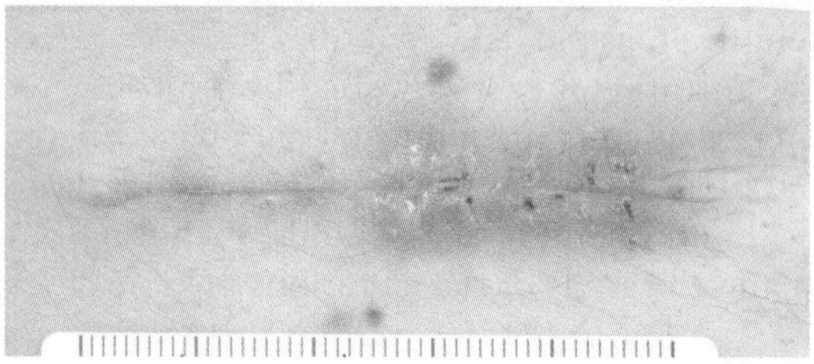

Figure 21.24 After bringing the sides of this wound together with deep sutures, closure on the left side was achieved with light-activated technology, whereas on the right side closure was carried out with sutures. This photo, taken at the end of two weeks, shows that healing was enhanced with light activation.

Green laser light can also be used to stimulate the body's repair mechanisms via a different process. A pink dye, called "rose bengal," is applied to the tissue, which is then exposed to the laser light for a few minutes. When the dye absorbs the light, it causes cross-linkages between collagen molecules in the tissue. The cross-linked molecules promote the closing of the tissue while reducing or eliminating the formation of scar tissue. Figure 21.24 shows the contrast between tissue receiving the normal treatment and tissue irradiated by lasers. The technique is also being studied for application to damaged peripheral nerves, blood vessels, and other tissues, such as incisions made in the cornea during eye surgery. ■

APPLYING PHYSICS 21.5 | The Sun and the Evolution of the Eye BIO

The center of sensitivity of our eyes coincides with the center of the wavelength distribution of the Sun. Is this an amazing coincidence?

EXPLANATION This fact is not a coincidence; rather, it's the result of biological evolution. Humans have evolved with vision most sensitive to wavelengths that are strongest from the Sun. If aliens from another planet ever arrived at Earth, their eyes would have the center of sensitivity at wavelengths different from ours. If their sun were a red dwarf, for example, the alien's eyes would be most sensitive to red light. ■

21.13 The Doppler Effect for Electromagnetic Waves

LEARNING OBJECTIVE

1. Describe the Doppler effect for electromagnetic waves and discuss its role in scientific discoveries.

As we saw in Section 14.6, sound waves exhibit the Doppler effect when the observer, the source, or both are moving relative to the medium of propagation.

(the secondary current) at an output voltage of 6.0 V (rms). (a) Should the transformer have more turns in the secondary compared with the primary, or fewer turns? (b) Find the current in the primary. (c) Find the number of turns in the secondary.

47. A transformer on a pole near a factory steps the voltage down from 3 600 V (rms) to 120 V (rms). The transformer is to deliver 1 000 kW to the factory at 90% efficiency. Find (a) the power delivered to the primary, (b) the current in the primary, and (c) the current in the secondary.

48. A transmission line that has a resistance per unit length of $4.50 \times 10^{-4}\ \Omega/\text{m}$ is to be used to transmit 5.00 MW over 400 miles (6.44×10^5 m). The output voltage of the generator is 4.50 kV (rms). (a) What is the line loss if a transformer is used to step up the voltage to 500 kV (rms)? (b) What fraction of the input power is lost to the line under these circumstances? (c) What difficulties would be encountered on attempting to transmit the 5.00 MW at the generator voltage of 4.50 kV (rms)?

21.10 Production of Electromagnetic Waves by an Antenna

21.11 Properties of Electromagnetic Waves

49. The U.S. Navy has long proposed the construction of extremely low frequency (ELF waves) communications systems; such waves could penetrate the oceans to reach distant submarines. Calculate the length of a quarter-wavelength antenna for a transmitter generating ELF waves of frequency 75 Hz. How practical is this antenna?

50. (a) The distance to Polaris, the North Star, is approximately 6.44×10^{18} m. If Polaris were to burn out today, how many years would it take to see it disappear? (b) How long does it take sunlight to reach Earth? (c) How long does it take a microwave signal to travel from Earth to the Moon and back? (The distance from Earth to the Moon is 3.84×10^5 km.)

51. QIC The Earth reflects approximately 38.0% of the incident sunlight from its clouds and surface. (a) Given that the intensity of solar radiation at the top of the atmosphere is 1 370 W/m^2, find the radiation pressure on the Earth, in pascals, at the location where the Sun is straight overhead. (b) State how this quantity compares with normal atmospheric pressure at the Earth's surface, which is 101 kPa.

52. Experimenters at the National Institute of Standards and Technology have made precise measurements of the speed of light using the fact that, in vacuum, the speed of electromagnetic waves is $c = 1/\sqrt{\mu_0\epsilon_0}$, where the constants $\mu_0 = 4\pi \times 10^{-7}\ \text{N} \cdot \text{s}^2/\text{C}^2$ and $\epsilon_0 = 8.854 \times 10^{-12}\ \text{C}^2/\text{N} \cdot \text{m}^2$. What value (to four significant figures) does this formula give for the speed of light in vacuum?

53. BIO Oxygenated hemoglobin absorbs weakly in the red (hence its red color) and strongly in the near infrared, whereas deoxygenated hemoglobin has the opposite absorption. This fact is used in a "pulse oximeter" to measure oxygen saturation in arterial blood. The device clips onto the end of a person's finger and has two light-emitting diodes—a red (660 nm) and an infrared (940 nm)—and a photocell that detects the amount of light transmitted through the finger at each wavelength. (a) Determine the frequency of each of these light sources. (b) If 67% of the energy of the red source is absorbed in the blood, by what factor does the amplitude of the electromagnetic wave change? *Hint:* The intensity of the wave is equal to the average power per unit area as given by Equation 21.28.

54. BIO **Operation of the pulse oximeter (see previous problem).** The transmission of light energy as it passes through a solution of light-absorbing molecules is described by the Beer–Lambert law

$$I = I_0 10^{-\epsilon CL} \quad \text{or} \quad \log_{10}\left(\frac{I}{I_0}\right) = -\epsilon CL$$

which gives the decrease in intensity I in terms of the distance L the light has traveled through a fluid with a concentration C of the light-absorbing molecule. The quantity ϵ is called the extinction coefficient, and its value depends on the frequency of the light. (It has units of m^2/mol.) Assume the extinction coefficient for 660-nm light passing through a solution of oxygenated hemoglobin is identical to the coefficient for 940-nm light passing through deoxygenated hemoglobin. Also assume 940-nm light has zero absorption ($\epsilon = 0$) in oxygenated hemoglobin and 660-nm light has zero absorption in deoxygenated hemoglobin. If 33% of the energy of the red source and 76% of the infrared energy is transmitted through the blood, what is the fraction of hemoglobin that is oxygenated?

55. The Sun delivers an average power of 1 370 W/m^2 to the top of Earth's atmosphere. Find the magnitudes of $\vec{\mathbf{E}}_{\text{max}}$ and $\vec{\mathbf{B}}_{\text{max}}$ for the electromagnetic waves at the top of the atmosphere.

56. QIC A laser beam is used to levitate a metal disk against the force of Earth's gravity. (a) Derive an equation giving the required intensity of light, I, in terms of the mass m of the disk, the gravitational acceleration g, the speed of light c, and the cross-sectional area of the disk A. Assume the disk is perfectly reflecting and the beam is directed perpendicular to the disk. (b) If the disk has mass 5.00 g and radius 4.00 cm, find the necessary light intensity. (c) Give two reasons why using light pressure as propulsion near Earth's surface is impractical.

57. A microwave oven is powered by an electron tube called a magnetron that generates electromagnetic waves of frequency 2.45 GHz. The microwaves enter the oven and are reflected by the walls. The standing-wave pattern produced in the oven can cook food unevenly, with hot spots in the food at antinodes and cool spots at nodes, so a turntable is often used to rotate the food and distribute the energy. If a microwave oven is used with a cooking dish in a fixed position, the antinodes can appear as burn marks on foods such as carrot strips or cheese. The separation distance between the burns is measured to be 6.00 cm. Calculate the speed of the microwaves from these data.

58. Consider a bright star in our night sky. Assume its distance from the Earth is 20.0 light-years (ly) and its power output is 4.00×10^{28} W, about 100 times that of the Sun. (a) Find the intensity of the starlight at the Earth. (b) Find the power of the starlight the Earth intercepts. One light-year is the distance traveled by light through a vacuum in one year.

21.12 The Spectrum of Electromagnetic Waves

59. What are the wavelengths of electromagnetic waves in free space that have frequencies of (a) 5.00×10^{19} Hz and (b) 4.00×10^{9} Hz?

60. BIO A diathermy machine, used in physiotherapy, generates electromagnetic radiation that gives the effect of "deep heat" when absorbed in tissue. One assigned frequency for diathermy is 27.33 MHz. What is the wavelength of this radiation?

61. What are the wavelength ranges in (a) the AM radio band (540–1 600 kHz) and (b) the FM radio band (88–108 MHz)?

62. An important news announcement is transmitted by radio waves to people who are 100 km away, sitting next to their radios, and by sound waves to people sitting across the newsroom, 3.0 m from the newscaster. Who receives the news first? Explain. Take the speed of sound in air to be 343 m/s.

63. Infrared spectra are used by chemists to help identify an unknown substance. Atoms in a molecule that are bound together by a particular bond vibrate at a predictable frequency, and light at that frequency is absorbed strongly by the atom. In the case of the C=O double bond, for example, the oxygen atom is bound to the carbon by a bond that has an effective spring constant of 2 800 N/m. If we assume the carbon atom remains stationary (it is attached to other atoms in the molecule), determine the resonant frequency of this bond and the wavelength of light that matches that frequency. Verify that this wavelength lies in the infrared region of the spectrum. (The mass of an oxygen atom is 2.66×10^{-26} kg.)

21.13 The Doppler Effect for Electromagnetic Waves

64. A spaceship is approaching a space station at a speed of 1.8×10^{5} m/s. The space station has a beacon that emits green light with a frequency of 6.0×10^{14} Hz. (a) What is the frequency of the beacon observed on the spaceship? (b) What is the change in frequency? (Carry five digits in these calculations.)

65. While driving at a constant speed of 80 km/h, you are passed by a car traveling at 120 km/h. If the frequency of light emitted by the taillights of the car that passes you is 4.3×10^{14} Hz, what frequency will you observe? What is the change in frequency?

66. A speeder tries to explain to the police that the yellow warning lights she was approaching on the side of the road looked green to her because of the Doppler shift. How fast would she have been traveling if yellow light of wavelength 580 nm had been shifted to green with a wavelength of 560 nm? *Note:* For speeds less than $0.03c$, Equation 21.32 will lead to a value for the observed frequency accurate to approximately two significant digits.

Additional Problems

67. A 25.0-mW laser beam of diameter 2.00 mm is reflected at normal incidence by a perfectly reflecting mirror. Calculate the radiation pressure on the mirror.

68. The intensity of solar radiation at the top of Earth's atmosphere is 1 370 W/m^3. Assuming 60% of the incoming solar energy reaches Earth's surface and assuming you absorb 50% of the incident energy, make an order-of-magnitude estimate of the amount of solar energy you absorb in a 60-minute sunbath.

69. A 200-Ω resistor is connected in series with a 5.0-μF capacitor and a 60-Hz, 120-V rms line. If electrical energy costs \$0.080/kWh, how much does it cost to leave this circuit connected for 24 h?

70. S In an *RLC* series circuit that includes a source of alternating current operating at fixed frequency and voltage, the resistance *R* is equal to the inductive reactance. If the plate separation of the parallel-plate capacitor is reduced to one-half its original value, the current in the circuit doubles. Find the initial capacitive reactance in terms of *R*.

71. As a way of determining the inductance of a coil used in a research project, a student first connects the coil to a 12.0-V battery and measures a current of 0.630 A. The student then connects the coil to a 24.0-V (rms), 60.0-Hz generator and measures an rms current of 0.570 A. What is the inductance?

72. (a) What capacitance will resonate with a one-turn loop of inductance 400 pH to give a radar wave of wavelength 3.0 cm? (b) If the capacitor has square parallel plates separated by 1.0 mm of air, what should the edge length of the plates be? (c) What is the common reactance of the loop and capacitor at resonance?

73. M A dish antenna with a diameter of 20.0 m receives (at normal incidence) a radio signal from a distant source, as shown in Figure P21.73. The radio signal is a continuous sinusoidal wave with amplitude $E_{max} = 0.20$ μV/m. Assume the antenna absorbs all the radiation that falls

on the dish. (a) What is the amplitude of the magnetic field in this wave? (b) What is the intensity of the radiation received by the antenna? (c) What is the power received by the antenna?

Figure P21.73

74. A particular inductor has appreciable resistance. When the inductor is connected to a 12-V battery, the current in the inductor is 3.0 A. When it is connected to an AC source with an rms output of 12 V and a frequency of 60 Hz, the current drops to 2.0 A. What are (a) the impedance at 60 Hz and (b) the inductance of the inductor?

75. One possible means of achieving space flight is to place a perfectly reflecting aluminized sheet into Earth's orbit and to use the light from the Sun to push this solar sail. Suppose such a sail, of area 6.00×10^4 m^2 and mass 6 000 kg, is placed in orbit facing the Sun. (a) What force is exerted on the sail? (b) What is the sail's acceleration? (c) How long does it take this sail to reach the Moon, 3.84×10^8 m away? Ignore all gravitational effects and assume a solar intensity of 1 340 W/m^2. *Hint:* The radiation pressure by a reflected wave is given by 2 (average power per unit area)/c.

76. BIO Q|C The U.S. Food and Drug Administration limits the radiation leakage of microwave ovens to no more than 5.0 mW/cm^2 at a distance of 2.0 in. A typical cell phone, which also transmits microwaves, has a peak output power of about 2.0 W. (a) Approximating the cell phone as a point source, calculate the radiation intensity of a cell phone at a distance of 2.0 in. How does the answer compare with the maximum allowable microwave oven leakage? (b) The distance from your ear to your brain is about 2 in. What would the radiation intensity in your brain be if you used a Bluetooth headset, keeping the phone in your pocket, 1.0 m away from your brain? Most headsets are so-called Class 2 devices with a maximum output power of 2.5 mW.

Light is bent (refracted) as it passes through water, with different wavelengths bending by different amounts, a phenomenon called dispersion. Together with reflection, these physical phenomena lead to the creation of a rainbow when light passes through small, suspended droplets of water.

Dave Walker/Shutterstock.com

22 Reflection and Refraction of Light

Light has a dual nature. In some experiments it acts like a particle, while in others it acts like a wave. In this and the next two chapters, we concentrate on the aspects of light that are best understood through the wave model. First we discuss the reflection of light at the boundary between two media and the refraction (bending) of light as it travels from one medium into another. We use these ideas to study the refraction of light as it passes through lenses and the reflection of light from mirrored surfaces. In Chapter 25, we describe how lenses and mirrors can be used to view objects with telescopes and microscopes and how lenses are used in photography. The ability to manipulate light has greatly enhanced our capacity to investigate and understand the nature of the Universe.

22.1 The Nature of Light

LEARNING OBJECTIVES

1. Discuss the dual nature of light as both a wave and as particles called photons.
2. State the equation yielding the energy of a photon.

Until the beginning of the 19th century, light was modeled as a stream of particles emitted by a source that stimulated the sense of sight on entering the eye. The chief architect of the particle theory of light was Newton. With this theory, he provided simple explanations of some known experimental facts concerning the nature of light, namely, the laws of reflection and refraction.

Most scientists accepted Newton's particle theory of light. During Newton's lifetime, however, another theory was proposed. In 1678 Dutch physicist and

medium 1 and medium 2. The frequency at which the wave fronts pass an observer at point *B* in medium 2 must equal the frequency at which they arrive at point *A*. If not, the wave fronts would either pile up at the boundary or be destroyed or created at the boundary. Because neither of these events occurs, the frequency must remain the same as a light ray passes from one medium into another.

Therefore, because the relation $v = f\lambda$ must be valid in both media and because $f_1 = f_2 = f$, we see that

$$v_1 = f\lambda_1 \text{ and } v_2 = f\lambda_2$$

Because $v_1 \neq v_2$, it follows that $\lambda_1 \neq \lambda_2$. A relationship between the index of refraction and the wavelength can be obtained by dividing these two equations and making use of the definition of the index of refraction given by Equation 22.4:

$$\frac{\lambda_1}{\lambda_2} = \frac{v_1}{v_2} = \frac{c/n_1}{c/n_2} = \frac{n_2}{n_1} \quad [22.5]$$

which gives

$$\lambda_1 n_1 = \lambda_2 n_2 \quad [22.6]$$

Let medium 1 be the vacuum so that $n_1 = 1$. It follows from Equation 22.6 that the index of refraction of any medium can be expressed as the ratio

$$n = \frac{\lambda_0}{\lambda_n} \quad [22.7]$$

where λ_0 is the wavelength of light in vacuum and λ_n is the wavelength in the medium having index of refraction n. Figure 22.9 is a schematic representation of this reduction in wavelength when light passes from a vacuum into a transparent medium.

We are now in a position to express Equation 22.3 in an alternate form. If we substitute Equation 22.5 into Equation 22.3, we get

$$n_1 \sin\theta_1 = n_2 \sin\theta_2 \quad [22.8]$$

◄ Snell's law of refraction

The experimental discovery of this relationship is usually credited to Willebrørd Snell (1591–1626) and is therefore known as **Snell's law of refraction**.

Tip 22.1 An Inverse Relationship

The index of refraction is *inversely* proportional to the wave speed. Therefore, as the wave speed v decreases, the index of refraction, *n, increases.*

Tip 22.2 The Frequency Remains the Same

The *frequency* of a wave does *not* change as the wave passes from one medium to another. Both the wave speed and the wavelength *do* change, but the frequency remains the same.

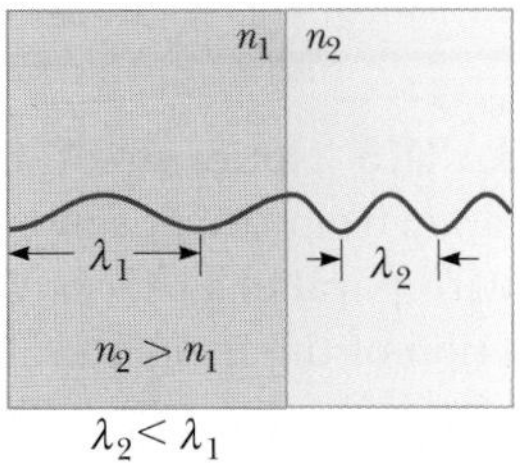

Figure 22.9 A schematic diagram of the *reduction* in wavelength when light travels from a medium with a low index of refraction to one with a higher index of refraction.

Quick Quiz

22.3 A material has an index of refraction that increases continuously from top to bottom. Of the three paths shown in Figure 22.10, which path will a light ray follow as it passes through the material?

22.4 As light travels from a vacuum ($n = 1$) to a medium such as glass ($n > 1$), which of the following properties remains the same, the (a) wavelength, (b) wave speed, or (c) frequency?

Figure 22.10 (Quick Quiz 22.3)

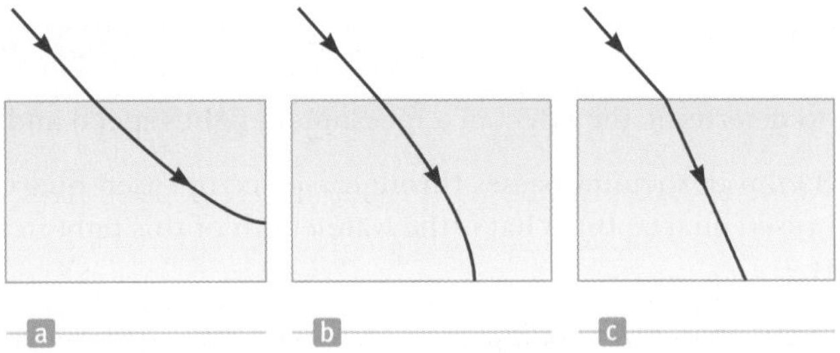

EXAMPLE 22.2 Angle of Refraction for Glass

GOAL Apply Snell's law.

PROBLEM A light ray of wavelength 589 nm (produced by a sodium lamp) traveling through air is incident on a smooth, flat slab of crown glass at an angle θ_1 of 30.0° to the normal, as sketched in Figure 22.11. **(a)** Find the angle of refraction, θ_2. **(b)** At what angle θ_3 does the ray leave the glass as it re-enters the air? **(c)** How does the answer for θ_3 change if the ray enters water below the slab instead of the air?

STRATEGY Substitute quantities into Snell's law and solve for the unknown angles of refraction.

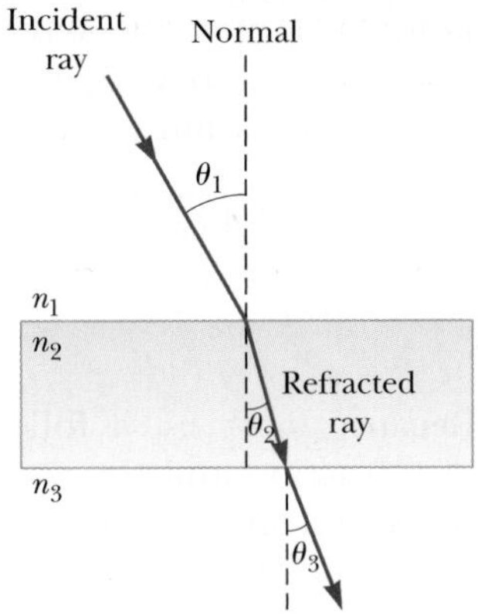

Figure 22.11 (Example 22.2) Refraction of light by glass.

SOLUTION

(a) Find the angle of refraction, θ_2.

Solve Snell's law (Eq. 22.8) for $\sin\theta_2$:

$$(1)\quad \sin\theta_2 = \frac{n_1}{n_2}\sin\theta_1$$

From Table 22.1, find $n_1 = 1.00$ for air and $n_2 = 1.52$ for crown glass. Substitute these values into Equation (1) and take the inverse sine of both sides:

$$\sin\theta_2 = \left(\frac{1.00}{1.52}\right)(\sin 30.0°) = 0.329$$

$$\theta_2 = \sin^{-1}(0.329) = \boxed{19.2°}$$

(b) At what angle θ_3 does the ray leave the glass as it re-enters the air?

Write Equation (1), replacing θ_3 with θ_2 and θ_1 with θ_2:

$$(2)\quad \sin\theta_3 = \frac{n_2}{n_3}\sin\theta_2 = \frac{1.52}{1.00}\sin(19.2°) = 0.500$$

Take the inverse sine of both sides to find θ_3:

$$\theta_3 = \sin^{-1}(0.500) = \boxed{30.0°}$$

(c) How does the answer for θ_3 change if the ray enters water below the slab instead of air?

Write Equation (2) and substitute a different value for n_3:

$$\sin\theta_3 = \frac{n_2}{n_3}\sin\theta_2 = \frac{1.52}{1.333}\sin(19.2°) = 0.375$$

$$\theta_3 = \sin^{-1}(0.375) = \boxed{22.0°}$$

REMARKS Notice that the light ray bends toward the normal when it enters a material of a higher index of refraction, and away from the normal when entering a material with a lower index of refraction. In passing through a slab of material with parallel surfaces, for example, from air to glass and back to air, the final direction of the ray is parallel to the direction of the incident ray. The only effect in that case is a lateral displacement of the light ray.

QUESTION 22.2 If the glass is replaced by a transparent material with smaller index of refraction, will the refraction angle θ_2 be (a) smaller, (b) larger, or (c) unchanged?

EXERCISE 22.2 Suppose a light ray in air ($n = 1.00$) enters a cube of material ($n = 2.50$) at a 45.0° angle with respect to the normal and then exits the bottom of the cube into water ($n = 1.333$). At what angle to the normal does the ray leave the slab?

ANSWER 32.0°

EXAMPLE 22.3 Light in Fused Quartz

GOAL Use the index of refraction to determine the effect of a medium on light's speed and wavelength.

PROBLEM Light of wavelength 589 nm in vacuum passes through a piece of fused quartz of index of refraction $n = 1.458$. **(a)** Find the speed of light in fused quartz. **(b)** What is the wavelength of this light in fused quartz? **(c)** What is the frequency of the light in fused quartz?

STRATEGY Substitute values into Equations 22.4 and 22.7.

SOLUTION

(a) Find the speed of light in fused quartz.

Obtain the speed from Equation 22.4:

$$v = \frac{c}{n} = \frac{3.00 \times 10^8 \text{ m/s}}{1.458} = \boxed{2.06 \times 10^8 \text{ m/s}}$$

(b) What is the wavelength of this light in fused quartz?

Use Equation 22.7 to calculate the wavelength:

$$\lambda_n = \frac{\lambda_0}{n} = \frac{589 \text{ nm}}{1.458} = \boxed{404 \text{ nm}}$$

(c) What is the frequency of the light in fused quartz?

The frequency in quartz is the same as in vacuum. Solve $c = f\lambda$ for the frequency:

$$f = \frac{c}{\lambda} = \frac{3.00 \times 10^8 \text{ m/s}}{589 \times 10^{-9} \text{ m}} = \boxed{5.09 \times 10^{14} \text{ Hz}}$$

REMARKS It's interesting to note that the speed of light in vacuum, 3.00×10^8 m/s, is an upper limit for the speed of material objects. In our treatment of relativity in Chapter 26, we will find that this upper limit is consistent with experimental observations. However, it's possible for a particle moving in a medium to have a speed that exceeds the speed of light in that medium. For example, it's theoretically possible for a particle to travel through fused quartz at a speed greater than 2.06×10^8 m/s, but it must still have a speed less than 3.00×10^8 m/s.

QUESTION 22.3 True or False: If light with wavelength λ in glass passes into water with index n_w, the new wavelength of the light is λ/n_w.

EXERCISE 22.3 Light with wavelength 589 nm passes through crystalline sodium chloride. In this medium, find (a) the speed of light, (b) the wavelength, and (c) the frequency of the light.

ANSWER (a) 1.94×10^8 m/s (b) 381 nm (c) 5.09×10^{14} Hz

EXAMPLE 22.4 Refraction of Laser Light in a Digital Videodisc (DVD)

GOAL Apply Snell's law together with geometric constraints.

PROBLEM A DVD is a video recording consisting of a spiral track about 1.0 μm wide with digital information. (See Fig. 22.12a.) The digital information consists of a series of pits that are "read" by a laser beam sharply focused on a track in the information layer. If the width a of the beam at the information layer must equal 1.0 μm to distinguish individual tracks and the width w of the beam as it enters the plastic is 0.700 0 mm, find the angle θ_1 at which the conical beam should enter the plastic. (See Fig. 22.12b.) Assume the plastic has a thickness t = 1.20 mm and an index of refraction n = 1.55. Note that this system is relatively immune to small dust particles degrading the video quality because particles would have to be as large as 0.700 mm to obscure the beam at the point where it enters the plastic.

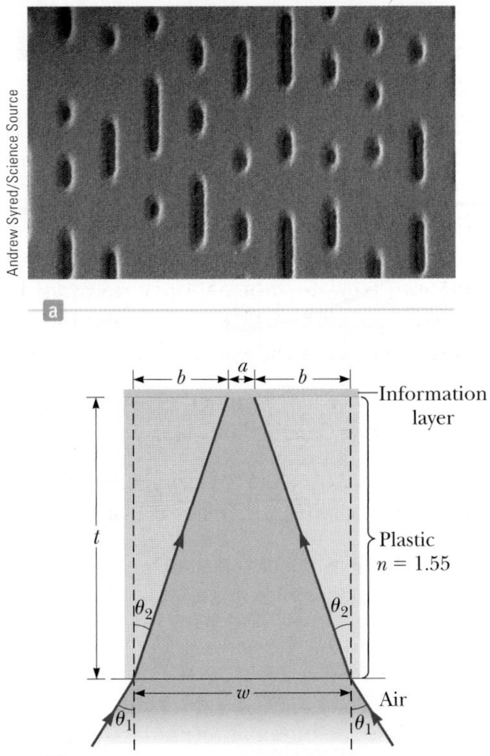

Figure 22.12 (Example 22.4) A micrograph of a DVD surface showing tracks and pits along each track. (b) Cross section of a cone-shaped laser beam used to read a DVD.

STRATEGY Use right-triangle trigonometry to determine the angle θ_2 and then apply Snell's law to obtain the angle θ_1.

(Continued)

SOLUTION

From the top and bottom of Figure 22.12b, obtain an equation relating w, b, and a:

$$w = 2b + a$$

Solve this equation for b and substitute given values:

$$b = \frac{w - a}{2} = \frac{700.0 \times 10^{-6}\,\text{m} - 1.0 \times 10^{-6}\,\text{m}}{2} = 349.5\ \mu\text{m}$$

Now use the tangent function to find θ_2:

$$\tan\theta_2 = \frac{b}{t} = \frac{349.5\ \mu\text{m}}{1.20 \times 10^3\ \mu\text{m}} \quad \rightarrow \quad \theta_2 = 16.2°$$

Finally, use Snell's law to find θ_1:

$$n_1 \sin\theta_1 = n_2 \sin\theta_2$$

$$\sin\theta_1 = \frac{n_2 \sin\theta_2}{n_1} = \frac{1.55 \sin 16.2°}{1.00} = 0.432$$

$$\theta_1 = \sin^{-1}(0.432) = 25.6°$$

REMARKS Despite its apparent complexity, the problem isn't that different from Example 22.2.

QUESTION 22.4 Suppose the plastic were replaced by a material with a higher index of refraction. How would the width of the beam at the information layer be affected? (a) It would remain the same. (b) It would decrease. (c) It would increase.

EXERCISE 22.4 Suppose you wish to redesign the system to decrease the initial width of the beam from 0.700 0 mm to 0.600 0 mm but leave the incident angle θ_1 and all other parameters the same as before, except the index of refraction for the plastic material (n_2) and the angle θ_2. What index of refraction should the plastic have?

ANSWER 1.79

22.4 Dispersion and Prisms

LEARNING OBJECTIVES

1. Define dispersion.
2. Describe how prisms work.

In Table 22.1 we presented values for the index of refraction of various materials. If we make careful measurements, however, we find that the index of refraction in anything but vacuum depends on the wavelength of light. The dependence of the index of refraction on wavelength is called **dispersion**. Figure 22.13 is a graphical representation of this variation in the index of refraction with wavelength. Because n is a function of wavelength, Snell's law indicates that **the angle of refraction made when light enters a material depends on the wavelength of the light.** As seen in the figure, the index of refraction for a material usually decreases with increasing wavelength. This means that violet light ($\lambda \cong 400$ nm) refracts more than red light ($\lambda \cong 650$ nm) when passing from air into a material.

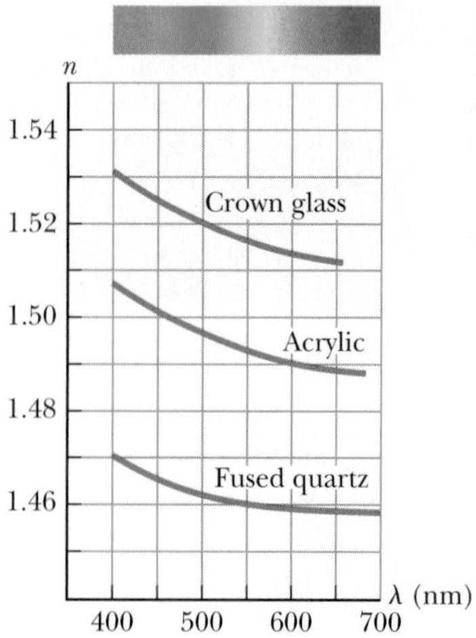

Figure 22.13 Variations of index of refraction in the visible spectrum with respect to vacuum wavelength for three materials.

To understand the effects of dispersion on light, consider what happens when light strikes a prism, as in Figure 22.14a. A ray of light of a single wavelength that is incident on the prism from the left emerges bent away from its original direction of travel by an angle δ, called the **angle of deviation**. Now suppose a beam of white light (a combination of all visible wavelengths) is incident on a prism. Because of dispersion, the different colors refract through different angles of deviation, as illustrated in Figure 22.14b. The rays that emerge from the second face of the prism spread out in a series of colors known as a visible **spectrum**, as shown in Figure 22.15. These colors, in order of decreasing wavelength, are red, orange, yellow, green, blue, and violet. Violet light deviates the most, red light the least, and the remaining colors in the visible spectrum fall between these extremes.

Tip 22.3 Dispersion

Light of shorter wavelength, such as violet light, refracts more than light of longer wavelengths, such as red light.

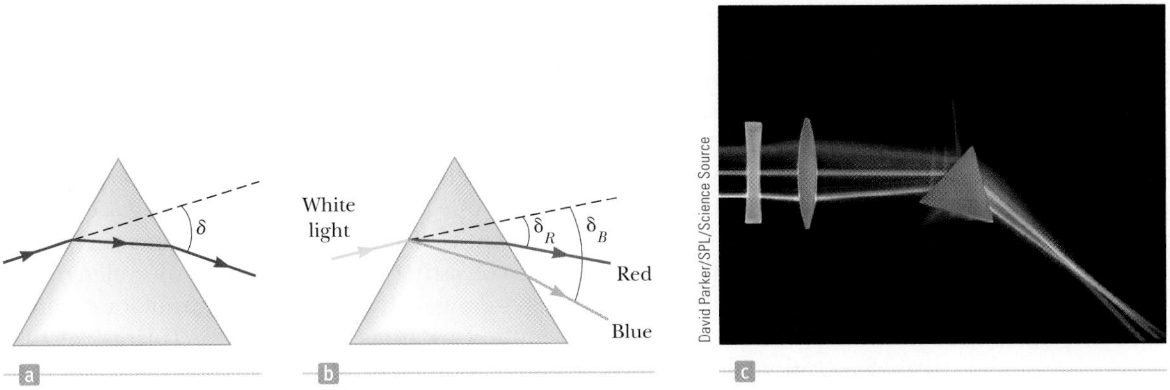

Figure 22.14 (a) A prism refracts a light ray and deviates the light through the angle δ. (b) When light is incident on a prism, the blue light is bent more than the red light. (c) Light of different colors passes through a prism and two lenses. Note that as the light passes through the prism, different wavelengths are refracted at different angles.

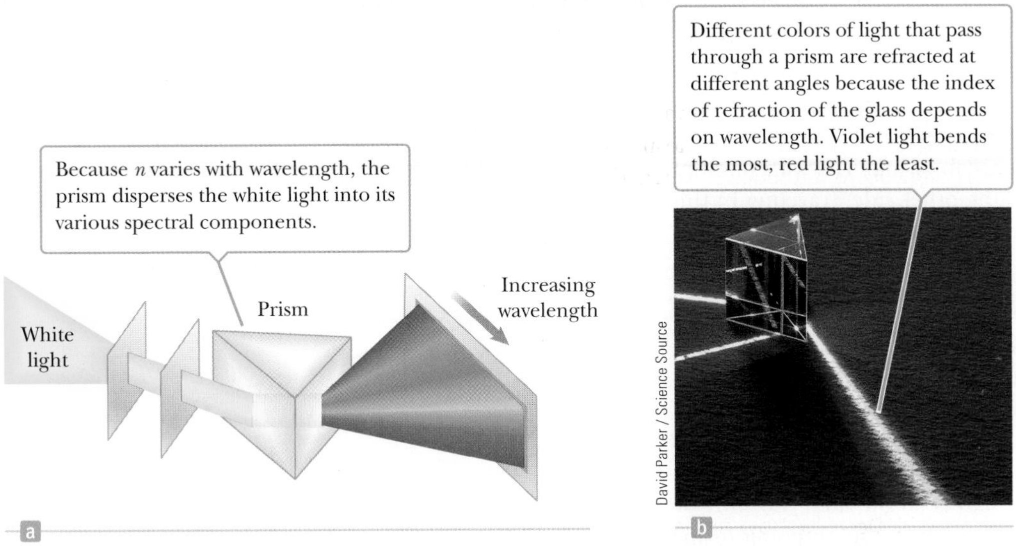

Figure 22.15 (a) Dispersion of white light by a prism. (b) White light enters a glass prism at the upper left.

Prisms are often used in an instrument known as a **prism spectrometer**, the essential elements of which are shown in Figure 22.16a. This instrument is commonly used to study the wavelengths emitted by a light source, such as a sodium vapor lamp. Light from the source is sent through a narrow, adjustable slit and lens to produce a parallel, or collimated, beam. The light then passes through the prism and is dispersed into a spectrum. The refracted light is observed through a telescope. The experimenter sees different colored images of the slit through the

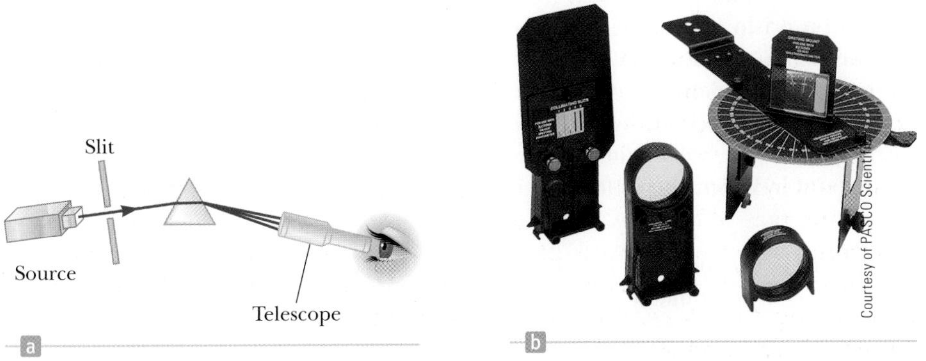

Figure 22.16 (a) A diagram of a prism spectroscope. The colors in the spectrum are viewed through a telescope. (b) A prism spectrometer with interchangeable components. A spectrometer is a spectroscope with a scale or detector that measures wavelengths.

eyepiece of the telescope. The telescope can be moved or the prism can be rotated to view the various wavelengths, which have different angles of deviation. Figure 22.16b shows one type of prism spectrometer used in undergraduate laboratories.

APPLICATION
Identifying Gases with a Spectrometer

All hot, low-pressure gases emit their own characteristic spectra, so one use of a prism spectrometer is to identify gases. For example, sodium emits only two wavelengths in the visible spectrum: two closely spaced yellow lines. (The bright line-like images of the slit seen in a spectroscope are called *spectral lines.*) A gas emitting these, and only these, colors can be identified as sodium. Likewise, mercury vapor has its own characteristic spectrum, consisting of four prominent wavelengths—orange, green, blue, and violet lines—along with some wavelengths of lower intensity. The particular wavelengths emitted by a gas serve as "fingerprints" of that gas. Spectral analysis, which is the measurement of the wavelengths emitted or absorbed by a substance, is a powerful general tool in many scientific areas. As examples, chemists and biologists use infrared spectroscopy to identify molecules, astronomers use visible-light spectroscopy to identify elements on distant stars, and geologists use spectral analysis to identify minerals.

APPLYING PHYSICS 22.3 | Dispersion

When a beam of light enters a glass prism, which has nonparallel sides, the rainbow of color exiting the prism is a testimonial to the dispersion occurring in the glass. Suppose a beam of light enters a slab of material with parallel sides. When the beam exits the other side, traveling in the same direction as the original beam, is there any evidence of dispersion?

EXPLANATION Due to dispersion, light at the violet end of the spectrum exhibits a larger angle of refraction on entering the glass than light at the red end. All colors of light return to their original direction of propagation as they refract back out into the air. As a result, the outgoing beam is white. The net shift in the position of the violet light along the edge of the slab is larger than the shift of the red light, however, so one edge of the outgoing beam has a bluish tinge to it (it appears blue rather than violet because the eye is not very sensitive to violet light), whereas the other edge has a reddish tinge. This effect is indicated in Figure 22.17. The colored edges of the outgoing beam of white light are evidence of dispersion. ■

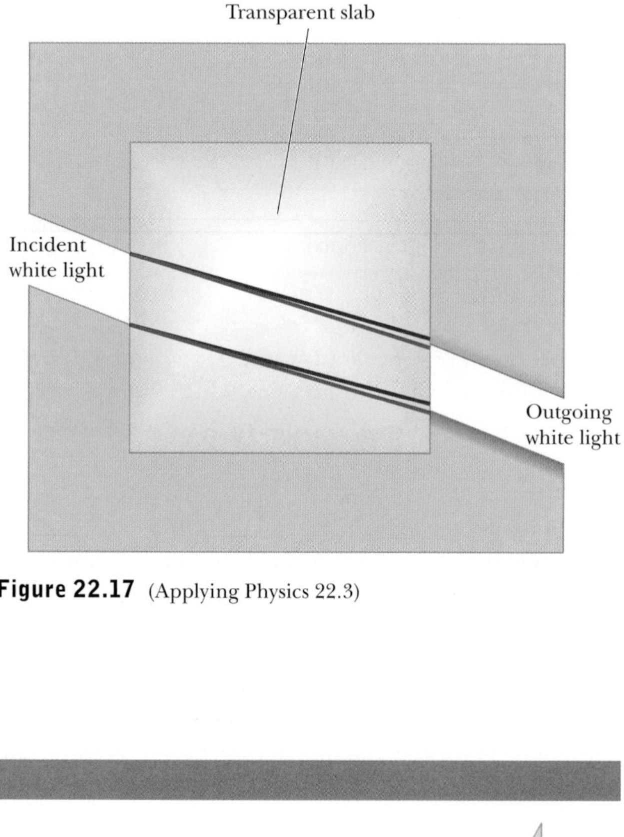

Figure 22.17 (Applying Physics 22.3)

EXAMPLE 22.5 | Light Through a Prism

GOAL Calculate the consequences of dispersion.

PROBLEM A beam of light is incident on a prism of a certain glass at an angle of $\theta_1 = 30.0°$, as shown in Figure 22.18. If the index of refraction of the glass for violet light is 1.80, find **(a)** θ_2, the angle of refraction at the air–glass interface, **(b)** ϕ_2, the angle of incidence at the glass–air interface, and **(c)** ϕ_1, the angle of refraction when the violet light exits the prism. **(d)** What is the value of Δy, the amount by which the violet light is displaced vertically?

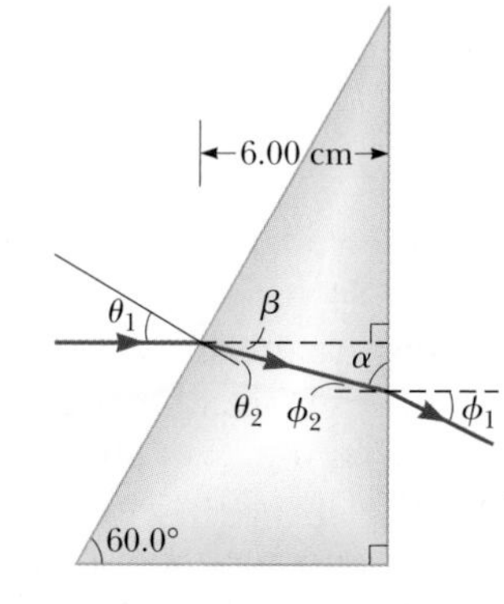

Figure 22.18 (Example 22.5)

STRATEGY This problem requires Snell's law to find the refraction angles and some elementary geometry and trigonometry based on Figure 22.18.

SOLUTION

(a) Find θ_2, the angle of refraction at the air–glass interface.

Use Snell's law to find the first angle of refraction:

$$n_1 \sin \theta_1 = \sin \theta_2 \rightarrow (1.00) \sin 30.0 = (1.80) \sin \theta_2$$

$$\theta_2 = \sin^{-1}\left(\frac{0.500}{1.80}\right) = \boxed{16.1°}$$

(b) Find ϕ_2, the angle of incidence at the glass–air interface.

Compute the angle β:

$$\beta = 30.0° - \theta_2 = 30.0° - 16.1° = 13.9°$$

Compute the angle α using the fact that the sum of the interior angles of a triangle equals 180°:

$$180° = 13.9° + 90° + \alpha \rightarrow \alpha = 76.1°$$

The incident angle ϕ_2 at the glass–air interface is complementary to α:

$$\phi_2 = 90° - \alpha = 90° - 76.1° = \boxed{13.9°}$$

(c) Find ϕ_1, the angle of refraction when the violet light exits the prism.

Apply Snell's law:

$$\phi_1 = \left(\frac{1}{n_1}\right) \sin^{-1}(n_2 \sin \phi_2)$$

$$= \left(\frac{1}{1.00}\right) \sin^{-1}[(1.80) \sin 13.9°] = \boxed{25.6°}$$

(d) What is the value of Δy, the amount by which the violet light is displaced vertically?

Use the tangent function to find the vertical displacement:

$$\tan \beta = \frac{\Delta y}{\Delta x} \rightarrow \Delta y = \Delta x \tan \beta$$

$$\Delta y = (6.00 \text{ cm}) \tan(13.9°) = \boxed{1.48 \text{ cm}}$$

REMARKS The same calculation for red light is left as an exercise. The violet light is bent more and displaced farther down the face of the prism. Notice that a theorem in geometry about parallel lines and the angles created by a transverse line give $\phi_2 = \beta$ immediately, which would have saved some calculation. In general, however, this tactic might not be available.

QUESTION 22.5 On passing through the prism, will yellow light bend through a larger angle or smaller angle than the violet light? (a) Yellow light bends through a larger angle. (b) Yellow light bends through a smaller angle. (c) The angles are the same.

EXERCISE 22.5 Repeat parts (a) through (d) of the example for red light passing through the prism, given that the index of refraction for red light is 1.72.

ANSWERS (a) 16.9° (b) 13.1° (c) 22.9° (d) 1.40 cm

22.5 The Rainbow

LEARNING OBJECTIVE

1. Describe how the dispersion of light forms rainbows.

The dispersion of light into a spectrum is demonstrated most vividly in nature through the formation of a rainbow, often seen by an observer positioned between the Sun and a rain shower. To understand how a rainbow is formed, consider Figure 22.19 (page 786). A ray of light passing overhead strikes a drop of water in the atmosphere and is refracted and reflected as follows: It is first refracted at the front surface of the drop, with the violet light deviating the most and the red light

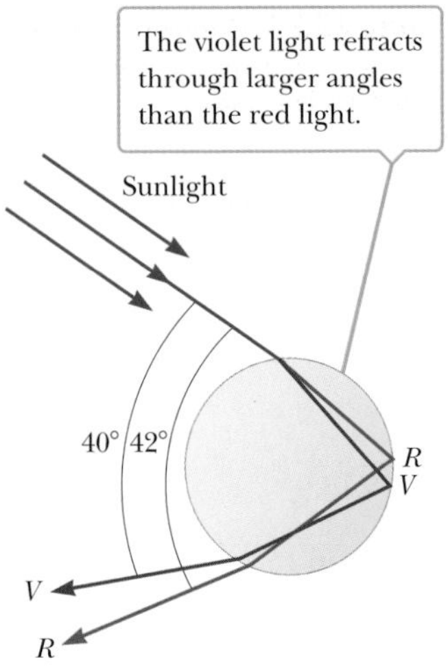

Figure 22.19 Refraction of sunlight by a spherical raindrop.

the least. At the back surface of the drop, the light is reflected and returns to the front surface, where it again undergoes refraction as it moves from water into air. The rays leave the drop so that the angle between the incident white light and the returning violet ray is 40° and the angle between the white light and the returning red ray is 42°. This small angular difference between the returning rays causes us to see the bow as explained in the next paragraph.

Now consider an observer viewing a rainbow, as in Figure 22.20a. If a raindrop high in the sky is being observed, the red light returning from the drop can reach the observer because it is deviated the most, but the violet light passes over the observer because it is deviated the least. Hence, the observer sees this drop as being red. Similarly, a drop lower in the sky would direct violet light toward the observer and appear to be violet. (The red light from this drop would strike the ground and not be seen.) The remaining colors of the spectrum would reach the observer from raindrops lying between these two extreme positions. Figure 22.20b shows a beautiful rainbow and a secondary rainbow with its colors reversed.

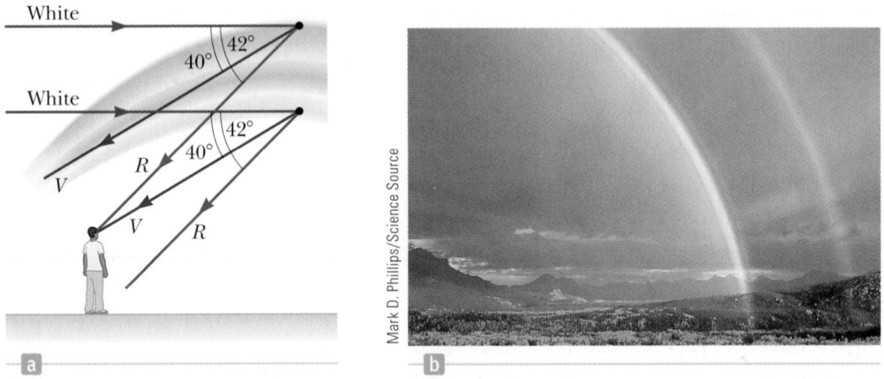

Figure 22.20 The formation of a rainbow seen by an observer standing with the Sun behind his back. (b) This photograph of a rainbow shows a distinct secondary rainbow with the colors reversed.

22.6 Huygens' Principle

LEARNING OBJECTIVE

1. State Huygens' principle, and apply it to the understanding of reflection and refraction.

The laws of reflection and refraction can be deduced using a geometric method proposed by Huygens in 1678. Huygens assumed light is a form of wave motion rather than a stream of particles. He had no knowledge of the nature of light or of its electromagnetic character. Nevertheless, his simplified wave model is adequate for understanding many practical aspects of the propagation of light.

Huygens' principle is a geometric construction for determining at some instant the position of a new wave front from knowledge of the wave front that preceded it. (A wave front is a surface passing through those points of a wave which have the same phase and amplitude. For instance, a wave front could be a surface passing through the crests of waves.) In Huygens' construction, **all points on a given wave front are taken as point sources for the production of spherical secondary waves, called wavelets, that propagate in the forward direction with speeds characteristic of waves in that medium. After some time has elapsed, the new position of the wave front is the surface tangent to the wavelets.**

Huygens' principle ▶

Figure 22.21 illustrates two simple examples of Huygens' construction. First, consider a plane wave moving through free space, as in Figure 22.21a. At $t = 0$, the wave front is indicated by the plane labeled AA'. In Huygens' construction, each point on this wave front is considered a point source. For clarity, only a few points

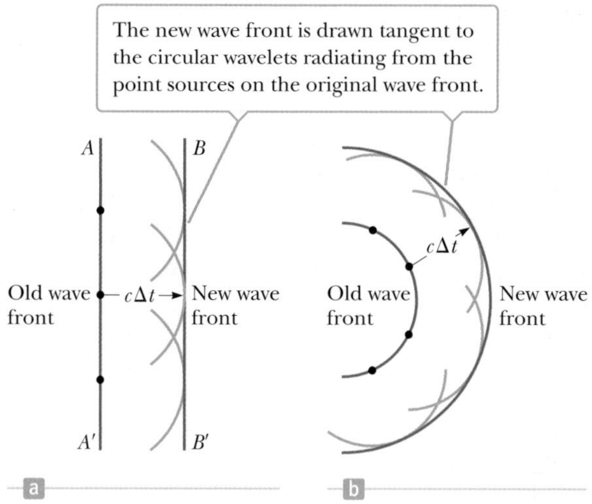

Figure 22.21 Huygens' constructions for (a) a plane wave propagating to the right and (b) a spherical wave.

on AA' are shown. With these points as sources for the wavelets, we draw circles of radius $c\,\Delta t$, where c is the speed of light in vacuum and Δt is the period of propagation from one wave front to the next. The surface drawn tangent to these wavelets is the plane BB', which is parallel to AA'. In a similar manner, Figure 22.21b shows Huygens' construction for an outgoing spherical wave.

Huygens' Principle Applied to Reflection and Refraction

The laws of reflection and refraction were stated earlier in the chapter without proof. We now derive these laws using Huygens' principle. Figure 22.22a illustrates the law of reflection. The line AA' represents a wave front of the incident light. As ray 3 travels from A' to C, ray 1 reflects from A and produces a spherical wavelet of radius AD. (Recall that the radius of a Huygens wavelet is $v\,\Delta t$.) Because the two wavelets having radii $A'C$ and AD are in the same medium, they have the same speed v, so $AD = A'C$. Meanwhile, the spherical wavelet centered at B has spread only half as far as the one centered at A because ray 2 strikes the surface later than ray 1.

From Huygens' principle, we find that the reflected wave front is CD, a line tangent to all the outgoing spherical wavelets. The remainder of our analysis depends on geometry, as summarized in Figure 22.22b. Note that the right triangles ADC and $AA'C$ are congruent because they have the same hypotenuse, AC, and because $AD = A'C$. From the figure, we have

$$\sin\theta_1 = \frac{A'C}{AC} \qquad \text{and} \qquad \sin\theta_1' = \frac{AD}{AC}$$

The right-hand sides are equal, so $\sin\theta_1 = \sin\theta_1'$, and it follows that $\theta_1 = \theta_1'$, which is the law of reflection.

Huygens' principle and Figure 22.23a (page 788) can be used to derive Snell's law of refraction. In the time interval Δt, ray 1 moves from A to B and ray 2 moves from A' to C. The radius of the outgoing spherical wavelet centered at A is equal

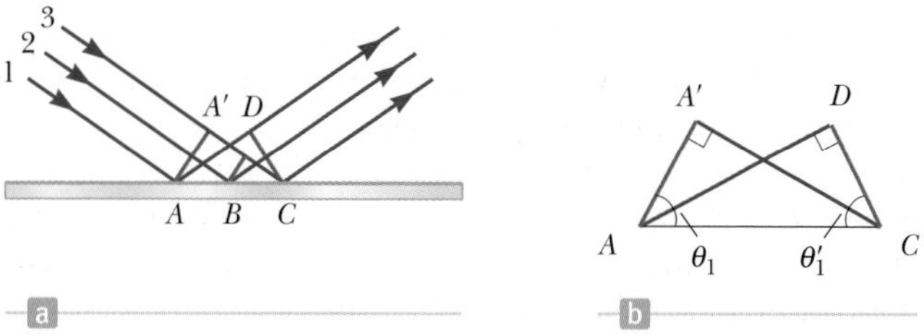

Figure 22.22 (a) Huygens' construction for proving the law of reflection. (b) Triangle ADC is congruent to triangle $AA'C$.

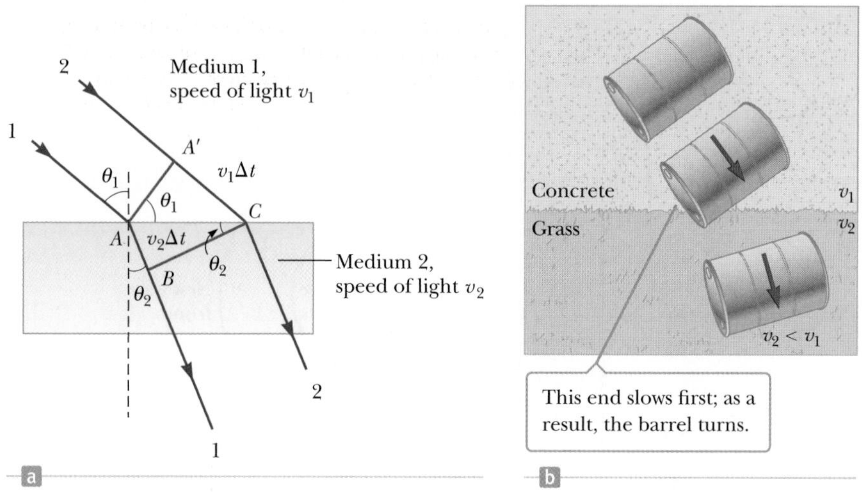

Figure 22.23 (a) Huygens' construction for proving the law of refraction. (b) Overhead view of a barrel rolling from concrete onto grass.

to $v_2\,\Delta t$. The distance $A'C$ is equal to $v_1\,\Delta t$. Geometric considerations show that angle $A'AC$ equals θ_1 and angle ACB equals θ_2. From triangles $AA'C$ and ACB, we find that

$$\sin\theta_1 = \frac{v_1\Delta t}{AC} \quad \text{and} \quad \sin\theta_2 = \frac{v_2\Delta t}{AC}$$

If we divide the first equation by the second, we get

$$\frac{\sin\theta_1}{\sin\theta_2} = \frac{v_1}{v_2}$$

From Equation 22.4, though, we know that $v_1 = c/n_1$ and $v_2 = c/n_2$. Therefore,

$$\frac{\sin\theta_1}{\sin\theta_2} = \frac{c/n_1}{c/n_2} = \frac{n_2}{n_1}$$

and it follows that

$$n_1 \sin\theta_1 = n_2 \sin\theta_2$$

which is the law of refraction.

A mechanical analog of refraction is shown in Figure 22.23b. When the left end of the rolling barrel reaches the grass, it slows down, while the right end remains on the concrete and moves at its original speed. This difference in speeds causes the barrel to pivot, changing the direction of its motion.

22.7 Total Internal Reflection

LEARNING OBJECTIVE

1. Define total internal reflection and apply it in basic physical contexts.

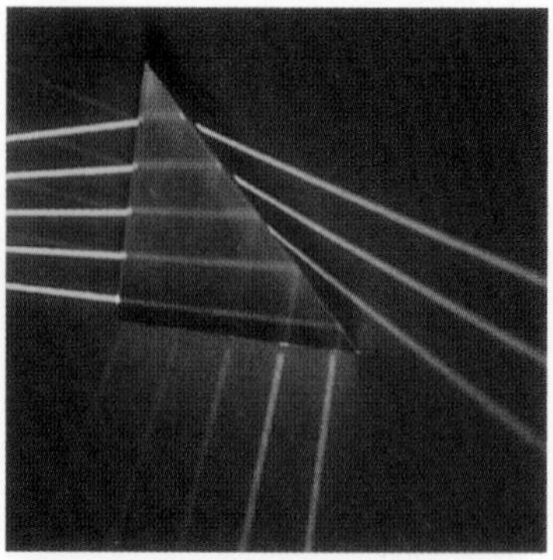

Courtesy of Henry Leap and Jim Lehman

This photograph shows nonparallel light rays entering a glass prism. The bottom two rays undergo total internal reflection at the longest side of the prism. The top three rays are refracted at the longest side as they leave the prism.

An interesting effect called *total internal reflection* can occur when light encounters the boundary between a medium with a *higher* index of refraction and one with a *lower* index of refraction. Consider a light beam traveling in medium 1 and meeting the boundary between medium 1 and medium 2, where n_1 is greater than n_2 (Fig. 22.24). Possible directions of the beam are indicated by rays 1 through 5. Note that the refracted rays are bent away from the normal because n_1 is greater than n_2. At some particular angle of incidence θ_c, called the **critical angle**, the refracted light ray moves parallel to the boundary so that $\theta_2 = 90°$ (Fig. 22.24b). *For angles of incidence greater than* θ_c, the beam is entirely reflected at the boundary, as is ray 5 in Figure 22.24a. This ray is reflected as though it had struck a perfectly reflecting surface. It and all rays like it obey the law of reflection: the angle of incidence equals the angle of reflection.

22.7 Total Internal Reflection

When light propagating in a medium with index of refraction n_1 is incident on the boundary of a region with index of refraction n_2, and $n_1 > n_2$, then total internal reflection can occur if the angle of incidence equals or exceeds a **critical angle** θ_c given by

$$\sin\theta_c = \frac{n_2}{n_1} \qquad (n_1 > n_2) \qquad [22.9]$$

Total internal reflection is used in the optical fibers that carry data at high speed around the world.

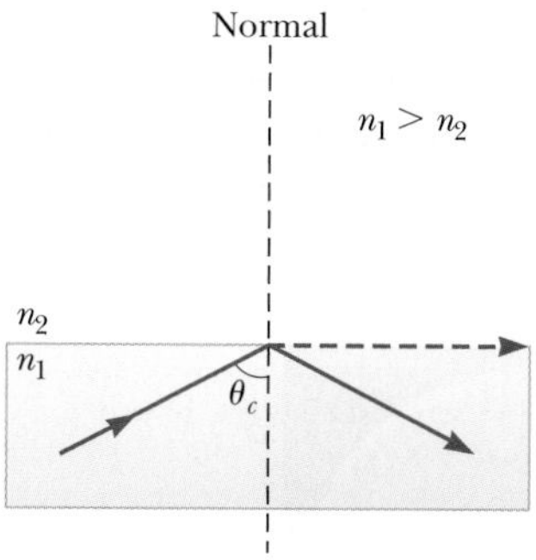

A ray undergoes total internal reflection at angles equal to or greater than the critical angle.

WARM-UP EXERCISES

ENHANCED WebAssign The warm-up exercises in this chapter may be assigned online in Enhanced WebAssign.

1. How many 800-nm photons does it take to have the same total energy as four 200-nm photons? (See Section 22.1.)
2. Two mirrors are set up at an angle of 90.0°. If light is incident on the first mirror at an angle of 30.0° from the normal, at what angle is it incident on the second mirror? (See Section 22.2.)
3. A ray of light in air is incident at an angle of 30.0° on a glass slide with index of refraction 1.73. (a) At what angle is the ray refracted? (b) If the wavelength of the light in vacuum is 564 nm, find its wavelength in the glass. (See Section 22.3.)
4. A source emits monochromatic light of wavelength 495 nm in air. When the light passes through a liquid, its wavelength reduces to 434 nm. (a) What is the liquid's index of refraction? (b) Find the speed of light in the liquid. (See Section 22.3.)
5. Carbon disulfide (n = 1.63) is poured into a container made of crown glass (n = 1.52). What is the critical angle for total internal reflection of a light ray in the liquid when it is incident on the liquid-to-glass surface? (See Section 22.7.)

CONCEPTUAL QUESTIONS

ENHANCED WebAssign The conceptual questions in this chapter may be assigned online in Enhanced WebAssign.

1. Why does the arc of a rainbow appear with red on top and violet on the bottom?
2. A ray of light is moving from a material having a high index of refraction into a material with a lower index of refraction. (a) Is the ray bent toward the normal or away from it? (b) If the wavelength is 600 nm in the material with the high index of refraction, is it greater, smaller, or the same in the material with the lower index of refraction? (c) How does the frequency change as the light moves between the two materials? Does it increase, decrease, or remain the same?
3. A light ray travels through three parallel slabs having different indices of refraction as in Figure CQ22.3. The rays shown are only the refracted rays. Rank the materials according to the size of their indices of refraction, from largest to smallest.

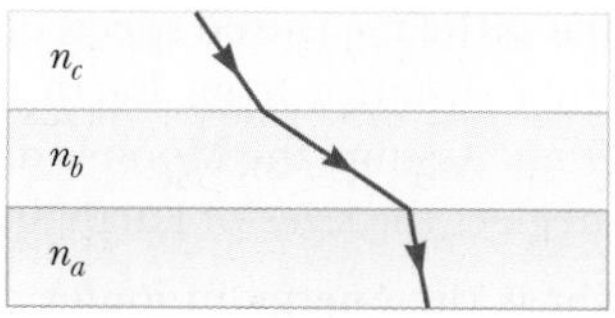

Figure CQ22.3

4. Under what conditions is a mirage formed? On a hot day, what are we seeing when we observe a mirage of a water puddle on the road?
5. Explain why a diamond loses most of its sparkle when submerged in carbon disulfide.
6. A type of mirage called a *pingo* is often observed in Alaska. Pingos occur when the light from a small hill passes to an observer by a path that takes the light over a body of water warmer than the air. What is seen is the hill and an inverted image directly below it. Explain how these mirages are formed.
7. In dispersive materials, the angle of refraction for a light ray depends on the wavelength of the light. Does the angle of reflection from the surface of the material depend on the wavelength? Why or why not?
8. The level of water in a clear, colorless glass can easily be observed with the naked eye. The level of liquid helium in a clear glass vessel is extremely difficult to see with the naked eye. Explain. *Hint:* The index of refraction of liquid helium is close to that of air.
9. Suppose you are told that only two colors of light (X and Y) are sent through a glass prism and that X is bent more than Y. Which color travels more slowly in the prism?

10. Is it possible to have total internal reflection for light incident from air on water? Explain.

11. Figure CQ22.11 shows a pencil partially immersed in a cup of water. Why does the pencil appear to be bent?

Figure CQ22.11

12. Try this simple experiment on your own. Take two opaque cups, place a coin at the bottom of each cup near the edge, and fill one cup with water. Next, view the cups at some angle from the side so that the coin in water is just visible as shown on the left in Figure CQ22.12. Notice that the coin in air is not visible as shown on the right in Figure CQ22.12. Explain this observation.

Figure CQ22.12

13. Why do astronomers looking at distant galaxies talk about looking backward in time?

14. Light can travel from air into water. Some possible paths for the light ray in the water are shown in Figure CQ22.14. Which path will the light most likely follow? (a) *A* (b) *B* (c) *C* (d) *D* (e) *E*

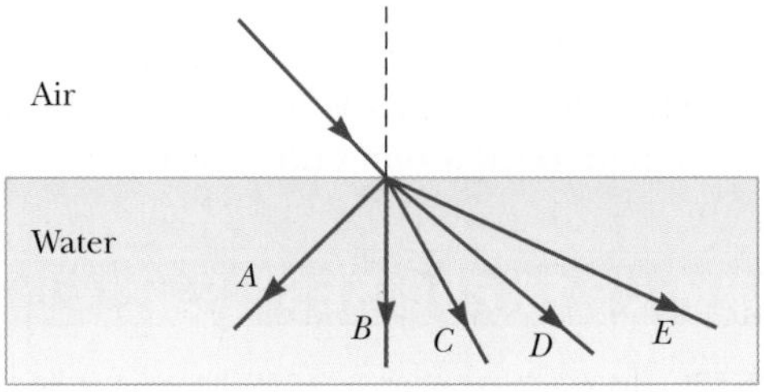

Figure CQ22.14

15. A light ray containing both blue and red wavelengths is incident at an angle on a slab of glass. Which of the sketches in Figure CQ22.15 represents the most likely outcome? (a) *A* (b) *B* (c) *C* (d) *D* (e) none of these

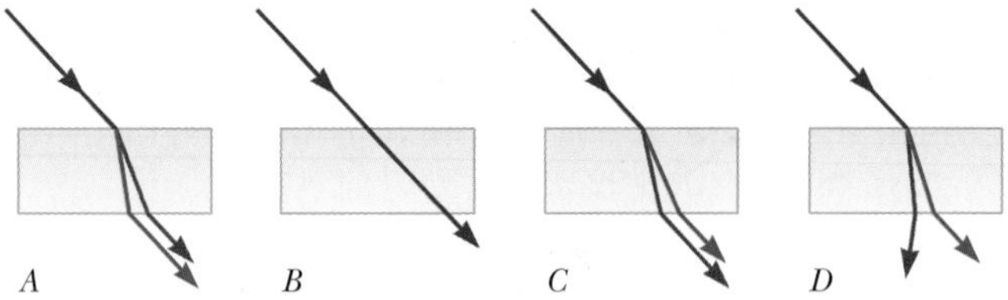

Figure CQ22.15

PROBLEMS

ENHANCED WebAssign The problems in this chapter may be assigned online in Enhanced WebAssign.

1. denotes straightforward problem; **2.** denotes intermediate problem;

3. denotes challenging problem

1. denotes full solution available in *Student Solutions Manual/Study Guide*

1. denotes problems most often assigned in Enhanced WebAssign

BIO denotes biomedical problems

GP denotes guided problems

M denotes Master It tutorial available in Enhanced WebAssign

Q|C denotes asking for quantitative and conceptual reasoning

S denotes symbolic reasoning problem

W denotes Watch It video solution available in Enhanced WebAssign

22.1 The Nature of Light

1. During the Apollo XI Moon landing, a retroreflecting panel was erected on the Moon's surface. The speed of light can be found by measuring the time it takes a laser beam to travel from Earth, reflect from the panel, and return to Earth. If this interval is found to be 2.51 s, what is the measured speed of light? Take the center-to-center distance from Earth to the Moon to be 3.84×10^8 m. Assume the Moon is directly overhead and do not neglect the sizes of Earth and the Moon.

2. Q|C (a) What is the energy in joules of an x-ray photon with wavelength 1.00×10^{-10} m? (b) Convert the energy to electron volts. (c) If more penetrating x-rays are desired, should the wavelength be increased or decreased? (d) Should the frequency be increased or decreased?

3. M Find the energy of (a) a photon having a frequency of 5.00×10^{17} Hz and (b) a photon having a wavelength of 3.00×10^{2} nm. Express your answers in units of electron volts, noting that 1 eV = 1.60×10^{-19} J.

4. Q|C (a) Calculate the wavelength of light in vacuum that has a frequency of 5.45×10^{14} Hz. (b) What is its wavelength in benzene? (c) Calculate the energy of one photon of such light in vacuum. Express the answer in electron volts. (d) Does the energy of the photon change when it enters the benzene? Explain.

5. Find the speed of light in (a) water, (b) crown glass, and (c) diamond.

6. S (a) Find a symbolic expression for the wavelength λ of a photon in terms of its energy E, Planck's constant h, and the speed of light c. (b) What does the equation say about the wavelengths of higher-energy photons?

7. A ray of light travels from air into another medium, making an angle of $\theta_1 = 45.0°$ with the normal as in Figure P22.7. Find the angle of refraction θ_2 if the second medium is (a) fused quartz, (b) carbon disulfide, and (c) water.

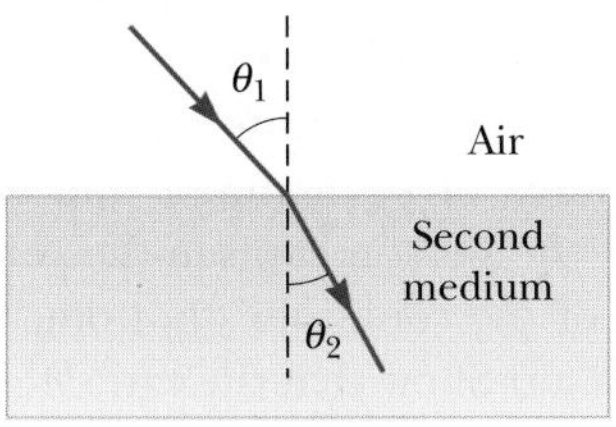

Figure P22.7

22.2 Reflection and Refraction

22.3 The Law of Refraction

8. W The two mirrors in Figure P22.8 meet at a right angle. The beam of light in the vertical plane P strikes mirror 1 as shown. (a) Determine the distance the reflected light beam travels before striking mirror 2. (b) In what direction does the light beam travel after being reflected from mirror 2?

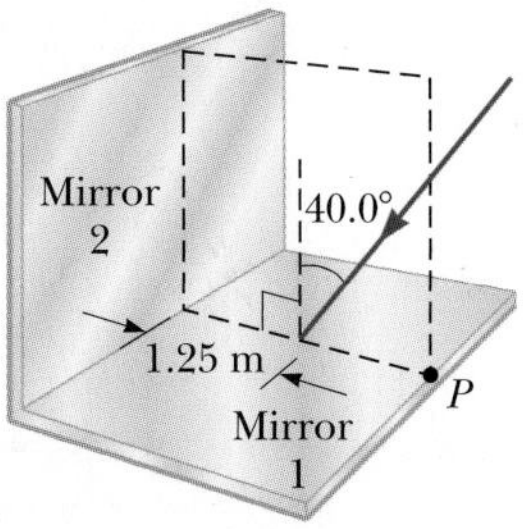

Figure P22.8

9. An underwater scuba diver sees the Sun at an apparent angle of 45.0° from the vertical. What is the actual direction of the Sun?

10. Two plane mirrors are at right angles to each other as shown by the side view in Figure P22.10. A light ray is incident on mirror 1 at an angle θ with the vertical. Using the law of reflection and geometry, show that after the ray is reflected off of both mirrors, the outgoing reflected ray is parallel to the incident ray.

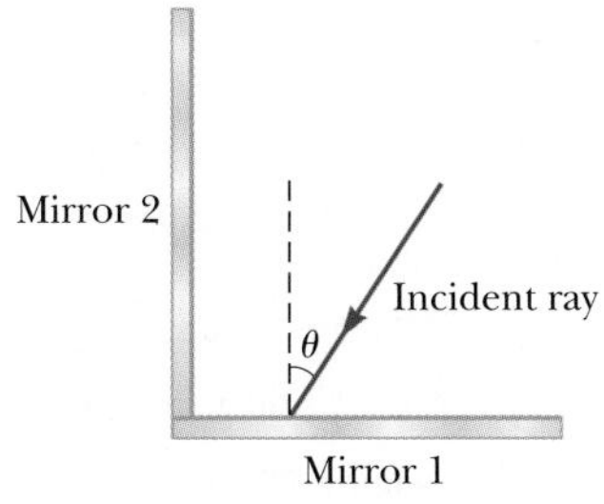

Figure P22.10

11. A laser beam is incident at an angle of 30.0° to the vertical onto a solution of corn syrup in water. If the beam is refracted to 19.24° to the vertical, (a) what is the index of refraction of the syrup solution? Suppose the light is red, with wavelength 632.8 nm in a vacuum. Find its (b) wavelength, (c) frequency, and (d) speed in the solution.

12. Light containing wavelengths of 400 nm, 500 nm, and 650 nm is incident from air on a block of crown glass at an angle of 25.0°. (a) Are all colors refracted alike, or is one color bent more than the others? (b) Calculate the angle of refraction in each case to verify your answer.

13. A ray of light is incident on the surface of a block of clear ice at an angle of 40.0° with the normal. Part of the light is reflected, and part is refracted. Find the angle between the reflected and refracted light.

14. Two plane mirrors are at an angle of $\theta_1 = 50.0°$ with each other as in the side view shown in Figure P22.14. If a horizontal ray is incident on mirror 1, at what angle θ_2 does the outgoing reflected ray make with the surface of mirror 2?

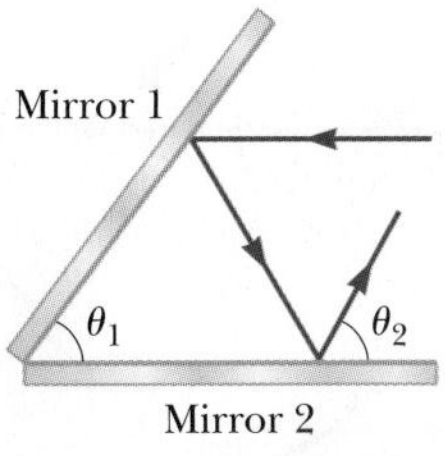

Figure P22.14

15. The light emitted by a helium–neon laser has a wavelength of 632.8 nm in air. As the light travels from air into zircon, find its (a) speed, (b) wavelength, and (c) frequency, all in the zircon.

16. Figure P22.16 shows a light ray traveling in a slab of crown glass surrounded by air. The ray is incident on the right surface at an angle of 55° with the normal and then reflects from points A, B, and C. (a) At which of these points does part of the ray enter the air? (b) If the glass slab is surrounded by carbon disulfide, at which point does part of the ray enter the carbon disulfide?

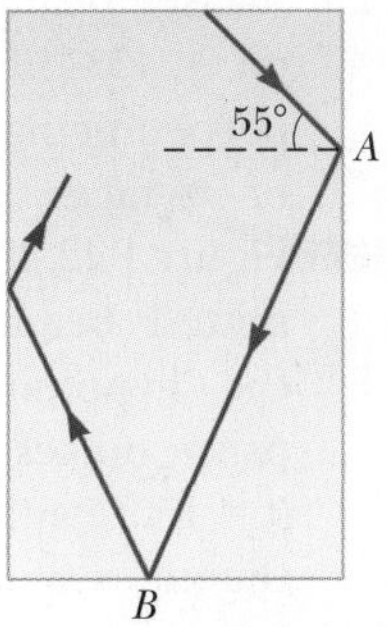

Figure P22.16

17. How many times will the incident beam shown in Figure P22.17 be reflected by each of the parallel mirrors?

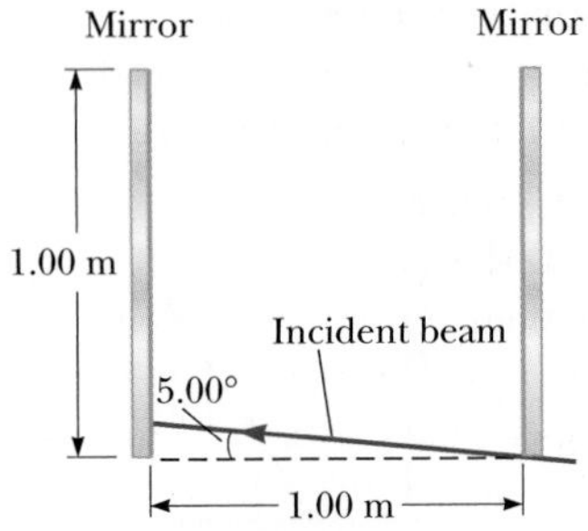

Figure P22.17

18. Q|C A ray of light strikes a flat, 2.00-cm-thick block of glass ($n = 1.50$) at an angle of 30.0° with respect to the normal (Fig. P22.18). (a) Find the angle of refraction at the top surface. (b) Find the angle of incidence at the bottom surface and the refracted angle. (c) Find the lateral distance d by which the light beam is shifted. (d) Calculate the speed of light in the glass and (e) the time required for the light to pass through the glass block. (f) Is the travel time through the block affected by the angle of incidence? Explain.

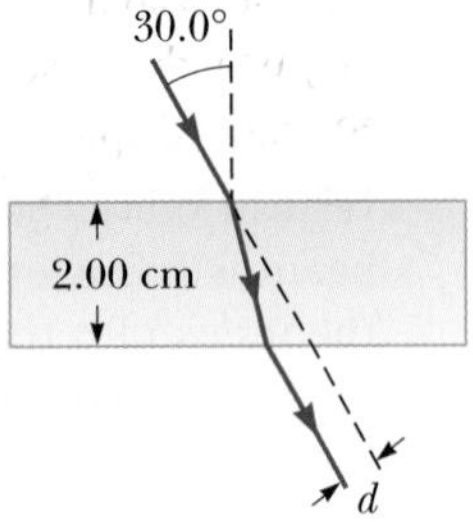

Figure P22.18

19. M The light beam shown in Figure P22.19 makes an angle of 20.0° with the normal line NN' in the linseed oil. Determine the angles θ and θ'. (The refractive index for linseed oil is 1.48.)

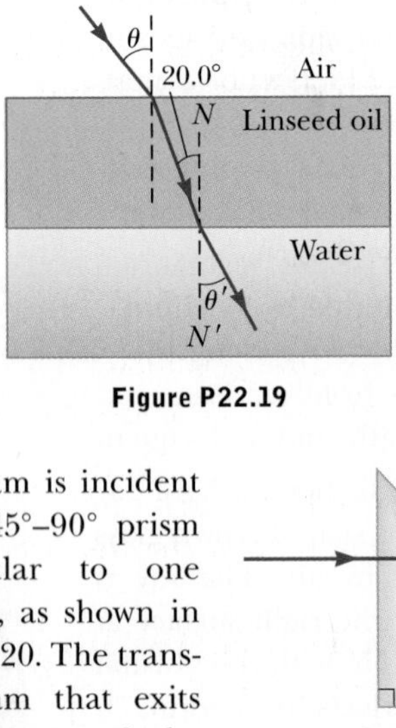

Figure P22.19

20. A laser beam is incident on a 45°–45°–90° prism perpendicular to one of its faces, as shown in Figure P22.20. The transmitted beam that exits the hypotenuse of the prism makes an angle of $\theta = 15.0°$ with the direction of the incident beam. Find the index of refraction of the prism.

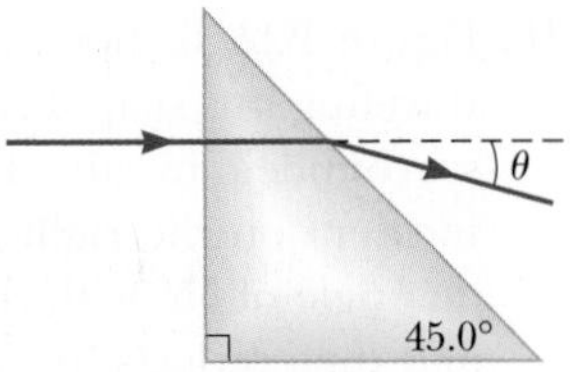

Figure P22.20

21. A block of crown glass is immersed in water as in Figure P22.21. A light ray is incident on the top face at an angle of $\theta_1 = 42.0°$ with the normal and exits the block at point P. (a) Find the vertical distance y from the top of the block to P. (b) Find the angle of refraction θ_2 of the light ray leaving the block at P.

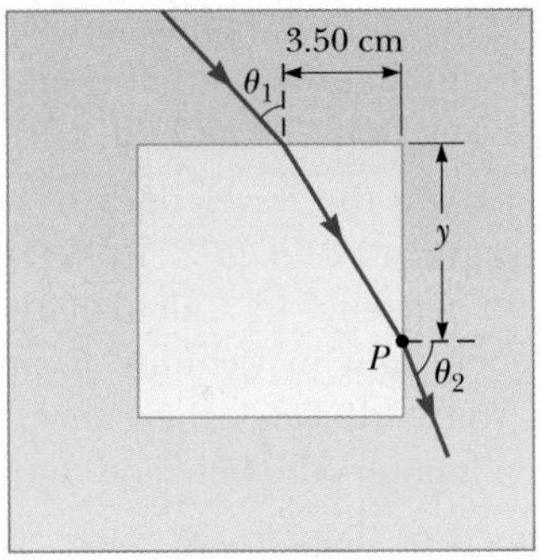

Figure P22.21

22. W BIO A narrow beam of ultrasonic waves reflects off the liver tumor in Figure P22.22. If the speed of the wave is 10.0% less in the liver than in the surrounding medium, determine the depth of the tumor.

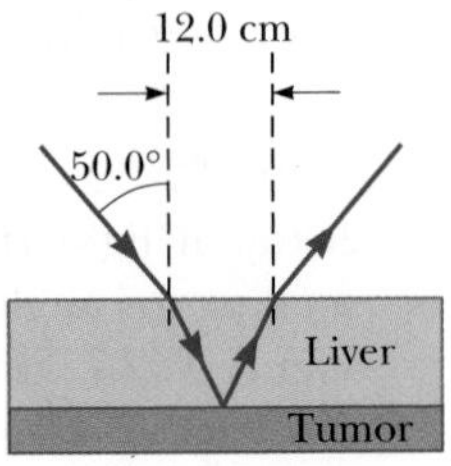

Figure P22.22

23. S A person looking into an empty container is able to see the far edge of the container's bottom, as shown in Figure P22.23a. The height of the container is h, and its width is d. When the container is completely filled with a fluid of index of refraction n and viewed from the same angle, the person can see the center of a coin at the middle of the container's bottom, as shown in Figure P22.23b. (a) Show that the ratio h/d is given by

$$\frac{h}{d} = \sqrt{\frac{n^2 - 1}{4 - n^2}}$$

(b) Assuming the container has a width of 8.00 cm and is filled with water, use the expression above to find the height of the container.

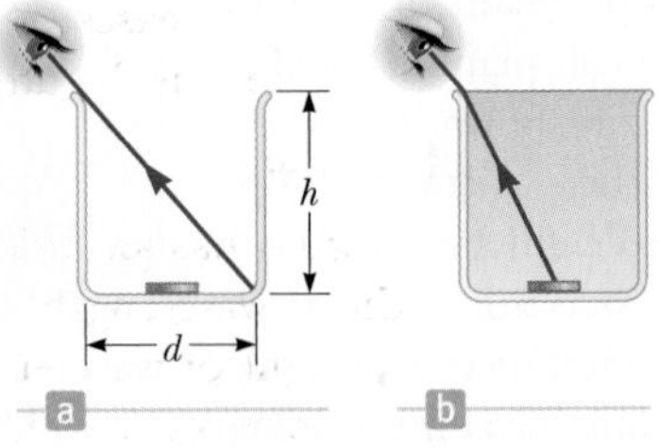

Figure P22.23

24. GP A submarine is 3.00×10^2 m horizontally from shore and 1.00×10^2 m beneath the surface of the water. A laser beam is sent from the submarine so that the beam strikes the surface of the water 2.10×10^2 m from the shore. A building stands on the shore, and

the laser beam hits a target at the top of the building. The goal is to find the height of the target above sea level. (a) Draw a diagram of the situation, identifying the two triangles that are important to finding the solution. (b) Find the angle of incidence of the beam striking the water–air interface. (c) Find the angle of refraction. (d) What angle does the refracted beam make with respect to the horizontal? (e) Find the height of the target above sea level.

25. **S** A beam of light both reflects and refracts at the surface between air and glass, as shown in Figure P22.25. If the index of refraction of the glass is n_g, find the angle of incidence, θ_1, in the air that would result in the reflected ray and the refracted ray being perpendicular to each other. *Hint:* Remember the identity $\sin(90° - \theta) = \cos\theta$.

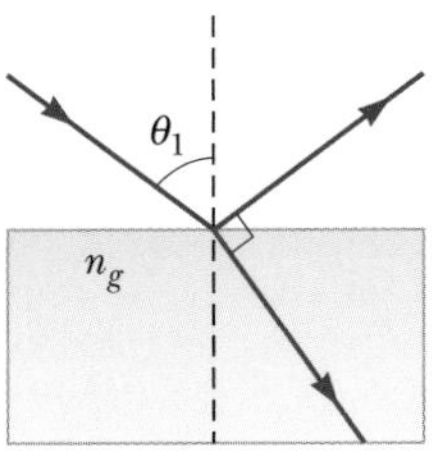

Figure P22.25

26. Figure P22.26 shows a light ray incident on a series of slabs having different refractive indices, where $n_1 < n_2 < n_3 < n_4$. Notice that the path of the ray steadily bends toward the normal. If the variation in n were continuous, the path would form a smooth curve. Use this idea and a ray diagram to explain why you can see the Sun at sunset after it has fallen below the horizon.

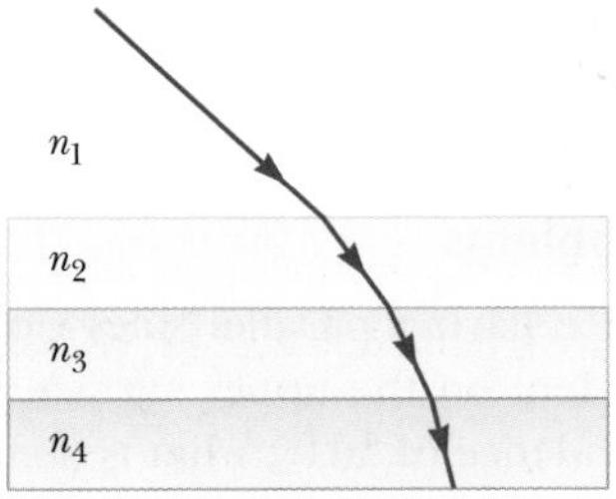

Figure P22.26

27. **M** An opaque cylindrical tank with an open top has a diameter of 3.00 m and is completely filled with water. When the afternoon Sun reaches an angle of 28.0° above the horizon, sunlight ceases to illuminate the bottom of the tank. How deep is the tank?

22.4 Dispersion and Prisms

28. **W** A certain kind of glass has an index of refraction of 1.650 for blue light of wavelength 430 nm and an index of 1.615 for red light of wavelength 680 nm. If a beam containing these two colors is incident at an angle of 30.00° on a piece of this glass, what is the angle between the two beams inside the glass?

29. The index of refraction for red light in water is 1.331 and that for blue light is 1.340. If a ray of white light enters the water at an angle of incidence of 83.00°, what are the underwater angles of refraction for the (a) blue and (b) red components of the light?

30. The index of refraction for crown glass is 1.512 at a wavelength of 660 nm (red), whereas its index of refraction is 1.530 at a wavelength of 410 nm (violet). If both wavelengths are incident on a slab of crown glass at the same angle of incidence, 60.0°, what is the angle of refraction for each wavelength?

31. A light beam containing red and violet wavelengths is incident on a slab of quartz at an angle of incidence of 50.00°. The index of refraction of quartz is 1.455 at 660 nm (red light), and its index of refraction is 1.468 at 410 nm (violet light). Find the dispersion of the slab, which is defined as the difference in the angles of refraction for the two wavelengths.

32. The index of refraction for violet light in silica flint glass is 1.66 and that for red light is 1.62. What is the angular dispersion of visible light passing through an equilateral prism of apex angle 60.0° if the angle of incidence is 50.0°? (See Fig. P22.32.)

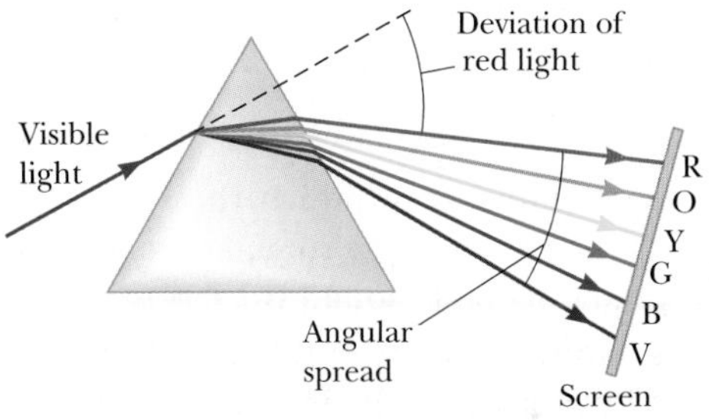

Figure P22.32

33. **M** A ray of light strikes the midpoint of one face of an equiangular (60°–60°–60°) glass prism ($n = 1.5$) at an angle of incidence of 30°. (a) Trace the path of the light ray through the glass and find the angles of incidence and refraction at each surface. (b) If a small fraction of light is also reflected at each surface, what are the angles of reflection at the surfaces?

22.7 Total Internal Reflection

34. For light of wavelength 589 nm, calculate the critical angles for the following substances when surrounded by air: (a) fused quartz, (b) polystyrene, and (c) sodium chloride.

35. Repeat Problem 34, but this time assume the quartz, polystyrene, and sodium chloride are surrounded by water.

36. **M** A beam of light is incident from air on the surface of a liquid. If the angle of incidence is 30.0° and the angle of refraction is 22.0°, find the critical angle for the liquid when surrounded by air.

37. A plastic light pipe has an index of refraction of 1.53. For total internal reflection, what is the minimum angle of incidence if the pipe is in (a) air and (b) water?

38. Determine the maximum angle θ for which the light rays incident on the end of the light pipe in Figure P22.38 are subject to total internal reflection along the walls of the pipe. Assume the light pipe has an index of refraction of 1.36 and the outside medium is air.

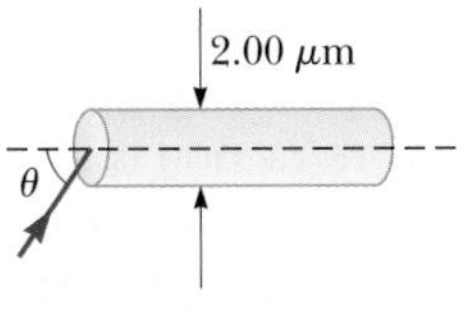

Figure P22.38

39. A light ray is incident normally to the long face (the hypotenuse) of a 45°–45°–90° prism surrounded by air, as shown in Figure 22.26b. Calculate the minimum index of refraction of the prism for which the ray will totally internally reflect at each of the two sides making the right angle.

40. QIC A beam of laser light with wavelength 612 nm is directed through a slab of glass having index of refraction 1.78. (a) For what minimum incident angle would a ray of light undergo total internal reflection? (b) If a layer of water is placed over the glass, what is the minimum angle of incidence on the glass–water interface that will result in total internal reflection at the water–air interface? (c) Does the thickness of the water layer or glass affect the result? (d) Does the index of refraction of the intervening layer affect the result?

41. A room contains air in which the speed of sound is 343 m/s. The walls of the room are made of concrete, in which the speed of sound is 1 850 m/s. (a) Find the critical angle for total internal reflection of sound at the concrete–air boundary. (b) In which medium must the sound be traveling in order to undergo total internal reflection? (c) "A bare concrete wall is a highly efficient mirror for sound." Give evidence for or against this statement.

42. GP QIC Consider a light ray traveling between air and a diamond cut in the shape shown in Figure P22.42. (a) Find the critical angle for total internal reflection for light in the diamond incident on the interface between the diamond and the outside air. (b) Consider the light ray incident normally on the top surface of the diamond as shown in Figure P22.42. Show that the light traveling toward point *P* in the diamond is totally reflected. (c) If the diamond is immersed in water, find the critical angle at the diamond–water interface. (d) When the diamond is immersed in water, does the light ray entering the top surface in Figure P22.42 undergo total internal reflection at *P*? Explain. (e) If the light ray entering the diamond remains vertical as shown in Figure P22.42, which way should the diamond in the water be rotated about an axis perpendicular to the page through *O* so that light will exit the diamond at *P*? (f) At what angle of rotation in part (e) will light first exit the diamond at point *P*?

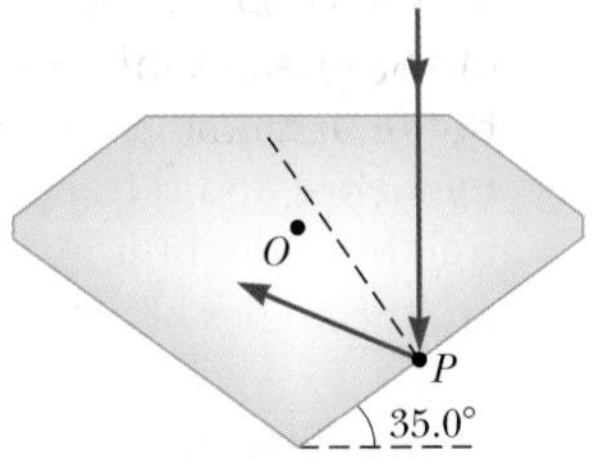

Figure P22.42

43. The light beam in Figure P22.43 strikes surface 2 at the critical angle. Determine the angle of incidence, θ_1.

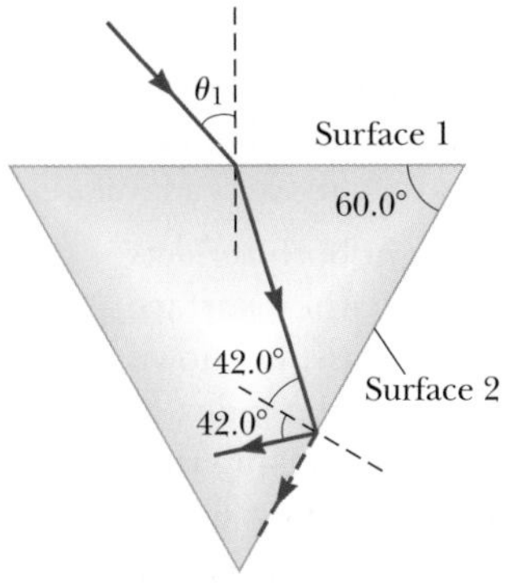

Figure P22.43

44. W A jewel thief hides a diamond by placing it on the bottom of a public swimming pool. He places a circular raft on the surface of the water directly above and centered over the diamond, as shown in Figure P22.44. If the surface of the water is calm and the pool is 2.00 m deep, find the minimum diameter of the raft that would prevent the diamond from being seen.

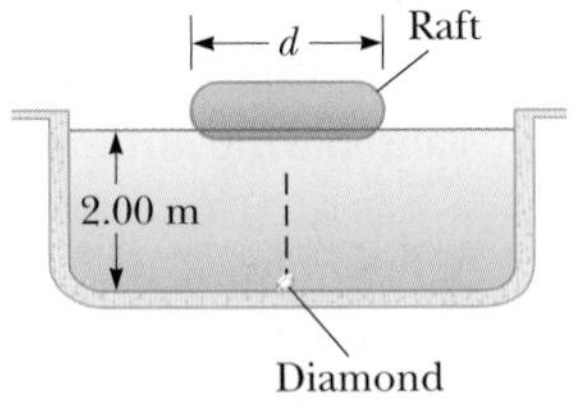

Figure P22.44

Additional Problems

45. A layer of ice having parallel sides floats on water. If light is incident on the upper surface of the ice at an angle of incidence of 30.0°, what is the angle of refraction in the water?

46. QIC A ray of light is incident at an angle 30.0° on a plane slab of flint glass surrounded by water. (a) Find the refraction angle. (b) Suppose the index of refraction of the surrounding medium can be adjusted, but the incident angle of the light remains the same. As the index of refraction of the medium approaches that of the glass, what happens to the refraction angle? (c) What happens to the refraction angle when the medium's index of refraction exceeds that of the glass?

47. When a man stands near the edge of an empty drainage ditch of depth 2.80 m, he can barely see the boundary between the opposite wall and bottom of the ditch as in Figure P22.47a. The distance from his eyes to the ground is 1.85 m. (a) What is the horizontal distance *d* from the man to the edge of the drainage ditch? (b) After the drainage ditch is filled with water as in Figure P22.47b, what is the maximum distance *x* the man can stand from the edge and still see the same boundary?

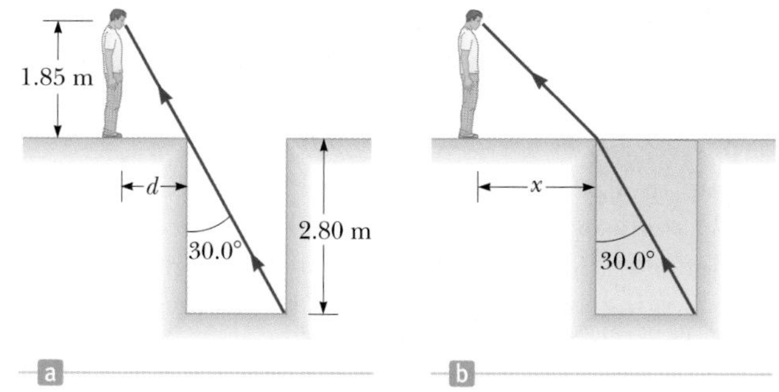

Figure P22.47

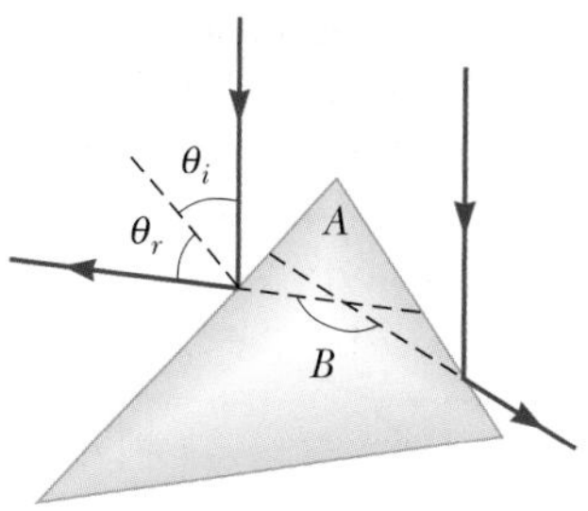

Figure P22.51

48. A light ray of wavelength 589 nm is incident at an angle θ on the top surface of a block of polystyrene surrounded by air, as shown in Figure P22.48. (a) Find the maximum value of θ for which the refracted ray will undergo total internal reflection at the left vertical face of the block. (b) Repeat the calculation for the case in which the polystyrene block is immersed in water. (c) What happens if the block is immersed in carbon disulfide?

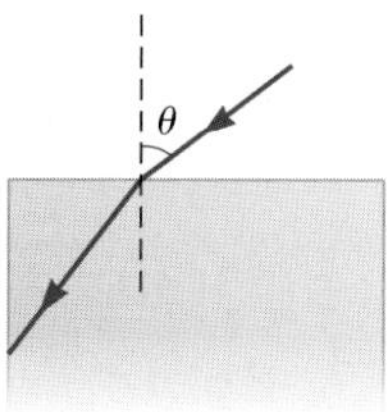

Figure P22.48

49. As shown in Figure P22.49, a light ray is incident normal to one face of a 30°–60°–90° block of flint glass (a prism) that is immersed in water. (a) Determine the exit angle θ_3 of the ray. (b) A substance is dissolved in the water to increase the index of refraction n_2. At what value of n_2 does total internal reflection cease at point P?

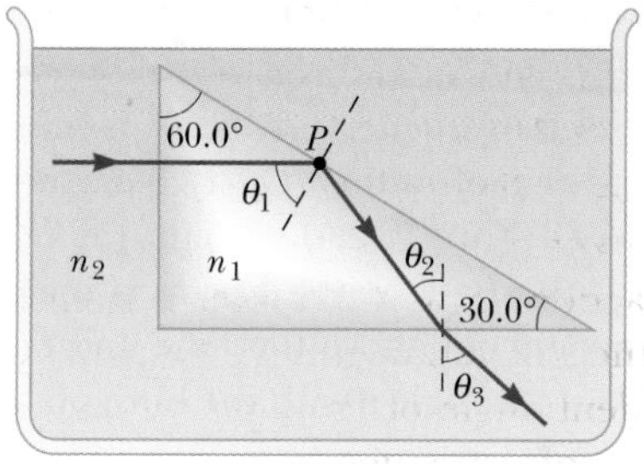

Figure P22.49

50. A narrow beam of light is incident from air onto a glass surface with index of refraction 1.56. Find the angle of incidence for which the corresponding angle of refraction is one-half the angle of incidence. *Hint:* You might want to use the trigonometric identity $\sin 2\theta = 2 \sin \theta \cos \theta$.

51. One technique for measuring the angle of a prism is shown in Figure P22.51. A parallel beam of light is directed onto the apex of the prism so that the beam reflects from opposite faces of the prism. Show that the angular separation of the two reflected beams is given by $B = 2A$.

52. An optical fiber with index of refraction n and diameter d is surrounded by air. Light is sent into the fiber along its axis, as shown in Figure P22.52. (a) Find the smallest outside radius R permitted for a bend in the fiber if no light is to escape. (b) Does the result for part (a) predict reasonable behavior as d approaches zero? As n increases? As n approaches unity? (c) Evaluate R, assuming the diameter of the fiber is 100 μm and its index of refraction is 1.40.

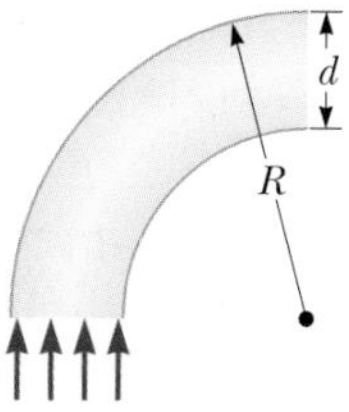

Figure P22.52

53. A piece of wire is bent through an angle θ. The bent wire is partially submerged in benzene (index of refraction = 1.50) so that, to a person looking along the dry part, the wire appears to be straight and makes an angle of 30.0° with the horizontal. Determine the value of θ.

54. A light ray traveling in air is incident on one face of a right-angle prism with index of refraction $n = 1.50$, as shown in Figure P22.54, and the ray follows the path shown in the figure. Assuming $\theta = 60.0°$ and the base of the prism is mirrored, determine the angle ϕ made by the outgoing ray with the normal to the right face of the prism.

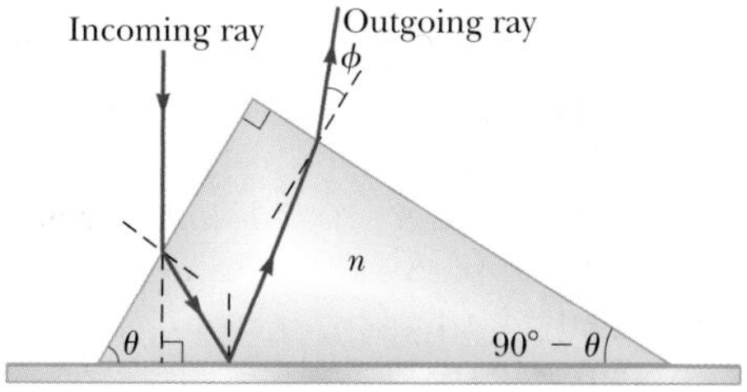

Figure P22.54

55. A transparent cylinder of radius $R = 2.00$ m has a mirrored surface on its right half, as shown in Figure P22.55 (page 800). A light ray traveling in air is incident on the left side of the cylinder. The incident light ray and the exiting light ray are parallel, and $d = 2.00$ m. Determine the index of refraction of the material.

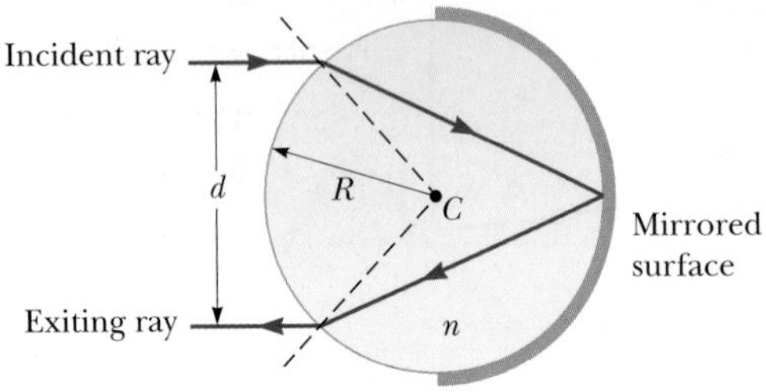

Figure P22.55

56. A laser beam strikes one end of a slab of material, as in Figure P22.56. The index of refraction of the slab is 1.48. Determine the number of internal reflections of the beam before it emerges from the opposite end of the slab.

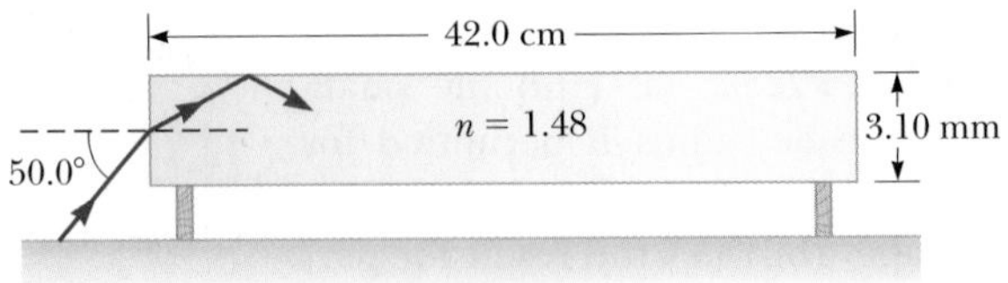

Figure P22.56

57. A light ray enters a rectangular block of plastic at an angle $\theta_1 = 45.0°$ and emerges at an angle $\theta_2 = 76.0°$, as shown in Figure P22.57. (a) Determine the index of refraction of the plastic. (b) If the light ray enters the plastic at a point $L = 50.0$ cm from the bottom edge, how long does it take the light ray to travel through the plastic?

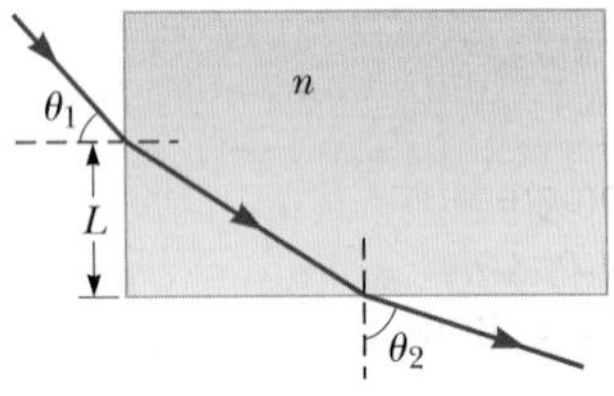

Figure P22.57

58. Students allow a narrow beam of laser light to strike a water surface. They arrange to measure the angle of refraction for selected angles of incidence and record the data shown in the following table:

Angle of Incidence (degrees)	Angle of Refraction (degrees)
10.0	7.5
20.0	15.1
30.0	22.3
40.0	28.7
50.0	35.2
60.0	40.3
70.0	45.3
80.0	47.7

Use the data to verify Snell's law of refraction by plotting the sine of the angle of incidence versus the sine of the angle of refraction. From the resulting plot, deduce the index of refraction of water.

59. Figure P22.59 shows the path of a beam of light through several layers with different indices of refraction. (a) If $\theta_1 = 30.0°$, what is the angle θ_2 of the emerging beam? (b) What must the incident angle θ_1 be to have total internal reflection at the surface between the medium with $n = 1.20$ and the medium with $n = 1.00$?

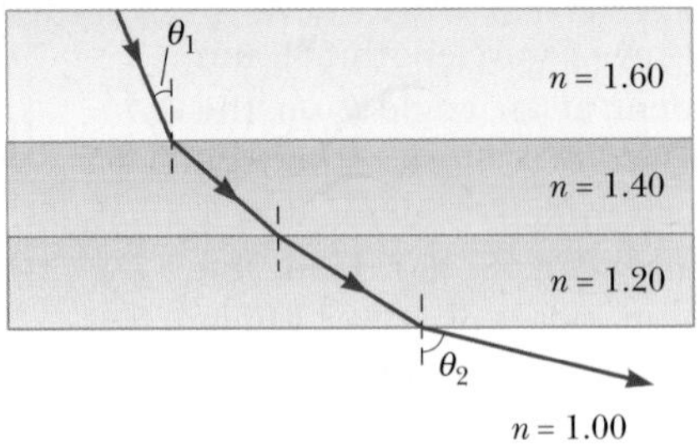

Figure P22.59

60. Three sheets of plastic have unknown indices of refraction. Sheet 1 is placed on top of sheet 2, and a laser beam is directed onto the sheets from above so that it strikes the interface at an angle of 26.5° with the normal. The refracted beam in sheet 2 makes an angle of 31.7° with the normal. The experiment is repeated with sheet 3 on top of sheet 2, and with the same angle of incidence, the refracted beam makes an angle of 36.7° with the normal. If the experiment is repeated again with sheet 1 on top of sheet 3, what is the expected angle of refraction in sheet 3? Assume the same angle of incidence.

61. A thick piece of Lucite ($n = 1.50$) is has the shape of a quarter circle of radius $R = 12.0$ cm as shown in the side view of Figure P22.61. A light ray traveling in air parallel to the base of the Lucite is incident at a distance $h = 6.00$ cm above the base and emerges out of the Lucite at an angle θ with the horizontal. Determine the value of θ.

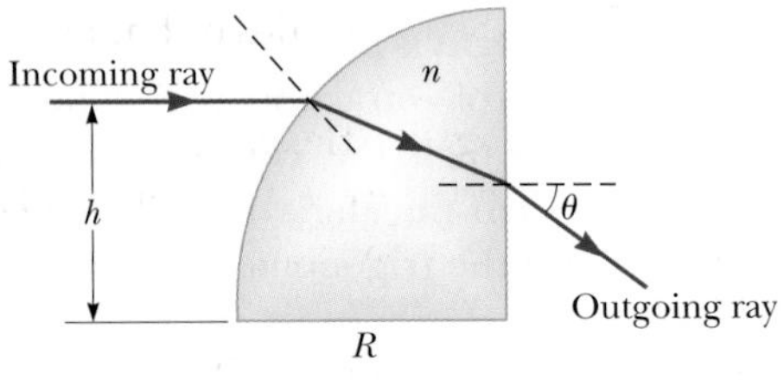

Figure P22.61

Gail Mooney/Masterfile Corporation

Funhouse mirrors distort images because the curved surfaces essentially change the angle of incidence of incoming rays, change that differs depending on the mirror's shape in a given location. In every case, however, the angle of reflection equals the angle of incidence.

Mirrors and Lenses 23

The development of the technology of mirrors and lenses led to a revolution in the progress of science. These devices, relatively simple to construct from cheap materials, led to microscopes and telescopes, extending human sight and opening up new pathways to knowledge, from microbes to distant planets.

This chapter covers the formation of images when plane and spherical light waves fall on plane and spherical surfaces. Images can be formed by reflection from mirrors or by refraction through lenses. In our study of mirrors and lenses, we continue to assume light travels in straight lines (the ray approximation), ignoring diffraction.

23.1 Flat Mirrors

LEARNING OBJECTIVE

1. Discuss and apply the properties of flat mirrors.

We begin by examining the flat mirror. Consider a point source of light placed at O in Figure 23.1, a distance p in front of a flat mirror. The distance p is called the **object distance**. Light rays leave the source and are reflected from the mirror. After reflection, the rays diverge (spread apart), but they appear to the viewer to come from a point I behind the mirror. Point I is called the **image** of the object at O. Regardless of the system under study, **images are formed at the point where rays of light actually intersect or where they appear to originate**. Because the rays in the figure appear to originate at I, which is a distance q behind the mirror, that is the location of the image. The distance q is called the **image distance**.

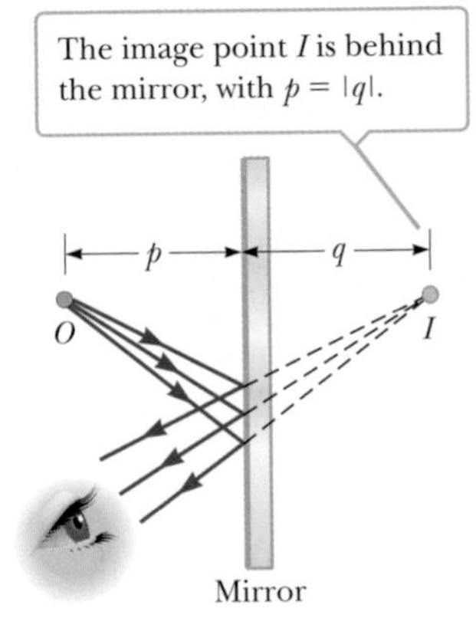

Figure 23.1 An image formed by reflection from a flat mirror. The image at point I is virtual. In Section 23.3, it will be shown that q must be taken as negative for virtual images: the object distance p, therefore, equals the absolute value of the image distance q.

Images are classified as real or virtual. In the formation of a *real image*, light actually passes through the image point. For a *virtual image*, light doesn't pass through the image point, but appears to come (diverge) from there. The image formed by the flat mirror in Figure 23.1 is a virtual image. In fact, the images seen in flat mirrors are always virtual (for real objects). Real images can be displayed on a screen (as at a movie), but virtual images cannot.

We examine some of the properties of the images formed by flat mirrors by using the simple geometric techniques. To find out where an image is formed, it's necessary to follow at least two rays of light as they reflect from the mirror as in Figure 23.2. One of those rays starts at P, follows the horizontal path PQ to the mirror, and reflects back on itself. The second ray follows the oblique path PR and reflects as shown. An observer to the left of the mirror would trace the two reflected rays back to the point from which they appear to have originated: point P'. A continuation of this process for points other than P on the object would result in a virtual image (drawn as a yellow arrow) to the right of the mirror. Because triangles PQR and $P'QR$ are identical, $PQ = P'Q$. Hence, we conclude that **the image formed by an object placed in front of a flat mirror is as far behind the mirror as the object is in front of the mirror.** Geometry also shows that the object height h equals the image height h'. The **lateral magnification** M is defined as

$$M \equiv \frac{\text{image height}}{\text{object height}} = \frac{h'}{h} \qquad [23.1]$$

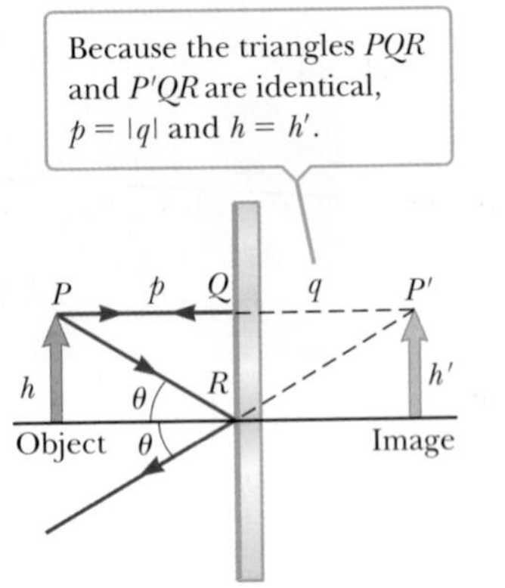

Figure 23.2 A geometric construction to locate the image of an object placed in front of a flat mirror.

Equation 23.1 is a general definition of the lateral magnification of any type of mirror. For a flat mirror, $M = 1$ because $h' = h$.

In summary, the image formed by a flat mirror has the following properties:

1. The image is as far behind the mirror as the object is in front.
2. The image is unmagnified, virtual, and upright. (By *upright*, we mean that if the object arrow points upward, as in Figure 23.2, so does the image arrow. The opposite of an upright image is an inverted image.)

Finally, note that a flat mirror produces an image having an *apparent* left–right reversal. You can see this reversal standing in front of a mirror and raising your right hand. Your image in the mirror raises the left hand. Likewise, your hair appears to be parted on the opposite side, and a mole on your right cheek appears to be on your image's left cheek.

Quick Quiz

23.1 In the overhead view of Figure 23.3, the image of the stone seen by observer 1 is at C. Where does observer 2 see the image: at A, at B, at C, at D, at E, or not at all?

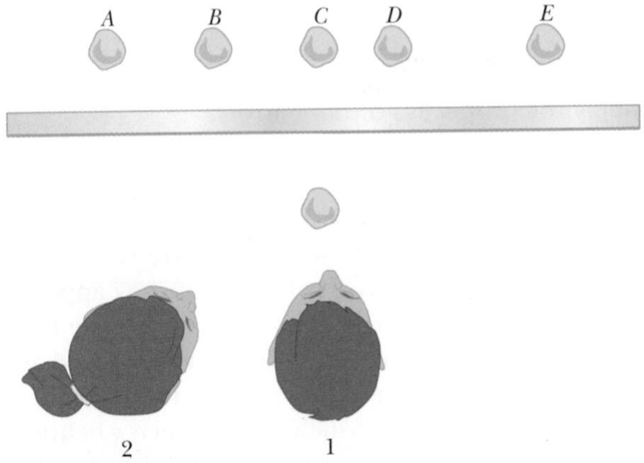

Figure 23.3 (Quick Quiz 23.1)

Tip 23.1 Magnification ≠ Enlargement

Note that the word *magnification*, as used in optics, doesn't always mean *enlargement* because the image could be smaller than the object.

■ EXAMPLE 23.1 "Mirror, Mirror, on the Wall"

GOAL Apply the properties of a flat mirror.

PROBLEM A man 1.80 m tall stands in front of a mirror and sees his full height, no more and no less. If his eyes are 0.14 m from the top of his head, what is the minimum height of the mirror?

STRATEGY Figure 23.4 shows two rays of light, one from the man's feet and the other from the top of his head, reflecting off the mirror and entering his eye. The ray from his feet just strikes the bottom of the mirror, so if the mirror were longer, it would be too long, and if shorter, the ray would not be reflected. The angle of incidence and the angle of reflection are equal, labeled θ. This means the two triangles, *ABD* and *DBC*, are identical because they are right triangles with a common side (*DB*) and two identical angles θ. Use this key fact and the small isosceles triangle *FEC* to solve the problem.

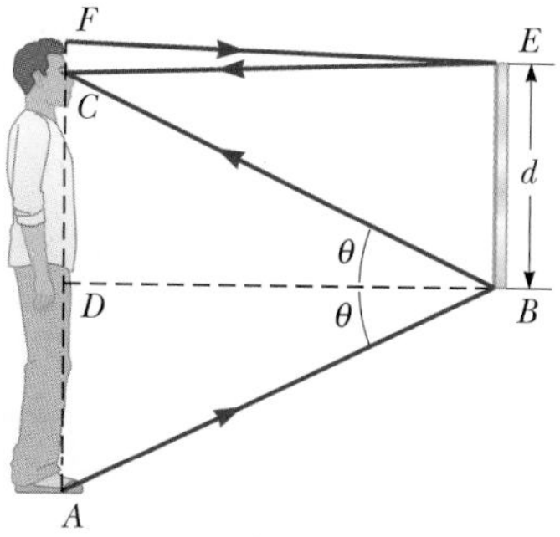

Figure 23.4 (Example 23.1)

SOLUTION

We need to find *BE*, which equals *d*. Relate this length to lengths on the man's body:

$$\textbf{(1)} \quad BE = DC + \tfrac{1}{2}CF$$

We need the lengths *DC* and *CF*. Set the sum of sides opposite the identical angles θ equal to *AC*:

$$\textbf{(2)} \quad AD + DC = AC = (1.80 - 0.14) = 1.66 \text{ m}$$

$AD = DC$, so substitute into Equation (2) and solve for *DC*:

$$AD + DC = 2DC = 1.66 \text{ m} \quad \rightarrow \quad DC = 0.83 \text{ m}$$

CF is given as 0.14 m. Substitute this value and *DC* into Equation (1):

$$BE = d = DC + \tfrac{1}{2}CF = 0.83 \text{ m} + \tfrac{1}{2}(0.14 \text{ m}) = \boxed{0.90 \text{ m}}$$

REMARKS The mirror must be exactly equal to half the height of the man for him to see only his full height and nothing more or less. Notice that the answer doesn't depend on his distance from the mirror.

QUESTION 23.1 Would a taller man be able to see his full height in the same mirror?

EXERCISE 23.1 How large should the mirror be if he wants to see only the upper third of his full height?

ANSWER 0.30 m

APPLICATION
Day and Night Settings for Rearview Mirrors

Most rearview mirrors in cars have a day setting and a night setting. The night setting greatly diminishes the intensity of the image so that lights from trailing cars will not blind the driver. To understand how such a mirror works, consider Figure 23.5. The mirror is a wedge of glass with a reflecting metallic coating on the back side. When the mirror is in the day setting, as in Figure 23.5a, light from an object behind the car strikes the mirror at point 1. Most of the light enters the

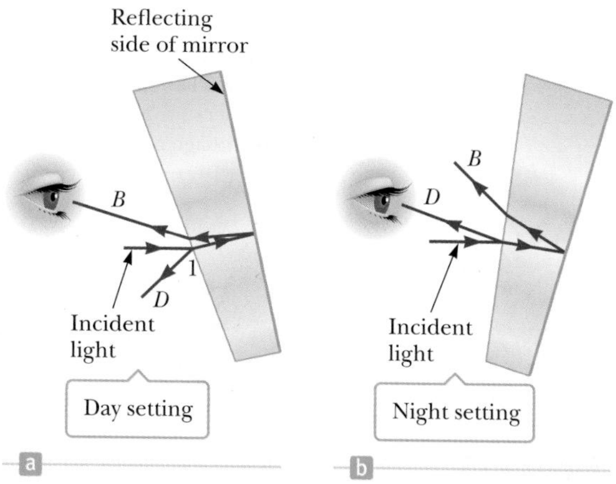

Figure 23.5 Cross-sectional views of a rearview mirror. (a) With the day setting, the silvered back surface of the mirror reflects a bright ray *B* into the driver's eyes. (b) With the night setting, the glass of the unsilvered front surface of the mirror reflects a dim ray *D* into the driver's eyes.

wedge, is refracted, and reflects from the back of the mirror to return to the front surface, where it is refracted again as it reenters the air as ray B (for *bright*). In addition, a small portion of the light is reflected at the front surface, as indicated by ray D (for *dim*). This dim reflected light is responsible for the image observed when the mirror is in the night setting, as in Figure 23.5b. Now the wedge is rotated so that the path followed by the bright light (ray B) doesn't lead to the eye. Instead, the dim light reflected from the front surface travels to the eye, and the brightness of trailing headlights doesn't become a hazard.

APPLYING PHYSICS 23.1 | Illusionist's Trick

The professor in the box shown in Figure 23.6 appears to be balancing himself on a few fingers with both of his feet elevated from the floor. He can maintain this position for a long time, and appears to defy gravity. How do you suppose this illusion was created?

EXPLANATION This trick is an example of an optical illusion, used by magicians, that makes use of a mirror. The box the professor is standing in is a cubical open frame that contains a flat, vertical mirror through a diagonal plane. The professor straddles the mirror so that one leg is in front of the mirror and the other leg is behind it, out of view. When he raises his front leg, that leg's reflection rises also, making it appear both his feet are off the ground, creating the illusion that he's floating in the air. In fact, he supports himself with the leg behind the mirror, which remains in contact with the ground. ■

Courtesy of Henry Leap and Jim Lehman

Figure 23.6 (Applying Physics 23.1)

23.2 Images Formed by Concave Mirrors

LEARNING OBJECTIVES

1. Use ray diagrams to understand relationships between the radius, focal length, object position, and image position for concave mirrors.
2. Define the focal length.
3. State the mirror equation.

A **spherical mirror**, as its name implies, has the shape of a segment of a sphere. Figure 23.7 shows a spherical mirror with a silvered inner, concave surface; this type of mirror is called a **concave mirror**. The mirror has radius of curvature R,

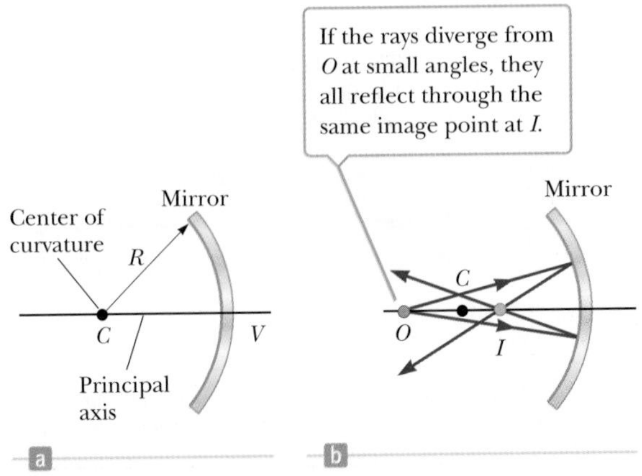

Figure 23.7 (a) A concave mirror of radius R. The center of curvature, C, is located on the principal axis. (b) A point object placed at O in front of a concave spherical mirror of radius R, where O is any point on the principal axis farther than R from the surface of the mirror, forms a real image at I.

and its center of curvature is at point C. Point V is the center of the spherical segment, and a line drawn from C to V is called the **principal axis** of the mirror.

Now consider a point source of light placed at point O in Figure 23.7b, on the principal axis and outside point C. Several diverging rays originating at O are shown. After reflecting from the mirror, these rays converge to meet at I, called the **image point**. The rays then continue and diverge from I as if there were an object there. As a result, a real image is formed. **Whenever reflected light actually passes through a point, the image formed there is real.**

We often assume all rays that diverge from the object make small angles with the principal axis. All such rays reflect through the image point, as in Figure 23.7b. Rays that make a large angle with the principal axis, as in Figure 23.8, converge to other points on the principal axis, producing a blurred image. This effect, called **spherical aberration**, is present to some extent with any spherical mirror and will be discussed in Section 23.7.

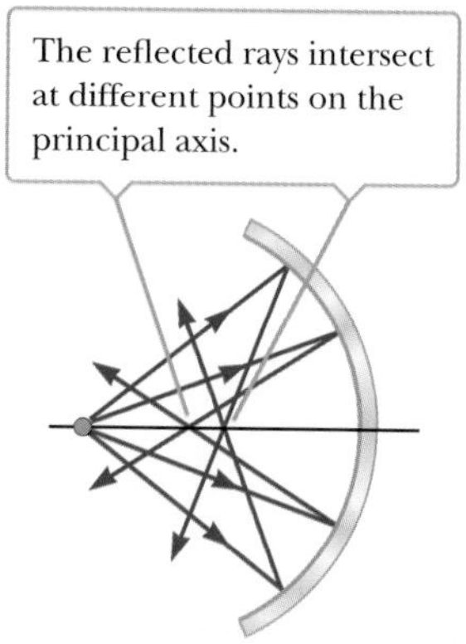

Figure 23.8 A spherical concave mirror exhibits *spherical aberration* when light rays make large angles with the principal axis.

We can use the geometry shown in Figure 23.9 to calculate the image distance q from the object distance p and radius of curvature R. By convention, these distances are measured from point V. The figure shows two rays of light leaving the tip of the object. One ray passes through the center of curvature, C, of the mirror, hitting the mirror head-on (perpendicular to the mirror surface) and reflecting back on itself. The second ray strikes the mirror at point V and reflects as shown, obeying the law of reflection. The image of the tip of the arrow is at the point where the two rays intersect. From the largest triangle in Figure 23.9, we see that $\tan\theta = h/p$; the light-blue triangle gives $\tan\theta = -h'/q$. The negative sign has been introduced to satisfy our convention that h' is negative when the image is inverted with respect to the object, as it is here. From Equation 23.1 and these results, we find that the magnification of the mirror is

$$M = \frac{h'}{h} = -\frac{q}{p} \qquad [23.2]$$

From two other triangles in the figure, we get

$$\tan\alpha = \frac{h}{p - R} \quad \text{and} \quad \tan\alpha = -\frac{h'}{R - q}$$

from which we find that

$$\frac{h'}{h} = -\frac{R - q}{p - R} \qquad [23.3]$$

If we compare Equation 23.2 with Equation 23.3, we see that

$$\frac{R - q}{p - R} = \frac{q}{p}$$

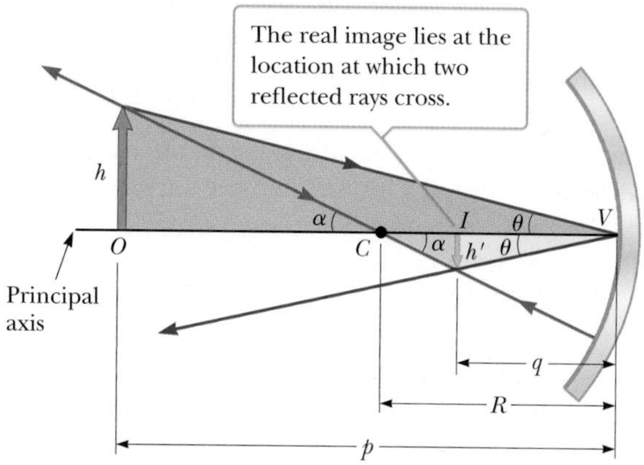

Figure 23.9 The image formed by a spherical concave mirror, where the object at O lies outside the center of curvature, C.

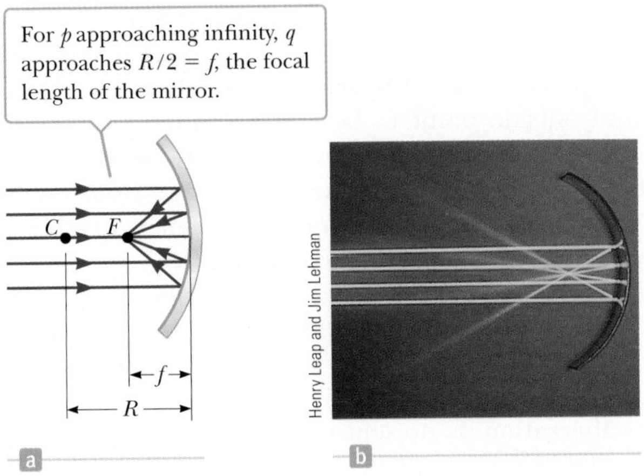

Figure 23.10 (a) Light rays from a distant object ($p = \infty$) reflect from a concave mirror through the focal point F. (b) A photograph of the reflection of parallel rays from a concave mirror.

Simple algebra reduces this equation to

Mirror equation ▶

$$\frac{1}{p} + \frac{1}{q} = \frac{2}{R} \quad [23.4]$$

This expression is called the **mirror equation.**

If the object is very far from the mirror—if the object distance p is great enough compared with R that p can be said to approach infinity—then $1/p \approx 0$, and we see from Equation 23.4 that $q \approx R/2$. In other words, when the object is very far from the mirror, **the image point is halfway between the center of curvature and the center of the mirror**, as in Figure 23.10a. The incoming rays are essentially parallel in that figure because the source is assumed to be very far from the mirror. In this special case we call the image point the **focal point** F and the image distance the **focal length** f, where

Focal length ▶

$$f = \frac{R}{2} \quad [23.5]$$

The mirror equation can therefore be expressed in terms of the focal length:

$$\frac{1}{p} + \frac{1}{q} = \frac{1}{f} \quad [23.6]$$

Note that rays from objects at infinity are always focused at the focal point.

Tip 23.2 Focal Point ≠ Focus Point

The focal point is *not* the point at which light rays focus to form an image. The focal point of a mirror is determined *solely* by its curvature; it doesn't depend on the location of any object.

23.3 Convex Mirrors and Sign Conventions

LEARNING OBJECTIVES

1. Understand the optical properties of convex mirrors.
2. Master the sign conventions for concave and convex mirrors.
3. Analyze concave and convex mirrors, finding image positions, magnifications, and stating whether images are upright or inverted, real or virtual.

Figure 23.11 shows the formation of an image by a **convex mirror**, which is silvered so that light is reflected from the outer, convex surface. It is sometimes called a **diverging mirror** because the rays from any point on the object diverge after reflection, as though they were coming from some point behind the mirror. The image in Figure 23.11 is virtual rather than real because it lies behind the mirror at the point the reflected rays appear to originate. In general, the image formed by a convex mirror is upright, virtual, and smaller than the object.

We won't derive any equations for convex spherical mirrors. If we did, we would find that the equations developed for concave mirrors can be used with convex

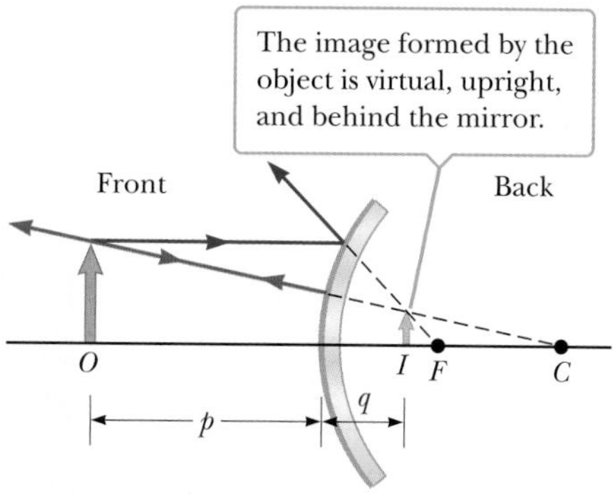

Figure 23.11 Formation of an image by a spherical, convex mirror.

mirrors if particular sign conventions are used. We call the region in which light rays move the *front side* of the mirror, and the other side, where virtual images are formed, the *back side*. For example, in Figures 23.9 and 23.11, the side to the left of the mirror is the front side and the side to the right is the back side. Figure 23.12 is helpful for understanding the rules for object and image distances, and Table 23.1 summarizes the sign conventions for all the necessary quantities. Notice that when the quantities p, q, and f (and R) are located where the light is—in front of the mirror—they are positive, whereas when they are located behind the mirror (where the light isn't), they are negative.

Tip 23.3 Positive Is Where the Light Is

The quantities p, q, and f are all positive when they are located where the light is—in front of the mirror as indicated in Figure 23.12.

Ray Diagrams for Mirrors

We can conveniently determine the positions and sizes of images formed by mirrors by constructing *ray diagrams* similar to the ones we have been using. This kind of graphical construction tells us the overall nature of the image and can be used to check parameters calculated from the mirror and magnification equations. Making a ray diagram requires knowing the position of the object and the location of the center of curvature. To locate the image, three rays are constructed (rather than only the two we have been constructing so far), as shown by the examples in Figure 23.13 (page 808). All three rays start from the same object point; for these examples, the tip of the arrow was chosen. For the concave mirrors in Figures 23.13a and 23.13b, the rays are drawn as follows:

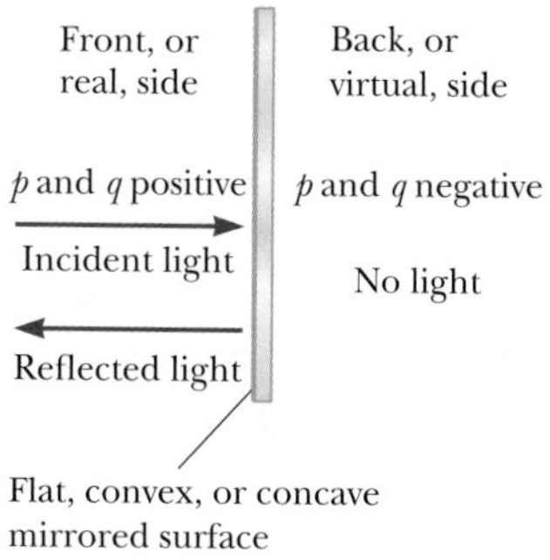

Figure 23.12 A diagram describing the signs of p and q for convex and concave mirrors.

1. Ray 1 is drawn parallel to the principal axis and is reflected back through the focal point F.
2. Ray 2 is drawn through the focal point and is reflected parallel to the principal axis.
3. Ray 3 is drawn through the center of curvature, C, and is reflected back on itself.

Note that rays actually go in all directions from the object; we choose to follow those moving in a direction that simplifies our drawing.

The intersection of any *two* of these rays at a point locates the image. The third ray serves as a check of our construction. The image point obtained in

Table 23.1 Sign Conventions for Mirrors

Quantity	Symbol	In Front	In Back	Upright Image	Inverted Image
Object location	p	+	−		
Image location	q	+	−		
Focal length	f	+	−		
Image height	h'			+	−
Magnification	M			+	−

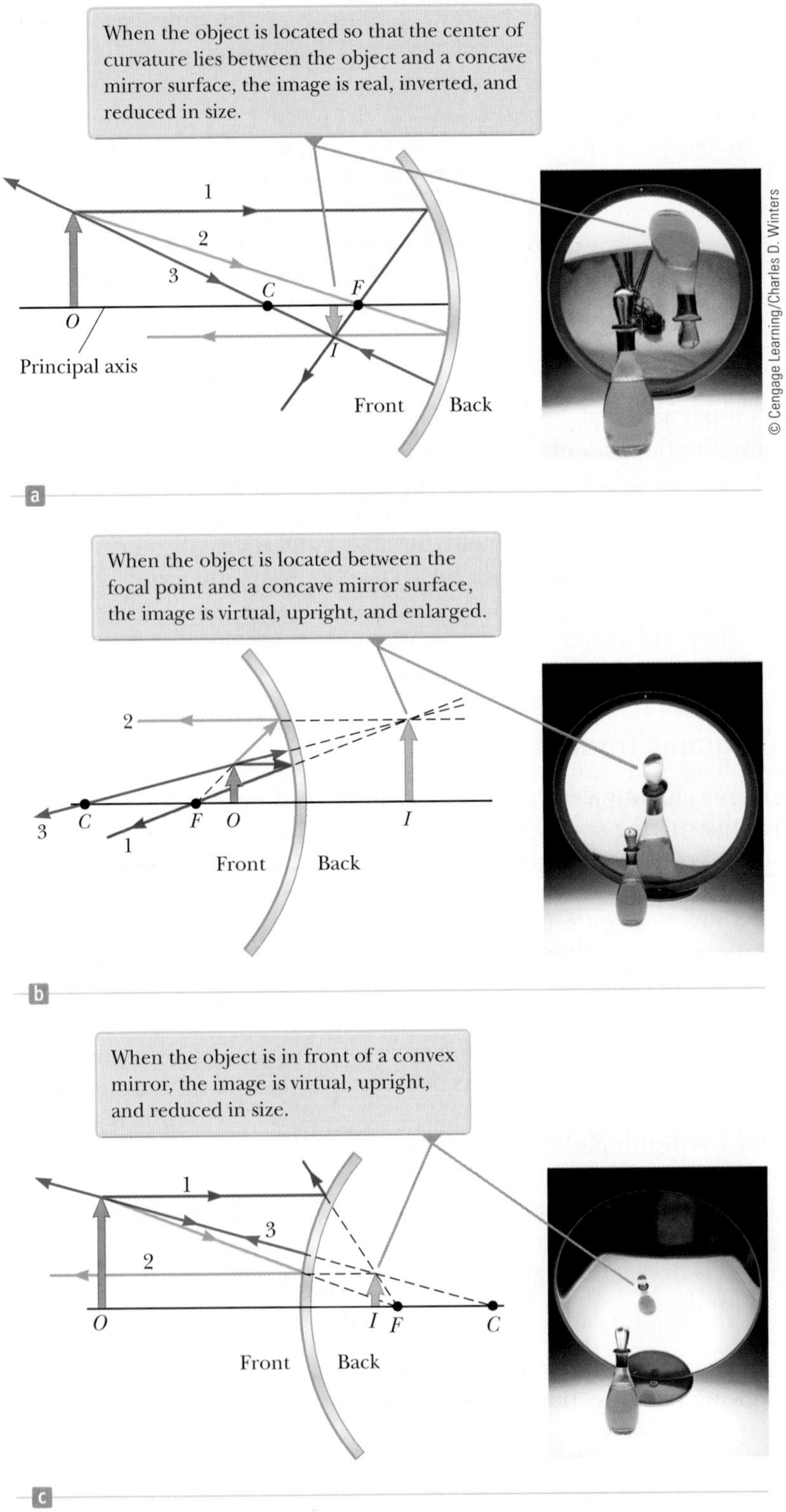

Figure 23.13 Ray diagrams for spherical mirrors and corresponding photographs of the images of bottles.

this fashion must always agree with the value of q calculated from the mirror formula.

In the case of a concave mirror, note what happens as the object is moved closer to the mirror. The real, inverted image in Figure 23.13a moves to the left as the object approaches the focal point. When the object is at the focal point, the image is infinitely far to the left. When the object lies between the focal point and the mirror surface, as in Figure 23.13b, however, the image is virtual and upright.

With the convex mirror shown in Figure 23.13c, the image of a real object is always virtual and upright. As the object distance increases, the virtual image shrinks and approaches the focal point as p approaches infinity. You should construct a ray diagram to verify these statements.

The image-forming characteristics of curved mirrors obviously determine their uses. For example, suppose you want to design a mirror that will help people shave or apply cosmetics. For this, you need a concave mirror that puts the user inside the focal point, such as the mirror in Figure 23.13b. With that mirror, the image is upright and greatly enlarged. In contrast, suppose the primary purpose of a mirror is to observe a large field of view. In that case you need a convex mirror such as the one in Figure 23.13c. The diminished size of the image means that a fairly large field of view is seen in the mirror. Mirrors like this one are often placed in stores to help employees watch for shoplifters. A second use of such a mirror is as a side-view mirror on a car (Fig. 23.14). This kind of mirror is usually placed on the passenger side of the car and carries the warning "Objects are closer than they appear." Without such warning, a driver might think she is looking into a flat mirror, which doesn't alter the size of the image. She could be fooled into believing that a truck is far away because it looks small, when it's actually a large semi very close behind her, but diminished in size because of the image formation characteristics of the convex mirror.

Junebug Clark/Science Source

Figure 23.14 A convex side-view mirror on a vehicle produces an upright image that is smaller than the object. The smaller image means that the object is closer than its apparent distance as observed in the mirror.

APPLYING PHYSICS 23.2 | Concave Versus Convex

A virtual image can be anywhere behind a concave mirror. Why is there a maximum distance at which the image can exist behind a *convex* mirror?

EXPLANATION Consider the concave mirror first and imagine two different light rays leaving a tiny object and striking the mirror. If the object is at the focal point, the light rays reflecting from the mirror will be parallel to the mirror axis. They can be interpreted as forming a virtual image infinitely far away behind the mirror. As the object is brought closer to the mirror, the reflected rays will diverge through larger and larger angles, resulting in their extensions converging closer and closer to the back of the mirror. When the object is brought right up to the mirror, the image is right behind the mirror. When the object is much closer to the mirror than the focal length, the mirrors acts like a flat mirror and the image is just as far behind the mirror as the object is in front of it. The image can therefore be anywhere from infinitely far away to right at the surface of the mirror. For the convex mirror, an object at infinity produces a virtual image at the focal point. As the object is brought closer, the reflected rays diverge more sharply and the image moves closer to the mirror. As a result, the virtual image is restricted to the region between the mirror and the focal point. ■

APPLYING PHYSICS 23.3 | Reversible Waves

Large trucks often have a sign on the back saying, "If you can't see my mirror, I can't see you." Explain this sign.

EXPLANATION The trucking companies are making use of the principle of the reversibility of light rays. For an image of you to be formed in the driver's mirror, there must be a pathway for rays of light to reach the mirror, allowing the driver to see your image. If you can't see the mirror, this pathway doesn't exist. ■

EXAMPLE 23.2 | Images Formed by a Concave Mirror

GOAL Calculate properties of a concave mirror.

PROBLEM Assume a certain concave, spherical mirror has a focal length of 10.0 cm. **(a)** Locate the image and find the magnification for an object distance of 25.0 cm. Determine whether the image is real or virtual, inverted or upright, and larger or smaller. Do the same for object distances of **(b)** 10.0 cm and **(c)** 5.00 cm.

STRATEGY For each part, substitute into the mirror and magnification equations. Part (b) involves a limiting process because the answers are infinite. Notice that when the magnification M is positive, the image is upright, and when M is negative, the image is inverted. Similarly, when q is positive the image is real, and when q is negative the image is virtual.

(Continued)

SOLUTION

(a) Find the image position for an object distance of 25.0 cm. Calculate the magnification and describe the image.

Use the mirror equation to find the image distance:

$$\frac{1}{p} + \frac{1}{q} = \frac{1}{f}$$

Substitute and solve for q. According to Table 23.1, p and f are positive.

$$\frac{1}{25.0 \text{ cm}} + \frac{1}{q} = \frac{1}{10.0 \text{ cm}}$$

$$q = \boxed{16.7 \text{ cm}}$$

Because q is positive, the image is in front of the mirror and is real. The magnification is given by substituting into Equation 23.2:

$$M = -\frac{q}{p} = -\frac{16.7 \text{ cm}}{25.0 \text{ cm}} = \boxed{-0.668}$$

The image is smaller than the object because $|M| < 1$, and it is inverted because M is negative. (See Fig. 23.13a.)

(b) Locate the image when the object distance is 10.0 cm. Calculate the magnification and describe the image.

The object is at the focal point. Substitute $p = 10.0$ cm and $f = 10.0$ cm into the mirror equation:

$$\frac{1}{10.0 \text{ cm}} + \frac{1}{q} = \frac{1}{10.0 \text{ cm}}$$

$$\frac{1}{q} = 0 \quad \rightarrow \quad \boxed{q = \infty}$$

Because $M = -q/p$, the magnification is also infinite.

(c) Locate the image when the object distance is 5.00 cm. Calculate the magnification and describe the image.

Once again, substitute into the mirror equation:

$$\frac{1}{5.00 \text{ cm}} + \frac{1}{q} = \frac{1}{10.0 \text{ cm}}$$

$$\frac{1}{q} = \frac{1}{10.0 \text{ cm}} - \frac{1}{5.00 \text{ cm}} = -\frac{1}{10.0 \text{ cm}}$$

$$q = \boxed{-10.0 \text{ cm}}$$

The image is virtual (behind the mirror) because q is negative. Use Equation 23.2 to calculate the magnification:

$$M = -\frac{q}{p} = -\left(\frac{-10.0 \text{ cm}}{5.00 \text{ cm}}\right) = \boxed{2.00}$$

The image is larger (magnified by a factor of 2) because $|M| > 1$, and upright because M is positive. (See Fig. 23.13b.)

REMARKS Note the characteristics of an image formed by a concave, spherical mirror. When the object is outside the focal point, the image is inverted and real; at the focal point, the image is formed at infinity; inside the focal point, the image is upright and virtual.

QUESTION 23.2 What location does the image approach as the object gets arbitrarily far away from the mirror? (a) infinity (b) the focal point (c) the radius of curvature of the mirror (d) the mirror itself

EXERCISE 23.2 If the object distance is 20.0 cm, find the image distance and the magnification of the mirror.

ANSWER $q = 20.0$ cm, $M = -1.00$

EXAMPLE 23.3 | Images Formed by a Convex Mirror

GOAL Calculate properties of a convex mirror.

PROBLEM An object 3.00 cm high is placed 20.0 cm from a convex mirror with a focal length of magnitude 8.00 cm. Find **(a)** the position of the image, **(b)** the magnification of the mirror, and **(c)** the height of the image.

STRATEGY This problem again requires only substitution into the mirror and magnification equations. Multiplying the object height by the magnification gives the image height.

SOLUTION

(a) Find the position of the image.

Because the mirror is convex, its focal length is negative. Substitute into the mirror equation:

$$\frac{1}{p} + \frac{1}{q} = \frac{1}{f}$$

$$\frac{1}{20.0\text{ cm}} + \frac{1}{q} = \frac{1}{-8.00\text{ cm}}$$

Solve for q:

$$q = -5.71\text{ cm}$$

(b) Find the magnification of the mirror.

Substitute into Equation 23.2:

$$M = -\frac{q}{p} = -\left(\frac{-5.71\text{ cm}}{20.0\text{ cm}}\right) = 0.286$$

(c) Find the height of the image.

Multiply the object height by the magnification:

$$h' = hM = (3.00\text{ cm})(0.286) = 0.858\text{ cm}$$

REMARKS The negative value of q indicates the image is virtual, or behind the mirror, as in Figure 23.13c. The image is upright because M is positive.

QUESTION 23.3 True or False: A convex mirror can produce only virtual images.

EXERCISE 23.3 Suppose the object is moved so that it is 4.00 cm from the same mirror. Repeat parts (a) through (c).

ANSWERS (a) −2.67 cm (b) 0.668 (c) 2.00 cm; the image is upright and virtual.

EXAMPLE 23.4 | The Face in the Mirror

GOAL Find a focal length from a magnification and an object distance.

PROBLEM When a woman stands with her face 40.0 cm from a cosmetic mirror, the upright image is twice as tall as her face. What is the focal length of the mirror?

STRATEGY To find f in this example, we must first find q, the image distance. Because the problem states that the image is upright, the magnification must be positive (in this case, $M = +2$), and because $M = -q/p$, we can determine q.

SOLUTION

Obtain q from the magnification equation:

$$M = -\frac{q}{p} = 2$$

$$q = -2p = -2(40.0\text{ cm}) = -80.0\text{ cm}$$

Because q is negative, the image is on the opposite side of the mirror and hence is virtual. Substitute q and p into the mirror equation and solve for f:

$$\frac{1}{40.0\text{ cm}} - \frac{1}{80.0\text{ cm}} = \frac{1}{f}$$

$$f = 80.0\text{ cm}$$

REMARKS The positive sign for the focal length tells us that the mirror is concave, a fact we already knew because the mirror magnified the object. (A convex mirror would have produced a smaller image.)

(Continued)

QUESTION 23.4 If she moves the mirror closer to her face, what happens to the image? (a) It becomes inverted and smaller. (b) It remains upright and becomes smaller. (c) It becomes inverted and larger. (d) It remains upright and becomes larger.

EXERCISE 23.4 Suppose a fun-house spherical mirror makes you appear to be one-third your normal height. If you are 1.20 m away from the mirror, find its focal length. Is the mirror concave or convex?

ANSWERS -0.600 m, convex

23.4 Images Formed by Refraction

LEARNING OBJECTIVES

1. State the generalized equations relating object distance, image distance, and radius of curvature for images formed by transparent spherical surfaces.
2. Master the sign conventions for refracting surfaces.
3. Analyze the images formed by refracting surfaces.

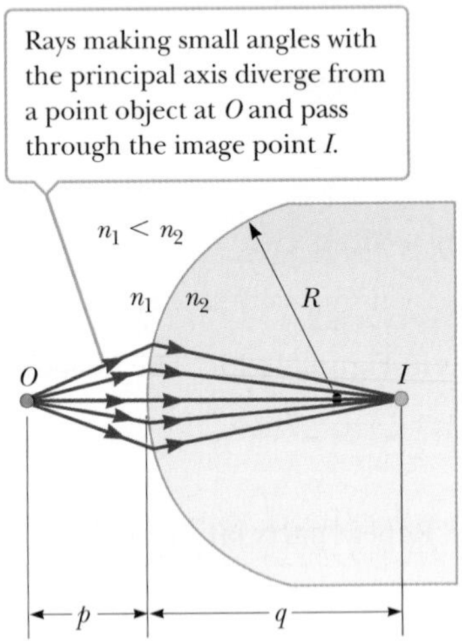

Figure 23.15 An image formed by refraction at a spherical surface.

In this section we describe how images are formed by refraction at a spherical surface. Consider two transparent media with indices of refraction n_1 and n_2, where the boundary between the two media is a spherical surface of radius R (Fig. 23.15). We assume the medium to the right has a higher index of refraction than the one to the left: $n_2 > n_1$. That would be the case for light entering a curved piece of glass from air or for light entering the water in a fishbowl from air. The rays originating at the object location O are refracted at the spherical surface and then converge to the image point I. We can begin with Snell's law of refraction and use simple geometric techniques to show that the object distance, image distance, and radius of curvature are related by the equation

$$\frac{n_1}{p} + \frac{n_2}{q} = \frac{n_2 - n_1}{R} \qquad [23.7]$$

Further, the magnification of a refracting surface is

$$M = \frac{h'}{h} = -\frac{n_1 q}{n_2 p} \qquad [23.8]$$

As with mirrors, certain sign conventions hold, depending on circumstances. First note that real images are formed by refraction on the side of the surface *opposite* the side from which the light comes, in contrast to mirrors, where real images are formed on the *same* side of the reflecting surface. This makes sense because light reflects off mirrors, so any real images must form on the same side the light comes from. With a transparent medium, the rays pass through and naturally form real images on the opposite side. We define the side of the surface where light rays originate as the front side. The other side is called the back side. Because of the difference in location of real images, the refraction sign conventions for q and R are the opposite of those for reflection. For example, p, q, and R are all positive in Figure 23.15. The sign conventions for spherical refracting surfaces are summarized in Table 23.2.

Table 23.2 Sign Conventions for Refracting Surfaces

Quantity	Symbol	In Front	In Back	Upright Image	Inverted Image
Object location	p	+	−		
Image location	q	−	+		
Radius	R	−	+		
Image height	h'			+	−

APPLYING PHYSICS 23.4 | Underwater Vision BIO

Why does a person with normal vision see a blurry image if the eyes are opened underwater with no goggles or diving mask in use?

EXPLANATION The eye presents a spherical refraction surface. The eye normally functions so that light entering from the air is refracted to form an image in the retina located at the back of the eyeball. The difference in the index of refraction between water and the eye is smaller than the difference in the index of refraction between air and the eye. Consequently, light entering the eye from the water doesn't undergo as much refraction as does light entering from the air, and the image is formed behind the retina. A diving mask or swimming goggles have no optical action of their own; they are simply flat pieces of glass or plastic in a rubber mount. They do, however, provide a region of air adjacent to the eyes so that the correct refraction relationship is established and images will be in focus. ■

Flat Refracting Surfaces

If the refracting surface is flat, then R approaches infinity and Equation 23.7 reduces to

$$\frac{n_1}{p} = -\frac{n_2}{q}$$

$$q = -\frac{n_2}{n_1}p \qquad [23.9]$$

From Equation 23.9, we see that the sign of q is opposite that of p. Consequently, **the image formed by a flat refracting surface is on the same side of the surface as the object.** This statement is illustrated in Figure 23.16 for the situation in which n_1 is greater than n_2, where a virtual image is formed between the object and the surface. Note that the refracted ray bends *away* from the normal in this case because $n_1 > n_2$.

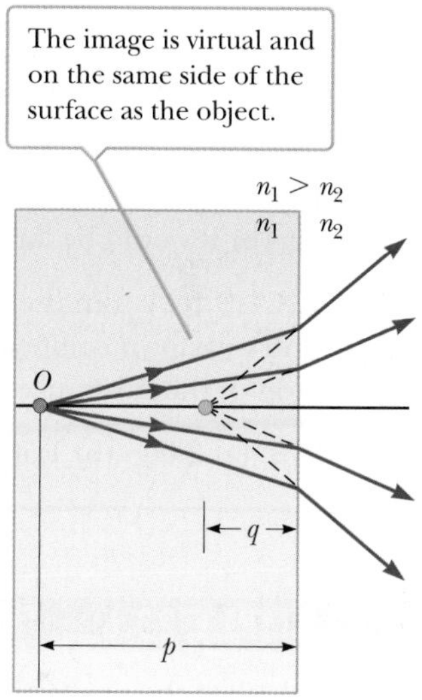

Figure 23.16 The image formed by a flat refracting surface.

Quick Quiz

23.2 A person spearfishing from a boat sees a fish located 3 m from the boat at an apparent depth of 1 m. To spear the fish, should the person aim (a) at, (b) above, or (c) below the image of the fish?

23.3 True or False: (a) The image of an object placed in front of a concave mirror is always upright. (b) The height of the image of an object placed in front of a concave mirror must be smaller than or equal to the height of the object. (c) The image of an object placed in front of a convex mirror is always upright and smaller than the object.

EXAMPLE 23.5 | Gaze into the Crystal Ball

GOAL Calculate the properties of an image created by a spherical lens.

PROBLEM A coin 2.00 cm in diameter is embedded in a solid glass ball of radius 30.0 cm (Fig. 23.17). The index of refraction of the ball is 1.50, and the coin is 20.0 cm from the surface. Find the position of the image of the coin and the height of the coin's image.

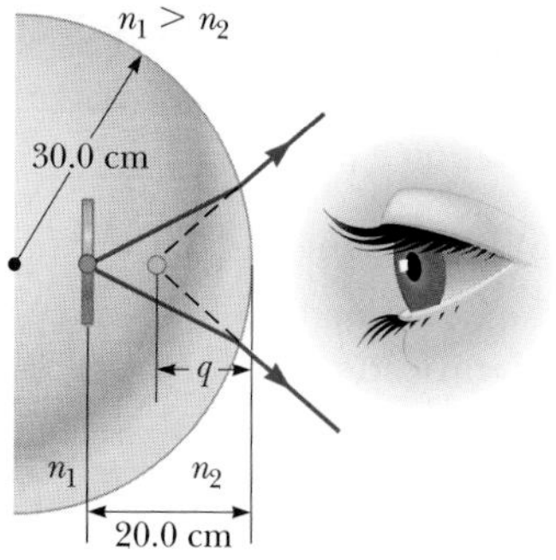

Figure 23.17 (Example 23.5) A coin embedded in a glass ball forms a virtual image between the coin and the surface of the glass.

STRATEGY Because the rays are moving from a medium of high index of refraction (the glass ball) to a medium of lower index of refraction (air), the rays originating at the coin are refracted away from the normal at the surface and diverge outward. The image is formed in the glass and is virtual. Substitute into Equations 23.7 and 23.8 for the image position and magnification, respectively.

(Continued)

SOLUTION

Apply Equation 23.7 and take $n_1 = 1.50$, $n_2 = 1.00$, $p = 20.0$ cm, and $R = -30.0$ cm:

$$\frac{n_1}{p} + \frac{n_2}{q} = \frac{n_2 - n_1}{R}$$

$$\frac{1.50}{20.0 \text{ cm}} + \frac{1.00}{q} = \frac{1.00 - 1.50}{-30.0 \text{ cm}}$$

Solve for q:

$$q = -17.1 \text{ cm}$$

To find the image height, use Equation 23.8 for the magnification:

$$M = -\frac{n_1 q}{n_2 p} = -\frac{1.50(-17.1 \text{ cm})}{1.00(20.0 \text{ cm})} = \frac{h'}{h}$$

$$h' = 1.28h = (1.28)(2.00 \text{ cm}) = 2.56 \text{ cm}$$

REMARKS The negative sign on q indicates that the image is in the same medium as the object (the side of incident light), in agreement with our ray diagram, and therefore must be virtual. The positive value for M means that the image is upright.

QUESTION 23.5 How would the final answer be affected if the ball and observer were immersed in water? (a) It would be smaller. (b) It would be larger. (c) There would be no change.

EXERCISE 23.5 A coin is embedded 20.0 cm from the surface of a similar ball of transparent substance having radius 30.0 cm and unknown composition. If the coin's image is virtual and located 15.0 cm from the surface, find the (a) index of refraction of the substance and (b) magnification.

ANSWERS (a) 2.00 (b) 1.50

EXAMPLE 23.6 The One That Got Away

GOAL Calculate the properties of an image created by a flat refractive surface.

PROBLEM A small fish is swimming at a depth d below the surface of a pond (Fig. 23.18). **(a)** What is the *apparent depth* of the fish as viewed from directly overhead? **(b)** If the fish is 12 cm long, how long is its image?

STRATEGY In this example the refracting surface is flat, so R is infinite. Hence, we can use Equation 23.9 to determine the location of the image, which is the apparent location of the fish.

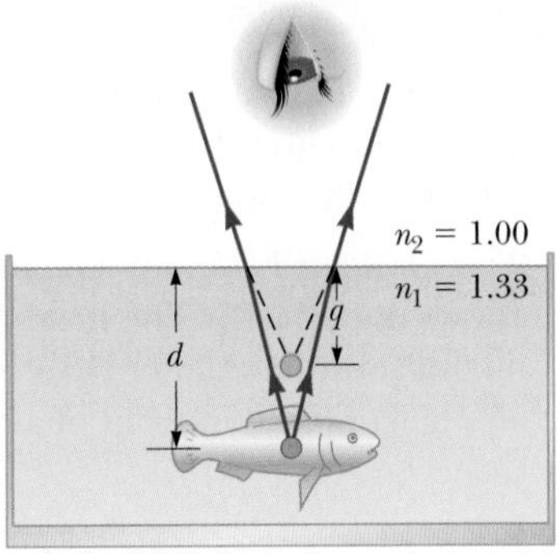

Figure 23.18 (Example 23.6) The apparent depth q of the fish is less than the true depth d.

SOLUTION

(a) Find the apparent depth of the fish.

Substitute $n_1 = 1.33$ for water and $p = d$ into Equation 23.9:

$$q = -\frac{n_2}{n_1}p = -\frac{1}{1.33}d = -0.752d$$

(b) What is the size of the fish's image?

Use Equation 23.9 to eliminate q from Equation 23.8, the magnification equation:

$$M = \frac{h'}{h} = -\frac{n_1 q}{n_2 p} = -\frac{n_1\left(-\dfrac{n_2}{n_1}p\right)}{n_2 p} = 1$$

$$h' = h = 12 \text{ cm}$$

REMARKS Again, because q is negative, the image is virtual, as indicated in Figure 23.18. The apparent depth is approximately three-fourths the actual depth. For instance, if $d = 4.0$ m, then $q = -3.0$ m.

QUESTION 23.6 Suppose a similar experiment is carried out with an object immersed in oil ($n = 1.5$) the same distance below the surface. How does the apparent depth of the object compare with its apparent depth when immersed in water? (a) The apparent depth is unchanged. (b) The apparent depth is larger. (c) The apparent depth is smaller.

EXERCISE 23.6 A spear fisherman estimates that a trout is 1.5 m below the water's surface. What is the actual depth of the fish?

ANSWER 2.0 m

23.5 Atmospheric Refraction

LEARNING OBJECTIVE

1. Discuss images formed by atmospheric refraction, such as mirages.

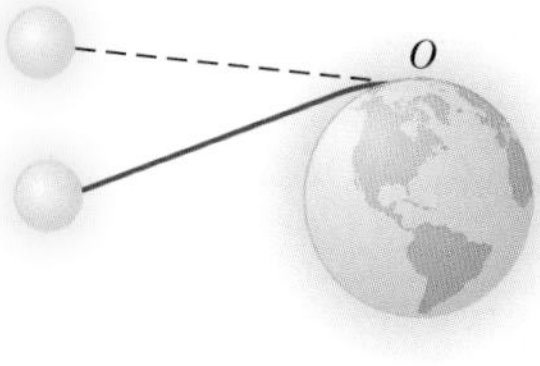

Figure 23.19 Because light is refracted by Earth's atmosphere, an observer at *O* sees the Sun even though it has fallen below the horizon.

Images formed by refraction in our atmosphere lead to some interesting phenomena. One such phenomenon that occurs daily is the visibility of the Sun at dusk even though it has passed below the horizon. Figure 23.19 shows why it occurs. Rays of light from the Sun strike Earth's atmosphere (represented by the shaded area around the planet) and are bent as they pass into a medium that has an index of refraction different from that of the almost empty space in which they have been traveling. The bending in this situation differs somewhat from the bending we have considered previously in that it is gradual and continuous as the light moves through the atmosphere toward an observer at point *O*. This is because the light moves through layers of air that have a continuously changing index of refraction. When the rays reach the observer, the eye follows them back along the direction from which they appear to have come (indicated by the dashed path in the figure). The end result is that the Sun appears to be above the horizon even after it has fallen below it.

The **mirage** is another phenomenon of nature produced by refraction in the atmosphere. A mirage can be observed when the ground is so hot that the air directly above it is warmer than the air at higher elevations. The desert is a region in which such circumstances prevail, but mirages are also seen on heated roadways during the summer. The layers of air at different heights above Earth have different densities and different refractive indices. The effect these differences can have is pictured in Figure 23.20a. The observer sees the sky and a cactus in two different ways. One group of light rays reaches the observer by the straight-line path *A*, and the eye traces these rays back to see the cactus in the normal fashion. In addition, a second group of rays travels along the curved path *B*. These rays are directed toward the ground and are then bent as a result of refraction. As a consequence,

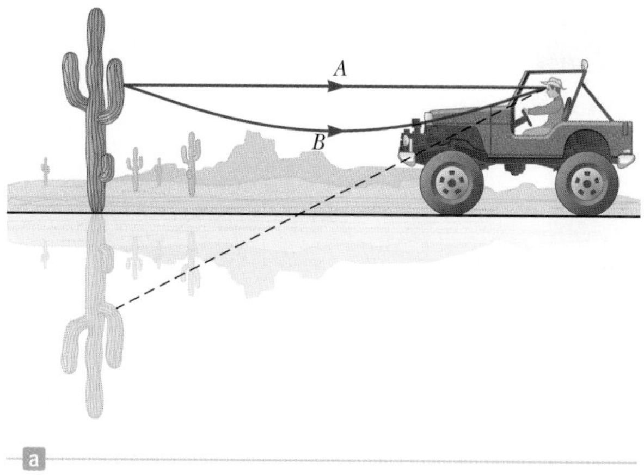

John M. Dunay IV, Fundamental Photographs, NYC

Figure 23.20 (a) A mirage is produced by the bending of light rays in the atmosphere when there are large temperature differences between the ground and the air. (b) Notice the reflection of the cars in this photograph of a mirage. The road looks like it's flooded with water, but it is actually dry.

the observer also sees an inverted image of the cactus and the background of the sky as he traces the rays back to the point at which they appear to have originated. Because both an upright image and an inverted image are seen when the image of a cactus or other object is observed in a reflecting pool of water, the observer unconsciously calls on this past experience and concludes that the sky is reflected by a pool of water in front of the cactus.

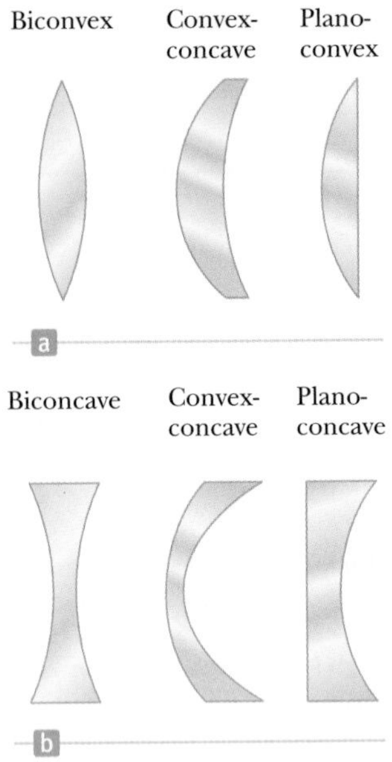

Figure 23.21 Various lens shapes. (a) Converging lenses have positive focal lengths and are thickest at the middle. (b) Diverging lenses have negative focal lengths and are thickest at the edges.

23.6 Thin Lenses

LEARNING OBJECTIVES

1. Discuss converging and diverging thin lenses and their properties.
2. Master the sign conventions for thin lenses.
3. Apply the lens-maker's equation to calculating the physical properties of thin lenses.
4. Analyze systems involving more than one thin lens or having a lens and a mirror.

A typical **thin lens** consists of a piece of glass or plastic, ground so that each of its two refracting surfaces is a segment of either a sphere or a plane. Lenses are commonly used to form images by refraction in optical instruments, such as cameras, telescopes, and microscopes. The equation that relates object and image distances for a lens is virtually identical to the mirror equation derived earlier, and the method used to derive it is also similar.

Figure 23.21 shows some representative shapes of lenses. Notice that we have placed these lenses in two groups. Those in Figure 23.21a are thicker at the center than at the rim, and those in Figure 23.21b are thinner at the center than at the rim. The lenses in the first group are examples of **converging lenses**, and those in the second group are **diverging lenses**. The reason for these names will become apparent shortly.

As we did for mirrors, it is convenient to define a point called the **focal point** for a lens. For example, in Figure 23.22a, a group of rays parallel to the axis passes

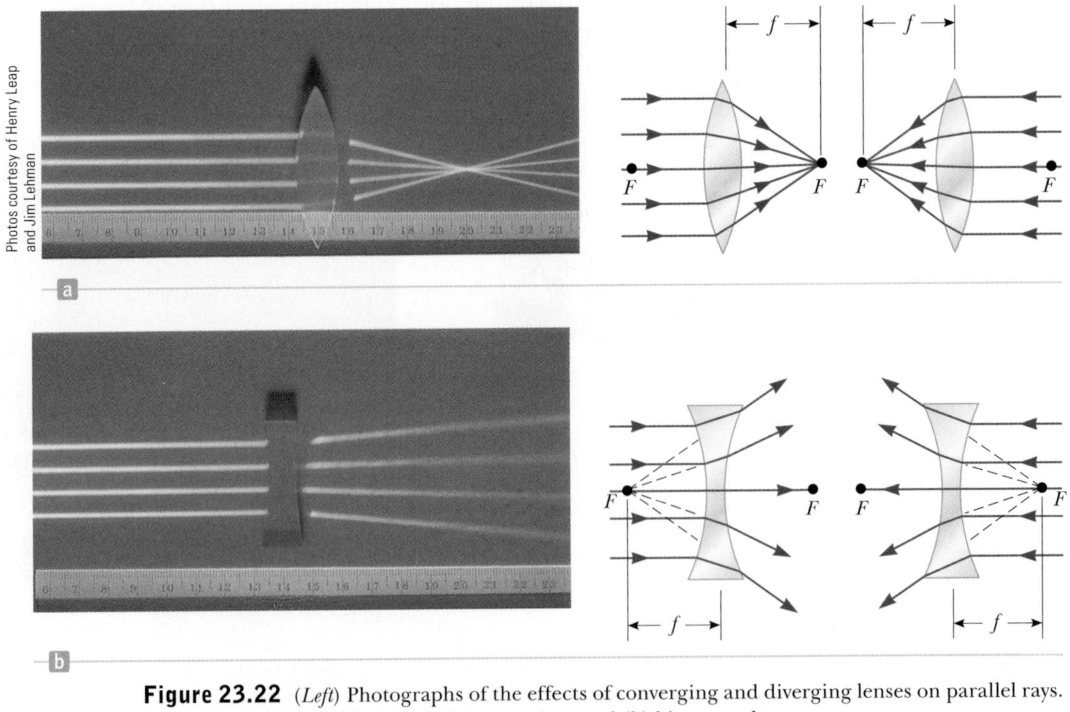

Figure 23.22 (*Left*) Photographs of the effects of converging and diverging lenses on parallel rays. (*Right*) The focal points of the (a) biconvex lens and (b) biconcave lens.

through the focal point F after being converged by the lens. The distance from the focal point to the lens is called the **focal length f. The focal length is the image distance that corresponds to an infinite object distance.** Recall that we are considering the lens to be very thin. As a result, it makes no difference whether we take the focal length to be the distance from the focal point to the surface of the lens or the distance from the focal point to the center of the lens because the difference between these two lengths is negligible. A thin lens has *two* focal points, as illustrated in Figure 23.22, one on each side of the lens. One focal point corresponds to parallel rays traveling from the left and the other corresponds to parallel rays traveling from the right.

Rays parallel to the axis diverge after passing through a lens of biconcave shape, shown in Figure 23.22b. In this case the focal point is defined to be the point where the diverged rays appear to originate, labeled F in the figure. Figures 23.22a and 23.22b indicate why the names *converging* and *diverging* are applied to these lenses.

Now consider a ray of light passing through the center of a lens. Such a ray is labeled ray 1 in Figure 23.23. For a thin lens, a ray passing through the center is undeflected. Ray 2 in the same figure is parallel to the principal axis of the lens (the horizontal axis passing through O), and as a result it passes through the focal point F after refraction. Rays 1 and 2 intersect at the point that is the tip of the image arrow.

We first note that the tangent of the angle α can be found by using the blue and gold shaded triangles in Figure 23.23:

$$\tan \alpha = \frac{h}{p} \qquad \text{or} \qquad \tan \alpha = -\frac{h'}{q}$$

From this result, we find that

$$M = \frac{h'}{h} = -\frac{q}{p} \qquad [23.10]$$

The equation for magnification by a lens is the same as the equation for magnification by a mirror. We also note from Figure 23.23 that

$$\tan \theta = \frac{PQ}{f} \qquad \text{or} \qquad \tan \theta = -\frac{h'}{q - f}$$

The height PQ used in the first of these equations, however, is the same as h, the height of the object. Therefore,

$$\frac{h}{f} = -\frac{h'}{q - f}$$

$$\frac{h'}{h} = -\frac{q - f}{f}$$

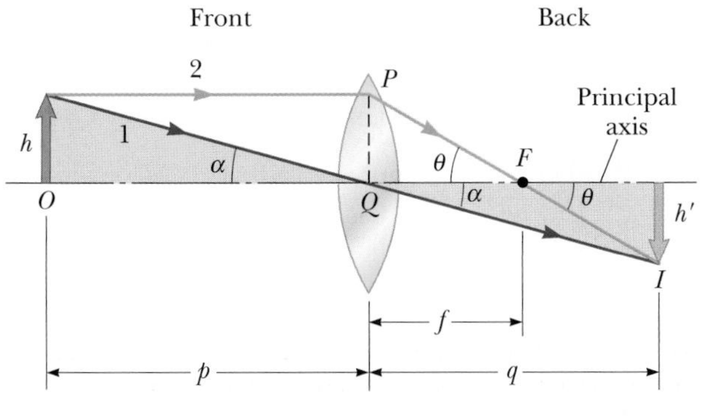

Figure 23.23 A geometric construction for developing the thin-lens equation.

Using the latter equation in combination with Equation 23.10 gives

$$\frac{q}{p} = \frac{q - f}{f}$$

which reduces to

Thin-lens equation ▶

$$\frac{1}{p} + \frac{1}{q} = \frac{1}{f} \qquad [23.11]$$

This equation, called the **thin-lens equation**, can be used with both converging and diverging lenses if we adhere to a set of sign conventions. Figure 23.24 is useful for obtaining the signs of p and q, and Table 23.3 gives the complete sign conventions for lenses. Note that **a converging lens has a positive focal length** under this convention and **a diverging lens has a negative focal length**. Hence, the names *positive* and *negative* are often given to these lenses.

The focal length for a lens in air is related to the curvatures of its front and back surfaces and to the index of refraction n of the lens material by

Lens-maker's equation ▶

$$\frac{1}{f} = (n - 1)\left(\frac{1}{R_1} - \frac{1}{R_2}\right) \qquad [23.12]$$

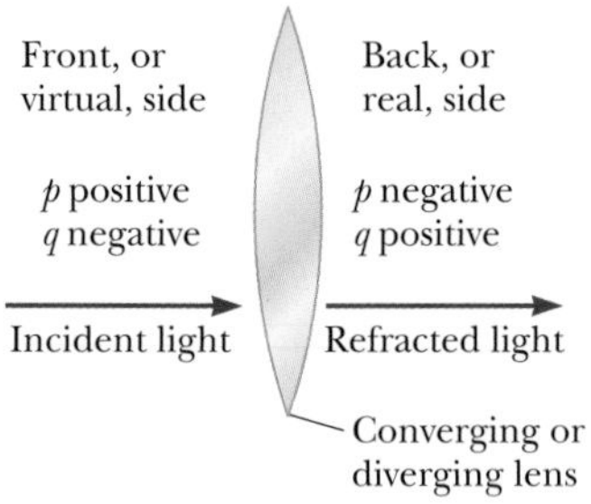

Figure 23.24 A diagram for obtaining the signs of p and q for a thin lens or a refracting surface.

where R_1 is the radius of curvature of the front surface of the lens and R_2 is the radius of curvature of the back surface. (As with mirrors, we arbitrarily call the side from which the light approaches the *front* of the lens.) Table 23.3 gives the sign conventions for R_1 and R_2. Equation 23.12, called the **lens-maker's equation**, enables us to calculate the focal length from the known properties of the lens.

Ray Diagrams for Thin Lenses

Ray diagrams are essential for understanding the overall image formation by a thin lens or a system of lenses. They should also help clarify the sign conventions already discussed. Figure 23.25 illustrates this method for three single-lens situations. To locate the image formed by a converging lens (Figs. 23.25a and 23.25b), the following three rays are drawn from the top of the object:

1. The first ray is drawn parallel to the principal axis. After being refracted by the lens, this ray passes through (or appears to come from) one of the focal points.
2. The second ray is drawn through the center of the lens. This ray continues in a straight line.
3. The third ray is drawn through the other focal point and emerges from the lens parallel to the principal axis.

A similar construction is used to locate the image formed by a diverging lens, as shown in Figure 23.25c. The point of intersection of *any two* of the rays in these diagrams can be used to locate the image. The third ray serves as a check on construction.

For the converging lens in Figure 23.25a, where the object is *outside* the front focal point ($p > f$), the ray diagram shows that the image is real and inverted.

Tip 23.4 Positive Is Again Where the Light Is

For lenses, p and q are positive where the light is, where the object or image is real. For real objects, the light originates with the object in front of the lens, so p is positive there as indicated in Figure 23.24. If the image forms in back of the lens, q is positive there, as well.

Tip 23.5 We *Choose* Only a Few Rays

Although our ray diagrams in Figure 23.25 only show three rays leaving an object, an infinite number of rays can be drawn between the object and its image.

Table 23.3 Sign Conventions for Thin Lenses

Quantity	Symbol	In Front	In Back	Convergent	Divergent
Object location	p	+	−		
Image location	q	−	+		
Lens radii	R_1, R_2	−	+		
Focal length	f			+	−

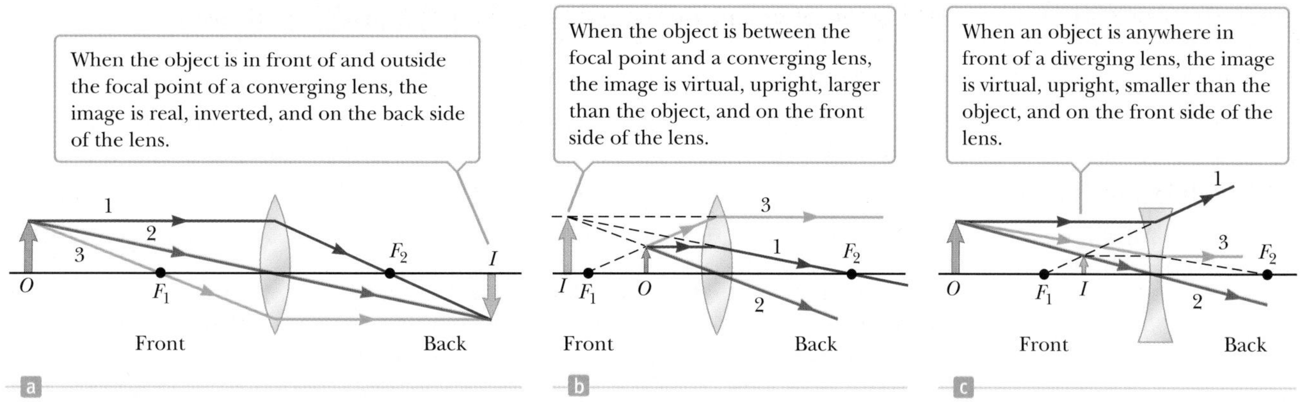

Figure 23.25 Ray diagrams for locating the image formed by a thin lens.

When the real object is *inside* the front focal point ($p < f$), as in Figure 23.25b, the image is virtual and upright. For the diverging lens of Figure 23.25c, the image is virtual and upright.

Quick Quiz

23.4 A clear plastic sandwich bag filled with water can act as a crude converging lens in air. If the bag is filled with air and placed under water, is the effective lens (a) converging or (b) diverging?

23.5 In Figure 23.25a the blue object arrow is replaced by one that is much taller than the lens. How many rays from the object will strike the lens?

23.6 An object is placed to the left of a converging lens. Which of the following statements are true, and which are false? (a) The image is always to the right of the lens. (b) The image can be upright or inverted. (c) The image is always smaller or the same size as the object.

Your success in working lens or mirror problems will be determined largely by whether you make sign errors when substituting into the lens or mirror equations. The only way to ensure you don't make sign errors is to become adept at using the sign conventions. The best way to do so is to work a multitude of problems on your own and construct confirming ray diagrams. Watching an instructor or reading the example problems is no substitute for practice.

APPLYING PHYSICS 23.5 | Vision and Diving Masks BIO

Diving masks often have a lens built into the glass faceplate for divers who don't have perfect vision. This lens allows the individual to dive without the necessity of glasses because the faceplate performs the necessary refraction to produce clear vision. Normal glasses have lenses that are curved on both the front and rear surfaces. The lenses in a diving-mask faceplate often have curved surfaces only on the inside of the glass. Why is this design desirable?

SOLUTION The main reason for curving only the inner surface of the lens in the diving-mask faceplate is to enable the diver to see clearly while underwater and in the air. If there were curved surfaces on both the front and the back of the diving lens, there would be two refractions. The lens could be designed so that these two refractions would give clear vision while the diver is in air. When the diver went underwater, however, the refraction between the water and the glass at the first interface would differ because the index of refraction of water is different from that of air. Consequently, the diver's vision wouldn't be clear underwater. ■

EXAMPLE 23.7 | Images Formed by a Converging Lens

GOAL Calculate geometric quantities associated with a converging lens.

PROBLEM A converging lens of focal length 10.0 cm forms images of an object situated at various distances. **(a)** If the object is placed 30.0 cm from the lens, locate the image, state whether it's real or virtual, and find its magnification. **(b)** Repeat the problem when the object is at 10.0 cm and **(c)** again when the object is 5.00 cm from the lens.

STRATEGY All three problems require only substitution into the thin-lens equation and the associated magnification equation, Equations 23.10 and 23.11, respectively. The conventions of Table 23.3 must be followed.

Figure 23.26 (Example 23.7)

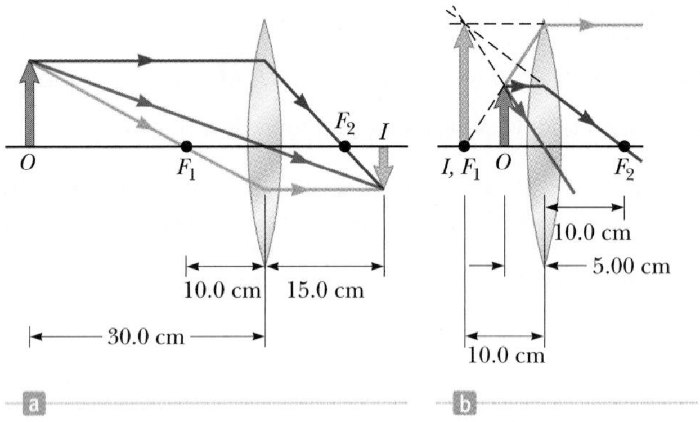

SOLUTION

(a) Find the image distance and describe the image when the object is placed at 30.0 cm.

The ray diagram is shown in Figure 23.26a. Substitute values into the thin-lens equation to locate the image:

$$\frac{1}{p} + \frac{1}{q} = \frac{1}{f}$$

$$\frac{1}{30.0 \text{ cm}} + \frac{1}{q} = \frac{1}{10.0 \text{ cm}}$$

Solve for q, the image distance. It's positive, so the image is real and on the far side of the lens:

$$q = \boxed{+15.0 \text{ cm}}$$

The magnification of the lens is obtained from Equation 23.10. M is negative and less than 1 in absolute value, so the image is inverted and smaller than the object:

$$M = -\frac{q}{p} = -\frac{15.0 \text{ cm}}{30.0 \text{ cm}} = \boxed{-0.500}$$

(b) Repeat the problem, when the object is placed at 10.0 cm.

Locate the image by substituting into the thin-lens equation:

$$\frac{1}{10.0 \text{ cm}} + \frac{1}{q} = \frac{1}{10.0 \text{ cm}} \quad \rightarrow \quad \frac{1}{q} = 0$$

This equation is satisfied only in the limit as q becomes infinite. Similarly, M becomes infinite, as well.

$$q \rightarrow \boxed{\infty}$$

(c) Repeat the problem when the object is placed 5.00 cm from the lens.

See the ray diagram shown in Figure 23.26b. Substitute into the thin-lens equation to locate the image:

$$\frac{1}{5.00 \text{ cm}} + \frac{1}{q} = \frac{1}{10.0 \text{ cm}}$$

Solve for q, which is negative, meaning the image is on the same side as the object and is virtual:

$$q = \boxed{-10.0 \text{ cm}}$$

Substitute the values of p and q into the magnification equation. M is positive and larger than 1, so the image is upright and double the object size:

$$M = -\frac{q}{p} = -\left(\frac{-10.0 \text{ cm}}{5.00 \text{ cm}}\right) = \boxed{+2.00}$$

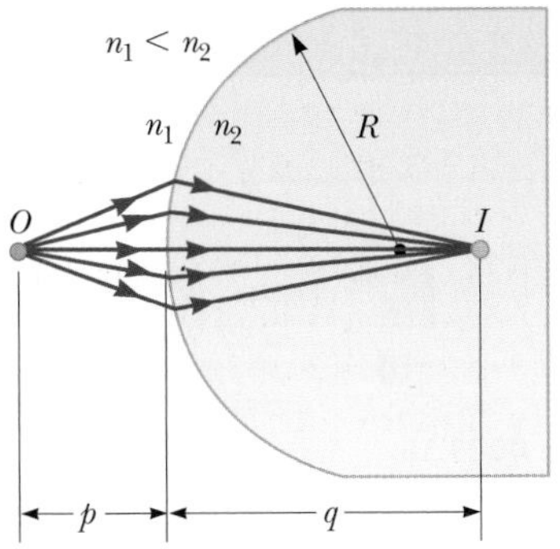

An image formed by refraction at a spherical surface. Rays making small angles with the principal axis diverge from a point object at *O* and pass through the image point *I*.

The **magnification of a refracting surface** is

$$M = \frac{h'}{h} = -\frac{n_1 q}{n_2 p} \qquad [23.8]$$

where the object is located in the medium with index of refraction n_1 and the image is formed in the medium with index of refraction n_2. Equations 23.7 and 23.8 are subject to the sign conventions of Table 23.2.

23.6 Thin Lenses

The **magnification of a thin lens** is

$$M = \frac{h'}{h} = -\frac{q}{p} \qquad [23.10]$$

The object and image distances of a thin lens are related by the **thin-lens equation**:

$$\frac{1}{p} + \frac{1}{q} = \frac{1}{f} \qquad [23.11]$$

Equations 23.10 and 23.11 are subject to the sign conventions of Table 23.3.

23.7 Lens and Mirror Aberrations

Aberrations are responsible for the formation of imperfect images by lenses and mirrors. **Spherical aberration** results from the focal points of light rays far from the principal axis of a spherical lens or mirror being different from those of rays passing through the center. **Chromatic aberration** arises because light rays of different wavelengths focus at different points when refracted by a lens.

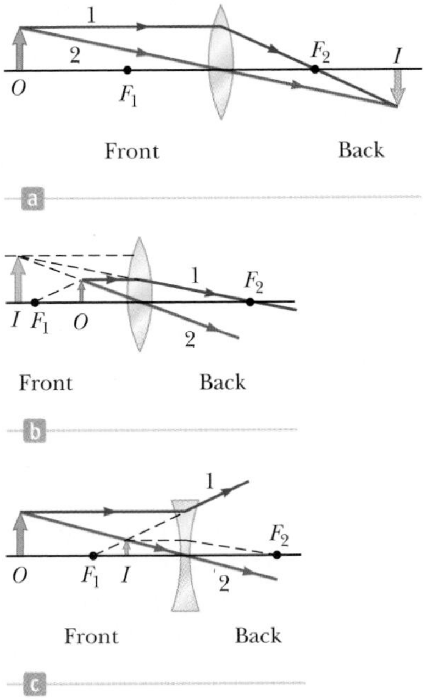

Ray diagrams for locating the image of an object. (a) The object is outside the focal point of a converging lens. (b) The object is inside the focal point of a converging lens. (c) The object is outside the focal point of a diverging lens.

WARM-UP EXERCISES

ENHANCED WebAssign The warm-up exercises in this chapter may be assigned online in Enhanced WebAssign.

1. A woman stands 1.50 m in front of a full-length flat mirror. Find (a) the position of her image, with proper sign, (b) the magnification of her image, and (c) whether it is real or virtual. (See Section 23.1.)
2. An object is 20.0 cm away from a concave mirror with focal length 15.0 cm. Find (a) the image distance, *q*, and (b) its magnification. State whether the image is (c) real or virtual, (d) upright or inverted. (See Section 23.3.)
3. An object is placed 16.0 cm away from a convex mirror with a focal length of magnitude 6.00 cm. (a) Is the sign of the focal length negative or positive? Find (b) the image distance *q*, and (c) the magnification. State whether the image is (d) real or virtual, (e) upright or inverted. (See Section 23.3.)
4. A coin is at the bottom of a pool of water 1.20 m deep. Find (a) the object distance, (b) the image distance, and (c) the magnification of the image. (d) Is the image real or virtual? (See Section 23.4.)
5. A gold coin is embedded in solid ball of clear plastic of radius 25.0 cm, 10.0 cm from the surface. A person looks directly at the coin along a line going through the coin and to the center of the ball, as in Example 23.5. If the plastic has index of refraction 1.70, find (a) the image distance *q*, and (b) the coin's magnification. (See Section 23.4.)
6. A thin, convergent lens has a focal length of 8.00 cm. If the object distance is 24.0 cm, find (a) the image distance, and (b) the magnification. State whether the image is (c) real or virtual, (d) upright or inverted. (See Section 23.6.)
7. A real object is 10.0 cm to the left of a thin, diverging lens having a focal length of magnitude 16.0 cm. (a) Is the sign of the focal length negative or positive? Find (b) the image distance, and (c) the magnification. State whether the image is (d) real or virtual, (e) upright or inverted. (See Section 23.6.)

CONCEPTUAL QUESTIONS

WebAssign The conceptual questions in this chapter may be assigned online in Enhanced WebAssign.

1. Tape a picture of yourself on a bathroom mirror. Stand several centimeters away from the mirror. Can you focus your eyes on *both* the picture taped to the mirror *and* your image in the mirror *at the same time*? So where is the image of yourself?

2. Why does a clear stream always appear to be shallower than it actually is?

3. A flat mirror creates a virtual image of your face. Suppose the flat mirror is combined with another optical element. Can the mirror form a real image in such a combination?

4. Explain why a mirror cannot give rise to chromatic aberration.

5. A common mirage is formed when the air gets gradually cooler as the height above the ground increases. What might happen if the air grows gradually warmer as the height increases? This often happens over bodies of water or snow-covered ground; the effect is called *looming*.

6. A virtual image is often described as an image through which light rays don't actually travel, as they do for a real image. Can a virtual image be photographed?

7. Suppose you want to use a converging lens to project the image of two trees onto a screen. One tree is a distance x from the lens; the other is at $2x$, as in Figure CQ23.7. You adjust the screen so that the near tree is in focus. If you now want the far tree to be in focus, do you move the screen toward or away from the lens?

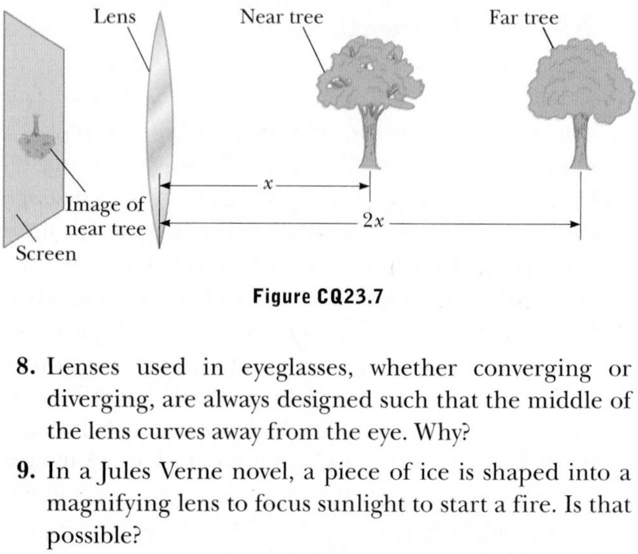

Figure CQ23.7

8. Lenses used in eyeglasses, whether converging or diverging, are always designed such that the middle of the lens curves away from the eye. Why?

9. In a Jules Verne novel, a piece of ice is shaped into a magnifying lens to focus sunlight to start a fire. Is that possible?

10. If a cylinder of solid glass or clear plastic is placed above the words LEAD OXIDE and viewed from the side, as shown in Figure CQ23.10, the word LEAD appears inverted, but the word OXIDE does not. Explain.

Figure CQ23.10

11. Can a converging lens be made to diverge light if placed in a liquid? How about a converging mirror?

12. Light from an object passes through a lens and forms a visible image on a screen. If the screen is removed, would you be able to see the image (a) if you remained in your present position and (b) if you could look at the lens along its axis, beyond the original position of the screen?

13. Why does the focal length of a mirror not depend on the mirror material when the focal length of a lens does depend on the lens material?

14. An inverted image of an object is viewed on a screen from the side facing a converging lens. An opaque card is then introduced covering only the upper half of the lens. What happens to the image on the screen? (a) Half the image would disappear. (b) The entire image would appear and remain unchanged. (c) Half the image would disappear and be dimmer. (d) The entire image would appear, but would be dimmer.

15. Why do some emergency vehicles have the symbol ƎƆИA⅃UBMA written on the front?

16. A person spear fishing from a boat sees a stationary fish a few meters away in a direction about 30° below the horizontal. To spear the fish, and assuming the spear does not change direction when it enters the water, should the person (a) aim above where he sees the fish, (b) aim below the fish, or (c) aim precisely at the fish?

17. An object, represented by a gray arrow, is placed in front of a plane mirror. Which of the diagrams in Figure CQ23.17 best describes the image, represented by the pink arrow?

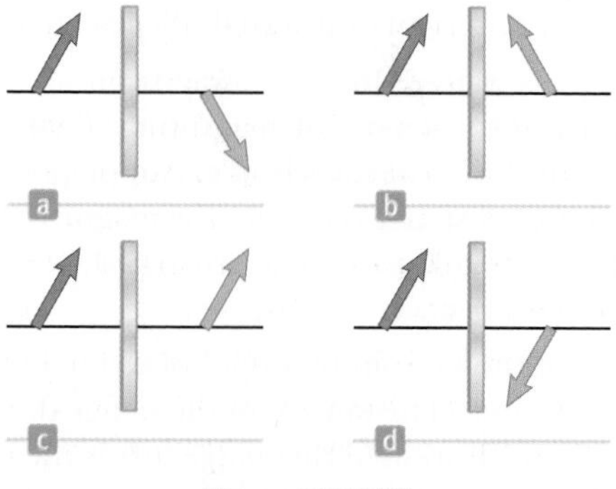

Figure CQ23.17

PROBLEMS

WebAssign The problems in this chapter may be assigned online in Enhanced WebAssign.

1. denotes straightforward problem; **2.** denotes intermediate problem;

3. denotes challenging problem

1. denotes full solution available in *Student Solutions Manual/Study Guide*

1. denotes problems most often assigned in Enhanced WebAssign

BIO denotes biomedical problems

GP denotes guided problems

M denotes Master It tutorial available in Enhanced WebAssign

Q|C denotes asking for quantitative and conceptual reasoning

S denotes symbolic reasoning problem

W denotes Watch It video solution available in Enhanced WebAssign

23.1 Flat Mirrors

1. (a) Does your bathroom mirror show you older or younger than your actual age? (b) Compute an order-of-magnitude estimate for the age difference, based on data you specify.

2. Q|C Two plane mirrors stand facing each other, 3.00 m apart, and a woman stands between them. The woman faces one of the mirrors from a distance of 1.00 m and holds her left arm out to the side of her body with the palm of her left hand facing the closer mirror. (a) What is the apparent position of the closest image of her left hand, measured perpendicularly from the surface of the mirror in front of her? (b) Does it show the palm of her hand or the back of her hand? (c) What is the position of the next image? (d) Does it show the palm of her hand or the back of her hand? (e) What is the position of the third closest image? (f) Does it show the palm of her hand or the back of her hand? (g) Which of the images are real and which are virtual?

3. A person walks into a room that has, on opposite walls, two plane mirrors producing multiple images. Find the distances from the person to the first three images seen in the left-hand mirror when the person is 5.00 ft from the mirror on the left wall and 10.0 ft from the mirror on the right wall.

4. In a church choir loft, two parallel walls are 5.30 m apart. The singers stand against the north wall. The organist faces the south wall, sitting 0.800 m away from it. So that she can see the choir, a flat mirror 0.600 m wide is mounted on the south wall, straight in front of the organist. What width of the north wall can she see? *Hint:* Draw a top-view diagram to justify your answer.

5. A periscope (Fig. P23.5) is useful for viewing objects that cannot be seen directly. It can be used in submarines and when watching golf matches or parades from behind a crowd of people. Suppose the object is a distance p_1 from the upper mirror and the centers of the two flat mirrors are separated by a distance h. (a) What is the distance of the final image from the lower mirror? (b) Is the final image real or virtual? (c) Is it upright or inverted? (d) What is its magnification? (e) Does it appear to be left–right reversed?

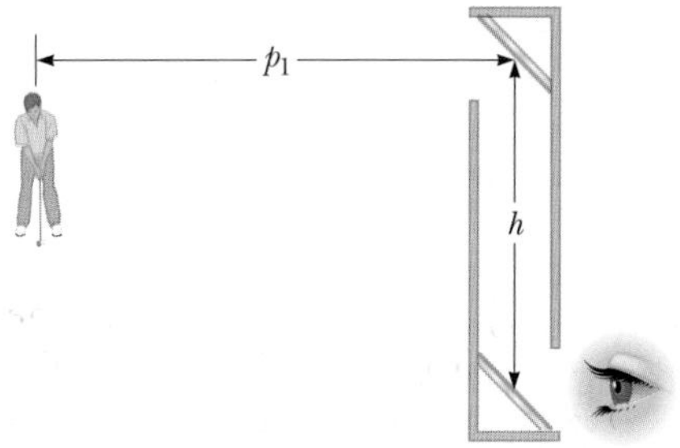

Figure P23.5

23.2 Images Formed by Concave Mirrors

23.3 Convex Mirrors and Sign Conventions

In the following problems, algebraic signs are not given. We leave it to you to determine the correct sign to use with each quantity, based on an analysis of the problem and the sign conventions in Table 23.1.

6. A dentist uses a mirror to examine a tooth that is 1.00 cm in front of the mirror. The image of the tooth is formed 10.0 cm behind the mirror. Determine (a) the mirror's radius of curvature and (b) the magnification of the image.

7. A convex spherical mirror, whose focal length has a magnitude of 15.0 cm, is to form an image 10.0 cm behind the mirror. (a) Where should the object be placed? (b) What is the magnification of the mirror?

8. BIO To fit a contact lens to a patient's eye, a *keratometer* can be used to measure the curvature of the cornea—the front surface of the eye. This instrument places an illuminated object of known size at a known distance p from the cornea, which then reflects some light from the object, forming an image of it. The magnification M of the image is measured by using a small viewing telescope that allows a comparison of the image formed by the cornea with a second calibrated image projected into the field of view by a prism arrangement. Determine the radius of curvature of the cornea when $p = 30.0$ cm and $M = 0.013\,0$.

9. A virtual image is formed 20.0 cm from a concave mirror having a radius of curvature of 40.0 cm. (a) Find the position of the object. (b) What is the magnification of the mirror?

10. Q|C While looking at her image in a cosmetic mirror, Dina notes that her face is highly magnified when she is close to the mirror, but as she backs away from the mirror, her image first becomes blurry, then disappears when she is about 30 cm from the mirror, and then inverts when she is beyond 30 cm. Based on these observations, what can she conclude about the properties of the mirror?

11. W A 2.00-cm-high object is placed 3.00 cm in front of a concave mirror. If the image is 5.00 cm high and virtual, what is the focal length of the mirror?

12. A dedicated sports car enthusiast polishes the inside and outside surfaces of a hubcap that is a section of a sphere. When he looks into one side of the hubcap, he sees an image of his face 30.0 cm in back of it. He then turns the hubcap over, keeping it the same distance from his face. He now sees an image of his face 10.0 cm in back of the hubcap. (a) How far is his face from the hubcap? (b) What is the magnitude of the radius of curvature of the hubcap?

13. A concave makeup mirror is designed so that a person 25 cm in front of it sees an upright image magnified by a factor of two. What is the radius of curvature of the mirror?

14. A concave mirror has a focal length of 30.0 cm. (a) What is its radius of curvature? Locate and describe the properties of the image when the object distance is (b) 100 cm and (c) 10.0 cm.

15. M A man standing 1.52 m in front of a shaving mirror produces an inverted image 18.0 cm in front of it. How close to the mirror should he stand if he wants to form an upright image of his chin that is twice the chin's actual size?

16. A convex mirror has a focal length of magnitude 8.0 cm. (a) If the image is virtual, what is the object location for which the magnitude of the image distance is one third the magnitude of the object distance? (b) Find the magnification of the image and state whether it is upright or inverted.

17. M At an intersection of hospital hallways, a convex spherical mirror is mounted high on a wall to help people avoid collisions. The magnitude of the mirror's radius of curvature is 0.550 m. (a) Locate the image of a patient located 10.0 m from the mirror. (b) Indicate whether the image is upright or inverted. (c) Determine the magnification of the image.

18. Q|C A concave mirror has a radius of curvature of 24.0 cm. (a) Determine the object position for which the resulting image is upright and larger than the object by a factor of 3.00. (b) Draw a ray diagram to determine the position of the image. (c) Is the image real or virtual?

19. Q|C A spherical mirror is to be used to form an image, five times as tall as an object, on a screen positioned 5.0 m from the mirror. (a) Describe the type of mirror required. (b) Where should the mirror be positioned relative to the object?

20. Q|C A ball is dropped from rest 3.00 m directly above the vertex of a concave mirror having a radius of 1.00 m and lying in a horizontal plane. (a) Describe the motion of the ball's image in the mirror. (b) At what time do the ball and its image coincide?

23.4 Images Formed by Refraction

21. A cubical block of ice 50.0 cm on an edge is placed on a level floor over a speck of dust. Locate the image of the speck, when viewed from directly above, if the index of refraction of ice is 1.309.

22. A goldfish is swimming inside a spherical bowl of water having an index of refraction $n = 1.333$. Suppose the goldfish is $p = 10.0$ cm from the wall of a bowl of *radius* $|R| = 15.0$ cm, as in Figure P23.22. Neglecting the refraction of light caused by the wall of the bowl, determine the apparent distance of the goldfish from the wall according to an observer outside the bowl.

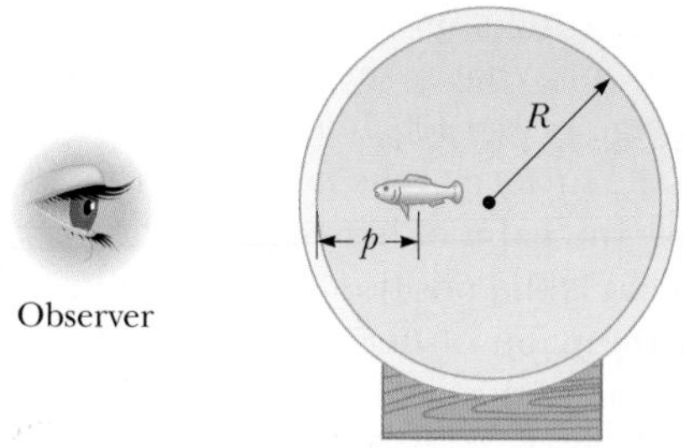

Figure P23.22

23. A paperweight is made of a solid glass hemisphere with index of refraction 1.50. The radius of the circular cross section is 4.0 cm. The hemisphere is placed on its flat surface, with the center directly over a 2.5-mm-long line drawn on a sheet of paper. What length of line is seen by someone looking vertically down on the hemisphere?

24. W The top of a swimming pool is at ground level. If the pool is 2.00 m deep, how far below ground level does the bottom of the pool appear to be located when (a) the pool is completely filled with water? (b) When it is filled halfway with water?

25. A transparent sphere of unknown composition is observed to form an image of the Sun on its surface opposite the Sun. What is the refractive index of the sphere material?

26. A flint glass plate ($n = 1.66$) rests on the bottom of an aquarium tank. The plate is 8.00 cm thick (vertical dimension) and covered with water ($n = 1.33$) to a depth of 12.0 cm. Calculate the apparent thickness of the plate as viewed from above the water. (Assume nearly normal incidence of light rays.)

27. Q|C A jellyfish is floating in a water-filled aquarium 1.00 m behind a flat pane of glass 6.00 cm thick and having an index of refraction of 1.50. (a) Where is the image of the jellyfish located? (b) Repeat the problem when the glass is so thin that its thickness can be neglected. (c) How does the thickness of the glass affect the answer to part (a)?

28. S Figure P23.28 shows a curved surface separating a material with index of refraction n_1 from a material with index n_2. The surface forms an image I of object O. The ray shown in red passes through the surface along a radial line. Its angles of incidence and refraction are both zero, so its direction does not change at the surface. For the ray shown in blue, the direction changes according to $n_1 \sin \theta_1 = n_2 \sin \theta_2$. For paraxial rays, we assume θ_1 and θ_2 are small, so we may write $n_1 \tan \theta_1 = n_2 \tan \theta_2$. The magnification is defined as $M = h'/h$. Prove that the magnification is given by $M = -n_1 q/n_2 p$.

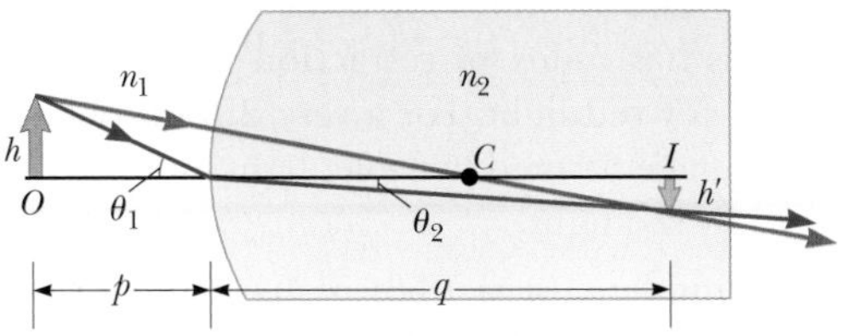

Figure P23.28

23.6 Thin Lenses

29. BIO A contact lens is made of plastic with an index of refraction of 1.50. The lens has an outer radius of curvature of +2.00 cm and an inner radius of curvature of +2.50 cm. What is the focal length of the lens?

30. An object is placed 50.0 cm from a screen. (a) Where should a converging lens of focal length 10.0 cm be placed to form an image on the screen? (b) Find the magnification of the lens.

31. A converging lens has a focal length of 10.0 cm. Locate the images for object distances of (a) 20.0 cm, (b) 10.0 cm, and (c) 5.00 cm, if they exist. For each case, state whether the image is real or virtual, upright or inverted, and find the magnification.

32. An object is placed 20.0 cm from a concave spherical mirror having a focal length of magnitude 40.0 cm. (a) Use graph paper to construct an accurate ray diagram for this situation. (b) From your ray diagram, determine the location of the image. (c) What is the magnification of the image? (d) Check your answers to parts (b) and (c) using the mirror equation.

33. A diverging lens has a focal length of magnitude 20.0 cm. (a) Locate the images for object distances of (i) 40.0 cm, (ii) 20.0 cm, and (iii) 10.0 cm. For each case, state whether the image is (b) real or virtual and (c) upright or inverted. (d) For each case, find the magnification.

34. Q|C A diverging lens has a focal length of 20.0 cm. Use graph paper to construct accurate ray diagrams for object distances of (a) 40.0 cm and (b) 10.0 cm. In each case determine the location of the image from the diagram and the image magnification, and state whether the image is upright or inverted. (c) Estimate the magnitude of uncertainty in locating the points in the graph. Are your answers and the uncertainty consistent with the algebraic answers found in Problem 33?

35. A transparent photographic slide is placed in front of a converging lens with a focal length of 2.44 cm. An image of the slide is formed 12.9 cm from the slide. How far is the lens from the slide if the image is (a) real? (b) Virtual?

36. W The nickel's image in Figure P23.36 has twice the diameter of the nickel when the lens is 2.84 cm from the nickel. Determine the focal length of the lens.

Figure P23.36

37. The projection lens in a certain slide projector is a single thin lens. A slide 24.0 mm high is to be projected so that its image fills a screen 1.80 m high. The slide-to-screen distance is 3.00 m. (a) Determine the focal length of the projection lens. (b) How far from the slide should the lens of the projector be placed to form the image on the screen?

38. M An object is located 20.0 cm to the left of a diverging lens having a focal length $f = -32.0$ cm. Determine (a) the location and (b) the magnification of the image. (c) Construct a ray diagram for this arrangement.

39. A converging lens is placed 30.0 cm to the right of a diverging lens of focal length 10.0 cm. A beam of parallel light enters the diverging lens from the left, and the beam is again parallel when it emerges from the converging lens. Calculate the focal length of the converging lens.

40. Q|C S (a) Use the thin-lens equation to derive an expression for q in terms of f and p. (b) Prove that for a real object and a diverging lens, the image must always be virtual. *Hint:* Set $f = -|f|$ and show that q must be less than zero under the given conditions. (c) For a real object and converging lens, what inequality involving p and f must hold if the image is to be real?

41. W Two converging lenses, each of focal length 15.0 cm, are placed 40.0 cm apart, and an object is placed 30.0 cm in front of the first lens. Where is the final image formed, and what is the magnification of the system?

42. Object O_1 is 15.0 cm to the left of a converging lens with a 10.0-cm focal length. A second lens is positioned 10.0 cm to the right of the first lens and is observed to form a virtual image at the position of the original object O_1. (a) What is the focal length of the second lens? (b) What is the overall magnification of this system? (c) What is the nature (i.e., real or virtual, upright or inverted) of the final image?

43. M A 1.00-cm-high object is placed 4.00 cm to the left of a converging lens of focal length 8.00 cm. A diverging lens of focal length −16.00 cm is 6.00 cm to the right of the converging lens. Find the position and height of the final image. Is the image inverted or upright? Real or virtual?

44. Two converging lenses having focal lengths of f_1 = 10.0 cm and f_2 = 20.0 cm are placed d = 50.0 cm apart, as shown in Figure P23.44. The final image is to be located between the lenses, at the position x = 31.0 cm indicated. (a) How far to the left of the first lens should the object be positioned? (b) What is the overall magnification of the system? (c) Is the final image upright or inverted?

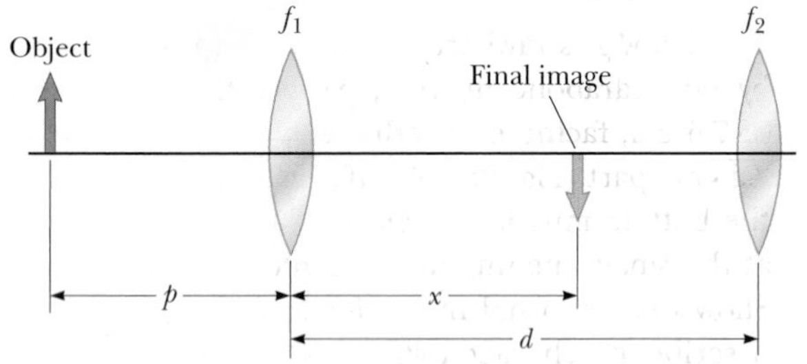

Figure P23.44

45. Lens L_1 in Figure P23.45 has a focal length of 15.0 cm and is located a fixed distance in front of the film plane of a camera. Lens L_2 has a focal length of 13.0 cm, and its distance d from the film plane can be varied from 5.00 cm to 10.0 cm. Determine the range of distances for which objects can be focused on the film.

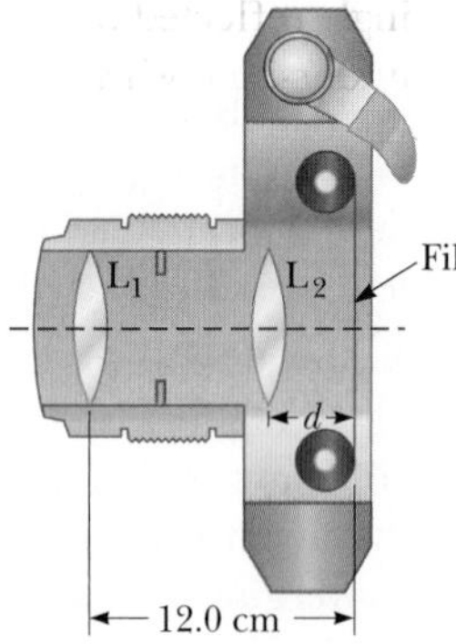

Figure P23.45

46. GP An object is placed 15.0 cm from a first converging lens of focal length 10.0 cm. A second converging lens with focal length 5.00 cm is placed 10.0 cm to the right of the first converging lens. (a) Find the position q_1 of the image formed by the first converging lens. (b) How far from the second lens is the image of the first lens? (c) What is the value of p_2, the object position for the second lens? (d) Find the position q_2 of the image formed by the second lens. (e) Calculate the magnification of the first lens. (f) Calculate the magnification of the second lens. (g) What is the total magnification for the system? (h) Is the final image real or virtual? Is it upright or inverted (compared to the original object for the lens system)?

Additional Problems

47. An object placed 10.0 cm from a concave spherical mirror produces a real image 8.00 cm from the mirror. If the object is moved to a new position 20.0 cm from the mirror, what is the position of the image? Is the final image real or virtual?

48. S A real object's distance from a converging lens is five times the focal length. (a) Determine the location of the image q in terms of the focal length f. (b) Find the magnification of the image. (c) Is the image real or virtual? Is it upright or inverted? Is the image on the same side of the lens as the object or on the opposite side?

49. The magnitudes of the radii of curvature are 32.5 cm and 42.5 cm for the two faces of a biconcave lens. The glass has index of refraction 1.53 for violet light and 1.51 for red light. For a very distant object, locate (a) the image formed by violet light and (b) the image formed by red light.

50. A diverging lens (n = 1.50) is shaped like that in Figure 23.25c. The radius of the first surface is 15.0 cm, and that of the second surface is 10.0 cm. (a) Find the focal length of the lens. Determine the positions of the images for object distances of (b) infinity, (c) $3|f|$, (d) $|f|$, and (e) $|f|/2$.

51. The lens and the mirror in Figure P23.51 are separated by 1.00 m and have focal lengths of +80.0 cm and −50.0 cm, respectively. If an object is placed 1.00 m to the left of the lens, where will the final image be located? State whether the image is upright or inverted, and determine the overall magnification.

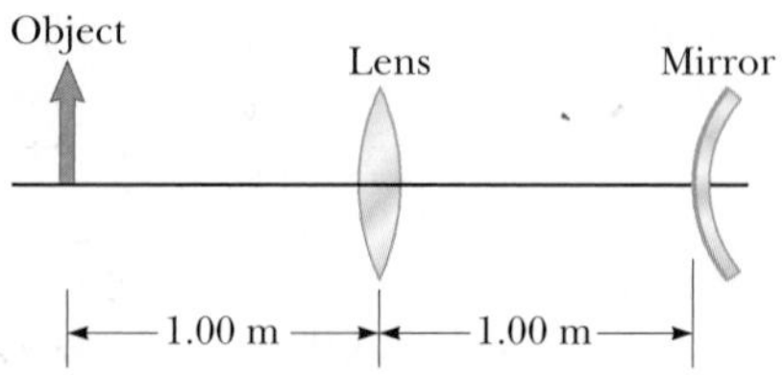

Figure P23.51

52. The object in Figure P23.52 is midway between the lens and the mirror, which are separated by a distance d = 25.0 cm. The magnitude of the mirror's radius of curvature is 20.0 cm, and the lens has a focal length of −16.7 cm. (a) Considering only the light that leaves the object and travels first toward the mirror, locate the final image formed by this system. (b) Is the image

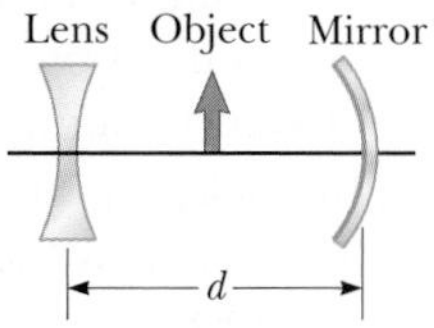

Figure P23.52

real or virtual? (c) Is it upright or inverted? (d) What is the overall magnification of the image?

53. A parallel beam of light enters a glass hemisphere perpendicular to the flat face, as shown in Figure P23.53. The radius of the hemisphere is $R = 6.00$ cm, and the index of refraction is $n = 1.56$. Determine the point at which the beam is focused. (Assume paraxial rays; i.e., assume all rays are located close to the principal axis.)

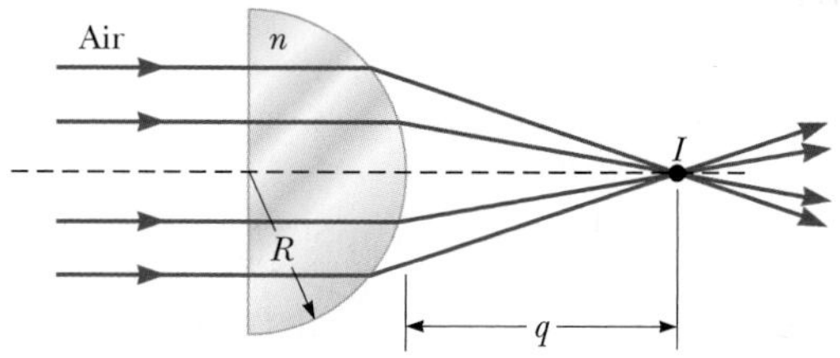

Figure P23.53

54. Two rays traveling parallel to the principal axis strike a large plano-convex lens having a refractive index of 1.60 (Fig. P23.54). If the convex face is spherical, a ray near the edge does not pass through the focal point (spherical aberration occurs). Assume this face has a radius of curvature of $R = 20.0$ cm and the two rays are at distances $h_1 = 0.500$ cm and $h_2 = 12.0$ cm from the principal axis. Find the difference Δx in the positions where each crosses the principal axis.

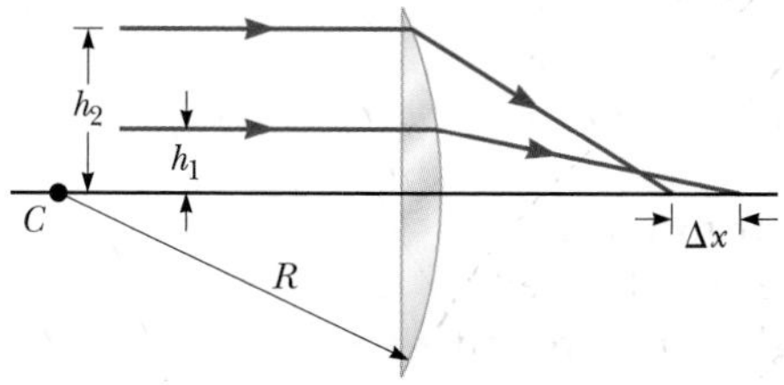

Figure P23.54

55. To work this problem, use the fact that the image formed by the first surface becomes the object for the second surface. Figure P23.55 shows a piece of glass with index of refraction $n = 1.50$ surrounded by air. The ends are hemispheres with radii $R_1 = 2.00$ cm and $R_2 = 4.00$ cm, and the centers of the hemispherical ends are separated by a distance of $d = 8.00$ cm. A point object is in air, a distance $p = 1.00$ cm from the left end of the glass. (a) Locate the image of the object due to refraction at the two spherical surfaces. (b) Is the image real or virtual?

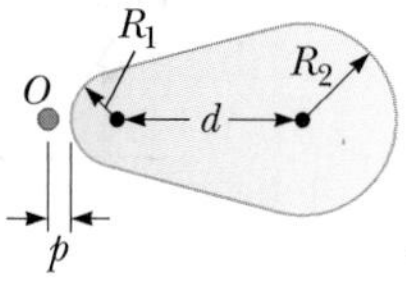

Figure P23.55

56. **S** Consider two thin lenses, one of focal length f_1 and the other of focal length f_2, placed in contact with each other, as shown in Figure P23.56. Apply the thin-lens equation to each of these lenses and combine the results to show that this combination of lenses behaves like a thin lens having a focal length f given by $1/f = 1/f_1 + 1/f_2$. Assume the thicknesses of the lenses can be ignored in comparison to the other distances involved.

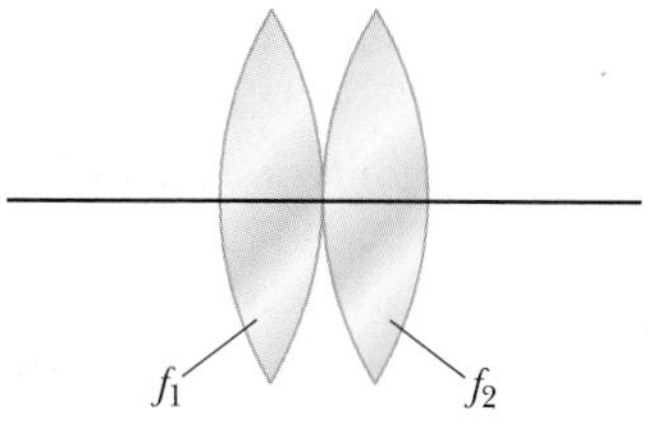

Figure P23.56

57. An object 2.00 cm high is placed 40.0 cm to the left of a converging lens having a focal length of 30.0 cm. A diverging lens having a focal length of −20.0 cm is placed 110 cm to the right of the converging lens. (a) Determine the final position and magnification of the final image. (b) Is the image upright or inverted? (c) Repeat parts (a) and (b) for the case in which the second lens is a converging lens having a focal length of +20.0 cm.

58. A "floating strawberry" illusion can be produced by two parabolic mirrors, each with a focal length of 7.5 cm, facing each other so that their centers are 7.5 cm apart (Fig. P23.58). If a strawberry is placed on the bottom mirror, an image of the strawberry forms at the small opening at the center of the top mirror. Show that the final image forms at that location and describe its characteristics. *Note:* A flashlight beam shone on these *images* has a very startling effect: Even at a glancing angle, the incoming light beam is seemingly reflected off the *images* of the strawberry! Do you understand why?

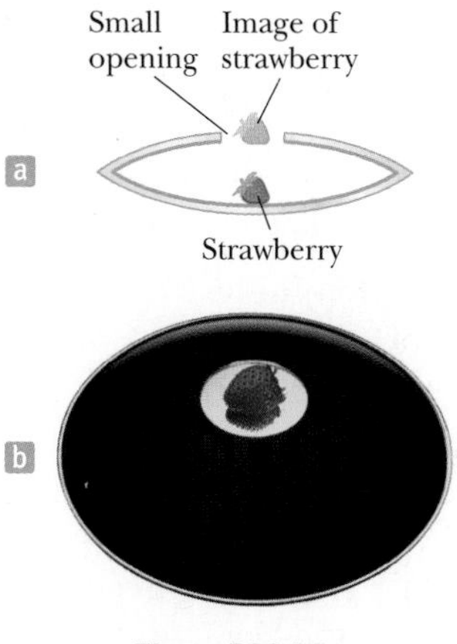

Michael Levin/Opti-Gone Associates

Figure P23.58

59. Figure P23.59 shows a converging lens with radii $R_1 = 9.00$ cm and $R_2 = -11.00$ cm, in front of a concave spherical mirror of radius $R = 8.00$ cm. The focal points (F_1 and F_2) for the thin lens and the center of

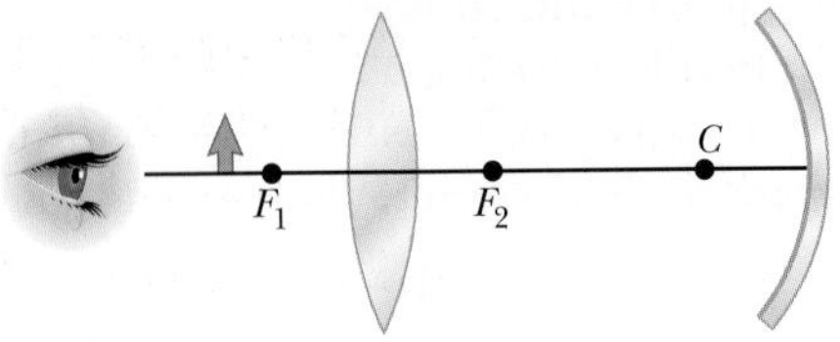

Figure P23.59

curvature (C) of the mirror are also shown. (a) If the focal points F_1 and F_2 are 5.00 cm from the vertex of the thin lens, what is the index of refraction of the lens? (b) If the lens and mirror are 20.0 cm apart and an object is placed 8.00 cm to the left of the lens, what is the position of the final image and its magnification as seen by the eye in the figure? (c) Is the final image inverted or upright? Explain.

60. **S** Find the object distances (in terms of f) for a thin converging lens of focal length f if (a) the image is real and the image distance is four times the focal length and (b) the image is virtual and the absolute value of the image distance is three times the focal length. (c) Calculate the magnification of the lens for cases (a) and (b).

61. The lens-maker's equation for a lens with index n_1 immersed in a medium with index n_2 takes the form

$$\frac{1}{f} = \left(\frac{n_1}{n_2} - 1\right)\left(\frac{1}{R_1} - \frac{1}{R_2}\right)$$

A thin diverging glass (index = 1.50) lens with $R_1 = -3.00$ m and $R_2 = -6.00$ m is surrounded by air. An arrow is placed 10.0 m to the left of the lens. (a) Determine the position of the image. Repeat part (a) with the arrow and lens immersed in (b) water (index = 1.33) (c) a medium with an index of refraction of 2.00. (d) How can a lens that is diverging in air be changed into a converging lens?

62. An observer to the right of the mirror–lens combination shown in Figure P23.62 sees two real images that are the same size and in the same location. One image is upright, and the other is inverted. Both images are 1.50 times larger than the object. The lens has a focal length of 10.0 cm. The lens and mirror are separated by 40.0 cm. Determine the focal length of the mirror. (Don't assume the figure is drawn to scale.)

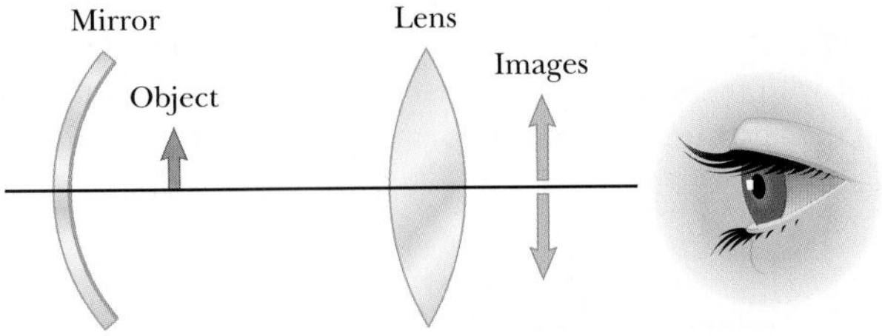

Figure P23.62

63. The lens-maker's equation applies to a lens immersed in a liquid if n in the equation is replaced by n_1/n_2. Here n_1 refers to the refractive index of the lens material and n_2 is that of the medium surrounding the lens. (a) A certain lens has focal length of 79.0 cm in air and a refractive index of 1.55. Find its focal length in water. (b) A certain mirror has focal length of 79.0 cm in air. Find its focal length in water.

64. A certain Christmas tree ornament is a silver sphere having a diameter of 8.50 cm. (a) If the size of an image created by reflection in the ornament is three-fourth's the reflected object's actual size, determine the object's location. (b) Use a principal-ray diagram to determine whether the image is upright or inverted.

65. **M** A glass sphere ($n = 1.50$) with a radius of 15.0 cm has a tiny air bubble 5.00 cm above its center. The sphere is viewed looking down along the extended radius containing the bubble. What is the apparent depth of the bubble below the surface of the sphere?

66. An object 10.0 cm tall is placed at the zero mark of a meterstick. A spherical mirror located at some point on the meterstick creates an image of the object that is upright, 4.00 cm tall, and located at the 42.0-cm mark of the meterstick. (a) Is the mirror convex or concave? (b) Where is the mirror? (c) What is the mirror's focal length?

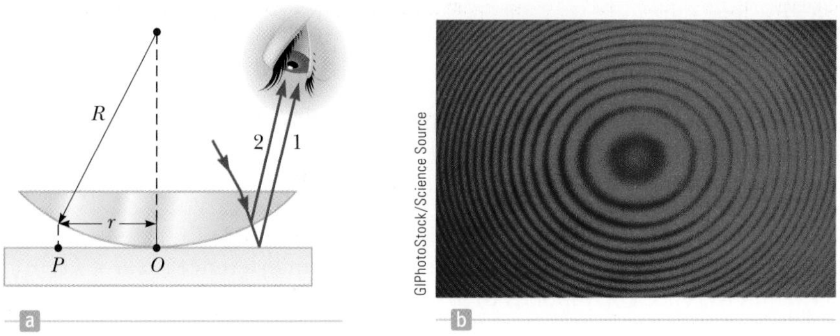

Figure 24.8 (a) The combination of rays reflected from the glass plate and the curved surface of the lens gives rise to an interference pattern known as Newton's rings. (b) A photograph of Newton's rings.

Newton's Rings

Another method for observing interference in light waves is to place a plano-convex lens on top of a flat glass surface, as in Figure 24.8a. With this arrangement, the air film between the glass surfaces varies in thickness from zero at the point of contact to some value t at P. If the radius of curvature R of the lens is much greater than the distance r and the system is viewed from above using light of wavelength λ, a pattern of light and dark rings is observed (Fig. 24.8b). These circular fringes, discovered by Newton, are called **Newton's rings**. The interference is due to the combination of ray 1, reflected from the plate, with ray 2, reflected from the lower surface of the lens. Ray 1 undergoes a phase change of 180° on reflection because it is reflected from a boundary leading into a medium of higher refractive index, whereas ray 2 undergoes no phase change because it is reflected from a medium of lower refractive index. Hence, the conditions for constructive and destructive interference are given by Equations 24.9 and 24.10, respectively, with $n = 1$ because the "film" is air. The contact point at O is dark, as seen in Figure 24.8b, because there is no path difference and the total phase change is due only to the 180° phase change upon reflection. Using the geometry shown in Figure 24.8a, we can obtain expressions for the radii of the bright and dark bands in terms of the radius of curvature R and vacuum wavelength λ. For example, the dark rings have radii of $r \approx \sqrt{m\lambda R/n}$.

Tip 24.2 The Two Tricks of Thin Films

Be sure to include *both* effects—path length and phase change—when you analyze an interference pattern from a thin film.

One important use of Newton's rings is in the testing of optical lenses. A circular pattern like that in Figure 24.8b is achieved only when the lens is ground to a perfectly spherical curvature. Variations from such symmetry produce distorted patterns that also give an indication of how the lens must be reground and repolished to remove imperfections.

APPLICATION

Checking for Imperfections in Optical Lenses

PROBLEM-SOLVING STRATEGY

Thin-Film Interference

The following steps are recommended in addressing thin-film interference problems:

1. Identify the thin film causing the interference, and the indices of refraction in the film and in the media on either side of it.
2. Determine the number of phase reversals: zero, one, or two.
3. Consult the following table, which contains Equations 24.9 and 24.10, and select the correct column for the problem in question:

Equation (m = 0, 1, . . .)	1 Phase Reversal	0 or 2 Phase Reversals
$2nt = (m + \frac{1}{2})\lambda$ **[24.9]**	Constructive	Destructive
$2nt = m\lambda$ **[24.10]**	Destructive	Constructive

4. Substitute values in the appropriate equations, as selected in the previous step.

■ EXAMPLE 24.2 Interference in a Soap Film

GOAL Study constructive interference effects in a thin film.

PROBLEM **(a)** Calculate the minimum thickness of a soap-bubble film ($n = 1.33$) that will result in constructive interference in the reflected light if the film is illuminated by light with wavelength 602 nm in free space. **(b)** Recalculate the minimum thickness for constructive interference when the soap-bubble film is on top of a glass slide with $n = 1.50$.

STRATEGY In part (a) there is only one inversion, so the condition for constructive interference is $2nt = \left(m + \frac{1}{2}\right)\lambda$. The minimum film thickness for constructive interference corresponds to $m = 0$ in this equation. Part (b) involves two inversions, so $2nt = m\lambda$ is required.

SOLUTION

(a) Calculate the minimum thickness of the soap-bubble film that will result in constructive interference.

Solve $2nt = \lambda/2$ for the thickness t and substitute:

$$t = \frac{\lambda}{4n} = \frac{602\text{ nm}}{4(1.33)} = \boxed{113\text{ nm}}$$

(b) Find the minimum soap-film thickness when the film is on top of a glass slide with $n = 1.50$.

Write the condition for constructive interference, when two inversions take place:

$$2nt = m\lambda$$

Solve for t and substitute:

$$t = \frac{m\lambda}{2n} = \frac{1 \cdot (602\text{ nm})}{2(1.33)} = \boxed{226\text{ nm}}$$

REMARKS The different colors in a soap bubble result from the thickness of the soap layer varying from one place to another. The swirling is caused by the changing thickness of the layer with time.

QUESTION 24.2 A soap film looks red in one area and violet in a nearby area. In which area is the soap film thicker?

EXERCISE 24.2 What other film thicknesses in part (a) will produce constructive interference?

ANSWERS 339 nm, 566 nm, 792 nm, and so on

■ EXAMPLE 24.3 Nonreflective Coatings for Solar Cells and Optical Lenses

GOAL Study destructive interference effects in a thin film when there are two inversions.

PROBLEM Semiconductors such as silicon are used to fabricate solar cells, devices that generate electric energy when exposed to sunlight. Solar cells are often coated with a transparent thin film, such as silicon monoxide (SiO; $n = 1.45$), to minimize reflective losses (Fig. 24.9). A silicon solar cell ($n = 3.50$) is coated with a thin film of silicon monoxide for this purpose. Assuming normal incidence, determine the minimum thickness of the film that will produce the least reflection at a wavelength of 552 nm.

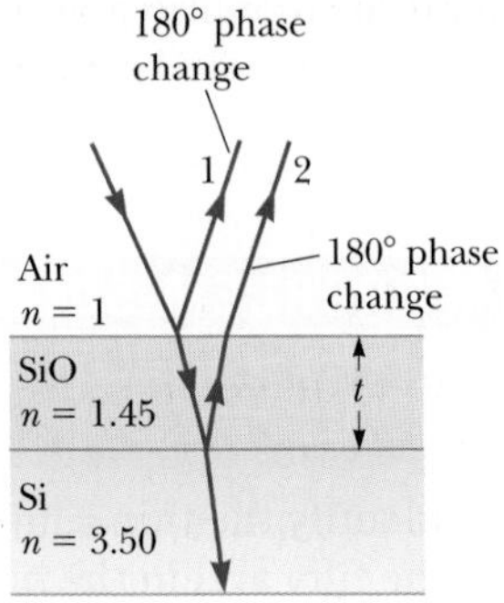

Figure 24.9 (Example 24.3) Reflective losses from a silicon solar cell are minimized by coating it with a thin film of silicon monoxide (SiO).

STRATEGY Reflection is least when rays 1 and 2 in Figure 24.9 meet the condition for destructive interference. Note that *both* rays undergo 180° phase changes on reflection. The condition for a reflection *minimum* is therefore $2nt = \lambda/2$.

SOLUTION

Solve $2nt = \lambda/2$ for t, the required thickness:

$$t = \frac{\lambda}{4n} = \frac{552\text{ nm}}{4(1.45)} = \boxed{95.2\text{ nm}}$$

REMARKS Typically, such coatings reduce the reflective loss from 30% (with no coating) to 10% (with a coating), thereby increasing the cell's efficiency because more light is available to create charge carriers in the cell. In reality the coating is never perfectly nonreflecting because the required thickness is wavelength dependent and the incident light covers a wide range of wavelengths.

QUESTION 24.3 To minimize reflection of a smaller wavelength, should the thickness of the coating be thicker or thinner?

EXERCISE 24.3 Glass lenses used in cameras and other optical instruments are usually coated with one or more transparent thin films, such as magnesium fluoride (MgF_2), to reduce or eliminate unwanted reflection. Carl Zeiss developed this method; his first coating was 1.00×10^2 nm thick, on glass. Using $n = 1.38$ for MgF_2, what visible wavelength would be eliminated by destructive interference in the reflected light?

ANSWER 552 nm

EXAMPLE 24.4 Interference in a Wedge-Shaped Film

GOAL Calculate interference effects when the film has variable thickness.

PROBLEM A pair of glass slides 10.0 cm long and with $n = 1.52$ are separated on one end by a hair, forming a triangular wedge of air, as illustrated in Figure 24.10. When coherent light from a helium–neon laser with wavelength 633 nm is incident on the film from above, 15.0 dark fringes per centimeter are observed. How thick is the hair?

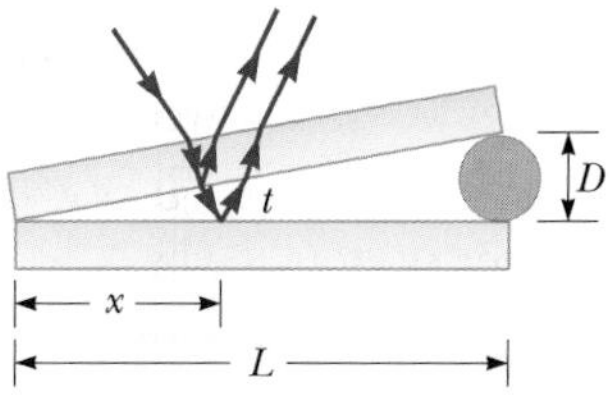

Figure 24.10 (Example 24.4) Interference bands in reflected light can be observed by illuminating a wedge-shaped film with monochromatic light. The dark areas in the interference pattern correspond to positions of destructive interference.

STRATEGY The interference pattern is created by the thin film of air having variable thickness. The pattern is a series of alternating bright and dark parallel bands. A dark band corresponds to destructive interference, and there is one phase reversal, so $2nt = m\lambda$ should be used. We can also use the similar triangles in Figure 24.10 to obtain the relation $t/x = D/L$. We can find the thickness for any m, and if the position x can also be found, this last equation gives the diameter of the hair, D.

SOLUTION

Solve the destructive-interference equation for the thickness of the film, t, with $n = 1$ for air:

$$t = \frac{m\lambda}{2}$$

If d is the distance from one dark band to the next, then the x-coordinate of the mth band is a multiple of d:

$$x = md$$

By dimensional analysis, d is just the inverse of the number of bands per centimeter.

$$d = \left(15.0\,\frac{\text{bands}}{\text{cm}}\right)^{-1} = 6.67 \times 10^{-2}\,\frac{\text{cm}}{\text{band}}$$

Now use similar triangles and substitute all the information:

$$\frac{t}{x} = \frac{m\lambda/2}{md} = \frac{\lambda}{2d} = \frac{D}{L}$$

Solve for D and substitute given values:

$$D = \frac{\lambda L}{2d} = \frac{(633 \times 10^{-9}\,\text{m})(0.100\,\text{m})}{2(6.67 \times 10^{-4}\,\text{m})} = \boxed{4.75 \times 10^{-5}\,\text{m}}$$

REMARKS Some may be concerned about interference caused by light bouncing off the top and bottom of, say, the upper glass slide. It's unlikely, however, that the thickness of the slide will be half an integer multiple of the wavelength of the helium–neon laser (for some very large value of m). In addition, in contrast to the air wedge, the thickness of the glass doesn't vary.

QUESTION 24.4 If the air wedge is filled with water, how is the distance between dark bands affected? Explain.

EXERCISE 24.4 The air wedge is replaced with water, with $n = 1.33$. Find the distance between dark bands when the helium–neon laser light hits the glass slides.

ANSWER 5.02×10^{-4} m

APPLYING PHYSICS 24.3 | Perfect Mirrors

When light hits a metallic mirror, electrons in the metal move in response to the electromagnetic fields, absorbing some of the light's energy. For many applications, such as directing high-intensity laser light, that reduction in intensity is undesirable. A dielectric mirror, on the other hand, is made of glass or plastic and doesn't conduct electricity. To improve reflectance, thin layers of different dielectric materials are stacked on the glass surface. If the thicknesses and dielectric constants are chosen properly, light reflected off one layer combines constructively with light reflected from the layer underneath, increasing the mirror's reflectance. Nearly perfect mirrors can be constructed using several thin dielectric layers. The mirrors can be designed to reflect a particular wavelength or a range of wavelengths.

Dielectric mirrors, first developed at MIT in 1998, have an enormous number of applications. One of the most important is the OmniGuide fiber, a hollow tube the size of a spaghetti noodle that can guide light without any significant loss of intensity. (Ordinary optical fibers tend to heat up.) Such fibers have up to forty concentric layers of plastic and glass and are highly flexible. Using an OmniGuide fiber, an intense laser light can be safely guided into the human body during surgery to remove tumors and other diseased tissues without harming the healthy surrounding tissue. ■

24.5 Using Interference to Read CDs and DVDs

LEARNING OBJECTIVES

1. Discuss the application of interference principles to the reading of CDs and DVDs.
2. Apply interference principles to CDs and DVDs.

APPLICATION
The Physics of CDs and DVDs

Compact discs (CDs) and digital videodiscs (DVDs) provide high-density storage of text, graphics, and movies; and high-quality sound recordings. The data on these discs are stored digitally as a series of zeros and ones, and these zeros and ones are read by laser light reflected from the disc. Strong reflections (constructive interference) from the disc are chosen to represent zeros, and weak reflections (destructive interference) represent ones.

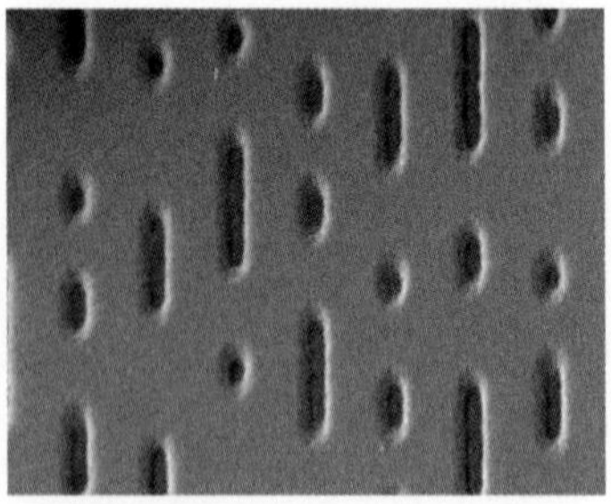
Andrew Syred / Science Source

Figure 24.11 A photomicrograph of adjacent tracks on a digital video disc (DVD). The information encoded in these pits and smooth areas is read by a laser beam.

To see in more detail how thin-film interference plays a crucial role in reading CDs and DVDs, consider Figure 24.11. This figure shows a photomicrograph of several DVD tracks, which consist of a sequence of pits (when viewed from the top or label side of the disc) of varying length formed in a reflecting-metal information layer. A cross-sectional view of a CD as shown in Figure 24.12 reveals that the pits appear as bumps to the laser beam, which shines on the metallic layer through a clear plastic coating from below.

As the disk rotates, the laser beam reflects off the sequence of bumps and lower areas into a photodetector, which converts the fluctuating reflected light intensity into an electrical string of zeros and ones. To make the light fluctuations more pronounced and easier to detect, the pit depth t is made equal to one-quarter of a

Figure 24.12 Cross section of a CD showing metallic pits of depth t and a laser beam detecting the edge of a pit.

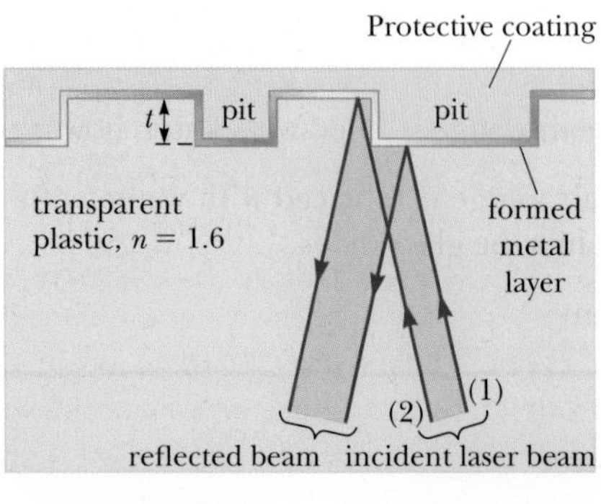

SOLUTION

(a) Calculate the intensity of the beam after it passes through the second polarizer.

The incident intensity is $I_b/2$. Apply Malus's law to the second polarizer:

$$I_2 = I_0 \cos^2\theta = \frac{I_b}{2}\cos^2(30.0°) = \frac{I_b}{2}\left(\frac{\sqrt{3}}{2}\right)^2 = \boxed{\frac{3}{8}I_b}$$

(b) Calculate the intensity of the beam after it passes through the third polarizer.

The incident intensity is now $3I_b/8$. Apply Malus's law to the third polarizer.

$$I_3 = I_2 \cos^2\theta = \frac{3}{8}I_b\cos^2(45.0°) = \frac{3}{8}I_b\left(\frac{\sqrt{2}}{2}\right)^2 = \boxed{\frac{3}{16}I_b}$$

REMARKS Notice that the angle used in part (b) was not 75.0°, but 75.0° − 30.0° = 45.0°. The angle is always with respect to the previous polarizer's transmission axis because the polarizing material physically determines what direction the transmitted electric fields can have.

QUESTION 24.8 At what angle relative to the previous polarizer must an additional polarizer be placed so as to completely block the light?

EXERCISE 24.8 The polarizers are rotated so that the second polarizer has a transmission axis of 40.0° with respect to the first polarizer, and the third polarizer has an angle of 90.0° with respect to the first. If I_b is the intensity of the original unpolarized light, what is the intensity of the beam after it passes through (a) the second polarizer and (b) the third polarizer? (c) What is the final transmitted intensity if the second polarizer is removed?

ANSWERS (a) $0.293I_b$ (b) $0.121I_b$ (c) 0

Polarization by Reflection

When an unpolarized light beam is reflected from a surface, the reflected light is completely polarized, partially polarized, or unpolarized, depending on the angle of incidence. If the angle of incidence is either 0° or 90° (a normal or grazing angle), the reflected beam is unpolarized. For angles of incidence between 0° and 90°, however, the reflected light is polarized to some extent. For one particular angle of incidence the reflected beam is completely polarized.

Suppose an unpolarized light beam is incident on a surface, as in Figure 24.28a. The beam can be described by two electric field components, one parallel to the surface (represented by dots) and the other perpendicular to the first component and to the direction of propagation (represented by orange arrows). It is found

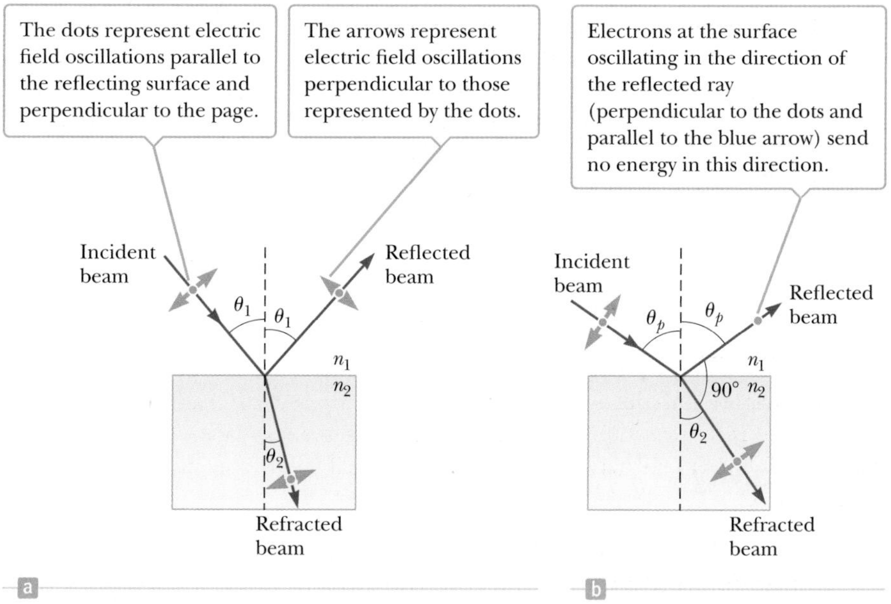

Figure 24.28 (a) When unpolarized light is incident on a reflecting surface, the reflected and refracted beams are partially polarized. (b) The reflected beam is completely polarized when the angle of incidence equals the polarizing angle θ_p, satisfying the equation $n = \tan\theta_p$.

that the parallel component reflects more strongly than the other components, and the result is a partially polarized beam. In addition, the refracted beam is also partially polarized.

Now suppose the angle of incidence, θ_1, is varied until the angle between the reflected and refracted beams is 90° (Fig. 24.28b). At this particular angle of incidence, called the **polarizing angle** θ_p, the reflected beam is completely polarized, with its electric field vector parallel to the surface, while the refracted beam is partially polarized.

An expression relating the polarizing angle to the index of refraction of the reflecting surface can be obtained by the use of Figure 24.28b. From this figure, we see that at the polarizing angle, $\theta_p + 90° + \theta_2 = 180°$, so $\theta_2 = 90° - \theta_p$. Using Snell's law and taking $n_1 = n_{air} = 1.00$ and $n_2 = n$ yields

$$n = \frac{\sin\theta_1}{\sin\theta_2} = \frac{\sin\theta_p}{\sin\theta_2}$$

Because $\sin\theta_2 = \sin(90° - \theta_p) = \cos\theta_p$, the expression for n can be written

Brewster's law ▶

$$n = \frac{\sin\theta_p}{\cos\theta_p} = \tan\theta_p \qquad [24.14]$$

Equation 24.14 is called **Brewster's law**, and the polarizing angle θ_p is sometimes called **Brewster's angle** after its discoverer, Sir David Brewster (1781–1868). For example, Brewster's angle for crown glass (where $n = 1.52$) has the value $\theta_p = \tan^{-1}(1.52) = 56.7°$. Because n varies with wavelength for a given substance, Brewster's angle is also a function of wavelength.

APPLICATION
Polaroid Sunglasses

Polarization by reflection is a common phenomenon. Sunlight reflected from water, glass, or snow is partially polarized. If the surface is horizontal, the electric field vector of the reflected light has a strong horizontal component. Sunglasses made of polarizing material reduce the glare, which *is* the reflected light. The transmission axes of the lenses are oriented vertically to absorb the strong horizontal component of the reflected light. Because the reflected light is mostly polarized, most of the glare can be eliminated without removing most of the normal light.

Polarization by Scattering

When light is incident on a system of particles, such as a gas, the electrons in the medium can absorb and reradiate part of the light. The absorption and reradiation of light by the medium, called **scattering**, is what causes sunlight reaching an observer on Earth from straight overhead to be polarized. You can observe this effect by looking directly up through a pair of sunglasses made of polarizing glass. Less light passes through at certain orientations of the lenses than at others.

Figure 24.29 illustrates how the sunlight becomes polarized. The left side of the figure shows an incident unpolarized beam of sunlight on the verge of striking an air molecule. When the beam strikes the air molecule, it sets the electrons of the molecule into vibration. These vibrating charges act like those in an antenna except that they vibrate in a complicated pattern. The horizontal part of the electric field vector in the incident wave causes the charges to vibrate horizontally, and the vertical part of the vector simultaneously causes them to vibrate vertically. A horizontally polarized wave is emitted by the electrons as a result of their horizontal motion, and a vertically polarized wave is emitted parallel to Earth as a result of their vertical motion.

Scientists have found that bees and homing pigeons use the polarization of sunlight as a navigational aid.

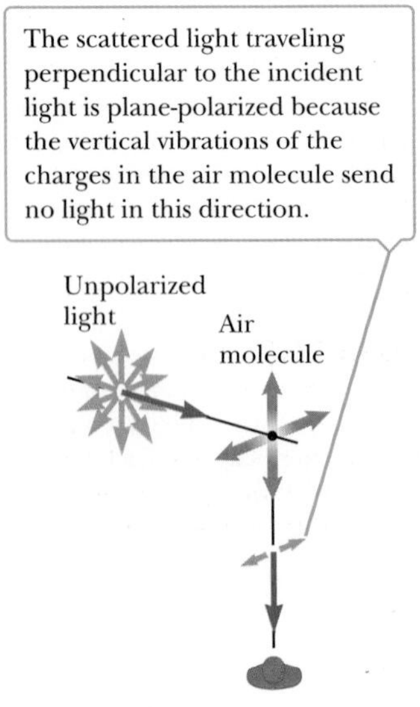

Figure 24.29 The scattering of unpolarized sunlight by air molecules.

Optical Activity

Many important practical applications of polarized light involve the use of certain materials that display the property of **optical activity**. A substance is said to be optically active if it rotates the plane of polarization of transmitted light. Suppose

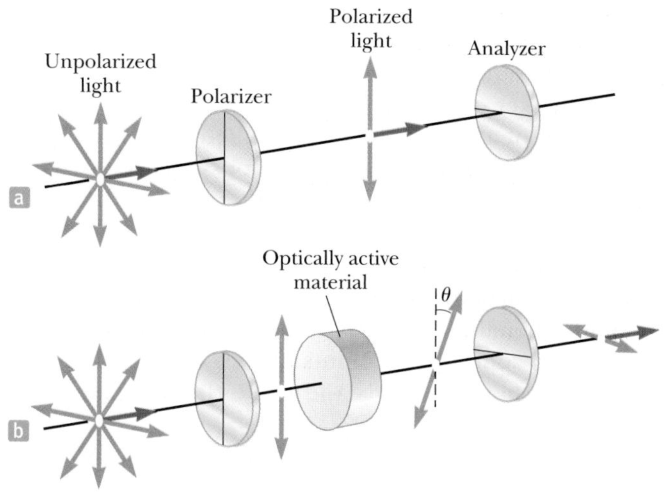

Figure 24.30 (a) When crossed polarizers are used, none of the polarized light can pass through the analyzer. (b) An optically active material rotates the direction of polarization through the angle θ, enabling some of the polarized light to pass through the analyzer.

unpolarized light is incident on a polarizer from the left, as in Figure 24.30a. The transmitted light is polarized vertically, as shown. If this light is then incident on an analyzer with its axis perpendicular to that of the polarizer, no light emerges from it. If an optically active material is placed between the polarizer and analyzer, as in Figure 24.30b, the material causes the direction of the polarized beam to rotate through the angle θ. As a result, some light is able to pass through the analyzer. The angle through which the light is rotated by the material can be found by rotating the polarizer until the light is again extinguished. It is found that the angle of rotation depends on the length of the sample and, if the substance is in solution, on the concentration. One optically active material is a solution of common sugar, dextrose. A standard method for determining the concentration of a sugar solution is to measure the rotation produced by a fixed length of the solution.

APPLICATION

Finding the Concentrations of Solutions by Means of Their Optical Activity

Optical activity occurs in a material because of an asymmetry in the shape of its constituent molecules. For example, some proteins are optically active because of their spiral shapes. Other materials, such as glass and plastic, become optically active when placed under stress. If polarized light is passed through an unstressed piece of plastic and then through an analyzer with an axis perpendicular to that of the polarizer, none of the polarized light is transmitted. If the plastic is placed under stress, however, the regions of greatest stress produce the largest angles of rotation of polarized light, and a series of light and dark bands are observed in the transmitted light. Engineers often use this property in the design of structures ranging from bridges to small tools. A plastic model is built and analyzed under different load conditions to determine positions of potential weakness and failure under stress. If the design is poor, patterns of light and dark bands will indicate the points of greatest weakness, and the design can be corrected at an early stage. Figure 24.31 shows examples of stress patterns in plastic.

Figure 24.31 A plastic model of an arch structure under load conditions observed between perpendicular polarizers. Such patterns are useful in the optimum design of architectural components.

APPLICATION
Liquid Crystal Displays (LCDs)

Liquid Crystals

An effect similar to rotation of the plane of polarization is used to create the familiar displays on pocket calculators, wristwatches, notebook computers, and so forth. The properties of a unique substance called a liquid crystal make these displays (called LCDs, for *liquid crystal displays*) possible. As its name implies, a **liquid crystal** is a substance with properties intermediate between those of a crystalline solid and those of a liquid; that is, the molecules of the substance are more orderly than those in a liquid, but less orderly than those in a pure crystalline solid. The forces that hold the molecules together in such a state are just barely strong enough to enable the substance to maintain a definite shape, so it is reasonable to call it a solid. Small inputs of mechanical or electrical energy, however, can disrupt these weak bonds and make the substance flow, rotate, or twist.

To see how liquid crystals can be used to create a display, consider Figure 24.32a. The liquid crystal is placed between two glass plates in the pattern shown, and electrical contacts, indicated by the thin lines, are made. When a voltage is applied across any segment in the display, that segment turns dark. In this fashion any number between 0 and 9 can be formed by the pattern, depending on the voltages applied to the seven segments.

To see why a segment can be changed from dark to light by the application of a voltage, consider Figure 24.32b, which shows the basic construction of a portion of the display. The liquid crystal is placed between two glass substrates that are packaged between two pieces of Polaroid material with their transmission axes perpendicular. A reflecting surface is placed behind one of the pieces of Polaroid. First consider what happens when light falls on this package and no voltages are applied to the liquid crystal, as shown in Figure 24.32b. Incoming light is polarized by the polarizer on the left and then falls on the liquid crystal. As the light passes through the crystal, its plane of polarization is rotated by 90°, allowing it to pass through the polarizer on the right. It reflects from the reflecting surface and

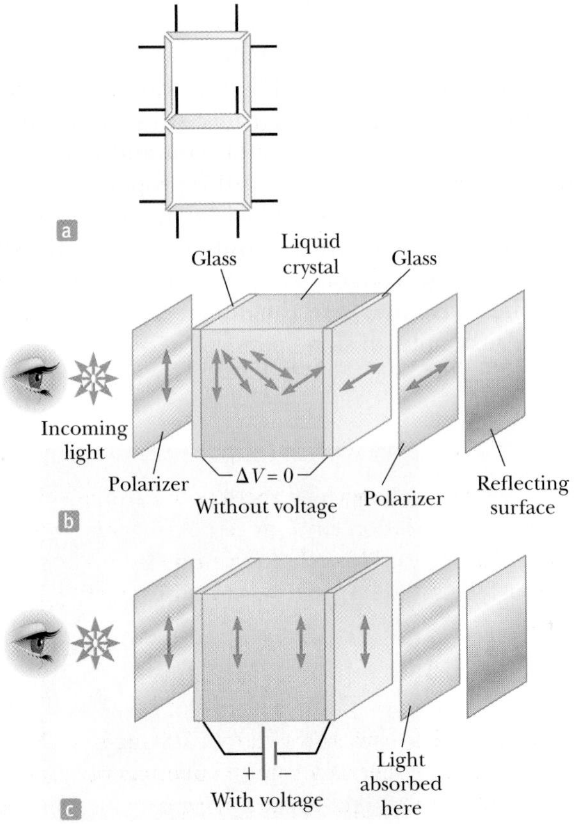

Figure 24.32 (a) The light-segment pattern of a liquid crystal display. (b) Rotation of a polarized light beam by a liquid crystal when the applied voltage is zero. (c) Molecules of the liquid crystal align with the electric field when a voltage is applied.

retraces its path through the crystal. Thus, an observer to the left of the crystal sees the segment as being bright. When a voltage is applied as in Figure 24.32c, the molecules of the liquid crystal don't rotate the plane of polarization of the light. In this case the light is absorbed by the polarizer on the right, and none is reflected back to the observer to the left of the crystal. As a result, the observer sees this segment as black. Changing the applied voltage to the crystal in a precise pattern at precise times can make the pattern tick off the seconds on a watch, display a letter on a computer display, and so forth.

SUMMARY

24.1 Conditions for Interference

Interference occurs when two or more light waves overlap at a given point. A sustained interference pattern is observed if (1) the sources are coherent (that is, they maintain a constant phase relationship with one another), (2) the sources have identical wavelengths, and (3) the superposition principle is applicable.

24.2 Young's Double-Slit Experiment

In **Young's double-slit experiment** two slits separated by distance d are illuminated by a single-wavelength light source. An interference pattern consisting of bright and dark fringes is observed on a screen a distance L from the slits. The condition for **bright fringes** (constructive interference) is

$$d \sin \theta_{\text{bright}} = m\lambda \qquad m = 0, \pm 1, \pm 2, \ldots \qquad [24.2]$$

The number m is called the **order number** of the fringe.

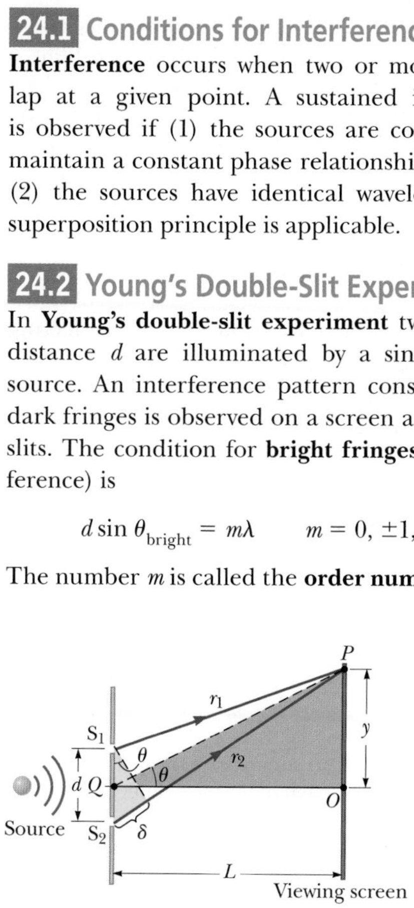

A geometric construction that describes Young's double-slit experiment. (This figure is not drawn to scale.)

The condition for **dark fringes** (destructive interference) is

$$d \sin \theta_{\text{dark}} = (m + \tfrac{1}{2})\lambda \qquad m = 0, \pm 1, \pm 2, \ldots \qquad [24.3]$$

The position y_m of the bright fringes on the screen can be determined by using the relation $\sin \theta \approx \tan \theta = y_m/L$, which is true for small angles. This relation can be substituted into Equations 24.2 and 24.3, yielding the location of the bright fringes:

$$y_{\text{bright}} = \frac{\lambda L}{d} m \qquad m = 0, \pm 1, \pm 2, \ldots \qquad [24.5]$$

A similar expression can be derived for the dark fringes. This equation can be used either to locate the maxima or to determine the wavelength of light by measuring y_m.

24.3 Change of Phase Due to Reflection

24.4 Interference in Thin Films

An electromagnetic wave undergoes a phase change of 180° on reflection from a medium with an index of refraction higher than that of the medium in which the wave is traveling. There is no change when the wave, traveling in a medium with higher index of refraction, reflects from a medium with a lower index of refraction.

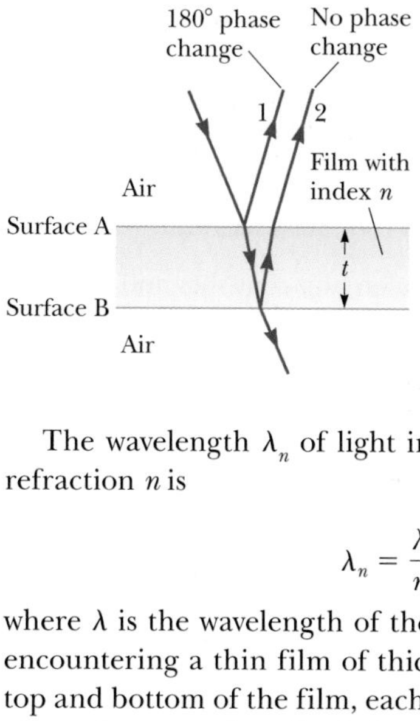

Interference in light reflected from a thin film is due to a combination of rays 1 and 2 reflected from the upper and lower surfaces of the film.

The wavelength λ_n of light in a medium with index of refraction n is

$$\lambda_n = \frac{\lambda}{n} \qquad [24.7]$$

where λ is the wavelength of the light in free space. Light encountering a thin film of thickness t will reflect off the top and bottom of the film, each ray undergoing a possible phase change as described above. The two rays recombine, and bright and dark fringes will be observed, with the conditions of interference given by the following table:

Equation ($m = 0, 1, \ldots$)		1 Phase Reversal	0 or 2 Phase Reversals
$2nt = (m + \tfrac{1}{2})\lambda$	[24.9]	Constructive	Destructive
$2nt = m\lambda$	[24.10]	Destructive	Constructive

24.6 Diffraction

24.7 Single-Slit Diffraction

Diffraction occurs when waves pass through small openings, around obstacles, or by sharp edges. The **diffraction pattern** produced by a single slit on a distant screen consists of a central bright maximum flanked by less bright fringes alternating with dark regions. The angles θ at which the diffraction pattern has zero intensity (regions of destructive interference) are described by

$$\sin\theta_{\text{dark}} = m\frac{\lambda}{a} \qquad m = \pm 1, \pm 2, \pm 3, \ldots \qquad [24.11]$$

where λ is the wavelength of the light incident on the slit and a is the width of the slit.

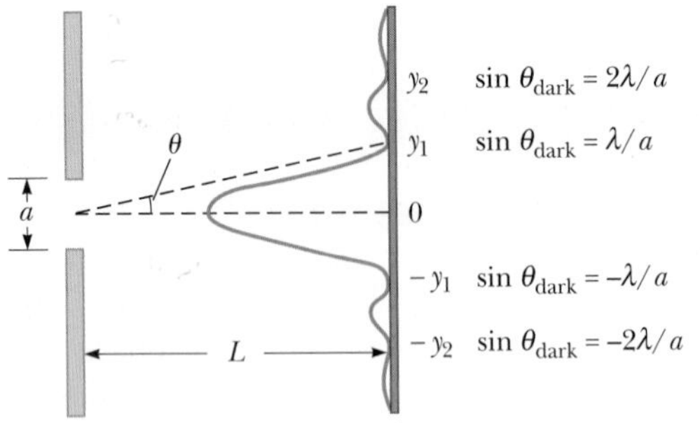

Positions of the minima for the Fraunhofer diffraction pattern of a single slit of width a. (This drawing is not to scale.)

24.8 The Diffraction Grating

A **diffraction grating** consists of many equally spaced, identical slits. The condition for **maximum intensity** in the interference pattern of a diffraction grating is

$$d\sin\theta_{\text{bright}} = m\lambda \qquad m = 0, \pm 1, \pm 2, \ldots \qquad [24.12]$$

where d is the spacing between adjacent slits and m is the order number of the diffraction pattern. A diffraction grating can be made by putting a large number of evenly spaced scratches on a glass slide. The number of such lines per centimeter is the inverse of the spacing d.

24.9 Polarization of Light Waves

Unpolarized light can be polarized by selective absorption, reflection, or scattering. A material can polarize light if it transmits waves having electric field vectors that vibrate in a plane parallel to a certain direction and absorbs waves with electric field vectors vibrating in directions perpendicular to that direction. When unpolarized light passes through a polarizing sheet, its intensity is reduced by half and the light becomes polarized. When this light passes through a second polarizing sheet with transmission axis at an angle of θ with respect to the transmission axis of the first sheet, the transmitted intensity is given by

$$I = I_0\cos^2\theta \qquad [24.13]$$

where I_0 is the intensity of the light after passing through the first polarizing sheet.

In general, light reflected from an amorphous material, such as glass, is partially polarized. Reflected light is completely polarized, with its electric field parallel to the surface, when the angle of incidence produces a 90° angle between the reflected and refracted beams. This angle of incidence, called the **polarizing angle** θ_p, satisfies **Brewster's law,** given by

$$n = \tan\theta_p \qquad [24.14]$$

where n is the index of refraction of the reflecting medium.

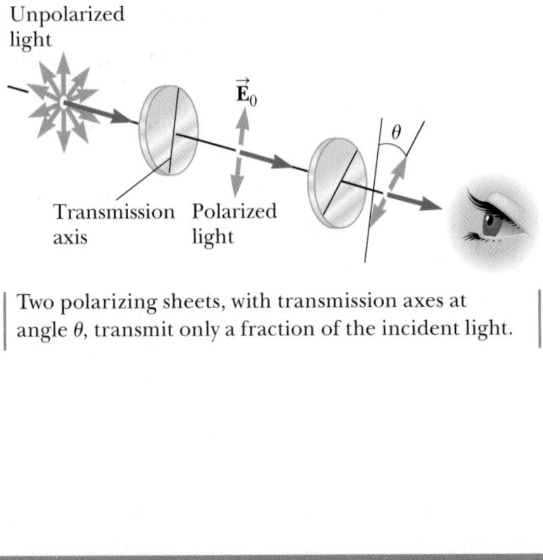

Two polarizing sheets, with transmission axes at angle θ, transmit only a fraction of the incident light.

WARM-UP EXERCISES

ENHANCED WebAssign The warm-up exercises in this chapter may be assigned online in Enhanced WebAssign.

1. **Physics Review** Two speakers, one at the origin and the other facing it at x = 1.00 m, are driven by the same oscillator at a frequency of 655 Hz. If the speed of sound is 343 m/s, find (a) the wavelength of sound waves produced by the speakers and (b) the location of a point of destructive interference closest to 0.500 m on the interval [0, 0.500 m]. (See Section 14.7.)

2. **Physics Review** (a) Calculate the frequency of monochromatic light having a wavelength in air of 5.30×10^2 nm. Find (b) the wavelength and (c) the frequency of the same light when inside diamond. (See Sections 21.12 and 22.3.)

3. In a Young's double-slit experiment, 633-nm laser light illuminates two slits separated by a distance of 3.50×10^{-5} m. (a) Find the angle at which the first order maximum occurs. (b) If the interference pattern is displayed on a screen 1.50 m away, determine the distance from the centerline to the first order maximum. Find also

(c) the angle at which the second-order ($m = 2$) minimum occurs and (d) the distance of that minimum from the centerline. (See Section 24.2.)

4. For a 4.76×10^{14}-Hz light wave, determine the wavelength (a) in space where the index of refraction is 1.00 and (b) in water where the index of refraction is 1.33. (See Section 24.4.)

5. Suppose the thin-film coating ($n = 1.17$) on an eyeglass lens ($n = 1.33$) is designed to eliminate reflection of 535-nm light. Determine (a) the number of phase inversions that occur, (b) the wavelength of light in the thin-film coating and (c) the minimum required coating thickness. Take the index of refraction for air to be 1.00. (See Section 24.4.)

6. Suppose 633-nm laser light illuminates a single slit of width 2.50×10^{-4} m. Determine the angle from the laser beam's path to the third dark fringe. (See Section 24.7.)

7. Light from a 543-nm laser beam is normally incident on a diffraction grating with 4.50×10^5 lines/m. Determine (a) the spacing between slits in the grating and (b) the angle to the third-order maximum. (See Section 24.8.)

8. An unpolarized beam of light with intensity of 50.0 W/m^2 passes through a polarizer with a vertical transmission axis. Determine the intensity of the light beam after passing through the polarizer. (See Section 24.9.)

9. A beam of light has an intensity of 70.0 W/m^2 and is polarized along the horizontal axis. Determine the intensity of the light beam after passing through a polarizer with a transmission axis oriented 30.0° from the horizontal. (See Section 24.9.)

10. Determine Brewster's angle for light reflected from water with an index of refraction equal to 1.33. (See Section 24.9.)

CONCEPTUAL QUESTIONS

ENHANCED WebAssign The conceptual questions in this chapter may be assigned online in Enhanced WebAssign.

1. Your automobile has two headlights. What sort of interference pattern do you expect to see from them? Why?

2. Holding your hand at arm's length, you can readily block direct sunlight from your eyes. Why can you not block sound from your ears this way?

3. Consider a dark fringe in an interference pattern at which almost no light energy is arriving. Light from both slits is arriving at this point, but the waves cancel. Where does the energy go?

4. If Young's double-slit experiment were performed under water, how would the observed interference pattern be affected?

5. In a laboratory accident, you spill two liquids onto water, neither of which mixes with the water. They both form thin films on the water surface. As the films spread and become very thin, you notice that one film becomes bright and the other black in reflected light. Why might that be?

6. If white light is used in Young's double-slit experiment rather than monochromatic light, how does the interference pattern change?

7. A lens with outer radius of curvature R and index of refraction n rests on a flat glass plate, and the combination is illuminated from white light from above. Is there a dark spot or a light spot at the center of the lens? What does it mean if the observed rings are noncircular?

8. Fingerprints left on a piece of glass such as a windowpane can show colored spectra like that from a diffraction grating. Why?

9. In everyday experience, why are radio waves polarized, whereas light is not?

10. Suppose reflected white light is used to observe a thin, transparent coating on glass as the coating material is gradually deposited by evaporation in a vacuum. Describe some color changes that might occur during the process of building up the thickness of the coating.

11. Would it be possible to place a nonreflective coating on an airplane to cancel radar waves of wavelength 3 cm?

12. Certain sunglasses use a polarizing material to reduce the intensity of light reflected from shiny surfaces, such as water or the hood of a car. What orientation of the transmission axis should the material have to be most effective?

13. Why is it so much easier to perform interference experiments with a laser than with an ordinary light source?

14. A soap film is held vertically in air and is viewed in reflected light as in Figure CQ24.14. Explain why the film appears to be dark at the top.

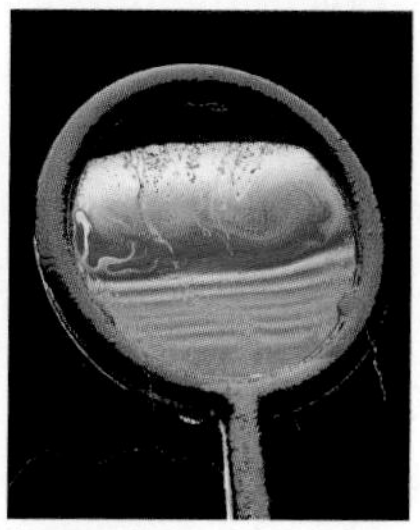

Figure CQ24.14

15. A plane monochromatic light wave is incident on a double-slit as illustrated in Figure 24.4. As the slit separation decreases, what happens to the separation between the interference fringes on the screen? (a) It decreases. (b) It increases. (c) It remains the same.

(d) It may increase or decrease, depending on the wavelength of the light. (e) More information is required.

16. A plane monochromatic light wave is incident on a double-slit as illustrated in Figure 24.4. If the viewing screen is moved away from the double slit, what happens to the separation between the interference fringes on the screen? (a) It increases. (b) It decreases. (c) It remains the same. (d) It may increase or decrease, depending on the wavelength of the light. (e) More information is required.

PROBLEMS

ENHANCED WebAssign The problems in this chapter may be assigned online in Enhanced WebAssign.

1. denotes straightforward problem; 2. denotes intermediate problem;
3. denotes challenging problem
[1.] denotes full solution available in *Student Solutions Manual/ Study Guide*
1. denotes problems most often assigned in Enhanced WebAssign

BIO denotes biomedical problems
GP denotes guided problems
M denotes Master It tutorial available in Enhanced WebAssign
Q|C denotes asking for quantitative and conceptual reasoning
S denotes symbolic reasoning problem
W denotes Watch It video solution available in Enhanced WebAssign

24.2 Young's Double-Slit Experiment

1. W A laser beam is incident on two slits with a separation of 0.200 mm, and a screen is placed 5.00 m from the slits. If the bright interference fringes on the screen are separated by 1.58 cm, what is the wavelength of the laser light?

2. In a Young's double-slit experiment, a set of parallel slits with a separation of 0.100 mm is illuminated by light having a wavelength of 589 nm, and the interference pattern is observed on a screen 4.00 m from the slits. (a) What is the difference in path lengths from each of the slits to the location of a third-order bright fringe on the screen? (b) What is the difference in path lengths from the two slits to the location of the third dark fringe on the screen, away from the center of the pattern?

[3.] A pair of narrow, parallel slits separated by 0.250 mm is illuminated by the green component from a mercury vapor lamp (λ = 546.1 nm). The interference pattern is observed on a screen 1.20 m from the plane of the parallel slits. Calculate the distance (a) from the central maximum to the first bright region on either side of the central maximum and (b) between the first and second dark bands in the interference pattern.

4. Light of wavelength 620 nm falls on a double slit, and the first bright fringe of the interference pattern is seen at an angle of 15.0° from the central maximum. Find the separation between the slits.

5. In a location where the speed of sound is 354 m/s, a 2 000-Hz sound wave impinges on two slits 30.0 cm apart. (a) At what angle is the first maximum located? (b) If the sound wave is replaced by 3.00-cm microwaves, what slit separation gives the same angle for the first maximum? (c) If the slit separation is 1.00 μm, what frequency of light gives the same first maximum angle?

6. A double slit separated by 0.058 0 mm is placed 1.50 m from a screen. (a) If yellow light of wavelength 588 nm strikes the double slit, what is the separation between the zeroth-order and first-order maxima on the screen? (b) If blue light of wavelength 412 nm strikes the double slit, what is the separation between the second-order and fourth-order maxima?

7. M Two radio antennas separated by d = 300 m, as shown in Figure P24.7, simultaneously broadcast identical signals at the same wavelength. A car travels due north along a straight line at position x = 1 000 m from the center point between the antennas, and its radio receives the signals. (a) If the car is at the position of the second maximum after that at point O when it has traveled a distance of y = 400 m northward, what is the wavelength of the signals? (b) How much farther must the car travel from this position to encounter the next minimum in reception? *Hint:* Do not use the small-angle approximation in this problem.

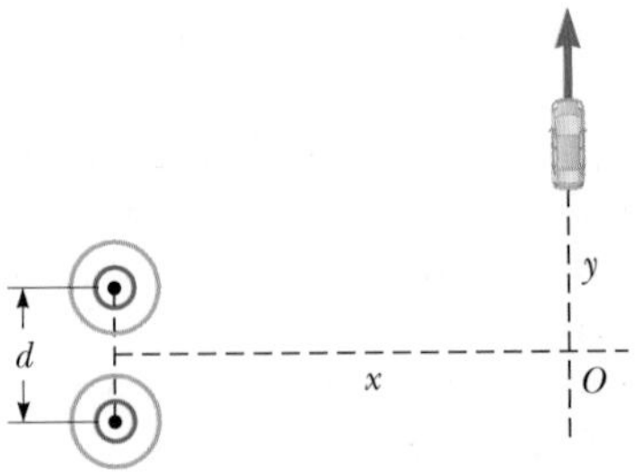

Figure P24.7

8. Light of wavelength 6.0×10^2 nm falls on a double slit, and the first bright fringe of the interference pattern is observed to make an angle of 12° with the horizontal. Find the separation between the slits.

9. Monochromatic light falls on a screen 1.75 m from two slits separated by 2.10 mm. The first- and second-order bright fringes are separated by 0.552 mm. What is the wavelength of the light?

10. A pair of slits, separated by 0.150 mm, is illuminated by light having a wavelength of λ = 643 nm. An

interference pattern is observed on a screen 140 cm from the slits. Consider a point on the screen located at y = 1.80 cm from the central maximum of this pattern. (a) What is the path difference δ for the two slits at the location y? (b) Express this path difference in terms of the wavelength. (c) Will the interference correspond to a maximum, a minimum, or an intermediate condition?

11. A riverside warehouse has two open doors, as in Figure P24.11. Its interior is lined with a sound-absorbing material. A boat on the river sounds its horn. To person A, the sound is loud and clear. To person B, the sound is barely audible. The principal wavelength of the sound waves is 3.00 m. Assuming person B is at the position of the first minimum, determine the distance between the doors, center to center.

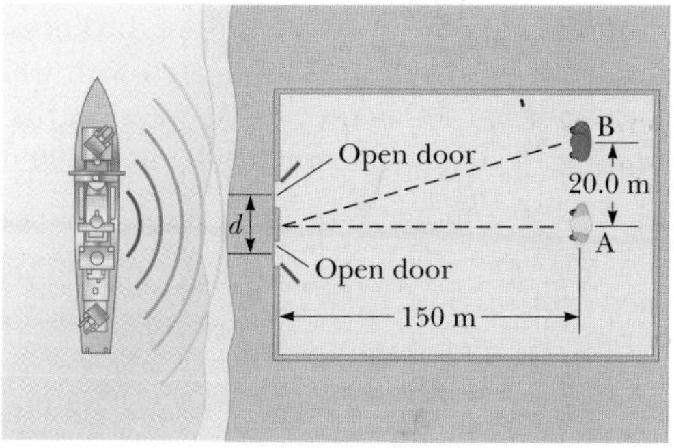

Figure P24.11

12. QIC A student sets up a double-slit experiment using monochromatic light of wavelength λ. The distance between the slits is equal to 25λ. (a) Find the angles at which the m = 1, 2, and 3 maxima occur on the viewing screen. (b) At what angles do the first three dark fringes occur? (c) Why are the answers so evenly spaced? Is the spacing even for all orders? Explain.

13. Radio waves from a star, of wavelength 250 m, reach a radio telescope by two separate paths, as shown in Figure P24.13. One is a direct path to the receiver, which is situated on the edge of a cliff by the ocean. The second is by reflection off the water. The first minimum of destructive interference occurs when the star is $\theta = 25.0°$ above the horizon. Find the height of the cliff. (Assume no phase change on reflection.)

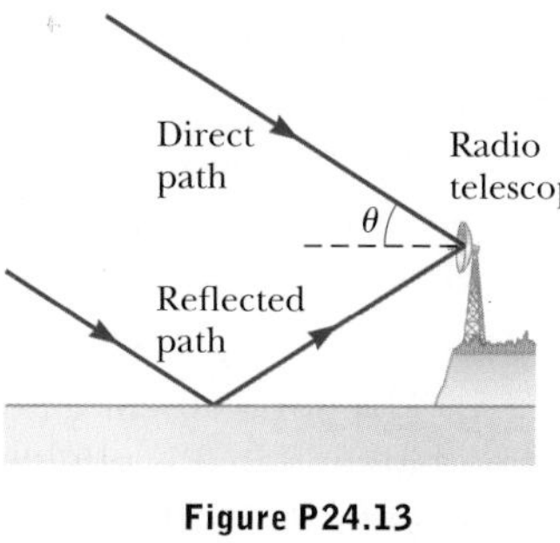

Figure P24.13

14. GP Monochromatic light of wavelength λ is incident on a pair of slits separated by 2.40×10^{-4} m, and forms an interference pattern on a screen placed 1.80 m away from the slits. The first-order bright fringe is 4.52 mm from the center of the central maximum. (a) Draw a picture, labeling the angle θ and the legs of the right triangle associated with the first-order bright fringe. (b) Compute the tangent of the angle θ associated with the first-order bright fringe. (c) Find the angle corresponding to the first-order bright fringe and compute the sine of that angle. Are the sine and tangent of the angle comparable in value? Does your answer always hold true? (d) Calculate the wavelength of the light. (e) Compute the angle of the fifth-order bright fringe. (f) Find its position on the screen.

15. Waves from a radio station have a wavelength of 300 m. They travel by two paths to a home receiver 20.0 km from the transmitter. One path is a direct path, and the second is by reflection from a mountain directly behind the home receiver. What is the minimum distance from the mountain to the receiver that produces destructive interference at the receiver? (Assume that no phase change occurs on reflection from the mountain.)

24.3 Change of Phase Due to Reflection

24.4 Interference in Thin Films

16. QIC A soap bubble (n = 1.33) having a wall thickness of 120 nm is floating in air. (a) What is the wavelength of the visible light that is most strongly reflected? (b) Explain how a bubble of different thickness could also strongly reflect light of this same wavelength. (c) Find the two smallest film thicknesses larger than the one given that can produce strongly reflected light of this same wavelength.

17. A thin layer of liquid methylene iodide (n = 1.756) is sandwiched between two flat, parallel plates of glass (n = 1.50). What is the minimum thickness of the liquid layer if normally incident light with $\lambda = 6.00 \times 10^2$ nm in air is to be strongly reflected?

18. S A thin film of oil (n = 1.25) is located on smooth, wet pavement. When viewed from a direction perpendicular to the pavement, the film reflects most strongly red light at 640 nm and reflects no green light at 512 nm. (a) What is the minimum thickness of the oil film? (b) Let m_1 correspond to the order of the constructive interference and m_2 to the order of the destructive interference. Obtain a relationship between m_1 and m_2 that is consistent with the given data.

19. A thin film of glass (n = 1.52) of thickness 0.420 μm is viewed under white light at near normal incidence. What wavelength of *visible* light is most strongly reflected by the film when surrounded by air?

20. W A transparent oil with index of refraction 1.29 spills on the surface of water (index of refraction 1.33), producing a maximum of reflection with normally incident orange light (wavelength 600 nm in air). Assuming the maximum occurs in the first order, determine the thickness of the oil slick.

21. M A possible means for making an airplane invisible to radar is to coat the plane with an antireflective polymer. If radar waves have a wavelength of 3.00 cm and the index of refraction of the polymer is $n = 1.50$, how thick would you make the coating?

22. QIC An oil film ($n = 1.45$) floating on water is illuminated by white light at normal incidence. The film is 2.80×10^2 nm thick. Find (a) the wavelength and color of the light in the visible spectrum most strongly reflected and (b) the wavelength and color of the light in the visible spectrum most strongly transmitted. Explain your reasoning.

23. QIC Astronomers observe the chromosphere of the Sun with a filter that passes the red hydrogen spectral line of wavelength 656.3 nm, called the H_α line. The filter consists of a transparent dielectric of thickness d held between two partially aluminized glass plates. The filter is kept at a constant temperature. (a) Find the minimum value of d that will produce maximum transmission of perpendicular H_α light if the dielectric has an index of refraction of 1.378. (b) If the temperature of the filter increases above the normal value increasing its thickness, what happens to the transmitted wavelength? (c) The dielectric will also pass what near-visible wavelength? One of the glass plates is colored red to absorb this light.

24. Two rectangular optically flat plates ($n = 1.52$) are in contact along one end and are separated along the other end by a 2.00-μm-thick spacer (Fig. P24.24). The top plate is illuminated by monochromatic light of wavelength 546.1 nm. Calculate the number of dark parallel bands crossing the top plate (including the dark band at zero thickness along the edge of contact between the plates).

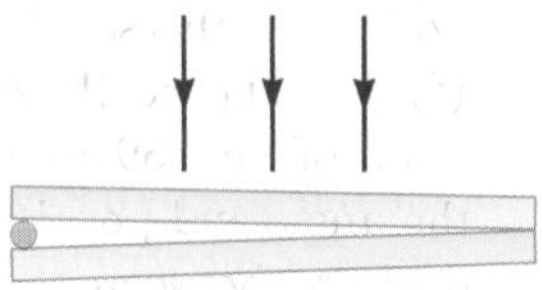

Figure P24.24
Problems 24 and 25.

25. An investigator finds a fiber at a crime scene that he wishes to use as evidence against a suspect. He gives the fiber to a technician to test the properties of the fiber. To measure the diameter of the fiber, the technician places it between two flat glass plates at their ends as in Figure P24.24. When the plates, of length 14.0 cm, are illuminated from above with light of wavelength 650 nm, she observes bright interference bands separated by 0.580 mm. What is the diameter of the fiber?

26. A plano-convex lens with radius of curvature $R = 3.0$ m is in contact with a flat plate of glass. A light source and the observer's eye are both close to the normal, as shown in Figure 24.8a. The radius of the 50th bright Newton's ring is found to be 9.8 mm. What is the wavelength of the light produced by the source?

27. A plano-convex lens rests with its curved side on a flat glass surface and is illuminated from above by light of wavelength 500 nm. (See Fig. 24.8.) A dark spot is observed at the center, surrounded by 19 concentric dark rings (with bright rings in between). How much thicker is the air wedge at the position of the 19th dark ring than at the center?

28. Nonreflective coatings on camera lenses reduce the loss of light at the surfaces of multilens systems and prevent internal reflections that might mar the image. Find the minimum thickness of a layer of magnesium fluoride ($n = 1.38$) on flint glass ($n = 1.66$) that will cause destructive interference of reflected light of wavelength 550 nm near the middle of the visible spectrum.

29. A thin film of glycerin ($n = 1.473$) of thickness 524 nm with air on both sides is illuminated with white light at near normal incidence. What wavelengths will be strongly reflected in the range 300 nm to 700 nm?

30. QIC A lens made of glass ($n_g = 1.52$) is coated with a thin film of MgF_2 ($n_s = 1.38$) of thickness t. Visible light is incident normally on the coated lens as in Figure P24.30. (a) For what minimum value of t will the reflected light of wavelength 540 nm (in air) be missing? (b) Are there other values of t that will minimize the reflected light at this wavelength? Explain.

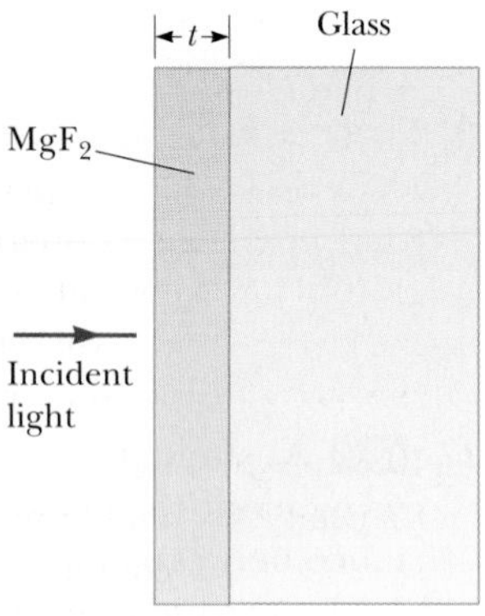

Figure P24.30

24.7 Single-Slit Diffraction

31. Light of wavelength 5.40×10^2 nm passes through a slit of width 0.200 mm. (a) Find the width of the central maximum on a screen located 1.50 m from the slit. (b) Determine the width of the first-order bright fringe.

32. W Light of wavelength 600 nm falls on a 0.40-mm-wide slit and forms a diffraction pattern on a screen 1.5 m away. (a) Find the position of the first dark band on each side of the central maximum. (b) Find the width of the central maximum.

33. M Light of wavelength 587.5 nm illuminates a slit of width 0.75 mm. (a) At what distance from the slit should a screen be placed if the first minimum in the diffraction pattern is to be 0.85 mm from the central maximum? (b) Calculate the width of the central maximum.

34. Microwaves of wavelength 5.00 cm enter a long, narrow window in a building that is otherwise essentially opaque to the incoming waves. If the window is 36.0 cm wide, what is the distance from the central maximum to the first-order minimum along a wall 6.50 m from the window?

35. A beam of monochromatic light is diffracted by a slit of width 0.600 mm. The diffraction pattern forms on a wall 1.30 m beyond the slit. The width of the central maximum is 2.00 mm. Calculate the wavelength of the light.

36. A screen is placed 50.0 cm from a single slit that is illuminated with light of wavelength 680 nm. If the distance between the first and third minima in the diffraction pattern is 3.00 mm, what is the width of the slit?

37. A slit of width 0.50 mm is illuminated with light of wavelength 500 nm, and a screen is placed 120 cm in front of the slit. Find the widths of the first and second maxima on each side of the central maximum.

38. The second-order dark fringe in a single-slit diffraction pattern is 1.40 mm from the center of the central maximum. Assuming the screen is 85.0 cm from a slit of width 0.800 mm and assuming monochromatic incident light, calculate the wavelength of the incident light.

24.8 The Diffraction Grating

39. Three discrete spectral lines occur at angles of 10.1°, 13.7°, and 14.8°, respectively, in the first-order spectrum of a diffraction-grating spectrometer. (a) If the grating has 3 660 slits/cm, what are the wavelengths of the light? (b) At what angles are these lines found in the second-order spectra?

40. Intense white light is incident on a diffraction grating that has 600 lines/mm. (a) What is the highest order in which the complete visible spectrum can be seen with this grating? (b) What is the angular separation between the violet edge (400 nm) and the red edge (700 nm) of the first-order spectrum produced by the grating?

41. M The hydrogen spectrum has a red line at 656 nm and a violet line at 434 nm. What angular separation between these two spectral lines is obtained with a diffraction grating that has 4 500 lines/cm?

42. Consider an array of parallel wires with uniform spacing of 1.30 cm between centers. In air at 20.0°C, ultrasound with a frequency of 37.2 kHz from a distant source is incident perpendicular to the array. (Take the speed of sound to be 343 m/s.) (a) Find the number of directions on the other side of the array in which there is a maximum of intensity. (b) Find the angle for each of these directions relative to the direction of the incident beam.

43. A helium–neon laser (λ = 632.8 nm) is used to calibrate a diffraction grating. If the first-order maximum occurs at 20.5°, what is the spacing between adjacent grooves in the grating?

44. W White light is spread out into its spectral components by a diffraction grating. If the grating has 2 000 lines per centimeter, at what angle does red light of wavelength 640 nm appear in the first-order spectrum?

45. Light from an argon laser strikes a diffraction grating that has 5 310 grooves per centimeter. The central and first-order principal maxima are separated by 0.488 m on a wall 1.72 m from the grating. Determine the wavelength of the laser light.

46. S Take red light at 700 nm and violet at 400 nm as the ends of the visible spectrum and consider the continuous spectrum of white light formed by a diffraction grating with a spacing of d meters between adjacent lines. Show that the interval $\theta_{v2} \leq \theta \leq \theta_{r2}$ of the continuous spectrum in second order must overlap the interval $\theta_{v3} \leq \theta \leq \theta_{r3}$ of the third-order spectrum. *Note:* θ_{v2} is the angle of the violet light in second order, and θ_{r2} is the angle made by red light in second order.

47. Sunlight is incident on a diffraction grating that has 2 750 lines/cm. The second-order spectrum over the visible range (400–700 nm) is to be limited to 1.75 cm along a screen that is a distance L from the grating. What is the required value of L?

48. GP A diffraction grating has 4.200×10^3 rulings per centimeter. The screen is 2.000 m from the grating. In parts (a) through (e), round each result to four digits, using the rounded values for subsequent calculations. (a) Compute the value of d, the distance between adjacent rulings. Express the answer in meters. (b) Calculate the angle of the second-order maximum made by the 589.0-nm wavelength of sodium. (c) Find the position of this second-order maximum on the screen. (d) Repeat parts (b) and (c) for the second-order maximum of the 589.6-nm wavelength of sodium, finding the angle and position on the screen. (e) What is the distance between the two second-order maxima on the screen? (f) Now find the distance between the two second-order maxima without rounding, carrying all digits in calculator memory. How many significant digits are in agreement? What can you conclude about rounding and the use of intermediate answers in this case?

49. Light of wavelength 500 nm is incident normally on a diffraction grating. If the third-order maximum of the diffraction pattern is observed at 32.0°, (a) what is the number of rulings per centimeter for the grating? (b) Determine the total number of primary maxima that can be observed in this situation.

50. Light containing two different wavelengths passes through a diffraction grating with 1 200 slits/cm. On a screen 15.0 cm from the grating, the third-order maximum of the shorter wavelength falls midway between the central maximum and the first side maximum for the longer wavelength. If the neighboring maxima of the longer wavelength are 8.44 mm apart on the screen, what are the wavelengths in the light? *Hint:* Use the small-angle approximation.

24.9 Polarization of Light Waves

51. The angle of incidence of a light beam in air onto a reflecting surface is continuously variable. The reflected ray is found to be completely polarized when the angle of incidence is 48.0°. (a) What is the index of refraction of the reflecting material? (b) If some of the incident light (at an angle of 48.0°) passes into the material below the surface, what is the angle of refraction?

52. Unpolarized light passes through two Polaroid sheets. The transmission axis of the analyzer makes an angle of 35.0° with the axis of the polarizer. (a) What fraction of the original unpolarized light is transmitted through the analyzer? (b) What fraction of the original light is absorbed by the analyzer?

53. The index of refraction of a glass plate is 1.52. What is the Brewster's angle when the plate is (a) in air and (b) in water? (See Problem 57.)

54. At what angle above the horizon is the Sun if light from it is completely polarized upon reflection from water?

55. A light beam is incident on a piece of fused quartz (n = 1.458) at the Brewster's angle. Find (a) the value of Brewster's angle and (b) the angle of refraction for the transmitted ray.

56. The critical angle for total internal reflection for sapphire surrounded by air is 34.4°. Calculate the Brewster's angle for sapphire if the light is incident from the air.

57. Equation 24.14 assumes the incident light is in air. If the light is incident from a medium of index n_1 onto a medium of index n_2, follow the procedure used to derive Equation 24.14 to show that $\tan \theta_p = n_2/n_1$.

58. Plane-polarized light is incident on a single polarizing disk, with the direction of E_0 parallel to the direction of the transmission axis. Through what angle should the disk be rotated so that the intensity in the transmitted beam is reduced by a factor of (a) 2.00, (b) 4.00, and (c) 6.00?

59. M Three polarizing plates whose planes are parallel are centered on a common axis. The directions of the transmission axes relative to the common vertical direction are shown in Figure P24.59. A linearly polarized beam of light with plane of polarization parallel to the vertical reference direction is incident from the left onto the first disk with intensity I_i = 10.0 units (arbitrary). Calculate the transmitted intensity I_f when θ_1 = 20.0°, θ_2 = 40.0°, and θ_3 = 60.0°. *Hint:* Make repeated use of Malus's law.

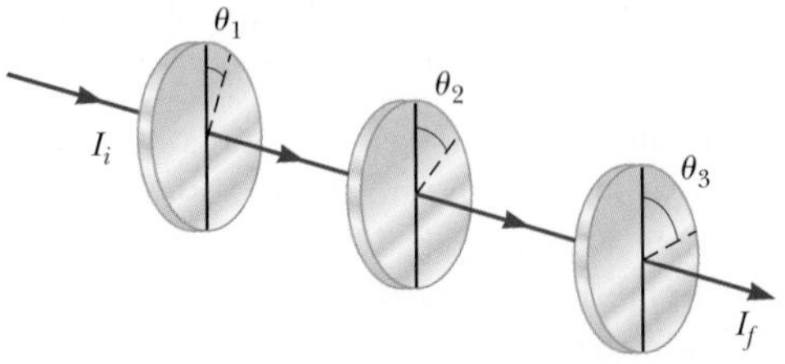

Figure P24.59 Problems 59 and 70.

60. Light of intensity I_0 and polarized parallel to the transmission axis of a polarizer is incident on an analyzer. (a) If the transmission axis of the analyzer makes an angle of 45° with the axis of the polarizer, what is the intensity of the transmitted light? (b) What should the angle between the transmission axes be to make $I/I_0 = 1/3$?

61. BIO Light with a wavelength in vacuum of 546.1 nm falls perpendicularly on a biological specimen that is 1.000 μm thick. The light splits into two beams polarized at right angles, for which the indices of refraction are 1.320 and 1.333, respectively. (a) Calculate the wavelength of each component of the light while it is traversing the specimen. (b) Calculate the phase difference between the two beams when they emerge from the specimen.

Additional Problems

62. Light from a helium–neon laser (λ = 632.8 nm) is incident on a single slit. What is the maximum width of the slit for which no diffraction minima are observed?

63. Q|C Laser light with a wavelength of 632.8 nm is directed through one slit or two slits and allowed to fall on a screen 2.60 m beyond. Figure P24.63 shows the pattern on the screen, with a centimeter ruler below it. Did the light pass through one slit or two slits? Explain how you can tell. If the answer is one slit, find its width. If the answer is two slits, find the distance between their centers.

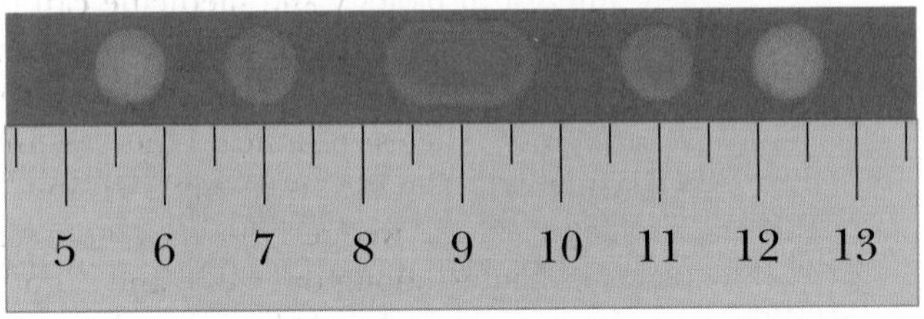

Figure P24.63

64. S In a Young's interference experiment, the two slits are separated by 0.150 mm and the incident light includes two wavelengths: λ_1 = 540 nm (green) and λ_2 = 450 nm (blue). The overlapping interference patterns are observed on a screen 1.40 m from the slits. (a) Find a relationship between the orders m_1 and m_2 that determines where a bright fringe of the green light coincides with a bright fringe of the blue light. (The order m_1 is associated with λ_1, and m_2 is associated with λ_2.) (b) Find the minimum values of m_1 and m_2 such that the overlapping of the bright fringes will occur and find the position of the overlap on the screen.

65. Light of wavelength 546 nm (the intense green line from a mercury source) produces a Young's interference pattern in which the second minimum from the central maximum is along a direction that makes an angle of 18.0 min of arc with the axis through the central maximum. What is the distance between the parallel slits?

66. The two speakers are placed 35.0 cm apart. A single oscillator makes the speakers vibrate in phase at a frequency of 2.00 kHz. At what angles, measured from the perpendicular bisector of the line joining the speakers, would a distant observer hear maximum sound intensity? Minimum sound intensity? (Take the speed of sound to be 340 m/s.)

67. Interference effects are produced at point P on a screen as a result of direct rays from a 500-nm source and reflected rays off a mirror, as shown in Figure P24.67. If the source is $L = 100$ m to the left of the screen and $h = 1.00$ cm above the mirror, find the distance y (in millimeters) to the first dark band above the mirror.

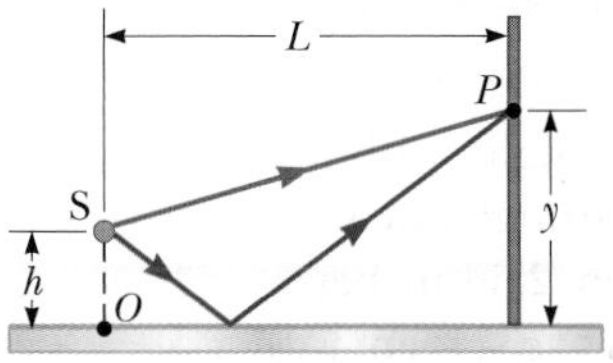

Figure P24.67

68. BIO Many cells are transparent and colorless. Structures of great interest in biology and medicine can be practically invisible to ordinary microscopy. An *interference microscope* reveals a difference in refractive index as a shift in interference fringes to indicate the size and shape of cell structures. The idea is exemplified in the following problem: An air wedge is formed between two glass plates in contact along one edge and slightly separated at the opposite edge. When the plates are illuminated with monochromatic light from above, the reflected light has 85 dark fringes. Calculate the number of dark fringes that appear if water ($n = 1.33$) replaces the air between the plates.

69. Figure P24.69 shows a radio-wave transmitter and a receiver, both $h = 50.0$ m above the ground and $d = 600$ m apart. The receiver can receive signals directly from the transmitter and indirectly from signals that bounce off the ground. If the ground is level between the transmitter and receiver and a $\lambda/2$ phase shift occurs upon reflection, determine the longest wavelengths that interfere (a) constructively and (b) destructively.

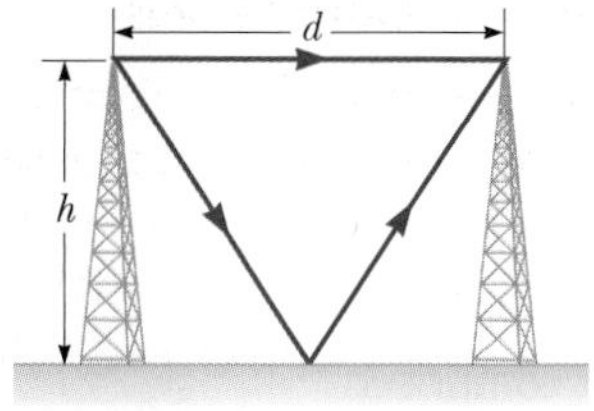

Figure P24.69

70. Three polarizers, centered on a common axis and with their planes parallel to one another, have transmission axes oriented at angles of θ_1, θ_2, and θ_3 from the vertical, as shown in Figure P24.59. Light of intensity I_i, polarized with its plane of polarization oriented vertically, is incident from the left onto the first polarizer. What is the ratio I_f/I_i of the final transmitted intensity to the incident intensity if (a) $\theta_1 = 45°$, $\theta_2 = 90°$, and $\theta_3 = 0°$? (b) $\theta_1 = 0°$, $\theta_2 = 45°$, and $\theta_3 = 90°$?

71. The transmitting antenna on a submarine is 5.00 m above the water when the ship surfaces. The captain wishes to transmit a message to a receiver on a 90.0-m-tall cliff at the ocean shore. If the signal is to be completely polarized by reflection off the ocean surface, how far must the ship be from the shore?

72. S A plano-convex lens (flat on one side, convex on the other) with index of refraction n rests with its curved side (radius of curvature R) on a flat glass surface of the same index of refraction with a film of index n_{film} between them. The lens is illuminated from above by light of wavelength λ. Show that the dark Newton rings that appear have radii of

$$r \approx \sqrt{m\lambda R/n_{\text{film}}}$$

where m is an integer.

73. A diffraction pattern is produced on a screen 140 cm from a single slit, using monochromatic light of wavelength 500 nm. The distance from the center of the central maximum to the first-order maximum is 3.00 mm. Calculate the slit width. *Hint:* Assume that the first-order maximum is halfway between the first- and second-order minima.

74. A flat piece of glass is supported horizontally above the flat end of a 10.0-cm-long metal rod that has its lower end rigidly fixed. The thin film of air between the rod and the glass is observed to be bright when illuminated by light of wavelength 500 nm. As the temperature is slowly increased by 25.0°C, the film changes from bright to dark and back to bright 200 times. What is the coefficient of linear expansion of the metal?

The twin Keck telescopes atop the summit of Hawaii's Mauna Kea volcano are the world's largest optical and infrared telescopes. Each telescope has a primary mirror 10 m in diameter comprised of 36 hexagonal segments that function together as a single piece of reflective glass. Far from city lights, high on a dormant volcano in dry, unpolluted air, the telescopes are ideally placed to probe the mysteries of the Universe.

Ed Darack/RGB Ventures LLC dba SuperStock/Alamy

25 Optical Instruments

We use devices made from lenses, mirrors, and other optical components every time we put on a pair of eyeglasses or contact lenses, take a photograph, look at the sky through a telescope, and so on. In this chapter we examine how optical instruments work. For the most part, our analyses involve the laws of reflection and refraction and the procedures of geometric optics. To explain certain phenomena, however, we must use the wave nature of light.

25.1 The Camera

LEARNING OBJECTIVES

1. **Describe the components and operation of a single-lens camera.**
2. **Define the *f*-number and discuss its effects on a camera's image.**

The single-lens photographic camera is a simple optical instrument having the features shown in Figure 25.1. It consists of an opaque box, a converging lens that produces a real image, and a photographic film behind the lens to receive the image. Digital cameras differ in that the image is formed on a charge-coupled device (CCD) or a complementary metal-oxide semiconductor (CMOS) sensor instead of on film. Both the CCD and the CMOS image sensors convert the image into digital form, which can then be stored in the camera's memory.

Focusing a camera is accomplished by varying the distance between the lens and sensor, with an adjustable bellows in antique cameras and other mechanisms in contemporary models. For proper focusing, which leads to sharp images, the lens-to-sensor distance depends on the object distance as well as on the focal length of the lens. The shutter, located behind the lens, is a mechanical device that is opened for selected time intervals. With this arrangement, moving objects can be photographed by using short exposure times and dark scenes (with low light levels) by using long exposure times. If this adjustment were not available, it would be impossible to take stop-action photographs. A rapidly moving vehicle, for example, could move far enough while the shutter was open to produce a blurred image. Another major cause of blurred images is movement of the *camera* while the shutter is open. To prevent such movement, you should mount the camera on a tripod or use short exposure times. Typical shutter speeds (that is, exposure times) are 1/30 s, 1/60 s, 1/125 s, and 1/250 s. Stationary objects are often shot with a shutter speed of 1/60 s.

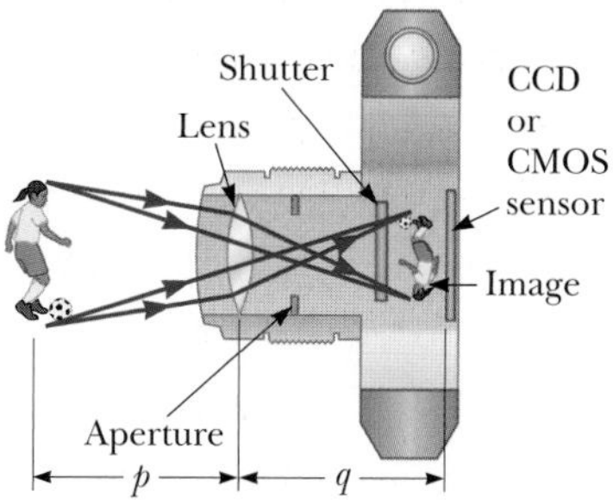

Figure 25.1 Cross-sectional view of a simple digital camera. The CCD or CMOS image sensor is the light-sensitive component of the camera. In a nondigital camera the light from the lens falls onto photographic film. In reality $p >> q$.

Most cameras also have an aperture of adjustable diameter to further control the intensity of the light reaching the sensor. When an aperture of small diameter is used, only light from the central portion of the lens reaches the sensor, reducing spherical aberration.

The intensity I of the light reaching the sensor is proportional to the area of the lens. Because this area in turn is proportional to the square of the lens diameter D, the intensity is also proportional to D^2. Light intensity is a measure of the rate at which energy is received by the sensor per unit area of the image. Because the area of the image is proportional to q^2 in Figure 25.1 and $q \approx f$ (when $p >> f$, so that p can be approximated as infinite), we conclude that the intensity is also proportional to $1/f^2$. Therefore, $I \propto D^2/f^2$. The brightness of the image formed on the sensor depends on the light intensity, so we see that it ultimately depends on both the focal length f and diameter D of the lens. The ratio f/D is called the ***f*-number** (or focal ratio) of a lens:

$$f\text{-number} \equiv \frac{f}{D} \qquad [25.1]$$

The *f*-number is often given as a description of the lens "speed." A lens with a low *f*-number is a "fast" lens. Extremely fast lenses, which have an *f*-number as low as approximately 1.2, are expensive because of the difficulty of keeping aberrations acceptably small with light rays passing through a large area of the lens. Camera lenses are often marked with a range of *f*-numbers, such as 1.4, 2, 2.8, 4, 5.6, 8, and 11. Any one of these settings can be selected by adjusting the aperture, which changes the value of D. Increasing the setting from one *f*-number to the next-higher value (for example, from 2.8 to 4) decreases the area of the aperture by a factor of 2. The lowest *f*-number setting on a camera corresponds to a wide-open aperture and the use of the maximum possible lens area.

Simple cameras usually have a fixed focal length and fixed aperture size, with an *f*-number of about 11. This high value for the *f*-number allows for a large **depth of field** and means that objects at a wide range of distances from the lens form reasonably sharp images on the sensor. In other words, the camera doesn't have to be focused. Most cameras with variable *f*-numbers adjust them automatically.

25.2 The Eye BIO

LEARNING OBJECTIVES

1. Identify the essential parts of the eye and discuss the eye's ability to adjust to varying light conditions.
2. Define near point and far point and discuss the focusing process of accommodation.
3. Describe several focusing conditions of the eye and the characteristics of optical lenses used to correct them.
4. Define the power of a lens.
5. Calculate the required power of corrective optical lenses.

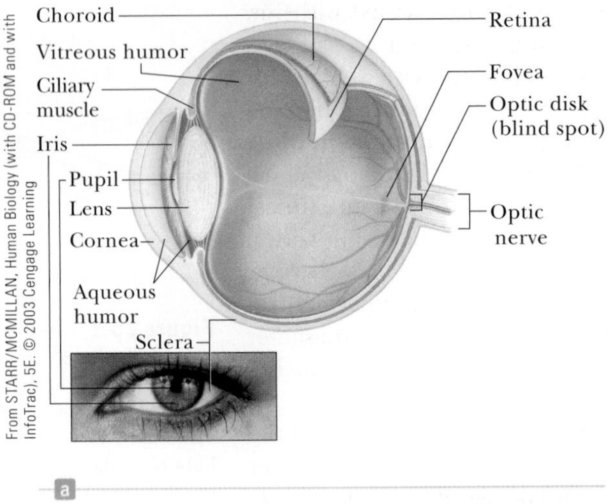

Figure 25.2 (a) Essential parts of the eye. Can you correlate the essential parts of the eye with those of the simple camera in Figure 25.1? (b) Close-up photograph of the human cornea.

Like a camera, a normal eye focuses light and produces a sharp image. The mechanisms by which the eye controls the amount of light admitted and adjusts to produce correctly focused images, however, are far more complex, intricate, and effective than those in even the most sophisticated camera. In all respects the eye is a physiological wonder.

Figure 25.2a shows the essential parts of the eye. Light entering the eye passes through a transparent structure called the *cornea,* behind which are a clear liquid (the *aqueous humor*), a variable aperture (the *pupil,* which is an opening in the *iris*), and the *crystalline lens.* Most of the refraction occurs at the outer surface of the eye, where the cornea is covered with a film of tears. Relatively little refraction occurs in the crystalline lens because the aqueous humor in contact with the lens has an average index of refraction close to that of the lens. The iris, which is the colored portion of the eye, is a muscular diaphragm that controls pupil size. The iris regulates the amount of light entering the eye by dilating the pupil in low-light conditions and contracting the pupil under conditions of bright light. The *f*-number of the eye ranges from about 2.8 to 16.

The cornea–lens system focuses light onto the back surface of the eye—the *retina*—which consists of millions of sensitive receptors called *rods* and *cones.* When stimulated by light, these structures send impulses to the brain via the optic nerve, converting them into our conscious view of the world. The process by which the brain performs this conversion is not well understood and is the subject of much speculation and research. Unlike film in a camera, the rods and cones chemically adjust their sensitivity according to the prevailing light conditions. This adjustment, which takes about 15 minutes, is responsible for the experience of "getting used to the dark" in such places as movie theaters. Iris aperture control, which takes less than a second, helps protect the retina from overload in the adjustment process.

The eye focuses on an object by varying the shape of the pliable crystalline lens through an amazing process called **accommodation**. An important component in accommodation is the *ciliary muscle,* which is situated in a circle around the rim of the lens. Thin filaments, called *zonules,* run from this muscle to the edge of the lens. When the eye is focused on a distant object, the ciliary muscle is relaxed, tightening the zonules that attach the ciliary muscle to the edge of the lens. The force of the zonules causes the lens to flatten, increasing its focal length. For an object distance of infinity, the focal length of the eye is equal to the fixed distance between lens and retina, about 1.7 cm. The eye focuses on nearby objects by tensing the ciliary muscle, which relaxes the zonules. This action allows the lens to bulge a bit and its focal length decreases, resulting in the image being focused on the retina. All these lens adjustments take place so swiftly that we are not even aware of the change. In this respect even the finest electronic camera is a toy compared with the eye.

There is a limit to accommodation because objects that are very close to the eye produce blurred images. The **near point** is the closest distance for which the lens can accommodate to focus light on the retina. This distance usually increases with age and has an average value of 25 cm. Typically, at age 10 the near point of the eye is about 18 cm. This increases to about 25 cm at age 20, 50 cm at age 40, and 500 cm or greater at age 60. The **far point** of the eye represents the farthest distance for which the lens of the relaxed eye can focus light on the retina. A person with normal vision is able to see very distant objects, such as the Moon, and so has a far point at infinity.

Conditions of the Eye

When the eye suffers a mismatch between the focusing power of the lens–cornea system and the length of the eye, so that light rays reach the retina before they converge to form an image, as in Figure 25.3a, the condition is known as **farsightedness** (or *hyperopia*). A farsighted person can usually see faraway objects clearly but not nearby objects. Although the near point of a normal eye is approximately 25 cm, the near point of a farsighted person is much farther than that. The eye of a farsighted person tries to focus by accommodation, by shortening its focal length. Accommodation works for distant objects, but because the focal length of the farsighted eye is longer than normal, the light from nearby objects can't be brought to a sharp focus before it reaches the retina, causing a blurred image. The condition can be corrected by placing a converging lens in front of the eye, as in Figure 25.3b. The lens refracts the incoming rays more toward the principal axis before entering the eye, allowing them to converge and focus on the retina.

Nearsightedness (or *myopia*) is another mismatch condition in which a person is able to focus on nearby objects, but not faraway objects. In the case of *axial myopia*, nearsightedness is caused by the lens being too far from the retina. It is also possible to have *refractive myopia*, in which the lens–cornea system is too powerful for the normal length of the eye. The far point of the nearsighted eye is not at infinity and may be less than 1 meter. The maximum focal length of the nearsighted eye is insufficient to produce a sharp image on the retina, and rays from a distant object converge to a focus in front of the retina. They then continue past that point, diverging before they finally reach the retina and produce a blurred image (Fig. 25.4a, page 874).

Nearsightedness can be corrected with a diverging lens, as shown in Figure 25.4b. The lens refracts the rays away from the principal axis before they enter the eye, allowing them to focus on the retina.

BIO APPLICATION
Using Optical Lenses to Correct for Defects

Beginning with middle age, most people lose some of their accommodation ability as the ciliary muscle weakens and the lens hardens. Unlike farsightedness, which is a

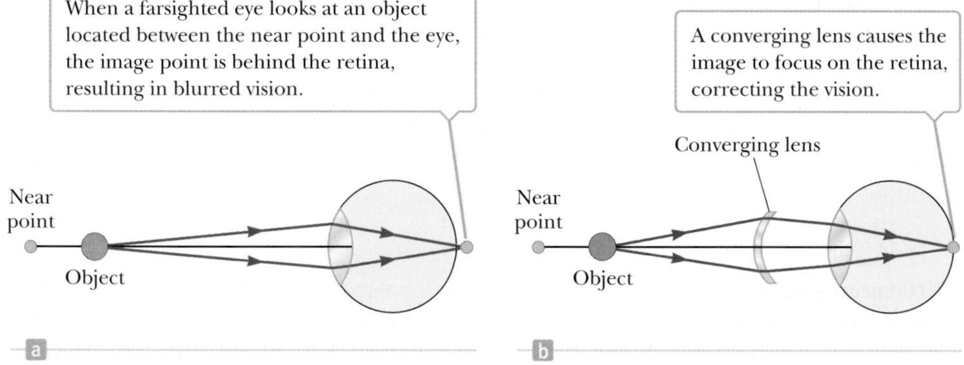

Figure 25.3 (a) An uncorrected farsighted eye. (b) A farsighted eye corrected with a converging lens. (The object is assumed to be very small in these figures.)

Figure 25.4 (a) An uncorrected nearsighted eye. (b) A nearsighted eye corrected with a diverging lens. (The object is assumed to be very small in these figures.)

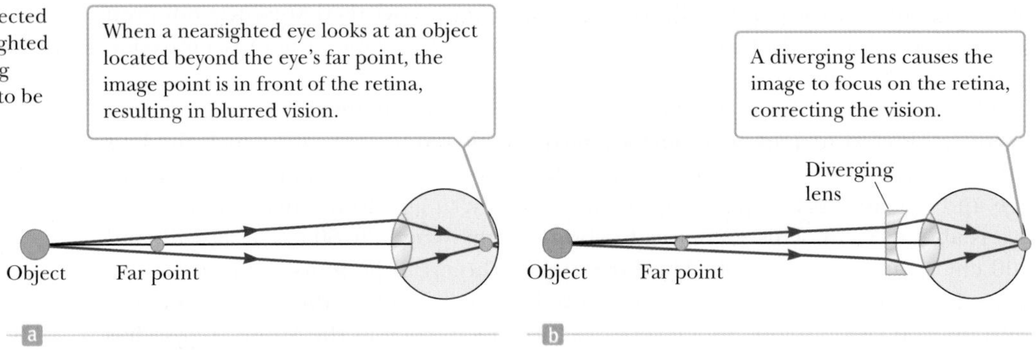

mismatch of focusing power and eye length, **presbyopia** (literally, "old-age vision") is due to a reduction in accommodation ability. This means the cornea and lens aren't able to bring nearby objects into focus on the retina. The symptoms are the same as with farsightedness, and the condition can be corrected with converging lenses.

In the eye defect known as **astigmatism**, light from a point source produces a line image on the retina. This condition arises when either the cornea or the lens (or both) is not perfectly symmetric. Astigmatism can be corrected with lenses having different curvatures in two mutually perpendicular directions.

Optometrists and ophthalmologists usually prescribe lenses measured in **diopters**:

> The **power** P of a lens in diopters equals the inverse of the focal length in meters: $P = 1/f$.

For example, a converging lens with a focal length of +20 cm has a power of +5.0 diopters, and a diverging lens with a focal length of −40 cm has a power of −2.5 diopters. (Although the symbol P is the same as for mechanical power, there is no relationship between the two concepts.)

The position of the lens relative to the eye causes differences in power, but they usually amount to less than one-quarter diopter, which isn't noticeable to most patients. As a result, practicing optometrists deal in increments of one-quarter diopter. Neglecting the eye–lens distance is equivalent to doing the calculation for a contact lens, which rests directly on the eye.

EXAMPLE 25.1 Prescribing a Corrective Lens for a Farsighted Patient BIO

GOAL Apply geometric optics to correct farsightedness.

PROBLEM The near point of a patient's eye is 50.0 cm. **(a)** What focal length must a corrective lens have to enable the eye to clearly see an object 25.0 cm away? Neglect the eye–lens distance. **(b)** What is the power of this lens? **(c)** Repeat the problem, taking into account that, for typical eyeglasses, the corrective lens is 2.00 cm in front of the eye.

STRATEGY This problem requires substitution into the thin-lens equation (Eq. 23.11) and then using the definition of lens power in terms of diopters. The object is at 25.0 cm, but the lens must form an image at the patient's near point, 50.0 cm, the closest point at which the patient's eye can see clearly. In part (c) 2.00 cm must be subtracted from both the object distance and the image distance to account for the position of the lens.

SOLUTION

(a) Find the focal length of the corrective lens, neglecting its distance from the eye.

Apply the thin-lens equation:

$$\frac{1}{p} + \frac{1}{q} = \frac{1}{f}$$

Substitute $p = 25.0$ cm and $q = -50.0$ cm (the latter is negative because the image must be virtual) on the same side of the lens as the object:

$$\frac{1}{25.0 \text{ cm}} + \frac{1}{-50.0 \text{ cm}} = \frac{1}{f}$$

Solve for f. The focal length is positive, corresponding to a converging lens.

$$f = 50.0 \text{ cm}$$

(b) What is the power of this lens?

The power is the reciprocal of the focal length in meters:

$$P = \frac{1}{f} = \frac{1}{0.500 \text{ m}} = +2.00 \text{ diopters}$$

(c) Repeat the problem, noting that the corrective lens is actually 2.00 cm in front of the eye.

Substitute the corrected values of p and q into the thin-lens equation:

$$\frac{1}{p} + \frac{1}{q} = \frac{1}{23.0 \text{ cm}} + \frac{1}{(-48.0 \text{ cm})} = \frac{1}{f}$$

$$f = 44.2 \text{ cm}$$

Compute the power:

$$P = \frac{1}{f} = \frac{1}{0.442 \text{ m}} = +2.26 \text{ diopters}$$

REMARKS Notice that the calculation in part (c), which doesn't neglect the eye–lens distance, results in a difference of 0.26 diopter.

QUESTION 25.1 True or False: The larger the distance to a near point, the larger the power of the required corrective lens.

EXERCISE 25.1 Suppose a lens is placed in a device that determines its power as 2.75 diopters. Find (a) the focal length of the lens and (b) the minimum distance at which a patient will be able to focus on an object if the patient's near point is 60.0 cm. Neglect the eye–lens distance.

ANSWERS (a) 36.4 cm (b) 22.7 cm

EXAMPLE 25.2 A Corrective Lens for Nearsightedness

GOAL Apply geometric optics to correct nearsightedness.

PROBLEM A particular nearsighted patient can't see objects clearly when they are beyond 25 cm (the far point of the eye). **(a)** What focal length should the prescribed contact lens have to correct this problem? **(b)** Find the power of the lens, in diopters. Neglect the distance between the eye and the corrective lens.

STRATEGY The purpose of the lens in this instance is to take objects at infinity and create an image of them at the patient's far point. Apply the thin-lens equation.

SOLUTION

(a) Find the focal length of the corrective lens.

Apply the thin-lens equation for an object at infinity and image at 25.0 cm:

$$\frac{1}{p} + \frac{1}{q} = \frac{1}{\infty} + \frac{1}{(-25.0 \text{ cm})} = \frac{1}{f}$$

$$f = -25.0 \text{ cm}$$

(b) Find the power of the lens in diopters.

$$P = \frac{1}{f} = \frac{1}{-0.250 \text{ m}} = -4.00 \text{ diopters}$$

REMARKS The focal length is negative, consistent with a diverging lens. Notice that the power is also negative and has the same numeric value as the sum on the left side of the thin-lens equation.

QUESTION 25.2 True or False: The shorter the distance to a patient's far point, the more negative the power of the required corrective lens.

EXERCISE 25.2 (a) What power lens would you prescribe for a patient with a far point of 35.0 cm? Neglect the eye–lens distance. (b) Repeat, assuming an eye-corrective lens distance of 2.00 cm.

ANSWERS (a) −2.86 diopters (b) −3.03 diopters

APPLYING PHYSICS 25.1 | Vision of the Invisible Man

A classic science fiction story, *The Invisible Man* by H. G. Wells, tells of a man who becomes invisible by changing the index of refraction of his body to that of air. Students who know how the eye works have criticized this story; they claim that the invisible man would be unable to see. On the basis of your knowledge of the eye, would he be able to see?

EXPLANATION He wouldn't be able to see. For the eye to see an object, incoming light must be refracted at the cornea and lens to form an image on the retina. If the cornea and lens have the same index of refraction as air, refraction can't occur and an image wouldn't be formed. ■

Quick Quiz

25.1 Two campers wish to start a fire during the day. One camper is nearsighted and one is farsighted. Whose glasses should be used to focus the Sun's rays onto some paper to start the fire? (a) either camper's (b) the nearsighted camper's (c) the farsighted camper's

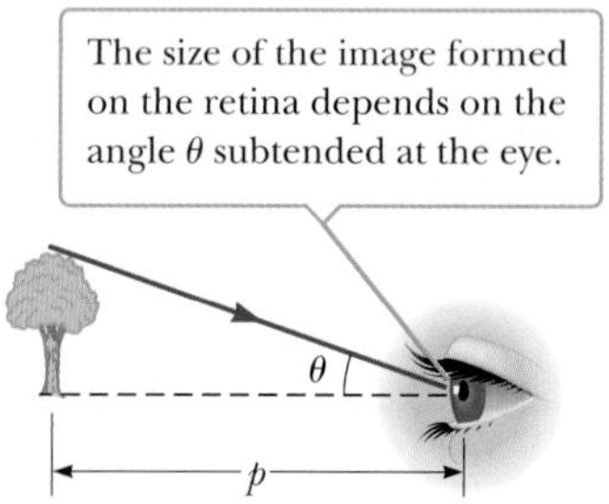

Figure 25.5 An observer looks at an object at distance p.

25.3 The Simple Magnifier

LEARNING OBJECTIVES

1. Define the angular magnification of a simple magnifier.
2. Apply geometric optics to systems involving a simple magnifier.

The **simple magnifier** is one of the most basic of all optical instruments because it consists only of a single converging lens. As the name implies, this device is used to increase the apparent size of an object. Suppose an object is viewed at some distance p from the eye, as in Figure 25.5. Clearly, the size of the image formed at the retina depends on the angle θ subtended by the object at the eye. As the object moves closer to the eye, θ increases and a larger image is observed. A normal eye, however, can't focus on an object closer than about 25 cm, the near point (Fig. 25.6a). (Try it!) Therefore, θ is a maximum at the near point.

To further increase the apparent angular size of an object, a converging lens can be placed in front of the eye with the object positioned at point O, just inside the focal point of the lens, as in Figure 25.6b. At this location, the lens forms a virtual, upright, and enlarged image, as shown. The lens allows the object to be viewed closer to the eye than is otherwise possible. We define the **angular magnification** m as the ratio of the angle subtended by a small object when the lens is in use (angle θ in Fig. 25.6b) to the angle subtended by the object placed at the near point with no lens in use (angle θ_0 in Fig. 25.6a):

Angular magnification with the object at the near point ▶

$$m \equiv \frac{\theta}{\theta_0} \qquad [25.2]$$

For the case in which the lens is held close to the eye, the angular magnification is a maximum when the image formed by the lens is at the near point of

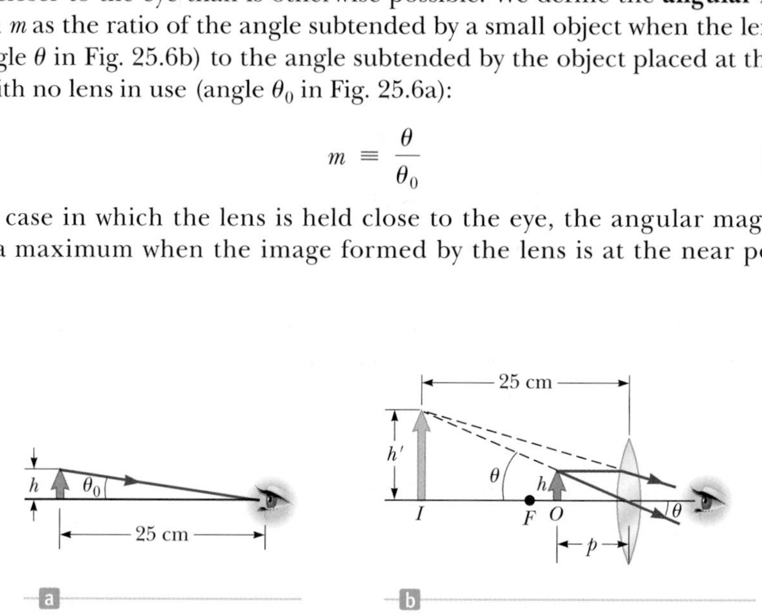

Figure 25.6 (a) An object placed at the near point (p = 25 cm) subtends an angle of $\theta_0 \approx h/25$ at the eye. (b) An object placed near the focal point of a converging lens produces a magnified image, which subtends an angle of $\theta \approx h'/25$ at the eye. Note that in this situation q = −25 cm.

the eye, which corresponds to $q = -25$ cm (see Fig. 25.6b). The object distance corresponding to this image distance can be calculated from the thin-lens equation:

$$\frac{1}{p} + \frac{1}{-25 \text{ cm}} = \frac{1}{f} \qquad [25.3]$$

$$p = \frac{25f}{25 + f}$$

Here, f is the focal length of the magnifier in centimeters. From Figures 25.6a and 25.6b, the small-angle approximation gives

$$\tan \theta_0 \approx \theta_0 \approx \frac{h}{25} \quad \text{and} \quad \tan \theta \approx \theta \approx \frac{h}{p} \qquad [25.4]$$

Equation 25.2 therefore becomes

$$m_{\text{max}} = \frac{\theta}{\theta_0} = \frac{h/p}{h/25} = \frac{25}{p} = \frac{25}{25f/(25 + f)}$$

so that

$$m_{\text{max}} = 1 + \frac{25 \text{ cm}}{f} \qquad [25.5]$$

The maximum angular magnification given by Equation 25.5 is the ratio of the angular size seen with the lens to the angular size seen without the lens, with the object at the near point of the eye. Although the normal eye can focus on an image formed anywhere between the near point and infinity, it's most relaxed when the image is at infinity (Sec. 25.2). For the image formed by the magnifying lens to appear at infinity, the object must be placed at the focal point of the lens so that $p = f$. In this case Equation 25.4 becomes

$$\theta_0 \approx \frac{h}{25} \quad \text{and} \quad \theta \approx \frac{h}{f}$$

and the angular magnification is

$$m = \frac{\theta}{\theta_0} = \frac{25 \text{ cm}}{f} \qquad [25.6]$$

With a single lens, it's possible to achieve angular magnifications up to about 4 without serious aberrations. Magnifications up to about 20 can be achieved by using one or two additional lenses to correct for aberrations.

EXAMPLE 25.3 | Magnification of a Lens

GOAL Compute magnifications of a lens when the image is at the near point and when it's at infinity.

PROBLEM **(a)** What is the maximum angular magnification of a lens with a focal length of 10.0 cm? **(b)** What is the angular magnification of this lens when the eye is relaxed? Assume an eye–lens distance of zero.

STRATEGY The maximum angular magnification occurs when the image formed by the lens is at the near point of the eye. Under these circumstances, Equation 25.5 gives us the maximum angular magnification. In part (b) the eye is relaxed only if the image is at infinity, so Equation 25.6 applies.

SOLUTION

(a) Find the maximum angular magnification of the lens.

Substitute into Equation 25.5:

$$m_{\text{max}} = 1 + \frac{25 \text{ cm}}{f} = 1 + \frac{25 \text{ cm}}{10.0 \text{ cm}} = 3.5$$

(Continued)

(b) Find the magnification of the lens when the eye is relaxed.

When the eye is relaxed, the image is at infinity, so substitute into Equation 25.6:

$$m = \frac{25\text{ cm}}{f} = \frac{25\text{ cm}}{10.0\text{ cm}} = 2.5$$

QUESTION 25.3 For greater magnification, should a lens with a larger or smaller focal length be selected?

EXERCISE 25.3 What focal length would be necessary if the lens were to have a maximum angular magnification of 4.0?

ANSWER 8.3 cm

25.4 The Compound Microscope

LEARNING OBJECTIVES

1. Discuss the optical components, characteristics, and limitations of a compound microscope.
2. Evaluate the magnification of a compound microscope.

A simple magnifier provides only limited assistance with inspection of the minute details of an object. Greater magnification can be achieved by combining two lenses in a device called a compound microscope, a schematic diagram of which is shown in Figure 25.7a. The instrument consists of two lenses: an objective with a very short focal length f_o (where $f_o < 1$ cm) and an ocular lens, or eyepiece, with a focal length f_e of a few centimeters. The two lenses are separated by a distance L that is much greater than either f_o or f_e.

The basic approach used to analyze the image formation properties of a microscope is that of two lenses in a row: the image formed by the first becomes the object for the second. The object O placed just outside the focal length of the objective forms a real, inverted image at I_1 that is at or just inside the focal point of the eyepiece. This image is much enlarged. (For clarity, the enlargement of I_1 is not shown in Fig. 25.7a.) The eyepiece, which serves as a simple magnifier, uses the

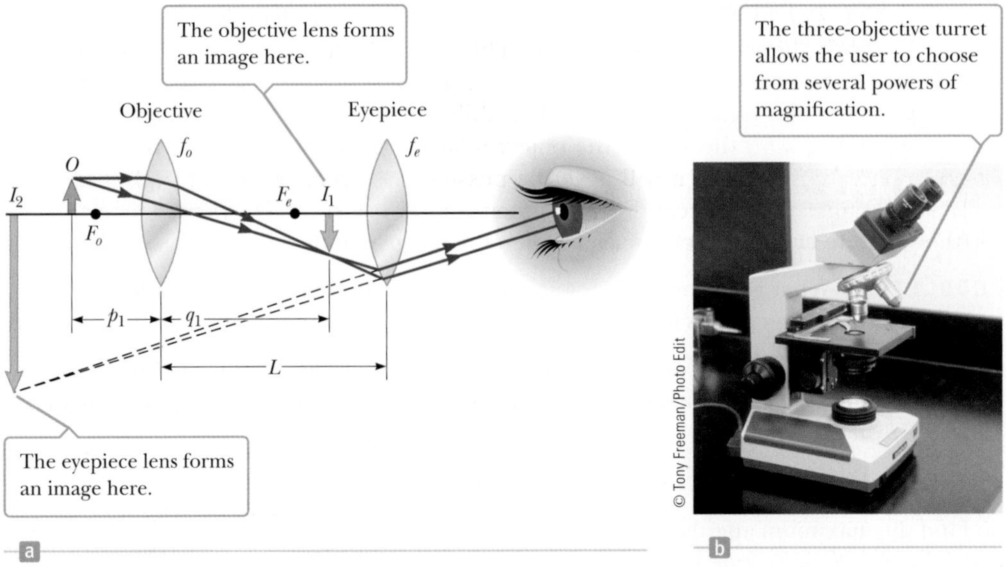

Figure 25.7 (a) A diagram of a compound microscope, which consists of an objective and an eyepiece, or ocular lens. (b) A compound microscope. Combinations of eyepieces with different focal lengths and different objectives can produce a wide range of magnifications.

image at I_1 as its object and produces an image at I_2. The image seen by the eye at I_2 is virtual, inverted, and very much enlarged.

The lateral magnification M_1 of the first image is $-q_1/p_1$. Note that q_1 is approximately equal to L because the object is placed close to the focal point of the objective lens, which ensures that the image formed will be far from the objective lens. Further, because the object is very close to the focal point of the objective lens, $p_1 \approx f_o$. Therefore, the lateral magnification of the objective is

$$M_1 = -\frac{q_1}{p_1} \approx -\frac{L}{f_o}$$

From Equation 25.6, the angular magnification of the eyepiece for an object (corresponding to the image at I_1) placed at the focal point is found to be

$$m_e = \frac{25\text{ cm}}{f_e}$$

The overall magnification of the compound microscope is defined as the product of the lateral and angular magnifications:

$$m = M_1 m_e = -\frac{L}{f_o}\left(\frac{25\text{ cm}}{f_e}\right) \qquad [25.7]$$

◀ Magnification of a microscope

The negative sign indicates that the image is inverted with respect to the object.

The microscope has extended our vision into the previously unknown realm of incredibly small objects, and the capabilities of this instrument have increased steadily with improved techniques in precision grinding of lenses. A natural question is whether there is any limit to how powerful a microscope could be. For example, could a microscope be made powerful enough to allow us to see an atom? The answer to this question is no, as long as visible light is used to illuminate the object. To be seen, the object under a microscope must be at least as large as a wavelength of light. An atom is many times smaller than the wavelength of visible light, so its mysteries must be probed via other techniques.

The wavelength dependence of the "seeing" ability of a wave can be illustrated by water waves set up in a bathtub in the following way. Imagine that you vibrate your hand in the water until waves with a wavelength of about 6 in. are moving along the surface. If you fix a small object, such as a toothpick, in the path of the waves, you will find that the waves are not appreciably disturbed by the toothpick, but continue along their path. Now suppose you fix a larger object, such as a toy sailboat, in the path of the waves. In this case the waves are considerably disturbed by the object. The toothpick was much smaller than the wavelength of the waves, and as a result the waves didn't "see" it. The toy sailboat, however, is about the same size as the wavelength of the waves and hence creates a disturbance. Light waves behave in this same general way. The ability of an optical microscope to view an object depends on the size of the object relative to the wavelength of the light used to observe it. Hence, it will never be possible to observe atoms or molecules with such a microscope because their dimensions are so small ($\approx$0.1 nm) relative to the wavelength of the light ($\approx$500 nm).

■ EXAMPLE 25.4 | Microscope Magnifications

GOAL Understand the critical factors involved in determining the magnifying power of a microscope.

PROBLEM A certain microscope has two interchangeable objectives. One has a focal length of 2.0 cm, and the other has a focal length of 0.20 cm. Also available are two eyepieces of focal lengths 2.5 cm and 5.0 cm. If the length of the microscope is 18 cm, compute the magnifications for the following combinations: the 2.0-cm objective and 5.0-cm eyepiece, the 2.0-cm objective and 2.5-cm eyepiece, and the 0.20-cm objective and 5.0-cm eyepiece.

STRATEGY The solution consists of substituting into Equation 25.7 for three different combinations of lenses.

(*Continued*)

SOLUTION

Apply Equation 25.7 and combine the 2.0-cm objective with the 5.0-cm eyepiece:

$$m = -\frac{L}{f_o}\left(\frac{25\text{ cm}}{f_e}\right) = -\frac{18\text{ cm}}{2.0\text{ cm}}\left(\frac{25\text{ cm}}{5.0\text{ cm}}\right) = \boxed{-45}$$

Combine the 2.0-cm objective with the 2.5-cm eyepiece:

$$m = -\frac{18\text{ cm}}{2.0\text{ cm}}\left(\frac{25\text{ cm}}{2.5\text{ cm}}\right) = \boxed{-9.0 \times 10^1}$$

Combine the 0.20-cm objective with the 5.0-cm eyepiece:

$$m = -\frac{18\text{ cm}}{0.20\text{ cm}}\left(\frac{25\text{ cm}}{5.0\text{ cm}}\right) = \boxed{-450}$$

REMARKS Much higher magnifications can be achieved, but the resolution starts to fall, resulting in fuzzy images that don't convey any details. (See Section 25.6 for further discussion of this point.)

QUESTION 25.4 True or False: A shorter focal length for either the eyepiece or objective lens will result in greater magnification.

EXERCISE 25.4 Combine the 0.20-cm objective with the 2.5-cm eyepiece and find the magnification.

ANSWER -9.0×10^2

25.5 The Telescope

LEARNING OBJECTIVES

1. Describe the optical components of reflecting and refracting telescopes and discuss the advantages and disadvantages of each.
2. Evaluate the angular magnification of a telescope.

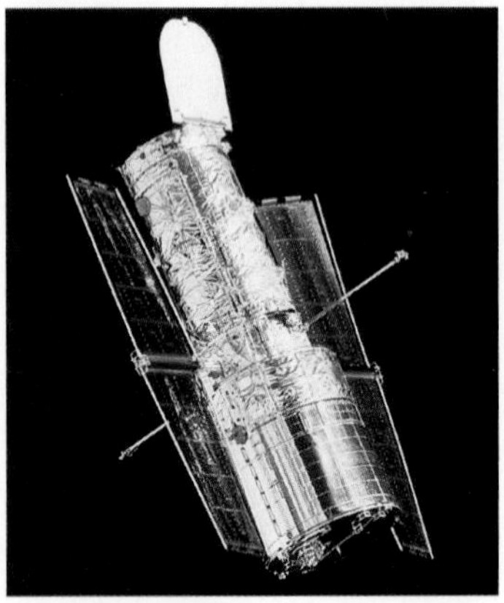

NASA

The Hubble Space Telescope enables us to see both further into space and further back in time than ever before.

There are two fundamentally different types of telescope, both designed to help us view distant objects such as the planets in our solar system: (1) the **refracting telescope**, which uses a combination of lenses to form an image, and (2) the **reflecting telescope**, which uses a curved mirror and a lens to form an image. Once again, we can analyze the telescope by considering it to be a system of two optical elements in a row. As before, the image formed by the first element becomes the object for the second.

In the refracting telescope two lenses are arranged so that the objective forms a real, inverted image of the distant object very near the focal point of the eyepiece (Fig. 25.8a). Further, the image at I_1 is formed at the focal point of the objective because the object is essentially at infinity. Hence, the two lenses are separated by the distance $f_o + f_e$, which corresponds to the length of the telescope's tube. Finally, at I_2, the eyepiece forms an enlarged image of the image at I_1.

The angular magnification of the telescope is given by θ/θ_o, where θ_o is the angle subtended by the object at the objective and θ is the angle subtended by the final image. From the triangles in Figure 25.8a, and for small angles, we have

$$\theta_o \approx \frac{h'}{f_e} \quad \text{and} \quad \theta_o \approx \frac{h'}{f_o}$$

Therefore, the angular magnification of the telescope can be expressed as

Angular magnification of a telescope ▶

$$m = \frac{\theta}{\theta_o} = \frac{h'/f_e}{h'/f_o} = \frac{f_o}{f_e} \qquad [25.8]$$

This equation says that the angular magnification of a telescope equals the ratio of the objective focal length to the eyepiece focal length. Here again, the angular magnification is the ratio of the angular size seen with the telescope to the angular size seen with the unaided eye.

In some applications—for instance, the observation of relatively nearby objects such as the Sun, the Moon, or planets—angular magnification is important. Stars, however, are so far away that they always appear as small points of light regardless

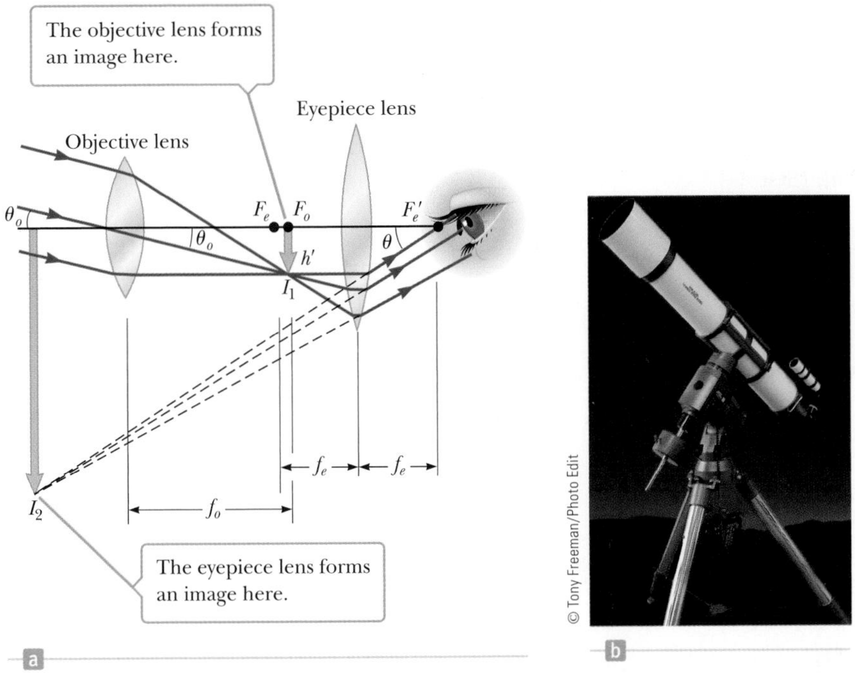

© Tony Freeman/Photo Edit

Figure 25.8 (a) A diagram of a refracting telescope, with the object at infinity. (b) A refracting telescope.

of how much angular magnification is used. The large research telescopes used to study very distant objects must have great diameters to gather as much light as possible. It's difficult and expensive to manufacture such large lenses for refracting telescopes. In addition, the heaviness of large lenses leads to sagging, which is another source of aberration.

These problems can be partially overcome by replacing the objective lens with a reflecting, concave mirror, usually having a parabolic shape so as to avoid spherical aberration. Figure 25.9 shows the design of a typical reflecting telescope. Incoming light rays pass down the barrel of the telescope and are reflected by a parabolic mirror at the base. These rays converge toward point A in the figure, where an image would be formed on a photographic plate or another detector. Before this image is formed, however, a small, flat mirror at M reflects the light toward an opening in the side of the tube that passes into an eyepiece. This design is said to have a *Newtonian focus*, after its developer. Note that in the reflecting telescope the light never passes through glass (except in the small eyepiece). As a result, problems associated with chromatic aberration are virtually eliminated.

The largest optical telescopes in the world are the two 10-m-diameter Keck reflectors on Mauna Kea in Hawaii. The largest single-mirrored reflecting telescope in the United States is the 5-m-diameter instrument on Mount Palomar in California. (See Fig. 25.10.) In contrast, the largest refracting telescope in the world, at the Yerkes Observatory in Williams Bay, Wisconsin, has a diameter of only 1 m.

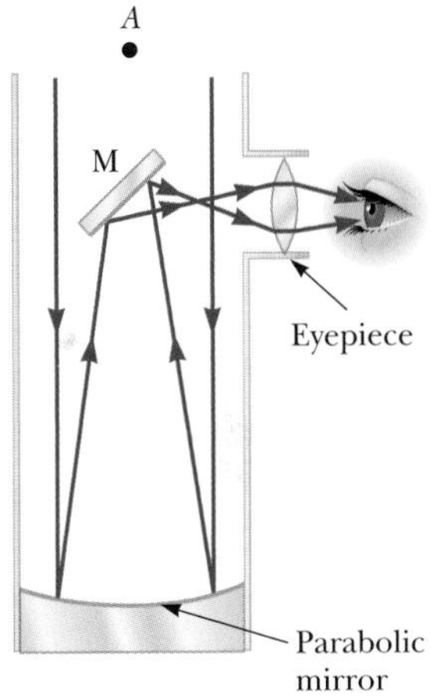

Figure 25.9 A reflecting telescope with a Newtonian focus.

Courtesy of Palomar Observatory/ California Institute of Technology

Figure 25.10 The Hale telescope at Mount Palomar Observatory. Just before taking the elevator up to the prime-focus cage, a first-time observer is always told, "Good viewing! And, if you should fall, try to miss the mirror."

EXAMPLE 25.5 Hubble Power

GOAL Understand magnification in telescopes.

PROBLEM The Hubble Space Telescope is 13.2 m long, but has a secondary mirror that increases its effective focal length to 57.8 m. (See Fig. 25.11.) The telescope doesn't have an eyepiece because various instruments, not a human eye, record the collected light. It can, however, produce images several thousand times larger than they would appear with the unaided human eye. What focal-length eyepiece used with the Hubble mirror system would produce a magnification of 8.00×10^3?

STRATEGY Equation 25.8 for telescope magnification can be solved for the eyepiece focal length. The equation for finding the angular magnification of a reflector is the same as that for a refractor.

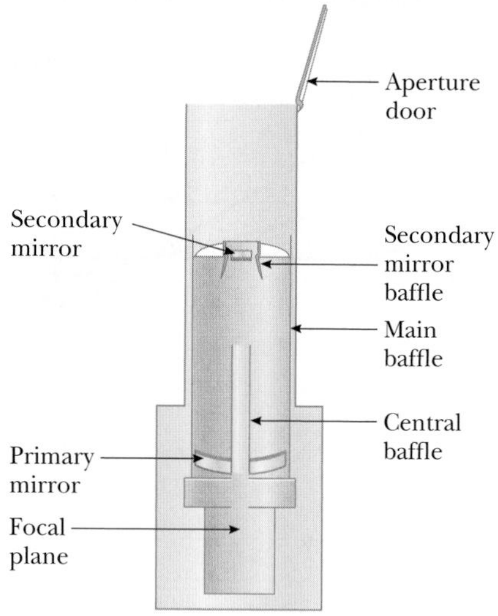

Figure 25.11 A schematic of the Hubble Space Telescope.

SOLUTION

Solve for f_e in Equation 25.8 and substitute values:

$$m = \frac{f_o}{f_e} \quad \rightarrow \quad f_e = \frac{f_o}{m} = \frac{57.8 \text{ m}}{8.00 \times 10^3} = 7.23 \times 10^{-3} \text{ m}$$

REMARKS The light-gathering power of a telescope and the length of the baseline over which light is gathered are in fact more important than a telescope's magnification, because these two factors contribute to the resolution of the image. A high-resolution image can always be magnified so its details can be examined. A low resolution image, however, is often fuzzy when magnified. (See Section 25.6.)

QUESTION 25.5 Can greater magnification of a telescope be achieved by increasing the focal length of the mirror? What effect will increasing the focal length of the eyepiece have on the magnification?

EXERCISE 25.5 The Hale telescope on Mount Palomar has a focal length of 16.8 m. Find the magnification of the telescope in conjunction with an eyepiece having a focal length of 5.00 mm.

ANSWER 3.36×10^3

25.6 Resolution of Single-Slit and Circular Apertures

LEARNING OBJECTIVES

1. State and discuss the physical origins of Rayleigh's criterion.
2. Determine the limiting angle for slit and circular apertures.
3. Evaluate the resolution limitations of optical instruments and the resolving power of a diffraction grating.

The ability of an optical system such as the eye, a microscope, or a telescope to distinguish between closely spaced objects is limited because of the wave nature of light. To understand this difficulty, consider Figure 25.12, which shows two light sources far from a narrow slit of width a. The sources can be taken as two point

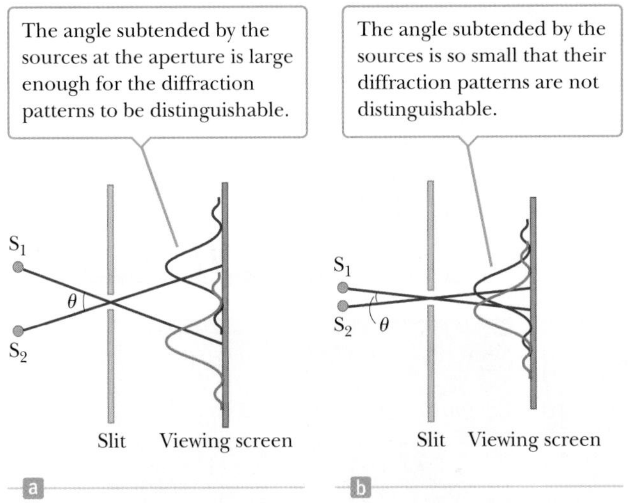

Figure 25.12 Two point sources far from a narrow slit each produce a diffraction pattern. (a) The sources are separated by a large angle. (b) The sources are separated by a small angle. (Notice that the angles are greatly exaggerated. The drawing is not to scale.)

sources S_1 and S_2 that are *not* coherent. For example, they could be two distant stars. If no diffraction occurred, two distinct bright spots (or images) would be observed on the screen at the right in the figure. Because of diffraction, however, each source is imaged as a bright central region flanked by weaker bright and dark rings. What is observed on the screen is the sum of two diffraction patterns, one from S_1 and the other from S_2.

If the two sources are separated so that their central maxima don't overlap, as in Figure 25.12a, their images can be distinguished and are said to be *resolved.* If the sources are close together, however, as in Figure 25.12b, the two central maxima may overlap and the images are *not resolved.* To decide whether two images are resolved, the following condition is often applied to their diffraction patterns:

◀ Rayleigh's criterion

When the central maximum of one image falls on the first minimum of another image, the images are said to be just resolved. This limiting condition of resolution is known as **Rayleigh's criterion**.

Figure 25.13 (page 884) shows diffraction patterns in three situations. In Fig. 25.13a, the sources are sufficiently separated, so the images are resolved. As the sources are brought closer together, as in Figure 25.13b, the central maximum of one image is centered on the first minimum of the other, so by Rayleigh's criterion, the images are just resolved. Finally, when the sources are very close to each other, their images are not resolved (Fig. 25.13c).

From Rayleigh's criterion, we can determine the minimum angular separation θ_{min} subtended by the source at the slit so that the images will be just resolved. In Chapter 24 we found that the first minimum in a single-slit diffraction pattern occurs at the angle that satisfies the relationship

$$\sin\theta = \frac{\lambda}{a}$$

where a is the width of the slit. According to Rayleigh's criterion, this expression gives the smallest angular separation for which the two images can be resolved. Because $\lambda << a$ in most situations, $\sin\theta$ is small and we can use the approximation $\sin\theta \approx \theta$. Therefore, the limiting angle of resolution for a slit of width a is

$$\theta_{min} \approx \frac{\lambda}{a} \quad [25.9]$$

◀ Limiting angle for a slit

where θ_{min} is in radians. Hence, the angle subtended by the two sources at the slit must be *greater* than λ/a if the images are to be resolved.

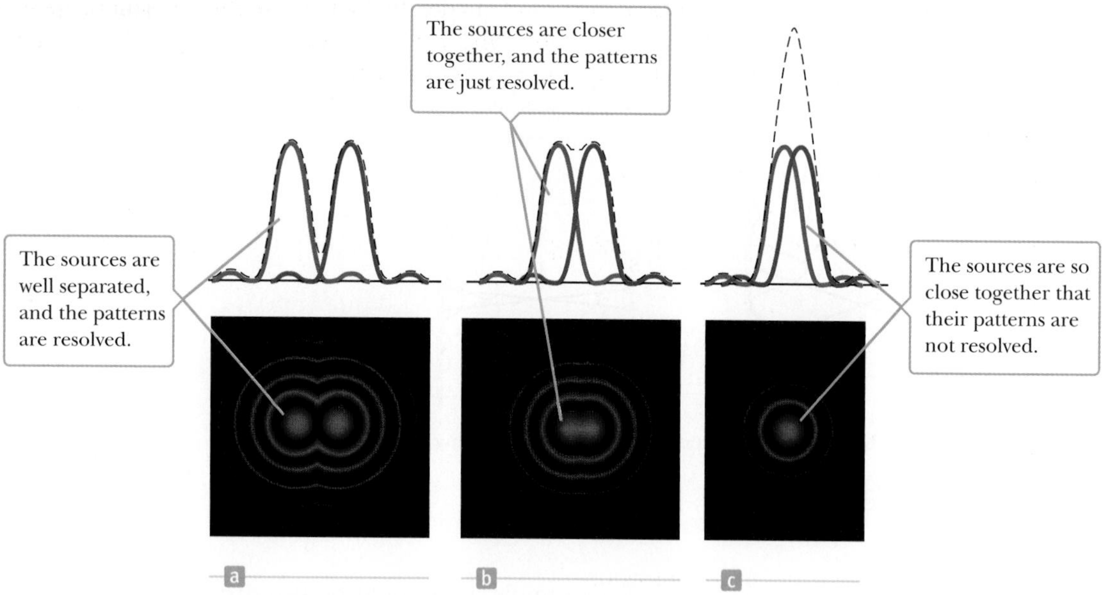

Figure 25.13 The diffraction patterns of two point sources (solid curves) and the resultant pattern (dashed curve) for three angular separations of the sources.

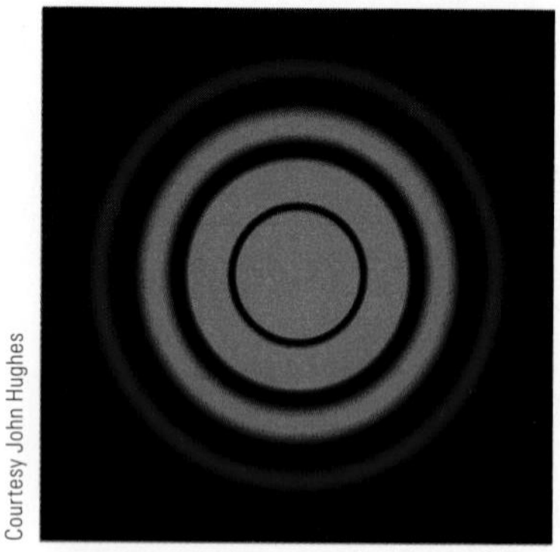

Figure 25.14 The diffraction pattern of a circular aperture consists of a central bright disk surrounded by concentric bright and dark rings.

Many optical systems use circular apertures rather than slits. The diffraction pattern of a circular aperture (Fig. 25.14) consists of a central circular bright region surrounded by progressively fainter rings. Analysis shows that the limiting angle of resolution of the circular aperture is

$$\theta_{\min} = 1.22\frac{\lambda}{D} \qquad [25.10]$$

where D is the diameter of the aperture. Note that Equation 25.10 is similar to Equation 25.9 except for the factor 1.22, which arises from a complex mathematical analysis of diffraction from a circular aperture.

Quick Quiz

25.2 Suppose you are observing a binary star with a telescope and are having difficulty resolving the two stars. Which color filter will better help resolve the stars? (a) blue (b) red (c) neither because colored filters have no effect on resolution

APPLYING PHYSICS 25.2 | Cat's Eyes BIO

Cats' eyes have vertical pupils in dim light. Which would cats be most successful at resolving at night, headlights on a distant car or vertically separated running lights on a distant boat's mast having the same separation as the car's headlights?

EXPLANATION The effective slit width in the vertical direction of the cat's eye is larger than that in the horizontal direction. Thus, it has more resolving power for lights separated in the vertical direction and would be more effective at resolving the mast lights on the boat. ■

EXAMPLE 25.6 | Resolution of a Microscope

GOAL Study limitations on the resolution of a microscope.

PROBLEM Sodium light of wavelength 589 nm is used to view an object under a microscope. The aperture of the objective has a diameter of 0.90 cm. **(a)** Find the limiting angle of resolution for this microscope. **(b)** Using visible light of any wavelength you desire, find the best limit of resolution for this microscope. **(c)** Water of index of refraction 1.33 now fills the space between the object and the objective. What effect would this water have on the resolving power of the microscope, using 589-nm light?

STRATEGY Parts (a) and (b) require substitution into Equation 25.10. Because the wavelength appears in the numerator, violet light, with the shortest visible wavelength, gives the maximum resolution. In part (c) the only difference is that the wavelength changes to λ/n, where n is the index of refraction of water.

SOLUTION

(a) Find the limiting angle of resolution for this microscope.

Substitute into Equation 25.10 to obtain the limiting angle of resolution:

$$\theta_{\min} = 1.22\frac{\lambda}{D} = 1.22\left(\frac{589 \times 10^{-9}\text{ m}}{0.90 \times 10^{-2}\text{ m}}\right)$$
$$= \boxed{8.0 \times 10^{-5}\text{ rad}}$$

(b) Calculate the microscope's best limit of resolution.

To obtain the best resolution, substitute the shortest visible wavelength available, which is violet light, of wavelength 4.0×10^2 nm:

$$\theta_{\min} = 1.22\frac{\lambda}{D} = 1.22\left(\frac{4.0 \times 10^{-7}\text{ m}}{0.90 \times 10^{-2}\text{ m}}\right)$$
$$= \boxed{5.4 \times 10^{-5}\text{ rad}}$$

(c) What effect does water between the object and the objective lens have on the resolution, with 589-nm light?

Calculate the wavelength of the sodium light in the water:

$$\lambda_w = \frac{\lambda_a}{n} = \frac{589\text{ nm}}{1.33} = 443\text{ nm}$$

Substitute this wavelength into Equation 25.10 to get the resolution:

$$\theta_{\min} = 1.22\left(\frac{443 \times 10^{-9}\text{ m}}{0.90 \times 10^{-2}\text{ m}}\right) = \boxed{6.0 \times 10^{-5}\text{ rad}}$$

REMARKS In each case any two points on the object subtending an angle of less than the limiting angle $\theta_{\min}$ at the objective cannot be distinguished in the image. Consequently, it may be possible to see a cell but then be unable to clearly see smaller structures within the cell. Obtaining an increase in resolution is the motivation behind placing a drop of oil on the slide for certain objective lenses.

QUESTION 25.6 Does having two eyes instead of one improve the human ability to resolve distant objects? In general, would more widely spaced eyes increase visual resolving power? Explain.

EXERCISE 25.6 Suppose oil with $n = 1.50$ fills the space between the object and the objective for this microscope. Calculate the limiting angle $\theta_{\min}$ for sodium light of wavelength 589 nm in air.

ANSWER 5.3×10^{-5} rad

EXAMPLE 25.7 Resolving Craters on the Moon

GOAL Calculate the resolution of a telescope.

PROBLEM The Hubble Space Telescope has an aperture of diameter 2.40 m. **(a)** What is its limiting angle of resolution at a wavelength of 6.00×10^2 nm? **(b)** What's the smallest crater it could resolve on the Moon? (The Moon's distance from Earth is 3.84×10^8 m.)

STRATEGY After substituting into Equation 25.10 to find the limiting angle, use $s = r\theta$ to compute the minimum size of crater that can be resolved.

SOLUTION

(a) What is the limiting angle of resolution at a wavelength of 6.00×10^2 nm?

Substitute $D = 2.40$ m and $\lambda = 6.00 \times 10^{-7}$ m into Equation 25.10:

$$\theta_{\min} = 1.22\frac{\lambda}{D} = 1.22\left(\frac{6.00 \times 10^{-7}\text{ m}}{2.40\text{ m}}\right)$$
$$= \boxed{3.05 \times 10^{-7}\text{ rad}}$$

(Continued)

(b) What's the smallest lunar crater the Hubble Space Telescope can resolve?

The two opposite sides of the crater must subtend the minimum angle. Use the arc length formula:

$$s = r\theta = (3.84 \times 10^8 \text{ m})(3.05 \times 10^{-7} \text{ rad}) = \boxed{117 \text{ m}}$$

REMARKS The distance is so great and the angle so small that using the arc length of a circle is justified because the circular arc is very nearly a straight line. The Hubble Space Telescope has produced several gigabytes of data every day since it first began operation.

QUESTION 25.7 Is the resolution of a telescope better at the red end of the visible spectrum or the violet end?

EXERCISE 25.7 The Hale telescope on Mount Palomar has a diameter of 5.08 m (200 in.). (a) Find the limiting angle of resolution for a wavelength of 6.00×10^2 nm. (b) Calculate the smallest crater diameter the telescope can resolve on the Moon. (c) The answers appear better than what the Hubble can achieve. Why are the answers misleading?

ANSWERS (a) 1.44×10^{-7} rad (b) 55.3 m (c) Although the numbers are better than Hubble's, the Hale telescope must contend with the effects of atmospheric turbulence, so the smaller space-based telescope actually obtains far better results.

It's interesting to compare the resolution of the Hale telescope with that of a large radio telescope, such as the system at Arecibo, Puerto Rico, which has a diameter of 1 000 ft (305 m). This telescope detects radio waves at a wavelength of 0.75 m. The corresponding minimum angle of resolution can be calculated as 3.0×10^{-3} rad (10 min 19 s of arc), which is more than 10 000 times larger than the calculated minimum angle for the Hale telescope.

With such relatively poor resolution, why is Arecibo considered a valuable astronomical instrument? Unlike its optical counterparts, Arecibo can see through clouds of dust. The center of our Milky Way galaxy is obscured by such dust clouds, which absorb and scatter visible light. Radio waves easily penetrate the clouds, so radio telescopes allow direct observations of the galactic core.

Resolving Power of the Diffraction Grating

The diffraction grating studied in Chapter 24 is most useful for making accurate wavelength measurements. Like the prism, it can be used to disperse a spectrum into its components. Of the two devices, the grating is better suited to distinguishing between two closely spaced wavelengths. We say that the grating spectrometer has a higher *resolution* than the prism spectrometer. If λ_1 and λ_2 are two nearly equal wavelengths between which the spectrometer can just barely distinguish, the **resolving power** of the grating is defined as

$$R \equiv \frac{\lambda}{\lambda_2 - \lambda_1} = \frac{\lambda}{\Delta\lambda} \qquad [25.11]$$

where $\lambda \approx \lambda_1 \approx \lambda_2$ and $\Delta\lambda = \lambda_2 - \lambda_1$. From this equation, it's clear that a grating with a high resolving power can distinguish small differences in wavelength. Further, if N lines of the grating are illuminated, it can be shown that the resolving power in the mth-order diffraction is given by

Resolving power ▶ of a grating

$$R = Nm \qquad [25.12]$$

So, the resolving power R increases with the order number m and is large for a grating with a great number of illuminated slits. Note that for $m = 0$, $R = 0$, which signifies that *all wavelengths are indistinguishable* for the zeroth-order maximum. (All wavelengths fall at the same point on the screen.) Consider, however, the second-order diffraction pattern of a grating that has 5 000 rulings illuminated by the light source. The resolving power of such a grating in second order is $R = 5\,000 \times 2 = 10\,000$. Therefore, the *minimum* wavelength separation between two spectral lines that can be just resolved, assuming a mean wavelength of 600 nm, is calculated from Equation 25.12 to be $\Delta\lambda = \lambda/R = 6 \times 10^{-2}$ nm. For the third-order principal maximum, $R = 15\,000$ and $\Delta\lambda = 4 \times 10^{-2}$ nm, and so on.

EXAMPLE 25.8 | Light from Sodium Atoms

GOAL Find the necessary resolving power to distinguish spectral lines.

PROBLEM Two bright lines in the spectrum of sodium have wavelengths of 589.00 nm and 589.59 nm, respectively. **(a)** What must the resolving power of a grating be so as to distinguish these wavelengths? **(b)** To resolve these lines in the second-order spectrum, how many lines of the grating must be illuminated?

STRATEGY This problem requires little more than substituting into Equations 25.11 and 25.12.

SOLUTION

(a) What must the resolving power of a grating be in order to distinguish the given wavelengths?

Substitute into Equation 25.11 to find R:

$$R = \frac{\lambda}{\Delta\lambda} = \frac{589.00\text{ nm}}{589.59\text{ nm} - 589.00\text{ nm}} = \frac{589\text{ nm}}{0.59\text{ nm}}$$

$$= 1.0 \times 10^3$$

(b) To resolve these lines in the second-order spectrum, how many lines of the grating must be illuminated?

Solve Equation 25.12 for N and substitute:

$$N = \frac{R}{m} = \frac{1.0 \times 10^3}{2} = 5.0 \times 10^2 \text{ lines}$$

REMARKS The ability to resolve spectral lines is particularly important in experimental atomic physics.

QUESTION 25.8 True or False: If two diffraction gratings differ only by the number of lines, the grating with the larger number of lines can yield a greater resolving power.

EXERCISE 25.8 Due to a phenomenon called electron spin, when the lines of a spectrum are examined at high resolution, each line is actually found to be two closely spaced lines called a doublet. An example is the doublet in the hydrogen spectrum having wavelengths of 656.272 nm and 656.285 nm. (a) What must be the resolving power of a grating so as to distinguish these wavelengths? (b) How many lines of the grating must be illuminated to resolve these lines in the third-order spectrum?

ANSWERS (a) 5.0×10^4 (b) 1.7×10^4 lines

25.7 The Michelson Interferometer

LEARNING OBJECTIVE

1. Describe the optical components and operating principles of the Michelson interferometer.

The Michelson interferometer is an optical instrument having great scientific importance. Invented by American physicist A. A. Michelson (1852–1931), it is an ingenious device that splits a light beam into two parts and then recombines them to form an interference pattern. The interferometer is used to make accurate length measurements.

Figure 25.15 (page 888) is a schematic diagram of an interferometer. A beam of light provided by a monochromatic source is split into two rays by a partially silvered mirror M inclined at an angle of 45° relative to the incident light beam. One ray is reflected vertically upward to mirror M_1, and the other ray is transmitted horizontally through mirror M to mirror M_2. Hence, the two rays travel separate paths, L_1 and L_2. After reflecting from mirrors M_1 and M_2, the two rays eventually recombine to produce an interference pattern, which can be viewed through a telescope. The glass plate P, equal in thickness to mirror M, is placed in the path of the horizontal ray to ensure that the two rays travel the same distance through glass.

The interference pattern for the two rays is determined by the difference in their path lengths. When the two rays are viewed as shown, the image of M_2 is at

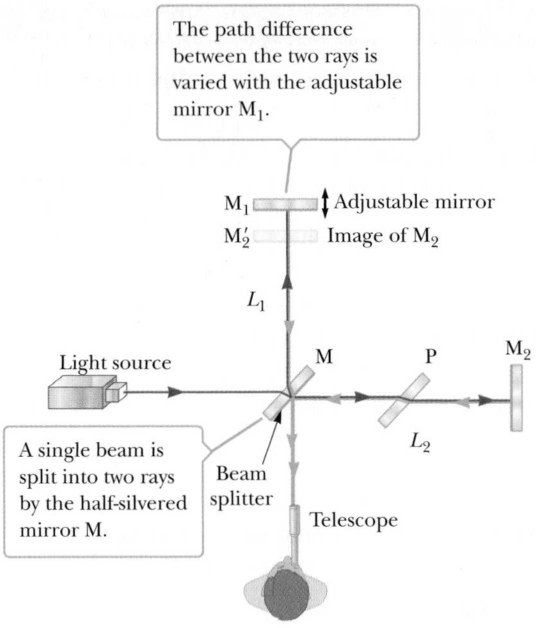

Figure 25.15 A diagram of the Michelson interferometer.

M_2', parallel to M_1. Hence, the space between M_2' and M_1 forms the equivalent of a parallel air film. The effective thickness of the air film is varied by using a finely threaded screw to move mirror M_1 in the direction indicated by the arrows in Figure 25.15. If one of the mirrors is tipped slightly with respect to the other, the thin film between the two is wedge shaped and an interference pattern consisting of parallel fringes is set up, as described in Example 24.4. Now suppose we focus on one of the dark lines with the crosshairs of a telescope. As mirror M_1 is moved to lengthen the path L_1, the thickness of the wedge increases. When the thickness increases by $\lambda/4$, the destructive interference that initially produced the dark fringe has changed to constructive interference, and we now observe a bright fringe at the location of the crosshairs. The term *fringe shift* is used to describe the change in a fringe from dark to light or from light to dark. Successive light and dark fringes are formed each time M_1 is moved a distance of $\lambda/4$. The wavelength of light can be measured by counting the number of fringe shifts for a measured displacement of M_1. Conversely, if the wavelength is accurately known (as with a laser beam), the mirror displacement can be determined to within a fraction of the wavelength. Because the interferometer can measure displacements precisely, it is often used to make highly accurate measurements of the dimensions of mechanical components.

If the mirrors are perfectly aligned rather than tipped with respect to each other, the path difference differs slightly for different angles of view. This arrangement results in an interference pattern that resembles Newton's rings. The pattern can be used in a fashion similar to that for tipped mirrors. An observer pays attention to the center spot in the interference pattern. For example, suppose the spot is initially dark, indicating that destructive interference is occurring. If M_1 is now moved a distance of $\lambda/4$, this central spot changes to a light region, corresponding to a fringe shift.

SUMMARY

25.1 The Camera

The light-concentrating power of a lens of focal length f and diameter D is determined by the ***f*-number**, defined as

$$f\text{-number} \equiv \frac{f}{D} \quad [25.1]$$

The smaller the f-number of a lens, the brighter the image formed.

25.2 The Eye

Hyperopia (farsightedness) is a defect of the eye that occurs either when the eyeball is too short or when the ciliary muscle cannot change the shape of the lens enough to form a properly focused image. **Myopia** (nearsightedness) occurs either when the eye is longer than normal or when the maximum focal length of the lens is insufficient to produce a clearly focused image on the retina.

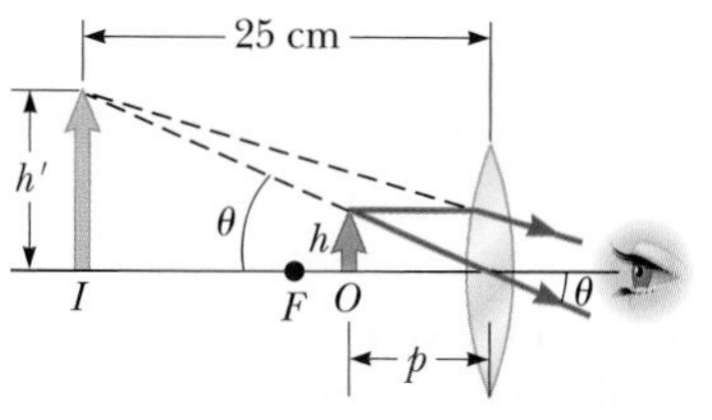

An object placed near the focal point of a converging lens produces a magnified image, which subtends an angle of $\theta \approx h'/25$ at the eye. Note that in this situation $q = -25$ cm.

The **power** of a lens in **diopters** is the inverse of the focal length in meters.

25.3 The Simple Magnifier

The **angular magnification of a lens** is defined as

$$m \equiv \frac{\theta}{\theta_0} \qquad [25.2]$$

where θ is the angle subtended by an object at the eye with a lens in use and θ_0 is the angle subtended by the object when it is placed at the near point of the eye and no lens is used. The **maximum angular magnification of a lens** is

$$m_{\text{max}} = 1 + \frac{25 \text{ cm}}{f} \qquad [25.5]$$

When the eye is relaxed, the angular magnification is

$$m = \frac{25 \text{ cm}}{f} \qquad [25.6]$$

25.4 The Compound Microscope

The overall **magnification of a compound microscope** of length L is the product of the magnification produced by the objective, of focal length f_o, and the magnification produced by the eyepiece, of focal length f_e:

$$m = -\frac{L}{f_o}\left(\frac{25 \text{ cm}}{f_e}\right) \qquad [25.7]$$

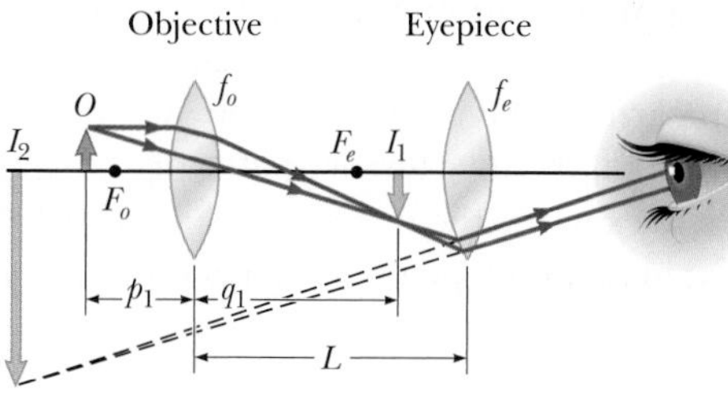

A ray diagram for a compound microscope, which consists of an objective and an eyepiece, or ocular lens.

25.5 The Telescope

The **angular magnification of a telescope** is

$$m = \frac{f_o}{f_e} \qquad [25.8]$$

where f_o is the focal length of the objective and f_e is the focal length of the eyepiece.

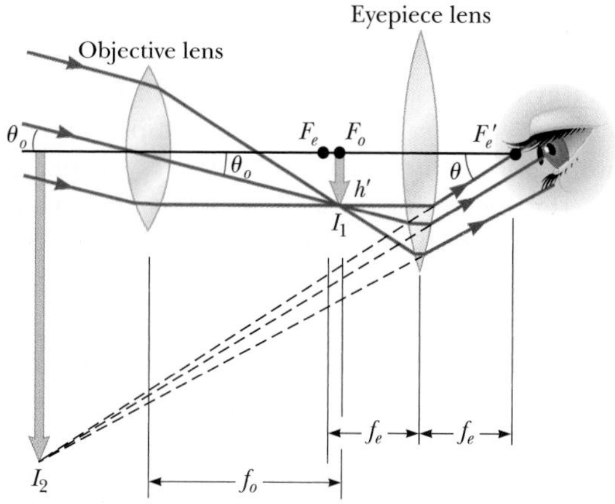

A ray diagram for a refracting telescope, with the object at infinity.

25.6 Resolution of Single-Slit and Circular Apertures

Two images are said to be **just resolved** when the central maximum of the diffraction pattern for one image falls on the first minimum of the other image. This limiting condition of resolution is known as **Rayleigh's criterion**. The limiting angle of resolution for a **slit** of width a is

$$\theta_{\text{min}} \approx \frac{\lambda}{a} \qquad [25.9]$$

The limiting angle of resolution of a **circular aperture** is

$$\theta_{\text{min}} = 1.22\frac{\lambda}{D} \qquad [25.10]$$

where D is the diameter of the aperture.

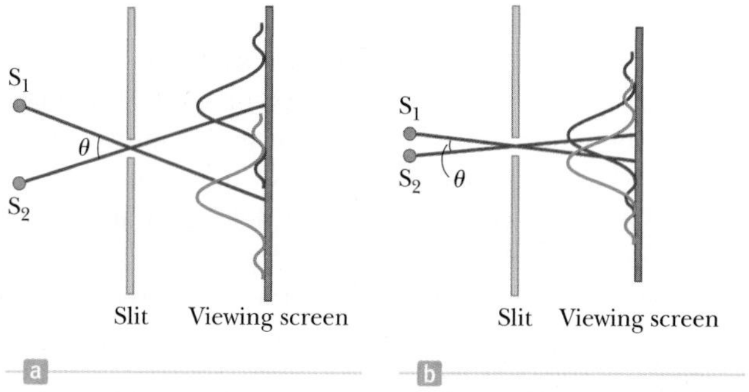

Two point sources far from a narrow slit each produce a diffraction pattern. (a) The sources are separated by a large angle. (b) The sources are separated by a small angle.

If λ_1 and λ_2 are two nearly equal wavelengths between which a grating spectrometer can just barely distinguish, the **resolving power** R of the grating is defined as

$$R \equiv \frac{\lambda}{\lambda_2 - \lambda_1} = \frac{\lambda}{\Delta\lambda} \qquad [25.11]$$

where $\lambda \approx \lambda_1 \approx \lambda_2$ and $\Delta\lambda = \lambda_2 - \lambda_1$. The **resolving power** of a diffraction grating in the mth order is

$$R = Nm \qquad [25.12]$$

where N is the number of illuminated rulings on the grating.

WARM-UP EXERCISES

ENHANCED WebAssign The warm-up exercises in this chapter may be assigned online in Enhanced WebAssign.

1. **Physics Review** A real object is 12.0 cm in front of a thin, convergent lens with a focal length of 10.0 cm. Determine (a) the distance from the lens to the image and (b) the image magnification. (c) Is the image upright or inverted? (d) Is the image real or virtual? (See Section 23.6.)
2. **Physics Review** A thin, diverging lens has a focal length of magnitude 15.0 cm. An object is placed 10.0 cm in front of the lens. Find (a) the image distance and (b) the image magnification. (c) Is the image upright or inverted? (d) Real or virtual? (See Section 23.6.)
3. A large telephoto camera lens has a focal length of 0.825 m and an *f*-number of 4. Determine the lens diameter. (See Section 25.1.)
4. A prescription lens has a focal length of 25.0 cm. What is the power of the lens in diopters? (See Section 25.2.)
5. A man can see no farther than 42.0 cm without corrective eyeglasses. (a) Is the man nearsighted or farsighted? Find (b) the focal length of the appropriate corrective lens and (c) the power of the lens in diopters. (See Section 25.2.)
6. A woman can see clearly only when objects are 48.0 cm or farther away from her. (a) Is she nearsighted or farsighted? Find (b) the focal length of the appropriate corrective lens and (c) the power of the lens in diopters. (See Section 25.2.)
7. A single-lens magnifier has a maximum angular magnification of 9.33. Determine (a) the lens's focal length (in cm) and (b) the magnification when used with a relaxed eye. (See Section 25.3.)
8. A compound microscope has objective and eyepiece lenses of focal lengths 0.80 cm and 4.0 cm, respectively. If the microscope length is 15 cm, what is the magnification of the microscope? (See Section 25.4.)
9. Determine the angular magnification of a small, handheld telescope if its objective lens has a focal length of 25.0 cm and its eyepiece has a focal length of 2.20 cm. (See Section 25.5.)
10. Determine the approximate limiting angle of resolution (in rad) for 645-nm light incident on a single slit having a width of 1.12×10^{-3} m. (See Section 25.6.)
11. Two stars have an angular separation of 7.50×10^{-8} rad when viewed in the night sky from Earth. Determine the minimum diameter of a telescope's circular aperture if the two stars are to be angularly resolved at a wavelength of 532 nm. (See Section 25.6.)
12. What minimum resolving power must a diffraction grating have to distinguish between two emissions near a wavelength of 632 nm and separated by 2.00 nm? (See Section 25.6.)

CONCEPTUAL QUESTIONS

ENHANCED WebAssign The conceptual questions in this chapter may be assigned online in Enhanced WebAssign.

1. A lens is used to examine an object across a room. Is the lens probably being used as a simple magnifier? Explain in terms of focal length, the image, and magnification.
2. A laser beam is incident at a shallow angle on a horizontal machinist's ruler that has a finely calibrated scale. The engraved rulings on the scale give rise to a diffraction pattern on a vertical screen. Discuss how you can use this technique to obtain a measure of the wavelength of the laser light.
3. BIO The optic nerve and the brain invert the image formed on the retina. Why don't we see everything upside down?
4. Suppose you are observing the interference pattern formed by a Michelson interferometer in a laboratory and a joking colleague holds a lit match in the light path of one arm of the interferometer. Will this match have an effect on the interference pattern?
5. If you want to examine the fine detail of an object with a magnifying glass with a power of +20.0 diopters, where should the object be placed so as to observe a magnified image of the object?
6. BIO Compare and contrast the eye and a camera. What parts of the camera correspond to the iris, the retina, and the cornea of the eye?

7. If you want to use a converging lens to set fire to a piece of paper, why should the light source be farther from the lens than its focal point?

8. Large telescopes are usually reflecting rather than refracting. List some reasons for this choice.

9. Explain why it is theoretically impossible to see an object as small as an atom regardless of the quality of the light microscope being used.

10. Which is most important in the use of a camera photoflash unit, the intensity of the light (the energy per unit area per unit time) or the product of the intensity and the time of the flash, assuming the time is less than the shutter speed?

11. BIO A patient has a near point of 1.25 m. Is she nearsighted or farsighted? Should the corrective lens be converging or diverging?

12. A lens with a certain power is used as a simple magnifier. If the power of the lens is doubled, does the angular magnification increase or decrease?

13. A laser produces a beam a few millimeters wide, with uniform intensity across its width. A hair is stretched vertically across the front of the laser to cross the beam. (a) How is the diffraction pattern it produces on a distant screen related to that of a vertical slit equal in width to the hair? (b) How could you determine the width of the hair from measurements of its diffraction pattern?

14. BIO During LASIK eye surgery (laser-assisted *in situ* keratomileusis), the shape of the cornea is modified by vaporizing some of its material. If the surgery is performed to correct for nearsightedness, how does the cornea need to be reshaped?

15. If you increase the aperture diameter of a camera by a factor of 3, how is the intensity of the light striking the film affected? (a) It increases by a factor of 3. (b) It decreases by a factor of 3. (c) It increases by a factor of 9. (d) It decreases by a factor of 9. (e) Increasing the aperture doesn't affect the intensity.

PROBLEMS

ENHANCED WebAssign The problems in this chapter may be assigned online in Enhanced WebAssign.

1. denotes straightforward problem; 2. denotes intermediate problem;
3. denotes challenging problem

[1.] denotes full solution available in *Student Solutions Manual/Study Guide*

1. denotes problems most often assigned in Enhanced WebAssign

BIO denotes biomedical problems

GP denotes guided problems

M denotes Master It tutorial available in Enhanced WebAssign

Q|C denotes asking for quantitative and conceptual reasoning

S denotes symbolic reasoning problem

W denotes Watch It video solution available in Enhanced WebAssign

25.1 The Camera

1. A lens has a focal length of 28 cm and a diameter of 4.0 cm. What is the *f*-number of the lens?

2. A certain camera has *f*-numbers that range from 1.2 to 22. If the focal length of the lens is 55 mm, what is the range of aperture diameters for the camera?

[3.] A photographic image of a building is 0.092 0 m high. The image was made with a lens with a focal length of 52.0 mm. If the lens was 100 m from the building when the photograph was made, determine the height of the building.

4. The image area of a typical 35 mm slide is 23.5 mm by 35.0 mm. If a camera's lens has a focal length of 55.0 mm and forms an image of the constellation Orion, which is 20° across, will the full image fit on a 35-mm slide?

5. A camera is being used with a correct exposure at $f/4$ and a shutter speed of $\frac{1}{15}$ s. In addition to the *f*-numbers listed in Section 25.1, this camera has *f*-numbers $f/1$, $f/1.4$, and $f/2$. To photograph a rapidly moving subject, the shutter speed is changed to $\frac{1}{125}$ s. Find the new *f*-number setting needed on this camera to maintain satisfactory exposure.

6. (a) Use conceptual arguments to show that the intensity of light (energy per unit area per unit time) reaching the film in a camera is proportional to the square of the reciprocal of the *f*-number as

$$I \propto \frac{1}{(f/D)^2}$$

(b) The correct exposure time for a camera set to $f/1.8$ is (1/500) s. Calculate the correct exposure time if the *f*-number is changed to $f/4$ under the same lighting conditions. *Note:* "$f/4$," on a camera, means "an *f*-number of 4."

[7.] A certain type of film requires an exposure time of 0.010 s with an $f/11$ lens setting. Another type of film requires twice the light energy to produce the same level of exposure. What *f*-number does the second type of film need with the 0.010-s exposure time?

8. A certain camera lens has a focal length of 175 mm. Its position can be adjusted to produce images when the lens is between 180 mm and 210 mm from the plane of the film. Over what range of object distances is the lens useful?

25.2 The Eye

9. BIO The near point of a person's eye is 60.0 cm. To see objects clearly at a distance of 25.0 cm, what should be the (a) focal length and (b) power of the appropriate corrective lens? (Neglect the distance from the lens to the eye.)

10. BIO GP A patient can't see objects closer than 40.0 cm and wishes to clearly see objects that are 20.0 cm from his eye. (a) Is the patient nearsighted or farsighted? (b) If the eye–lens distance is 2.00 cm, what is the minimum object distance p from the lens? (c) What image position with respect to the lens will allow the patient to see the object? (d) Is the image real or virtual? Is the image distance q positive or negative? (e) Calculate the required focal length. (f) Find the power of the lens in diopters. (g) If a contact lens is to be prescribed instead, find p, q, and f, and the power of the lens.

11. BIO M The accommodation limits for Nearsighted Nick's eyes are 18.0 cm and 80.0 cm. When he wears his glasses, he is able to see faraway objects clearly. At what minimum distance is he able to see objects clearly?

12. BIO W A certain child's near point is 10.0 cm; her far point (with eyes relaxed) is 125 cm. Each eye lens is 2.00 cm from the retina. (a) Between what limits, measured in diopters, does the power of this lens–cornea combination vary? (b) Calculate the power of the eyeglass lens the child should use for relaxed distance vision. Is the lens converging or diverging?

13. BIO An individual is nearsighted; his near point is 13.0 cm and his far point is 50.0 cm. (a) What lens power is needed to correct his nearsightedness? (b) When the lenses are in use, what is this person's near point?

14. BIO Q|C A patient has a near point of 45.0 cm and far point of 85.0 cm. (a) Can a single lens correct the patient's vision? Explain the patient's options. (b) Calculate the power lens needed to correct the near point so that the patient can see objects 25.0 cm away. Neglect the eye–lens distance. (c) Calculate the power lens needed to correct the patient's far point, again neglecting the eye–lens distance.

15. BIO An artificial lens is implanted in a person's eye to replace a diseased lens. The distance between the artificial lens and the retina is 2.80 cm. In the absence of the lens, an image of a distant object (formed by refraction at the cornea) falls 5.33 cm behind the implanted lens. The lens is designed to put the image of the distant object on the retina. What is the power of the implanted lens? *Hint:* Consider the image formed by the cornea to be a virtual object.

16. BIO A person is to be fitted with bifocals. She can see clearly when the object is between 30 cm and 1.5 m from the eye. (a) The upper portions of the bifocals (Fig. P25.16) should be designed to enable her to see distant objects clearly. What power should they have? (b) The lower portions of the bifocals should enable her to see objects located 25 cm in front of the eye. What power should they have?

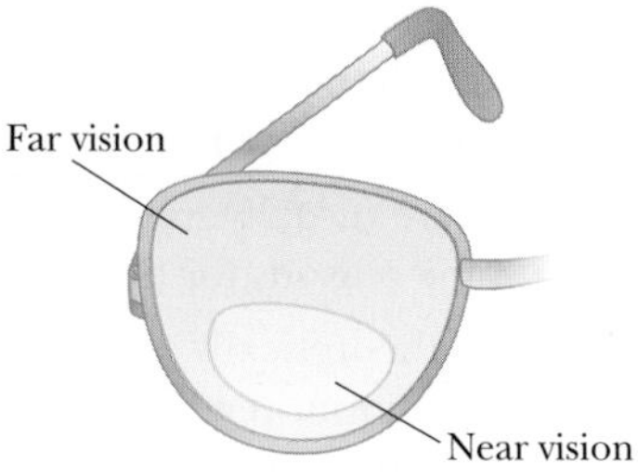

Figure P25.16

17. BIO A nearsighted woman can't see objects clearly beyond 40.0 cm (her far point). If she has no astigmatism and contact lenses are prescribed, what power and type of lens are required to correct her vision?

18. BIO Q|C A person sees clearly wearing eyeglasses that have a power of −4.00 diopters when the lenses are 2.00 cm in front of the eyes. (a) What is the focal length of the lens? (b) Is the person nearsighted or farsighted? (c) If the person wants to switch to contact lenses placed directly on the eyes, what lens power should be prescribed?

25.3 The Simple Magnifier

19. A stamp collector uses a lens with 7.5-cm focal length as a simple magnifier. The virtual image is produced at the normal near point (25 cm). (a) How far from the lens should the stamp be placed? (b) What is the expected angular magnification?

20. A lens that has a focal length of 5.00 cm is used as a magnifying glass. (a) To obtain maximum magnification and an image that can be seen clearly by a normal eye, where should the object be placed? (b) What is the angular magnification?

21. W A biology student uses a simple magnifier to examine the structural features of an insect's wing. The wing is held 3.50 cm in front of the lens, and the image is formed 25.0 cm from the eye. (a) What is the focal length of the lens? (b) What angular magnification is achieved?

22. A jeweler's lens of focal length 5.0 cm is used as a magnifier. With the lens held near the eye, determine (a) the angular magnification when the object is at the focal point of the lens and (b) the angular magnification when the image formed by the lens is at the near point of the eye (25 cm). (c) What is the object distance giving the maximum magnification?

23. A leaf of length h is positioned 71.0 cm in front of a converging lens with a focal length of 39.0 cm. An observer views the image of the leaf from a position 1.26 m behind the lens, as shown in Figure P25.23. (a) What is the magnitude of the lateral magnification (the ratio of the image size to the object size) produced

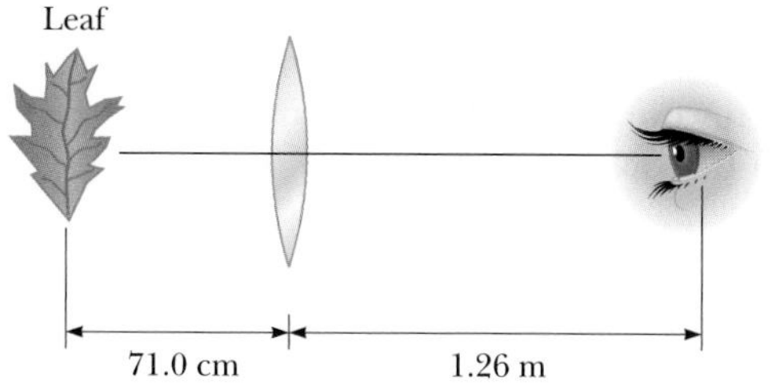

Figure P25.23

by the lens? (b) What angular magnification is achieved by viewing the image of the leaf rather than viewing the leaf directly?

24. (a) What is the maximum angular magnification of an eyeglass lens having a focal length of 18.0 cm when used as a simple magnifier? (b) What is the magnification of this lens when the eye is relaxed?

25.4 The Compound Microscope

25.5 The Telescope

25. The desired overall magnification of a compound microscope is 140×. The objective alone produces a lateral magnification of 12×. Determine the required focal length of the eyepiece.

26. The distance between the eyepiece and the objective lens in a certain compound microscope is 20.0 cm. The focal length of the objective is 0.500 cm, and that of the eyepiece is 1.70 cm. Find the overall magnification of the microscope.

27. Find the magnification of a telescope that uses a 2.75-diopter objective lens and a 35.0-diopter eyepiece.

28. BIO A microscope has an objective lens with a focal length of 16.22 mm and an eyepiece with a focal length of 9.50 mm. With the length of the barrel set at 29.0 cm, the diameter of a red blood cell's image subtends an angle of 1.43 mrad with the eye. If the final image distance is 29.0 cm from the eyepiece, what is the actual diameter of the red blood cell? *Hint:* To solve this question, go back to basics and use the thin-lens equation.

29. A certain telescope has an objective mirror with an aperture diameter of 200 mm and a focal length of 2 000 mm. It captures the image of a nebula on photographic film at its prime focus with an exposure time of 1.50 min. To produce the same light energy per unit area on the film, what is the required exposure time to photograph the same nebula with a smaller telescope that has an objective with a 60.0-mm diameter and a 900-mm focal length?

30. S (a) Find an equation for the length L of a refracting telescope in terms of the focal length of the objective f_o and the magnification m. (b) A knob adjusts the eyepiece forward and backward. Suppose the telescope is in focus with an eyepiece giving a magnification of 50.0. By what distance must the eyepiece be adjusted when the eyepiece is replaced, with a resulting magnification of 1.00×10^2? Must the eyepiece be adjusted backward or forward? Assume the objective lens has a focal length of 2.00 m.

31. Suppose an astronomical telescope is being designed to have an angular magnification of 34.0. If the focal length of the objective lens being used is 86.0 cm, find (a) the required focal length of the eyepiece and (b) the distance between the two lenses for a relaxed eye. *Hint:* For a relaxed eye, the image formed by the objective lens is at the focal point of the eyepiece.

32. A certain telescope has an objective of focal length 1 500 cm. If the Moon is used as an object, a 1.0-cm-long image formed by the objective corresponds to what distance, in miles, on the Moon? Assume 3.8×10^8 m for the Earth–Moon distance.

33. S Astronomers often take photographs with the objective lens or mirror of a telescope alone, without an eyepiece. (a) Show that the image size h' for a telescope used in this manner is given by $h' = fh/(f - p)$, where h is the object size, f is the objective focal length, and p is the object distance. (b) Simplify the expression in part (a) if the object distance is much greater than the objective focal length. (c) The "wingspan" of the International Space Station is 108.6 m, the overall width of its solar panel configuration. When it is orbiting at an altitude of 407 km, find the width of the image formed by a telescope objective of focal length 4.00 m.

34. BIO W An elderly sailor is shipwrecked on a desert island, but manages to save his eyeglasses. The lens for one eye has a power of +1.20 diopters, and the other lens has a power of +9.00 diopters. (a) What is the magnifying power of the telescope he can construct with these lenses? (b) How far apart are the lenses when the telescope is adjusted for minimum eyestrain?

35. M A person decides to use an old pair of eyeglasses to make some optical instruments. He knows that the near point in his left eye is 50.0 cm and the near point in his right eye is 100 cm. (a) What is the maximum angular magnification he can produce in a telescope? (b) If he places the lenses 10.0 cm apart, what is the maximum overall magnification he can produce in a microscope? *Hint:* Go back to basics and use the thin-lens equation to solve part (b).

36. Q|C Galileo devised a simple terrestrial telescope that produces an upright image. It consists of a converging objective lens and a diverging eyepiece at opposite ends of the telescope tube. For distant objects, the tube length is the objective focal length less the absolute value of the eyepiece focal length. (a) Does the user of the telescope see a real or virtual image? (b) Where is the final image? (c) If a telescope is to be constructed with a tube of length 10.0 cm and a magnification of 3.00, what are the focal lengths of the objective and eyepiece?

25.6 Resolution of Single-Slit and Circular Apertures

37. A converging lens with a diameter of 30.0 cm forms an image of a satellite passing overhead. The satellite has two green lights (wavelength 500 nm) spaced 1.00 m apart. If the lights can just be resolved according to the Rayleigh criterion, what is the altitude of the satellite?

38. While flying at an altitude of 9.50 km, you look out the window at various objects on the ground. If your ability to distinguish two objects is limited only by diffraction, find the smallest separation between two objects on the ground that are distinguishable. Assume your pupil has a diameter of 4.0 mm and take $\lambda = 575$ nm.

39. M To increase the resolving power of a microscope, the object and the objective are immersed in oil ($n = 1.5$). If the limiting angle of resolution without the oil is 0.60 μrad, what is the limiting angle of resolution with the oil? *Hint:* The oil changes the wavelength of the light.

40. BIO (a) Calculate the limiting angle of resolution for the eye, assuming a pupil diameter of 2.00 mm, a wavelength of 500 nm *in air*, and an index of refraction for the eye of 1.33. (b) What is the maximum distance from the eye at which two points separated by 1.00 cm could be resolved?

41. A vehicle with headlights separated by 2.00 m approaches an observer holding an infrared detector sensitive to radiation of wavelength 885 nm. What aperture diameter is required in the detector if the two headlights are to be resolved at a distance of 10.0 km?

42. Two stars located 23 light-years from Earth are barely resolved using a reflecting telescope having a mirror of diameter 68 cm. Assuming $\lambda = 575$ nm and assuming that the resolution is limited only by diffraction, find the separation between the stars.

43. Suppose a 5.00-m-diameter telescope were constructed on the Moon, where the absence of atmospheric distortion would permit excellent viewing. If observations were made using 500-nm light, what minimum separation between two objects could just be resolved on Mars at closest approach (when Mars is 8.0×10^7 km from the Moon)?

44. W A spy satellite circles Earth at an altitude of 200 km and carries out surveillance with a special high-resolution telescopic camera having a lens diameter of 35 cm. If the angular resolution of this camera is limited by diffraction, estimate the separation of two small objects on Earth's surface that are just resolved in yellow-green light ($\lambda = 550$ nm).

45. A 15.0-cm-long grating has 6 000 slits per centimeter. Can two lines of wavelengths 600.000 nm and 600.003 nm be separated with this grating? Explain.

46. The H_α line in hydrogen has a wavelength of 656.20 nm. This line differs in wavelength from the corresponding spectral line in deuterium (the heavy stable isotope of hydrogen) by 0.18 nm. (a) Determine the minimum number of lines a grating must have to resolve these two wavelengths in the first order. (b) Repeat part (a) for the second order.

25.7 The Michelson Interferometer

47. Light of wavelength 550 nm is used to calibrate a Michelson interferometer. With the use of a micrometer screw, the platform on which one mirror is mounted is moved 0.180 mm. How many fringe shifts are counted?

48. Q|C Monochromatic light is beamed into a Michelson interferometer. The movable mirror is displaced 0.382 mm, causing the central spot in the interferometer pattern to change from bright to dark and back to bright $N = 1\,700$ times. (a) Determine the wavelength of the light. What color is it? (b) If monochromatic red light is used instead and the mirror is moved the same distance, would N be larger or smaller? Explain.

49. BIO An interferometer is used to measure the length of a bacterium. The wavelength of the light used is 650 nm. As one arm of the interferometer is moved from one end of the cell to the other, 310 fringe shifts are counted. How long is the bacterium?

50. Mirror M_1 in Figure 25.15 is displaced a distance ΔL. During this displacement, 250 fringe shifts are counted. The light being used has a wavelength of 632.8 nm. Calculate the displacement ΔL.

51. A thin sheet of transparent material has an index of refraction of 1.40 and is 15.0 μm thick. When it is inserted in the light path along one arm of an interferometer, how many fringe shifts occur in the pattern? Assume the wavelength (in a vacuum) of the light used is 600 nm. *Hint:* The wavelength will change within the material.

52. M The Michelson interferometer can be used to measure the index of refraction of a gas by placing an evacuated transparent tube in the light path along one arm of the device. Fringe shifts occur as the gas is slowly added to the tube. Assume 600-nm light is used, the tube is 5.00 cm long, and 160 fringe shifts occur as the pressure of the gas in the tube increases to atmospheric pressure. What is the index of refraction of the gas? *Hint:* The fringe shifts occur because the wavelength of the light changes inside the gas-filled tube.

Additional Problems

53. The Yerkes refracting telescope has a 1.00-m-diameter objective lens of focal length 20.0 m. Assume it is used with an eyepiece of focal length 2.50 cm. (a) Determine the magnification of the planet Mars as seen through the telescope. (b) Are the observed Martian polar caps right side up or upside down?

54. Estimate the minimum angle subtended at the eye of a hawk flying at an altitude of 50 m necessary to recognize a mouse on the ground.

55. An American standard analog television picture (non-HDTV), also known as NTSC, is composed of approximately 485 visible horizontal lines of varying light intensity. Assume your ability to resolve the lines is limited only by the Rayleigh criterion, the pupils of your eyes are 5.00 mm in diameter, and the average wavelength of the light coming from the screen is 550 nm. Calculate the ratio of the minimum viewing distance to the vertical dimension of the picture such that you will not be able to resolve the lines.

56. BIO A person with a nearsighted eye has near and far points of 16 cm and 25 cm, respectively. (a) Assuming a lens is placed 2.0 cm from the eye, what power must the lens have to correct this condition? (b) Suppose contact lenses placed directly on the cornea are used to correct the person's eyesight. What is the power of the lens required in this case, and what is the new near point? *Hint:* The contact lens and the eyeglass lens require slightly different powers because they are at different distances from the eye.

57. BIO The near point of an eye is 75.0 cm. (a) What should be the power of a corrective lens prescribed to enable the eye to see an object clearly at 25.0 cm? (b) If, using the corrective lens, the person can see an object clearly at 26.0 cm but not at 25.0 cm, by how many diopters did the lens grinder miss the prescription?

58. BIO If a typical eyeball is 2.00 cm long and has a pupil opening that can range from about 2.00 mm to 6.00 mm, what are (a) the focal length of the eye when it is focused on objects 1.00 m away, (b) the smallest *f*-number of the eye when it is focused on objects 1.00 m away, and (c) the largest *f*-number of the eye when it is focused on objects 1.00 m away?

59. BIO M A cataract-impaired lens in an eye may be surgically removed and replaced by a manufactured lens. The focal length required for the new lens is determined by the lens-to-retina distance, which is measured by a sonarlike device, and by the requirement that the implant provide for correct distance vision. (a) If the distance from lens to retina is 22.4 mm, calculate the power of the implanted lens in diopters. (b) Since there is no accommodation and the implant allows for correct distance vision, a corrective lens for close work or reading must be used. Assume a reading distance of 33.0 cm, and calculate the power of the lens in the reading glasses.

60. BIO If the aqueous humor of the eye has an index of refraction of 1.34 and the distance from the vertex of the cornea to the retina is 2.00 cm, what is the radius of curvature of the cornea for which distant objects will be focused on the retina? (For simplicity, assume all refraction occurs in the aqueous humor.)

61. A Boy Scout starts a fire by using a lens from his eyeglasses to focus sunlight on kindling 5.0 cm from the lens. The Boy Scout has a near point of 15 cm. When the lens is used as a simple magnifier, (a) what is the maximum magnification that can be achieved and (b) what is the magnification when the eye is relaxed? *Caution:* The equations derived in the text for a simple magnifier assume a "normal" eye.

62. A laboratory (astronomical) telescope is used to view a scale that is 300 cm from the objective, which has a focal length of 20.0 cm; the eyepiece has a focal length of 2.00 cm. Calculate the angular magnification when the telescope is adjusted for minimum eyestrain. *Note:* The object is not at infinity, so the simple expression $m = f_o/f_e$ is not sufficiently accurate for this problem. Also, assume small angles, so that $\tan\theta \approx \theta$.

Albert Einstein revolutionized modern physics. He explained the random motion of pollen grains, which proved the existence of atoms, and the photoelectric effect, which showed that light was a particle as well as a wave. With Satyendra Nath Bose he predicted a new form of matter, the Bose-Einstein condensate, which was discovered in the laboratory forty years after his death. His theory of special relativity made clear the foundations of space and time, and established a key relationship between mass and energy. His theory of gravitation, general relativity, led to a deeper understanding of planetary motion, the structure and evolution of stars, and the expanding universe. The equation in this photo, loosely translated, says that the average curvature of spacetime is zero in empty space.

26 Relativity

Most of our everyday experiences and observations have to do with objects that move at speeds much less than the speed of light. Newtonian mechanics was formulated to describe the motion of such objects, and its formalism is quite successful in describing a wide range of phenomena that occur at low speeds. It fails, however, when applied to particles having speeds approaching that of light.

Experimentally, for example, it's possible to accelerate an electron to a speed of $0.99c$ (where c is the speed of light) by using a potential difference of several million volts. According to Newtonian mechanics, if the potential difference is increased by a factor of 4, the electron's kinetic energy is four times greater and its speed should double to $1.98c$. Experiments, however, show that the speed of the electron—as well as the speed of any other particle that has mass—always remains *less* than the speed of light, regardless of the size of the accelerating voltage.

The existence of a universal speed limit has far-reaching consequences. It means that the usual concepts of force, momentum, and energy no longer apply for rapidly moving objects. A less obvious consequence is that observers moving at different speeds will measure different time intervals and displacements between the same two events. Relating the measurements made by different observers is the subject of relativity.

26.1 Galilean Relativity

LEARNING OBJECTIVE

1. State and discuss the principle of Galilean relativity.

To describe a physical event, it's necessary to choose a *frame of reference.* When you perform an experiment in a laboratory, for example, you select a coordinate system, or frame of reference, that is at rest with respect to the laboratory. Suppose, instead, you choose to do an experiment in the back of a truck moving at a constant velocity $\vec{v}$. You can then select a moving frame of reference that's at rest with respect to the truck. If you found Newton's first law to be valid in that frame, would an observer at rest with respect to the Earth agree with you?

According to the principle of Galilean relativity, **the laws of mechanics must be the same in all inertial frames of reference**. Inertial frames of reference are those in which Newton's laws are valid. In these frames, objects move in straight lines at constant speed unless acted on by a nonzero net force, thus the name "inertial frame" because objects observed from these frames obey Newton's first law, the law of inertia. For the situation described in the previous paragraph, the laboratory coordinate system and the coordinate system of the moving car are both inertial frames of reference. Consequently, if the laws of mechanics are found to be true in the laboratory, then the person in the car must also observe the same laws.

Consider a truck in motion, moving with a constant velocity, as in Figure 26.1a. If a passenger in the truck throws a ball straight up in the air, the passenger observes that the ball moves in a vertical path. The motion of the ball is precisely the same as it would be if the ball were thrown while at rest on Earth. The law of gravity and the equations of motion under constant acceleration are obeyed whether the truck is at rest or in uniform motion.

Now consider the same experiment when viewed by another observer at rest on Earth. This stationary observer views the path of the ball in the truck to be a parabola, as in Figure 26.1b. Further, according to this observer, the ball has a velocity to the right equal to the velocity of the truck. Although the two observers disagree on the shape of the ball's path, both agree that the motion of the ball obeys the law of gravity and Newton's laws of motion, and they even agree on how long the ball is in the air. We draw the following important conclusion: **There is no preferred frame of reference for describing the laws of mechanics.**

Figure 26.1 Two observers watch the path of a thrown ball and obtain different results.

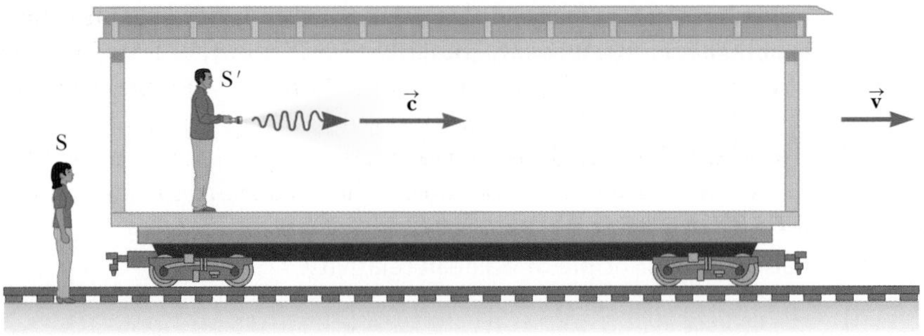

Figure 26.2 A pulse of light is sent out by a person in a moving boxcar. According to Galilean relativity, the speed of the pulse should be $\vec{c} + \vec{v}$ relative to a stationary observer.

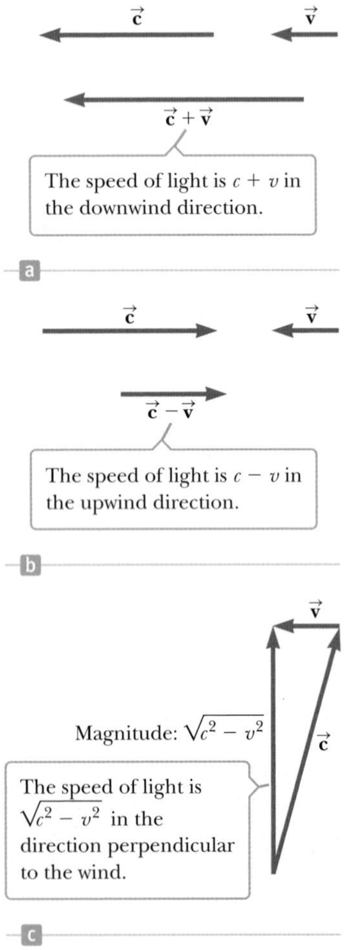

Figure 26.3 If the speed of the ether wind relative to Earth is v and c is the speed of light relative to the ether, the speed of light relative to Earth is (a) $c + v$ in the down-wind direction, (b) $c - v$ in the upwind direction, and (c) $\sqrt{c^2 - v^2}$ in the direction perpendicular to the wind. The Michelson–Morley experiment, however, disproved the ether wind hypothesis, leading to Einstein's postulate that the speed of light in vacuum has the same value regardless of the motion of an inertial observer.

26.2 The Speed of Light

LEARNING OBJECTIVES

1. Discuss the speed of light in the context of Galilean relativity and their predictions.
2. Explain the discredited concept of the luminiferous ether.
3. Describe the Michelson-Morley experiment and its consequences for the speed of light as measured by different observers, and what the experiment implies about the ether.

It's natural to ask whether the concept of Galilean relativity in mechanics also applies to experiments in electricity, magnetism, optics, and other areas. Experiments indicate that the answer is no. Further, if we assume the laws of electricity and magnetism are the same in all inertial frames, a paradox concerning the speed of light immediately arises. According to electromagnetic theory, the speed of light always has the fixed value of $2.997\,924\,58 \times 10^8$ m/s in free space. According to Galilean relativity, however, the speed of the pulse relative to the stationary observer S outside the boxcar in Figure 26.2 should be $c + v$. Hence, Galilean relativity is inconsistent with Maxwell's well-tested theory of electromagnetism.

Electromagnetic theory predicts that light waves must propagate through free space with a speed equal to the speed of light. The theory doesn't require the presence of a medium for wave propagation, however. This is in contrast to other types of waves, such as water or sound waves, that do require a medium to support the disturbances. In the 19th century, physicists thought that electromagnetic waves also required a medium to propagate. They proposed that such a medium existed and gave it the name **luminiferous ether**. The ether was assumed to be present everywhere, even in empty space, and light waves were viewed as ether oscillations. Further, the ether would have to be a massless but rigid medium with no effect on the motion of planets or other objects. These concepts are indeed strange. In addition, it was found that the troublesome laws of electricity and magnetism would take on their simplest forms in a special frame of reference at *rest* with respect to the ether. This frame was called the *absolute frame*. The laws of electricity and magnetism would be valid in this absolute frame, but they would have to be modified in any reference frame moving with respect to the absolute frame.

As a result of the importance attached to the ether and the absolute frame, it became of considerable interest in physics to prove by experiment that they existed. Because it was considered likely that Earth was in motion through the ether, from the view of an experimenter on Earth, there was an "ether wind" blowing through the laboratory. A direct method for detecting the ether wind would use an apparatus fixed to Earth to measure the wind's influence on the speed of light. If v is the speed of the ether relative to Earth, then the speed of light should have its maximum value, $c + v$, when propagating downwind, as shown in Figure 26.3a. Likewise,

the other in the spaceship. The observer at rest on Earth (and also assumed to be at rest with respect to the two stars) measures the distance between the stars to be L_p. According to this observer, the time it takes the spaceship to complete the voyage is $\Delta t = L_p/v$. Because of time dilation, the space traveler, using his spaceship clock, measures a smaller time of travel: $\Delta t_p = \Delta t/\gamma$. The space traveler claims to be at rest and sees the destination star moving toward the spaceship with speed v. Because the space traveler reaches the star in time Δt_p, he concludes that the distance L between the stars is shorter than L_p. The distance measured by the space traveler is

$$L = v\,\Delta t_p = v\frac{\Delta t}{\gamma}$$

Because $L_p = v\,\Delta t$, it follows that

$$L = \frac{L_p}{\gamma} = L_p\sqrt{1 - v^2/c^2} \qquad [26.4]$$

According to this result, illustrated in Figure 26.9, if an observer at rest with respect to an object measures its length to be L_p, an observer moving at a speed v relative to the object will find it to be shorter than its proper length by the factor $\sqrt{1 - v^2/c^2}$. Note that **length contraction takes place only along the direction of motion.**

Time-dilation and length contraction effects have interesting applications for future space travel to distant stars. For the star to be reached in a fraction of a human lifetime, the trip must be taken at very high speeds. According to an Earth-bound observer, the time for a spacecraft to reach the destination star will be dilated compared with the time interval measured by travelers. As discussed in the treatment of the twin paradox, the travelers will be younger than their twins when they return to Earth. Therefore, by the time the travelers reach the star, they will have aged by some number of years, while their partners back on Earth will have aged a larger number of years, the exact ratio depending on the speed of the spacecraft. At a spacecraft speed of $0.94c$, this ratio is about 3:1.

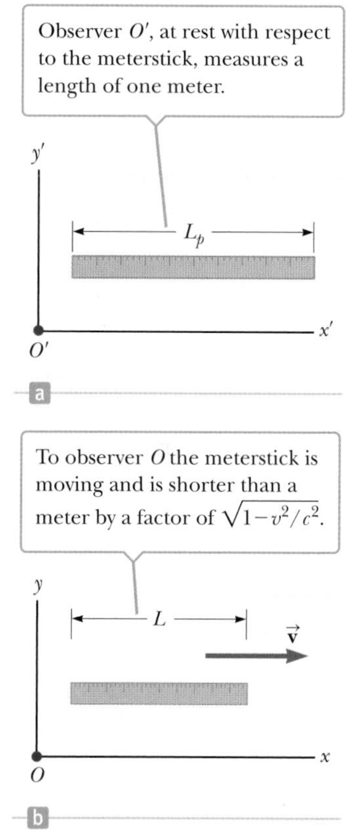

Figure 26.9 The length of a meterstick is measured by two observers.

Quick Quiz

26.4 You are packing for a trip to another star, and on your journey you will be traveling at a speed of $0.99c$. Can you sleep in a smaller cabin than usual, because you will be shorter when you lie down? Explain your answer.

26.5 You observe a rocket moving away from you. **(i)** Compared with its length when it was at rest on the ground, will you measure its length to be (a) shorter, (b) longer, or (c) the same? Compared to the passage of time measured by the watch on your wrist, is the passage of time on the rocket's clock (d) faster, (e) slower, or (f) the same? **(ii)** Answer the same questions from part (i) if the rocket turns around and comes toward you.

EXAMPLE 26.2 Speedy Plunge

GOAL Apply the concept of length contraction to a distance.

PROBLEM **(a)** An observer on Earth sees a spaceship at an altitude of 4 350 km moving downward toward Earth with a speed of $0.970c$. What is the distance from the spaceship to Earth as measured by the spaceship's captain? **(b)** After firing her engines, the captain measures her ship's altitude as 267 km, whereas the observer on Earth measures it to be 625 km. What is the speed of the spaceship at this instant?

(Continued)

STRATEGY To the captain, Earth is rushing toward her ship at $0.970c$; hence, the distance between her ship and Earth is contracted. Substitution into Equation 26.9 yields the answer. In part (b) use the same equation, substituting the distances and solving for the speed.

SOLUTION

(a) Find the distance from the ship to Earth as measured by the captain.

Substitute into Equation 26.4, getting the altitude as measured by the captain in the ship:

$$L = L_p\sqrt{1 - v^2/c^2} = (4\,350 \text{ km})\sqrt{1 - (0.970c)^2/c^2}$$

$$= \boxed{1.06 \times 10^3 \text{ km}}$$

(b) What is the subsequent speed of the spaceship if the Earth observer measures the distance from the ship to Earth as 625 km and the captain measures it as 267 km?

Apply the length-contraction equation:

$$L = L_p\sqrt{1 - v^2/c^2}$$

Square both sides of this equation and solve for v:

$$L^2 = L_p^2(1 - v^2/c^2) \quad \rightarrow \quad 1 - v^2/c^2 = \left(\frac{L}{L_p}\right)^2$$

$$v = c\sqrt{1 - (L/L_p)^2} = c\sqrt{1 - (267 \text{ km}/625 \text{ km})^2}$$

$$v = \boxed{0.904c}$$

REMARKS The proper length is always the length measured by an observer at rest with respect to that length.

QUESTION 26.2 As a spaceship approaches an observer at nearly the speed of light, the captain directs a beam of yellow light at the observer. What would the observer report upon seeing the light? (a) Its wavelength would be shifted toward the red end of the spectrum. (b) Its wavelength would correspond to yellow light. (c) Its wavelength would be shifted toward the blue end of the spectrum.

EXERCISE 26.2 Suppose the observer on the ship measures the distance from Earth as 50.0 km, whereas the observer on Earth measures the distance as 125 km. At what speed is the spacecraft approaching Earth?

ANSWER $0.917c$

Length contraction occurs only in the direction of the observer's motion. No contraction occurs perpendicular to that direction. For example, a spaceship at rest relative to an observer may have the shape of an equilateral triangle, but if it passes the observer at relativistic speed in a direction parallel to its base, the base will shorten while the height remains the same. Hence the observer will report that the craft has the form of an isosceles triangle. An observer traveling with the ship will still observe it to be equilateral.

26.5 Relativistic Momentum

LEARNING OBJECTIVES

1. Generalize the concept of momentum to relativity.
2. Through calculation, compare the classical and relativistic concepts of momentum.

Properly describing the motion of particles within the framework of special relativity requires generalizing Newton's laws of motion and the definitions of momentum and energy. These generalized definitions reduce to the classical (nonrelativistic) definitions when v is much less than c.

First, recall that conservation of momentum states that when two objects collide, the total momentum of the system remains constant, assuming the objects are isolated, reacting only with each other. When analyzing such collisions from rapidly moving inertial frames, however, it is found that momentum is not conserved if the classical definition of momentum, $p = mv$, is used. To have momentum conservation in all inertial frames—even those moving at an appreciable fraction of c—the definition of momentum must be modified to read

$$p \equiv \frac{mv}{\sqrt{1 - v^2/c^2}} = \gamma m v \qquad [26.5]$$

◂ Relativistic momentum

where v is the speed of the particle and m is its mass as measured by an observer at rest with respect to the particle. Note that when v is much less than c, the denominator of Equation 26.5 approaches 1, so that p approaches mv. Therefore, the relativistic equation for momentum reduces to the classical expression when v is small compared with c.

■ EXAMPLE 26.3 The Relativistic Momentum of an Electron

GOAL Contrast the classical and relativistic definitions of momentum.

PROBLEM An electron, which has a mass of 9.11×10^{-31} kg, moves with a speed of $0.750c$. Find the classical (nonrelativistic) momentum and compare it with its relativistic counterpart p_{rel}.

STRATEGY Substitute into the classical definition to get the classical momentum, then multiply by the gamma factor to obtain the relativistic version.

SOLUTION

First, compute the classical (nonrelativistic) momentum with $v = 0.750c$:

$$p = mv = (9.11 \times 10^{-31}\text{ kg})(0.750 \times 3.00 \times 10^8\text{ m/s})$$
$$= 2.05 \times 10^{-22}\text{ kg}\cdot\text{m/s}$$

Multiply this result by γ to obtain the relativistic momentum:

$$p_{\text{rel}} = \frac{mv}{\sqrt{1 - v^2/c^2}} = \frac{2.05 \times 10^{-22}\text{ kg}\cdot\text{m/s}}{\sqrt{1 - (0.750c/c)^2}}$$
$$= 3.10 \times 10^{-22}\text{ kg}\cdot\text{m/s}$$

REMARKS The (correct) relativistic result is 50% greater than the classical result. In subsequent calculations no notational distinction will be made between classical and relativistic momentum. For problems involving relative speeds of $0.2c$, the answer using the classical expression is about 2% below the correct answer.

QUESTION 26.3 A particle with initial momentum p_i doubles its speed. How does its final momentum p_f compare with its initial momentum? (a) $p_f > 2p_i$ (b) $p_f = 2p_i$ (c) $p_f < 2p_i$

EXERCISE 26.3 Repeat the calculation for a proton traveling at $0.600c$.

ANSWERS $p = 3.01 \times 10^{-19}\text{ kg}\cdot\text{m/s}$, $p_{\text{rel}} = 3.76 \times 10^{-19}\text{ kg}\cdot\text{m/s}$

26.6 Relative Velocity in Special Relativity

LEARNING OBJECTIVES

1. Contrast the expression for relative velocity in Galilean relativity to the corresponding expression in special relativity.
2. Calculate one-dimensional relative velocities in special relativity.

In Galilean relativity, if a man in a spaceship traveling at velocity v shines a laser straight ahead, the light beam would be expected to travel at velocity $c + v$ relative to an observer on Earth. The null result of the Michelson–Morley experiment, however, indicates the laser light will still travel at speed c relative to that same observer, although the light's frequency increases. (When light's frequency increases, the light is said to be "blue-shifted." When the frequency decreases, the light is said to be "red-shifted.") Evidently, some new formula must be derived to allow the comparison of velocities by observers moving at high relative speeds.

The procedure is very similar to that used in Chapter 3 for nonrelativistic relative velocity. Here, an observer in a spaceship will be labeled B and the Earth observer E, with B moving in the positive x-direction at speed v_{BE} with respect to the Earth observer E. The goal is to find the relationship between their independent measurements of an object A. Given v_{AE}, the velocity of A according to observer E, what is v_{AB}, the velocity of A relative to observer B? According to Galilean relativity, the answer derived in Chapter 3 is

Relative velocity in Galilean relativity ▶

$$v_{AB} = v_{AE} - v_{BE} \qquad [26.6]$$

For velocities that are about 10% of the speed of light and greater, relativistic effects start becoming appreciable, so the relativistic expression for relative velocity should be used[2]:

Relative velocity in special relativity ▶

$$v_{AB} = \frac{v_{AE} - v_{BE}}{1 - \dfrac{v_{AE}v_{BE}}{c^2}} \qquad [26.7]$$

Notice that when either v_{AE} or v_{BE} is much smaller than c, this expression agrees with the Galilean (nonrelativistic) relationship, as it should. Equation 26.7 is useful for determining velocities measured by B in the moving frame of reference when the velocity measured by observer E in the rest frame is known. On the other hand, when the velocity in question is measured by observer B and the task is to find the velocity measured by observer E, Equation 26.7 must be algebraically solved for v_{AE}. The resulting expression is

Relativistic addition of velocities ▶

$$v_{AE} = \frac{v_{AB} + v_{BE}}{1 + \dfrac{v_{AB}v_{BE}}{c^2}} \qquad [26.8]$$

Ignoring the expression in the denominator, Equation 26.8 has the expected form: if observer B is moving with respect to observer E at velocity v_{BE} and fires a projectile at velocity v_{AB} relative to himself, then adding v_{AB} and v_{BE} should give the velocity of the projectile v_{AE} as measured by the Earth observer. Special relativity contributes the denominator of Equation 26.8, an equation often called the relativistic addition of velocities.

Using Equation 26.8, the speed of an object projected forward from a moving vehicle as measured by an observer on Earth can be calculated. Suppose, for example, that observer B is moving at v_{BE} with respect to the Earth observer and directs the beam of a laser in front of his rapidly moving spacecraft. Here v_{AB} is the velocity of light relative to observer B on the spacecraft. The speed of the light v_{AE} as measured by the Earth observer is, therefore,

$$v_{AE} = \frac{v_{AB} + v_{BE}}{1 + \dfrac{v_{AB}v_{BE}}{c^2}} = \frac{c + v_{BE}}{1 + \dfrac{cv_{BE}}{c^2}} = \frac{c\left(1 + \dfrac{v_{BE}}{c}\right)}{1 + \dfrac{v_{BE}}{c}} = c$$

This calculation shows that the derived velocity transformation is consistent with the experimental result proving the speed of light is the same for all observers.

[2] The derivation of Equation 26.7 requires the use of Lorentz transformations and will not be presented in this textbook.

This is the distribution of the cosmic microwave background as observed by the Planck space-based telescope, leftover blackbody radiation created 380 000 years after the Big Bang. The temperature of this thermal radiation is about 2.7 K. The false color image represents differences from that temperature by about one part in one hundred thousand, giving clues to structure formation in the early universe that led to the stars, galaxies, and clusters of galaxies observed in the universe today.

Quantum Physics 27

Although many problems were resolved by the theory of relativity in the early part of the 20th century, many others remained unsolved. Attempts to explain the behavior of matter on the atomic level with the laws of classical physics were consistently unsuccessful. Various phenomena, such as the electromagnetic radiation emitted by a heated object (blackbody radiation), the emission of electrons by illuminated metals (the photoelectric effect), and the emission of sharp spectral lines by gas atoms in an electric discharge tube couldn't be understood within the framework of classical physics. Between 1900 and 1930, however, a modern version of mechanics called *quantum mechanics* or *wave mechanics* was highly successful in explaining the behavior of atoms, molecules, and nuclei.

The earliest ideas of quantum theory were introduced by Planck, and most of the subsequent mathematical developments, interpretations, and improvements were made by a number of distinguished physicists, including Einstein, Bohr, Schrödinger, de Broglie, Heisenberg, Born, and Dirac. In this chapter we introduce the underlying ideas of quantum theory and the wave-particle nature of matter, and discuss some simple applications of quantum theory, including the photoelectric effect, the Compton effect, and x-rays.

27.1 Blackbody Radiation and Planck's Hypothesis

LEARNING OBJECTIVES

1. Define thermal radiation and discuss its physical origin.
2. Define blackbody and explain the reason for the definition.

3. Sketch a typical curve for the intensity of blackbody radiation as a function of wavelength.
4. State Wien's displacement law. Explain its origin from the graph of the intensity of blackbody radiation versus wavelength and how the graph changes with temperature.
5. State the key assumption in Planck's theory of blackbody radiation: the quantization of energy.

Tip 27.1 Expect the Unexpected

Our life experiences take place in the macroscopic world, where quantum effects are not evident. Quantum effects can be even more bizarre than relativistic effects. As Nobel Prize–winning physicist Richard Feynman once said, "Nobody understands quantum mechanics."

An object at any temperature emits electromagnetic radiation, called **thermal radiation**. Stefan's law, discussed in Section 11.5, describes the total power radiated. The spectrum of the radiation depends on the temperature and properties of the object. At low temperatures, the wavelengths of the thermal radiation are mainly in the infrared region and hence not observable by the eye. As the temperature of an object increases, the object eventually begins to glow red. At sufficiently high temperatures, it appears to be white, as in the glow of the hot tungsten filament of a lightbulb. A careful study of thermal radiation shows that it consists of a continuous distribution of wavelengths from the infrared, visible, and ultraviolet portions of the spectrum.

From a classical viewpoint, thermal radiation originates from accelerated charged particles near the surface of an object; such charges emit radiation, much as small antennas do. The thermally agitated charges can have a distribution of frequencies, which accounts for the continuous spectrum of radiation emitted by the object. By the end of the 19th century, it had become apparent that the classical theory of thermal radiation was inadequate. The basic problem was in understanding the observed distribution energy as a function of wavelength in the radiation emitted by a blackbody. By definition, a **blackbody** is an ideal system that absorbs *all* radiation incident on it. A good approximation of a blackbody is a small hole leading to the inside of a hollow object, as shown in Figure 27.1. The nature of the radiation emitted through the small hole leading to the cavity depends *only on the temperature* of the cavity walls and not at all on the material composition of the object, its shape, or other factors.

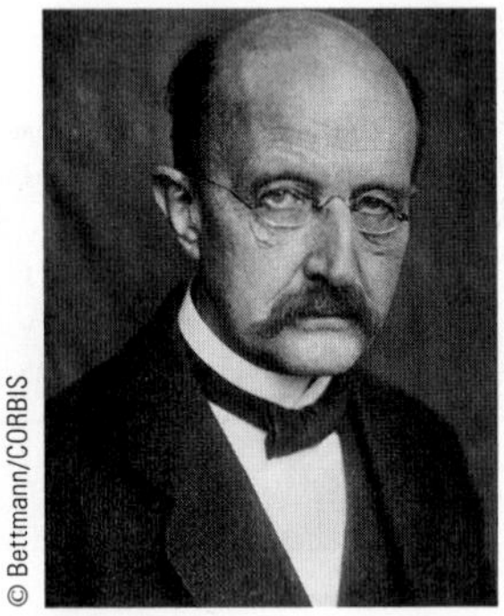

Max Planck
German Physicist (1858–1947)
Planck introduced the concept of a "quantum of action" (Planck's constant h) in an attempt to explain the spectral distribution of blackbody radiation, which formed the foundations for quantum theory. In 1918 he was awarded the Nobel Prize in Physics for this discovery of the quantized nature of energy.

Experimental data for the distribution of energy in blackbody radiation at three temperatures are shown in Figure 27.2. The radiated energy varies with wavelength and temperature. As the temperature of the blackbody increases, the total amount of energy (area under the curve) it emits increases. Also, with increasing temperature, the peak of the distribution shifts to shorter wavelengths. This shift obeys the following relationship, called **Wien's displacement law**,

$$\lambda_{max} T = 0.289\,8 \times 10^{-2}\ \text{m} \cdot \text{K} \tag{27.1}$$

where λ_{max} is the wavelength at which the curve peaks and T is the absolute temperature of the object emitting the radiation.

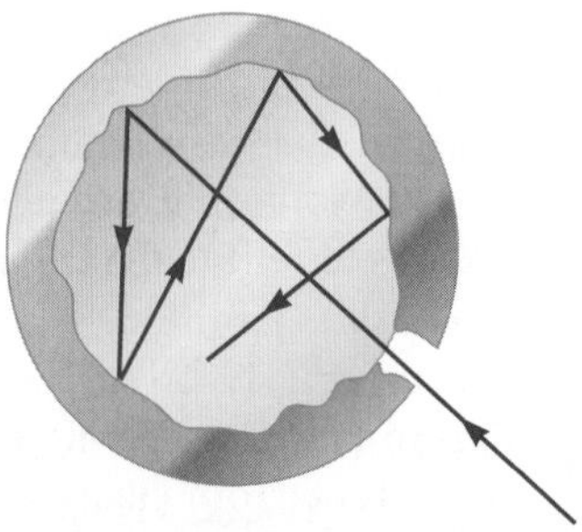

Figure 27.1 An opening in the cavity of a body is a good approximation of a blackbody. As light enters the cavity through the small opening, part is reflected and part is absorbed on each reflection from the interior walls. After many reflections, essentially all the incident energy is absorbed.

APPLYING PHYSICS 27.1 | Star Colors

If you look carefully at stars in the night sky, you can distinguish three main colors: red, white, and blue. What causes these particular colors?

EXPLANATION These colors result from the different surface temperatures of stars. A relatively cool star, with a surface temperature of 3 000 K, has a radiation curve similar to the middle curve in Figure 27.2. The peak in this curve is above the visible wavelengths, 0.4 μm to 0.7 μm, beyond the wavelength of red light, so significantly more radiation is emitted within the visible range at the red end than the blue end of the spectrum. Consequently, the star appears reddish in color, similar to the red glow from the burner of an electric range. ■

A hotter star has a radiation curve more like the upper curve in Figure 27.2. In this case the star emits significant radiation throughout the visible range, and the combination of all colors causes the star to look white. Such is the case with our own Sun, with a surface temperature of 5 800 K. For much hotter stars, the peak can be shifted so far below the visible range that significantly more blue radiation is emitted than red, so the star appears bluish in color. Stars cooler than the Sun tend to have orange or red colors. The surface temperature of a star can be obtained by first finding the wavelength corresponding to the maximum in the intensity versus wavelength curve, then substituting that wavelength into Wien's law. For example, if the wavelength were 2.30 $\times 10^{-7}$ m, the surface temperature would be given by

$$T = \frac{0.289\,8 \times 10^{-2}\ \text{m}\cdot\text{K}}{\lambda_{\text{max}}} = \frac{0.289\,8 \times 10^{-2}\ \text{m}\cdot\text{K}}{2.30 \times 10^{-7}\ \text{m}} = 1.26 \times 10^{4}\ \text{K}$$

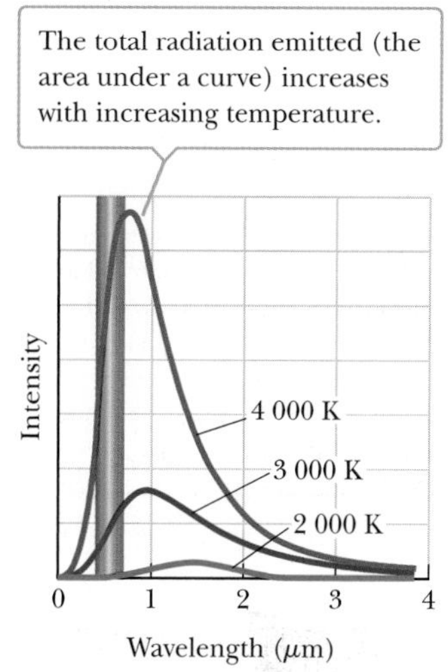

Figure 27.2 Intensity of blackbody radiation versus wavelength at three temperatures. The visible range of wavelengths is between 0.4 μm and 0.7 μm. At approximately 6 000 K, the peak is in the center of the visible wavelengths, and the object appears white.

Attempts to use classical ideas to explain the shapes of the curves shown in Figure 27.2 failed. Figure 27.3 shows an experimental plot of the blackbody radiation spectrum (blue curve), together with the theoretical picture of what this curve should look like based on classical theories (rust-colored curve). At long wavelengths, classical theory is in good agreement with the experimental data. At short wavelengths, however, major disagreement exists between classical theory and experiment. As λ approaches zero, classical theory erroneously predicts that the intensity should go to infinity, when the experimental data show it should approach zero.

In 1900 Planck developed a formula for blackbody radiation that was in complete agreement with experiments at all wavelengths, leading to a curve shown by the blue line in Figure 27.3. Planck hypothesized that blackbody radiation was produced by submicroscopic charged oscillators, which he called *resonators.* He assumed the walls of a glowing cavity were composed of billions of these resonators, although their exact nature was unknown. The resonators were allowed to have only certain discrete energies E_n, given by

$$E_n = nhf \qquad [27.2]$$

where n is a positive integer called a **quantum number**, f is the frequency of vibration of the resonator, and h is a constant known as **Planck's constant**, which has the value

$$h = 6.626 \times 10^{-34}\ \text{J}\cdot\text{s} \qquad [27.3]$$

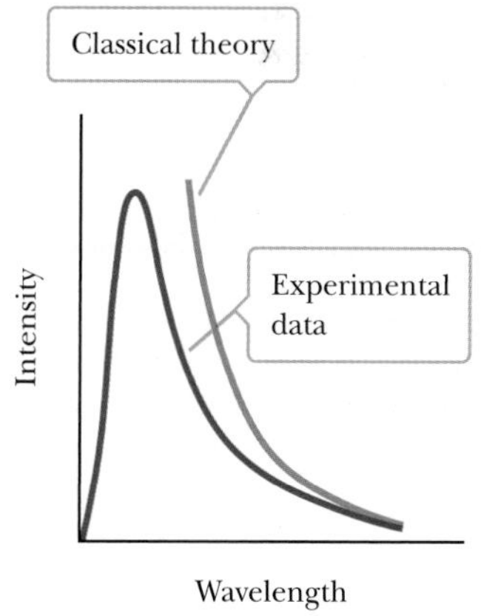

Figure 27.3 Comparison of experimental data with the classical theory of blackbody radiation. Planck's theory matches the experimental data perfectly.

Because the energy of each resonator can have only discrete values given by Equation 27.2, we say the energy is *quantized.* Each discrete energy value represents a different *quantum state,* with each value of n representing a specific quantum state. (When the resonator is in the $n = 1$ quantum state, its energy is hf; when it is in the $n = 2$ quantum state, its energy is $2hf$; and so on.)

The key point in Planck's theory is the assumption of quantized energy states. This radical departure from classical physics is the "quantum leap" that led to a totally new understanding of nature. It's shocking: it's like saying a pitched baseball can have only a fixed number of different speeds, and no speeds in between those fixed values. The fact that energy can assume only certain, discrete values instead of any one of a continuum of values is the single most important difference between quantum theory and the classical theories of Newton and Maxwell.

27.2 The Photoelectric Effect and the Particle Theory of Light

LEARNING OBJECTIVES

1. Describe the photoelectric effect.
2. Discuss non-classical features of the photoelectric effect and their explanations using the photon theory of light.
3. Apply the photoelectric effect equation to quantum systems.

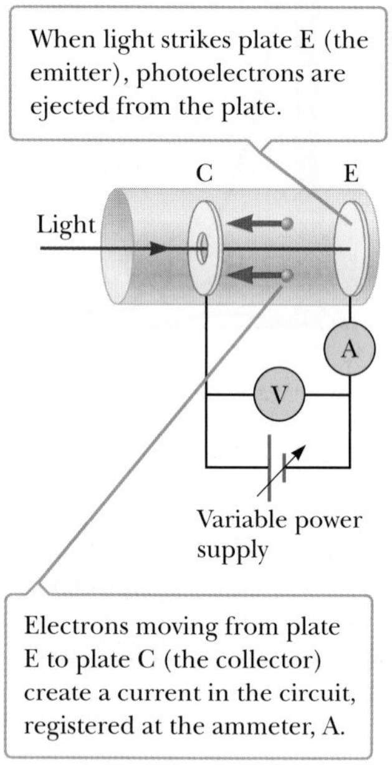

Figure 27.4 A circuit diagram for studying the photoelectric effect.

In the latter part of the 19th century, experiments showed that light incident on certain metallic surfaces caused the emission of electrons from the surfaces. This phenomenon is known as the **photoelectric effect**, and the emitted electrons are called **photoelectrons**. The first discovery of this phenomenon was made by Hertz, who was also the first to produce the electromagnetic waves predicted by Maxwell.

Figure 27.4 is a schematic diagram of a photoelectric effect apparatus. An evacuated glass tube known as a photocell contains a metal plate E (the emitter) connected to the negative terminal of a variable power supply. Another metal plate, C (the collector), is maintained at a positive potential by the power supply. When the tube is kept in the dark, the ammeter reads zero, indicating that there is no current in the circuit. When plate E is illuminated by light having a wavelength shorter than some particular wavelength that depends on the material used to make plate E, however, a current is detected by the ammeter, indicating a flow of charges across the gap between E and C. This current arises from photoelectrons emitted from the negative plate E and collected at the positive plate C.

Figure 27.5 is a plot of the photoelectric current versus the potential difference ΔV between E and C for two light intensities. At large values of ΔV, the current reaches a maximum value. In addition, the current increases as the incident light intensity increases, as you might expect. Finally, when ΔV is negative—that is, when the power supply in the circuit is reversed to make E positive and C negative—the current drops to a low value because most of the emitted photoelectrons are repelled by the now negative plate C. In this situation, only those electrons having a kinetic energy greater than the magnitude of $e\,\Delta V$ reach C, where e is the charge on the electron.

When ΔV is equal to or more negative than $-\Delta V_s$, the **stopping potential**, no electrons reach C, and the current is zero. The stopping potential is *independent* of the radiation intensity. The maximum kinetic energy of the photoelectrons is related to the stopping potential through the relationship

$$KE_{\max} = e\Delta V_s \qquad [27.4]$$

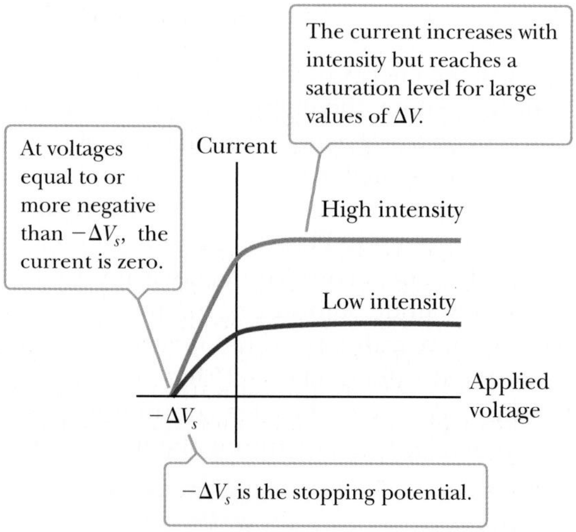

Figure 27.5 Photoelectric current versus applied potential difference for two light intensities.

Several features of the photoelectric effect can't be explained with classical physics or with the wave theory of light:

- No electrons are emitted if the incident light frequency falls below some **cut-off frequency** f_c, also called the threshold frequency, which is characteristic of the material being illuminated. This fact is inconsistent with the wave theory, which predicts that the photoelectric effect should occur at *any* frequency, provided the light intensity is sufficiently high.
- The maximum kinetic energy of the photoelectrons is independent of light intensity. According to wave theory, light of higher intensity should carry more energy into the metal per unit time and therefore eject photoelectrons having higher kinetic energies.
- The maximum kinetic energy of the photoelectrons increases with increasing light frequency. The wave theory predicts no relationship between photoelectron energy and incident light frequency.
- Electrons are emitted from the surface almost instantaneously (less than 10^{-9} s after the surface is illuminated), even at low light intensities. Classically, we expect the photoelectrons to require some time to absorb the incident radiation before they acquire enough kinetic energy to escape from the metal.

A successful explanation of the photoelectric effect was given by Einstein in 1905, the same year he published his special theory of relativity. As part of a general paper on electromagnetic radiation, for which he received the Nobel Prize in Physics in 1921, Einstein extended Planck's concept of quantization to electromagnetic waves. He suggested that a tiny packet of light energy or **photon** would be emitted when a quantized oscillator made a jump from an energy state $E_n = nhf$ to the next lower state $E_{n-1} = (n - 1)hf$. Conservation of energy would require the decrease in oscillator energy, hf, to be equal to the photon's energy E, so that

$$E = hf \qquad [27.5]$$

◀ Energy of a photon

where h is Planck's constant and f is the frequency of the light, which is equal to the frequency of Planck's oscillator.

The key point here is that the light energy lost by the emitter, hf, stays sharply localized in a tiny packet or particle called a photon. In Einstein's model a photon is so localized that it can give *all* its energy hf to a single electron in the metal. According to Einstein, the maximum kinetic energy for these liberated photoelectrons is

$$KE_{\text{max}} = hf - \phi \qquad [27.6]$$

◀ Photoelectric effect equation

where ϕ is called the **work function** of the metal. The work function, which represents the minimum energy with which an electron is bound in the metal, is on the order of a few electron volts. Table 27.1 lists work functions for various metals.

With the photon theory of light, we can explain the previously mentioned features of the photoelectric effect that cannot be understood using concepts of classical physics:

- Photoelectrons are created by absorption of a single photon, so the energy of that photon must be greater than or equal to the work function, else no photoelectrons will be produced. This explains the cutoff frequency.
- From Equation 27.6, KE_{max} depends only on the frequency of the light and the value of the work function. Light intensity is immaterial because absorption of a single photon is responsible for the electron's change in kinetic energy.
- Equation 27.6 is linear in the frequency, so KE_{max} increases with increasing frequency.

Table 27.1 Work Functions of Selected Metals

Metal	ϕ (eV)
Ag	4.73
Al	4.08
Cu	4.70
Fe	4.50
Na	2.46
Pb	4.14
Pt	6.35
Zn	4.31

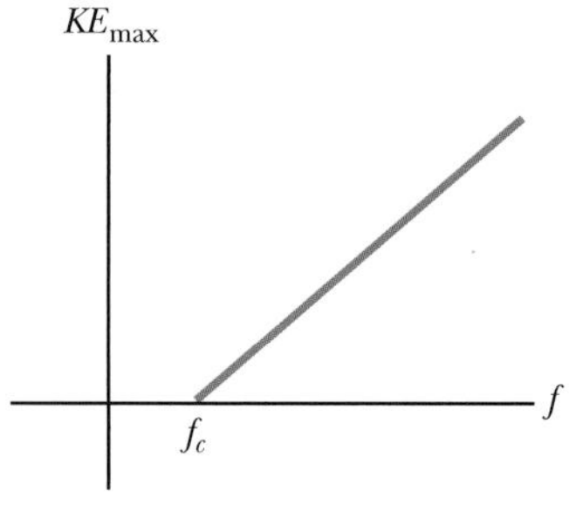

Figure 27.6 A sketch of KE_{max} versus the frequency of incident light for photoelectrons in a typical photoelectric effect experiment. Photons with frequency less than f_c don't have sufficient energy to eject an electron from the metal.

- Electrons are emitted almost instantaneously, regardless of intensity, because the light energy is concentrated in packets rather than spread out in waves. If the frequency is high enough, no time is needed for the electron to gradually acquire sufficient energy to escape the metal.

Experimentally, a linear relationship is observed between f and KE_{max}, as sketched in Figure 27.6. The intercept on the horizontal axis, corresponding to $KE_{max} = 0$, gives the cutoff frequency below which no photoelectrons are emitted, regardless of light intensity. The cutoff *wavelength* λ_c can be derived from Equation 27.6:

$$KE_{max} = hf_c - \phi = 0 \quad \rightarrow \quad h\frac{c}{\lambda_c} - \phi = 0$$

$$\lambda_c = \frac{hc}{\phi} \qquad [27.7]$$

where c is the speed of light. Wavelengths *greater* than λ_c incident on a material with work function ϕ don't result in the emission of photoelectrons.

EXAMPLE 27.1 Photoelectrons from Sodium

GOAL Understand the quantization of light and its role in the photoelectric effect.

PROBLEM A sodium surface is illuminated with light of wavelength 0.300 μm. The work function for sodium is 2.46 eV. Calculate **(a)** the energy of each photon in electron volts, **(b)** the maximum kinetic energy of the ejected photoelectrons, and **(c)** the cutoff wavelength for sodium.

STRATEGY Parts (a), (b), and (c) require substitution of values into Equations 27.5, 27.6, and 27.7, respectively.

SOLUTION

(a) Calculate the energy of each photon.

Obtain the frequency from the wavelength:

$$c = f\lambda \quad \rightarrow \quad f = \frac{c}{\lambda} = \frac{3.00 \times 10^8 \text{ m/s}}{0.300 \times 10^{-6} \text{ m}}$$

$$f = 1.00 \times 10^{15} \text{ Hz}$$

Use Equation 27.5 to calculate the photon's energy:

$$E = hf = (6.63 \times 10^{-34} \text{ J}\cdot\text{s})(1.00 \times 10^{15} \text{ Hz})$$
$$= 6.63 \times 10^{-19} \text{ J}$$
$$= (6.63 \times 10^{-19} \text{ J})\left(\frac{1.00 \text{ eV}}{1.60 \times 10^{-19} \text{ J}}\right) = \boxed{4.14 \text{ eV}}$$

(b) Find the maximum kinetic energy of the photoelectrons.

Substitute into Equation 27.6:

$$KE_{max} = hf - \phi = 4.14 \text{ eV} - 2.46 \text{ eV} = \boxed{1.68 \text{ eV}}$$

(c) Compute the cutoff wavelength.

Convert ϕ from electron volts to joules:

$$\phi = 2.46 \text{ eV} = (2.46 \text{ eV})(1.60 \times 10^{-19} \text{ J/eV})$$
$$= 3.94 \times 10^{-19} \text{ J}$$

Find the cutoff wavelength using Equation 27.7:

$$\lambda_c = \frac{hc}{\phi} = \frac{(6.63 \times 10^{-34} \text{ J}\cdot\text{s})(3.00 \times 10^8 \text{ m/s})}{3.94 \times 10^{-19} \text{ J}}$$
$$= 5.05 \times 10^{-7} \text{ m} = \boxed{505 \text{ nm}}$$

REMARKS The cutoff wavelength is in the yellow-green region of the visible spectrum.

QUESTION 27.1 True or False: Suppose in a given photoelectric experiment the frequency of light is larger than the cutoff frequency. The magnitude of the stopping potential times the electron charge, then, is larger than the energy of the incident photons.

EXERCISE 27.1 (a) What minimum-frequency light will eject photoelectrons from a copper surface? (b) If this frequency is tripled, find the maximum kinetic energy (in eV) of the resulting photoelectrons.

ANSWERS (a) 1.13×10^{15} Hz (b) 9.40 eV

Photocells

The photoelectric effect has many interesting applications using a device called the *photocell.* The photocell shown in Figure 27.4 produces a current in the circuit when light of sufficiently high frequency falls on the cell, but it doesn't allow a current in the dark. This device is used in streetlights: a photoelectric control unit in the base of the light activates a switch that turns off the streetlight when ambient light strikes it. Many garage-door systems and elevators use a light beam and a photocell as a safety feature in their design. When the light beam strikes the photocell, the electric current generated is sufficiently large to maintain a closed circuit. When an object or a person blocks the light beam, the current is interrupted, which signals the door to open.

APPLICATION
Photocells

27.3 X-Rays

LEARNING OBJECTIVES

1. Discuss the production of x-rays by electron bombardment of a metal. State their typical wavelengths.
2. Discuss the intensity of a typical x-ray spectrum and its dependence on wavelength. Explain the discrete and continuous portions of the spectrum.
3. Define threshold voltage and bremsstrahlung.
4. Derive the shortest wavelength that can be produced by electron bombardment of a given metal.

X-rays were discovered in 1895 by Wilhelm Röntgen and much later identified as electromagnetic waves, following a suggestion by Max von Laue in 1912. X-rays have higher frequencies than ultraviolet radiation and can penetrate most materials with relative ease. Typical x-ray wavelengths are about 0.1 nm, which is on the order of the atomic spacing in a solid. As a result, they can be diffracted by the regular atomic spacings in a crystal lattice, which act as a diffraction grating. The x-ray diffraction pattern of a well-ordered protein crystal is shown in Figure 27.7.

X-rays are produced when high-speed electrons are suddenly slowed down, such as when a metal target is struck by electrons that have been accelerated through a potential difference of several thousand volts. Figure 27.8a shows a schematic

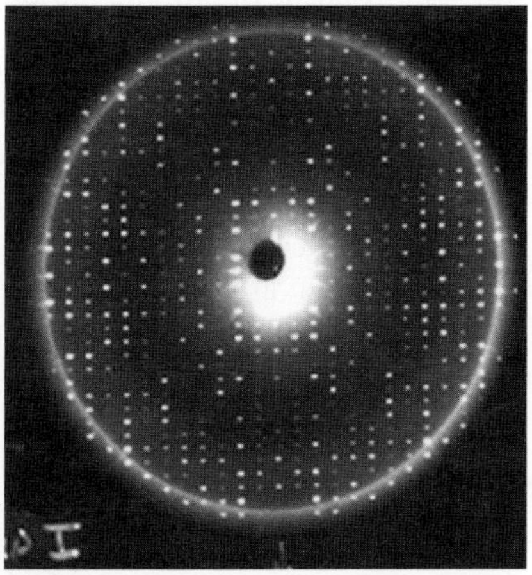
Courtesy of NASA MARSHALL SPACE FLIGHT CENTER/0101745

Figure 27.7 An x-ray diffraction pattern of a well-ordered protein crystal can be analyzed mathematically with powerful computers, leading to a determination of the structure of the protein. Once the structure is known, molecules can be designed that fit the active site of the protein. Such molecules can be used in developing therapeutic drugs that deactivate a given protein without affecting other biological systems.

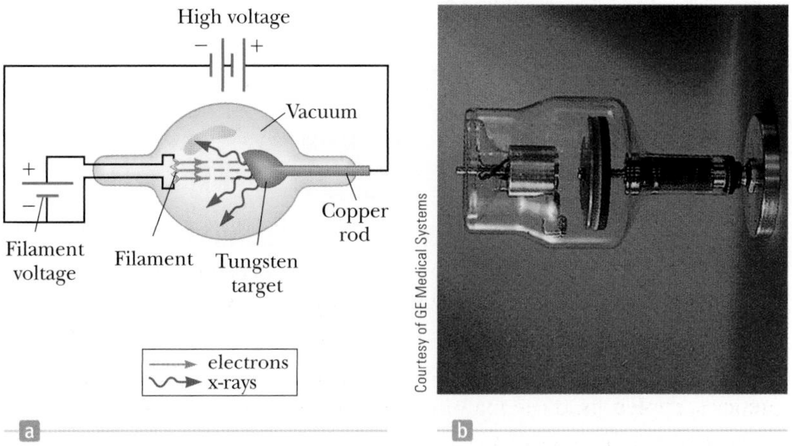

Courtesy of GE Medical Systems

Figure 27.8 (a) Diagram of an x-ray tube. (b) Photograph of an x-ray tube.

diagram of an x-ray tube. A current in the filament causes electrons to be emitted, and these freed electrons are accelerated toward a dense metal target, such as tungsten, which is held at a higher potential than the filament.

Figure 27.9 represents a plot of x-ray intensity versus wavelength for the spectrum of radiation emitted by an x-ray tube. Note that the spectrum has two distinct components. One component is a continuous broad spectrum that depends on the voltage applied to the tube. Superimposed on this component is a series of sharp, intense lines that depend on the nature of the target material. To observe these sharp lines, which represent radiation emitted by the target atoms as their electrons undergo rearrangements, the accelerating voltage must exceed a certain value, called the **threshold voltage**. We discuss threshold voltage further in Chapter 28. The continuous radiation is sometimes called **bremsstrahlung**, a German word meaning "braking radiation," because electrons emit radiation when they undergo an acceleration inside the target.

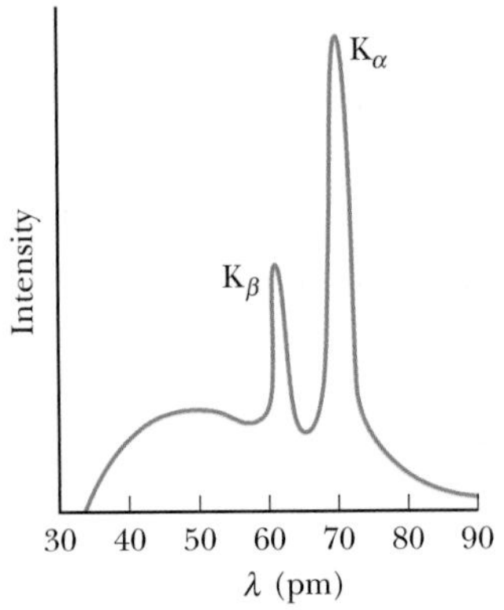

Figure 27.9 The x-ray spectrum of a metal target consists of a broad continuous spectrum plus a number of sharp lines, which are due to *characteristic x-rays*. The data shown were obtained when 35-keV electrons bombarded a molybdenum target. Note that 1 pm = 10^{-12} m = 10^{-3} nm.

Figure 27.10 illustrates how x-rays are produced when an electron passes near a charged target nucleus. As the electron passes close to a positively charged nucleus in the target material, it is deflected from its path because of its electrical attraction to the nucleus; hence, it undergoes an acceleration. An analysis from classical physics shows that any charged particle will emit electromagnetic radiation when it is accelerated. (An example of this phenomenon is the production of electromagnetic waves by accelerated charges in a radio antenna, as described in Chapter 21.) According to quantum theory, this radiation must appear in the form of photons. Because the radiated photon shown in Figure 27.10 carries energy, the electron must lose kinetic energy because of its encounter with the target nucleus. An extreme example consists of the electron losing all of its energy in a single collision. In this case the initial energy of the electron ($e\Delta V$) is transformed completely into the energy of the photon (hf_{max}). In equation form,

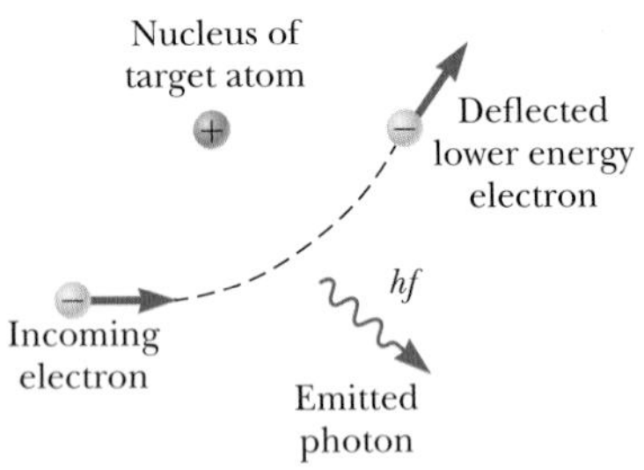

Figure 27.10 An electron passing near a charged target atom experiences an acceleration, and a photon is emitted in the process.

$$e\Delta V = hf_{\text{max}} = \frac{hc}{\lambda_{\text{min}}} \qquad [27.8]$$

where $e\Delta V$ is the energy of the electron after it has been accelerated through a potential difference of ΔV volts and e is the charge on the electron. This equation says that the shortest wavelength radiation that can be produced is

$$\lambda_{\text{min}} = \frac{hc}{e\Delta V} \qquad [27.9]$$

Not all the radiation produced has this particular wavelength because many of the electrons aren't stopped in a single collision. The result is the production of the continuous spectrum of wavelengths.

APPLICATION
Using X-Rays to Study the Work of Master Painters

Interesting insights into the process of painting and revising a masterpiece are being revealed by x-rays. Long-wavelength x-rays are absorbed in varying degrees by some paints, such as those having lead, cadmium, chromium, or cobalt as a base. The x-ray interactions with the paints give contrast because the different elements in the paints have different electron densities. Also, thicker layers will absorb more than thin layers. To examine a painting by an old master, a film is placed behind it while it is x-rayed from the front. Ghost outlines of earlier paintings and earlier forms of the final masterpiece are sometimes revealed when the film is developed.

27.4 Diffraction of X-Rays by Crystals

LEARNING OBJECTIVES

1. Discuss x-ray diffraction and state Bragg's law.
2. Apply Bragg's law to crystalline systems.

In Chapter 24 we described how a diffraction grating could be used to measure the wavelength of light. In principle the wavelength of *any* electromagnetic wave can be measured if a grating having a suitable line spacing can be found. The spacing between lines must be approximately equal to the wavelength of the radiation to

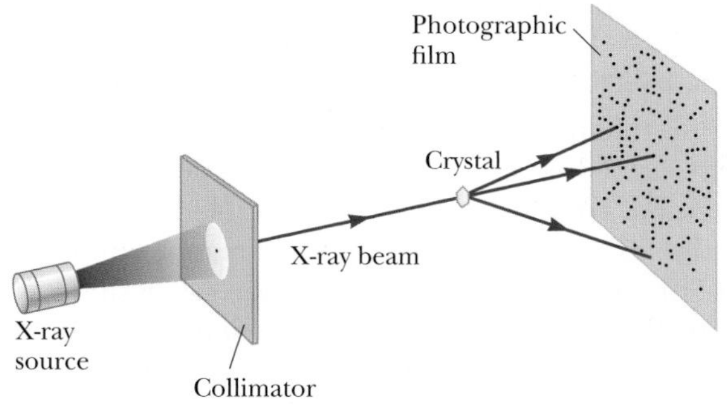

Figure 27.11 Schematic diagram of the technique used to observe the diffraction of x-rays by a single crystal. The array of spots formed on the film by the diffracted beams is called a Laue pattern. (See Fig. 27.7.)

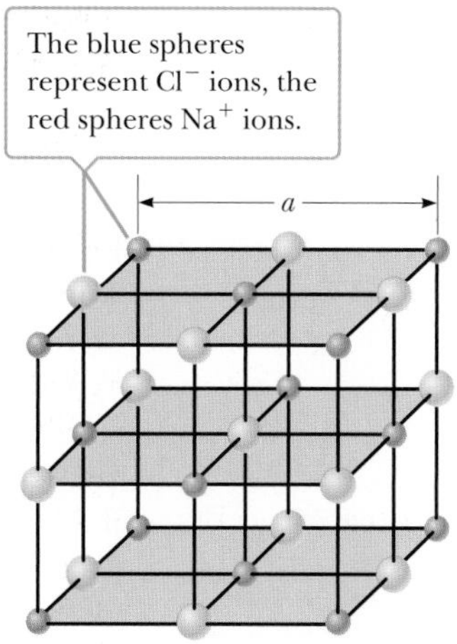

Figure 27.12 A model of the cubic crystalline structure of sodium chloride. The length of the cube edge is $a = 0.563$ nm.

be measured. X-rays are electromagnetic waves with wavelengths on the order of 0.1 nm. It would be impossible to construct a grating with such a small spacing. As noted in the previous section, however, the regular array of atoms in a crystal could act as a three-dimensional grating for observing the diffraction of x-rays.

One experimental arrangement for observing x-ray diffraction is shown in Figure 27.11. A narrow beam of x-rays with a continuous wavelength range is incident on a crystal such as sodium chloride. The diffracted radiation is very intense in certain directions, corresponding to constructive interference from waves reflected from layers of atoms in the crystal. The diffracted radiation is detected by a photographic film and forms an array of spots known as a *Laue pattern*. The crystal structure is determined by analyzing the positions and intensities of the various spots in the pattern.

The arrangement of atoms in a crystal of NaCl is shown in Figure 27.12. The smaller red spheres represent Na^+ ions, and the larger blue spheres represent Cl^- ions. The spacing between successive Na^+ (or Cl^-) ions in this cubic structure, denoted by the symbol a in Figure 27.12, is approximately 0.563 nm.

A careful examination of the NaCl structure shows that the ions lie in various planes. The shaded areas in Figure 27.12 represent one example, in which the atoms lie in equally spaced planes. Now suppose an x-ray beam is incident at grazing angle θ on one of the planes, as in Figure 27.13. The beam can be reflected from both the upper and lower plane of atoms. The geometric construction in Figure 27.13, however, shows that the beam reflected from the lower surface travels farther than the beam reflected from the upper surface by a distance of $2d \sin \theta$. The two portions of the reflected beam will combine to produce constructive interference when this path difference equals some integral multiple of the wavelength λ. The condition for constructive interference is given by

$$2d \sin \theta = m\lambda \qquad m = 1, 2, 3, \ldots \qquad [27.10]$$

◀ Bragg's law

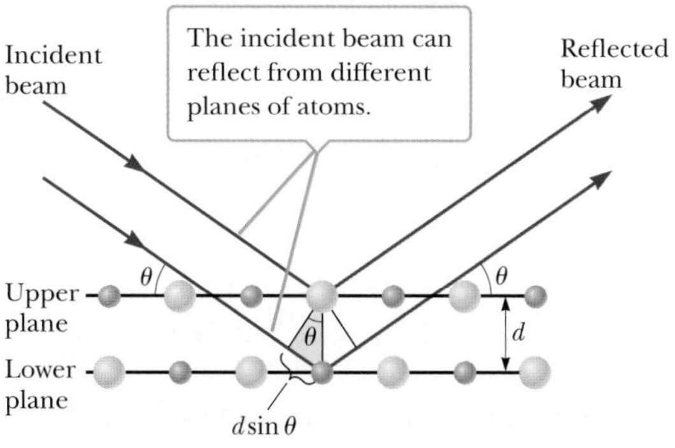

Figure 27.13 A two-dimensional depiction of the reflection of an x-ray beam from two parallel crystalline planes separated by a distance d. The beam reflected from the lower plane travels farther than the one reflected from the upper plane by an amount equal to $2d \sin \theta$.

This condition is known as **Bragg's law**, after W. L. Bragg (1890–1971), who first derived the relationship. If the wavelength and diffraction angle are measured, Equation 27.10 can be used to calculate the spacing between atomic planes.

The technique of x-ray diffraction has been used to determine the atomic arrangement of complex organic molecules such as proteins. Proteins are large molecules containing thousands of atoms that help regulate chemical processes in cells. Some proteins are amazing catalysts, speeding up the slow room-temperature reactions in cells by 17 orders of magnitude. To understand this incredible biochemical reactivity, it is important to determine the structure of these intricate molecules.

The main technique used to determine the molecular structure of proteins, DNA, and RNA is x-ray diffraction using x-rays of wavelength of about 1 Å (1 Å = 0.1 nm = 1×10^{-10} m). This technique allows the experimenter to "see" individual atoms that are separated by about this distance in molecules. Because the biochemical x-ray diffraction sample is prepared in crystal form, the *geometry* (position of the bright spots in space) of the diffraction pattern is determined by the regular three-dimensional crystal lattice arrangement of molecules in the sample. The *intensities* of the bright diffraction spots are determined by the atoms and their electronic distributions in the fundamental building block of the crystal: the unit cell. Using complicated computational techniques, investigators can essentially deduce the molecular structure by matching the observed intensities of diffracted beams with a series of assumed atomic positions that determine the atomic structure and electron density of the molecule.

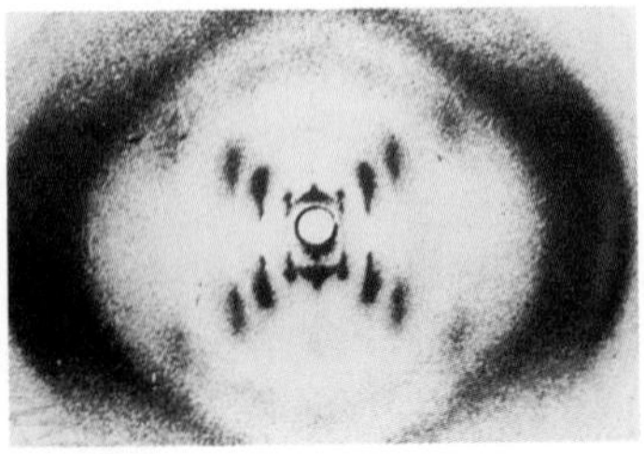

Figure 27.14 An x-ray diffraction photograph of DNA taken by Rosalind Franklin. The cross pattern of spots was a clue that DNA has a helical structure.

Crystallizing a large molecule such as DNA and obtaining its x-ray diffraction pattern is highly challenging. In the case of DNA, obtaining sufficiently pure crystals is especially difficult because there exist two crystalline forms, A and B, which arise in mixed form during preparation. These two forms result in diffraction patterns that can't be easily deciphered. In 1951 Rosalind Franklin, a researcher at King's College in London, developed an ingenious method of separating the two forms and managed to obtain excellent x-ray diffraction images of pure crystalline DNA in B-form. With these images she determined that the helical shape of DNA consisted of two interwoven strands, with the sugar-phosphate backbone on the outside of the molecule, refuting prior models that had the backbone on the inside. One of her images is shown in Figure 27.14.

Figure 27.15 The double-helix structure of DNA.

James Watson and Francis Crick used Franklin's work to uncover further details of the molecule and its function in heredity, particularly in regards to the internal structure. Attached to each sugar-phosphate unit of each strand is one of four base molecules: adenine, cytosine, guanine, or thymine. The bases are arranged sequentially along the strand, with patterns in the sequence of bases acting as codes for proteins that carry out various functions for a given organism. The bases in one strand bind to those in the other strand, forming a double helix. A model of the double helix is shown in Figure 27.15. In 1962 Watson, Crick, and a colleague of Franklin's, Maurice Wilkins, received the Nobel Prize for Physiology and Medicine for their work on understanding the structure and function of DNA. Franklin would have shared the prize, but died in 1958 of cancer at the age of thirty-eight. (The prize is not awarded posthumously.)

EXAMPLE 27.2 | X-Ray Diffraction from Calcite

GOAL Understand Bragg's law and apply it to a crystal.

PROBLEM If the spacing between certain planes in a crystal of calcite ($CaCO_3$) is 0.314 nm, find the grazing angles at which first- and third-order interference will occur for x-rays of wavelength 0.070 0 nm.

STRATEGY Solve Bragg's law for sin θ and substitute, using the inverse-sine function to obtain the angle.

SOLUTION

Find the grazing angle corresponding to $m = 1$, for first-order interference:

$$\sin \theta = \frac{m\lambda}{2d} = \frac{(0.070\,0 \text{ nm})}{2(0.314 \text{ nm})} = 0.111$$

$$\theta = \sin^{-1}(0.111) = \boxed{6.37°}$$

Repeat the calculation for third-order interference ($m = 3$):

$$\sin\theta = \frac{m\lambda}{2d} = \frac{3(0.070\,0 \text{ nm})}{2(0.314 \text{ nm})} = 0.334$$

$$\theta = \sin^{-1}(0.334) = 19.5°$$

REMARKS Notice there is little difference between this kind of problem and a Young's slit experiment.

QUESTION 27.2 True or False: A smaller grazing angle implies a smaller distance between planes in the crystal lattice.

EXERCISE 27.2 X-rays of wavelength 0.060 0 nm are scattered from a crystal with a grazing angle of 11.7°. Assume $m = 1$ for this process. Calculate the spacing between the crystal planes.

ANSWER 0.148 nm

27.5 The Compton Effect

LEARNING OBJECTIVES

1. Describe the Compton effect in terms of conservation of energy and momentum.
2. Apply the Compton effect to systems of photons and electrons.

Further justification for the photon nature of light came from an experiment conducted by Arthur H. Compton in 1923. In his experiment Compton directed an x-ray beam of wavelength λ_0 toward a block of graphite. He found that the scattered x-rays had a slightly longer wavelength λ than the incident x-rays and hence the energies of the scattered rays were lower. The amount of energy reduction depended on the angle at which the x-rays were scattered. The change in wavelength $\Delta\lambda$ between a scattered x-ray and an incident x-ray is called the **Compton shift**.

To explain this effect, Compton assumed that if a photon behaves like a particle, its collision with other particles is similar to a collision between two billiard balls. Hence, the x-ray photon carries both measurable *energy* and *momentum*, and these two quantities must be conserved in a collision. If the incident photon collides with an electron initially at rest, as in Figure 27.16, the photon transfers some of its energy and momentum to the electron. As a consequence, the energy and frequency of the scattered photon are lowered and its wavelength increases. Applying relativistic energy and momentum conservation to the collision described in Figure 27.16, the shift in wavelength of the scattered photon is given by

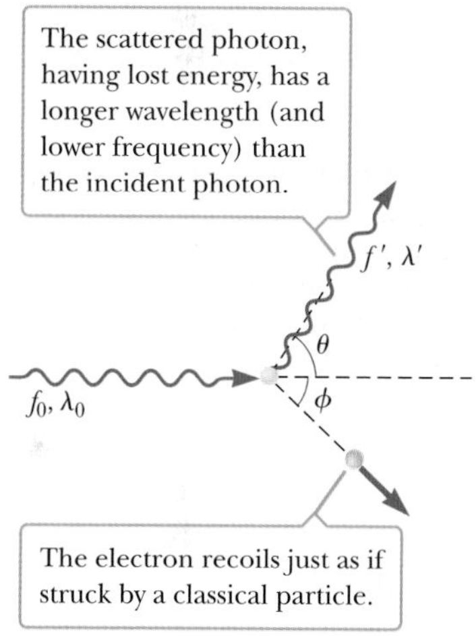

Figure 27.16 Diagram representing Compton scattering of a photon by an electron.

$$\Delta\lambda = \lambda - \lambda_0 = \frac{h}{m_e c}(1 - \cos\theta) \qquad [27.11]$$

◀ The Compton shift formula

© Mary Evans Picture Library/Alamy

Arthur Holly Compton
American Physicist (1892–1962)
Compton attended Wooster College and Princeton University. He became director of the laboratory at the University of Chicago, where experimental work concerned with sustained chain reactions was conducted. This work was of central importance to the construction of the first atomic bomb. Because of his discovery of the Compton effect and his work with cosmic rays, he shared the 1927 Nobel Prize in Physics with Charles Wilson.

where m_e is the mass of the electron and θ is the angle between the directions of the scattered and incident photons. The quantity h/m_ec is called the **Compton wavelength** and has a value of 0.002 43 nm. The Compton wavelength is very small relative to the wavelengths of visible light, so the shift in wavelength would be difficult to detect if visible light were used. Further, note that the Compton shift depends on the scattering angle θ and not on the wavelength. Experimental results for x-rays scattered from various targets obey Equation 27.11 and strongly support the photon concept.

Quick Quiz

27.1 True or False: When a photon scatters off an electron, the photon loses energy.

27.2 An x-ray photon is scattered by an electron. Does the frequency of the scattered photon relative to that of the incident photon (a) increase, (b) decrease, or (c) remain the same?

27.3 A photon of energy E_0 strikes a free electron, with the scattered photon of energy E moving in the direction opposite that of the incident photon. In this Compton effect interaction, what is the resulting kinetic energy of the electron? (a) E_0 (b) E (c) $E_0 - E$ (d) $E_0 + E$

EXAMPLE 27.3 | Scattering X-Rays

GOAL Understand Compton scattering and its effect on the photon's energy.

PROBLEM X-rays of wavelength $\lambda_i = 0.200\ 000$ nm are scattered from a block of material. The scattered x-rays are observed at an angle of 45.0° to the incident beam. **(a)** Calculate the wavelength of the x-rays scattered at this angle. **(b)** Compute the fractional change in the energy of a photon in the collision.

STRATEGY To find the wavelength of the scattered x-ray photons, substitute into Equation 27.11 to obtain the wavelength shift, then add the result to the initial wavelength, λ_i. In part (b), calculating the fractional change in energy involves calculating the energy of the x-ray photon before and after, using $E = hf = hc/\lambda$. Taking the difference and dividing by the initial energy yields the desired fractional change in energy. Here, however, a symbolic expression is derived that relates energy terms and wavelengths.

SOLUTION

(a) Calculate the wavelength of the x-rays.

Substitute into Equation 27.11 to obtain the wavelength shift:

$$\Delta\lambda = \frac{h}{m_e c}(1 - \cos\theta)$$

$$= \frac{6.63 \times 10^{-34}\ \text{J}\cdot\text{s}}{(9.11 \times 10^{-31}\ \text{kg})(3.00 \times 10^{8}\ \text{m/s})}(1 - \cos 45.0°)$$

$$= 7.11 \times 10^{-13}\ \text{m} = 0.000\ 711\ \text{nm}$$

Add this shift to the original wavelength to obtain the wavelength of the scattered photon:

$$\lambda_f = \Delta\lambda + \lambda_i = \boxed{0.200\ 711\ \text{nm}}$$

(b) Find the fraction of energy lost by the photon in the collision.

Rewrite the energy E in terms of wavelength, using $c = f\lambda$:

$$E = hf = h\frac{c}{\lambda}$$

Compute $\Delta E/E$ using this expression:

$$\frac{\Delta E}{E} = \frac{E_f - E_i}{E_i} = \frac{hc/\lambda_f - hc/\lambda_i}{hc/\lambda_i}$$

Cancel hc and rearrange terms:

$$\frac{\Delta E}{E} = \frac{1/\lambda_f - 1/\lambda_i}{1/\lambda_i} = \frac{\lambda_i}{\lambda_f} - 1 = \frac{\lambda_i - \lambda_f}{\lambda_f} = -\frac{\Delta\lambda}{\lambda_f}$$

Substitute values from part (a):

$$\frac{\Delta E}{E} = -\frac{0.000\ 711\ \text{nm}}{0.200\ 711\ \text{nm}} = \boxed{-3.54 \times 10^{-3}}$$

REMARKS It is also possible to find this answer by substituting into the energy expression at an earlier stage, but the algebraic derivation is more elegant and instructive because it shows how changes in energy are related to changes in wavelength.

QUESTION 27.3 The incident photon loses energy. Where does it go?

EXERCISE 27.3 Repeat the example for a photon with wavelength 3.00×10^{-2} nm that scatters at an angle of 60.0°.

ANSWERS (a) 3.12×10^{-2} nm (b) $\Delta E/E = -3.88 \times 10^{-2}$

27.6 The Dual Nature of Light and Matter

LEARNING OBJECTIVES

1. State de Broglie's hypothesis and discuss the dual nature of light and matter.
2. Apply de Broglie's hypothesis to quantum and classical objects.

Phenomena such as the photoelectric effect and the Compton effect offer evidence that when light (or other forms of electromagnetic radiation) and matter interact, the light behaves as if it were composed of particles having energy hf and momentum h/λ. In other contexts, however, light acts like a wave, exhibiting interference and diffraction effects. This apparent duality can be partly explained by considering the energies of photons in different contexts. For example, photons with frequencies in the radio wavelengths carry very little energy, and it may take 10^{10} such photons to create a signal in an antenna. These photons therefore act together like a wave to create the effect. Gamma rays, on the other hand, are so energetic that a single gamma ray photon can be detected.

In his doctoral dissertation in 1924, Louis de Broglie postulated that **because photons have wave and particle characteristics, perhaps all forms of matter have both properties**. This highly revolutionary idea had no experimental confirmation at that time. According to de Broglie, electrons, just like light, have a dual particle–wave nature.

◀ De Broglie's hypothesis

In Chapter 26 we found that the relationship between energy and momentum for a photon, which has a rest energy of zero, is $p = E/c$. We also know from Equation 27.5 that the energy of a photon is

$$E = hf = \frac{hc}{\lambda} \qquad [27.12]$$

Consequently, the momentum of a photon can be expressed as

$$p = \frac{E}{c} = \frac{hc}{c\lambda} = \frac{h}{\lambda} \qquad [27.13]$$

◀ Momentum of a photon

From this equation, we see that the photon wavelength can be specified by its momentum, or $\lambda = h/p$. De Broglie suggested that *all* material particles with momentum p should have a characteristic wavelength $\lambda = h/p$. Because the momentum of a particle of mass m and speed v is $mv = p$, the **de Broglie wavelength** of a particle is

$$\lambda = \frac{h}{p} = \frac{h}{mv} \qquad [27.14]$$

◀ De Broglie wavelength

Further, de Broglie postulated that the frequencies of matter waves (waves associated with particles having nonzero rest energy) obey the Einstein relationship for photons, $E = hf$, so that

$$f = \frac{E}{h} \qquad [27.15]$$

◀ Frequency of matter waves

The dual nature of matter is quite apparent in Equations 27.14 and 27.15 because each contains both particle concepts (mv and E) and wave concepts (λ and f). The fact that these relationships had been established experimentally for photons made the de Broglie hypothesis that much easier to accept. The Davisson–Germer experiment in 1927 confirmed de Broglie's hypothesis by showing that electrons

Louis de Broglie
French Physicist (1892–1987)
De Broglie attended the Sorbonne in Paris, where he changed his major from history to theoretical physics. He was awarded the Nobel Prize in Physics in 1929 for his discovery of the wave nature of electrons.

scattering off crystals form a diffraction pattern. The regularly spaced planes of atoms in crystalline regions of a nickel target act as a diffraction grating for electron matter waves.

Quick Quiz

27.4 True or False: As the momentum of a particle of mass m increases, its wavelength increases.

27.5 A nonrelativistic electron and a nonrelativistic proton are moving and have the same de Broglie wavelength. Which of the following are also the same for the two particles? (a) speed (b) kinetic energy (c) momentum (d) frequency

EXAMPLE 27.4 The Electron Versus the Baseball

GOAL Apply the de Broglie hypothesis to a quantum and a classical object.

PROBLEM **(a)** Compare the de Broglie wavelength for an electron ($m_e = 9.11 \times 10^{-31}$ kg) moving at a speed equal to 1.00×10^7 m/s with that of a baseball of mass 0.145 kg pitched at 45.0 m/s. **(b)** Compare these wavelengths with that of an electron traveling at $0.999c$.

STRATEGY This problem is a matter of substitution into Equation 27.14 for the de Broglie wavelength. In part (b) the relativistic momentum must be used.

SOLUTION

(a) Compare the de Broglie wavelengths of the electron and the baseball.

Substitute data for the electron into Equation 27.14:

$$\lambda_e = \frac{h}{m_e v} = \frac{6.63 \times 10^{-34}\ \text{J}\cdot\text{s}}{(9.11 \times 10^{-31}\ \text{kg})(1.00 \times 10^7\ \text{m/s})}$$

$$= \boxed{7.28 \times 10^{-11}\ \text{m}}$$

Repeat the calculation with the baseball data:

$$\lambda_b = \frac{h}{m_b v} = \frac{6.63 \times 10^{-34}\ \text{J}\cdot\text{s}}{(0.145\ \text{kg})(45.0\ \text{m/s})} = \boxed{1.02 \times 10^{-34}\ \text{m}}$$

(b) Find the wavelength for an electron traveling at $0.999c$.

Replace the momentum in Equation 27.14 with the relativistic momentum:

$$\lambda_e = \frac{h}{m_e v/\sqrt{1 - v^2/c^2}} = \frac{h\sqrt{1 - v^2/c^2}}{m_e v}$$

Substitute:

$$\lambda_e = \frac{(6.63 \times 10^{-34}\ \text{J}\cdot\text{s})\sqrt{1 - (0.999c)^2/c^2}}{(9.11 \times 10^{-31}\ \text{kg})(0.999 \cdot 3.00 \times 10^8\ \text{m/s})}$$

$$= \boxed{1.09 \times 10^{-13}\ \text{m}}$$

REMARKS The electron wavelength corresponds to that of x-rays in the electromagnetic spectrum. The baseball, by contrast, has a wavelength much smaller than any aperture through which the baseball could possibly pass, so we couldn't observe any of its diffraction effects. It is generally true that the wave properties of large-scale objects can't be observed. Notice that even at extreme relativistic speeds, the electron wavelength is still far larger than the baseball's.

QUESTION 27.4 How does doubling the speed of a particle affect its wavelength? Is your answer always true? Explain.

EXERCISE 27.4 Find the de Broglie wavelength of a proton ($m_p = 1.67 \times 10^{-27}$ kg) moving at 1.00×10^7 m/s.

ANSWER 3.97×10^{-14} m

Application: The Electron Microscope

BIO APPLICATION
Electron Microscopes

A practical device that relies on the wave characteristics of electrons is the **electron microscope**. A *transmission* electron microscope, used for viewing flat, thin samples, is shown in Figure 27.17. In many respects it is similar to an optical microscope, but the electron microscope has a much greater resolving power because it can accelerate electrons to very high kinetic energies, giving them very short wavelengths. No microscope can resolve details that are significantly smaller than the wavelength of the radiation used to illuminate the object. Typically, the wavelengths of electrons in an electron microscope are smaller than the visible wavelengths by a factor of about 10^{-5}.

The electron beam in an electron microscope is controlled by electrostatic or magnetic deflection, which acts on the electrons to focus the beam to an image. Due to limitations in the electromagnetic lenses used, however, the improvement in resolution over light microscopes is only about a factor of 1 000, two orders

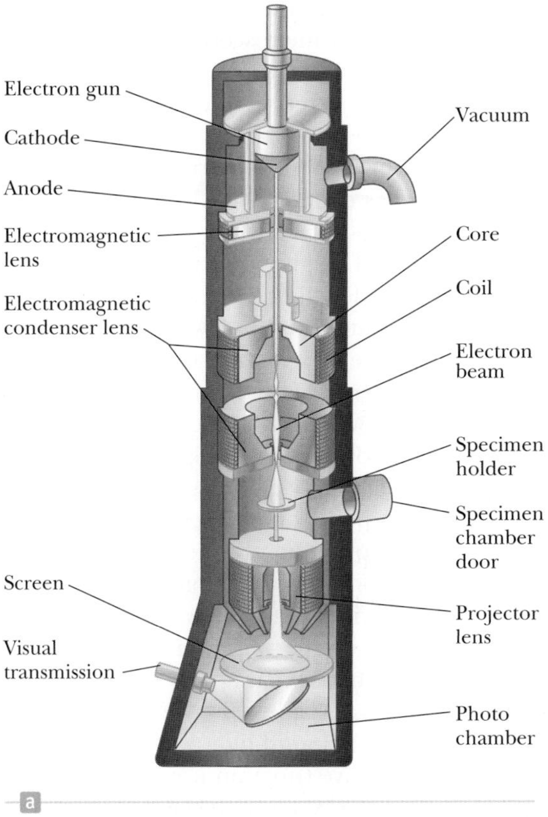

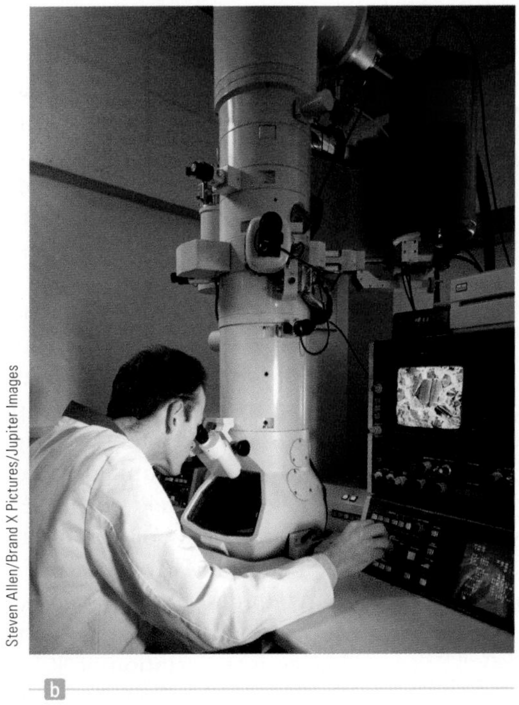

Steven Allen/Brand X Pictures/Jupiter Images

Figure 27.17 (a) Diagram of a transmission electron microscope for viewing a thin, sectioned sample. The "lenses" that control the electron beam are magnetic deflection coils. (b) An electron microscope.

of magnitude smaller than that implied by the electron wavelength. Rather than examining the image through an eyepiece as in an optical microscope, the viewer looks at an image formed on a fluorescent screen. (The viewing screen must be fluorescent because otherwise the image produced wouldn't be visible.)

APPLYING PHYSICS 27.2 | X-Ray Microscopes?

Electron microscopes (Fig. 27.17) take advantage of the wave nature of particles. Electrons are accelerated to high speeds, giving them a short de Broglie wavelength. Imagine an electron microscope using electrons with a de Broglie wavelength of 0.2 nm. Why don't we design a microscope using 0.2-nm *photons* to do the same thing?

EXPLANATION Because electrons are charged particles, they interact electrically with the sample in the microscope and scatter according to the shape and density of various portions of the sample, providing a means of viewing the sample. Photons of wavelength 0.2 nm are uncharged and in the x-ray region of the spectrum. They tend to simply pass through the thin sample without interacting. ■

27.7 The Wave Function

LEARNING OBJECTIVE

1. Discuss the Schrödinger wave equation and the wave function.

Erwin Schrödinger
Austrian Theoretical Physicist (1887–1961)
Schrödinger is best known as the creator of wave mechanics, a less cumbersome theory than the equivalent matrix mechanics developed by Werner Heisenberg. In 1933 Schrödinger left Germany and eventually settled at the Dublin Institute of Advanced Study, where he spent 17 happy, creative years working on problems in general relativity, cosmology, and the application of quantum physics to biology. In 1956 he returned home to Austria and his beloved Tyrolean mountains, where he died in 1961.

De Broglie's revolutionary idea that particles should have a wave nature soon moved out of the realm of skepticism to the point where it was viewed as a necessary concept in understanding the subatomic world. In 1926 Austrian–German physicist Erwin Schrödinger proposed a wave equation that described how matter waves change in space and time. The Schrödinger wave equation represents a key element in the theory of quantum mechanics. It's as important in quantum mechanics as Newton's laws in classical mechanics. Schrödinger's equation has been successfully applied to the hydrogen atom and to many other microscopic systems.

Solving Schrödinger's equation (beyond the level of this course) determines a quantity Ψ called the **wave function**. Each particle is represented by a wave function Ψ that depends on both position and time. Once Ψ is found, Ψ^2 gives us information on the **probability** (per unit volume) of finding the particle in any given region. To understand this, we return to Young's experiment involving coherent light passing through a double slit.

First, recall from Chapter 21 that the intensity of a light beam is proportional to the square of the electric field strength E associated with the beam: $I \propto E^2$. According to the wave model of light, there are certain points on the viewing screen where the net electric field is zero as a result of destructive interference of waves from the two slits. Because E is zero at these points, the intensity is also zero and the screen is dark there. Likewise, at points on the screen where constructive interference occurs, E is large, as is the intensity; hence, these locations are bright.

Now consider the same experiment when light is viewed as having a particle nature. The number of photons reaching a point on the screen per second increases as the intensity (brightness) increases. Consequently, the number of photons that strike a unit area on the screen each second is proportional to the square of the electric field, or $N \propto E^2$. From a probabilistic point of view, a photon has a high probability of striking the screen at a point where the intensity (and E^2) is high and a low probability of striking the screen where the intensity is low.

When describing particles rather than photons, Ψ rather than E plays the role of the amplitude. Using an analogy with the description of light, we make the following interpretation of Ψ for particles: If Ψ is a wave function used to describe a single particle, the value of Ψ^2 at some location at a given time is proportional to the probability per unit volume of finding the particle at that location at that time. Adding all the values of Ψ^2 in a given region gives the probability of finding the particle in that region.

an explanation of the spectra of hydrogen that includes some features of the currently accepted theory. His model of the hydrogen atom included the following basic assumptions:

1. The electron moves in circular orbits about the proton under the influence of the Coulomb force of attraction, as in Figure 28.5. The Coulomb force produces the electron's centripetal acceleration.
2. Only certain electron orbits are stable and allowed. In these orbits no energy in the form of electromagnetic radiation is emitted, so the total energy of the atom remains constant.
3. Radiation is emitted by the hydrogen atom when the electron "jumps" from a more energetic initial state to a less energetic state. The "jump" can't be visualized or treated classically. The frequency f of the radiation emitted in the jump is related to the change in the atom's energy, given by

$$E_i - E_f = hf \qquad [28.4]$$

where E_i is the energy of the initial state, E_f is the energy of the final state, h is Planck's constant, and $E_i > E_f$. The frequency of the radiation is *independent of the frequency of the electron's orbital motion.*

4. The circumference of an electron's orbit must contain an integral number of de Broglie wavelengths,

$$2\pi r = n\lambda \qquad n = 1, 2, 3, \ldots$$

(See Fig. 28.6.) Because the de Broglie wavelength of an electron is given by $\lambda = h/m_e v$, we can write the preceding equation as

$$m_e v r = n\hbar \qquad n = 1, 2, 3, \ldots \qquad [28.5]$$

where $\hbar = h/2\pi$.

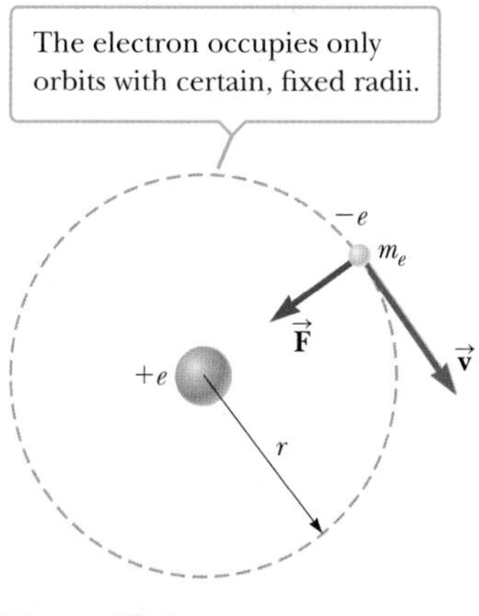

Figure 28.5 Diagram representing Bohr's model of the hydrogen atom.

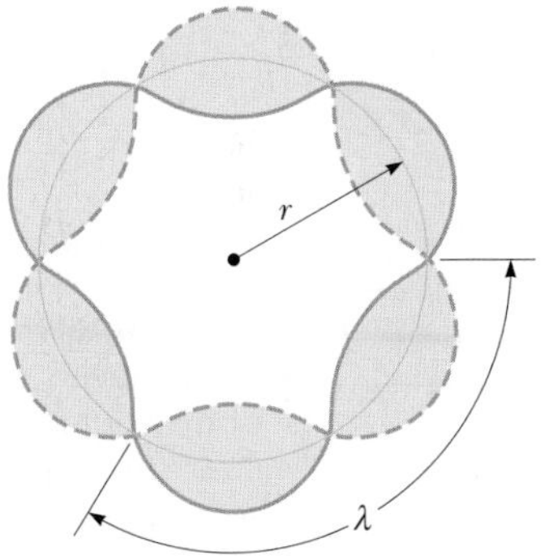

Figure 28.6 Standing-wave pattern for an electron wave in a stable orbit of hydrogen. There are three full wavelengths in this orbit.

With these four assumptions, we can calculate the allowed energies and emission wavelengths of the hydrogen atom using the model pictured in Figure 28.5, in which the electron travels in a circular orbit of radius r with an orbital speed v. The electrical potential energy of the atom is

$$PE = k_e \frac{q_1 q_2}{r} = k_e \frac{(-e)(e)}{r} = -k_e \frac{e^2}{r}$$

where k_e is the Coulomb constant. Assuming the nucleus is at rest, the total energy E of the atom is the sum of the kinetic and potential energy:

$$E = KE + PE = \tfrac{1}{2} m_e v^2 - k_e \frac{e^2}{r} \qquad [28.6]$$

By Newton's second law, the electric force of attraction on the electron, $k_e e^2/r^2$, must equal $m_e a_r$, where $a_r = v^2/r$ is the centripetal acceleration of the electron, so

$$m_e \frac{v^2}{r} = k_e \frac{e^2}{r^2} \qquad [28.7]$$

Multiply both sides of this equation by $r/2$ to get an expression for the kinetic energy:

$$\tfrac{1}{2} m_e v^2 = \frac{k_e e^2}{2r} \qquad [28.8]$$

Combining this result with Equation 28.6 gives an expression for the energy of the atom,

$$E = -\frac{k_e e^2}{2r} \qquad [28.9]$$

where the negative value of the energy indicates that the electron is bound to the proton.

Bain Collection, Prints & Photographs Division, Library of Congress, LC-DIG-ggbain-35303

Niels Bohr
Danish Physicist (1885–1962)
Bohr was an active participant in the early development of quantum mechanics and provided much of its philosophical framework. During the 1920s and 1930s he headed the Institute for Advanced Studies in Copenhagen, where many of the world's best physicists came to exchange ideas. Bohr was awarded the 1922 Nobel Prize in Physics for his investigation of the structure of atoms and of the radiation emanating from them.

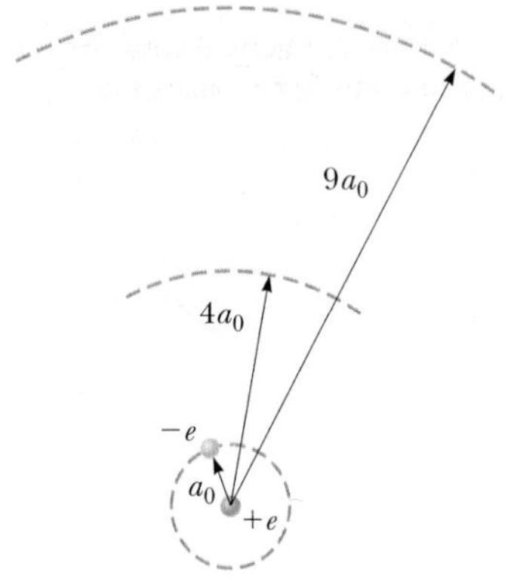

Figure 28.7 The first three circular orbits predicted by the Bohr model of the hydrogen atom. The electron is shown in the ground state orbit.

An expression for r can be obtained by solving Equations 28.5 and 28.7 for v^2 and equating the results:

$$v^2 = \frac{n^2\hbar}{m_e^{\,2}r^2} = \frac{k_e e^2}{m_e r}$$

$$r_n = \frac{n^2\hbar^2}{m_e k_e e^2} \qquad n = 1, 2, 3, \ldots \qquad [28.10]$$

This equation is based on the assumption that the **electron can exist only in certain allowed orbits determined by the integer *n*.**

The orbit with the smallest radius, called the **Bohr radius**, a_0, corresponds to $n = 1$ and has the value

$$a_0 = \frac{\hbar^2}{mk_e e^2} = 0.052\ 9\ \text{nm} \qquad [28.11]$$

A general expression for the radius of any orbit in the hydrogen atom is obtained by substituting Equation 28.11 into Equation 28.10:

$$r_n = n^2 a_0 = n^2(0.052\ 9\ \text{nm}) \qquad [28.12]$$

The first three Bohr orbits for hydrogen are shown in Figure 28.7.

Equation 28.10 can then be substituted into Equation 28.9 to give the following expression for the energies of the quantum states:

The energy levels of hydrogen ▶

$$E_n = -\frac{m_e k_e^{\,2} e^4}{2\hbar^2}\left(\frac{1}{n^2}\right) \qquad n = 1, 2, 3, \ldots \qquad [28.13]$$

If we substitute numerical values into Equation 28.13, we obtain

$$E_n = -\frac{13.6}{n^2}\ \text{eV} \qquad [28.14]$$

The lowest-energy state, or **ground state**, corresponds to $n = 1$ and has an energy $E_1 = -m_e k_e^{\,2} e^4/2\hbar^2 = -13.6$ eV. The next state, corresponding to $n = 2$, has an energy $E_2 = E_1/4 = -3.40$ eV, and so on. An energy level diagram showing the energies of these stationary states and the corresponding quantum numbers is given in Figure 28.8. The uppermost level shown, corresponding to $E = 0$ and $n \to \infty$, represents the state for which the electron is completely removed from the atom. In this state the electron's *KE* and *PE* are both zero, which means that the electron is at rest infinitely far away from the proton. The minimum energy required to ionize the atom—that is, to completely remove the electron—is called the **ionization energy**. The ionization energy for hydrogen is 13.6 eV.

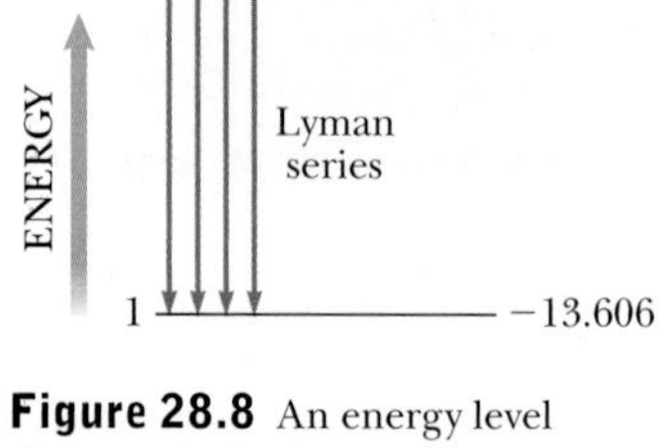

Figure 28.8 An energy level diagram for hydrogen. Quantum numbers are given on the left, and energies (in electron volts) are given on the right. Vertical arrows represent the four lowest-energy transitions for each of the spectral series shown. The colored arrows for the Balmer series indicate that this series results in visible light.

Equations 28.4 and 28.13 and the third Bohr postulate show that if the electron jumps from one orbit with quantum number n_i to a second orbit with quantum number n_f, it emits a photon of frequency f given by

$$f = \frac{E_i - E_f}{\hbar} = \frac{m_e k_e^{\,2} e^4}{4\pi\,\hbar^3}\left(\frac{1}{n_f^{\,2}} - \frac{1}{n_i^{\,2}}\right) \qquad [28.15]$$

where $n_f < n_i$.

To convert this equation into one analogous to the Rydberg equation, substitute $f = c/\lambda$ and divide both sides by c, obtaining

$$\frac{1}{\lambda} = R_{\text{H}}\left(\frac{1}{n_f^{\,2}} - \frac{1}{n_i^{\,2}}\right) \qquad [28.16]$$

where

$$R_{\text{H}} = \frac{m_e k_e^{\,2} e^4}{4\pi c\hbar^3} \qquad [28.17]$$

Substituting the known values of m_e, k_e, e, c, and $\hbar$ verifies that this theoretical value for the Rydberg constant is in excellent agreement with the experimentally derived

28.7 Atomic Transitions and Lasers

LEARNING OBJECTIVES

1. Define the stimulated absorption of photons by atoms.
2. Define what is meant by excited states of atomic electrons.
3. Define the spontaneous emission of photons, and contrast it with the stimulated emission of photons.
4. Define population inversion and describe how a laser works on the atomic level.

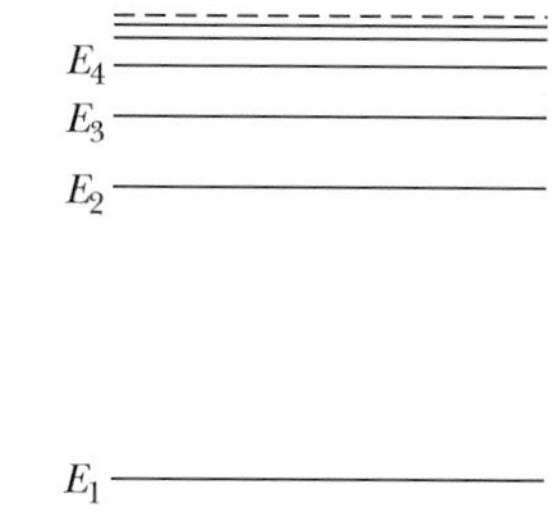

Figure 28.16 Energy level diagram of an atom with various allowed states. The lowest-energy state, E_1, is the ground state. All others are excited states.

We have seen that an atom will emit radiation only at certain frequencies that correspond to the energy separation between the various allowed states. Consider an atom with many allowed energy states, labeled E_1, E_2, E_3, . . . , as in Figure 28.16. When light is incident on the atom, only those photons with energy hf matching the energy separation ΔE between two levels can be absorbed. A schematic diagram representing this **stimulated absorption process** is shown in Figure 28.17. At ordinary temperatures, most of the atoms in a sample are in the ground state. If a vessel containing many atoms of a gas is illuminated with a light beam containing all possible photon frequencies (that is, a continuous spectrum), only those photons of energies $E_2 - E_1$, $E_3 - E_1$, $E_4 - E_1$, and so on can be absorbed. As a result of this absorption, some atoms are raised to various allowed higher-energy levels, called **excited states.**

Once an atom is in an excited state, there is a constant probability that it will jump back to a lower level by emitting a photon, as shown in Figure 28.18. This process is known as **spontaneous emission**. Typically, an atom will remain in an excited state for only about 10^{-8} s.

A third process that is important in lasers, **stimulated emission**, was predicted by Einstein in 1917. Suppose an atom is in the excited state E_2, as in Figure 28.19 (page 964), and a photon with energy $hf = E_2 - E_1$ is incident on it. The incoming photon increases the probability that the excited atom will return to the ground state and thereby emit a second photon having the same energy hf. Note that two identical photons result from stimulated emission: the incident photon and the emitted photon. *The emitted photon is exactly in phase with the incident photon.* These photons can stimulate other atoms to emit photons in a chain of similar processes.

The intense, coherent (in-phase) light in a laser (*l*ight *a*mplification by *s*timulated *e*mission of *r*adiation) is a result of stimulated emission. In a laser, voltages can be used to put more electrons in excited states than in the ground state. This process is called **population inversion**. The excited state of the system must be a *metastable state,* which means that its lifetime must be relatively long. When that

Scientist checking the performance of an experimental laser-cutting device mounted on a robot arm. The laser is being used to cut through a metal plate.

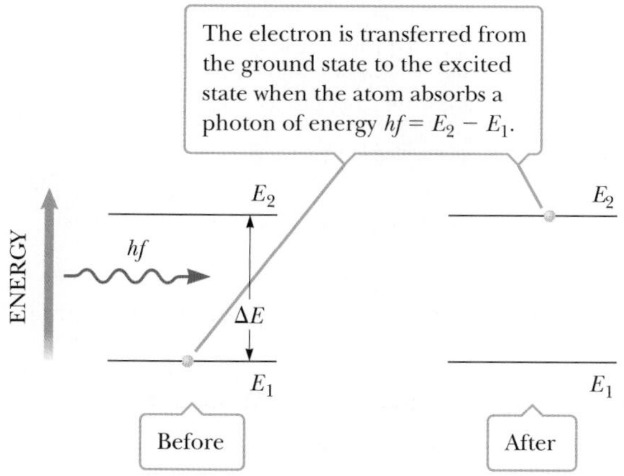

Figure 28.17 Diagram representing the process of *stimulated absorption* of a photon by an atom.

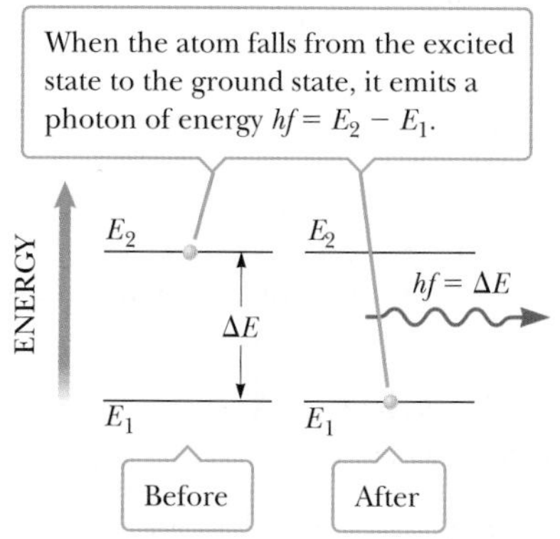

Figure 28.18 Diagram representing the process of *spontaneous emission* of a photon by an atom that is initially in the excited state E_2.

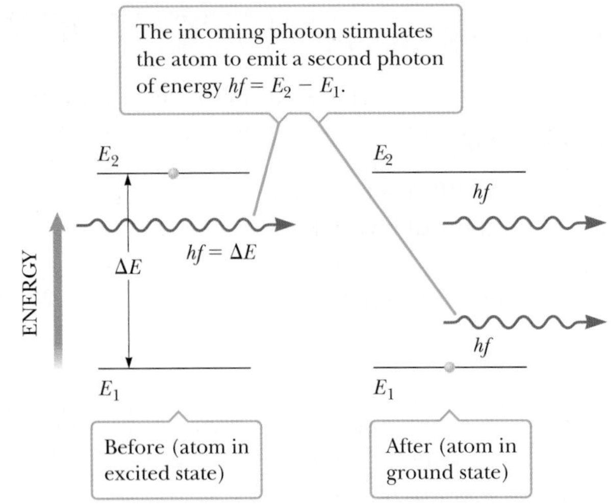

Figure 28.19 Diagram representing the process of *stimulated emission* of a photon by an incoming photon of energy *hf*. Initially, the atom is in the excited state.

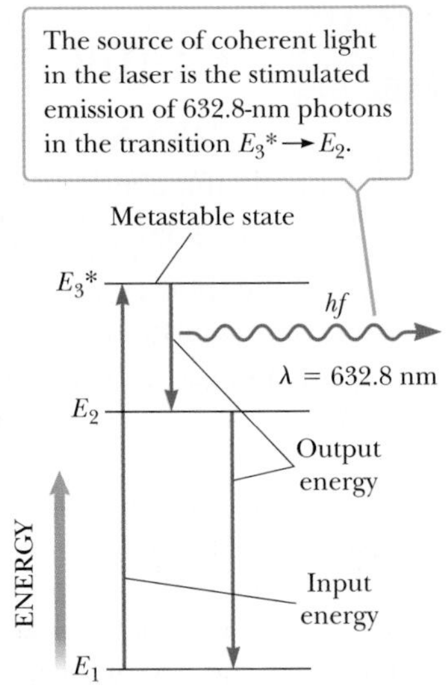

Figure 28.20 Energy level diagram for the neon atom in a helium–neon laser.

is the case, stimulated emission will occur before spontaneous emission. Finally, the photons produced must be retained in the system for a while so that they can stimulate the production of still more photons. This step can be done with mirrors, one of which is partly transparent.

Figure 28.20 is an energy level diagram for the neon atom in a helium–neon gas laser. The mixture of helium and neon is confined to a glass tube sealed at the ends by mirrors. A high voltage applied to the tube causes electrons to sweep through it, colliding with the atoms of the gas and raising them into excited states. Neon atoms are excited to state E_3^* through this process and also as a result of collisions with excited helium atoms. When a neon atom makes a transition to state E_2, it stimulates emission by neighboring excited atoms. The result is the production of coherent light at a wavelength of 632.8 nm. Figure 28.21 summarizes the steps in the production of a laser beam.

APPLICATION
Laser Technology

Lasers that cover wavelengths in the infrared, visible, and ultraviolet regions of the spectrum are now available. Applications include the surgical "welding" of detached retinas, "lasik" surgery, precision surveying and length measurement, a potential source for inducing nuclear fusion reactions, precision cutting of metals and other materials, and telephone communication along optical fibers.

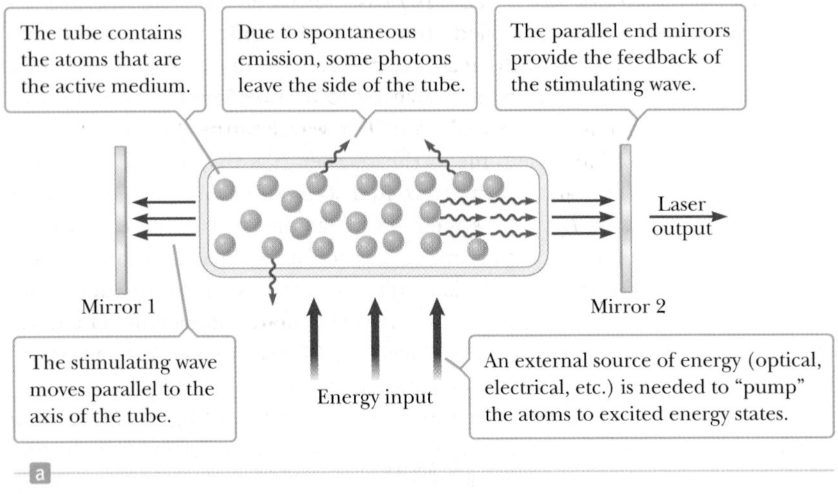

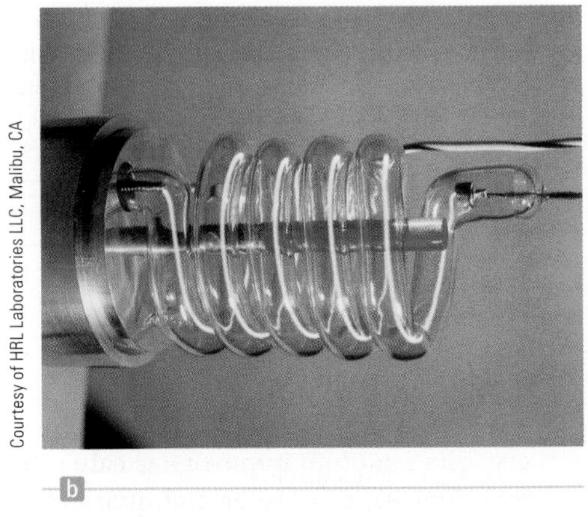

Figure 28.21 (a) Steps in the production of a laser beam. (b) Photograph of the first ruby laser, showing the flash lamp surrounding the ruby rod.

SUMMARY

28.3 The Bohr Model

The **Bohr model** of the atom is successful in describing the spectra of atomic hydrogen and hydrogen-like ions. One basic assumption of the model is that the electron can exist only in certain orbits such that its angular momentum mvr is an integral multiple of $\hbar$, where $\hbar$ is Planck's constant divided by 2π. Assuming circular orbits and a Coulomb force of attraction between electron and proton, the energies of the quantum states for hydrogen are

$$E_n = -\frac{m_e k_e^2 e^4}{2\hbar^2}\left(\frac{1}{n^2}\right) \qquad n = 1, 2, 3, \ldots \qquad [28.13]$$

where k_e is the Coulomb constant, e is the charge on the electron, and n is an integer called a **quantum number**.

If the electron in the hydrogen atom jumps from an orbit having quantum number n_i to an orbit having quantum number n_f, it emits a photon of frequency f, given by

$$f = \frac{E_i - E_f}{h} = \frac{m_e k_e^2 e^4}{4\pi\hbar^3}\left(\frac{1}{n_f^2} - \frac{1}{n_i^2}\right) \qquad [28.15]$$

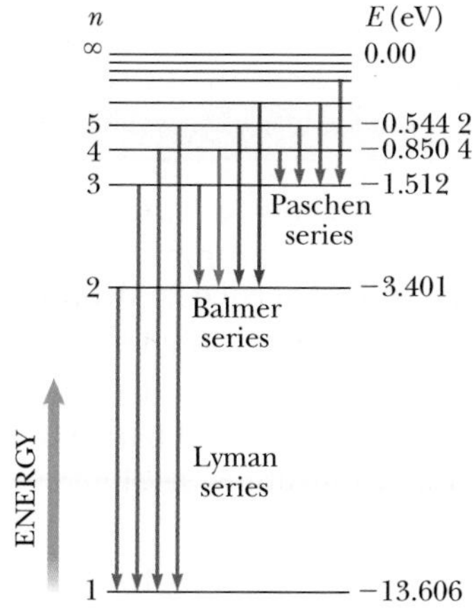

Energy level diagram for hydrogen. Vertical arrows represent the lowest-energy transitions for each of the spectral series shown.

Bohr's **correspondence principle** states that quantum mechanics is in agreement with classical physics when the quantum numbers for a system are very large.

The Bohr theory can be generalized to hydrogen-like atoms, such as singly ionized helium or doubly ionized lithium. This modification consists of replacing e^2 by Ze^2 wherever it occurs.

28.4 Quantum Mechanics and the Hydrogen Atom

One of the many successes of quantum mechanics is that the quantum numbers n, ℓ, and m_ℓ associated with atomic structure arise directly from the mathematics of the theory. The quantum number n is called the **principal quantum number**, ℓ is the **orbital quantum number**, and m_ℓ is the **orbital magnetic quantum number.** These quantum numbers can take only certain values: $1 \le n < \infty$ in integer steps, $0 \le \ell \le n - 1$, and $-\ell \le m_\ell \le \ell$. In addition, a fourth quantum number, called the **spin magnetic quantum number** m_s, is needed to explain a fine doubling of lines in atomic spectra, with $m_s = \pm\frac{1}{2}$.

28.5 The Exclusion Principle and the Periodic Table

An understanding of the periodic table of the elements became possible when Pauli formulated the **exclusion principle**, which states that no two electrons in the same atom can have the same values for the set of quantum numbers n, ℓ, m_ℓ, and m_s. A particular set of these quantum numbers is called a quantum state. The exclusion principle explains how different energy levels in atoms are populated. Once one subshell is filled, the next electron goes into the vacant subshell that is lowest in energy. Atoms with similar configurations in their outermost shell have similar chemical properties and are found in the same column of the periodic table.

28.6 Characteristic X-Rays

Characteristic x-rays are produced when a bombarding electron collides with an electron in an inner shell of an atom with sufficient energy to remove the electron from the atom. The vacancy is filled when an electron from a higher level drops down into the level containing the vacancy, emitting a photon in the x-ray part of the spectrum in the process.

28.7 Atomic Transitions and Lasers

When an atom is irradiated by light of all different wavelengths, it will only absorb wavelengths equal to the difference in energy of two of its energy levels. This phenomenon, called **stimulated absorption**, places an atom's electrons into **excited states.** Atoms in an excited state have a probability of returning to a lower level of excitation by **spontaneous emission**. The wavelengths that can be emitted are the same as the wavelengths that can be absorbed. If an atom is in an excited state and a photon with energy $hf = E_2 - E_1$ is incident on it, the probability of emission of a second photon of this energy is greatly enhanced. The emitted photon is exactly in phase with the incident photon. This process is called **stimulated emission**. The emitted and original photon can then stimulate more emission, creating an amplifying effect.

Lasers are monochromatic, coherent light sources that work on the principle of **stimulated emission** of radiation from a system of atoms.

WARM-UP EXERCISES

ENHANCED WebAssign The warm-up exercises in this chapter may be assigned online in Enhanced WebAssign.

1. **Physics Review** Determine the (a) energy in joules and (b) wavelength in nm of a photon with energy of 4.05 eV. (See Section 27.2.)
2. Determine the energies in eV of the (a) third and (b) fourth energy levels of the hydrogen atom. (See Section 28.3.)
3. An electron in the $n = 5$ energy level of hydrogen undergoes a transition to the $n = 3$ energy level. Determine (a) the energy in eV, (b) the energy in joules, and (c) the frequency of the emitted photon. (See Section 28.3.)
4. The so-called Lyman-α photon is the lowest energy photon in the Lyman series of hydrogen and results from an electron transitioning from the $n = 2$ to the $n = 1$ energy level. Determine (a) the energy in eV, (b) in joules, and (c) the wavelength in nm of the Lyman-α line. (See Section 28.3.)
5. Determine the orbital radius in nm of an electron in hydrogen's (a) $n = 2$, and (c) $n = 4$ energy levels. (See Section 28.3.)
6. Singly-ionized helium (He^+) is a hydrogen-like atom. Determine the energy in eV required to raise a He^+ electron from the $n = 1$ to the $n = 2$ energy level. (See Section 28.3.)
7. Hydrogen's single electron can occupy any of the atom's distinct quantum states. Determine the number of distinct quantum states in the (a) $n = 1$, (b) $n = 2$, and (c) $n = 3$ energy levels. (See Section 28.4.)
8. For an electron in a $3d$ state, determine (a) the principle quantum number and (b) the orbital quantum number. (c) How many different magnetic quantum numbers are possible for electrons in that state? (d) What total number of electrons could occupy that state? (See Section 28.4.)
9. (a) Identify the number of electrons in the ground-state outer shell of atomic oxygen (atomic number 8). (b) How many electrons are in the ground-state outer shell of aluminum? (See Section 28.5 and Table 28.4.)
10. Estimate the energy in eV of a K shell electron in lead ($Z = 82$) (See Section 28.7.)

CONCEPTUAL QUESTIONS

ENHANCED WebAssign The conceptual questions in this chapter may be assigned online in Enhanced WebAssign.

1. In the hydrogen atom, the quantum number n can increase without limit. Because of this fact, does the frequency of possible spectral lines from hydrogen also increase without limit?
2. Does the light emitted by a neon sign constitute a continuous spectrum or only a few colors? Defend your answer.
3. In an x-ray tube, if the energy with which the electrons strike the metal target is increased, the wavelengths of the characteristic x-rays do not change. Why not?
4. An energy of about 21 eV is required to excite an electron in a helium atom from the $1s$ state to the $2s$ state. The same transition for the He^+ ion requires approximately twice as much energy. Explain.
5. Is it possible for a spectrum from an x-ray tube to show the continuous spectrum of x-rays without the presence of the characteristic x-rays?
6. Suppose the electron in the hydrogen atom obeyed classical mechanics rather than quantum mechanics. Why should such a hypothetical atom emit a continuous spectrum rather than the observed line spectrum?
7. When a hologram is produced, why must the system (including light source, object, beam splitter, and so on) be held motionless within a quarter of the light's wavelength?
8. Why are three quantum numbers needed to describe the state of a one-electron atom (ignoring spin)?
9. Describe how the structure of atoms would differ if the Pauli exclusion principle were not valid. What consequences would follow, both at the atomic level and in the world at large?
10. Can the electron in the ground state of hydrogen absorb a photon of energy less than 13.6 eV? Can it absorb a photon of energy greater than 13.6 eV? Explain.
11. Why do lithium, potassium, and sodium exhibit similar chemical properties?
12. List some ways in which quantum mechanics altered our view of the atom pictured by the Bohr theory.
13. It is easy to understand how two electrons (one with spin up, one with spin down) can fill the $1s$ shell for a helium atom. How is it possible that eight more electrons can fit into the $2s$, $2p$ level to complete the $1s^22s^22p^6$ shell for a neon atom?
14. The ionization energies for Li, Na, K, Rb, and Cs are 5.390, 5.138, 4.339, 4.176, and 3.893 eV, respectively. Explain why these values are to be expected in terms of the atomic structures.
15. Why is stimulated emission so important in the operation of a laser?

PROBLEMS

ENHANCED WebAssign The problems in this chapter may be assigned online in Enhanced WebAssign.

1. denotes straightforward problem; 2. denotes intermediate problem;
3. denotes challenging problem

[1.] denotes full solution available in *Student Solutions Manual/ Study Guide*

1. denotes problems most often assigned in Enhanced WebAssign

BIO denotes biomedical problems

GP denotes guided problems

M denotes Master It tutorial available in Enhanced WebAssign

Q|C denotes asking for quantitative and conceptual reasoning

S denotes symbolic reasoning problem

W denotes Watch It video solution available in Enhanced WebAssign

28.1 Early Models of the Atom

28.2 Atomic Spectra

1. Q|C The wavelengths of the Lyman series for hydrogen are given by

$$\frac{1}{\lambda} = R_H\left(1 - \frac{1}{n^2}\right) \qquad n = 2, 3, 4, \ldots$$

(a) Calculate the wavelengths of the first three lines in this series. (b) Identify the region of the electromagnetic spectrum in which these lines appear.

2. Q|C The wavelengths of the Paschen series for hydrogen are given by

$$\frac{1}{\lambda} = R_H\left(\frac{1}{3^2} - \frac{1}{n^2}\right) \qquad n = 4, 5, 6, \ldots$$

(a) Calculate the wavelengths of the first three lines in this series. (b) Identify the region of the electromagnetic spectrum in which these lines appear.

3. The "size" of the *atom* in Rutherford's model is about 1.0×10^{-10} m. (a) Determine the attractive electrostatic force between an electron and a proton separated by this distance. (b) Determine (in eV) the electrostatic potential energy of the atom.

4. An isolated atom of a certain element emits light of wavelength 520 nm when the atom falls from its fifth excited state into its second excited state. The atom emits a photon of wavelength 410 nm when it drops from its sixth excited state into its second excited state. Find the wavelength of the light radiated when the atom makes a transition from its sixth to its fifth excited state.

5. Q|C The "size" of the atom in Rutherford's model is about 1.0×10^{-10} m. (a) Determine the speed of an electron moving about the proton using the attractive electrostatic force between an electron and a proton separated by this distance. (b) Does this speed suggest that Einsteinian relativity must be considered in studying the atom? (c) Compute the de Broglie wavelength of the electron as it moves about the proton. (d) Does this wavelength suggest that wave effects, such as diffraction and interference, must be considered in studying the atom?

6. W In a Rutherford scattering experiment, an α-particle (charge $= +2e$) heads directly toward a gold nucleus (charge $= +79e$). The α-particle had a kinetic energy of 5.0 MeV when very far ($r \to \infty$) from the nucleus. Assuming the gold nucleus to be fixed in space, determine the distance of closest approach. *Hint:* Use conservation of energy with $PE = k_e q_1 q_2 / r$.

28.3 The Bohr Model

7. M A hydrogen atom is in its first excited state ($n = 2$). Using the Bohr theory of the atom, calculate (a) the radius of the orbit, (b) the linear momentum of the electron, (c) the angular momentum of the electron, (d) the kinetic energy, (e) the potential energy, and (f) the total energy.

8. W For a hydrogen atom in its ground state, use the Bohr model to compute (a) the orbital speed of the electron, (b) the kinetic energy of the electron, and (c) the electrical potential energy of the atom.

9. S Show that the speed of the electron in the nth Bohr orbit in hydrogen is given by

$$v_n = \frac{k_e e^2}{n\hbar}$$

10. A photon is emitted when a hydrogen atom undergoes a transition from the $n = 5$ state to the $n = 3$ state. Calculate (a) the wavelength, (b) the frequency, and (c) the energy (in electron volts) of the emitted photon.

11. A hydrogen atom emits a photon of wavelength 656 nm. From what energy orbit to what lower-energy orbit did the electron jump?

12. Following are four possible transitions for a hydrogen atom

I. $n_i = 2;\ n_f = 5$ II. $n_i = 5;\ n_f = 3$

III. $n_i = 7;\ n_f = 4$ IV. $n_i = 4;\ n_f = 7$

(a) Which transition will emit the shortest-wavelength photon? (b) For which transition will the atom gain the most energy? (c) For which transition(s) does the atom lose energy?

13. M What is the energy of a photon that, when absorbed by a hydrogen atom, could cause an electronic transition from (a) the $n = 2$ state to the $n = 5$ state and (b) the $n = 4$ state to the $n = 6$ state?

14. W A hydrogen atom initially in its ground state ($n = 1$) absorbs a photon and ends up in the state for which

$n = 3$. (a) What is the energy of the absorbed photon? (b) If the atom eventually returns to the ground state, what photon energies could the atom emit?

15. The Balmer series for the hydrogen atom corresponds to electronic transitions that terminate in the state with quantum number $n = 2$ as shown in Figure P28.15. Consider the photon of longest wavelength corresponding to a transition shown in the figure. Determine (a) its energy and (b) its wavelength. Consider the spectral line of shortest wavelength corresponding to a transition shown in the figure. Find (c) its photon energy and (d) its wavelength. (e) What is the shortest possible wavelength in the Balmer series?

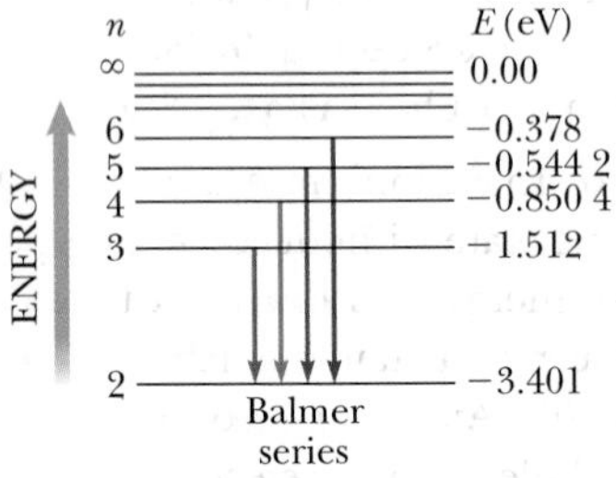

Figure P28.15

16. **S** A particle of charge q and mass m, moving with a constant speed v, perpendicular to a constant magnetic field B, follows a circular path. If in this case the angular momentum about the center of this circle is quantized so that $mvr = 2n\hbar$, show that the allowed radii for the particle are

$$r_n = \sqrt{\frac{2n\hbar}{qB}}$$

where $n = 1, 2, 3, \ldots$.

17. **M** (a) If an electron makes a transition from the $n = 4$ Bohr orbit to the $n = 2$ orbit, determine the wavelength of the photon created in the process. (b) Assuming that the atom was initially at rest, determine the recoil speed of the hydrogen atom when this photon is emitted.

18. Consider a large number of hydrogen atoms, with electrons all initially in the $n = 4$ state. (a) How many different wavelengths would be observed in the emission spectrum of these atoms? (b) What is the longest wavelength that could be observed? (c) To which series does the wavelength found in (b) belong?

19. A photon with energy 2.28 eV is absorbed by a hydrogen atom. Find (a) the minimum n for a hydrogen atom that can be ionized by such a photon and (b) the speed of the electron released from the state in part (a) when it is far from the nucleus.

20. (a) Calculate the angular momentum of the Moon due to its orbital motion about Earth. In your calculation use 3.84×10^8 m as the average Earth–Moon distance and 2.36×10^6 s as the period of the Moon in its orbit. (b) If the angular momentum of the Moon obeys Bohr's quantization rule ($L = n\hbar$), determine the value of the quantum number n. (c) By what fraction would the Earth–Moon radius have to be increased to increase the quantum number by 1?

21. An electron is in the second excited orbit of hydrogen, corresponding to $n = 3$. Find (a) the radius of the orbit and (b) the wavelength of the electron in this orbit.

22. **S** (a) Write an expression relating the kinetic energy KE of the electron and the potential energy PE in the Bohr model of the hydrogen atom. (b) Suppose a hydrogen atom absorbs a photon of energy E, resulting in the transfer of the electron to a higher-energy level. Express the resulting change in the potential energy of the system in terms of E. (c) What is the change in the electron's kinetic energy during this process?

23. The orbital radii of a hydrogen-like atom is given by the equation

$$r_n = \frac{n^2\hbar^2}{Zm_e k_e e^2}$$

What is the radius of the first Bohr orbit in (a) He^+, (b) Li^{2+}, and (c) Be^{3+}?

24. **GP** Consider a Bohr model of doubly ionized lithium. (a) Write an expression similar to Equation 28.14 for the energy levels of the sole remaining electron. (b) Find the energy corresponding to $n = 4$. (c) Find the energy corresponding to $n = 2$. (d) Calculate the energy of the photon emitted when the electron transits from the fourth energy level to the second energy level. Express the answer both in electron volts and in joules. (e) Find the frequency and wavelength of the emitted photon. (f) In what part of the spectrum is the emitted light?

25. A general expression for the energy levels of one-electron atoms and ions is

$$E_n = -\frac{\mu k_e^2 q_1^2 q_2^2}{2\hbar^2 n^2}$$

Here μ is the reduced mass of the atom, given by $\mu = m_1 m_2/(m_1 + m_2)$, where m_1 is the mass of the electron and m_2 is the mass of the nucleus; k_e is the Coulomb constant; and q_1 and q_2 are the charges of the electron and the nucleus, respectively. The wavelength for the $n = 3$ to $n = 2$ transition of the hydrogen atom is 656.3 nm (visible red light). What are the wavelengths for this same transition in (a) positronium, which consists of an electron and a positron, and (b) singly ionized helium? *Note:* A positron is a positively charged electron.

26. **S** Using the concept of standing waves, de Broglie was able to derive Bohr's stationary orbit postulate. He assumed a confined electron could exist only in states where its de Broglie waves form standing-wave patterns, as in Figure 28.6. Consider a particle confined in a box of length L to be equivalent to a string of length L and

fixed at both ends. Apply de Broglie's concept to show that (a) the linear momentum of this particle is quantized with $p = mv = nh/2L$ and (b) the allowed states correspond to particle energies of $E_n = n^2E_0$, where $E_0 = h^2/(8mL^2)$.

28.4 Quantum Mechanics and the Hydrogen Atom

27. List the possible sets of quantum numbers for electrons in the $3d$ subshell.

28. When the principal quantum number is $n = 4$, how many different values of (a) ℓ and (b) m_ℓ are possible?

29. The ρ-meson has a charge of $-e$, a spin quantum number of 1, and a mass 1 507 times that of the electron. If the electrons in atoms were replaced by ρ-mesons, list the possible sets of quantum numbers for ρ-mesons in the $3d$ subshell.

28.5 The Exclusion Principle and the Periodic Table

30. W (a) Write out the electronic configuration of the ground state for nitrogen ($Z = 7$). (b) Write out the values for the possible set of quantum numbers n, ℓ, m_ℓ, and m_s for the electrons in nitrogen.

31. A certain element has its outermost electron in a $3p$ subshell. It has valence +3 because it has three more electrons than a certain noble gas. What element is it?

32. Two electrons in the same atom have $n = 3$ and $\ell = 1$. (a) List the quantum numbers for the possible states of the atom. (b) How many states would be possible if the exclusion principle did not apply to the atom?

33. Zirconium ($Z = 40$) has two electrons in an incomplete d subshell. (a) What are the values of n and ℓ for each electron? (b) What are all possible values of m_ℓ and m_s? (c) What is the electron configuration in the ground state of zirconium?

28.6 Characteristic X-Rays

34. A tungsten target is struck by electrons that have been accelerated from rest through a 40.0-kV potential difference. Find the shortest wavelength of the radiation emitted.

35. A bismuth target is struck by electrons, and x-rays are emitted. Estimate (a) the M- to L-shell transitional energy for bismuth and (b) the wavelength of the x-ray emitted when an electron falls from the M shell to the L shell.

36. When an electron drops from the M shell ($n = 3$) to a vacancy in the K shell ($n = 1$), the measured wavelength of the emitted x-ray is found to be 0.101 nm. Identify the element.

37. M The K series of the discrete spectrum of tungsten contains wavelengths of 0.018 5 nm, 0.020 9 nm, and 0.021 5 nm. The K-shell ionization energy is 69.5 keV. Determine the ionization energies of the L, M, and N shells.

Additional Problems

38. In a hydrogen atom, what is the principal quantum number of the electron orbit with a radius closest to 1.0 μm?

39. (a) How much energy is required to cause an electron in hydrogen to move from the $n = 1$ state to the $n = 2$ state? (b) If the electrons gain this energy by collision between hydrogen atoms in a high-temperature gas, find the minimum temperature of the heated hydrogen gas. The thermal energy of the heated atoms is given by $3k_BT/2$, where k_B is the Boltzmann constant.

40. A pulsed ruby laser emits light at 694.3 nm. For a 14.0-ps pulse containing 3.00 J of energy, find (a) the physical length of the pulse as it travels through space and (b) the number of photons in it. (c) If the beam has a circular cross section 0.600 cm in diameter, what is the number of photons per cubic millimeter?

41. An electron in chromium moves from the $n = 2$ state to the $n = 1$ state without emitting a photon. Instead, the excess energy is transferred to an outer electron (one in the $n = 4$ state), which is then ejected by the atom. In this Auger (pronounced "ohjay") process, the ejected electron is referred to as an Auger electron. (a) Find the change in energy associated with the transition from $n = 2$ into the vacant $n = 1$ state using Bohr theory. Assume only one electron in the K shell is shielding part of the nuclear charge. (b) Find the energy needed to ionize an $n = 4$ electron, assuming 22 electrons shield the nucleus. (c) Find the kinetic energy of the ejected (Auger) electron. (All answers should be in electron volts.)

42. QIC S As the Earth moves around the Sun, its orbits are quantized. (a) Follow the steps of Bohr's analysis of the hydrogen atom to show that the allowed radii of the Earth's orbit are given by

$$r_n = \frac{n^2\hbar^2}{GM_SM_E^2}$$

where n is an integer quantum number, M_S is the mass of the Sun, and M_E is the mass of the Earth. (b) Calculate the numerical value of n for the Sun–Earth system. (c) Find the distance between the orbit for quantum number n and the next orbit out from the Sun corresponding to the quantum number $n + 1$. (d) Discuss the significance of your results from parts (b) and (c).

43. BIO M A laser used in eye surgery emits a 3.00-mJ pulse in 1.00 ns, focused to a spot 30.0 μm in diameter on the retina. (a) Find (in SI units) the power per unit area at the retina. (This quantity is called the *irradiance*.) (b) What energy is delivered per pulse to an area of molecular size (say, a circular area 0.600 nm in diameter)?

44. An electron has a de Broglie wavelength equal to the diameter of a hydrogen atom in its ground state. (a) What is the kinetic energy of the electron? (b) How does this energy compare with the magnitude of the ground-state energy of the hydrogen atom?

45. **S** Use Bohr's model of the hydrogen atom to show that when the atom makes a transition from the state n to the state $n - 1$, the frequency of the emitted light is given by

$$f = \frac{2\pi^2 m k_e^2 e^4}{h^3}\left[\frac{2n - 1}{(n - 1)^2 n^2}\right]$$

46. Suppose the ionization energy of an atom is 4.100 eV. In this same atom, we observe emission lines that have wavelengths of 310.0 nm, 400.0 nm, and 1 378 nm. Use this information to construct the energy level diagram with the least number of levels. Assume that the higher energy levels are closer together.

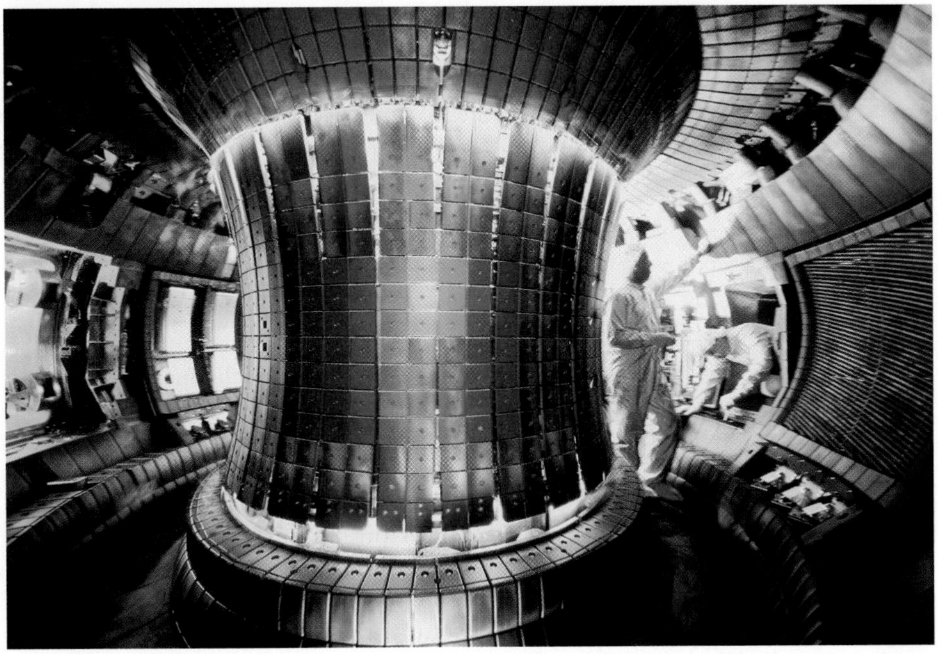

Peter Ginter/Science Faction/Getty Images

Technicians prepare the vacuum chamber of the ASDEX Upgrade Fusion Reactor (Axially Symmetric Divertor Experiment), where plasma is heated to over sixty million degrees, leading to the fusion of deuterium and tritium and the release of energy. Such experimental tokamaks may lead to working fusion reactors.

Nuclear Physics 29

In this chapter we discuss the properties and structure of the atomic nucleus. We start by describing the basic properties of nuclei and follow with a discussion of the phenomenon of radioactivity. Finally, we explore nuclear reactions and the various processes by which nuclei decay.

29.1 Some Properties of Nuclei

LEARNING OBJECTIVES

1. Define a nuclei's atomic, neutron, and mass numbers and describe the notation used to represent them.
2. Define the unified mass unit and the term isotope.
3. Discuss the size of nuclei and the concept of nuclear stability.

All nuclei are composed of two types of particles: protons and neutrons. The only exception is the ordinary hydrogen nucleus, which is a single proton. In describing some of the properties of nuclei, such as their charge, mass, and radius, we make use of the following quantities:

- the **atomic number** Z, which equals the number of protons in the nucleus
- the **neutron number** N, which equals the number of neutrons in the nucleus
- the **mass number** A, which equals the number of nucleons in the nucleus (*nucleon* is a generic term used to refer to either a proton or a neutron)

The symbol we use to represent nuclei is ${}^{A}_{Z}\text{X}$, where X represents the chemical symbol for the element. For example, ${}^{27}_{13}\text{Al}$ has the mass number 27 and the atomic number 13; therefore, it contains 13 protons and 14 neutrons. When no confusion

Ernest Rutherford
New Zealand Physicist (1871–1937)
Rutherford was awarded the 1908 Nobel Prize in Chemistry for studying radioactivity and for discovering that atoms can be broken apart by alpha rays. "On consideration, I realized that this scattering backward must be the result of a single collision, and when I made calculations I saw that it was impossible to get anything of that order of magnitude unless you took a system in which the greater part of the mass of the atom was concentrated in a minute nucleus. It was then that I had the idea of an atom with a minute massive center carrying a charge."

Table 29.1 Masses of the Proton, Neutron, and Electron in Various Units

	Mass		
Particle	**kg**	**u**	**MeV/c^2**
Proton	$1.672\,6 \times 10^{-27}$	1.007 276	938.28
Neutron	$1.675\,0 \times 10^{-27}$	1.008 665	939.57
Electron	9.109×10^{-31}	5.486×10^{-4}	0.511

is likely to arise, we often omit the subscript Z because the chemical symbol can always be used to determine Z.

The nuclei of all atoms of a particular element must contain the same number of protons, but they may contain different numbers of neutrons. Nuclei that are related in this way are called **isotopes. The isotopes of an element have the same *Z* value, but different *N* and *A* values.** The natural abundances of isotopes can differ substantially. For example, ${}^{11}_{6}\text{C}$, ${}^{12}_{6}\text{C}$, ${}^{13}_{6}\text{C}$, and ${}^{14}_{6}\text{C}$ are four isotopes of carbon. The natural abundance of the ${}^{12}_{6}\text{C}$ isotope is about 98.9%, whereas that of the ${}^{13}_{6}\text{C}$ isotope is only about 1.1%. Some isotopes don't occur naturally, but can be produced in the laboratory through nuclear reactions. Even the simplest element, hydrogen, has isotopes: ${}^{1}_{1}\text{H}$, hydrogen; ${}^{2}_{1}\text{H}$, deuterium; and ${}^{3}_{1}\text{H}$, tritium.

Charge and Mass

The proton carries a single positive charge $+e = 1.602\,177\,33 \times 10^{-19}$ C, the electron carries a single negative charge $-e$, and the neutron is electrically neutral. Because the neutron has no charge, it's difficult to detect. The proton is about 1 836 times as massive as the electron, and the masses of the proton and the neutron are almost equal (Table 29.1).

Definition of the unified mass unit, u ▶

For atomic masses, it is convenient to define the **unified mass unit, u,** in such a way that the mass of one atom of the isotope ${}^{12}\text{C}$ is exactly 12 u, where 1 u = $1.660\,559 \times 10^{-27}$ kg. The proton and neutron each have a mass of about 1 u, and the electron has a mass that is only a small fraction of an atomic mass unit.

Because the rest energy of a particle is given by $E_R = mc^2$, it is often convenient to express the particle's mass in terms of its energy equivalent. For one atomic mass unit, we have an energy equivalent of

$$E_R = mc^2 = (1.660\,559 \times 10^{-27}\ \text{kg})(2.997\,92 \times 10^8\ \text{m/s})^2$$
$$= 1.492\,431 \times 10^{-10}\ \text{J} = 931.494\ \text{MeV}$$

In calculations nuclear physicists often express *mass* in terms of the unit MeV/c^2, where

$$1\ \text{u} = 931.494\ \text{MeV}/c^2$$

The Size of Nuclei

The size and structure of nuclei were first investigated in the scattering experiments of Rutherford, discussed in Section 28.1. Using the principle of conservation of energy, Rutherford found an expression for how close an alpha particle moving directly toward the nucleus can come to the nucleus before being turned around by Coulomb repulsion.

In such a head-on collision, the kinetic energy of the incoming alpha particle must be converted completely to electrical potential energy when the particle stops at the point of closest approach and turns around (Fig. 29.1). If we equate the initial kinetic energy of the alpha particle to the maximum electrical potential energy of the system (alpha particle plus target nucleus), we have

$$\tfrac{1}{2}mv^2 = k_e\frac{q_1q_2}{r} = k_e\frac{(2e)(Ze)}{d}$$

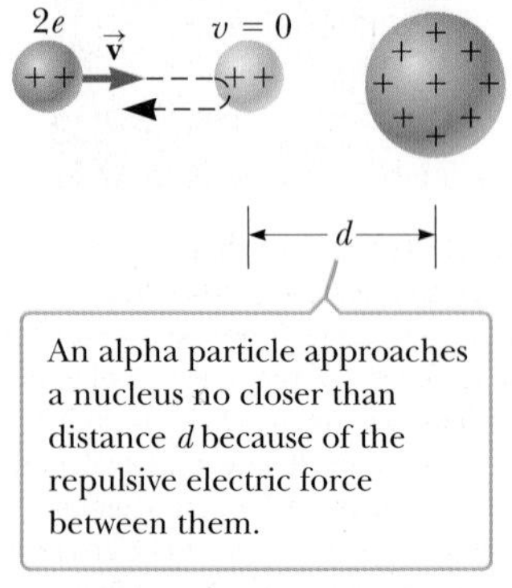

Figure 29.1 An alpha particle on a head-on collision course with a nucleus of charge *Ze*.

If a star of mass $2 \times 1.99 \times 10^{30}$ kg collapsed into neutrons ($m_n = 1.67 \times 10^{-27}$ kg), what would its radius be? Assume $r = r_0 A^{1/3}$.

29.2 Binding Energy

9. Find the average binding energy per nucleon of (a) $^{24}_{12}$Mg and (b) $^{85}_{37}$Rb.

10. Calculate the binding energy per nucleon for (a) ^{2}H, (b) ^{4}He, (c) ^{56}Fe, and (d) ^{238}U.

11. A pair of nuclei for which $Z_1 = N_2$ and $Z_2 = N_1$ are called *mirror isobars*. (The atomic and neutron numbers are interchangeable.) Binding-energy measurements on such pairs can be used to obtain evidence of the charge independence of nuclear forces. Charge independence means that the proton–proton, proton–neutron, and neutron–neutron forces are approximately equal. Calculate the difference in binding energy for the two mirror nuclei $^{15}_{8}$O and $^{15}_{7}$N.

12. W The peak of the stability curve occurs at ^{56}Fe, which is why iron is prominent in the spectrum of the Sun and stars. Show that ^{56}Fe has a higher binding energy per nucleon than its neighbors ^{55}Mn and ^{59}Co. Compare your results with Figure 29.4.

13. Q|C Two nuclei having the same mass number are known as *isobars*. (a) Calculate the difference in binding energy per nucleon for the isobars $^{23}_{11}$Na and $^{23}_{12}$Mg. (b) How do you account for this difference? (The mass of $^{23}_{12}$Mg = 22.994 127 u.)

14. Calculate the binding energy of the last neutron in the $^{43}_{20}$Ca nucleus. *Hint:* You should compare the mass of $^{43}_{20}$Ca with the mass of $^{42}_{20}$Ca plus the mass of a neutron. The mass of $^{42}_{20}$Ca = 41.958 622 u, whereas the mass of $^{43}_{20}$Ca = 42.958 770 u.

29.3 Radioactivity

15. Radon gas has a half-life of 3.83 days. If 3.00 g of radon gas is present at time $t = 0$, what mass of radon will remain after 1.50 days have passed?

16. BIO A drug tagged with $^{99}_{43}$Tc (half-life = 6.05 h) is prepared for a patient. If the original activity of the sample was 1.1×10^4 Bq, what is its activity after it has been on the shelf for 2.0 h?

17. GP The half-life of ^{131}I is 8.04 days. (a) Convert the half-life to seconds. (b) Calculate the decay constant for this isotope. (c) Convert 0.500 μCi to the SI unit the becquerel. (d) Find the number of ^{131}I nuclei necessary to produce a sample with an activity of 0.500 μCi. (e) Suppose the activity of a certain ^{131}I sample is 6.40 mCi at a given time. Find the number of half-lives the sample goes through in 40.2 d and the activity at the end of that period.

18. Q|C Tritium has a half-life of 12.33 years. What fraction of the nuclei in a tritium sample will remain (a) after 5.00 yr? (b) After 10.0 yr? (c) After 123.3 yr? (d) According to Equation 29.4a, an infinite amount of time is required for the entire sample to decay. Discuss whether that is realistic.

19. After 2.00 days, the activity of a sample of an unknown type radioactive material has decreased to 84.2% of the initial activity. What is the half-life of this material?

20. BIO W After a plant or animal dies, its ^{14}C content decreases with a half-life of 5 730 yr. If an archaeologist finds an ancient fire pit containing partially consumed firewood and the ^{14}C content of the wood is only 12.5% that of an equal carbon sample from a present-day tree, what is the age of the ancient site?

21. M A freshly prepared sample of a certain radioactive isotope has an activity of 10.0 mCi. After 4.00 h, the activity is 8.00 mCi. (a) Find the decay constant and half-life of the isotope. (b) How many atoms of the isotope were contained in the freshly prepared sample? (c) What is the sample's activity 30 h after it is prepared?

22. A building has become accidentally contaminated with radioactivity. The longest-lived material in the building is strontium-90. (The atomic mass of $^{90}_{38}$Sr is 89.907 7 u.) If the building initially contained 5.0 kg of this substance and the safe level is less than 10.0 counts/min, how long will the building be unsafe?

29.4 The Decay Processes

23. Identify the missing nuclides in the following decays:
(a) $^{212}_{83}\text{Bi} \rightarrow ? + ^{4}_{2}\text{He}$
(b) $^{95}_{36}\text{Kr} \rightarrow ? + e^- + \bar{\nu}$
(c) $? \rightarrow ^{4}_{2}\text{He} + ^{140}_{58}\text{Ce}$

24. Complete the following radioactive decay formulas:
(a) $^{12}_{5}\text{B} \rightarrow ? + e^- + \bar{\nu}$
(b) $^{234}_{90}\text{Th} \rightarrow ^{230}_{88}\text{Ra} + ?$
(c) $? \rightarrow ^{14}_{7}\text{N} + e^- + \bar{\nu}$

25. The mass of ^{56}Fe is 55.934 9 u, and the mass of ^{56}Co is 55.939 9 u. Which isotope decays into the other and by what process?

26. Find the energy released in the alpha decay of $^{238}_{92}$U. The following mass value will be useful: $^{234}_{90}$Th has a mass of 234.043 583 u.

27. Determine which of the following suggested decays can occur spontaneously:
(a) $^{40}_{20}\text{Ca} \rightarrow e^+ + ^{40}_{19}\text{K}$ (b) $^{144}_{60}\text{Nd} \rightarrow ^{4}_{2}\text{He} + ^{140}_{58}\text{Ce}$

28. W $^{66}_{28}$Ni (mass = 65.929 1 u) undergoes beta decay to $^{66}_{29}$Cu (mass = 65.928 9 u). (a) Write the complete decay formula for this process. (b) Find the maximum kinetic energy of the emerging electrons.

29. A ^{3}H (tritium) nucleus beta decays into ^{3}He by creating an electron and an antineutrino according to the reaction

$$^{3}_{1}\text{H} \rightarrow ^{3}_{2}\text{He} + e^- + \bar{\nu}$$

Use Appendix B to determine the total energy released in this reaction.

30. A piece of charcoal used for cooking is found at the remains of an ancient campsite. A 1.00-kg sample of carbon from the wood has an activity equal to 5.00×10^2 decays per minute. Find the age of the charcoal. *Hint:* Living material has an activity equal to 15.0 decays/minute per gram of carbon present.

31. M A wooden artifact is found in an ancient tomb. Its carbon-14 ($^{14}_{6}C$) activity is measured to be 60.0% of that in a fresh sample of wood from the same region. Assuming the same amount of ^{14}C was initially present in the artifact as is now contained in the fresh sample, determine the age of the artifact.

29.6 Nuclear Reactions

32. A beam of 6.61-MeV protons is incident on a target of $^{27}_{13}Al$. Those protons that collide with the target produce the reaction

$$p + {}^{27}_{13}Al \rightarrow {}^{27}_{14}Si + n$$

($^{27}_{14}Si$ has a mass of 26.986 721 u.) Neglecting any recoil of the product nucleus, determine the kinetic energy of the emerging neutrons.

33. Identify the unknown particles X and X′ in the following nuclear reactions:
(a) $X + {}^{4}_{2}He \rightarrow {}^{24}_{12}Mg + {}^{1}_{0}n$
(b) ${}^{235}_{92}U + {}^{1}_{0}n \rightarrow {}^{90}_{38}Sr + X + 2{}^{1}_{0}n$
(c) $2{}^{1}_{1}H \rightarrow {}^{2}_{1}H + X + X'$

34. GP One method of producing neutrons for experimental use is to bombard $^{7}_{3}Li$ with protons. The neutrons are emitted according to the reaction

$${}^{1}_{1}H + {}^{7}_{3}Li \rightarrow {}^{7}_{4}Be + {}^{1}_{0}n$$

(a) Calculate the mass in atomic mass units of the particles on the left side of the equation. (b) Calculate the mass (in atomic mass units) of the particles on the right side of the equation. (c) Subtract the answer for part (b) from that for part (a) and convert the result to mega electron volts, obtaining the Q value for this reaction. (d) Assuming lithium is initially at rest, the proton is moving at velocity v, and the resulting beryllium and neutron are both moving at velocity V after the collision, write an expression describing conservation of momentum for this reaction in terms of the masses m_p, m_{Be}, m_n, and the velocities. (e) Write an expression relating the kinetic energies of particles before and after together with Q. (f) What minimum kinetic energy must the incident proton have if this reaction is to occur?

35. (a) Suppose $^{10}_{5}B$ is struck by an alpha particle, releasing a proton and a product nucleus in the reaction. What is the product nucleus? (b) An alpha particle and a product nucleus are produced when $^{13}_{6}C$ is struck by a proton. What is the product nucleus?

36. Q|C Consider two reactions:

$$(1)\ n + {}^{2}_{1}H \rightarrow {}^{3}_{1}H$$

$$(2)\ {}^{1}_{1}H + {}^{2}_{1}H \rightarrow {}^{3}_{2}He$$

(a) Compute the Q values for these reactions. Identify whether each reaction is exothermic or endothermic. (b) Which reaction results in more released energy? Why? (c) Assuming the difference is primarily due to the work done by the electric force, calculate the distance between the two protons in helium-3.

37. Natural gold has only one isotope, $^{197}_{79}Au$. If gold is bombarded with slow neutrons, e^- particles are emitted. (a) Write the appropriate reaction equation. (b) Calculate the maximum energy of the emitted beta particles. The mass of $^{198}_{80}Hg$ is 197.966 75 u.

38. Complete the following nuclear reactions:
(a) $? + {}^{14}_{7}N \rightarrow {}^{1}_{1}H + {}^{17}_{8}O$ (b) ${}^{7}_{3}Li + {}^{1}_{1}H \rightarrow {}^{4}_{2}He + ?$

39. (a) Determine the product of the reaction ${}^{7}_{3}Li + {}^{4}_{2}He \rightarrow ? + n$ (b) What is the Q value of the reaction?

29.7 Medical Applications of Radiation

40. BIO In terms of biological damage, how many rad of heavy ions are equivalent to 100 rad of x-rays?

41. BIO M A person whose mass is 75.0 kg is exposed to a whole-body dose of 25.0 rad. How many joules of energy are deposited in the person's body?

42. BIO A 200-rad dose of radiation is administered to a patient in an effort to combat a cancerous growth. Assuming all the energy deposited is absorbed by the growth, (a) calculate the amount of energy delivered per unit mass. (b) Assuming the growth has a mass of 0.25 kg and a specific heat equal to that of water, calculate its temperature rise.

43. A "clever" technician decides to heat some water for his coffee with an x-ray machine. If the machine produces 10 rad/s, how long will it take to raise the temperature of a cup of water by 50°C? Ignore heat losses during this time.

44. BIO An x-ray technician works 5 days per week, 50 weeks per year. Assume the technician takes an average of eight x-rays per day and receives a dose of 5.0 rem/yr as a result. (a) Estimate the dose in rem per x-ray taken. (b) How does this result compare with the amount of low-level background radiation the technician is exposed to?

45. BIO A patient swallows a radiopharmaceutical tagged with phosphorus-32 ($^{32}_{15}P$), a β^- emitter with a half-life of 14.3 days. The average kinetic energy of the emitted electrons is 700 keV. If the initial activity of the sample is 1.31 MBq, determine (a) the number of electrons emitted in a 10-day period, (b) the total energy deposited in the body during the 10 days, and (c) the absorbed dose if the electrons are completely absorbed in 100 g of tissue.

46. BIO A particular radioactive source produces 100 mrad of 2-MeV gamma rays per hour at a distance of 1.0 m. (a) How long could a person stand at this distance before accumulating an intolerable dose of 1 rem? (b) Assuming the gamma radiation is emitted

uniformly in all directions, at what distance would a person receive a dose of 10 mrad/h from this source?

Additional Problems

47. A radioactive sample contains 3.50 μg of pure ^{11}C, which has a half-life of 20.4 min. (a) How many moles of ^{11}C are present initially? (b) Determine the number of nuclei present initially. What is the activity of the sample (c) initially and (d) after 8.00 h?

48. Find the threshold energy that the incident neutron must have to produce the reaction: $^1_0\text{n} + ^4_2\text{He} \rightarrow ^2_1\text{H} + ^3_1\text{H}$.

49. M A 200.0-mCi sample of a radioactive isotope is purchased by a medical supply house. If the sample has a half-life of 14.0 days, how long will it keep before its activity is reduced to 20.0 mCi?

50. The ^{14}C isotope undergoes beta decay according to the process given by Equation 29.15. Find the Q value for this process.

51. Q|C In a piece of rock from the Moon, the ^{87}Rb content is assayed to be 1.82×10^{10} atoms per gram of material and the ^{87}Sr content is found to be 1.07×10^9 atoms per gram. (The relevant decay is $^{87}\text{Rb} \rightarrow ^{87}\text{Sr} + e^-$. The half-life of the decay is 4.8×10^{10} yr.) (a) Determine the age of the rock. (b) Could the material in the rock actually be much older? (c) What assumption is implicit in using the radioactive-dating method?

52. Many radioisotopes have important industrial, medical, and research applications. One of these is ^{60}Co, which has a half-life of 5.2 years and decays by the emission of a beta particle (energy 0.31 MeV) and two gamma photons (energies 1.17 MeV and 1.33 MeV). A scientist wishes to prepare a ^{60}Co sealed source that will have an activity of at least 10 Ci after 30 months of use. What is the minimum initial mass of ^{60}Co required?

53. BIO A medical laboratory stock solution is prepared with an initial activity due to ^{24}Na of 2.5 mCi/mL, and 10.0 mL of the stock solution is diluted at $t_0 = 0$ to a working solution whose total volume is 250 mL. After 48 h, a 5.0-mL sample of the working solution is monitored with a counter. What is the measured activity? *Note:* 1 mL = 1 milliliter.

54. After the sudden release of radioactivity from the Chernobyl nuclear reactor accident in 1986, the radioactivity of milk in Poland rose to 2 000 Bq/L due to iodine-131, with a half-life of 8.04 days. Radioactive iodine is particularly hazardous because the thyroid gland concentrates iodine. The Chernobyl accident caused a measurable increase in thyroid cancers among children in Belarus. (a) For comparison, find the activity of milk due to potassium. Assume 1 liter of milk contains 2.00 g of potassium, of which 0.011 7% is the isotope ^{40}K, which has a half-life of 1.28×10^9 yr. (b) After what length of time would the activity due to iodine fall below that due to potassium?

55. The theory of nuclear astrophysics is that all the heavy elements like uranium are formed in the interior of massive stars. These stars eventually explode, releasing the elements into space. If we assume that at the time of explosion there were equal amounts of ^{235}U and ^{238}U, how long ago were the elements that formed our Earth released, given that the present $^{235}\text{U}/^{238}\text{U}$ ratio is 0.007? (The half-lives of ^{235}U and ^{238}U are 0.70×10^9 yr and 4.47×10^9 yr, respectively.)

56. A by-product of some fission reactors is the isotope $^{239}_{94}$Pu, which is an alpha emitter with a half-life of 24 000 years:

$$^{239}_{94}\text{Pu} \rightarrow ^{235}_{92}\text{U} + ^4_2\text{He}$$

Consider a sample of 1.0 kg of pure $^{239}_{94}$Pu at $t = 0$. Calculate (a) the number of $^{239}_{94}$Pu nuclei present at $t = 0$ and (b) the initial activity of the sample. (c) How long does the sample have to be stored if a "safe" activity level is 0.10 Bq?

The Compact Muon Solenoid (or CMS), a detector at the Large Hadron Collider (LHC) located near Geneva, Switzerland at CERN, is designed to search for the Higgs boson, particles associated with dark matter, and extra dimensions beyond the usual four. The CMS discovered a Higgs-like boson in July of 2012. The worker in the photo is performing winter maintenance.

George Grassie/ZUMAPRESS/Newscom

30 Nuclear Energy and Elementary Particles

In this concluding chapter we discuss the two means by which energy can be derived from nuclear reactions: fission, in which a nucleus of large mass number splits into two smaller nuclei, and fusion, in which two light nuclei fuse to form a heavier nucleus. In either case, there is a release of large amounts of energy that can be used destructively through bombs or constructively through the production of electric power. We end our study of physics by examining the known subatomic particles and the fundamental interactions that govern their behavior. We also discuss the current theory of elementary particles, which states that all matter in nature is constructed from only two families of particles: quarks and leptons. Finally, we describe how such models help us understand the evolution of the Universe.

30.1 Nuclear Fission

LEARNING OBJECTIVES

1. Describe nuclear fission and the sequence of events in a typical fission process.
2. Define a self-sustained nuclear reaction, the reproduction constant K, and the moderator, and explain their roles in reactor design.
3. Apply fission concepts to unstable nuclei.

Nuclear fission occurs when a heavy nucleus, such as ^{235}U, splits, or fissions, into two smaller nuclei. In such a reaction **the total mass of the products is less than the original mass of the heavy nucleus**.

The fission of ^{235}U by slow (low-energy) neutrons can be represented by the sequence of events

$$^{1}_{0}\text{n} + ^{235}_{92}\text{U} \rightarrow ^{236}_{92}\text{U}^* \rightarrow \text{X} + \text{Y} + \text{neutrons} \qquad [30.1]$$

where $^{236}U^*$ is an intermediate state that lasts only for about 10^{-12} s before splitting into nuclei X and Y, called **fission fragments**. Many combinations of X and Y satisfy the requirements of conservation of energy and charge. In the fission of uranium, about 90 different daughter nuclei can be formed. The process also results in the production of several (typically two or three) neutrons per fission event. On the average, 2.47 neutrons are released per event.

A typical reaction of this type is

$$ {}^{1}_{0}n + {}^{235}_{92}U \rightarrow {}^{141}_{56}Ba + {}^{92}_{36}Kr + 3{}^{1}_{0}n \qquad [30.2]$$

The fission fragments, barium and krypton, and the released neutrons have a great deal of kinetic energy following the fission event. Notice that the sum of the mass numbers, or number of nucleons, on the left (1 + 235 = 236) is the same as the total number of nucleons on the right (141 + 92 + 3 = 236). The total number of protons (92) is also the same on both sides. The energy Q released through the disintegration in Equation 30.2 can be easily calculated using the data in Appendix B. The details of this calculation can be found in Chapter 26 (Example 26.5), with an answer of $Q = 200.422$ MeV.

The breakup of the uranium nucleus can be compared to what happens to a drop of water when excess energy is added to it. All the atoms in the drop have energy, but not enough to break up the drop. If enough energy is added to set the drop vibrating, however, it will undergo elongation and compression until the amplitude of vibration becomes large enough to cause the drop to break apart. In the uranium nucleus a similar process occurs (Fig. 30.1). The sequence of events is as follows:

◀ Sequence of events in a nuclear fission process

1. The ^{235}U nucleus captures a thermal (slow-moving) neutron.
2. The capture results in the formation of $^{236}U^*$, and the excess energy of this nucleus causes it to undergo violent oscillations.
3. The $^{236}U^*$ nucleus becomes highly elongated, and the force of repulsion between protons in the two halves of the dumbbell-shaped nucleus tends to increase the distortion.
4. The nucleus splits into two fragments, emitting several neutrons in the process.

Typically, the amount of energy released by the fission of a single heavy radioactive atom is about one hundred million times the energy released in the combustion of one molecule of the octane used in gasoline engines.

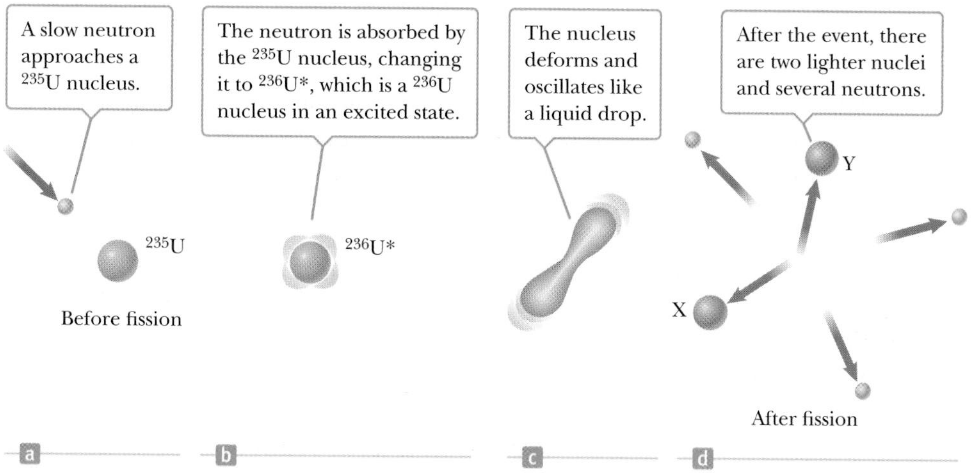

Figure 30.1 A nuclear fission event as described by the liquid-drop model of the nucleus.

APPLYING PHYSICS 30.1 | Unstable Products

If a heavy nucleus were to fission into only two product nuclei, they would be very unstable. Why?

EXPLANATION According to Figure 29.3, the ratio of the number of neutrons to the number of protons increases with Z. As a result, when a heavy nucleus splits in a fission reaction to two lighter nuclei, the lighter nuclei tend to have too many neutrons. The result is instability because the nuclei return to the curve in Figure 29.3 by decay processes that reduce the number of neutrons. ■

EXAMPLE 30.1 | A Fission-Powered World

GOAL Relate raw material to energy output.

PROBLEM **(a)** Calculate the total energy released if 1.00 kg of ^{235}U undergoes fission, taking the disintegration energy per event to be $Q = 208$ MeV. **(b)** How many kilograms of ^{235}U would be needed to satisfy the world's annual energy consumption (about 4×10^{20} J)?

STRATEGY In part (a), use the concept of a mole and Avogadro's number to obtain the total number of nuclei. Multiplying by the energy per reaction then gives the total energy released. Part (b) requires some light algebra.

SOLUTION

(a) Calculate the total energy released from 1.00 kg of ^{235}U.

Find the total number of nuclei in 1.00 kg of uranium:

$$N = \left(\frac{6.02 \times 10^{23}\ \text{nuclei/mol}}{235\ \text{g/mol}}\right)(1.00 \times 10^3\ \text{g})$$

$$= 2.56 \times 10^{24}\ \text{nuclei}$$

Multiply N by the energy yield per nucleus, obtaining the total disintegration energy:

$$E = NQ = (2.56 \times 10^{24}\ \text{nuclei})\left(208\ \frac{\text{MeV}}{\text{nucleus}}\right)$$

$$= \boxed{5.32 \times 10^{26}\ \text{MeV}}$$

(b) How many kilograms would provide for the world's annual energy needs?

Set the energy per kilogram, E_{kg}, times the number of kilograms, N_{kg}, equal to the total annual energy consumption. Solve for N_{kg}:

$$E_{\text{kg}}N_{\text{kg}} = E_{\text{tot}}$$

$$N_{\text{kg}} = \frac{E_{\text{tot}}}{E_{\text{kg}}} = \frac{4 \times 10^{20}\ \text{J}}{(5.32 \times 10^{32}\ \text{eV/kg})(1.60 \times 10^{-19}\ \text{J/eV})}$$

$$= \boxed{5 \times 10^6\ \text{kg}}$$

REMARKS The calculation implicitly assumes perfect conversion to usable power, which is never the case in real systems. There are sufficient easily recoverable reserves of uranium to supply the entire world's energy needs for about seven years at current levels of usage. Breeder reactor technology can greatly extend those reserves.

QUESTION 30.1 Estimate the average mass of ^{235}U needed to provide power for one family for one year.

EXERCISE 30.1 How long can 1 kg of uranium-235 keep a 100-watt lightbulb burning if all its released energy is converted to electrical energy?

ANSWER $\sim$ 30 000 yr

Nuclear Reactors

The neutrons emitted when ^{235}U undergoes fission can in turn trigger other nuclei to undergo fission, with the possibility of a chain reaction (Fig. 30.2). Calculations show that if the chain reaction isn't controlled, it will proceed too rapidly and possibly result in the sudden release of an enormous amount of energy (an explosion), even from only 1 g of ^{235}U. If the energy in 1 kg of ^{235}U were released, it would equal that released by the detonation of about 20 000 tons of

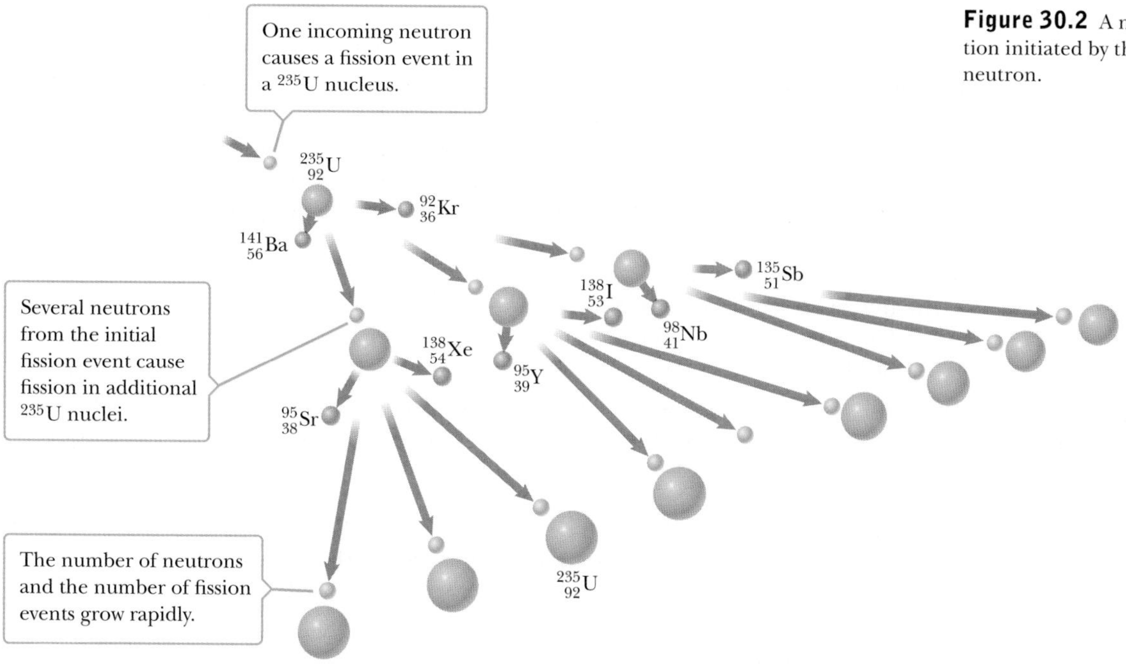

Figure 30.2 A nuclear chain reaction initiated by the capture of a neutron.

TNT! An uncontrolled fission reaction, of course, is the principle behind the first nuclear bomb.

A nuclear reactor is a system designed to maintain what is called a **self-sustained chain reaction**, first achieved in 1942 by a team led by Enrico Fermi. Most reactors in operation today also use uranium as fuel. Natural uranium contains only about 0.7% of the ^{235}U isotope, with the remaining 99.3% being the ^{238}U isotope. This fact is important to the operation of a reactor because ^{238}U almost never undergoes fission. Instead, it tends to absorb neutrons, producing neptunium and plutonium. For this reason, reactor fuels must be artificially enriched so that they contain several percent of the ^{235}U isotope.

On average, about 2.5 neutrons are emitted in each fission event of ^{235}U. To achieve a self-sustained chain reaction, one of these neutrons must be captured by another ^{235}U nucleus and cause it to undergo fission. A useful parameter for describing the level of reactor operation is the **reproduction constant *K*, defined as the average number of neutrons from each fission event that will cause another event**.

A self-sustained chain reaction is achieved when $K = 1$. Under this condition, the reactor is said to be **critical**. When K is less than 1, the reactor is subcritical and the reaction dies out. When K is greater than 1, the reactor is said to be supercritical and a runaway reaction occurs. In a nuclear reactor used to furnish power to a utility company, it is necessary to maintain a K value close to 1.

The basic design of a nuclear reactor is shown in Figure 30.3. The fuel elements consist of enriched uranium. The size of the reactor is important in reducing neutron leakage: a large reactor has a smaller surface-to-volume ratio and smaller leakage than a smaller reactor.

It's also important to regulate the neutron energies because slow neutrons are far more likely to cause fissions than fast neutrons in ^{235}U. Further, ^{238}U doesn't absorb slow neutrons. For the chain reaction to continue, the neutrons must, therefore, be slowed down. This slowing is accomplished by surrounding the fuel with a substance called a **moderator**, such as graphite (carbon) or heavy water (D_2O). Most modern reactors use heavy water. Collisions in the moderator slow the neutrons and enhance the fissioning of ^{235}U.

The power output of a fission reactor is controlled by the control rods depicted in Figure 30.3. These rods are made of materials like cadmium that readily absorb neutrons.

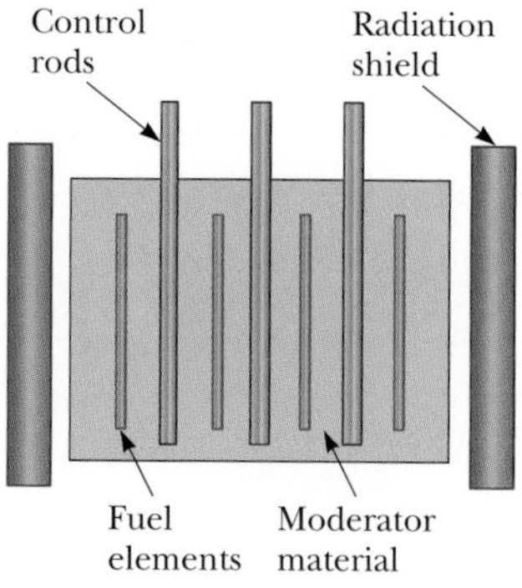

Figure 30.3 Cross section of a reactor core showing the control rods, fuel elements containing enriched fuel, and moderating material, all surrounded by a radiation shield.

APPLICATION
Nuclear Reactor Design

Fissions in a nuclear reactor heat molten sodium (or water, depending on the system), which is pumped through a heat exchanger. There, the thermal energy is transferred to water in a secondary system. The water is converted to steam, which drives a turbine-generator to create electric power.

Fission reactors are extremely safe. According to the *Oak Ridge National Laboratory Review*, "The health risk of living within 8 km (5 miles) of a nuclear reactor for 50 years is no greater than the risk of smoking 1.4 cigarettes, drinking 0.5 L of wine, traveling 240 km by car, flying 9 600 km by jet, or having one chest x-ray in a hospital. Each activity, by itself, is estimated to increase a person's chance of dying in any given year by one in a million."

The safety issues associated with nuclear power reactors are complex and often emotional and overblown. All sources of energy have associated risks. Coal, for example, exposes workers to health hazards (including radioactive radon) and produces atmospheric pollution (including greenhouse gases and highly radioactive ash). Solar panels introduce toxic substances into the environment such as cadmium and silicon tetrachloride, and in addition occupy significant land areas relative their power production capabilities. In each case the risks must be weighed against the benefits and the availability and cost of the energy source.

Given concerns about greenhouse gasses and global climate change, studying alternative forms of energy production is extremely important. Among green energy alternatives, nuclear fission is the most robust, providing reliable power with little down time. Despite advances in technology and reductions in cost, solar and wind power remain very expensive and intermittent.

The known sources of uranium ore are sufficient to supply all humanity's power requirements for several years at the current rate of usage if burned in conventional reactors. (See Problem 8.) Breeder reactors, on the other hand, convert nuclear waste into nuclear fuel and can also change thorium to an isotope of uranium suitable for a reactor. Using that technology, current reserves would likely be sufficient for well over a thousand years, even without the discovery and exploitation of new sources of uranium or thorium. The largest untapped source is the 4.5 billion metric tons of uranium dissolved in Earth's oceans. Several methods of extracting this uranium economically are currently under study. The dissolved uranium is steadily replenished by Earth's rivers and could potentially provide reactor fuel almost indefinitely. (See Problem 10.)

30.2 Nuclear Fusion

LEARNING OBJECTIVES

1. Describe the process of nuclear fusion and the proton-proton cycle in stars.
2. Discuss the potential application of fusion to power reactors and state Lawson's criterion.
3. Apply fusion concepts to nuclei.

When two light nuclei combine to form a heavier nucleus, the process is called nuclear fusion. Because the mass of the final nucleus is less than the sum of the masses of the original nuclei, there is a loss of mass, accompanied by a release of energy. Although fusion power plants have not yet been developed, a worldwide effort is under way to harness the energy from fusion reactions in the laboratory.

Fusion in the Sun

All stars generate their energy through fusion processes. About 90% of stars, including the Sun, fuse hydrogen, whereas some older stars fuse helium or other heavier elements. The energy produced by fusion increases the pressure inside the star and prevents its collapse due to gravity.

Two conditions must be met before fusion reactions in the star can sustain its energy needs. First, the temperature must be high enough (about 10^7 K for hydrogen) to allow the kinetic energy of the positively charged hydrogen nuclei to overcome their mutual Coulomb repulsion as they collide. Second, the density of nuclei must be high enough to ensure a high rate of collision.

It's interesting to note that a quantum effect is key in making sunshine. Temperatures inside stars like the Sun are not high enough to allow colliding protons to overcome the Coulomb repulsion. In a certain percentage of collisions, however, the nuclei pass through the barrier anyway, an example of *quantum tunneling*.

The **proton–proton cycle** is a series of three nuclear reactions that are believed to be the stages in the liberation of energy in the Sun and other stars rich in hydrogen. An overall view of the proton–proton cycle is that four protons combine to form an alpha particle and two positrons, with the release of 25 MeV of energy in the process.

The specific steps in the proton–proton cycle are

$$^1_1\mathrm{H} + {}^1_1\mathrm{H} \rightarrow {}^2_1\mathrm{D} + \mathrm{e}^+ + \nu$$

and

$$^1_1\mathrm{H} + {}^2_1\mathrm{D} \rightarrow {}^3_2\mathrm{He} + \gamma \qquad [30.3]$$

where D stands for deuterium, the isotope of hydrogen having one proton and one neutron in the nucleus. (It can also be written as $^2_1\mathrm{H}$.) The second reaction is followed by either hydrogen-helium fusion or helium-helium fusion:

$$^1_1\mathrm{H} + {}^3_2\mathrm{He} \rightarrow {}^4_2\mathrm{He} + \mathrm{e}^+ + \nu$$

or

$$^3_2\mathrm{He} + {}^3_2\mathrm{He} \rightarrow {}^4_2\mathrm{He} + 2({}^1_1\mathrm{H})$$

The energy liberated is carried primarily by gamma rays, positrons, and neutrinos, as can be seen from the reactions. The gamma rays are soon absorbed by the dense gas, raising its temperature. The positrons combine with electrons to produce gamma rays, which in turn are also absorbed by the gas within a few centimeters. The neutrinos, however, almost never interact with matter; hence, they escape from the star, carrying about 2% of the energy generated with them. These energy-liberating fusion reactions are called **thermonuclear fusion reactions**. The hydrogen (fusion) bomb, first exploded in 1952, is an example of an uncontrolled thermonuclear fusion reaction.

Fusion Reactors

APPLICATION
Fusion Reactors

A great deal of effort is under way to develop a sustained and controllable fusion power reactor. Controlled fusion is often called the ultimate energy source because of the availability of water, its fuel source. For example, if deuterium, the isotope of hydrogen consisting of a proton and a neutron, were used as the fuel, 0.06 g of it could be extracted from 1 gal of water at a cost of about four cents. Hence, the fuel costs of even an inefficient reactor would be almost insignificant. An additional advantage of fusion reactors is that comparatively few radioactive by-products are formed. As noted in Equation 30.3, the end product of the fusion of hydrogen nuclei is safe, nonradioactive helium. Unfortunately, a thermonuclear reactor that can deliver a net power output over a reasonable time interval is not yet a reality, and many problems must be solved before a successful device is constructed.

The fusion reactions that appear most promising in the construction of a fusion power reactor involve deuterium (D) and tritium (T), which are isotopes of hydrogen. These reactions are

$$^2_1\mathrm{D} + {}^2_1\mathrm{D} \rightarrow {}^3_2\mathrm{He} + {}^1_0\mathrm{n} \qquad Q = 3.27\ \mathrm{MeV}$$

$$^2_1\mathrm{D} + {}^2_1\mathrm{D} \rightarrow {}^3_1\mathrm{T} + {}^1_1\mathrm{H} \qquad Q = 4.03\ \mathrm{MeV} \qquad [30.4]$$

and

$$^2_1\text{D} + ^3_1\text{T} \rightarrow ^4_2\text{He} + ^1_0\text{n} \qquad Q = 17.59 \text{ MeV}$$

where the Q values refer to the amount of energy released per reaction. As noted earlier, deuterium is available in almost unlimited quantities from our lakes and oceans and is very inexpensive to extract. Tritium, however, is radioactive ($T_{1/2}$ = 12.3 yr) and undergoes beta decay to ^{3}He. For this reason, tritium doesn't occur naturally to any great extent and must be artificially produced.

The fundamental challenge of nuclear fusion power is to give the nuclei enough kinetic energy to overcome the repulsive Coulomb force between them at close proximity. This step can be accomplished by heating the fuel to extremely high temperatures (about 10^8 K, far greater than the interior temperature of the Sun). Such high temperatures are not easy to obtain in a laboratory or a power plant. At these high temperatures, the atoms are ionized and the system then consists of a collection of electrons and nuclei, commonly referred to as a *plasma*.

In addition to the high temperature requirements, two other critical factors determine whether or not a thermonuclear reactor will function: the **plasma ion density** n and the **plasma confinement time** τ, the time the interacting ions are maintained at a temperature equal to or greater than that required for the reaction to proceed. The density and confinement time must both be large enough to ensure that more fusion energy will be released than is required to heat the plasma.

Lawson's criterion ▶

Lawson's criterion states that a net power output in a fusion reactor is possible under the following conditions:

$$n\tau \geq 10^{14} \text{ s/cm}^3 \qquad \text{Deuterium–tritium interaction} \qquad [30.5]$$

$$n\tau \geq 10^{16} \text{ s/cm}^3 \qquad \text{Deuterium–deuterium interaction}$$

The problem of plasma confinement time has yet to be solved. How can a plasma be confined at a temperature of 10^8 K for times on the order of 1 s? Most fusion experiments use magnetic field confinement to contain a plasma. One device, called a **tokamak**, has a doughnut-shaped geometry (a toroid), as shown in Figure 30.4. This device uses a combination of two magnetic fields to confine the plasma inside the doughnut. A strong magnetic field is produced by the current in the windings, and a weaker magnetic field is produced by the current in the toroid. The resulting magnetic field lines are helical, as shown in the figure. In this configuration the field lines spiral around the plasma and prevent it from touching the walls of the vacuum chamber.

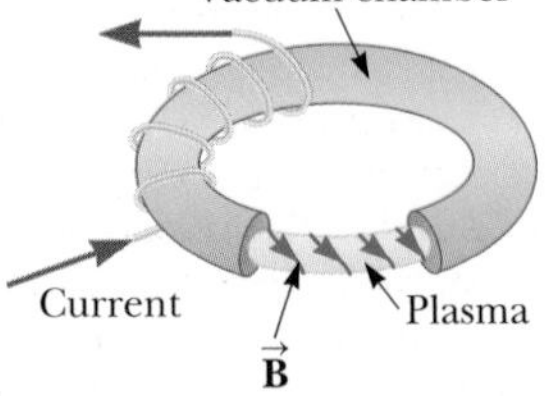

Figure 30.4 Diagram of a tokamak used in the magnetic confinement scheme. The plasma is trapped within the spiraling magnetic field lines as shown.

In inertial confinement fusion, the fuel, typically deuterium and tritium, is put in the form of a small pellet. Directly or indirectly, powerful lasers deliver energy rapidly to the pellet, exploding off the outer layers and imploding the rest of the pellet, heating and compressing it. Shock waves form and meet at the center, greatly increasing the density and pressure and causing fusion reactions. The released energy then causes further fusion reactions. Fusion can also take place in a device the size of a TV set and in fact was invented by Philo Farnsworth, one of the pioneers of electronic television. In this method, called inertial electrostatic confinement, positively charged particles are rapidly attracted toward a negatively charged grid. Some of the positive particles then collide and fuse.

■ EXAMPLE 30.2 | Astrofuel on the Moon

GOAL Calculate the energy released in a fusion reaction.

PROBLEM Find the energy released in the reaction of helium-3 with deuterium:

$$^3_2\text{He} + ^2_1\text{D} \rightarrow ^4_2\text{He} + ^1_1\text{H}$$

STRATEGY The energy released is the difference between the mass energy of the reactants and the products.

Conservation of Strangeness

The K, Λ, and Σ particles exhibit unusual properties in their production and decay and hence are called *strange particles.*

One unusual property of strange particles is that they are always produced in pairs. For example, when a pion collides with a proton, two neutral strange particles are produced with high probability (Fig. 30.6) following the reaction:

$$\pi^- + p^+ \rightarrow K^0 + \Lambda^0$$

On the other hand, the reaction $\pi^- + p^+ \rightarrow K^0 + n$ has never occurred, even though it violates no known conservation laws and the energy of the pion is sufficient to initiate the reaction.

The second peculiar feature of strange particles is that although they are produced by the strong interaction at a high rate, they don't decay into particles that interact via the strong force at a very high rate. Instead, they decay very slowly, which is characteristic of the weak interaction. Their half-lives are in the range of 10^{-10} s to 10^{-8} s; most other particles that interact via the strong force have much shorter lifetimes on the order of 10^{-23} s.

To explain these unusual properties of strange particles, a law called *conservation of strangeness* was introduced, together with a new quantum number S called **strangeness**. The strangeness numbers for some particles are given in Table 30.2. The production of strange particles in pairs is explained by assigning $S = +1$ to one of the particles and $S = -1$ to the other. All nonstrange particles are assigned strangeness $S = 0$. The **law of conservation of strangeness** states that whenever a nuclear reaction or decay occurs, the sum of the strangeness numbers before the process must equal the sum of the strangeness numbers after the process.

The slow decay of strange particles can be explained by assuming the strong and electromagnetic interactions obey the law of conservation of strangeness, whereas the weak interaction does not. Because the decay reaction involves the loss of one strange particle, it violates strangeness conservation and hence proceeds slowly via the weak interaction.

In checking reactions for proper strangeness conservation, the same procedure as with baryon number conservation and lepton number conservation is followed. Using Table 30.2, count the strangeness on each side. If the two results are equal, the reaction conserves strangeness.

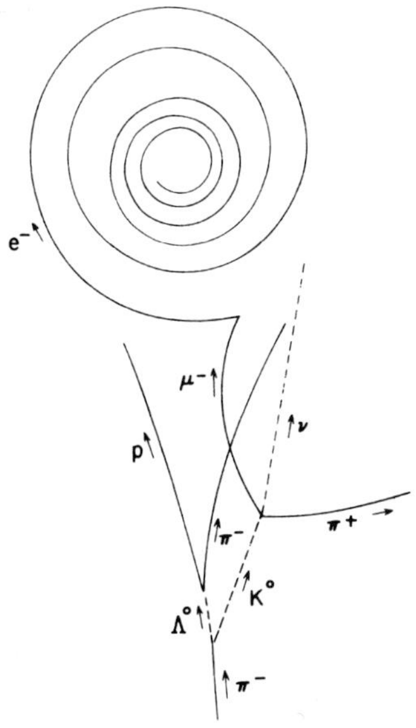

Figure 30.6 This drawing represents tracks of many events obtained by analyzing a bubble-chamber photograph. The strange particles Λ^0 and K^0 are formed (at the bottom) as the π^- interacts with a proton according to the interaction $\pi^- + p \rightarrow \Lambda^0 + K^0$. (Note that the neutral particles leave no tracks, as indicated by the dashed lines.) The Λ^0 and K^0 then decay according to the interactions $\Lambda^0 \rightarrow \pi + p$ and $K^0 \rightarrow \pi + \mu^- + \nu_\mu$.

APPLYING PHYSICS 30.2 | Breaking Conservation Laws

A student claims to have observed a decay of an electron into two neutrinos traveling in opposite directions. What conservation laws would be violated by this decay?

EXPLANATION Several conservation laws would be violated. Conservation of electric charge would be violated because the negative charge of the electron has disappeared. Conservation of electron-lepton number would also be violated because there is one lepton before the decay and two afterward. If both neutrinos were electron neutrinos, electron-lepton number conservation would be violated in the final state. If one of the product neutrinos were other than an electron neutrino, however, another lepton conservation law would be violated because there were no other leptons in the initial state. Other conservation laws would be obeyed by this decay. Energy can be conserved; the rest energy of the electron appears as the kinetic energy (and possibly some small rest energy) of the neutrinos. The opposite directions of the two neutrinos' velocities allow for the conservation of momentum. Conservation of baryon number and conservation of other lepton numbers would also be upheld in this decay. ■

30.7 The Eightfold Way

LEARNING OBJECTIVE

1. Describe the eightfold way and its similarities with the periodic table.

Quantities such as spin, baryon number, lepton number, and strangeness are labels we associate with particles. Many classification schemes that group particles into families based on such labels have been proposed. First, consider the first eight baryons listed in Table 30.2, all having a spin of $\frac{1}{2}$. The family consists of the proton, the neutron, and six other particles. If we plot their strangeness versus their charge using a sloping coordinate system, as in Figure 30.7a, a fascinating pattern emerges: six of the baryons form a hexagon, and the remaining two are at the hexagon's center. (Particles with spin quantum number $\frac{1}{2}$ or $\frac{3}{2}$ are called *fermions.*)

Now consider the family of mesons listed in Table 30.2 with spins of zero. (Particles with spin quantum number 0 or 1 are called *bosons.*) If we count both particles and antiparticles, there are nine such mesons. Figure 30.7b is a plot of strangeness versus charge for this family. Again, a fascinating hexagonal pattern emerges. In this case the particles on the perimeter of the hexagon lie opposite their antiparticles, and the remaining three (which form their own antiparticles) are at its center. These and related symmetric patterns, called the **eightfold way**, were proposed independently in 1961 by Murray Gell-Mann and Yuval Ne'eman.

Murray Gell-Mann
American Physicist (b. 1929)
Gell-Mann was awarded the 1969 Nobel Prize in Physics for his theoretical studies dealing with subatomic particles.

The groups of baryons and mesons can be displayed in many other symmetric patterns within the framework of the eightfold way. For example, the family of spin-$\frac{3}{2}$ baryons contains ten particles arranged in a pattern like the tenpins in a bowling alley. After the pattern was proposed, one of the particles was missing; it had yet to be discovered. Gell-Mann predicted that the missing particle, which he called the *omega minus* (Ω^-), should have a spin of $\frac{3}{2}$, a charge of -1, a strangeness of -3, and a mass of about 1 680 MeV/c^2. Shortly thereafter, in 1964, scientists at the Brookhaven National Laboratory found the missing particle through careful analyses of bubble-chamber photographs and confirmed all its predicted properties.

The patterns of the eightfold way in the field of particle physics have much in common with the periodic table. Whenever a vacancy (a missing particle or element) occurs in the organized patterns, experimentalists have a guide for their investigations.

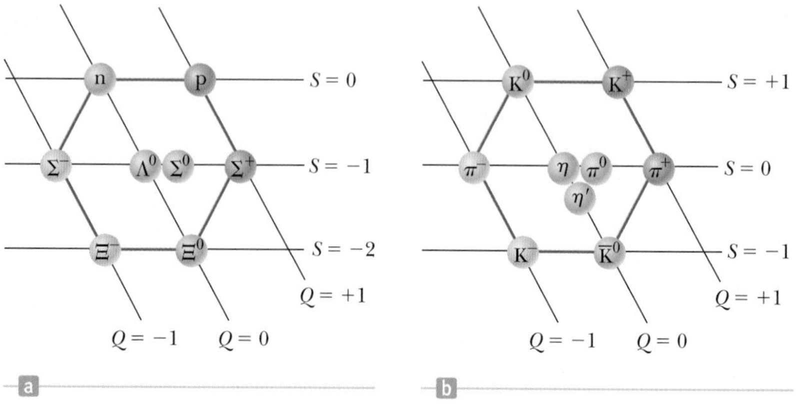

Figure 30.7 (a) The hexagonal eightfold-way pattern for the eight spin-$\frac{1}{2}$ baryons. This strangeness versus charge plot uses a horizontal axis for the strangeness values S, but a sloping axis for the charge number Q. (b) The eightfold-way pattern for the nine spin-zero mesons.

C. Factoring

The following are some useful formulas for factoring an equation:

$$ax + ay + az = a(x + y + z) \qquad \text{common factor}$$

$$a^2 + 2ab + b^2 = (a + b)^2 \qquad \text{perfect square}$$

$$a^2 - b^2 = (a + b)(a - b) \qquad \text{difference of squares}$$

D. Quadratic Equations

The general form of a quadratic equation is

$$ax^2 + bx + c = 0 \qquad \textbf{[A.7]}$$

where x is the unknown quantity and a, b, and c are numerical factors referred to as coefficients of the equation. This equation has two roots, given by

$$x = \frac{-b \pm \sqrt{b^2 - 4ac}}{2a} \qquad \textbf{[A.8]}$$

If $b^2 - 4ac > 0$, the roots are real.

EXAMPLE

The equation $x^2 + 5x + 4 = 0$ has the following roots corresponding to the two signs of the square-root term:

$$x = \frac{-5 \pm \sqrt{5^2 - (4)(1)(4)}}{2(1)} = \frac{-5 \pm \sqrt{9}}{2} = \frac{-5 \pm 3}{2}$$

$$x_1 = \frac{-5 + 3}{2} = -1 \qquad x_2 = \frac{-5 - 3}{2} = -4$$

where x_1 refers to the root corresponding to the positive sign and x_2 refers to the root corresponding to the negative sign.

EXAMPLE

A ball is projected upwards at 16.0 m/s. Use the quadratic formula to determine the time necessary for it to reach a height of 8.00 m above the point of release.

SOLUTION From the discussion of ballistics in Chapter 2, we can write

$$\textbf{(1)} \quad x = \tfrac{1}{2}at^2 + v_0 t + x_0$$

The acceleration is due to gravity, given by $a = -9.80\ \text{m/s}^2$; the initial velocity is $v_0 = 16.0$ m/s; and the initial position is the point of release, taken to be $x_0 = 0$. Substitute these values into Equation (1) and set $x = 8.00$ m, arriving at

$$x = -4.90t^2 + 16.00t = 8.00$$

where units have been suppressed for mathematical clarity. Rearrange this expression into the standard form of Equation A.7:

$$-4.90t^2 + 16.00t - 8.00 = 0$$

The equation is quadratic in the time, t. We have $a = -4.9$, $b = 16$, and $c = -8.00$. Substitute these values into Equation A.8:

$$t = \frac{-16.0 \pm \sqrt{16^2 - 4(-4.90)(-8.00)}}{2(-4.90)} = \frac{-16.0 \pm \sqrt{99.2}}{-9.80}$$

$$= 1.63 \mp \frac{\sqrt{99.2}}{9.80} = 0.614\ \text{s}, 2.65\ \text{s}$$

Both solutions are valid in this case, one corresponding to reaching the point of interest on the way up and the other to reaching it on the way back down.

Exercises

Solve the following quadratic equations:

	Answers	
1. $x^2 + 2x - 3 = 0$	$x_1 = 1$	$x_2 = -3$
2. $2x^2 - 5x + 2 = 0$	$x_1 = 2$	$x_2 = \frac{1}{2}$
3. $2x^2 - 4x - 9 = 0$	$x_1 = 1 + \sqrt{22}/2$	$x_2 = 1 - \sqrt{22}/2$

4. Repeat the ballistics example for a height of 10.0 m above the point of release.
Answer: $t_1 = 0.842$ s $t_2 = 2.42$ s

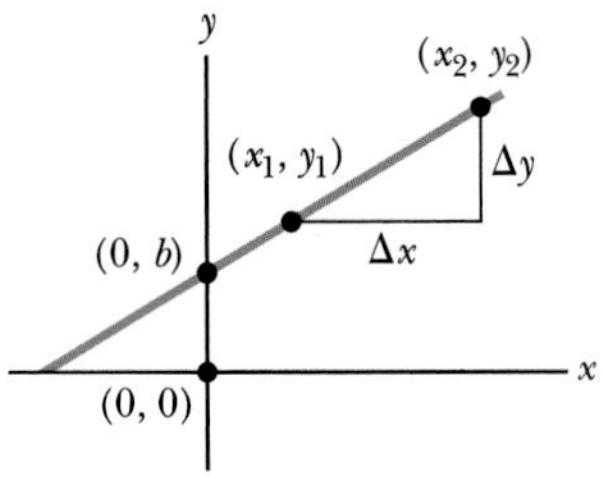

Figure A.1

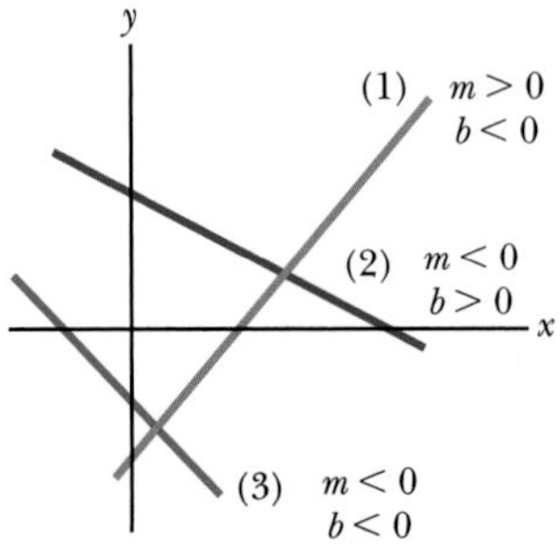

Figure A.2

E. Linear Equations

A linear equation has the general form

$$y = mx + b \quad \text{[A.9]}$$

where m and b are constants. This kind of equation is called linear because the graph of y versus x is a straight line, as shown in Figure A.1. The constant b, called the y-intercept, represents the value of y at which the straight line intersects the y-axis. The constant m is equal to the slope of the straight line. If any two points on the straight line are specified by the coordinates (x_1, y_1) and (x_2, y_2), as in Figure A.1, the slope of the straight line can be expressed as

$$\text{Slope} = \frac{y_2 - y_1}{x_2 - x_1} = \frac{\Delta y}{\Delta x} \quad \text{[A.10]}$$

Note that m and b can have either positive or negative values. If $m > 0$, the straight line has a positive slope, as in Figure A.1. If $m < 0$, the straight line has a negative slope. In Figure A.1, both m and b are positive. Three other possible situations are shown in Figure A.2.

EXAMPLE

Suppose the electrical resistance of a metal wire is 5.00 Ω at a temperature of 20.0°C and 6.14 Ω at 80.0°C. Assuming the resistance changes linearly, what is the resistance of the wire at 60.0°C?

SOLUTION Find the equation of the line describing the resistance R and then substitute the new temperature into it. Two points on the graph of resistance versus temperature, (20.0°C, 5.00 Ω) and (80.0°C, 6.14 Ω), allow computation of the slope:

$$\textbf{(1)} \quad m = \frac{\Delta R}{\Delta T} = \frac{6.14\ \Omega - 5.00\ \Omega}{80.0°\text{C} - 20.0°\text{C}} = 1.90 \times 10^{-2}\ \Omega/°\text{C}$$

Now use the point-slope formulation of a line, with this slope and (20.0°C, 5.00 Ω):

$$\textbf{(2)} \quad R - R_0 = m(T - T_0)$$

$$\textbf{(3)} \quad R - 5.00\ \Omega = (1.90 \times 10^{-2}\ \Omega/°\text{C})(T - 20.0°\text{C})$$

Finally, substitute $T = 60.0°$ into Equation (3) and solve for R, getting $R = 5.76\ \Omega$.

Exercises

1. Draw graphs of the following straight lines:
(a) $y = 5x + 3$ **(b)** $y = -2x + 4$ **(c)** $y = -3x - 6$

2. Find the slopes of the straight lines described in Exercise 1.
Answers: (a) 5 **(b)** -2 **(c)** -3

3. Find the slopes of the straight lines that pass through the following sets of points: **(a)** $(0, -4)$ and $(4, 2)$ **(b)** $(0, 0)$ and $(2, -5)$ **(c)** $(-5, 2)$ and $(4, -2)$
Answers: (a) $3/2$ **(b)** $-5/2$ **(c)** $-4/9$

4. Suppose an experiment measures the following displacements (in meters) from equilibrium of a vertical spring due to attaching weights (in Newtons): (0.025 0 m, 22.0 N), (0.075 0 m, 66.0 N). Find the spring constant, which is the slope of the line in the graph of weight versus displacement.
Answer: 880 N/m

F. Solving Simultaneous Linear Equations

Consider the equation $3x + 5y = 15$, which has two unknowns, x and y. Such an equation doesn't have a unique solution. For example, note that ($x = 0$, $y = 3$), ($x = 5$, $y = 0$), and ($x = 2$, $y = 9/5$) are all solutions to this equation.

If a problem has two unknowns, a unique solution is possible only if we have two equations. In general, if a problem has n unknowns, its solution requires n equations. To solve two simultaneous equations involving two unknowns, x and y, we solve one of the equations for x in terms of y and substitute this expression into the other equation.

EXAMPLE

Solve the following two simultaneous equations:

$$\textbf{(1)}\quad 5x + y = -8 \qquad \textbf{(2)}\quad 2x - 2y = 4$$

SOLUTION From Equation (2), we find that $x = y + 2$. Substitution of this value into Equation (1) gives

$$5(y + 2) + y = -8$$
$$6y = -18$$
$$y = -3$$
$$x = y + 2 = -1$$

ALTERNATE SOLUTION Multiply each term in Equation (1) by the factor 2 and add the result to Equation (2):

$$10x + 2y = -16$$
$$2x - 2y = 4$$
$$12x = -12$$
$$x = -1$$
$$y = x - 2 = -3$$

Two linear equations containing two unknowns can also be solved by a graphical method. If the straight lines corresponding to the two equations are plotted in a conventional coordinate system, the intersection of the two lines represents the solution. For example, consider the two equations

$$x - y = 2$$
$$x - 2y = -1$$

These equations are plotted in Figure A.3. The intersection of the two lines has the coordinates $x = 5$, $y = 3$, which represents the solution to the equations. You should check this solution by the analytical technique discussed above.

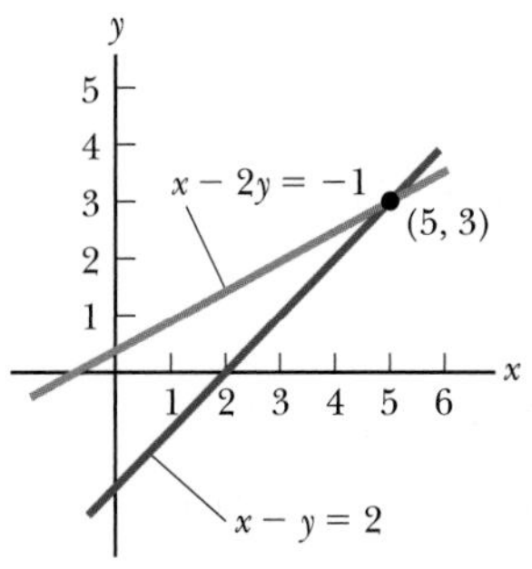

Figure A.3

EXAMPLE

A block of mass $m = 2.00$ kg travels in the positive x-direction at $v_i = 5.00$ m/s, while a second block, of mass $M = 4.00$ kg and leading the first block, travels in the positive x-direction at 2.00 m/s. The surface is frictionless. What are the blocks' final velocities if the collision is perfectly elastic?

(Continued)

SOLUTION As can be seen in Chapter 6, a perfectly elastic collision involves equations for the momentum and energy. With algebra, the energy equation, which is quadratic in v, can be recast as a linear equation. The two equations are given by

$$\textbf{(1)}\quad mv_i + MV_i = mv_f + MV_f$$

$$\textbf{(2)}\quad v_i - V_i = -(v_f - V_f)$$

Substitute the known quantities $v_i = 5.00$ m/s and $V_i = 2.00$ m/s into Equations (1) and (2):

$$\textbf{(3)}\quad 18 = 2v_f + 4V_f$$

$$\textbf{(4)}\quad 3 = -v_f + V_f$$

Multiply Equation (4) by 2 and add to Equation (3):

$$18 = 2v_f + 4V_f$$

$$6 = -2v_f + 2V_f$$

$$24 = 6V_f \quad \rightarrow \quad V_f = \boxed{4.00 \text{ m/s}}$$

Substituting the solution for V_f back into Equation (4) yields $v_f = \boxed{1.00 \text{ m/s}}$.

Exercises

Solve the following pairs of simultaneous equations involving two unknowns:

	Answers
1. $x + y = 8$ $x - y = 2$	$x = 5,\ y = 3$
2. $98 - T = 10a$ $T - 49 = 5a$	$T = 65.3,\ a = 3.27$
3. $6x + 2y = 6$ $8x - 4y = 28$	$x = 2,\ y = -3$

G. Logarithms and Exponentials

Suppose a quantity x is expressed as a power of some quantity a:

$$x = a^y \qquad \textbf{[A.11]}$$

The number a is called the **base** number. The **logarithm** of x with respect to the base a is equal to the exponent to which the base must be raised so as to satisfy the expression $x = a^y$:

$$y = \log_a x \qquad \textbf{[A.12]}$$

Conversely, the **antilogarithm** of y is the number x:

$$x = \text{antilog}_a\, y \qquad \textbf{[A.13]}$$

The antilog expression is in fact identical to the exponential expression in Equation A.11, which is preferable for practical purposes.

In practice, the two bases most often used are base 10, called the *common* logarithm base, and base $e = 2.718\ldots$, called the *natural* logarithm base. When common logarithms are used,

$$y = \log_{10} x \qquad (\text{or } x = 10^y) \qquad \textbf{[A.14]}$$

When natural logarithms are used,

$$y = \ln x \qquad (\text{or } x = e^y) \qquad \textbf{[A.15]}$$

For example, $\log_{10} 52 = 1.716$, so $\text{antilog}_{10} 1.716 = 10^{1.716} = 52$. Likewise, $\ln_e 52 = 3.951$, so $\text{antiln}_e 3.951 = e^{3.951} = 52$.

In general, note that you can convert between base 10 and base e with the equality

$$\ln x = (2.302\,585)\log_{10} x \qquad \textbf{[A.16]}$$

Finally, some useful properties of logarithms are

$$\log(ab) = \log a + \log b \qquad \ln e = 1$$

$$\log(a/b) = \log a - \log b \qquad \ln e^a = a$$

$$\log(a^n) = n \log a \qquad \ln\left(\frac{1}{a}\right) = -\ln a$$

Logarithms in college physics are used most notably in the definition of decibel level. Sound intensity varies across several orders of magnitude, making it awkward to compare different intensities. Decibel level converts these intensities to a more manageable logarithmic scale.

EXAMPLE (Logs)

Suppose a jet testing its engines produces a sound intensity of $I = 0.750$ W at a given location in an airplane hangar. What decibel level corresponds to this sound intensity?

SOLUTION Decibel level β is defined by

$$\beta = 10 \log\left(\frac{I}{I_0}\right)$$

where $I_0 = 1 \times 10^{-12}$ W/m^2 is the standard reference intensity. Substitute the given information:

$$\beta = 10 \log\left(\frac{0.750 \text{ W/m}^2}{10^{-12} \text{ W/m}^2}\right) = 119 \text{ dB}$$

EXAMPLE (Antilogs)

A collection of four identical machines creates a decibel level of $\beta = 87.0$ dB in a machine shop. What sound intensity would be created by only one such machine?

SOLUTION We use the equation of decibel level to find the total sound intensity of the four machines, and then we divide by 4. From Equation (1):

$$87.0 \text{ dB} = 10 \log\left(\frac{I}{10^{-12} \text{ W/m}^2}\right)$$

Divide both sides by 10 and take the antilog of both sides, which means, equivalently, to exponentiate:

$$10^{8.7} = 10^{\log(I/10^{-12})} = \frac{I}{10^{-12}}$$

$$I = 10^{-12} \cdot 10^{8.7} = 10^{-3.3} = 5.01 \times 10^{-4} \text{ W/m}^2$$

There are four machines, so this result must be divided by 4 to get the intensity of one machine:

$$I = 1.25 \times 10^{-4} \text{ W/m}^2$$

EXAMPLE (Exponentials)

The half-life of tritium is 12.33 years. (Tritium is the heaviest isotope of hydrogen, with a nucleus consisting of a proton and two neutrons.) If a sample contains 3.0 g of tritium initially, how much remains after 20.0 years?

SOLUTION The equation giving the number of nuclei of a radioactive substance as a function of time is

$$N = N_0\left(\frac{1}{2}\right)^n$$

(Continued)

where N is the number of nuclei remaining, N_0 is the initial number of nuclei, and n is the number of half-lives. Note that this equation is an exponential expression with a base of $\frac{1}{2}$. The number of half-lives is given by $n = t/T_{1/2} = 20.0\text{ yr}/12.33\text{ yr} = 1.62$. The fractional amount of tritium that remains after 20.0 yr is therefore

$$\frac{N}{N_0} = \left(\frac{1}{2}\right)^{1.62} = 0.325$$

Hence, of the original 3.00 g of tritium, $0.325 \times 3.00\text{ g} = 0.975\text{ g}$ remains.

A.4 Geometry

Table A.2 gives the areas and volumes for several geometric shapes used throughout this text. These areas and volumes are important in numerous physics applications. A good example is the concept of pressure P, which is the force per unit area. As an equation, it is written $P = F/A$. Areas must also be calculated in problems involving the volume rate of fluid flow through a pipe using the equation of continuity, the tensile stress exerted on a cable by a weight, the rate of thermal energy transfer through a barrier, and the density of current through a wire. There are numerous other applications. Volumes are important in computing the buoyant force exerted by water on a submerged object, in calculating densities, and in determining the bulk stress of fluid or gas on an object, which affects its volume. Again, there are numerous other applications.

Table A.2 Useful Information for Geometry

Shape	Area or Volume	Shape	Area or Volume
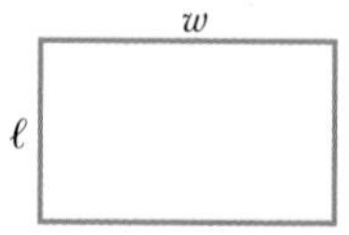 Rectangle	Area $= \ell w$	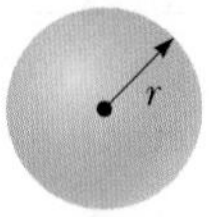 Sphere	Surface area $= 4\pi r^2$ Volume $= \frac{4}{3}\pi r^3$
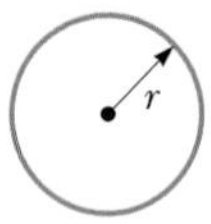 Circle	Area $= \pi r^2$ Circumference $= 2\pi r$	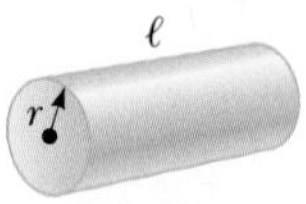 Cylinder	Lateral surface area $= 2\pi r\ell$ Volume $= \pi r^2\ell$
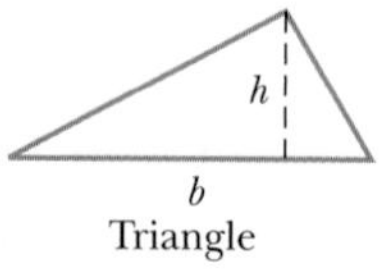 Triangle	Area $= \frac{1}{2}bh$	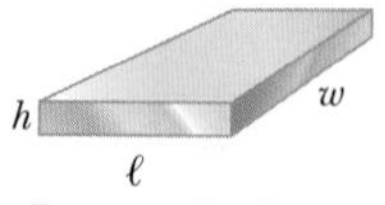 Rectangular box	Surface area $= 2(\ell h + \ell w + hw)$ Volume $= \ell wh$

A.5 Trigonometry

Some of the most basic facts concerning trigonometry are presented in Chapter 1, and we encourage you to study the material presented there if you are having trouble with this branch of mathematics. The most important trigonometric concepts include the Pythagorean theorem:

$$\Delta s^2 = \Delta x^2 + \Delta y^2 \tag{A.17}$$

This equation states that the square distance along the hypotenuse of a right triangle equals the sum of the squares of the legs. It can also be used to find distances between points in Cartesian coordinates and the length of a vector, where Δx is replaced by the x-component of the vector and Δy is replaced by the y-component of the vector. If the vector $\vec{\mathbf{A}}$ has components A_x and A_y, the magnitude A of the vector satisfies

$$A^2 = A_x^2 + A_y^2 \tag{A.18}$$

which has a form completely analogous to the form of the Pythagorean theorem. Also highly useful are the cosine and sine functions because they relate the length of a vector to its x- and y-components:

$$A_x = A\cos\theta \tag{A.19}$$
$$A_y = A\sin\theta \tag{A.20}$$

The direction θ of a vector in a plane can be determined by use of the tangent function:

$$\tan\theta = \frac{A_y}{A_x} \tag{A.21}$$

A relative of the Pythagorean theorem is also frequently useful:

$$\sin^2\theta + \cos^2\theta = 1 \tag{A.22}$$

Details on the above concepts can be found in the extensive discussions in Chapters 1 and 3. The following are some other useful trigonometric identities:

$$\sin\theta = \cos(90° - \theta)$$

$$\cos\theta = \sin(90° - \theta)$$

$$\sin 2\theta = 2\sin\theta\cos\theta$$

$$\cos 2\theta = \cos^2\theta - \sin^2\theta$$

$$\sin(\theta \pm \phi) = \sin\theta\cos\phi \pm \cos\theta\sin\phi$$

$$\cos(\theta \pm \phi) = \cos\theta\cos\phi \pm \sin\theta\sin\phi$$

The following relationships apply to any triangle, as shown in Figure A.4:

$$\alpha + \beta + \gamma = 180°$$

$$a^2 = b^2 + c^2 - 2bc\cos\alpha$$

$$b^2 = a^2 + c^2 - 2ac\cos\beta \qquad \text{law of cosines}$$

$$c^2 = a^2 + b^2 - 2ab\cos\gamma$$

$$\frac{a}{\sin\alpha} = \frac{b}{\sin\beta} = \frac{c}{\sin\gamma} \qquad \text{law of sines}$$

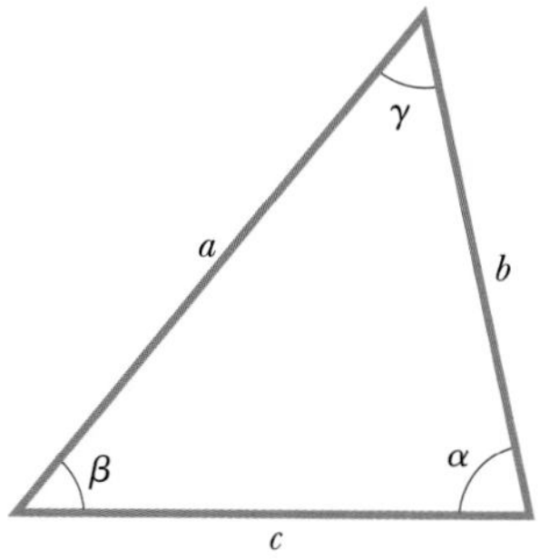

Figure A.4

Appendix B An Abbreviated Table of Isotopes

Atomic Number Z	Element	Symbol	Chemical Atomic Mass (u)	Mass Number (* Indicates Radioactive) *A*	Atomic Mass (u)	Percent Abundance	Half-Life (If Radioactive) $T_{1/2}$
0	(Neutron)	n		1*	1.008 665		10.4 min
1	Hydrogen	H	1.007 94	1	1.007 825	99.988 5	
	Deuterium	D		2	2.014 102	0.011 5	
	Tritium	T		3*	3.016 049		12.33 yr
2	Helium	He	4.002 602	3	3.016 029	0.000 137	
				4	4.002 603	99.999 863	
3	Lithium	Li	6.941	6	6.015 122	7.5	
				7	7.016 004	92.5	
4	Beryllium	Be	9.012 182	7*	7.016 929		53.3 days
				9	9.012 182	100	
5	Boron	B	10.811	10	10.012 937	19.9	
				11	11.009 306	80.1	
6	Carbon	C	12.010 7	10*	10.016 853		19.3 s
				11*	11.011 434		20.4 min
				12	12.000 000	98.93	
				13	13.003 355	1.07	
				14*	14.003 242		5 730 yr
7	Nitrogen	N	14.006 7	13*	13.005 739		9.96 min
				14	14.003 074	99.632	
				15	15.000 109	0.368	
8	Oxygen	O	15.999 4	15*	15.003 065		122 s
				16	15.994 915	99.757	
				18	17.999 160	0.205	
9	Fluorine	F	18.998 403 2	19	18.998 403	100	
10	Neon	Ne	20.179 7	20	19.992 440	90.48	
				22	21.991 385	9.25	
11	Sodium	Na	22.989 77	22*	21.994 437		2.61 yr
				23	22.989 770	100	
				24*	23.990 963		14.96 h
12	Magnesium	Mg	24.305 0	24	23.985 042	78.99	
				25	24.985 837	10.00	
				26	25.982 593	11.01	
13	Aluminum	Al	26.981 538	27	26.981 539	100	
14	Silicon	Si	28.085 5	28	27.976 926	92.229 7	
15	Phosphorus	P	30.973 761	31	30.973 762	100	
				32*	31.973 907		14.26 days
16	Sulfur	S	32.066	32	31.972 071	94.93	
				35*	34.969 032		87.5 days
17	Chlorine	Cl	35.452 7	35	34.968 853	75.78	
				37	36.965 903	24.22	
18	Argon	Ar	39.948	40	39.962 383	99.600 3	
19	Potassium	K	39.098 3	39	38.963 707	93.258 1	
				40*	39.963 999	0.011 7	1.28×10^9 yr
20	Calcium	Ca	40.078	40	39.962 591	96.941	
21	Scandium	Sc	44.955 910	45	44.955 910	100	
22	Titanium	Ti	47.867	48	47.947 947	73.72	

Atomic Number Z	Element	Symbol	Chemical Atomic Mass (u)	Mass Number (* Indicates Radioactive) A	Atomic Mass (u)	Percent Abundance	Half-Life (If Radioactive) $T_{1/2}$
23	Vanadium	V	50.941 5	51	50.943 964	99.750	
24	Chromium	Cr	51.996 1	52	51.940 512	83.789	
25	Manganese	Mn	54.938 049	55	54.938 050	100	
26	Iron	Fe	55.845	56	55.934 942	91.754	
27	Cobalt	Co	58.933 200	59	58.933 200	100	
				60*	59.933 822		5.27 yr
28	Nickel	Ni	58.693 4	58	57.935 348	68.076 9	
				60	59.930 790	26.223 1	
29	Copper	Cu	63.546	63	62.929 601	69.17	
				65	64.927 794	30.83	
30	Zinc	Zn	65.39	64	63.929 147	48.63	
				66	65.926 037	27.90	
				68	67.924 848	18.75	
31	Gallium	Ga	69.723	69	68.925 581	60.108	
				71	70.924 705	39.892	
32	Germanium	Ge	72.61	70	69.924 250	20.84	
				72	71.922 076	27.54	
				74	73.921 178	36.28	
33	Arsenic	As	74.921 60	75	74.921 596	100	
34	Selenium	Se	78.96	78	77.917 310	23.77	
				80	79.916 522	49.61	
35	Bromine	Br	79.904	79	78.918 338	50.69	
				81	80.916 291	49.31	
36	Krypton	Kr	83.80	82	81.913 485	11.58	
				83	82.914 136	11.49	
				84	83.911 507	57.00	
				86	85.910 610	17.30	
37	Rubidium	Rb	85.467 8	85	84.911 789	72.17	
				87*	86.909 184	27.83	4.75×10^{10} yr
38	Strontium	Sr	87.62	86	85.909 262	9.86	
				88	87.905 614	82.58	
				90*	89.907 738		29.1 yr
39	Yttrium	Y	88.905 85	89	88.905 848	100	
40	Zirconium	Zr	91.224	90	89.904 704	51.45	
				91	90.905 645	11.22	
				92	91.905 040	17.15	
				94	93.906 316	17.38	
41	Niobium	Nb	92.906 38	93	92.906 378	100	
42	Molybdenum	Mo	95.94	92	91.906 810	14.84	
				95	94.905 842	15.92	
				96	95.904 679	16.68	
				98	97.905 408	24.13	

(*Continued*)

Atomic Number Z	Element	Symbol	Chemical Atomic Mass (u)	Mass Number (* Indicates Radioactive) A	Atomic Mass (u)	Percent Abundance	Half-Life (If Radioactive) $T_{1/2}$
43	Technetium	Tc		98*	97.907 216		4.2×10^6 yr
				99*	98.906 255		2.1×10^5 yr
44	Ruthenium	Ru	101.07	99	98.905 939	12.76	
				100	99.904 220	12.60	
				101	100.905 582	17.06	
				102	101.904 350	31.55	
				104	103.905 430	18.62	
45	Rhodium	Rh	102.905 50	103	102.905 504	100	
46	Palladium	Pd	106.42	104	103.904 035	11.14	
				105	104.905 084	22.33	
				106	105.903 483	27.33	
				108	107.903 894	26.46	
				110	109.905 152	11.72	
47	Silver	Ag	107.868 2	107	106.905 093	51.839	
				109	108.904 756	48.161	
48	Cadmium	Cd	112.411	110	109.903 006	12.49	
				111	110.904 182	12.80	
				112	111.902 757	24.13	
				113*	112.904 401	12.22	9.3×10^{15} yr
				114	113.903 358	28.73	
49	Indium	In	114.818	115*	114.903 878	95.71	4.4×10^{14} yr
50	Tin	Sn	118.710	116	115.901 744	14.54	
				118	117.901 606	24.22	
				120	119.902 197	32.58	
51	Antimony	Sb	121.760	121	120.903 818	57.21	
				123	122.904 216	42.79	
52	Tellurium	Te	127.60	126	125.903 306	18.84	
				128*	127.904 461	31.74	$> 8 \times 10^{24}$ yr
				130*	129.906 223	34.08	$\leq 1.25 \times 10^{21}$ yr
53	Iodine	I	126.904 47	127	126.904 468	100	
				129*	128.904 988		1.6×10^7 yr
54	Xenon	Xe	131.29	129	128.904 780	26.44	
				131	130.905 082	21.18	
				132	131.904 145	26.89	
				134	133.905 394	10.44	
				136*	135.907 220	8.87	$\geq 2.36 \times 10^{21}$ yr
55	Cesium	Cs	132.905 45	133	132.905 447	100	
56	Barium	Ba	137.327	137	136.905 821	11.232	
				138	137.905 241	71.698	
57	Lanthanum	La	138.905 5	139	138.906 349	99.910	
58	Cerium	Ce	140.116	140	139.905 434	88.450	
				142*	141.909 240	11.114	$> 5 \times 10^{16}$ yr
59	Praseodymium	Pr	140.907 65	141	140.907 648	100	
60	Neodymium	Nd	144.24	142	141.907 719	27.2	
				144*	143.910 083	23.8	2.3×10^{15} yr
				146	145.913 112	17.2	

Atomic Number Z	Element	Symbol	Chemical Atomic Mass (u)	Mass Number (* Indicates Radioactive) A	Atomic Mass (u)	Percent Abundance	Half-Life (If Radioactive) $T_{1/2}$
61	Promethium	Pm		145*	144.912 744		17.7 yr
62	Samarium	Sm	150.36	147*	146.914 893	14.99	1.06×10^{11} yr
				149*	148.917 180	13.82	$> 2 \times 10^{15}$ yr
				152	151.919 728	26.75	
				154	153.922 205	22.75	
63	Europium	Eu	151.964	151	150.919 846	47.81	
				153	152.921 226	52.19	
64	Gadolinium	Gd	157.25	156	155.922 120	20.47	
				158	157.924 100	24.84	
				160	159.927 051	21.86	
65	Terbium	Tb	158.925 34	159	158.925 343	100	
66	Dysprosium	Dy	162.50	162	161.926 796	25.51	
				163	162.928 728	24.90	
				164	163.929 171	28.18	
67	Holmium	Ho	164.930 32	165	164.930 320	100	
68	Erbium	Er	167.6	166	165.930 290	33.61	
				167	166.932 045	22.93	
				168	167.932 368	26.78	
69	Thulium	Tm	168.934 21	169	168.934 211	100	
70	Ytterbium	Yb	173.04	172	171.936 378	21.83	
				173	172.938 207	16.13	
				174	173.938 858	31.83	
71	Lutecium	Lu	174.967	175	174.940 768	97.41	
72	Hafnium	Hf	178.49	177	176.943 220	18.60	
				178	177.943 698	27.28	
				179	178.945 815	13.62	
				180	179.946 549	35.08	
73	Tantalum	Ta	180.947 9	181	180.947 996	99.988	
74	Tungsten (Wolfram)	W	183.84	182	181.948 206	26.50	
				183	182.950 224	14.31	
				184*	183.950 933	30.64	$> 3 \times 10^{17}$ yr
				186	185.954 362	28.43	
75	Rhenium	Re	186.207	185	184.952 956	37.40	
				187*	186.955 751	62.60	4.4×10^{10} yr
76	Osmium	Os	190.23	188	187.955 836	13.24	
				189	188.958 145	16.15	
				190	189.958 445	26.26	
				192	191.961 479	40.78	
77	Iridium	Ir	192.217	191	190.960 591	37.3	
				193	192.962 924	62.7	
78	Platinum	Pt	195.078	194	193.962 664	32.967	
				195	194.964 774	33.832	
				196	195.964 935	25.242	
79	Gold	Au	196.966 55	197	196.966 552	100	
80	Mercury	Hg	200.59	199	198.968 262	16.87	
				200	199.968 309	23.10	
				201	200.970 285	13.18	
				202	201.970 626	29.86	

(Continued)

Atomic Number Z	Element	Symbol	Chemical Atomic Mass (u)	Mass Number (* Indicates Radioactive) A	Atomic Mass (u)	Percent Abundance	Half-Life (If Radioactive) $T_{1/2}$
81	Thallium	Tl	204.383 3	203	202.972 329	29.524	
				205	204.974 412	70.476	
		(Th C″)		208*	207.982 005		3.053 min
		(Ra C″)		210*	209.990 066		1.30 min
82	Lead	Pb	207.2	204*	203.973 029	1.4	$\geq 1.4 \times 10^{17}$ yr
				206	205.974 449	24.1	
				207	206.975 881	22.1	
				208	207.976 636	52.4	
		(Ra D)		210*	209.984 173		22.3 yr
		(Ac B)		211*	210.988 732		36.1 min
		(Th B)		212*	211.991 888		10.64 h
		(Ra B)		214*	213.999 798		26.8 min
83	Bismuth	Bi	208.980 38	209	208.980 383	100	
		(Th C)		211*	210.987 258		2.14 min
84	Polonium	Po					
		(Ra F)		210*	209.982 857		138.38 days
		(Ra C′)		214*	213.995 186		164 μs
85	Astatine	At		218*	218.008 682		1.6 s
86	Radon	Rn		222*	222.017 570		3.823 days
87	Francium	Fr					
		(Ac K)		223*	223.019 731		22 min
88	Radium	Ra		226*	226.025 403		1 600 yr
		(Ms Th_1)		228*	228.031 064		5.75 yr
89	Actinium	Ac		227*	227.027 747		21.77 yr
90	Thorium	Th	232.038 1				
		(Rd Th)		228*	228.028 731		1.913 yr
		(Th)		232*	232.038 050	100	1.40×10^{10} yr
91	Protactinium	Pa	231.035 88	231*	231.035 879		32.760 yr
92	Uranium	U	238.028 9	232*	232.037 146		69 yr
				233*	233.039 628		1.59×10^{5} yr
		(Ac U)		235*	235.043 923	0.720 0	7.04×10^{8} yr
				236*	236.045 562		2.34×10^{7} yr
		(UI)		238*	238.050 783	99.274 5	4.47×10^{9} yr
93	Neptunium	Np		237*	237.048 167		2.14×10^{6} yr
94	Plutonium	Pu		239*	239.052 156		2.412×10^{4} yr
				242*	242.058 737		3.73×10^{6} yr
				244*	244.064 198		8.1×10^{7} yr

Sources: Chemical atomic masses are from T. B. Coplen, "Atomic Weights of the Elements 1999," a technical report to the International Union of Pure and Applied Chemistry, and published in *Pure and Applied Chemistry,* 73(4), 667–683, 2001. Atomic masses of the isotopes are from G. Audi and A. H. Wapstra, "The 1995 Update to the Atomic Mass Evaluation," *Nuclear Physics,* A595, vol. 4, 409–480, December 25, 1995. Percent abundance values are from K. J. R. Rosman and P. D. P. Taylor, "Isotopic Compositions of the Elements 1999," a technical report to the International Union of Pure and Applied Chemistry, and published in *Pure and Applied Chemistry,* 70(1), 217–236, 1998.

Some Useful Tables

Appendix C

Table C.1 Mathematical Symbols Used in the Text and Their Meaning

Symbol	Meaning
$=$	is equal to
$\neq$	is not equal to
$\equiv$	is defined as
$\propto$	is proportional to
$>$	is greater than
$<$	is less than
$\gg$	is much greater than
$\ll$	is much less than
$\approx$	is approximately equal to
$\sim$	is on the order of magnitude of
Δx	change in x or uncertainty in x
Σx_i	sum of all quantities x_i
$\|x\|$	absolute value of x (always a positive quantity)

Table C.2 Standard Symbols for Units

Symbol	Unit	Symbol	Unit
A	ampere	kcal	kilocalorie
Å	angstrom	kg	kilogram
atm	atmosphere	km	kilometer
Bq	bequerel	kmol	kilomole
Btu	British thermal unit	L	liter
C	coulomb	lb	pound
°C	degree Celsius	ly	lightyear
cal	calorie	m	meter
cm	centimeter	min	minute
Ci	curie	mol	mole
d	day	N	newton
deg	degree (angle)	nm	nanometer
eV	electronvolt	Pa	pascal
°F	degree Fahrenheit	rad	radian
F	farad	rev	revolution
ft	foot	s	second
G	Gauss	T	tesla
g	gram	u	atomic mass unit
H	henry	V	volt
h	hour	W	watt
hp	horsepower	Wb	weber
Hz	hertz	yr	year
in.	inch	μm	micrometer
J	joule	Ω	ohm
K	kelvin		

Table C.3 The Greek Alphabet

Alpha	A	α	Nu	N	ν
Beta	B	β	Xi	Ξ	ξ
Gamma	Γ	γ	Omicron	O	o
Delta	Δ	δ	Pi	Π	π
Epsilon	E	ϵ	Rho	P	ρ
Zeta	Z	ζ	Sigma	Σ	σ
Eta	H	η	Tau	T	τ
Theta	Θ	θ	Upsilon	Υ	υ
Iota	I	ι	Phi	Φ	ϕ
Kappa	K	κ	Chi	X	χ
Lambda	Λ	λ	Psi	Ψ	ψ
Mu	M	μ	Omega	Ω	ω

Table C.4 Physical Data Often Used[a]

Average Earth–Moon distance	3.84×10^8 m
Average Earth–Sun distance	1.496×10^{11} m
Equatorial radius of Earth	6.38×10^6 m
Density of air (20°C and 1 atm)	1.20 kg/m^3
Density of water (20°C and 1 atm)	1.00×10^3 kg/m^3
Free-fall acceleration	9.80 m/s^2
Mass of Earth	5.98×10^{24} kg
Mass of Moon	7.36×10^{22} kg
Mass of Sun	1.99×10^{30} kg
Standard atmospheric pressure	1.013×10^5 Pa

[a] These are the values of the constants as used in the text.

Table C.5 Some Fundamental Constants

Quantity	Symbol	Value[a]
Atomic mass unit	u	$1.660\,538\,782\,(83) \times 10^{-27}$ kg $931.494\,028\,(23)$ MeV/c^2
Avogadro's number	N_A	$6.022\,141\,79\,(30) \times 10^{23}$ particles/mol
Bohr magneton	$\mu_B = \dfrac{e\hbar}{2m_e}$	$9.274\,009\,15\,(23) \times 10^{-24}$ J/T
Bohr radius	$a_0 = \dfrac{\hbar^2}{m_e e^2 k_e}$	$5.291\,772\,085\,9\,(36) \times 10^{-11}$ m
Boltzmann's constant	$k_B = \dfrac{R}{N_A}$	$1.380\,650\,4\,(24) \times 10^{-23}$ J/K
Compton wavelength	$\lambda_C = \dfrac{h}{m_e c}$	$2.426\,310\,217\,5\,(33) \times 10^{-12}$ m
Coulomb constant	$k_e = \dfrac{1}{4\pi\epsilon_0}$	$8.987\,551\,788\ldots \times 10^9$ N·m²/C² (exact)
Deuteron mass	m_d	$3.343\,583\,20\,(17) \times 10^{-27}$ kg $2.013\,553\,212\,724\,(78)$ u
Electron mass	m_e	$9.109\,382\,15\,(45) \times 10^{-31}$ kg $5.485\,799\,094\,3\,(23) \times 10^{-4}$ u $0.510\,998\,910\,(13)$ MeV/c^2
Electron volt	eV	$1.602\,176\,487\,(40) \times 10^{-19}$ J
Elementary charge	e	$1.602\,176\,487\,(40) \times 10^{-19}$ C
Gas constant	R	$8.314\,472\,(15)$ J/mol·K
Gravitational constant	G	$6.674\,28\,(67) \times 10^{-11}$ N·m²/kg²
Neutron mass	m_n	$1.674\,927\,211\,(84) \times 10^{-27}$ kg $1.008\,664\,915\,97\,(43)$ u $939.565\,346\,(23)$ MeV/c^2
Nuclear magneton	$\mu_n = \dfrac{e\hbar}{2m_p}$	$5.050\,783\,24\,(13) \times 10^{-27}$ J/T
Permeability of free space	μ_0	$4\pi \times 10^{-7}$ T·m/A (exact)
Permittivity of free space	$\epsilon_0 = \dfrac{1}{\mu_0 c^2}$	$8.854\,187\,817\ldots \times 10^{-12}$ C²/N·m² (exact)
Planck's constant	h	$6.626\,068\,96\,(33) \times 10^{-34}$ J·s
	$\hbar = \dfrac{h}{2\pi}$	$1.054\,571\,628\,(53) \times 10^{-34}$ J·s
Proton mass	m_p	$1.672\,621\,637\,(83) \times 10^{-27}$ kg $1.007\,276\,466\,77\,(10)$ u $938.272\,013\,(23)$ MeV/c^2
Rydberg constant	R_H	$1.097\,373\,156\,852\,7\,(73) \times 10^7$ m^{-1}
Speed of light in vacuum	c	$2.997\,924\,58 \times 10^8$ m/s (exact)

Note: These constants are the values recommended in 2006 by CODATA, based on a least-squares adjustment of data from different measurements. For a more complete list, see P. J. Mohr, B. N. Taylor, and D. B. Newell, "CODATA Recommended Values of the Fundamental Physical Constants: 2006." *Rev. Mod. Phys.* **80**:2, 633–730, 2008.

[a]The numbers in parentheses represent the uncertainties of the last two digits.

Appendix D

SI Units

Table D.1 SI Base Units

	SI Base Unit	
Base Quantity	**Name**	**Symbol**
Length	meter	m
Mass	kilogram	kg
Time	second	s
Electric current	ampere	A
Temperature	kelvin	K
Amount of substance	mole	mol
Luminous intensity	candela	cd

Table D.2 Derived SI Units

Quantity	Name	Symbol	Expression in Terms of Base Units	Expression in Terms of Other SI Units
Plane angle	radian	rad	m/m	
Frequency	hertz	Hz	s^{-1}	
Force	newton	N	$kg \cdot m/s^2$	J/m
Pressure	pascal	Pa	$kg/m \cdot s^2$	N/m^2
Energy: work	joule	J	$kg \cdot m^2/s^2$	$N \cdot m$
Power	watt	W	$kg \cdot m^2/s^3$	J/s
Electric charge	coulomb	C	$A \cdot s$	
Electric potential (emf)	volt	V	$kg \cdot m^2/A \cdot s^3$	W/A, J/C
Capacitance	farad	F	$A^2 \cdot s^4/kg \cdot m^2$	C/V
Electric resistance	ohm	Ω	$kg \cdot m^2/A^2 \cdot s^3$	V/A
Magnetic flux	weber	Wb	$kg \cdot m^2/A \cdot s^2$	$V \cdot s$, $T \cdot m^2$
Magnetic field intensity	tesla	T	$kg/A \cdot s^2$	Wb/m^2
Inductance	henry	H	$kg \cdot m^2/A^2 \cdot s^2$	Wb/A

Answers to Quick Quizzes, Example Questions, Odd-Numbered Warm-Up Exercises, Conceptual Questions, and Problems

CHAPTER 1

Example Questions

1. False
2. True
3. $2.6 \times 10^2\ \text{m}^2$
4. $28.0\ \text{m/s} = \left(28.0\frac{\text{m}}{\text{s}}\right)\left(\frac{2.24\ \text{mi/h}}{1.00\ \text{m/s}}\right) = 62.7\ \text{mi/h}$

   The answer is slightly different because the different conversion factors were rounded, leading to small, unpredictable differences in the final answers.
5. $\left(\frac{60.0\ \text{min}}{1.00\ \text{h}}\right)^2$
6. An answer of 10^{12} cells is within an order of magnitude of the given answer, corresponding to slightly different choices in the volume estimations. Consequently, 10^{12} cells is also a reasonable estimate. An estimate of 10^9 cells would be suspect because (working backwards) it would imply cells that could be seen with the naked eye!
7. $\sim 4 \times 10^{11}$
8. $\sim 10^{12}$
9. Working backwards, $r = 4.998$, which further rounds to 5.00, whereas $\theta = 37.03°$, which further rounds to 37.0°. The slight differences are caused by rounding.
10. Yes. The cosine function divided into the distance to the building will give the length of the hypotenuse of the triangle in question.
11. 540 km

Warm-Up Exercises

1. (a) $5.680\,17 \times 10^5$ (b) 3.09×10^{-4}
3. $25.2\ \frac{\text{km}}{\text{min}^2}$
5. $132\ \text{m}^2$
7. 58
9. 22 m

Conceptual Questions

1. (a) ~ 0.1 m (b) ~ 1 m (c) Between 10 m and 100 m (d) ~ 10 m (e) ~ 100 m
3. $\sim 10^9$ s
5. (a) $\sim 10^6$ beats (b) $\sim 10^9$ beats
7. The length of a hand varies from person to person, so it isn't a useful standard of length.
9. (a) A dimensionally correct equation isn't necessarily true. For example, the equation 2 dogs = 5 dogs is dimensionally correct, but isn't true. (b) If an equation is not dimensionally correct, it cannot be true.
11. An estimate, even if imprecise by an order of magnitude, greatly reduces the range of plausible answers to a given question. The estimate gives guidance as to what corrective measures might be feasible. For example, if you estimate that 40 000 people in a country will die unless they have food assistance and if this number is reliable within an order of magnitude, you know that at most 400 000 people will need provisions.
13. (a) yes (b) no (c) yes (d) no (e) yes

Problems

3. (b) $A_{\text{cylinder}} = \pi R^2$, $A_{\text{rectangular plate}} = \text{length} \times \text{width}$
5. $\text{m}^3/(\text{kg} \cdot \text{s}^2)$
7. $52\ \text{m}^2$
9. (a) three significant figures (b) four significant figures (c) three significant figures (d) two significant figures
11. (a) $35.7\ \text{cm}^2$ (b) $99.2\ \text{cm}^3$ (c) $17.7\ \text{cm}^2$, $99.5\ \text{cm}^3$ (d) In the rounding process, small amounts are either added to or subtracted from an answer to satisfy the rules of significant figures. For a given rounding, different small adjustments are made, introducing a certain amount of randomness in the last significant digit of the final answer.
13. $5.9 \times 10^3\ \text{cm}^3$
15. 2×10^8 fathoms
17. $0.204\ \text{m}^3$
19. Yes. The speed of 38.0 m/s converts to a speed of 85.0 mi/h, so the driver exceeds the speed limit by 10 mi/h.
21. (a) 6.81 cm (b) $5.83 \times 10^2\ \text{cm}^2$ (c) $1.32 \times 10^3\ \text{cm}^3$
23. 6.71×10^8 mi/h
25. $3.08 \times 10^4\ \text{m}^3$
27. 9.82 cm
29. $\sim 10^8$ breaths
31. (a) $\sim 10^{14}$. Other answers are possible. This answer assumes an intestinal length of about 7 m and average diameter of about 4 cm. There are probably more bacteria in your intestinal tract than there are cells in your body! (b) The large number means that they are probably not dangerous. Intestinal bacteria help digest food and provide important nutrients. Humans and bacteria enjoy a mutually beneficial symbiotic relationship.
33. $\sim 10^7$ rev
35. (2.0 m, 1.4 m)
37. $r = 2.2$ m, $\theta = 27°$
39. 5.69 m
41. (a) 6.71 m (b) 0.894 (c) 0.746
43. 3.41 m
45. (a) 3.00 (b) 3.00 (c) 0.800 (d) 0.800 (e) 1.33
47. 5.00/7.00; the angle itself is 35.5°
49. 70.0 m
51. (a) MLT^{-2} (b) $\text{kg} \cdot \text{m/s}^2$
53. (a) 1.0×10^3 kg (b) $m_{\text{cell}} = 5.2 \times 10^{-16}$ kg (c) $m_{\text{kidney}} = 0.27$ kg (d) $m_{\text{fly}} = 1.3 \times 10^{-5}$ kg
55. The value of k, a dimensionless constant, cannot be found by dimensional analysis.
57. 152 μm
59. 1×10^{10} gal/yr
61. (a) 3.16×10^7 s (b) Between 10^{10} yr and 10^{11} yr
63. 1×10^5

CHAPTER 2

Quick Quizzes

1. (a) 200 yd (b) 0 (c) 0 (d) 8 yd/s
2. (a) False (b) True (c) True
3. The velocity vs. time graph (a) has a constant slope, indicating a constant acceleration, which is represented by the acceleration vs. time graph (e).

Graph (b) represents an object with increasing speed, but as time progresses, the lines drawn tangent to the curve have increasing slopes. Since the acceleration is equal to the slope of the tangent line, the acceleration must be increasing, and the acceleration vs. time graph that best indicates this behavior is (d).

Graph (c) depicts an object which first has a velocity that increases at a constant rate, which means that the object's acceleration is constant. The velocity then stops changing, which means that the acceleration of the object is zero. This behavior is best matched by graph (f).

4. (b)
5. (a) blue graph (b) red graph (c) green graph
6. (e)
7. (c)
8. (a) and (f)

Example Questions

1. No. The object may not be traveling in a straight line. If the initial and final positions are in the same place, for example, the displacement is zero regardless of the total distance traveled during the given time.
2. No. A vertical line in a position vs. time graph would mean that an object had somehow traversed all points along the given path instantaneously, which is physically impossible.
3. No. A vertical tangent line would correspond to an infinite acceleration, which is physically impossible.
4. 35.0 m/s
5. The graphical solution is the intersection of a straight line and a parabola.
6. The coasting displacement would double to 143 m, with a total displacement of 715 m.
7. The acceleration is zero wherever the tangent to the velocity versus time graph is horizontal. Visually, that occurs from −50 s to 0 s and then again at approximately 180 s, 320 s, and 330 s.
8. The upward jump would slightly increase the ball's initial velocity, slightly increasing the maximum height.
9. 6
10. The engine should be fired again at 235 m.

Warm-Up Exercises

1. $t = 4.10, -1.10$
3. (a) 10.3 s (b) 3.20×10^2 m
5. 30.5 m/s
7. 1.00 s

Conceptual Questions

1. Yes. If the velocity of the particle is nonzero, the particle is in motion. If the acceleration is zero, the velocity of the particle is unchanging or is constant.
3. Yes. If this occurs, the acceleration of the car is opposite to the direction of motion, and the car will be slowing down.
5. No. Car B may be traveling at a lower velocity but have a greater acceleration at that instant.
7. (a) Yes. (b) Yes.
9. (a) The car is moving to the east and speeding up.
(b) The car is moving to the east but slowing down.
(c) The car is moving to the east at constant speed.
(d) The car is moving to the west but slowing down.
(e) The car is moving to the west and speeding up.
(f) The car is moving to the west at constant speed.
(g) The car starts from rest and begins to speed up toward the east.
(h) The car starts from rest and begins to speed up toward the west.
11. (b)
13. (d)

Problems

1. ≈ 0.02 s
3. (a) 52.9 km/h (b) 90.0 km
5. (a) Boat A wins by 60 km (b) 0
7. (a) 180 km (b) 63.4 km/h
9. Yes. Taking the initial speed to be zero, the final speed as 75 m/s, and $a = 1.3$ m/s², we find that the distance the plane travels before takeoff is $x = v_{\text{to}}^2/2a = 2.2$ km, which is less than the length of the runway.
11. (a) 4.4 m/s² (b) 27 m
13. (a) 2.80 h (b) 218 km
15. 274 km/h
17. (a) 5.00 m/s (b) −2.50 m/s (c) 0 (d) 5.00 m/s
19. 0.18 mi west of the flagpole
21. $|\bar{a}| = 1.34 \times 10^4$ m/s²
23. 3.7 s
25. (a) 70.0 mi/h · s = 31.3 m/s² = 3.19g (b) 321 ft = 97.8 m
27. −16.0 cm/s²
29. (a) 6.61 m/s (b) −0.448 m/s²
31. (a) 2.32 m/s² (b) 14.4 s
33. (a) 8.14 m/s² (b) 1.23 s (c) Yes. For uniform acceleration, the velocity is a linear function of time.
35. There will be a collision only if the car and the van meet at the same place at some time. Writing expressions for the position versus time for each vehicle and equating the two gives a quadratic equation in t whose solution is either 11.4 s or 13.6 s. The first solution, 11.4 s, is the time of the collision. The collision occurs when the van is 212 m from the original position of Sue's car.
37. 200 m
39. (a) 1.5 m/s (b) 32 m
41. 958 m
43. (a) 8.2 s (b) 1.3×10^2 m
45. (a) 31.9 m (b) 2.55 s (c) 2.55 s (d) −25.0 m/s
47. (a) −12.7 m/s (b) 38.2 m
49. Hardwood floor: $a = 2.0 \times 10^3$ m/s², $\Delta t = 1.4$ ms; carpeted floor: $a = 3.9 \times 10^2$ m/s², $\Delta t = 7.1$ ms
51. (a) It is a freely falling object, so its acceleration is −9.80 m/s² (downward) while in flight. (b) 0 (c) 9.80 m/s (d) 4.90 m
53. (a) It continues upward, slowing under the influence of gravity until it reaches a maximum height, and then it falls to Earth. (b) 308 m (c) 8.51 s (d) 16.4 s
55. 15.0 s
57. (a) -3.50×10^5 m/s² (b) 2.86×10^{-4} s
59. (a) 10.0 m/s upward (b) 4.68 m/s downward
61. (a) 4.0 m/s (b) 1.0 ms (c) 0.82 m
63. 0.60 s
65. (a) 0.231 m (b) 0.364 m (c) 0.399 m (d) 0.175 m
67. (a) 1.21 s after the first ball is dropped (b) 7.18 m below the window
69. No. In the time interval equal to David's reaction time, the $1 bill (a freely falling object) falls a distance of $gt^2/2 \cong$ 20 cm, which is about twice the distance between the top of the bill and its center.
71. (a) $t_1 = 5.0$ s, $t_2 = 85$ s (b) 200 ft/s (c) 18 500 ft from starting point (d) 10 s after starting to slow down (total trip time = 100 s)
73. (a) 5.5×10^3 ft (b) 3.7×10^2 ft/s (c) The plane would travel only 0.002 ft in the time it takes the light from the bolt to reach the eye.
75. (a) 7.82 m (b) 0.782 s

CHAPTER 3

Quick Quizzes

1. (c)

6. Because the wavelength becomes smaller in water, the angles to the first maxima become smaller, resulting in a smaller central maximum.
7. The separation between principal maxima will be larger.
8. The additional polarizer must make an angle of 90° with respect to the previous polarizer.

Warm-Up Exercises

1. (a) 0.524 m (b) 0.369 m
3. (a) 1.04° (b) 2.71×10^{-2} m (c) 2.59° (d) 6.79×10^{-2} m
5. (a) two (b) 4.57×10^{-7} m (c) 1.14×10^{-7} m
7. (a) 2.22×10^{-6} m (b) 47.1°
9. 52.5 W/m^2

Conceptual Questions

1. You will *not* see an interference pattern from the automobile headlights, for two reasons. The first is that the headlights are not coherent sources and are therefore incapable of producing sustained interference. Also, the headlights are so far apart in comparison to the wavelengths emitted that, even if they were made into coherent sources, the interference maxima and minima would be too closely spaced to be observable.
3. The result of the double slit is to redistribute the energy arriving at the screen. Although there is no energy at the location of a dark fringe, there is four times as much energy at the location of a bright fringe as there would be with only a single narrow slit. The total amount of energy arriving at the screen is twice as much as with a single slit, as it must be according to the law of conservation of energy.
5. One of the materials has a higher index of refraction than water, and the other has a lower index. The material with the higher index will appear black as it approaches zero thickness. There will be a 180° phase change for the light reflected from the upper surface, but no such phase change for the light reflected from the lower surface, because the index of refraction for water on the other side is lower than that of the film. Thus, the two reflections will be out of phase and will interfere destructively. The material with index of refraction lower than water will have a phase change for the light reflected from both the upper and the lower surface, so that the reflections from the zero-thickness film will be back in phase and the film will appear bright.
7. Because the light reflecting at the lower surface of the film undergoes a 180° phase change, while light reflecting from the upper surface of the film does not undergo such a change, the central spot (where the film has near zero thickness) will be dark. If the observed rings are not circular, the curved surface of the lens does not have a true spherical shape.
9. For regional communication at the Earth's surface, radio waves are typically broadcast from currents oscillating in tall vertical towers. These waves have vertical planes of polarization. Light originates from the vibrations of atoms or electronic transitions within atoms, which represent oscillations in all possible directions. Thus, light generally is not polarized.
11. Yes. In order to do this, first measure the radar reflectivity of the metal of your airplane. Then choose a light, durable material that has approximately half the radar reflectivity of the metal in your plane. Measure its index of refraction, and place onto the metal a coating equal in thickness to one-quarter of 3 cm, divided by that index. Sell the plane quickly, and then you can sell the supposed enemy new radars operating at 1.5 cm, which the coated metal will reflect with extra-high efficiency.
13. If you wish to perform an interference experiment, you need monochromatic coherent light. To obtain it, you must first pass light from an ordinary source through a prism or diffraction grating to disperse different colors into different directions. Using a single narrow slit, select a single color and make that light diffract to cover both slits for a Young's experiment. The procedure is much simpler with a laser because its output is already monochromatic and coherent.
15. (b)

Problems

1. 632 nm
3. (a) 2.6 mm (b) 2.62 mm
5. (a) 36.2° (b) 5.08 cm (c) 5.08×10^{14} Hz
7. (a) 55.7 m (b) 124 m
9. 662 nm
11. 11.3 m
13. 148 m
15. 75.0 m
17. 85.4 nm
19. 511 nm
21. 0.500 cm
23. (a) 238 nm (b) λ will increase (c) 328 nm
25. 78.4 μm
27. 4.75 μm
29. 617 nm, 441 nm, 343 nm
31. (a) 8.10 mm (b) 4.05 mm
33. (a) 1.1 m (b) 1.7 mm
35. 462 nm
37. 1.20 mm, 1.20 mm
39. (a) 479 nm, 647 nm, 698 nm (b) 20.5°, 28.3°, 30.7°
41. 5.91° in first order; 13.2° in second order; and 26.5° in third order
43. 1.81 μm
45. 514 nm
47. 9.17 cm
49. (a) 3.53×10^3 grooves/cm (b) 11 maxima
51. (a) 1.11 (b) 42.0°
53. (a) 56.7° (b) 48.8°
55. (a) 55.6° (b) 34.4°
59. 6.89 units
61. (a) 413.7 nm, 409.7 nm (b) 8.6°
63. One slit 0.12 mm wide. The central maximum is twice as wide as the other maxima.
65. 0.156 mm
67. 2.50 mm
69. (a) 16.6 m (b) 8.28 m
71. 127 m
73. 0.350 mm

CHAPTER 25

Quick Quizzes

1. (c)
2. (a)

Example Questions

1. True
2. True
3. A smaller focal length gives a greater magnification and should be selected.
4. True
5. Yes. Increasing the focal length of the mirror increases the magnification. Increasing the focal length of the eyepiece decreases the magnification.
6. More widely spaced eyes increase visual resolving power by effectively increasing the aperture size, *D*, in Equation 25.10. The limiting angle of resolution is thereby decreased, meaning finer details of distant objects can be resolved.

7. Resolution is better at the violet end of the visible spectrum.
8. True

Warm-Up Exercises

1. (a) 60.0 cm (b) –5.00 (c) inverted (d) real
3. 0.206 m
5. (a) nearsighted (b) –42.0 cm (c) –2.38 diopters
7. (a) 3.00 cm (b) 8.33
9. 11.4
11. 8.65 m

Conceptual Questions

1. The observer is *not* using the lens as a simple magnifier. For a lens to be used as a simple magnifier, the object distance must be less than the focal length of the lens. Also, a simple magnifier produces a virtual image at the normal near point of the eye, or at an image distance of about $q = -25$ cm. With a large object distance and a relatively short image distance, the magnitude of the magnification by the lens would be considerably less than one. Most likely, the lens in this example is part of a lens combination being used as a telescope.
3. The image formed on the retina by the lens and cornea is already inverted.
5. For a lens to operate as a simple magnifier, the object should be located just inside the focal point of the lens. If the power of the lens is +20.0 diopters, its focal length is
   $f = (1.00 \text{ m})/P = (1.00 \text{ m})/{+20.0} = 0.050\,0 \text{ m} = 5.00 \text{ cm}$
   The object should be placed slightly less than 5.00 cm in front of the lens.
7. You want a real image formed at the location of the paper. To form such an image, the object distance must be greater than the focal length of the lens.
9. In order for someone to see an object through a microscope, the wavelength of the light in the microscope must be smaller than the size of the object. An atom is much smaller than the wavelength of light in the visible spectrum, so an atom can never be seen with the use of visible light.
11. farsighted; converging
13. (a) The diffraction pattern of a hair is the same as the diffraction pattern produced by a single slit of the same width. (b) The central maximum is flanked by minima. Measure the width $2y$ of the central maximum between the minima bracketing it. Because the angle is small, you can use

$$\sin \theta_{\text{dark}} \approx \tan \theta_{\text{dark}}$$

$$m\lambda/a \approx y/L$$

   to find the width a of the hair.
15. (c)

Problems

1. 7.0
3. 177 m
5. f/1.4
7. 8.0
9. (a) 42.9 cm (b)+2.33 diopters
11. 23.2 cm
13. (a) −2.00 diopters (b) 17.6 cm
15. +17.0 diopters
17. −2.50 diopters, a diverging lens
19. (a) 5.8 cm (b) $m = 4.3$
21. (a) 4.07 cm (b) $m = +7.14$
23. (a) $|M| = 1.22$ (b) $\theta/\theta_0 = 6.08$
25. 2.1 cm
27. 12.7
29. 3.38 min
31. (a) 2.53 cm (b) 88.5 cm
33. (b) $-fh/p$ (c) −1.07 mm
35. (a) 1.50 (b) 1.9
37. 492 km
39. 0.40 μrad
41. 5.40 mm
43. 9.8 km
45. No. A resolving power of 2.0×10^5 is needed, and that available is only 1.8×10^5.
47. 1.31×10^3
49. 50.4 μm
51. 40
53. (a) 8.00×10^2 (b) The image is inverted.
55. 15.4
57. (a) +2.67 diopters (b) 0.16 diopter too low
59. (a) +44.6 diopters (b) 3.03 diopters
61. (a) $m = 4.0$ (b) $m = 3.0$

CHAPTER 26

Quick Quizzes

1. False: the speed of light is c for all observers.
2. (a)
3. False
4. No. From your perspective you're at rest with respect to the cabin, so you will measure yourself as having your normal length, and will require a normal-sized cabin.
5. **(i)** (a), (e) **(ii)** (a), (e)
6. False
7. (a) False (b) False (c) True (d) False

Example Questions

1. 9.61 s
2. (c)
3. (a)
4. $-0.946c$
5. (a)
6. Very little of the mass is converted to other forms of energy in these reactions because the total number of neutrons and protons doesn't change. The energy liberated is only the energy associated with their interactions.

Warm-Up Exercises

1. −122 m/s
3. 5.74 m/s
5. (a) 1.36 (b) 2.82×10^5 km
7. (a) $0.750c$ (b) $0.667c$
9. (a) 0.334 MeV (b) 6.02×10^{-13} J

Conceptual Questions

1. An ellipsoid. The dimension in the direction of motion would be measured to be less than D.
3. No. The principle of relativity implies that nothing can travel faster than the speed of light in a *vacuum*, which is equal to 3.00×10^8 m/s.
5. As the object approaches the speed of light, its kinetic energy grows without limit. It would take an infinite investment of work to accelerate the object to the speed of light.
7. For a wonderful fictional exploration of this question, get a "Mr. Tompkins" book by George Gamow. All of the relativity effects would be obvious in our lives. Time dilation and length contraction would both occur. Driving home in a hurry, you would push on the gas pedal not to increase your speed very much, but to make the blocks shorter. Big Doppler shifts in wave frequencies would make red lights look green as you approached and make car horns and radios useless. High-speed transportation would be very expensive, requiring huge fuel purchases, as well as dangerous, since a speeding car could knock down a building. When you got home, hungry for lunch, you would find that you had missed

dinner; there would be a five-day delay in transit when you watch a live TV program originating in Australia. Finally, we would not be able to see the Milky Way, since the fireball of the Big Bang would surround us at the distance of Rigel or Deneb.

9. A photon transports energy. The relativistic equivalence of mass and energy is enough to give it momentum.

11. Suppose a railroad train is moving past you. One way to measure its length is this: you mark the tracks at the cowcatcher forming the front of the moving engine at 9:00:00 AM, while your assistant marks the tracks at the back of the caboose at the same time. Then you find the distance between the marks on the tracks with a tape measure. You and your assistant must make the marks simultaneously in your frame of reference, for otherwise the motion of the train would make its length different from the distance between marks.

Problems

1. (a) 1.38 yr (b) 1.31 light-years
3. $0.103c$
5. (a) 1.3×10^{-7} s (b) 38 m (c) 7.6 m
7. (a) 1.55×10^{-5} s (b) 7.09 (c) 2.19×10^{-6} s (d) 649 m (e) From the third observer's point of view, the muon is traveling faster, so according to the third observer, the muon's lifetime is longer than that measured by the observer at rest with respect to Earth.
9. $0.950c$
11. Yes, with 19 m to spare
13. (a) $\frac{2\sqrt{2}}{3}c = 0.943c = 2.83 \times 10^8$ m/s (b) The result would be the same.
15. $0.285c$
17. $0.54c$ to the right
19. $0.357c$
21. $0.998c$ toward the right
23. (a) $0.999\,997c$ (b) 3.74×10^5 MeV
25. $0.786c$
27. (a) 8.65×10^{22} J (b) 9.61×10^5 kg
31. (a) 3.91×10^4 (b) 7.67 cm
33. (a) $0.979c$ (b) $0.065\,2c$ (c) $0.914c$
35. $0.917c$
37. (a) $v/c = 1 - 1.13 \times 10^{-10}$ (b) 5.99×10^{27} J (c) \$$2.16 \times 10^{20}$
39. (a) $0.800c$ (b) 7.50×10^3 s (c) 1.44×10^{12} m (d) $0.385c$
41. $0.80c$
43. (a) $v = 0.141c$ (b) $v = 0.436c$
45. (a) 7.0 μs (b) 1.1×10^4 muons
47. 5.5 yr; Goslo is older
49. (a) 17.4 m (b) 3.30° with respect to the direction of motion

CHAPTER 27

Quick Quizzes

1. True
2. (b)
3. (c)
4. False
5. (c)

Example Questions

1. False
2. False
3. Some of the photon's energy is transferred to the electron.
4. Doubling the speed of a particle doubles its momentum, reducing the particle's wavelength by a factor of one-half. This answer is no longer true when the doubled speed is relativistic.
5. True

Warm-Up Exercises

1. (a) 7.69×10^{14} Hz (b) 4.29×10^{14} Hz
3. 1.96×10^6 m/s
5. 2.46 eV
7. 0.124 nm
9. 4.85×10^{-3} nm
11. 2.32×10^4 m/s

Conceptual Questions

1. The shape of an object is normally determined by observing the light reflecting from its surface. In a kiln, the object will be very hot and will be glowing red. The emitted radiation is far stronger than the reflected radiation, and the thermal radiation emitted is only slightly dependent on the material from which the object is made. Unlike reflected light, the emitted light comes from all surfaces with equal intensity, so contrast is lost and the shape of the object is harder to discern.

3. The "blackness" of a blackbody refers to its ideal property of absorbing all radiation incident on it. If an observed room temperature object in everyday life absorbs all radiation, we describe it as (visibly) black. The black appearance, however, is due to the fact that our eyes are sensitive only to visible light. If we could detect infrared light with our eyes, we would see the object emitting radiation. If the temperature of the blackbody is raised, Wien's law tells us that the emitted radiation will move into the visible range of the spectrum. Thus, the blackbody could appear as red, white, or blue, depending on its temperature.

5. All objects do radiate energy, but at room temperature this energy is primarily in the infrared region of the electromagnetic spectrum, which our eyes cannot detect. (Pit vipers have sensory organs that *are* sensitive to infrared radiation; thus, they can seek out their warm-blooded prey in what we would consider absolute darkness.)

7. We can picture higher frequency light as a stream of photons of higher energy. In a collision, one photon can give all of its energy to a single electron. The kinetic energy of such an electron is measured by the stopping potential. The reverse voltage (stopping voltage) required to stop the current is proportional to the frequency of the incoming light. More intense light consists of more photons striking a unit area each second, but atoms are so small that one emitted electron never gets a "kick" from more than one photon. Increasing the intensity of the light will generally increase the size of the current, but it will not change the energy of the individual electrons that are ejected. Thus, the stopping potential remains constant.

9. Wave theory predicts that the photoelectric effect should occur at any frequency, provided that the light intensity is high enough. However, as seen in photoelectric experiments, the light must have sufficiently high frequency for the effect to occur.

11. (a) Electrons are emitted only if the photon frequency is greater than the cutoff frequency.

13. No. Suppose that the incident light frequency at which you first observed the photoelectric effect is above the cutoff frequency of the first metal, but less than the cutoff frequency of the second metal. In that case, the photoelectric effect would not be observed at all in the second metal.

15. An electron has both classical-wave and classical-particle characteristics. In single- and double-slit diffraction and interference experiments, electrons behave like classical waves. An electron has mass and charge. It carries kinetic energy and momentum in parcels of definite size, as classical particles do. At the same time it has a particular wavelength and frequency. Since an electron displays characteristics of both classical waves and classical particles, it is neither a classical wave nor a classical particle. It is customary to call it a *quantum particle,* but another invented term, such as "wavicle," could serve equally well.

Problems

1. (a) $\approx$ 3 000 K (b) $\approx$ 20 000 K
3. 9.47 μm, which is in the infrared region of the spectrum
5. (a) 2.49×10^{-5} eV (b) 2.49 eV (c) 249 eV
7. (a) 2.57 eV (b) 1.28×10^{-5} eV (c) 1.91×10^{-7} eV
9. (a) 2.24 eV (b) 555 nm (c) 5.41×10^{14} Hz
11. (a) 1.02×10^{-18} J (b) 1.53×10^{15} Hz (c) 196 nm (d) 2.15 eV (e) 2.15 V
13. 4.8×10^{14} Hz, 2.0 eV
15. 1.2×10^{2} V and 1.2×10^{7} V, respectively
17. 17.8 kV
19. 0.078 nm
21. 0.281 nm
23. 70.0°
25. (a) 1.18×10^{-23} kg·m/s (b) 478 eV
27. (a) 1.46 km/s (b) 7.28×10^{-11} m
29. 3.58×10^{-13} m
31. (a) 15 keV (b) 1.2×10^{2} keV
33. $\sim 10^{6}$ m/s
35. Within 1.16 mm for the electron, 5.28×10^{-32} m for the bullet
37. 3×10^{-29} J
39. (a) 2.21×10^{-32} kg·m/s (b) 1.00×10^{10} Hz (c) 4.14×10^{-5} eV
41. 2.5×10^{20} photons
43. (a) 5 200 K (b) Clearly, a firefly is not at this temperature, so this is not blackbody radiation.
45. 1.36 eV
47. 2.00 eV
49. (a) 0.022 0c (b) 0.999 2c

CHAPTER 28

Quick Quizzes

1. (b)
2. (a) 5 (b) 9 (c) 25
3. (d)

Example Questions

1. The energy associated with the quantum number n increases with increasing quantum number n, going to zero in the limit of arbitrarily large n. A transition from a very high energy level to the ground state therefore results in the emission of photons approaching an energy of 13.6 eV, the same as the ionization energy.
2. The energy difference in the helium atom will be four times that of the same transition in hydrogen. Energy levels in hydrogen-like atoms are proportional to Z^2, where Z is the atomic number.
3. The quantum numbers n and ℓ are never negative.
4. No. The M shell is at a higher energy; hence, transitions from the M shell to the K shell will always result in more energetic photons than any transition from the L shell to the K shell.

Warm-Up Exercises

1. (a) 6.48×10^{-19} J (b) 307 nm
3. (a) 0.967 eV (b) 1.55×10^{-19} J (c) 2.33×10^{14} Hz
5. (a) 0.212 nm (b) 0.846 nm
7. (a) 2 (b) 8 (c) 18
9. (a) 6 (b) 3

Conceptual Questions

1. If the energy of the hydrogen atom were proportional to n (or any power of n), then the energy would become infinite as n grew to infinity. But the energy of the atom is inversely proportional to n^2. Thus, as n grows to infinity, the energy of the atom approaches a value that is above the ground state by a finite amount, namely, the ionization energy 13.6 eV. As the electron falls from one bound state to another, its energy loss is always less than the ionization energy. The energy and frequency of any emitted photon are finite.
3. The characteristic x-rays originate from transitions within the atoms of the target, such as an electron from the L shell making a transition to a vacancy in the K shell. The vacancy is caused when an accelerated electron in the x-ray tube supplies energy to the K shell electron to eject it from the atom. If the energy of the bombarding electrons were to be increased, the K shell electron will be ejected from the atom with more remaining kinetic energy. But the energy difference between the K and L shell has not changed, so the emitted x-ray has exactly the same wavelength.
5. A continuous spectrum without characteristic x-rays is possible. At a low accelerating potential difference for the electron, the electron may not have enough energy to eject an electron from a target atom. As a result, there will be no characteristic x-rays. The change in speed of the electron as it enters the target will result in the continuous spectrum.
7. The hologram is an interference pattern between light scattered from the object and the reference beam. If anything moves by a distance comparable to the wavelength of the light (or more), the pattern will wash out. The effect is just like making the slits vibrate in Young's experiment, to make the interference fringes vibrate wildly so that a photograph of the screen displays only the average intensity everywhere.
9. If the Pauli exclusion principle were not valid, the elements and their chemical behavior would be grossly different because every electron would end up in the lowest energy level of the atom. All matter would therefore be nearly alike in its chemistry and composition, since the shell structures of each element would be identical. Most materials would have a much higher density, and the spectra of atoms and molecules would be very simple, resulting in the existence of less color in the world.
11. The three elements have similar electronic configurations, with filled inner shells plus a single electron in an s orbital. Because atoms typically interact through their unfilled outer shells, and since the outer shells of these atoms are similar, the chemical interactions of the three atoms are also similar.
13. Each of the eight electrons must have at least one quantum number different from each of the others. They can differ (in m_s) by being spin-up or spin-down. They can differ (in ℓ) in angular momentum and in the general shape of the wave function. Those electrons with $\ell = 1$ can differ (in m_ℓ) in orientation of angular momentum.
15. Stimulated emission is the reason laser light is coherent and tends to travel in a well-defined parallel beam. When a photon passing by an excited atom stimulates that atom to emit a photon, the emitted photon is in phase with the original photon and travels in the same direction. As this process is repeated many times, an intense, parallel beam of coherent light is produced. Without stimulated emission, the excited atoms would return to the ground state by emitting photons at random times and in random directions. The resulting light would not have the useful properties of laser light.

Problems

1. (a) 121.5 nm, 102.5 nm, 97.20 nm (b) Far ultraviolet
3. (a) 2.3×10^{-8} N (b) −14 eV

5. (a) 1.6×10^6 m/s (b) No. $v/c = 5.3 \times 10^{-3} << 1$ (c) 0.45 nm (d) Yes. The wavelength is roughly the same size as the atom.
7. (a) 0.212 nm (b) 9.95×10^{-25} kg·m/s (c) 2.11×10^{-34} J·s (d) 3.40 eV (e) −6.80 eV (f) −3.40 eV
11. $E = -1.51$ eV ($n = 3$) to $E = -3.40$ eV ($n = 2$)
13. (a) 2.86 eV (b) 0.472 eV
15. (a) 1.89 eV (b) 658 nm (c) 3.02 eV (d) 412 nm (e) 366 nm
17. (a) 488 nm (b) 0.814 m/s
19. (a) 3 (b) 520 km/s
21. (a) 0.476 nm (b) 0.997 nm
23. (a) 0.026 5 nm (b) 0.017 6 nm (c) 0.013 2 nm
25. (a) 1.31 μm (b) 164 nm
27.

n	ℓ	m_ℓ	m_s
3	2	2	$\frac{1}{2}$
3	2	2	$-\frac{1}{2}$
3	2	1	$\frac{1}{2}$
3	2	1	$-\frac{1}{2}$
3	2	0	$\frac{1}{2}$
3	2	0	$-\frac{1}{2}$
3	2	−1	$\frac{1}{2}$
3	2	−1	$-\frac{1}{2}$
3	2	−2	$\frac{1}{2}$
3	2	−2	$-\frac{1}{2}$

29. Fifteen possible states, as summarized in the following table:

n	3	3	3	3	3	3	3	3	3	3	3	3	3	3	3
ℓ	2	2	2	2	2	2	2	2	2	2	2	2	2	2	2
m_ℓ	2	2	2	1	1	1	0	0	0	−1	−1	−1	−2	−2	−2
m_s	1	0	−1	1	0	−1	1	0	−1	1	0	−1	1	0	−1

31. aluminum
33. (a) $n = 4$ and $\ell = 2$ (b) $m_\ell = (0, \pm1, \pm2)$, $m_s = \pm1/2$ (c) $1s^22s^22p^63s^23p^63d^{10}4s^24p^64d^25s^2 = [\text{Kr}]\ 4d^25s^2$
35. (a) 14 keV (b) 8.9×10^{-11} m
37. L shell: 11.7 keV; M shell: 10.0 keV; N shell: 2.30 keV
39. (a) 10.2 eV (b) 7.88×10^4 K
41. (a) 5.40 keV (b) 3.40 eV (c) 5.40 keV
43. (a) 4.24×10^{15} W/m^2 (b) 1.20×10^{-12} J

CHAPTER 29

Quick Quizzes

1. False
2. (c)
3. (c)
4. (a) and (b)
5. (b)

Example Questions

1. Tritium has the greater binding energy. Unlike tritium, helium has two protons that exert a repulsive electrostatic force on each other. The helium-3 nucleus is therefore not as tightly bound as the tritium nucleus.
2. Doubling the initial mass of radioactive material doubles the initial activity. Doubling the mass has no effect on the half-life.
3. 7.794×10^{-13} J, 4.69×10^{11} J
4. The binding energy of carbon-14 should be greater than nitrogen-14 because nitrogen has more protons in its nucleus. The mutual repulsion of the protons means that the nitrogen nucleus would require less energy to break up than the carbon nucleus.
5. No
6. Reactions between helium and beryllium can be found in the Sun.

Warm-Up Exercises

1. (a) 0.657 J (b) 4.11×10^{12} MeV
3. 2.7×10^{-15} m
5. 1.49×10^4 MeV/c^2
7. 3.12×10^{-17} s^{-1}
9. (a) 230 (b) 88

Conceptual Questions

1. An alpha particle contains two protons and two neutrons. Because a hydrogen nucleus contains only one proton, it cannot emit an alpha particle.
3. Isotopes of a given element correspond to nuclei with different numbers of neutrons. This will result in a variety of different physical properties for the nuclei, including the obvious one of mass. The chemical behavior, however, is governed by the element's electrons. All isotopes of a given element have the same number of electrons and, therefore, the same chemical behavior.
5. Nucleus Y will be more unstable. The nucleus with the higher binding energy requires more energy to be disassembled into its constituent parts and has less available energy to release in a decay.
7. In alpha decay, there are only two final particles: the alpha particle and the daughter nucleus. There are also two conservation principles: of energy and of momentum. As a result, the alpha particle must be ejected with a discrete energy to satisfy both conservation principles. However, beta decay is a three-particle decay: the beta particle, the neutrino (or antineutron), and the daughter nucleus. As a result, the energy and momentum can be shared in a variety of ways among the three particles while still satisfying the two conservation principles. This allows a continuous range of energies for the beta particle.
11. The larger rest energy of the neutron means that a free proton in space will not spontaneously decay into a neutron and a positron. When the proton is in the nucleus, however, the important question is that of the total rest energy of the nucleus. If it is energetically favorable for the nucleus to have one less proton and one more neutron, then the decay process will occur to achieve this lower energy.
13. Carbon dating cannot generally be used to estimate the age of a rock, because the rock was not alive to receive carbon, and hence radioactive carbon-14, from the environment. Only the ages of objects that were once alive can be estimated with carbon dating.

Problems

1. (a) 1.5 fm (b) 4.7 fm (c) 7.0 fm (d) 7.4 fm
3. 1.8×10^2 m
5. (a) 27.6 N (b) 4.16×10^{27} m/s^2 (c) 1.73 MeV
7. (a) 1.9×10^7 m/s (b) 7.1 MeV
9. (a) 8.26 MeV/nucleon (b) 8.70 MeV/nucleon
11. 3.54 MeV
13. (a) 0.210 MeV/nucleon greater for $^{23}_{11}$Na (b) attributable to less proton repulsion
15. 2.29 g
17. (a) 6.95×10^5 s (b) 9.97×10^{-7} s^{-1} (c) 1.9×10^4 decays/s (d) 1.9×10^{10} nuclei (e) 5, 0.200 mCi
19. 8.06 days
21. (a) 5.58×10^{-2} h^{-1}, 12.4 h (b) 2.39×10^{13} nuclei (c) 1.9 mCi
23. (a) $^{208}_{81}$Tl (b) $^{95}_{37}$Rb (c) $^{144}_{60}$Nd
25. e^+ decay, $^{56}_{27}\text{Co} \rightarrow {}^{56}_{26}\text{Fe} + e^+ + \nu$
27. (a) cannot occur spontaneously (b) can occur spontaneously
29. 18.6 keV
31. 4.22×10^3 yr
33. (a) $^{21}_{10}$Ne (b) $^{144}_{54}$Xe (c) X = e^+, X′ = ν
35. (a) $^{13}_{6}$C (b) $^{10}_{5}$B

37. (a) $^{197}_{79}Au + n \rightarrow {}^{198}_{80}Hg + e^- + \bar{\nu}$ (b) 7.89 MeV
39. (a) The product nucleus is $^{10}_{5}B$. (b) −2.79 MeV
41. 18.8 J
43. 24 d
45. (a) 8.97×10^{11} electrons (b) 0.100 J (c) 100 rad
47. (a) 3.18×10^{-7} mol (b) 1.91×10^{17} nuclei (c) 1.08×10^{14} Bq (d) 8.92×10^{6} Bq
49. 46.5 d
51. (a) 4.0×10^{9} yr (b) It could be no older. (c) The rock could be younger if some ^{87}Sr were initially present.
53. 54 μCi
55. 5.9 billion years

CHAPTER 30

Quick Quizzes

1. (a)
2. (b)

Example Questions

1. 1 mg/yr
2. 1.77×10^{12} J
3. True
4. No. That reaction violates conservation of baryon number.

Warm-Up Exercises

1. (a) 7.11×10^{13} J (b) 7.90×10^{-4} kg
3. (a) 3.33×10^{-11} J (b) 3.70×10^{-28} kg
5. 6.67×10^{15} cm^{-3}
7. 0
9. 1

Conceptual Questions

1. The experiment described is a nice analogy to the Rutherford scattering experiment. In the Rutherford experiment, alpha particles were scattered from atoms and the scattering was consistent with a small structure in the atom containing the positive charge.

3. The largest quark charge is $2e/3$, so a combination of only two particles, a quark and an antiquark forming a meson, could not have an electric charge of $+2e$. Only particles containing three quarks, each with a charge of $2e/3$, can combine to produce a total charge of $2e$.

5. Until about 700 000 years after the Big Bang, the temperature of the Universe was high enough for any atoms that formed to be ionized by ambient radiation. Once the average radiation energy dropped below the hydrogen ionization energy of 13.6 eV, hydrogen atoms could form and remain as neutral atoms for relatively long period of time.

7. In the quark model, all hadrons are composed of smaller units called quarks. Quarks have a fractional electric charge and a baryon number of $\frac{1}{3}$. There are six flavors of quarks: up (u), down (d), strange (s), charmed (c), top (t), and bottom (b). All baryons contain three quarks, and all mesons contain one quark and one antiquark. Section 30.8 has a more detailed discussion of the quark model.

9. Baryons and mesons are hadrons, interacting primarily through the strong force. They are not elementary particles, being composed of either three quarks (baryons) or a quark and an antiquark (mesons). Baryons have a nonzero baryon number with a spin of either $\frac{1}{2}$ or $\frac{3}{2}$. Mesons have a baryon number of zero and a spin of either 0 or 1.

11. All stable particles other than protons and neutrons have baryon number zero. Since the baryon number must be conserved, and the final states of the kaon decay contain no protons or neutrons, the baryon number of all kaons must be *zero*.

Problems

1. 1.1×10^{16} fissions
3. 126 MeV
5. (a) 16.2 kg (b) 117 g
7. (a) $3/a$ (b) $3.72/a$ (c) The surface-to-volume ratio is lowest for the sphere; therefore, the sphere has the better shape for minimum leakage.
9. 1.01 g
11. (a) $^{8}_{4}Be$ (b) $^{12}_{6}C$ (c) 7.27 MeV
13. 4.03 MeV
15. 14.1 MeV
17. (a) 4.53×10^{23} Hz (b) 0.662 fm
19. 67.5 MeV, 67.5 MeV/c, 1.63×10^{22} Hz
21. (a) conservation of electron-lepton number and conservation of muon-lepton number (b) conservation of charge (c) conservation of baryon number (d) conservation of baryon number (e) conservation of charge
23. (a) Violates conservation of baryon number (not allowed). (b) This reaction may occur via the strong interaction. (c) This reaction may occur via the weak interaction only. (d) This reaction may occur by the weak interaction only. (e) This reaction may occur via the electromagnetic interaction.
25. (a) not conserved (b) conserved (c) conserved (d) not conserved (e) not conserved (f) not conserved
27. 3.34×10^{26} electrons, 9.36×10^{26} up quarks, 8.70×10^{26} down quarks
29. (a) Σ^+ (b) π^- (c) K^0 (d) Ξ^-
31. a neutron, udd
33. 18.8 MeV
35. (a) K^+ (b) Ξ^0 (c) π^0
37. 26
39. (b) 12 days
41. 29.8 MeV
43. 3.60×10^{38} protons/s

Index

Note: Page numbers followed by *f* and *t* refer to figures and tables respectively.

A

D

E

F

J

K

N

T

U

PHYSICAL CONSTANTS

Quantity	Symbol	Value	SI unit
Avogadro's number	N_A	6.02×10^{23}	particles/mol
Bohr radius	a_0	5.29×10^{-11}	m
Boltzmann's constant	k_B	1.38×10^{-23}	J/K
Coulomb constant, $1/4\pi\epsilon_0$	k_e	8.99×10^{9}	$N \cdot m^2/C^2$
Electron Compton wavelength	$h/m_e c$	2.43×10^{-12}	m
Electron mass	m_e	9.11×10^{-31}	kg
		5.49×10^{-4}	u
		0.511 MeV/c^2	
Elementary charge	e	1.60×10^{-19}	C
Gravitational constant	G	6.67×10^{-11}	$N \cdot m^2/kg^2$
Mass of Earth	M_E	5.98×10^{24}	kg
Mass of Moon	M_M	7.36×10^{22}	kg
Molar volume of ideal gas at STP	V	22.4	L/mol
		2.24×10^{-2}	m^3/mol
Neutron mass	m_n	$1.674\,93 \times 10^{-27}$	kg
		1.008 665	u
		939.565 MeV/c^2	
Permeability of free space	μ_0	1.26×10^{-6}	$T \cdot m/A$
		($4\pi \times 10^{-7}$ exactly)	
Permittivity of free space	ϵ_0	8.85×10^{-12}	$C^2/N \cdot m^2$
Planck's constant	h	6.63×10^{-34}	$J \cdot s$
	$\hbar = h/2\pi$	1.05×10^{-34}	$J \cdot s$
Proton mass	m_p	$1.672\,62 \times 10^{-27}$	kg
		1.007 276	u
		938.272 MeV/c^2	
Radius of Earth (at equator)	R_E	6.38×10^{6}	m
Radius of Moon	R_M	1.74×10^{6}	m
Rydberg constant	R_H	1.10×10^{7}	m^{-1}
Speed of light in vacuum	c	3.00×10^{8}	m/s
Standard free-fall acceleration	g	9.80	m/s^2
Stefan-Boltzmann constant	σ	5.67×10^{-8}	$W/m^2 \cdot K^4$
Universal gas constant	R	8.31	$J/mol \cdot K$

The values presented in this table are those used in computations in the text. Generally, the physical constants are known to much better precision.